T0327511

INTRODUCTION TO AEROSPACE ENGINEERING WITH A FLIGHT TEST PERSPECTIVE

Aerospace Series

INTRODUCTION TO AEROSPACE ENGINEERING WITH A FLIGHT TEST PERSPECTIVE

Stephen Corda

WILEY

Library of Congress Cataloging-in-Publication Data

Names: Corda, Stephen, 1958- author.
Title: Introduction to aerospace engineering with a flight test perspective /
  Stephen Corda.
Description: Chichester, West Sussex, United Kingdom : John Wiley & Sons,
  Inc., 2017. | Includes bibliographical references and index.
Identifiers: LCCN 2016039084| ISBN 9781118953365 (cloth) | ISBN 9781118953372
  (epub) | ISBN 9781118953389 (Adobe PDF)
Subjects: LCSH: Aerospace engineering–Textbooks. | Airplanes–Flight
  testing–Textbooks.
Classification: LCC TL546 .C6915 2017 | DDC 629.1–dc23 LC record available at https://lccn.loc.gov/2016039084

A catalogue record for this book is available from the British Library.

Cover image: Jupiterimages/Gettyimages

Set in 10/12pt, TimesLTStd by SPi Global, Chennai, India.

10 9 8 7 6 5 4 3 2 1

To Glenn, Gerard, and Marta.
"Blue skies and tailwinds."

Table of Contents

About the Author

Stephen Corda was born in Newburg, New York on February 3, 1958. He attended the University of Maryland, graduating in 1980 with a Bachelor of Science degree in Aerospace Engineering. He continued with graduate studies in hypersonic aerodynamics at the University of Maryland, graduating with a Master of Science degree in 1982 and a PhD in 1988, both in Aerospace Engineering. From 1983 to 1984, he attended the Von Karman Institute for Fluid Dynamics in Rhode-Saint-Genese, Belgium, graduating with an Aeronautics Diploma in 1984. In 1988, he accepted a position with the Hypersonic Propulsion Group at The Johns Hopkins University Applied Physics Laboratory, Laurel, Maryland. From 1990 to 2001 and then from 2004 to 2006, he served in various positions at the NASA Dryden (now Armstrong) Flight Research Center, Edwards, California, including project engineer, flight test engineer, project manager, and Chief of the Propulsion and Performance Branch. He was an instructor and flight test engineer in the Performance Branch at the U.S. Air Force Test Pilot School, Edwards, California from 2001 to 2002. From 2003 to 2004, he was an Assistant Professor in the Aerospace Engineering Department, U.S. Naval Academy, Annapolis, Maryland. He was Associate Professor and Chairman of the Aviation Systems and Flight Research Program at the University of Tennessee Space Institute, Tullahoma, Tennessee, from 2006 to 2012. Returning to the Mojave Desert in 2012, he was the engineering manager of Flight Sciences at Virgin Galactic-The Spaceship Company until 2014. Dr. Corda is currently a private engineering consultant.

Series Preface

The field of aerospace is multi-disciplinary and wide-ranging, covering a large variety of products, disciplines, and domains, not merely in engineering but in many related supporting activities. These combine to enable the aerospace industry to produce innovative and technologically advanced vehicles. The wealth of knowledge and experience that has been gained by expert practitioners in the various aerospace fields needs to be passed on to others working in the industry and also to researchers, teachers, and students in universities.

The *Aerospace Series* aims to be a practical, topical, and relevant series of books aimed at people working in the aerospace industry, including engineering professionals and operators, engineers in academia, and allied professionals such commercial and legal executives. The topics are wide-ranging, covering design and development, manufacture, operation and support of aircraft, and topics such as infrastructure operations and current advances in research and technology.

The design of future aircraft will depend not only on a deep understanding of the fundamental scientific disciplines that provide the foundations for aerospace engineering, but also on the test techniques that enable verification and validation of novel designs.

This book, *Introduction to Aerospace Engineering with a Flight Test Perspective*, provides a comprehensive introduction to the fundamentals of aerodynamics, propulsion, performance, and stability and control, as required for the design of fixed-wing aircraft. It is a welcome addition to the Wiley Aerospace Series, complementing many of the other books in the Series. Of particular note is the inclusion of various ground and flight testing techniques that relate to the various sections of the book, an area that is rarely documented in textbooks.

Peter Belobaba, Jonathan Cooper and Allan Seabridge

Preface

This book is an introductory level text in aerospace engineering with a unique perspective. Flight test, where dreams of aircraft and space vehicles take to the sky, is the bottom line in the application of aerospace engineering theories and principles. Designing and flying the real machines is often the reason that these theories and principles were developed in the first place. This book provides a solid foundation in many of the fundamentals of aerospace engineering, while illuminating many aspects of real-world flight. Fundamental aerospace engineering subjects that are covered include aerodynamics, propulsion, performance, and stability and control.

The test perspective provides an applied, hands-on engineering flavor to the book. The reader comes away with engineering insights about how to do many different types of aerospace testing, topics that are seldom covered or integrated into a university aerospace engineering curriculum. These topics are essential to becoming a well-rounded aerospace engineer, regardless of what discipline or role one may have in aerospace.

The text is suitable for use in an introductory, undergraduate course in aerospace engineering. The addition of the sections dealing with testing provides the opportunity to introduce these important subjects, especially for those aerospace programs that do not have a dedicated flight test course. In addition, the text may be used to support a dedicated flight test course.

The text can also serve working engineers who seek to broaden their aerospace engineering "toolbox", to include some of the fundamentals of flight testing. The text can be helpful to those engaged in flight test, as a convenient reference source in fundamental aerospace engineering theory and applied flight test practice. The flight test perspective can also provide the non-engineer, aviation professional, or enthusiast with a deeper understanding of aerospace and flight test. However, the text should not be used as a "how to" manual for the non-professional to attempt their own flight testing.

Sections entitled Flight Test Techniques (FTTs) and Ground Test Techniques (GTTs) present test methods used in applying the aerospace engineering theories and concepts discussed in the previous sections. Rather than presenting a step-by-step list of procedures, the FTTs are described in a unique manner, by placing the reader "in the cockpit" of different aircraft, giving them an exciting perspective for learning about flight test concepts, test techniques, and in-flight data collection. A collateral benefit of this approach is that the reader learns about several different types of aircraft. This approach is a unique and interesting way to learn about aerospace engineering and flight testing, short of actually flying the real airplanes!

Other useful resources may be found online at the companion Wiley website associated with the text (www.wiley.com\go\corda\aerospace_engg_flight_test_persp), where you will find a collection of technical papers and information, which are referenced in the text. These are organized by chapter and by reference to flight test techniques. Instructors may also access complete solutions to all of the homework problems on the website.

In many ways, a textbook is autobiographical in nature, drawing on the author's personal career experiences, interactions, and lessons learned. The material for this book is derived from the author's experience as an aerospace engineer, flight test engineer, flight research pilot, and educator at the NASA Dryden (now Armstrong) Flight Research Center, The Johns Hopkins University Applied Physics Laboratory, the US Air Force Test Pilot School, the US Naval Academy, the University of Tennessee Space Institute, and Virgin Galactic – The Spaceship Company. I owe a great debt to the many engineers, scientists, students, technicians, managers, administrators, and test pilots that I have worked and flown with over my career, who have helped me to learn my trade.

In preparing the manuscript of the textbook, I am very grateful to the many folks who contributed material, especially the following:

Katie Bell, Lycoming Engines
Jennifer Bowman, Orbital ATK
Richard Ferriere, www.richard.ferriere.free.fr
Guillaume Grossir and Sebastien Paris, Von Karman Institute for Fluid Dynamics
Phil Hays, www.okieboat.com
Christian Hochheim, Extra Aircraft
Kate Igoe, National Air & Space Museum
Bernardo Malfitano, www.understandingairplanes.com
Paul Niewald, The Boeing Company
Ray Watkins, http://1000aircraftphotos.com/Contributions/WatkinsRay/WatkinsRay.htm
Jessika Wichner, German Aerospace Center, DLR

A special thank-you goes to Jim Ross (NASA *Photo One*) and his photo team at the NASA Armstrong Flight Research Center, who provided exceptional photos of NASA aircraft. Dr Saeed Farokhi, professor of aerospace engineering at the University of Kansas and a fellow author of a Wiley engineering textbook, provided valuable guidance and insight into the publishing process. Finally, I must thank Dr John D. Anderson, Jr., for his unwavering support in my early years as an aerospace engineering undergraduate and graduate student at the University of Maryland, and for all of the wonderful textbooks that he has written, from which I learned many of the fundamentals of aerospace engineering.

If you have comments, questions, suggestions, or corrections for the next edition of the book, please email them to scdaos@gmail.com.

Stephen Corda
Rosamond, California

About the Companion website

This book is accompanied by a companion website:

www.wiley.com/go/corda/aerospace_engg_flight_test_persp

The website includes:

- Chapters 1 - 6 PDF files
- Homework problems solution manual
- Flight Test Techniques papers

1

First Flights

The first controlled flight of a heavier-than-air airplane, 17 December 1903. (Source: *W. Wright, O. Wright, and J. Daniels, 1903, US Library of Congress.*)

"Wilbur, having used his turn in the unsuccessful attempt on the 14th, the right to the first trial now belonged to me. After running the motor a few minutes to heat it up, I released the wire that held the machine to the track, and the machine started forward in the wind. Wilbur ran at the side of the machine, holding the wing to balance it on the track. Unlike the start on the 14th, made in a calm, the machine, facing a 27-mile wind, started very slowly. Wilbur was able to stay with it till it lifted from the track after a forty-foot run. One of the Life Saving men snapped the camera for us, taking a picture just as the machine had reached the end of the track and had risen to a height

Introduction to Aerospace Engineering with a Flight Test Perspective, First Edition. Stephen Corda.
© 2017 John Wiley & Sons Ltd. Published 2017 by John Wiley & Sons Ltd.
Companion Website: www.wiley.com/go/corda/aerospace_engg_flight_test_persp

of about two feet.[1] The slow forward speed of the machine over the ground is clearly shown in the picture by Wilbur's attitude. He stayed along beside the machine without any effort.

The course of the flight up and down was exceedingly erratic, partly due to the irregularity of the air, and partly to lack of experience in handling this machine. The control of the front rudder was difficult on account of its being balanced too near the center. This gave it a tendency to turn itself when started; so that it turned too far on one side and then too far on the other. As a result the machine would rise suddenly to about ten feet, and then as suddenly dart for the ground. A sudden dart when a little over a hundred feet from the end of the track, or a little over 120 ft from the point at which it rose into the air, ended the flight. As the velocity of the wind was over 35 ft per second and the speed of the machine over the ground against this wind ten feet per second, the speed of the machine relative to the air was over 45 ft per second, and the length of the flight was equivalent to a flight of 540 feet made in calm air. This flight lasted only 12 seconds, but it was nevertheless the first in the history of the world in which a machine carrying a man had raised itself by its own power into the air in full flight, had sailed forward without reduction of speed and had finally landed at a point as high as that from which it started."

> Orville Wright writing about the first successful flight of a heavier-than-air flying machine from Kill Devil Hills, North Carolina, on 17 December, 1903[2]

1.1 Introduction

The history of aerospace engineering is full of firsts, such as the first balloon flight, the first airplane flight, the first helicopter flight, the first artificial satellite flight, the first manned spacecraft flight, and many others. In this first chapter, these many firsts are discussed in the context of the aerospace engineering involved in making these historic events happen. The first flight of a new vehicle design is a significant achievement and milestone. It is usually the culmination of years of hard work by many people, including engineers, technicians, managers, pilots, and other support personnel. First flights often represent firsts in the application of new aerospace engineering concepts or theories that are being validated by the actual flight.

As an aerospace engineer, you have the opportunity to contribute to the first flight of a new aircraft, a new spacecraft, or a new technology. Aerospace engineers are involved in all facets of the design, analysis, research, development, and testing of aerospace vehicles. This encompasses many different aerospace engineering discipline specialties, including aerodynamics, propulsion, performance, stability, control, structures, systems, and others. Several of these fundamental disciplines of aerospace engineering are introduced in this text. The aerospace engineer tests the vehicle, on the ground and in flight, to verify that it can perform as predicted and to improve its operating

---

[1] John Thomas Daniels, Jr. (1873–1948) snapped the iconic photograph of the Wright brothers' historic first flight. Daniels was a member of the Kill Devil Hills, North Carolina, Life Saving Station, which relied on volunteers to respond to the frequent shipwrecks in this barrier island area. The Wright's *Flyer* made four flights on 17 December 1903, three of which were photographed. After the fourth and final flight, a gust of wind caught the airplane and Daniels grabbed a wing strut, attempting to hold the airplane down. He was caught between the biplane wings when the *Flyer* flipped over in the wind. Although the *Flyer I* was destroyed, Daniels was unhurt, and he would later recount that he had "survived the first airplane crash".

[2] Orville Wright, *How We Made the First Flight* (1986) Federal Aviation Administration, Office of Public Affairs, Aviation Education Program, US Department of Transportation.

characteristics. Flight testing is usually the final test to be performed on the complete vehicle or system.

In many areas of engineering and technology, there is sometimes a perception that there is "nothing left to be done", or that "there is nothing left to be invented". The impressive successes of our aerospace past may appear, to some, to dim the prospects for future innovations. Aerospace engineers have indeed designed, built, and flown some of the most innovative, complex, and amazing machines known to humanity. However, there is still ample room for creativity and innovation in the design of aerospace vehicles, and opportunities for technological breakthroughs to make the skies and stars far more accessible. By the end of this textbook, you will have greatly increased your knowledge of aerospace engineering, but you will also be humbled by how much more there is to be discovered.

1.1.1   Organization of the Book

Aerospace engineering encompasses the fields of aeronautical and astronautical engineering. As a broad generalization, the aeronautical field tends to deal with vehicles that fly through the sensible atmosphere, that is, *aircraft*. Astronautics deals with vehicles that operate in the airless space environment, that is, *spacecraft*. Aerospace engineering is, in many ways, a merging of these two fields, and includes aircraft, spacecraft, and other vehicles that operate in both the air and space environments. In the coming sections, we get more precise with the definitions of the various types of aerospace vehicles, such as aircraft and spacecraft.

The material in the text is organized in an academic building-block fashion as shown in Figure 1.1. In Chapter 1, we start by defining and discussing some of the many different types of aircraft and spacecraft. Many first flights of these different types of aerospace vehicles are described, providing insights and perspectives into the development and evolution of aerospace engineering. The terms *aircraft* and *spacecraft* are clearly defined, along with definitions of the various parts, components, and assemblies that make up various examples of these types of vehicles. The reader also makes a literary "first flight" in a modern, supersonic jet airplane, which introduces many of the areas to be discussed in the coming chapters.

In Chapter 2, several *introductory concepts* in aerospace engineering and flight test are discussed. This chapter gives the reader some of the basic concepts and terminology, in aerospace

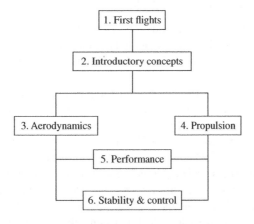

Figure 1.1   Academic building blocks followed in the text.

engineering and flight test, from which to learn the material in the subsequent chapters. Some basic mathematical ideas, definitions, and concepts are reviewed, which starts to fill our engineering toolbox with the basic tools required to analyze and design aerospace vehicles. Basic aerospace engineering concepts, relating to the flight of aerospace vehicles, are introduced, including aircraft axis systems, free-body diagrams, the regimes of flight, and the flight envelope. Basic flight test concepts are introduced, including the different types of flight test, the flight test process, the players involved, and the use of flight test techniques.

The fundamental disciplines of *aerodynamics* and *propulsion* are discussed in Chapters 3 and 4, respectively. The study of aerodynamics, in Chapter 3, provides the theories and tools required to analyze the flow of air over aerospace vehicles, the flow that produces aerodynamic forces such as lift and drag. We discover how and why these aerodynamic forces are created, and how this affects the design of aerodynamic surfaces such as airfoils and wings. In studying propulsion in Chapter 4, we learn about the devices that generate the thrust force to propel aerospace vehicles both in the atmosphere and in space. We develop a deeper understanding of how thrust is produced, regardless of the type of machinery that is used.

The study of *performance*, in Chapter 5, builds upon an understanding of aerodynamics and propulsion, as shown graphically in Figure 1.1. Performance deals with the linear motion of the vehicle caused by the aerodynamic forces (lift and drag) and propulsive force (thrust) acting upon it. Performance seeks to determine how fast, how high, how far, and how long a vehicle can fly.

In Chapter 6, the study of *stability and control* also builds upon the fundamental disciplines of aerodynamics and propulsion. Stability and control deals with the angular motion of the vehicle caused by the aerodynamic and propulsive moments acting on it. We investigate the vehicle's stability when disturbed from its equilibrium condition and seek to understand the impacts of various vehicle configurations and geometries. We also look at the means by which the vehicle can be controlled throughout its flight regime.

Many examples of ground and flight testing are integrated throughout the text, in sections entitled *Ground Test Techniques* and *Flight Test Techniques*. The flight test techniques are described in a unique manner, by placing the reader "in the cockpit" of different aircraft as the test pilot or flight test engineer. The reader obtains an intimate knowledge of the engineering concepts, test techniques, and in-flight data collection by "flying" the flight test techniques. A collateral benefit of this approach is that the reader is familiarized with several different types of real aircraft.

1.1.2 FTT: Your Familiarization Flight

This is the first of many flight test techniques (FTTs) that are "flown" in the text. The FTT is a precise and standardized method, used to efficiently collect data during flight test, research, and evaluation of aerospace vehicles. The FTT process is discussed in more detail in a later section of this chapter.

This first FTT introduces you to aerospace engineering in an exciting way, by taking a flight in a supersonic jet aircraft. A flight test engineer (FTE) often flies a *familiarization flight* in an aircraft prior to performing test flights, especially if this is an aircraft that is new to the FTE. As the name implies, this flight serves to *familiarize* the FTE with the aircraft and the flight environment. The areas of familiarization usually include the aircraft's performance, flying qualities, cockpit environment, avionics, or other special test equipment and instrumentation. The present FTT provides a general description of a familiarization flight, but the primary objective is to introduce you to a wide range of aerospace engineering and test concepts that are explored in later chapters. Your familiarization flight will raise many technical questions about aerospace engineering and flight test, and this provides motivation to seek answers in the chapters to come.

Figure 1.2 McDonnell Douglas F/A-18B *Hornet* supersonic fighter. (Source: *Courtesy of the author.*)

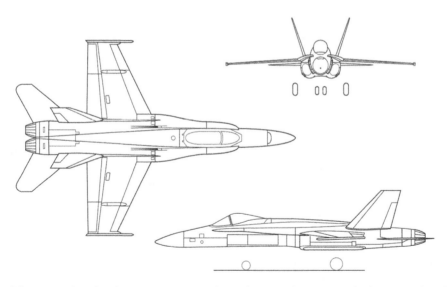

Figure 1.3 Three-view drawing of the McDonnell Douglas F/A-18A *Hornet* (single-seat version shown). (Source: *NASA.*)

For your familiarization flight, you will be flying the McDonnell Douglas (now Boeing) F/A-18B *Hornet* supersonic jet aircraft, shown in Figure 1.2. The F/A-18B is a two-seat, twin-engine, supersonic fighter jet aircraft, designed for launching from and landing on an aircraft carrier. Almost all aerospace vehicles are designated with letters and numbers, which we will decipher in a later section. A three-view drawing of the F/A-18A is shown in Figure 1.3. You will get very familiar

Table 1.1 Selected specifications of the McDonnell Douglas F/A-18B *Hornet*.

Item	Specification
Primary function	All-weather, supersonic fighter/attack jet aircraft
Manufacturer	McDonnell Douglas Aircraft, St Louis, Missouri
First flight	18 November 1978
Crew	1 pilot + 1 instructor pilot or flight test engineer
Powerplant	$2 \times$ F404-GE-400 afterburning turbofan engine
Thrust, MIL (ea. engine)	10,700 lb (47,600 N), military power
Thrust, MAX (ea. engine)	17,700 lb (78,700 N), maximum afterburner
Empty weight	~25,000 lb (11,300 kg)
Maximum takeoff weight	51,900 lb (23,500 kg)
Length	56 ft (17.1 m)
Height	15 ft 4 in (4.67 m)
Wingspan	37 ft 6 in (11.4 m)
Wing area	400 ft^2 (37.2 m^2)
Airfoil, wing root	NACA 65A005 modified
Airfoil, wingtip	NACA 65A003.5 modified
Maximum speed	1190 mph (1915 km/h), Mach 1.7+
Service ceiling	>50,000 ft (>15,240 m)
Load factor limits	+7.5 g, −3.0 g

with these types of drawings of aerospace vehicles, where typically side, top, and front views of the vehicle are depicted. Selected specifications of the F-18 *Hornet* are given in Table 1.1. The chapters to come will help you understand all of the technical details in these specifications, such as what defines a "low bypass turbofan jet engine with an afterburner" or why wing area, maximum weights, or load factor limits are important.

Before you can go flying in an F-18, you need to be properly dressed. You don an olive-green flight suit, black flight boots, and an anti-G suit, an outer garment that fits snugly over the lower half of your body. Inflatable bladders, sewn into the anti-G suit, inflate with pressurized air to prevent blood from pooling in your lower extremities, keeping the blood in your head, so that you do not lose consciousness when the aircraft is maneuvering at high load factors or g's. You slip your arms into a parachute harness that buckles around your chest and both legs. You are wearing the harness for the parachute, but not the actual parachute, as you will buckle this harness into your ejection seat, which contains your emergency parachute in the headrest.

With your flight helmet, oxygen mask, and kneeboard, a small clipboard-type writing surface, in your helmet bag, you walk out to the airport ramp, where the jet is parked. As you walk up to the aircraft, you note its general configuration. The aircraft has a slender fuselage with a low-mounted, thin wing, aft-mounted horizontal tail, twin vertical tails, and tricycle landing gear, comprising two main wheels, extending from either side of the fuselage, and a fuselage nosewheel. You observe that the landing gear looks quite sturdy, designed for harsh aircraft carrier landings. The jet is powered by twin engines, with semicircular air inlets on each side of the fuselage and side-by-side exhaust nozzles at the aft end of the fuselage. The two aviators sit in a tandem configuration, beneath a long "bubble" canopy that is hinged behind the aft cockpit. Your test pilot will be seated in the front cockpit and you will be in the aft cockpit.

You approach the aircraft from its left side, next to the cockpit, as shown in Figure 1.4. Before you climb into the cockpit, you perform a walk-around of the jet to learn a little more about it. Underneath the left wing, near the fuselage, you look into the left engine inlet, which is a semicircular

Figure 1.4 F/A-18B *Hornet* walk-around, left wing. (Source: *Courtesy of NASA/Lauren Hughes.*)

opening. This inlet feeds air to the turbofan jet engine. Later, we will learn about why the inlet is shaped in this way and how the air mass flow, which is ingested through the inlet, is related to the production of thrust. Looking underneath the fuselage, you see a large cylindrical fuel tank with pointed ends, hung underneath the centerline of the fuselage. Of course, you know that the fuel quantity carried aboard the aircraft dictates how far and how long the aircraft can fly. We will see that the range and endurance is a function of more than just the fuel quantity; it is also a function of key parameters related to the aerodynamics and propulsion of the vehicle. We will also learn about how to obtain range and endurance through flight testing.

You move towards the leading edge of the left wing. You observe that the wing is thin, with a somewhat sharp leading edge, and that the wing leading edge is swept backward. There is a large hinged flap surface at the inboard wing trailing edge. We will explore the aerodynamics of three-dimensional wings and their two-dimensional cross-sectional shapes, known as airfoils. We will learn why airfoils and wings are shaped differently for flight at different speeds, including why wings are swept back. We will discuss how hinged flaps increase the lift of a wing. Fundamentally, we will discuss how a wing produces aerodynamic lift, and will discuss the many ways of quantifying the lift and drag of an aircraft, through analysis, ground test, or flight test.

Now you are at the rear of the aircraft, looking at the two engine nozzles, as shown in Figure 1.5. The nozzles have interlocking metal petals that can expand and contract to change the nozzle exit area. We will examine how the flow properties change with area in subsonic and supersonic nozzle flows. We will learn how to calculate the velocity, pressure, and temperature of the gas flowing through the nozzle. You look down the afterburner of the jet engine, which appears to be an almost empty duct. We will discuss the various components of the jet engine, including the afterburner, and will explain their functions. The jet engine is an amazing engineering achievement. We will explore its beginnings, and the engineers who invented it. We also learn about how engines are tested in the ground and flight environments.

Coming around the right, aft end of the airplane, as shown in Figure 1.6, you look at the horizontal and vertical tail surfaces. We will learn why these surfaces are critical to the stability and control of the aircraft. We will see that the locations and sizes of these surfaces are important parameters in defining the aircraft's stability in flight, and will also learn about the control forces associated with

Figure 1.5   F/A-18B *Hornet* walk-around, engine nozzles. (Source: *Courtesy of NASA/Lauren Hughes.*)

Figure 1.6   F/A-18B *Hornet* walk-around, aft, right empennage, and right wing flaps. (Source: *Courtesy of NASA/Lauren Hughes.*)

deflection of these surfaces in an air stream. We will discuss several different flight test techniques used to quantify an aircraft's stability. Near the nose of the airplane, you notice several L-shaped tubes mounted on the lower side of the fuselage. We will learn about these Pitot tubes, which are used to measure the F-18's airspeed, and will investigate how they work in subsonic and supersonic flight. We will also see that flight testing is required to calibrate these probes to obtain accurate airspeed information. You come to the aircraft nose, which has a pointed shape. We will explore the aerodynamics of two and three-dimensional bodies, such as this nose shape. We will also touch on the interesting phenomena that occur when these types of pointed shapes are at high angles of attack.

Returning to the left side of the fuselage, with your walk-around complete, you meet up with your test pilot. It is time to get into the airplane and go flying. The pilot climbs the ladder into the front cockpit and you follow, making your way into the back cockpit. You buckle your lap belt, plug in your G-suit hose, and connect the ejection seat shoulder belts to your harness. Next, you don your flight helmet, connect your oxygen mask hose, plug in your communications cable, and slip on your flying gloves. You strap your kneeboard on your right thigh so that you will be able to take some notes during your flight. Now that you are strapped in and have connected all of your gear, you have a chance to relax and look around. There is a center control stick, two rudder pedals at your feet, and two throttle levers by your left side. The instrument panel in front of you has three square display screens, surrounded by buttons, and an array of other circular, analog instruments (Figure 1.7).

You hear the test pilot in your helmet earphones, asking if you can hear him and if you are ready for engine start. You reply affirmatively, and a few seconds later, you hear the whir of the engines coming alive. After engine start, the canopy lowers, and the pilot performs various checks, including checks of the flight control system. After these checks, the pilot taxis the jet to the end of the runway. The pilot performs the pre-takeoff checks, then tells you to arm your ejection seat, and asks if you are ready for takeoff. You say you are ready to go. The pilot contacts the control tower and requests a takeoff clearance with an unrestricted climb. The tower grants both requests, and the pilot taxis the jet to the centerline of the runway.

The pilot pushes the throttles forward into full afterburner and the jet accelerates forward. In what seems like a very short distance, the F-18 is airborne. We will learn how to calculate the takeoff

Figure 1.7   F/A-18B *Hornet* front cockpit. (Source: *NASA*.)

distance and define the parameters that affect this calculation. We will also learn how to measure the takeoff distance in flight test. The pilot keeps the jet low to the ground, continuing to gain airspeed, and then pulls the jet up to what seems like a near vertical climb. You feel pushed down heavily into your seat. Looking at the g-meter, you see that you are pulling about 4 g's, making you feel four times heavier than your normal weight. We will see how the load factor affects the turn radius of this type of pull-up maneuver and we will calculate the radius of this vertical turn. We will discuss climb performance and define how to calculate the rate and angle of climb. We will also investigate climb performance from an energy perspective, where we account for kinetic and potential energies of the vehicle. We will make energy plots that define the performance capabilities of the aircraft, and we will discuss the flight test techniques used to quantify climb performance.

Looking at the altitude indication, you see that the numbers are increasing rapidly. At an altitude of about 14,000 ft (4270 m), the pilot rolls and pulls out of the vertical climb, so that you are upside down, and then rolls the aircraft upright to wings-level flight. Reducing the engine power, the pilot stabilizes the aircraft at a constant airspeed and altitude to let you catch your breath for a moment. Looking at the cockpit instruments, you see that you are at an airspeed of about 220 knots (253 mph, 407 km/h) and a Mach number of 0.6. We will discover that there are many different kinds of airspeeds and look at the reason for these different definitions. We will learn about Mach number, how it is defined, what it physically means, and why it is so important in high-speed aerodynamics. We will see that in this steady-state flight condition, there are four forces acting on the aircraft, which are in balance, and we will learn that this steady-state trim condition is an important starting point for most of the flight test techniques.

The pilot climbs the F-18 higher, leveling off at an altitude of 30,000 ft (9140 m). The airspeed indicates 350 knots (403 mph, 643.7 km/h), the outside air temperature (OAT) is a frigid −48°F (412°R, 228.7 K), and the Mach number reads about 0.6. We will learn about how the atmosphere changes, from sea level to high altitudes, and how this affects the calculations of aircraft performance. We will develop models of the atmosphere that will be used in our analyses. The pilot advances the throttles into full afterburner, and the F-18 accelerates in level flight. You watch the Mach indicator, waiting for it to indicate that you have broken the sound barrier and are flying at supersonic speed. We will discuss what is meant by the "sound barrier" and how it was "broken" for the first time. Looking out at the wing, you see something that looks like blurry light-and-dark lines or bands, dancing on the wing surface. You glance at the airspeed indication and it shows 530 knots (609.9 mph, 981.6 km/h). These are shock waves forming on the wing as the jet reaches transonic speeds. We will explain why these form on the wing and at what flight speeds. We will discuss the implications of these shock waves on the aerodynamics of the aircraft. We will learn that there are techniques to visualize these flow structures in flight.

The jet continues the level acceleration, and the Mach meter is indicating about Mach 0.96, when it jumps to Mach 1.1. We will explain the aerodynamic cause of this jump in the Mach indication. You have been on the ground and heard the sonic boom of a jet flying overhead at supersonic speed, yet you heard no sonic boom as your F-18 went supersonic. We will learn about sonic booms and some of the research that has been conducted to understand them. The F-18 continues to accelerate, reaching about Mach 1.3 in level flight. You are now flying at supersonic speed, traveling a distance of about one mile every four seconds. We will learn about high-speed supersonic flow and how it is fundamentally different from low-speed, subsonic flow. We will delve into discussions about even higher speed hypersonic flow, where the Mach number is greater than about five, and the flow physics is distinctly different.

The pilot pulls the throttles back, and the F-18 decelerates to subsonic speed. The pilot asks if you are ready to do some maneuvering. Of course, you say that you are more than ready. First, the pilot does some high-g, level turns, so that you can acclimatize to higher g-loadings. You successively fly level turns at 2g, then 4g, then 6g. With each successive turn, the high load factors push you down further into your seat. We will learn about level turn performance capabilities, the important parameters involved, and the associated flight test techniques. We will also learn about the flight envelope of the aircraft, as related to the airspeed and load factors that are within the aircraft's capabilities.

Now, the pilot asks if you want to fly, so you grab hold of the control stick and get a "feel" for the F-18. We will discuss why the handling qualities of an aircraft are important and how they are evaluated. The pilot tells you to do some rolls, so you push the stick full over to the left and the jet rolls around the horizon in a blur. We will discuss the aircraft stability in all three of its axes, and determine what characteristics determine whether its motion is stable or unstable about these axes. The pilot tells you to pull back the throttles to slow the jet down so that you can do some low-speed flight. You pull the control stick back, raising the nose of the aircraft, and increasing the angle-of-attack. You continue slowly pulling the nose up and, at an angle-of-attack of about 20°, the aircraft starts to gently rock from side-to-side. Pulling back a little more and the wing rock increases and the nose wanders a bit from side to side. At about 25° angle-of-attack, you cannot pull the stick back any further. Looking at the altimeter, you see that you are descending at a high rate. The pilot tells you to recover by returning the control stick to its center or neutral position. After you do this, the nose attitude decreases rapidly and the aircraft is back flying in level flight. We will learn about the aerodynamics associated with high angle-of-attack flight and stall. We will also discuss the aerodynamics and issues involved with aircraft spins. The various ground and flight test techniques to learn about stalls and spins will be covered.

It is time to head back to the airport. The pilot takes the flight controls, rolls the F-18 inverted, and then pulls down in a maneuver called a split-S. As the aircraft is coming through the vertical, you have a great view of the ground below, as shown in Figure 1.8. While this is a fun maneuver to lose altitude quickly, we will see that it can also be used to obtain aerodynamic data about the aircraft. The pilot enters the landing pattern, lowers the landing gear, and slows for the landing approach. Similar to takeoff performance, we will discuss the important parameters, associated with landing performance, and determine how to calculate the landing distance. We will see how the type of runway surface and other factors affects this distance. The F-18 touches down and rolls

Figure 1.8 F-18 familiarization flight, view from the aft cockpit, on the backside of a split-S. (Source: *Courtesy of NASA/Jim Ross.*)

to a stop on the runway. You have had a successful familiarization flight during which we have identified many areas to be discussed and explored in the chapters ahead.

1.2 Aircraft

In the broadest sense, the term "aircraft" refers to all types of vehicles that fly within our Earth's sensible atmosphere. The Federal Aviation Administration (FAA), in its Federal Aviation Regulations [6], defines an *aircraft* as "a device that is used or intended to be used for flight in the air". Aircraft support their weight with the force derived from either static or dynamic sources. For example, a lighter-than-air balloon supports its weight with static buoyancy, while a heavier-than-air airplane generates aerodynamic lift, which balances its weight, due to the dynamic reaction of air flowing over its wings.

1.2.1 Classification of Aircraft

There are many different types of aircraft, and a wide variety of ways that one could classify these different types. We could classify the different types of aircraft based on their geometric configuration, the type of propulsion, the mission or function, or other factors. Perhaps, a reasonable first distinction that we can make is between aircraft that are lighter-than-air and those that are heavier-than-air. A classification of aircraft, based on this starting point, is shown in Figure 1.9.

Lighter-than-air aircraft include airships and balloons. We can further subdivide heavier-than-aircraft into *powered* and *unpowered* aircraft, that is, aircraft with and without one or more propulsive devices or engines. Unpowered, heavier-than-air aircraft include gliders or sailplanes. Powered, heavier-than-air aircraft can be subdivided into *airplanes, rotorcraft,* and *ornithopters,* where the distinction between these different types of aircraft is based on their type of lift production. Airplanes have a *fixed wing,* which produces lift due to the air flowing over it. Rotorcraft encompass all heavier-than-air aircraft that generate lift from rotating wings or spinning *rotor blades.* Rotorcraft can be further divided into *autogyro* and *helicopter.* The autogyro has unpowered, free-spinning rotor blades, which require forward motion for lift production, whereas the helicopter has powered rotors that can produce lift even without forward speed. Ornithopters use *flapping wings* to generate both lift and thrust, similar to a bird. Many early would-be inventors

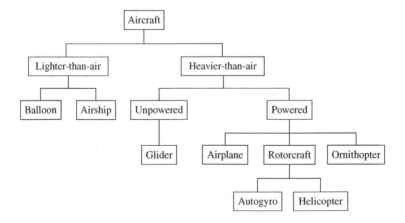

Figure 1.9 Classification of different types of aircraft.

of the first heavier-than-air airplane attempted to fly this type of flapping wing machine, but without success. We generally follow the classifications given in Figure 1.9 to describe aircraft in the following sections. We start our discussion of aircraft with the fixed-wing airplane.

1.2.2   The Airplane

Since most of us have grown up in a time where airplanes are commonplace, it is difficult to imagine that we do not know what an airplane is "supposed to look like". However, if we were living in the late 1800s, prior to the first successful flight of a heavier-than-air machine, we would probably be influenced by nature, and think that airplane flight should mimic bird flight. Some of the early aviation enthusiasts took this to the extreme, attempting to construct flyable ornithopters. Other early aviation pioneers made careful observations of bird flight, trying to understand nature's secrets about flight. There are now many variants on what an airplane looks like, but there are several common fundamental engineering aspects to heavier-than-air flight that have made it successful. We will see examples of this in the design and successful flight of the first airplane.

1.2.2.1   The First Airplane

At the beginning of this chapter, the iconic photograph of the first controlled, sustained flight of a heavier-than-air, powered airplane was presented. This first flight was the culmination of years of hard work by two brothers from Dayton, Ohio: Orville (1871–1948) and Wilbur (1867–1912) Wright. The brothers followed a logical and systematic approach in the design, construction, and flight test of their powered airplane. They critically reviewed much of the existing technical information and data relevant to aeronautical theory and aircraft design. In several important areas, the Wright brothers determined that the state-of-the-art information and data was not adequate or was incorrect, so they performed their own, independent analyses and tests to obtain what they needed. An example of this is the designs of the airfoil shapes for their wings and propellers, which were based on data that they collected using a wind tunnel of their own design. They also developed their own aircraft internal combustion engine, with the help of expert machinist, Charlie Taylor. The Wright brothers' determination to ensure that their airplane design was based on sound technical data was fundamental to their success.

The Wright brothers were also methodical and systematic in their approach to flying and flight testing. Between 1900 and 1903, they performed extensive flight testing with gliders of their own design. Starting first with unmanned, kite-like gliders (Figure 1.10), they systematically progressed to manned glider flights (Figure 1.11). The Wright brothers designed, built, and flew their first manned glider at Kitty Hawk, North Carolina, in 1900 with disappointing results. They test flew another glider design in 1901, but this second manned glider also flew poorly. It was not until their third glider design in 1903 that the Wright brothers were satisfied with how the glider flew. These glider design iterations systematically improved the performance and flying qualities of their unpowered airplanes and these lessons learned were incorporated into their 1903 powered airplane design.

The glider flying had another very important purpose, in addition to collecting flight data to improve their designs. By flying these many glider flights, the Wright brothers were learning *how to fly*. They gained extensive piloting experience in how to *control* their aircraft in the new three-dimensional world of flying. They understood that not only must a successful heavier-than-air vehicle lift its own weight, but it must also be *controllable*. They designed their aircraft to be controllable by the pilot in all three axes, with independent control effectors in pitch, roll, and yaw.

Figure 1.10   Unmanned, kite-like gliders from 1901 (left) and 1902 (right). (Source: *Wright Brothers, 1901 and 1902, US Library of Congress, PD-old-100.*)

Figure 1.11   Flight of a Wright brothers manned glider, October 24, 1902. Note the single vertical rudder on this glider. (Source: *O. Wright, 1902, US Library of Congress, PD-old-100.*)

Their airplane design had an elevator for pitch control, a rudder for yaw control, and for roll control, they used a scheme of *warping* or twisting of the wings.

The Wright brothers spent a considerable amount of time observing the flight of birds, and in particular the flights of buzzards. Their observations of bird flight gave them valuable insights into how to control a flying vehicle. They observed that as the birds soared and turned, the shape of their wings changed. Realizing that this wing twisting or warping was critical to the roll control of the maneuvering birds, the Wright brothers incorporated the wing warping concept into their airplane designs, and finally into the design of the first successful heavier-than-air airplane.

It is interesting to read the Wright brothers' description of their invention of a heavier-than-air flying machine in their original patent, as shown below. Note, that they make particular mention of the stability and control aspects of their airplane.

> Be it known that we, Orville Wright and Wilbur Wright, citizens of the United States, residing in the city of Dayton, county of Montgomery, and State of Ohio, have invented certain new and useful Improvements in Flying-Machines, of which the following is a specification. Our invention relates to that class of flying machines in which the weight is sustained by the reactions resulting when one or more aeroplanes are moved through the air edge-wise at a small angle of incidence, either by the application of mechanical power or by the utilization of the force of gravity. The objects of our invention are to provide means for maintaining or restoring the equilibrium or lateral balance of the apparatus, to provide means for guiding the machine both vertically and horizontally, and to provide a structure combining lightness, strength, convenience of construction, and certain other advantages which will hereinafter appear.

US patent 821,393, "Flying-Machine"
Application filed March 23, 1903
Patent granted May 22, 1906

The Wright brothers' successful, first powered airplane, the *Flyer I*, was a canard[3] configuration biplane, with a forward-mounted, all-moving horizontal, biplane elevator and an aft-mounted, vertical, twin rudder. (The all-moving nature of the elevator is clearly seen in the photograph of the *Flyer I*'s first flight, shown at the beginning of the chapter.). The airplane structure was a spruce and ash wooden framework, covered with finely woven muslin cotton fabric. The wing bracing wires were 15-gauge bicycle spoke wire. The airplane had a single, four-cylinder, gasoline-fueled piston engine, capable of producing about 12 horsepower (8.9 kW). Less than half a gallon of gasoline fuel was carried onboard the airplane. There was no engine throttle, the pilot could only open or close the fuel line that supplied the engine. The engine drove two contra-rotating, pusher propellers through a chain-drive transmission system. The propellers rotated at an average speed of about 350 revolutions per minute (rpm). The 170 lb (77 kg) engine was mounted on the right wing. To counterbalance the engine weight, the pilot was placed on the left wing. Since the typical pilot weight of about 145 lb (66 kg) was less than the engine weight, the right wing was about 4″ (10 cm) longer than the left.

Unusual by today's standards, the pilot lay prone on his stomach, with his hips in a padded wooden cradle, facing towards the front-mounted elevator. The wing-warping roll control and rudder-deflection yaw control were interconnected, such that sliding of the hip cradle sideways caused the wings to warp and the rudders to deflect. A wooden lever in the pilot's left hand controlled the aircraft pitch by changing both the angle of the elevator and the camber or shape of the elevator airfoil section. If the pilot pulled back on the lever, the elevator angle and camber were increased, thereby increasing its lift. If the pilot pushed the lever forward, the elevator angle and camber were decreased, resulting in less elevator lift. (Airfoil camber is discussed in Chapter 3.)

---

[3] The word *canard* is literally translated from French as "duck". It is speculated that the aeronautical usage came from the French public's comparison of a 1906 airplane, designed and flown by Brazilian aviation pioneer Alberto Santos-Dumont, to a duck. This 1906 airplane, named the *No. 14-bis*, was a biplane with a forward-mounted elevator. Santos-Dumont first flew the *No. 14-bis* on 13 September 1906, but this powered hop was only 23 feet (7 m) in distance. He is credited with flying the first public flight of a heavier-than-air airplane in Europe, when he flew the *No. 14-bis* on 23 October 1906, traveling a distance of about 200 feet (61 m).

Table 1.2   Selected specifications of the 1903 Wright *Flyer I*.

Item	Specification
Primary function	First heavier-than-air flying machine
Manufacturer	Orville and Wilbur Wright, Dayton, Ohio
First flight	17 December 1903
Crew	One pilot
Powerplant	In-line, 4-cylinder, water-cooled piston engine
Engine power	12 hp (8.9 kW) at 1020 rpm
Fuel capacity	0.2 gal (0.65 l) of gasoline
Propellers	Two 2-bladed, 8 ft (2.4 m) diameter
Empty weight	605 lb (274 kg)
Gross weight	750 lb (341 kg)
Length	21 ft 1 in (6.43 m)
Height	9 ft 4 in (2.8 m)
Wingspan	40 ft 4 in (12.3 m)
Wing area	510 ft^2 (47.4 m^2) (upper and lower wings)
Wing loading	1.47 lb/ft^2 (7.18 kg$_f$/m^2)
Maximum speed	30 mph (48.3 km/h)
Stall speed	22 mph (35 km/h)
Ceiling	30 ft (9.0 m)

The *Flyer I* used a 60 ft (18.3 m) launch rail for takeoff. The aircraft was restrained, sitting on the rail, until the pilot was ready for takeoff. He then released the restraining rope and the aircraft started its takeoff roll along the rail, riding on two modified bicycle wheel hubs. The aircraft had wooden skids for landing on the sandy ground. The *Flyer I* had a maximum airspeed of about 30 mph (48 km/h) and a maximum altitude of about 30 ft (9.0 m). Selected specifications of the Wright *Flyer I* are given in Table 1.2.

After winning a coin toss, Wilbur Wright attempted the first flight of the *Flyer I* on 14 December 1903. The launch rail was placed on an incline, giving the aircraft a downhill, gravity-assisted takeoff roll. Taking off in a light wind, Wilbur pulled the *Flyer I* off the launch rail, but almost immediately stalled the aircraft, causing it to return to earth in about three seconds. This "powered hop", with a gravity-assisted takeoff, could not be considered a first, controlled flight of a heavier-than-air airplane. The aircraft sustained some minor damage, which took three days to repair.

On 17 December 1903, it was Orville's turn to attempt the first flight. Since the winds were blowing at more than 20 mph (32.2 km/h), the launch rail was placed on level ground and pointed into the wind. At 10:35 am, Orville Wright made the first controlled, powered flight in a heavier-than-air airplane, with the flight lasting about 12 seconds, landing 120 ft (37 m) from the point of takeoff. The Wright brothers made four flights that day, with the final flight lasting almost a full minute. A summary of the initial flights of the *Flyer I* on 14 and 17 December 1903 is given in Table 1.3. After the successful flights of 17 December, the Wright brothers sent a telegram to their father, telling him about their accomplishment (Figure 1.12). Soon after the fourth landing, a gust of wind picked up the *Flyer I* and it tumbled end-over-end across the rough and sandy terrain. The *Flyer I* was destroyed and never flew again. Quite fittingly, a part of the *Flyer I* would soar again, when a piece of its wing fabric and a piece of wood from one of its propellers were carried inside a spacesuit pocket of Neil Armstrong when he stepped onto the surface of the moon on 20 July 1969.

Table 1.3   Wright brothers' flights of 14 and 17 December 1903.

Flight No.	Date	Flight Time	Ground Distance	Pilot
1	14 Dec	3 sec	112 ft (34.1 m)	Wilbur
2	17 Dec	12 sec	120 ft (36.6 m)	Orville
3	17 Dec	13 sec	175 ft (53.3 m)	Wilbur
4	17 Dec	15 sec	200 ft (61.0 m)	Orville
5	17 Dec	59 sec	852 ft (260 m)	Wilbur

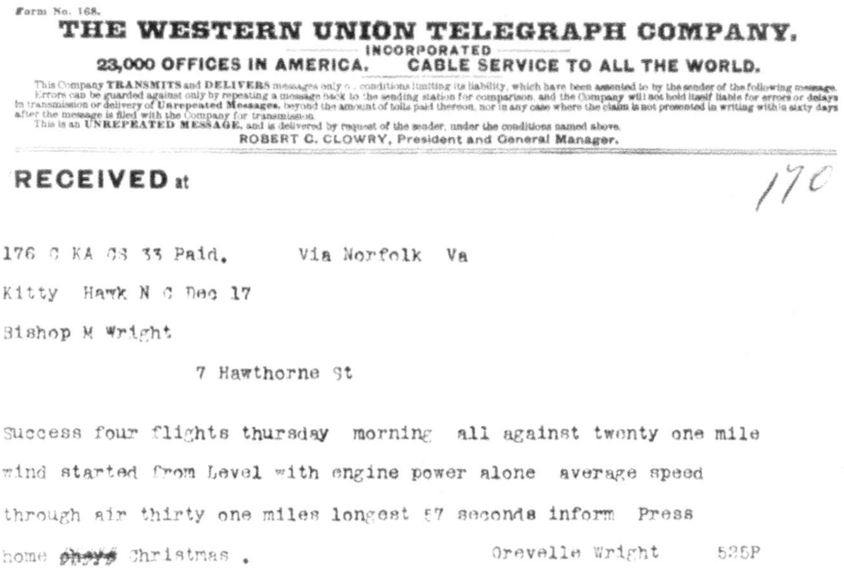

Form No. 168.

THE WESTERN UNION TELEGRAPH COMPANY.
INCORPORATED
23,000 OFFICES IN AMERICA.   CABLE SERVICE TO ALL THE WORLD.

This Company TRANSMITS and DELIVERS messages only n . conditions limiting its liability, which have been assented to by the sender of the following message.
Errors can be guarded against only by repeating a message back to the sending station for comparison, and the Company will not hold itself liable for errors or delays
in transmission or delivery of Unrepeated Messages, beyond the amount of tolls paid thereon, nor in any case where the claim is not presented in writing within sixty days
after the message is filed with the Company for transmission.
This is an UNREPEATED MESSAGE, and is delivered by request of the sender, under the conditions named above.
ROBERT C. CLOWRY, President and General Manager.

RECEIVED at   *170*

176 C KA GS 33 Paid.   Via Norfolk   Va

Kitty Hawk N C Dec 17

Bishop M Wright

            7 Hawthorne St

Success four flights thursday   morning   all against twenty one mile

wind started from Level with engine power alone   average speed

through air thirty one miles longest 57 seconds inform   Press

home ~~phays~~ Christmas .   Orevelle Wright   525P

Figure 1.12   Telegram from Orville Wright on 17 December 1903 after a successful day of flying. The stated speed through the air of 31 mph is the sum of the ground speed and wind speed. (Source: *PD-old-100.*)

1.2.2.2   Parts of an Airplane

In this section, the major parts of a fixed-wing airplane are described. There are many different aircraft configurations, as discussed in the next section. For our present purpose, we reference a somewhat standard aircraft configuration, with a single fuselage, a single wing attached to the fuselage, podded engines mounted underneath the wings, and horizontal and vertical tail surfaces mounted to the fuselage, aft of the wing, as shown in Figure 1.13. This configuration is in wide use today for commercial, military, and general aviation applications. The following discussion is generally applicable to other aircraft configurations, discussed in the next section. The major components of an airplane are the *fuselage*, *main wing*, *empennage*, *engines*, and *landing gear*. The fuselage contains the cockpit, passenger, and cargo compartments. The main wing extends from either side of the fuselage and often has integral fuel tanks within it. The *empennage*[4] is the tail area of the airplane, comprising the horizontal and vertical *stabilizers* and the associated moving control surfaces: the *elevators* and *rudders*, respectively. If the airplane is a powered airplane, there is one or

---

[4] The word *empennage* comes from the French *empenner*, to feather an arrow, where *empennage* refers to the feathers of an arrow.

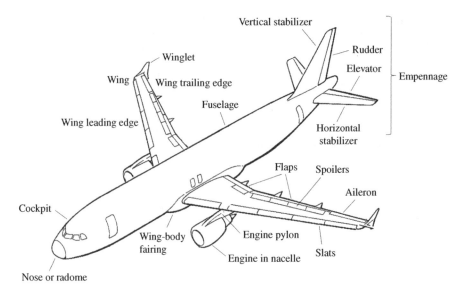

Figure 1.13   Parts of the conventional configuration airplane.

more wing or fuselage-mounted engines. The powerplant may be a reciprocating-engine–propeller combination or a jet engine. The engines may be *podded*, with the engine pods or *nacelles* mounted above or below the wings or on the sides of the fuselage. The engines may be *buried* in the fuselage, with an inlet or intake opening towards the front of the fuselage and exhaust openings at the aft end. The landing gear is composed of wheels with tires attached to struts, extending from the fuselage, wings, or engine pods. Often, the landing gear configuration consists of two main gear assemblies under the wings and a nose gear at the front of the fuselage, although other configurations are possible.

The elevators and the rudder on the empennage, and the ailerons on the wings comprise the primary flight control system. Each of these control system surfaces provides an incremental aerodynamic force that creates a moment to rotate the aircraft about its center of gravity (CG) in the desired direction. As shown in Figure 1.14, these control surfaces enable rotation of the airplane in three dimensions, where the elevator, ailerons, and rudder provide pitch, roll, and yaw rotations, respectively. Elevators are flap-like devices located at the trailing edges of the horizontal stabilizers. Some aircraft, typically military fighter aircraft, have all-moving horizontal stabilizers, called *stabilators* or *stabs*, instead of a combination of stabilizers and elevators.

The ailerons on the left and right wings deflect in opposite directions; that is, when the right aileron deflects upward, the left aileron deflects downward and vice versa. The downward deflected aileron results in additional lift on one side of the wing, while the upward deflected aileron results in decreased lift on the other side of the wing, creating the rolling moment. The additional lift produced by the downward deflected aileron also results in additional drag. This additional drag produces a yawing moment in a direction opposite or *adverse* to the desired direction of roll, and therefore is called *adverse yaw*. To counter this adverse yaw, the rudder is deflected to produce an opposing yawing moment, resulting in what is termed a *coordinated turn*.

High-speed aircraft also have secondary or auxiliary flight controls, which include devices on the wings called flaps, slats, and spoilers. Flaps are high-lift devices, located at the inboard wing trailing edge sections. When deflected or lowered, the flaps provide increased lift at lower airspeeds, enabling steeper landing approach glide paths without an increase in the approach airspeed. Slats,

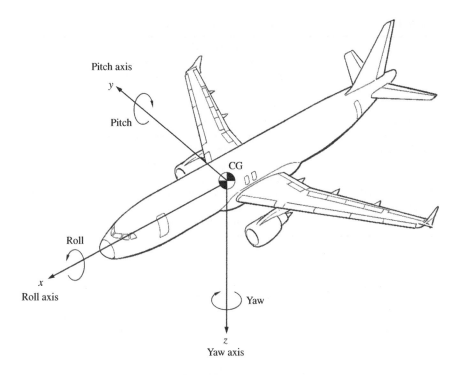

Figure 1.14   Airplane axes and rotations.

which are extended from the wing leading edge, are also high-lift devices that increase the wing lift at low speeds. There are several different types of wing flaps and slats, of varying mechanical complexity and aerodynamic effectiveness, which are discussed in Chapter 3. Spoilers, which extend upward from the wing upper surface, reduce or "spoil" the lift, to assist the airplane in slowing down and descending. They are also deployed after landing, to "dump" the wing lift and transfer the airplane's weight from the wings to the landing gear, which improves braking. Spoilers can also be used as a means of airplane roll control, when deployed differentially (extending from one wing and not the other).

1.2.2.3   Airplane Configurations

Airplanes come in all shapes and sizes. Usually, the configuration of an airplane is driven, or at least strongly influenced, by its mission requirements. For example, a commercial airliner has a large fuselage cabin area due to the requirement to transport passengers. A military fighter jet may have a highly swept wing to allow it to fly supersonically (we will see why this is so in Chapter 3.). A utility airplane that must be able to take off and land on snow might have skis for landing gear. These are a few examples of the types of aircraft configurations that may be driven by the mission requirements. It may be possible to satisfy the mission requirements with a variety of design solutions, limited only by the imagination and creativity of the airplane designer, and influenced by advancements in technology. A listing of possible airplane configurations for different components is given in Table 1.4. This listing is not meant to be exhaustive, but rather to illustrate the many possibilities in airplane designs. We briefly discuss several of these configuration options, citing real airplane design examples along the way, to better appreciate the possibilities.

Table 1.4   Sampling of possible airplane configurations.

Area	Possible airplane configurations
Fuselage type	Single fuselage, twin fuselage, twin boom
Number of wings	Monoplane, biplane, triplane
Wing location	Low-wing, mid-wing, high-wing
Wing type	Straight, aft-swept, forward-swept
Horizontal tail	Aft-mounted, forward-mounted (canard), tailless
Vertical tail	Single or twin vertical fin
Propulsion	Reciprocating piston, gas turbine (jet), rocket
Number of engines	Single or multi-engine
Engine(s) location	Above or below wing, fuselage side-mounted, internal
Landing gear type	Wheel, skid, float, ski
Landing gear	Tricycle, tail wheel, bicycle

Figure 1.15   North American F-82 *Twin Mustang* twin fuselage airplane. Note that there is a pilot in each fuselage cockpit. (Source: *US Air Force.*)

It is quite common for airplanes to have a single fuselage, whereas twin fuselage airplane designs are somewhat rare. A twin fuselage aircraft may offer some advantages for some applications. The twin fuselage airplane may have reduced design and development time and costs, if an existing single-fuselage airplane can be used as a baseline. This was the case for the North American F-82 *Twin Mustang*, developed near the end of World War II (see photo in Figure 1.15, drawing in Figure 1.16). Based on the single-fuselage XP-51 Mustang (see Figure 3.72), the F-82 was designed as a very long range fighter escort aircraft, with a nominal range of over 2000 miles (3200 km). The F-82 twin fuselages were from the single-fuselage P-51, which was stretched by 57″ (1.45 m), allowing for the installation of additional fuel tanks. Both cockpits were retained from the single-fuselage airplanes, so that a pilot in either cockpit could fly the airplane, which was advantageous for very long duration flights. The F-82 saw combat during the Korean War,

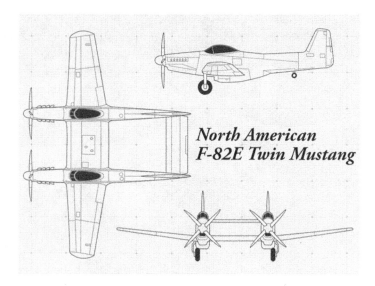

Figure 1.16 Multiple-view drawing of North American F-82 *Twin Mustang*. (Source: *NASA*.)

being the first fighter to shoot down a North Korean aircraft. The F-82 *Twin Mustang* still holds the record for the longest, non-stop flight by a propeller-driven fighter airplane, when it flew from Hawaii to New York, a distance of 5051 miles (8128 km), in 14 hours 32 minutes on 27 February 1947.

The twin fuselage configuration has found an application for airplanes that carry a large, center-line payload, such as the Virgin Galactic *White Knight Two*, which carries the *Spaceship Two* (see photo in Figure 1.17, drawing in Figure 1.18). The twin-fuselage *White Knight Two* is the first stage of a two-stage space launch system, with the *Spaceship Two* being the second stage. Only the right fuselage of the *White Knight Two* is configured to carry pilots and passengers, but, conceivably, the left fuselage could be designed to do so also. All three fuselages, the two *White Knight Two* and single *Spaceship Two* fuselages, are similar in design. This is an interesting design philosophy, whereby the *White Knight Two* is configured to be flown like the *Spaceship Two*, with a similar cockpit arrangement, equipment, and pilot sight picture. This allows for training and proficiency flying in the *White Knight Two* airplane which simulates, at least, the glide, approach, and landing phases of the *Spaceship Two*.

Similar to the twin fuselage configuration, an airplane may have twin longitudinal *booms* that extend from the main wing to the tail. The twin boom configuration may be advantageous for powerplant integration or for ease of access to aft fuselage cargo doors. The twin booms also provide additional volume for carrying fuel or equipment. The Cessna 337 *Skymaster* is an example of a twin-boom, twin-engine airplane that has been used as a general aviation and military utility aircraft (see photo in Figure 1.19, drawing in Figure 1.20). The twin booms allow both engines to be mounted on the fuselage centerline, with one in a *puller* or *tractor* configuration (forward-mounted engine) and the other in a *pusher* configuration (aft-mounted engine). An advantage of having both engines along the airplane centerline, versus mounted on either side of the fuselage, is that lateral-directional control is not degraded in the event of an engine failure, i.e. there is no yawing tendency with the power loss of one engine.

An airplane with tailwheel landing gear, also sometimes called *conventional landing gear*, is the Extra 300 airplane (see photo in Figure 1.21, drawing in Figure 1.22). The Extra 300 is a two-place,

Figure 1.17 Virgin Galactic *White Knight Two* and *Spaceship Two*. (Source: © *Virgin Galactic/Mark Greenberg, "SS2 and VMS Eve" https://en.wikipedia.org/wiki/File:SS2_and_VMS_Eve.jpg, CC-BY-SA-3.0. License at https://creativecommons.org/licenses/by-sa/3.0/legalcode.*)

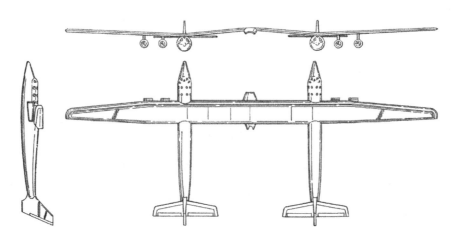

Figure 1.18 Three-view drawing of the Virgin Galactic *White Knight Two* (*Spaceship Two* not attached). (Source: *US Design Patent D612,719 S1, US Patent and Trademark Office, July 25, 2008.*)

single engine, high-performance, aerobatic, general aviation airplane with an all-composite, carbon fiber main wing. The tailwheel configuration is needed to provide ground clearance for the large-diameter propeller at the front of the airplane. The wing is attached to the middle of the fuselage, hence, it is termed a *mid-wing* configuration. The North American *Twin Mustang* is a *low-wing* monoplane and the Cessna *Skymaster* is a *high-wing* monoplane, where the main wing is attached to the bottom and top of the fuselage, respectively.

An example of a forward-swept wing configuration is the Grumman X-29 experimental, supersonic research aircraft (see photo in Figure 1.23, drawing in Figure 1.24). The X-29

Figure 1.19 Cessna 337 *Skymaster* twin-engine airplane with twin booms. (Source: © *User: Kogo, "Cessna Skymaster O-2" https://en.wikipedia.org/wiki/File:Cessna_Skymaster_O-2_5.jpg, GFDL 1.2. License at https://commons.wikimedia.org/wiki/Commons:GNU_Free_Documentation_License,_version_1.2.*)

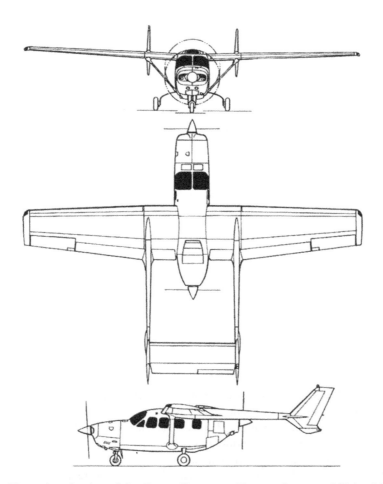

Figure 1.20 Three-view drawing of the Cessna *Skymaster*. (Source: *Courtesy of Richard Ferriere, with permission.*)

Figure 1.21   Extra 300 single-engine, mid-wing, tailwheel airplane. (Source: *Courtesy of the author.*)

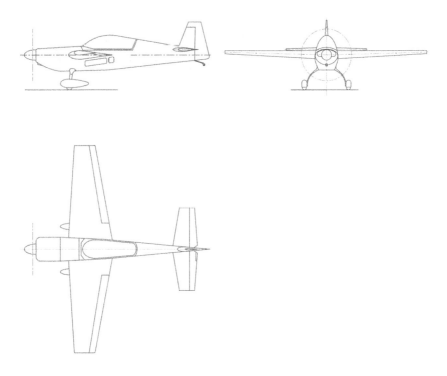

Figure 1.22   Three-view drawing of Extra 300. (Source: *Courtesy of Extra Aircraft, Germany, with permission.*)

investigated forward-swept wing maneuverability and other advanced technologies. Two X-29 aircraft were built, with test flights conducted by NASA and the US Air Force. The single-seat X-29 had a forward-swept main wing and trapezoidal-shaped canard surfaces forward of the wing. Forward-swept wings are susceptible to divergent aeroelastic twisting, so the X-29 wing was fabricated with advanced composite materials, which could provide the required structural stiffness with low weight. The forward-swept wing X-29 was inherently unstable, requiring a

Figure 1.23   Grumman X-29 forward-swept wing research aircraft. (Source: *NASA*.)

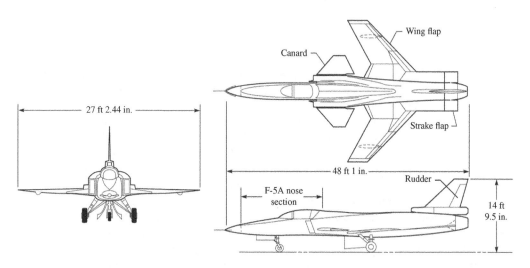

Figure 1.24   Three-view drawing of the Grumman X-29 forward-swept wing aircraft. (Source: *NASA*.)

state-of-the-art "fly-by-wire" flight control system, where the aircraft was constantly flown and stabilized by computers. A single General Electric F404 turbofan jet engine powered the X-29, enabling a top speed of Mach 1.8 at 33,000 ft (10,000 m). The first flight of the X-29 was on 14 December 1984. The two X-29 aircraft completed 422 research test flights over a period from 1984 to 1991.

Most of the airplane configurations that we have discussed so far are single-wing or *monoplane* configurations. An example of an airplane with two main wings, a *biplane*, is the Russian Antonov An-2 *Colt* (see photo in Figure 1.25, drawing in Figure 1.26). The two wings need not have the same dimensions. In fact, a biplane's wings can differ in size, airfoil shape, wing sweep, or other characteristics. The An-2 is a large, rugged, single-engine aircraft designed to perform a variety of utility

Figure 1.25 Antonov An-2 *Colt* single-engine, biplane with ski landing gear. (Source: © *Sergey Ryabt-sev, "Randonezh Antonov An-2R" https://en.wikipedia.org/wiki/File:Antonov_An-2R_on_ski_Ryabtsev .jpg, GFDL-1.2, License at https://commons.wikimedia.org/wiki/Commons:GNU_Free_Documentation_ License,_version_1.2.*)

tasks such as cargo hauling, crop dusting, water bombing (for fighting forest fires), parachute drop, glider towing, or military troop or civilian passenger transport. Designed by the Antonov Design Bureau, Kiev, Ukraine in 1946, the An-2 was produced for the next 45 years. Because of its sturdy construction, relatively simple systems, low speed capabilities, and large payload capacity, the An-2 has become a popular "bush" plane for flying people and cargo in and out of remote, unimproved areas. Known as a short takeoff and landing, or STOL, airplane, the An-2 can takeoff in less than about 600 ft (180 m) and, due to its extremely low stall speed of less than 30 mph (48 km/h), it needs only about 700 ft (210 m) to land. The An-2 shown in Figure 1.25 has conventional landing gear, but with skis for operation on snow-covered terrain replacing the tires.

All of the airplane configurations that we have discussed so far have distinct fuselage, wing, and tail components. The *flying wing* is a *tailless* airplane configuration, where the fuselage and wing are blended together. The flying wing concept is not new. Flying wing prototype aircraft were built and flown as early as the 1940s. Several flying wing designs were also built and flown in the early 20th century. The Northrop B-2 *Spirit* "stealth bomber" is a modern example of a flying wing airplane (see photo in Figure 1.27, drawing in Figure 1.28). Its two jet engines are "buried" in the blended wing-fuselage to mask their heat signature, enhancing its stealth capability. While there are significant aerodynamic advantages, especially in terms of reduced drag, for a tailless flying wing configuration, the stability and control issues require some special considerations. The advent of "fly-by-wire" flight control technology has made these design issues much easier to manage. We discuss the interesting stability and control considerations of flying wings further in Chapter 6.

1.2.3 Rotorcraft: the Helicopter

Thus far, we have discussed only *fixed-wing* aircraft. We now discuss *rotary-wing* aircraft or *rotor-craft*, where the lift-producing surfaces are rotating. The rotating wings, more properly called *rotor blades*, are attached to the *rotor hub* at the top of a *rotor mast* above the aircraft. The rotor

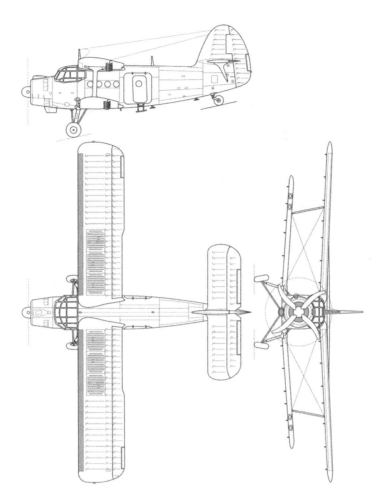

Figure 1.26 Multiple-view drawing of the Antonov An-2 *Colt*. (Source: *Kaboldy, "Antonov An-2 3-View" https://en.wikipedia.org/wiki/File:Antonov_An-2_3view.svg, CC-BY-SA-3.0, https://creativecommons.org/ licenses/by-sa/3.0/legalcode*.)

blades, hub, and mast collectively are simply called the *rotor*. Rotorcraft include helicopters and autogyros. Helicopters are heavier-than-air flying machines that can take off and land vertically, translate in any direction, including backwards, and remain stationary in the air or *hover*. They have engine-driven rotor blades that produce both lift and thrust. The lift produced does not depend on the forward speed, so the helicopter can take off and land with zero forward velocity. The rotor can still produce lift if the engine is not running, as long as there is forward speed to keep the rotor blades spinning or auto-rotating. In this manner, the helicopter can glide like a fixed-wing airplane.

Two models of helicopters are shown in Figure 1.29, the Sikorsky UH-60 *Black Hawk* helicopter and the Bell OH-58 *Kiowa* light helicopter. The UH-60 is a twin-engine, single-rotor, 4-bladed military helicopter, designed for utility and transport operations. It carries a crew of two, and up to 11 passengers. The rotor of the UH-60 has a diameter of 53 ft 8 in (16.36 m). The UH-60 has a cruise speed of about 170 mph (294 km/h) and can climb to a maximum altitude of about 20,000 ft (6100 m). A three-view drawing of the UH-60 is shown in Figure 1.30. The Bell OH-58 *Kiowa* is a single-rotor, 2-bladed, military helicopter, designed for light utility and transport. The *Kiowa* is

Figure 1.27   Northrop Grumman B-2 *Spirit* flying wing airplane. (Source: *US Air Force.*)

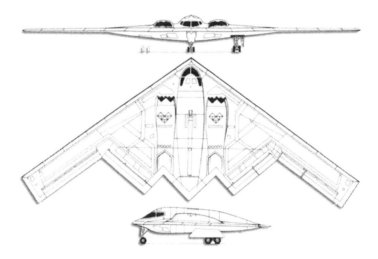

Figure 1.28   Three-view drawing of the Northrop Grumman B-2 *Spirit*. (Source: *PD-USGov-Military.*)

the military version of the popular Model 206A *Jet Ranger* civilian helicopter. The OH-58 carries a crew of one or two pilots with the civilian version capable of carrying up to four passengers. The rotor diameter of the OH-58 is 35 ft (10.7 m). The OH-58 has a cruise speed of 127 mph (204 km/h) and a maximum ceiling of about 15,000 ft (4600 m).

Autogyros, also known as gyrocopters or gyroplanes, have unpowered, free-spinning rotor blades that require forward motion to produce lift. Thrust is provided by an independent engine–propeller combination mounted in the fuselage as in a fixed-wing airplane. A short, fixed wing attached to the fuselage may also generate lift. The autogyro has many attributes of a helicopter, but since it requires forward motion to generate lift, it cannot take off and land vertically, fly backwards, or hover in still air. For completeness, we make the distinction between helicopters and autogyros, but our discussion focuses on the helicopter, as it is the predominant rotorcraft in use today.

Figure 1.29   Two models of helicopters, the twin-engine, Sikorsky UH-60 *Black Hawk* medium-lift heli-copter and the single-engine Bell OH-58 *Kiowa* light helicopter. (Source: *NASA*.)

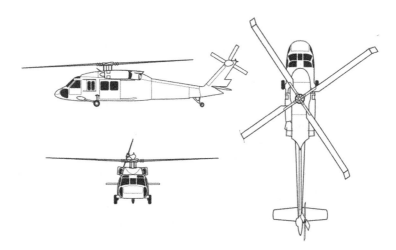

Figure 1.30   Three-view drawing of the Sikorsky UH-60 *Black Hawk* helicopter. (Source: © *User: Fox 52, "UH-60 Orthographical Image" https://en.wikipedia.org/wiki/File:UH-60_orthographical_image.svg, CC-BY-SA-4.0. License at https://creativecommons.org/licenses/by-sa/4.0/legalcode.*)

1.2.3.1   The First Rotorcraft

Early inspiration for the design of rotorcraft may have come from nature. There are several examples of rotating winged seeds in nature that glide through the air as a means of dispersion. These flying seeds or *samaras* (a fruit with a wing) have been the interest of past aeronautical enthusiasts or inventors and current aeronautical engineers. In 1808, aeronautical pioneer, Sir George Cayley (1773–1857) wrote about the sycamore seed, as follows.

I was much struck with the beautiful contrivance of the chat of the sycamore tree. It is an oval seed furnished with one thin wing, which one would at first imagine would not impede its fall but only guide the seed downward, like the feathers of an arrow. But it is so formed and balanced that it no sooner is blown from the tree than it instantly creates a rotative motion preserving the seed for the centre, and the … wing keeps it nearly horizontal, meeting the air in a very small angle like the bird's wing.

The aerodynamics of this natural rotating wing has been studied extensively, including through the application of modern computational techniques, attempting to unlock the secrets of another example of nature's optimization of flight.

The notion of a flying, rotating wing dates back to an ancient Chinese rotating toy, which was essentially a feather acting as a propeller, attached to the end of a stick. By applying rotation to the stick between the palms of one's hands and releasing the toy, the rotating feather propeller would generate lift, making the toy fly for a short time. In 1483, Leonardo da Vinci conceived of a human-carrying rotorcraft he called an "aerial screw". The da Vinci aerial screw concept did not address several critical issues in the design of a practical rotorcraft, which would not be solved for many centuries.

One of these issues was the development of a propulsion system, to rotate the blades with sufficient power and yet light enough in weight to lift the rotorcraft into the air. This propulsion issue was shared by the designers of heavier-than-air fixed-wing airplanes and would not be solved until the early 20th century, with the advent of the internal combustion engine. Another issue, unique to rotary-wing aircraft, had to do with the reaction torque developed by a rotating wing. The torque imparted by the engine to the rotor blade shaft also results in a reaction torque, which tends to want to rotate the vehicle in the opposite direction of the blade motion. This reaction-torque must be countered by some means, so that the vehicle does not rotate when the rotor blades are spinning. Other issues to be solved included the high vibration environment due to the large, spinning rotor, which can lead to mechanical failure and structural metal fatigue. Many of these issues are still being actively worked on today to improve helicopter designs.

During the early 20th century, there were many attempts at building and flying a vehicle capable of vertical flight. On 13 November 1907, about four years after the Wright brothers' first successful flight of a heavier-than-air fixed-wing airplane, a French bicycle maker, Paul Cornu (1881–1944), flew a helicopter of his own design to a vertical height of 1 ft (30 cm) and hovered for 20 seconds, making this the first free flight of a heavier-than-air rotorcraft. Cornu's helicopter had two 20 ft (6.1 m) diameter rotors with large low aspect ratio blades mounted on spinning spoked wheels, as shown in Figure 1.31. The opposite rotation of the rotors, located at opposite ends of the vehicle, served to counter the reaction-torque. A 24 horsepower, gasoline-fueled internal combustion

Figure 1.31 Paul Cornu's rotorcraft. (Source: *PD-USGov.*)

engine, powered the rotors. In the several flights of the Cornu helicopter, it achieved a maximum vertical height of only about 6 ft (1.8 m), never rising above the region of aerodynamic ground effect, where there is increased lift and decreased drag. (Aerodynamic ground effect is discussed in Chapter 3.)

The first practical helicopter design is perhaps the experimental Vought Sikorsky VS-300 helicopter, designed by helicopter pioneer Igor Sikorsky. The first flight of the VS-300 was on 14 September 1939 in Stratford, Connecticut, piloted by Igor Sikorsky himself. While the first flight of the VS-300 was only a few vertical inches in the air and lasted only 10 seconds, this experimental helicopter prototype, and subsequent variants, would set rotorcraft speed and endurance records (Figure 1.32). The VS-30 had a three-bladed, 28 ft (8.5 m) diameter main rotor, powered by a 75 hp Lycoming engine, and weighed 1325 lb (601 kg). For the first time, the reaction-torque of the main rotor was countered using an anti-torque tail rotor, an additional spinning rotor mounted vertically at the end of the fuselage (Figure 1.33). The single main rotor coupled with an anti-torque tail rotor configuration became the predominant helicopter configuration, the predecessor of the modern-day helicopter.

1.2.3.2 The Helicopter

The major components of a typical, modern helicopter with a single main rotor and anti-torque tail rotor are shown in Figure 1.34. Most, if not all, of the major components are attached to or contained within the structural airframe, including the cockpit, passenger or cargo cabin, engine, fuel tanks, transmission, and landing gear. The landing gear may be skids, fixed or retractable wheels, or amphibious floats. The powerplant may be an internal combustion engine or a turboshaft engine. There may be a single engine or dual engines for additional power and redundancy. The main rotor, comprising the blades, hub, and mast, and the tail rotor are connected to the engine through the transmission, where gearboxes reduce the engine's rotational speed, allowing them to rotate at the required lower speed.

Helicopter main rotor systems are usually of a single or dual rotor configuration. As we have discussed, the single rotor configuration requires an anti-torque mechanism, such as a tail rotor. In a dual rotor system, the rotors spin in opposite directions, which cancels the rotor torque.

Figure 1.32   The VS-300 helicopter, piloted by its designer, Igor Sikorsky. (Source: *PD-USGov.*)

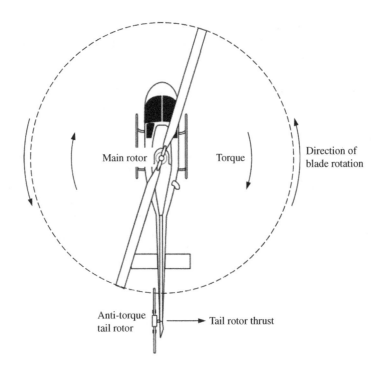

Figure 1.33   Helicopter with a single main rotor and anti-torque tail rotor.

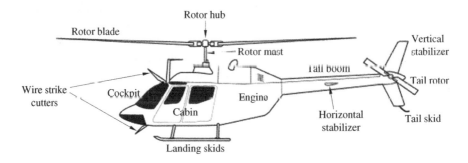

Figure 1.34   Components of the modern helicopter.

The Boeing CH-47 *Chinook*, shown in Figure 1.35, is an example of a twin engine, heavy-lift helicopter with dual tandem rotors.

The main rotor blades are attached to the top of the rotor mast at the rotor hub. A rotor system, whether single or dual, is classified as a *fully articulated*, *semi-rigid*, or *rigid* rotor system, based on the method of attachment of the blades to the hub and the way that the blades move relative to the rotor plane of rotation. Of the three rotor systems, a fully articulated rotor system has the most degrees of freedom for blade movement. With this system, each rotor blade can move independently in three directions relative to the plane of rotation: up or down, called *blade flap*, and fore or aft, called *blade lead or lag*, respectively, and in rotation about the blade spanwise axis, that is, a rotation that changes the blade pitch angle, called *blade feathering*. The blades are attached to the hub using three independent mechanical hinges, appropriately called the flapping hinge, the lead/lag hinge,

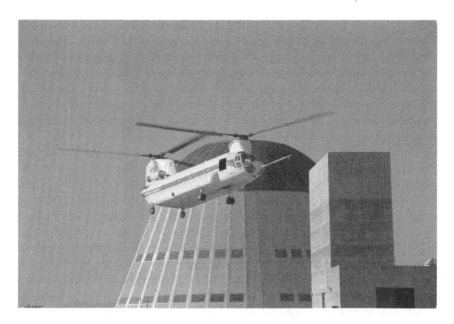

Figure 1.35   Boeing CH-47 *Chinook* twin engine, dual tandem rotor heavy-lift helicopter. (Source: *NASA.*)

and the feathering hinge. Fully articulated rotor systems are used on helicopters with more than two main rotor blades.

Blade flapping and lead/lag motion is needed to balance the unequal lift being produced across the rotor disk. In forward flight, the rotation of the blades results in an increase in lift for the rotor blade that is advancing into the relative wind and a decrease in lift for the retreating blade. Blade feathering controls the amount of lift that is produced by changing the blade pitch or angle-of-attack. Increasing or decreasing the blade pitch increases or decreases the lift, respectively.

With the semi-rigid rotor system, the rotor blades have two degrees of motion relative to the rotor plane of motion, flapping and feathering. The rotor blades are rigidly attached to the rotor hub, but the hub attachment to the mast is such that it can have a see-saw or teetering motion relative to the plane of rotation. This teetering motion allows the rotor blades to flap, but since the blades are rigidly attached to the hub, the blades on either side of the hub flap as a unit. This means that for a typical two-blade semi-rigid rotor system, when the blade on one side goes down, the blade on the opposite side goes up. Blade feathering is the same as in the fully articulated system, using a feathering hinge. Semi-rigid rotor systems are usually found on helicopters with two main rotor blades.

In the rigid rotor system, the rotor blades are rigidly attached to the hub and the hub is rigidly attached to the mast, such that the blades have a single degree of motion relative to the rotor plane of motion, that of feathering. Mechanically, the rigid system is much simpler than the other systems, since there are no flapping and lead/lag hinges and mechanisms. Any aerodynamically induced flapping and lead/lag motions of the blades must be absorbed by the blades and hub, making the structural design of these components more complex. A rigid rotor system may also have higher vibration characteristics than the other types of systems.

Returning to the single main rotor configuration, let us investigate other types of anti-torque devices. In addition to the conventional tail rotor, other types of anti-torque devices may be used, such as a Fenestron or NOTAR® system. The Fenestron design, also called a fantail, is essentially a tail rotor with multiple blades shrouded within a circular duct. While a conventional tail rotor may have two to five rotor blades, a fantail may have as many as 8–13 blades. The fantail blades are also shorter in length or span, and spin at a higher rotational speed than conventional tail rotor blades. The shrouded fantail acts like a ducted fan, which is more aerodynamically efficient than an exposed tail rotor. Vibration and noise are also reduced with the fantail. The shrouding has some safety advantages, protecting the rotor from striking foreign objects in flight, such as trees or power lines, and reducing risk to personnel on the ground. A disadvantage of the fantail is the added weight due to the structure around the rotor.

NOTAR® is an acronym for NO TAil Rotor. The NOTAR® system is based on a combination of an aerodynamic phenomenon, known as the Coanda effect, and direct jet thrust. A fan, located at the forward end of the tail boom, produces a low pressure, high volume flow of ambient air that is expelled through two longitudinal slots on the right side of the tail boom. These horizontal air jets create a low pressure area that causes the downwash flow from the main rotor to curve around the circular cross-section of the boom. This circulation control around the boom, created by the air jets, is known as the Coanda effect. The accelerated flow around the right side of the tail boom results in an aerodynamic lift force in a direction that counteracts the main rotor torque. In hovering flight, this circulation control system provides up to 60% of the required anti-torque. Additional anti-torque is provided by a rotating, direct jet thruster that is fed by the fan air in the boom. Vertical stabilizers provide additional directional control in forward flight. Advantages of the NOTAR® system include the elimination of tail rotor mechanisms and transmissions, and the safety benefit of not having a tail rotor with regards to tail strike.

Several desirable features of rotary-wing and fixed-wing aircraft are brought together in the *tilt-rotor* aircraft, such as the Boeing V-22 *Osprey* (Figure 1.36). The tilt-rotor aircraft combines the rotorcraft capabilities of vertical takeoff, hover, and landing with the benefits of a fixed-wing aircraft, such as improved speed, range, and fuel efficiency, as compared with a pure rotorcraft. The tilt-rotor has two counter-rotating main rotors or propellers that are mounted on engine nacelles at the ends of a short wing. The nacelles can be rotated in flight between horizontal and vertical

Figure 1.36   Boeing V-22 *Osprey* tilt-rotor aircraft. (Source: *US Air Force.*)

positions. With the nacelles in their vertical position, the tilt-rotor can operate like a helicopter with two counter-rotating main rotors. With the nacelles in the horizontal position, the tilt-rotor flies like a fixed-wing, twin-engine airplane with two large propellers. As can be seen in Figure 1.36, the 38 ft (11.6 m) diameter, rotating blades are a compromise between a helicopter and an airplane. With a cruising speed of about 240 knots (444 km/h), a maximum altitude of about 25,000 ft (7600 m), and a capability to takeoff vertically at a weight of about 53,000 pounds (24,040 kg), the V-22 tilt-rotor combines the benefits of rotary and fixed-wing aircraft.

1.2.4   Lighter-Than-Air Aircraft: Balloon and Airship

As the name implies, lighter-than-air vehicles are aircraft that utilize gases that are less dense than atmospheric air. The gas may be less dense because it is heated, as in a hot air balloon, or because it has an inherently lower density than air, such as helium or hydrogen in a gas balloon or airship. Lighter-than-air aircraft obtain their lift primarily from buoyancy, rather than from aerodynamic lift. We discuss two types of lighter-than-air aircraft, the *balloon* and the *airship*. The distinction between a balloon and an airship has to do with the ability to propel and steer the vehicle. A balloon does not have a propulsion system, while the airship has a means of propulsion, and is steerable.

Buoyancy is based on Archimedes' principle, which states that an object, submerged in a fluid, is acted upon by a buoyant force with a magnitude equal to the weight of the fluid displaced by the object, and in a direction that is opposite to the weight of the object. The fluid can be a liquid or a gas, so Archimedes' principle is applicable to a ship or submarine in the ocean or a balloon or airship in the air. Assuming that the fluid is air, the buoyancy force, F_b, can be written as

$$F_b = W_a = m_a g = \rho_a \mathcal{V} g \qquad (1.1)$$

where W_a, m_a, and ρ_a are the weight, mass, and density, respectively, of the air displaced by the object of volume, $\mathcal{V}$, and g is the acceleration due to gravity. Now imagine that we have a hollow object, with a volume $\mathcal{V}$, which we can fill with a substance of density, ρ_g. The weight of the substance in the object is given by

$$W_g = \rho_g \mathcal{V} g \qquad (1.2)$$

(For simplicity, we ignore the weight of the hollow object that contains the substance of density ρ_g.)

If we fill the object with a substance that has the same density as air, $\rho_g = \rho_a$, the weight of the object equals the buoyancy force, and the object remains stationary in the air, as shown in Figure 1.37a. If we fill the object with a substance that has a density greater than air, $\rho_g > \rho_a$, the object's weight is greater than the buoyancy force and the object sinks. If we fill the object with a substance that has a density less than air, $\rho_g < \rho_a$, the object's weight is less than the buoyancy force and the object rises. As common sense would dictate, if we fill the hollow object with lead, it sinks, and if we fill it with helium or hydrogen, the object rises. We could also fill the object with air, at a higher temperature than the external, ambient air, so that the air inside the object is at a lower density. This then is the fundamental physics behind the buoyancy of the balloon and airship.

1.2.4.1   The First Balloon

Balloons are perhaps the earliest form of manned flying vehicles. The first recorded, manned flight of a *hot air balloon* occurred on November 21, 1783 in Paris, France. The balloon, with aeronauts Jean Francis Pilatre de Rozier and the Marquis d'Arlandes onboard (a person who operates or

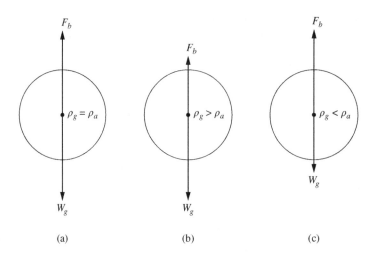

Figure 1.37 Buoyancy, (a) stationary, $\rho_g = \rho_a$, (b) sinking, $\rho_g > \rho_a$, and (c) rising, $\rho_g < \rho_a$.

travels in a balloon or airship is called an *aeronaut*), flew for about 25 minutes over the city of Paris, rising to an altitude of about 3000 ft (914 m) and covering a distance of about 5 miles (8.1 km). This was the first free flight by mankind in an aerial vehicle. The balloon was built by Joseph and Etienne Montgolfier of France, who were to play a major role in the future development of balloon flight. A firebox, suspended underneath an opening at the bottom of the balloon, held a fire that filled the balloon with hot air. The balloon aeronauts stood on a platform, encircling the bottom of the balloon, from which they could add fuel to and tend the fire in the firebox.

With a background in paper manufacturing, the Montgolfier brothers were supposedly inspired by seeing scraps of paper, in their paper mill, being lifted aloft by smoke from a fire. Based on these observations, they believed that the smoke was a new, undiscovered gas that was less dense than air, which they dubbed "Montgolfier gas". They believed that a thicker smoke contained more of this "Montgolfier gas", so they sometimes burned unusual materials, such as rotted meat and shoes, to produce as thick a smoke as possible for their balloons. They did not realize that the smoke was simply heated air and was therefore less dense than unheated air. The brothers used a trial-and-error method, rather than one based on an understanding of the physics, in developing their balloons.

An aspect of the Montgolfier brothers' balloon development, which was insightful and may have contributed to their success, was their incremental design and flight test approach. They started with the flight of a smaller scale, 10 m (32.8 ft) diameter, unmanned hot air balloon that was tethered to the ground, and built up to flights of larger, manned balloons. The Montgolfier brothers' flight test approach was also admirable from a risk reduction standpoint. Prior to risking a balloon flight with people onboard, they flew a balloon carrying three farm animals, a sheep (aptly named Montauciel, French for "climb to the heavens"), a duck, and a rooster, to assess the effects of balloon flight on living creatures. There was logic to their selection of these three particular animals. The sheep was thought to have a physiology that was similar to a human being, thus it was selected to assess the physiological effects of altitude. Since the duck was capable of flight at the balloon altitudes, it was used to assess any non-physiological effects of balloon flight. The rooster was a non-flight capable bird, so it was used to assess altitude effects in comparison with the duck. On 9 September 1783, the sheep, duck, and rooster made history as the first living creatures to fly in a balloon. Their balloon flight lasted about 8 minutes, ascending to an altitude of about 1500 ft (460 m) and landing safely about 2 miles (3.2 km) from their launch point.

The next incremental step was the flight of a 75 ft (22.9 m) tall, 55 ft (16.8 m) diameter, tethered balloon with a man onboard. On October 15, 1783, Etienne Montgolfier was the first person to ascend in a tethered balloon followed, later that day, by Jean Francis Pilatre de Rozier, who rode the tethered balloon to a height of about 80 ft (24.4 m), the length of the tethered line attached to the balloon. Just a little over a month later, the first free flight of a balloon was completed by de Rozier and the Marquis d'Arlandes in a Montgolfier balloon.

Ten days after the first manned hot air balloon flight, aeronauts Jacques Alexander Charles and Nicholas Louis Robert flew the first manned flight of a *gas balloon* on 1 December 1783, also in Paris, France. Charles and Robert ascended to an altitude of about 1800 ft (550 m), covered a distance of about 25 miles (40.2 km), and were airborne for about 2 hours. The rubber-coated, silk balloon was filled with flammable hydrogen gas, an attractive choice from a buoyancy standpoint, but a poor choice from a flight safety perspective. Hydrogen gas would be used in balloons (and airships) well into the 20th century – with many instances of catastrophic events due to its high flammability – until being replaced by helium gas. In fact, de Rozier, of hot air balloon fame, died when his hydrogen gas balloon exploded while attempting to cross the English Channel in 1785. De Rozier's balloon was a hybrid gas-and-hot air balloon, essentially a hot air balloon with an internal hydrogen gas chamber. Sadly, in addition to being one of the first persons to fly, de Rozier was also the first air crash fatality. The first manned balloon flight in the USA was in a hydrogen gas balloon, piloted by the Frenchman Jean-Pierre Blanchard on 9 January 1793. Blanchard's balloon lifted off from Philadelphia, Pennsylvania, climbed to an altitude of about 5800 ft (1770 m) and landed in New Jersey.

In addition to opening up a new era in flight, balloons also found military applications. Tethered balloons were used as military observation platforms by the French in the late 18th century and by the armies during the US Civil War. Balloons were also used for artillery spotting in World War I.

1.2.4.2   The Balloon

The two types of balloons that we have been discussing, the hot air balloon and the gas balloon, are different based on the source of the lighter-than-air substance that provides the buoyancy. As its name implies, the hot air balloon is filled with air at a higher temperature, hence, a lower density, than the external, ambient air. The gas balloon is filled with an unheated gas, with a lower density than air, such as hydrogen, helium, or ammonia.

Balloons do not have a means of propulsion, so they literally drift with the wind. By adjusting the balloon's buoyancy, the balloon pilot can cause the balloon to rise or sink, moving the balloon vertically into different wind currents and thereby having some, albeit limited, control of horizontal motion. Both types of balloons have a fabric *envelope* that is filled with the lifting gas, a *basket* or *payload* suspended underneath the envelope, and a means of adjusting the buoyancy in flight. The basket is used to carry people, while the payload could be any type of equipment or instrumentation that is carried aloft.

The major components of a conventional, modern hot air balloon are shown in Figure 1.38. The envelope of the modern hot air balloon is constructed of lightweight, synthetic fabric panels that are sewn together in banana peel shaped vertical rows, called gores. The fabric is structurally reinforced with horizontal and vertical load tapes. In a conventional hot air balloon, the envelope has a teardrop shape, but it can have a variety of other shapes. Hot air can be vented from the envelope, either through a deflation port located at the top of the envelope or through other vents in the side of the envelope. Venting of hot air is one means of buoyancy control for the balloon pilot. The envelope side vents can also be used to turn the balloon about its vertical axis, providing some control of the basket position relative to the direction of motion, which may be useful to the

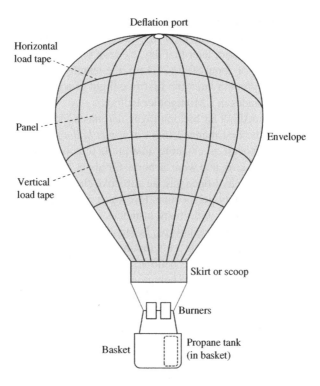

Figure 1.38   The hot air balloon.

pilot in landing the balloon. The burners, mounted beneath the envelope, are used to heat the air inside the envelope. Unlike the first manned balloon flight, which used damp straw, old rags, and rotting meat as fuel for their firebox, modern balloons use liquid propane, which is stored in tanks inside the basket. The opening at the bottom of the envelope, called the skirt or scoop, is coated with a fire resistant material to prevent the burner flames from igniting the envelope. By controlling the firing of the burner, the balloon pilot can control the temperature of the hot air in the envelope and hence the buoyancy of the balloon. The basket or gondola is suspended beneath the envelope using stainless steel or Kevlar composite cables. The basket is commonly made of wicker, metal, or fabric, covering a metal frame. Flight instruments and avionics, such as an altimeter, variometer or rate-of-climb indicator, radio, and transponder, are mounted in the basket. In the example problem below, we gain an appreciation for the size of a hot air balloon required to carry a reasonable weight, which includes the weight of the envelope, heating system, basket, aeronauts, and hot air inside the envelope.

The early hot air balloons had the obvious disadvantages of literally carrying a fire aloft and needing to carry the heavy load of firewood or other combustibles to fuel the fire. In fact, the first hot air balloon flight by de Rozier and d'Arlandes was cut short due to their concern that the balloon was starting to catch fire. Once balloon designers figured out how to adequately seal balloons to prevent the leakage of the buoyant gas, the gas balloon soon became preferred over the hot air balloon. However, the burners of modern-day hot air balloons are much more efficient and safer, making hot air balloons the current preference for sport ballooning.

The major components of a typical gas balloon are shown in Figure 1.39. Similar to a hot air balloon, the gas balloon has an envelope that is inflated with the buoyant gas. The gas balloon envelope is typically spherical in shape and made of a thin, gas-tight synthetic material. Typical

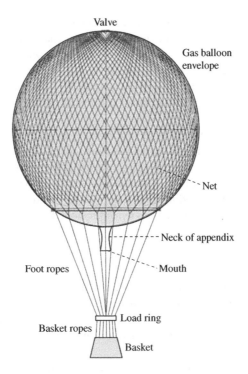

Figure 1.39   Major components of the gas balloon.

lifting gases include helium, hydrogen, and ammonia. A net surrounds the gas envelope and is connected via ropes to the load ring, from which the gondola or basket is suspended. The net serves to spread the load of the payload evenly over the surface of the envelope. A valve is located at the top of the envelope, which can be opened by the pilot, allowing gas to escape, to control the rate of ascent. In addition to this vent valve, the ascent and descent of gas balloons are controlled by throwing ballast bags, filled with sand or water, overboard. A tube at the bottom of the envelope, called the appendix, is used to fill the balloon and serves as an outlet to relieve the buildup of gas pressure inside the envelope due to temperature increases. There is also a rip panel on the envelope, which can be opened to rapidly deflate the balloon on the ground, in a high wind condition, or in an emergency situation. The basket or gondola is similar to that used for hot air balloons.

Gas balloons are used for sport ballooning, but less so than hot air balloons, due to their increased complexity and the high cost of the lifting gas. The maximum altitude capability of gas balloons is much greater than hot air balloons. Gas balloons can ascend to near-space altitudes of over 120,000 ft (37 km), above 99.5% of the earth's atmosphere. For this reason, high-altitude gas balloons are used extensively for scientific research.

Scientific gas balloons are used for a myriad of research and observation purposes, including studies of the weather, the upper atmosphere, and deep space. The envelope volume of the gas balloon expands significantly as it ascends and the external, ambient air pressure decreases. When fully expanded, these specialized gas balloons can be as large as 400 ft (120 m) in height and 460 ft (140 m) in diameter, with a volume of 40 million ft^3 (1.1 million m^3). The gas envelope skin of these massive balloons is made of a thin polyethylene film, with a thickness of only 0.8 mil (one mil is one thousandth of an inch) or 20 microns. With a maximum payload capability of about 8000 pounds (3629 kg), a scientific balloon can reach an altitude of 120,000 ft (37 km). They are also used as a means of lifting a test object, such as a parachute or vehicle, to an altitude where it can be released to study aerodynamics, flight dynamics, or other characteristics.

Unlike the gas balloons that expand as they ascend, the *superpressure gas balloon* is designed to maintain a constant volume at all altitudes. The gas envelope of a superpressure balloon is constructed of a high-strength polyester film that can bear the high loads as the gas pressure changes. Superpressure balloons can stay aloft for months, making ideal long endurance, high altitude scientific platforms.

The hybrid balloon combines features of the hot air and gas balloons. The hybrid balloon generates its buoyancy from a combination of heated gas from a burner and the carriage of an unheated, lighter-than-air gas such as helium or hydrogen. De Rozier attempted to cross the English Channel in a hybrid hot air–hydrogen gas balloon. Since de Rozier's time, hybrid balloons have been used for several long distance flights, including a solo, around-the-world flight by Steve Fossett in 2002. Fossett's circumnavigation in a hot air–helium hybrid balloon took over 14 days.

1.2.4.3   The Airship

An airship is distinguished from a balloon by both its ability to propel itself through the air, typically using internal combustion engines driving propellers, and also its ability to be steered. Early airships were called dirigible balloons, after the French *dirigible*, meaning "capable of being directed or steerable". Developed in the early 20th century, airships were the first powered aircraft that had flight controls for steering.

Similar to a balloon, an airship obtains its lift from a buoyant gas that is contained within a gas envelope. The buoyant gas used in airships is the same as that used in gas balloons, with the inert helium being used in most modern airships. Early airships were filled with hydrogen gas, with the same disastrous results as experienced with hydrogen-filled balloons. Typically, an airship's envelope has an axisymmetric, streamlined shape that contains the buoyant gas in separate gas bags or cells. Airships can be classified as rigid or non-rigid, based on the construction of the envelope, as shown in Figure 1.40.

The rigid airship has an envelope comprising a structural frame with a fabric outer covering. In early airships, the structural frame was made of wood, covered by cotton cloth fabric, similar to the construction of early airplane airframes. Modern airships use a metal, typically aluminum, framework and synthetic materials for the covering. The rigid structure maintains the shape of the airship and carries the structural loads of the vehicle. The gas bags or cells are mounted inside the rigid envelope. While the gas bags or cells are filled with a pressurized, buoyant gas, the rigid envelope is typically a non-pressurized structure.

A non-rigid airship, also called a *blimp*, is somewhat similar to a gas balloon in that the internal pressure of the buoyant gas maintains the shape of the envelope. Gas bags inside the envelope are filled with the buoyant gas, while other gas bags called *ballonets* are filled with air at sea level pressure. The combination of these gas bags is used to maintain the shape of the non-rigid airship hull. To compensate for the change in size of the buoyant gas bags, due to changes in pressure with altitude, air is forced into or out of the ballonets. Air is pumped into the ballonets using auxiliary blowers and is released from the ballonets using vents.

A third type of airship, the semi-rigid airship, is a combination of the rigid and non-rigid designs. Its envelope shape is maintained by the internal gas bags, but there is also a supporting structure, such as a "backbone" keel, much like a ship.

All of these airship types usually have a gondola beneath the envelope structure, where the aircrew, passengers, and cargo are carried. On larger, rigid airships, passenger and cargo compartments can be located inside the rigid envelope structure. This is not possible for non-rigid airships. Larger

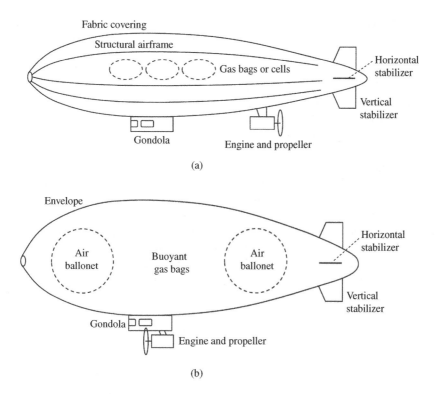

Figure 1.40 Components of airships, (a) rigid airship and (b) non-rigid airship.

airships typically may have multiple *power cars* or *engine cars* – separate nacelles where the internal combustion engine and propeller are installed. These engine cars may be mounted from the gondola structure or from other locations on the envelope. The propeller mounting may be of the *pusher* (facing aft) or *tractor* (facing forward) configuration. The multiple engines allow for the application of asymmetric thrust, which is used to help steer the airship. Movable, horizontal and vertical, control surfaces, located at the tail of the airship, are also used for steering and to control the attitude of the airship. To descend, the airship can vent gas and to climb, it can drop ballast. Longitudinal trimming of the airship can be accomplished through weight shifting, by pumping water or gas fore and aft inside the vehicle.

Takeoff, landing, and general ground handling of an airship requires unique facilities and a large ground crew. An airship can take off or ascend much like a balloon, but it can also use its engines to assist in the lift-off. The landing is made by slowly descending towards a ground crew, dropping ropes for them to grab, and being anchored to the ground. A mooring mast may also be used, where the nose or bow of the airship is attached or moored to the mast. As might be imagined, a large number of people are required in the ground crew for ground handling of the airship. The process is more difficult in gusty or high wind conditions.

Perhaps the most famous rigid airships were those built in the early 20th century by the German Zeppelin Company. One of the most famous Zeppelin passenger airships was the LZ-129 *Hindenburg* (LZ stood for Luftschiff Zeppelin, German for "Airship Zeppelin" and "129" is the airship designation number). The *Hindenburg* was 803.8 ft (245.0 m) in length and 135.1 ft (41.2 m) in diameter. The giant airship contained over 7 million ft^3 (200,000 m^3) of hydrogen gas. Powered by four Daimler-Benz 16-cylinder diesel engines, the *Hindenburg* had a cruise speed of 76 mph (125 km/h) and a typical cruise altitude of only 650 ft (198 m). The *Hindenburg* had a flight crew of 39 men, an additional dozen chefs and stewards, and a doctor on board. Luxury accommodation for 72 passengers included private cabins, promenade observation areas, a dining room, a lounge with a grand baby piano, and a smoking room, which was kept at a higher than ambient pressure to prevent any leaking hydrogen gas from entering. For the passengers, flying in a Zeppelin passenger airship was much like taking a voyage on a luxury cruise ship. These large passenger airships were the first commercial airliners, able to cross long distances, including routinely flying across the Atlantic Ocean.

Even though these Zeppelins were filled with highly flammable hydrogen gas, they had a very good safety record and flew all over the world carrying passengers and cargo for almost a decade. However, on 6 May 1937, the *Hindenburg* caught fire while attempting to dock at the Naval Air Station in Lakehurst, New Jersey, reducing the huge airship to a skeleton and ashes in less than a minute. Although never definitively proven, the leading theory for the cause of the fire was the ignition of leaking hydrogen gas by a static electric spark. This high profile disaster, along with a series of other airship accidents, contributed to the decline and ultimate demise of the airship as a viable means of commercial air travel. Replacing the flammable hydrogen gas with helium made the airship safer for flight, but the advancements in fixed-wing airplanes soon made the airship obsolete.

Airships are still used today, but mostly for applications where flying "low and slow" is desired, such as aerial advertising, tourism, remote sensing, and aerial observation. There has been renewed interest in using airships as long duration, very high altitude, scientific and commercial platforms, similar to high altitude balloons. An advantage of the airship is its ability to maintain a constant location over a point on the earth, similar to a stationary satellite. Scientific applications of airships include astronomical or weather observations. Commercial uses include acting as telecommunications platforms.

An example of a modern airship is the Zeppelin NT, shown in Figure 1.41. The Zeppelin NT is a semi-rigid, helium-filled airship with a gas volume of 290,450 ft^3 (8255 m^3). With a length of 246 ft (75 m) and a diameter of 46 ft (14.2 m), the Zeppelin NT has a gross weight of about 23,500 lb (10,700 kg). It carries a crew of 2 and 12 passengers at speeds up to 77 mph (125 km/h) and altitudes up to about 8500 ft (2600 m). The airship is powered by four 200 hp (149 kW) Lycoming IO-360, air-cooled, piston engines.

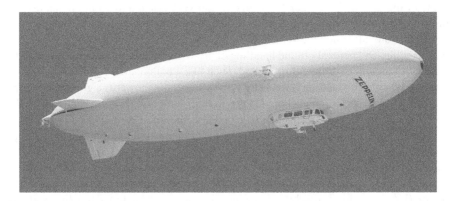

Figure 1.41   A modern airship, the Zeppelin NT, 2010. (Source: *User: Stefan-Xp, "Zeppelin NT" https:// commons.wikimedia.org/wiki/File:Zeppelin_NT.jpg, CC-BY-SA-3.0. License at https://creativecommons .org/licenses/by-sa/3.0/legalcode.*)

Another related aircraft, the *hybrid airship*, combines elements of the lighter-than-air airship and the heavier-than-air, fixed-wing airplane. The hybrid airship obtains its lift from a combination of aerostatic (buoyant) lift and aerodynamic lift. By virtue of its more aerodynamic shape and higher cruise airspeeds, as compared with a conventional airship, the aerodynamic lift of a hybrid airship can approach 50% of the total lift.

1.2.5   The Unmanned Aerial Vehicle

As the name implies, an *unmanned aerial vehicle* (UAV) is an aircraft without a pilot inside the vehicle. As such, any of the aircraft types shown in Figure 1.9 can be a UAV. The UAV may be flown remotely by a pilot on the ground or may operate autonomously, independent of real-time human intervention. A UAV that is flown remotely by a pilot is also sometimes referred to as a remotely piloted aircraft. For autonomous operation, a set of instructions or a flight plan may be programmed into computers onboard the UAV, or these instructions may be transmitted to the UAV via a communications or data link from the ground. This includes the capability to perform autonomous takeoffs and landings. The autonomous landing of a UAV on an aircraft carrier has even been demonstrated. The *unmanned aircraft system*, or UAS, refers to the complete system involved with the UAV operation, including the flight vehicle, ground control station, instrumentation, telemetry, communication, navigation equipment, and other support equipment.

Since it does not carry a human pilot, a major advantage of a UAV is its ability to fly into hostile environments, such as military combat, a forest fire, or a hurricane, without the potential for loss of human life. UAVs can also be used to reach remote areas that are otherwise inaccessible, such as during disaster relief. Unmanned aerial vehicles have seen an amazing proliferation in recent years, finding applications in a wide range of applications, including military, scientific, and commercial uses. Military UAVs perform reconnaissance, surveillance, and combat missions. A military UAV that carries weapons is designated as an unmanned combat air vehicle (UCAV). Equipped with the appropriate sensors and instruments, scientific applications of UAVs include earth remote sensing, meteorological sensing, geophysical mapping, and archaeological surveying, just to name a few. Some of the commercial applications of UAVs include oil, gas, and mineral exploration, aerial surveying of crops, livestock monitoring, wildlife mapping, pipeline and power line inspection, forest fire detection, motion picture filming, and potentially in the near future, parcel delivery.

UAVs come in all shapes and sizes, ranging from vehicles with conventional airplane configurations and weights to biologically inspired *micro air vehicles* that are the size and weight of insects. The General Atomics *Predator* UAV, shown in Figure 1.42, has a length of 36 ft (11 m), wingspan of 86 ft (26 m), and maximum weight of about 7000 lb (3200 kg). The much larger Northrop Grumman RQ-4 *Global Hawk* UAV, shown in Figure 1.43, has a length of 44 ft (13.5 m), wingspan of 116 ft (35.4 m), and maximum weight of about 25,600 lb (12,600 kg).

Both the *Predator* and the *Global Hawk* can be classified as *long endurance* UAVs with the ability to fly at very high altitudes for a very long time. The *Predator* can fly at over 50,000 ft (15,000 m) for over 30 hours, and the *Global Hawk* can fly at over 60,000 ft (18,000 m) for over 30 hours. Long endurance capability is another advantage of the UAV in performing surveillance, reconnaissance, and remote sensing. While the *Predator* is turboprop powered and the *Global Hawk* has a turbofan jet engine, other UAV propulsion system options, such as solar power, may enable *ultra-long endurance* UAVs to stay aloft for days, weeks, or longer. Some development work has been performed on the design of UAVs that could stay aloft for such long periods of time that they would act as airborne communication or earth-sensing "satellites".

Contrast the size, weight, shape, and capabilities of the UAVs such as the *Predator* and *Global Hawk* with the *micro air vehicle* (Figure 1.44), which can be the size and weight of an insect.

Figure 1.42 The unmanned aerial vehicle, the General Atomics MQ-9 *Predator* UAV. (Source: *NASA*.)

Figure 1.43 A large, jet engine-powered UAV, the Northrop Grumman RQ-4 *Global Hawk*. (Source: *NASA*.)

While some micro air vehicles have conventional airplane configurations, many are *biologically inspired* designs, based on insects, birds, or other small flying creatures. They often mimic the flight characteristics of the living creature being copied, such as using flapping wings to generate lift and thrust. The extremely small size of micro air vehicles make them ideal for applications where it is

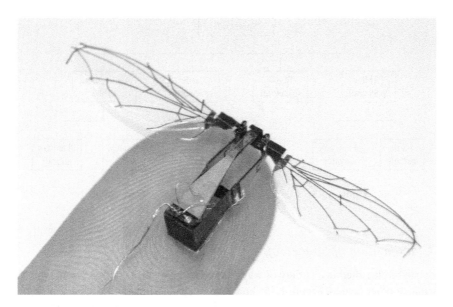

Figure 1.44   A biologically inspired micro air vehicle. (Source: *Air Force Research Laboratory, US Air Force.*)

desirable for the vehicle to be unnoticed, such as for clandestine surveillance, or where they need to operate in a very confined space, such as search and rescue in the rubble of a building that has collapsed. However, due to their small size, the integration of desired systems and components, to provide power, navigation, control, and sensing, is quite a challenge.

1.3   Spacecraft

Spacecraft are vehicles that are designed to operate in space, where there is no sensible atmosphere, including in Earth orbit and in the vacuum of space. Spacecraft are, in many ways, fundamentally different from aircraft. Since aircraft operate in the sensible atmosphere, their design is usually dominated by aerodynamic considerations, such as maximizing lift and minimizing drag. Spacecraft that operate solely in airless space do not typically have these aerodynamic design constraints. In fact, they may appear to be completely non-aerodynamic, that is, there may be components and parts that protrude from the spacecraft in every direction. However, spacecraft that need to return from space through the atmosphere require similar aerodynamic designs considerations to fixed-wing airplanes. Spacecraft must operate in the harsh environment of the vacuum of space, where there are extremes in temperature, space radiation, and possible impacts from micrometeorites.

Similar to aircraft, spacecraft come in a wide variety of shapes and sizes. In this section, we discuss several different types of spacecraft, classified primarily based on their function or mission. In keeping with the general theme of this chapter, we look at several unmanned and manned spacecraft that were "firsts", including the first artificial satellite and the first manned spacecraft. In addition, we discuss a unique type of "one-person spacecraft", the spacesuits worn be astronauts. All spacecraft require some kind of launch vehicle or launch system to get them into space. At the end of this section, the rocket booster is discussed, still the only feasible launch system that we currently have to get spacecraft into space. Other non-rocket-based launch systems are mentioned, which may provide space access in the future.

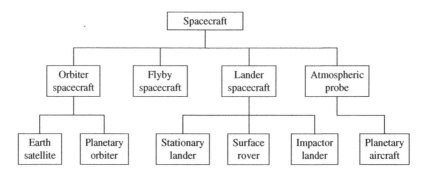

Figure 1.45   Classification of different types of spacecraft.

1.3.1   Classification of Spacecraft

Spacecraft can be classified in a variety of ways, based on geometric configuration, type of propulsion, mission, or other factors. Figure 1.45 shows one such classification of spacecraft types, based primarily on the mission or function of the vehicle. Using this scheme, the spacecraft are classified as an *orbiter, flyby spacecraft, lander,* or *atmospheric probe.* Brief descriptions of these different types of spacecraft are given below.

1.3.1.1   Orbiter Spacecraft

Orbiter spacecraft are launched and placed into orbit around the earth or another planet. Earth orbiter spacecraft, also called artificial satellites, have a variety of missions and functions, including those involving communications, navigation, meteorology, reconnaissance, and scientific research. Spacecraft that are sent to orbit another planet are typically used for scientific research. These planetary orbiter spacecraft may serve as a communications or data relay platform for a lander that is sent to the planet's surface. We discuss several orbiter spacecraft, including different types of unmanned, artificial earth satellites and manned orbiting vehicles in later sections.

1.3.1.2   Flyby Spacecraft

Unlike orbiting spacecraft, flyby spacecraft follow a trajectory that is not a closed orbit around a planet. They may be sent on interplanetary journeys that take many years. As their name implies, this type of spacecraft is designed to fly by a planet or other celestial object of scientific interest, collecting data with their array of sensors and instruments. Some flyby trajectories can be cleverly designed to bring the spacecraft in close proximity to a series of planets.

Four of the most successful flyby spacecraft are the *Pioneer 10* and *11* and *Voyager 1* and *2* spacecraft, all launched in the 1970s. These spacecraft were sent on trajectories to fly by the solar system's outer planets and then continue, beyond our solar system. *Pioneer 10* was the first spacecraft to fly by the gas giant planet, Jupiter. *Pioneer 11* was the second spacecraft to visit Jupiter and the first to fly by the ringed, gas giant, Saturn. *Voyager 1* completed flybys of Jupiter and Saturn and then, in 2012, became the first manmade object to leave our solar system. It also discovered the first volcanic activity on another world, the active volcanoes on the Jupiter moon, Io. In addition to flybys of Jupiter and Saturn, *Voyager 2* was the first spacecraft to fly by Uranus and Neptune. A more detailed discussion of the *Pioneer 10* flyby spacecraft is given in a later section.

1.3.1.3   Lander Spacecraft

Lander spacecraft are designed to land on the surface of another planet or other celestial body. The landing may be a "soft" landing, where the spacecraft survives and is able to perform other functions on the surface, or it may be a high velocity impact, where the spacecraft obtains data during its descent, but is not designed to survive the impact. At the time of writing, lander spacecraft have successfully impacted or soft landed on the Moon, Mercury, Mars, Venus, Saturn's moon Titan, and a few asteroids and comets.

The first type of lander spacecraft that was flown into space was the *impactor* lander. This is, perhaps, the simplest type of lander spacecraft, since it does not have the design complexity to soft land on a surface. Data is collected about the planet or celestial body, including the atmosphere, if present, as the impactor descends to the surface. The Russian impactor spacecraft, *Luna 2*, was the first manmade object to "land" on another celestial body, impacting the Moon on 14 September 1959. With an estimated impact speed of 3.3 km/s (7382 mph), the vehicle was certainly destroyed upon impact. *Luna 2* was the second in a series of lunar explorer spacecraft sent to the Moon by Russia. The first intended Russian impactor lander, *Luna 1*, was unsuccessful, due to an error in its trajectory, causing it to miss hitting the Moon by about 6000 km (3700 miles).

As shown in Figure 1.46, the 390 kg (860 lb) *Luna 2* spacecraft was spherical in shape with pro-truding antennas. Sensors and instrumentation on board the spacecraft included radiation detectors, micrometeorite detectors, and a magnetometer to detect the Moon's magnetic field. Data from these sensors confirmed that the Moon does not have a radiation belt or any significant magnetic field. On its way to the Moon, *Luna 2* released an amount of sodium gas into space, which created a

Figure 1.46   An impactor spacecraft, Soviet *Luna 2*, the first manmade object to land on another celestial body, the Moon, 1959. (Source: *NASA*.)

comet-like, bright orange gas trail that was visible from Earth. The purpose of this gas release was to study the behavior of a gas in outer space.

A variation of the impactor lander is the *penetrator* spacecraft, which is designed to survive the tremendous forces of the impact and penetrate into the surface. It then makes measurements that are telemetered back to Earth, typically by relaying the data to a "mothership" spacecraft in orbit.

The *stationary lander* and the *surface rover* make soft landings on a planet or other body. The stationary lander remains at its landing spot, while the surface rover is able to move about the surface. The rover has the advantage of being able to move about the surface, allowing it to explore a larger area than the stationary lander, but this comes at a higher risk of damage in navigating the terrain and surface obstacles. Both the stationary lander and the surface rover may have semi-autonomous functions, such as unfolding solar arrays or antennas, but they are often sent commands by controllers on Earth. The movement of a surface rover on another planet is precisely choreographed by controllers on Earth to ensure the safety and success of the vehicle's movements. We discuss the *Curiosity* Mars surface rover in a later section.

If there is sufficient atmosphere to produce significant frictional heating, shielding may be required to protect the lander spacecraft from high heating during its entry and descent. The landing must be soft enough so that the spacecraft is undamaged and able to perform its mission on the surface, typically scientific data collection. Parachutes, rockets, or both, may be used to decelerate the lander during its descent. Rockets may be fired, right before touchdown, to reduce the landing impact velocity. Touchdown of the spacecraft may be on mechanical landing gear or inflatable cushions or bags, to absorb the final landing loads. For landing on bodies with very low gravity, a harpoon-type device may fire an anchor cable into the surface to hold the spacecraft onto the surface.

This was the scheme used for the European Space Agency (ESA) *Philae*, a small robotic lander that made the first soft landing on the surface of a comet, *67P/Churyumov–Gerasimenko*, on 12 November 2014 (Figure 1.47). The *Philae* lander had anchoring harpoons on its belly that would fire downward into the comet, when the spacecraft touched down on the gravity-less comet. The

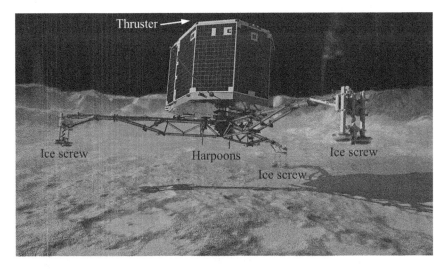

Figure 1.47   Depiction of *Philae* landing on Comet *67P Churyumov–Gerasimenko*. (Source: *Adapted from DLR German Aerospace Center, "Rosetta's Philae Touchdown" https://en.wikipedia.org/wiki/File: Rosetta%27s_Philae_touchdown.jpg, CC-BY-SA-3.0. License at https://creativecommons.org/licenses/by-sa/ 3.0/legalcode.*)

lander would then fire a thruster, on top of the spacecraft, gently pushing it onto the surface, while ice screws were drilled into the surface from its three landing footpads. Unfortunately, the landing impact was softer than planned, so that the anchoring harpoons did not fire. Without the anchors in place, the *Philae* lander bounced off the comet surface a few times, but luckily settled down to a permanent landing without being damaged.

The first spacecraft to soft land on another celestial body was the Russian *Luna 9*, which landed on the Moon on 3 February 1966. This was followed by the United States *Surveyor 1*, which soft landed on the Moon four months later, on 2 June 1966. Both landers answered a question that was in debate prior to a spacecraft actually landing on the Moon: would a spacecraft sink deeply into the dust on the surface of the Moon, perhaps even burying the lander? This was a question that was of some concern for future plans to land people on the Moon. The lunar surface supported the weight of the landers, definitively putting this issue to rest. Both *Luna 9* and *Surveyor 1* were stationary landers, and both used retrorockets to reduce their rate of descent for a soft landing. One of the major objectives achieved by *Surveyor 1* was the validation of the technologies required to soft land on the moon, paving the way for manned landers.

The Lunar Excursion Module (LEM) and the Lunar Roving Vehicle (LRV) were manned lander spacecraft that landed on the moon during the *Apollo* program (Figure 1.48). The LEM was a stationary lander that carried two astronauts from Moon orbit to the lunar surface. It was a two-stage spacecraft, with a descent stage and an ascent stage. The ascent stage returned the astronauts to rendezvous in lunar orbit with the *Apollo* Command Module spacecraft. There were six successful landings of the LEM on the moon between 1969 and 1972. The LRV was a surface rover that was carried to the lunar surface by the LEM. It was an open-frame, electrically powered, four-wheeled, car-like vehicle with side-by-side seating for two astronauts. The LRV was used on the moon for the last three *Apollo* missions.

Figure 1.48   *Apollo 16* Lunar Excursion Module and Lunar Roving Vehicle on the moon, 1972. (Source: *NASA.*)

1.3.1.4 Atmospheric Probe

Some spacecraft carry smaller, specially instrumented atmospheric probe spacecraft that are released to enter the atmosphere of another planet. These atmospheric probes are used to collect scientific data about the planet and its atmosphere as they descend. Entering the atmosphere at hypersonic speeds, they typically encounter large aerodynamic forces and high heating. Deceleration and thermal protection may be required using an aeroshell, a rigid shell structure that detaches from the beneath the probe. High drag devices, such as parachutes, may also be deployed to decelerate the probe and allow more time for data collection during its descent. Data from the probe is typically telemetered to the orbiter "mothership" spacecraft, where it is relayed back to Earth. Often, atmospheric probes are not designed to survive to landing on the planet surface, burning up in their fiery descent.

The *Pioneer 13* spacecraft, also known as the *Pioneer Venus Multiprobe*, carried four atmospheric probes to the planet Venus in the late 1970s. The probes were spherically shaped, pressure vessels with aeroshells. One probe was 1.5 m (4.9 ft) in diameter and the other three were smaller, at 0.8 m (2.6 ft) in diameter (Figure 1.49). The aeroshell of the large probe detached to provide thermal protection, and a parachute was used to decelerate the probe. The smaller probes did not have parachutes and their aeroshells remained attached. The probes were not designed to survive to landing on Venus, but amazingly, one of the small probes continued to transmit signals for over an hour after impacting the surface. The main spacecraft body or *bus* (to be discussed in the next section) was also used as an atmospheric probe, even though it did not have a heat shield or parachute.

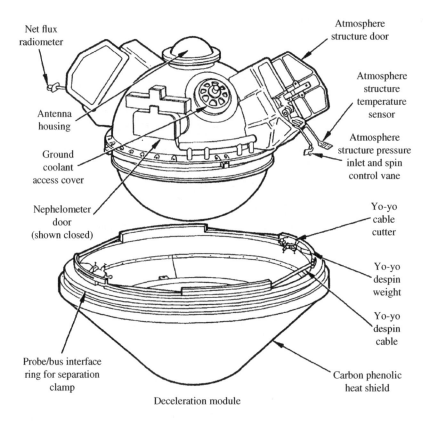

Figure 1.49 Diagram of *Pioneer 13* small probe. (Source: *NASA.*)

When the 2.5 m (8.2 ft) diameter, 290 kg (639 lb) cylindrical spacecraft was no longer able to stay in orbit around Venus, it entered the atmosphere and transmitted scientific data until it burned up at an altitude of about 110 km (68.4 miles).

1.3.1.5 Planetary Aircraft

Another type of atmospheric spacecraft is the planetary aircraft. This concept involves the deployment of an unmanned aircraft in the atmosphere of a planet, where it flies in the atmosphere and makes scientific measurements and observations, similar to an airborne science-type aircraft on Earth. An advantage of the planetary aircraft is that it can explore a larger area than other types of lander spacecraft. It may even be possible for the planetary aircraft to land and take off on the planet, giving it the capability to explore several different landing sites. While many of these concepts involve unmanned aerial vehicles, an ultimate possibility might be manned planetary aircraft, allowing astronauts to travel about and explore a planet from the air and cover large distances over the planet surface. A wide variety of planetary aircraft have been proposed, including gliders, powered-airplanes, rotorcraft, airships, and balloons. Propulsion concepts for these aircraft have included rocket power and a combustion, electric battery, or solar-powered engine-propeller combination.

Aircraft flight on another planet involves some unique design challenges. First, the aircraft must be packaged to fit within the volume constraints of the rocket booster that is launched from Earth, which usually involves folding of the aircraft's structural components. Upon arriving at the planet of interest, the aircraft must be deployed and unfolded in the planetary atmosphere. The aircraft may also need to be capable of transitioning from a near-vertical, free-fall type deployment to horizontal flight. Other complexities include autonomous operation of the aircraft flight control and navigation systems, propulsion, sensors, telemetry, and other systems. The atmospheres of other planets are dramatically different, in terms of density, pressure, temperature, composition, and other factors, than Earth's atmosphere. This must be taken into consideration in the aerodynamic design of the aircraft and its flying qualities. For instance, since the atmosphere of Mars is much less dense than on Earth, flight near the surface of Mars corresponds to flight at over 100,000 ft (30 km) on Earth.

The rocket pioneer, Wernher von Braun, proposed one of the earliest concepts of using large manned gliders to land on Mars. In the 1970s, the use of unmanned airplanes was investigated for the exploration of Mars. There was renewed interest in flying an airplane on Mars in 2003, which would have coincided with the centennial of the Wright brothers' first flight. There has also been recent interest in using aerial vehicles to explore the atmosphere of Venus and Titan, a moon of Saturn, with various concepts ranging from inflatable airships to solar-powered airplanes.

One of the designs for a Mars airplane, that has been extensively studied, is the NASA Aerial Regional-scale Environmental Survey (ARES) Mars airplane (Figure 1.50). The 150 kg (330 lb) ARES design had a wingspan of 6.25 m (20.5 ft) and was powered by a hydrazine-fueled, liquid rocket engine. It had a predicted cruise speed of 145 m/s (476 ft/s, 324 mph) and range of 680 km (423 miles). Although the 2003 Mars airplane mission was cancelled, several key enabling technologies were demonstrated. In September 2002, a 50%-scale prototype of the ARES Mars airplane was flight tested by releasing it from a high-altitude balloon. Since the atmosphere of Mars is much thinner than on Earth, the release and flight of the ARES airplane at an altitude of about 100,000 ft (30 km) on Earth simulated flight near the surface of Mars. The airplane was packaged, as it would be for a Mars mission, thus the successful unfolding and deployment of the airplane's wings and tail in the test, verified this aspect of the design. After being released, aerodynamic and stability and control data were obtained for the ARES airplane at high altitudes, simulating flight on Mars.

Figure 1.50   Artist's concept of NASA ARES Mars airplane flying over Mars. (Source: *NASA*.)

1.3.2   Parts of a Spacecraft

A spacecraft can often be separated into the *bus* and the *payload*. The bus is the structural framework of the spacecraft, onto which subsystem components and the payload are mounted. Typically, the spacecraft bus contains subsystems that support the operation of the payload. The support functions may include electric power and distribution, communications and telemetry, thermal control, attitude sensing and control, and in-orbit propulsion. The payload and spacecraft bus may be separate units or they may be a combined system. The spacecraft bus infrastructure is commonly used for earth satellites, especially communications satellites. The payload contains the systems to perform the primary mission of the spacecraft, which may be specialized instrumentation, sensors, or other scientific equipment. The payload could also be the people who are being transported into space.

The spacecraft structure is typically made of high-strength, lightweight materials such as aluminum, titanium, beryllium, or composite materials. The spacecraft structure must be designed for a variety of conditions, including the high g-loads experienced during launch and orbit insertion, and extremes in the thermal, pressure, and acoustic environments. There may be tight tolerances on the deformation of the structure caused by these environmental factors, driven by the critical alignment of sensors, antennas, or other components. The structure must be as light as possible, as there are usually weight limits for the launch vehicle being used.

Most spacecraft have systems for power generation, communication, thermal control, vehicle attitude control, and propulsion. Spacecraft power can be provided by a variety of sources, including batteries, solar panels, fuel cells, and even nuclear power sources. The choice of a power source usually depends on the power requirements and the mission duration. Batteries or fuel cells are often satisfactory for missions lasting days or a few weeks. Typical types of batteries include silver zinc, lithium-ion, nickel-cadmium, and nickel-hydrogen. Fuel cells produce electricity through the chemical reaction of a fuel and an oxidizer, often hydrogen and oxygen. For longer duration missions of months or years, such as interplanetary voyages, nuclear and solar power generation is usually required. An example of a nuclear power generator is the radioisotope thermoelectric generator (RTG). The RTG operation is based on the natural decay of a radioactive material, such as plutonium; therefore, radiation shielding may be needed to protect other spacecraft systems. Heat given off by the decay of the isotope is converted into electricity. Solar cells are a reliable means of power generation, but they require the deployment and pointing of solar arrays. With any of these types of power sources, systems for power distribution, power regulation, and energy storage are typically required.

Communications equipment is required to transmit information, data, and commands between the spacecraft and the ground users or controllers on Earth. The communications signal is often a radio signal, but may be a laser signal. The communications signal that is sent *to* the spacecraft is called the *uplink*, while the signal *from* the spacecraft is called the *downlink*. Receivers, transmitters, and antennas are basic components of the communications system. Spacecraft typically have redundant sets of communications equipment and antennas.

Spacecraft often have high-gain and low-gain antennas. The gain refers to the amount of radio power that can be collected by the antenna and sent to the receiver. The higher the gain, the higher the rate of data transmission of radio signals that can be sent and received. The antenna gain can be increased by making the radio signal collecting area of the antenna larger. This is why most high-gain spacecraft antennas are large, parabolic-shaped, dish antennas. Because of their high data rate, spacecraft primarily use their high-gain antennas for communications with Earth. However, high-gain antennas are also highly directional, that is, they send and receive radio signals in a narrow radio beam width. (Hence, high-gain antennas are also called directional antennas.) A stronger, focused radio signal can be sent and received from a high-gain antenna, but the pointing of this signal is more difficult. Typically, a spacecraft's high-gain antenna must be pointed to Earth within a fraction of a degree to send and receive radio signals. Hence, a high-gain antenna is more susceptible to signal loss and loss of communications. By contrast, low-gain antennas can send and receive radio signals over a wide coverage area, albeit at a lower data rate. The wide beam width of the low-gain antenna makes it less susceptible to signal loss, so they are usually used as a backup for the high-gain antenna, especially in situations where the radio signal must be reacquired after loss of signal with the high-gain antenna.

Spacecraft are exposed to a severe thermal environment in vacuum, with extremes in temperature that are dependent on the amount of sun exposure. In space, the sun-lit side of a spacecraft may have a surface temperature of over 120°C (248°F, 393 K), while the shaded side may feel a frigid −200°C (−328°F, 73.2 K). The thermal control system must maintain the spacecraft's temperature within the operating limits of the onboard systems and components. It must prevent electronics from overheating and must keep mechanical, moving parts from freezing. Thermal control may be accomplished using passive or active methods. Passive techniques include the use of insulation, blankets, surface coatings (for instance, simply painting a surface white or black), and mirrors. Active thermal control includes the use of electrical heaters, fluid-filled radiators, and louvers. The louvers act much like window blinds, opening and closing to regulate the dissipation of heat from inside the spacecraft.

Stabilization of the spacecraft is usually required in terms of attitude control in three dimensions. External forces and torques acting on the spacecraft may disturb it from a desired attitude or orientation. An attitude control system is required to sense and correct any changes from the desired orientation. It may also be necessary to intentionally change the attitude or orientation of the spacecraft for pointing of sensors, communications antennas, or solar arrays, thermal management, or docking with another structure or spacecraft. A reaction control system, composed of multiple small rocket thrusters, may be used to make attitude corrections. Another system that may be used for spacecraft stabilization is the spinning *reaction* or *momentum wheel*. By changing the speed of the spinning momentum wheel, angular momentum can be traded between the spacecraft and the wheel. For example, if a rotation of the spacecraft is desired in a certain direction, the wheel is spun up in the opposite direction. By conservation of angular momentum, the desired spacecraft rotation is obtained.

A spacecraft may have a propulsion system that consists of tankage, propellant, thrusters, and the associated feed system plumbing and valves. The propellant may be a compressed gas, such as nitrogen, a liquid monopropellant, such as hydrazine, or a solid fuel. Spacecraft propulsion may be needed for attitude or altitude changes or corrections. A spacecraft may have a separate propulsion unit or rocket motor, called an *apogee boost motor,* or *kick stage*, used to boost the vehicle into its final orbit.

Finally, spacecraft that are intended for human spaceflight, or space habitation, require life support systems. Life-sustaining oxygen must be supplied, along with the removal of carbon dioxide and other harmful contaminants. The life support systems must maintain the spacecraft cabin environment at an adequate pressure, temperature, and humidity. Protection from harmful space radiation and impacts from micrometeorites must also be provided. Systems must be available to handle human waste products.

To illustrate the parts of a spacecraft, we examine the NASA *Magellan* spacecraft, shown in Figure 1.51. *Magellan* was launched on 4 May 1989, in the cargo bay of the Space Shuttle *Atlantis*, the first interplanetary spacecraft to be carried aboard a Shuttle. *Magellan* was sent to explore the planet Venus, including obtaining radar imaging of its surface. The spacecraft was 6.4 m in height (21 ft), 4.6 m (15.1 ft) in diameter, 10 m (32.8 ft) across from tip to tip of the solar panels, and had a total mass of 3453 kg (7612 lb), including 2414 kg of propellants (5322 lb). To reduce costs, the spacecraft was built from spare parts from other spacecraft programs, including the *Voyager*, *Galileo*, *Ulysses*, and *Mariner 9* programs. As shown in Figure 1.51, the main components of the *Magellan* spacecraft were the bus, forward equipment module, antennas, solar panels, attitude control module, propulsion module, and rocket engine module.

The *Magellan* spacecraft was built around a 10-sided, aluminum bus (a spare bus from the *Voyager* program), as shown in Figure 1.52. The bus was a bolted-together, aluminum structure

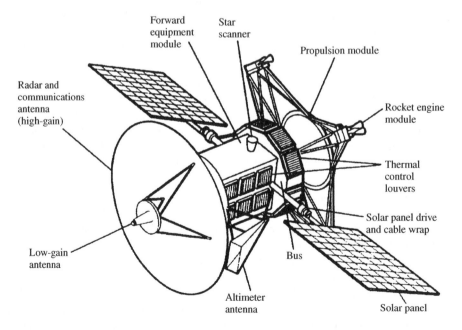

Figure 1.51   Parts of the *Magellan* spacecraft. (Source: *NASA*.)

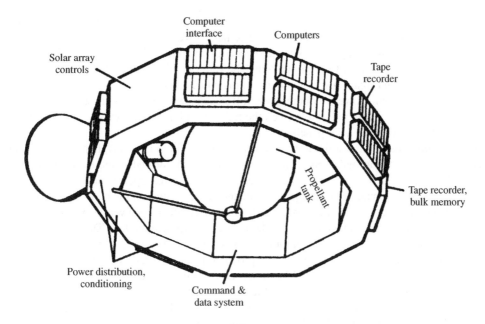

Figure 1.52   *Magellan* spacecraft bus. (Source: *NASA*.)

with ten independent compartments, designed to hold electronic components. Housed in the bus were the spacecraft's flight computers, power distribution and conditioning components, solar array controls, tape recorders, and command and data subsystem. The spherical, propellant tank for the spacecraft's propulsion module was mounted in the center of the bus.

The flight computers (from the *Galileo* program) controlled the command and data subsystem, as well as the spacecraft's attitude. The command and data subsystem stored the commands sent to the spacecraft, by controllers on Earth, and could autonomously control the spacecraft, if contact with Earth was lost. Scientific and radar mapping data were also stored on the command and data subsystem. Commands and data were stored on two digital tape recorders, which could store up to 225 megabytes of data, a meager amount of data storage by today's standards. The forward equipment module, a box-like structure, sat atop the bus. The radar electronics, telecommunications equipment, and batteries were mounted in the forward equipment module.

Spacecraft thermal control was accomplished with a combination of passive and active techniques. Multi-layered thermal blankets were wrapped around the electronic compartments. In addition to providing thermal insulation, the blankets had an outer coating that reflected solar radiation. Thermal control louvers were mounted on the face of each electronic compartment of the bus and on two sides of the forward equipment module.

The spacecraft had four different antennas to support communications and radar mapping functions. As shown in Figure 1.51, a 3.7 m (12.1 ft) diameter, high-gain, parabolic dish antenna (from the *Voyager* program) was mounted at the top of the spacecraft stack. This large dish antenna was the primary antenna for communications with Earth and for the radar mapping. A low-gain antenna (also from the *Voyager* program) was mounted in the center of this high-gain dish. A cone-shaped, medium gain antenna (from the *Mariner 9* program) was mounted to the top side of the bus. Both of these antennas augmented the high-gain antenna. The altimeter antenna, mounted on the side of the forward equipment module, was used for radar mapping.

The spacecraft electrical power was a 28 V system that was supplied by two large, square solar panels or two nickel-cadmium batteries. The solar panels supplied 1200 W of power at the start of the mission, but this gradually decreased over time as the efficiency of the panels degraded. Each solar panel measured 2.5 m (8.2 ft) on a side. The panels were hinged for stowage in the Space Shuttle cargo bay and were deployed once the spacecraft was released. The panels could also rotate so that they could be oriented towards the Sun. The rotation of the panels was controlled by the solar array controls and solar sensors at the tips of the panels. The batteries, located in the forward equipment module, could power all of the spacecraft systems when it was not in sunlight. They also provided the required additional power when the radar mapping system was in use. The batteries could be recharged with the solar panels.

The *Magellan* spacecraft's attitude was controlled using 36 cm (14 in) diameter momentum wheels, mounted in the attitude control module (Figure 1.53), located in the forward equipment module. The spacecraft's rotation rate was sensed by a set of gyroscopes, which sent their data to an attitude control computer. The computer commanded the rotation of the momentum wheels, as required, to correct the attitude of the spacecraft. The spacecraft attitude was also determined, to a high accuracy, with a star scanner, located in the forward equipment module. The star scanner data was used to correct the small errors accumulated by the drift of the gyroscopes.

Propulsion for the *Magellan* spacecraft was provided by an Inertial Upper Stage, a propulsion module, and a rocket engine module. When released from the Space Shuttle, in low Earth orbit, the *Magellan* spacecraft was attached to the Inertial Upper Stage, a two-stage solid-fueled rocket booster. The Inertial Upper Stage rocket was fired to send *Magellan* on its interplanetary trajectory from low Earth orbit to Venus. The *Magellan* rocket engine module (Figure 1.54) was a Star 48B solid rocket motor used for orbit insertion at Venus. The Star-48B motor (designed and used to

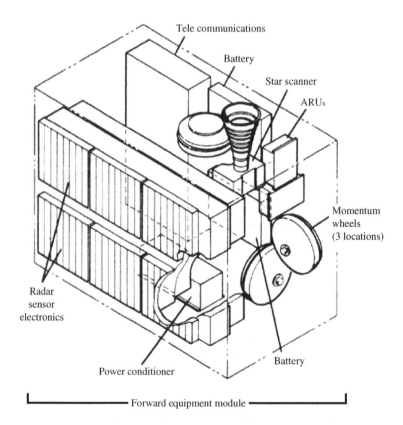

Figure 1.53 *Magellan* spacecraft attitude control module. (Source: *NASA*.)

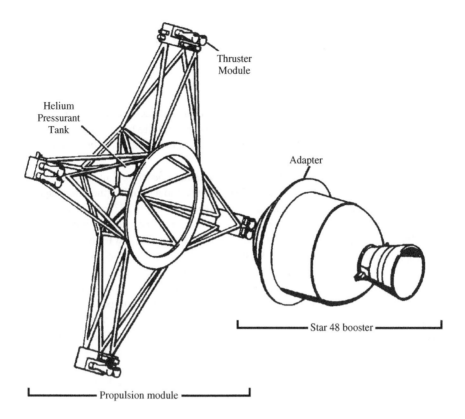

Figure 1.54 *Magellan* spacecraft propulsion and rocket engine modules. (Source: *NASA*.)

raise communications satellites from low Earth orbit to geosynchronous orbit) contained 2014 kg (4440 lb) of solid propellant and delivered a thrust of about 89,000 N (320,000 lb).

The *Magellan* propulsion module (Figure 1.54) was a four-armed truss structure with six liquid-propellant thrusters on the tip of each arm. The 24 thrusters were used for spacecraft attitude control, trajectory and orbit corrections, and reaction wheel desaturation. (The reaction wheels build up excess momentum from external torques, such as from solar pressure on the solar panels. The process of removing this excess momentum is called *desaturation*.) Each cluster of six thrusters comprised two 100 lb (445 N) thrusters, one 5 lb (22.2 N) thruster, and three, tiny 0.2 lb (0.89 N) thrusters. The 100 lb thrusters, pointed aft, were used for trajectory corrections, large corrections to the Venus orbit, and spacecraft stabilization during the Star-48B orbit insertion burn. The 5 lb thrusters prevented the spacecraft from rolling during these maneuvers. The 0.2 lb thrusters were used for reaction-wheel desaturation and small maneuver corrections. All 24 thrusters were fueled from a single 71 cm (28 in) diameter, titanium, propellant tank, located in the middle of the spacecraft bus, filled with 133 kg (293 lb) of monopropellant hydrazine. Additional pressure for the hydrazine system was provided by a small, helium-filled pressurant tank, mounted to the propulsion module struts, as needed.

1.3.3   Unmanned Spacecraft

Unmanned spacecraft include a wide variety of vehicles that operate in the space environment, performing a wide variety of missions. Different types of unmanned spacecraft include satellites that orbit the earth, probes that journey to other celestial bodies, and landers and rovers that operate on other planets and moons. We start with a discussion of satellites.

1.3.3.1   The First Earth Artificial Satellite

Spacecraft that are made to orbit the earth are called *satellites*, sometimes called *artificial* satellites as opposed to natural satellites, such as the Moon. For a spacecraft to be placed into *low earth orbit*, which we define as an altitude of a few hundred miles, it must attain an *orbital velocity* of about 17,000 mph (27,000 km/h).

The first artificial earth satellite was the *Sputnik 1* (meaning *Satellite 1* in Russian), launched by the Soviet Union on 4 October 1957 (Figure 1.55). *Sputnik* had a simple spherical shape, about the size of a beach ball, 22.8 in (58 cm) in diameter, and weighed 184 lb (83 kg). It was launched into an elliptical, low earth orbit, with an apogee (point in the orbit that is the farthest from the earth) of 584 miles (940 km) and a perigee (point in the orbit that is the closest to the earth) of 143 miles (230 km), completing an orbit about every 96 minutes (also known as the *orbital period*).

Sputnik broadcast radio pulses, using four external antennas, which were detected by radio receivers on the ground. In addition to demonstrating the ability to launch an artificial satellite into low earth orbit, *Sputnik* provided scientific data about the earth's upper atmosphere. The drag on the satellite in its orbit provided data about the atmospheric density, while the attenuation of the broadcast radio signals provided information about the ionosphere. The satellite remained in orbit for about three months, before the atmospheric drag brought it into the thicker regions of the atmosphere, where it burned up. This first satellite was visible in the night sky, as it raced around the earth every 96 minutes, broadcasting its radio pulses. The *space age* had begun.

In the USA, plans were in progress to develop the first American earth satellite, but the *Sputnik* launch caught all of America by surprise. There was a firestorm of political, scientific, and emotional turmoil over the concern that the Soviet Union was technologically ahead of the USA and that they would use this advantage to control space. There was also much concern that the Soviet Union had the capability to launch intercontinental ballistic missiles, armed with nuclear weapons, across the globe. Thus, *Sputnik* had not only started the space age, but had also set off the space race

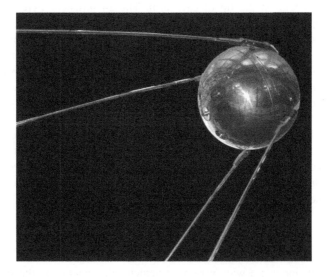

Figure 1.55   *Sputnik 1*, the first artificial satellite. (Source: *NASA*.)

between the USA and the Soviet Union. As a side note, a significant event that occurred, due to this new space race, was the creation of the National Aeronautics and Space Administration (NASA) on 1 October 1958, from its predecessor, the National Advisory Committee for Aeronautics (NACA).

Prior to *Sputnik*, the American effort to launch an artificial satellite into low earth orbit resided with the *Vanguard* program, led by the Naval Research Laboratory in Washington, DC. The *Vanguard* program was ultimately successful in placing several artificial satellites into earth orbit, but they were not the first to place an American satellite into orbit. *Vanguard 1* was successfully launched and placed into earth orbit on 17 March 1958, the fourth manmade satellite to orbit the earth, after *Sputnik 1*, *Sputnik 2* (on 3 November 1957), and *Explorer 1* (discussed below). The *Vanguard 1*, a spherically shaped satellite, was 6.4 inches (16.4 cm) in diameter and weighed 3.2 lb (1.5 kg). It was placed into an elliptical orbit with an apogee of 2387 miles (3841 km), perigee of 409 miles (659 km), and an orbital period of 132.8 minutes. Despite its small size, the *Vanguard 1* satellite had some impressive scientific achievements, including obtaining data to prove that the earth is not a perfect sphere, but rather, more "pear-shaped". It was also the first solar-powered satellite. Amazingly, the *Vanguard 1* is still in orbit today, albeit non-functioning, making it the oldest manmade satellite still in earth orbit. It is expected to remain in earth orbit into the 22nd century.

After *Sputnik*, the USA initiated the *Explorer* project to place an artificial satellite into low earth orbit. The *Explorer 1* satellite was the top stage of its rocket launch vehicle. The aft end of the satellite was the burnt-out fourth stage of the rocket and the forward end was the satellite instrumentation section (Figure 1.56). The *Explorer 1* had a total weight of 30.7 lb (13.9 kg) with the instrumentation section weighing 18.4 lb (8.35 kg). As shown in Figure 1.56, the instrumentation section had substantial instrumentation, containing a nose cone temperature probe, a cosmic ray detector, an internal temperature sensor, micrometeorite erosion gauges, external temperature sensors, a micrometeorite ultrasonic microphone detector, and low-power and high-power data transmitters. There were two different types of antennas on the satellite, two fiberglass slot antennas on the

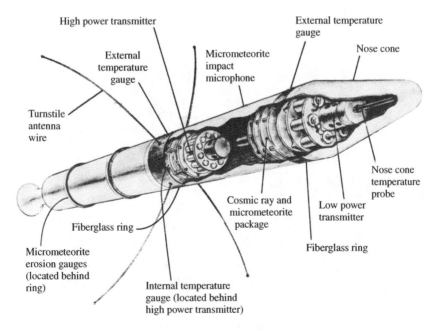

Figure 1.56   *Explorer 1*, the first American artificial satellite. (Source: *US Army Redstone Arsenal*.)

cylindrical satellite body and four flexible "whip" wire antennas located around the circumference in a turnstile arrangement. The satellite was spun about its longitudinal axis at 750 revolutions per minute, to keep the flexible wire antennas extended.

The *Explorer 1* satellite was built by the California Institute of Technology's Jet Propulsion Laboratory (JPL), under the direction of Dr William H. Pickering. The satellite instrumentation was designed and built by Dr James A. van Allen of the State University of Iowa. The satellite's *Jupiter-C* launch vehicle, a modified *Redstone* ballistic missile, was designed by a team led by Dr Wernher von Braun at the Army Redstone Arsenal in Huntsville, Alabama. Von Braun and many of his team had come to the USA after World War II, having led much of the German V-2 rocket development. He would later lead the major efforts to develop the *Saturn V* rocket engines that would take man to the Moon. The success of the *Explorer 1* satellite is a tribute to the combined contributions of these three scientists and the teams that they led (Figure 1.57). The modest size of the *Explorer 1* satellite can be appreciated in this photo.

A little less than four months after the success of *Sputnik 1*, on 31 January 1958, the USA successfully placed *Explorer 1* into low earth orbit. The small satellite went around the earth in an elliptical orbit with an apogee of 1575 miles (2535 km), a perigee of 224 miles (361 km), and

Figure 1.57 Dr William H. Pickering, Dr James A. van Allen, and Dr Wernher von Braun hold up a full-scale model of *Explorer 1*. (Source: *NASA*.)

an orbital period of 114.9 minutes. *Explorer 1* is credited with collecting the scientific data that led James van Allen to the discovery of the radiation belt that surrounds the earth and now bears his name.

Orbiting earth satellites have truly transformed our world by providing us with a variety of previously unavailable capabilities and perspectives. Earth satellites have revolutionized many areas, including weather prediction, earth observation, communications, and navigation. Satellites are placed into a variety of earth orbits, including very high orbits, called *geostationary orbits*, where they can remain over the same spot on the earth. Since the success of these early satellites, literally thousands of artificial satellites have been placed in earth orbit (some estimates place the number at over 6500). Of these, perhaps several hundred are currently operational, many have reentered the atmosphere and burned up, while others are no longer operational and have become space debris. In fact, there is so much space debris or space "junk" orbiting the earth that it has to be tracked so that collisions can be avoided with operational spacecraft, including manned spacecraft.

1.3.3.2 Detailed Descriptions of a Few Types of Unmanned Spacecraft

In this section, brief examples are given of a few types of unmanned spacecraft, a communications satellite, a scientific satellite, a deep space probe, and a planetary rover. These different types of unmanned spacecraft have distinctly different missions, which drive their individual designs and configurations. These different spacecraft look quite different from each other because of their quite different missions.

A Communications Satellite: the Iridium Constellation

Communications satellites have become commonplace in today's society, providing global coverage for the television, radio, and telecommunications industries. The concept of using satellites for global communications was proposed well before the launch of the first artificial satellite in 1957. In October 1945, Arthur C. Clarke published a short article entitled "Extra-Terrestrial Relays – Can Rocket Stations Give Worldwide Radio Coverage?" which proposed the use of artificial earth satellites for the worldwide relay of television signals. The satellite communications relay concept involves sending signals from earth to the orbiting satellite, which then relays the signal to other points on the globe. The first artificial satellite, dedicated to the communications relay, was the NASA *Echo 1* high-altitude balloon, launched to 1000 miles (1609 km) above the earth in 1960. *Echo 1* was a 100 ft (30.5 m) diameter balloon with a mirror-like, metallic surface, which passively reflected communications radio signals. Later communications satellites would have active repeaters, which could store and actively transmit radio signals and data.

An *Iridium* communications satellite is shown in Figure 1.58. This communications satellite is one of a large number of similar spacecraft that form a satellite *constellation*. The *Iridium* constellation comprises 66 satellites that provide satellite telephone communications coverage over the entire earth's surface. Each satellite is in low earth orbit at an altitude of about 485 miles (781 km). The constellation name came from the original plan to have a total of 77 satellites, matching the number of electrons orbiting the atom of the element iridium. The *Iridium* satellites orbit around the earth in six *orbital planes*, spaced 30° apart, with 11 satellites in each plane.

The *Iridium* satellites take advantage of the spacecraft bus design philosophy, making the manufacturing and assembly of so many satellites more efficient, much like the assembly line concept used in the automotive industry. The *Iridium* satellite payload consists of the communications components, such as transponders and antenna. The satellite bus contains the power, communications, and attitude control systems. As shown in Figure 1.58, the satellite has large solar panels as part of

Figure 1.58   An artificial earth satellite, the *Iridium* communications satellite. (Source: *User: Ideonexus, "Iridium Satellite" https://en.wikipedia.org/wiki/File:Iridium_satellite.jpg, CC-BY-SA-2.0. License at https:// creativecommons.org/licenses/by-sa/2.0/legalcode.*)

the solar power system. Due to the configuration of the solar panels, the reflected sunlight makes the satellite visible at times from the earth, even in the daytime, an event called an "Iridium flare", or more generally, "satellite glint". Batteries are also used to store power when the Sun is blocked, such as during a solar eclipse.

A Scientific Satellite: the Hubble Space Telescope
Earth-borne telescopes are handicapped by having to look through the earth's atmosphere, which distorts the images captured by the telescopes. The apparent "twinkling" of stars at night is due to this atmospheric distortion. In addition, the atmosphere absorbs some wavelengths of radiation, such as ultraviolet, gamma, and X-ray radiation, making it difficult for earth-based sensors to observe these types of radiation from astronomical bodies. Placing a telescope in space, eliminates the atmospheric distortion and absorption of visible light and other wavelengths of radiation.

Space-borne telescopes were proposed well before the advent of earth satellites. The German rocket scientist, Hermann Oberth (1894–1989), wrote about a space telescope in his 1923 book, "Die Rakete zu den Plantraumen" ("By rocket into planetary space"). Later in 1946, the American astrophysicist, Lyman Spitzer, Jr, (1914–1997) proposed the idea of an orbiting astronomical observatory in his paper "Astronomical advantages of an extraterrestrial observatory". Spitzer was instrumental in the advocacy of space telescopes throughout his career.

One of the largest space telescopes to be placed into earth orbit is the *Hubble Space Telescope* (HST) (Figure 1.59), with a length of 13.2 m (43.3 ft), diameter of 4.3 m (13.8 ft), and a mass of 11,110 kg (24,500 lb). Named after the American astronomer Edwin Hubble (1889–1953), the HST is a joint venture between NASA and the European Space Agency (ESA). Carried into space by the Space Shuttle *Discovery* (STS-31) on 24 April 1990, the *Hubble* was placed into a near-circular orbit, 569 km (354 miles) above the earth.

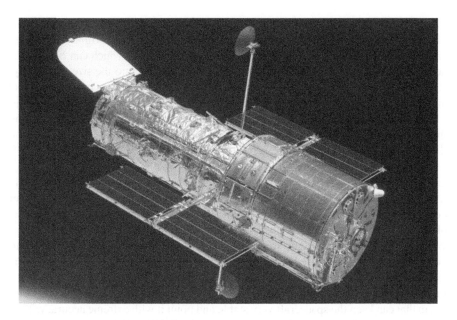

Figure 1.59   A scientific satellite, the *Hubble* space telescope in low earth orbit. (Source: *NASA*.)

The HST is one of the four large space-based telescopes in NASA's Great Observatories program. Each of these space telescopes is designed for a specific region of the electromagnetic spectrum. The *Hubble* space telescope is designed to observe primarily in the visible light spectrum. The *Compton* gamma ray observatory and the *Chandra* X-ray observatory are designed to observe in the gamma ray and X-ray radiation bands, respectively. The *Spitzer* space telescope is designed for infrared observations. All of these space telescopes are still in earth orbit, except the *Compton* gamma ray observatory, which had to be de-orbited when one of its stabilizing gyroscopes failed in 2000. Most of the *Compton* observatory burned up when entering the atmosphere, with any remaining parts falling into the Pacific Ocean.

The *Hubble* spacecraft is essentially a long telescope tube consisting of two mirrors, a support truss structure, an aperture door, and sensing instruments and equipment. The telescope is a type of Cassegrain reflector telescope, known as a Rictchey–Chreiten Cassegrain. As shown in Figure 1.60, the aperture door at one end of the telescope is opened and light passes down the tube to the primary mirror. The concave, primary mirror reflects the light onto a smaller, convex, secondary mirror,

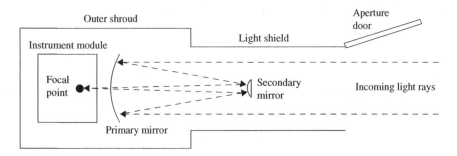

Figure 1.60   Schematic of HST Cassegrain reflector telescope operation.

which then focuses the light back, through a hole in the center of the primary mirror to the focal point where the telescope's sensing instruments are located. The HST primary mirror is 2.4 m (7.9 ft) in diameter, small in comparison to ground-based telescopes, which can be 10 m (32.8 ft) in diameter. A larger mirror can collect more light, which is critical to the size and clarity of the objects that can be "seen" by a telescope. (A telescope works by collecting as much light as possible, not by magnifying the size of an object. In fact, the HST has no magnifying lenses at all, just two mirrors for light collection.) Since the HST can collect undistorted light, in its perch high above the atmosphere, it can provide much improved optical resolution over ground-based systems, despite its smaller mirror. The HST mirrors are kept at a constant temperature of about 21°C (70°F) to prevent warping which would distort images.

Instruments, onboard the *Hubble*, include an infrared camera and spectrometer, an optical survey camera, a wide field of view optical camera, an ultraviolet spectrograph, and an optical spectrometer. The HST four main sensing instruments observe in the near infrared, visible light, and near ultraviolet wavelengths. Power for the instruments and equipment onboard the spacecraft is generated by two large solar panels, 7.56 m (24.8 ft) long by 2.45 m (8.0 ft) wide. Power stored in six nickel-hydrogen batteries is used when the HST is in the earth's shadow for about half an hour during each orbit.

Capturing high resolution images of distant objects requires a sophisticated stability and guidance system that can keep the spacecraft very stable and point it with extreme accuracy. The HST's attitude is adjusted using a set of spinning reactions wheels. The HST spacecraft does not have any propulsion systems for attitude or orbit adjustments. During its five servicing missions, the Space Shuttle boosted the HST's orbit, as required, to compensate for orbit degradation from atmospheric drag.

After the HST was in orbit, the images it sent back to earth were better than those that could be obtained from ground-based telescopes, but they were of lower quality than was expected; in fact, they were somewhat blurry. It was determined that there was a serious flaw in the telescope's primary mirror, resulting in the image distortion. After an investigation, it was discovered that when the mirror was fabricated, it had been ground (a process where the mirror is shaped by removing glass with abrasives) to the wrong shape. The error in the shape of the mirror was extremely small, about 2200 nanometers (0.0000866 in) or about 1/50th the thickness of a sheet of paper, but this was enough to make the HST images blurry. Luckily, the HST was actually designed to be serviced and maintained in space with the aid of the Space Shuttle. This original intent was for the repair and updating of instruments and components, not a major fix of the primary mirror. In a series of five complex Space Shuttle missions, astronauts successfully installed fixes for the HST mirrors, correcting its "blurry vision".

The *Hubble* has proved to be a very productive astronomical observatory, having made over a million observations since its insertion into orbit, by some estimates. Observations from *Hubble* have led to significant scientific discoveries in many areas of astronomy and cosmology. With its increased optical resolution, the *Hubble* can look farther back in time to observe the events closer to the creation of the universe. *Hubble* helped scientists discover dark energy, a mysterious force theorized to be accelerating the expansion of the universe. The *Hubble* data has led scientists to revise the age of universe from about 10–20 billion years to about 13–14 billion years. The *Hubble* Space Telescope is expected to continue providing exciting astronomical observations until at least 2020. Its successor, the James Webb Space Telescope, is planned to be launched into space sometime in 2018.

A Deep Space Flyby Spacecraft: Pioneer 10

Launched from Cape Canaveral, Florida on 3 March 1972, *Pioneer 10* (Figure 1.61) was the first flyby spacecraft to be sent into deep space. It was the first spacecraft to go beyond the orbit of

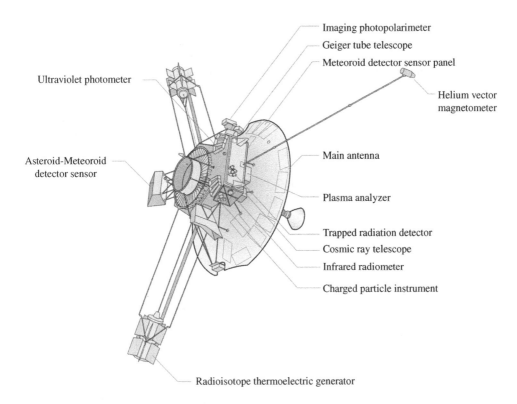

Ultraviolet photometer

Imaging photopolarimeter

Geiger tube telescope

Meteoroid detector sensor panel

Helium vector
magnetometer

Asteroid-Meteoroid
detector sensor

Main antenna

Plasma analyzer

Trapped radiation detector

Cosmic ray telescope

Infrared radiometer

Charged particle instrument

Radioisotope thermoelectric generator

Figure 1.61   A deep space flyby spacecraft, Pioneer 10. (Source: *NASA*.)

Mars and through the asteroid belt, beyond Mars and Jupiter. The primary mission of *Pioneer 10* was to explore the gas giant planet Jupiter. *Pioneer 10* obtained scientific information about Jupiter and several of its moons, including infrared, visible light, and ultraviolet images, measurements of the planet's atmosphere and radiation environment, and data about the bodies' masses. The probe passed within about 130,000 km (81,000 miles) of Jupiter in December 1973, a distance equal to about three diameters of the planet. The probe continued its journey past Jupiter, continuing to send back scientific data about deep space and became the first manmade object to leave our solar system.

As shown in Figure 1.61, a prominent feature of the *Pioneer 10* spacecraft was its large, 27.4 m (90 ft) parabolic, dish antenna. At the base of the antenna, the main body of the spacecraft was comprised of a 36 cm (14 in) deep hexagonal bus structure with each side 76 cm (30 in) in length. Eight of the probe's eleven science instruments were housed within an equipment compartment inside the bus. Insulating blankets, made of aluminized Mylar and Kapton, provided passive thermal protection of the components in the compartment. An active thermal control system, using movable louvers, dissipated excess heat that was generated by the electrical components. The total weight of the probe at launch was 258 kg (569 lb).

The spacecraft was stabilized by spinning the vehicle around the axis of the dish antenna, a technique known as spin stabilization. Six hydrazine-fueled, rocket thrusters provided spacecraft attitude and orientation control. Each small rocket motor generated about 4.5 N (1.0 lb) of thrust. The rocket fuel was liquid hydrazine, a highly toxic, highly flammable, clear liquid monopropellant, commonly used in satellite reaction control systems. A monopropellant does not require a separate fuel and oxidizer as it can generate a hot thrust-producing gas by itself. The spacecraft was launched with about 36 kg (79 lb) of hydrazine fuel, stored in a single, 42 cm (17 in) diameter spherical tank. Two of the thrusters were used to maintain the spacecraft spin-rate at a constant 4.8 rpm. Two Sun sensors and one star sensor were used to keep the spacecraft properly oriented.

The spacecraft had two 8 W transceivers (a combination of a transmitter and receiver) for redundancy. Several antennas were connected to the transceivers, including the large dish, high-gain, narrow-beam, dish antenna and smaller omni-directional and lower-gain antennas. Data from the probe were transmitted back to earth using the transceivers and antennas at a maximum transmission rate of only 256 bits per second. This data transmission rate was degraded as the probe traveled further away from earth. Commands were transmitted to the probe from controllers on earth.

Spacecraft electrical power was supplied by four radioisotope thermoelectric generators (RTGs) that used plutonium-238 fuel. The RTGs were mounted on two 3 m (9.8 ft) support rods (Figure 1.61) to keep the radioactive fuel away from the other spacecraft equipment and instruments. The four RTGs could supply a total of 155 W of power, which decreased to about 140 W as the plutonium fuel decayed. The power system was designed to provide 100 W of power, required for all of the spacecraft systems, for two years. This design goal was far exceeded as the spacecraft continued to be at least partially operational until 2003. Eventually, *Pioneer 10* depleted its electrical power so that it could no longer transmit radio messages back to earth.

A gold-anodized, aluminum plaque was attached to the spacecraft, which was designed to provide information about the civilization on earth, in the event the probe was found by extraterrestrial beings. The plaque had unique diagrams and symbols, including a depiction of a human male and female, with the right hand of the male raised in a gesture of good will. The raised hand was also meant to inform an extraterrestrial being that humans have opposable thumbs. The human figures were drawn to scale relative to a diagram of the *Pioneer 10* spacecraft, making it possible to deduce the size of humans. A kind of interstellar map was provided showing the position of the Sun relative to the center of the galaxy. The trajectory of the spacecraft was depicted on a diagram of our solar system, tracing its path from earth to Jupiter and beyond the solar system. There were also binary numbers etched on the plaque, which provided the height of the female and the distance of the earth from the Sun. The binary numbers were defined on the plaque to be in units of the spin state transition of a hydrogen electron, which can be interpreted as a unit of length, with a wavelength of 21 cm (8.3 in), or a unit of time, with a frequency of 1420 MHz. Hydrogen was selected because it is the most abundant element in the universe.

The last signal received from *Pioneer 10* was on 23 January 2003 when it was 12 billion km (7.5 billion miles) or 80 AU from Earth (an AU is an astronomical unit equal to the mean distance from the earth to the Sun). At this distance, it takes over 11 hours for a radio signal to reach the earth. In 2012, the spacecraft was over 100 AU (15 billion km, 9.3 billion miles) from Earth. At this enormous distance, it takes about 14 hours for the light from the Sun to reach the spacecraft. *Pioneer 10* is flying towards the star *Aldebaran*, a giant, orange star in the zodiac constellation *Tarus* (the Bull). Its interstellar journey to *Aldebaran* will take over two million years.

A Surface Rover: the Curiosity Mars Rover

The exploration of other planets has always sparked the imagination of mankind. The successful landing of a vehicle on another world is perhaps one of the most difficult engineering feats in aerospace engineering. Some planetary landers are unmanned spacecraft that land in a specific location and remain stationary, collecting scientific data and information with an array of sensors and instruments, perhaps including mechanical arms to collect surface samples. Other unmanned planetary rovers have the ability to move about the surface, driving around on a set of wheels, much like a remotely controlled robotic automobile.

One such planetary rover is the car-sized, *Curiosity* robotic rover (Figure 1.62), sent to explore the planet Mars. *Curiosity* was a part of the NASA Mars Science Laboratory mission, launched

Figure 1.62   Self-portrait of a surface rover, the *Curiosity* rover on the surface of Mars. (Source: *NASA*.)

on 26 November 2011. The *Curiosity* rover successfully landed in the Gale Crater on Mars on 6 August 2012. *Curiosity* is the fourth robotic rover that has been sent by NASA to explore the surface of Mars. The rover has collected scientific data about the weather and geology of Mars, including searching for clues as to whether the Martian environment was ever suitable for microbial life.

With a length of 2.9 m (9.5 ft), width of 2.7 m (8.9 ft), and height of 2.2 m (7.2 ft), the *Curiosity* rover has a mass of about 900 kg (1980 lb), significantly heavier than previous Mars rovers. Due to its greater mass, a new descent and landing technique was required. A main parachute and retro-rockets were used to decelerate the spacecraft until it was close to the Martian surface. The new landing technique utilized a "sky crane" upper portion of the spacecraft, which lowered the rover to the surface, suspended from the sky crane by cables. After the rover touched down, the sky crane vehicle cut the cables and flew away from the rover, crash-landing away from the rover.

Components and instrumentation on the *Curiosity* rover are shown in Figure 1.63. The rover has six, 50 cm (20 in) diameter, independently actuated wheels, each with its own motor. The front and rear sets of wheels can be independently steered, allowing for a tight turning radius. With a ground

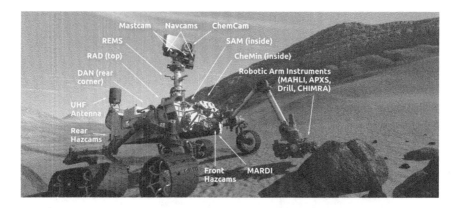

Figure 1.63   Components and instruments on the *Curiosity* Mars rover. (Source: *NASA*.)

clearance of 60 cm (24 in), the six-wheeled rover can roll over obstacles as high as 75 cm (29 in), climb slopes up to 12.5°, tilt in any direction up to 50°, and has an average speed of about 90 m per hour (300 ft/h).

The rover has a three-joint, robotic arm with a cross-shaped turret at its end, which functions much like a human hand. The turret can hold various types of tools, enabling *Curiosity* to perform geological tasks such as drilling into rocks, grinding samples, and digging in the Martian soil. Two instruments are located at the tip of the arm, the alpha particle X-ray spectrometer (APXS) and the Mars hand lens imager (MAHLI). The arm can position these instruments close to features on the Martian surface allowing for an X-ray spectrographic analysis or microscopic imaging.

Operation of the robotic rover is controlled by commands sent by a ground station on earth. *Curiosity* has two identical, radiation-hardened computers, a primary and a backup, to support the rover's robotic functions. An internal, three-axis inertial measurement unit (IMU) is used to calculate the rover's position.

A plutonium-fueled radioisotope thermoelectric generator (RTG) provides about 125 W of electrical power for the rover. The 4.8 kg (11 lb) of radioactive plutonium-238 onboard the rover supplied power for at least 687 (earth) days, equal to a Martian year. The rover also has two rechargeable, lithium-ion batteries that can provide additional electrical power. The rover and its electronic components and instruments are temperature controlled to survive the extremes in temperature on the Martian surface, ranging from about −127 to 40°C (−197 to 104°F). The passive heat from the RTGs and electric heaters are used to keep the equipment at the appropriate temperatures.

Curiosity has redundant communications capabilities, including three antennas and multiple communications links. The rover can send and receive signals directly back to Earth using a direct X-band communications link or relay signals via other spacecraft in Mars orbit using an ultra-high frequency (UHF) communications link. Antennas include a steerable, high-gain antenna and non-steerable, omnidirectional, low-gain antennas. Due to the long distance involved, it takes over 14 minutes for a communications signal to travel between Mars and Earth.

The rover has 17 cameras, eight hazard avoidance cameras, four navigation cameras, four science cameras, and one descent imager. Mounted on the front and rear sections of the rover, the four pairs of hazard avoidance cameras are used to detect terrain hazards such as large rocks and trenches. These cameras allow the rover to move autonomously, but usually the imagery data is used by Earth ground controllers to plan the rover's path. The four navigation cameras are two pairs of stereo cameras that are mounted on a vertical mast at the front of the rover. They provide three-dimensional, panoramic images that are used in conjunction with the hazard avoidance cameras to support navigation of the rover on the Martian surface. The science cameras included the MastCam, a pair of cameras mounted on the vertical mast, which take three-dimensional, stereo, color images and video from as high as seven feet above the surface. The laser-induced remote sensing for chemistry and micro-imaging camera, or ChemCam, also located on the mast, can fire a laser at a rock or soil sample up to 7 m (23 ft) away and vaporize a pinhead-sized spot on the sample. A spectrograph analyzes the vapor to determine its composition and provide information as to whether the sample is worthy of more close-up, intense scrutiny by the rover. The fourth science camera is the Mars hand lens imager (MHLI), a special type of "magnifying glass" or microscope, which can see objects smaller than the diameter of a human hair. The Mars descent imager (MARDI) provided high-resolution imagery during *Curiosity*'s descent and landing on Mars.

One of the major goals of the *Curiosity* mission is to search for signs of past life on Mars. *Curiosity* is a true science laboratory on Mars, with the capability to analyze atmosphere and surface samples using its onboard test equipment. Using its robotic arm, *Curiosity* can insert a Martian rock or soil sample into the "sample analysis at Mars" (SAM) instrument or chemistry and mineralogy (CheMin) X-ray diffraction and fluorescence instrument. The SAM instrument is used to detect organic, carbon-based molecules, a possible precursor to the chemical building blocks of life. The CheMin instrument is used to detect the minerals in the samples, which could provide information as to the past presence of water on Mars. The dynamic albedo of neutrons (DAN) instrument also searched for signs of past water on Mars. At the time of writing, the *Curiosity* rover is still operational on the surface of Mars.

1.3.4   Manned Spacecraft

Placing human beings in space is no easy task. The technological and economic challenges are enormous. To date, only three nations have successfully flown manned spacecraft, the Soviet Union, the USA, and China. The Soviet Union was the first nation to put a man in space on 12 April 1961, with the United States second, less than a month later. The Chinese are relative newcomers to manned spaceflight, flying their first manned mission in 2003. The early manned spacecraft were all capsule configurations. The airplane-like Space Shuttle was flown in the 1980s by the United States for three decades, until its retirement in 2011. The Soviets and Chinese have used a capsule-type spacecraft since the start of their space programs, although the Russians worked on a Shuttle-like vehicle, the *Buran*, for a short time. The United States is returning to a capsule configuration with the development of the *Orion* spacecraft.

In addition to manned spacecraft that can transport people from the earth into space, there are manned space structures or space stations that are placed into low earth orbit. A space station is designed to remain in orbit for long periods of time, with rotating crews of astronauts living onboard for many months. The long duration stays on space stations have provided a large amount of scientific information about the effects of long-term spaceflight on the human body. A wide range of other scientific research is typically performed on a space station, including studies in astronomy, atmospheric science, and biology. A space station usually has a modular structure, comprising connected pressurized modules, which serve as work spaces, laboratories, and living quarters. Other connected structures include large solar panel arrays to provide electrical power, airlocks allowing astronauts to exit and enter the station to and from space, and structures to allow docking of spacecraft.

The first space station was the Russian *Salyut*, launched in 1971. The International Space Station (ISS) (Figure 1.64) is currently in low earth orbit. The ISS is a large, modular structure that is visible to the naked eye in the night sky. With an apogee of 330 km (205 miles) and a perigee of

Figure 1.64   The International Space Station (2010). (Source: *NASA*.)

435 km (270 miles), the ISS completes an orbit of the earth about every 90 minutes. The ISS has had a continuous human presence onboard since November 2000. Currently, the Chinese also have a small, single module station in orbit, the *Tiangong 1*. Prior to the ISS, there were eight earlier space stations that were placed in earth orbit by Russia and the United States.

1.3.4.1   The First Manned Spacecraft

Less than four years after the successful launch of the first artificial satellite into earth orbit, the Soviet Union achieved another space first, launching the first human being into space. On 12 April 1961, the first manned spacecraft, *Vostok 1* (Figure 1.65), entered low earth orbit with 27-year-old Soviet cosmonaut Yuri Gagarin onboard. Gagarin and the *Vostok 1* completed a single earth orbit, making it the shortest manned orbital flight in history, with a flight time from launch to landing of 108 minutes.

The entire *Vostok 1* flight was controlled by either automatic systems or by ground control, even though there were manual controls that could be operated by the cosmonaut. Since this was the first time that a human being had been exposed to the space environment, including the effects of weightlessness, it was unknown whether there would be adverse reactions on the human body, incapacitating the cosmonaut. Therefore, it was decided that automatic systems or ground control would be the safest option. In fact, the manual controls were locked during the mission, requiring a code to unlock them. The unlock code was sealed in an envelope, to be opened by Gagarin in the event of an emergency.

The *Vostok 1* spacecraft consisted of a spherical capsule attached to a service module. The single cosmonaut sat in an aircraft-type ejection seat in the spherical capsule, which had three small porthole windows (Figure 1.65). The spherical capsule was 2.3 m (7.5 ft) in diameter and weighed about 2400 kg (5300 lb). The service module contained the batteries for electrical power, consumables for life support, instrumentation and telemetry systems, the spacecraft attitude control system, and the retrorocket propulsion system, to slow the vehicle for return to earth. The spherical capsule was

Figure 1.65   Russian *Vostok 1*, the first manned spacecraft on display at the RKK Energiya Museum, Moscow, Russia. The capsule hatch, on the ground at left, is replaced by a clear window. The ejection seat is on the right. (Source: © *D.R. Siefkin, "Gagarin Capsule" https://en.wikipedia.org/wiki/File:Gagarin_Capsule .jpg, CC-BY-SA-3.0. License at https://creativecommons.org/licenses/by-sa/3.0/legalcode.*)

separated from the service module for the descent back to earth, but the cosmonaut did not remain in the capsule for the landing. At an altitude of about 7 km (23,000 ft), Gagarin ejected from the capsule and parachuted to the ground. The capsule used parachutes for its final descent and was recovered on the ground.

The spherical descent capsule was covered with an ablative material to protect it from the intense temperatures generated during entry to the earth's atmosphere. Ablation is a method of thermal protection where a material coating is allowed to vaporize or melt away. Heat is absorbed in the chemical transformations and phase changes of the material during ablation and is carried away from the vehicle by the flow of the vaporized material into the freestream flow. The black, charred appearance of the ablative material on the *Vostok* capsule, after experiencing the intense thermal environment of atmospheric entry, is evident in Figure 1.65.

Just 25 days after Gagarin's flight in *Vostok 1*, the United States launched the world's second manned space vehicle, the *Mercury Freedom 7* spacecraft, on 5 May 1961 (Figure 1.66). The first American in space was Alan B. Sheppard, Jr, who completed a sub-orbital flight, lasting 15 minutes and 28 seconds, to an altitude of over 187 km (116 miles). This was followed by another sub-orbital flight with Virgil I, Grissom piloting the *Mercury* spacecraft, *Liberty Bell 7*, on 21 July 1961. John H. Glenn, Jr became the first American to orbit the earth on 20 February 1962 with the flight of the *Mercury Friendship 7* spacecraft. Glenn and the *Friendship 7* completed three orbits of the earth with the flight lasting 4 hours and 55 minutes. The orbit of *Friendship 7* had a perigee of 98.8 miles (159 km) and an apogee of 165 miles (265 km). (These three astronauts were in the first group of seven astronauts selected by NASA. They each named their *Mercury* spacecraft, adding the number "7", at the end of the name, to represent their group of seven.)

Figure 1.66   Launch of the *Mercury-Redstone* rocket with astronaut Alan Shepard, aboard *Freedom 7*, the first American in space. (Source: *NASA*.)

The *Mercury* spacecraft was a one-man capsule with a truncated cone-shaped main body and cylindrical upper body (Figure 1.67). The spacecraft had a length of 7.2 ft (2.2 m), a maximum diameter at the base of the cone-shaped body of 6.2 ft (1.9 m), and a weight of about 2400 pounds (1090 kg). The astronaut sat with his back to the capsule base, looking forward towards the cylinder top, with a small window on the angled cone surface. Unlike, the *Vostok* capsule, there was no ejection seat. The capsule's recovery drogue, main, and reserve parachutes were packed in the forward cylindrical section.

A 17 ft (5.2 m) long escape tower was mounted on the cylinder top of the capsule. Solid rocket motors, mounted at the top of the escape tower, could be fired to pull the capsule clear of the rocket booster in the event of a launch emergency. The escape tower was jettisoned after the spacecraft had reached a safe altitude. An ablative heat shield was mounted to the base of the capsule's truncated cone shape. A retrorocket package was attached to the heat shield with metal straps.

To return to earth, the spacecraft was oriented "backwards", so that firing the retrorockets would slow the spacecraft, causing it to descend from orbital altitude. After this entry burn, the retrorocket

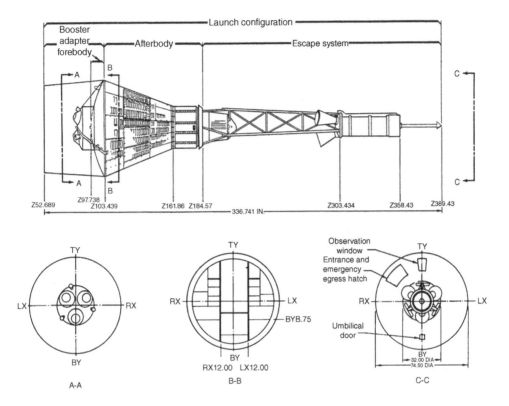

Figure 1.67   The *Mercury* spacecraft components and dimensions. (Source: *NASA*.)

package was jettisoned. The spacecraft was thus oriented with its blunt heat shield entering the atmosphere, protecting the capsule and its backwards-sitting astronaut occupant. The heat shield reached a maximum temperature of about 3000°F (1900 K) during entry. Prior to John Glenn's entry, there were some indications that the spacecraft's heat shield had come loose. Therefore, the retrorocket package was not jettisoned for entry, with the thinking that the metal straps would help retain the heat shield in place. After landing and recovery of the spacecraft, it was determined that a warning light had erroneously indicated a problem with the heat shield.

A drogue parachute was deployed at an altitude of about 21,000 ft (6400 m) to stabilize the spacecraft, followed by deployment of the main parachute at about 10,000 ft (3050 m) to slow the final descent for a landing in the Atlantic Ocean. A landing bag was inflated behind the heat shield to cushion the water impact. A fleet of US Navy ships was used in the recovery operation of the spacecraft in the ocean.

It is interesting to compare the different design approaches for the first manned spacecraft developed by the Soviet Union and the United States, as shown in Figure 1.68. Fundamentally, both spacecraft had a common design requirement, that of placing a man in space and returning him safely, but the design solutions were quite different. The *Vostok* was a two-module system, comprising a spherical entry capsule and a service module, while the *Mercury* was a single-module system with a single capsule. One of the most striking differences between the *Vostok* and *Mercury* was the shape of the spacecraft. The *Vostok* spacecraft had the shape of a simple sphere, while the *Mercury* spacecraft had the shape of a truncated cone topped with a short cylinder. Given its non-aerodynamic shape, an aerodynamic *shroud* covered the *Vostok* spacecraft at the top of the

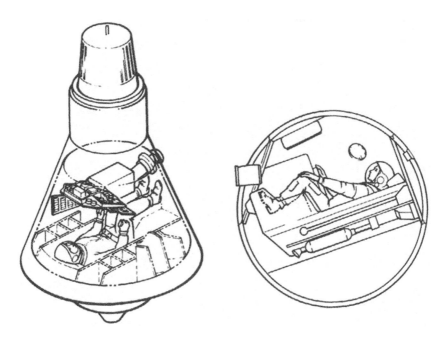

Figure 1.68 Comparison of the *Mercury* and *Vostok* spacecraft. (Source: *From Ezel, NASA SP-4209, 1978.*)

rocket booster during launch into space. The *Mercury* capsule's shape was amenable to being placed at the top of the rocket booster with no shroud. In contrast, both vehicles used a blunt-body shape for entry into the atmosphere from orbit. The *Sputnik* used a spherical shape and the Mercury entered backwards, with its blunt base facing into the flow. As discussed in Chapter 3, a blunt-body shape is optimum for minimum heat transfer at hypersonic speed.

The *Vostok* sphere is an optimum shape for a pressure vessel with maximum interior volume and minimum structural mass. The aerodynamics of the sphere was also well-known at the time. It was certainly known that the spherical *Vostok* would generate considerable drag during entry to slow the craft and it would not generate any aerodynamic lift. The *Mercury* spacecraft's shape produced a small amount of aerodynamic lift, enough to enable some control over the entry trajectory. The lift-to-drag ratio is a measure of a vehicle's aerodynamic efficiency, its ability to produce lift in relation to the drag. The *Mercury* capsule's lift-to-drag ratio was about 0.2–03, a relatively small number, but still a positive number that enabled limited trajectory control. In fact, an objective of the early *Mercury* flights was to determine if an astronaut could actively control the spacecraft's entry trajectory in a weightless environment. This highlights another difference between the *Vostok* and *Mercury*, having to do with spacecraft control. The *Vostok* entry was entirely controlled by automatic systems and by ground controllers. *Mercury*, on the other hand, allowed manual control of the vehicle by the astronaut.

Both the *Vostok* and *Mercury* had a retrorocket system to slow the spacecraft for entry. The *Vostok* retrorocket system was installed on the service module while the *Mercury* system was attached to the capsule heat shield. Both spacecraft had small thrusters for attitude control in three dimensions. After the retrorocket de-orbit burn, both spacecraft were oriented so that the occupant was facing backwards as they entered the earth's atmosphere. The Soviet engineers came up with a clever way to control the spacecraft attitude during entry. The center of mass of the capsule was offset from the centroid of the sphere, so that the spherical capsule would orient itself in the proper attitude for the entry. This passive attitude control system did not require an active reaction control-type attitude control system, as used in the *Mercury* spacecraft. The *Vostok* spacecraft was of aluminum construction with an ablative coating around the sphere. The *Mercury* had an ablative heat shield protecting the capsule and a nickel alloy pressure vessel with an outer shell of titanium. The *Vostok* cosmonauts experienced a deceleration of up to ten times the force of gravity, or 10 g during entry, while the *Mercury* astronauts felt about 8 g. The difference in the deceleration force is due to the difference between the *Vostok* ballistic entry and the *Mercury* entry with lift. The *Vostok* cosmonaut ejected from the spacecraft and parachuted down to the ground, while the *Mercury* astronaut landed in the water inside the capsule.

In summary, it is interesting and enlightening to compare the first two spacecraft that placed human beings in space. Designed by two different countries, with different design philosophies and different technical capabilities, the two spacecraft represent two different solutions to the design requirement of placing a man in space and returning him safely to earth. Both the *Vostok* and the *Mercury* spacecraft were successful in meeting this design requirement. The *Vostok* spacecraft design would be used for a total of eight spaceflights with six of those being manned missions. The *Mercury* spacecraft would fly a total of 16 times, with six of those being manned spaceflights. The legacy of the *Vostok* and *Mercury* designs would be imprinted on future Soviet and American spacecraft designs, respectively, for many years to come.

1.3.4.2 The Spacesuit: A One-Person Spacecraft

We include in this section about spacecraft, a description of the spacesuit, which in many ways, can be considered a one-person spacecraft. The spacesuit must perform many of the same functions as a spacecraft in maintaining a safe and habitable environment for a human being in space. Spacesuits are worn by astronauts for extravehicular activity (EVA), commonly referred to as a "spacewalk". In modern times, spacewalks have become almost commonplace, with the hundreds of hours of EVA time that it has taken to construct and maintain the International Space Station (ISS). The first spacewalk occurred on 18 March 1965, when Russian cosmonaut Alexei Leonov exited his *Voskhod 2* space capsule and floated in space for 12 minutes and 9 seconds. He was kept from floating too far away from his capsule by a 5 m (16 ft) umbilical, which supplied his spacesuit with oxygen. Less than three months later, astronaut Edward White performed the first American spacewalk on 3 June 1965 during the *Gemini 4* space mission (Figure 1.69). White's EVA lasted 23 minutes. In addition to EVA in earth orbit, spacesuits have been used for EVA on the moon by the *Apollo* astronauts.

Similar to spacecraft, the design of spacesuits has evolved over time, driven by changing requirements, design improvements gained by real-world experience, and advancements in technology. We examine one particular spacesuit to get some idea as to what is involved in the design of these complex systems. The NASA extravehicular mobility unit (EMU) is the EVA spacesuit system that was used for the NASA Space Shuttle program and that is currently in use for the International Space Station (ISS) program (Figure 1.70). The EMU is composed of a pressure garment, called the space suit assembly (SSA), and an integrated life support system. The SSA is attached to the hard upper torso (HUT), a fiberglass and aluminum outer shell that is the primary structural unit of the EMU. The portable life support system (PLSS) and secondary oxygen package (SOP) are attached to the HUT. The SSA, PLSS, and SOP are covered by the thermal micro-meteoroid garment (TMG), an outer garment that is a thermal and impact barrier.

The SSA is a multi-layered garment, pressurized to 29.6 kPa (4.29 lb/in^2) with 100% oxygen. The SSA layers consist of a pressure bladder, a pressure-restraint layer, and a thermal insulation layer. The innermost layer is a urethane-coated nylon pressure bladder for the spacesuit. A

Figure 1.69   The first American spacewalk, performed by Edward White, on June 3, 1965. (Source: *NASA*.)

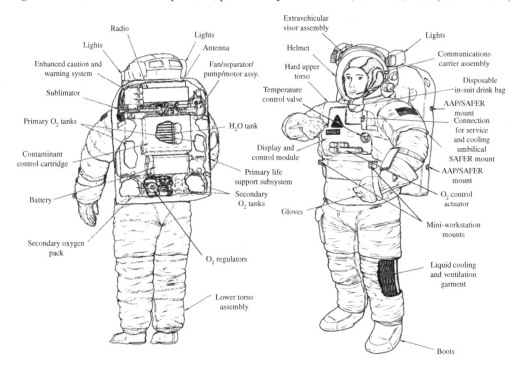

Figure 1.70   NASA extravehicular mobility unit (EMU). (Source: *NASA*.)

pressure restraint garment covers the pressure bladder and carries the spacesuit pressure loads. The outermost covering of the SSA consists of five thermal insulation layers of tear-resistant, neoprene-coated nylon and aluminized Mylar.

Underneath all of the SSA layers, the astronaut wears a liquid cooling and ventilation garment (LCVG), a kind of long underwear that keeps the wearer cool and dry. The LCVG has 90 m (300 ft) of narrow tubes sewn into it that circulate chilled water to remove excess body heat. Vents in the LCVG serve to draw sweat away from the body.

The life support system has a primary system, the Portable Life Support System (PLSS), and a backup, secondary oxygen package (SOP). The PLSS backpack houses breathing oxygen tanks, an oxygen circulation fan, carbon dioxide "scrubbers", water cooling equipment, a two-way radio, and a silver-zinc battery power source. The SOP has backup oxygen and water cooling tanks. The PLSS can provide life support for up to seven hours, but the actual duration is a function of the astronaut's metabolic rate (how much energy is being expended, impacting the suit cooling and how fast the consumables are being used) and the solar thermal environment (impacting the suit cooling). The PLSS operation is monitored and controlled using the enhanced caution and warning system (ECWS) and the display and control module (DCM), respectively, mounted on the front of the HUT. Since the front of the DCM cannot be seen while in the spacesuit, the astronaut wears a wrist mirror to read the displays which are labeled backwards so as to be legible in a mirror. The astronaut's heart rate and other vital suit parameters, such as carbon dioxide level, are telemetered to the ground for monitoring during EVA.

The spacesuit helmet is a clear plastic, polycarbonate bubble. The visor assembly covers the bubble and contains the visor, a movable sunshade, lights, and camera. The visor is coated with a thin layer of gold to provide protection from cosmic rays. Communications headphones and microphones are in the communications carrier assembly (CCA), also called the "Snoopy cap", a fabric head cover worn under the helmet. Inside the helmet, the astronaut can drink from a water bag via a plastic tube.

The gloves are one of the more complex items of the spacesuit, as they must provide adequate protection and insulation while remaining as dexterous as possible. The gloves must integrate the same layers of protection and insulation as in the SSA, yet they must allow finger motion and tactility sufficient to perform manipulation and handling tasks, such as the use of tools and operation of DCM suit controls. Since the fingers tend to get the coldest in space, the gloves have fingertip heaters.

The weight of the complete spacesuit, including the life support system and consumables, is about 120 pounds (54 kg). The spacesuit is a complex system that incorporates many of the features of a manned spacecraft, such as life support functions and protection from the space environment.

1.3.5   Space Access Systems and Vehicles

For spacecraft to operate in space, they must first get into space. In this section, we examine several different systems used to place spacecraft into space. Currently, the use of expendable rockets is the predominant means of access to space. We discuss the birth of the liquid-fueled rocket, occurring independently in the USA and Germany. Other access to space systems are explored, including air-launched systems and non-rocket-based systems.

According to the ancient Greek novel by Lucian in the 2nd century AD, a sailing ship is swept upward by a waterspout to 350 miles above the earth. After seven days atop the waterspout, the ship is deposited on the Moon where there are strange, extraterrestrial creatures from the Moon and Sun. This fictional story is the first recorded description of travel to outer space, albeit using a technically implausible launch system.

Figure 1.71   Space cannon from Jules Verne's *From the Earth to the Moon*. (Source: *PD-old-100.*)

Jules Verne, in his 1865 novel *From the Earth to the Moon*, imagined an enormous cannon to launch a projectile, with three people inside, to the Moon (Figure 1.71). Verne's fictional space cannon was 900 ft (270 m) long and had a bore diameter (interior diameter of the cannon barrel) of 9 ft (2.7 m). Made of cast iron, the cannon was 6 ft (1.8 m) thick and weighed more than 68,000 tons (60 Mkg). Due to its immense size and weight, the cannon was built directly into the ground, pointing vertically straight up. Coincidentally, Verne selected Tampa, Florida as his launch site, not far from the current Cape Canaveral rocket launch complex. Verne did make some technical calculations for the design of his space cannon, but in reality the barrel of his cannon was much too short for the projectile to achieve escape velocity from the earth. A longer barrel would enable the high pressure gases from the firing charge to push on the projectile for a longer time, producing a higher exit velocity. In addition, the acceleration loads on the people inside the projectile would not have been survivable, estimated to be over 20,000 g. However, the idea of a space cannon may have some technical feasibility, at least for unmanned payloads.

1.3.5.1   Expendable Rocket-Based Launch Systems

Currently, the only method that we have for placing spacecraft into space is using rockets. Usually the rocket is launched from a land-based pad, although there are a few sea-based launch platforms. Mobile launch platforms include submarines and airplanes, but these can only accommodate smaller rocket systems.

A rocket has a self-contained propulsion system, carrying both the fuel and oxidizer required for combustion. Since rocket propulsion does not rely on atmospheric air for combustion, rockets can function beyond the sensible atmosphere into outer space. There are two primary types of rockets: *liquid-fueled* and *solid propellant*. In the liquid-propellant rocket, the fuel and oxidizer are stored in separate tanks in the vehicle. The fuel and oxidizer are combined into a solid mixture in the solid-fuel rocket.

Most rockets used today for space access are *expendable rockets*, that is, they are used for one launch and discarded, usually by allowing them to fall back into the atmosphere and burn up or fall into the open ocean. To be a little more precise with our terminology, a *rocket booster* includes everything in the rocket (rocket engine, propellant, tankage, propellant feed systems, structure, etc.) except the payload, which may be a spacecraft.

Two of the largest liquid-fueled rocket boosters ever built, the American *Saturn V* and the Soviet N1, are shown in Figure 1.72. Both were expendable, man-rated, multi-stage, heavy-lift launch vehicles, designed to transport human beings beyond earth orbit. Both boosters had three stages, each stage containing its own rocket engines, tankage, propellants, and propellant feed system. The concept of *staging* is used to reduce the weight of the rocket as it ascends by discarding the mass of each successive stage as its propellants are consumed.

The *Saturn V* booster was developed to transport human beings to the Moon as part of the *Apollo* program in the late 1960s and early 1970s. The *Saturn V* was successfully launched 13 times from Cape Canaveral, Florida, including six moon landing missions. The five F-1 liquid-fuel rocket engines in the first stage of the *Saturn V* produced a combined thrust of over 7.6 million lb (33.8 MN) at lift-off. The 363 ft (111 m) height of the *Saturn V* booster was greater than the length of an American football field. The total lift-off weight of the *Saturn V* was over 6.5 million lb (2900 tonnes). The booster was capable of placing very large payloads into low earth orbit, including the heaviest payload ever launched of 260,000 pounds (118,000 kg or 118 tonnes).

The N1 booster was the Soviet counterpart to the American *Saturn V*. The N1 was not quite as long as the *Saturn V*, standing 344 ft (105 m) tall. Its lift-off weight of about 6 million lb (2700 tonnes) was also lower than that of the *Saturn V*. The N1 was designed to deliver up to 200,000 pounds (90,000 kg) of load to low earth orbit. The first stage of the N1 housed 30 NK-15 liquid-fuel rocket engines, making the first stage propellant feed system very complex. The 30 rocket engines produced a combined lift-off thrust of 11.3 million lb (50.3 MN), making it the most powerful rocket stage ever built. Only four unmanned attempts were made to launch the N1 booster and all failed catastrophically. The N1 never made it into space, with its longest flight of 107 seconds never reaching first stage separation. While the N1 may not have been a success, future Soviet efforts with heavy-lift, expendable rocket boosters were successful in their manned spaceflight program.

1.3.5.2   The First Unmanned Liquid-Propellant Rocket

Robert Hutchings Goddard (1882–1945) was born at a time when airplanes and spacecraft were by no mean, commonplace. He was 11 years old when the Wright brothers made their historic first flight of a heavier-than-air airplane. When Goddard was 17, he climbed a cherry tree in his backyard to prune its dead branches. Resting in its branches, he gazed up at the sky and imagined what could be. He later wrote, in an autobiographical account, of the inspiration that he felt while gazing up at the sky in that cherry tree.

Figure 1.72 Models comparing sizes of United States *Saturn V* and Soviet N1 "moon rockets". A scale model of a person is shown at the bottom, between the boosters. (Source: *From Portree, NASA-RP-1357, 1995.*)

> On the afternoon of October 19, 1899, I climbed a tall cherry tree and, armed with a saw which I still have, and a hatchet, started to trim the dead limbs from the cherry tree. It was one of the quiet, colorful afternoons of sheer beauty which we have in October in New England, and as I looked towards the fields at the east, I imagined how wonderful it would be to make some device which had even the possibility of ascending to Mars. I was a different boy when I descended the tree from when I ascended for existence at last seemed very purposive.

The purpose to which Goddard alluded was his decision to dedicate his life to making spaceflight a reality. For the rest of his life, Goddard privately observed that day, 19 October, as the anniversary day of his inspiration.

In 1907, he began his lifelong research and testing of rockets in earnest, as an undergraduate physics student at the Worcester Polytechnic Institute, Massachusetts. After earning his PhD in physics from Clark University in Worcester, Massachusetts, he accepted a research fellowship at Princeton University in 1912. In addition to possessing exceptional academic skills, Goddard was a prolific inventor, obtaining 214 patents over his career (some of these were granted after his death).

Goddard was granted two significant patents in 1914; one described a solid-fueled, multi-stage rocket and the other, a rocket fueled with gasoline and liquid nitrous oxide, or in other words, a liquid-fueled rocket. In the fall of 1914, Goddard returned to Clark University, where he conducted experiments with different types of solid propellant rockets, much of this at his own expense. He performed *static ground tests* of these rocket engines, carefully measuring their thrust and efficiency. (A static ground test is a test of the rocket engine that is performed with the engine securely mounted in a test stand on the ground. The engine is fired *statically*, meaning that it cannot move. The engine is instrumented so that data is collected about the operation of the engine and its systems. A load cell, or other force-measuring device, is often attached to the rocket, so that the thrust force is measured. Today, static ground tests of rocket engines are a standard, engineering practice.)

In 1916, Goddard built a vacuum tube apparatus to show that the efficiency of a rocket increased with decreasing external pressure (Figure 1.73). The rocket was placed at the top of the long vertical

Figure 1.73   Robert Goddard with vacuum tube device, which he used to prove that a rocket could produce thrust in the vacuum of space, June 1916. (Source: *NASA*.)

tube, with its exhaust firing into the tube. The oval portion of the tube apparatus served to reduce the rebound of the rocket exhaust gas. It was a misconception of Goddard's time that a rocket would not produce thrust in the vacuum of space. According to this misconception, Newton's third law of "action with equal and opposite reaction" required that the rocket exhaust gas have something to "push against". It was thought that the vacuum of space could not provide the "equal and opposite" reaction to the rocket exhaust gas, to propel the rocket. The error in this reasoning is that the equal and opposite reaction is from the rocket, not the vacuum, reacting to the exhaust gas. Using the vacuum tube apparatus, Goddard was the first to prove that a rocket produces thrust in the vacuum of space. (Goddard obtained several patents, an "apparatus for vacuum tube transportation" and "vacuum tube transportation systems", which were visionary ideas of very high-speed transportation using magnetic levitation, or *maglev*, vehicles traveling inside vacuum tubes.)

During this time, Goddard made some significant advances in the design of rockets. He had the rare ability to transform his theoretical understanding into practical engineering, that is, real flight hardware. To increase the thrust and efficiency of his rockets, he understood several key design requirements. He realized that the rocket exhaust velocity and the rocket's propellant mass fraction, the mass of the propellant relative to the rocket's total mass, must both be as high as possible. To increase the rocket exhaust velocity, Goddard used a converging-diverging exhaust nozzle, called a *de Laval nozzle*. Using this exhaust nozzle, he was able to accelerate the flow exiting the nozzle to supersonic speeds, as high as Mach 7.

In the rocket designs of his day, the fuel and combustion chambers were combined, requiring a large, thick-walled, heavy chamber to withstand the high pressures and temperatures of combustion. To increase the rocket's propellant mass fraction, he separated the fuel chambers or tanks from the combustion chamber. By separating the fuel and combustion chambers, only a smaller combustion chamber was required to withstand the high pressures and temperatures, while the propellant tanks could be made as lightweight as possible. At the time, Goddard also realized that liquid propellants have much higher energy content per unit mass than solid propellants, but he resisted using these as he thought the handling of extremely cold, or *cryogenic*, propellants, such as liquid oxygen, was not practical.

Based on the substantial progress that he was making and the fact that he was unable to continue self-funding all of his research, Goddard started submitting research funding proposal to sponsors such as the Smithsonian Institution and others. In his 1916 proposal to the Smithsonian Institution, he included a paper he had written that detailed his solid propellant rocket experiments, his mathematical theories of rocket propulsion, and his vision of using rockets to explore the Earth's atmosphere and beyond. He provided a quantitative analysis of launching a rocket to the Moon with a payload of flash powder, which would explode upon impact on the Moon. He calculated that a multi-stage rocket, with an initial launch mass of 6436 lb (2919 kg), could deliver a payload of 2.67 lb (1.21 kg) of flash powder to the surface of the moon. He calculated that this amount of flash powder would make a flash "just visible" from an Earth-bound, high power telescope, confirming the rocket's impact on the Moon. The Smithsonian was impressed with Goddard's proposal and, in 1917, awarded him a $5000 grant.

Later, in 1919, the Smithsonian published Goddard's paper as Publication No. 2540 of the Smithsonian Miscellaneous Collections, entitled "A method of reaching extreme altitudes" [11]. This publication has become one of the most significant scientific contributions to the development of rocket propulsion. Unfortunately, at the time of its publication, it was not viewed as such. The public at large still thought that the idea of space travel was a fanciful dream without much scientific basis. In fact, the US government and US military did not see much use for either spaceflight or Goddard's rockets. To make matters worse, the press ridiculed Goddard's ideas of flying rockets to the Moon. This led to Goddard's distrust of the press and his penchant for working in secret, sentiments that he would have for the rest of his life.

By 1921, Goddard was conducting experiments with liquid propellants. He successfully ground tested the first liquid propellant rocket engine, using gasoline and liquid oxygen, in November 1923. He initially pursued a *pump-fed* engine design, where mechanical pumps are used to move the propellants from their tanks to the combustion chamber, but this proved to be problematic. Goddard abandoned the pump-fed system and decided to use a *pressure-fed* system, where a high-pressure, inert gas, such as nitrogen, is used to "push" the propellants out of their tanks and into the combustion chamber. On 6 December 1925, Goddard conducted a static test of a liquid-propellant rocket engine, with a pressure-fed system, in a laboratory at Clark University in Worcester, Massachusetts. The rocket engine fired for 27 seconds, lifting its own weight in the test stand, proving the feasibility of a liquid-propellant rocket engine.

The configuration of Goddard's first liquid-propellant rocket is shown in Figure 1.74. The combustion chamber and exhaust nozzle were located at the top of the rocket and the two cylindrical propellant tanks were at the bottom. Goddard chose this arrangement in his early rocket designs because he thought that the rocket would be more stable in flight with this configuration. A conical-shaped exhaust shield, with an asbestos fabric covering, protected the liquid oxygen

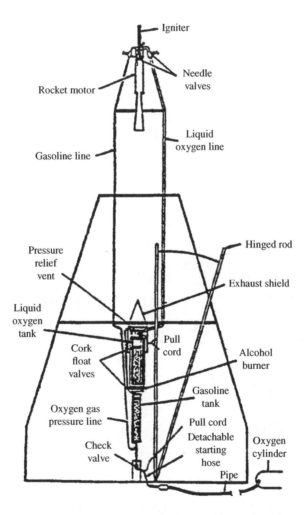

Figure 1.74   Goddard's first pressure-fed liquid-propellant rocket. (Source: *NASA*.)

tank from the hot exhaust. The gasoline and liquid oxygen lines ran from their tanks to the combustion chamber along the left and right sides of the rocket, respectively. These propellant feed lines also served as structural supports for the rocket. The rocket motor ignitor was mounted at the very top of the rocket, above the combustion chamber. The high-pressure gas generated by the boil-off of the liquid oxygen was the pressurant gas that was used to move the propellants to the combustion chamber. Goddard adopted today's conventional rocket configuration, with the combustion chamber and nozzle at the base of the rocket, in his later designs. This early Goddard design lacked any aerodynamic fairings over the rocket nose or body and had no fins for stability.

In 1926, Goddard was ready to conduct flight tests of his liquid-fueled rockets. He moved his operation to a farm owned by a distant relative, about two miles from Clark University. It offered a remote setting for conducting rocket launches, with less chance of an errant rocket crashing and hurting people or property. Despite its remote location, the neighbors still complained about the noise from the rockets. On 26 March 1926, the first flight of a liquid-propellant rocket occurred. The 10 ft (3.0 m) long, gasoline and liquid oxygen-fueled rocket (Figure 1.75) reached a maximum altitude of about 41 ft (12.5 m) and ended its 2.5 second flight in a cabbage field, about 184 ft (56.1 m) from its launch point. Since his youth, Goddard had avidly written in his daily diary. On

Figure 1.75   Robert Goddard with the first liquid-propellant rocket (Goddard is holding the trapezoidal launch frame). (Source: *NASA*.)

this auspicious day of the first flight of a liquid-propellant rocket, he rather succinctly wrote the following.

March 16, 1926

Went to Auburn with S[achs] in am. E[sther] and Mr. Roope[5] came out at 1 pm. Tried rocket at 2:30. It rose 41 ft, & went 184 ft, in 2.5 secs, after the lower half of nozzle had burned off.

The next day, Goddard wrote about the previous day's first liquid-fueled rocket flight with a bit more fanfare.

March 17, 1926

The first flight with a rocket using liquid propellants was made yesterday at Aunt Effie's farm[6] in Auburn. The day was clear and comparatively quiet. The anemometer on the Physics lab was turning leisurely when Mr. Sachs and I left in the morning, and was turning as leisurely when we returned at 5:30 pm. Even though the release was pulled, the rocket did not rise at first, but the flame came out, and there was a steady roar. After a number of seconds it rose, slowly until it cleared the frame, and then at express train speed, curving over to the left, and striking the ice and snow, still going at a rapid rate. It looked almost magical as it rose, without any appreciably greater noise or flame, as if it said "I've been here long enough; I think I'll be going somewhere else, if you don't mind." Esther said that it looked like a fairy or an aesthetic dancer, as it started off. The sky was clear, for the most part, with large shadowy white clouds, but late in the afternoon there was a large pink cloud in the west, over which the sun shone. One of the surprising things was the absence of smoke, the lack of very loud roar, and the smallness of the flame.

In 1930, Goddard moved his rocket flight test operation to remote Roswell, New Mexico, where he had plenty of open space and a clear, dry climate conducive to year-round testing. Today, New Mexico is the location of the White Sands Missile Range, the US Army's rocket test range. Covering almost 3200 square miles (8200 km^2), it is the largest military installation in the United States.

Goddard continued to design and build larger and more powerful rockets, integrating design improvements and technology advancements with each new model. Soon, Goddard's rockets resembled the configuration that is common today. Goddard's A-series rocket had an aerodynamic nose cone, a cylindrical body with a smooth aluminum skin that covered the internal tanks, a nozzle at the base of the rocket, and thin highly swept tail fins. An A-series rocket was the first rocket to fly faster than the speed of sound on 8 March 1935. Goddard was responsible for significant advancements in rocket guidance and control, including inventing a steering system using gyroscope-controlled, movable vanes in the exhaust and another using a movable or *gimbaled* exhaust nozzle, forerunners of systems in use today. He continually developed innovations in rocket propulsion technology, including eventually building turbopumps for pump-fed propellant feed systems. He was the first to launch a scientific payload in a rocket, consisting of a barometer, a thermometer, and a camera.

---

[5] In addition to Goddard, three other people witnessed the first flight of a liquid-propellant rocket: Esther Goddard, his wife and photographer, Henry Sachs, the crew chief, and Percy Roope, an assistant professor of physics at Clark University.

[6] "Aunt Effie" was Effie Ward, a distant relative of Robert Goddard. Her farm was in a rural area, about two miles from Clark University in Worcester, Massachusetts. This site of the first flights of a liquid-fueled rocket is now part of a golf course.

Between 1926 and 1941, Goddard launched 34 rockets, reaching altitudes of about 8500 ft (2600 m) and speeds of about 550 mph (885 km/h). He performed many more static ground tests than flight tests, using a methodical, engineering approach to improve the designs. Many of the ground and flight tests ended with failures of the engine, nozzle, guidance system, or other components, but Goddard was never deterred. He always believed that important lessons were learned from any test results, a noteworthy perspective for anyone involved with ground or flight testing.

Robert Goddard was a true visionary, recognizing the incredible potential of rockets for atmospheric research, ballistic missiles, and space travel. Unfortunately, the US government, the US military, and the public were blind to his vision. Another aerospace visionary, Charles Lindbergh, took great interest in Goddard's work and personally helped with obtaining funds for his rocket research. Overall, Goddard's research and testing of rockets received meager funding and support throughout his career and he received little recognition for his work. Today, Robert Goddard is rightfully considered the father of rocket propulsion. Much like the Wright brothers' first flight, Goddard's first flight was a fledgling step of remarkable consequence. It would ultimately shape the future of rocketry and affect all of humanity. Four short years after Goddard's first flight, the first man to set foot on the Moon was born, destined to fly 238,900 miles (384,500 km) from the earth to the Moon atop a 363 ft (111 m) tall, 6.5 million lb (2900 tonnes) *Saturn V* rocket, a rocket that could trace its roots back to Robert Goddard's first liquid-fuel rocket.

1.3.5.3   The First Rocket to Reach Space

Robert Goddard was not alone in his quest to make rocket flight a reality. Across the Atlantic Ocean, Dr Wernher von Braun and his team of German rocket scientists and engineers were busy at the Peenemunde Army Research Center, on a small Baltic Sea island off the northern coast of Germany, developing the world's first long-range, guided ballistic missile with a liquid-propellant rocket engine. The German scientists and engineers were aware of Goddard's work and monitored it closely. Some believe that Goddard's work made a significant impact on the German rocket designs. The German efforts culminated in the development of the V-2 rocket, which was used as a weapon against the Allied forces during World War II. Over 3000 V-2 rockets were launched by the Germans against Allied targets during the war, many in London, England. On 3 October 1942, a V-2 rocket, launched from Peenemunde, soared to an altitude of 190 km (118 miles, 623,000 ft), becoming the first manmade object to reach the edge of space.

The German developed V-2 liquid-propellant rocket was substantially larger than Goddard's rockets. The V-2 rocket airframe had a length of 14 m (46 ft), diameter of 1.65 m (5 ft 5 in), and wingspan of 3.56 m (11 ft 8 in). With a total launch weight of 12,500 kg (27,600 lb), the V-2 carried 3800 kg (8400 lb) of fuel, comprising a 75% ethanol-25% water mixture, 4900 kg (10,800 lb) of liquid oxygen, and a 1000 kg warhead (2200 lb). The V-2 could reach speeds of 5700 km/h (3540 mph) and altitudes of over 200 km (124 miles, 656,000 ft). The missile had a maximum range of about 320 km (200 miles).

The V-2 rocket was technologically advanced, for its time. In addition to the liquid-propellant rocket engine technology, the rocket incorporated advancements in the areas of supersonic aerodynamics, stability, control, guidance, and navigation. The various components of the V-2 rocket are shown in Figure 1.76. The liquid rocket engine had a pump-fed propellant system, where the fuel and oxidizer pumps were driven by a steam turbine. The steam to drive the turbine was generated by the combustion of hydrogen peroxide with a sodium permanganate catalyst. The propellant tanks were made of a lightweight aluminum–magnesium alloy. The alcohol–water mixture fuel was also

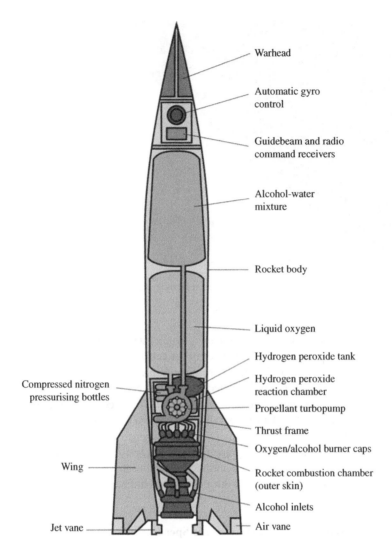

Figure 1.76 Components of the V-2 rocket. (Source: *User: PD-Fastfission, https://en.wikipedia.org/wiki/ File:V-2_rocket_diagram_(with_English_labels).svg.*)

used as a coolant for the combustion chamber and nozzle. The fuel was pumped behind the combustion chamber walls, cooling the chamber and heating the fuel. The heated fuel was then sprayed into the combustion chamber. Fuel was also sprayed inside the nozzle, providing film cooling of the nozzle walls. Steering control of the rocket was provided by a combination of movable rudders (air vanes) on the tail fins (wings) and movable guide vanes (jet vanes) in the rocket engine exhaust. An automatic gyroscope control system was used for vehicle stabilization. Early V-2 rockets used a simple type of analogue computer for guidance and navigation, but later versions used a ground-transmitted, radio signal guide beam.

After World War II, many of the German scientists and engineers that had been involved with the V-2 development, including von Braun, were brought to the United States. Many of these rocket scientists and engineers were located in Huntsville, Alabama, laying the foundations for what would eventually become the US Army Redstone Arsenal and the NASA Marshall

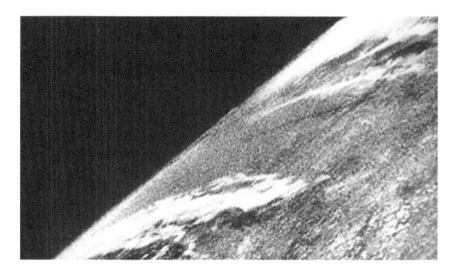

Figure 1.77   The first photograph of the Earth taken from space, 24 October 1946. (Source: *US Army.*)

Spaceflight Center. Accompanying the German personnel was significant amounts of V-2 rocket hardware and components, including complete rocket vehicles. These captured V-2 rockets were studied and launched on test flights by the US from the White Sands Missile Range in New Mexico.

On 24 October 1946, a V-2 rocket, launched from White Sands, carried a small 35-mm motion picture camera, taking black-and-white photographs every 1.5 seconds during its flight. The photographs were not telemetered back to earth; rather the camera film was recovered from a protective case after the rocket crashed back to earth. The rocket reached an altitude of 65 miles (105 km, 340,000 ft), where it took the first photograph of the Earth from space (Figure 1.77). Today, photographs of the Earth from space are quite commonplace, but this first grainy picture of the Earth from space gave us the first perspective of our planet from the new frontier of space.

1.3.5.4   The First Vehicle to Fly at Hypersonic Speed

In the 1940s, the United States developed the *WAC Corporal*, the first *sounding rocket*, specifically designed for upper atmospheric research. Designed and built by the Douglas Aircraft Company and the California Institute of Technology's Guggenheim Aeronautical Laboratory, the *WAC Corporal* was 7 ft, 11 in (2.4 m) in length, 12 in (30 cm) in diameter, weighed about 760 lb (340 kg), and carried a payload with a weight of about 25 lb (11 kg). The *WAC Corporal* was a two-stage rocket system with a solid-propellant first stage and a liquid propellant second stage. The first stage was a 5 ft (1.5 m) long *Tiny Tim* solid rocket booster with three stabilizing fins, capable of producing about 50,000 lb (11,000 N) of thrust for 0.6 s. The second stage had an Aerojet liquid-propellant motor delivering about 1500 lb (340 N) of thrust for 47 s. The rocket was not stabilized and was unguided (some references stated that *WAC* was an acronym for "without attitude control"). Later versions of the *WAC Corporal* incorporated stabilization and guidance systems. Upon reaching its peak altitude, the nosecone of the rocket separated and fell back to earth for recovery of scientific instruments and recording equipment. The first flight of a fully operational *WAC Corporal* occurred on 11 October 1945 at the White Sands proving ground, New Mexico (later to be renamed the White Sands Missile Range). The rocket reached at altitude of about 230,000 ft (44 miles, 70 km). On 22 May 1946, a White Sands-launched *WAC Corporal* reached an altitude of 50 miles (80 km, 264,000 ft), making this the first sub-orbital flight of a manmade object. (An altitude of 50 miles is the altitude boundary for space used by the US Air Force.)

Later, the liquid-propellant, upper stage of the *WAC Corporal* was mated to a much larger V-2 rocket and re-named the *Bumper-WAC* rocket. The V-2 rocket stage was about 45 ft (14 m) in length and had a thrust of about 55,000 lb (245,000 N) at launch. After launch, the V-2 rocket motor burned for only about one minute, before the *WAC Corporal* second stage was ignited, which burned for about 45 s. The *Bumper-WAC* was so-named because of the "bump" in altitude provided by the V-2 rocket.

A total of eight flights of the *Bumper-WAC* rocket were conducted, six from White Sands proving ground, New Mexico and two from Cape Canaveral, Florida. The two Cape Canaveral launches were the first two ever conducted from this fledgling rocket launch complex (Figure 1.78). The first *Bumper-WAC* flight from White Sands was on 13 May 1948, reaching a maximum altitude of about 80 miles (129 km, 422,000 ft) and a maximum speed of about 2740 mph (4400 km/h, 4020 ft/s). The firsts for this little liquid-propellant rocket were to continue. On 24 February 1949, the fifth *Bumper-WAC* flight from White Sands reached an altitude of 244 miles (390 km) and a maximum speed of 5150 miles per hour (8290 km/h), making it the first manmade object to fly at hypersonic speed, in excess of Mach 5.

Figure 1.78 Launch of the *Bumper-WAC Corporal* from Cape Canaveral, Florida, 24 July 1950. (Source: *NASA*.)

1.3.5.5   Reusable Rocket-Based Launch Systems

As one might imagine, the use of expendable rocket launch systems is very expensive and ineffi-
cient, given the fact that the rocket boosters can only be used once. Imagine the cost of air travel,
if a commercial airliner was discarded after a single flight! Of course, space travel is a bit more
complex than airline travel, but not having to manufacture new boosters for every flight could provide a cost and efficiency benefit. There have been several studies of reusable booster systems,
where the separated booster stages are returned to earth so they can be reused. These studies have
included *fly-back boosters*, where the booster is flown back as a glider or under power, using an
air-breathing jet engine. There have been several recent successes by commercial space companies
with returning a first stage booster for potential reuse. These have involved guiding the booster back
to a landing pad, where the booster's main engines are relighted to slow the descent rate, landing
gear are extended, and the booster is landed vertically. Below, we discuss a partially reusable,
rocket-based launch system, the Space Shuttle.

The NASA Space Shuttle, shown in Figure 1.79, is an example of a partially reusable,
rocket-based launch system. The Space Shuttle program, more formally called the Space
Transportation System (STS), followed the *Apollo* program as the access to space system for the
United States. The Space Shuttle was a man-rated launch system used to transport astronauts to
low earth orbit for three decades. The first flight of the Space Shuttle into space was on 12 April
1981. There were a total of 135 Space Shuttle missions, until the program ended in 2011. In total,
six Orbiter Vehicles were built, the *Enterprise*, *Columbia*, *Challenger*, *Discovery*, *Atlantis*, and

Figure 1.79   The first Space Shuttle launch, STS-1, 12 April 1981, the 20th anniversary of Yuri Gagarin's
flight. (Source: *NASA*.)

Endeavor. The *Enterprise* was a non-space-worthy vehicle that was used for the approach and landing glide tests prior to any space flights. Tragically, two Space Shuttles were lost in accidents, *Challenger* during launch in 1986 (STS-25) and *Columbia* during entry in 2003 (STS-113).

The Space Shuttle launch system comprised the winged Orbiter Vehicle (OV), a large External Tank (ET), and two solid rocket boosters (SRBs). The Orbiter housed the various flight crew decks, a large cargo payload bay, Orbital Maneuvering System (OMS) rocket motors, and three Space Shuttle Main Engines (SSMEs). The Orbiter was attached to the External Tank, which contained the liquid hydrogen and liquid oxygen propellants that fueled the Orbiter's three SSMEs. The Solid Rocket Boosters were mounted on either side of the External Tank. The Orbiter Vehicle, along with the SSMEs, and the SRBs were the reusable components of the Space Transportation System, while the External Tank was not reusable.

The Space Shuttle was vertically launched from Cape Canaveral, Florida. The complete Space Shuttle launch stack (Orbiter, External Tank, and Solid Rocket Boosters) was 184 ft (56 m) tall with a gross lift-off weight of about 4.4 million lb (2000 tonnes). The total lift-off thrust produced by the SSMEs and SRBs was about 6.78 million lb (30.2 MN). The Shuttle was a two-stage rocket booster system with the first stage SRBs being jettisoned about two minutes after launch, at an altitude of about 150,000 ft (46,000 m). The External Tank continued to supply fuel and oxidizer to the SSMEs until main engine cut-off (MECO), just prior to orbit insertion. The SSMEs burned for about 8 minutes from lift-off to MECO. The ET was jettisoned after MECO, falling back to earth into the ocean. The Orbiter entered low earth orbit and could perform orbital maneuvers using its OMS engines. After completion of the in-orbit mission, the Orbiter used its OMS engine to slow down and enter the atmosphere as a hypersonic glider. It glided to a horizontal landing on a very long paved runway, at the NASA Kennedy Space Center, Florida or Edwards Air Force Base, California.

The Orbiter Vehicle was a space plane, designed to launch like a rocket and land like an airplane. The Orbiter flight envelope encompassed altitudes from sea level to 330 miles (530 km) and speeds from 213 mph (343 km/h) to Mach 25. At hypersonic speeds, the Orbiter lift-to-drag ratio was about 1, increasing to about 2 for supersonic flight, and about 4.5 for subsonic flight. (This is an example of the decrease in the lift-to-drag ratio as the Mach number increases and the fact that hypersonic lift-to-drag ratios are small.) The Orbiter had a length of 122 ft (37 m), height of 56.6 ft (17.2 m) to the top of its vertical tail, wingspan of 78.1 ft (23.8 m), and gross lift-off weight of about 240,000 lb (110,000 kg).

The Orbiter had a somewhat conventional high-speed airplane configuration, with a highly swept double-delta wing and single vertical tail. Flight control surfaces included elevons, mounted at the wing trailing edges, provided pitch and roll control, and a rudder at the trailing edge of the vertical tail provided yaw control. The rudder was of a split design, such that it could deflect a surface in both the left and right directions to act as a speed brake for landing. The aft end of the Orbiter housed the three SSMEs and two OMS engines, mounted in pods on either side of the vertical tail. Each SSME had a sea level thrust of 393,800 lb (1.75 MN), with a combined thrust of over 1.18 million lb (5.3 MN), and a specific impulse of 455 s. (Specific impulse is a measure of the efficiency of a propulsive device as given by the ratio of the thrust produced to the propellant consumed, to be discussed in Chapter 4.) The Orbiter also had a reaction control system (RCS) comprising 44 small liquid-fuel rocket thrusters that were distributed at the forward and aft ends of the vehicle. The RCS provided attitude control and maneuvering of the Orbiter in pitch, roll, and yaw while in orbit and during entry.

There were three flight deck areas in the Orbiter crew compartment, the flight deck where two pilots and two Mission Specialists were seated, the mid-deck where additional crew were seated, and a utility area where consumables, such as air and water were located. Typical Shuttle missions had a crew of seven astronauts, but up to 11 people could be accommodated in an emergency.

A unique feature of the Orbiter was its capability to carry a large payload in its 59 ft long (18 m) by 15 ft wide (4.6 m), fuselage cargo bay. Two long cargo bay doors ran the length of the cargo bay, allowing the deployment of large payloads. Typical payload weights that could be carried to orbit were about 50,000 lb (22,700 kg). In addition to transporting payloads to orbit, the Orbiter could also capture payloads from orbit and return them to earth. The Orbiter could land with payloads weighing up to 32,000 lb (14,400 kg).

To survive the 3000°F (3460°R, 1922 K) temperature of atmospheric entry, the Orbiter was covered with a thermal protection system (TPS). The type of TPS on different parts of the Orbiter varied depending on the heat load. TPS material included reinforced carbon-carbon for high heat load areas and various types of lightweight ceramic and composite tiles for lower heat load areas. Unlike the ablative heat shields used in the previously discussed space capsules, the Orbiter TPS was reusable, although it did require careful maintenance and repair between flights. The TPS was very lightweight, especially as compared to ablative materials, but it was also fragile, requiring careful handling.

The Solid Rocket Boosters were the largest solid rocket motors ever flown, each producing a peak thrust of over 3 million lb (13.3 MN). The SRBs provided over 70% of the total thrust at lift-off and during the first stage ascent. With a length of 149.2 ft (45.5 m) and diameter of 12.2 ft (3.7 m), each SRB weighed about 1.3 million lb (590 tonnes). The solid fuel in the SRB was ammonium perchlorate composite propellant, a mixture of ammonium perchlorate oxidizer and aluminum fuel. Other ingredients in the solid fuel included iron oxide catalysts, polymer binders (to hold the solid fuel together), and epoxy curing agents. The SRB had a sea level specific impulse of about 240 s. After being jettisoned, the SRBs descended back to earth under parachutes, falling into the ocean. The SRBs were recovered by ship and were refurbished for use on another launch.

The External Tank was the largest and heaviest component of the Space Shuttle, with a length of 153.8 ft (46.9 m), diameter of 27.6 ft (8.4 m), and lift-off weight of about 1.67 million lb (756 tonnes). The ET contained the liquid hydrogen and liquid oxygen propellants to fuel the SSMEs. The ET was covered with a thermal protection system (TPS) primarily composed of a spray-on foam insulation. Thermal protection and insulation was required to prevent aero-thermodynamic heating of the cryogenic propellants and to prevent the cryogenic propellants from liquefying the air next to the metal propellant tanks. The weight of the TPS on the ET was about 4800 pounds (2180 kg). The ET is jettisoned 10 s after MECO. The majority of the tank disintegrates in the atmosphere, with the remaining pieces falling into the ocean.

1.3.5.6 Air-Launched Space Access Systems

So far, we have discussed multi-stage, access to space systems where all of the stages utilize rocket-power. Another option that has been developed is an *air-launched system* where the first stage is an airplane rather than a rocket. The second stage is typically some kind of rocket-powered vehicle that is dropped from the carrier aircraft or "mothership", as it is sometimes called. A variety of schemes have been investigated, using different types of carrier aircraft and different configurations for the attachment of the second stage vehicle to the carrier aircraft. The carrier aircraft could be a fighter-type jet, business jet, transport aircraft, or a new purpose-built aircraft. Of course, an existing aircraft design requires modifications to accommodate carriage of a second-stage vehicle.

An example of an air-launched system for small, unmanned spacecraft is the Orbital Sciences L-1011 and *Pegasus* launch system. The *Pegasus* rocket booster is carried aloft, underneath a Lockheed L-1011 jet (Figure 1.80), to a nominal release altitude of 40,000 ft (12,000 m). The booster has three, solid propellant rocket motor stages. The first stage of the booster has a small delta wing that provides lift to help the vehicle transition from the horizontal launch attitude to the desired

Figure 1.80   Orbital Sciences ASB-11 *Pegasus* launch vehicle dropped from Lockheed L-1011 carrier aircraft. (Source: *NASA*.)

ascent climb angle. The *Pegasus* is capable of placing small payloads of about 1000 lb (450 kg) into low earth orbit.

The Virgin Galactic *White Knight Two* mothership is a specially designed carrier aircraft for the *Spaceship Two* rocket-powered, sub-orbital, manned spacecraft (Figure 1.17). The *Spaceship Two* vehicle is attached to a pylon between the unique dual fuselage configuration of the *White Knight Two*. After release from the *White Knight Two* at about 47,000 ft (14,000 m), the *Spaceship Two* is designed to carry up to six people on a sub-orbital trajectory into space, reaching an apogee of about 100 km (62 miles) before gliding back for a horizontal landing on a runway.

All of the air-launched systems to date have release conditions at subsonic speeds and altitudes below about 50,000 ft (15,000 m). There may be benefits, in terms of payload weight delivered to space, if the first stage can reach a higher energy state prior to release of the second stage. Several concepts have been studied where the first stage vehicle is capable of releasing a second stage vehicle at supersonic speeds and much higher altitudes, but none of these have been attempted in actual flight.

1.3.5.7   Non-Rocket-Based Space Access Systems

We now consider a few non-rocket concepts for space access, returning first to the idea of the space cannon. As improbable as Jules Verne's space cannon may have seemed, it does have some technical feasibility. In the 1950s, the United States and Canada investigated the use of "superguns" to launch research probes into the upper atmosphere and small unmanned satellites into orbit. Significant progress was also made in developing the electronics and instrumentation that could survive the huge acceleration forces associated with gun launching. Much of these efforts culminated in the High Altitude Research Project (HARP) during the 1960s, a joint project of the United States and Canada, to develop a gun-launched system to place a satellite into orbit.

Several HARP superguns were built, constructed from decommissioned US Navy 16 in (400 mm) battleship gun barrels. The "16 in" specification denotes the inner diameter or *bore* of the gun barrel and also defines the maximum outer diameter of the projectile that can be fired. Each 16 in gun barrel was about 60 ft (18 m) long and ultimately, two of these barrels were welded together to form 120 ft (36.6 m) long superguns that weighed about 100 tons (90,700 kg).

The island of Barbados was selected as the first HARP gun launch site due to its proximity to the equator and its general remoteness. Placing a space launch site near the equator is beneficial in terms of launching in the direction of the earth's rotation, thereby imparting the earth's rotational velocity to the vehicle. Hundreds of instrumented projectiles were launched from the Barbados supergun, many reaching sub-orbital altitudes (Figure 1.81). On 18 November 1966, a supergun in Yuma, Arizona fired a 180 kg (400 lb) projectile to a record altitude of 180 km (590,000 ft,

Figure 1.81   The HARP supergun being fired on the island of Barbados. (Source: *US Department of Defense.*)

110 miles). The program was cancelled shortly after this, prior to achieving the goal of placing a satellite in orbit with a supergun or space cannon.

Another imaginative, non-rocket-based space access concept is the *space elevator*. In 1895, the Russian rocket scientist, Konstantin Tsiolkovsky, postulated the building of a tower that would reach from the surface of the earth to space. A huge technical hurdle to this concept is the enormous compressive weight that has to be borne by the structure, just as for a very high building. In contrast to this, the space elevator is based upon a *tensile* structure, where the weight of the system is borne from above by a counterweight in space. The components of a notional space elevator are shown in Figure 1.82. The space elevator cable or tether extends from the counterweight in space to the surface of the earth near the equator. The tether is in tension due to the circular motion of the counterweight. The tether remains vertically centered over the same position on the earth as the earth rotates. An elevator or climber, attached to the tether, mechanically ascends and descends along the cable, to and from space. Advances in materials technology are required to actually build a tether that is strong and light enough to make the space elevator concept a reality. One area that is being pursued is the use of high-strength, lightweight carbon nanotubes, but much more work must be done to build the large structures required, based on this technology.

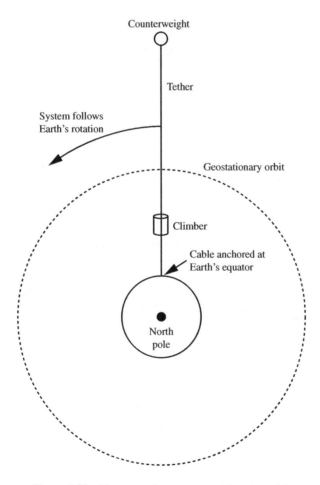

Figure 1.82   The space elevator concept (not to scale).

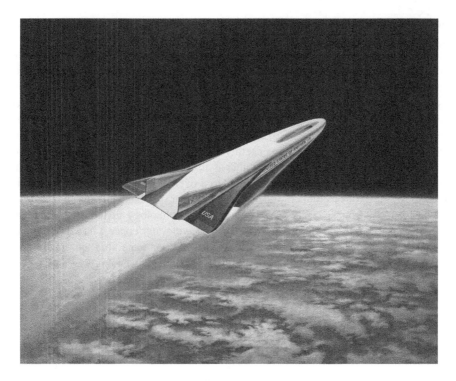

Figure 1.83   The X-30 National Aerospace Plane single-stage-to-orbit concept, 1990. (Source: *NASA.*)

The epitome of reusability and efficiency in space access is perhaps the concept of a single-stage-to-orbit (SSTO) vehicle. An SSTO vehicle would fly into space without the need for rocket stages or expendable boosters and then return to earth as a hypersonic airplane. The ideal SSTO vehicle would be completely reusable, much like an ordinary airplane, able to fly back into space after re-fueling, with a minimum of refurbishment and maintenance.

During the 1990s, the US National Aerospace Plane (NASP) program was a national effort to design and flight test the X-30 SSTO vehicle (Figure 1.83). Utilizing supersonic combustion ramjet, or *scramjet*, engines, the X-30 would use the air in the atmosphere as its propulsion system oxidizer, rather than carry it onboard like a conventional rocket. Reaching hypersonic speeds within the denser parts of the atmosphere, the X-30 would be exposed to extremely high heat loads, making thermal management and thermal protection a difficult design problem. While the X-30 program was eventually cancelled, there have been several significant technological advancements in the development of hypersonic vehicles and hypersonic propulsion that may one day make access to space, using an SSTO vehicle, a reality.

References

1. Anderson, J.D., Jr., *Introduction to Flight*, 4th edition, McGraw-Hill, Boston, Massachusetts, 2000.
2. Crouch, T.D., *A Dream of Wings: Americans and the Airplane*, W.W. Norton & Company, February 2002.
3. Crouch, T.D., *The Bishop's Boys*, W.W. Norton & Company, April 2003.
4. Federal Aviation Administration, US Department of Transportation, *Methods Techniques and Practices – Aircraft Inspection and Repair*, Advisory Circular (AC) **43**.13-1 (as revised), Oklahoma City, Oklahoma, September 8, 1998.
5. Federal Aviation Administration, US Department of Transportation, *Balloon Flying Handbook*, FAA-H-8083-11A, Oklahoma City, Oklahoma, 2008.

6. Federal Aviation Administration, US Department of Transportation, Code of Federal Regulations, Federal Aviation Regulations, 14 CFR Part 1, Definitions and Abbreviations, 2014.

7. Federal Aviation Administration, US Department of Transportation, Code of Federal Regulations, Federal Aviation Regulations, 14 CFR Part 401, Organization and Definitions, 2014.

8. Federal Aviation Administration, US Department of Transportation, *Instrument Flying Handbook*, FAA-H-8083-15B, Oklahoma City, Oklahoma, 2012.

9. Federal Aviation Administration, US Department of Transportation, *Rotorcraft Flying Handbook*, FAA-H-8083-21, Oklahoma City, Oklahoma, 2000.

10. Gessow, A. and Meyers, G.C., Jr., *Aerodynamics of the Helicopter*, Frederick Ungar Publishing Company, New York, New York 1952.

11. Goddard, R.H., *Rockets: Comprising "A Method of Reaching Extreme Altitudes" and "Liquid-Propellant Rocket Development"*, Facsimile Edition, American Institute of Aeronautics and Astronautics, Reston, Virginia, 2002.

12. Griffin, M.D. and French, J.R., *Space Vehicle Design*, 2nd edition, AIAA Education Series, American Institute of Aeronautics and Astronautics, Inc., Reston, Virginia, 2004.

13. Hurt, H.H., Jr, *Aerodynamics for Naval Aviators*, US Navy NAVWEPS 00-80T-80, US Government Printing Office, Washington, DC, January 1965.

14. Hybrid Air Vehicles, Ltd., http://www.hybridairvehicles.com/

15. Jackson, P. (ed.), *Jane's All the World's Aircraft: 2002–2003*, Jane's Information Group Limited, Coulsdon, Surrey, United Kingdom, 2002.

16. Page, B.R., "The Rocket Experiments of Robert H. Goddard," *The Physics Teacher*, November 1991, pp. 490–496.

17. Leishmann, J.G., *Principles of Helicopter Aerodynamics*, 1st edition, Cambridge University Press, New York, New York, 2000.

18. Lilienthal, O., *Birdflight as the Basis of Aviation*, translated from the 2nd edition by A.W. Isenthal, Markowski International Publishers, Hummelstown, Pennsylvania, 2001.

19. Young, H.D. and Freedman, R.A., *University Physics*, 11th edition, Addison Wesley, San Francisco, California, 2004.

2

Introductory Concepts

Entrance to the US Air Force Test Pilot School, Edwards, California, where the fundamentals of flight test are learned.[1] (Source: *US Air Force*.)

2.1   Introduction

In this chapter, several basic concepts in aerospace engineering and flight test are introduced. Some basic concepts in mathematics and physics that are relevant to our study of aerospace vehicles, are

[1] The Lockheed NF-104A points skyward at the entrance to the US Air Force Test Pilot School (USAFTPS). The NF-104A was a modified F-104A *Starfighter* jet that was used as a low-cost space plane trainer at the Aerospace Research Pilots School, the forerunner to USAFTPS in the 1960s. Modifications included the addition of a small rocket engine and a reaction control system for flight in the upper atmosphere. A typical NF-104A flight profile was a level acceleration at 35,000 ft to Mach 1.9 using its J79 jet engine, ignition of the rocket engine, then a 3.5 g pull-up at Mach 2.1, to enter a very steep climb. The J79 engine was shut down at about 85,000 ft, with the rocket power lasting about 100 seconds. The aircraft followed a ballistic arc, reaching altitudes well over 100,000 ft. The NF-104A set a record altitude of 120,800 ft on 6 December 1963. After coasting down from its peak altitude to denser air, the jet engine was restarted and the NF-104A made a normal landing.

Introduction to Aerospace Engineering with a Flight Test Perspective, First Edition. Stephen Corda.
© 2017 John Wiley & Sons Ltd. Published 2017 by John Wiley & Sons Ltd.
Companion Website: www.wiley.com/go/corda/aerospace_engg_flight_test_persp

also briefly reviewed. Basic aerospace engineering concepts and terminology, such as the regimes of flight, vehicle axes systems, free body diagrams, angle-of-attack, Mach number, and others, are introduced. Some of the fundamental concepts of flight testing are introduced, including the definition of flight test, the flight test process, flight test hazards and safety, and the flight test technique.

2.2 Introductory Mathematical Concepts

Mathematics is the language of engineering. To be a competent engineer, one must be able to "speak" and understand the language of mathematics. Many quantitative aspects of aerospace engineering theories and principles must be explained through mathematics, using equations and numbers. To become "fluent" in the language of mathematics, one must use it regularly, through application in both theoretical and real world problems. Unlike pure mathematicians, engineers typically use mathematics as a *tool*, to perform engineering analysis and design. Remember that the mathematics does have *physical* meaning, which is embodied in the physics that is captured by its equations and numbers. We start with a review units and unit conversions, a topic that may seem mundane, but it is critically important in real-world engineering. The topic of measurement and numerical uncertainty is covered, an area of high importance in engineering, especially as applied to ground and flight testing. Finally, a few aspects of scalar and vector quantities are reviewed.

2.2.1 Units and Unit Systems

In engineering, we deal with quantities that are described by *units* or a combination of units. For instance, we can specify that for a Boeing 787 *Dreamliner* (Figure 2.1), the wingspan is 196.0 feet (60.0 meters) and the maximum takeoff weight is 502,500 lb (227,930 kilograms). Whether we are performing an engineering calculation or taking data during a flight test, it is critical that we use the proper units to quantify the numbers. However, in doing this, there are quite a few choices as to which units to use. For example, we can state that the *Dreamliner* cruise airspeed is 490 knots or 564 miles per hour or 907 kilometers per hour or 827 feet per second or 252 meters per second. The

Figure 2.1 Boeing 787 *Dreamliner*. (Source: *H. Michael Miley, "Boeing 787 Dreamliner Arrival Air-Venture 2011" https://commons.wikimedia.org/wiki/File:Boeing_787_Dreamliner_arrival_Airventure_2011 .jpg, CC-BY-SA-3.0. License at https://creativecommons.org/licenses/by-sa/2.0/legalcode.*)

specific units that are chosen may depend on the situation. In the *Dreamliner* airspeed example, it would be appropriate to use units of knots for a cockpit airspeed indicator while units of feet per second or meters per second may be more appropriate for an aircraft performance calculation.

We first discuss the two primary unit systems in use today and the basis for these systems. We then discuss consistent units, in regards to dimensional consistency and conversion factors. Even though the discussion of units may seem mundane, the proper use of units is extremely important in engineering. Two real-world examples of the importance of units are discussed at the end of this section.

2.2.1.1 Unit Systems

The two systems of units that are in widespread use in engineering today are the *English System* and the *International System*. The term *English System* has some ambiguity, as it may refer to the system of units used in the United Kingdom, known as *British Imperial Units*, or in the United States, sometimes referred to as *United States customary units*. The United States system was developed from the British system, so they are very similar, although there are some distinct differences. Whenever we use the terms *English System* and *English units* in this text, we are referring to the system and units that are used in the United States.

The International System is commonly referred to as the *metric system* and abbreviated by *SI*, from the French translation, *Systeme International*. The International System is used worldwide and increasingly so in the United States. The SI system is the internationally agreed reference from which all other units (and unit systems) are now defined. It is still quite common to have quantities expressed in English units for certain engineering disciplines, such as thermodynamics and air-breathing propulsion. There is also a vast amount of technical literature that was written using English units.

As far as is practical, whenever dimensional quantities are discussed in the book, the engineering units are given in both the International System and the English System. This "bilingual" display of units is done to help the reader obtain an intuitive engineering "feel" for both systems of units, a critical skill in the bilingual scientific and engineering world.

The SI system is founded upon seven *base units*, defined for seven *base quantities*, as given in Table 2.1. While we are usually interested in the base quantities of length, mass, time, and temperature, all of the base units have been provided in Table 2.1 for completeness. The definitions of the SI base units, also given in Table 2.1, are obtained from the most accurate and reproducible measurements that are possible. It should be noted that while the base quantities are assumed to be mutually independent, their respective base units are in fact interdependent, based upon their definitions. For example, the base quantity of length is independent of the other base quantities, but the definition of its base unit, the meter, is dependent on another base unit, the second. This interdependency is true for all of the other base units, except the kelvin. All other units in the unit system are derived from the base units. These *derived units* are expressed as products of powers of the base units, as shown in Table 2.2 for selected SI derived units and quantities.

The British Imperial base units, from which English units are derived, are officially defined in terms of SI units, with a length of one inch equal to exactly 2.54 cm and a force of one pound equal to exactly 4.448221615260 newton. The British Imperial unit of time is the second, the same as in the SI system. Even though English units are derived from British units, we still refer to a set of fundamental or "base" units for the English system, as they set the foundation or basis for the units used in this system. The fundamental units of the English system for length, mass, time, and temperature are the *foot, slug, second*, and *degree Rankine*, respectively. As stated previously, we

Table 2.1 SI base units and definitions.

Base unit	Unit symbol	Base quantity	Definition of base unit
meter	m	length	Distance travelled by light in a vacuum in 1/299,792,458 of a second.
kilogram	kg	mass	Mass equal to the international prototype of the kilogram (a cylinder of platinum-iridium alloy).
second	s	time	Based on an atomic clock, which uses the transition between the two lowest energy levels of the cesium 133 atom. One second is the duration of 9,192,631,770 cycles of the radiation that causes this transition between levels.
ampere	A	electric current	Constant current between two straight parallel conductors of infinite length, placed 1 m apart in vacuum, that would produce a force equal to 2×10^{-7} N per meter of length between these conductors.
kelvin	K	temperature	Temperature that is 1/273.16 of the thermodynamic temperature of the triple point of water.
mole	mol	amount of substance	Amount of substance which contains as many elementary entities as there are atoms in 0.012 kg of carbon 12.
candela	cd	luminous intensity	Luminous intensity, in a given direction, of a source that emits monochromatic radiation of frequency 540×10^{12} hertz and that has a radiant intensity in that direction of 1/683 watt per steradian*.

*A steradian is the solid angle subtended at the center of a unit sphere by a unit area on its surface.

Table 2.2 Selected SI derived quantities and units.

Derived unit	Symbol	Derived quantity	Derived unit in terms of other units	Derived unit in terms of base units
radian	rad	plane angle	W	—
newton	N	force	—	$m \cdot kg \cdot s^{-2}$
joule	J	work, energy	N·m	$m^2 \cdot kg \cdot s^{-2}$
pascal	Pa	pressure, stress	N/m^2	$m^{-1} \cdot kg \cdot s^{-2}$
watt	W	power	J/s	$m^2 \cdot kg \cdot s^{-3}$
hertz	Hz	frequency	—	s^{-1}
degree Celsius	°C	temperature	—	K
coulomb	C	electric charge	—	$s \cdot A$
volt	V	electric potential difference	W/A	$m^2 \cdot kg \cdot s^{-3} \cdot A^{-1}$
farad	F	capacitance	C/V	$m^2 \cdot kg^{-1} \cdot s^4 \cdot A^2$
ohm	—	electric resistance	V/A	$m^2 \cdot kg \cdot s^{-3} \cdot A^{-2}$
weber	Wb	magnetic flux	V·s	$m^2 \cdot kg \cdot s^{-2} \cdot A^{-1}$
tesla	T	magnetic flux density	Wb/m^2	$kg \cdot s^{-2} \cdot A^{-1}$
henry	H	inductance	Wb/A	$m^2 \cdot kg \cdot s^{-2} \cdot A^{-2}$

use both the English and SI systems of units throughout the book, and it is important to understand the basis of both unit systems.

2.2.1.2 Dimensional Consistency

When performing calculations, we must ensure that the equations that we use are *dimensionally consistent*, that is, quantities must have common units to be summed or equated. For instance, we cannot add a length of 12 m to a temperature 100 K. The units must also be consistent with the parameter that we are interested in. If we are calculating the temperature in the combustor of a jet engine, then the result should be in degrees Rankine or kelvins, or some other unit of temperature.

Dimensional consistency is especially important as analyses get more complex, where calculations may involve many different parameters and many unit conversions. It is always good practice to carry all of the units through, in writing, to the end of a calculation to ensure dimensional consistency. If the units are dimensionally inconsistent, then an error has been made somewhere in the calculation.

2.2.1.3 Consistent Sets of Units and Unit Conversions

Dimensional consistency should not be confused with a *consistent set of units*, which refers to a unit set that does not require any conversion factors in calculations or in mathematical expressions of fundamental physics. The base units of the SI and English systems are each a consistent set of units. For instance, if we look at the consistent English units for the weight, W, of a mass, m, in a gravitational field with an acceleration of gravity, g, we have

$$[W] = [m][g] = \text{slug} \times \frac{\text{ft}}{\text{s}^2} = \text{lb}_f \tag{2.1}$$

where the resultant weight is in the consistent units of pounds force, lb_f. If we use inconsistent units of pound mass, lb_m, we get

$$[W] = [m][g] = \text{lb}_m \times \frac{\text{ft}}{\text{s}^2} = \frac{\text{lb}_m \text{ ft}}{\text{s}^2} \tag{2.2}$$

To obtain the consistent units of pounds force, we need to convert the pounds mass to slugs, where 32.2 lb_m is equal to 1 slug. We would then need to write the weight equation as

$$W = \frac{1}{g_c} mg \tag{2.3}$$

where g_c is the conversion factor, equal to

$$g_c = 32.2 \, \frac{\text{lb}_m}{\text{slug}} \tag{2.4}$$

Equation (2.4) is telling us that a mass of one slug is 32.2 times larger than a pound mass. Using inconsistent units, Equation (2.3) is then given by

$$[W] = \frac{1}{g_c}[m][g] = \frac{1}{g_c} \left(\text{lb}_m \times \frac{\text{ft}}{\text{s}^2} \right) = \text{lb}_f \tag{2.5}$$

In SI units, the conversion factor is

$$g_c = 9.81 \, \frac{\text{kg}_f}{\text{N}} \tag{2.6}$$

If we use this conversion factor for our weight in SI units, we obtain the inconsistent SI unit for weight of kilogram force, kg_f.

$$[W] = \frac{1}{g_c}[m][g] = \frac{1}{g_c}\left(kg \times \frac{m}{s^2}\right) = kg_f \tag{2.7}$$

From a certain perspective, the use of inconsistent units in Equations (2.5) and (2.7) makes some sense. In each case, the unit weight is the same as the unit of mass. In Equation (2.5), a pound of weight equals a pound of mass, and in Equation (2.7) a kilogram of weight equals a kilogram of mass. However, from an engineering perspective this makes things more complicated and prone to errors, usually by a factor of 32.2 or 9.81. The bottom line is that we should always strive to use consistent units and avoid the addition of these inconsistent unit conversion factors in our equations. That said, be aware that inconsistent units, such as the pound mass, lb_m, are still found frequently in engineering, especially in the fields thermodynamics and propulsion.

As a clarification of unit symbols used in this text, we simply use lb, rather than lb_f, to denote a pound force, lb_m to denote a pound mass, and kg to denote a kilogram mass. We will not find much occasion to use the unit of kilogram force, kg_f.

In regards to temperature, the kelvin, in the SI system, and the degree Rankine, in the English system, represent consistent units based on *absolute temperature scales*. In an absolute temperature scale, the "bottom" of the scale corresponds to absolute zero, where, theoretically, all molecular translational motion ceases. Therefore, zero kelvin, 0 K, and zero degrees Rankine, 0°R, are equivalent, corresponding to absolute zero.

We often deal with temperatures in units of degrees Fahrenheit and degrees Celsius, which are not based on absolute temperature scales. Conversions from units of degrees Celsius and degrees Fahrenheit to consistent units are given below.

$$K = °C + 273.15 \tag{2.8}$$

$$°R = °F + 459.67 \tag{2.9}$$

Based on these equations, we see that $0 K = -273.15°C$ and $0°R = -459.67°F$. Conversely, $0°C = 273.15 K$ and $0°F = 459.67°R$. Conversions between Celsius and Fahrenheit are as follows.

$$°C = \frac{5}{9}(°F - 32) \tag{2.10}$$

$$°F = \frac{9}{5}°C + 32 \tag{2.11}$$

Again, be prepared to properly convert temperature to consistent units, as the inconsistent units of Centigrade and Fahrenheit are still widely used.

Finally, as a natural consequence of dealing with all of these different sets of units, we must be able to perform *unit conversions* within the same unit system and between different unit systems. In performing calculations, we usually want to convert to the set of units that are appropriate for the quantities being calculated. For example, if we are calculating the time it takes to fly from Los Angeles to London, it would be more appropriate for us to use units of hours instead of seconds. If we are recording the airspeed of an aircraft, we are probably writing down knots or miles per hour from an airspeed indicator, rather than feet per second. We often find it necessary to perform many unit conversions in a calculation. (A list of useful unit conversion factors is given in Appendix B.) Sometimes, it is beneficial to convert all or most of the given quantities into consistent units prior to starting the calculation. Nevertheless, it is best to carry your units through the entire calculation to minimize errors and to help check the validity of the results.

The following two examples illustrate the importance of units.

Example 2.1 The Importance of Units: The "Gimli" Glider *On 23 July 1983, Air Canada Flight 143, a twin-engine Boeing 767 commercial airliner, departed Ottawa, Canada with a planned destination of Edmonton, Canada. About an hour into the flight, at a cruising altitude of 41,000 ft (12,500 m), both turbofan engines "flamed out". The Boeing 767 airliner had inexplicably run out of fuel. Luckily, there was a decommissioned Canadian air force base within gliding distance of the aircraft, in Gimli, Manitoba, Canada. The airliner successfully landed at the Gimli airfield, collapsing the nose landing gear, which could only be partially extended due to the power loss of both engines. After the landing, the powerless airliner was dubbed the "Gimli Glider". So, how did this advanced jet airliner simply run out of fuel? As with most aviation incidents and accidents, there was a chain of events that led up to this potentially catastrophic event.*

The aircraft needed to be fueled for the flight from Ottawa to Edmonton, but unlike "filling up" your automobile, an airliner is not necessarily just "filled up". If the airliner carries more fuel than is required for a flight, it is carrying excess weight, which results in a performance and cost penalty. So, the required fuel quantity, which includes an extra amount of fuel called a reserve, is calculated for each flight. A total of 22,300 kg of fuel was required for the Ottawa to Edmonton flight. The night before the flight, the Boeing 767's computerized fuel indication and monitoring system had failed. Rather than using this computerized system, the amount of fuel, to be added, was calculated manually. It was determined that the aircraft had 7682 liters of fuel in its tanks prior to adding any fuel. This fuel quantity was converted from liters to kilograms, using a conversion factor of 1.77 kg/liter, as follows.

$$7682 \ liters \times 1.77 \frac{kg}{liter} = 13{,}597 \ kg$$

Subtracting this amount of fuel in the aircraft fuel tanks from the amount of fuel required for the flight, the fuel quantity to be added was calculated as

$$22{,}300 \ kg - 13{,}597 \ kg = 8703 \ kg$$

Since the fuel was added to the aircraft from a fuel truck, which dispensed fuel in liters rather than kilograms, the fuel quantity to be added was converted from kilograms to liters, as

$$\frac{8703 \ kg}{1.77 \frac{kg}{liter}} = 4907 \ liters$$

The problem with the above calculations is that an incorrect value for the liters-kilogram conversion was used. The correct conversion is 0.8 kg/liter rather than 1.77 kg/liter. Using the correct conversion, the actual fuel quantity in the aircraft tanks prior to adding fuel was

$$7682 \ liters \times 0.8 \ \frac{kg}{liter} = 6146 \ kg$$

The amount of fuel that needed to be added should have been

$$22{,}300 \ kg - 6146 \ kg = 16{,}154 \ kg$$

Thus, the amount of fuel in liters that should have been added was

$$\frac{8703 \ kg}{0.8 \ \frac{kg}{liter}} = 10{,}879 \ liters$$

Using the incorrect conversion, the aircraft took off with a total of only 12,589 liters of fuel in its fuel tanks, rather than the required 22,300 liters.

The reason for the use of the incorrect conversion value was not just a matter of an erroneous number, but was also due to an error in units. The Boeing 767 was a new addition to the Air Canada fleet, bringing with it several advancements in computerized control of the aircraft systems. However, these advancements came with some changes in the normal procedures that had been followed with the other aircraft in the Air Canada fleet. The Boeing 767 was the first aircraft in their fleet which measured fuel in SI units of kilograms, rather than English units of pounds. Prior to the introduction of the Boeing 767, the fuel quantities for Air Canada aircraft were converted from pounds to liters, using the correct conversion factor of 1.77 lb/liter. Hence, this same value was erroneously used for the calculation of fuel, in kilograms, for the new Boeing 767.

Example 2.2 The Importance of Units: The Mars Climate Orbiter *The Mars Climate Orbiter (MCO), shown in Figure 2.2, and the Mars Polar Lander (MPL) were part of a series of NASA spacecraft missions to explore Mars in the late 1990s. The MCO spacecraft bus (the bus is the spacecraft platform or modular infrastructure upon which the payload or experiments and instrumentation are mounted) dimensions were approximately 2.1 m (6.9 ft) tall, 1.6 m (5.2 ft) wide, and 2.0 m (6.6 ft) deep with a launch weight of 338 kg (745 lb). The fully extended solar panel array measured 5.5 m (18 ft) in length. The MCO and MPL total mission cost was $327.6 million which included $193.1 million for spacecraft development, $91.7 million for launch services, and $42.8 million for operations.*

The MCO spacecraft was launched from Cape Canaveral, Florida, aboard a Delta II launch vehicle, on 11 December 1998. The MPL was launched from Cape Canaveral on 3 January 1999

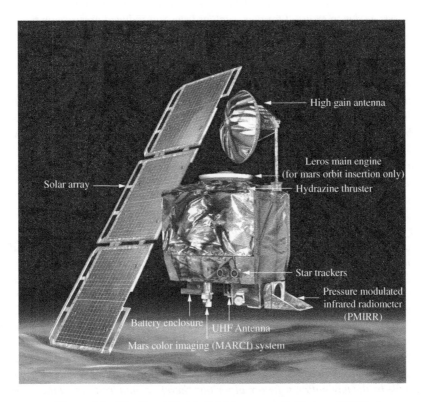

Figure 2.2 The Mars Climate Orbiter spacecraft. (Source: *NASA*.)

aboard a similar Delta II rocket. After a nine and a half month, 416 million mile (669 million km) journey to Mars, the MCO was to enter Mars orbit and remain there to collect long-term atmospheric and weather data and to serve as a communications relay for the MPL, which would land on the surface of Mars.

On arrival to Mars, the MCO was to perform an orbital insertion burn (fire its main engine to decelerate) and enter an elliptical orbit about the planet. The spacecraft would then use aerobraking, "dipping" in and out of the Martian atmosphere, creating atmospheric drag that would slow the vehicle and circularize its orbit. On 23 September 1999, shortly after entering the Martian atmosphere on a much lower than planned trajectory, all communication with the MCO was lost. Communication with the MCO was never reestablished and the spacecraft was presumed lost.

An MCO Mishap Investigation Board (MIB) was established to investigate the loss of the spacecraft. It was discovered that the spacecraft had entered Mars orbit with a periapsis (the lowest altitude of its orbit) of 57 km, when this lowest altitude should have been a much higher 226 km. At the lower entry altitude, the spacecraft encountered a much denser region of the Martian atmosphere, resulting in either intolerable atmospheric drag and the subsequent destruction of the spacecraft, as it descended further into the atmosphere, or "skipping" of the vehicle out of the atmosphere, sending it into an orbit around the Sun. The lowest survivable altitude was determined to be about 80 km. So, why was the periapsis 170 km lower than expected, resulting in the loss of the MCO?

To fully understand the cause of the mishap, we need to know a little more about the MCO spacecraft attitude and trajectory control. During its nine-month journey to Mars, the spacecraft's attitude and trajectory was controlled using eight small hydrazine monopropellant thrusters and three reaction wheels. As is typical of reaction wheel systems, excess momentum built up in the MCO wheel system due to external torques, such as from Sun-induced pressure on the solar panel array. To remove this excess angular momentum, the MCO thrusters were fired periodically during its nine-month spaceflight, in what are called angular momentum desaturation (AMD) maneuvers. The required attitude and trajectory corrections were calculated using ground-based computer software over the course of the nine-month spaceflight. Two relevant pieces of software used for the corrections, to be discussed in the root cause of the mishap, included the software that calculated the thruster forces (software file "Small Forces") and the software that used these thruster force numbers to calculate the spacecraft attitude corrections (software file "Angular Momentum Desaturation").

Given this background information, we now review the single root cause of the MCO mishap, as determined by the MCO Mishap Investigation Board, cited below.

> *The MCO MIB has determined that the root cause for the loss of the MCO spacecraft was the failure to use metric units in the coding of a ground software file, "Small Forces," used in trajectory models. Specifically, thruster performance data in English units instead of metric units was used in the software application code titled SM_FORCES (small forces). The output from the SM_FORCES application code as required by a MSOP (Mars Surveyor Operations Project) Software Interface Specification (SIS) was to be in metric units of newton-seconds (N-s). Instead, the data was reported in English units of pound-seconds (lbf-s). The Angular Momentum Desaturation (AMD) file contained the output data from the SM_FORCES software. The SIS, which was not followed, defines both the format and units of the AMD file generated by ground-based computers. Subsequent processing of the data from the AMD file by the navigation software algorithm therefore, underestimated the effect on the spacecraft trajectory by a factor of 4.45, which is the required conversion factor*

from force in pounds to newtons. An erroneous trajectory was computed using this incorrect data.

Excerpt from MCO MIB Phase I Report, *10 November 1999 ([31])*

This unit conversion error resulted in several small errors that accumulated, over the nine-month trip of the Mars Climate Orbiter, into a larger, ultimately catastrophic, error in the trajectory. The bottom line is that a "simple" unit conversion error, not converting the thruster force from English units to metric units, led to the demise of a multi-million dollar space probe. Needless to say, careful attention to units is important!

2.2.2   Measurement and Numerical Uncertainty

In performing a test or a numerical analysis, there is always uncertainty or error in measurements or calculations. To really understand and properly interpret results, this uncertainty or error must be quantified. In this section, we first discuss uncertainty in general. Then we define accuracy and precision, two important concepts that help us to quantify the uncertainty or error. Finally, we discuss significant figures, a way to specify the uncertainty in our numbers. This section is only meant to be a brief introduction to the important topic of measurement and numerical uncertainty. See [7] for more details concerning this topic.

2.2.2.1   Measurement Uncertainty

Suppose that we collect flight data, as represented by the data points (circular symbols) in Figure 2.3a. Now, suppose that we let two persons analytically model this flight data. One person develops a linear model and the other person fits the data to a non-linear curve, as shown. Which model is correct and better represents the flight data? Now, let us assume that we can quantify the *uncertainty* or error in the data measurements and we place *error bars* on the data that represent this uncertainty, as shown in Figure 2.3b. It is important to realize that the error bands represent the range of values that may be the "true values" of the data. The circular symbols are not necessarily closer to these true values. We now see that it is impossible for us to determine which analytical model is better. In fact, we cannot even determine if the physics behind the data is linear or

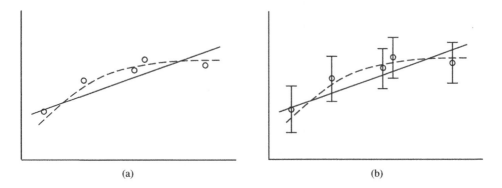

(a)   (b)

Figure 2.3   Interpreting data, (a) without uncertainty bands, (b) with uncertainty bands.

non-linear. Thus, we see, with this simple example, how important it is that we understand and quantify the uncertainty or error in measurements.

This also gives us some insight into the importance of *how* we should make measurements in our data collection. An *uncertainty analysis* may be required, where the sources of uncertainty or error are identified and quantified for a test. The propagation of the uncertainties in individual variables, through the data reduction to a final result, could also be determined. Although the details of a formal uncertainty analysis is beyond the scope of this book, such analyses are a powerful tool in the efficient planning and design of a test. By understanding the uncertainties or errors in the proposed data collection, informed decisions can be made concerning instrumentation requirements, including sensor calibrations and the measurement techniques to be used. This leads us to a discussion about accuracy and precision.

2.2.2.2 Accuracy and Precision

Accuracy and precision are sometimes confused as meaning the same thing, but they are distinctly different. *Accuracy* is defined as how close the measured or calculated value is to the true value. Therefore, the degree of inaccuracy in a measurement or calculation is the difference between the measured or calculated value and the true value. The *precision* is defined as the degree to which the same results can be repeated or reproduced from the same measurements, under the same measurement conditions.

The accuracy and precision are related to the *total measurement error*, which is the sum of the *systematic error* and the *random error*. The accuracy is related to the systematic error, which is the constant component of the total measurement error, often referred to as the measurement *bias*. The precision is related to the random error, which is the random component of the total measurement error, sometimes called the repeatability or precision error. If we made a number of measurements of a quantity and plotted the frequency of occurrence of the measured values, as in Figure 2.4, the mean value of all of the measurements would be different from the true value by a constant amount, which is the bias or accuracy of the measurement. The random or precision error of our many measurements would be distributed about the mean value as shown.

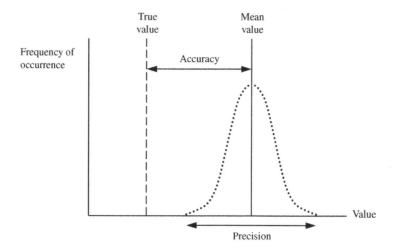

Figure 2.4 Accuracy and precision definitions.

The accuracy and precision can be independent of each other. For instance, we could increase the measurement precision by perhaps changing the measurement technique, but the fixed bias or accuracy would be unchanged. Similarly, we could decrease the bias, by perhaps performing a better instrumentation *calibration*, but the precision would remain unchanged, assuming the same measurement technique is used. Calibration involves comparing the instrument to a *standard* with a known uncertainty or error, so that we can quantify the instrument bias.

Let us look at a simple example to help us understand how we can have accuracy and precision in a measurement. Imagine that we shoot at a target, such that the bull's eye of the target is the true value. Our shooting results are shown as a function of accuracy versus precision in Figure 2.5. The target in the lower left of Figure 2.5 shows a bullet pattern that is not close to the "bull's eye" and not closely packed together, hence, it is a bullet pattern with low accuracy and low precision. The bullet pattern in the upper left is more closely packed together, but it is still not close to the bull's eye, hence, it has high precision and low accuracy. The bullet pattern in the lower right is closer to the bull's eye, but it is not closely packed together, hence, it has high accuracy and low precision. Finally, the bullet pattern in the upper right is both close to the bull's eye and closely packed together, hence, it has high accuracy and high precision. This is certainly the best result for target shooting, but it is also the best for measurements in general, that is, we usually want our measurements to be both accurate and precise.

Now, let us relate our understanding of accuracy and precision to a test situation. Let us replace the targets in Figure 2.5 with analogue gauges, which are displaying the airspeed of an aircraft, as shown in Figure 2.6. Let us assess the accuracy and precision with which we can read the airspeed from each of these gauges. The lower left dial depicts a gauge that is not calibrated and has large divisions in its scale, hence, we read the airspeed from this gauge with low accuracy and low precision. The gauge in the upper left is not calibrated, but it has finer divisions in its scale, hence, we read the airspeed with low accuracy and high precision. The lower right gauge is calibrated, but it has a scale with large divisions, hence, we read the airspeed with high accuracy and low

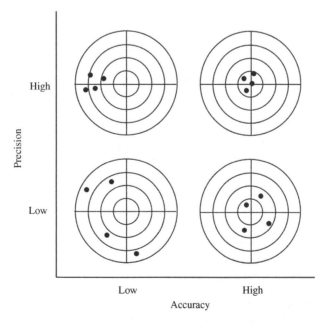

Figure 2.5   Accuracy and precision in shooting at a target.

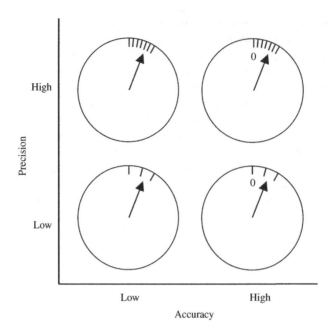

Figure 2.6   Accuracy and precision in reading an analogue gauge.

precision. Finally, the upper right gauge is calibrated and has finer divisions in its scale, allowing us to read the airspeed with both high accuracy and high precision.

2.2.2.3   Significant Figures

As a final topic concerning uncertainty, we make brief mention about *significant figures* of a number. This topic applies to any numbers that we deal with, whether they are measured in a test or are being manipulated in a post-test data reduction or analysis. This fundamental topic is often covered early in one's scientific or engineering education, but it is then often "forgotten", especially with the use of calculators and computers that can spit out an unlimited number of digits.

The numerical uncertainty is indicated by how many meaningful digits, or *significant figures*, there are in a value. The order of magnitude of the last digit in the significant figures is the uncertainty. For instance, if we measure the airspeed of an airplane as 243.7 miles per hour, we have four significant figures, where we are certain of the first three digits and the fourth digit is uncertain. Therefore, the uncertainty in our airspeed value is on the order of 0.1 mile per hour.

We must also be careful to maintain the proper number of significant figures in our calculations. When numbers are multiplied or divided, the result should have the same number of significant figure as the number with the fewest number of significant figures in the calculation. So, if we multiply an airspeed of 243.7 miles per hour by a time of 1.45 hours, the result for distance should be 353 miles (even though our calculator can supply a result of 353.365 miles).

When numbers are added or subtracted, the result should have the same uncertainty as the highest uncertainty in the numbers being added or subtracted. So, if we add an airspeed of 10 miles per hour to an airspeed of 243.7 miles per hour, the resulting sum should be 253 miles per hour, since our value of 10 miles per hour has the greatest uncertainty, on the order of 1 mile per hour. We have been careful not to write the result as 253.0 miles per hour, as this would imply an uncertainty on the order of 0.1 mile per hour, which would be incorrect.

Fractions and integers are considered to have an infinite number of significant figures. For instance, in the equation $y = \frac{1}{2}x^2$, the fraction is *exactly* equal to one divided by two, with an infinite number of significant figures ($0.50000\ldots$) and the exponent of x is exactly equal to 2, with an infinite number of significant figures ($2.0000\ldots$).

Always remember that despite the infinite number of digits that are available to us with calculators and computers, we should not infer a greater certainty in our numbers than is really there. This helps us to avoid drawing incorrect conclusions about test data or numerical analyses.

Example 2.3 Tomahawk Cruise Missile Drag Uncertainty Analysis *The present example, from [5], illustrates the use of an uncertainty analysis in the assessment of the aerodynamic drag of a flight vehicle. The lift and drag are two of the most important aerodynamic parameters that determine the performance and flying qualities of a flight vehicle. Since it is not possible to directly measure the lift and drag in flight, other basic parameters must be measured to calculate these forces. The propagation of the errors and uncertainties in these basic measurements results in an uncertainty in the lift and drag. This example shows the results of an uncertainty analysis, applied to the calculation of the total vehicle drag obtained from several flight test measured parameters.*

The uncertainty analysis was applied to the AGM-109 Tomahawk *air-launched cruise missile (ALCM), designed and built by the General Dynamics Corporation in the late 1970s (Figure 2.7). The AGM-109 missile had an 18.25 ft (5.563 m) long, cylindrical fuselage, with a circular cross-section, and a gross weight of 2553 lb (1158 kg). The missile had a cruciform tail (four tail fins in a cross pattern) and a small, straight wing (zero wing sweep) with an area of 12 ft^2 (1.1 m^2). Propulsion was provided by a Williams F107 turbofan jet engine. After being launched from a military aircraft, the AGM-109 unfolded its wing and flew to its target at subsonic speeds using its turbofan jet engine. Flight testing of the AGM-109 was conducted as part of a US military "fly-off" competition among various cruise missile designs in the late 1970s.*

Figure 2.7 Raytheon BGM-109 *Tomahawk* cruise missile, similar to the AGM-109 cruise missile, proposed in the 1970s. The AGM-109 did not win the "fly-off" competition and was never produced. (Source: *US Navy.*)

Table 2.3 Effect of a 1% change in the independent, measured parameters on the AGM-109 *Tomahawk* drag coefficient.

Independent, measured parameter	Change in drag coefficient
Engine airflow calibration	1.2%
Engine core speed	10.9%
Engine fan speed	1.3%
Engine thrust calibration	4.0%
Indicated air temperature	0.8%
Inlet total pressure calibration	3.0%
Nozzle area	2.5%
Sea level temperature	4.0%
Static pressure	1.0%
Wing area	1.0%

(Source: *Data from* [5].)

The in-flight drag of the AGM-109 cruise missile could not be measured directly. Instead, other basic parameters (termed the independent *parameters) were measured in flight, and the drag (termed the* dependent *parameter) was calculated from these independent parameters. The independent measurement parameters included those that defined the flight condition (air temperature and pressure conditions), geometry of the vehicle (wing area), and engine performance (engine airflow, engine fan speed, nozzle area, and core speed). Thus, the dependent parameter (drag) could be mathematically expressed as a function of these independent parameters. The uncertainty analysis used a numerical technique that perturbed the independent variables, in a functional expression for the drag, to estimate the uncertainty in the drag.*

Table 2.3 provides selected results from the uncertainty analysis, where the effects of changes in the independent measurement parameters on the dependent variable (the drag coefficient[2]) are shown for a subsonic flight condition. The independent parameters in the uncertainty analysis included the in-flight measured parameters and the parameters associated with certain instrumentation calibrations. In calculating the drag coefficient, each independent measurement parameter was changed by 1%, while keeping all of the other independent parameters constant. The table shows the resulting percentage change in the drag coefficient due to a 1% change in each independent measurement parameter. For instance, a 1% change in the measurement of the sea level temperature resulted in a 4.0% change in the drag coefficient.

The results of the uncertainty analysis provided several important insights about calculating the in-flight drag for the vehicle. It identified which measurements were the most critical, and the major sources of error, in calculating the drag. Based on Table 2.3, the drag calculation is most sensitive to the measurement of the jet engine core speed.[3] If the core speed measurement is in error by just one percent, then the resulting error in the drag computation is about 11%. This understanding of the measurement sensitivities helps determine where to invest time, effort, and funding in making test measurements, for example which measurements should be made more carefully or require higher accuracy sensors. Conversely, the uncertainty analysis also provides insight into

---

[2] The non-dimensional drag coefficient, C_D, is defined as the drag, D, divided by the product of the dynamic pressure, q, and a reference area, S, usually the wing area. The drag coefficient and other aerodynamic coefficients are discussed in Chapter 3.

[3] The engine core speed is the rotational speed of the turbomachinery in the central core of the engine. The faster the core speed, the more air is being sucked into the engine and the higher the thrust. This is discussed in Chapter 4.

which measurements are not as critical, hence, requiring less attention. By including the instrument calibrations, the uncertainty analysis indicates which calibrations are most critical. Table 2.3 indicates that the calibrations associated with the engine thrust measurement and the inlet total pressure are more critical for the drag computation. Thus, it would be worthwhile to expend more effort in performing these calibrations to obtain a more accurate value of the drag.

2.2.3   Scalars and Vectors

In this section, we briefly review a few aspects of scalar and vector quantities. A *vector* quantity possesses both magnitude and direction. This is in contrast to *scalar* quantities that are defined only by their magnitude. Velocity and force are examples of vector quantities, while density and temperature are scalars. In the text, vectors are represented with a single letter with an arrow above them, such as the vector $\vec{A}$. Scalars are represented by the letter symbol alone.

2.2.3.1   The Unit Vector and Vector Magnitude

We use vector notation to mathematically express the various laws of nature and conservation principles. When written in this vector form, these mathematical expressions of physical laws and principles are independent of any specific coordinate system. This form of the equations, independent of a coordinate system, is known as the *invariant form*. This invariant form allows us to obtain an understanding of the physics, without any restrictions or complications of using a specific coordinate system. When we are ready to solve problems, we apply these invariant forms of the equations to an appropriate coordinate system, typically selecting one that best fits the geometry of the problem.

A vector can be represented by its components. The specification of the vector components depends on the coordinate system used. The magnitude of a vector, $\vec{A}$, is defined as the square root of the sum of the square of the components and is denoted by absolute symbol bars as $|\vec{A}|$ or simply as A. The magnitude of a vector quantity is a scalar.

A unit vector is defined as the vector divided by its magnitude. Thus, the unit vector of $\vec{A}$, denoted by $\hat{e}_A$, is given by

$$\hat{e}_A = \frac{\vec{A}}{|\vec{A}|} \tag{2.12}$$

The unit vector has a unit length. Hence, the unit vector is sometimes called the normalized vector. The unit vector points in the direction of the original vector, hence, $\hat{e}_A$ points in the direction of $\vec{A}$. The unit vector is sometimes called the *direction vector*. We can rewrite Equation (2.12) as

$$\vec{A} = A\,\hat{e}_A \tag{2.13}$$

where A is the scalar magnitude of the $\vec{A}$. Thus, we see that we can represent a vector as the product of its magnitude and its unit vector.

2.2.3.2   Vectors in the Cartesian Coordinate System

For much of our discussions in future chapters, we need to reference our vehicle orientation, position, or motion with respect to a *coordinate system*. Depending on the problem of interest, the origin of the coordinate system could be attached to the vehicle or to the earth. If attached to the vehicle, the origin of the coordinate system is typically located at the vehicle center of gravity. If attached

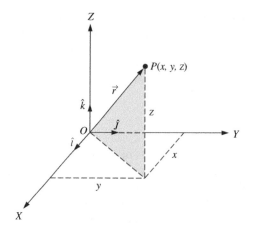

Figure 2.8   Cartesian coordinate system.

to the earth, the coordinate system origin could be located at the *surface* of the earth, typical of aircraft motion problems, or at the *center* of the earth, typical of spacecraft motion problems.

In the Cartesian coordinate system, a point in space, P, is specified by three coordinates, (x, y, z), measured with respect to three mutually perpendicular axes, X, Y, Z, as shown in Figure 2.8. The vector $\overrightarrow{OP}$, from the origin O of the coordinate system to the point P, defines the position vector, $\vec{r}$, to the point P.

$$\overrightarrow{OP} = \vec{r} = (x, y, z) = x\,\hat{\imath} + y\,\hat{\jmath} + z\,\hat{k} \tag{2.14}$$

where x, y, and z are the scalar components of the vector $\vec{r}$, and $\hat{\imath}, \hat{\jmath}$, and $\hat{k}$ are the unit vectors along the X, Y, and Z axes, respectively. The magnitude of the position vector is given by

$$r \equiv |\vec{r}| = \sqrt{x^2 + y^2 + z^2} \tag{2.15}$$

If we have a vector quantity, such as the velocity of an aircraft, $\vec{V}$, we can define $\vec{u}, \vec{v}$, and $\vec{w}$ as the velocity components of $\vec{V}$ in the X, Y, and Z directions, respectively, as shown in Figure 2.9. Hence, we can represent the velocity vector in terms of these components as

$$\vec{V} = u\,\hat{\imath} + v\,\hat{\jmath} + w\,\hat{k} \tag{2.16}$$

The magnitude of the velocity vector is given by

$$V \equiv |\vec{V}| = \sqrt{u^2 + v^2 + w^2} \tag{2.17}$$

Remember that these equations apply to any vector quantity, not just the velocity vector that we chose as an example.

2.3   Introductory Aerospace Engineering Concepts

In this section, several basic aerospace engineering concepts, definitions, and nomenclature are introduced. Some of these concepts may be familiar from basic physics, but with a new focus on aerospace applications. Others concepts are new and specific to aerospace engineering or flight test. We start by defining several aerospace vehicle axis systems and the associated conditions that

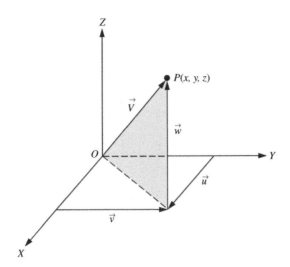

Figure 2.9 Velocity vector and velocity components in Cartesian coordinates.

define the vehicle's orientation and motion. An idealized, point mass model of an aerospace vehicle is discussed, which is useful in analyzing the vehicle state by applying Newton's laws of motion. The speed of sound and Mach number are introduced and used to discuss the different regimes of flight. Finally, several aerospace concepts are introduced that are captured by a defining diagram or chart, these being the flight envelope, the aircraft load factor versus airspeed plot, and the aircraft weight and balance plot.

2.3.1 Aircraft Body Axes

There are several different axis systems that may be used to define the orientation or attitude of an aerospace vehicle. The selection of a particular system usually depends on the type of problem that is being analyzed. In this text, we usually deal with a three-dimensional coordinate system that is rigidly attached to the aircraft, called the *body axis* system.

The origin of the body axis system is located at the aircraft center of gravity, commonly referred to as the "CG", as shown in Figure 2.10. The x_b-axis points out through the aircraft nose along a defined reference line, which may be a line through the fuselage or wing (usually the wing *chord*, to be defined in Chapter 3). The y_b-axis points out of the aircraft's right wing and is positive in that direction. Using the right-hand rule, the z_b-axis points through the bottom of the aircraft and is positive in that direction. The x_b-z_b plane is a *symmetry plane* that "cuts" the aircraft into two symmetrical halves. The x_b, y_b, and z_b axes are also referred to as the *longitudinal*, *lateral*, and *vertical* axes, respectively.

The body axes are "bolted" to the aircraft and do not change their orientation, relative to the aircraft, as the aircraft translates and rotates in three-dimensional space. The x_b- and y_b-axes always points out of the aircraft nose and right wing, respectively, regardless of the aircraft orientation. Aircraft moments of inertia and products of inertia are referenced to the body axis system, since they then remain constant, regardless of changes in the aircraft orientation. The body axis system is often the frame of reference for the pilot, since the pilot is attached to this coordinate system as the aircraft translates and rotates.

The motion of the aircraft can be described about the three body axes. In general, an aircraft has *six degrees of freedom*, three linear translations and three angular rotations. An aircraft can translate

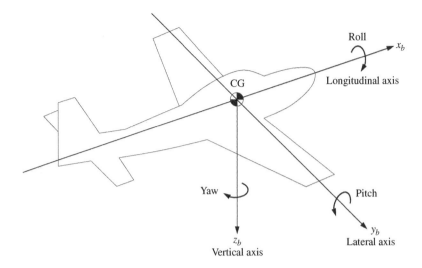

Figure 2.10   Aircraft body axis system.

forward (and aft, if it is a helicopter or airship) along the longitudinal axis, move to its right or left along the lateral axis, and move up or down along the vertical axis. As shown in Figure 2.10, the aircraft can rotate about each of the three body axes. Rotation about the longitudinal axis is called *roll*, hence this axis is called the *roll axis*. Rotation about the lateral axis is called *pitch*, hence this axis is called the *pitch axis*. Rotation about the vertical axis is called *yaw*, hence this axis is called the *yaw axis*.

2.3.2   Angle-of-Attack and Angle-of-Sideslip

Consider an aircraft that is flying at a velocity, V_∞, as shown in Figure 2.11, where the nose of the aircraft may not be pointing in the direction of the velocity vector. The orientation of the aircraft can be defined with respect to the velocity vector in terms of two angles, the *angle-of-attack*, α, and the *angle-of-sideslip*, β, as shown.

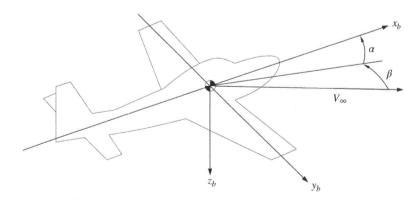

Figure 2.11   Orientation of an aircraft with respect to the velocity vector, V_∞.

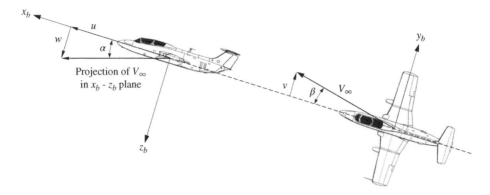

Figure 2.12   Definition of angle-of-attack, α, and angle-of-sideslip, β. (Source: *Adapted from Dynamics of Flight: Stability and Control, B. Etkin and L.D. Reid, Fig. 1.7, p. 16, (1996), [7], with permission from John Wiley & Sons, Inc.*)

The aircraft angle-of-attack, α, measured in the x_b-z_b plane, is defend as the angle between the aircraft longitudinal axis (the x_b-axis) and the projection of the velocity vector in the x_b-z_b plane, as shown in Figure 2.12. The projection of the velocity vector is defined by its components, u and w, along the x_b- and z_b-axes, respectively. Positive angle-of-attack is measured up from the velocity vector to the aircraft reference line. Later, in Chapter 3, we define an angle-of-attack specific to the airfoil section of a wing.

The aircraft angle-of-sideslip β, is the angle between the aircraft x_b-z_b symmetry plane and the velocity vector. The sideslip angle is not measured in the x_b-y_b plane, since the velocity vector is not necessarily in this plane. If the angle-of-sideslip is zero, then the angle-of-attack is simply the angle between the aircraft longitudinal axis and the total velocity vector. Positive angle-of-sideslip is with the aircraft nose pointing to the left with respect to the velocity vector. Positive sideslip angle is also referred to as "wind in the right ear" as this is what the pilot would feel in an open cockpit airplane with the nose pointing left relative to the velocity vector.

Let us define the velocity vector, $\vec{V}_\infty$, as

$$\vec{V}_\infty = u\,\hat{\imath} + v\,\hat{\jmath} + w\,\hat{k} \tag{2.18}$$

where u, v, and w are the components of the velocity in the x_b, y_b, and z_b axis directions, respectively, and $\hat{\imath}$, $\hat{\jmath}$, and $\hat{k}$ are the unit vectors along these axes, respectively. The magnitude of the velocity, V, is given by

$$V_\infty = \sqrt{u^2 + v^2 + w^2} \tag{2.19}$$

Using these definitions of the velocity, the angle-of-attack is defined as

$$\alpha = \tan^{-1}\frac{w}{u} \tag{2.20}$$

and the angle-of-sideslip is defined as

$$\beta = \sin^{-1}\frac{v}{V_\infty} \tag{2.21}$$

The angle-of-attack and angle-of-sideslip are two important parameters that are frequently used in describing the aircraft orientation, especially in the areas of aerodynamics and stability and control.

Example 2.4 Calculation of Angle-of-attack and Angle-of-sideslip *The components of velocity of an aircraft, in the body axis system, are u = 173.8 kt (nautical mile per hour), v = 1.27 kt, and w = 13.2 kt. Calculate the magnitude of the velocity, the angle-of-attack, and the angle-of-sideslip.*

Solution

From Equation (2.19), the velocity magnitude is

$$V_\infty = \sqrt{u^2 + v^2 + w^2} = \sqrt{(173.8\,kt)^2 + (1.27\,kt)^2 + (13.2\,kt)^2} = 174\,kt$$

From Equation (2.20), the angle-of-attack is

$$\alpha = tan^{-1}\frac{w}{u} = tan^{-1}\left(\frac{13.2\,kt}{173.8\,kt}\right) = 4.34°$$

From Equation (2.21), the angle-of-sideslip is

$$\beta = sin^{-1}\frac{v}{V_\infty} = sin^{-1}\left(\frac{1.27\,kt}{174\,kt}\right) = 0.418°$$

2.3.3 Aircraft Stability Axes

Similar to the aircraft body axis system, the aircraft *stability axis* system, composed of the x_s, y_s, and z_s axes, is attached to the aircraft center of gravity, as depicted in Figure 2.13. The stability y_s-axis points out through the right wing of the aircraft and is coincident with the body y_b-axis. To obtain the stability axes from the body axes, the x- and z-axes are rotated around the y-axis, through the angle-of-attack, α, so that the x_s-axis is aligned with the projection of the velocity vector in the x-z plane, as shown in Figure 2.13. This alignment of the stability axes makes the aircraft lift parallel to the stability z_s-axis and the aircraft drag parallel to the stability x_s-axis. (The lift and drag are defined as perpendicular to and parallel to the velocity vector, respectively.) This alignment of the aerodynamic forces with the stability axes is useful in the in-flight determination of lift and drag, as discussed in Chapter 3.

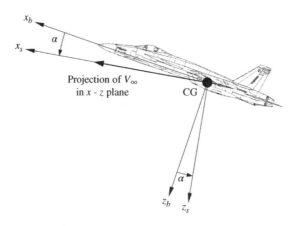

Figure 2.13 Aircraft stability axis system.

2.3.4   *Aircraft Location Numbering System*

Another aircraft coordinate system, which is commonly used to specify the location of aircraft structural or other components, is the *aircraft location numbering system*. Use of this system is usually started during the aircraft design process and is maintained throughout the life of the aircraft. Location numbering is typically used on aircraft technical drawings and in maintenance and operational manuals. Aircraft location numbering is based on systems that were originally developed for ships and boats; hence, it has retained some maritime-influenced terminology. An example of aircraft location numbering is shown in Figure 2.14. In the USA, the location numbers are typically denoted in units of inches.

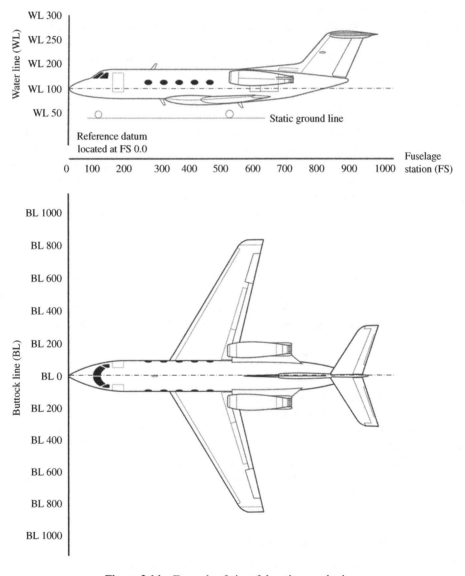

Figure 2.14   Example of aircraft location numbering.

The *fuselage station* (FS) number is the longitudinal distance measured from a *reference datum* or *zero station* (FS 0.0), which may be located in front of the airplane. A positive fuselage station number is measured aft from the reference datum. The FS 0.0 is sometimes selected ahead of the aircraft to allow the aircraft to "grow" or "shrink" during the design process or the course of its operational life. Using this scheme, no fuselage station numbers have a negative value and components that are not moved, retain a consistent station number. For instance, if during the design process, the nose of an aircraft increases in length while the wing location is not changed, the nose FS number decreases to a smaller, positive value, but the wing FS numbers are unchanged.

The *buttock line* or *butt line* (BL) number is the lateral distance from the reference datum, measured positive outward to each wingtip. The butt line zero station, BL 0.0, is almost always at the centerline of the aircraft, which is usually a plane of symmetry. In Figure 2.14, the left and right wingtips are both located at about BL 850.

The *water line* (WL) number is the vertical distance measured, positive upward, from a zero station (WL 0.0), which is a major, longitudinal structural member in the fuselage or the ground plane beneath the aircraft. In Figure 2.14, the top of the vertical tail is located at about WL 275.

2.3.5 The Free-Body Diagram and the Four Forces

When we consider the flight of aircraft, it is sometimes convenient to think about the aircraft as a point mass, that is, it is assumed that all of the vehicle's mass is concentrated at a single point, called the center of mass or center of gravity (CG). This point mass assumption allow us to analyze the aircraft motion as a *free-body problem*, that is, the motion of a single point mass that is *free* of its surroundings, acted upon by distinct forces. The *free-body diagram* depicts this point mass representation of the vehicle, with vectors drawn to show the magnitude and direction of the forces acting on it.

In many instances, we apply Newton's first and second laws of motion to a free-body diagram in order to analyze the vehicle state or motion. Newton's first law states that a body remains in an equilibrium state, either at rest (zero velocity) or in motion at constant velocity. Newton's second law deals with the non-equilibrium state, where the sum of the net force acting on a body is equal to the time rate of change of the body's momentum, $m\vec{V}$. For many of the situations of interest to us, we assume that the mass of the body is constant, so that Newton's second law becomes

$$\sum \vec{F} = \frac{d}{dt}(m\vec{V}) = m\frac{d\vec{V}}{dt} = m\vec{a} \tag{2.22}$$

where $\vec{a}$ is the time rate of change of the velocity, or the acceleration.

In the text, we consider two cases for the motion of an aircraft, unaccelerated motion with a straight-line flight path and accelerated motion with a curved flight path. Unaccelerated flight is associated with the climb, cruise, and descent flight conditions and accelerated flight is associated with takeoff, landing, and turning flight. In Chapter 5, we analyze the vehicle performance by applying Newton's laws to the vehicle translational motion. In Chapter 6, we analyze the vehicle stability and control by applying Newton's laws to the vehicle's curved or rotational motion.

2.3.5.1 Wings-Level, Unaccelerated Flight

For the case of unaccelerated flight, the acceleration is zero, the velocity is constant, and Equation (2.22) is simply

$$\sum \vec{F} = 0 \tag{2.23}$$

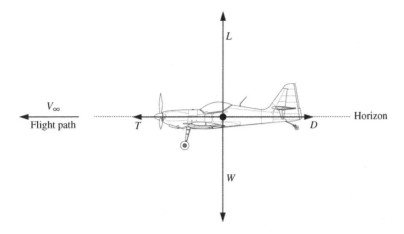

Figure 2.15   The four forces acting on an aircraft in level, unaccelerated flight.

Let us now draw a free-body diagram for an aircraft flying in level, unaccelerated flight at a constant altitude and constant airspeed. The aircraft's flight path is horizontal to the surface of the earth as is its velocity, V_∞. There are four distinct forces acting on the aircraft, lift, L, drag, D, thrust, T, and weight, W, as shown in Figure 2.15. The lift is perpendicular to the velocity vector and the drag is parallel to the velocity vector. The thrust acts along a vector defined by the propulsion system, defined by a thrust vector angle, α_T, relative to the velocity vector. For simplicity, it is often assumed that the thrust vector angle is zero so that the thrust is parallel to the velocity vector, as shown in Figure 2.15. The weight acts in a direction towards the center of the earth, along the gravity vector. Ignoring the curvature of the earth, the weight acts downward.

If we assume that the aircraft in Figure 2.15 is in steady level flight, that is, flying at constant altitude and constant airspeed, we can apply Equation (2.23) to the forces in the directions perpendicular and parallel to the velocity vector to obtain

$$\sum F_\perp = L - W = 0 \tag{2.24}$$

$$\sum F_\parallel = T - D = 0 \tag{2.25}$$

where $F_\perp$ and $F_\parallel$ are the components of force perpendicular to and parallel to the flight path, respectively. This simply gives us the obvious result that for steady, constant velocity flight, the lift equals the weight and the thrust equals the drag.

$$L = W \tag{2.26}$$

$$D = T \tag{2.27}$$

Despite its simplicity, this equilibrium case will be useful in future analyses. If we divide Equation (2.27) by (2.26), we obtain an expression relating the *thrust-to-weight ratio*, (T/W), to the *lift-to-drag ratio*, (L/D), for an aircraft in steady level flight.

$$\frac{T}{W} = \frac{1}{L/D} \tag{2.28}$$

These two non-dimensional ratios are important parameters in many aspects of aircraft aerodynamics, performance, and design. Broadly speaking, the thrust-to-weight ratio is a propulsion related parameter and the lift-to-drag ratio is an aerodynamics related parameter. The thrust is dependent on the propulsion system and the weight is a function of the aircraft structure, payload, and fuel. The thrust can be changed during a flight, for example by the pilot selecting a different throttle setting. The aircraft total weight changes during a flight, usually decreasing, due to fuel consumption. Therefore, the thrust-to-weight ratio varies continuously during a flight, having different values at different phases of flight, such as takeoff, cruise, or landing. The thrust-to-drag ratio is a measure of the propulsion system's capability to accelerate the aircraft mass. We can see this by applying Newton's second law to the propulsion system thrust force ($T = ma$) and the aircraft weight ($W = mg$), as follows

$$\frac{T}{W} = \frac{ma}{mg} = \frac{a}{g} \tag{2.29}$$

where m is the aircraft mass, g is the acceleration due to gravity, and a is the aircraft acceleration. Equation (2.29) shows that the thrust-to-weight ratio is directly proportional to the aircraft's acceleration. A higher thrust-to-weight ratio indicates that an aircraft has a higher acceleration or climb capability. If the thrust-to-weight ratio is greater than one, then the vehicle is capable of accelerating in a vertical climb. High performance fighter aircraft may have this capability, while it is a requirement for a vertical takeoff rocket vehicle.

Thrust-to-weight ratios for various types of vehicles are shown in Table 2.4. Aircraft thrust-to-weight ratios are usually specified for the maximum static thrust that is produced at sea level divided by the aircraft maximum takeoff weight. Thrust-to-weight ratios are also quoted for a jet or rocket engine alone, as a measure of the engine's acceleration capability without an airframe.

The lift-to-drag ratio is a measure of the aerodynamic efficiency of an aircraft. The more aerodynamic lift that an aircraft can produce in relation to the aerodynamic drag, the more aerodynamically efficient it is. The lift and drag are strongly influenced by the size and design of the aircraft wing. The lift and drag, and hence the lift-to-drag ratio, vary with the airspeed. We are often interested in the maximum value of the lift-to-drag ratio, denoted as $(L/D)_{max}$. Values of the lift-to-drag ratio are given for various types of vehicles in Table 2.4. Examine these values closely, as it is worthwhile to obtain a "feel" for the L/D of different types of vehicles.

2.3.5.2 Climbing, Unaccelerated Flight

Consider now the case of climbing, unaccelerated flight, where the aircraft is in a constant airspeed climb, as shown in Figure 2.16. The *flight path angle*, γ, is defined as the angle between the aircraft's

Table 2.4 Lift-to-drag and thrust-to-weight ratios for various types of aerospace vehicles.

Type of aerospace vehicle	Lift-to-drag ratio, L/D	Thrust-to-weight ratio, T/W
Wright *Flyer I*	8.3	—
General aviation airplane	7–15	—
High performance glider	40–60	—
Commercial airliner	15–25	0.25–0.4
Military fighter airplane	4–15	0.6–1.1
Helicopter	4–5	—
Space capsule (*Apollo* capsule)	~0.35 (reentry)	—
Lifting space plane (Space Shuttle)	4–5 (subsonic glide)	1.5 (lift-off)

Values are for cruise flight conditions unless otherwise noted.

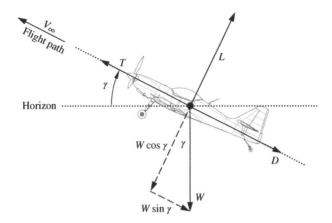

Figure 2.16   The four forces acting on an aircraft in climbing, unaccelerated flight.

velocity vector and the horizon. The same four forces of lift, drag, weight, and thrust act on the aircraft at its center of gravity. Again, it is assumed that the thrust vector angle is zero, aligning the thrust with the velocity vector. As shown in Figure 2.16, the weight vector points vertically downward, making an angle γ with respect to the direction perpendicular to the flight path.

Even though the aircraft is climbing, it is not accelerating or decelerating, hence, Equation (2.23) is still valid. Summing the forces perpendicular and parallel to the flight path, we have

$$\sum F_\perp = L - W \cos \gamma = 0 \tag{2.30}$$

$$\sum F_\| = T - D - W \sin \gamma = 0 \tag{2.31}$$

Solving for the lift and drag, we have

$$L = W \cos \gamma \tag{2.32}$$

$$D = T - W \sin \gamma \tag{2.33}$$

Solving Equation (2.33) for thrust, we have

$$T = D + W \sin \gamma \tag{2.34}$$

Equation (2.32) states that, for a constant airspeed climb, the lift must equal the component of weight perpendicular to the flight path. Equation (2.34) states that, for a steady climb, the thrust must equal the drag plus a component of the weight in the direction opposite to the flight path. Comparing Equation (2.34) with Equation (2.27) for level flight, we see that, as expected, more thrust is required for a constant airspeed climb than for level flight at constant airspeed, by an additional amount equal to $W \sin \gamma$. Note the case of level, unaccelerated flight simply corresponds to the case of zero flight path angle.

2.3.5.3   Descending, Unaccelerated Flight

Consider now the case of steady, unaccelerated flight, where the aircraft is in a constant airspeed descent, as shown in Figure 2.16. The *flight path angle* is now negative, $-\gamma$, since the flight path is below the horizon. The angle, θ, is defined as the magnitude of the negative flight path angle. The

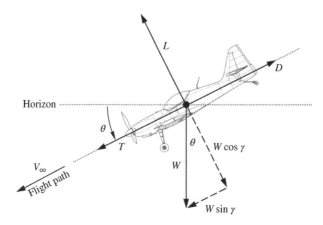

Figure 2.17   The four forces acting on an aircraft in descending, unaccelerated flight.

same four forces of lift, drag, weight, and thrust act on the aircraft at its center of gravity. Again, it is assumed that the thrust vector angle is zero, aligning the thrust with the velocity vector. As shown in Figure 2.17, the weight vector points vertically downward, making an angle, θ with respect to the direction perpendicular to the flight path.

Even though the aircraft is descending, it is not accelerating or decelerating, and Equation (2.23) is valid. Summing the forces perpendicular and parallel to the flight path, we have

$$\sum F_\perp = L - W \cos \theta = 0 \tag{2.35}$$

$$\sum F_\parallel = T - D + W \sin \theta = 0 \tag{2.36}$$

Solving for the lift and drag, we have

$$L = W \cos \theta \tag{2.37}$$

$$D = T + W \sin \theta \tag{2.38}$$

Solving Equation (2.38) for thrust, we have

$$T = D - W \sin \theta \tag{2.39}$$

Equation (2.37) states that, for a constant airspeed descent, the lift must equal the component of weight perpendicular to the flight path. Equation (2.39) states that, for a steady descent, the thrust equals the drag minus a component of the weight in the direction of the flight path. Thus, we see that less thrust is required for descending flight at constant airspeed than for constant airspeed, level, or climbing flight, since there is the component of weight, $W \sin \theta$, acting in the direction of the thrust. Note again, that the case of level, unaccelerated flight simply corresponds to that of zero flight path angle.

In summary, by using the point mass assumption for the vehicle, drawing a free-body diagram, and applying Newton's second law of motion, we can develop the equations relating the forces to the state of the vehicle. We apply this procedure to many vehicle motion problems in the future, especially in analyzing vehicle aerodynamics, performance, and stability and control.

2.3.6 FTT: the Trim Shot

It is fitting that the *trim shot* is one of the first flight test techniques (FTTs) that is introduced, as this flight condition is the starting point for many, if not most, of the other flight test techniques. The trim shot is an equilibrium flight condition where the aircraft is in steady, unaccelerated flight. The trim shot is perhaps the simplest flight test technique in concept and one of the most critical in proper execution. If this setup maneuver is not performed properly, it may be difficult, if not impossible, to obtain high quality flight test data.

In a trimmed state, all of the forces acting on the vehicle are stabilized and the vehicle moments are zero. For an aircraft with conventional flight controls, a trimmed state is obtained by setting the elevator, ailerons, and rudder at their trimmed positions, thereby reducing the pitching, rolling, and yawing moments, respectively, to zero. Typically, the control surfaces are held fixed at the required trim positions by the use of trimming devices, such as control surface position actuators or trim tabs. At this trimmed condition, the control forces felt by the pilot are zero. The point at which this trim condition is established is called the trim point or *trim shot*. (Although the vehicle is in an equilibrium state at the trim point, this does not guarantee that the vehicle remains in equilibrium, if it disturbed from this trim point. However, we defer this for a later discussion about static and dynamic stability in Chapter 6.)

While a trim shot can be set up in turning, climbing, or descending flight, we are often interested in the trim shot with the aircraft at constant altitude, constant airspeed, and constant attitude. However, these three parameters are not independent. For instance, if the attitude of an aircraft, trimmed at a constant altitude and constant airspeed, is disturbed slightly from its trim state by atmospheric turbulence, its airspeed changes (assuming the altitude remains constant).

In setting up the trim shot, the fundamental flight technique of *attitude flying* is used. Attitude flying relates to flying the aircraft by reference to the "outside world" or horizon. By using this outside reference, small changes in the aircraft's attitude can be perceived well before the cockpit instruments display these changes. The altimeter (which measures altitude) and the airspeed indicator are pressure-sensing instruments that are susceptible to lag in displaying their readings, due to the tubing lengths required to measure the pressure and other instrument factors. Attitude flying enables the pilot to perceive small attitude changes and apply small corrections to return the aircraft to its trim attitude, before the airspeed changes from the desired trim condition.

To set up a trim shot, the aircraft is first stabilized at a constant altitude with a constant power setting. It may take several tens of seconds or even minutes for the engines to completely stabilize. Attitude flying is then used to establish the desired airspeed or Mach number by setting the appropriate aircraft attitude. Another power adjustment may be required to arrest a rate of climb or descent and maintain constant altitude. The control forces are held to maintain the aircraft attitude and are trimmed to zero force by setting the appropriate trim devices. Lateral and directional controls (ailerons and rudder) are used to maintain a wings-level attitude and a constant heading, respectively. The accuracy of the trim shot can be checked by releasing the flight controls and verifying that the desired trim conditions of constant altitude, airspeed, and attitude are stabilized. Patience must be exercised in setting up the trim shot, to allow all of the flight condition variables to stabilize.

The trim shot is demonstrated by flying the Extra 300 aircraft (see photograph in Figure 1.21 and three-view drawing in Figure 1.22). The Extra 300 is a single-engine, mid-wing, high-performance, two-place aerobatic airplane. It is powered by a single Lycoming AEIO-540-L1B5 normally aspirated, air-cooled, horizontally opposed, six-cylinder piston engine producing 300 hp (224 kW). The engine is designed for aerobatic flight with an oil system capable of supplying engine oil in

Table 2.5   Selected specifications of the Extra 300.

Item	Specification
Primary function	General aviation, advanced aerobatics
Manufacturer	Extra Aircraft, Germany
First flight	6 May 1988
Crew	1 pilot + 1 passenger
Powerplant	Lycoming AEIO-540-L1B5 six-cylinder engine
Engine power	300 hp (224 kW) at 2700 rpm
Empty weight	1643 lb (745.3 kg)
Maximum gross weight	2095 lb (950.3 kg)
Length	23.4 ft (7.12 m)
Height	8.60 ft (2.62 m)
Wingspan	26.25 ft (8.0 m)
Wing area	115.2 ft^2 (10.7 m^2)
Wing loading	16.7 lb/ft^2 (81.3 kg$_f$/m^2)
Airfoil, wing root	MA15S (symmetric, 15% thickness)
Airfoil, wingtip	MA12S (symmetric, 12% thickness)
Maximum cruising speed	158 knots (182 mph, 293 km/h)
Service ceiling	17,000 ft (5200 m)
Load factor limits	+10.0 g, −10.0 g

sustained inverted flight. The first flight of the Extra 300 was on 6 May 1988. Selected specifications of the Extra 300 are provided in Table 2.5.

You will be flying the aircraft solo from the aft cockpit, as required to stay with the center of gravity limits for solo flight. You will perform the trim shot FTT as a setup for another maneuver, a maximum rate aileron roll, where the aircraft will complete a full, 360° roll about its longitudinal axis. Your desired trim shot flight conditions for this maneuver are an airspeed and altitude of 155 knots (178 mph, 287 km/h) and 7000 ft (2100 m), respectively.

As shown in Figure 2.18, the aft cockpit flight controls include a center-mounted stick for pitch and roll control and rudder pedals for yaw control. A pitch trim lever, located on the right side of the cockpit, can be rotated up or down to reduce the control stick pitch forces to zero. A forward and aft moving throttle lever on the left side of the cockpit controls engine power. Cockpit instrumentation includes round analog indicators for airspeed and altitude and an electronic flight information system (EFIS) display. Flight test data is supplied to the EFIS by three onboard data sources, a global positioning system (GPS) receiver, an engine information system, and an attitude and heading reference system (AHRS). The system can provide aircraft airspeed, altitude, attitude, position data, engine information, three-dimensional linear accelerations and angular rates, and other data. The flight test data is collected at a rate of 10 samples per second, or a sample rate of 10 Hz. The primary flight test data to be collected for the trim shot FTT are airspeed, altitude, pitch and roll angles, and roll rate. The cockpit instrumentation does not include an artificial horizon, an instrument used to determine the aircraft's pitch and roll attitude. For the setup of the trim shot in flight, you will rely on basic *attitude flying*, where you will set the aircraft attitude with respect to the outside horizon.

Now that you have reviewed the instrumentation and data system, you are ready to go flying. You decide to perform your test in the early morning, when there is less chance of atmospheric turbulence. The still air will make it easier to set up a stable trim shot, and the quality of the test data will be better. You climb into the Extra 300 aft cockpit, secure your safety belts, start the Lycoming engine, and taxi to the runway for takeoff. Applying full engine power, you take off and

Figure 2.18 Aft cockpit and instrument panel of the Extra 300 aircraft (Source: *Courtesy of the author.*)

start the climb. After your climb is established, you set climb power, slightly less than full power. At 7000 ft, you push forward slightly on the control stick to lower the aircraft nose and set up for the trim shot.

You adjust the nose attitude, in relation to the horizon, to maintain your constant altitude. After giving the engine 15 seconds to stabilize, you confirm that the altitude and airspeed are stable. The altitude is constant at 7000 ft and the airspeed is stable at 160 knots (184 mph, 296 km/h), faster than your desired target airspeed. Using attitude flying, you make a small nose attitude adjustment, slightly raising the nose relative to the horizon to decrease the airspeed and make a small power reduction, from the previously set climb power, to prevent the aircraft from climbing. All of these adjustments are small, giving time between them to allow the flight condition to stabilize. While you are patiently waiting for everything to stabilize, you are careful to maintain a wings-level attitude and a constant heading, with very small adjustments in roll and yaw, using lateral stick and rudder inputs, respectively. After some time, it looks like the aircraft is stable in airspeed, altitude, and attitude. You are holding some aft pitch control force to maintain this trim shot, so you slowly move the pitch trim lever to reduce this force to zero. Finally, you check your trim shot by gingerly releasing your grip on the control stick and verify that the flight condition is stable. The airspeed and altitude are stable at 155 knots and 7000 ft, respectively. You maintain this trim shot for several seconds to ensure that nothing is changing. It has taken some patience and precise attitude flying, but you have succeeded in setting up a stable trim shot for your test maneuver.

To execute the maximum rate roll maneuver, you apply a rapid, full left lateral stick input, trying to avoid any pitch input. The Extra 300 has full span ailerons, extending almost the full length of each wing, so the roll is almost a blur as the aircraft rolls 360° around its longitudinal axis. After the aircraft rolls through inverted flight and back towards upright, you rapidly center the control stick to stop the roll. The test maneuver is complete and you descend for landing. After landing, you download the flight test data from the data system. You are pleased with your trim shot, but the data will really tell how stable it was.

The data from the trim shot and roll maneuver are plotted in Figure 2.19. The parameters are plotted on the vertical axes as a function of time, commonly known as a "strip chart" format. Starting from the top of the figure, the plotted parameters are airspeed, altitude, roll rate, pitch angle, and roll angle. The pitch angle is the angle between the aircraft longitudinal axis and the horizon, indicating where the aircraft nose is pointing relative to the horizon. The roll angle is the angle of bank of the wings with respect to the horizon. The trim shot segment of the data is to the left of the vertical dashed line and the roll maneuver is to the right, as indicated.

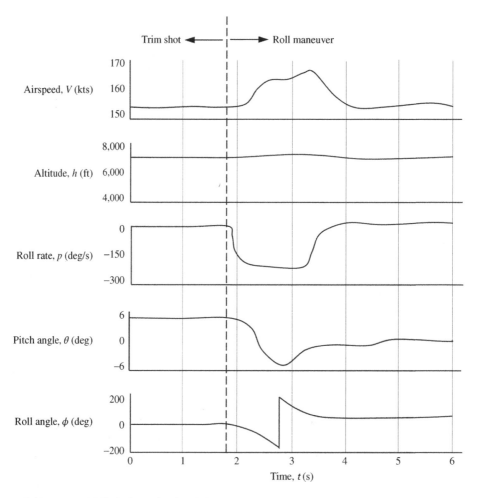

Figure 2.19 Extra 300 flight data; trim shot followed by a maximum rate aileron roll. (Source: *Figure created by author based on data trends in* [28], *with permission of Christopher Ludwig.*)

Examining the data for the trim shot, the most important few seconds, preceding the maneuver, are shown. The altitude is stable at 7000 ft and the airspeed is about 154 knots, about 1 knot less than the target of 155 knots. The pitch angle is stable at about 5°. Since the test maneuver was an aileron roll, it was critical that the trim shot start at a zero roll angle and with zero roll rate. If these were non-zero at the start of the maneuver, it would be difficult to accurately determine the aircraft roll performance. This again emphasizes the importance of the trim shot. From the data, the wings are level, as indicated by a roll angle of zero and the roll rate is zero. Based on the data, the trim shot looks correct and stable, prior to initiating the roll maneuver.

The roll maneuver is completed in less than two seconds. The roll angle is seen to go from −180 to +180 degrees, indicating a 360-degree roll. The altitude remains constant throughout the roll, while the airspeed increases, indicating that the nose was dropping in the maneuver, as also verified by the decrease in pitch attitude. A parameter of particular interest in assessing roll performance is the roll rate. As is seen by the data, the maximum roll rate was about 250 degrees per second. This maximum roll rate was obtained for a little less than one second of the two second-duration roll, due to the finite amount of time to achieve the roll rate and to recover from the roll. This review of the aileron roll shows the degree of analysis detail that can be obtained from a few, properly selected measurement parameters.

2.3.7 Mach Number and the Regimes of Flight

The maximum airspeed of the first successful heavier-than-air airplane, the Wright Flyer, was about 12 miles per hour (19 km/h). Just a short 20 years after this first flight, airplane speeds were topping 200 miles per hour (320 km/h) in the Schneider Cup air races of the 1920s. By the 1940s, World War II fighter airplanes were exceeding 400 miles per hour (640 km/h) in level flight. Then on 14 October 1947, the Bell X-1 rocket plane flew into history as the first airplane to fly faster than the speed of sound. Routine supersonic flight was soon a reality. On 12 April 1961, Yuri Gagarin was the first person to fly at hypersonic speeds, entering the earth's atmosphere in the Russian *Vostok 1* spacecraft at over 17,000 miles per hour (27,400 km/h) or almost 25 times the speed of sound. Returning from the moon, the *Apollo* spacecraft reached 36 times the speed of sound entering the earth's atmosphere, the fastest that a manned aerospace vehicle has ever flown.

In these examples, as the airspeed increased, the magnitude of the velocity was referenced to the speed of sound. The speed of sound is literally the speed that sound waves travel through the air. (Of course, sound waves can travel through mediums other than air, such as water or other gases, and there is a corresponding speed of sound for any given medium.) In Chapter 3, we show that the speed of sound is a function of the gas properties and the gas temperature. However, for now, let us get a "feel" for the magnitude of the speed of sound in air. At sea level and a temperature of 59°F (519°R, 288 K), the speed of sound in air is 661.6 knots (761.3 mph, 1116.6 ft/s, 340.2 m/s, 1225 km/h). This may seem quite fast, but it makes sense when we think about it. When someone speaks to you from across the room, their words are traveling to your ears via sound waves. You can hear the words instantaneously, so it makes sense that the sound waves are traveling at well over 700 mph.

The ratio of the velocity of the flow, V, to the speed of sound, a, is a dimensionless number called the *Mach number*, M, defined as

$$M \equiv \frac{V}{a} = \frac{\text{airspeed}}{\text{speed of sound}} \tag{2.40}$$

Let us think about the Mach number in terms of the motion of the air molecules. The flow velocity is a *directed* motion of the air molecules, that is, the air molecules are all moving in the same direction. The speed of sound is related to the *random* motion of the air molecules, which is a function of the air temperature. The higher the temperature, the more "excited" the air molecules become, increasing their random motion. Therefore, the Mach number can be physically interpreted as the ratio of the directed motion to the random thermal motion of the air molecules.

So, why do we often use the Mach number to quantify how fast we are flying rather than just using the velocity? The answer is that by using the Mach number, we are not only quantifying the magnitude of the flow velocity, but we are also saying something about the physical characteristics of the flow. As we change the Mach number of a flow, there are distinct changes in the physical nature of the flow. Keep this in mind, as we define the various flow regimes based upon specific physical phenomena and characteristics of the flow.

To explore the various flight regimes, let's think about what happened during your F-18 familiarization flight. When you were sitting in the cockpit on the runway at sea level, assume that the air temperature was 59°F (519°R, 288 K). The engines were running, so there were sound waves from the engine noise traveling away from the aircraft in all directions at the speed of sound, roughly 760 miles per hour (1200 km/h). This situation is shown in Figure 2.20 for $M = 0$, where the sound waves form concentric circles emanating from center.

When the F-18 leveled off at 30,000 ft (9100 m), you checked several of the cockpit instruments. The airspeed indicator read 350 knots (403 mph, 644 km/h), the outside air temperature (OAT) was −48°F (412°R, 229 K), and the Mach indication was about 0.6. As shown in Chapter 3, the speed of sound is proportional to the square root of the temperature, therefore the speed of sound at 30,000 ft, a_{30K}, is given by

$$\frac{a_{30K}}{a_{SL}} = \sqrt{\frac{T_{30K}}{T_{SL}}} = \sqrt{\frac{412°\text{R}}{519°\text{R}}} = 0.891 \tag{2.41}$$

$$a_{30K} = 0.891\, a_{SL} = 0.891(661.6\,\text{kt}) = 589\,\text{kt} \tag{2.42}$$

where T_{30K} is the air temperature at 30,000 ft, T_{SL} is the air temperature at sea level, and a_{SL} is the speed of sound at sea level. Using Equation (2.40), the Mach number at 30,000 ft, M_{30K}, was

$$M_{30K} = \frac{V_{30K}}{a_{30K}} = \frac{350\,\text{kt}}{589\,\text{kt}} = 0.594 \tag{2.43}$$

which agrees with the Mach indication that you read. The aircraft was in the *subsonic* flow regime, where the velocity is less than the speed of sound and the Mach number is less than one. The sound waves from the engine were still moving at the speed of sound, but since the aircraft had a forward velocity, the sound waves "bunched up" in front of the aircraft and spread out behind the aircraft, as shown in Figure 2.20 for $M < 1$. In the subsonic flow regime, the flow properties, such

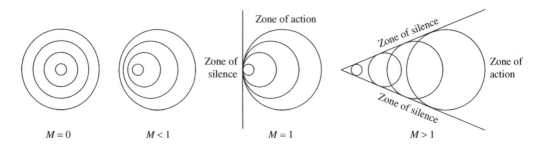

Figure 2.20 Sound waves patterns corresponding to different Mach numbers.

as the air temperature and pressure, change smoothly and continuously throughout the flow. Later, in Chapter 3, we will see that the air density of the flow remains constant, or nearly so, in much of the subsonic regime and this is called *incompressible flow*.

Returning to the F-18 flight, the aircraft was leveled off at 30,000 ft, stabilized for a moment, then a level acceleration was performed. Looking through the side of the canopy during the acceleration, you saw some light and dark shadowy lines "dancing" on the wing of the aircraft. Glancing at the airspeed indicator at this moment, you noted that the airspeed was about 530 knots (610 mph, 982 km/h). Let us calculate the Mach number corresponding to this airspeed and altitude. Since the acceleration was performed at a constant altitude of 30,000 ft, the speed of sound is still 589 knots, so that the Mach number is

$$M_{30K} = \frac{V_{30K}}{a_{30K}} = \frac{530\,\text{kt}}{589\,\text{kt}} = 0.900 \tag{2.44}$$

During the level acceleration, the Mach number increased from about 0.6 to 0.9, from the *subsonic* to the *transonic* flow regime. We can define the flow as subsonic below about Mach 0.8 and transonic starting from about Mach 0.8 up to about Mach 1.2. In the transonic flow regime, there is both subsonic, and for the first time, supersonic flow present. *Supersonic* flow is defined as flow where the Mach number is greater than one. In the transonic *mixed flow* region, there are localized "pockets" of the flow that accelerate from a subsonic Mach number to slightly beyond Mach 1, to supersonic flow. The supersonic flow, in these pockets of flow, has to readjust to the overall subsonic flow in an abrupt fashion, through what is called a *shock wave*. The Mach number decreases discontinuously from above Mach 1 to less than 1 through the shock wave. Other thermodynamic flow properties (pressure, temperature, density, etc.) also change discontinuously through the shock wave. In the other parts of the flow, where there are no supersonic pockets of flow, the flow properties change continuously, since the flow is subsonic. The sound wave pattern around the aircraft in transonic flow is similar to that in subsonic flow, except for the small pockets of flow, as we have just described.

As the level acceleration continued in the F-18, the aircraft reached Mach 1, meaning that it was traveling at an airspeed equal to the speed of sound. The aircraft was moving at the same speed as the sound waves from the engine, so that these waves could not travel forward of the aircraft and remained stationary with respect to the aircraft, as shown in Figure 2.20 for $M = 1$. The wavefronts of the sound waves overlap or coalesce to form a near perpendicular, dividing line between the upstream region, where the sound waves cannot travel, called the *zone of silence*, and the downstream region, where the sound can still be heard, called the *zone of action*.

As the F-18 accelerated past about Mach 1.2, it entered the *supersonic* flow regime. The aircraft was traveling faster than the sound waves emanating from the engine, so that these waves start to "fall behind", as shown in Figure 2.20 for $M > 1$. The wavefronts from the sound waves start forming a conical shock wave around the aircraft, where again sound waves cannot travel upstream past this conical boundary. In the supersonic flow regime, the flow upstream of the shock wave is entirely supersonic, with a Mach number greater than one. The flow properties, such as the pressure and temperature, change discontinuously through shock waves. Unlike subsonic flow, the air density cannot be considered constant, rather the air is *compressible* in supersonic flow.

Table 2.6 summarizes the different flow regimes that have been discussed, as a function of the Mach number. If you could have flown at much higher Mach numbers in the F-18, you would have reached the *hypersonic* flow regime beyond about Mach 5. Here, the shock waves form at a steeper angle, with respect to the flow direction and there are larger jumps in the flow properties across these stronger shock waves. At such a high Mach number, the flow has a large amount of kinetic energy. Hypersonic flows are synonymous with high temperature flows, which are discussed further in Chapter 3.

Table 2.6   Classification of flight regimes based on Mach number.

Flight regime	Mach number range	Physical flow features
Subsonic	$M < 0.8$	Smoothly changing flow properties Constant density flow (incompressible flow) Acoustic disturbances (sound waves) can propagate upstream
Transonic	$0.8 < M < 1.2$	Subsonic and supersonic flow present Local pocket(s) of supersonic flow, terminating in a shock wave
Supersonic	$1.2 < M < 5$	Shock waves and expansion waves are present in flow Discontinuous flow properties across shock waves Flow density is not constant (compressible flow) Acoustic disturbances (sound waves) cannot propagate upstream
Hypersonic	$M > 5$	Shock waves are closer to a body than for supersonic flow Very high heat transfer High temperature, chemically reacting flows

Keep in mind that the Mach numbers that bound the different flow regimes are only approximate. The Mach number where the effects of the different flow regimes are realized may vary, depending on factors such as the vehicle geometry. For instance, a slender body does not disturb the flow as much as a non-slender, thicker body, so that the onset of transonic shock waves on a slender body occurs at a slightly higher Mach number than for a non-slender body.

As a final note, we look at Figure 2.21, a photograph of a bullet in a supersonic flow, obtained using a flow visualization technique that makes the shock waves visible. This was the first photograph ever obtained of shock waves in a supersonic flow. The photograph was taken by the 19th century Austrian physicist Ernst Mach (1838–1916), after whom the Mach number is named. Ernst Mach pioneered many of the principles of supersonic flow and developed optical techniques to visualize these flows. The shock wave is clearly visible at the front of the bullet, trailing downstream at an angle. Weaker waves are seen trailing from the body of the bullet and from the turbulent wake behind the bullet.

2.3.8   The Flight Envelope

The *flight envelope* depicts the steady wings-level flight regime of an aircraft on an altitude versus Mach number or airspeed plot, as shown in Figure 2.22. It bounds the airspeed and altitude range of the aircraft, from its minimum to its maximum airspeed and from sea level to the maximum attainable altitude. The flight envelope is defined for specific aircraft conditions of weight, load factor, configuration, and power setting. For these specified conditions, the aircraft can maintain a constant altitude and constant airspeed at any point within, and on the boundaries of, the flight envelope. The aircraft is in equilibrium at all points in the flight envelope, where the lift equals the weight and the thrust equals the drag. Typically, a flight envelope corresponds to the aircraft at its maximum gross weight and a load factor of one, commonly referred to as 1 g flight. The aircraft is usually assumed to be in a clean configuration – for example landing gear retracted, flaps up, and no external stores – but a flight envelope may be defined for other configurations. The flight envelope provides significant information about the capabilities and limitations of an aircraft.

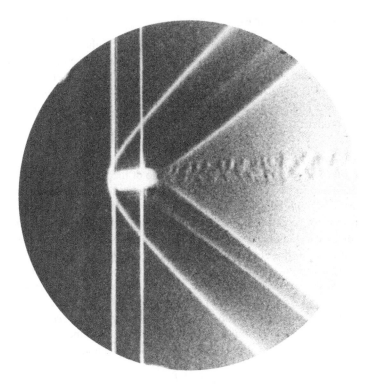

Figure 2.21 Photograph of a supersonic bullet taken by Ernst Mach. The flow is from left to right. The two bright vertical lines are "trip" wires used to trigger the photographic light source. (Source: *Ernst Mach, 1888, PD-old-70.*)

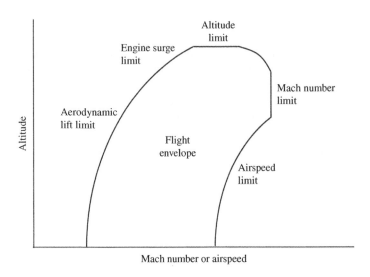

Figure 2.22 The flight envelope.

2.3.8.1   Flight Envelope Boundaries

The boundaries of the flight envelope define many of the limiting flight characteristics of an aircraft. These aircraft limits include the aerodynamic lift limit, jet engine surge limit, maximum altitude, maximum Mach number, and maximum airspeed (Figure 2.22). The flight envelope boundaries are dictated by several factors, some of the most important being the aircraft aerodynamics (lift and drag), propulsion (thrust and engine operation), and structural capabilities (static loads, dynamic loads, and material properties).

Aerodynamic Lift Limit
The *aerodynamic lift limit* boundary is along the left side of the flight envelope. The points along this boundary correspond to the minimum, level flight airspeed at each altitude. It is also defined as the *aerodynamic stall boundary*. If the airspeed is decreased any further, the aircraft (primarily the aircraft's wing) cannot generate sufficient lift to balance the weight, and the aircraft stalls. The loss of lift is usually due to massive separation of the air flow over the wing, destroying the lift.

The airspeed along the lift limit line is defined as the 1 g, aircraft *stall speed*, V_s. The stall speed at the intersection of the lift limit line and the horizontal (airspeed) axis represents the aircraft aerodynamic stall speed at sea level. (The stall speed may also be defined by other, non-aerodynamic considerations, as is discussed in Chapter 3.) As shown in Figure 2.22, the stall speed increases with increasing altitude. (To be precise, this is correct for the *true airspeed*, the speed of the aircraft relative to the air mass. The various types of airspeeds are explained in Chapter 3.) To show that the stall airspeed increases with altitude, we first introduce the non-dimensional *lift coefficient*, C_L, as

$$C_L = \frac{L}{q_\infty S} \tag{2.45}$$

where L is the aircraft total lift, q_∞ is the freestream dynamic pressure, and S is the wing planform area. The *dynamic pressure* is defined as

$$q_\infty \equiv \frac{1}{2}\rho_\infty V_\infty^2 \tag{2.46}$$

where ρ_∞ and V_∞ are the freestream density and freestream velocity, respectively. The lift coefficient increases linearly with increasing aircraft angle-of-attack, up to a maximum value, $C_{L,max}$, at the stall angle-of-attack, α_s.

Along the lift limit line, the aircraft is in steady level flight and has not yet stalled, such that the lift equals the weight. The airspeed, angle-of-attack, and lift coefficient along the lift limit line are the stall speed, V_s, the stall angle-of-attack, α_s, and the maximum lift coefficient, $C_{L,max}$, respectively. Therefore, using Equations (2.45) and (2.46), the lift along the lift limit line is given by

$$L = W = q_\infty S C_{L,max} = \frac{1}{2}\rho_\infty V_s^2 S C_{L,max} \tag{2.47}$$

Solving Equation (2.47) for the stall speed, V_s, we obtain

$$V_s = \sqrt{\frac{2W}{\rho_\infty S C_{L,max}}} \tag{2.48}$$

The aircraft weight, W, wing reference area, S, and maximum lift coefficient, $C_{L,max}$, are constants in Equation (2.48), independent of airspeed or altitude. The freestream density, ρ_∞, in Equation (2.48), decreases in magnitude with increasing altitude. Therefore, as the altitude increases, the stall speed, V_s, decreases with the inverse, square root of the freestream density.

Jet Engine Surge Limit

Another possible limit along the upper, left boundary of the flight envelope is the *jet engine surge limit*. This limit is somewhat analogous to the aerodynamic lift limit, since it is associated with aerodynamic stall. However, the surge limit has to do with stall in a jet engine, rather than on an aircraft wing. The engine stall starts with the aerodynamic stall of jet engine compressor blades,[4] which disrupts the air flow entering the jet engine. The high pressure air downstream of the compressor may flow upstream through the compressor and exit the engine inlet. This can occur in an explosive and dramatic manner, with flames shooting out of the engine inlet. The engine is more susceptible to surge at high altitudes and low airspeeds, hence the location of the surge limit boundary at the upper left corner of the flight envelope.

Altitude Limit

The top-most boundary of the flight envelope is the aircraft *altitude limit* for steady level flight. The aircraft cannot climb higher than this altitude limit and sustain steady level flight. It may be possible for the aircraft to *zoom climb* above the altitude limit, exchanging its kinetic energy (airspeed) for potential energy (altitude), but it cannot maintain steady level flight at this higher, zoom altitude.

The altitude limit is sometimes specified as either an *absolute altitude ceiling* or *service ceiling*. The absolute ceiling is the highest altitude that the aircraft can maintain steady level flight. The service ceiling is defined as the altitude where the aircraft rate of climb is a specified value. For piston-powered aircraft, the service ceiling is defined as the altitude where the rate of climb is 100 ft/min (30.5 m/min). For jet-engine powered aircraft, the service ceiling is defined as the altitude where the rate of climb is 500 ft/min (152 m/min). The service ceiling for jet-powered military aircraft is sometimes referred to as the *combat ceiling*.

The altitude limit is set by several factors, including the maximum thrust available, the minimum wing loading (defined as the aircraft weight divided by the wing area), or the cabin pressure limit. As an aircraft climbs, the air density decreases with increasing altitude, reducing the lift. To compensate for the decrease in air density, the aircraft must increase its airspeed so that the lift remains equal to weight. The angle-of-attack can be increased, thereby increasing the lift coefficient, but once the aircraft reaches its maximum angle-of-attack and maximum lift coefficient, the only option available is to increase airspeed. In addition to obtaining more lift at higher airspeed, the drag also increases. For an aircraft in steady level flight with thrust equal to drag, the thrust must be increased to compensate for the increase in drag. The engine thrust is also affected by the decrease in air density with increasing altitude, as this decreases the air mass flow entering the engine, which decreases the thrust. At the altitude limit, there is insufficient thrust to equal the drag, and the aircraft cannot maintain steady level flight. At this point, the aircraft is thrust limited and cannot climb any higher.

Assuming that the aircraft is not thrust limited, another factor that may affect the altitude limit is the wing loading, defined as the ratio of the aircraft weight to the wing area, W/S. To compensate for the decrease in air density with increasing altitude, a larger wing area is required, to provide more lift. For a constant aircraft weight, the wing loading decreases as the wing area increases. As the wing area increases, the wing structural weight increases. Simply put, the wing gets larger and heavier to fly higher. At some altitude, the required wing area results in a structural weight that is prohibitive. For some aircraft, this altitude may set a limit corresponding to a maximum wing area and a minimum value of the wing loading. There are several other factors, which may set the wing area, to be discussed in Chapter 3.

---

[4] The compressor is a fan-like, rotating disk, at the front of the jet engine, composed of many short, fin-like "wings" or blades with airfoil-shaped cross-sections. The air flow entering the engine is compressed to high pressure as it passes through the compressor, before entering the engine combustor. Jet engines and compressors are discussed in Chapter 4.

The altitude constraint due to the passenger cabin structural strength relates to a physiological or human limit. Since the ambient air pressure decreases with increasing altitude, aircraft cabin pressurization is required to maintain a life-sustaining breathing environment at high altitudes. Typically, commercial airliners maintain passenger cabin altitudes of about 7000–8000 ft (2100–2400 m) when flying at altitudes of about 40,000 ft (12,000 m). The cabin pressurization capability is set by the structural limit of the cabin design, which is based on the pressure differential between the inside and outside of the cabin. At some altitude, the ambient pressure is so low that the pressure differential would exceed the cabin structural limits if the cabin were pressurized to required values to maintain life. This altitude may set a limit based on the cabin pressure (differential) limit.

Airspeed Limit

The lower, right boundary of the flight envelope is the *airspeed limit*. This boundary is the maximum airspeed that the aircraft can obtain in steady level flight. Typically, the limiting factor that sets this boundary is the maximum thrust available from the aircraft propulsion system. Once the maximum thrust is equal to the aircraft total drag, the aircraft cannot accelerate and it has reached its maximum airspeed in steady level flight. The aircraft may be capable of exceeding this boundary by descending or diving, but this higher airspeed is transitory, as the aircraft cannot sustain this flight condition. Although this boundary represents the maximum airspeed that can be obtained in steady level flight, the normal, maximum cruise speed of the aircraft is defined as a slightly lower airspeed to provide a safety margin.

The maximum airspeed boundary may also be influenced by other considerations, especially *flutter*, a structural dynamic coupling of the aerodynamic flow and the elastic motion of the aircraft wing or control surfaces. The onset of flutter is often unpredictable and catastrophic, potentially leading to failure of the wing or control surfaces. Usually, an aircraft is flight tested to an airspeed well beyond the steady level flight airspeed boundary, by diving the aircraft to demonstrate that it is free of flutter issues. This flutter boundary is normally a function of the dynamic pressure rather than airspeed, so that the maximum airspeed boundary may follow a line of constant dynamic pressure.

Mach Number Limit

The *Mach number limit* is the furthermost right boundary on the flight envelope, and pertains to aircraft that are capable of supersonic flight. This boundary corresponds to the maximum Mach number that is sustainable in steady level flight. A higher Mach number could be reached by diving the aircraft, but this would be a transitory, unsteady condition since the aircraft could not sustain this flight condition. The Mach number limit may be set by aerodynamic, heating, or engine operation considerations.

Aerodynamically, the limit may be driven by the formation of shock waves on the wing. A shock wave is a thin region of flow, across which there is a large, discontinuous increase in the flow pressure. This pressure increase results in aerodynamic flow separation on the wing and significant loss of lift. There may also be significant aircraft controllability issues associated with either the flow separation or the location of the shock waves.

Heating considerations involve exceeding the material temperature limits of the aircraft external airframe or internal engine components. At supersonic Mach numbers, the aircraft airframe skin may reach high temperatures due to skin friction heating, especially in regions where the flow stagnates or goes to near zero velocity, such as at the aircraft nose and the leading edges of wings, tails, and engine inlets. Thus, the material temperature limits may dictate the maximum sustained flight Mach number. However, the engine materials are typically more limiting than the airframe, as the temperature of the flow through the engine is further increased from the freestream value by the engine compression and combustion processes. The engine turbine materials are usually the limiting factor in setting the maximum flight Mach number.

Another Mach number and engine related issue, which may restrict the maximum Mach number, concerns the interaction of shock waves with the engine inlet geometry. This interaction can lead to unsteady shock wave oscillations, called *inlet buzz*, which can lead to damage or failure of the engine structure. Avoiding the onset of this phenomenon may set the Mach number limit at high altitudes.

For some very high altitude flying aircraft, the aerodynamic lift limit boundary and the maximum Mach number boundary nearly converge at the maximum altitude. This may make it difficult or hazardous to fly the aircraft at this high altitude limit, as a small decrease in airspeed results in an aerodynamic stall, and a small increase in airspeed exceeds the maximum Mach number limit. This corner of the flight envelope is therefore referred to as the *coffin corner* (not shown in Figure 2.22).

2.3.8.2 Flight Envelope Examples

In this section, the flight envelopes of several different types of aircraft are presented, including a general aviation airplane, a commercial airliner, a supersonic military jet fighter, and a very high altitude, very high Mach number airplane. The wide variations in flight envelopes are demonstrated by these different aircraft types. A comparison of flight envelopes is also provided at the end of this section.

Example 2.5 General Aviation Airplane Flight Envelope: Beechcraft A36 *The Beechcraft A36 Bonanza is a six-place high-performance single-engine aircraft designed and manufactured by the Beechcraft Aircraft Corporation, Wichita, Kansas (Figure 2.23). The A36 has a conventional*

Figure 2.23 Beechcraft A36 *Bonanza* general aviation aircraft. (Source: *Alan Lebeda, "Beechcraft A36 Bonanza" https://commons.wikimedia.org/wiki/File:Beech_A36_Bonanza_36_AN1890204.jpg, GFDL-1.2. License at https://commons.wikimedia.org/wiki/Commons:GNU_Free_Documentation_License,_version_1 .2.)*

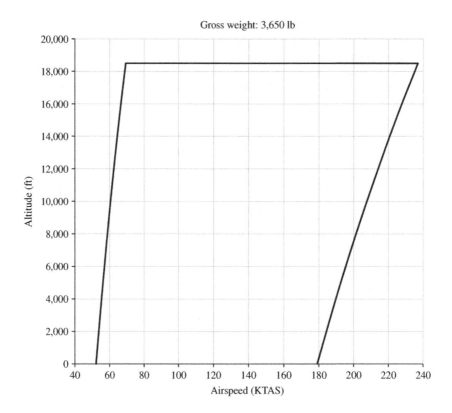

Figure 2.24   Beechcraft A36 *Bonanza* flight envelope.

configuration with a low-mounted main wing, straight horizontal stabilizers, single vertical tail, and retractable tricycle landing gear. The Bonanza *has a wingspan of 33.5 ft (10.2 m), height of 8.58 ft (2.62 m), and length of 27.5 ft (8.38 m). The aircraft is powered by a Continental IO-550-B horizontally opposed six-cylinder fuel injected air-cooled piston engine with approximately 330 hp (246 kW), turning an 84.0 inch (2.13 m) diameter propeller. The aircraft has a maximum takeoff weight of approximately 3650 lb (1660 kg). The Beechcraft A36* Bonanza *was introduced in 1970.*

The flight envelope of the Beechcraft A36 Bonanza *is shown in Figure 2.24, plotted as true airspeed (in units of knots true airspeed or KTAS) versus altitude. The flight envelope is for the A36 at its maximum gross weight of 3650 lb. Similar to most piston-powered, general aviation aircraft, the A36 has a simple flight envelope, bounded only by an aerodynamic lift limit, maximum altitude, and maximum airspeed. Starting at the lower left-hand corner of the flight envelope, the 1 g, sea level stall speed of the A36 is 52 KTAS (60 mph, 96 km/h), increasing to about 69 KTAS (79 mph, 128 km/h) at the maximum altitude of 18,500 ft (5640 m). The maximum airspeed at sea level is 237 KTAS (273 mph, 439 km/h). As with many general aviation aircraft, the A36 flight envelope is not very large, relative to other types of aircraft.*

Example 2.6 Commercial Airliner Airplane Flight Envelope: Boeing 767 *The Boeing 767 is a wide-body twin-engine jet airliner, designed and manufactured by Boeing Commercial Airplanes, Everett, Washington (Figure 2.25). The airliner has a conventional configuration with a low-mounted wing, straight horizontal stabilizers, single vertical tail, and retractable tricycle landing gear. The Boeing 767 has a wingspan of 156.1 ft (47.6 m), height of 52.0 ft (15.8 m) and*

Figure 2.25   Boeing 767 prototype airliner flying over Mount Rainier, Washington. (Source: *Seattle Municipal Archives, "Boeing 767 Over Mount Rainier, circa 1980s" https://en.wikipedia.org/wiki/File:Boeing_767_over_Mount_Rainier,_circa_1980s.jpg, CC-BY-2.0. License at https://creativecommons.org/licenses/by-sa/2.0/legalcode.*)

length of 180.25 ft (54.9 m). The aircraft is powered by two Pratt & Whitney PW4056 high bypass ratio turbofan engines, with the engines mounted in nacelles hung from pylons underneath the wings. Each engine provides an uninstalled, sea level, static thrust of 63,300 lb (282 kN). The aircraft has a maximum takeoff weight of approximately 412,000 lb (186,900 kg). The first flight of the Boeing 767 was on 26 September 1981.

The flight envelope of the Boeing 767 is shown in Figure 2.26, plotted as true airspeed (in units of knots true airspeed or KTAS) versus altitude. The flight envelope is for the Boeing 767 with two Pratt & Whitney PW4056 turbofan engines and a gross weight of 412,000 lb. As might be expected, the flight envelope of a subsonic, commercial airliner covers a broader range of altitudes and airspeeds than a general aviation aircraft. The flight envelope has aerodynamic lift limit, maximum altitude, maximum Mach number, and maximum airspeed boundaries.

The 1 g, sea level stall speed is 133 KTAS (153 mph, 246 km/h), increasing to about 280 KTAS (322 mph, 519 km/h) at the maximum altitude of 43,000 ft (13,100 m). The passenger cabin pressurization limits the maximum altitude. The Boeing 767 has a maximum operating airspeed, V_{MO}, of 360 KTAS (414 mph, 667 km/h) at sea level, increasing to about 518 KTAS (596 mph, 959 km/h) at an altitude of 26,000 ft (7900 m). Although not obvious in Figure 2.26, the flight envelope is limited to a maximum operating Mach number, M_{MO}, of 0.86 from 26,000 ft up to its maximum altitude. The maximum cruise airspeed of the Boeing 767 is less than the M_{MO}, to provide a safety margin.

Example 2.7 Supersonic Military jet Flight Envelope: McDonnell Douglas F-15 *The F-15 Eagle is an air-superiority military jet fighter aircraft designed and built by McDonnell Douglas Aircraft Company (now The Boeing Company), St Louis, Missouri (Figure 2.27). The aircraft has a high-mounted, swept main wing with a modified delta shape, twin vertical tails, all-moving horizontal stabilators, and twin turbofan jet engines. The F-15 airplane has a wingspan of 42.8 ft*

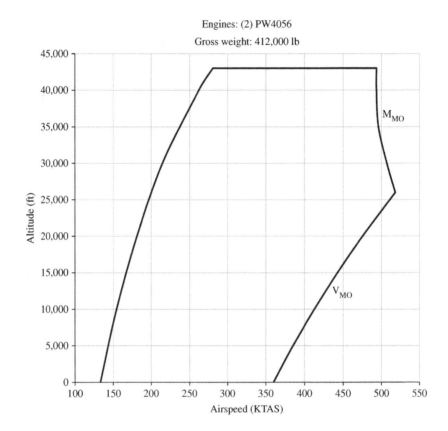

Figure 2.26   Boeing 767-300 flight envelope.

(13.0 m), height of 18.7 ft (5.7 m), and length of 63.7 ft (19.4 m). The aircraft is powered by two Pratt & Whitney F100-PW-100 turbofan engines. Each engine produces an uninstalled, sea level static thrust of approximately 25,000 lb (111 kN) in full afterburner. The aircraft has a fully fueled take-off weight of approximately 42,000 lb (19,000 kg) and a landing weight of approximately 32,000 lb (14,500 kg). The aircraft has aerial refueling capability for extended duration flight. The first flight of the McDonnell Douglas F-15A was on 27 July 1972.

The flight envelope of the F-15 is shown in Figure 2.28, on a Mach number versus altitude plot. Often, two power settings are shown on a flight envelope for a military jet with an afterburning jet engine, that of military thrust *(maximum thrust without use of the afterburner) and* maximum thrust *(maximum thrust with afterburner operating). This is shown in Figure 2.28, where the smaller flight envelope boundary (dotted line) is for military thrust and the larger, complete flight envelope boundary (solid line) is for maximum thrust. The military thrust flight envelope indicates that the aircraft is barely able to exceed Mach 1 without afterburner, and has a ceiling of about 50,000 ft (15,000 m). When afterburner is used, the maximum thrust flight envelope gives a standard day, maximum Mach number of about 2.2 at an altitude of 36,000 ft (11,000 m) and a ceiling of about 60,000 ft (18,000 m). The low speed portion of the F-15 flight envelope is bounded by a lift limit boundary, where the lift boundary minimum Mach number increases with increasing altitude as shown, reaching high Mach numbers at high altitudes.*

Example 2.8 High Mach Number, High Altitude Aircraft Flight Envelope: Lockheed SR-71 *The SR-71 Blackbird, designed and manufactured by the Lockheed Advanced Development*

Figure 2.27 Mcdonnell douglas F-15 *Eagle* supersonic jet fighter aircraft (F-15B two-seat version shown). (Source: *NASA.*)

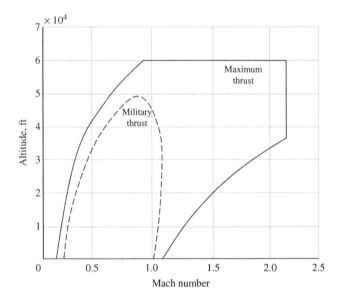

Figure 2.28 McDonnell Douglas F-15A *Eagle* flight envelope. (Source: *Adapted from Vachon, M.J., Moes, T.R, and Corda, S., "Local Flow Conditions for Propulsion Experiments on the F-15B Propulsion Flight Test Fixture," NASA TM-2005-213670, November 2005, Fig. 3.*)

Company (commonly called the "Skunk Works"), Palmdale, California, is an ultra-high-altitude, supersonic reconnaissance aircraft (Figure 2.29). The SR-71 has a wingspan of 55.6 ft (16.9 m), height of 18.5 ft (5.64 m), and length of 107.4 ft (32.74 m). The aircraft has a long narrow fuselage, a large delta wing, two large engine nacelles mounted in the wings, and twin canted all-moving rudders. Lifting surfaces called "chines" extend along the sides of the fuselage, from the aircraft nose to the intersection of the wing and fuselage. The aircraft has a tandem, two-place cockpit configuration with flight controls in the forward cockpit only. The SR-71 has titanium construction and is painted black to increase radiative heat transfer for flight at the high temperatures

Figure 2.29   Lockheed SR-71 *Blackbird* reconnaissance aircraft flying near Mt Whitney, California. (Source: *NASA*.)

associated with high Mach number supersonic flight: hence its designation as the Blackbird. *The aircraft is powered by two Pratt & Whitney PW J58 turbojet engines. A prominent feature of the propulsion system is the cone-shaped spikes at the entrance of each nacelle. Each engine produces an uninstalled, sea level static thrust of approximately 34,000 lb (151 kN) in full afterburner. The aircraft has a fully fueled takeoff weight of approximately 143,000 lb (65,000 kg). The aircraft has aerial refueling capability for extended duration flight. The first flight of the Lockheed SR-71* Blackbird *was on 22 December 1964.*

The flight envelope of the Lockheed SR-71 is shown in Figure 2.30, on a Mach number versus altitude plot. Even though the SR-71 was a supersonic aircraft, its flight envelope is markedly different from the F-15A. The SR-71 flight envelope looks narrow in contrast to the more full flight envelope of the F-15. The SR-71 design was very focused on its mission to fly at triple-sonic Mach numbers at high altitude. In this sense, the aircraft was somewhat point designed *for this specific flight condition, rather than being designed to fly in a broader flight envelope.*

The minimum airspeed boundary on the left side of the flight envelope is specified in KEAS, for knots equivalent airspeed. *(This is yet another type of airspeed, which is typically used in very high speed aircraft. The structural loads correlate with the square of the equivalent airspeed, which is an important consideration for very high Mach number aircraft, such as the SR-71.) The minimum airspeed below 25,000 ft (7600 m) is 145 KEAS and increases to 300 KEAS above 25,000 ft and below Mach 1. Above Mach 1, the minimum airspeed increases to 310 KEAS. The maximum altitude is above 80,000 ft (24,000 m), but this can only be reached when flying at the maximum Mach number of 3.2. Unlike some flight envelopes where flight on the Mach limit boundary may not be advisable, the SR-71 was designed for sustained cruise at its limit Mach number and altitude. The Mach number limit decreases with decreasing altitude as shown, with a sea-level maximum*

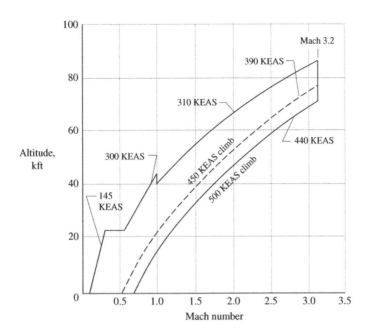

Figure 2.30 The SR-71 *Blackbird* flight envelope. (Source: *S. Corda, et al., "The SR-71 Test Bed Aircraft: A Facility for High-Speed Flight Research," NASA TP-2000-209023, June 2000, Fig. 3.*)

Mach number of about 0.76. In actuality, this maximum Mach number boundary corresponds to a maximum equivalent airspeed limit of 500 KEAS (575 mph, 926 km/h), a limit set by aircraft structural considerations.

2.3.8.3   Comparison of Flight Envelopes for Different Aircraft Types

It is worthwhile to compare the flight envelopes, for the different types of aircraft that have been discussed. These flight envelopes are compared on a Mach number versus altitude plot in Figure 2.31. The maximum performance from the flight envelopes of these different types of aircraft is shown in Table 2.7. The flight envelope range shown, from sea level to about 90,000 ft (27,000 m) and up to about Mach 3.5 encompasses the manned aircraft that have been designed and built for sustained flight within the atmosphere.

Note how the shapes of the flight envelopes change with airspeed and altitude, especially in moving from subsonic to supersonic aircraft. It is quite amazing that a supersonic jet aircraft's flight envelope can have the breadth in airspeed and Mach number, as shown. However, beyond

Table 2.7   Maximum performance from flight envelopes of different aircraft types.

Aircraft	Maximum altitude	Maximum airspeed	Maximum Mach no.
Wright Flyer I	30 ft (9.1 m)	26 KTAS (48 km/h)	0.04
Beechcraft A36	18,500 ft (5600 m)	237 KTAS (439 km/h)	0.37
Boeing 767-300	43,000 ft (13,000 m)	518 KTAS (959 km/h)	0.86
McDonnell Douglas F-15	60,000 ft (18,000 m)	1434 KTAS (2656 km/h)	2.5
Lockheed SR-71	>80,000 ft (>24,400 m)	1854 KTAS (3434 km/h)	3.2

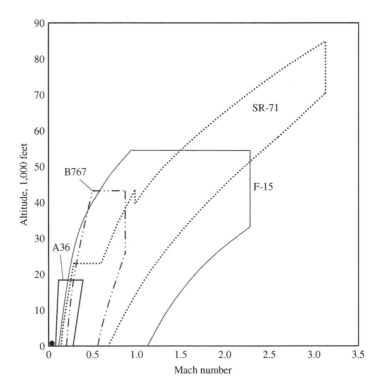

Figure 2.31   Flight envelope comparison of different types of aircraft.

about Mach 2.5, it becomes more and more difficult to design an aircraft that possesses a broad flight envelope. As with the SR-71, very high Mach number vehicles become *point design* aircraft that have little design margin for flying efficiently at other flight conditions than their design point.

As a final note, looking closely at Figure 2.31, there is a small black dot in the lower left corner. This represents the flight envelope of the Wright Flyer I, which had a maximum speed of about 30 mph (48 km/h) and a maximum altitude of about 30 ft (9.0 m), corresponding to a maximum Mach number of about 0.04. With this starting point, the flight envelopes in Figure 2.31 can be viewed from a historical perspective of the evolution of aircraft. Expanding in both airspeed and altitude, aircraft have evolved from piston-powered, low subsonic speed airplanes, to high subsonic speed, commercial airliners, to supersonic jet aircraft, and to the fastest manned aircraft that has ever been built.

2.3.9   The V-n Diagram

We now introduce a diagram that depicts the structural load limits for an aircraft as a function of airspeed and load factor. We first define the non-dimensional load factor, n, as

$$n \equiv \frac{L}{W} \tag{2.49}$$

where L and W are the aircraft lift and weight, respectively. The flight envelope, discussed in the previous section, applies to an aircraft in level flight, where the lift equals the weight, and thus the load factor, from Equation (2.49), equals one. In this situation, the inertia force acting on the aircraft is simply equal to its mass times the acceleration due to gravity. We commonly refer to this as flight at 1 g or a load factor of one. The load factor, n, is non-dimensional, but when we refer to a load factor, we often specify it in terms of g's. For instance, if an aircraft is flying such that the lift is three times greater than the weight, the aircraft is flying at 3 g or a load factor of 3. The aircraft load factor can be greater than or less than one if the aircraft is maneuvering, such as in a horizontal turn, pull-up, or pushover. Wind gusts encountered by the aircraft can also increase or decrease load factor, sometimes significantly.

The *V-g* or *V-n diagram* defines an allowable structural envelope that is a function of the load factor, n, and airspeed, V (Figure 2.32). Flight at airspeeds and load factors within the boundaries of the V-n diagram are within the structural operating limits of an aircraft. The limits shown on a V-n diagram are the aerodynamic (stall) limit on the left boundary, positive and negative structural load limits on the upper and lower boundaries, respectively, and a maximum airspeed limit on the right boundary. Each V-n diagram is valid for a specific type or model of an aircraft at a specific gross weight and altitude, in a specific aircraft configuration, and for a specific type of loading. The aircraft configuration definition includes the position of the flaps, the landing gear position, etc. The type of loading is defined as either symmetrical or asymmetrical (rolling) loading.

There are two types of structural load factor boundaries that are shown on the V-n diagram, the *limit load factor* and the *ultimate load factor*. The limit load factor is the maximum load factor that an aircraft can be subjected to without any permanent structural deformation of the primary structure. Above the limit load factor, the aircraft primary structure may be permanently deformed or damaged, perhaps resulting in an unsafe flight condition. If the ultimate load factor is exceeded, the aircraft primary structure will fail. There are positive and negative boundaries for both the limit load factor and the ultimate load factor, corresponding to positive and negative g's, respectively. As shown in the figure, the maximum negative limit and negative ultimate load factors are typically smaller in magnitude than the positive values for aircraft that are designed primarily to fly in a positive g flight regime. For some types of aerobatic aircraft, the maximum negative limit and negative ultimate load factors can be equal in magnitude to the positive values.

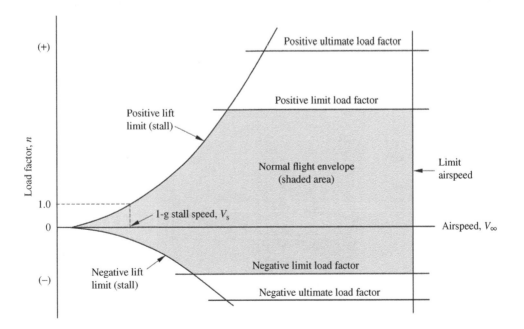

Figure 2.32   The V-n diagram.

The left side of the V-n diagram is the aerodynamic lift boundary of an aircraft. The airspeed along the left boundary is the aircraft aerodynamic stall speed as a function of the load factor. The aircraft is unable to maintain level flight at any airspeed–load factor combinations above this line. The intersection of the lift limit line with a line corresponding to a load factor of one provides the 1 g stall speed of an aircraft, $V_{s,1g}$. There is a lift limit line for positive and negative load factors. The negative load factors corresponds to negative lift flight, that is, the lift being produced due to negative angle-of-attack on the wing or inverted flight. For a given load factor magnitude, the negative load factor stall speed is typically higher than the positive load factor stall speed. This is primarily due to the shape of the wing airfoil section, which is designed to be more efficient at producing lift at positive, rather than negative angle-of-attack. The shape of the lift limit lines is not linear; rather they vary with the square of the airspeed. For an aircraft in steady flight at a load factor n, the lift is given by

$$L = nW = \frac{1}{2}\rho_\infty V_\infty^2 S C_L \tag{2.50}$$

The stall speed at a load factor n, $V_{s,n}$, is given by

$$V_{s,n} = \sqrt{\frac{2nW}{\rho_\infty S C_{L,max}}} = \sqrt{n}\, V_{s,1g} \tag{2.51}$$

where the 1 g stall speed, $V_{s,1g}$, is defined by Equation (2.48). Hence, the stall speed at a load factor n, varies with the square root of the load factor multiplied by the 1 g stall speed.

The lift limit line and the positive limit load factor line intersect at a point called the *maneuver point*. The airspeed corresponding to the maneuver point is the called the *corner speed*, V_A. The maneuver point and the corner speed have important implications relative to the turn performance and structural limitations of an aircraft. Below the corner speed, an aircraft stalls before it reaches the limit load. Above the corner speed, the limit load can be reached. The right side boundary of the V-n diagram is the limit airspeed, above which the aircraft will sustain structural damage or failure of the primary structure. The damage or failure may be due to exceeding structural loads in a critical gust situation, structural dynamic phenomena, or compressibility effects.

We now look at the V-n diagrams of a subsonic aircraft, the Beechcraft T-34A *Mentor* (Figure 2.33), and a supersonic aircraft, the Lockheed F-104 *Starfighter* (Figure 2.34). The

Figure 2.33 Beechcraft T-34A *Mentor*. (Source: *Courtesy of the author.*)

Figure 2.34 Lockheed F-104 *Starfighter* Mach 2 supersonic interceptor aircraft. (Source: *NASA.*)

Beechcraft T-34A aircraft is a single-engine, two-place ex-military trainer with a straight, low-mounted wing, retractable landing gear, tandem cockpit, and bubble canopy. The T-34 is powered by a single, 225 hp (168 kW), air-cooled, six-cylinder piston engine. The first flight of the T-34A was in 1948. The T-34A is still flying today as a civilian general aviation airplane.

Designed and manufactured by the Lockheed Skunk Works, the F-104 is a Mach 2-class, supersonic jet aircraft designed for high dash speeds to intercept enemy aircraft. The F-104 was the first military jet capable of sustained flight at Mach 2. The F-104 has a slender, pointed fuselage, a mid-mounted, low aspect ratio wing with a trapezoidal planform, and an aft-mounted T-tail. Powered by a single General Electric J79 turbojet engine with afterburner, the lightweight F-104 has excellent climb and acceleration capabilities. The F-104 set many world speed and altitude records during its time in service. The first flight of the Lockheed F-104 *Starfighter* was on 17 February 1956.

The V-n diagrams of the subsonic T-34A and the supersonic F-104 are shown in Figure 2.35 and Figure 2.36, respectively. Load factor is plotted versus indicated airspeed in these figures. Figure 2.35, for the T-34, is based on an aircraft gross weight of 2900 lb (1315 kg) or less. The stall speed of the T-34 is 53 knots (61 mph, 98 km/h) at 1 g and increases with increasing load factor. The T-34 has a positive limit load factor of 6 g and a positive ultimate load factor of 9 g. The maximum dive airspeed of the T-34 is 243 knots (280 mph, 450 km/h). For the F-104, the 1 g stall speed is close to 200 knots (230 mph, 370 km/h). The positive limit and negative limit load factors are 7.33 g and −3.0 g, respectively. The F-104 maximum dive speed is 750 knots (860 mph, 1390 km/h) at sea level. The F-104 is limited to a maximum Mach number of 2.0 at an altitude of 40,000 ft (12,200 m) and above.

Example 2.9 Stall Speed at Elevated Load Factor *A Beechcraft T-34 has a 1 g stall speed of 53 knots (1 knot equals 1 nautical mile, nm, per hour). Calculate the maximum lift coefficient, $C_{L,max}$, and the stall speed at a load factor of 3. Assume a weight, W, of 2900 lb, a wing area, S, of 177.6 ft^2, and sea level conditions.*

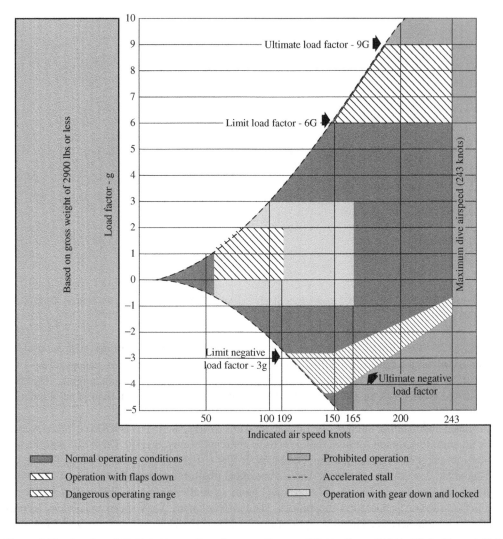

Figure 2.35 Beechcraft T-34A *Mentor* V-n diagram. (Source: *US Air Force, T-34A Flight Manual. T.O. IT-34A-1S, 1 July 1960.*)

Solution

Convert the 1 g stall speed into consistent units.

$$V_{s,1g} = 53 \, \frac{nm}{h} \times \frac{6076\,ft}{1\,nm} \times \frac{1\,h}{3600\,s} = 89.45 \, \frac{ft}{s}$$

Using Equation (2.48), the maximum lift coefficient is obtained from the 1 g stall speed as

$$V_{s,1g} = \sqrt{\frac{2W}{\rho_\infty S C_{L,max}}}$$

$$C_{L,max} = \frac{2W}{\rho_\infty S (V_{s,1g})^2}$$

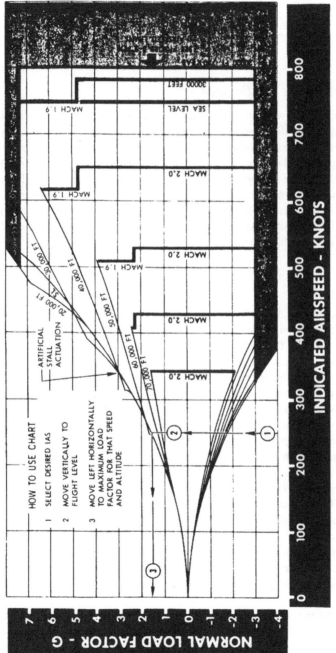

Figure 2.36 Lockheed F-104 *Starfighter* V-n diagram. (Source: *US Air Force, F/RF/TF-104G Flight Manual, T.O. 1F-104G-1, 31 March 1975.*)

$$C_{L,max} = \frac{2(2900\,lb)}{\left(0.002377\,\frac{slug}{ft^3}\right)(177.6\,ft^2)\left(89.45\,\frac{ft}{s}\right)^2} = 1.717$$

Using Equation (2.51), the stall speed at a load factor of 3 is

$$V_{s,n} = \sqrt{n}\,V_{s,1g} = \sqrt{3}\,(53\,knots) = 91.8\,knots$$

2.3.10   Aircraft Weight and Balance

In this section, we discuss two important parameters in the design and operation of aircraft, the total aircraft weight and the distribution of the weight, commonly referred to as aircraft *weight* and *balance*. These parameters can significantly affect the aircraft performance, stability, control, and structural loads. Operating the aircraft outside of its weight and balance limits is a safety of flight issue. Operating an aircraft at a weight greater than the approved maximum weight can significantly degrade performance and compromise structural integrity. Aircraft stability and control may be adversely affected if an aircraft is flown with its weight distributed such that its center of gravity is outside of its balance limits. Operating the aircraft outside of its approved weight and balance envelope can have catastrophic consequences.

2.3.10.1   Aircraft Weight

In discussing aircraft weight, it is helpful to first define some weight-related terminology. The *gross weight* is the total aircraft weight in a specific ground or flight condition, and includes the airframe structure, engines, systems, fuel, oil, people, baggage, equipment, etc. The aircraft gross weight generally decreases during a flight due to fuel consumption. The *maximum allowable gross weight* is the heaviest weight that the aircraft can have in any ground or flight condition. It is typically defined during the design and testing process, based on performance, stability, control, and structural considerations. The *maximum takeoff weight* is the heaviest weight that the aircraft can have during takeoff. The maximum takeoff weight may be less than the maximum gross weight due to operational or performance considerations, such as the necessity to have a lower weight for acceptable takeoff climb performance. The *maximum ramp or taxi weight*, the heaviest weight the aircraft may have on the ground, may be greater than the maximum takeoff weight. The ramp weight must be reduced to the takeoff weight prior to takeoff and this is accomplished through fuel burn during start-up and taxi. However, neither the maximum ramp weight nor the maximum takeoff weight may exceed the maximum allowable gross weight. The *maximum landing weight* is the heaviest weight that the aircraft may have when landing. The maximum landing weight may be less than the maximum takeoff weight due to structural limitations on the landing gear, or associated structure, when landing loads are taken into account. For large aircraft, such as commercial airliners, the maximum landing weight may be over 100,000 lb (45,000 kg) less than the maximum takeoff weight.

The aircraft *basic empty weight* is defined as the weight of airframe, engines, fixed equipment, unusable fuel and oil (fuel and oil that cannot be drained from the aircraft or consumed by the engine), and usable oil. Typically, the basic empty weight is obtained by weighing an aircraft on mechanical scales. The aircraft basic empty weight may change due to structural modifications to an aircraft or due to the addition or removal of equipment. A new empty weight is obtained either through re-weighing or through analytical calculation. The Federal Aviation Administration (FAA) advises that a new empty weight should be obtained if the weight increase is more than one pound

(0.450 kg) for an aircraft with an empty weight less than 5000 lb (2268 kg), more than two pounds (0.900 kg) for an aircraft with an empty weight between 5000 and 50,000 lb (22,700 kg), and more than 5 lb (2.27 kg) for an aircraft with an empty weight greater than 50,000 lb. Periodic re-weighing may be performed as a good practice for several reasons, such as correcting analytically based calculations, which did not account for the weight of wiring and attachment hardware, and because aircraft gain weight over time, due to the accumulation of dirt, oil, and other contaminants that cannot be removed. For some types of aircraft, such as commercial airliners, FAA regulations may require re-weighing every 36 months.

If the aircraft basic empty weight is subtracted from the maximum gross weight, the *useful load* is obtained, which includes the weight of the people (aircrew and passengers), baggage, and usable fuel. The aircraft designer seeks to minimize the basic empty weight to maximize the useful load. The *fuel load* is the weight of the usable fuel that is carried in the aircraft. The *zero fuel weight* is equal to the aircraft weight minus the fuel load.

The aircraft weight has a major influence on the performance of the aircraft. As you might imagine, there is a negative impact on performance as the aircraft weight increases. In the extreme, if the weight gets too large, such that the weight is greater than the lift, the aircraft will not be capable of lifting off the ground at all. Some of the aircraft performance impacts of a higher weight include longer takeoff and landing distances, reduced rate of climb, lower maximum altitude, less range, and higher stalling airspeed. If the aircraft weight exceeds the maximum gross weight, it is possible that the performance may be degraded to the point that safe flight is not possible.

For aircraft that have met a civil certification or military specification standard, this is usually accomplished for the aircraft at its maximum gross weight. This means that the structural analyses, ground loads tests, and flight tests were all performed at the aircraft's maximum gross weight. If the aircraft weight exceeds the maximum gross weight, it is possible that the aircraft could sustain a catastrophic structural failure even if it is flown within the boundaries for which it has been certified or cleared. The weight exceedance could also result in progressive structural damage, such as fatigue cracking that will eventually lead to a catastrophic structural failure even when the aircraft is flown within its approved flight loads envelope.

2.3.10.2 Aircraft Balance and Center of Gravity Location

The center of gravity (CG) is defined as a single point where it can be assumed that the vehicle mass is concentrated. In regards to aircraft balance, the center of gravity is a point about which the aircraft is balanced about its longitudinal, lateral, and vertical axes. We are primarily interested in the CG location on the longitudinal axis, although the CG location on the lateral or vertical axes is important under certain circumstances. As shown in Figure 2.37, the aircraft would be longitudinally balanced if we placed a fulcrum at the center of gravity. (In Figure 2.37, the CG location is commonly symbolized by a circle that is divided into four equal quadrants, with two of the quadrants filled in.) The primary flight control used to longitudinally balance the aircraft is the elevator.

The CG location is dependent on the weight distribution of the fixed structure, systems, equipment, etc. and on the loading of non-fixed items, such as fuel, people, baggage and equipment. Fuel burn in flight can also cause the CG location to change. For aircraft with conventional configurations, the lateral CG location is mostly affected by the distribution of weight in or on the wings. This could be internal fuel that is carried in the wings or, for military aircraft, external stores hung underneath the wings.

The longitudinal CG location is usually measured as a distance from a *datum*, a reference line that is usually specified by the aircraft manufacturer. The datum is usually located at or near the

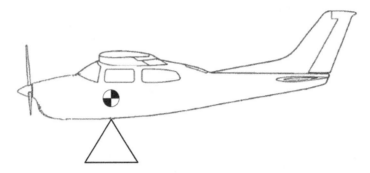

Figure 2.37   Aircraft balanced on a fulcrum at its center of gravity.

aircraft nose, but sometimes it is specified as a distance forward of the aircraft, not associated with the physical structure of the aircraft at all. The longitudinal distance, in inches, from the datum is called the *fuselage station* (FS). The fuselage station is a positive number aft of the datum and a negative number forward of the datum. For example, FS 30 denotes the fuselage station that is 30 in (76 cm) aft of the datum. The datum itself is FS 0.

The aircraft longitudinal CG location must be between the forward and aft *CG limits* for safe flight. The forward and aft CG limits are determined by the aircraft design, and are verified through flight test. These limits may vary with aircraft gross weight, aircraft configuration, such as flap or landing gear position, or type of flight operation, such as aerobatic flight.

Similar to the effects of weight, aircraft balance, or more specifically, the location of the center of gravity, can have a major influence on the aircraft performance, stability, control, and structural loads. The aircraft designer often sets the forward CG limit based on the desired landing characteristics of the aircraft. If the CG is too far forward, the aircraft tends to be "nose heavy" making it more difficult for the pilot to set the nose in the proper position for takeoff or landing. A far forward CG can result in larger elevator control forces, higher stall speeds, and higher structural loads on the nose landing gear when landing. The aft CG limit is usually set by the aircraft longitudinal stability requirements. An aircraft becomes less stable as the CG location moves aft. If the CG is too far aft, the aircraft becomes less stable in most flight conditions. It may also seriously degrade or eliminate the capability to recover from stall or departure situations. A too far aft CG location can result in very light elevator control forces, making the aircraft more susceptible to inadvertent structural overstress. As discussed in Chapter 5, the CG position relative to the aircraft's center of lift is critical. Normally, the CG must be forward of the center of lift for stability.

2.3.10.3   Weight and Balance Computation

We now discuss the determination of the aircraft weight and the location of the longitudinal center of gravity, commonly called the computation of *aircraft weight and balance*. The weight and CG limits, as described in the previous sections, are usually depicted on a weight versus station number chart, called a *center of gravity chart*, as shown in Figure 2.38. The computed aircraft weight and CG location are plotted on this chart to determine if the aircraft weight and balance are within the specified limits.

The aircraft gross weight is the sum of the basic empty weight and the weight of all items that are loaded into or onto the aircraft. Loaded items include the people (aircrew and passengers), baggage, usable fuel, and external stores, if attached. For safe operation of the aircraft, the computed aircraft gross weight must be less than the specified maximum gross weight.

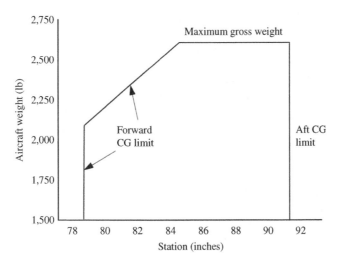

Figure 2.38 Aircraft center of gravity chart.

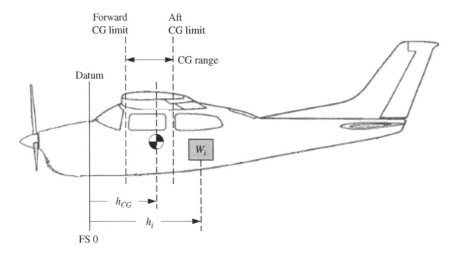

Figure 2.39 Aircraft center of gravity limits.

The aircraft center of gravity location is computed as follows. Each item that is loaded in the aircraft is located a certain distance from the datum, h_i, as shown in Figure 2.39. The product of the item weight and this distance is a moment about the datum (FS 0). The distance from the datum is called a moment arm, or simply, an *arm*. The aircraft center of gravity location, h_{CG}, is obtained by dividing the sum of the moments created by each item weight, $\sum_i M_i$, by the sum of the item weights, $\sum_i W_i$, as given by

$$h_{CG} = \frac{\sum_i M_i}{\sum_i W_i} \tag{2.52}$$

The weight of each item is known or is obtained through weighing. The value of each item's arm is obtained by measuring the distance of the item's location from the datum.

These distances are usually specified by the aircraft manufacturer to the aircraft operator in the operating manuals, such as the distances to the seats, fuel tanks, baggage compartments, etc. The basic empty weight of the aircraft has an arm and a moment associated with it, and these values are also supplied by the aircraft manufacturer or are obtained through weighing. The computation of an aircraft weight and balance is illustrated in the example below.

Example 2.10 Calculation of Aircraft Weight and Balance *A Beechcraft A36 Bonanza, six-place, high-performance, general aviation aircraft (Figure 2.23) has a basic empty weight of 2230 lb (1012 kg), a maximum takeoff weight of 3600 lb (1632.9 kg) and a maximum landing weight of 3100 lb (1406.1 kg). The moment and arm corresponding to the empty weight is supplied by the aircraft manufacturer as 171,130.2 in-lb (19,335.1 N-m) and 76.74 in (194.9 cm), respectively. At the maximum takeoff weight, the forward and aft center of gravity limits are 81.0 in (205.7 cm) and 87.7 in (222.8 cm), respectively, aft of the datum. At the maximum landing weight, the forward and aft center of gravity limits are 74.0 in (188.0 cm) and 87.7 in (222.8 cm), respectively, aft of the datum.*

A 165 lb (74.8 kg) pilot and a 182 lb (82.6 kg) co-pilot are seated in the cockpit at FS 79. Two 180 lb (81.6 kg) passengers are seated in the cabin at FS 117.5. The aircraft has 74 gal (280 liters) of usable fuel in the wing tanks at FS 75. (See Figure 2.40 for fuselage station numbers.)

If the Bonanza takes off, flies for 3 hours, then lands, calculate the zero fuel weight, the takeoff weight and center of gravity location, and the landing weight and center of gravity location. Assume that the aviation gasoline has a weight of 6.0 lb (2.72 kg) per gallon with a fuel burn rate of 15.1 gallons per hour (57.2 liters/h). Also, determine whether the takeoff and landing weights and center of gravity locations are within the allowable limits. Draw a weight versus center of gravity diagram showing all of the limits and the takeoff and landing conditions.

Solution

The following calculations are made in the order given below.

1. *The zero fuel weight, $W_{zero\,fuel}$, is the sum of the basic empty weight and the weights of the pilot, co-pilot, and passengers.*

$$W_{zero\,fuel} = W_{empty} + W_{pilot} + W_{co-pilot} + W_{passengers}$$

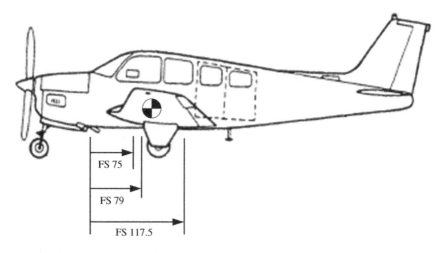

Figure 2.40 Beechcraft A-36 *Bonanza* fuselage station numbers (not to scale).

Inserting numerical values, the zero fuel weight is

$$W_{zero\,fuel} = 2230\,lb + 165\,lb + 182\,lb + 2 \times 180\,lb = 2937\,lb$$

2. *The fuel load is converted from gallons to pounds using the conversion of 6.0 lb per gallon.*

$$W_{fuel} = 74\,gal \times 6.0\frac{lb}{gal} = 444\,lb$$

3. *The moments, M, corresponding to the pilot, co-pilot, passengers, and fuel are the product of their respective weight, W, and arm, h (fuselage station).*

$$M_{pilot} = (Wh)_{pilot} = (165\,lb)(79.00\,in) = 13{,}035\,in \cdot lb$$

$$M_{co-pilot} = (Wh)_{co-pilot} = (182\,lb)(79.00\,in) = 14{,}378\,in \cdot lb$$

$$M_{passengers} = (Wh)_{passengers} = (2 \times 180\,lb)(117.50\,in) = 42{,}300\,in \cdot lb$$

$$M_{fuel} = (Wh)_{fuel} = (444\,lb)(75.0\,in) = 33{,}300\,in \cdot lb$$

4. *The takeoff weight is the sum of the zero fuel weight and the fuel load.*

$$W_{takeoff} = W_{zero\,fuel} + W_{fuel} = 2937\,lb + 444.0\,lb = 3381\,lb$$

5. *The takeoff moment is the sum of the moments due to the empty weight, pilot, co-pilot, passengers, and fuel load.*

$$M_{takeoff} = M_{empty} + M_{pilot} + M_{co-pilot} + M_{passengers} + M_{fuel}$$

$$M_{takeoff} = 171,130.2 + 13{,}035 + 14{,}378 + 42{,}300 + 33{,}300 = 274{,}143\,in \cdot lb$$

6. *Using Equation (2.52), the takeoff center of gravity is the takeoff moment divided by the takeoff weight.*

$$h_{CG,\,takeoff} = \frac{M_{takeoff}}{W_{takeoff}} = \frac{274{,}143\,in \cdot lb}{3381\,lb} = 81.08\,in$$

7. *The fuel burn, in gallons, is the flight time (3 hours) multiplied by the fuel burn rate (15.1 gallons/h). The fuel burn weight is then converted from gallons to pounds.*

$$W_{fuel\,burn} = (3\,h \times 15.1\,gal) \times 6.0\frac{lb}{gal} = 271.8\,lb$$

8. *The fuel burn moment is the fuel burn weight multiplied by the fuel arm.*

$$M_{fuel\,burn} = (Wh)_{fuel\,burn} = (271.8\,lb)(75.0\,in) = 20{,}385\,in \cdot lb$$

9. *The landing weight is the takeoff weight minus the fuel burn weight.*

$$W_{land} = W_{takeoff} - W_{fuel\,burn} = 3381\,lb - 271.8\,lb = 3109.2\,lb$$

10. *The landing moment is the takeoff moment minus the fuel burn moment.*

$$M_{land} = M_{takeoff} - M_{fuel\,burn}$$

$$M_{land} = 274{,}143\,in \cdot lb - 20{,}385\,in \cdot lb = 253{,}758\,in \cdot lb$$

11. *Using Equation (2.52), the landing center of gravity is the landing moment divided by the landing weight.*

$$h_{CG,land} = \frac{M_{land}}{W_{land}} = \frac{253{,}758\,in \cdot lb}{3109.2\,lb} = 81.62\,in$$

The results of these calculations, along with the data supplied for the empty weight, are used to fill in Table 2.8. The zero fuel weight is 2937.0 lb (1332.2 kg). The takeoff weight and center of gravity location are 3381.0 lb (1553.6 kg) and 81.08 inches (205.9 cm), respectively. The landing weight and center of gravity location are 3109.2 lb (1410.3 kg) and 81.62 inches (207.3 cm), respectively.

The weight and center of gravity limits provided in the problem statement are used to develop the center of gravity chart shown in Figure 2.41. The takeoff and landing conditions are placed on the chart. Both conditions are within the acceptable weight and CG limits for safe flight.

Table 2.8 Beechcraft A-36 *Bonanza* weight and balance data.

Item	Weight, W (lb)	Arm, h (in)	Moment/100, $M/100$ (in-lb)
Empty weight	2230.0	76.74	1711.30
Pilot	165.0	79.00	130.35
Co-pilot	182.0	79.00	143.78
Passengers (2)	360.0	117.5	423.00
Zero fuel weight	**2937.0**	–	–
Fuel load	444.0	75.00	333.00
Takeoff weight	**3381.0**	**81.08**	2741.43
Fuel Burn	271.8	75.00	203.85
Landing weight	**3109.2**	**81.62**	2537.58

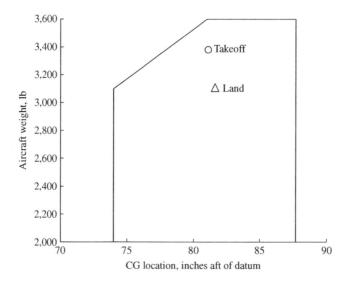

Figure 2.41 Beechcraft A-36 *Bonanza* center of gravity chart.

2.3.11 Aerospace Vehicle Designations and Naming

For the great number of aerospace vehicles that have been designed and built, there is a virtual "alphabet soup" of letters and numbers that are used to designate these different vehicles. It can be difficult to interpret and understand what all of these letters and numbers mean. For instance, what are the meanings of the designations NKC-135A, VH-3D, or ASB-11A? What kind of vehicles are these? As might be expected, there is a logical designation system for aerospace vehicles, which has evolved over time for both military and non-military vehicles. A brief overview is provided of some of the vehicle designations currently in use. Hopefully, this will be useful in interpreting the many designations that the reader may come across in this book and in the engineering workplace. Keep in mind that there are still some exceptions to these current guidelines, and older designations may still be in use.

The US Department of Defense has defined a formal designation system for military aerospace vehicles, including aircraft, unmanned vehicles, missiles, rockets, space probes, satellites, etc. All military aerospace vehicles are assigned a *Mission Design Series (MDS)* designation that consists of letters and numbers, symbolizing characteristics of the aerospace vehicle. Many of the vehicles also have a *popular name*, which usually has some relevance to the type of vehicle or its mission. There are even guidelines for the selection of popular names, to include using no more than two short words and choosing a name that characterizes the vehicle mission and operational capabilities. In the following, the designation system for two classes of aerospace vehicles are summarized: (1) aircraft and (2) guided missiles, rockets, probes, boosters, and satellites. These designation systems are best explained by "decoding" examples, as given below.

2.3.11.1 US Military Aircraft Designations

Our first example is the Boeing KC-135A *Stratotanker* (Figure 2.42). The KC-135 is a US Air Force aerial refueling aircraft, commonly called a *tanker*. The design of this aircraft was based on Boeing's first commercial jet airliner, the Boeing 707. The popular name, *Stratotanker*, is fitting, describing a jet *tanker* that flies at high altitude, in the stratosphere.

Figure 2.42 US Air Force KC-135 *Stratotanker* with its aerial refueling boom extended. (Source: *US Air Force*.)

We "decode" a special version of the KC-135, designated as the NKC-135A. The first letter is the *status prefix* symbol, designating a non-standard vehicle, that is a test, modification, experimental, or prototype design. The status prefix "N" designates a vehicle that is in a permanent, special test function. The second and third letters are the *modified mission* symbol and *basic mission* symbol, respectively. The basic mission symbol identifies the primary function or capability of the aircraft. The modified mission symbol identifies modifications to the basic mission. For our example, the basic mission symbol "C" identifies the aircraft as a transport, and the symbol "K" shows it has a modified mission as a tanker. The number "135" in our designation is the *design number*, designating the manufacturer's 135th airplane design. Finally, the letter after the number is the *series*, identifying the first production model of a design and then advancing one letter for each later model. In our case, the "A" indicates the first model of the 135 airplane design.

To summarize for the aircraft designation NKC-135A, we have

N	status prefix	permanently operating in special test capacity
K	modified mission	tanker
C	basic mission	transport
135	design number	135th design
A	series	1st version of this design

Table 2.9 provides a complete list of US Military designator symbols and descriptions for aircraft. As shown in this table, some aircraft also have a *vehicle type* designation, required only for certain vehicles, such as glider, helicopter, vertical takeoff and landing (VTOL) vehicle, missile, or space vehicle. There is a vehicle type symbol "D" which applies to the ground control equipment for unmanned aerial vehicles, rather than an actual vehicle. An example of the use of a vehicle type designation is the Sikorsky VH-3D *Sea King* helicopter, as "decoded" below.

	status prefix	(none)
V	modified mission	staff
	basic mission	(none)
H	vehicle type	helicopter
3	design number	3rd design
D	series	4th version of this design

The *Sea King* was designed as an antisubmarine warfare helicopter, with the "VH" model serving in its modified mission role as the US Presidential helicopter.

Two aircraft identifying numbers, which are not included in the MDS designator, are *serial numbers* and *block numbers*. Serial numbers uniquely identify a specific vehicle. The numbering system for serial numbers varies with the different military services. Block numbers identify a manufacturer's production group of aircraft with the same configuration, within a specific design series. Block number assignments are usually in multiples of five: 1, 5, 10, 15, etc., but intermediate numbers are also sometimes used. For example, the US Air Force F-16 *Fighting Falcon* has been assigned quite a few block numbers in its modification history. The Block 1 F-16 was an early production model with a black nose cone. Changes for Block 5 aircraft included a low-visibility grey nose cone and the addition of fuselage and tail fin rain water drainage holes. Structural material changes from titanium to aluminum and new material bonding techniques were incorporated into F-16 Block 10 aircraft. Block 15 F-16s had 30% larger horizontal stabilizers, improved radar, and other improvements. The F-16 block numbers now reach 60, with its continuing evolution and improvements.

Table 2.9 US Military designator symbols and descriptions for aircraft (from "Designating and Naming Military Aerospace Vehicles," AFJI 16-401, NAVAIRINST 8800.3B, AR 70-50, 14 March 2005).

Status prefix	Modified mission	Basic mission	Vehicle type
G – Permanently grounded	A – Attack	A – Attack	D – UAV control segment
J – Special test (temporary)	C – Transport	B – Bomber	G – Glider
N – Special test (permanent)	D – Director	C – Transport	H – Helicopter
X – Experimental	E – Special electronic installation	E – Special electronic installation	Q – Unmanned aerial vehicle
Y – Prototype			
Z – Planning	F – Fighter	F – Fighter	S – Spaceplane
	H – Search and rescue/medevac	L – Laser	V – Vertical takeoff and landing (VTOL)/short takeoff and landing (STOL)
	K – Tanker	O – Observation	Z – Lighter-than-air vehicle
	L – Cold weather	P – Patrol	
	M – Multi-mission	R – Reconnaissance	
	O – Observation	S – Antisubmarine	
	P – Patrol	T – Trainer	
	Q – Drone	U – Utility	
	R – Reconnaissance	X – Research	
	S – Antisubmarine		
	T – Trainer		
	U – Utility		
	V – Staff		
	W – Weather		

2.3.11.2   Designations for Missiles, Rockets, Space Probes, Boosters, and Satellites

Let us now look at the DoD designation system for aerospace vehicles other than aircraft, that is, for guided missiles, rockets, probes, boosters, and satellites. We "decode" the designation ASB-11A. The first letter specifies the *launch environment*, that is, where the vehicle is launched from, where the "A" designation specifies an *air launched* vehicle. The second letter specifies the *basic mission* or the primary function or capability of the vehicle. *Space support* is the basic mission for our example, indicated by the second letter designation "S". The *vehicle type* is specified by the third letter, where "B" signifies a *booster* in our example. Similar to the aircraft designation, the *design number* and series number are given by the number "11" and the final letter "A", respectively. To summarize, the ASB-11A "decodes" as follows.

	status prefix	(none)
A	launch environment	air launched
S	basic mission	space support
B	vehicle type	booster
11	design number	11th design
A	series	1st version of this design

Table 2.10 US Military designator symbols and descriptions for guided missiles, rockets, probes, boosters, and satellites (from "Designating and Naming Military Aerospace Vehicles," AFJI 16-401, NAVAIRINST 8800.3B, AR 70-50, 14 March 2005).

Status prefix	Launch environment	Basic mission	Vehicle type
C – Captive	A – Air	C – Transport	B – Booster
D – Dummy	B – Multiple	D – Decoy	M – Guided Missile
J – Special test (temporary)	C – Container	E – Electronic/communication	N – Probe
N – Special test (permanent)	F – Individual	G – Surface attack	R – Rocket
X – Experimental	G – Surface	I – Aerial/space intercept	S – Satellite
Y – Prototype	H – Silo stored	L – Launch	
Z – Planning		detection/surveillance	
	L – Silo launched	M – Scientific/calibration	
	M – Mobile	N – Navigation	
	P – Soft pad	Q – Drone	
	R – Ship	S – Space support	
	S – Space	T – Training	
	U – Underwater	U – Underwater attack	
		W – Weather	

ASB-11A is the designation of the Orbital Sciences *Pegasus* air-launched, space support booster, previously described in Section 1.3.5.6 (Figure 1.80).

Table 2.10 provides a complete list of US Military designator symbols and descriptions for guided missiles, rockets, probes, and satellites. Similar to the aircraft designations, there is an optional *status prefix* first letter. (The ASB-11A Pegasus launch vehicle example does not have a status prefix.).

2.3.11.3 Designations for Non-Military Aircraft

Designation and naming for non-military aircraft, including commercial and general aviation aircraft, is less standardized, at least among all the different US and international aircraft manufacturers. In the past, there were some regulated two-letter codes that were used by US aircraft manufacturers for non-military aircraft, but this is no longer necessarily in use. Today, each individual aircraft manufacturer seems to have its own type of designation and numbering system.

For many, if not most, non-military aircraft, the designation is usually a letter followed by a number. The letter often indicates the aircraft manufacturer, such as "A" for Airbus, "B" for Boeing, "C" for Cessna, etc., and the number usually indicates a model number. For example, "B747" signifies a Boeing model 747 aircraft, a wide-body commercial jet airliner and cargo transport. Some companies use a two-letter code, which may be a vestige of the old, regulated two-letter designation system, an example being the "PA-31" for the Piper model 31, a cabin-class, twin-engine, general aviation aircraft.

The model number is often followed by a dash and a series number for the specified model, for example "B747-400" being the 400 series of the Boeing 747 model. There is sometimes a suffix letter added to the model or series number to signify a different version. For example, the "F" in the B747-400F designation is the freighter version of the Boeing B747-400 and the "P" in the PA-31P-350 designation is the pressurized verison of the Piper PA-31-350.

The non-military aircraft designation should not be confused with the *international aircraft registration prefix code*. These codes are unique letters and numbers that precede the aircraft registration number. All civil aircraft in the world must be registered in accordance with regulations established by the International Civil Aviation Organization (ICAO). The aircraft registration prefix code for the US is "N", so that a US registered civil aircraft might have a registration number N1234. In the US, the aircraft registration number is commonly called an *N number*, because of the prefix code, or a *tail number*, because the registration number is often displayed near the tail of the aircraft.

2.4 Introductory Flight Test Concepts

In this section, some of the fundamental concepts of flight testing are introduced. A more precise definition of *flight test* is given and several different types of flight testing are described. As with most scientific endeavors, flight test involves a methodical and systematic approach, using well-defined processes and techniques. The flight test process is described, first in terms of its philosophical basis in the classical scientific method and then in terms of the details of its specific application. The fundamental test techniques, called *flight test techniques* or *FTTs*, and the typical methods used for data collection are described. Flight test typically involves a team of people with different areas of expertise. The roles and responsibilities of the people involved in flight test are explained. Finally, flight test safety and risk assessment are discussed, along with the typical approaches used to manage this risk.

2.4.1 What is a Flight Test?

Based on popular treatments, flight test could be perceived as a cavalier and death-defying endeavor. Nothing is further from the truth. When performed properly, flight test is a precisely planned and meticulously executed scientific approach, where the hazards are minimized as much as possible. Make no mistake; there are sometimes risks involved or potential hazards. However, the goal of all flight tests should be to minimize and mitigate these risks or hazards as much as possible. If the flight test goes very well, it may be predictable and uneventful, perhaps even considered mundane at times by some. However, even the most mundane test has the excitement of turning theories and equations into real flight.

Aerospace vehicles, whether they are aircraft or spacecraft, are some of the most complex systems and machines that man can build. Despite all of our modern theories and computing capabilities, accurately predicting all of the flight characteristics of a new aerospace vehicle is not possible. Flight test is needed to determine the *real* characteristics of the vehicle in the *real* flight environment. Flight test is often the last or final step in the aerospace design and development process. It can provide the real data to be compared with the design predictions and analyses. The actual characteristics and performance of the vehicle and its systems are determined through flight test. These flight-determined characteristics are compared with the predictions, to determine whether the design objectives have been met or whether design changes are warranted.

Flight test is usually a team endeavor. The flight test team comprises pilots, engineers, technicians, and other support personnel. Success depends on the combined efforts and contributions of a diverse group of people, with a broad range of technical, management, and operational skills. The technical team must usually be interdisciplinary, requiring expertise in aerodynamics, stability, control, structures, instrumentation, avionics, and other areas.

Flight test often requires specialized sensors, instrumentation, and equipment to measure desired parameters. These parameters may be measurements that are not available using the standard aircraft instruments and they may be sampled at a higher data rate than achievable with standard instruments. Flight test equipment or "boxes" that are installed in a vehicle are typically painted orange to distinguish them from other non-flight test items and to make them highly visible. The measured flight test parameters are usually recorded onboard the vehicle and may also be transmitted or telemetered to a ground station, where they are recorded or displayed to ground observers. The data may also be displayed onboard the vehicle to the pilot or flight test engineers.

2.4.1.1 Types of Flight Testing

There are various types of flight testing, with different objectives and approaches. It may be commonly thought that the first flight of a new type of aircraft *is* the definition of flight testing, but flight test encompasses much more. Some of the different types of civilian flight testing include experimental, engineering, production, systems, and maintenance flight test. Flight test may also be performed for pure scientific research.

Experimental flight test includes the first flights of a new or prototype aircraft type and the determination or expansion of its flight envelope. It may also include testing of a new aircraft model or of an existing aircraft design that has been significantly modified. Experimental flight test seeks to define the unknown performance, flying qualities, systems operation, or other characteristics of a new aircraft or model.

Engineering flight test is performed on an aircraft within its existing flight envelope. The flight and system characteristics are expected to be the same as already known. This type of flight test may include functional and reliability testing of systems and components. Engineering flight test includes the testing required to certify an aircraft under government regulations. US civilian certification standards are specified by the Federal Aviation Administration for different categories of aerospace vehicles. For fixed-wing airplanes, the most often used standards are the Code of Federal Regulations (CFR) Part 23, "Airworthiness Standards for Normal, Utility, Acrobatic and Commuter Airplanes," and Part 25, "Airworthiness Standards for Transport Category Airplanes."

Production flight test is of aircraft being produced by an aircraft manufacturer. Prior to being issued a government airworthiness certificate, production flight test is performed on a newly built aircraft to ensure conformity to the approved or certified design. These flight tests are also performed within the defined flight envelope of the existing design.

Systems flight test is conducted to assess the systems onboard the aircraft that have been newly installed, updated, or modified. It is assumed that the aircraft flight characteristics are not affected by the new or changed systems, but this may not always be the case. For example, the installation of a large antenna or sensor on the outer mold line of the aircraft may significantly change the aerodynamics, performance, or flying qualities. In-flight testing of avionics, such as communications and navigation equipment, is a common systems flight test. Flights required for aircraft certification include systems flight tests.

Maintenance flight test or *functional check flights* (FCFs) are flights that are conducted after maintenance or new installation has been performed on the aircraft. These flights serve to verify the normal performance and flying qualities of the aircraft and the correct operation of its systems.

Types of flight testing that are specific to the US Military include developmental test & evaluation (DT&E) and operational test & evaluation (OT&E). Both DT&E and OT&E can be performed on a new aircraft type or an existing aircraft that has been modified. However, DT&E flight testing precede OT&E, as explained below. Similar to the aircraft certification requirements in the civilian world, there are US Military Specifications (MIL-SPECs) and US Military Standards (MIL-STDs) for the flight characteristics and systems operations of a military aircraft.

Developmental test & evaluation is performed to quantify the flight dynamics of an aircraft, that is, "how it flies". Typically, this includes flight test evaluation of the vehicle's performance and flying qualities. These flight evaluations are usually performed using what is called *open loop* testing. Here, the pilot provides an input to the vehicle control system and the resulting vehicle output or response is measured, without the pilot disturbing this response. The aircraft systems, such as avionics, autopilots, cockpit displays, radar, and sensors are also usually evaluated in DT&E flight testing.

In *Operational test & evaluation* flight testing, the vehicle is flown in a manner that duplicates how it will be flown operationally, that is, for its intended use or mission. The test pilot flies the aircraft as the non-test pilot, in its "everyday" use, will fly it. Maintenance of the aircraft is also conducted, as it will be maintained operationally. OT&E testing evaluates the vehicle's suitability, reliability, and maintainability for its intended use or mission. As an example, if an aircraft is designed for a cargo mission, OT&E flight test involves flying the aircraft along representative flight profiles, with representative cargo payloads, and operating to and from representative types and lengths of runways. The aircraft's operation may also be tested and evaluated for representative climates. For instance, if the aircraft is going to be normally operated in a hot, humid climate or an artic climate, OT&E flight testing will be performed in these kinds of environments. In contrast to DT&E open-loop flight test, OT&E tends to be *closed loop* flight testing, where the pilot puts in an input, the aircraft responds, and the pilot stays "in the loop" by applying additional inputs, based on the response.

While we are making clear distinctions between different types of flight testing, in reality, there is often overlap between these various kinds of testing. For instance, in performing Certification flight testing for a new engine installation in an aircraft, DT&E performance and flying qualities flight testing may need to be completed, in addition to OT&E flight testing, to determine how well the aircraft can still perform the desired mission.

2.4.1.2   Who Does Flight Testing?

Flight testing and flight research are performed by a wide variety of entities, including large and small commercial aerospace companies, the military, and the government. The various organizations may work together as a team in performing the flight testing or research. Flight testing may even be performed by individuals, as when the builder of an experimental, homebuilt aircraft performs the testing of his or her newly constructed aircraft. Often, organizations have their own, dedicated flight test groups or departments, staffed by pilots, engineers, managers, technicians, and other support personnel who are specifically trained to perform flight operations and testing. The organizations may also have non-test aircraft to support flight testing and to be flown for aircrew proficiency.

Government agencies often act in an oversight or regulatory role, ensuring that the vehicle being tested ultimately meets the government regulations or standards. Some of the major government aviation regulatory authorities in the world that oversee flight testing include the Federal Aviation Administration (FAA) in the United States, the Civil Aviation Authority (CAA) in the United Kingdom, and the European Aviation Safety Agency (EASA) in Europe.

The US military has several major flight test facilities or complexes in the USA. These include the US Air Force flight test centers at Edwards, California, and Eglin, Florida and the US Navy facilities at Patuxent River, Maryland and China Lake, California. The test locations are large, remote areas that are conducive to flight testing of high-speed military aircraft and weapons systems. Non-military companies and government organizations also use these test ranges by arrangement with the military. NASA operates the Armstrong Flight Research Center as a tenant on Edwards Air Force Base, California.

Where does one learn to do flight testing? Early in aviation, there was little formal flight training, much less, any formal flight test training. Flight testing was often done by trial-and-error, hopefully with the result that the flight tester survived their flight to learn from their mistakes. Eventually, it was recognized that formal training in flight test would be beneficial, and several formal schools and training programs were established. Today, many aerospace companies train their flight testers "in-house", developing their flight test skills through a mix of formal academic training and practical experience, under the tutelage of veteran flight test personnel. There are also several formal flight test training facilities and test pilot schools worldwide, where student test pilots and student flight test engineers attend a formal, typically year-long curriculum of academics and flight training. These include military, industry, and civilian operated training facilities. The major flight test schools and training institutions include:

- Divisão de Formação em Ensaios em Voo, Brazilian Air Force Test Pilot School, São José dos Campos, Brazil
- Empire Test Pilot School, Boscombe Down, Wiltshire, England
- Aircraft and Systems Testing Establishment (ASTE), Indian Air Force Test Pilot School, Bangalore, India
- International Test Pilot School (civilian), Woodford, England, and London, Ontario, Canada
- L'Ecole du Personnel Navigant d'Essais et de Reception (EPNER), French Test Pilot School, Istres, France
- National Test Pilot School (civilian), Mojave, California, USA
- Russian Ministry of Aviation Industry Test Pilot School, Zhukovsky, Moscow Olast, Russia
- US Air Force Test Pilot School, Edwards, California, USA
- US Naval Test Pilot School, Patuxent River, Maryland, USA.

Typically, flight test training includes the basic areas of performance, flying qualities, systems testing, and test management. All of these subject areas draw heavily upon the technical aspects of aerospace engineering. The management of flight test projects is also typically covered. Academic theory is usually taught in a classroom environment, which is then coupled with practical application of the theory in flight. The flight exercises are typically flown in a variety of aircraft, to give the students exposure to a wide range of different aircraft. Student test pilots and student flight test engineers often work as a team in preparing for, flying, and analyzing test flights. In addition to flying real aircraft, flight simulators are also used, due to their unique capabilities and cost effectiveness.

2.4.1.3 The X-Planes

Flight test has had a rich history in the USA, starting with the first test flights of the Wright *Flyer I* by the Wright brothers. As the flight testing of aerospace vehicles became more formalized, a special designation was created for the flight test vehicles that were the first of their kind or unique in other ways. Starting with the Bell X-1,[5] the "X-plane" designation has become synonymous with US experimental aerospace vehicles that expand the frontiers of air and space.

Many of the early X-planes were experimental, rocket-powered aircraft, expanding the airspeed and altitude boundaries of high-speed flight. However, the X-plane designation includes a wide

---

[5] The Bell X-1 was initially designated the XS-1, the "S" signifying "supersonic". The "S" was deleted from the designation early in the project.

range of aerospace vehicles, such as low-speed propeller-driven airplanes, unmanned aerial vehicles, vertical takeoff and landing vehicles, unmanned missile test beds, space access vehicles, and prototypes of advanced aircraft, missiles, and spacecraft. Despite their differences, the flight test of X-planes share the common goal of advancing the research and technology boundaries of aerospace engineering. The US X-planes, from the Bell X-1 to the present, are listed in Table 2.11. We refer back to many of the significant accomplishments of the various X-planes throughout the text.

2.4.2   The Flight Test Process

Similar to other scientific fields of inquiry, the flight test process is based on the fundamental tenants of the classical *scientific method*. With its roots dating back to the 17th century, the scientific method is a systematic process that is used to investigate phenomena and acquire scientific knowledge. One way of applying the scientific method to flight test is shown in Figure 2.43, comprising the following steps.

1. Formulation or application of a theory or hypothesis
2. Application of the theory to make predictions
3. Performance of an experiment (flight test), using flight test techniques, to compare with the predictions
4. Analysis of the data from the experiment (flight test)
5. Formulation of conclusions from the analysis to modify the theory or hypothesis, as appropriate

In the first step of the process, a new theory may be formulated or an existing theory may be used. The objective of the test may be to prove the new theory or to validate the predictions based upon the existing theory. For example, a new type of propulsion system may be flight tested to prove the theoretical basis of the system. Alternatively, an existing aircraft performance theory may be used to predict takeoff performance, and then a flight test is conducted to validate the prediction. Whether the theory is new or not, pre-test predictions are usually made to compare with the test data.

The flight test is the "experiment" to be performed for data collection. Similar to the setup of an experiment for any scientific endeavor, a flight test must be carefully designed and planned to successfully collect the desired data. The actual in-flight data collection is often accomplished by using standard *flight test techniques*, to be discussed in the next section. Once the test is complete and the data has been collected, the data must be analyzed, which may include comparisons with predictions. Results from the data analysis are used to draw conclusions about the theory or objectives of the test. The theory may then be revised, based on the results and conclusions from the test.

Applying the above philosophy, the detailed steps of the flight test process might look like the example shown in Figure 2.44. The example process shown is not meant to be applicable to all flight test situations, rather, it serves as a template from which to discuss several of the important elements that are typically included in the process. The first step in the process is the definition of the test objectives and requirements. For example, the objectives might be to certify an aircraft to government regulatory standards, and the requirements might be those specified in the government regulations. After the objectives and requirements are well understood, a test plan is written that includes details of how the testing is to be performed to meet the objectives and requirements. These details typically include the statement of the objectives, an aircraft description and configuration, flight test techniques and maneuvers to be used, roles and responsibilities of the personnel involved,

Table 2.11 The X-planes.

X No.	First flight	Goals or accomplishments
X-1	25 Jan 1946	First manned supersonic flight, reaching Mach 1.06 at 45,000 ft on 14 October 1947
X-1A-E	24 Jul 1951	Continuation of X-1 high-speed flight research
X-2	27 Jun 1952	High-speed flight research aircraft with swept wing flight, 1st to exceed Mach 3
X-3	20 Oct 1952	Mach 2 research aircraft, but never flew faster than Mach 0.95
X-4	15 Dec 1948	Tailless (no horizontal tail) research aircraft designed for high subsonic speed flight
X-5	20 Jun 1951	First variable-sweep wing aircraft
X-6	None	Evaluation of nuclear propulsion using a modified B-36 aircraft (not built)
X-7	26 Apr 1951	Testbed for ramjet propulsion, fastest flight Mach 4.3
X-8	24 Apr 1947	Upper air research sounding rocket (highest to 800,000 ft), led to *Aerobee* rocket
X-9	28 Apr 1949	Testbed for air-to-surface missile technology
X-10	14 Oct 1953	Aerodynamics and systems testbed for intercontinental cruise missile technology
X-11	None	Proposed test vehicle for original *Atlas* intercontinental ballistic cruise missile concept
X-12	None	Proposed test vehicle for original *Atlas* intercontinental ballistic cruise missile concept
X-13	10 Dec 1955	Vertical takeoff and landing (VTOL) flight research with a jet aircraft
X-14	17 Feb 1957	VTOL flight research using vectored thrust, data used to design the *Harrier* prototype
X-15	8 Jun 1959	Hypersonic flight research, first manned hypersonic aircraft flight, flew to Mach 6.70
X-16	None	Designed to be a high-altitude long range reconnaissance aircraft (not built)
X-17	17 Apr 1956	Multi-stage rocket used for hypersonic entry research up to Mach 14.4
X-18	20 Nov 1959	First tilt-wing vertical takeoff and landing (VTOL) aircraft
X-19	20 Nov 1963	VTOL flight research using tandem, tilt-rotor concept (similar to V-22 *Osprey*)
X-20	None	Hypersonic "space plane" design, called the *Dyna-Soar* (not built)
X-21	18 Apr 1963	Northrop laminar boundary layer control test aircraft
X-22	17 Mar 1966	V/STOL aircraft with dual tandem ducted-propellers and variable-stability system
X-23	21 Dec 1966	Lifting-body, maneuvering reentry test vehicle
X-24	17 Apr 1969	Rocket-powered lifting body, explored low-speed flight and landing of lifting bodies
X-25	5 Jun 1968	"Gyro-chute" concept for emergency egress capability using an ultralight gyrocopter
X-26	3 Jul 1962	Schweitzer 2-32 sailplane used as a Navy trainer and as a stealth observation platform
X-27	None	Lockheed design for an advanced lightweight fighter to replace the F-104 (not built)
X-28	12 Aug 1970	Prototype of small, single-engine seaplane for reconnaissance use in Southeast Asia
X-29	14 Dec 1984	Forward swept wing flight research

Table 2.11   (*continued*)

X No.	First flight	Goals or accomplishments
X-30	None	Hypersonic, scramjet-powered, single-stage-to-orbit (SSTO) vehicle (not built)
X-31	11 Oct 1990	High angle-of-attack flight research, including post-stall region, using vectored thrust
X-32	18 Sep 2000	Boeing concept demonstrator aircraft for Joint Strike Fighter competition
X-33	None	Single-stage-to-orbit vehicle with linear aerospike rocket motor (not built)
X-34	None	Reusable access to space testbed; captive-carry flight in June 1999
X-35	24 Oct 2000	Lockheed concept demonstrator aircraft for Joint Strike Fighter competition
X-36	17 May 1997	Boeing remotely piloted, 28% scale vehicle with no vertical or horizontal tails
X-37	22 Apr 2010	Orbital space plane to demonstrate reusable space technologies, operated by USAF
X-38	12 Mar 1998	Concept demonstrator of a crew rescue vehicle for the International Space Station
X-39	None	Reserved for use by USAF for sub-scale unmanned demonstrators
X-40	11 Aug 1998	80% scale version of proposed Space Maneuver Vehicle, became the X-37
X-41	Unknown	Classified DARPA common aero vehicle (CAV) maneuvering reentry vehicle
X-42	Unknown	Experimental, expendable upper stage, designed to boost payloads into orbit
X-43	2 Jun 2001	Air-launched, unmanned, hydrogen-fueled, scramjet testbed; flew to Mach 9.68
X-44	None	Tailless research aircraft concept (not built)
X-45	22 May 2002	Tailless, thrust-vectoring unmanned combat air vehicle (UCAV) demonstrator
X-46	None	US Navy unmanned combat air vehicle (UCAV-N) demonstrator, (cancelled)
X-47	24 Feb 2003	Tailless, diamond-shaped wing planform UCAV demonstrator
X-48	20 Jul 2007	Unmanned, sub-scale, blended wing body (BWB) testbed
X-49	29 Jul 2007	Compound fixed wing airplane-helicopter with vectored thrust ducted propeller design
X-50	24 Nov 2003	Canard rotor wing demonstrator vehicle
X-51	26 May 2010	Air-launched, unmanned, hydrocarbon-fueled scramjet testbed; flew to >Mach 5
X-52	None	Number skipped
X-53	8 Dec 2006	Active aeroelastic wing (AAW) technology demonstrator using highly modified F-18
X-54	Pending	Reserved for Gulfstream/NASA supersonic business jet demonstrator
X-55	2 Jun 2009	Advanced composite cargo aircraft testbed
X-56	26 Jul 2013	UAV to study high altitude, long endurance flight technologies
X-57+	Unknown	Unknown

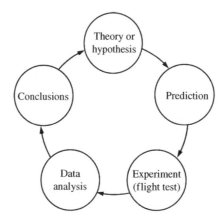

Figure 2.43   The scientific method applied to flight test.

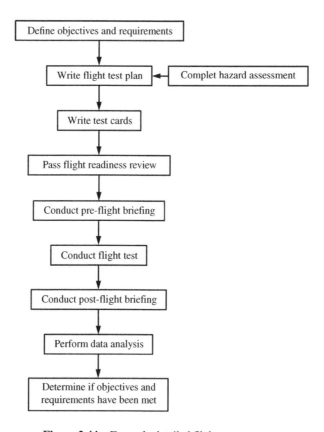

Figure 2.44   Example detailed flight test process.

measurements and instrumentation required, ground testing to be performed prior to flight, and any other test requirements. A critical element of the flight test process is the assessment of the potential hazards in performing the test. This hazard assessment may be included in the written test plan or it may be a separate document. The safety and risk assessment aspects of the flight test planning process are covered in a later section. After the test plan has been written, reviewed, and approved, test cards are written that specify the step-by-step procedures to be used to set up and perform the test maneuvers. Test cards are discussed in more detail in a later section. Prior to the test, there is usually a technical and safety review of the readiness to proceed to flight, often called a flight readiness review. This final, formal review is usually presented to technical, safety, and management personnel who are not associated with the flight test, so as to provide an independent, objective assessment and approval as to whether the flight test team is ready to proceed to flight. Once this readiness review is passed, the flight test team is complete with the preparation phase of the test and is ready to move on to test execution.

On, or very near to, the day of the flight, the test team meets to brief the planned flight to ensure that everyone understands the test objectives, the flight test techniques that will be flown, test data that is required, and any flight restrictions or limits. The test cards for the day's flight are talked through, as they will be flown. Finally, it is time to go flying, and the test flight is performed, following the briefed plan. The well-known adage for this process is to "plan the flight and fly the plan". After the flight is completed, the test team holds a post-flight briefing to review the flight, discuss what went well and what did not go as planned, and to identify any issues or discrepancies. The flight test data is analyzed by the engineering analysts to ultimately determine if the objectives and requirements have been met.

2.4.3   Flight Test Techniques

Test methods called *flight test techniques* (FTT) are often used to obtain flight test data. Many standard FTTs have been developed to obtain flight test data in areas such as performance, flying qualities, structures, and systems. Most FTTs are conceptually straightforward, but flying the FTT with the required precision can be a demanding task for the test pilot. We will learn many of the fundamental FTTs used in flight test in the coming chapters. Table 2.12 lists the flight test techniques that are described in the text, along with the aircraft that is used in discussing the FTT. In addition to the FTTs, several ground test techniques (GTTs) are introduced throughout the text. These are test and analysis techniques that are performed on the ground, often prior to proceeding to flight tests. The GTTs, discussed in the text, are also is given in Table 2.12.

2.4.3.1   Flight Profile

It is often useful to organize a flight on an altitude versus time plot, called a *flight profile*, as shown in Figure 2.45. The flight profile is composed of *test points*, where flight test techniques or data collection are performed at specific flight conditions. The flight conditions are typically specified as an altitude and Mach number or airspeed, but they may include angle-of-attack, angle-of-sideslip, load factor, aircraft configuration, or other required conditions. The flight profile plot provides an overview of the complete flight. It identifies the individual test points, FTTs, and the *transition* between test points. The transition information is helpful for the test pilot to set up the succeeding test points. Ideally, the test point transitions have been optimized, in the test planning, to match energy levels between test points as much as possible. For example, an energy losing test point maneuver, such as a high-g, descending turn, might follow a test point where the energy is high, such as a high Mach number, high altitude maneuver.

Table 2.12 Ground and flight test techniques discussed in text.

Discipline	Ground or flight test technique (GTT or FTT)	Aircraft used in description
Fundamentals	Familiarization flight	McDonnell Douglas F/A-18B *Hornet*
	"Trim shot"	Extra 300
Aerodynamics	In-flight flow visualization	NASA F/A-18 HARV
	Drag cleanup	GTT
	Wind tunnel testing	GTT
	Computation fluid dynamics	GTT
	Lift and drag in steady, gliding flight	North American XP-51B *Mustang*
	Aerodynamic modeling	Boeing F/A-18E *Super Hornet*
	Visualizing shock waves in flight	Various
	Stall, departure, and spin	Christen *Eagle II*
	Hypersonic flight testing	North American X-15
Propulsion	Engine test cell and test stand	GTT
	Flying engine testbed	Various
	In-flight thrust measurement	Convair F-106B *Delta Dart*
Performance	Altitude and airspeed calibration	Northrop T-38A *Talon*
	Cruise performance	Ryan NYP *Spirit of St. Louis*
	Climb performance	Cessna 172 *Cutlass*
	Energy	Lockheed F-104G *Starfighter*
	Turn performance	Lockheed F-16 *Fighting Falcon*
	Takeoff performance	North American XB-70 *Valkyrie*
Stability & Control	Longitudinal static stability	Piper PA32 *Saratoga*
	Lateral-directional static stability	NASA M2-F1 lifting body
	Longitudinal dynamic stability	Piper PA31 *Navajo*
	Variable-stability aircraft	Various
	First flight	New or modified vehicle

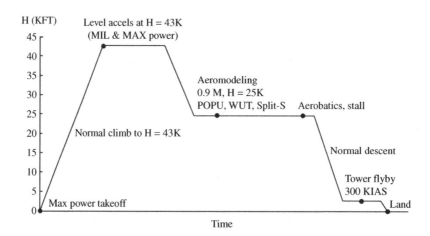

Figure 2.45 Example flight profile.

2.4.3.2 Flight Test Cards

The details of each test point and FTT are usually written down on a set of *flight test cards*. The test cards provide step-by-step procedures for each test point. Details about setting up the test point, the maneuvers to be flown, and the post-maneuver prompt (i.e. what test point is coming next) are usually included. In addition to the individual test point cards, there are usually some overview cards that describe the big picture of the test, often including a card showing the flight profile, as described in the previous section. The overview cards usually include other items that are applicable to all the test points, such as aircrew assignments, communications frequencies, airspace information, aircraft and test limits, aircraft weight and balance information, performance charts, or other pertinent information.

In constructing test cards, the format selected depends on the test situation and on personal preferences. For example, textual directions, tables, pictures, or diagrams are different formats that may be appropriate. The physical size of a test card can be small or large, depending on the size of the cockpit or cabin where it is utilized. Typical elements of a test card include the test point name and number, aircraft type, and card number. Items specific to the test point might include the aircraft configuration, limits, data band and tolerance, setup of instrumentation or data system requirements, and setup or trim shot flight conditions. There is usually space on the card to record data or write comments. The order in which data is recorded should be prioritized on the card: the most important data that is required should be written down first. The test card should specify data responsibilities, such as specifying that the pilot make a verbal call at given time intervals or the flight test engineer starts a data system and records certain data. An example flight test card is shown in Figure 2.46. The test cards are a record of the events and data from a test flight. They should not be altered or rewritten after a flight, as this might taint the original information or data on the card.

2.4.3.3   Flight Test Data Collection

Almost always, a flight test technique involves the collection of data. The data may be quantitative, such as obtaining performance or stability data about an aircraft. Alternatively, the data may be qualitative, such as a subjective pilot opinion of how capable an aircraft is at performing a mission task, such as aerial refueling or formation flying.

Typically, *data bands* and *tolerances* are specified for the data collection. For many FTTs, especially those that require maneuvering, it is not possible to maintain the aircraft at perfectly constant flight conditions. A data band of ± 1000 ft (300 m), above and below a specific altitude, may be used for a constant altitude test point, since the atmospheric properties are essentially constant over this altitude range. Tolerances for holding flight conditions, such as airspeeds or load factors, are set by the accuracy required in the data. Tighter tolerances usually translate into more pilot effort in flying the aircraft precisely.

The first step in the flight data collection process is the definition of what data needs to be collected. This definition flows down from the test requirements in the form of the data that is needed to perform the data analyses. In addition to this data – the direct inputs to analyze a specific aircraft characteristic – other data is usually required to validate the test conditions. For example, in an aircraft climb performance test, primary data such as airspeed, altitude, engine power, and time are required for the performance analysis. In addition, data such as the angle-of-sideslip may be desired to ensure that the climb was performed with zero sideslip. The sideslip data may not feed directly into the climb performance analysis, but it validates the quality of the data.

Of course, the desired number of data parameters may exceed the number available. This constraint is usually due to the number of sensors on the vehicle and the architecture and size of the data acquisition system. Another factor to be considered is the data *sampling rate*, the data samples per second that are collected. The required data sampling rate is a function of the frequency of the physics that is being measured and other data acquisition related requirements.

F-16B S/N:	F-16 PERF FLT	DATE:

MAX PWR LEVEL ACCEL
43 K ±500 FT / 0.85 M → M_{MAX}

LIMITS: 600 KCAS, 1.6 M
CONFIG: S/B IN, MAX PWR

A   TRIM SHOT @ 41 K, 0.8 M (15 sec)

B   DAS: On (REC LT ON)

C   LVL ACCEL -- MAX PWR
       0.85 M → M_{MAX}
       ALT ± 100 FT, VVI±100 FPM

Mi	ΔHi	VVI	FUEL	TIME
0.85	____	____	____	____
0.90	____	____		____
0.95	____	____		____
1.0	____	____		____
1.05	____	____		____
1.1	____	____		____
____	____	____		____

D   DAS: OFF (REC LT OFF)

E   COMMENTS:

NEXT: DESCENT TO 27.5K	5B

Figure 2.46   Example of a flight test data card for a maximum power level acceleration in an F-16.

Flight test data may be collected in a variety of ways, ranging from simple hand recording of information and data in flight to the use of a sophisticated data acquisition system (DAS). There are many types of data acquisition systems available, from simple inexpensive systems to complex costly systems. A DAS can be commercial off-the-shelf (COTS) equipment or it can be a custom-made system. Modern data acquisition systems are capable of measuring and recording thousands of parameters. DAS data may be recorded onboard the aircraft or may be telemetered to a ground station, or both. Telemetry systems require a transmitter and antenna on the aircraft, as well as receiving equipment and an antenna on the ground. A benefit of data telemetry is that personnel in a ground control station or control room can monitor the flight data in real time.

At the other end of the data collection spectrum is the use of "hand-held" data, where personnel in the aircraft manually record flight data by hand. Hand-held data is the simplest source of flight test data, requiring only a pencil and paper in its most basic form (of course, an electronic tablet, laptop computer, or other electronic device could be used for hand recording). For testing where budget or schedule precludes use of a data acquisition system, hand-held data may be the only option. However, even when data is electronically recorded, hand-held data is useful in providing a real-time record of the test and as a convenient way to reconstruct the test events after the flight. The quantitative, hand-held data can serve as a "backup" to the DAS or telemetered data. There may be instances when the electronically recorded data is unavailable or incomplete, making the hand-held data extremely valuable. Hand-held data may be quantitative or qualitative. Numerical readings may be taken from cockpit gauges and instruments, providing quantitative data collection. Qualitative data may consist of observations, comments, or descriptions which provide valuable insights of the test.

2.4.4   Roles of Test Pilot, Flight Test Engineer, and Flight Test Analyst

Typically, safe and successful flight testing requires the contributions of a multi-disciplinary team of pilots, engineers, technicians, and other support personnel. Each person contributes specific expertise and talents to the test team, so that the team as a whole possesses the required expertise and skills to successfully and safely conduct the flight test. For instance, a typical test team may require specialists in various operational, engineering, and management disciplines, such as aircraft piloting, maintenance, instrumentation, aerodynamics, stability, control, or project management. In this section, we briefly define the roles of the test pilot, flight test engineer, and flight test analyst.

2.4.4.1   Test Pilot

The test pilot flies the test aircraft and performs the test maneuvers and evaluations. They are proficient in the operation of the test aircraft and all of its systems. In addition to being able to fly the test aircraft for normal operations, they have the training and skills to accurately perform flight test techniques and test procedures in the aircraft. They have developed observational skills to perceive and analyze aircraft flight characteristics or handling qualities, from an engineering test perspective. Typically, the test pilot has flight experience in a wide variety of aircraft types, giving them the capability of adapting to a diverse range of new, and perhaps unexpected, flight characteristics of a test aircraft. Ideally, a test pilot has an engineering or scientific education or background, with an in-depth understanding of the theories, test techniques, and execution of flight test.

2.4.4.2   Flight Test Engineer

The flight test engineer (FTE) is typically a professionally trained engineer in aerospace, electrical, mechanical, or other engineering discipline. The FTE is significantly involved with the test planning and coordination among the various engineering disciplines and management. They must be knowledgeable with all of the tasks and techniques to be flown by the test pilot. Often, the FTE is responsible for the preparation and revision of the test cards. The FTE may need to serve as an aircrew member in the test aircraft, or as a discipline engineer in the ground station or control room. As an aircrew member, the FTE works closely with the test pilot in flying the test cards, pacing the test point items, and collecting hand-held data, as appropriate.

2.4.4.3   Flight Test Analyst

The flight test analyst is typically an engineering specialist, with expertise in a particular discipline, such as aerodynamics, performance, stability, control, instrumentation, avionics, or structures. The analyst may perform or analyze data from ground tests, such as wind tunnel tests or systems functional checks. The analyst may perform detailed analyses or computations prior to the flight test. Prior to the flight test, they typically provide test objectives and requirements for their discipline, and contribute to the development of the required test points and maneuvers. During the test, the analyst may sit at a console in a ground station or control room, monitoring the real-time data and pacing the test points for the given discipline. After the flight, the analyst reviews, analyzes, and reports on the flight data that was collected.

2.4.5   Flight Test Safety and Risk Assessment

All flight testing has some degree of risk associated with it. As the saying goes, the only way to reduce the flight test risk to zero is to stay on the ground. However, the risk associated with any flight test can and should be minimized, by applying good judgement and by performing extensive pre-test planning. One of the primary goals of any flight test is to maximize the amount of data collected while minimizing the risk. A flight test mishap can have significant impacts. Damage to or complete loss of the test vehicle can lead to lengthy delays or even cancellation of a project. Worst still, injury to people or loss of human life is devastating to everyone involved.

A cornerstone of flight test safety is the concept of an *incremental buildup* in the test program. The incremental buildup is a methodical process where the testing proceeds from the *known* to the *unknown*. In terms of the aircraft flight envelope, the buildup process involves an *envelope expansion* where the test progresses from test points at the lowest risk flight conditions to the higher risk areas of the flight envelope. Typically, this means starting at the center or "heart" of the subsonic flight envelope and moving outward, toward the edges of the envelope. Moving to the edges of the flight envelope may be an increase or a decrease in airspeed, altitude, dynamic pressure, or other flight condition parameter. For instance, envelope expansion to obtain aerodynamic stall data should start at higher airspeeds, and progress to slower and slower airspeeds. Envelope expansion for flutter testing should start at normal airspeeds and low dynamic pressures, and progress to high airspeeds and high dynamic pressure, possibly in diving flight. There should be no surprises in the buildup process, as the results of successive test points should track the test predictions and trends. If they do not, testing should be halted, and the unexpected results must be investigated and understood before proceeding.

Flight test safety begins in the planning phase of the tests. All possible mishap scenarios, hazards, and risks should be considered. Lessons learned from previous tests, especially similar types of tests, should be reviewed. There are methods by which the level of risk is quantified to some degree. Before we discuss some of these, let us get a bit more precise about some definitions of hazards and risk.

A *hazard* is defined as a condition, event, object, or circumstance that could lead to an unplanned or undesired event, such as an accident. It is an existing or potential condition that could cause injury, illness, or death to human beings or damage to, or loss of, the vehicle. Once the hazards have been identified, the *risks* associated with these hazards must be assessed, in a formal process called *risk assessment*. Hazard identification and risk assessment originated in the nuclear and chemical industries, where it helped to make these industries much safer.

Risk is the future impact of a hazard that is not controlled or eliminated. In assessing the *level of risk*, one must determine the *severity* and likelihood, or *probability*, of a hazard leading to an undesired outcome. The risk increases if either the severity or probability increases. There is some degree of uncertainty in all risk assessment, as it is often difficult to accurately predict the hazard severity or probability.

The risk assessment process starts with the identification of the hazards. There are a number of ways that hazards are identified, ranging from qualitative processes, such as brainstorming, to more data-driven techniques, such as failure modes and effects analysis (FMEA). The FMEA is a systematic review of the test vehicle components, assemblies, and subsystems to identify possible failure modes, their underlying causes, and their subsequent consequences or effects. Here again, a review of hazards, incidents, or accidents from past tests can provide valuable information and insights into the hazard identification process. Brainstorming is typically an unstructured, unbounded discussion among experts, usually those associated with the test. They utilize imaginative thinking to come up with hazard scenarios and what-ifs that could be unsafe. The structured what-if technique (SWIFT) may be used, involving a multi-disciplinary team of experts that use brainstorming applied to a systems-level description of the test equipment or vehicle. Identified hazards are often organized and documented in the form of threat hazard reports (THAs). The THA is a brief, one or two page form that succinctly states the hazard, its causes and potential outcomes, any risk controls or mitigations, and a final risk assessment of the hazard's severity and probability.

A tool used to help quantify the risk, associated with the hazards, is the *risk assessment matrix*, shown as Table 2.13. Each hazard is located in the matrix, based on its *severity* and *probability* of occurrence. The definition of the mishap severity categories and probability levels are somewhat subjective. An example of definitions for the mishap severity categories is given in Table 2.14. Note the subjectivity in the severity criteria definitions, especially in regards to the monetary values associated with the mishap severity. Table 2.15 shows an example of definitions for the mishap probability levels, where there is again subjectivity in the definition of the quantities.

The population of the risk assessment matrix with the identified hazards is open to interpretation and may be somewhat subjective, although quantitative data, based on historical trends or failure analyses can help to remove some of this subjectivity. The combinations of severity and probability are categorized in terms of level of risk, ranging from high (red) to low (green). Almost always,

Table 2.13 Typical flight test risk assessment matrix.

Severity / Probability	Catastrophic (1)	Critical (2)	Marginal (3)	Negligible (4)
Frequent (A)	High (red)	High (red)	Serious (orange)	Medium (yellow)
Probable (B)	High (red)	High (red)	Serious (orange)	Medium (yellow)
Occasional (C)	High (red)	Serious (orange)	Medium (yellow)	Low (green)
Remote (D)	Serious (orange)	Medium (yellow)	Medium (yellow)	Low (green)
Improbable (E)	Medium (yellow)	Medium (yellow)	Medium (yellow)	Low (green)

Table 2.14 Example mishap severity category descriptions.

Severity	Category	Criteria
Catastrophic	1	Loss of life or permanent total disability, or material loss or damage in excess of $1M.
Critical	2	Permanent partial disability or injury resulting in hospitalization, or material loss or damage exceeding $200K but less than $1M.
Marginal	3	Injury resulting in one or more lost work day(s), or material loss or damage exceeding $10K but less than $200K.
Negligible	4	Injury not resulting in a lost workday, or material loss or damage exceeding $2K but less than $10K.

Table 2.15 Example mishap probability level descriptions.

Probability	Level	Criteria
Frequent	A	Likely to occur often. Probability of occurrence greater than 10^{-1}.
Probable	B	Will occur several times. Probability of occurrence less than 10^{-1} but greater than 10^{-2}.
Occasional	C	Likely to occur sometime. Probability of occurrence less than 10^{-2} but greater than 10^{-3}.
Remote	D	Unlikely but possible to occur. Probability of occurrence less than 10^{-3} but greater than 10^{-6}.
Improbable	E	So unlikely, can be assumed may not occur. Probability of occurrence less than 10^{-6}.

any hazards that have been assessed to be in the high (red) risk level are deemed to be of too high a risk to perform the test. This may also be true for any hazards in the serious (orange) range. Hazard mitigations are used to reduce the risk to *acceptable risk* levels. The *acceptable risk* is the threshold below which the risk is tolerated and testing can be performed. Above the acceptable risk level, testing cannot be conducted. Usually, this threshold of acceptable risk is set by the senior management of an organization or by accepted company policies or procedures.

To reduce the risk level of a hazard, *risk mitigations* are applied. Risk mitigation might be accomplished by changing the design of a component or of the complete vehicle, although this may be time and cost prohibitive, especially if the component or vehicle is already built. Often, the risk mitigation may be a *minimizing procedure*, where the risk is reduced or managed by implementation of a new procedure. The buildup flight test approach is considered a minimizing procedure. Other minimizing procedures might include setting test or vehicle limits, such as limiting airspeed or angle-of-attack to lower values if the flight characteristics are unknown at the higher values.

Despite its inherent subjectivity, the risk assessment matrix is a valuable tool in assessing the risk levels associated with a test. By capturing the severities and probabilities of all of the test hazards in the single matrix, a better appreciation is obtained of the overall test risk.

An example of the application of a risk assessment matrix is shown in Table 2.16, where the risks associated with "everyday life" have been assessed. An everyday life hazard of getting a paper cut has been assessed with a risk level of 4A, with a severity category of negligible and a probability level of frequent. A shark attack hazard has a 1E risk level rating, with a catastrophic severity and improbable likelihood of occurrence. There may be subjectivity in the placement of these hazards in the matrix, as one could have a differing opinion as to the frequency that one gets a paper cut or whether a shark attack is necessarily fatal.

Table 2.16 Example risk assessment matrix for "everyday life".

Severity / Probability	Catastrophic 1	Critical 2	Marginal 3	Negligible 4
Frequent A				Paper cut
Probable B				
Occasional C			Trip and fall	
Remote D		Automobile accident		
Improbable E	Shark attack			

Table 2.17 Example risk assessment matrix for a hypothetical flight test.

Severity / Probability	Catastrophic 1	Critical 2	Marginal 3	Negligible 4
Frequent A				
Probable B				Avionics overheating
Occasional C			Hard landing	
Remote D	Loss of control			
Improbable E				

An example risk assessment matrix for a hypothetical flight test is shown in Table 2.17. Suppose that this hypothetical test is of a new aircraft, where there are known cooling issues with the avionics, such that overheating of the equipment is probable. However, mitigations are in place to easily identify the issue and shut the equipment down before any damage is done, resulting in a risk of level 4B. Perhaps it is predicted that this new aircraft may be a bit difficult to land, hence the possibility exists of an occasional hard landing. The landing gear may have been built to be especially sturdy to account for this, so the hard landing risk is given a level of 3C. Finally, maneuvers may be planned during the test program such that loss of control of the aircraft could occur, which could result in a catastrophic loss of the aircraft. However, test limits have been established far from the predicted loss of control boundaries and a careful, incremental test buildup is followed, such that the probability of loss of control is remote, giving this risk a level of 1E.

References

1. *Aerodynamics*, USAF Test Pilot School, Edwards AFB, California, January 2000.
2. *Aircraft Performance*, USAF Test Pilot School, Edwards AFB, California, January 2000.
3. Anderson, J.D., Jr, *Introduction to Flight*, 4th edition, McGraw-Hill, Boston, Massachusetts, 2000.
4. Anderson, J.D., Jr, *Fundamentals of Aerodynamics*, 3rd edition, McGraw-Hill, New York, 2001.
5. Brown, S.W. and Bradley, D., "Uncertainty Analysis for Flight Test Performance Calculations," AIAA 1st Flight Test Conference, Las Vegas, Nevada, 11–13 November 1981.
6. Clarke, A.C., "Extraterrestrial Relays – Can Rocket Stations Give Worldwide Radio Coverage?" *Wireless World*, October, 1945, pp. 305–308.
7. Coleman, H.W. and Steele, W.G., *Experimentation and Uncertainty Analysis for Engineers*, 2nd edition, John Wiley & Sons, Inc., New York, 1999.
8. Department of the Air Force, "Designating and Naming Defense Military Aerospace Vehicles," AFJI 16-401, NAVAIRINST 8800.3A, AR 70-50, 14 March 2005, http://www.e-publishing.af.mil.
9. Department of Defense, "Model Designation of Military Aerospace Vehicles," DoD 4120.15-L, 12 May 2004.
10. Federal Aviation Administration, US Department of Transportation, *Methods Techniques and Practices – Aircraft Inspection and Repair*, Advisory Circular (AC) **43**.13-1 (as revised), Oklahoma City, Oklahoma, 8 September 1998.
11. Federal Aviation Administration, US Department of Transportation, *Balloon Flying Handbook*, FAA-H-8083-11A, Oklahoma City, Oklahoma, 2008.
12. Federal Aviation Administration, US Department of Transportation, Code of Federal Regulations, Federal Aviation Regulations, 14 CFR Part 1, Definitions and Abbreviations, 2014.
13. Federal Aviation Administration, US Department of Transportation, Code of Federal Regulations, Federal Aviation Regulations, 14 CFR Part 401, Organization and Definitions, 2014.
14. Federal Aviation Administration, US Department of Transportation, *Instrument Flying Handbook*, FAA-H-8083-15B, Oklahoma City, Oklahoma, 2012.
15. Federal Aviation Administration, US Department of Transportation, *Risk Management Handbook*, FAA-H-8083-2, Oklahoma City, Oklahoma, 2009.
16. Federal Aviation Administration, US Department of Transportation, *Rotorcraft Flying Handbook*, FAA-H-8083-21, Oklahoma City, Oklahoma, 2000.

17. Gallagher, G.L., Higgins, L.B., Khinoo, L.A., and Pierce, P.W., *Fixed Wing Performance*, USNTPS-FTM-NO. 108, US Naval Test Pilot School, Patuxent River, Maryland, 30 September 1992.

18. Gessow, A. and Meyers, G.C., Jr, *Aerodynamics of the Helicopter*, Frederick Ungar Publishing Company, New York, New York 1952.

19. Griffin, M.D. and French, J.R., *Space Vehicle Design*, 2nd edition, AIAA Education Series, American Institute of Aeronautics and Astronautics, Inc., Reston, Virginia, 2004.

20. Hurt, H.H., Jr, *Aerodynamics for Naval Aviators,* US Navy NAVWEPS 00-80T-80, US Government Printing Office, Washington, DC, January 1965.

21. Hybrid Air Vehicles, Ltd., http://www.hybridairvehicles.com/

22. Jackson, P. (ed.), *Jane's All the World's Aircraft: 2002–2003*, Jane's Information Group Limited, Coulsdon, Surrey, United Kingdom, 2002.

23. Jenkins, D.R., Landis, T., and Miller, J., *"American X-Vehicles: An Inventory, X-1 to X-50" NASA SP-2003-4531*, US Government Printing Office, Washington, DC, June 2003.

24. Karamcheti, K., *Principles of Ideal-Fluid Aerodynamics*, 2nd edition, Robert E. Krieger Publishing Company, Huntington, New York, 1980.

25. Leishmann, J.G., *Principles of Helicopter Aerodynamics*, 1st edition, Cambridge University Press, New York, New York, 2000.

26. Lilienthal, O., *Birdflight as the Basis of Aviation*, translated from the 2nd edition by A.W. Isenthal, Markowski International Publishers, Hummelstown, Pennsylvania, 2001.

27. Lucian of Samosata, *True History*, translated by Francis Hicks, A.H. Bullen, London, England, 1894.

28. Ludwig, C.G., "Flight Test Evaluation of a Low-Cost, Compact, and Reconfigurable Airborne Data Acquisition System Based on Commercial Off-the-Shelf Hardware," Master's Thesis, University of Tennessee, 2009.

29. "Mars Climate Orbiter Press Kit," NASA-JPL, September 1999.

30. "Mars Climate Orbiter Fact Sheet," NASA-JPL, http://mars.jpl.nasa.gov/msp98/orbiter/fact.html, June 2014.

31. "Mars Climate Orbiter Mishap Investigation Board Phase I Report," NASA, November 10, 1999.

32. *"History of the X-Plane Program"* NASA, http://history.nasa.gov/x1/appendixa1.html, March 2015.

33. "The NIST Reference on Constants, Units, and Uncertainty," National Institute of Standards and Technology (NIST), Physical Measurement Laboratory, http://physics.nist.gov/cuu/Units/units.html, last update June 2011.

34. Stolicker, F.N., *Introduction to Flight Test Engineering*, AGARD-AG-300, Vol. **14**, AGARD, France, 1995.

35. Talay, T.A., *Introduction to the Aerodynamics of Flight, NASA SP-367*, US Government Printing Office, Washington, D.C., 1975.

36. Taylor, B.N. and Thompson, A., editors, "The International System of Units (SI)," NIST Special Publication 330 (SP330), National Institute of Standards and Technology, Gaithersburg, Maryland, March 2008.

37. Ward, D.T., and Strganac, T.W., *Introduction to Flight Test Engineering*, 2nd edition, Kendall Hunt Publishing Company, Dubuque, Iowa, 2001.

38. Young, H.D. and Freedman, R.A., *University Physics*, 11th edition, Addison Wesley, San Francisco, California, 2004.

39. Young, J.O., *Meeting the Challenge of Supersonic Flight* Air Force Flight Test Center History Office, Edwards AFB, California, 1997.

Problems

1. The components of velocity of an aircraft, in the body axis system, are $u = 120.7$ mph, $v = 3.12$ ft/s, and $w = 11.63$ ft/s. Calculate the magnitude of the velocity, the angle-of-attack, and the angle-of-sideslip.

2. A flight test aircraft is flying at a velocity of 228.1 kt. The angle-of-attack and angle-of-sideslip are measured as 3.28° and 1.27°, respectively. Calculate the velocity components in the body axis coordinate system.

3. A helicopter is in a constant airspeed vertical climb. Draw a free-body diagram for the helicopter at this flight condition. Obtain an expression for the thrust-to-weight ratio of this vertically climbing helicopter.

4. A hybrid airship is in a constant airspeed descent with its engine not producing thrust. Draw a free-body diagram for the airship at this flight condition. Obtain an expression for the lift-to-drag ratio of this descending hybrid airship.

5. A Northrop T-38 jet is in a constant airspeed climb with a flight path angle of 30°. Obtain an expression for the flight path angle in terms of the lift, drag, and thrust.

6. A Northrop T-38 *Talon* is flying in steady level flight. The aircraft weight is 10,060 lb and the engines are producing a total thrust of 3935 lb. The T-38 wing reference area, S, is 170 ft^2. Calculate the lift-to-drag ratio, the thrust-to-weight ratio, and the wing loading at this flight condition.

7. An aircraft is flying at an altitude of 13,100 m and an airspeed, V_∞, of 670.3 km/h. The air temperature at this altitude is 216.5 K. Calculate the speed of sound, a_∞, in units of m/s and the Mach number, M_∞.

8. An aircraft is flying at an altitude of 27,000 ft and a Mach number, M_∞, of 0.58. The air temperature at this altitude is 47.3°F. Calculate the speed of sound, a_∞, in units of ft/s and the airspeed, V_∞, in units of mph.

9. A Cessna 172, small general aviation, four-seat aircraft is flying in straight-and-level flight at a constant airspeed of 121 knots and an altitude of 5000 ft, where the freestream air density, ρ_∞, is 0.0020482 slug/ft^3. The aircraft lift coefficient at this flight condition is 0.3174. Calculate the dynamic pressure, angle-of-attack, and the weight of the aircraft at this flight condition. (As shown in Chap. 2, we can relate the lift coefficient, C_L, to the angle-of-attack, α, using the approximate linear relation $C_L = 2\pi\alpha$, where the angle-of-attack is in radians.) The following table of specifications is provided for the aircraft. Given the aircraft specifications and flight condition, comment on whether the values for the lift coefficient and angle-of-attack "make sense".

Parameter	Value
Aircraft length	27 ft 2 in
Wing span	36 ft 1 in
Wing area	174 ft^2
Maximum takeoff weight	2550 lb
Total fuel capacity	56 gal
Maximum cruise speed	124 kt

10. A Cessna 172, small general aviation, four-seat aircraft is flying in straight-and-level flight at a constant airspeed of 62 knots and an altitude of 3150 ft, where the freestream air density, ρ_∞, is 0.0021657 slug/ft^3. The aircraft weight at this flight condition is 2510 lb. Calculate the dynamic pressure, lift coefficient, and angle-of-attack of the aircraft at this flight condition. (The lift coefficient, C_L, can be related to the angle-of-attack, α, using the approximate linear relation $C_L = 2\pi\alpha$, where the angle-of-attack is in radians.). Use the table of aircraft specifications provided in Prob. 9. Given the aircraft specifications and flight condition, comment on whether the values for the lift coefficient and angle-of-attack "make sense".

11. Draw the V-n diagram for a general aviation trainer aircraft with the following specifications. Assume sea level conditions.

Parameter	Value
Weight	1670 lb
Wing area	159.5 ft^2
1 g stall speed	43 kt
Never exceed airspeed	149 kt
Positive limit load factor	+4.4 g
Negative limit load factor	−1.76 g

12. For the Beechcraft *Bonanza* in Example 2.10, a 176 lb pilot and a 147 lb co-pilot are seated in the cockpit at FS 79. The aircraft has 57 gallons (280 liters) of usable fuel in the wing tanks at FS 75. If the *Bonanza* takes off, flies for 1.2 hours, then lands, calculate the zero fuel weight, the takeoff weight and center of gravity location, and the landing weight and center of gravity location. Assume that the aviation gasoline has a weight of 6.0 lb (2.72 kg) per gallon with a fuel burn rate of 15.1 gallons per hour (57.2 liters/h). Also, determine whether the takeoff and landing weights and center of gravity locations are within the allowable limits.

3

Aerodynamics

The Flight of Icarus by Flemish Baroque painter Jacob P. Gowy. (Source: *Jacob P. Gowy, PD-100-old.*)

(Daedalus) laid down lines of feathers, beginning with the smallest, following the shorter with longer ones … Then he fastened them together with thread at the middle, and bees' wax at the base, and, when he had arranged them, he flexed each one into a gentle curve, so that they imitated real bird's wings.

Introduction to Aerospace Engineering with a Flight Test Perspective, First Edition. Stephen Corda.
© 2017 John Wiley & Sons Ltd. Published 2017 by John Wiley & Sons Ltd.
Companion Website: www.wiley.com/go/corda/aerospace_engg_flight_test_persp

When he had put the last touches to what he had begun, the artificer balanced his own body between the two wings and hovered in the moving air. He instructed the boy as well, saying 'Let me warn you, Icarus, to take the middle way, in case the moisture weighs down your wings if you fly too low, or if you go too high, the sun scorches them. At the same time as he laid down the rules of flight, he fitted the newly created wings on the boy's shoulders ... He gave a never to be repeated kiss to his son, and lifting upwards on his wings ... He urged the boy to follow, and showed him the dangerous art of flying, moving his own wings, and then looking back at his son.

... the boy began to delight in his daring flight, and abandoning his guide, drawn by desire for the heavens, soared higher. His nearness to the devouring sun softened the fragrant wax that held the wings and the wax melted, he flailed with bare arms, but losing his oar-like wings, could not ride the air. Even as his mouth was crying his father's name, it vanished into the dark blue sea.

Ovid, Metamorphoses Book 8, *Daedalus and Icarus*[1,2]

3.1   Introduction

Aerodynamics is sometimes considered the *science of flight*. Perhaps more so than the other aerospace engineering disciplines, aerodynamics captures the essence of heavier-than-air flight, that of a body moving through the air and generating *lift* greater than its weight. This lift does not come free, as the movement of the body through the air creates a *drag* force, which opposes the body's motion. To overcome the drag force, a means of *propulsion* is usually required, this being the topic of the next chapter. Often, the goal of aerodynamics is to design shapes that create large amounts of lift with as low a drag as possible. Aerodynamic shapes are considered by some to be quite beautiful and elegant, with their gently sloping surfaces, almost an art form, in themselves.

More technically speaking, aerodynamics is the science of the flow of air and its interaction with bodies such as wings, automobiles, buildings, or airplanes. Aerodynamic flows may be external flows, such as over the wing of an airplane or the hood of a car, or internal flows, such as inside a jet engine or wind tunnel. The flow of air over a body produces forces and moments on the body. Aerodynamics often focuses specifically on the *lift* and *drag* forces produced on the body.

Aerodynamics is a subset of *fluid dynamics*, which encompasses the motion of *liquids* and *gases*, in general. Other subsets of fluid dynamics include *hydrodynamics*, the study of the flow of water or liquids, and *gas dynamics*, the study of gas flows. Fluid dynamics is a broad subject with a diverse range of applications in engineering, including astrophysics, meteorology, oceanography, and others. In addition to the flight of airplanes and missiles through the atmosphere, fluid dynamics applies to the motion of ships and submarines in the ocean, the swimming of microscopic organisms in a pond, the flow of blood in our circulatory system, and the entry of a space probe in a planetary atmosphere, to name just a few. In this chapter, we explore many aspects of aerodynamics, related to flows over aerospace vehicles. We will see that the fundamental characteristics of the flow change as the speed of the vehicle increases. First, we learn about the fundamental properties of a fluid.

---

[1] From Ovid, Metamorphoses, (2004), translated by A.S. Kline, Borders Classics.

[2] In Greek mythology, Daedalus and his son, Icarus, are imprisoned on the Greek island of Crete. Daedalus constructs wings of feathers, held together with wax, to escape the island by flight. He teaches Icarus to fly, but warns him of the dangers of flying too low, where the sea's dampness will make the wings too heavy for flight, or flying too high, where the sun's heat will melt the wax. They fly off the island together and Icarus, in his exuberance of flight, soars higher and higher, until the sun melts the wax and he tragically falls into the sea and drowns.

3.2   Fundamental Physical Properties of a Fluid

We all know that a solid is fundamentally different from a fluid, whether it is a liquid or gas. If we place a fluid into a container, it "spreads out" and fills the container, whereas a solid does not. From a molecular perspective, the atoms and molecules of a solid are packed closely together, while the spacing is much larger for a gas. The atoms and molecules of a solid form a rigid geometric structure that is held together by strong intermolecular forces. In a fluid, the influence of intermolecular forces is weak, allowing the relative motion of the molecules, resulting in *fluidity*.

We need to precisely define several properties of a fluid to help us in our quantitative discussions of aerodynamics. These properties are grouped together as *thermodynamic*, *kinematic*, or *transport* properties. Kinematic properties have to do with the fluid flow or the *flow in motion*. Kinematic properties of a fluid flow include the linear and angular velocities, linear and angular accelerations, and the strain rate. (There are additional kinematic flow properties that could be included, but these are beyond the scope of the text.) Thermodynamic properties are related to the *thermodynamic state* of the fluid. Pressure, temperature, and density are familiar thermodynamic properties of a fluid. Transport properties have to do with the movement or *transport of mass, momentum,* and *heat* in a fluid. The three transport properties are the coefficients of diffusion, viscosity, and thermal conductivity, which are related to the transport of mass, momentum, and heat, respectively. We briefly discuss each of these types of flow properties below. First, we define a *fluid element*, a concept that is used frequently in discussing fluid properties.

3.2.1   The Fluid Element

The *fluid element* is not a physical property of the fluid; it is a concept or model that helps us to visualize and discuss fluid motion. A fluid element is a hypothetical cube of fluid, with infinitesimal mass, m, and with infinitesimally small dimensions Δx, Δy, and Δz, as shown in Figure 3.1.

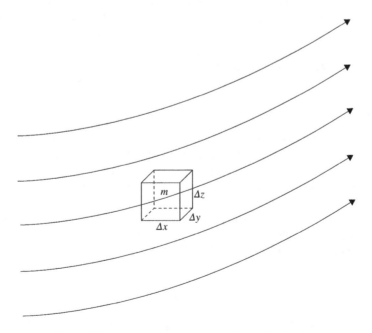

Figure 3.1   A fluid element moving along with the flow.

The fluid element is subjected to forces that act on its entire mass, called *body forces*, and forces that act only at its surface, called *surface forces*. The force due to gravity, or the fluid element's weight, is an example of a typical body force. The body forces are proportional to the mass or volume of the fluid element. Examples of surface forces are the pressure and viscous shear stress acting over a surface area of the fluid element. The force due to the pressure acts in a direction normal to the surface area, while the viscous force acts tangentially to the area.

The fluid element moves along with the flow and has six degrees of freedom in three-dimensional space: three linear translations and three angular rotations. This does not mean that the fluid element is always moving and rotating in all six ways, as this depends on the nature of the flow, however, it has the freedom to do so. The motion of the fluid element is governed by Newton's laws of motion, more specifically, by Newton's second law of motion.

3.2.2   Thermodynamic Properties of a Fluid

Consider an airplane sitting on the ground preparing for takeoff. The properties, such as the pressure, temperature, and density, at a point on the wing are related to each other using the laws of thermodynamics. After the airplane takes off and air is flowing over the wing, the laws of thermodynamics still relate these properties at the point on the wing. In addition to the pressure, temperature, and density, other thermodynamic flow properties include the enthalpy, entropy, internal energy, specific heat, bulk modulus, and the coefficient of thermal expansion. The *thermodynamic state* is uniquely defined by specification of any two of these thermodynamic properties. All of the other properties are obtained from any two independent thermodynamic properties, using an appropriate *equation of state*, a subject of a later section.

Consider, once again, the wing that is flying through the air. If we were moving along with the wing and measured the thermodynamic flow properties, such as the pressure, density, and temperature, we would be measuring the *static* pressure, density, and temperature. These *static conditions* are due to the random motion of the molecules, rather than due to the directed motion or velocity of the wing. Relating this concept to the fluid element, moving along with the flow, in Figure 3.1, the static conditions are those that are measured as we move along with the fluid element. Later, we define another type of condition, called *stagnation* or *total* conditions, which are defined when the flow is brought to zero velocity without any losses.

3.2.2.1   Pressure

We think of the air around us as a continuous, uniform gas. In reality, the air is made up of discrete molecules of nitrogen, oxygen, and other trace gas species. Even in the still air around you, these molecules are literally bouncing all over the place, colliding with each other, the walls of the room, the book that you are reading, and even you. The atmospheric pressure on our bodies is due to the impact of these molecules against our skin.

The pressure is defined as the normal force per unit area due to the time rate of change of the momentum of the gas molecules impacting our skin, or some other surface. The pressure, p, over an area, A, is defined as the force, F, divided by the area, as given by

$$p = \frac{F}{A} \tag{3.1}$$

Consider, once again, the fluid element shown in Figure 3.1 and one side of the cube of surface area, dA. In the limit, as the surface area, dA, goes to zero, the pressure is given by

$$p = \lim_{dA \to 0} \left(\frac{dF}{dA} \right) \qquad (3.2)$$

where this now represents the pressure at a point in the flow. Thus, the pressure is defined as a *point property*, which can vary from point to point in the flow. As such, a finite area need not be identified to specify a pressure at a point in the flow. Typical consistent units of pressure, in English and SI units, are lb/ft² and N/m² (also defined as the pascal, Pa), respectively.

The pressure, defined above, is the *static pressure* in a flow. It is the pressure that you would feel or measure if you were moving along with the flow, due to the random motion and collisions of the fluid molecules.

3.2.2.2 Specific Volume and Density

Consider, once again, the fluid element shown in Figure 3.1. We can define its mass, m, as its weight, W, divided by the acceleration due to gravity, g.

$$m = \frac{W}{g} \qquad (3.3)$$

We are sometimes interested in quantities *per unit mass*. For instance, we sometimes use the volume per unit mass, v, or *specific volume*, simply defined as the volume divided by the mass.

$$v = \frac{\mathcal{V}}{m} \qquad (3.4)$$

Quantities per unit mass are called *specific* quantities.

The density, ρ, of the fluid element is defined as its mass, m, per unit volume, $\mathcal{V}$, given by

$$\rho = \frac{m}{\mathcal{V}} \qquad (3.5)$$

In the limit, as the volume, $d\mathcal{V}$, of the fluid element of incremental mass, dm, goes to zero, the density is given by

$$\rho = \lim_{d\mathcal{V} \to 0} \left(\frac{dm}{d\mathcal{V}} \right) \qquad (3.6)$$

where this represents the density at a point in the flow. Therefore, the density can also be defined as a point property in the flow that varies from point to point. As such, a finite volume need not be identified to specify a density at a point in the flow. Typical consistent units of density, in English and SI units, are slugs/ft³ and kg/m³, respectively. Similar to the static pressure, this is the static density that is measured when moving along with the flow, due to the random motion of the molecules.

The specific volume, v, is defined as the inverse of the density.

$$v = \frac{1}{\rho} \qquad (3.7)$$

3.2.2.3 Temperature

Imagine a container filled with air at room temperature. The air in the container is composed of molecules of different species of gases, such as nitrogen, oxygen, and others species. The gas molecules are in random motion inside the container, bouncing around, hitting each other and the walls of the container. If we place a burner underneath the container, the molecules get more "excited" and start moving faster inside the container; the *translational energy* of the molecules has increased. The temperature, T, is a measure of this translational energy of the gas molecules.

The increase in the translational energy also increases the random kinetic energy of the gas molecules, thus the temperature can also be viewed as a measure of the random kinetic energy of the molecules. If the temperature is raised, the gas molecules get more excited and their random kinetic energy increases. Similarly, if the temperature is lowered, the random kinetic energy of the gas molecules decreases. More precisely, the average kinetic energy of the fluid molecules, KE, is related to the fluid temperature, T, by

$$KE = \frac{3}{2}kT \tag{3.8}$$

where k is the Boltzmann constant, which has a value of 1.38×10^{-23} J/K (5.65×10^{-24} ft · lb/°R). Equation (3.8) is rigorously derived using the *kinetic theory of gases*. Typical consistent units of temperature, in English and SI units, are degrees Rankine, °R, and kelvins, K, respectively. Similar to the static pressure, this is the *static temperature* that is measured when moving along with the flow, due to the random motion of the molecules.

3.2.2.4   Standard Conditions

We are often interested in the thermodynamic properties of air at a specific reference condition at mean sea level, called *standard sea level conditions (SSL)*, or simply *standard conditions*. The values of selected properties of air at standard conditions are given in Table 3.1. These standard conditions are derived from a *standard atmosphere*, which is developed in Chapter 5.

Recalling basic thermodynamics, the thermodynamic state is uniquely defined by specifying two independent thermodynamic variables, such as the pressure and temperature or the pressure and density. The other thermodynamic properties are obtained from the two specified properties by applying an appropriate equation of state. Therefore, the standard conditions for air are defined by specifying the standard temperature and pressure or two other independent thermodynamic variables.

Table 3.1 includes the value of the speed of sound at standard conditions, due to the importance of this quantity in aerodynamics. Other selected *transport properties* of air at standard conditions are also given, to be discussed in an upcoming section. Get familiar with the values of air at standard conditions given in Table 3.1, as we see and use them often in aerospace engineering. They serve as a reference condition for other conditions that we investigate and analyze.

3.2.3   Kinematic Properties of a Flow

The flow properties of velocity and acceleration are probably quite familiar. We all have a "feel" for what it means when we say the wind is blowing at 20 mph (32.2 km/h). However, this is a reference

Table 3.1   Values of selected properties of air at standard conditions.

Property	Symbol	SI units	English units
Density	ρ_{SSL}	1.225 kg/m^3	0.002377 slug/ft^3
Pressure	p_{SSL}	101,325 N/m^2	2116 lb/ft^2
Temperature	T_{SSL}	288 K (15 °C)	519 °R (59 °F)
Speed of sound	a_{SSL}	340.2 m/s	1116.6 ft/s
Dynamic viscosity	μ_{SSL}	17.89×10^{-6} kg/(m · s)	0.3737×10^{-6} slug/(ft · s)
Thermal conductivity	k_{SSL}	0.02533 J/(m · s · K)	4.067×10^{-6} Btu/(ft · s · °R)

to a wind *speed*, not a velocity. Velocity is a vector term that has a magnitude and a direction. We could more precisely reference a wind velocity of 20 mph *from the North*, to give it a direction as well as a magnitude. Normally, we use a coordinate system to define vectors. We can define the general linear velocity vector, $\vec{V}$, for a steady flow in Cartesian coordinates, (x, y, z), as

$$\vec{V} = \vec{V}(x, y, z) = \vec{V}_x + \vec{V}_y + \vec{V}_z = u\hat{\imath} + v\hat{\jmath} + w\hat{k} \tag{3.9}$$

where $\vec{V}_x$, $\vec{V}_y$, and $\vec{V}_z$ are the x, y, and z velocity vectors, respectively, u, v, and w, are the x, y, and z scalar components of the velocity, and $\hat{\imath}$, $\hat{\jmath}$, and $\hat{k}$ are the unit vectors in the x, y, and z directions.

Returning to our fluid element concept, we can imagine a fluid element moving through a flow field and, as it does, its velocity changes in magnitude and direction. As the fluid element passes through a point in the flow with coordinates (x_1, y_1, z_1), from Equation (3.9), its velocity is given by

$$\vec{V}_1 = u_1\hat{\imath} + v_1\hat{\jmath} + w_1\hat{k} \tag{3.10}$$

When the fluid element moves to a new point with coordinates (x_2, y_2, z_2), its velocity is

$$\vec{V}_2 = u_2\hat{\imath} + v_2\hat{\jmath} + w_2\hat{k} \tag{3.11}$$

Therefore, we see that Equation (3.9) defines a *velocity field* of the flow, where the velocity can vary from point to point in the flow. Based on this, the velocity is called a *point property* of the flow. The magnitude of the velocity at any point in the flow is the square root of the sum of the squares of the components, as given by

$$|\vec{V}| = \sqrt{u^2 + v^2 + w^2} \tag{3.12}$$

The other kinematic properties of angular velocity, linear acceleration, and angular acceleration are defined in a similar manner as just outlined for the linear velocity.

3.2.4 Streamlines, Pathlines, and Flow Visualization

If we "watch" a particular fluid element as it moves through an unsteady flow, we can trace its path as a function of space and time. This path is called the fluid element's *trajectory* or *pathline*. If the flow does not change with time – that is, the flow is steady – the fluid elements follows a fixed path in space that does not vary with time. For a steady flow, the pathlines of the fluid elements are called *streamlines*. Streamlines are curves that are tangential to the local velocity vector everywhere along their lengths, as shown in Figure 3.2. The streamlines (dotted lines) are tangential to the local velocity vectors, $\vec{V}(x, y)$, everywhere in the flow. Since the streamlines are tangential to the velocity, there is no flow *perpendicular* to a streamline. Streamlines are a useful way to *visualize* an aerodynamic flow, which can tell us a lot about the flow over a body.

Since air and many other fluids are transparent, it is usually not possible to see the flow streamlines or pathlines. *Flow visualization techniques* are used to make the flow visible, such as by introducing smoke, colored dyes, or other markers into the flow or onto a body surface. Flow visualization may be categorized as either *on-surface* or *off-surface* techniques.

On-surface methods make flow patterns and streamlines visible on the surface of a body. They may also be used to quantify surface pressures or temperatures. Typical on-surface flow visualization techniques include oil flow, pressure- or temperature-sensitive liquid crystals, optical imaging, and mechanical tufts. These colored agents respond to the surface shear stress and mark the flow pattern over the surface. These types of surface visualization are especially useful in identifying regions of separated flow.

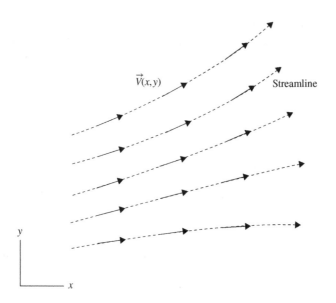

Figure 3.2   Streamlines are tangential to the local velocity $\vec{V}(x, y)$ everywhere in the flow.

Off-surface or tracer flow visualization methods use smoke, dyes, microspheres, or other very low mass particles, injected into the flow to trace the fluid streamlines or pathlines. These tracer techniques are based on the assumption that the particles follow the streamlines or pathlines of the flow. For more accurate flow tracing, the density of the particles should match the flow density. The tracer particles may be visible to the naked eye or they may be illuminated with lasers or other light sources to make them visible or even fluorescent. Optical methods are also used for off-surface flow visualization and are usually non-intrusive since the flow can be viewed from a distance.

The flow over a model of an F-18 aircraft in a water tunnel is shown in Figure 3.3, made visible using colored dye that is injected into the flow from several small ports on the model surface. (A water tunnel is similar to a wind tunnel, to be discussed in Section 3.7.4, except that a model is immersed in a flow of water rather than air.) Sometimes, sheets of laser light are used to illuminate the particles, making planes of the flow field highly visible.

In addition to the qualitative definition of the flow field, tracer methods can provide quantitative flow information by tracking the injected particles. Techniques, such as particle image velocimetry, have become powerful, non-intrusive ways of obtaining velocity data in a flow.

A caveat must be added to the earlier statement that air flow is usually invisible. There are examples of *natural flow visualization*, where the flow over a vehicle is made visible by natural phenomena, such as condensation of water vapor or optical effects of sunlight. An excellent pictorial description of natural flow visualization on aircraft is found in [19]. Details of in-flight flow visualization methods are discussed in the next section.

3.2.5   FTT: In-Flight Flow Visualization

The ability to visualize the flow over a body in flight is extremely useful in understanding the aerodynamics of the flow or interpreting quantitative measurements. As the saying goes, "a picture is worth a thousand words." The present flight test technique discusses a few of the flow visualization techniques available for use in flight. In-flight flow visualization techniques are not limited to subsonic flight, and several can be used in transonic and supersonic flight test.

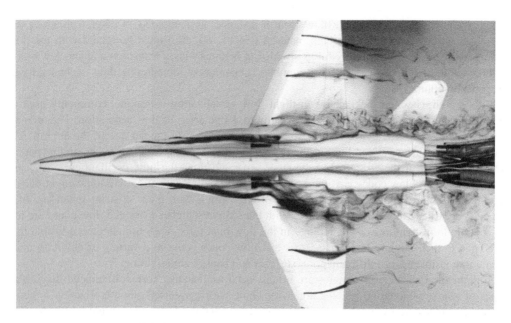

Figure 3.3   Flow over a model of an F-18 aircraft, visualized using colored dye in a water tunnel. (Source: *NASA*.)

Flow visualization techniques have been used for many years in ground test facilities, including wind tunnels and water tunnels, to visualize a variety of flow features. Many of the flow visualization techniques, which have been used on the ground, have been applied to the flight environment. The installation of flow visualization equipment in an aircraft can be more difficult than in a ground facility, due to space limitations, power requirements, inaccessibility to certain aircraft areas, or other constraints. The operation of flow visualization systems must also be compatible with the flight environment, such as low pressures and low temperatures at high altitudes or high vibrations associated with propulsion systems, or air turbulence. A means of photographically observing or recording the flow visualization in flight is usually desired. This is accomplished with cameras mounted on the test aircraft or on another aircraft observing the test aircraft. Several different types of in-flight flow visualization methods are described below.

Dyed oils are usually painted on an aircraft surface prior to flight. In flight, the oil flows on the surface, literally "painting" a picture of the surface flow. The oil thins, thickens, or puddles in different regions of the flow, depending on the local flow phenomena. For example, a thick, puddled line of oil may indicate the presence of a shock wave, or pooling of oil in an area may indicate a region of separated flow. The oil viscosity is adjusted to provide the desired flow properties at the flight conditions, such as different airspeeds or altitudes. The oil dye color is selected to provide the best contrast and definition against the particular surface color.

Liquid crystals are materials that change their reflective color as a function of shear forces or temperature. The patterns of changing color on a surface are used to identify flow phenomena such as boundary layer transition or shock waves. The liquid crystals are mixed with a solvent and sprayed onto the aircraft surface in a thin film prior to flight. When exposed to the shear forces or temperatures in flight, the color change of the liquid crystals is photographically recorded. The liquid crystals are calibrated to obtained quantitative data of temperature fields.

An example of in-flight optical imaging for flow visualization is the use of infrared imaging for the measurement of surface temperatures. This type of imaging is non-intrusive, since the imaging equipment can view the flow from a distance, as compared to other techniques that have been discussed, where a foreign material (smoke, oil, etc.) must be mixed with the flow. Infrared imaging has been successful in visualizing boundary layer transition and shock waves in flight, even at higher supersonic Mach numbers. Quantitative temperature data is obtained with proper calibration.

Tufts are an easy and inexpensive in-flight flow visualization technique, commonly used on aircraft for surface flow visualization, especially to define areas of flow separation. The tufts are usually made of colored nylon cord or wool, which are taped to the surface of the aircraft. The tufts follow the surface streamlines of the flow, providing a picture of the flow streamline pattern over a large area of the aircraft. In regions of separated flow, the tufts oscillate erratically and reverse direction. Flow cones, a variant of this type of flow visualization, are lightweight narrow rigid hollow conical shapes, usually made of plastic, with a short piece of string extending from the cone apex that is taped to the surface. With their slightly higher mass, the flow cones are less susceptible to instabilities or "whipping" in the flow than tufts. They are also more visible than tufts, due to their slightly large size. Tufts and flow cones come in a variety of colors to make them more visible against different color surfaces and background lighting.

In-flight smoke generator systems are usually used to visualize vortex-dominated flows. Two types of smoke generating systems are pyrotechnic cartridge-based and vaporization systems. Smoke cartridges are pyrotechnically ignited canisters that produce a dense, non-toxic chemical smoke, which is ducted to the flow region of interest. The smoke is entrained into the flow, especially vortical-type flows, tracing the path of the vortex. Since these are pyrotechnic devices, safety precautions must be taken to prevent premature detonation, overpressurization, fire, or other in-flight hazards. Different colored smoke is used to provide the best visualization and contrast in flight. Vaporization smoke systems generate smoke by vaporizing a chemical, such as propylene glycol, with electric heaters.

Flow visualization techniques used on the NASA F/A-18 High-Alpha Research Vehicle (HARV) are described next. The NASA F-18 High-Alpha Research Vehicle (HARV) was a pre-production model of the single-place, twin-engine, McDonnell Douglas F-18 *Hornet* aircraft, modified for its role as a test aircraft, including the addition of paddle-type thrust vectoring vanes in the engine exhaust. The F-18 HARV explored high angle-of-attack and thrust vectoring flight during a test program that lasted from April 1987 to September 1996. The aircraft flew 385 research flights, demonstrating stabilized flight at high angles-of-attack between 65 and 70°.

Figure 3.4 shows the surface streamlines over the nose of the F-18 HARV after a high angle-of-attack flight. The flow visualization fluid was emitted from multiple, flush-surface orifices in the aircraft forebody and leading edge extension (LEX) during flight. (The LEX is the flat, somewhat triangular-shaped surface, to the right of the cockpit in Figure 3.4. It is an extension from the wing leading edge, providing additional lift.) The fluid used for the flights was a mixture of propylene glycol monomethyl ether (PGME) and a toluene-based red dye. After flowing along the streamlines, the PGME in the fluid evaporated, leaving the red dye "painted" on the surface. The PGME evaporation and setting of the dye took about 75–90 s, which required the pilot to maintain the test condition for that time. Note how the surface streamlines wrap around from underneath the aircraft nose. A darker streak of fluid is seen on the fuselage leading edge extension, indicating a strong vortex flowing from the LEX, as described in more detail below.

The F-18 HARV is shown flying at 20° angle-of-attack in Figure 3.5, where smoke was used to visualize the vortex flow from the fuselage leading edge extension. The smoke was generated by 12 specially formulated pyrotechnic cartridges, which ducted smoke to exhaust ports on the aircraft

Figure 3.4 Surface flow visualization on the nose of the F-18 HARV after a high angle-of-attack flight. (Source: *NASA*.)

surface. Two smoke cartridges were fired at a time, generating about 30 s of steady smoke for flow visualization. The vortex is a horizontal "tornado" of flow that trails downstream of the LEX. The smoke shows a great amount of detail about the path and flow structure of the vortex. Near the aircraft's vertical tail, the tightly wound vortex filament "bursts" and dissipates.

The flow over the fuselage and wing surfaces was visualized using tufts, made of 5″ (12.7 cm) long pieces of either wool or nylon cord that were taped to the surface, and flow cones, 3″ (7.6 cm) long plastic cones covered with reflective tape. Many of the surface flow details are discerned from the tuft patterns shown in Figure 3.5. The tufts indicate that the surface streamlines on the top of the center fuselage, behind the cockpit, flow straight back. In the proximity of the LEX vortex, the tufts turn sideways as they are caught up in the vortical flow structure. On the wing,

Figure 3.5   F-18 HARV in-flight flow visualization using tufts, flow cones, and smoke. (Source: *NASA.*)

the tufts do not flow straight back, rather they appear to be pointing in several different directions, indicating large regions of separated flow. Several of the tufts, near the wing leading edge, have a "U" shape, indicating the surface flow has reversed direction. The tufts show that the surface flow on the ailerons, outboard on the wing, is also separated.

3.2.6   Transport Properties of a Fluid

Consider a rocket in flight, traveling at a velocity U_∞. The rocket engine has a hot exhaust flow that exits the engine's nozzle at a velocity U_1 and temperature T_1, as shown in Figure 3.6. There is a *transport of mass*, *momentum*, and *heat (energy)* from the rocket engine exhaust flow into the freestream flow, where each of these *transport phenomena* are related to the gradient of a flow property.

The exhaust is composed of various species of combustion gases, which are not present in the freestream air flow, so there is a gradient of the species mass concentrations, c_i, between the rocket exhaust and the air flow, which is related to the mass transport. The exhaust velocity is much higher than the freestream flow velocity, so there is a gradient in the velocity between the exhaust flow and freestream flow that is related to the momentum transport. The exhaust flow is at a much higher temperature than the external freestream flow temperature, so there is a

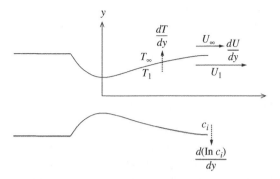

Figure 3.6   Transport of mass, momentum, and heat in a rocket exhaust flow.

temperature gradient between the exhaust flow and freestream flow, which is related to the transfer of heat.

These three transport phenomena are present in all fluid flows, but the degree to which they are important varies significantly. Flows where any of these transport phenomena significantly affect the flow are called *viscous flows*. Flows where these transport phenomena have a small or negligible impact are called *inviscid flows*. For much of the remainder of this chapter, we deal with inviscid flows, but we learn more about viscous flows at the end of this chapter. We now define these three transport phenomena more precisely.

3.2.6.1   Mass Transport

The transport of mass is governed by *Fick's law of diffusion*, given by

$$\frac{\dot{m}_i}{A} = -\rho_i D \frac{d}{dy}(\ln c_i) \tag{3.13}$$

where $\dot{m}_i$ is the mass transport or flux of species i, ρ_i is the density of species i, $d(\ln c_i)/dy$ is the gradient in the mass species concentration of species i, and D is the coefficient of mass diffusivity, a transport property of the fluid.

3.2.6.2   Momentum Transport

The transport of momentum is given by

$$\tau_{xy} = \mu \frac{dU}{dy} \tag{3.14}$$

where τ_{xy} is the shear stress, dU/dy is the velocity gradient, and μ is the coefficient of absolute viscosity (also called the dynamic viscosity), a transport property of the fluid. (In the text, we refer to μ as simply the *coefficient of viscosity*.)

3.2.6.3   Heat Transport

The transport of heat is governed by *Fourier's Law*, given by

$$\dot{q}_y = -k \frac{dT}{dy} \tag{3.15}$$

where $\dot{q}_y$ is the rate of heat flow per unit area or heat flux, dT/dy is the temperature gradient, and k is the thermal conductivity, a transport property of the fluid. Since the heat flux is in the opposite direction to the temperature gradient (from high to low temperature), there is a negative sign in front of the temperature gradient term in Equation (3.15).

3.2.6.4   Coefficient of Viscosity and Sutherland's Law

We are primarily concerned with the coefficient of viscosity, μ, in this text, so it is worthwhile to see how it is calculated. The coefficient of viscosity is a function of pressure and temperature. Often, sufficient accuracy is obtained by assuming that the viscosity is only a function of temperature. A widely used approximation for the coefficient of viscosity is *Sutherland's law*, given by

$$\frac{\mu}{\mu_{ref}} \approx \left(\frac{T}{T_{ref}}\right)^{3/2} \left(\frac{T_{ref} + S}{T + S}\right) \tag{3.16}$$

where μ_{ref} and T_{ref} are reference values, S is Sutherland's constant, a temperature that is dependent on the type of gas, and T is the temperature for which μ is sought. From Sutherland's Law, it is seen that the coefficient of viscosity increases with increasing temperature.

The values of μ_{ref}, T_{ref}, and S for air are given in Table 3.2. The value of μ, calculated using Sutherland's Law, has an error of about $\pm 2\%$ from the true value, for a temperature range of between $300\,°R$ ($-160\,°F$, $167\,K$) and $3420\,°R$ ($2960\,°F$, $1900\,K$), which is quite satisfactory for the majority of the aerodynamics problems. A sample calculation for μ, using Sutherland's law, is given in the example problem below.

We sometimes use another viscosity term, the kinematic viscosity, ν, defined in terms of the coefficient of viscosity, μ, and density, ρ, as

$$\nu = \frac{\mu}{\rho} \tag{3.17}$$

Example 3.1 Coefficient of Viscosity for Air at Standard Temperature *Calculate the coefficient of viscosity for air at standard conditions.*

Solution

From Table 3.1, the standard temperature is $519\,°R$ or $288\,K$. Using Equation (3.16) and Table 3.2, the coefficient of viscosity of air at standard temperature in English units is

$$\mu = \mu_{ref} \left(\frac{T}{T_{ref}}\right)^{3/2} \left(\frac{T_{ref} + S}{T + S}\right)$$

Table 3.2 Sutherland's law reference values and Sutherland's constant for air.

Parameter	Symbol	SI units	English units
Reference viscosity	μ_{ref}	$17.16 \times 10^{-6} \dfrac{kg}{m \cdot s}$	$3.584 \times 10^{-7} \dfrac{slug}{ft \cdot s}$
Reference temperature	T_{ref}	$273.15\,K$	$491.6\,°R$
Sutherland's constant	S	$110.6\,K$	$199\,°R$

$$\mu = \left(3.584 \times 10^{-7} \frac{slug}{ft \cdot s}\right) \left(\frac{519°R}{491.6°R}\right)^{3/2} \left(\frac{491.6°R + 199°R}{519°R + 199°R}\right)$$

$$\mu = \left(3.584 \times 10^{-7} \frac{slug}{ft \cdot s}\right)(1.043) = 3.739 \times 10^{-7} \frac{slug}{ft \cdot s}$$

In SI units, the coefficient of air at standard temperature is

$$\mu = \left(17.16 \times 10^{-6} \frac{kg}{m \cdot s}\right) \left(\frac{288\,K}{273.15\,K}\right)^{3/2} \left(\frac{273.15\,K + 110.6\,K}{288\,K + 110.6\,K}\right)$$

$$\mu = \left(17.16 \times 10^{-6} \frac{kg}{m \cdot s}\right)(1.042) = 17.89 \times 10^{-6} \frac{kg}{m \cdot s}$$

3.3 Types of Aerodynamic Flows

Aerodynamic flows may be classified based upon the dominant flow physics. This often leads to simplifying assumptions about the flow that may make the analysis of the flow much easier. We now define several categories of aerodynamic flows and describe their dominant physics or characteristics.

3.3.1 Continuum and Non-Continuum Flows

Fluids, like all matter, are composed of molecules. Theoretically, one could analyze fluid motion based on the motion of the individual molecules. This approach would be a formidable task and is usually not desired or required. This molecular approach is required for the analysis of some unique aerodynamic situations, such as low density flows at very high altitudes. However, for most aerodynamics problems, we want to know the *macroscopic* or *bulk* properties of the fluid in motion or at rest, so it is not necessary to consider the state of each individual fluid molecule. Therefore, the flow is usually treated as a continuous distribution of matter, called *continuum flow*, rather than as a flow of discrete molecules, called *non-continuum* or *free molecular flow*.

Continuum flow is a good assumption for most flows of air in "normal" conditions, that is pressures and temperatures that are not too high or too low. (We are more specific as to the definition of these "normal conditions" a little later.) For instance, the number of air molecules inside a minute cube of continuum air, with sides of length 0.001 mm (0.00004 in), is about 27 million molecules. Based on this, an assumption that the flow is a continuous medium is appropriate.

We can be a bit more quantitative about the definitions of continuum and free molecular flows by considering how closely packed the molecules are to each other. A measure of this molecular density is the mean distance that a molecule travels before colliding with another molecule, called the *mean free path*, λ. If the mean free path is much smaller than the *characteristic dimension* of interest, L, then the flow is considered a continuum flow. The characteristic dimension is related to the scale of the geometry of interest, so this could be the length in the flow direction, of a fuselage, wing, or other geometry. The ratio of the mean free path to the characteristic dimension is a non-dimensional parameter called the *Knudsen number*, Kn, defined as

$$Kn = \frac{\lambda}{L} \tag{3.18}$$

If the Knudsen number is much smaller than one, the flow is assumed to be a continuum flow. If the Knudsen number is on the order of or greater than one, then free molecular flow must be assumed. However, there is no distinct boundary between continuum and free molecular flow. For example, the flow does not sharply transition from continuum to free molecular flow when the Knudsen number increases from 0.999 to 1.001. Rather, certain physical effects start to get more important as the flow transitions from one or the other type of flow. The transition region, where the flow may have physical characteristics of both continuum and free molecular flow, is called the *low density region*, with the associated flow being called *low density flow*.

For air at standard sea level conditions of 1 atmosphere (2116 lb/ft^2, 101,325 N/m^2) and 59 °F (519 °R, 288 K), the mean free path is equal to about 66 nm (6.6×10^{-8} m, 2.6×10^{-10} in), a very small distance. For instance, if the characteristic dimension is the 3.7 m (12 ft) length of a missile body, the Knudsen number is

$$Kn = \frac{\lambda}{L} = \frac{6.63 \times 10^{-8}\,\text{m}}{3.7\,\text{m}} = 1.79 \times 10^{-8} \tag{3.19}$$

Since this Knudsen number is many orders of magnitude smaller than one, the assumption of a continuum flow, for the missile flying at standard conditions, is a good one. So, when does the value of the Knudsen number approach one, such that the assumption of continuum flow is no longer valid? This can happen if the mean free path gets large or the scale of our body gets small.

As altitude increases and the air gets thinner, the distance between molecules increases, but the mean free path remains a very small distance. Even at an altitude of 50 miles (80 km, 260,000 ft), considered by some as the boundary of space, the mean free path is equal to about 0.005 m (0.2 in). Calculating the Knudsen number for the missile at this altitude, we have

$$Kn = \frac{\lambda}{L} = \frac{0.005\,\text{m}}{3.7\,\text{m}} = 1.3 \times 10^{-3} \tag{3.20}$$

which is some five orders of magnitude larger than at sea level, but still much less than one. As we get even higher, the molecular spacing increases to the point where the Knudsen number approaches one, and the flow can no longer be considered a continuum. The fluid dynamics of free molecular flow is quite different from that of a continuum flow, since the effects of each molecular collision and interaction must be considered. Conceptually, this is somewhat like thinking of the fluid as made up of a collection of "billiard ball molecules" that impact a body as it moves through the flow. The fluid dynamic analysis of free molecular flow is performed using *kinetic theory*, which is beyond the scope of our discussions.

3.3.2 Steady and Unsteady Flows

An important distinction affecting the physics and theoretical analysis of the flow of air is whether or not the flow is changing with time. Flows that do not vary with time are called *steady flows*, whereas flows that change with time are called *unsteady flows*. While this is an obvious and simple distinction, the difference between steady and unsteady flows has a profound impact, not only on the analysis of the flows, but also on the magnitude of the resultant aerodynamic forces on a body.

For a steady flow, the flow properties such as pressure, temperature, density, and velocity, are constant and do not change with time. This does not mean that the flow properties do not change with location in the flow. For example, the surface pressure on an aircraft wing changes from the leading edge to the trailing edge, but if the flow is steady, the pressures at these points on the wing are invariant with time.

In some instances, a flow may at first be unsteady as it transitions from some initial condition to a final, steady-state condition. In other cases, the flow may remain unsteady regardless of how much time has passed. Examples of steady and unsteady flows are shown in Figure 3.7. The flow

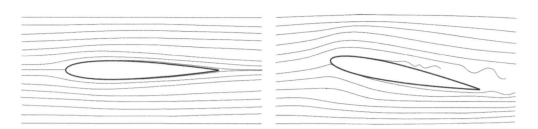

Figure 3.7   Steady (left) and unsteady flows (right) over an airfoil.

is steady over a wing at a small angle-of-attack, well below the stall angle. The streamlines are smooth and steady over the wing's upper and lower surfaces. The pressure distribution over the wing is also steady, resulting in steady values of the lift and drag. In contrast, the flow over a wing at an angle-of-attack exceeding the stall angle is an unsteady flow. Typically, the flow separates over the top surface of the wing and the flow is turbulent and unsteady in this region. The separated flow produces an unsteady and oscillatory shedding of turbulent vortices that trail downstream of the wing, resulting in an unsteady pressure distribution on the wing. The magnitudes of the lift and drag are therefore also changing with time. The separated flow lift is much smaller and the unsteady drag is much larger than their steady values. Separated flows almost always result in unsteady flow.

3.3.3   Incompressible and Compressible Flows

In the previous chapter, the different regimes of flight were introduced. Subsonic, transonic, supersonic, and hypersonic flight were categorized, primarily based on a range of Mach number. While this is a meaningful way to differentiate between the various flow regimes, the change in density or *compressibility* of the gas is also a useful way to differentiate and understand these different flow regimes.

Let us examine the *compressibility* of substances a little more closely. From our everyday experience, we know that gases are much more compressible than liquids. Imagine that we have a cylindrical container with a movable, piston-type lid, as shown in Figure 3.8. If the container is filled with a gas, we can push the lid down into the container and compress the gas. If the cylinder is filled with a liquid, it is virtually impossible to push the lid down and compress the liquid.

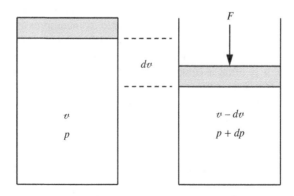

Figure 3.8   Compressibility of a gas.

Now suppose that we fill the container with a unit mass of a gas, such that the gas has a specific volume, v, and is at a pressure, p, as shown on the left side of Figure 3.8. If we push the piston down, the specific volume of the gas decreases by an incremental amount, dv, and the gas pressure increases by an incremental amount dp, as shown on the right side of Figure 3.8. The volume of the gas has been decreased to $v - dv$ and the gas pressure has been increased to $p + dp$. The *compressibility* of the gas, τ, is defined as the fractional change in the volume per unit change in pressure, given by

$$\tau = -\frac{1}{v}\left(\frac{dv}{dp}\right) \tag{3.21}$$

(The symbol τ is used in aerodynamics for both compressibility and shear stress, so care must be taken to ensure the proper meaning.)

The compressibility is a property of a fluid and therefore varies with the type of fluid. The compressibility of water at standard pressure (1 atm, $101{,}325\,\text{N/m}^2$, $2116\,\text{lb/ft}^2$) is about $5 \times 10^{-10}\,\text{m}^2\text{/N}$. Air, with a compressibility of about $1 \times 10^{-5}\,\text{m}^2\text{/N}$ at standard pressure, is about 20,000 times more compressible than water.

Since the specific volume is equal to the inverse of the density, we can rewrite Equation (3.21) as

$$\tau = -\frac{1}{v}\left(\frac{dv}{dp}\right) = -\rho\left[\frac{d\left(\rho^{-1}\right)}{dp}\right] = -\rho(-\rho^{-2})\left(\frac{d\rho}{dp}\right) = \frac{1}{\rho}\left(\frac{d\rho}{dp}\right) \tag{3.22}$$

Solving for the change in density, $d\rho$, we obtain

$$d\rho = \rho\,\tau\,dp \tag{3.23}$$

which gives the change in density for a given change in pressure, dp, as a function of the compressibility, τ.

A pressure change or pressure *gradient* in a flow causes the fluid to move from the high pressure region to the low pressure region. According to Equation (3.23), this pressure change, dp, also produces a density change, $d\rho$. The density change is very small for a liquid, since its compressibility is small, while it is much larger for a gas, since its compressibility is much larger. For an *incompressible* flow, with a compressibility that is theoretically zero, the change in density, $d\rho$, is zero. Hence, an incompressible flow is a *constant density* flow. In contrast, a compressible flow is a *variable density* flow. In reality, all substances are compressible to some degree, but to all intents and purposes, a liquid is considered *incompressible* while a gas is *compressible*.

Returning to Equation (3.23), the density change is small, even for a gas, if the pressure change is small. The magnitude of the pressure change corresponds to the speed of the flow, such that a small pressure change results in a low speed flow and a large pressure gradient produces a high speed flow. Based on this, a low-speed flow is considered as incompressible. Figure 3.9 is a plot of the percent change in the density of air versus the flow Mach number. For a flow Mach number less than 0.3, there is less than a 5% change in the air density. Based on this somewhat arbitrary small change in density, it is often assumed that air flows with a Mach number less than about 0.3 are incompressible. If the flow Mach number is greater than 0.3, then a compressible flow assumption should be used. (Later, we derive the equation for density as a function of Mach number that generated this plot.)

3.3.4   Inviscid and Viscous Flows

Earlier, a *viscous flow* was defined as one where mass diffusion, viscosity, or thermal conduction significantly affects the flow. When these transport phenomena are small or negligible, the flow is

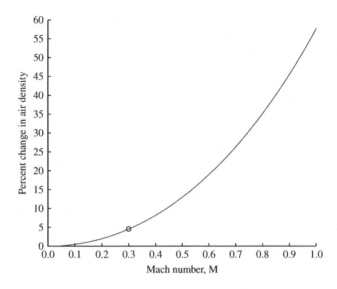

Figure 3.9   Percent change in air density versus flow Mach number.

considered *inviscid*. So, what determines whether the influences of these transport phenomena on the flow are significant or not? In aerodynamics, we are often interested in the resultant forces and moments on a body, such as the lift, drag, and pitching moment. With this in mind, we might say that the influence of these transport phenomena are not important if they do not significantly affect the resultant forces and moments on the body. In other words, the flow may be treated as inviscid if the resultant forces and moments are not significantly different if mass diffusion, viscosity, and thermal conduction are ignored.

Consider the flow of air over an airfoil, as sketched in Figure 3.10. The freestream air flow has a uniform velocity, V_∞, upstream of the airfoil. As the air molecules approach the airfoil, the molecules flowing directly over the surface of the airfoil are slowed down by skin friction. In fact, right at the airfoil surface, the flow velocity is zero due to the friction. The air molecules that are further away from the airfoil surface feel the slowing effect of the skin friction less, until, far from

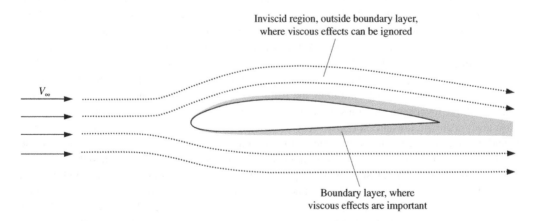

Figure 3.10   Inviscid and viscous regions of the flow around a body.

the airfoil, there is essentially no influence of skin friction on the flow. The velocity of the air molecules, far from the airfoil, is simply the freestream velocity, V_∞. The thin region, adjacent to the airfoil, where skin friction slows the freestream flow, is called the *boundary layer*. Since skin friction is a viscous effect, the boundary layer is a *viscous region*, where viscous effects influence the flow. In addition to skin friction, the boundary layer is also a region where the other viscous effects of mass diffusion and thermal conduction may be significant. Outside of the boundary layer, the viscous effects do not influence the flow, so we can consider this region of the flow as *inviscid*.

The boundary layer concept was the idea of the German engineer and physicist, Ludwig Prandtl (1875–1953), considered by many as the father of modern fluid dynamics. In 1904, Prandtl presented a paper entitled "Uber Flussigkeitsbewegung bei sehr kleiner Reibung" ("Fluid flow with very little friction") which introduced, for the first time, the boundary layer concept and its relation to flow separation, drag, and aerodynamic stall. Prandtl would go on to develop pioneering theories in wing design, supersonic compressible flows, and many other areas of fluid dynamics. We will encounter his name several more times throughout this chapter.

The viscous nature of the boundary layer varies dramatically, depending on several factors related to the flow properties or body geometry. The flow in the viscous region may be smooth and orderly, called *laminar* flow, or it may be chaotic, disorderly, random, and unsteady, called *turbulent* flow. The flow may also be somewhere in between laminar and turbulent, or in a *transitional* flow regime. These different types of viscous flow are seen in the thermal convective flow above a candle, as visualized in Figure 3.11. Later, an aerodynamic parameter called the *Reynolds number* is introduced that is used to distinguish between laminar, turbulent, and transitional flows. As shown in Figure 3.11, a flow transitions from laminar to turbulent flow as the Reynolds number increases.

For many of the flows having to do with the flight of aircraft in the sensible atmosphere, the flow is primarily turbulent. In fact, the existence of fully laminar flow over an aircraft in flight is a rarity, but it can be found. For instance, the flow may be fully laminar if the speed is sufficiently low, the density is sufficiently low (as for very high altitudes), or the length scale of the body is sufficiently small. These three assumptions (small ρ, L, and V) result in low Reynolds numbers, which is identified with laminar flow.

In addition to the thin boundary layer next to the surface of a body, some types of flows are dominated by viscous effects. Separated flows, such as the stalled flow over a wing or the wake flow behind a bluff-based body, are viscous dominated flows. These separated flows tend to be unsteady and turbulent. The size and location of separated flow regions are dependent on the nature of the viscous boundary layer.

In reality, all flows are viscous in nature, but the analysis of viscous flows is much more complex than inviscid flows. Prandtl's boundary layer concept allows us to separate the analysis of the flow into two regions, a thin, boundary layer region close to the body, where viscous effects are important, and an inviscid region, outside the boundary layer, where viscous effects are ignored. This was one of the breakthroughs of Prandtl's theory. The viscous skin friction and heat conduction effects are confined to the thin boundary layer, while these effects are negligible in the outer, inviscid region, which greatly simplifies the mathematical analysis of fluid flows. In general, the solution of the governing equations of fluid flow over a body is extremely complicated, and closed form analytical solutions are not possible, except for a very few specific, simple geometries. The boundary layer concept allows the much simpler solution of the inviscid flow field, outside the boundary layer, and the solution of a simplified set of equations, appropriately called the boundary layer equations, in the thin, viscous boundary layer region. But how far away from the body do you have to be before inviscid flow can be assumed? How far from the body does the viscous boundary layer extend? Or, in other words, just how "thin" is the boundary layer? Later, when we discuss viscous flows in more detail, we introduce the equations to quantitatively answer these questions. Based on what we have discussed so far, the geometry of the body influences whether viscous

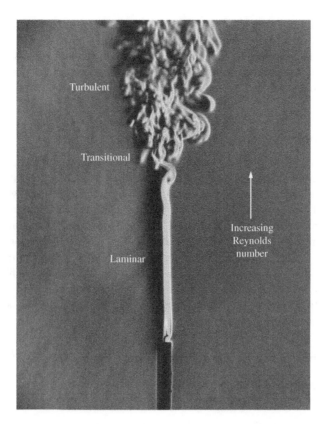

Figure 3.11 Schlieren photograph of the thermal convection plume rising from a candle, showing laminar, transitional, and turbulent flow. (Source: *Adapted from Gary Settles, "Laminar-Turbulent Transition" https://commons.wikimedia.org/wiki/File:Laminar-turbulent_transition.jpg, CC-BY-SA-3.0. License at https://creativecommons.org/licenses/by-sa/3.0/legalcode.*)

effects are important. However, what other factors influence whether or not the flow is considered viscous or inviscid? Does the speed of the flow or the type of fluid matter? We address these and other matters concerning viscous flows, later in this chapter.

3.4   Similarity Parameters

In this section, several non-dimensional parameters that are important in fluid dynamics and heat transfer are introduced. Many of the parameters that are particularly significant in aerodynamics are emphasized, and many are introduced by definition, with further explanation and discussion provided in later sections.

Most of the parameters that are introduced are known as *similarity parameters*. These are parameters that allow us to compare the flows over different sized bodies, based on similarities in their geometries and flow physics. The concept of flow similarity is discussed in detail in the ground test technique about wind tunnel testing.

Many of the parameters that have a length scale associated with them are based on a *characteristic length*. The characteristic length is typically the primary dimension of the vehicle or body in the flow direction. For example, the characteristic length of a simple flat plate is the length of the flat plate in the flow direction. For an airplane wing, the characteristic length is typically the chord length, defined as the straight-line distance from the leading edge to the trailing edge of the wing's airfoil shape.

The first two similarity parameters that are discussed, the Mach number and the Reynolds number, are perhaps the most important in aerodynamics. Aerodynamic flows scale with these two parameters throughout all of the flow regimes, with the exception of hypersonic flight.

3.4.1 Mach Number

In Chapter 2, the Mach number was introduced and shown to be very useful in categorizing the different regimes of flight. The Mach number, M, was defined as the ratio of the airspeed, V, to the speed of sound, a. We also made a qualitative argument that the Mach number is the ratio of the directed motion of the flow relative to the random thermal motion. Now, let us think about the Mach number from the perspective of the forces involved. When an aerospace vehicle flies through the atmosphere, forces are generated on the vehicle. These forces depend on the vehicle's geometry, attitude, altitude, and velocity. They are also a function of the properties of the air, including its viscosity and elasticity or compressibility. As the vehicle flies through the air, it affects a large volume of air. Let us assume that the volume of air affected is equal to L^3, where L is the characteristic length of the vehicle. Assuming that a vehicle is flying at a velocity V, the inertia force imparted to the air is equal to

$$\text{inertia force} = \text{mass} \times \text{acceleration} \sim (\rho L^3)\left(\frac{V}{t}\right) \sim \frac{\rho L^3 V}{L/V} \sim \rho L^2 V^2 \tag{3.24}$$

where ρ is the air density and t is time.

The elasticity force of the air is given by

$$\text{elasticity force} = \text{pressure force} \times \text{area} \sim pL^2 \tag{3.25}$$

where p is the air pressure and L^2 is the area of air acted upon by the pressure. In our earlier discussion about Mach number, it was stated that the speed of sound is proportional to the square root of the air temperature, T. Later in this chapter, it is shown that the air temperature is proportional to the air pressure divided by the air density through an equation of state. Therefore, the speed of sound is related to the pressure and density as follows.

$$a^2 \sim RT \sim \frac{p}{\rho} \tag{3.26}$$

where R is the specific gas constant for air. Using Equation (3.26) in (3.25), the elasticity force is given by

$$\text{elasticity force} \sim \rho a^2 L^2 \tag{3.27}$$

Using Equations (3.24) and (3.27), the ratio of the inertia force to the elasticity force is given by

$$\frac{\text{inertia force}}{\text{elasticity force}} = \frac{\rho L^2 V^2}{\rho a^2 L^2} = \frac{V^2}{a^2} = M^2 \tag{3.28}$$

Equation (3.28) states that another physical interpretation of the Mach number is the ratio of the inertia force to the elasticity force, which is related to the compressibility of the gas. Hence, the Mach number is a governing parameter for compressible flows.

Summarizing these various interpretations for the Mach number, we have

$$M = \frac{V}{a} = \frac{\text{airspeed}}{\text{speed of sound}} \sim \frac{\text{directed motion}}{\text{random (thermal) motion}} \sim \frac{\text{inertia force}}{\text{elasticity force}} \tag{3.29}$$

3.4.2   Reynolds Number

The viscous force in a fluid may be defined as

$$\text{Viscous force} \sim \mu VL \tag{3.30}$$

where μ is the coefficient of viscosity, V is the fluid velocity, and L is the characteristic length. To confirm that this is a force, we check the units in Equation (3.30) as

$$[\mu VL] = \left(\frac{\text{slug}}{\text{ft} \cdot \text{s}}\right)\left(\frac{\text{ft}}{\text{s}}\right)(\text{ft}) = \frac{\text{slug} \cdot \text{ft}}{\text{s}^2} = \text{lb} \tag{3.31}$$

which gives units of force.

Using Equations (3.24) and (3.30), the Reynolds number, Re, is equal to the ratio of the inertia force to the viscous force.

$$Re = \frac{\text{inertia force}}{\text{viscous force}} = \frac{\rho L^2 V^2}{\mu VL} = \frac{\rho VL}{\mu} \tag{3.32}$$

where ρ and V are the freestream density and velocity, respectively, and L is the characteristic length of the body or vehicle.

The inertia forces are those that are due to the velocity and momentum in the flow. The viscous forces are due to shear stress acting over the surface of the body. If the Reynolds number is small ($Re \ll 1$), then the viscous forces are significant in relation to the inertial forces and cannot be ignored. If the Reynolds number is large ($Re \gg 1$), then the inertial forces are dominant over the viscous forces. For a high Reynolds number flow, the influence of viscosity is small, except in a region close to the surface of bodies, called the *boundary layer*. Hence, a major portion of the flow, outside this viscous boundary layer region, may be considered as inviscid when the Reynolds number is large. This is a significant simplification that is useful for many aerodynamic analysis.

The Reynolds number is named in honor of the British engineer, Osborne Reynolds (1842–1912), who conducted landmark experiments in the fluid dynamics of laminar and turbulent flows. In his famous pipe flow experiments, a colored dye was injected in the center of a transparent pipe, through which water was flowing at a constant velocity (Figure 3.12). Reynolds observed that as

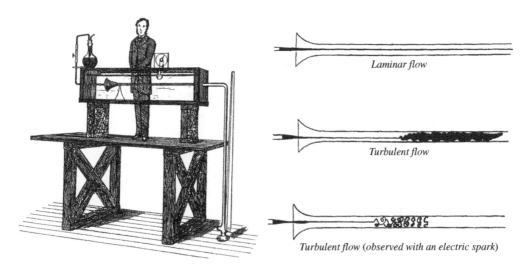

Figure 3.12   Osborne Reynolds' pipe experiment (left) and his drawings of dye patterns (right). (Source: *O. Reynolds, 1883, PD-old-100.*)

the velocity of the dye was increased, the pattern of the dye changed dramatically, from a smooth, orderly path, distinct from the water flow, to a random chaotic mixing of the dye with the water. He was able to correlate this change or transition, from the orderly laminar flow of the dye to a disorderly turbulent flow, with his now famous Reynolds number. Reynolds published his findings in an 1883 paper, entitled "An experimental investigation of the circumstances which determine whether the motion of water in parallel channels shall be direct or sinuous and of the law of resistance in parallel channels."

While the correlation of Reynolds number with the nature of the flow is valid, in practice it is often difficult to assign very specific values of the transition Reynolds number that apply to all flow situations and geometries. However, we can specify some approximate values of the Reynolds number that are generally applicable. If the Reynolds number is less than about 100,000, the flow is most likely laminar. If the Reynolds number is greater than about 500,000, the flow is probably turbulent. In between these values, the flow is likely of a transitional nature, transitioning from laminar to turbulent flow. Summarizing these statements, we have

$$Re < \sim 100,000 : \text{Laminar flow} \tag{3.33}$$

$$\sim 100,000 < Re < \sim 500,000 : \text{Transitional flow} \tag{3.34}$$

$$Re > \sim 500,000 : \text{Turbulent flow} \tag{3.35}$$

Figure 3.13 shows the ranges of Reynolds number (based on the vehicle or body length) and airspeed for different types of aerospace vehicles and other interesting flying things, such as insects and birds. The Reynolds number and speed are plotted on log scales, so that the increments along each axis represent an order of magnitude increase in the quantity. The Reynolds number of dust moving in the air is very small, less than 10. For insects, the Reynolds numbers range from about 100 to less than 10,000. Birds, which are larger than insects and fly faster, have a Reynolds number range from about 10,000 to several hundred thousand ($\sim 10^5$). The Reynolds number of full-scale aircraft are in the millions ($\sim 10^6$) to tens of millions ($\sim 10^7$).

The Reynolds number is sometimes presented in another useful form, where the specification of a characteristic length is not required, called the Reynolds number per unit length or *unit Reynolds*

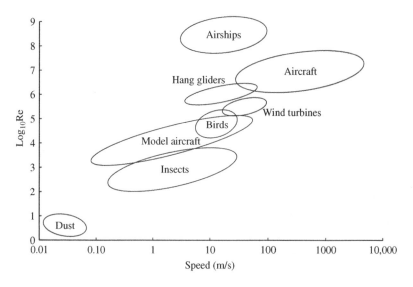

Figure 3.13 Ranges of Reynolds number and airspeed for various types of vehicles.

number, defined as

$$\frac{Re}{m} \text{ or } \frac{Re}{ft} = \frac{\rho V}{\mu} \tag{3.36}$$

The units of the freestream conditions in Equation (3.36) are specified in SI or English units, as desired, to give the Reynolds per meter or per foot, respectively.

3.4.3 Pressure Coefficient

The pressure acting over the surface area of a body makes an important contribution to the aerodynamic force on the body. The pressure is a dimensional quantity with units of force per unit area, such as N/m^2 in SI units and lb/ft^2 in English units. We define a *dimensionless pressure coefficient* as

$$C_p \equiv \frac{p - p_\infty}{q_\infty} \sim \frac{\text{static pressure}}{\text{dynamic pressure}} \tag{3.37}$$

where p is the local pressure, p_∞ is the freestream static pressure, and q_∞ is the freestream dynamic pressure. Equation (3.37) indicates the magnitude of the difference between the local pressure and the freestream pressure, relative to the magnitude of the dynamic pressure. A pressure coefficient equal to zero indicates that the local pressure is equal to the freestream pressure. If the pressure coefficient is positive, the local pressure is higher than the freestream pressure and if it is negative, it is lower than the freestream pressure. The magnitude of the pressure coefficient provides an indication of how much higher or lower the pressure difference is, relative to the freestream pressure.

The pressure coefficient is a true similarity parameter, in that its value is independent of the size of a body such that two geometrically similar bodies, of different size, have the same pressure coefficient distributions. Thus, one could measure the dimensional pressure distribution on the surface of a wind tunnel model and, by then calculating the pressure coefficient distribution, determine the pressure distribution on a full-scale vehicle.

3.4.4 Force and Moment Coefficients

We are often interested in the aerodynamic forces and moments acting on an aircraft or other body in flight. The lift, drag, and pitching moment are three of the most important of these that we encounter. It is useful to have a non-dimensional form of these aerodynamic forces and moments.

The *force coefficient*, C_F, is defined as the aerodynamic force, F, non-dimensionalized by the freestream dynamic pressure, q_∞, multiplied by a reference area, S_{ref}, which is typically the wing planform area for an airplane.

$$C_F = \frac{F}{q_\infty S_{ref}} = \frac{F}{\frac{1}{2}\rho_\infty V_\infty^2 S_{ref}} \sim \frac{\text{aerodynamic force}}{\text{dynamic force}} \tag{3.38}$$

Similarly, the *moment coefficient* is defined as

$$C_M = \frac{M}{\frac{1}{2}\rho_\infty V_\infty^2 c_{ref} S_{ref}} = \frac{M}{q_\infty c_{ref} S_{ref}} \sim \frac{\text{aerodynamic moment}}{\text{dynamic moment}} \tag{3.39}$$

where c_{ref} is the moment reference length, which is typically the wing chord length for an airplane. The non-dimensional aerodynamic coefficients are a function of Reynolds number and Mach number, making them particularly useful in comparing different geometries and flows.

3.4.5 Ratio of Specific Heats

The ratio of the specific heat at constant pressure, c_p, to the specific heat at constant volume, c_v, is another important similarity parameter for compressible flows. This ratio of specific heats is given the Greek symbol γ and is defined as

$$\gamma = \frac{c_p}{c_v} \sim \frac{\text{enthalpy}}{\text{internal energy}} \tag{3.40}$$

Physically, the ratio of specific heats is the ratio of the flow enthalpy to the internal energy. The ratio of specific heats is assumed a constant for air, and many other gases, at normal conditions. The value of γ for air at normal conditions is 1.4. The parameter γ is found in many of the compressible flow equations, alongside the Mach number.

3.4.6 Prandtl Number

The Prandtl number, Pr, is defined as

$$Pr = \frac{c_p \mu}{k} \sim \frac{\text{momentum diffusion rate}}{\text{thermal diffusion rate}} \tag{3.41}$$

where c_p is the specific heat at constant pressure, μ is the coefficient of viscosity, and k is the thermal conductivity. It is named after the German physicist, Ludwig Prandtl. Unlike the Reynolds number, there is no length scale associated with it. The Prandtl number is a function of the fluid properties only. As such, tabulated values of the Prandtl number are usually found alongside other fluid properties, such as the coefficient of viscosity, μ, and the thermal conductivity, k.

Physically, the Prandtl number is the ratio of the momentum diffusion rate to the thermal diffusion rate. Momentum is diffused or spread in a fluid due to velocity gradients, primarily in boundary layers. Hence, momentum diffusion is also referred to as viscous diffusion, since it has to do with the viscous boundary layer. Thermal diffusion is the spread of thermal energy or heat in a fluid. Similar to the velocity boundary layer, there can exist a thermal boundary layer, where there is a gradient of temperature near the surface of a body, which drives the thermal diffusion.

If the Prandtl number is much less than one, the thermal diffusion rate dominates, so that the heat diffuses or spreads in the fluid more rapidly than the momentum. If the Prandtl number is much greater than one, the spread of momentum is more rapid than the spread of heat. In terms of heat transfer, a Prandtl number much less than one means that conduction is more dominant than convection. A Prandtl number much greater than one indicates that convection is the dominant form of heat transfer over conduction.

Let us get a better feel for values of the Prandtl number. For air at standard, sea level conditions, the Prandtl number is calculated as

$$Pr = \frac{c_p \mu_\infty}{k_\infty} = \frac{(1006\,\text{J/kg} \cdot \text{K})(17.89 \times 10^{-6}\,\text{kg/m} \cdot \text{s})}{0.02533\,\text{J/m} \cdot \text{s} \cdot \text{K}} = 0.7105 \tag{3.42}$$

Thus, in air, the thermal diffusion rate is slightly dominant over the momentum diffusion rate. The Prandtl numbers for various types of substances are given in Table 3.3. In liquid metals, such as mercury, conduction is the primary mode of heat transfer, while in oils, convection is the dominant heat transfer mode.

3.4.7 Other Similarity Parameters

While the Mach number, Reynolds number, and Prandtl number are the primary similarity parameters that we deal with extensively in the present text, there are several other similarity parameters

Table 3.3 Typical values and ranges of the Prandtl number for various substances.

Substance	Prandtl number
Air at standard conditions	0.71
Liquid metals	0.001–0.03
Gases	0.7–1.0
Water	1–10
Oils	50–2,000
Glycerin	2000–100,000

that are of interest in other areas of fluid mechanics. Some of these other similarity parameters are briefly introduced in the present section, which may be useful in future applications.

3.4.7.1 Froude Number

The Froude number, Fr, is defined as

$$Fr = \sqrt{\frac{V^2}{gL}} \sim \sqrt{\frac{\text{inertia forces}}{\text{gravity forces}}} \tag{3.43}$$

where V is the flow velocity, g is the acceleration due to gravity, and L is the characteristic length of the geometry of interest. Physically, the Froude number is the ratio of the inertial forces in the flow to the gravity forces.

The Froude number is important for flows where the gravity forces are significant. In hydrodynamics, it is a scaling parameter for similarly shaped objects of different sizes that are submerged in water. The water wave patterns generated by two similarly shaped objects of different scale are the same if the Froude number is the same. The Froude number, as applied to hydrodynamics, is somewhat analogous to the Mach number, as applied to air flows.

The Froude number is named after the English engineer, William Froude (1810–1879), who specialized in hydrodynamics and naval architecture, the design of ships, boats, and other marine vessels. Froude developed laws for the resistance of ship hulls in water and contributed to the prediction of the stability of ships.

3.4.7.2 Grashof Number

The Grashof number, Gr, is defined as

$$Gr = \frac{\rho^2 g \beta (T_s - T_\infty) L}{\mu^2} \sim \frac{\text{buoyancy forces}}{\text{viscous forces}} \tag{3.44}$$

where ρ is the fluid density, g is the acceleration due to gravity, β is the volumetric thermal expansion coefficient, T_s is the surface temperature, T_∞ is the bulk temperature of the fluid, L is the characteristic length of the geometry of interest, and μ is the coefficient of viscosity of the fluid. Physically, the Grashof number is the ratio of the buoyancy forces to the viscous forces acting on a fluid.

The Grashof number is a similarity parameter for *free convective heat transfer*. Convection is the transfer of heat due to mass motion of a fluid, such as air. In free or natural convection, the fluid motion and transfer of heat is driven by buoyancy changes in the fluid due to changes in fluid density, which are caused by temperature changes. Warm air is more buoyant than cool air, which causes it to rise, resulting in the free convective transfer of heat. The weather is a result of free convection in the atmosphere due to spatial changes in temperature.

For Grashof numbers below about 10^8 (based on a length of a vertical flat plate), the boundary layer in free convection is laminar. For Grashof numbers between 10^8 and 10^9, the boundary layer is transitional. The free convective boundary layer is turbulent (for a vertical flat plate) at Grashof numbers above about 10^9 (based on the plate length). The Grashof number, as applied to free convective boundary layers, is somewhat analogous to the Reynolds number, as applied to viscous boundary layers.

The Grashof number is named in honor of the German engineer, Franz Grashof (1826–1893), who developed early formulas for steam flow and contributed to heat transfer theories of free convection.

3.4.7.3 Knudsen Number

The Knudsen number, Kn, is defined as

$$Kn = \frac{\lambda}{L} \sim \frac{\text{mean free path length}}{\text{characteristic length}} \tag{3.45}$$

where λ is the molecular mean free path length and L is the characteristic length of the geometry of interest. The mean free path is defined as the average distance that a molecule travels before it collides with another molecule. This distance is very small, equal to about 66.3 nm (6.63×10^{-8} m, 2.61×10^{-10} in) for air at standard sea level conditions. The Knudsen number is used to distinguish between continuum and free molecular flow (discussed in Section 3.3.1). The continuum flow assumption is valid for a flow with a Knudsen number much less than one, while a Knudsen number on the order of, or greater than, one, corresponds to free molecular flow.

The Knudsen number is named in honor of the Danish physicist, Martin Knudsen (1871–1949), who devoted much of his scientific career to the study of the kinetic theory of gases. He also performed research in physical oceanography and the properties of seawater.

3.4.7.4 Stanton Number

The Stanton number, C_H, is defined as

$$C_H = \frac{\dot{q}}{\rho_\infty V_\infty c_p (T_0 - T_w)} = \frac{\text{heat flux to the surface}}{\text{convected heat flux}} \tag{3.46}$$

where $\dot{q}$ is the heat transfer rate per unit area or heat flux, c_p is the specific heat at constant pressure, T_0 is the flow total temperature, T_w is the wall or surface temperature, and ρ_∞ and V_∞ are the freestream density and velocity, respectively. (The symbol St is often used for the Stanton number, but we use the symbol C_H to avoid confusion with the Strouhal number, also given the symbol St.) The temperature difference, $T_0 - T_w$, is sometimes referred to as the driving temperature potential, since its magnitude drives the heat flux.

The Stanton number is a dimensionless heat transfer number applied to flows with *forced convective heat transfer*. Convection is the transfer of heat due to mass motion of a fluid, such as air. If the fluid motion is induced by some external means, such as a fan, the wind, or vehicle motion, the process is forced convection. This is in contrast to free or natural convection, discussed earlier, where the Grashof number applies. The forced convective heat transfer is a function of the density of the flow, the flow velocity, and the difference between the flow total temperature and the wall temperature, as seen in Equation (3.46). The Stanton number is applicable to the study of aerodynamic heating, which is discussed in Section 3.13.7.

The Stanton number is named after the British engineer Thomas E. Stanton (1865–1931), who studied engineering under Osborne Reynolds and took his first employment in Reynolds' laboratory. Stanton's primary area of interest was in viscous fluid flow, researching problems involving friction and heat transfer. After the Wright Brothers' first flight in Europe in 1908, Stanton studied airplane and airship design, and heat transfer problems associated with air-cooled airplane engines.

3.4.7.5 Strouhal Number

The Strouhal number, St, is defined as

$$St = \frac{fL}{V} \sim \frac{\text{local acceleration}}{\text{convective acceleration}} \sim \frac{\text{oscillation}}{\text{mean flow speed}} \tag{3.47}$$

where f is the vortex shedding frequency, L is the characteristic length of the geometry of interest, and V is the flow velocity. Physically, the Strouhal number is the ratio of the inertial forces due to the unsteady flow (the local acceleration) to the inertial forces due to changes in the steady flow velocity from point to point in the flow (the convective acceleration). One could also interpret the Strouhal number as a measure of oscillations in the flow relative to the mean flow speed.

Many fluid dynamic flows are unsteady and oscillatory in nature. If one observes water flowing over a rock in a river, an unsteady, oscillating pattern of swirling flow is seen streaming in the wake of the rock. The flow behind a bluff body, such as the rock, exhibits an unsteady, oscillating flow pattern called *vortex shedding*, as shown for a cylinder in Figure 3.14. The vortices are shed alternately from either side of the bluff body at a frequency that is a function of the Reynolds number, based on the body's diameter. The double row of alternating vortices, that are shed behind a two-dimensional bluff body, such as a cylinder, is called a *von Karman vortex street*. Vortices are shed from a non-bluff body, that is, an aerodynamically *streamlined* body, if it presents itself to the flow at a high angle. This is the case for a streamlined wing or fuselage at high angle-of-attack.

Sometimes the frequency is such that the shedding flow is audible, as when power or transmission lines are heard to "sing", due to wind blowing over their cylindrical cross-section. The Strouhal number is named in honor of the Czech experimental physicist, Vincent Strouhal (1850–1922),

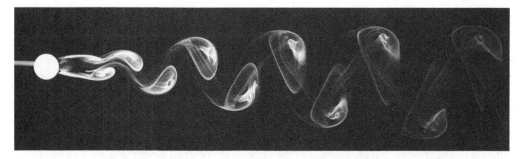

Figure 3.14 Vortex shedding in the wake of a cylinder (flow is from left to right). (Source: *Jurgen Wagner, "Karman Eddy Small Re" https://en.wikipedia.org/wiki/File:Karmansche_Wirbelstr_kleine_Re.JPG, CC-BY-SA-4.0. License at https://creativecommons.org/licenses/by-sa/4.0/legalcode.*)

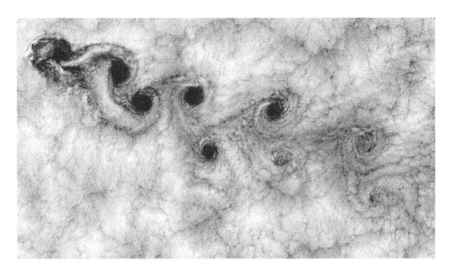

Figure 3.15   Satellite image of clouds show a von Karman street from wind blowing over islands off the coast of Chile (wind direction is from upper left to lower right of picture).   (Source: *NASA*.)

who studied this shedding of vortices from wires in the wind. The ancient Greeks were aware of this phenomenon, inventing a musical instrument called the Aeolian harp (named after the Greek god of the wind, Aeolus). The ancient Greek Aeolian harp is composed of strings stretched across the length of a wooden box with a sounding board. The harp was placed in an open window where the wind would blow across the strings and create musical tones. Large musical sculptures, based on the Aeolian harp, are found today, mounted on rooftops or on windy landscapes. There are many other examples of vortex shedding flows in nature, including the flow behind a swimming fish, the flow around skyscrapers, and the flow of cloud around an island, as shown in Figure 3.15.

The wind flow over a transmission line may also make the line itself start oscillating, a tell-tale sign that shedding vortices can also create alternating forces on a body. These wind-induced oscillation forces can lead to structural fatigue or other destructive forces on transmission lines or wires, buildings, bridges, towers, chimneys, and other types of slender (length much greater than their cross-sectional width) structures exposed to high wind. Resonance between the vortex shedding frequencies and the natural frequencies of the structure must be avoided, as this could lead to catastrophic structural failure. Wind tunnel testing is commonly performed on models of new structures, such as high-rise buildings and suspension bridges, to ensure that the designs are safe with regards to wind-induced oscillation forces.

Mechanical devices are also used to dampen the oscillations and vibrations caused by the wind on structures. An example is the Stockbridge damper – also called a dog-bone damper because of its shape – that is used to dampen the wind-induced oscillations of lines, wires, or cables (Figure 3.16). The Stockbridge damper has a dumbbell shape with two masses attached to a short cable or flexible rod that is attached to the line, cable, or wire. Patented in 1928 by George H. Stockbridge, an American engineer working for the Southern California Edison electrical company, the original device used pieces of concrete as the masses attached to the ends of a short length of cable (Figure 3.16 left). Modern versions of the Stockbridge damper may use different materials and have slightly different configurations, but are essentially the same as the original design (Figure 3.16 right).

As given by Equation (3.47), the Strouhal number, St, is indicative of the vortex shedding frequency, f, for a body of height L, in a flow with a velocity V. The Strouhal number is a function of the Reynolds number for a given geometry, as shown for the flow over a circular cylinder in

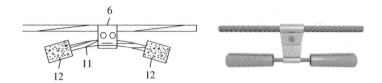

Figure 3.16 Stockbridge damper, original patent drawing (left) and modern device (right). (Source: *Left: US Patent, 1928, PD-US-Patent, Right: User: BillC, "Stockbridge Damper" https://en.wikipedia.org/wiki/ File:Stockbridge_damper_POV.jpg, CC-BY-SA-3.0. License at https://creativecommons.org/licenses/by-sa/3 .0/legalcode.*)

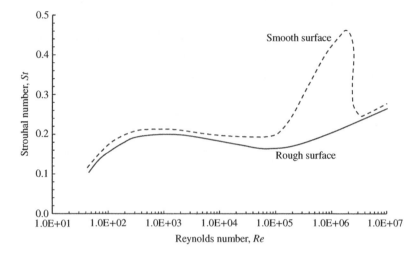

Figure 3.17 Strouhal number as a function of Reynolds number for the flow over a circular cylinder.

Figure 3.17. As shown in this figure, the Strouhal number has a constant value of about 0.2 over a large range of Reynolds numbers, from about 100 to 100,000. At higher Reynolds numbers, above 100,000, the Strouhal number varies considerably, depending on the smoothness or roughness of the cylinder surface. Keep in mind that a given Strouhal number corresponds to a specific shedding frequency of the vortices from the cylinder. Thus, the vortex shedding frequency for a cylinder is fairly constant below a Reynolds number of 100,00, and varies above this value.

The Strouhal number is useful in correlating the propulsive efficiency of animals that use flapping motion for locomotion, such as birds and fish. The Strouhal number for the up and down motion of a wing or tail is defined using Equation (3.47), where f is the frequency of the flapping, L is the vertical distance made by the tip of the "flapper" (e.g. wingtip or tip of a fish tail), and V is the forward speed of the animal. For flapping motion, the product in the numerator of the Strouhal number, fL, is interpreted as the vertical velocity of the wingtip or tail tip. Thus, the Strouhal number for flapping motion represents the ratio of how much the animal moves its flapper up and down relative to its forward speed. Given this definition of the Strouhal number for flapping motion, the propulsive efficiency is optimum for Strouhal number between 0.2 and 0.4. This applies to the cruising flight or swimming of a wide range of animals, including birds, bats, dolphins, sharks, and whales. This interesting correlation with the Strouhal number is useful to designers of small, unmanned aerial vehicles that utilize flapping locomotion, in optimizing the flapping frequency and amplitude for a desired forward speed.

Table 3.4 Summary of selected similarity parameters in fluid mechanics and heat transfer.

Symbol	Similarity parameter	Definition	Physical meaning	Flow application
C_F	Force coefficient	$\dfrac{F}{q_\infty S_{ref}}$	$\dfrac{\text{aerodynamic force}}{\text{dynamic forces}}$	All flows
C_H	Stanton number	$\dfrac{\dot{q}}{\rho_\infty V_\infty (h_0 - h_w)}$	$\dfrac{\text{heat flux to surface}}{\text{convective heat flux}}$	Forced convective heat transfer
C_M	Moment coefficient	$\dfrac{M}{q_\infty c_{ref} S_{ref}}$	$\dfrac{\text{aerodynamic moment}}{\text{dynamic moment}}$	All flows
C_p	Pressure coefficient	$\dfrac{p - p_\infty}{q_\infty}$	$\dfrac{\text{static pressure}}{\text{dynamic pressure}}$	All flows
Fr	Froude number	$\dfrac{V}{\sqrt{gL}}$	$\dfrac{\text{inertia forces}}{\text{gravity (body) forces}}$	Free-surface flows
Gr	Grashof number	$\dfrac{\rho^2 g \beta (T_s - T_\infty) L}{\mu^2}$	$\dfrac{\text{buoyancy forces}}{\text{viscous forces}}$	Natural convection
Kn	Knudsen number	$\dfrac{\lambda}{L}$	$\dfrac{\text{mean free path length}}{\text{characteristic length}}$	Free molecular flows
M	Mach number	$\dfrac{V}{a}$	$\dfrac{\text{airspeed}}{\text{speed of sound}}$	Compressible flows
Pr	Prandtl number	$\dfrac{c_p \mu}{k}$	$\dfrac{\text{viscous dissipation}}{\text{thermal dissipation}}$	Heat convection
Re	Reynolds number	$\dfrac{\rho V L}{\mu}$	$\dfrac{\text{inertia forces}}{\text{viscous forces}}$	Viscous or compressible flows
St	Strouhal number	$\dfrac{fL}{V}$	$\dfrac{\text{oscillation}}{\text{mean flow speed}}$	Flow-excited vibrations
γ	Ratio of specific heats	$\dfrac{c_p}{c_v}$	$\dfrac{\text{enthalpy}}{\text{internal energy}}$	Compressible flows

3.4.8   Summary of Similarity Parameters

A summary of the non-dimensional parameters that have been introduced in this section is provided in Table 3.4. These particular parameters were chosen because of their relevance to problems in fluid mechanics and heat transfer. Many of these parameters are used throughout the text, and Table 3.4 may be referenced for quick definitions.

Example 3.2 Calculation of Reynolds Number *The Lockheed SR-71 aircraft (see Figure 2.29) is flying at Mach 3.0 at an altitude of 80,000 ft. The speed of sound at 80,000 ft is 395.9 ft/s. Calculate the Reynolds number based on its wing reference chord length of 37.70 ft.*

Solution

The Reynolds number is calculated using Equation (3.32). The freestream density, velocity, and viscosity must first be obtained. From Appendix C, for an altitude of 80,000 ft, the freestream density, ρ_∞, is 8.683×10^{-5} slug/ft^3 and the freestream temperature, T_∞, is 390.0 °R.

The freestream velocity is

$$V_\infty = M_\infty a_\infty = (3.0)\left(395.9\frac{ft}{s}\right) = 1{,}188\frac{ft}{s}$$

Using Equation (3.16) and the values in Table 3.2, the freestream viscosity is calculated as

$$\frac{\mu}{\mu_{ref}} = \left(\frac{T_\infty}{T_{ref}}\right)^{3/2}\left(\frac{T_{ref}+S}{T_\infty+S}\right) = \left(\frac{390.0°R}{491.6°R}\right)^{3/2}\left(\frac{491.6°R+199°R}{390.0°R+199°R}\right) = 0.8285$$

$$\mu = (3.584\times10^{-7}\,slug/ft\cdot s)(0.8285) = 2.969\times10^{-7}\,slug/ft\cdot s$$

The Reynolds number based on the SR-71 wing reference chord length may now be calculated as

$$Re_L = \frac{\rho_\infty V_\infty L}{\mu_\infty} = \frac{\left(8.683\times10^{-5}\,\frac{slug}{ft^3}\right)\left(1{,}188\,\frac{ft}{s}\right)(37.70\,ft)}{2.969\times10^{-7}\,\frac{slug}{ft\cdot s}} = 1.310\times10^7$$

3.5 A Brief Review of Thermodynamics

In our study of aerodynamics and propulsion, we need a foundation in the basic concepts and terminology of thermodynamics. It is assumed that the student has some familiarity with thermodynamics, either as introduced in a basic physics course or through a dedicated course on the subject. The following provides a brief review of some basic thermodynamic concepts and terms.

Thermodynamics deals with energy and its transformations into heat and work. These transformations affect the thermodynamic properties of aerodynamic and propulsive flows, such as pressure, temperature, and density. We define the thermodynamic terms, energy, work, and heat, more precisely in the sections below. We also introduce the concept of entropy, which is essential in understanding the laws of thermodynamics. In fact, thermodynamics is viewed as the science of energy and entropy, embodied by the first and second laws of thermodynamics, respectively.

3.5.1 Thermodynamic System and State

Some basic thermodynamic concepts, needed to develop the laws of thermodynamics, are presented in this section. These include the concepts of a thermodynamic system, state, and process.

3.5.1.1 Thermodynamic System

When we discuss the transfer of energy, we usually refer to this transfer as *to* or *from* a *thermodynamic system*, defined as a specific quantity of matter, of fixed mass, which is convenient to identify as a cohesive unit. Everything external to what we have defined as our system is the *surroundings*. The system is separated from the surroundings by the *system boundaries*, which may be fixed or movable.

For example, consider a sealed, flexible balloon that is on the ground at sea level. The balloon is filled with a gas at sea level pressure and temperature. We can identify the system as the fixed mass of gas inside the balloon. The flexible balloon is the system boundary, which separates the gas inside the balloon (system) from the ambient air (surroundings). Now suppose that we let the balloon ascend to an altitude of 10,000 ft, where the ambient air pressure is less than at sea level.

The balloon expands until the internal gas pressure equals the ambient pressure at 10,000 ft. The system boundary has moved outward, but the fixed mass of gas, which we identified as the system, does not change.

We are usually concerned with the transfer of energy between the system and its surrounding, through the system boundaries. One must be careful and precise in clearly defining the system, boundaries, and surroundings in order to accurately quantify the energy exchanges.

3.5.1.2   Properties of a System, Thermodynamic State, and Processes

Consider again the system composed of the gas in the balloon. We previously defined several fundamental physical properties of a substance, such as a fluid, including its mass, pressure, temperature, density, and specific volume. We can similarly specify the properties for a system, such as the pressure and temperature of the balloon gas system. Specification of any two independent thermodynamic properties, such as the pressure and temperature, uniquely defines the *thermodynamic state* of the system. Thus, if we specify the gas pressure and temperature of the balloon when it is on the ground, the thermodynamic state of the system is uniquely defined. The state of the system changes when the balloon is at an altitude of 10,000 ft, as the gas properties are changed.

The properties of the system are categorized as being either *intensive* or *extensive* properties. Intensive properties are independent of the system mass, while extensive properties are a function of the system mass. Pressure, temperature, and density are examples of intensive properties. Imagine that we could cut the balloon in half such that each half contained half of the original mass of gas (assuming that the gas does not escape). The gas temperature in each half would be the same as that before the balloon was divided, so the temperature is independent of the mass. The mass of gas is certainly changed, confirming that mass is an extensive property. Other examples of extensive properties are total volume and specific volume.

Assuming the balloon is stationary, either on the ground or at 10,000 ft, the system properties remain constant and the state is said to be in *equilibrium*. If the system temperature does not change, the system is in *thermal equilibrium*. If the system pressure does not change, the system is in *mechanical equilibrium*. If all of the system properties are constant, such that the thermodynamic state is constant, the system is in *thermodynamic equilibrium*. In many instances, we can approximate the change from an initial to a final state of a system with a series of *quasi-equilibrium* states. Here, it is assumed that the state is changed in infinitesimally small increments, such that the deviation from thermodynamic equilibrium for each incremental state is also infinitesimal.

Whenever one or more of the properties of a system changes, the result is a change in the state of the system. How the properties change from an initial state to a final one is called a *process*. There are several processes of interest where one property in the system remains constant. Examples include the *isobaric* or constant pressure process, the *isochoric* or constant volume process, and the *isothermal* or constant temperature process. Other processes that are useful in aerodynamics and propulsion include the *adiabatic* process, where there is zero heat transfer in or out of the system, the *reversible* process, a constant entropy process where there are no dissipative losses such as due to friction, and the *isentropic* process which is a process that is both adiabatic and reversible. To be clear, a reversible process may involve heat transfer in or out of the system, while an isentropic process is a reversible process with zero heat transfer. Table 3.5 provides a summary of the various processes of interest that have been mentioned.

3.5.1.3   Processes on p-$\mathcal{V}$ and T-s Diagrams

An informative way to follow a thermodynamic process is to plot the changes in the state of the working fluid on pressure–volume, p–$\mathcal{V}$, and temperature–entropy, T–s, diagrams, as shown in

Table 3.5 Summary of processes.

Process	Type of process	Description in terms of e, w, q, s
Isobaric	Constant pressure	$w = pdv$
Isochoric	Constant volume	$w = 0$
Isothermal	Constant temperature	$de = 0^*$
Adiabatic	Zero heat transfer	$q = 0$
Reversible	Constant entropy (no dissipative losses)	$s = $ constant
Isentropic	Adiabatic and reversible	$q = 0, s = $ constant

* This holds for an *ideal gas*, discussed in the following section.

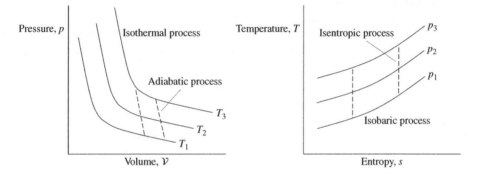

Figure 3.18 Pressure-volume and temperature-entropy diagrams.

Figure 3.18. By drawing lines of constant temperature on the p–V diagram, we can identify a constant temperature or isothermal process. An adiabatic process has zero heat transfer and follows the dashed lines in Figure 3.18, where the temperature changes with pressure and volume. For both of these types of process, we can trace the changes in the state of the gas, to include the temperature, pressure, and volume. We can draw lines of constant pressure on the T–s diagram, as shown in Figure 3.18, defining a constant pressure or isobaric process. An isentropic process follows the vertical dashed lines of constant entropy, in Figure 3.18.

The p–V and T–s diagrams are used to trace many processes of interest in aerodynamics and propulsion. Often, we are interested in a series of connected processes where the state of the gas changes with each process. In some cases, the gas returns to its original state after a series of processes, having thus undergone a *cycle*. This is the case for many propulsive devices where the gas follows a propulsive cycle.

3.5.2 Connecting the Thermodynamic State: The Equation of State

In the previous section, it was stated that the thermodynamic state is uniquely defined by specifying any two independent, thermodynamic properties. The other thermodynamic state properties are related to these two independent properties through an *equation of state*. In the following, we introduce several equations of state that are based on different physical models of the gas. We start with the simplest equation of state, based on the assumption of an *ideal gas*, which is first defined.

3.5.2.1 The Ideal Gas

Consider two physical aspects of a gas, the size of the gas molecules and how the gas molecules interact with each other. First, let us assume that the gas is composed of a large number of identical molecules that are in random motion. We assume that these molecules move in straight-line paths until they collide with another molecule or a surface. The average distance that a molecule travels between collisions is the *mean free path*, λ, which for air at standard sea level conditions (Table 3.1), is equal to 66.3 nm (6.63×10^{-8} m, 2.61×10^{-10} in). This is a very small distance, but it is several orders of magnitude greater than the size of an "air" molecule, since the diameter of a nitrogen molecule or oxygen molecule is on the order of 3×10^{-10} m (1.2×10^{-12} in). Based on this molecular diameter, the volume occupied by a molecule is about 1.4×10^{-29} m^3 (8.5×10^{-25} in^3). Therefore, for our gas molecules moving in three-dimensional space, the distance traveled between collisions is large as compared to the size of the molecules. Based on this, we can assume that the *molecular volume is negligible*.

Concerning molecular collisions, let us assume that these collisions are perfectly elastic, that is, the gas molecules do not lose or gain energy in collisions. The average translational kinetic energy of a gas molecule, KE_{avg}, is shown to be directly proportional to the gas temperature, T, as given by

$$KE_{avg} = \frac{3}{2}kT \tag{3.48}$$

where the constant of proportionality, k, is the Boltzmann constant, equal to 1.38×10^{-23} J/K (5.65×10^{-24} ft·lb/°R). For air at a temperature of 27 °C (80.6 °F, 300 K), the translational kinetic energy of an "air" molecule is

$$
\begin{aligned}
KE_{avg} = \frac{3}{2}kT &= \frac{3}{2}\left(1.38 \times 10^{-23}\,\frac{J}{K}\right)(300\,K) \\
&= 6.21 \times 10^{-21}\,J
\end{aligned}
\tag{3.49}
$$

The molecular collisions are one form of intermolecular interaction, but there is also an interaction due to the electrical nature of the charged particles of the atoms, such as electrons and protons, which comprise the molecules. This intermolecular interaction results in a repulsive force when the molecules are very close together and becomes a weak attractive force as the molecular distance increases. The energies associated with these intermolecular forces vary considerably with the type of gas, but they are orders of magnitude lower than the kinetic energies of the molecules. Therefore, we assume that the molecular collisions are the *only* way that the molecules interact with each other, ignoring the intermolecular forces. Therefore, it is assumed that the *intermolecular forces are negligible*.

Based on the arguments that we have just made about the size, interactions, and energies of gas molecules, we define an *ideal gas* as composed of non-interacting, "point" particles where the *molecules have no volume* and there are *no intermolecular forces*. The ideal gas model is a *theoretical* model and there are limitations to the assumptions in the model. At very high gas densities, when the gas molecules are packed closely together, the assumption that the mean free path is much larger than the molecular volume breaks down. At very low temperatures, the assumption that the kinetic energy of the molecular motion is much larger than the energy associated with the intermolecular forces breaks down. However, at normal pressures and temperatures, results for an ideal gas are within about ±5% of those for a *real gas*, where there are no ideal gas assumptions. The ideal gas assumption is appropriate for most of our external aerodynamic flows of interest, to include subsonic, transonic, and supersonic flows over aircraft, wings, or other bodies. It is also used for internal aerodynamic and propulsive flows, to include flows inside engines, diffusers, ducts, or nozzles.

3.5.2.2   The Ideal Gas Equation of State

If we observe the behavior of an ideal gas as we change the pressure, temperature, or volume, we find that the gas obeys certain physical laws. If we hold the temperature constant, we find that the pressure varies with the inverse of the volume. For example, if we decrease the gas volume by one half while keeping the temperature constant, the gas pressure doubles in magnitude. We can express this observation that the pressure, p, is inversely proportional to the volume, $\mathcal{V}$, at a constant temperature, as

$$p = \frac{C_1}{\mathcal{V}} \tag{3.50}$$

where C_1 is a constant. Rearranging Equation (3.50), we have

$$p\mathcal{V} = C_1 \tag{3.51}$$

which is known as *Boyle's law*.

If we now change the temperature of an ideal gas, we observe that the pressure increases with increasing temperature and decreases with decreasing temperature. This is expressed as the gas pressure being proportional to the temperature, as given by

$$p = C_2 T \tag{3.52}$$

where C_2 is a constant.

If we now change the amount of an ideal gas, or more precisely, if we change the number of moles of the gas, we observe that the volume, $\mathcal{V}$, is proportional to the number of moles, n. (Recall from basic chemistry, that a mole is an amount of a substance that contains a specific number of molecules, this specific number being Avogadro's number, N_A, which has a value of $6.02214199 \times 10^{23}$ molecules/mol.) We can express this last observation as

$$\mathcal{V} = C_3 n \tag{3.53}$$

where C_3 is a constant. There is often confusion with the concept of moles, but Equation (3.53) is simply stating that the gas volume is proportional to the *amount* of gas, with the amount given in terms of moles of gas.

Using Equations (3.52) and (3.53), Boyle's law is given by

$$p\mathcal{V} = (C_2 T)(C_3 n) = C_4 n T \tag{3.54}$$

where C_4 is a constant, defined as the *universal gas constant*, $\mathcal{R}$. The universal gas constant has the same value for all gases. In SI and English units, the universal gas constant is

$$\mathcal{R} = 8{,}314 \frac{J}{(kg \cdot mol)\,K} = 49{,}709 \frac{ft\,lb}{(slug \cdot mol)\,°R} = 1{,}545 \frac{ft\,lb}{(lb_m \cdot mol)\,°R} \tag{3.55}$$

where the non-consistent units of $(lb_m \cdot mol)$ has been included, as this value of the universal gas constant is commonly encountered in the technical literature. To be clear, the (kg·mol) and (slug·mol) are units in themselves, they are not a kilogram or a slug multiplied by a mole. The (kg·mol) and (slug·mol) denote that amount of a substance with a mass equal to the molecular weight in kilograms or slugs, respectively. For instance, one (kg·mol) of air has a mass of 28.96 kg, the molecular weight of air in kilograms. The number of molecules of a substance in a mole, whether it is a (kg·mol), (slug·mol), or other type of mole, is equal to Avogadro's number.

To summarize, we have taken the three physical observations of the behavior of an ideal gas, embodied by Equations (3.51), (3.52), and (3.53), and combined these equations into a single, rather simple equation, given by

$$pV = nRT \tag{3.56}$$

Since Equation (3.56) is based upon the assumption of an ideal gas, it is called the *ideal gas equation*.

We can define a *specific gas constant*, R, in terms of the universal gas constant, $\mathcal{R}$, and the molecular weight of the gas, $\mathcal{M}$, as given by

$$R = \frac{\mathcal{R}}{\mathcal{M}} \tag{3.57}$$

Given the molecular weight of air is 28.96 kg/(kg·mol) or 28.96 slug/(slug·mol), Equation (3.57) is used to calculate the specific gas constant for air, in SI and English units, as

$$R_{air} = \frac{\mathcal{R}}{\mathcal{M}_{air}} = \frac{8314 \frac{J}{(kg \cdot mol)\,K}}{28.96 \frac{kg}{(kg \cdot mol)}} = 287 \frac{J}{kg\,K} \tag{3.58}$$

$$R_{air} = \frac{\mathcal{R}}{\mathcal{M}_{air}} = \frac{49{,}709 \frac{ft\,lb}{(slug \cdot mol)\,°R}}{28.96 \frac{slug}{(slug \cdot mol)}} = 1716 \frac{ft\,lb}{slug\,°R} \tag{3.59}$$

The molecular weight is also sometimes called the *molar mass* and is defined as the total mass, m, of a substance divided by the number of moles, n.

$$\mathcal{M} = \frac{m}{n} \tag{3.60}$$

Inserting Equations (3.57) and (3.60) into the ideal gas equation, Equation (3.56) we obtain

$$p = \frac{n}{V}RT = \frac{n}{V}R\mathcal{M}T = \frac{m}{V}RT$$

$$p = \rho RT \tag{3.61}$$

Equation (3.61) is a form of the ideal gas equation of state that relates the pressure, density, and temperature of an ideal gas.

Equations (3.56) and (3.61) give us a simple way to calculate the properties of an ideal gas, given any two independent, thermodynamic variables. Remember that these equations are valid for an idealized model of a gas, where the gas molecules are assumed to have no volume and to be far apart, so that the intermolecular forces are negligible. However, for many, if not most, of the aerodynamic applications discussed in this book, the ideal gas equation of state is suitable. That said, let us take a look at when the ideal gas assumption may not be valid.

3.5.2.3   Deviation from Ideal Gas Behavior

We have defined an ideal gas as one where the volume of the gas molecules is zero and the intermolecular force between the gas molecules is neglected. We now examine the thermodynamic conditions where these assumptions are valid and invalid. To this end, we introduce the *compressibility factor*, Z, defined as

$$Z = \frac{pV}{nRT} = \frac{p}{\rho RT} \tag{3.62}$$

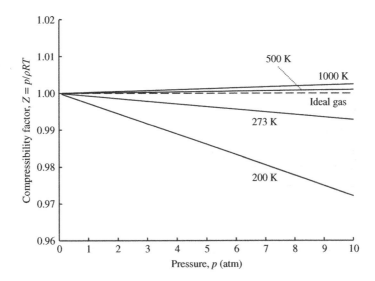

Figure 3.19 Deviation from ideal gas behavior as a function of temperature and pressure.

Figure 3.19 shows the variation of the compressibility factor, Z, as a function of the gas pressure and temperature. For an ideal gas, $Z = 1$, and Equation (3.62) reduces to the ideal gas equation of state. We see that the assumptions of zero molecular volume and negligible intermolecular forces are truly valid only at zero pressure. The gas deviates from ideal gas behavior when $Z \neq 1$, with the greatest deviations occurring at low temperatures and high pressures. At low temperatures, the gas molecules have less kinetic energy to overcome the effects of intermolecular forces. At high pressures, the gas molecules are packed more closely together, making the effects of finite molecular volume and intermolecular forces more significant. In aerospace applications, low temperatures are typically associated with low pressures and high pressures are associated with high temperatures. The low temperature, low pressure regime corresponds to flight at very high altitude. The high pressure, high temperature regime corresponds to flight at hypersonic Mach numbers.

For flight that is not at these extremes of altitude and Mach number, the compressibility factor is near unity, and the ideal gas equation of state is valid. As shown in Figure 3.20, the deviation from ideal gas behavior is less than about 1% for pressures below about 10 atm ($21{,}000 \, \text{lb/ft}^2$, $1.0 \times 10^6 \, \text{N/m}^2$) and temperatures above about 270 K ($-3 \, ^\circ\text{C}$, $26 \, ^\circ\text{F}$). At these higher temperatures, the gas molecules have higher average kinetic energy to overcome the intermolecular forces. At the lower pressures, the molecular spacing is greater, so that the intermolecular forces are not significant. Based on this pressure-temperature range, we see that the ideal gas law has a wide applicability to many of our gas dynamics problems. The compressibility factor approaches one for all gases at low pressures.

3.5.2.4 Other Equations of State

While the ideal gas equation of state has a broad range of application in fluid flows, it is not the only equation of state that can or should be used for all flow situations. In defining any equation of state, we are expressing a relationship among the thermodynamic state variables, such as the pressure, temperature, or density of the system at a given thermodynamic state. This state could be for the flow of air over a wing or for stagnant hydrogen gas in a rocket fuel tank. In many instances,

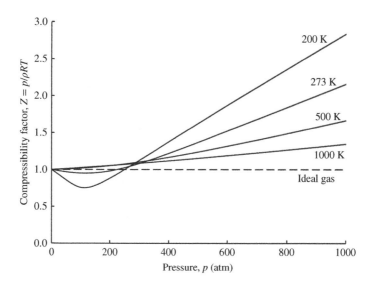

Figure 3.20 Deviation from ideal gas behavior as a function of temperature and pressure (up to 10 atm).

the relationship among the thermodynamic state variables is an actual equation, such as the ideal gas equation.

Sometimes, the relationship is too complex to obtain a closed form equation. For example, when dealing with flows where the gas is chemically reacting, such as the hypersonic flow over a Space Shuttle entering the atmosphere at Mach 25 or the flow of hot combustion gases through a rocket engine, it may not be possible or practical to define an analytical equation that adequately defines the state variables in the flow. For these cases, the thermodynamic state variables may be related through graphs or tables of thermodynamic properties, where we can look up the numbers for the state variables corresponding to the known condition. This look-up could be performed using a numerical scheme or software package on a computer, where an input is provided of the given state conditions, say the pressure and temperature, and the output is the density. The use of these more complex methods is beyond the scope of this book, but we introduce another equation of state that includes additional physics and is therefore slightly more complex than the ideal gas equation of state.

Let us seek to improve the fidelity of the equation of state by addressing its two main assumptions: the gas molecules have zero volume and intermolecular forces are negligible. We distinguish between an ideal gas, a gas that obeys the ideal equation of state, and a *real gas*, which has the true gas behavior and for which the ideal gas equation of state does not apply.

Imagine that we fill a large cylinder of volume, $\mathcal{V}$, with a gas at normal pressure. The volume of the gas molecules is negligible relative to the volume of the cylinder. Now imagine that we compress the gas in the cylinder with a piston, so that the gas pressure is very high and the cylinder volume is reduced to a small fraction of its original volume. At this high pressure condition, the volume occupied by the gas molecules is a significant fraction of the total cylinder volume, so it cannot be neglected. To correct for the fact that the gas molecules are taking up some of the volume of the cylinder, we can correct the volume term in the ideal equation of state, Equation (3.56), as follows

$$\mathcal{V}_{ideal} = \mathcal{V}_{real} - nb \tag{3.63}$$

where $\mathcal{V}_{ideal}$ is the ideal gas volume, $\mathcal{V}_{real}$ is the real gas volume, and nb is the total molecular volume equal to the product of the number of moles of gas, n, and the volume of a single gas molecule, b. This term corrects for the fact that, at high pressures, the real gas volume is larger than predicted by the ideal gas equation.

Now, let us address the issue of the intermolecular forces in a gas. In the ideal gas model, it is assumed that the gas molecules travel in straight paths, uninfluenced by the forces between molecules. In a real gas, the intermolecular forces cause the gas molecules to move in curved rather than straight paths. Since the gas pressure is related to intermolecular collisions, the real gas, curved path collisions result in a lower pressure than the direct, straight path collisions of an ideal gas. Therefore, to correct for the fact that the real gas pressure is lower than predicted by an ideal gas, a correction term is added to the pressure term in the ideal gas equation of state, as follows

$$p_{ideal} = p_{real} + \frac{an^2}{\mathcal{V}^2} \tag{3.64}$$

where p_{ideal} is the ideal gas pressure, p_{real} is the real gas pressure, and $(an^2/\mathcal{V}^2)$ is the pressure correction term with the influence of the intermolecular forces being captured in the gas specific constant, a. The magnitude of the constant a is proportional to the strength of the intermolecular forces, i.e. a is large for substances with strong intermolecular forces.

If we take the corrections for the volume and pressure from Equations (3.63) and (3.64), respectively, and insert them into the ideal equation of state, Equation (3.56), we obtain

$$\left(p + \frac{an^2}{\mathcal{V}^2}\right)(\mathcal{V} - nb) = n\mathcal{R}T \tag{3.65}$$

This equation is known as the *Van der Waals equation of state*, named after the Dutch physicist, Johannes van der Waals (1837–1923), who derived the equation in 1873 and who later was awarded the 1910 Nobel prize in physics for his work on the equation of state. The Van der Waals equation incorporates correction terms for the pressure and volume to account for the intermolecular forces and finite molecular volume, respectively.

At standard pressure and temperature, the correction terms in Equation (3.65) are very small, that is

$$\frac{an^2}{\mathcal{V}^2} \ll p \tag{3.66}$$

$$nb \ll \mathcal{V} \tag{3.67}$$

so that the Van der Waals equation of state reduces to the ideal gas equation of state.

The constants, a and b, in Van der Waals equation of state are properties of a particular gas, which are obtained by experiment. This means that the Van der Waals equation applies to the specific gas corresponding to these particular gas constants. This is in contrast to the ideal gas equation of state, which applies to any gas. Therefore, we see that while the Van der Waals equation of state provides higher fidelity than the ideal gas equation of state, at some conditions, it does this with a bit more complexity and a loss of some general applicability.

Example 3.3 The Ideal Gas Equation of State *At a point in a flow, the density and temperature are 1.134 kg/m^3 and 322.6 K, respectively. Calculate the pressure and specific volume at this point in the flow.*

Solution

Using the ideal gas equation of state, given by Equation (3.61), the pressure is

$$p = \rho RT = \left(1.134 \frac{kg}{m^3}\right)\left(287 \frac{J}{kg\,K}\right)(322.6\,K) = 1.050 \times 10^5 \frac{N}{m^2}$$

The specific volume is given by

$$v = \frac{1}{\rho} = \frac{1}{1.134 \frac{kg}{m^3}} = 0.8818 \frac{m^3}{kg}$$

Example 3.4 Validity of the Ideal Gas Equation of State *Use the Van der Waals equation of state to show that the ideal gas equation of state is valid for air at a pressure and temperature of 101,325 N/m² and 300 K, respectively. The Van der Waals constants for air are*

$$a = 1.358 \times 10^5 \frac{N\,m^4}{(kg \cdot mol)^2}$$

$$b = 3.64 \times 10^{-2} \frac{m^3}{kg \cdot mol}$$

Solution

To show that the ideal gas equation of state is valid for this condition, we show that the correction terms in the Van der Waals equation of state are negligible. Using the ideal gas equation of state, Equation (3.56), we calculate the molar volume or volume per mole as

$$p\mathcal{V} = n\mathcal{R}T$$

$$\frac{\mathcal{V}}{n} = \frac{\mathcal{R}T}{p} = \frac{\left(8314 \frac{J}{(kg\cdot mol)\,K}\right)(300\,K)}{1.01325 \times 10^5 \frac{N}{m^2}} = 24.62 \frac{m^3}{kg \cdot mol}$$

The Van der Waals equation of state is given by Equation (3.65) as

$$\left(p + \frac{an^2}{\mathcal{V}^2}\right)(\mathcal{V} - nb) = n\mathcal{R}T$$

We want to show that the correction terms for the pressure and specific volume are negligible, that is

$$\frac{an^2}{\mathcal{V}^2} = \frac{a}{(\mathcal{V}/n)^2} \ll p$$

$$nb \ll \mathcal{V} \; or \; b \ll \frac{\mathcal{V}}{n}$$

which reduces Van der Waals equation to the perfect gas equation of state. The pressure correction term is

$$\frac{an^2}{\mathcal{V}^2} = \frac{a}{(\mathcal{V}/n)^2} = \frac{1.358 \times 10^5 \frac{N\,m^4}{(kg\cdot mol)^2}}{\left(24.62 \frac{m^3}{kg\cdot mol}\right)^2} = 0.2241 \frac{N}{m^2}$$

which is negligible compared to the magnitude of the pressure, as given below.

$$\frac{an^2}{\mathcal{V}^2} = 0.2241 \frac{N}{m^2} \ll p = 1.01325 \times 10^5 \frac{N}{m^2}$$

The volume correction term is similarly negligible.

$$b = 3.64 \times 10^{-2} \frac{m^3}{kg \cdot mol} \ll \frac{\mathcal{V}}{n} = 24.62 \frac{m^3}{kg \cdot mol}$$

Therefore, the ideal gas equation of state is valid for air at this condition. This is the expected result, as the given pressure and temperature are very near standard conditions, where the ideal gas equation of state is expected to be valid.

3.5.3   Additional Thermodynamic Properties: Internal Energy, Enthalpy, and Entropy

We now define several additional, important thermodynamic properties related to energy, namely internal energy, enthalpy, and entropy. It is important that we clearly understand the physical meaning of each of these terms, the differences between them, and the sign conventions used, as appropriate.

3.5.3.1   Internal Energy

Consider again the fixed mass of gas inside a sealed balloon, which was defined earlier as a thermodynamic system. The gas is composed of molecules that are in random motion inside the balloon. Each gas molecule has a translational kinetic energy due to its translational motion. Each gas molecule also has additional energies associated with the atomic and electronic structure of the molecule. These additional energies are neglected, except for very high temperature conditions, as in hypersonic flight. Therefore, we define the *internal energy*, E, of the system as the sum of the energies associated with the motion of all of the gas molecules. The internal energy is a thermodynamic property of a system, just as are the pressure, temperature, density, and specific volume, with a unique value dependent on the given state of the system. Later, we show that for an ideal gas, the internal energy is a function of the temperature only.

The system internal energy does not include the kinetic and potential energies associated with the motion and position of the system relative to the surroundings. If our balloon is traveling at 100 mph at an altitude of 10,000 ft, the associated kinetic and potential energies of the balloon do not affect the internal energy of the molecules *inside* the balloon. In other words, the energy due to the interaction of the system with the surroundings does not affect the energy associated with the interaction of the molecules with each other.

As we have stated, the internal energy, E, depends on the sum of the energies of a given mass of gas molecules. If we were to change the mass of the system, the internal energy would also change. Therefore, the internal energy is an *extensive* property, meaning that it depends on the mass of the system. We often deal with internal energy per unit mass, e, sometimes referred to as the *specific internal energy*, defined as

$$e \equiv \frac{E}{m} \tag{3.68}$$

The internal energy per unit mass is an *intensive* property. We often simply use the term *internal energy* when referring to the specific internal energy and the term *total internal energy* for the extensive property, E.

3.5.3.2   Enthalpy

We now introduce another thermodynamic property that is useful in aerodynamics and propulsion. This property is the enthalpy, where the total enthalpy, H, is defined as

$$H \equiv E + p\mathcal{V} \tag{3.69}$$

where E is the total internal energy, p is the pressure, and $\mathcal{V}$ is the volume.

We may also define an enthalpy per unit mass, or specific enthalpy, h, as

$$h = \frac{H}{m} \equiv e + pv \tag{3.70}$$

where e is the internal energy and v is the specific volume. The total enthalpy is an extensive property, while the specific enthalpy is an intensive property.

If we are dealing with an ideal gas, where the ideal equation of state is valid, we may write Equation (3.70) as

$$h \equiv e + RT \tag{3.71}$$

where R is the specific gas constant and T is the temperature. Since the internal energy is a function of temperature only for an ideal gas (to be shown later) and R is a constant for a specific gas, Equation (3.71) tells us that the enthalpy, of an ideal gas, is a function of temperature only.

3.5.3.3 Entropy

The concept of entropy was introduced in the 19th century by Rudolph Clasius (1822–1888), a German physicist who made significant contributions to the development of the science of thermodynamics. He is often credited with the formulation of the first and second laws of thermodynamics.

Entropy deals with the "waste" of energy or its unavailability to do useful work. Specifically, entropy is the thermal energy of a system, per unit temperature, that is "wasted" or not available to do useful work. If we think of work as having to do with the transformation of energy due to *ordered* molecular motion, then molecular *disorder* tends towards energy transfer that does no useful work. From this perspective, entropy is viewed as a measure of the *disorder* or *randomness* of a system.

The total entropy, S, is an extensive property of a substance, dependent on the mass of the system. As with work and heat, we can define the intensive property, specific entropy, s, or entropy per unit mass, as

$$s \equiv \frac{S}{m} \tag{3.72}$$

The total entropy and specific entropy have units of energy and energy per unit mass, respectively. Entropy is a point property, dependent only on the end states of a property.

We are almost always dealing with *changes* in the entropy of a system. This makes the assignment of a zero for entropy arbitrary in many cases. For instance, specifying the entropy of water to be zero at a pressure of one atmosphere and a temperature of $0\,°C$, provides a reference state from which to calculate the value of the entropy at different states, that is, different pressures and temperature. These entropy values are based on an arbitrary reference state, but this does not matter if all we are concerned with is the *change* in entropy from one state to another. The third law of thermodynamics provides a basis for a state of zero entropy at zero absolute temperature, but this is not critical to our applications, where we are only dealing with changes in entropy.

3.5.4 *Work and Heat*

We now discuss two forms of energy transfer for a system, work and heat. These quantities are defined as *boundary* phenomena because both occur at the boundaries of a system, when energy is transferred across the boundary. They are both *transient* phenomena, occurring when a system undergoes a change in state, resulting in work or heat crossing a system boundary. Finally, both work and heat are *path* functions, which depend on the way the system gets from the initial to the final state. We elaborate more on these aspects of work and heat in the following sections.

The units of work and heat are the units of energy, joules (J) in SI units and foot-pounds (ft·lb) or British thermal units (Btu) in English units. Work is defined as a force multiplied by a distance. The SI unit of work, the joule, is defined as a force of one newton multiplied by a distance of one meter. The joule is defined in terms of SI base units as

$$1\,J = 1\,N \cdot m = 1\,\frac{kg \cdot m^2}{s^2} \tag{3.73}$$

In English units, work is defined as a force of one pound multiplied by a distance of one foot. Energy and work have the same units, so that the units of energy, and sometimes work, are expressed in the inconsistent English units of the British thermal unit. The Btu is defined as the amount of energy required to raise the temperature of one pound of water by one degree Fahrenheit. The unit conversions for the Btu are as follows.

$$1\,Btu = 778\,ft \cdot lb = 1055\,J \tag{3.74}$$

3.5.4.1 Work

In mechanics, work, W, is defined as a force, F, acting through the distance, x, where the displacement is in the same direction as the force. Assuming a constant force and an incremental displacement, dx, that is integrated over some path, we have

$$W = \int F\,dx \tag{3.75}$$

Let us return to the balloon gas system, and consider the work associated with the system when it ascends from sea level to a high altitude. If the balloon rises a small increment in altitude, the balloon's radius increases by a small displacement, Δr, as shown in Figure 3.21. Assuming that these changes are small, we can consider the pressure difference, between the interior and the exterior of the balloon during this altitude change, as constant. Let us denote this constant pressure difference simply as p. Now consider an elemental area of the balloon surface area, dA, upon which

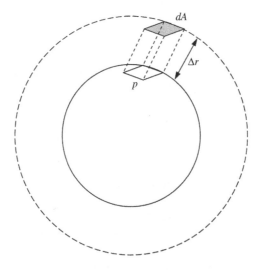

Figure 3.21   Work done in expanding a balloon.

this pressure acts. The force, F, on this elemental area is simply the force times the area, $p\,dA$. The increment of work, ΔW, done by the system is this force times the distance Δr that the incremental area has moved due to the balloon expansion. (For simplicity, we have ignored the surface tension inherent in the flexible surface of the balloon, which would contribute to the work.) Therefore, the increment in work is given by

$$\Delta W = F \cdot \Delta r = (p\,dA) \cdot \Delta r \tag{3.76}$$

Integrating over the entire surface area of the balloon, the total increment in work, δW, is given by

$$\delta W = \int_A (p\,dA)\,\Delta r = p \int_A \Delta r\,dA = p\,d\mathcal{V} \tag{3.77}$$

where the integral of $\Delta r\,dA$ is the change in volume, $d\mathcal{V}$, of the balloon due to the small displacement Δr. The change in volume, $d\mathcal{V}$, is a positive quantity since the gas is expanding, which increases the volume.

Now, we must make a clarification about the sign convention for work. It is desirable to use a frame of reference based on the system in regards to the direction of the energy transfer. When energy is transferred *to* the system, that is, work is done *on the system* by its surroundings, we define the sign of work as positive. When energy is transferred *from* the system, that is, work is done *by the system* against its surroundings, we define the sign of work as negative. Positive work corresponds to *compression* of the system and energy *entering* the system, while negative work corresponds to *expansion* of the system and energy *exiting* the system.

In our balloon expansion example, work is being done *by* the system because the gas is pushing the balloon surface outward. Therefore, the sign of work should be negative, so that Equation (3.77) is rewritten as

$$-\delta W = p\,d\mathcal{V} \quad \text{or} \quad \delta W = -p\,d\mathcal{V} \tag{3.78}$$

Dividing by the system mass, we can write Equation (3.78) in terms of the specific work and specific volume.

$$\delta w = -p\,dv \tag{3.79}$$

The total work done, $W_{1\to 2}$, for the finite volume change of the balloon gas, from the initial volume on the ground, $\mathcal{V}_1$, to a final volume at 10,000 ft, $\mathcal{V}_2$, is given by

$$\int_1^2 \delta W = W_{1\to 2} = -\int_{\mathcal{V}_1}^{\mathcal{V}_2} p\,d\mathcal{V} \tag{3.80}$$

The pressure in Equation (3.80) may not be a constant, for the finite volume change from state 1 to 2. Therefore, to evaluate the integral in Equation (3.80), we need to know how the pressure varies as a function of the volume. The integral, in Equation (3.80), is the area under the pressure curve function on a pressure–volume, or p–$\mathcal{V}$, diagram, as shown in Figure 3.22.

The work done for the finite volume change from $\mathcal{V}_1$ to $\mathcal{V}_2$ is the area under the $p = f(\mathcal{V})$ curve in Figure 3.22. By inspection of this figure, we see that if we had selected another curve or *path* from the state 1 (where the volume is $\mathcal{V}_1$) to state 2 (where the volume is $\mathcal{V}_2$), the area under the curve would be different, and hence the work done would have a different value. From this, we conclude that the work done is dependent not only on the end states (state 1 and 2) of the process, but also depends on the path that is followed to get from state 1 to 2.

Functions such as this are called *path functions*, which depend on the path taken to get from the initial to the final state. This is in contrast to *point functions* which depend only on the end states and not on the path. Thermodynamic properties, such as pressure, temperature, volume, internal energy, and entropy are examples of point functions that depend only on the given state and not

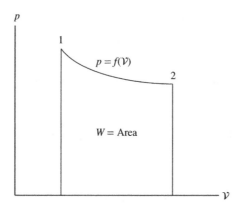

Figure 3.22   Pressure–volume diagram and work.

on the path that was taken to get to that state. There is also a mathematical distinction between path functions and point functions. The differentials of point functions are the more familiar *exact differentials*, while path functions are *inexact differentials*. For a function, F, exact differentials are denoted as dF, while inexact differentials are denoted as δF. Integration of an exact differential, such as volume, $d\mathcal{V}$, is given by

$$\int_1^2 d\mathcal{V} = \mathcal{V}_2 - \mathcal{V}_1 \tag{3.81}$$

We cannot integrate an inexact differential in this manner. In fact, all we can do is symbolize the integration as having followed a path, as in the work done from state 1 to 2, given by

$$\int_1^2 \delta W = W_{1\to 2} \tag{3.82}$$

We often deal with the work per unit mass, w, given by

$$w \equiv \frac{W}{m} \tag{3.83}$$

where m is the mass of the system.

The time rate of change of work is the power, P, defined as

$$P = \dot{W} \equiv \frac{\delta W}{dt} \tag{3.84}$$

Example 3.5 Work Done in an Isothermal Process of an Ideal Gas *Consider a sealed, flexible balloon with a diameter of 6 m, filled with helium gas ($R = 2077$ J/kg-K) at a pressure and temperature of 689 kPa and 288 K, respectively. If the balloon expands to a diameter of 9 m, what is the final pressure and the work done by the gas in the expansion, assuming the process is isothermal? Assume that the helium is an ideal gas.*

Solution

The thermodynamic system is the helium gas in the balloon. Since we are assuming that the helium is an ideal gas, we can use the ideal gas equation of state, given by

$$p = \rho RT = \frac{m}{\mathcal{V}} RT$$

We can solve for the constant mass of gas in the system, using the conditions for State 1.

$$m = \frac{p_1 V_1}{RT}$$

We use the equation for the volume of a sphere to calculate the initial (State 1) gas volume, corresponding to a radius of 6 m.

$$V_1 = \frac{4}{3}\pi \left(\frac{d_1}{2}\right)^3 = \frac{4}{3}\pi \left(\frac{6m}{2}\right)^3 = 113.1\,m^3$$

The system mass is

$$m = \frac{p_1 V_1}{RT} = \frac{\left(689 \times 10^3\,\frac{N}{m^2}\right)(113.1\,m^3)}{\left(2077\,\frac{J}{kg \cdot K}\right)(288\,K)} = 45.57\,kg$$

Since the system mass remains constant and we are assuming that the process is isothermal (constant temperature), we can calculate the pressure for State 2 as

$$p_2 = \frac{m}{V_2}RT$$

The volume for State 2 is

$$V_2 = \frac{4}{3}\pi \left(\frac{d_2}{2}\right)^3 = \frac{4}{3}\pi \left(\frac{9m}{2}\right)^3 = 381.7\,m^3$$

Therefore, the pressure for State 2 is

$$p_2 = \frac{m}{V_2}RT = \frac{45.57\,kg}{381.7\,m^3}\left(2077\,\frac{J}{kg \cdot K}\right)(288\,K) = 71{,}414\,\frac{N}{m^2}$$

The work done by the gas in the expansion is given by Equation (3.80).

$$W_{1 \rightarrow 2} = -\int_{V_1}^{V_2} p\,dV$$

Substituting in the ideal equation of state for the pressure, and recalling that the system mass and temperature are constant, we have an expression for the work done for an ideal gas in an isothermal process.

$$W_{1 \rightarrow 2} = -\int_{V_1}^{V_2} p\,dV = -\int_{V_1}^{V_2} \left(\frac{m}{V}RT\right)dV = -mRT \int_{V_1}^{V_2} \frac{dV}{V} = -mRT \ln\frac{V_2}{V_1}$$

This equation is valid whether the process is an isothermal expansion or an isothermal compression. Substituting in our values, we have the total work done by the gas due to the isothermal expansion.

$$W_{1 \rightarrow 2} = -mRT \ln\frac{V_2}{V_1} = -(45.57\,kg)\left(2{,}077\,\frac{J}{kg \cdot K}\right)(288\,K)\ln\frac{(381.7\,m^3)}{(113.1\,m^3)} = -3.316 \times 10^7\,J$$

The negative sign for the result indicates that work is being done by the system against the surroundings (the gas is expanding) and energy is leaving the system.

3.5.4.2 Heat

Similar to work, heat, Q, is a form of energy that is transferred across a system boundary. We know from common experience that if we place a hot object next to a cold object, there is heat transferred from the hot object to the cold one, until the temperatures equilibrate. Heat is the *form of energy that is transferred* due to the temperature difference between the system and its surroundings or another system. The energy transfer itself is called *heat flow* or *heat transfer*.

A system cannot possess work or heat, rather it is a transient phenomenon that is observed when a system undergoes a change in state. Once the change in the system state is complete, the system does not contain any work or heat. This concept may seem more intuitive when one thinks of *heat* as *heat transferred*, since heat is transferred across a system boundary due to a temperature difference. This should not be confused with the common usage of the term *heat transfer*, which is usually referring to the heat transfer *rate*, with units of energy per unit time.

Heat is a path function, similar to work. The heat transferred is a function of the end states and is also dependent on the path followed between the end states. We denote the inexact differential of heat as δQ. This inexact differential of heat should not be interpreted as a *change in the amount* of heat, as a system does not contain an amount of heat that is subject to change. Rather, δQ represents an increment of the energy in transit to or from a system. When there is no heat transferred in a change of state, $Q = 0$, the process is called *adiabatic*. This differs from an *isothermal* process, where the temperature remains constant, but the heat transfer may not be zero.

The sign convention used for heat is the same as that used for work, in regards to the direction of energy transfer. Heat is positive for energy transferred *to* the system (heat added to the system) and negative for energy transferred *from* the system (heat removed from the system).

We often deal with the heat per unit mass, q, given by

$$q \equiv \frac{Q}{m} \tag{3.85}$$

with the inexact differential of the heat per unit mass given by δq.

The time rate of change of heat, or heat transfer rate, $\dot{Q}$, is defined as

$$\dot{Q} \equiv \frac{\delta Q}{dt} \tag{3.86}$$

3.5.5 The Laws of Thermodynamics

The science of thermodynamics is based upon experimental observations of phenomena in the natural world. The fundamental laws of thermodynamics come from these observations. These are fundamental laws of science based on empirical observations, meaning that they are not derivable from first principles. There are four fundamental laws of thermodynamics, the zeroth, first, second, and third laws. We discuss the first and second laws of thermodynamics in detail below, as these two laws play an important role in our future development of theories and equations in aerodynamics and propulsion. The first and second laws lead to the properties of internal energy and entropy, respectively. We make brief mention here of the zeroth and third laws of thermodynamics for completeness.

3.5.5.1 The Zeroth and Third Laws of Thermodynamics

The zeroth law of thermodynamics deals with the thermal equilibrium of bodies (or systems). It states that when two bodies have the same temperature as a third body, then the temperatures of the two bodies are also the same. This law seems rather obvious, but since this observation is not derivable from any other basic principles, it must be put forth as a fundamental law. When we use a thermometer to measure a temperature, we are applying the zeroth law. We calibrate and mark a temperature scale on a thermometer, based on having the thermometer in thermal equilibrium with a system of known temperature. For example, we could insert a thermometer in an ice bath with a temperature of $0\,°\text{C}$ and place a zero mark on the thermometer. If we then use the thermometer to measure the temperature of another object and the thermometer reads zero, then the zeroth law tells us that the object's temperature must be equivalent to that of the ice bath or $0\,°\text{C}$.

The third law of thermodynamics has to do with the definition of an absolute temperature scale, where zero on the Kelvin temperature scale, known as *absolute zero*, corresponds to the minimum internal energy of a system. The third law does not state that all molecular motion ceases at absolute zero. The third law does provide a basis for the measurement of entropy, whereby the entropy of substances is defined at absolute zero. The third law states that a perfect crystal has an entropy equal to zero at a temperature of absolute zero. A perfect crystal may be thought of as a structure with the maximum degree of order or minimum degree of disorder.

3.5.5.2 The First Law of Thermodynamics

The first law of thermodynamics is a statement of the conservation of energy for a system of fixed mass. It states that the change in the internal energy of the system, de, is equal to the sum of the heat added to the system, δq and the work done on the system, δw. In equation form, the first law of thermodynamics is given by

$$de = \delta q + \delta w \tag{3.87}$$

We can write another useful form of the first law, using the definition of work in terms of pressure and specific volume, Equation (3.79), as

$$de = \delta q - p dv \tag{3.88}$$

(We are following the sign convention for heat and work as previously defined.) In Equation (3.88), heat and work are path dependent, while internal energy is not. During a thermodynamic process, the change in the internal energy of a system, Δe, depends only on the initial and final states, and not on the path between these states. We can write the first law for a thermodynamic process from an initial state 1 to a final state 2, as

$$\Delta e = e_2 - e_1 = w_{1 \to 2} + q_{1 \to 2} \tag{3.89}$$

where e_1 and e_2 are the internal energies at the initial and final states, respectively, and $w_{1 \to 2}$ and $q_{1 \to 2}$ are the work done and heat transfer from state 1 to 2, respectively.

For our aerodynamics and propulsion applications, we find it useful to rewrite the first law, in terms of other thermodynamics variables such as pressure, volume, and enthalpy. We start with the definition of enthalpy, Equation (3.70), rewritten here for convenience

$$h = e + pv \tag{3.90}$$

Differentiating Equation (3.90), we obtain

$$dh = de + p dv + v dp \tag{3.91}$$

Rearranging, we have

$$de = dh - p dv - v dp \tag{3.92}$$

Using Equation (3.88), we have

$$de = \delta q - pd\upsilon = dh - pd\upsilon - \upsilon dp \tag{3.93}$$

$$\delta q = dh - \upsilon dp \tag{3.94}$$

Equation (3.94) is another form of the first law of thermodynamics, relating the heat to the enthalpy, pressure, and specific volume.

For a constant pressure process, $dp = 0$, the first law of thermodynamics is simply

$$\delta q = dh \tag{3.95}$$

where the change in heat equals the change in enthalpy.

3.5.5.3   The Second Law of Thermodynamics

When we discussed heat and the transfer of energy between two objects at different temperatures, we stated that the heat flowed from the hot object to the cold one. This implied a *direction* for the heat transfer, from hot to cold. However, can the heat transfer be from the cold object to the hot one? The first law of thermodynamics does not dictate anything about the *direction* of the energy transfer. We know from common experience, though, that a hot cup of coffee cools down until its temperature is equilibrated with its surroundings. However, the hot cup of coffee does not get hotter by virtue of the cooler surroundings transferring energy to the hot coffee.

The second law of thermodynamics states the direction that a process must follow and the impossibility of the process in the opposite direction. In essence, the second law states that a process must proceed in a direction such that the disorder of the system and surroundings remains constant or increases. We have previously related the disorder of a system to the thermodynamic property entropy, and we use the concept of entropy to quantitatively describe the second law of thermodynamics.

Let us again return to our gas-filled balloon. Assume that the balloon is warmed by the Sun such that a small amount of heat, δq, is added to the gas in the balloon, causing it to expand by an infinitesimal amount. Let us assume that the balloon expands at just the right rate such that the gas temperature remains constant. Since the internal energy is a function of the temperature of an ideal gas, the internal energy remains constant and the change in the internal energy is zero. Applying the first law of thermodynamics, Equation (3.87), to this isothermal process, we have

$$de = 0 = \delta w + \delta q \tag{3.96}$$

Equation (3.96) says that all of the heat added to the system must be converted to work and vice versa. There are no losses in the energy conversions back and forth between heat and work. Therefore, this isothermal process of an ideal gas is a *reversible* process.

Using the definition of work and the ideal equation of state, we can rewrite Equation (3.96) as

$$\delta q = -\delta w = pd\upsilon = (\rho RT)d\upsilon = RT\frac{d\upsilon}{\upsilon} \tag{3.97}$$

Rearranging, we have

$$\frac{d\upsilon}{\upsilon} = \left(\frac{1}{R}\right)\frac{\delta q}{T} \tag{3.98}$$

where $\delta q/T$ represents an amount of heat added to the system at the temperature, T, and $d\upsilon/\upsilon$ represents the fractional increase in the gas volume due to the expansion of the balloon.

After the balloon expands, the gas molecules are moving in a slightly larger volume, therefore we can say that their randomness and disorder in position has increased. Therefore, we can connect the increase in volume, dv/v, with the increase in the degree of randomness or disorder of the system. So, Equation (3.98) is telling us that the increase in randomness or disorder of a system is proportional to the heat added to the system divided by the system temperature (the $1/R$ term may be considered a constant of proportionality). Previously, we defined the entropy as a measure of the randomness or disorder in a system, so let us now define the change in entropy, ds, for a reversible process as

$$ds \equiv \left(\frac{\delta q}{T} \right)_{rev} \qquad (3.99)$$

We can calculate the change in entropy of a system for an isothermal reversible process, from an initial state 1 to a final state 2, by integrating Equation (3.99) as given by

$$\Delta s = s_2 - s_1 = \int_1^2 \left(\frac{\delta q}{T} \right)_{rev} \qquad (3.100)$$

For irreversible processes, there are finite system losses, such that the entropy increases. Therefore, we can generalize Equation (3.99) for an irreversible process as

$$ds \geq \frac{\delta q}{T} \qquad (3.101)$$

where we have now indicated, by the inequality sign, that the entropy would increase for an irreversible process. The equality sign would still be valid for the case of a reversible process. Equation (3.101) is a statement of the second law of thermodynamics, indicating that a thermodynamic process must proceed in a direction such that entropy is increased, or at best remain constant if there are no losses. According to Equation (3.101), it is impossible for the process to proceed in direction such that the entropy decreases. However, what if heat is removed from the system, such that δq is negative? Would not the change in entropy be negative also, resulting in a decrease in entropy? The entropy would tend to decrease due to the heat transfer, but because of the real losses in an irreversible process, the entropy *of the system* would still increase.

3.5.6 Specific Heats of an Ideal Gas

Thinking again about our gas-filled balloon, if we add heat, δq, to the gas in the balloon, we would expect the gas temperature to increase by some amount, dT. (This is true, unless we did the heat addition process in a special way, such as an isothermal process, where the balloon is expanding at just the right rate to maintain a constant temperature.) For a system of fixed mass, we can say that the heat transferred, δq, is directly proportional to the temperature change, dT, as given by

$$\delta q = c \, dT \qquad (3.102)$$

where we define c as the *specific heat*, which has different values depending on the type of process and the type of material in the system.

Heat is a path function, so that the heat added to achieve the same temperature rise also differs depending on the path or process used. Let us consider two specific types of processes, the *isochoric* (constant volume) process and the *isobaric* (constant pressure) process. Both processes are of interest in aerodynamics and propulsion. (An isothermal or constant temperature process would be a meaningless exercise, as dT and δq are both zero.)

Back to our balloon gas example, imagine that we add heat to the gas, but constrain the volume so that the temperature change, dT, follows a constant volume process. We wish to apply the first law of thermodynamics, Equation (3.88), repeated below, to the process.

$$de = \delta q - pdv \tag{3.103}$$

Since there is no volume change, dv, the first law becomes

$$de = (\delta q)_v \tag{3.104}$$

where the subscript v on δq denotes heat addition at constant volume. Equation (3.104) tells us, that for this temperature increase at constant volume, the change in the gas internal energy, de, is simply equal to the heat added during the constant volume process, $(\delta q)_v$.

Now, let us assume that we perform the same temperature change, dT, using a constant pressure process. According to the ideal gas equation of state, if the gas temperature increases, the gas volume must also increase to maintain constant pressure. Therefore, application of the first law of thermodynamics to the constant pressure heat addition process yields

$$de = (\delta q)_p - pdv \tag{3.105}$$

where $(\delta q)_p$ is the heat added during the constant volume process and dv is the change in the gas volume to maintain the constant pressure, p.

Shortly, we show that the change in the internal energy, de, is directly proportional to the change in the temperature, dT. Therefore, the changes in internal energy, de, in Equations (3.104) and (3.105) are equivalent, since it was assumed that the temperature change, dT, was the same in both processes, giving us

$$(\delta q)_v = (\delta q)_p - pdv \tag{3.106}$$

For the left and right sides of this equation to be equal, the heat added at constant pressure must be greater than the heat added at constant volume, $(\delta q)_p > (\delta q)_v$, to make up for the work done by the gas in the constant pressure expansion. Thus, we have shown that the heat added to the system is indeed path dependent, making the specific heat, in Equation (3.102), also path dependent. We can write Equation (3.102) for a constant volume process, as

$$\delta q = c_v dT \tag{3.107}$$

where c_v is the specific heat at constant volume. Similarly, for a constant pressure process, we have

$$\delta q = c_p dT \tag{3.108}$$

where c_p is the specific heat at constant pressure. We can also think of these specific heats as the heat added to the system, per unit temperature, at either constant volume or constant pressure.

The specific heat varies depending on the type of material in the system. We know from common experience, that the quantity of heat needed to obtain a given temperature change depends on the type of material. For instance, more heat needs to be added to 1 kg of water to increase its temperature by 1 °C than to a 1 kg piece of steel. To obtain the same temperature rise, the heat added to the gas in our balloon differs depending on the type of gas. Therefore, the specific heat in Equation (3.102) depends on the type of material in the system.

In aerodynamics and propulsion, we generally deal with gases, and often with air in particular. Table 3.6 provides the numerical values of the specific heats at constant volume and at constant pressure for air and other common gases. We almost always deal with a gas that obeys the ideal or perfect gas equation of state. An ideal or perfect gas, where the specific heats at constant volume and

Table 3.6 Specific heats, specific gas constants, and ratio of specific heats of some common gases at 1 atm and 293 K (68 °F, 20 °C).

Gas	Specific heat at constant volume, c_p J/kg·K (ft·lb/slug·°R)	Specific heat at constant pressure, c_v J/kg·K (ft·lb/slug·°R)	Specific gas constant, $R = c_p - c_v$ J/kg·K (ft·lb/slug·°R)	Ratio of specific heats, $\gamma = c_p/c_v$
Air	1006 (6020.7)	719 (4303.1)	287 (1717.6)	1.399
Helium (He)	5190 (31,061)	3113 (18,630.6)	2077 (12,430)	1.667
Hydrogen (H$_2$)	14,320 (85,702)	10,200 (61,044.7)	4120 (24,657.3)	1.404
Nitrogen (N$_2$)	1039 (6218.2)	742 (4440.7)	297 (1777.5)	1.400
Oxygen (O$_2$)	75916 (5482.1)	656 (3926.0)	260 (1556.1)	1.396
Carbon dioxide (CO$_2$)	840 (5027.2)	651 (3896.1)	189 (1131.1)	1.290

at constant pressure are assumed to be constant, is called a *calorically perfect gas*. For all practical purposes, we assume a calorically perfect gas for all of our future aerodynamic discussions.

For gases, the specific heat at constant pressure is always greater than the specific heat at constant volume. This falls out from the analysis we just completed concerning heat addition for a constant volume and a constant pressure process, where it was shown that $(\delta q)_p > (\delta q)_v$. By substituting Equations (3.107) and (3.108) into this inequality, we get $c_p > c_v$. From Table 3.6, c_p is 40% greater than c_v for air. This means that the heat addition at constant pressure is 40% greater than that at constant volume, to obtain the same temperature change.

Also shown in Table 3.6 is the dimensionless *ratio of specific heats*, defined as

$$\gamma = \frac{c_p}{c_v} \tag{3.109}$$

We see this dimensionless parameter frequently in aerodynamics and propulsion. The ratio of specific heats, γ, is always greater than one for a gas, since c_p is always greater than c_v for a gas.

Looking closely at Table 3.6, we notice an interesting correlation between the type of gas and the ratio of specific heats. Helium is a monatomic gas (composed of a single atom), with a ratio of specific heats equal to 1.67. Hydrogen, nitrogen, and oxygen are diatomic gases (composed of two atoms), with a ratio of specific heats of about 1.4. Carbon dioxide is a polyatomic gas (composed of three atoms), with a ratio of specific heats of 1.3. Using kinetic theory, the ratio of specific heats is shown to have about the same value for the same type of gas. For monatomic gases, $\gamma \cong 5/3$, for diatomic gases, $\gamma \cong 7/5$, and for polyatomic gases (composed of three or more atoms), $\gamma \cong 4/3$. Air has a ratio of specific heats of 1.40, since it is composed mostly of diatomic nitrogen and oxygen.

We can now use our new definitions c_v and c_p to obtain some useful relations for internal energy and enthalpy. First, consider the equation for the first law of thermodynamics that we obtained for a constant volume process of an ideal gas, Equation (3.104). If we substitute the definition of specific heat at constant volume, Equation (3.107), into this equation, we obtain a relation between the internal energy, the specific heat at constant volume, and the temperature change.

$$de = c_v \, dT \tag{3.110}$$

Assuming that c_v is a constant, we can integrate Equation (3.110) from an initial state 1 to a final state 2, as

$$\int_1^2 de = c_v \int_1^2 dT \tag{3.111}$$

Setting both the internal energy and temperature to zero at state 1, we obtain

$$e = c_v T \tag{3.112}$$

Equation (3.112) states that the internal energy is a function of temperature only, as we have mentioned in past discussions. This equation defines the specific heat in terms of the properties of the gas, that is, the internal energy and temperature. Therefore, we have shown that c_v is also a property of the gas. Equation (3.112) does not contain any terms that relate to a thermodynamic process of any kind. It does contain a term, the specific heat at constant volume, which is defined for a specific process, but the validity of the equation is not restricted to this process. Equation (3.112) is valid for any process involving an ideal gas.

Now, consider the equation for the first law of thermodynamics, in terms of enthalpy, for a constant pressure process, Equation (3.95). If we substitute the definition of specific heat at constant pressure, Equation (3.108), into this equation, we obtain a relation between the enthalpy, the specific heat at constant pressure, and the temperature change.

$$dh = c_p\, dT \tag{3.113}$$

Assuming that c_p is a constant, we can integrate Equation (3.113) from an initial state 1 to a final state 2, as

$$\int_1^2 dh = c_p \int_1^2 dT \tag{3.114}$$

Setting both the internal energy and temperature to zero at state 1, we obtain

$$h = c_p T \tag{3.115}$$

Equation (3.115) tells us that the enthalpy is a function of temperature only. Similar to the specific heat at constant volume, we have shown that c_p is also a property of the gas, since the enthalpy and temperature are properties of the gas. Similar to Equation (3.112), Equation (3.115) is valid for any process involving an ideal gas.

Finally, let us return to the definition of enthalpy, as given by Equation (3.71), repeated below.

$$h = e + RT \tag{3.116}$$

Inserting the definitions of the internal energy and enthalpy in terms of the specific heats and temperature, Equations (3.112) and (3.115), into this equation, we obtain

$$c_p T = c_v T + RT \tag{3.117}$$

Simplifying, we have

$$c_p = c_v + R \tag{3.118}$$

Solving for the specific gas constant, we have

$$R = c_p - c_v \tag{3.119}$$

Equation (3.119) shows that the specific gas constant of an ideal gas is the difference between the specific heat at constant pressure and the specific heat at constant volume. This equation is also another confirmation that c_p is greater than c_v. Values of the specific gas constant for several gases are shown in Table 3.6.

Other useful equations for the specific heats, in terms of the ratio of specific heats and the specific gas constant, are obtained as follows. Dividing Equation (3.118) by the specific heat at constant volume, c_v, we have

$$\frac{c_p}{c_v} = \frac{c_v}{c_v} + \frac{R}{c_v} \tag{3.120}$$

$$\gamma = 1 + \frac{R}{c_v} \tag{3.121}$$

Solving for the specific heat at constant volume, we have

$$c_v = \frac{R}{\gamma - 1} \tag{3.122}$$

Inserting Equation (3.122) into Equation (3.109), we have

$$\gamma = \frac{c_p}{c_v} = \frac{c_p}{R/(\gamma - 1)} = (\gamma - 1)\frac{c_p}{R} \tag{3.123}$$

Solving for the specific heat at constant pressure, we have

$$c_p = \frac{\gamma R}{\gamma - 1} \tag{3.124}$$

3.5.7   Isentropic Flow

Earlier, we defined the adiabatic process, where there is no heat transfer in or out of the system, and the isentropic process, where there is no heat transfer (adiabatic) and no losses (reversible). At first glance, these may seem like overly restrictive processes with little application to real-world aerodynamic flows. However, let us consider the flow of air over an aircraft, wing, or other body. There is no heat being added or taken away from the external flow of air over the body, so the flow is adiabatic. There could be irreversible losses in the flow due to skin friction in viscous boundary layers, adjacent to the body surfaces or due to shock waves in a supersonic flow. However, if we are not concerned with the flow in the boundary layers or through shock waves, in supersonic flow, there are no irreversible losses in the flow. This applies to the flow upstream and downstream of a shock wave, as the losses are incurred in passing through the shock wave. The upstream and downstream flows are reversible and of constant entropy, although the entropies ahead of and behind the shock are not equal.

Thus, we see that the external aerodynamic flow over a body may be considered adiabatic and reversible, or isentropic, as long as we are focused on the regions outside of loss-producing mechanisms, such as boundary layers and shock waves. This applies to a wide range of external aerodynamic flows over many types of geometries of interest. The assumption of isentropic flow can also apply to internal flows of interest, such as through wind tunnels, engine inlets, rocket nozzles, and other internal flow geometries. Exceptions in propulsive flows include the addition of heat in a duct, such as an engine or rocket combustion chamber.

We now derive several relationships, applicable to isentropic processes, which we find useful in our analysis of the many types of aerodynamic flows that have been mentioned. We start with the second law of thermodynamics for a reversible process, Equation (3.99), repeated below,

$$ds \equiv \left(\frac{\delta q}{T}\right)_{rev} \tag{3.125}$$

and the first law of thermodynamics, Equation (3.94), in terms of the enthalpy, repeated below.

$$\delta q = dh - v\,dp \tag{3.126}$$

Inserting Equation (3.126) into (3.125), we have (where the "reversible" subscript has been dropped)

$$ds = \frac{dh - v\,dp}{T} = \frac{dh}{T} - \frac{v}{T}dp \tag{3.127}$$

Using the definition of enthalpy in terms of the specific heat at constant pressure, Equation (3.113), and the perfect gas equation of state, Equation (3.61), we have

$$ds = c_p\frac{dT}{T} - R\frac{dp}{p} \tag{3.128}$$

Applying the isentropic assumption, $ds = 0$, Equation (3.128) is rearranged as

$$\frac{dp}{p} = \frac{c_p}{R}\frac{dT}{T} \tag{3.129}$$

Inserting Equation (3.124) into (3.129), we have

$$\frac{dp}{p} = \frac{\gamma}{\gamma - 1}\frac{dT}{T} \tag{3.130}$$

Equation (3.132) relates a change in pressure to a change in temperature for an isentropic process.

For an isentropic process that proceeds from a state 1 to a state 2, we can integrate Equation (3.130) as

$$\int_{p_1}^{p_2}\frac{dp}{p} = \frac{\gamma}{\gamma - 1}\int_{T_1}^{T_2}\frac{dT}{T} \tag{3.131}$$

$$\ln\frac{p_2}{p_1} = \frac{\gamma}{\gamma - 1}\ln\frac{T_2}{T_1} \tag{3.132}$$

Simplifying Equation (3.132), we have an expression relating the pressures and temperatures at states 1 and 2 for an isentropic flow process

$$\frac{p_2}{p_1} = \left(\frac{T_2}{T_1}\right)^{\gamma/(\gamma-1)} \tag{3.133}$$

Using the perfect gas equation of state, Equation (3.61), we can obtain an expression relating the pressure to the density for an isentropic process. Inserting Equation (3.61) into (3.133), we have

$$\frac{p_2}{p_1} = \left(\frac{T_2}{T_1}\right)^{\gamma/(\gamma-1)} = \left(\frac{p_2/\rho_2 R}{p_1/\rho_1 R}\right)^{\gamma/(\gamma-1)} = \left(\frac{p_2}{p_1}\right)^{\gamma/(\gamma-1)}\left(\frac{\rho_2}{\rho_1}\right)^{-\gamma/(\gamma-1)} \tag{3.134}$$

$$\frac{p_2}{p_1}\left(\frac{p_2}{p_1}\right)^{-\gamma/(\gamma-1)} = \left(\frac{p_2}{p_1}\right)^{-1/(\gamma-1)} = \left(\frac{\rho_2}{\rho_1}\right)^{-\gamma/(\gamma-1)} \tag{3.135}$$

$$\frac{p_2}{p_1} = \left(\frac{\rho_2}{\rho_1}\right)^{\gamma} \tag{3.136}$$

Rearranging, we have

$$\frac{p_1}{\rho_1^{\gamma}} = \frac{p_2}{\rho_2^{\gamma}} \tag{3.137}$$

or, more generally

$$\frac{p}{\rho^\gamma} = \text{constant} \tag{3.138}$$

In summary, using Equations (3.133) and (3.136), we can relate the pressure, density, and temperature for an isentropic process as

$$\frac{p_2}{p_1} = \left(\frac{\rho_2}{\rho_1}\right)^\gamma = \left(\frac{T_2}{T_1}\right)^{\gamma/(\gamma-1)} \tag{3.139}$$

As a corollary to our discussion of an isentropic process, we introduce the concept of *stagnation* or *total properties*. In Section 3.2.2, *static* flow properties were defined as those that are felt or measured when moving along with the flow. Static properties are due to the random motion of the fluid molecules and not dependent on the directed motion or velocity of the flow. Now, suppose we have a flow where the static temperature, static pressure, static density, and Mach number are T, p, ρ, and M, respectively. If we were to *adiabatically* decelerate the flow from its Mach number, M, to zero velocity, we measure a new fluid temperature, defined as the *total* or *stagnation temperature*, T_t. Similarly, if the flow is *isentropically* brought to rest, that is, brought to zero velocity in an adiabatic and reversible process, we measure a new fluid pressure and density, defined as the *total* or *stagnation* pressure and density, p_t and ρ_t, respectively. Unlike the static conditions, the stagnation or total conditions are a function of directed motion of the flow or the flow velocity (or Mach number). Keep in mind that the stagnation or total conditions are *reference conditions*, for any point in the flow. We can define the stagnation or total conditions for any point in the flow based on the static conditions and Mach number at that point.

As a final note, because of how Equation (3.139) was derived, consistent units must be used when applying this equation for the temperature ratio. Since the unit conversion of temperature, from consistent to inconsistent units, involves addition of a constant, an erroneous result is obtained if inconsistent units are used. This is illustrated in the example problem below.

Example 3.6 Isentropic Flow Over a Wing *An airplane is flying at an altitude where the freestream pressure, p_∞, density, ρ_∞, and temperature, T_∞, are 1400 lb/ft², 0.001701 slug/ft², and 479.8 °R, respectively. The pressure, at a point on the wing, p_{wing}, is measured to be 1132 lb/ft². Assuming isentropic flow, calculate the density and temperature at this point on the wing.*

Solution

The pressure, density, and temperature are related using Equation (3.139) for an isentropic process, given by

$$\frac{p_{wing}}{p_\infty} = \left(\frac{\rho_{wing}}{\rho_\infty}\right)^\gamma = \left(\frac{T_{wing}}{T_\infty}\right)^{\gamma/(\gamma-1)}$$

Solving for the density on the wing, we have

$$\rho_{wing} = \rho_\infty \left(\frac{p_{wing}}{p_\infty}\right)^{1/\gamma} = \left(0.001701 \frac{slug}{ft^3}\right)\left(\frac{1132}{1400}\right)^{1/1.4}$$

$$\rho_{wing} = \left(0.001701 \frac{slug}{ft^3}\right)(0.8592) = 0.001461 \frac{slug}{ft^3}$$

Solving for the temperature on the wing, we have

$$T_{wing} = T_\infty \left(\frac{p_{wing}}{p_\infty} \right)^{(\gamma-1)/\gamma} = (479.8°R)\left(\frac{1132}{1400} \right)^{(1.4-1)/1.4}$$

$$T_{wing} = (479.8°R)(0.9411) = 451.5°R$$

Let us now calculate the temperature, on the wing, using inconsistent units of temperature. We first convert the freestream temperature from Rankine to Fahrenheit.

$$T_\infty = 479.8°R - 459 = 20.8°F$$

Solving for the temperature on the wing, we have

$$T_{wing} = T_\infty \left(\frac{p_{wing}}{p_\infty} \right)^{(\gamma-1)/\gamma} = (20.8°F)\left(\frac{1132}{1400} \right)^{(1.4-1)/1.4}$$

$$T_{wing} = (20.8°F)(0.9411) = 19.57°F$$

Converting back to Rankine, we have

$$T_{wing} = 19.57°F + 459 = 498.1°R$$

which is an erroneous result.

3.6 Fundamental Equations of Fluid Motion

In this section, we derive a set of fundamental equations that mathematically describe the motion of a fluid, such as air. In some respects, it is truly amazing that we can write down a few mathematical expressions and solve them for the details of a fluid in motion, extracting information about the flow velocity, pressure, temperature, density, or other flow variables. These equations can be applied to the flow of a fluid over an aircraft, submarine, automobile, building, or other object and solved for these flow variables. The equations of fluid motion can also be applied to the flows inside a wind tunnel, rocket nozzle, jet engine, or other internal flow. The flow variables are integrated over an external or internal surface to obtain the forces and moments acting on the surface.

In the following, the three fundamental equations of fluid motion are derived, the continuity, momentum, and energy equations. The physics of fluid motion are embodied in the mathematics of these equations. The continuity equation is an expression of the conservation of mass in fluid motion, the momentum equation is an expression of Newton's second law, and the energy equation is an expression of the conservation of energy.

3.6.1 Conservation of Mass: The Continuity Equation

Consider the steady, inviscid flow of air from a station 1 to a downstream station 2, as shown in Figure 3.23, where several streamlines are drawn. The outermost streamlines are the boundaries of a *streamtube* of flow, which could be considered as the inviscid flow along the walls of a duct. The streamtube cross-sectional area, velocity and density at station 1 are A_1, V_1, and ρ_1, respectively, and A_2, V_2, and ρ_2, respectively, at station 2. Here, the assumption is made that the flow properties are uniform across the cross-sectional area, such that the flow properties are changing in only one dimension, the direction of the flow.

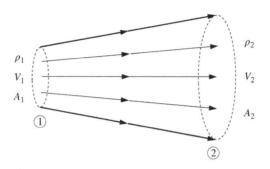

Figure 3.23   Mass flow through a streamtube.

The mass of air, flowing through any cross-section of the streamtube, per unit time, is defined as the mass flow rate, $\dot{m}$, with units of kg/s or slugs/s, in SI and English units respectively. The mass flow rate entering the streamtube at station 1, $\dot{m}_1$, is given by

$$\dot{m}_1 = \rho_1 A_1 V_1 \tag{3.140}$$

Similarly, the mass flow rate exiting the streamtube at station 2, $\dot{m}_2$, is given by

$$\dot{m}_2 = \rho_2 A_2 V_2 \tag{3.141}$$

Since the streamtube is bounded by the outermost streamlines (or walls), through which there is no mass flow, the mass flow rate entering and exiting the streamtube must be equal. Thus, we have

$$\dot{m} = \dot{m}_1 = \dot{m}_2 = \rho_1 A_1 V_1 = \rho_2 A_2 V_2 \tag{3.142}$$

or

$$\rho A V = \text{constant} \tag{3.143}$$

Equation (3.142) is an expression of the conservation of mass for a streamtube of flow. We could apply this equation between any two stations in a flow field that are bounded by the same streamlines.

If the flow is assumed to be incompressible, then $\rho_1 = \rho_2$, so that Equation (3.142) becomes

$$A_1 V_1 = A_2 V_2 \tag{3.144}$$

or, in general

$$A V = \text{constant} \tag{3.145}$$

Equation (3.144) is an expression of conservation of mass for an incompressible flow. Solving for the velocity at the exit of the streamtube, V_2, we have

$$V_2 = \frac{A_1}{A_2} V_1 \tag{3.146}$$

Equation (3.146) provides a relationship between the velocities at the entrance and exit of a stream-tube, or solid-wall duct, as a ratio of the cross-sectional areas at the exit and entrance. Based on the principle of conservation of mass, we see that, for an incompressible flow, the flow velocity increases with increasing area and decreases with decreasing area. Summarizing this, we have

$$V_2 > V_1 \text{ if } \frac{A_2}{A_1} < 1 \tag{3.147}$$

$$V_2 < V_1 \text{ if } \frac{A_2}{A_1} > 1 \tag{3.148}$$

Example 3.7   Calculation of Mass Flow Rate *A streamtube of air has a cross-section area, A, equal to 4.5 ft². The density, ρ, and velocity, V, at this cross-sectional area, are 0.00224 slug/ft³ and 100 ft/s, respectively. Calculate the mass flow rate of air, through this cross-sectional area.*

Solution

The mass flow rate is given by

$$\dot{m} = \rho A V = \left(0.00224 \frac{slug}{ft^3}\right)(4.5\,ft^2)\left(100\frac{ft}{s}\right) = 1.008\frac{slug}{s}$$

3.6.2   Newton's Second Law: The Momentum Equation

The momentum equation is an embodiment of Newton's second law of motion, where the sum of the forces, $\vec{F}$, acting on a body is equal to the mass, m, of the body multiplied by the rate of change of the body's velocity, $\vec{V}$.

$$\sum \vec{F} = m\frac{d\vec{V}}{dt} = m\vec{a} \tag{3.149}$$

Let us now apply Newton's second law, as given by Equation (3.149), to a fluid flow. Consider an infinitesimal fluid element, with dimensions dx, dy, and dz, moving in a fluid flow with a velocity $\vec{V}$, as shown in Figure 3.24. The fluid element is moving in a three-dimensional space, so the fluid element's velocity is given by

$$\vec{V} = u\hat{i} + v\hat{j} + w\hat{k} \tag{3.150}$$

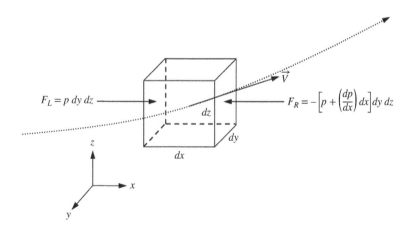

Figure 3.24   Pressure-derived forces in the *x*-direction on a fluid element.

where u, v, and w are the x, y, and z components of the velocity, respectively, and i, j, and k are the unit vectors along these Cartesian coordinate axes.

Forces due to pressure, shear stress (friction), and gravity act upon the fluid element. We assume an inviscid flow, that is, a flow without viscous effects or friction, so that the viscous forces are neglected. The effect of the force due to gravity, or the body force, on the motion of the fluid element is considered negligible. Thus, the significant forces affecting the motion of the fluid element are due to the pressure.

Let us now apply Equation (3.149) to our fluid element. Equation (3.149) is a vector equation, which for the x-direction, is given by

$$F_x = ma_x \tag{3.151}$$

where F_x is the total force in the x-direction, m is the fluid element mass, and a_x is the acceleration of the fluid element in the x-direction.

Examining Figure 3.24, the force due to the pressure on the left face of the fluid element, F_L, is the pressure, p, acting over the area of the face, $dy\,dz$.

$$F_L = p\,dy\,dz \tag{3.152}$$

The pressure changes throughout the flow field, so that the pressure on the right face of the fluid element is given by

$$p_R = p + \left(\frac{dp}{dx}\right) dx \tag{3.153}$$

where dp/dx is the change in pressure with distance in the x-direction. Therefore, the pressure force on the right face of the fluid element, F_R, is given by

$$F_R = -\left[p + \left(\frac{dp}{dx}\right) dx\right] dy\,dz \tag{3.154}$$

which is in the negative x-direction. Using Equations (3.152) and (3.154), the total force in the x-direction due to the pressure, F_x, is

$$F_x = F_L + F_R = p\,dy\,dz - \left[p + \left(\frac{dp}{dx}\right) dx\right] dy\,dz = -\left(\frac{dp}{dx}\right) dx\,dy\,dz \tag{3.155}$$

The mass of the fluid element, m, is given by

$$m = \rho(dx\,dy\,dz) \tag{3.156}$$

where ρ is the density of the air in the fluid element and $(dx\,dy\,dz)$ is the fluid element volume.

The acceleration of the fluid element in the x-direction is given by

$$a_x = \frac{du}{dt} = \frac{du}{dx}\frac{dx}{dt} = \frac{du}{dx}u \tag{3.157}$$

Substituting Equations (3.155), (3.156), and (3.157) into Equation (3.151), we have

$$-\frac{dp}{dx}(dx\,dy\,dz) = \rho(dx\,dy\,dz)u\frac{du}{dx} \tag{3.158}$$

Equation (3.158) is rewritten as

$$\frac{dp}{dx}dx = -\rho u\,du = -\rho\frac{1}{2}d(u^2) \tag{3.159}$$

Similarly, Newton's second law, as given by (3.149), is applied in the y- and z-directions, respectively, to obtain the following equations.

$$\frac{dp}{dy}dy = -\rho v\, dv = -\rho\frac{1}{2}d(v^2)$$ (3.160)

$$\frac{dp}{dz}dz = -\rho w\, dw = -\rho\frac{1}{2}d(w^2)$$ (3.161)

Summing Equations (3.159), (3.160), and (3.161), we have

$$\frac{dp}{dx}dx + \frac{dp}{dy}dy + \frac{dp}{dz}dz = -\rho\frac{1}{2}d(u^2) - \rho\frac{1}{2}d(v^2) - \rho\frac{1}{2}d(w^2)$$ (3.162)

Collecting the terms in the right-hand side of Equation (3.162), we have

$$\frac{dp}{dx}dx + \frac{dp}{dy}dy + \frac{dp}{dz}dz = -\rho\frac{1}{2}d(u^2 + v^2 + w^2)$$ (3.163)

The left-hand side of Equation (3.162) is simply the differential of the pressure, dp. The sum of the squares of the component velocities on the right-hand side of the equation is simply the square of the total velocity, so that

$$dp = -\rho\frac{1}{2}d(V^2)$$ (3.164)

Taking the derivative of V^2 on the right-hand side of Equation (3.164), we have

$$dp = -\rho V\, dV$$ (3.165)

Equation (3.165) is known as *Euler's equation*, which relates the change in pressure, dp, to the change in velocity, dV, along a streamline. Recall the initial assumptions that went into the derivation of this equation: that the flow is inviscid with no body forces. We did not make any assumption about whether the fluid density was constant or varied throughout the flow. Therefore, Equation (3.165) is valid for incompressible or compressible inviscid flows.

We now seek to integrate Equation (3.165) for a steady, compressible flow. If we make the further assumption of an isentropic (adiabatic and reversible) flow, the pressure is related to the density using Equation (3.138) as

$$\frac{p}{\rho^\gamma} = constant \equiv C$$ (3.166)

Solving for the density, we have

$$\rho = \left(\frac{p}{C}\right)^{1/\gamma}$$ (3.167)

Substituting Equation (3.167) into Euler's equation, Equation (3.165), we have

$$dp = -\left(\frac{p}{C}\right)^{1/\gamma} V\, dV$$ (3.168)

Rearranging, we have

$$\left(\frac{p}{C}\right)^{-1/\gamma} dp = -V\, dV$$ (3.169)

Integrating from a station 1, where the pressure and velocity are p_1 and V_1, respectively, to a downstream station 2, where the pressure and velocity are p_2 and V_2, respectively, we have

$$\int_{p_1}^{p_2} \left(\frac{p}{C}\right)^{-1/\gamma} dp = -\int_{V_1}^{V_2} V dV \tag{3.170}$$

$$C^{1/\gamma} \left(\frac{\gamma}{\gamma-1}\right)(p)^{\frac{\gamma-1}{\gamma}}\Bigg]_1^2 = -\left(\frac{V_2^2}{2} - \frac{V_1^2}{2}\right) \tag{3.171}$$

$$C^{1/\gamma} \left(\frac{\gamma}{\gamma-1}\right)\left[(p_2)^{\frac{\gamma-1}{\gamma}} - (p_1)^{\frac{\gamma-1}{\gamma}}\right] = -\left(\frac{V_2^2}{2} - \frac{V_1^2}{2}\right) \tag{3.172}$$

Assuming that the flow conditions at station 1 are known, the constant, C, may be evaluated as

$$C = \frac{p_1}{\rho_1^\gamma} \tag{3.173}$$

Substituting this constant into Equation (3.172), we have

$$\left(\frac{p_1}{\rho_1^\gamma}\right)^{1/\gamma}\left(\frac{\gamma}{\gamma-1}\right)\left[(p_2)^{\frac{\gamma-1}{\gamma}} - (p_1)^{\frac{\gamma-1}{\gamma}}\right] = -\left(\frac{V_2^2}{2} - \frac{V_1^2}{2}\right) \tag{3.174}$$

Equation (3.174) is the *Bernoulli equation for compressible flow*, relating the pressure and velocity along a streamline for an inviscid, compressible, isentropic flow. An analogous, yet much simpler, equation can be derived for incompressible flow.

It should be noted that the isentropic relation, Equation (3.166), also provides an equation that relates the flow properties at station 1 to those at station 2, as given by

$$\frac{p_1}{\rho_1^\gamma} = \frac{p_2}{\rho_2^\gamma} \tag{3.175}$$

This equation may be used, instead of Equation (3.174), in the solution of the flow properties in a compressible flow.

With the assumption of incompressible flow, the density is constant, which greatly simplifies the integration of Euler's equation. Integrating Equation (3.165) along a streamline in incompressible flow, we have

$$\int_{p_1}^{p_2} dp = -\rho \int_{V_1}^{V_2} V dV \tag{3.176}$$

$$p_2 - p_1 = -\rho \left(\frac{V_2^2}{2} - \frac{V_1^2}{2}\right) \tag{3.177}$$

Rearranging, we have

$$p_1 + \frac{1}{2}\rho V_1^2 = p_2 + \frac{1}{2}\rho V_2^2 \tag{3.178}$$

A general expression for Equation (3.178) may be written as

$$p + \frac{1}{2}\rho V^2 = p + q = \text{constant along a streamline in incompressible flow} \tag{3.179}$$

which states that the sum of the static pressure, p, and the dynamic pressure, q, are constant along a streamline.

Equation (3.179) is *Bernoulli's equation for incompressible flow* – commonly referred to as simply *Bernoulli's equation* – which relates the pressure and velocity along a streamline for an inviscid, incompressible flow. From Bernoulli's equation, we see that, along a streamline in an incompressible flow, the pressure increases with decreasing velocity and the pressure decreases with increasing velocity.

We have already stated that Equation (3.179) holds for inviscid flow, where there are no losses due to viscosity. If there are no other loss-producing mechanisms in the flow, such as shock waves, an isentropic flow may be assumed, and the constant in Equation (3.179) applies to the complete flow field, not just along a streamline. Therefore, for an isentropic flow, we can write Equation (3.179) as

$$p + \frac{1}{2}\rho V^2 = p_t = \text{constant everywhere in isentropic, incompressible flow} \qquad (3.180)$$

where p_t is the total pressure that is obtained by isentropically bringing the flow to rest (zero velocity). Thus, we see that the total pressure, the sum of the static pressure and dynamic pressure, is constant throughout an isentropic flow.

Example 3.8 Bernoulli's Equation *Consider the inviscid, incompressible, isentropic flow over the wing of an airplane at sea level. At a point in the flow, upstream of the wing, the pressure, p_1, and velocity, V_1, are 2116 lb/ft² and 123.2 mph, respectively. At a point on the wing, the pressure, p_2, is measured to be 1754 lb/ft². Calculate the total pressure of the flow, p_t, and the velocity at the point on the wing, V_2.*

Solution

Since the flow is inviscid, incompressible, and isentropic, Equation (3.180) is valid everywhere in the flow. The total pressure of the flow is given by

$$p_t = p_1 + \frac{1}{2}\rho V_1^2$$

Convert the velocity to consistent units.

$$V_1 = 123.2 \frac{mi}{h} \times \frac{5280\,ft}{1\,mi} \times \frac{1\,h}{3600\,s} = 180.7 \frac{ft}{s}$$

The total pressure is

$$p_t = 1{,}834 \frac{lb}{ft^2} + \frac{1}{2}\left(0.002377 \frac{slug}{ft^3}\right)\left(180.7 \frac{ft}{s}\right)^2 = 1873 \frac{lb}{ft^2}$$

Since the total pressure is constant throughout the flow, the velocity at the point on the wing, V_2, may be found using

$$p_t = p_2 + \frac{1}{2}\rho V_2^2$$

Solving for the velocity at point 2, we have

$$V_2 = \sqrt{2\left(\frac{p_t - p_2}{\rho}\right)} = \sqrt{2\left(\frac{1873 \frac{lb}{ft^2} - 1754 \frac{lb}{ft^2}}{0.002377 \frac{slug}{ft^3}}\right)} = 316.4 \frac{ft}{s} = 215.7 \frac{mi}{h}$$

3.6.3 Conservation of Energy: The Energy Equation

The energy equation of fluid flow is based upon the first law of thermodynamics. As discussed in Section 3.5.5.2, the first law of thermodynamics is a statement of the conservation of energy for a system of fixed mass. In equation form, the first law is written as Equation (3.87), repeated below.

$$de = \delta q + \delta w \qquad (3.181)$$

where the change in the internal energy of the system, de, is equal to the sum of the heat added to the system, δq, and the work done by the system, δw. We derived another form of the first law of thermodynamics, relating the heat transfer to the enthalpy, h, pressure, p, and specific volume, v, as Equation (3.94), repeated below.

$$\delta q = dh - v dp \qquad (3.182)$$

If the flow is assumed to be adiabatic, then $\delta q = 0$ (we address the energy equation for non-adiabatic flows in Chapter 4), and we have

$$dh - v dp = 0 \qquad (3.183)$$

Inserting Euler's equation for the change in pressure, dp, Equation (3.165), into Equation (3.183), we have

$$dh - v(-\rho V dV) = dh - \frac{1}{\rho}(-\rho V dV) = dh + V dV = 0 \qquad (3.184)$$

Integrating Equation (3.184) along a streamline from a point 1 to a point 2, we have

$$\int_{h_1}^{h_2} dh + \int_{V_1}^{V_2} V dV = 0 \qquad (3.185)$$

$$h_2 - h_1 + \frac{V_2^2}{2} - \frac{V_1^2}{2} = 0 \qquad (3.186)$$

or

$$h_2 + \frac{V_2^2}{2} = h_1 + \frac{V_1^2}{2} \qquad (3.187)$$

Equation (3.187) is written as

$$h + \frac{V^2}{2} = \text{constant along a streamline} \qquad (3.188)$$

which relates the enthalpy, h, and velocity, V, at any two points along a streamline. If all of the streamlines in the flow start upstream from a uniform flow, then the constant in Equation (3.188) is the same for all of the streamlines in the flow. Therefore, Equation (3.188) holds for anywhere in the flow, so that

$$h + \frac{V^2}{2} = h_t = \text{constant everywhere in flow field} \qquad (3.189)$$

where h_t is the total enthalpy.

Using the definition of the enthalpy in terms of the specific heat at constant pressure, c_p, and the temperature, T, given by Equation (3.113), we can rewrite Equation (3.189) as

$$c_p T + \frac{V^2}{2} = c_p T_t = \text{constant everywhere in flow field} \qquad (3.190)$$

For any two points in the flow, we have

$$c_p T_1 + \frac{V_1^2}{2} = c_p T_2 + \frac{V_2^2}{2} = c_p T_t \qquad (3.191)$$

where we have now related the temperature and velocity at any two points in the flow field to each other and to the total temperature, T_t. Equations (3.189), (3.190), and (3.191) are forms of the energy equation for steady, inviscid, adiabatic flow. Remember that these are still mathematical expressions of the first law of thermodynamics and the conservation of energy as applied to fluid flow.

Example 3.9   Calculation of Total Temperature and Total Enthalpy *The velocity and temperature in an inviscid, adiabatic flow are 415 ft/s and 519°R, respectively. Calculate the total temperature and total enthalpy of the flow.*

Solution

The static temperature and velocity are related to the total temperature and total enthalpy by

$$c_p T + \frac{V^2}{2} = c_p T_t = h_t$$

Solving for the total temperature, we have

$$T_t = T + \frac{V^2}{2c_p} = 519°R + \frac{\left(415\frac{ft}{s}\right)^2}{2\left(6020.7\frac{ft \cdot lb}{slug \cdot °R}\right)} = 533.3°R$$

The total enthalpy is given by

$$h_t = c_p T_t = \left(6020.7\frac{ft \cdot lb}{slug \cdot °R}\right)(533.3°R) = 3.211 \times 10^6 \frac{ft \cdot lb}{slug}$$

3.6.4   *Summary of the Governing Equations of Fluid Flow*

We have derived a set of governing equations for a fluid flow, composed of a continuity equation, momentum equation, and energy equation. The continuity equation is a mathematical expression for the conservation of mass, the momentum equation embodies Newton's second law, and the energy equation is an expression for the conservation of energy. For steady, inviscid, compressible flow, we have three equations and four unknowns: the pressure, p, density, ρ, velocity, V, and temperature, T. Since there are more unknowns than equations, an additional equation is needed to solve this set of governing equations for the flow properties. This final equation is the equation of state, as discussed in Section 3.5.2, relating the state variables, such as the pressure, temperature, and density to each other. Hence, for a steady, inviscid, compressible, adiabatic flow, the continuity, momentum, energy, and state equations are given by

Continuity:	$\rho A V = \text{constant}$	(3.192)
Momentum:	$dp = -\rho V \, dV$	(3.193)
Energy:	$c_p dT + V dV = 0$	(3.194)
Equation of state:	$p = \rho R T$	(3.195)

We have developed the equations for the steady flow of an inviscid, compressible, isentropic fluid through a streamtube of known area variation, where the flow properties vary in the flow direction only. The continuity, momentum, energy, and state equations for the flow properties at Station 1 and 2 of the streamtube are given by

Continuity:
$$\rho_1 A_1 V_1 = \rho_2 A_2 V_2 \tag{3.196}$$

Momentum:
$$\left(\frac{p_1}{\rho_1^\gamma}\right)^{1/\gamma} \left(\frac{\gamma}{\gamma-1}\right) \left[(p_2)^{\frac{\gamma-1}{\gamma}} - (p_1)^{\frac{\gamma-1}{\gamma}}\right] = -\left(\frac{V_2^2}{2} - \frac{V_1^2}{2}\right) \tag{3.197}$$

or

Isentropic relation:
$$\frac{p_1}{\rho_1^\gamma} = \frac{p_2}{\rho_2^\gamma} \tag{3.198}$$

Energy:
$$c_p T_1 + \frac{V_1^2}{2} = c_p T_2 + \frac{V_2^2}{2} \tag{3.199}$$

Equations of state:
$$p_1 = \rho_1 R T_1 \tag{3.200}$$

$$p_2 = \rho_2 R T_2 \tag{3.201}$$

If the additional restriction of incompressible flow is made, the continuity and momentum equations are written, between two points in a streamtube, as

Continuity:
$$A_1 V_1 = A_2 V_2 \tag{3.202}$$

Momentum:
$$p_1 + \frac{1}{2}\rho V_1^2 = p_2 + \frac{1}{2}\rho V_2^2 \tag{3.203}$$

With the assumptions of incompressible flow (density, ρ, constant and known) and a known streamtube geometry (the distribution of area, A, known), the continuity and momentum equations form a set of two equations with two unknowns (p, V). These two equations are solved for the two unknowns, independent of the energy equation. Hence, for incompressible flow, the energy equation is said to be *decoupled* from the continuity and momentum equations. From a physical perspective, the application of the laws of thermodynamics is not required for the solution of incompressible flow.

For many of our upcoming aerodynamic discussions about airfoils and wings, a subsonic, inviscid, incompressible flow is assumed. This may seem restrictive at first, but a large proportion of aerodynamic theory and aircraft design is encompassed by these assumptions. This also follows the historical development of aerodynamics as it tried to keep pace with the increasing performance capabilities of aircraft. Later in this chapter, we address several areas where these flow assumptions are not valid and new theories and approaches must be introduced.

3.7 Aerodynamic Forces and Moments

Aerodynamic forces are created from the motion of a body through the air. You may have experienced this by simply holding your hand outside the window of a moving car. The air flowing over your hand results in forces and moments that may push your hand back or lift it up. Intuitively, you know that these aerodynamic forces and moments change with the speed of the car and the orientation of your hand relative to the airstream. The aerodynamic forces and moments on a body, such as your hand or an aircraft wing, are fundamentally due to two sources, the *pressure* distribution and *skin friction* distribution over the surface, as shown in Figure 3.25 for an airfoil or wing section. The

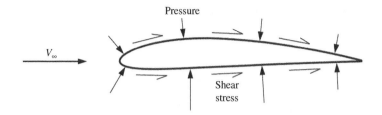

Figure 3.25 Pressure and shear stress distributions on an airfoil surface.

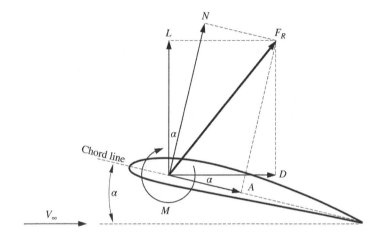

Figure 3.26 Aerodynamic forces on an airfoil.

pressure distribution acts in a direction normal or perpendicular to the surface, and the skin friction acts tangentially to the surface. The pressure and shear stress distributions are integrated over the surface of a body to obtain the forces and moments due to pressure and skin friction, respectively.

Consider an airfoil or wing section at an angle-of-attack, α, in an airflow with a freestream velocity V_∞, where the angle-of-attack is defined as the angle between the airfoil chord line and the freestream velocity, as shown in Figure 3.26. There is a total resultant aerodynamic force, F_R, and moment, M, on the airfoil due to the flow of air over the airfoil. Typically, the resultant force is resolved into components with respect to either the airfoil chord line or the freestream flow direction. The components of the resultant force perpendicular and parallel to the chord line are the normal force, N, and axial force, A, respectively. The lift, L, and drag, D, are the components of the resultant force that are perpendicular and parallel to the freestream velocity, respectively. The moment, M, may be taken about any point, but often the moment is taken about the airfoil's quarter-chord point, $M_{c/4}$.

As discussed in Section 3.4.4, it is often convenient to define non-dimensional coefficients for these forces, where the force is normalized by the product of the freestream dynamic pressure, q_∞, and a reference area, S, which is usually the planform area for a wing. Following Equation (3.38), the normal and axial force coefficients, C_N and C_A, respectively, are defined as

$$C_N \equiv \frac{N}{q_\infty S} \tag{3.204}$$

$$C_A \equiv \frac{A}{q_\infty S} \tag{3.205}$$

The lift and drag coefficients, C_L and C_D, respectively, are defined as

$$C_L \equiv \frac{L}{q_\infty S} \tag{3.206}$$

$$C_D \equiv \frac{D}{q_\infty S} \tag{3.207}$$

The moment coefficient, C_M, is defined as

$$C_M \equiv \frac{M}{q_\infty S c} \tag{3.208}$$

where c is the airfoil chord length.

A specific nomenclature is used for aerodynamics coefficients, depending on whether the coefficient refers to a two-dimensional or three-dimensional body. For a three-dimensional body, uppercase letters are used for the coefficient, so that the lift coefficient of a wing is denoted by C_L. Lowercase letters are used for the coefficients when referencing a two-dimensional body, so that the lift coefficient of an airfoil is denoted by c_l. The lift, drag, and moment coefficients for a two-dimensional body, such as an airfoil, are given by

$$c_l \equiv \frac{L'}{q_\infty c} \tag{3.209}$$

$$c_d \equiv \frac{D'}{q_\infty c} \tag{3.210}$$

$$c_m \equiv \frac{M'}{q_\infty c^2} \tag{3.211}$$

where the prime superscript denotes the force or moment per unit span.

The concept of resolving the resultant aerodynamic force on a body into components that are perpendicular and parallel to the freestream flow, or the lift and drag, respectively, was first documented by the Englishman Sir George Cayley (1773–1857) in 1799. He engraved a silver disk with a diagram showing the separation of the aerodynamic force into its lift and drag components for the flow over an inclined surface, representative of a wing at an angle-of-attack. On the other side of the disk, there is an engraved depiction of an airplane design of Cayley's, with a curved wing, aft-mounted tail, and a set of "flappers" used for propulsion. The significance of this airplane configuration is that it is the first recorded drawing of a fixed-wing airplane where the means of producing lift, the curved wing, is separated from the means of producing thrust, the "flappers". This separation of lift and thrust was a breakthrough in airplane design, leaving behind the many concepts that futilely attempted to mimic bird flight, where wings produced both lift and thrust. Cayley's fixed-wing airplane design was also the first to include a stabilizing tail, which is discussed further in Chapter 6. Cayley's engraved silver disk documented significant advancements in both aerodynamic theory and aircraft design.

How do we actually measure the lift and drag on a wing or a complete airplane? At the end of this section, we discuss several ground and flight test techniques that are used to obtain values for the lift and drag. We explore wind tunnel testing, computational techniques, and several types of flight tests. These fundamentally different methods or approaches are the cornerstones of aerodynamic analysis and design. The exact mix of these three approaches, that is, how much of each method is used in the analysis and design of a vehicle, may vary considerably with the specific application, but using all of the elements often leads to successful results. The fact that we spend so much time discussing these techniques underscores the importance of these aerodynamic parameters in aerospace engineering and vehicle design.

3.7.1 Lift

The concept of lift is an amazing thing. Who has not marveled at the sight of a massive airliner, weighing over a million pounds, made of aluminum and steel, full of hundreds of people, lift itself into "thin air" and fly away? Or who has not been amazed at the soaring flight of an eagle or hawk, with wings spread wide, seemingly suspended in the air? Despite their very large weight difference, these two examples rely on the same principles of aerodynamic lift to give them flight. In this section, we discuss several theories to explain the amazing phenomenon of aerodynamic lift.

Now is a good time for you, especially if you are a new student to aerospace engineering, to take a moment and think about your own ideas about aerodynamic lift. Take a piece of paper and write, in your own words, a short, fundamental explanation of aerodynamic lift. Draw a picture of an airfoil and sketch out your concept of aerodynamic lift. This is a good exercise in critical thinking before we start discussing several different theories. Then, after you have read the present section, you can compare your ideas and notions about lift with the theories that have been presented.

We now present several different theories of aerodynamic lift that have been put forward over the years. It is shown that some of these theories are not proper applications of the laws of physics and therefore are just plain wrong. Other theories have grains of truth in them, but they do not provide a fundamental explanation of how lift is produced. In the end, we arrive at a fundamental explanation that obeys the physical laws of nature, and which matches our experimental observations of aerodynamic lift.

3.7.1.1 Theories of Lift: Action and Reaction

This theory of lift is based on Newton's third law of *action and reaction*, that for any action, there is an equal and opposite reaction. In this theory, it is assumed that the airfoil exerts a downward force on the air as it flows past; this is the action. As a reaction, the air exerts an equal and opposite force on the airfoil, as shown in Figure 3.27. This equal and opposite reaction force is resolved into lift, L and drag, D, components, parallel and perpendicular to the freestream velocity, V_∞, respectively.

A problem with this theory is that it assumes that all of the lift production occurs on the airfoil lower surface only. In reality, the airfoil upper surface plays the major role in producing aerodynamic lift, especially the region near the leading edge of the upper surface. Using this theory, two airfoils with the same lower surface geometry, but with completely different upper

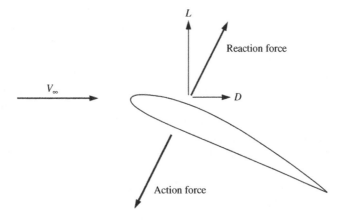

Figure 3.27   Action-reaction theory of lift.

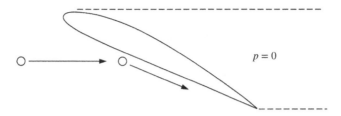

Figure 3.28   Newtonian theory of lift.

surface geometries would produce the same lift. We know from experimental evidence that this is not the case.

3.7.1.2   Theories of Lift: Newtonian Theory

The *Newtonian theory* of lift was postulated by Isaac Newton in his 1687 *Principia Mathematica* (Latin for *Mathematical Principles of Natural Philosophy*). In this theory, the flow of air is considered as a uniform stream of non-interacting particles. When the particles strike the bottom of an airfoil, as shown in Figure 3.28, it is assumed that they transfer their normal momentum to the airfoil, resulting in a force on the airfoil. According to Newton's second law, the force on the airfoil is equal to the time rate of change of the flow momentum normal to the airfoil bottom. The particles are assumed to "slide" along the body of the airfoil, preserving their tangential momentum.

Similar to the previous theory of lift, Newtonian theory neglects the role of the airfoil upper surface, in terms of its interaction with the air flow. The region shielded from the flow by the airfoil, called the shadow region, is unaffected by the air stream of particles. The pressure in the shadow region is assumed to be zero. Newton developed this theory for the prediction of low speed aerodynamic flows, but it is highly inaccurate for this speed regime. It is interesting that Newtonian theory has proven to be useful in providing approximate predictions of inviscid, hypersonic flows.

3.7.1.3   Theories of Lift: Equal Transit Time

This theory assumes that the air flow, reaching the airfoil leading edge, divides into fluid particles that flow over the top and bottom of the airfoil. It is assumed that two fluid particles, starting at the same location on the airfoil leading edge, with one flowing over the top surface and one under the bottom surface, reach the trailing edge at the same time. Hence, it is assumed that the time it takes for the fluid particle to travel over the top surface is the same as for the fluid particle traveling under the bottom surface, or that they have *equal transit times* from the leading edge to the trailing edge. Since the distance over the upper surface is longer than under the lower surface, the velocity of the upper surface fluid particle must be greater than that of the lower surface particle, for the two particles to meet at the trailing edge at the same time. Then, using Bernoulli's equation, as given by Equation (3.179), it is assumed that the higher velocity flow over the upper surface results in a lower pressure on the upper surface and the lower velocity flow under the lower surface results in higher pressure on the lower surface. The pressure difference, due to the velocity difference, results in the airfoil lift.

The fundamental error in this theory is that the assumption of equal transit times for the fluid particles, traveling over the upper and lower surfaces, is not correct. It has been experimentally proven that for two fluid particles, which start at the same leading edge location, the fluid particle traveling over the upper surface reaches the trailing edge before the particle traveling under the

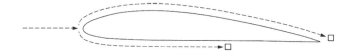

Figure 3.29 Equal transit time theory of lift, fluid elements do not rejoin at trailing edge.

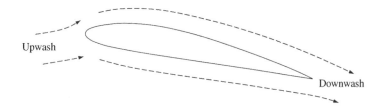

Upwash

Downwash

Figure 3.30 Flow deflection theory of lift.

lower surface, as shown in Figure 3.29. Hence, the foundation of the equal transit theory is incorrect. It is true that the flow velocity over the airfoil upper surface is greater than that under the lower surface, and that the pressure is lower on top versus the bottom, but not for the reasons given in this theory.

3.7.1.4 Theories of Lift: Flow Deflection

This theory of lift assumes that the airfoil changes the direction of the flow, upstream and downstream of the airfoil. This *flow deflection* by the airfoil is an *upwash* at the airfoil leading edge and a *downwash* at the trailing edge, as shown in Figure 3.30. The upwash has a velocity component in the upward direction and the downwash has a velocity component in the downward direction. It is assumed that the downward component of velocity is greater than the upward component, so that the net effect is that the airfoil imparts a downward component of momentum to the air. Using Newton's second law, this downward momentum is equal to a downward force, thus the airfoil is pushing the air flow down. The equal and opposite reaction, by Newton's third law, is the air pushing back on the airfoil, creating lift.

The physical aspects of this theory are correct, in that the air flow is deflected by the airfoil, resulting in an upwash and a downwash. The net downward momentum does result in a downward component of velocity, which pushes the flow downward. The flow deflection is a *result* of this force, rather than its *cause*. We address the fundamental source of the force in the next section.

3.7.1.5 Theories of Lift: Pressure and Shear Stress Distributions

The motion of an airfoil through the air creates distributions of pressure and shear stress over its surface, as shown in Figure 3.31a. By integrating the pressure and shear stress distributions over the entire airfoil surface, the resultant aerodynamic force is obtained, as shown in Figure 3.31b. The resultant force is resolved into the lift, L, and drag, D, perpendicular and parallel to the freestream flow velocity, V_∞, respectively. Thus, the pressure and shear stress distributions are certainly fundamental to the definition of the lift.

If the shear stress is ignored, assuming that it contributes primarily to the drag, the lift is generated as a result of a net pressure difference between the upper and lower airfoil surfaces. To provide a positive (upward) lift force, the net pressure over the upper surface must be less than the lower surface pressure. Assuming that the pressure distribution is fundamental to the creation of lift, the

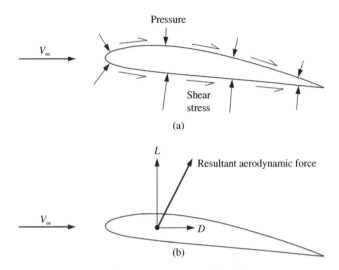

Figure 3.31 Motion of an airfoil through the air creates (a) pressure and shear stress distributions with (b) resultant forces.

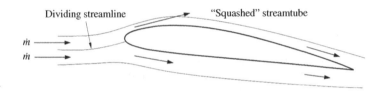

Figure 3.32 "Squashed" streamtube theory of lift.

theory of lift may be reduced to providing an explanation regarding the creation of the surface pressure distribution, specifically why the upper surface pressure is less than that on the lower surface.

3.7.1.6 Theories of Lift: "Squashed" Streamtubes

This theory is based upon the principles from which the governing equations of fluid flow are derived. Specifically, it is based on the conservation of mass, and Newton's second law. Consider an airfoil in a uniform flow of air, where we focus on two streamtubes of air of equal mass flow rate, $\dot{m}$, as shown in Figure 3.32. Far upstream of the airfoil, the two streamtubes also have equal cross-sectional areas. We have selected the two streamtubes, such that when they reach the airfoil leading edge, the shared streamline between them is the *dividing streamline* between the flow that goes over the top and under the bottom of the airfoil. Due to the airfoil shape and angle-of-attack, the air in the streamtube that flows over the top surface "sees" the airfoil as more of an obstruction that the lower streamtube. Both streamtubes have the same mass of fluid that is "pressing" against their boundaries, from above for the upper surface streamtube and from below for the lower surface streamtube. Hence, the upper surface streamtube is "squashed" in flowing over the upper surface, more than experienced by the lower surface streamtube.

The "squashing" of the streamtube results in reduction of the streamtube cross-section area. By conservation of mass, as embodied by Equation (3.143) where $\rho A V = $ constant, the velocity

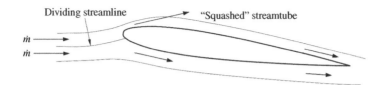

Figure 3.33 "Squashed" streamtube for a symmetric airfoil at angle-of-attack.

increases with a decrease in area. Since the area decrease is greater for the upper surface streamtube, the velocity increase is greater than for the lower surface streamtube. The greatest velocity increase occurs for the upper surface streamtube near the nose of the airfoil, where the streamtube is most constricted.

The velocity increases correspond to decreases in the static pressure. This is true for an incompressible flow, where the pressure varies according to Bernoulli's equation, as given by Equation (3.179), and for compressible flow, where the pressure varies according to Euler's equation, given by Equation (3.165). Recall that both of these equations are embodiments of Newton's second law of motion. Hence, the higher velocities over the airfoil upper surface result in lower pressures than those from the higher velocities over the airfoil lower surface. The pressure difference, between the airfoil upper and lower surfaces, results in the lift force.

If the airfoil were symmetric (same shape on the top and bottom) and at zero angle-of-attack, the streamtubes flowing over the airfoil top and bottom surfaces experience the same degree of obstruction, hence their area, velocity, and pressure changes are the same, resulting in zero lift. However, if the symmetric airfoil is placed at an angle-of-attack, the dividing streamline is below the symmetric nose of the airfoil (Figure 3.33) so that the upper surface streamtube of air must flow around this nose and over the top, hence it "sees" more of an obstruction than the lower surface streamtube.

The aerodynamic flow over an airfoil must obey the principles that govern fluid flow, specifically conservation of mass and Newton's second law. This theory of lift is based on these principles and provides an explanation of lift that is consistent with these principles. Experimental evidence also supports the flow descriptions used to explain this theory.

3.7.1.7 Theories of Lift: Circulation

The final theory of lift that we discuss is the *circulation theory of lift*. In the late 19th and early 20th centuries, the circulation theory of lift was independently developed by three different aerodynamicists in three different countries, Frederick W. Lanchester (1868–1946) in England, M. Wilhelm Kutta (1867–1944) in Germany, and Nikolai Y. Joukowsky (1847–1921) in Russia. This theory was a breakthrough in theoretical aerodynamics, providing a mathematical tool for the calculation of lift. However, the circulation theory of lift is not an explanation of lift; rather it is a mathematical foundation for the calculation of lift of an airfoil in incompressible flow. However, even though we have already decided that it is not an explanation of lift, it is worthwhile delving into some of its details.

The key element of this theory is the connection between *circulation* and lift. Circulation is mathematically defined as the line integral around a closed curve C, in a vector field. By convention, the circulation is positive for the counterclockwise direction around the curve C. (If you were to walk around the curve, the inside of the closed circuit would be on your left for positive circulation.) Applying this definition of circulation to a closed curve C, around an airfoil, in a velocity vector

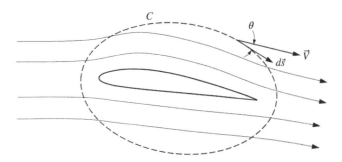

Figure 3.34   Circulation around an airfoil.

field, as shown in Figure 3.34, the circulation about the airfoil, Γ, is given by

$$\Gamma = \oint_C \vec{V} \cdot d\vec{s} = \oint_C V \cos \theta ds \qquad (3.212)$$

where V is the velocity at a point on the curve C, ds is an incremental distance along C, and θ is the angle between $\vec{V}$ and $d\vec{s}$.

The circulation about an airfoil is established when it initially starts in motion. When the airfoil moves forward, the higher-pressure air, flowing over the airfoil lower surface, tries to curl upward around the trailing edge. However, the flow is unable to turn the corner around the sharp trailing edge and a counterclockwise rotating vortex is created. This *starting vortex* is shed from the trailing edge as the airfoil continues forward. Based on conservation of angular momentum, a clockwise circulation around the airfoil is created, which is equal in strength and opposite in direction to the counterclockwise starting vortex. This establishment of circulation is shown in Figure 3.35.

The flow around the airfoil in a uniform freestream flow may be decomposed into a uniform flow added to the circulation around the airfoil, as shown in Figure 3.36. The velocity around the airfoil is obtained by the vector addition of the uniform flow and the circulation, which results in higher velocities over the top of the airfoil and lower velocities under the lower surface. For an incompressible flow, Bernoulli's equation, as given by Equation (3.179), indicates that the pressure decreases with higher velocity and increases with lower velocity. For compressible flow, Euler's equation, given by Equation (3.165), indicates this same pressure–velocity trend. Hence, the pressure difference, between the airfoil upper and lower surfaces, results in a lift force.

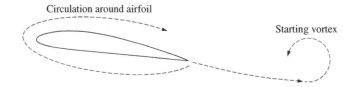

Figure 3.35   Establishment of circulation around an airfoil.

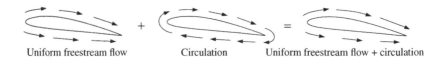

Figure 3.36   Uniform freestream flow and circulation.

It can be shown that the airfoil circulation, Γ, is related to the lift per unit span, L', by the *Kutta–Joukowsky theorem*, given by

$$L' = \rho_\infty V_\infty \Gamma \qquad (3.213)$$

where ρ_∞ and V_∞ are the freestream density and velocity, respectively. To calculate the lift, the circulation Γ must be known. *Thin airfoil theory* was developed after the circulation theory of lift as a means of analytically determining the circulation and hence the lift. In the next section, we explore some of the interesting applications of circulation by German engineer Anton Flettner.

3.7.1.8 Anton Flettner and His Spinning Cylinders

The Kutta–Joukowsky theorem may be applied to spinning objects, such as spinning cylinders or spheres, in a freestream flow, such that an aerodynamic force is generated. The spinning object provides the circulation, which when combined with a uniform flow, produces an aerodynamic force. This effect, called the *Magnus effect*, is seen in the curving trajectory of a baseball or golf ball, which is imparted with spin when thrown or hit, respectively. The imparted spin on the ball creates a circulation, which when combined with the free flight uniform flow, creates a force that curves the flight trajectory of the ball.

The practical application of the force-producing, spinning cylinder is perhaps best exemplified by the work of Anton Flettner (1885–1961), a German engineer and inventor, who made significant contributions to the design of airplanes, rotorcraft, and boats. In the 1920s and 1930s, Flettner applied the concept of the Magnus effect to a full-scale ship and an airplane. As shown in Figure 3.37, the Flettner rotor ship was a schooner with two, 50-foot (15.2 m) tall cylinders mounted on the deck. A motor was used to spin each cylinder, or "rotor sail", creating a force perpendicular to the oncoming wind. The sailing performance of Flettner's rotor ship, christened the *Baden-Baden*, was impressive, matching or bettering that of conventional schooners in moderate to heavy wind. The *Baden-Baden* even made a successful crossing of the Atlantic Ocean, from Germany to New York, in 1926. A larger, commercial rotor ship, the *Barbara*, was also constructed, with three tall vertical rotor sails. While demonstrating impressive performance, the rotor ships

Figure 3.37 Flettner rotor ship. (Source: *G.G. Bain, 1924, US Library of Congress, PD-Bain.*)

Figure 3.38   Flettner rotor airplane. (Source: *San Diego Air & Space Museum, PD-old, no known copyright restrictions.*)

were not as economical as diesel engine-powered vessels and thus were not a commercial success. The *Baden-Baden* was lost at sea in a storm in 1931. In modern times, several different types of rotor ships have been designed and built, some motivated by rising fuel costs or environmental concerns.

In the 1930s, Flettner applied spinning cylinders to aviation, developing the Flettner rotor airplane, shown in Figure 3.38. The rotating cylinders were mounted horizontally on either side of the airplane fuselage, similar to conventional wings. The spinning cylinders created a lift force, perpendicular to the direction of motion. While it is known that a Flettner rotor airplane was built, it is not known whether it ever actually flew.

In addition to the rotor ship and rotor airplane, Anton Flettner made other significant contributions to the marine and aviation fields. During World War I, Flettner invented a marine rudder system and a trim tab control system for aircraft. During World War II, he established an aircraft company in Germany that produced the Flettner Fl 282 helicopter for the German Luftwaffe. After the war, Flettner came to the USA and established the Flettner Aircraft Corporation in New York. While Flettner's aircraft company was not commercially successful, he did find financial success with his invention of the Flettner rotary ventilator. A type of spinning, centrifugal exhauster that promotes air cooling, the Flettner ventilator was widely used on boats, trucks, buses, and vans. Modern versions of the Flettner ventilator are still sold and in use today.

3.7.2   Drag

Drag is the component of the aerodynamic force that opposes a flight vehicle's motion. It acts in a direction opposite to the vehicle's velocity vector. Drag is the penalty for forward motion through the air. Thrust must match the drag in steady, level flight, or be greater than the drag to accelerate or climb. Usually, the goal of the aerodynamicist is to reduce drag to a minimum. Decreasing the drag increases the lift-to-drag ratio, a measure of the aerodynamic efficiency of a flight vehicle. As with any aerodynamic force, drag on a body is composed of the components due to the integrated pressure and skin friction distributions. The pressure acts normal to the body surface, while the

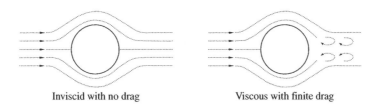

Inviscid with no drag   Viscous with finite drag

Figure 3.39   Inviscid and viscous, incompressible flow over a circular cylinder.

skin friction acts tangentially on the surface. The relative amounts of drag due to pressure and due to skin friction depend on several parameters that are a function of the body geometry and flow conditions. The present section defines the drag makeup for two- and three-dimensional bodies.

3.7.2.1   Drag of Two-Dimensional Shapes

We first consider the drag for two-dimensional shapes, such as circular cylinders and airfoils. (An airfoil is a two-dimensional cross-section of a three-dimensional lifting surface, such as an aircraft wing or helicopter rotor blade. Airfoil geometry is discussed in Section 3.8.) The uniform, incompressible flow over a circular cylinder is shown in Figure 3.39, assuming inviscid and viscous flows. For the inviscid case, where there is no friction, the air molecules flow smoothly around the cylinder, forming a perfectly symmetric streamline pattern. Since the flow is symmetric, the pressure distributions on the forward and aft surfaces of the cylinder are the same. Since the pressure distributions on the front and aft surfaces are the same, the forward and aft-acting forces are exactly balanced and there is zero drag. The same argument may be made for the pressure distributions on the upper and lower cylinder surfaces, which results in zero lift.

In reality, we know that if we place a cylinder in a uniform, incompressible flow, there is a drag force on the body. This contradiction between the theoretical result of zero drag for a body in an incompressible, inviscid flow and the finite drag, based on real-world experience, is known as *d'Alembert's paradox*, named after the French mathematician and philosopher, Jean le Rond d'Alembert (1717–1783), who documented this quandary in 1744. He and other fluid dynamicists of the 18th and 19th centuries were perplexed and confounded by this contradiction. It was not until the boundary layer theory of Ludwig Prandtl that it was realized that viscosity is required to obtain a non-zero drag.

The flow situation with viscosity is shown for the cylinder on the right side of Figure 3.39. With the retarding action of skin friction, the flow separates on the aft portion of the cylinder, creating a separated wake behind the cylinder. The addition of skin friction results in two drag-producing mechanisms, *skin friction drag* in the boundary layer next to the body surface and *pressure drag due to flow separation*. Skin friction drag is due to the tangential shearing stresses at the body surface, which slows the flow to zero velocity at the surface. The friction drag force is proportional to the velocity gradient in the boundary layer. The pressure drag is due to the reduced pressure in the separated wake, resulting in a pressure distribution over the cylinder that produces a net drag force.

The cylinder has a large separated wake, such that the pressure drag is large relative to the skin friction drag. Bodies that have a large pressure drag in relation to their skin friction drag are termed *bluff bodies* and are characterized by large separated wake regions. In contrast, geometries that have small wakes such that their pressure drag is small relative to their skin friction drag are called *streamlined bodies*. Perhaps the most common streamlined, two-dimensional geometry in aviation is the airfoil. Uniform, subsonic flow over an airfoil is shown in Figure 3.40 for inviscid and viscous

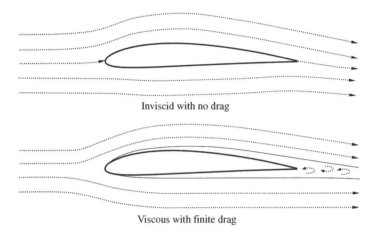

Figure 3.40   Inviscid and viscous, incompressible flow over an airfoil.

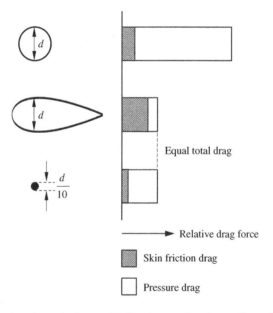

Figure 3.41   Aerodynamic drag on bluff and streamlined, two-dimensional shapes.

flows. The inviscid flow over the streamlined airfoil is similar to the inviscid flow over the bluff body cylinder in that the resultant net drag force is still zero. Hence, we see that d'Alembert's paradox holds for any shaped body, whether a bluff or streamlined shape, in an inviscid flow. The geometry need not be symmetric in any way for the drag to be zero in an inviscid flow. (This is shown in the example problem at the end of this section.)

A comparison of the relative drag of bluff and streamlined, two-dimensional shapes is shown in Figure 3.41. A bluff body cylinder, with the same diameter, d, as a more streamlined shape, has a higher total drag. The pressure drag due to flow separation makes up a high percentage of the total drag of the cylinder, while the skin friction drag comprises the majority of the drag for the streamlined shape. Another bluff body cylinder shape is shown, with a diameter of one tenth that of the other shapes. Despite its small size, this bluff body has an equal drag as the streamlined shape, due to its large pressure drag. These simple examples highlight the importance of *streamlining* for the reduction of pressure drag on flight vehicles.

To summarize, for any two-dimensional shape, in an incompressible, viscous flow, the total drag, D, is the sum of the pressure drag due to flow separation, also called the *form drag*, D_p, and the *skin friction drag*, D_f, given by

$$D = D_p + D_f \tag{3.214}$$

In terms of the drag coefficient, Equation (3.214) is written as

$$c_d = c_{d,p} + c_{d,f} \tag{3.215}$$

where c_d is the total drag coefficient, $c_{d,p}$ is the pressure drag or form drag coefficient, and $c_{d,f}$ is the skin friction drag coefficient. Lowercase letters are used to indicate the drag of a two-dimensional shape.

The sum of the form drag and the skin friction drag is defined as the *profile drag*. The profile drag is approximately constant with subsonic Mach number, but it is a function of the Reynolds number, which is indicative of whether the boundary layer is laminar or turbulent. Subsonic drag data for airfoils is typically compiled in terms of the profile drag coefficient.

Thus far, we have discussed only the case of an incompressible flow. For higher speed, transonic and supersonic flows, there is an additional drag term due to the presence of shock waves, called the *wave drag*. (These high-speed flows and shock waves are discussed in Section 3.11.) The pressure increases discontinuously across a shock wave, leading to two sources of shock wave-induced pressure drag: (1) the pressure increase in the flow direction, which causes boundary layer separation, leading to pressure drag due to flow separation and (2) the pressure rise across the shock wave, which results in a pressure distribution with an integrated net force in the drag direction. Therefore, the total drag of a two-dimensional shape, in a compressible, viscous flow, may be written as

$$D = D_p + D_f + D_w \tag{3.216}$$

where the possibility of wave drag, D_w, has been included. In coefficient form, we have

$$c_d = c_{d,p} + c_{d,f} + c_{d,w} \tag{3.217}$$

where $c_{d,w}$ is the wave drag coefficient.

3.7.2.2   Drag of Three-Dimensional (Finite) Wings

The drag of three-dimensional or finite wings is composed of the profile drag of the corresponding two-dimensional airfoil shape, plus an additional drag due to the production of lift. For a wing, the drag due to lift is a pressure drag caused by the wingtip vortices. These vortical flows induce a flow over the three-dimensional wing, which changes the pressure distribution, and results in a net drag force. The drag due to lift is usually referred to as the *induced* or *vortex drag*. For a wing, the induced drag coefficient, $C_{D,i}$, is given by

$$C_{D,i} = \frac{C_L^2}{\pi e AR} \tag{3.218}$$

where C_L is the wing lift coefficient, e is the *Oswald efficiency factor* or simply, the span efficiency factor, and AR is the wing aspect ratio. The span efficiency factor is a parameter related to the wing planform shape. As discussed in Section 3.9, a wing with an elliptical planform shape is the most efficient and has a span efficiency factor equal to one. Other wing planform shapes are less efficient, with span efficiency factors less than one, typically between about 0.85 and 0.95. The wing aspect ratio is equal to the wingspan squared divided by the wing area. The aspect ratio also relates to the wing's efficiency, with higher aspect ratio wings being more efficient at producing lift. With these brief explanations of the terms in Equation (3.218) (to be elaborated on in Section 3.9), we can comment on their effects on the induced drag. It is expected that there would be more drag if the lift is greater, and Equation (3.218) verifies this, as the induced drag increases with the square of the lift coefficient. The induced drag also increases with a decrease in the wing efficiency, embodied by either the span efficiency factor or aspect ratio.

Thus, the total drag coefficient of a finite a wing, C_D, is given by

$$C_D = c_d + C_{D,i} + C_{D,w} \tag{3.219}$$

where c_d is the two-dimensional, airfoil profile drag, given by Equation (3.215), $C_{D,i}$ is the induced drag, and $C_{D,w}$ is the wave drag. Uppercase letters are used when denoting drag for a three-dimensional body, while lowercase letters are still used for the two-dimensional case. The breakdown of aerodynamic drag for a three-dimensional wing or body is shown graphically in Figure 3.42. We develop a deeper understanding of the different types of drag, depicted in Figure 3.42, in future sections of this chapter.

3.7.2.3 Drag of the Complete Aircraft

We now look at the drag of a complete aircraft. The *total drag of a complete aircraft*, D, is defined as the sum of the *parasite drag*, D_e, the *wave drag*, D_w, and the lift-dependent *induced drag*, D_i, of the wing, as follows.

$$D = D_e + D_w + D_i \tag{3.220}$$

or in coefficient form

$$C_D = C_{D,e} + C_{D,w} + C_{D,i} \tag{3.221}$$

The parasite drag is composed of the profile drag of the complete aircraft, which includes the form drag and the skin friction drag of the fuselage, wings, tail, landing gear, engine nacelles, etc.

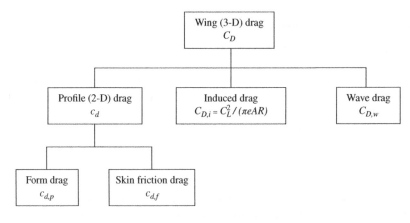

Figure 3.42   Breakdown of aerodynamic drag for a wing.

It also includes some other types of drag, which we capture in the *miscellaneous drag coefficient*, $C_{D,misc}$. Thus, the parasite drag coefficient is given by

$$C_{D,e} = C_{D,p} + C_{D,f} + C_{D,misc} \qquad (3.222)$$

The parasite drag coefficient, as used in Equations (3.221) and (3.222), includes the lift-independent and lift-dependent or induced components of the pressure drag and skin friction drag. The miscellaneous drag is assumed to be independent of the lift. For example, the pressure drag term is separated into a pressure drag that is independent of lift and a pressure drag that is dependent on lift. The same separation of lift-independent and lift-dependent drag is made for the skin friction drag. At a constant Mach number, the aircraft lift changes with angle-of-attack, so that the lift-dependent pressure drag and skin friction drag vary with the aircraft angle-of-attack. This change due to lift is small for the skin friction drag and more significant for the pressure drag. The same argument may be made for the wave drag coefficient, separating it into lift-independent and induced drag terms.

We seek to separate out the lift-independent and lift-dependent drag terms in Equation (3.221). Let us rewrite Equation (3.221) such that the parasite drag and the wave drag coefficients are the sum of lift-independent drag terms, $C_{D,e,0}$ and $C_{D,w,0}$, respectively, and induced drag terms, $C_{D,e,i}$ and $C_{D,w,i}$, respectively.

$$C_D = (C_{D,e,0} + C_{D,e,i}) + (C_{D,w,0} + C_{D,w,i}) + C_{D,wing,i} \qquad (3.223)$$

The lift-independent parasite drag, $C_{D,e,0}$, is given by

$$C_{D,e,0} = C_{D,p,0} + C_{D,f,0} + C_{D,misc} \qquad (3.224)$$

where $C_{D,p,0}$ is the lift-independent pressure drag, $C_{D,f,0}$ is the lift-independent skin friction drag, and $C_{D,misc}$ is the lift-independent wave drag.

Grouping the lift-independent and induced drag terms together, we have

$$C_D = (C_{D,e,0} + C_{D,w,0}) + (C_{D,e,i} + C_{D,w,i} + C_{D,wing,i}) \qquad (3.225)$$

Each of the induced drag coefficient terms has a form similar to Equation (3.218), that is, a constant multiplied with the square of the aircraft lift coefficient. Thus, we can rewrite Equation (3.225) as

$$C_D = (C_{D,e,0} + C_{D,w,0}) + \left(k_e C_L^2 + k_w C_L^2 + \frac{1}{\pi e AR} C_L^2 \right) \qquad (3.226)$$

where k_e and k_w are the induced drag constants for the parasite and wave drags, respectively. Defining a new induced drag constant, K, as

$$K \equiv k_e + k_w + \frac{1}{\pi e AR} \qquad (3.227)$$

and the aircraft *zero-lift, parasite drag coefficient*, $C_{D,0}$, as

$$C_{D,0} \equiv C_{D,e,0} + C_{D,w,0} \qquad (3.228)$$

Using Equations (3.227) and (3.228), Equation (3.226) becomes

$$C_D = C_{D,0} + K C_L^2 = C_{D,0} + C_{D,i} \qquad (3.229)$$

where $C_{D,i}$ is the *induced drag coefficient of the complete aircraft*, which includes the induced drag contributions from the parasite drag, wave drag, and induced drag of the wing. Often, the

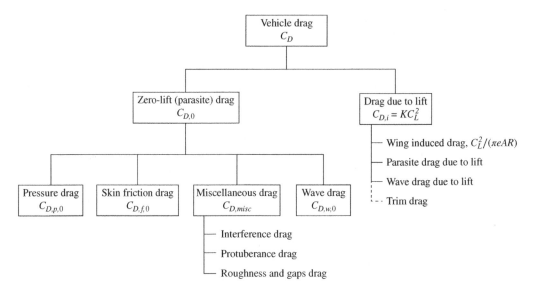

Figure 3.43   Breakdown of drag for the complete aircraft.

induced drag contributions from the parasite drag and the wave drag are much smaller than the induced drag of the wing, such as during cruise flight, so that Equation (3.230) is simplified to

$$C_D = C_{D,0} + \frac{C_L^2}{\pi e AR} \tag{3.230}$$

The total aircraft drag coefficient is shown graphically in Figure 3.43. There are several miscellaneous drag items that are included in the total drag, which are discussed below. There is also a trim drag due to lift that is discussed below.

Interference Drag

An aircraft is composed of many components, including a wing, tail, fuselage, engines, and landing gear. To obtain the total drag of the complete aircraft, one might assume that the drag of each component could be calculated and then summed, but often the sum of the drags of the individual components underestimates the total aircraft drag. This error in the drag estimation is due to the aerodynamic effect that each component may have on the other components. For example, the air flow over the wing affects the flow around the fuselage, tail, and engine nacelles (for wing-mounted engines), which changes the surface pressure distributions and the integrated aerodynamic forces on these components. One may say that the flow over a given component *interferes* with the flow over the other components, thus this type of drag is called *interference drag*. The interference drag may be on the order of 5–10% of the total aircraft drag.

The interference drag can be reduced with the addition of aerodynamic *fairings* and *fillets*. These are typically lightweight, non-structural parts that do not carry any aircraft loads. Fairings are typically streamline-shaped coverings of all or much of a component, such as the wheel and tire of a landing gear. Fillets smooth out the intersection where components come together, such as the intersection of the wing and fuselage, so that the air flow is not slowed or separated, causing drag. Fairings and fillets may also cover gaps and spaces between components, so that the air flow is not trapped or stagnated in regions.

While interference drag usually increases the total aircraft drag, there are instances where aerodynamic interference decreases the total drag. For example, by adding a wingtip fuel tank to a wing, the combined drag of the wing and tip tank is less than the sum of the individual drags. This is due to the favorable interference of the tip tank with the wing, where the tank acts like an end plate on the wing, reducing its induced drag.

The interference drag for an aircraft may be obtained through wind tunnel testing or calculated using computational fluid dynamics. Often, the effects of different fairings and fillets may be assessed parametrically to determine the configuration with the least interference drag.

Protuberance Drag

Protuberances are a variety of objects that protrude into the air flow, including antennas, navigation lights, or air data sensors, such as Pitot tubes and temperature probes. Protuberances also include protrusions or irregularities of the aircraft outer mold line due to steps in skin panel joints or non-flush head rivets in metal structures. The amount of protuberance drag is often related to the design and manufacturing detail or the "fit and finish" of the aircraft. It is possible to make the protuberance drag very small, but this requires significant time and expense in the manufacturing and assembly of the aircraft. The protuberance drag may make up several percent, perhaps as high as 10%, of the total aircraft parasite drag.

Drag due to Roughness and Gaps

As with the protuberance drag, the drag due to roughness and gaps is related to the outer mold line or surface of the aircraft. The surface roughness affects the state of the boundary layer, where transition from laminar to turbulent flow can result in increased skin friction drag. Roughness may promote flow separation, leading to additional pressure drag. Gaps in the aircraft structure may be due to specified tolerances or misalignment of skin panels. The gaps may trip boundary layers and promote turbulent flow over the surface, resulting in higher skin friction drag. There may also be flow leakage into gaps in high-pressure areas, leading to flow momentum losses and leakage drag. The gaps may be in a low-pressure area, resulting in air exhausting from inside the aircraft to the external flow, which may promote flow separation and additional drag. Similar to the protuberance drag, the drag penalty due to roughness and gaps can be reduced with improved "fit and finish" of the aircraft. The drag due to roughness and gaps may be several percent of the total parasite drag of the aircraft.

Some aircraft designs define *aerosmoothness* requirements for the manufacturing, construction, assembly, external finish, or control surface rigging of the aircraft. Aerosmoothness documents typically specify tolerances for surface smoothness, skin panel gaps or misalignments, mechanical fasteners, control surface rigging, etc. Manufacturing or rigging procedures may be given for achieving the specified tolerances.

Trim Drag

The trim drag is a lift-induced drag primarily due to the horizontal tail. Lift is produced by the horizontal tail, usually in the downward direction, to balance the pitching moment produced by the wing. The trim drag includes the induced drag of the horizontal tail and any additional induced drag of the wing, if it produces additional lift to counter the downward tail force (recall that the aircraft weight must be supported by the sum of the lift produced by the wing and the tail). As the aircraft center of gravity moves aft, more tail lift is required, which increases the trim drag. Interestingly, it is possible to obtain a negative trim drag (a force in the thrust direction) in some cases, for far forward center of gravity positions where less wing lift is required, which reduces the wing induced drag. The trim drag is usually small, making up only about 1–2% of total drag of airplane

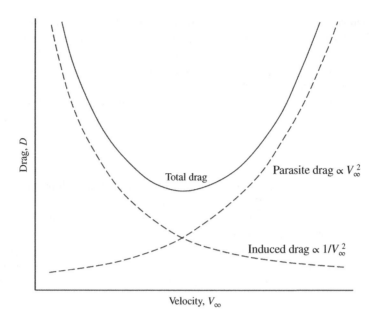

Figure 3.44   Drag versus airspeed.

in a cruise flight condition. It was not included in the development of the aircraft induced drag, as given by Equation (3.226), hence it is shown as connected to the induced drag by a dotted line in Figure 3.43.

3.7.2.4   Variation of Drag with Airspeed and Mach Number

Now that we have developed the total aircraft drag components as shown in Figure 3.43, let us now examine some general trends of some of these drag terms. The variation of the drag with airspeed, excluding wave drag, is shown in Figure 3.44. The parasite drag increases with the square of the velocity, while the induced drag decreases with one over the velocity squared. The sum of the parasite and induced drags, the total drag has a parabolic variation with velocity, with a minimum total drag at a specific airspeed.

The variation of the zero-lift or parasite drag coefficient with Mach number is shown in Figure 3.45. The coefficient remains approximately constant at subsonic Mach numbers. At transonic Mach numbers, the parasite drag coefficient increases rapidly to a maximum at low supersonic Mach numbers, then decreases with increasing supersonic Mach number. The sharp transonic drag rise is discussed in Section 3.10.

It is worthwhile to examine some parasite drag coefficients for an actual aircraft. Table 3.7 presents parasite drag coefficient values for the Gates (now Bombardier) *Learjet* business jet, shown in Figure 3.46. The parasite drag coefficients are specified for various items, along with the percent of the total drag that is accounted for by the item. The interference drag and drag due to roughness and gaps account for over a fifth (20%) of the total aircraft drag.

While examining this drag coefficient listing, note that sometimes the difference or change in the drag coefficient is expressed in terms of *drag counts*. One drag count is defined as a change in the drag coefficient of 0.0001. Thus, the difference in the parasite drag coefficient between the horizontal and vertical tails, in Table 3.7, is 0.0005 or 5 drag counts.

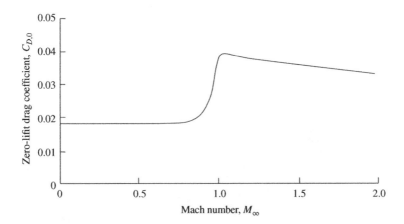

Figure 3.45   Zero-lift drag coefficient versus Mach number.

Table 3.7   Parasite drag breakdown for Gates *Learjet*

Item	Parasite drag coefficient[*] $C_{D,0}$	Percent of total drag
Wing	0.0053	23.45
Fuselage	0.0063	27.88
Tip tanks	0.0021	9.29
Tip tank fins	0.0001	0.44
Engine nacelles	0.0012	5.31
Engine pylons	0.0003	1.33
Horizontal tail	0.0016	7.08
Vertical tail	0.0011	4.86
Interference	0.0031	13.72
Roughness and gaps	0.0015	6.64
Total	0.0226	100.00

[*]Drag coefficient based on wing planform area.
(Source: *R. Ross and R.D. Neal, "Learjet Model 25 Drag Analysis," in Proceedings of the NASA, Industry, University, General Aviation Drag Reduction Workshop, NASA CR-145627, 1975.*)

Example 3.10   D'Alembert's Paradox *An airfoil is in a Mach 0.2 flow. Calculate the drag on the body if the flow is assumed to be incompressible and inviscid.*

Solution

To obtain the force on the airfoil, we apply Newton's second law in the form of the momentum equation for steady, incompressible, inviscid fluid flow, as was provided in Section 3.6. A control volume is drawn around the airfoil as shown in Figure 3.47. The inflow and outflow boundaries, station ∞ and station e, respectively, have the same cross-sectional area, A. They are located far from the airfoil so that the flow streamlines passing through these boundaries are uniform and equal to the freestream conditions of Mach number, velocity, pressure, and density.

The flow streamlines are perpendicular to the upstream and downstream boundaries and parallel to the x-axis. The upper and lower boundaries of the control volume are parallel to the freestream flow, such that there is no flow passing through these boundaries and the properties along these

Figure 3.46   Gates *Learjet* business jet. (Source: *Courtesy of the author.*)

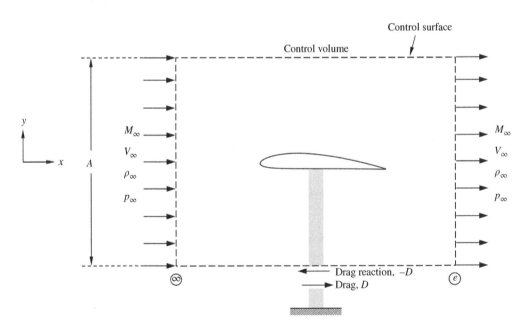

Figure 3.47   Control volume around body in a subsonic, inviscid flow.

boundaries are the freestream conditions. The airfoil is supported by a strut that cuts through the control surface of the control volume, such that the drag force reaction is obtained at the lower boundary.

The x-component of the momentum equation for steady, inviscid flow, in integral form, is given by

$$\sum F_x = \int_S \rho u_x (\vec{u} \cdot \hat{n}) \, dS$$

where F_x is the force component in the x-direction, ρ_∞ is the fluid density, $\vec{u}$ is the fluid velocity, u_x is the x-component of the fluid velocity, and $\hat{n}$ is the unit normal to the control surface, S. The left-hand side this equation is the sum of the forces in the x-direction and the right-hand side of this equation is the net rate of flow of momentum or momentum flux, in the x-direction, in through the inflow boundary, station ∞, and out through the outflow boundary, station e.

Applying the equation above to the control volume, we have

$$-D = \rho_e u_e (u_e)A - \rho_\infty V_\infty (V_\infty)A = \rho_\infty V_\infty (V_\infty)A - \rho_\infty V_\infty (V_\infty)A = 0$$

Thus, zero drag is predicted for the body in an incompressible, inviscid flow. Jean le Rond d'Alembert also applied momentum theory to this problem and arrived at the same conclusion, leading to the paradox that bears his name.

3.7.3   GTT: Drag Cleanup

Reduction of the total aircraft drag is usually a goal of the aerodynamicist. Lower drag translates into higher aerodynamic efficiency and increased performance for the aircraft, embodied by lower fuel consumption, increased range, higher cruise airspeed, higher climb rate, or increased glide capability. The present ground test technique is a systematic method to identify and quantify the sources of parasite drag for a complete aircraft. As discussed in the previous section, the parasite drag is increased due to protuberances, roughness, gaps, and flow leakage. While the increase in parasite drag due to each individual item may be negligible, the collective increase in the parasite drag can be significant. Once the individual drag-producing items are identified, the aerodynamicist can apply modifications, usually in the form of geometry changes, surface smoothing, or gap sealing, to reduce the overall vehicle parasite drag, hence this GTT is called *drag cleanup*.

Drag cleanup investigations have historically been carried out in large wind tunnels, where an aircraft is mounted on a force balance system such that the total aircraft aerodynamic forces and moments are measured directly. The aircraft is first tested in a *clean configuration*, where all protrusions have been removed or faired, and all gaps, openings, and external leaks have been sealed. The measured drag in this configuration provides a minimum, baseline value for the aircraft. The aircraft is then modified, item by item, back to its actual, operational configuration, by adding protuberances, gaps, roughness, etc., and the change in drag is systematically measured after each item is added.

While many drag cleanup studies have been performed in wind tunnels, it may be possible to apply the modern techniques of computational fluid dynamics (CFD) to these investigations. The use of CFD may make drag cleanup studies possible for large aircraft that cannot fit into existing wind tunnel facilities. Of course, the numerical schemes and computational geometries used for the CFD must have sufficient fidelity to quantify the small drag changes for individual drag cleanup items. This may be a more complex and costly endeavor than might be thought at first glance.

The drag cleanup investigation of a full-scale, twin-engine, general aviation airplane, in the NASA Langley 30×60 ft wind tunnel, is shown in Figure 3.48. The airplane tested was a modified Piper *Seneca I* twin-engine airplane. Wool tufts were attached to the aircraft fuselage and wing to visualize the flow over the aircraft, as seen in Figure 3.48. The tufts were used to identify regions of separated flow and areas of flow leakage into or out of the structure. Two sources of significant excess parasite drag were identified for the aircraft in climbing flight with the tufts, (1) premature flow separation near the juncture of the wing and fuselage and (2) flow leakage around the wing flaps and spoilers. These areas were "cleaned up" by installing a new wing–fuselage fillet to eliminate the separated flow, addition of vortex generators on the wing upper surface inboard of the engine nacelles, and sealing of the flap and spoiler leak paths. These modifications significantly reduced the aircraft parasite drag without affecting the cruise drag.

Figure 3.48   Drag cleanup wind tunnel investigation of a full-scale, twin-engine, general aviation airplane. (Source: *NASA*.)

Several modifications were made to reduce the protuberance drag. These included rigging of the wing spoilers to fit flush with the wing upper surface and redesign of the 16 fuel tank inspection hatches, underneath the wing, for a flush fit. Aerodynamic fairings were also added on the underside of the fuselage, to the flap tracks (the channels that the flaps move along) and to the round head rivets on the wing.

Results from the drag cleanup were a small decrease in the parasite drag of 5 drag counts (change in drag coefficient of 0.0005) at cruise flight and a significant decrease of 100 counts (0.0100) for the climb flight condition. It was determined that the addition of the wing–fuselage fairing and the sealing of the spoiler flow leakage made the largest contributions to the parasite drag reduction.

Drag cleanup studies are usually a worthwhile endeavor that can lead to significant reductions in parasite drag and increases in aircraft performance. It is a testament to the adage that many "little things" can add up to be significant. The effort spent on removing or streamlining protuberances, sealing gaps, and smoothing surfaces can significantly reduce the aircraft total parasite drag.

3.7.4   GTT: Wind Tunnel Testing

In this section, we discuss the ground test technique of using the ground-based wind tunnel to obtain aerodynamic data. The wind tunnel has been an indispensable engineering tool in the design and development of aerospace vehicles, which has been used to unlock many of the fundamental secrets in aerodynamics.

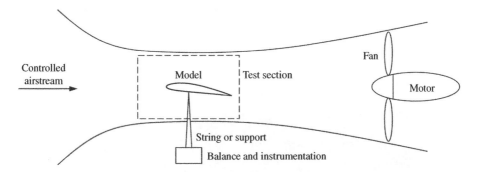

Figure 3.49 Wind tunnel concept. (Source: *Adapted from Baals and Corliss, NASA SP-440, 1981, [14].*)

3.7.4.1 Wind Tunnel Description

A wind tunnel is a ground-based facility that can produce a high-speed flow of air, or other gas, to simulate the flow over a body in flight. In a wind tunnel, a flow is created over a stationary body, while in flight, the body moves through a stationary fluid. However, this difference in the frame of reference does not matter in obtaining the aerodynamic forces. However, it is usually much easier to study and make measurements on a stationary body, versus a moving one. As shown in Figure 3.49, the basic concept of a wind tunnel involves a controlled airstream flowing through a converging section, where the flow velocity is increased. The flow enters the test section where the wind tunnel model is mounted. The flow is drawn through the wind tunnel by a motor-driven fan, at the exit of the duct. The fan is placed at the exit of the tunnel, drawing in the flow, rather than at the entrance, to avoid ingesting the swirling, turbulent wake from the fan into the test section. A majority of wind tunnels are designed for aerodynamic testing. Other special-use wind tunnels include propulsion tunnels for testing operating jet engines, icing tunnels with refrigeration systems that can simulate aircraft icing, low turbulence or "quiet" tunnels with very low air turbulence, vertical tunnels for testing of spinning aircraft models, free-flight tunnels where aircraft models are free-flown in the test section, and others.

A model is mounted in the *test section*, typically on one or more strut-type mounts called *stings*, as shown in Figure 3.50. The angle-of-attack and angle-of-sideslip of the model are adjusted by moving the sting or rotating the model on the sting. Often, forces and moments are measured on the model using an *internal* or *external force balance*, where the force-measuring sensors are located inside or outside of the model, respectively. Wind tunnel models may have numerous flush orifices or *pressure taps* on their surface, small holes that are plumbed via flexible tubing to pressure sensors. These surface pressure measurements are integrated to obtain the model forces and moments. They may also provide detailed, local pressure information such as the locations of separated flow regions or shock waves.

A wide variety of *flow visualization* techniques are also used in wind tunnel testing. The flow on the model surface can be visualized using oils, paints, tufts, or materials that sublime or evaporate on the surface. These techniques are especially useful for characterizing the flow patterns on the model, including regions of separated flow. Techniques used to visualize the flow around a body include injection of smoke, helium-filled bubbles, and fog into the air stream upstream of the body. More sophisticated, laser-based techniques can obtain quantitative, velocity field data around the wind tunnel model.

Figure 3.50   Model mounted on a sting in a wind tunnel test section. (Source: *NASA*.)

3.7.4.2   Geometric, Kinematic, and Dynamic Similarity

So how can measurements on a small model, mounted in a duct with air blowing over it, predict the aerodynamic characteristics of a full-scale vehicle in flight? The answer lies in the application of *similarity*. The first, perhaps obvious similarity requirement is that the model and the full-scale aircraft are *geometrically similar*. The two geometries must have the same external shapes or *outer mold lines*. The wind tunnel model is typically geometrically scaled from the full-scale vehicle.

The second similarity requirement, called *kinematic similarity*, has to do with similarity in time between the wind tunnel flow and flight. The paths of the moving fluid particles in the wind tunnel and in flight must be the same as a function of time. In other words, the flow streamlines over the wind tunnel model must be similar to the flow streamlines in flight.

The third requirement is for *dynamic similarity*, which involves matching the physics that determine the forces on the wind tunnel model and the full-scale vehicle. The forces could be influenced by viscous effects or by effects of the compressibility of the air. The similarity parameter that relates the viscous forces to the inertia forces is the Reynolds number. The similarity parameter that relates the inertia forces to the elasticity (air compressibility) forces is the Mach number. Thus, dynamic similarity requires that the wind tunnel flow have the same Reynolds number and Mach number as flight. The non-dimensional pressure coefficients, force coefficients, and moment coefficients are then the same for both flows. The non-dimensional coefficients are calculated from measurements of the dimensional forces and moments on the sub-scale model in the wind tunnel. The dimensional forces, such as the lift and drag, for the full-scale vehicle are then obtained by multiplying the lift and drag coefficients, respectively, by the flight dynamic pressure and full-scale reference area.

It is seldom possible to match exactly both the Mach number and Reynolds number between the wind tunnel and flight. A choice must then be made as to which parameter is more important to match. This decision is often dependent on the flow regime. For high-speed flight, where compressibility effects may predominate, it may be more important to match Mach number. For low-speed, subsonic flight, viscous effects may dominate over those due to compressibility, so matching Reynolds number may be more important. Often, it is possible to match both similarity parameters well enough so that the critical physics of the flow is captured.

Let us look more closely at the two situations, where it is more critical to match either the Mach number or the Reynolds number between the wind tunnel and flight. For both of these cases, it is assumed that the air flowing in the wind tunnel has the same properties as in flight, that is, the freestream pressure, p_∞, temperature, T_∞, viscosity, μ_∞, and ratio of specific heats, γ, are the same. In addition, it is assumed that the wind tunnel and flight are dynamically similar flows.

The lift on a full-scale aircraft wing in flight, L_f, is expressed as

$$L_f = q_\infty S_f C_{L,f} = \frac{1}{2}\gamma p_\infty M_\infty^2 S_f C_{L,f} \tag{3.231}$$

where S_f is the wing reference area of the aircraft, $C_{L,f}$ is the wing lift coefficient in flight, and q_∞ and M_∞ are the freestream dynamic pressure and Mach number, respectively.

Now, assume that a wind tunnel test is conducted of a scale model of the aircraft, where the air properties and Mach number are matched. The lift on the aircraft model wing, measured in the wind tunnel, is given by

$$L_w = q_\infty S_w C_{L,w} = \frac{1}{2}\gamma p_\infty M_\infty^2 S_w C_{L,w} \tag{3.232}$$

where S_w is the wing reference area of the aircraft model wing and $C_{L,w}$ is the lift coefficient obtained from the wind tunnel test. Since it is assumed that the flows in the wind tunnel and flight are dynamically similar, the lift coefficients in flight and in the wind tunnel are the same, $C_{L,f} = C_{L,w}$. Therefore, dividing Equation (3.231) by (3.232) and solving for the wing lift of the full-scale aircraft in flight, we have

$$L_f = \left(\frac{S_f}{S_w}\right) L_w \tag{3.233}$$

Thus, Equation (3.233) tells us that if the Mach number is matched between the wind tunnel and flight, the lift of the aircraft in flight scales with the lift of the wind tunnel model by the ratio of their reference areas.

Let us now look at the case where matching the Reynolds number is more important. The lift on the full-scale aircraft wing in flight is given by

$$L_f = q_\infty S_f C_{L,f} = \frac{1}{2}\rho_\infty V_\infty^2 b_f c_f C_{L,f} \tag{3.234}$$

where ρ_∞ is the freestream density, and it is assumed that the wing reference area, S_f, is that of a rectangular wing of span, b_f, and chord, c_f. Multiplying and dividing by several ratios equal to one, Equation (3.234) becomes

$$L_f = \frac{1}{2}\rho_\infty \left(\frac{\rho_\infty}{\rho_\infty}\right) V_\infty^2 \left(\frac{\mu_\infty}{\mu_\infty}\right)^2 b_f c_f \left(\frac{c_f}{c_f}\right) C_{L,f} = \frac{1}{2}\left(\frac{\mu_\infty^2}{\rho_\infty}\right)\left(\frac{\rho_\infty^2 V_\infty^2 c_f^2}{\mu_\infty^2}\right)\left(\frac{b_f}{c_f}\right) C_{L,f} \quad (3.235)$$

The terms in Equation (3.235) have been grouped such that the Reynolds number based on the wing chord length, Re_c, is identified, so that we have

$$L_f = \frac{1}{2}\left(\frac{\mu_\infty^2}{\rho_\infty}\right)\left(\frac{b_f}{c_f}\right) Re_c^2 C_{L,f} = \frac{1}{2}\left(\frac{\mu_\infty^2 R T_\infty}{p_\infty}\right)\left(\frac{b_f}{c_f}\right) Re_c^2 C_{L,f} \quad (3.236)$$

where the perfect gas equation of state has been used to replace the freestream density by the freestream pressure, p_∞, freestream temperature, T_∞, and specific gas constant, R.

Similarly, for the lift from the wind tunnel test of a scale model of the aircraft, where the air properties and Reynolds number, Re_c, match the flight conditions, Equations (3.235) and (3.236) are written as

$$L_f = \frac{1}{2}\rho_\infty \left(\frac{\rho_\infty}{\rho_\infty}\right) V_\infty^2 \left(\frac{\mu_\infty}{\mu_\infty}\right)^2 b_w c_w \left(\frac{c_w}{c_w}\right) C_{L,w} = \frac{1}{2}\left(\frac{\mu_\infty^2}{\rho_\infty}\right)\left(\frac{\rho_\infty^2 V_\infty^2 c_w^2}{\mu_\infty^2}\right)\left(\frac{b_w}{c_w}\right) C_{L,w} \quad (3.237)$$

$$L_w = \frac{1}{2}\left(\frac{\mu_\infty^2}{\rho_\infty}\right)\left(\frac{b_w}{c_w}\right) Re_c^2 C_{L,w} = \frac{1}{2}\left(\frac{\mu_\infty^2 R T_\infty}{p_\infty}\right)\left(\frac{b_w}{c_w}\right) Re_c^2 C_{L,w} \quad (3.238)$$

where $C_{L,w}$ is the lift coefficient from the wind tunnel and b_w and c_w are the wingspan and wing chord, respectively, of the wind tunnel model. The Reynolds number, Re_c, is now based on the model wing chord length, c_w. If the wind tunnel and flight are dynamically similar flows, we again have $C_{L,f} = C_{L,w}$.

As the wing planform area is scaled down in size, the ratio of the wingspan to the chord, b/c, remains constant. (This important ratio is the wing *aspect ratio*, which equals b/c for a rectangular wing, to be discussed later in this chapter.) Therefore, we have

$$\frac{b_w}{c_w} = \frac{b_f}{c_f} \quad (3.239)$$

Given that the air properties, wing aspect ratios, Reynolds numbers, and lift coefficients are the same between the wind tunnel and flight, Equations (3.236) and (3.238) gives us the simple result that the lift in flight equals the lift in the wind tunnel, $L_w = L_f$.

3.7.4.3 Subsonic Wind Tunnel Velocity–Area Relation

The basic components and cross-sectional area variations of a subsonic and a supersonic wind tunnel are shown in Figure 3.51. The *plenum* or reservoir is the source of the air or gas for the wind tunnel. The plenum draws in air from the atmosphere, while a reservoir is a vessel that stores the gas. The subsonic wind tunnel has a converging section or *contraction* downstream of the plenum or reservoir, where the tunnel cross-sectional area decreases to the *test section* area. Downstream of the subsonic test section, the area increases in the *diffuser* to the exhaust exit area. For the supersonic wind tunnel, the flow exiting the plenum or reservoir goes through a converging-diverging nozzle, which exhausts into the test section. The flow is exhausted downstream of the supersonic test section.

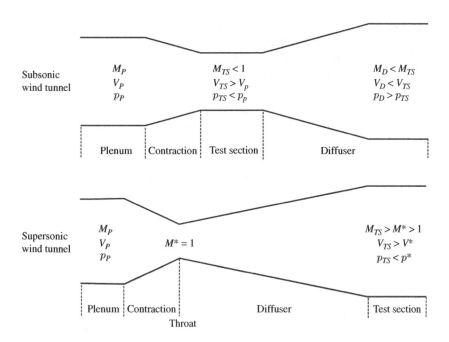

Figure 3.51   Subsonic and supersonic wind tunnel design.

Let us now examine how the flow velocity and other properties change as the air or gas flows through the wind tunnel. Since the wind tunnel has non-porous walls, the mass flow of gas, $\dot{m}_p$, that comes from the plenum (or reservoir) remains constant through the tunnel. Thus, for the subsonic wind tunnel, we can write

$$\dot{m}_p = \dot{m}_t = \dot{m}_d \tag{3.240}$$

where $\dot{m}_t$ and $\dot{m}_d$ are the mass flows through the test section and diffuser exit, respectively. The mass flow is defined as

$$\dot{m} = \rho A V \tag{3.241}$$

where ρ is the gas density, A is the cross-sectional area, and V is the flow velocity. The mass flow has units of slugs/s or kg/s, in English or SI units, respectively. Substituting Equation (3.241) into (3.240), we have

$$(\rho A V)_p = (\rho A V)_t = (\rho A V)_d \tag{3.242}$$

At low subsonic speed, the flow can assumed to be incompressible, so that the gas density is constant throughout the wind tunnel. Therefore, Equation (3.242) becomes

$$(A V)_p = (A V)_t = (A V)_d \tag{3.243}$$

Solving Equation (3.243) for the test section velocity, V_t, we have

$$V_t = \left(\frac{A_p}{A_t}\right) V_p = \frac{V_p}{A_t/A_p} \tag{3.244}$$

The area decreases or *contracts* between the plenum and the test section, so that the duct *contraction ratio*, A_t/A_p, is less than one, resulting in an increase of the velocity and Mach number from the plenum to the test section. For a subsonic flow, a decrease in the area cannot increase the Mach number beyond Mach one. As discussed in a later section, there is a unique relationship between the Mach number and area ratio for subsonic and supersonic flows. By properly selecting the contraction ratio, the desired test section Mach number is obtained.

Using Equation (3.243), the velocity at the diffuser exit, V_d, is given by

$$V_d = \left(\frac{A_t}{A_d}\right) V_t = \frac{V_p}{A_d/A_t} \qquad (3.245)$$

Here, we see that the area expansion of the diffuser, such that $(A_d/A_t) > 1$, results in a decrease in the velocity and Mach number through the diffuser duct. A similar analysis can be performed for the supersonic wind tunnel, but this discussion is deferred until the discussion of supersonic nozzles, later in this chapter.

3.7.4.4 Types of Wind Tunnels

The two basic types of winds tunnels are the *open circuit* and *closed circuit* designs, as shown in Figure 3.52. There are many variations of these basic types, designed for special applications and speed ranges. As shown in Figure 3.52, the flow follows a straight path through an open circuit wind tunnel. A motor-driven fan is used to draw in atmospheric air through a section composed of a honeycomb or wire screen, which serves to guide and straighten the flow. A contraction section or effuser is used to increase the flow velocity or Mach number to the desired test section conditions. The test section may be open to the atmosphere, called an open jet or Eiffel wind tunnel, or it may be totally enclosed. The flow is slowed in the diffuser and exits through the motor or fan section, which is also open to the atmosphere. Open circuit wind tunnels are limited to subsonic test section Mach numbers.

In the closed circuit wind tunnel, also called a closed return wind tunnel (it is also called a Prandtl or Gottingen wind tunnel, after the aerodynamicist Ludwig Prandtl and after the German city where its use was pioneered), the airflow recirculates continuously inside the tunnel ducting. Created by a motor-driven fan, the flow is turned in the closed circuit by turning or guide vanes. A contraction section is still used to increase the flow velocity or Mach number to the desired conditions at the test section. Subsonic or supersonic flow conditions can be achieved with a closed circuit wind tunnel. Closed circuit tunnels are more complex than open circuit tunnels, but the test section flow quality is usually better.

For the open circuit and closed circuit tunnels discussed so far, one or more motor-driven fans produce the flow. These are examples of *continuous flow wind tunnels*, where the flow can be maintained for a long time. Some larger closed circuit wind tunnels may have run time limitations due

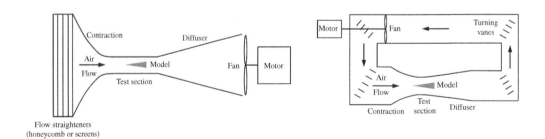

Figure 3.52   Open circuit (left) and closed circuit (right) wind tunnels.

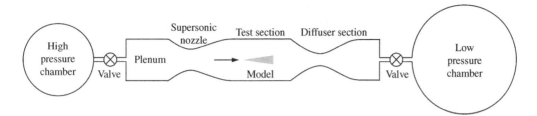

Figure 3.53   Supersonic blow down wind tunnel.

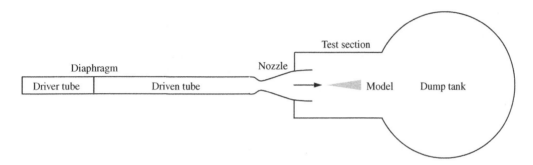

Figure 3.54   The impulse wind tunnel, a hypersonic shock tube.

to the large power requirements of the motor. In contrast, the *intermittent flow* or *blow down wind tunnel* can operate for only a short duration, since the flow is created by exhausting a high-pressure gas from a reservoir into the test section. The general arrangement of a closed circuit, supersonic, blow down wind tunnel is shown in Figure 3.53. The high-pressure air, or other gas, is stored in a reservoir upstream of the test section. A fast-acting valve is opened, allowing the high-pressure gas to exhaust through a convergent-divergent supersonic nozzle and the test section into a low-pressure chamber. Once the pressures in the high- and low-pressure chambers equalize, the flow stops. The flow time is limited by the capacities of the high- or low-pressure chambers, but supersonic run times on the order of several seconds is typical. The set up for another test requires pumping up of the gas in high-pressure reservoir and evacuation of the pressure in the low-pressure chamber.

Another type of intermittent wind tunnel is the *impulse tunnel* designed to simulate hypersonic Mach number flows. In these types of tunnels, a high pressure, high temperature gas is created which is then expanded through a convergent-divergent nozzle into the test section. In the shock tube, shown in Figure 3.54, a high-pressure gas in the driver section is separated from the low-pressure gas in the driven section by a metal diaphragm. When the diaphragm is burst, the high-pressure gas drives a shock wave into the driven section, which compresses the gas, greatly increasing its pressure and temperature. The high-pressure high-temperature gas is expanded through the nozzle, creating a hypersonic flow over the test model. Typical run times for impulse tunnels are on the order of milliseconds, with some on the order of microseconds. With these incredibly short test times, sophisticated, high-speed instrumentation must be used to collect data.

3.7.4.5   Examples of Wind Tunnels

In 1871, the first enclosed wind tunnel was designed and built by a British marine engineer, Francis Herbert Wenham (1824–1908) and a British scientific instrument maker, John Browning

(1831–1925). Despite the rudimentary nature of their wind tunnel, they obtained some noteworthy aerodynamic results. They measured lift-to-drag ratios on wings and discovered the significance of the wing aspect ratio. From their measurements, they concluded that higher lift-to-drag ratios were possible with higher aspect ratio wings. After the invention of the wind tunnel, significant strides were made in the empirical design of wings and other aerodynamic shapes. In this section, several examples are presented of wind tunnels, from devices or facilities used in early aviation to modern test complexes. This is not a comprehensive listing, but it provides some insight into the wide range of wind tunnels that are possible.

The Whirling Arm

Although not a wind tunnel, one of the earliest devices used to simulate the flow over a flying object was the *whirling arm*. This apparatus typically consisted of a test object that was placed at the end of an arm, which rotated, in a horizontal plane, being spun on a spindle by a falling weight. The whirling arm was first used by the British mathematician and military engineer, Benjamin Robins (1707–1751), to compare the air resistance of various blunt shapes (Figure 3.55). Robins' whirling arm could produce a flow of only a few feet per second. However, based on his whirling arm data, Robins concluded that the existing theories for aerodynamic drag, including those developed by Sir Isaac Newton, were incorrect. Of course, the whirling arm had some major deficiencies as an aerodynamic tool. After repeated rotations through the same air mass, the model and the air were both rotating, leading to the inability to define the true flow velocity seen by the model. The force measurements on the model were also not very accurate, as it was difficult to make these measurements on a spinning arm. Still, the whirling arm apparatus was used by many early scientists, engineers, and aviation enthusiasts, including Otto Lilienthal and Sir George Cayley.

Wright Brothers' Wind Tunnel

The Wright brothers designed a simple, open circuit wind tunnel, which was instrumental to the success of their airplane design (Figure 3.56). At the end of 1901, they had determined that the aerodynamic data, upon which they based their glider designs, was the reason why the gliders performed so poorly in flight. Some of this data was from Lilienthal's whirling arm measurements, which they now concluded was not accurate.

The Wright brothers' wind tunnel was essentially a long wooden box with a fan at the front end, upstream of the test section, with the flow exhausting to the atmosphere. They placed flow straighteners behind the fan to help make the flow more uniform. There was a viewing window at

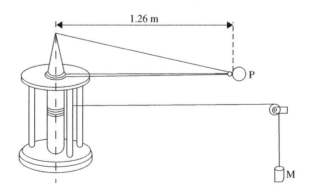

Figure 3.55 Whirling arm apparatus invented by Benjamin Robins, 1746. (Source: *Baals and Corliss, NASA SP-440, 1981, [14].*)

Figure 3.56 Replica of the Wright brothers' wind tunnel. (Source: *M.L. Watts, "Wright Brothers Wind Tunnel Replica" https://commons.wikimedia.org/wiki/File:Wright_Brothers_Wind_Tunnel_Replica .jpg, CC-BY-SA-3.0. License at https://creativecommons.org/licenses/by-sa/3.0/legalcode.*)

the back of tunnel, above the test section, used to view the model and force measurement devices. They built and tested hundreds of wing models in their wind tunnel, measuring lift and drag with a force balance of their own design. They also conducted aerodynamic studies for the design of their propellers. Following a systematic, scientific process, they conducted parametric studies of the various shapes, changing only one variable at a time. By the end of 1901, the Wright brothers had developed a detailed aerodynamic design database for aircraft wings. They used this database for the design of their future aircraft, including the successful 1903 *Flyer*.

Variable Density Tunnel
One of the wind tunnels that has had a major impact on the design of aircraft was the variable density tunnel (VDT) at the NACA Memorial Aeronautical Laboratory, Hampton, Virginia (now the NASA Langley Research Center). The VDT was designed by German aerodynamicist Max Munk (1890–1986), at the NACA aeronautical laboratory in the early 1920s. The VDT was a closed circuit design where the entire tunnel was essentially encased within a large, welded steel pressure vessel (Figure 3.57). By increasing the pressure up to 21 atmospheres (44,400 lb/ft^2, 309 lb/in^2, 2217 kPa), the VDT could simulate high Reynolds number flows with sub-scale models. (Looking back at the definition of the Reynolds number, Equation (3.32), if the length of the wind tunnel model wing chord is one tenth of the full-scale aircraft wing chord, the Reynolds number is matched if the wind tunnel density is increased by a factor of ten. A ten-fold increase of the density is achieved with a ten-fold increase of the pressure.) The VDT was the first wind tunnel that could match flight Reynolds numbers, allowing it to produce much more accurate aerodynamic data than ever before.

The VDT pressure vessel was constructed by a shipbuilding company, the Newport News Shipbuilding and Dry Dock Company, in nearby Newport News, Virginia. The 2 ¼″ thick (54 mm) steel tank was 34.5 ft (10.5 m) in length, 15 ft (4.6 m) in diameter, and weighed 85 tons (77,000 kg). The test section was 5 ft (1.5 m) in diameter. The VDT closed circuit design had a clever annular

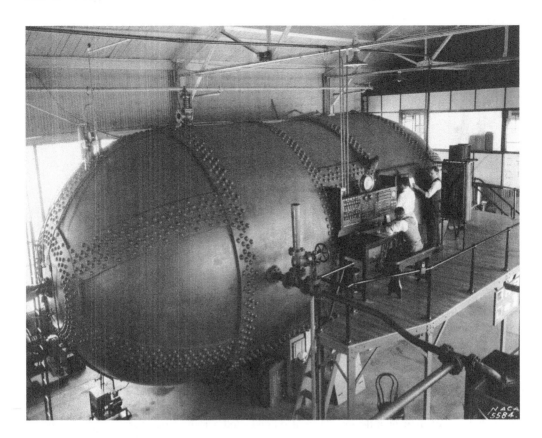

Figure 3.57   NACA variable density wind tunnel. (Source: *NASA*.)

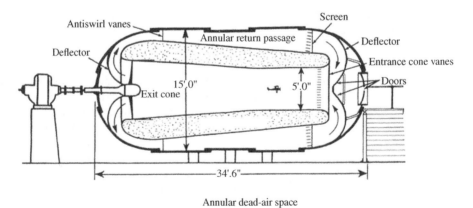

Figure 3.58   Diagram of the NACA variable density wind tunnel. (Source: *Baals and Corliss, NASA SP-440, 1981, [14].*)

return passage, which minimized the required pressure vessel volume (Figure 3.58). Power to the fan was supplied by a 250 hp (186 kW) motor. A maximum flow speed of about 50 mph (80 km/h) could be reached in the test section.

The VDT was extensively used for airfoil design between the early 1920s through the 1940s. Airfoil data from the VDT played an important role in aircraft design, especially for

World War II aircraft designs. A landmark NACA publication in 1933, NACA Report No. 46, cataloged the airfoil data for 78 airfoil shapes that were used for these aircraft designs [40]. Given its significant role in the history of aircraft development, the site of the VDT is now a National Historic Landmark.

Slotted-Wall Wind Tunnel

Prior to the first manned supersonic flight, aeronautical engineers had subsonic and supersonic wind tunnels at their disposal to obtain aerodynamic data. However, at this time, there were no wind tunnels capable of obtaining transonic aerodynamic data. During transonic or supersonic wind tunnel testing, shock waves emanate from the model and the sting in the test section. At transonic and low supersonic Mach numbers, the shock wave angles are very steep, perhaps nearly perpendicular or normal to the freestream flow direction. These shock waves reflect off the wind tunnel walls and may bounce back onto the model, sting, or subsonic wake of the model, contaminating aerodynamic measurements on the model. The solution to this problem came from NACA aerodynamicists who developed a wind tunnel with slots cut into the wind tunnel test section walls. The slots were parallel to the freestream flow direction and served to suck away the shock waves from the model, preventing them from reflecting back onto the model or its wake. The slotted wall test section of the NASA Langley 16-foot high-speed wind tunnel is shown in Figure 3.59. The development of the slotted-throat transonic wind tunnel was a significant accomplishment that paved the way for use of the wind tunnel in fundamental research of transonic aerodynamics. This accomplishment was so significant, that the 1951 Collier Trophy[3] was awarded to the NACA for the development of the slotted-throat transonic wind tunnel.

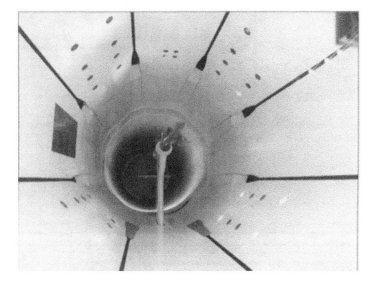

Figure 3.59 NASA Langley 16-foot transonic, slotted wall wind tunnel test section. (Source: *E.C. Ezel and L.N. Ezel, "The Partnership: A History of the Apollo-Soyuz Test Project," NASA SP-4209, 1978.*)

---

[3] The Robert J. Collier Trophy is awarded annually by the National Aeronautic Association "for the greatest achievement in aeronautics or astronautics in America, with respect to improving the performance, efficiency, and safety of air or space vehicles, the value of which has been thoroughly demonstrated by actual use during the preceding year". The Collier Trophy is on permanent display at the Smithsonian National Air and Space Museum, Washington, DC.

Hypersonic Gun Tunnel

An example of a hypersonic impulse wind tunnel is the *Longshot* free-piston or gun tunnel at the Von Karman Institute for Fluid Dynamics, Rhode-Saint-Genese, Belgium (Figure 3.60). The Longshot tunnel can simulate hypersonic flows between Mach 14 and 20 at very high unit Reynolds numbers up to about 14 million per meter (4.3 million per foot), the highest unit Reynolds number obtainable, at these hypersonic Mach numbers, in the world. This enables the *Longshot* to match both Mach number and Reynolds number at hypersonic reentry flight conditions.

The *Longshot* has a 6 m-long (20 ft-long) driver tube with a 12.5 cm (4.9 in) inner diameter that is filled with nitrogen gas (Figure 3.61). The driven tube is 27 m (89 ft) long with an inner diameter of 7.5 cm (3.0 in) and is filled with a gas, either nitrogen or carbon dioxide, at atmospheric pressure. The gas in the driver tube is pumped up to very high pressures of 300–1000 bars

Figure 3.60 *Longshot* hypersonic gun tunnel, Von Karman Institute, Belgium. (Source: *Photo courtesy of Sebastien Paris, Von Karman Institute for Fluid Dynamics, with permission.*)

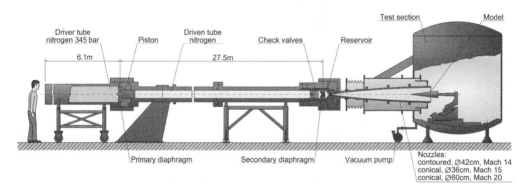

Figure 3.61 Schematic of the Von Karman Institute *Longshot* hypersonic gun tunnel. (Source: *Image courtesy of Guillaume Grossir, Von Karman Institute for Fluid Dynamics, with permission.*)

(4300–14,500 psi, 30–100 MPa). A primary diaphragm separates the driver section from the driven section. A secondary diaphragm separates the driven section from a contoured nozzle that exhausts into a 16 m^3 (565 ft^3) evacuated vessel or "dump tank". The wind tunnel model is mounted in the vacuum vessel, facing towards the exit of the nozzle. A piston, weighing 1.5–9 kg (3.3–20 lb), is located at the upstream end of the driven tube. By rupturing the primary diaphragm, the piston is exposed to the very high pressure of the driver section. The high pressure driver gas shoots the piston into the driven section, accelerating it to 600 m/s (2000 ft/s, 1300 mph). (The name *gun tunnel* is derived from the bullet-like motion of the piston.) The piston compresses the gas in the driven section to a high pressure and temperature. When the secondary diaphragm is ruptured, the high pressure, driven tube gas flows through the nozzle, establishing the hypersonic flow at the nozzle exit for about 5–10 ms. Due to the short run time, data is collected at high sample rates, up to about 50 kHz or 50,000 data samples per second.

Vertical Spin Tunnel
The vertical spin tunnel is a specialized facility, where models are tested in a vertical air flow. Free-flying aircraft models are tested to obtain data about departure, spin, and other out-of-control flight characteristics (see Section 3.12.4 for details about departure and spin). The models are dynamically scaled, matching the geometric dimensions, inertias, and mass distribution characteristics of the full-scale aircraft. Recovery characteristics of an aircraft configuration are investigated by using remotely actuated control surfaces on the free-flight models. Free fall and dynamic stability characteristics of spacecraft can be obtained in the vertical tunnel.

The NASA 20-foot Vertical Spin Tunnel, located at the NASA Langley Research Center, Hampton, Virginia, is shown in Figure 3.62. The vertical tunnel is a closed circuit, annular return wind tunnel that operates at atmospheric conditions. The fixed-pitch, three-bladed fan, at the top of the tunnel, draws the air flow upward through the test section, as shown in Figure 3.62. The velocity of the air flow, through the test section, can be varied from zero to approximately 85 ft/s (58 mph, 26 m/s, Mach 0.08). Test models are inserted into the test section by hand, as shown in Figure 3.63, or released from a cable, suspending the model above the tunnel. An angular velocity or spin is imparted to the model by hand, to establish the model in a spin. As the model drops down the test section, the vertical air velocity is increased until the model is stabilized and suspended in the vertical flow. Data is collected by an array of cameras and with instrumentation mounted in the model. Models are also mounted on a rotary balance, where they are set at different angles-of-attack and rotated at various spin rotation rates. The balance measures aerodynamic forces and moments on the model while it is spinning.

National Full-Scale Aerodynamics Complex
The wind tunnel with the largest test section in the world is the National Full-Scale Aerodynamics Complex (NFAC) located at the NASA Ames Research Center, Moffett Field, California (Figure 3.64). The NFAC has two large test sections, a 40 × 80 ft (12 × 24 m) test section, and an 80 × 120 ft (24 × 37 m) test section, that are both connected to a common fan-drive system (Figure 3.65). By moving a system of wall closure louvers, the air from the fans is drawn into either test section. The 40 × 80 ft wind tunnel is a closed circuit design and the 80 × 120 ft wind tunnel is an open circuit design. The large test sections are capable of testing full-scale aircraft as shown in Figure 3.66. A full size Boeing 737 airliner can be tested in the 80 × 120 ft wind tunnel. Test speeds of up to 300 knots (340 mph, 560 km/h) are obtainable in the 40 × 80 ft tunnel and up to 110 knots (130 mph, 200 km/h) in the 80 × 120 ft tunnel. The NFAC fan-drive system consists of six 40-foot diameter (12 m), variable-pitch fans, each with 15 laminated wood blades. A 22,500 hp (16,800 kW) electric motor powers each fan. It takes 106 MW (142,000 hp) of electricity to run

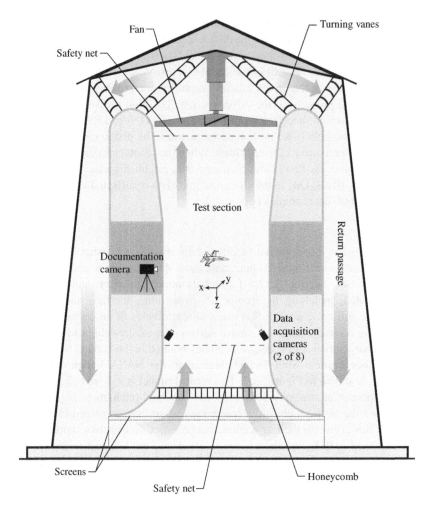

Fan

Turning vanes

Safety net

Test section

Documentation
camera

Return passage

Data
acquisition
cameras
(2 of 8)

$x \leftarrow$ y
z

Screens

Safety net

Honeycomb

Figure 3.62   NASA 20-foot Vertical Spin Tunnel. (Source: *NASA*.)

the six electric motors. The fan-drive provides an air mass flow of up to 60 tons per second (54,000 kg/sec).

Example 3.11   Matching Mach Number or Reynolds Number in Wind Tunnel Tests *The rectangular wing of a small ultralight aircraft has a span of 20 ft (6.1 m) and a chord length of 3.5 ft (1.1 m). A model of the aircraft, with a wing chord length of 0.7 ft (0.21 m), is tested in a wind tunnel at a flow velocity of 170 mph (273.6 km/h) and a lift of 17.8 lb (79.2 N) is measured. Assuming that the Mach number is matched between the wind tunnel and flight, what is the corresponding lift in flight? Assuming that the Reynolds number is matched between the wind tunnel and flight, what is the corresponding flight airspeed?*

Solution

For matching Mach numbers between the wind tunnel and flight, the lift in flight is related to the lift measured in the wind tunnel by Equation (3.233). The lift in flight is calculated as

$$L_f = \left(\frac{S_f}{S_w} \right) L_w = \left(\frac{b_f c_f}{b_w c_w} \right) L_w$$

Figure 3.63   Aircraft model thrown into test section of NASA 20-foot Vertical Spin Tunnel. (Source: *NASA.*)

The aspect ratio, AR, of the full-scale wing is

$$AR = \frac{b_f}{c_f} = \frac{20\,ft}{3.5\,ft} = 5.71$$

Since the wing aspect ratio of the full-scale aircraft and the wind tunnel model are the same, the model wingspan is given by

$$b_w = (AR)(c_w) = (5.71)(0.7\,ft) = 4.0\,ft$$

Inserting numerical values into the equation for the lift in flight, we have

$$L_f = \left(\frac{20\,ft \times 3.5\,ft}{4\,ft \times 0.7\,ft} \right)(17.8\,lb) = 445.0\,lb$$

Thus, the 17.8 lb of lift measured in the wind tunnel model corresponds to 445.0 lb of lift in flight.
If the Reynolds numbers are matched between the wind tunnel and flight, we can write

$$Re_f = \frac{\rho_\infty V_f c_f}{\mu_\infty} = Re_w = \frac{\rho_\infty V_w c_w}{\mu_\infty}$$

where Re_f is the Reynolds number in flight and Re_w is the wind tunnel Reynolds number. Assuming that the freestream density and viscosity are the same in flight and in the wind tunnel, we have

$$V_f c_f = V_w c_w$$

Figure 3.64   The National Full-Scale Aerodynamics Complex (NFAC), Moffett Field, California. (Source: *NASA.*)

Solving for the airspeed in flight, we have

$$V_f = V_w \frac{c_w}{c_d} = (170\,mph) \left(\frac{0.7\,ft}{4\,ft} \right) = 29.75\,mph$$

Thus, the wind tunnel test of the model at 170 mph corresponds to an airspeed of 29.75 mph of the full-scale aircraft in flight.

3.7.5   GTT: Computational Fluid Dynamics

We now discuss a ground test technique for calculating the forces and moments on a vehicle that is quite different from using a wind tunnel test facility to make measurements on a physical model. The present method can be thought of as a "numerical wind tunnel", where computers are used to numerically simulate aerodynamic flows around bodies with a technique called *computational fluid dynamics* or *CFD*. In addition to wind tunnel testing and flight testing, CFD represents another key element of aerospace engineering analysis and design.

The compressible flow over a body is defined mathematically by a set of non-linear, partial differential equations with the proper boundary conditions. However, the exact, closed form, analytical solution of these governing equations for the flow over a complex geometry, such as a complete aircraft, is impossible. So, how can we solve this complex set of differential equations? Let us look at a way to transform the partial differential equations into a form that we can solve.

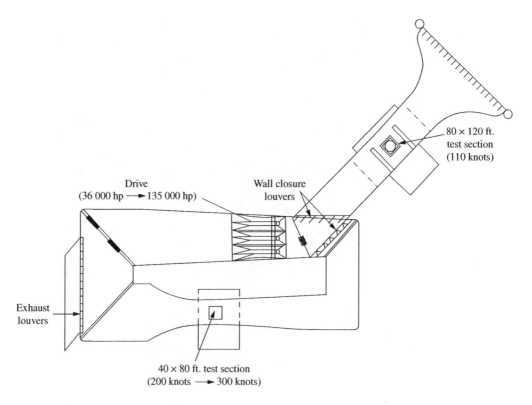

80 × 120 ft.
test section
(110 knots)

Drive
(36 000 hp ⟶ 135 000 hp)

Wall closure
louvers

Exhaust
louvers

40 × 80 ft. test section
(200 knots ⟶ 300 knots)

Figure 3.65   NFAC 40 × 80 ft closed circuit and 80 × 120 ft open circuit wind tunnels. (Source: *NASA.*)

Imagine that there is a flow of air over an F-18 airplane and, somewhere in this flow field, the pressure is defined as $p_{i,j}$, where the point (i, j) denotes a location (x, y) somewhere in the two-dimensional space, as shown in Figure 3.67. Now, imagine that we want to know how the pressure has changed in the x-direction, at a downstream point $(i + 1, j)$. The pressure at the point $(i + 1, j)$ is written as a Taylor series expanded about the point (i, j) as

$$p_{i+1,j} = p_{i,j} + \left(\frac{\partial p}{\partial x}\right)_{i,j} \Delta x + \left(\frac{\partial^2 p}{\partial x^2}\right)_{i,j} \frac{(\Delta x)^2}{2} + \left(\frac{\partial^3 p}{\partial x^3}\right)_{i,j} \frac{(\Delta x)^3}{6} + \cdots \tag{3.246}$$

Equation (3.246) is an exact expression for $p_{i+1,j}$ for an infinite number of higher-order terms or as the distance between the points, Δx, approaches zero. Solving for the first partial derivate term in Equation (3.246), we have

$$\left(\frac{\partial p}{\partial x}\right)_{i,j} = \frac{p_{i+1,j} - p_{i,j}}{\Delta x} - \text{HOT} \tag{3.247}$$

where Δx is the x distance between the point (i, j) and $(i + 1, j)$ and HOT denotes the higher-order terms. If the higher-order terms are neglected, the first partial derivative is approximated as

$$\left(\frac{\partial p}{\partial x}\right)_{i,j} \approx \frac{p_{i+1,j} - p_{i,j}}{\Delta x} \tag{3.248}$$

The partial derivative of the pressure is approximated by an algebraic difference quotient that is evaluated using the discrete, point values of the pressure at the points (i, j) and $(i + 1, j)$.

Figure 3.66 An F-18 aircraft mounted in the NFAC 80×120 ft test section. (Source: *NASA.*)

Using the same philosophy, we can transform the partial differential equations, governing the air flow over the F-18 aircraft, into algebraic equations. These transformed equations are an approximate form of the governing equations of fluid flow, but they still contain the physics embodied by the governing equations. Now, imagine that we place a grid or mesh of points all through the flow field and on the surface of the aircraft. The algebraic forms of the governing equations are now solved numerically by calculating the partial differential terms as algebraic difference quotients, as shown by Equation (3.248). This transformation of the governing fluid flow partial differential equations into a set of algebraic equations that are solved at discrete, grid points throughout the flow is known, in CFD parlance, as *discretization*. Various numerical techniques are used to calculate or *advance* the flow field solution in time or space. Our simple example is related to the CFD *finite-difference* numerical method, where the algebraic quotients are known as finite differences. The result is a set of numbers, at all of the grid points in the flow, for the flow properties, such as pressure, temperature, velocity, Mach number, etc.

An example of a CFD mesh over two-dimensional airfoils is shown in Figure 3.68. A close-up view of the CFD mesh around complex three-dimensional bodies, an airplane wing–body, engine

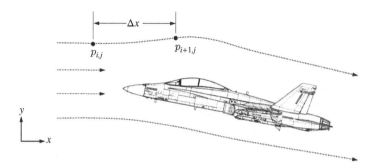

Figure 3.67 Two CFD grid points in the flow field over an F-18.

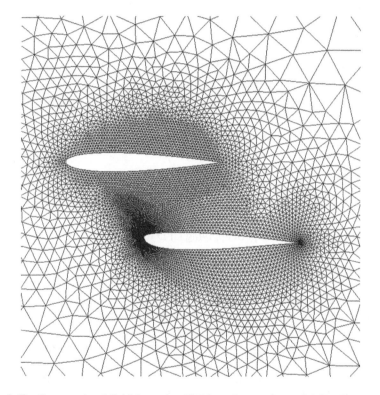

Figure 3.68 Computational fluid dynamics (CFD) mesh around two airfoils. (Source: *NASA*.)

nacelle, and pylon, is shown in Figure 3.69. There are many more grid points close to the airfoils in Figure 3.69 and on the surface of the three-dimensional bodies in Figure 3.69. The Taylor series approximations approach the exact equations as the grid spacing approaches zero. By making the distance between grid points, or *grid spacing*, smaller, the accuracy of the numerical solution improves. Thus, a higher *grid density* is needed to more accurately resolve the physics near the surface where there are larger gradients of the flow properties, such as the velocity. The mesh geometries are not simple rectangles or cubes, rather, they are a complex pattern of polygons. These complex mesh geometries allow a more accurate resolution of the two- or three-dimensional geometries to be analyzed. (You may have noticed that we are italicizing many of the CFD terms

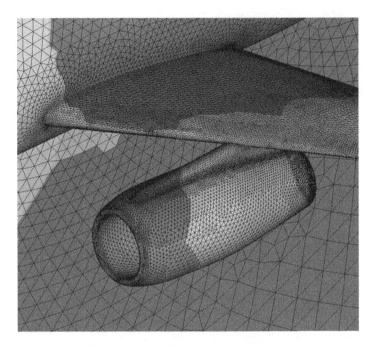

Figure 3.69   Close-up view of the computational fluid dynamics (CFD) mesh around an airplane wing–body, engine nacelle, and pylon. (Source: *NASA*.)

that we have been encountering. This accentuates the fact that the specialized field of CFD has a language and terminology of its own.) Even though these meshes look much more complex than our simple example, the basic concept of applying CFD to these flows is the same.

You may ask the next logical question: why not cover the entire flow field and body surface with a very fine mesh of points, with almost infinitesimal grid spacing, to obtain a high accuracy solution? Well, we now come to the "hardware" or "facility" aspect of CFD. A CFD mesh over complex geometries typically requires millions of grid points. The limiting factor now becomes the computer storage and speed that are available to perform the tremendous number of computations over that many grid points. Even with modern supercomputers, there is a limit to the mesh density that can be used, and the time to complete so many computations may not be practicable. Another factor to consider is that a high-density mesh may not be required in some regions where the flow gradients are small, such as close to the freestream boundaries where the flow is more uniform. The creation of the appropriate mesh for a CFD problem is a critical aspect of achieving an accurate solution and this is an active area of CFD research. Many software packages are specifically designed to perform just the meshing part of the CFD problem. Of course, the type of numerical method used has a major impact on the time it takes to compute a solution and the accuracy of that solution. There are a wide variety of numerical methods or *CFD solvers* that are used, the choice of which is dependent on many factors, such as the flow regime, geometry, and desired accuracy. The development and improvement of CFD solvers is also a vigorous area of CFD research.

The CFD simulation of the flow over a helicopter rotor blade is shown in Figure 3.70. The complex vortical flows over the rotor blade section are clearly evident. A CFD simulation of the hypersonic flow over the *Orion* capsule vehicle is shown in Figure 3.71. A word of caution must be voiced here. The results of CFD can provide the user with a tremendous amount of "data", which is displayed with impressive graphics, but one must always be mindful that the accuracy of these

Figure 3.70 CFD simulation of the flow over a helicopter rotor blade airfoil section. (Source: *NASA*.)

Figure 3.71 CFD simulation of the hypersonic flow over the *Orion* capsule vehicle at about Mach 23. (Source: *NASA*.)

results is dependent on the physics contained in the governing equations upon which the CFD is based. If the physics of interest is not properly captured by the fundamental governing equations and the appropriate boundary conditions, the results of the CFD are probably not valid, despite the ability of the CFD to produce a multitude of numerical results and impressive graphics.

Computational fluid dynamics has become an incredibly powerful analysis and design tool in aerospace engineering. In fact, it has permeated many other fields of study where the flow of a gas or liquid is of interest. It is applicable to both *external flows*, such as the flows over aircraft, rockets, submarines, automobiles, and buildings, and *internal flows*, such as the flows inside jet engines, rocket nozzles, and even wind tunnels. CFD gives engineers the ability to analyze some types of flows that cannot be duplicated in a wind tunnel. The capabilities of CFD in aerospace engineering have grown exponentially, providing an amazing aerodynamic analysis and design capability within reach of almost anyone with a computer, even a modest personal computer.

3.7.6   FTT: Lift and Drag in Steady, Gliding Flight

Use of a wind tunnel to obtain aircraft aerodynamic data typically involves the testing of a sub-scale model at modest subsonic speeds, as full-scale testing is usually not possible for large aircraft at high flight speeds. While wind tunnel testing can be a viable method of predicting the aerodynamic characteristics of an aircraft, there are potential issues that can degrade the accuracy of the measurements. These issues include aerodynamic *interference* from the model strut–support system or the wind tunnel walls. This interference can alter the pressure distribution over the model, thus leading to incorrect lift and drag measurements. Another major issue is the matching of the wind tunnel and flight Reynolds number, which affects the type of boundary layer and the subsequent skin friction and pressure drag.

It would be ideal if one could obtain aerodynamic data on the full-scale aircraft in actual flight. This may not be an easy or practical task for all aircraft, especially if the aerodynamic data is desired over a large portion of the vehicle's flight envelope. However, there are several flight test techniques (FTTs) that are used to obtain lift and drag data for a full-scale aircraft in flight. In the present section, you will fly a simple FTT that is used to obtain the lift and drag of an aircraft in steady, gliding flight, where the aircraft is in a constant airspeed descent with zero thrust. The addition of thrust makes the data collection and analysis more difficult, due to the need to accurately model or measure the in-flight thrust and correct the drag data for the influence of the propulsion system. This is possible, and we fly several other aeromodeling FTTs to obtain lift and drag of an aircraft in flight, accounting for thrust, later in this chapter.

To learn this FTT, you will be flying the North American Aviation (NAA) P-51B *Mustang*, a single-seat, long-range fighter and bomber escort aircraft used during World War II. The P-51 was the first airplane to incorporate a laminar flow airfoil, with the hope of promoting low-drag laminar flow over the wing. The first flight of the P-51 Mustang was on 26 October 1940. Specifically, you will be flying the XP-51B test airplane, used by the NACA for flight research in the 1940s and 1950s, shown in Figure 3.72. A three-view drawing of the XP-51B is shown in Figure 3.73, and selected specifications are given in Table 3.8. The XP-51B was a low wing monoplane with retractable landing gear, powered by a 1500 hp Packard V-1650, liquid-cooled, supercharged, V-12 piston engine, a version of the British Rolls Royce *Merlin* engine.

However, you will be flying this airplane in gliding flight with zero thrust, so you will not be taking advantage of the tremendous horsepower available from this engine. You will not even have the engine running for this FTT and, in fact, the four-bladed, 11 ft 2 in (3.40 m) diameter, Hamilton Standard propeller blades are removed from the airplane. The engine is not running and the propeller is removed because of the difficulties of accurately measuring the thrust force

Figure 3.72   NACA XP-51B *Mustang* test aircraft. (Source: *D.D. Baals and W.R. Corliss, NASA SP-440, 1981, [14].*)

due to the engine exhaust or propeller in flight. Without an accurate measurement or prediction of the in-flight thrust, it is impossible to determine the aircraft drag within any acceptable degree of accuracy. Without a running engine, a hydraulic pump and electrical batteries are installed to power the wing flaps and landing gear.

This somewhat unusual glide flight test of a powered airplane, with the propeller removed, was conducted by the NACA in 1945 [54]. At that time, the accuracy of the drag data obtained from wind tunnels was in question, especially at higher subsonic Mach numbers. The objective of the NACA XP-51 glide flight tests was to obtain high quality aerodynamic data for a full-scale airplane to correlate with wind tunnel data. Wind tunnel data was collected for a 1/3-scale model of the XP-51, with the propeller removed, in the NASA Ames 16 ft (4.9 m) wind tunnel (the wind tunnel test section has a diameter of 16 ft). With the engine shut down and without a propeller, the full-scale XP-51B test airplane was in an optimum configuration for the conduct of gliding flights and for comparison to the wind tunnel data. Of course, the flight safety risk of performing a "dead stick" landing of an airplane without power must be accepted. This risk was mitigated by performing the flight test at Muroc Dry Lake (now Edwards Air Force Base, California), where there are large expanses of dry lakebeds to land an unpowered glider airplane.

However, since our test airplane has no engine, how do we get the aircraft to altitude so that we can perform a glide? We use the same techniques as used by sailplane pilots, that is, the aircraft will be towed to altitude by another airplane. In the NACA tests, the XP-51 was towed aloft behind a Northrop P-61 *Black Widow*, a twin-engine aircraft that was originally designed to intercept enemy bombers, specifically at night, using radar.

We are ready to perform a glide test flight. Strapped into the pilot seat of the XP-51B, you are connected to the P-61 by two long tow cables and start climbing to the planned test release altitude of 28,000 ft (8500 m), as shown in Figure 3.74. The climb is amazingly quiet – not surprising, since the 1500 hp Packard engine is not running. Upon reaching 28,000 ft, the *Black Widow* tow plane

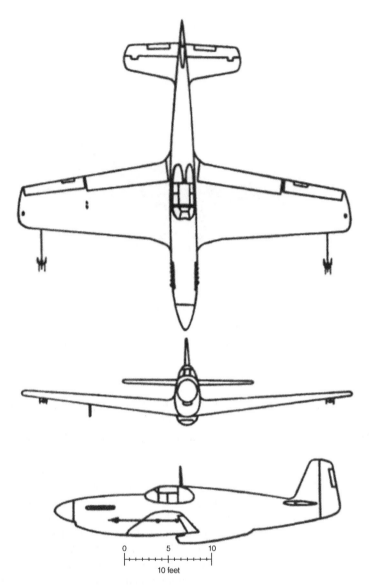

Figure 3.73   Three-view drawing of the NACA XP-51B test aircraft. (Source: *Nissen, et al., NACA ACR 4 K02, 1945, [55].*)

levels off and you prepare to start the glide test. You pull the tow cable release handle, detaching the cable from the airplane's nose with a soft "clunk" and you are now a glider. However, you have no time to dwell on this, as you have data to collect. You firmly grasp the control stick and set the aircraft's pitch attitude slightly below the horizon, capturing a constant airspeed. We will set the airspeed, V_∞, to a constant value to establish steady, gliding flight. We can perform glides at different constant airspeeds to obtain data over a range of airspeeds and Mach numbers.

The aircraft is in steady, gliding flight, as shown in Figure 3.75. Even though the aircraft is descending, the forces acting on it are balanced, so that the aircraft is not accelerating and the airspeed is constant. As shown by the free-body diagram in Figure 3.75, the aircraft is at an

Table 3.8 Selected specifications of NACA XP-51B as flown in test configuration.

Item	Specification
Primary function	Long range fighter, converted to flight research
Manufacturer	North American Aviation, Los Angeles, California
Crew	1 pilot
Weight, as tested	7335 lb (3327 kg)
Wing span	37.03 ft (11.29 m)
Wing reference area	233.2 ft^2 (21.66 m^2)
Wing loading	31.4 lb/ft^2 (153 kg$_f$/m^2)
Wing aspect ratio	5.815
Wing airfoil section	NACA low drag, laminar flow

Figure 3.74 The XP-51B test airplane being towed aloft by the P-61. (Source: *E.P. Hartman, "Adventures in Research: A History of the Ames Research Center, 1940–1965," NASA SP-4302, 1970.*)

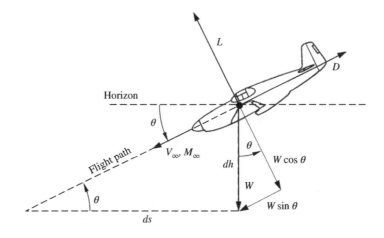

Figure 3.75 XP-51B in steady, gliding flight.

airspeed, V_∞, angle-of-attack, α, and flight path angle, $-\gamma$, defined as the angle between the horizon and the freestream velocity. The angle, θ, is the magnitude of the negative flight path angle. The forces acting on the aircraft are the lift, L, drag, D, and weight, W.

By applying Newton's second law to the aircraft and summing forces in the direction perpendicular to the velocity vector, we have

$$\sum F_{\perp \vec{v}} = ma_\perp = L - W \cos\theta = 0 \tag{3.249}$$

where the acceleration is zero because the aircraft is in steady, unaccelerated flight.

Summing forces in the direction parallel to the velocity, we have

$$\sum F_{\| \vec{v}} = ma_\| = D - W \sin\theta = 0 \tag{3.250}$$

where again the acceleration is zero.

Solving for the lift and drag in Equations (3.249) and (3.250), respectively, we have

$$D = W \sin\theta \tag{3.251}$$

$$L = W \cos\theta \tag{3.252}$$

Using the definitions for the lift and drag coefficients, C_L and C_D, respectively, we solve Equations (3.251) and (3.252) for these coefficients, obtaining

$$C_L = \frac{W}{q_\infty S} \cos\theta \tag{3.253}$$

$$C_D = \frac{W}{q_\infty S} \sin\theta \tag{3.254}$$

Equations (3.253) and (3.254) are expressions for the lift and drag coefficients of the aircraft in steady, gliding flight as a function of the aircraft's weight, W, the flight dynamic pressure, q_∞, the wing planform area, S, and the flight path angle, θ.

For the test aircraft, the test weight, W, and the wing reference area, S, are known quantities (see Table 3.8), which remain constant during the glide. We assume that the freestream density, ρ_∞, is a constant over the altitude band that data is collected. The flight dynamic pressure, q_∞, is calculated as

$$q_\infty = \frac{1}{2} \rho_\infty V_\infty^2 \tag{3.255}$$

Dividing Equation (3.253) by (3.254), we obtain an expression for the lift-to-drag ratio, L/D.

$$\frac{L}{D} = \frac{W \cos\theta}{W \sin\theta} = \frac{1}{\tan\theta} \tag{3.256}$$

Thus, the lift-to-drag ratio for steady, gliding flight is a function only of the flight path angle, θ.

Rearranging Equation (3.256) to solve for the flight path angle, θ, we have

$$\theta = \tan^{-1}\left(\frac{1}{L/D}\right) \tag{3.257}$$

The gliding flight path angle, θ, is inversely proportional to the lift-to-drag ratio L/D, which means that the glide angle is shallower for larger values of L/D and steeper for smaller values of L/D. The minimum glide angle, θ_{min}, is at the maximum lift-to-drag ratio, $(L/D)_{max}$, of the aircraft.

$$\theta_{min} = \tan^{-1}\left[\frac{1}{(L/D)_{max}}\right] \tag{3.258}$$

As shown in Figure 3.75, the flight path angle can also be defined in terms of an incremental decrease in the altitude, dh, and an incremental change in the forward movement of the aircraft, ds. Dividing these incremental distances by the time increment over which the distances are measured, we have

$$\tan\theta = \frac{dh}{ds} = \frac{dh/dt}{ds/dt} = \frac{dh/dt}{V_\infty} \tag{3.259}$$

where V_∞ is the aircraft total velocity and dh/dt is the aircraft rate of descent. Therefore, combining Equations (3.257) and (3.259), we have

$$\theta = \tan^{-1}\left(\frac{1}{L/D}\right) = \tan^{-1}\left(\frac{dh/dt}{V_\infty}\right) \tag{3.260}$$

Therefore, the flight path angle, θ, is obtained by measuring the aircraft rate of descent, dh/dt. Based on this, you start recording the altitude every 15 s to obtain a time history of altitude versus time, to calculate the rate of descent, which will then be used to obtain the flight path angle using Equation (3.260). Once the flight path angle is known, the lift coefficient, drag coefficient, and lift-to-drag ratio can be calculated, using Equations (3.253), (3.254), and (3.256), respectively.

In the NACA glide tests, the lift and drag were obtained from measurements of the longitudinal deceleration using a sensitive accelerometer, installed in the aircraft, rather than the method just described. A comparison of the drag data from a NACA glide flight and the NACA wind tunnel test is shown in Figure 3.76. The flight data is in good agreement with the wind tunnel data over the range of Mach numbers, including higher Mach numbers, where compressibility effects may be present.

The calculation of the lift and drag, using the method presented in this FTT, is less accurate than the NACA method using accelerometer measurements. In practice, the calculation of the rate of descent, from the slope of the altitude versus time curve, is very sensitive to inaccuracies in the data and may lead to erroneous results. However, the present technique has been described in order to discuss details about lift and drag related to steady, gliding flight. The following example problem provides a numerical calculation for the lift and drag in steady, gliding flight based on the FTT.

Example 3.12 XP-51 Glide Flight Test *You have completed a glide test flight in the XP-51 with the aircraft configured as given in Table 3.8. The glide was flown at a constant airspeed of*

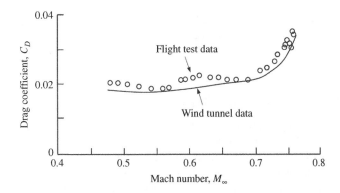

Figure 3.76 Comparison of XP-51 drag data from glide flight and wind tunnel. (Source: *Data reproduced from Nissen, et al., NACA ACR 4 K02, 1945, [55].*)

315 knots (Mach 0.523) from an altitude of 26,000 ft to 24,000 ft (you may assume atmospheric conditions corresponding to a constant altitude of 25,000 ft). Altitude versus time data was recorded for the glide and, from this data, the rate of descent, dh/dt, was determined to be 2990 ft/min. For this glide, calculate the aircraft glide path angle, lift and drag coefficients, and lift-to-drag ratio.

Solution

First, we convert the airspeed or freestream velocity, V_∞, from knots into consistent units.

$$V_\infty = 315 \frac{nmi}{h} \times \frac{6076 \, ft/nmi}{3600 \, s/h} = 531.7 \frac{ft}{s}$$

The glide path angle, θ, is given by Equation (3.260) as

$$\theta = \tan^{-1}\left(\frac{dh/dt}{V_\infty}\right) = \tan^{-1}\left(\frac{2990 \frac{ft}{min} \times \frac{1}{60} \frac{min}{s}}{531.7 \, ft/s}\right) = 5.354 \text{ deg}$$

The lift-to-drag ratio, L/D, is given Equation (3.256) as

$$\frac{L}{D} = \frac{1}{\tan\theta} = \frac{1}{\tan(5.354°)} = 10.67$$

From Appendix C, at an altitude of 25,000 ft, the freestream density, ρ_∞, is 0.0010663 slug/ft³. The freestream dynamic pressure, q_∞, is then

$$q_\infty = \frac{1}{2}\rho_\infty V_\infty^2 = \frac{1}{2}\left(0.0010663 \frac{slug}{ft^3}\right)\left(531.7 \frac{ft}{s}\right)^3 = 150.5 \frac{lb}{ft^2}$$

From Table 3.8, the aircraft weight, W, is 7335 lb and the wing reference area, S, is 233.2 ft². The lift coefficient, C_L, is given by Equation (3.253) as

$$C_L = \frac{W}{q_\infty S}\cos\theta = \frac{7335 \, lb}{\left(150.5 \frac{lb}{ft^2}\right)(233.2 ft^2)}\cos(5.354°) = 0.2081$$

The drag coefficient, C_D, is given by Equation (3.254) as

$$C_D = \frac{W}{q_\infty S}\sin\theta = \frac{7335 \, lb}{\left(150.5 \frac{lb}{ft^2}\right)(233.2 ft^2)}\sin(5.354°) = 0.01950$$

3.8 Two-Dimensional Lifting Shapes: Airfoils

In this section, we get more precise in describing two-dimensional geometries that produce significantly more lift than drag, known as *airfoils* or *wing sections*. An airfoil shape is the two-dimensional cross-section, parallel to the flow direction, of a three-dimensional wing. While a simple flat plate produces lift when oriented at an angle to the freestream flow, an airfoil section is often designed with curvature and thickness to produce aerodynamic lift more efficiently and effectively.

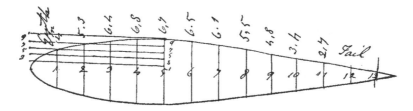

Figure 3.77 Sir George Cayley's sketch of the cross-section of a common trout, which is similar to the NACA 63A016 modern low-drag airfoil. (Source: *Hodgson, John Edmund, Aeronautical and Miscellaneous Notebook (ca. 1799–1826) of Sir George Cayley, with an Appendix Comprising a List of the Cayley Papers, W. Heffer & Sons, Ltd., Cambridge, 1933, Newcomen Society Extra Publication No. 3.*)

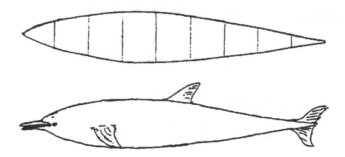

Figure 3.78 Sir George Cayley's sketch of a low drag body, based on the shape of a dolphin. (Source: *[33], Fig. 23b.*)

Nature has optimized the shapes of different animals for efficient locomotion through air or water. There are many examples where airfoil shapes mimic the aerodynamic, streamlined shapes of animals. Early aeronautical engineers and scientists understood the importance of low drag shapes in the design of the first air vehicles. British aeronautical pioneer, Sir George Cayley, looked to the bodies that nature has optimized. He sought to find low drag bodies that he called "solids of least resistance." Before considering the aerodynamic shapes of birds, he investigated the shapes of fish and marine mammals.

In 1809, Sir George Cayley made careful measurements of the cross-sectional shape of a fish, the common trout, as shown in Figure 3.77. The renowned aerodynamicist, Theodore von Karman, commented that Cayley's drawing of the trout cross-section closely matches the profile of the NACA 63A016 modern low-drag airfoil shape.

Cayley designed another solid of least resistance based on the shape of a dolphin, as shown in Figure 3.78. The dolphin shape has been of interest to aerodynamicists and hydrodynamicists for many years, because of its efficient locomotion through ocean waters. It is no surprise that the fins and tail flukes of the bottlenose dolphin have airfoil-like cross-sections and its body profile is similar to that of a symmetric NACA 0018 airfoil, as shown in Figure 3.79.

In the early 1800s, the Englishman, Horatio F. Phillips (1845–1924), performed some of the earliest studies of airfoil shapes. He designed a wind tunnel that used the induction of steam to create a flow of air through a wooden duct, to collect experimental data of different airfoil shapes. He patented a series of airfoil shapes that he called aerocurves, based on his wind tunnel tests, as shown in Figure 3.80. Phillips' wind tunnel data confirmed his belief that the curvature or camber

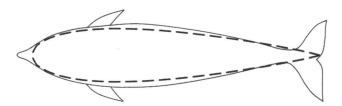

Figure 3.79 Body of a bottlenose dolphin (*Tursiops Truncatus*) compared with a NACA 0018 airfoil section (dotted line).

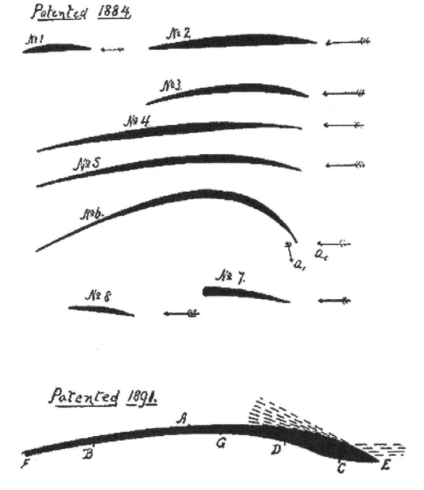

Figure 3.80 Aerocurves patented by Horatio Phillips, 1884 and 1891 (flow is from right to left). (Source: *Horatio F. Phillips, PD-old-70.*)

of an airfoil increased its lifting capability over a non-curved surface. He also proved that the lift was primarily generated by the upper or *suction surface* of an airfoil.

Phillips applied his airfoil test data to the construction of several full-scale aircraft. A characteristic feature of his aircraft designs was the multiple, stacked wings, with the appearance of a set

Figure 3.81 1904 Horatio Phillips' multiplane that made a 50 foot long "powered hop", 1908. (Source: *J.D. Fullerton, PD-old-70.*)

of Venetian blinds. His first man-carrying, powered aircraft had 21 stacked wings, an aft-mounted tail, and a tractor propeller, as shown in Figure 3.81. In 1904, this aircraft made a short uncontrolled, "powered hop" of about 50 ft (15 m), where it left the ground briefly. On 6 April 1907, a Phillips-designed multiplane, powered by a 22 hp (16 kW) engine, flew about 500 ft (150 m).

In the early days of aviation, many airfoils were designed by copying the shapes of bird wings. A contemporary of the Wright brothers, the German glider pioneer, Otto Lilienthal (1848–1896), made extensive studies of soaring birds, especially storks, as shown in Figure 3.82. The significant influence of biologically inspired flight on his glider designs is evident in his book *Birdflight as the Basis of Aviation* [29].

Lilienthal carefully measured bird wings and tested these shapes using a whirling arm apparatus. He recognized that the airfoil curvature or camber was a key design feature for obtaining high lift. Between 1891 and 1896, Lilienthal used his airfoil data to design a series of weight-shift controlled gliders, similar to hang gliders of today. He and his brother, Gustav, piloted the gliders on over 2000 flights, launching from hills near Berlin, Germany (Figure 3.83). The gliders flew for distances over 800 ft (240 m). Tragically, on 9 August 1896, Otto Lilienthal lost control while flying one of his gliders, falling from a height of about 50 ft (15 m). Although he survived the crash, his neck was fractured and he died the next day. His dying words to his brother were "Opfer müssen gebracht werden" ("Sacrifices must be made").

Even after rudimentary wind tunnels were employed, airfoil design was often a trial-and-error, empirical process, lacking any systematic design methodology based on aerodynamic theory. In the 1930s and 40s, aerodynamics groups at Gottingen in Germany, at the Royal Air Force Establishment in England, and at the NACA in the USA, initiated systematic studies of families of airfoil shapes, sets of shapes with common geometric characteristics. These investigations established several databases of airfoil shapes, which are still used today. These parametric studies also led

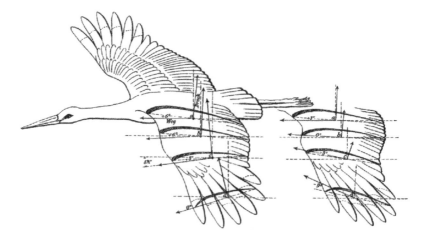

Figure 3.82 Otto Lilienthal's drawings of the White Stork wing and airfoil sections, 1889. (Source: *Otto Lilienthal, PD-old-100.*)

Figure 3.83 Otto Lilienthal in gliding flight, ca. 1895. (Source: *Anonymous, PD-old-100.*)

to an understanding of how to design airfoils, using a more systematic approach. Several compendiums of airfoil shapes and data can be found in the literature, an excellent example being the NACA collection [1]. (We discuss more details about some of the families of airfoil sections in Section 3.8.2.)

An evolution of airfoil shapes is shown in Figure 3.84, from airfoils used in early aviation to modern airfoils. The early airfoils used by the Wright Brothers and Bleriot, a well-known French aircraft designer in the early 1900s, were highly cambered and thin, similar to the cross-sections of a bird's wing. Through trial-and-error experimentation, it was discovered that a rounded leading edge and a sharp trailing edge were beneficial, but the theoretical basis for this was not known at the time. The growing recognition of the importance of airfoil thickness is evident in the design evolution. As discussed earlier, the airfoils designed by the groups at the RAF, Gottingen, and the

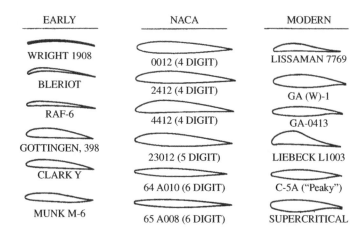

Figure 3.84   Evolution of airfoil sections.   (Source: *NASA, PD-USGov-NASA.*)

NACA incorporated many of these airfoil design features. The Gottingen 398 and Clark Y airfoils were very successful and were used by the NACA as a basis for some of their airfoil designs. Then NACA designed several families of airfoil shapes, using a four-, five-, and six-digit numbering system that is explained in Section 3.8.2. The modern airfoils have significantly different shapes that are departures from the "classic" NACA airfoils. Many of these modern airfoils are "custom designed" for specific flight applications. For example, the Lissamon 7769 airfoil is a low Reynolds number airfoil, designed for the low-speed flight of human-powered aircraft. The GA(W)-1 and GA-0413 airfoils are advanced designs by NASA for general aviation aircraft. The Liebeck L1003, known as a *laminar, rooftop airfoil*, is designed to produce high lift. The C-5A airfoil is specially designed for transonic flight of the Lockheed C-5A *Galaxy* transport aircraft. Many of the modern, high-speed airfoils are *supercritical airfoils*, designed to delay the onset of transonic drag, to be discussed in Section 3.11.6.

3.8.1   Airfoil Construction and Nomenclature

A systematic approach towards airfoil design is shown in Figure 3.85. Starting at Step 1 at the top of the figure, the airfoil designer first chooses the desired airfoil length, drawing a straight *chord line* from the leading to the trailing edge. The airfoil curvature is set by defining the *mean camber line*, as shown in Step 2. In Steps 3 and 4, a *thickness envelope* is wrapped around the mean camber line to form the airfoil upper and lower surfaces. The same thickness is added above and below the mean camber line, such that the mean camber line defines the midpoints between the upper and lower surfaces. The resulting final airfoil shape is shown in Step 5. If the mean camber line, defined in Step 2, is positioned above the chord line, the airfoil is said to have *positive camber* (Figure 3.86). If the mean camber line is below the chord line, the airfoil has *negative camber*. If the mean camber line is coincident with the chord line, the result is a *symmetric airfoil* (Figure 3.86). By using this systematic design approach, a family of airfoil shapes can be generated. For instance, a family of cambered airfoils of varying thickness may be generated by starting with a common mean camber line and wrapping envelopes of increasing thickness around the camber line.

As is probably evident by now, there is specific nomenclature associated with airfoils (Figure 3.87). The *leading* and *trailing edges* are the furthest forward and rearward points of the airfoil, respectively. The leading edge may have an associated radius, defined as the radius of

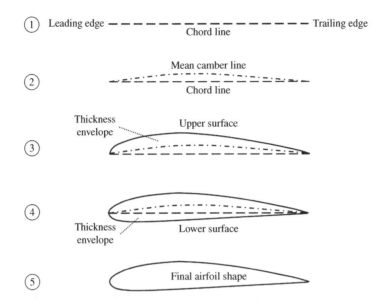

Figure 3.85   Airfoil construction. (Source: *Adapted from Talay, NASA SP 367, 1975, [65].*)

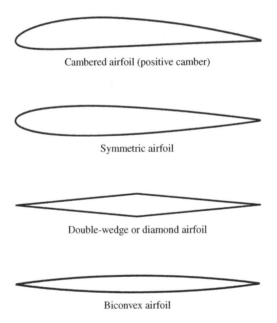

Figure 3.86   Various airfoil shapes.

the circle that fits between the upper and lower surfaces. Airfoils with sharp leading edges, such as the supersonic diamond or biconvex airfoils, have a zero leading edge radius (Figure 3.86). (The biconvex airfoil is formed by two opposing circular arcs.) The straight line connecting the leading and trailing edges is the *chord line*, often simply called the *chord*. The *mean camber line*, sometimes designated as MCL, also extends between the leading and trailing edges, and is

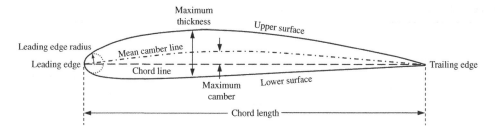

Figure 3.87 Airfoil nomenclature.

equidistant from the upper and lower surfaces. The airfoil *camber* is the distance between the mean camber line and the chord line, measured perpendicular to the chord line. The maximum camber is the maximum distance between the MCL and the chord line, located at some horizontal distance from the leading edge. The airfoil lifting and pitching moment characteristics are strongly influenced by the shape of the mean camber line and the maximum camber. The distance between the upper and lower surfaces is the airfoil *thickness*, which varies from the leading to the trailing edges. The maximum thickness is the maximum distance between the upper and lower surfaces, located at some horizontal distance from the leading edge.

3.8.2 Airfoil Numbering Systems

The NACA developed a numbering system to define the different airfoils, and families of airfoils, that they designed. Most aerospace organizations or companies also use some type of airfoil numbering system for the designs that they produce. It is useful to have an understanding of the NACA airfoil numbering system, because it provides insight into the aerodynamic characteristics of the airfoils, and because the NACA series of airfoils are still in wide use on modern-day aircraft.

In 1937, the NACA reported on the characteristics of 78 airfoil sections [40] that were tested in the Langley variable density tunnel (see Section 3.7.4.5 for details of this wind tunnel). These airfoils were a related family of airfoils, derived from a common thickness distribution, based on the Gottingen 398 and Clark Y airfoils. The 78 airfoils, derived and tested in this report, formed the first *NACA four-digit series*. The numbering system for the four-digit series is based on the geometry of the airfoil section, as given in Table 3.9. For example, the NACA 4412 airfoil has a maximum camber equal to 4% of the chord length, c; that is $0.04c$, located a distance of $0.4c$ from the airfoil leading edge, and a thickness equal to 12% of the chord length, or $0.12c$. As another example, consider the NACA 0018 airfoil. This airfoil's numbering indicates that it has zero camber and 18% thickness. Thus, we see that a four-digit series airfoil with "00" as its first two digits indicates an airfoil with zero camber or a symmetric airfoil.

Table 3.9 NACA four-digit series airfoil numbering.

Digit	Definition	Dimensions
1	maximum camber	percent of chord
2	location of maximum camber	tenths of chord from leading edge
3, 4	maximum thickness	percent of chord

Table 3.10   NACA five-digit series airfoil numbering.

Digit	Definition	Dimensions
1	design lift coefficient ($1.5 \times$ first digit)	tenths
2, 3	location of maximum camber ($0.5 \times$ second and third digits)	percent of chord from leading edge
4, 5	maximum thickness	percent of chord

The next series of airfoil shapes developed by the NACA was the *NACA five-digit series*. The numbering system for the five-digit series is based on a combination of the theoretical aerodynamic characteristics and geometry of the airfoil, as given in Table 3.10. As an example of a five-digit series airfoil, the NACA 23015 airfoil has a design lift coefficient of 0.3,[4] has its maximum camber located at 15% chord, or $0.15c$, and has a thickness of 15% of its chord length, or $0.15c$.

The original NACA *6-series* airfoils were designed in the 1940s with a goal of designing low drag airfoils. Many of these 6-series airfoils were designed to promote extensive laminar flow over the forward portions of the airfoils, hence these airfoils are sometimes called *laminar flow airfoils*. Obtaining laminar flow, versus turbulent flow, over the airfoil surface results in lower skin friction drag, as discussed in Section 3.12.3. The 6-series airfoils were designed for low drag over a range of lift coefficients (synonymous with a range of angles-of-attack).

The six-digit series is designated with five or six digits. The six-digit series numbering system is detailed in Table 3.11. There are several variations of the 6-series number system, so only a few of the predominant examples are described. A few examples will help to explain the 6-series numbering system.

The NACA 63_3-218 is a 6-series airfoil with the minimum pressure located at $0.3c$ from the leading edge, the low drag range of lift coefficient around the design lift coefficient equal to 0.3, a design lift coefficient of 0.2, and a maximum thickness of 18% of the chord or $0.18c$.

The NACA 64A204 airfoil is a 6A series airfoil, with the letter "A" replacing the dash as in the previous example, where the letter denotes a modification to the 6-series. The NACA 64A204 airfoil has a minimum pressure location of $0.4c$ from the leading edge, a design lift coefficient of

Table 3.11   NACA 6-series airfoil numbering.

Digit	Definition	Dimensions
1	series designation	none
2	location of minimum pressure (basic symmetrical section at zero lift)	tenths of chord from leading edge
3 (subscript)	low drag range around design $C_L{}^*$	tenths (of C_L)
dash or letter	spacer (dash) or modification letter, e.g. "A"	none
4	design lift coefficient	tenths
5, 6	maximum thickness	percent of chord

*If third digit omitted, low drag range is <0.1. See Section 3.8.5.2 for an explanation of low drag range.

---

[4] The *design lift coefficient* is defined as the lift coefficient where the airfoil has its best lift-to-drag ratio. Flight at the design lift coefficient usually corresponds to flight at or near minimum drag. There is a range of lift coefficients, around the design lift coefficient, where the drag is at or near minimum, termed the *low drag range* for the 6-digit airfoil series.

Table 3.12 Airfoil shapes used on selected aircraft.

Aircraft	Airfoil	
	Wing Root	Wing Tip
Bell X-1E	NACA 64A004	NACA 64A004
Boeing F/A-18 *Hornet*	NACA 65A005 mod	NACA65A003.5
Beechcraft A36 *Bonanza*	NACA 23016.5	NACA 23012
Bell 206L *Ranger* (rotor blades)	NACA 0012 mod (11.3%)	NACA 0012 mod (11.3%)
Cessna 337 *Skymaster*	NACA 2412	NACA 2409
Lockheed F-104 *Starfighter*	Biconvex 3.36%	Biconvex 3.36%
Lockheed-Martin F-16 *Fighting Falcon*	NACA 64A204	NACA 64A204
Lockheed U-2 *Dragon Lady*	NACA 63A409	NACA 63A406
McDonnell Douglas F-15 *Eagle*	NACA 64A006.6	NACA 64A203
North American X-15	NACA 66-005 mod	NACA 66-005 mod
North American XP-51 *Mustang*	NAA/NACA 45-100	NAA/NACA 45-100
North American XB-70A *Valkyrie*	hexagonal section	hexagonal section
Northrop T-38 *Talon*	NACA 65A004.8	NACA 65A004.8
Ryan NYP *Spirit of St. Louis*	Clark Y	Clark Y
Supermarine *Spitfire*	NACA 0013.5	NACA 0013.5

0.2, and a maximum thickness of 4% or $0.04c$. Since the third digit is omitted for this airfoil, its low drag range of lift coefficient is less than 0.1.

A listing of the NACA and other types of airfoil shapes used on selected aircraft is shown in Table 3.12. Many of these selected aircraft are discussed in the text. NACA airfoils are used on the rotor blades of helicopters, as shown for the Bell 206L *Ranger* helicopter, which uses modified NACA 0012 airfoil sections on its rotor blades. Note also that different airfoil shapes are often used at the wing root and the wing tip. As discussed in Section 3.9, different airfoil sections may be used along the wingspan to tailor the lift distribution, especially for low-speed or stalling flight conditions.

3.8.3 Airfoil Lift, Drag, and Pitching Moment

Consider an airfoil, of chord length c, at an angle-of-attack, α, in a freestream flow of velocity V_∞, as shown in Figure 3.88a. The forces and moment are the result of the pressure, p, and shear stress, τ, distributions integrated over the body surface, as shown in Figure 3.88b. The resultant force, F_R, may be resolved into the lift, L, and drag, D, perpendicular and parallel to the freestream velocity, respectively. It is possible to place the resultant force at a single point along the airfoil chord line, called the *center of pressure*, such that there is no net moment, shown as the distance x_{cp} from the airfoil leading edge in Figure 3.88c. The center of pressure is located at the centroid of the pressure distribution over the airfoil. If the angle-of-attack of the airfoil is changed, the pressure and shear stress distributions (and the lift and drag) change, thus the location of the center of pressure changes.

The resultant force at the center of pressure can be represented by an equivalent resultant force and a moment at any other location along the chord line. The lift, drag, and moment are often referenced at a point located one quarter of the chord length, $c/4$, aft of the leading edge (Figure 3.88d) or at the leading edge (Figure 3.88e). The pitching moments at these locations are referred to as the moment about the quarter chord point, $M_{c/4}$, and the moment about the leading edge, M_{LE}. Both of these moments are usually a function of angle-of-attack.

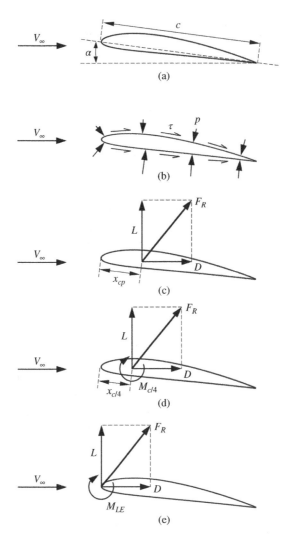

Figure 3.88   Airfoil forces and moment (a) airfoil at angle-of-attack in freestream flow, (b) pressure and shear stress distribution, (c) lift and drag with no moment, (d) lift, drag, and moment about the quarter-chord point, and (e) lift, drag, and moment about the leading edge.

There is one point along the chord line, called the *aerodynamic center*, where the pitching moment is independent of angle-of-attack. There is still a lift, drag, and moment, M_{ac}, at the aerodynamic center. To be clear, the aerodynamic center is not the same as the center of pressure, where there is no moment. The location of the center of pressure is usually aft of the aerodynamic center, although for subsonic flow, the aerodynamic center is usually very close to the quarter chord point. The aerodynamic center shifts aft, towards the airfoil mid-chord point, $c/2$, as the flow becomes supersonic.

3.8.4   Pressure Coefficient

The non-dimensional pressure coefficient, C_p, was introduced in Section 3.4.3. In this section, the use of the pressure coefficient is useful in understanding the surface pressure distributions on airfoils. Using the definition of the dynamic pressure, q_∞, given by Equation (2.46), we expand the definition of the pressure coefficient, given by Equation (3.37), as

$$C_p \equiv \frac{p - p_\infty}{q_\infty} = \frac{p - p_\infty}{\frac{1}{2}\rho_\infty V_\infty^2} \tag{3.261}$$

where p is the local surface pressure and p_∞, q_∞, ρ_∞, and V_∞ are the freestream values of pressure, dynamic pressure, density, and velocity, respectively.

Using the perfect gas equation of state, Equation (3.61) and the definition of Mach number, Equation (2.40), in Equation (3.37), the dynamic pressure is written in terms of the freestream pressure, p_∞, freestream Mach number, M_∞, and specific gas constant, γ, as

$$q_\infty = \frac{1}{2}\rho_\infty V_\infty^2 = \frac{1}{2}\left(\frac{p_\infty}{RT_\infty}\right)(Ma_\infty)^2 = \frac{1}{2}\left(\frac{p_\infty}{RT_\infty}\right)M_\infty^2(\gamma RT_\infty) = \frac{1}{2}\gamma p_\infty M_\infty^2 \tag{3.262}$$

Inserting Equation (3.262) into Equation (3.261), the pressure coefficient is written as

$$C_p = \frac{p - p_\infty}{\frac{1}{2}\gamma p_\infty M_\infty^2} = \frac{2}{\gamma M_\infty^2}\left(\frac{p}{p_\infty} - 1\right) \tag{3.263}$$

Equations (3.261) and (3.263) show that the pressure coefficient provides the magnitude of the local pressure, at a point on the body surface relative to the freestream pressure. If the surface pressure, p, equals the freestream pressure, p_∞, the pressure coefficient is zero. If the surface pressure is greater than the freestream pressure, $p > p_\infty$, the pressure coefficient is a positive number. A negative pressure coefficient indicates that the local surface pressure is below the freestream pressure, $p < p_\infty$.

The surface pressure coefficient, C_p, along the upper and lower surfaces of an airfoil, is plotted versus the airfoil non-dimensional chord line, x/c, in Figure 3.89. By convention, negative C_p is plotted above the abscissa and positive C_p is plotted below. The pressure coefficient on the airfoil upper surface is negative, indicating pressures below the freestream pressure. The pressure coefficient on the airfoil lower surface is mostly positive, indicating pressures higher than freestream pressure. The pressure coefficient at the airfoil leading edge stagnation point is positive, indicating local pressures above freestream pressure. As the air flows over the upper surface, the surface pressures decreases rapidly to below freestream pressure, making the pressure coefficient negative. The surface pressure increases along the airfoil upper surface, but remains below freestream pressure until close to the trailing edge. The pressure coefficient becomes positive near the airfoil trailing edge, indicating the surface pressure has risen to slightly above freestream pressure. Proceeding aft, from the stagnation point, along the airfoil lower surface, the pressure is higher than freestream pressure, with a positive pressure coefficient.

It can be shown that for small angles-of-attack and neglecting skin friction, the airfoil lift coefficient, c_l, is given by

$$c_l = \int_0^1 (C_{p,l} - C_{p,u})d\left(\frac{x}{c}\right) \tag{3.264}$$

where $C_{p,l}$ and $C_{p,u}$ are the pressure coefficients on the lower and upper surfaces, respectively. The integral in Equation (3.264) is the area bounded by the pressure coefficient curves on the upper and lower surfaces.

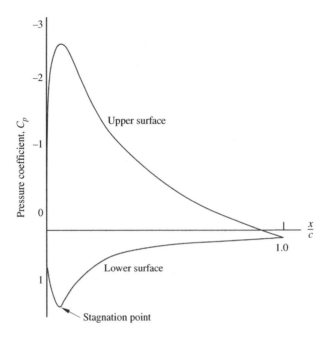

Figure 3.89   Airfoil surface pressure coefficient distribution.

3.8.5 Airfoil Lift, Drag, and Moment Curves

The aerodynamic data for airfoils is usually presented by a series of plots in a standard format. These include the *lift curve*, a plot of the lift coefficient, c_l, as a function of angle-of-attack, α, the *drag curve*, a plot of the drag coefficient, c_d, versus the lift coefficient or the angle-of-attack, and the *pitching moment curve*, a plot of pitching moment coefficient, c_m, as a function of either lift coefficient or angle-of-attack. Recall that lowercase letters are used for the coefficients since they are for two-dimensional airfoil sections.

In this section, generic forms of these various airfoil plots are presented and discussed for low subsonic flow. Experimental data for a wide array of airfoils may be found in the literature, which includes coordinate data of the airfoil shapes and the aerodynamic lift, drag, and moment plots. The NACA has published many technical reports of airfoil data, an excellent compilation of which can be found in [1] and [2].

The aerodynamic coefficients of an airfoil are a function of the airfoil section shape, angle-of-attack, Reynolds number, surface roughness, and Mach number. For low subsonic speeds, the airfoil aerodynamic characteristics are independent of Mach number. Effects of Mach number are significant at higher subsonic and supersonic Mach numbers, when compressible effects become important. The state of the boundary layer on the airfoil surface, whether it is laminar, turbulent, or transitional, is a function of the Reynolds number and surface roughness. For Reynolds numbers greater than about 100,000 and rough surfaces, the boundary layer is typically turbulent.

3.8.5.1 Airfoil Lift Curve

The lift curve for a cambered airfoil is shown in Figure 3.90. Since the airfoil has camber, the lift coefficient is positive at zero angle-of-attack. For a symmetric airfoil, the lift is zero at zero

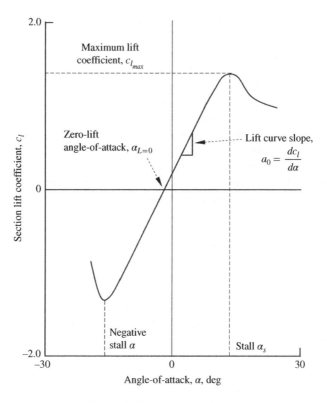

Figure 3.90   Airfoil lift curve.

angle-of-attack. The angle-of-attack where the lift coefficient is equal to zero is defined as the zero-lift angle-of-attack, $\alpha_{L=0}$. The zero-lift angle-of-attack is zero for a symmetric airfoil and a negative angle-of-attack for a cambered airfoil.

The lift curve is linear for much of the angle-of-attack range, usually from zero to about 12–15° positive angle-of-attack, depending on the airfoil shape. For negative angles-of-attack, the linear range is the same as the positive range for symmetric airfoils and does not extend as far as the positive range for cambered airfoils. The linear lift range typically covers the normal flight operating angles-of-attack of an aircraft. The slope of the lift coefficient versus angle-of-attack curve is known as the *lift curve slope*, a_0, defined as

$$a_0 = \frac{dc_l}{d\alpha} \equiv c_{l_\alpha} \tag{3.265}$$

where we have introduced a nomenclature for the derivative of a coefficient as the coefficient with a subscript.

Based on the classical theoretical analysis of thin airfoils at small angles-of-attack, called *thin airfoil theory*, the lift curve slope of a symmetric or a cambered airfoil is given by

$$a_0 = \frac{dc_l}{d\alpha} = 2\pi \text{ per radian} = 0.1097 \text{ per degree} \tag{3.266}$$

For a symmetric airfoil, thin airfoil theory predicts that the lift coefficient is given by

$$c_l = 2\pi\alpha \tag{3.267}$$

where the angle-of-attack, α, is in radians. For example, if an airfoil is at an angle-of-attack of 8°, thin airfoil theory predicts a lift coefficient of 0.877.

At the end of the linear range of the lift curve, the lift coefficient reaches a maximum, known as the maximum lift coefficient, $c_{l,max}$, and then decreases. The highest $c_{l,max}$ that can be obtained with conventional airfoils is about 1.8–1.9. Significant increases in the maximum lift coefficient can be obtained with high-lift devices (to be discussed in Section 3.9.3).

The angle-of-attack corresponding to the maximum lift coefficient is the stall angle-of-attack, α_s. There is a corresponding maximum negative lift coefficient and negative stall angle-of-attack. Beyond the stall angle-of-attack, the flow over the airfoil is no longer smooth and orderly, rather there are typically significant regions of separated flow. This causes a drastic loss of lift, as shown by the decrease in the lift coefficient after stall. The shape and slope of the lift curve after the stall is highly dependent on the airfoil shape. For a symmetric airfoil, the lift curve is symmetric about the $c_l = 0$ vertical axis, even beyond the stall angle-of-attack.

The lift curve is relatively insensitive to changes in Reynolds number, until the angle-of-attack is near or above the stall angle-of-attack. At these high angles-of-attack, the state of the boundary layer, as determined by the Reynolds numbers, is important for whether the flow stays attached or separates. As discussed in Section 3.12 concerning viscous flows, separation is delayed by a turbulent boundary layer due to the higher average kinetic energy in the boundary layer. A higher stall angle-of-attack can be reached with a turbulent boundary layer, hence the lift curve extends to this higher angle-of-attack at larger Reynolds number. The two major effects of Reynolds number on the lift curve are shown in Figure 3.91. As the Reynolds number is increased, the section lift coefficient, $c_{l,max}$, and stall angle-of-attack, α_s, both increase.

Up to now, we have been a bit "loose" with our definition of angle-of-attack. The airfoil angle-of-attack that we have been using thus far is the angle between the chord line and the freestream velocity, which is called the *geometric angle-of-attack* and simply denoted as α. The lift curve in Figure 3.90 plots the lift coefficient versus the geometric angle-of-attack. The relationship between the geometric angle-of-attack and the lift coefficient is shown in Figure 3.92. The airfoil can be set at a zero-lift angle-of-attack, $\alpha_{L=0}$, such that its lift is zero. For a cambered

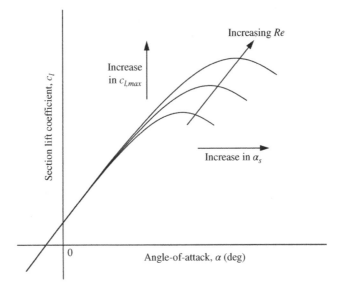

Figure 3.91 Effect of Reynolds number on the airfoil section lift curve.

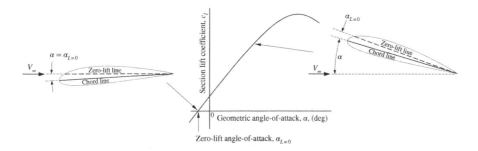

Figure 3.92   Lift curve versus geometric angle-of-attack.

airfoil, the zero-lift angle-of-attack is at a negative geometric angle-of-attack, as shown on the left in Figure 3.92. We can define a *zero-lift line*, drawn through the airfoil section that is parallel to the freestream velocity when the airfoil is at its zero-lift angle-of-attack. For the airfoil at an arbitrary, positive angle-of-attack, as shown on the right in Figure 3.92, the geometric angle-of-attack is the angle, α, between the chord line and the freestream velocity. The zero-lift line, which was drawn when the airfoil was set at the zero-lift angle-of-attack, lies above the chord line at an angle equal to the zero-lift angle-of-attack.

We now define the *absolute angle-of-attack*, α_a, as the angle between the freestream velocity and the zero-lift line. The absolute angle-of-attack can be calculated as

$$\alpha_a = \alpha + \alpha_{L=0} \tag{3.268}$$

where the absolute magnitude of the zero-lift angle-of-attack is used in this equaiton. If we plot the lift coefficient versus the absolute angle-of-attack, the lift curve is shifted to the right by an amount equal to the zero lift-angle-of-attack, as shown in Figure 3.93. The lift curve passes through the origin and, by definition, the absolute angle-of-attack is zero at zero lift. For an airfoil at an arbitrary, positive angle-of-attack, the absolute angle-of-attack is α_a, the angle between the zero-lift line and the freestream velocity, as shown on the right in Figure 3.93. The absolute angle-of-attack is a useful concept in aerodynamics applications and other areas, such as stability and control, as we will see in Chapter 6.

3.8.5.2   Airfoil Drag Curve

The cambered airfoil section drag coefficient, c_d, is plotted versus the section lift coefficient, c_l, in Figure 3.94. Recall from Section 3.7.2.1, that the two-dimensional drag coefficient for an airfoil is

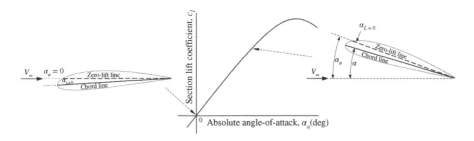

Figure 3.93   Lift curve versus absolute angle-of-attack.

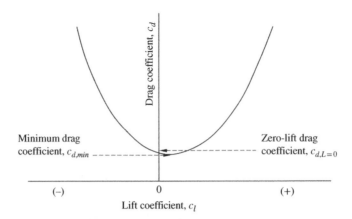

Figure 3.94   Airfoil drag coefficient versus lift coefficient.

called the profile drag, denoted by $c_{d,0}$. We simply use c_d, omitting the additional "0" subscript, in discussing the airfoil profile drag in this section. The airfoil profile drag coefficient is due to the sum of the airfoil skin friction drag and the pressure drag due to flow separation or form drag.

The drag coefficient is plotted for lift coefficients in the linear lift curve range, as described in the previous section. Since the lift coefficient is linearly related to the angle-of-attack in this linear range, essentially the same drag curve is obtained if the drag coefficient is plotted versus angle-of-attack instead of lift coefficient.

The profile drag is mostly due to skin friction at the small lift coefficients (or low angles-of-attack), while the pressure drag dominates at higher lift coefficients (larger angles-of-attack). The cambered airfoil drag coefficient has a minimum value, $c_{d,min}$, at a small positive angle-of-attack. The minimum drag coefficient of a symmetric airfoil is at zero angle-of-attack. The zero-lift drag coefficient, $c_{d,L=0}$, is the drag coefficient corresponding to the zero lift coefficient.

For airfoils designed to promote laminar flow, the drag curves look slightly different at small lift coefficients, as shown in Figure 3.95. There is a *laminar drag bucket*, over a *low drag range*

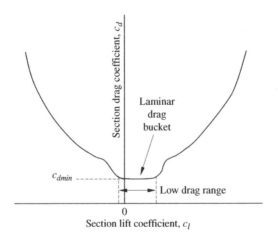

Figure 3.95   Laminar airfoil drag coefficient versus lift coefficient.

of lift coefficients, where the drag coefficient "dips" down to lower values. The drag coefficient is at or near its minimum value over this low drag range. This low drag range corresponds to small lift coefficients where the angle-of-attack is low. The laminar flow, promoted by the airfoil design, results in lower skin friction drag than a turbulent boundary layer, and, at low angles-of-attack, the pressure drag due to flow separation is also low. The low skin friction and pressure drag result in the "drag bucket", over the low angle-of-attack and low lift coefficient range, shown in Figure 3.95. However, as the angle-of-attack is increased, the laminar boundary layer is more susceptible to flow separation than a turbulent boundary layer, resulting in flow separation and an increase in pressure drag. Thus, at the higher angles-of-attack (or higher lift coefficients), the total drag coefficient is significantly higher than in the drag bucket.

For the general drag curve, an increase in Reynolds number results in transition of laminar boundary layers to turbulent flow. The higher energy, turbulent boundary layers are more resistant to flow separation than the laminar boundary layers, thus the pressure drag and the total drag are lower. This trend is shown in Figure 3.96, where the total drag curve shift towards lower drag coefficients as the Reynolds number is increased.

3.8.5.3 Airfoil Pitching Moment Curve

The airfoil section pitching moment, c_m, curve is shown in Figure 3.97. The pitching moment plotted is usually either the moment about the airfoil aerodynamic center, $c_{m,ac}$, or about the quarter chord point, $c_{m,c/4}$. The moment curve is plotted versus angle-of-attack, over the linear lift range and into the stall region. The moment coefficient is linear over the linear lift range angle-of-attack, as seen in Figure 3.97, and becomes non-linear in the stall region. The slope of the moment curve in the linear range, $c_{m,\alpha}$, is an important parameter in aircraft longitudinal stability, as discussed in Chapter 6. The moment curve is relatively insensitive to changes in the Reynolds number, until the angle-of-attack is large, for the same physical reasons as discussed for the lift curve.

3.8.6 Data for Selected Symmetric and Cambered Airfoils

Aerodynamic data for airfoil sections is readily available from many different sources, notably from NACA reports of systematic wind tunnel tests of sections, as discussed earlier. For example,

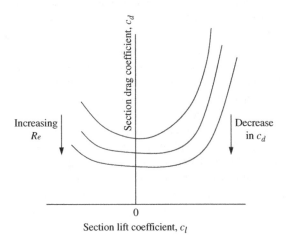

Figure 3.96 Effect of Reynolds number on the airfoil section drag curve.

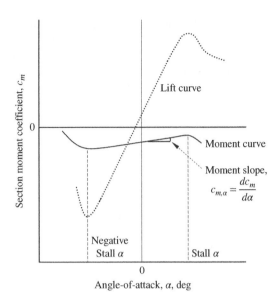

Figure 3.97   Airfoil pitching moment curve.

[1], [2], [27], and [40] provide a large amount of experimental data for a wide range of airfoil section geometries in subsonic and supersonic flows. Modern computational techniques have also enabled the rapid calculation of airfoil section characteristics for virtually any shape. Several online resources are also available that provide a database of airfoil section properties, determined from experimental and analytical means.

The present section provides a limited set of data for selected symmetric and cambered airfoils, obtained from [40]. Symmetric airfoil data is provided for the NACA 0012 and 0015 airfoils (Figures 3.98 and 3.99). Cambered airfoil data is given for the NACA 2412 and 4412 airfoils (Figures 3.100 and 3.101). This data is presented with the purpose of illustrating the interpretation and use of these types of charts. For each data chart shown in the figures, two different sets of data are shown.

In the left-side plot of each figure, the section aerodynamic characteristics were obtained for a rectangular wing with an aspect ratio of 6, corresponding to a wingspan of 30 in (76 cm) and chord length of 5 in (12.7 cm). The lift coefficient, c_l, drag coefficient, c_d, lift-to-drag ratio, L/D, and center of pressure location, x_{cp}, are plotted versus an angle-of-attack from -8 to $+32°$. (The uppercase letters are used for the section properties in these figures, while we maintain the usage of lowercase letters to denote airfoil properties. In the figures, the location of the center of pressure is simply marked as c.p.) The geometry of the airfoil is given at the top of the left chart, graphically and in a table of the section coordinates of the upper and lower surfaces. The data applies to low-speed incompressible flow, since the data was obtained at a nominal speed of 68 ft/s (47 mph, 75 km/h). The data was obtained in the NACA variable density tunnel, as was discussed in Section 3.7.4.5, at a nominal Reynolds number (RN in chart), based on chord length, of 3 million. One may wonder how such a high Reynolds number was obtained in the wind tunnel test, given the small chord size and low test velocity. The answer is the use of the variable density tunnel, where the tests were conducted at a high nominal pressure of 20 atm (300 lb/in^2, 42,000 lb/ft^2, 2,000,000 Pa). The right-side

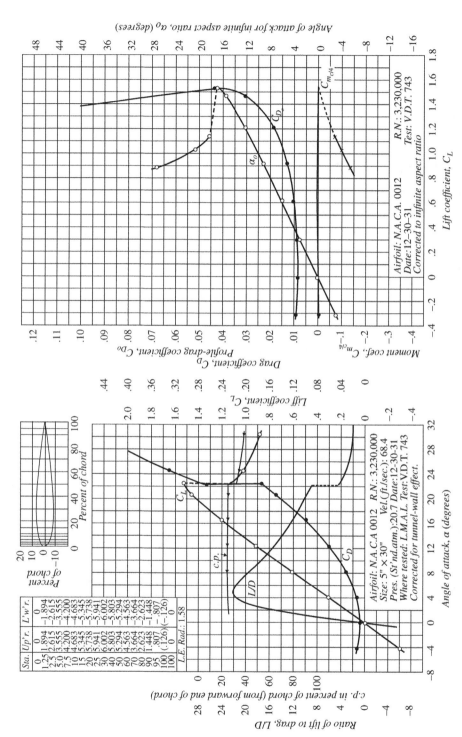

Figure 3.98 NACA 0012 airfoil data. (Source: Jacobs, E.N., Ward, K.E., and Pinkerton, R.M., NACA Report No. 460, 1935, [40].)

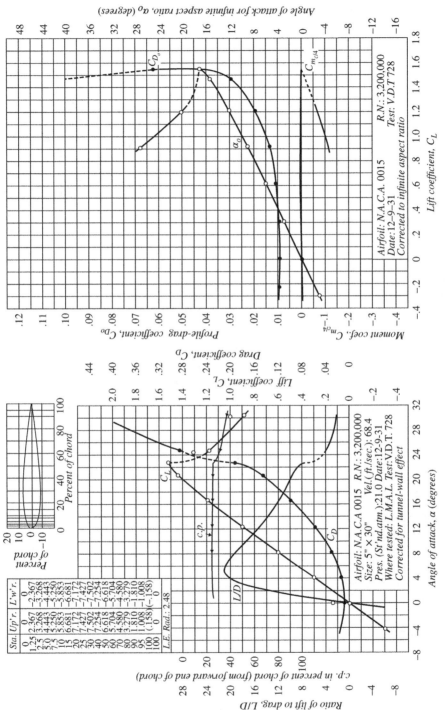

Figure 3.99　NACA 0015 airfoil data.　(Source: Jacobs, E.N., Ward, K.E., and Pinkerton, R.M., NACA Report No. 460, 1935, [401].)

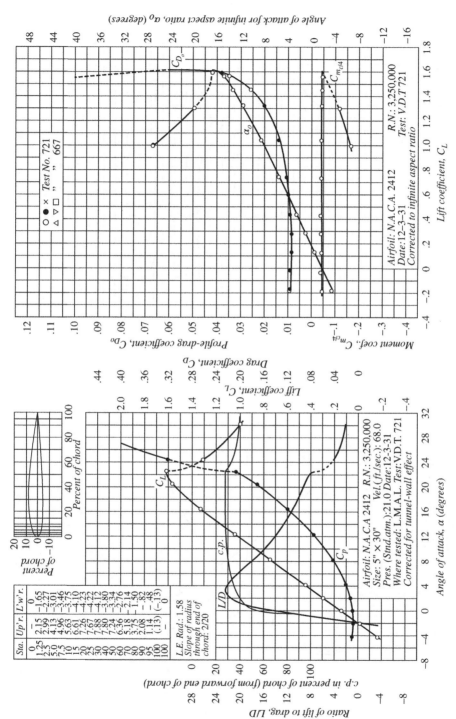

Figure 3.100 NACA 2412 airfoil data. (Source: Jacobs, E.N., Ward, K.E., and Pinkerton, R.M., NACA Report No. 460, 1935, [40].)

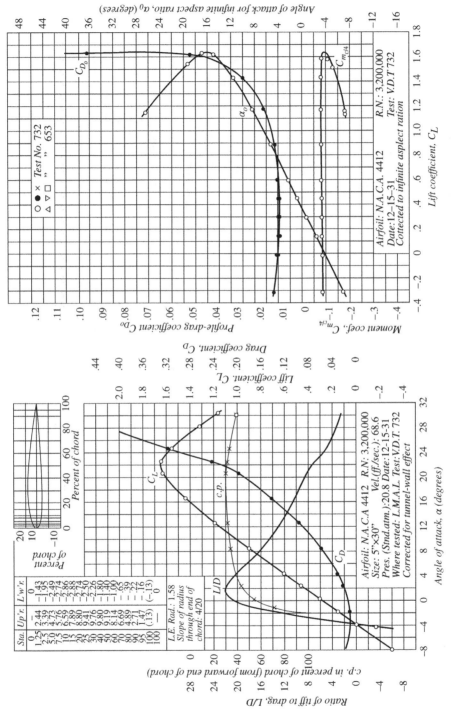

Figure 3.101   NACA 4412 airfoil data.   (Source: Jacobs, E.N., Ward, K.E., and Pinkerton, R.M., NACA Report No. 460, 1935, [40].)

plot of each figure provides aerodynamic data that has been corrected to infinite aspect ratio. The profile drag coefficient, moment coefficient about the quarter chord point, and angle-of-attack are plotted versus the lift coefficient. The data was obtained at the same test conditions as stated for the left-side data. The example problem, at the end of this section, further illustrates the use of these data charts.

Example 3.13   Airfoil Data *Consider the data for the NACA 4412 airfoil section, shown in Figure 3.101. Identify the following, (1) the maximum lift-to-drag ratio, $(L/D)_{max}$, (2) the maximum lift coefficient, $c_{l,max}$, (3) the stall angle-of-attack, α_{stall}, (4) the minimum drag coefficient, $c_{d,min}$, (5) the minimum profile drag coefficient, $c_{d,0,min}$, and (6) the pitching moment coefficient, $c_{m,c/4}$, at zero degrees angle-of-attack.*

Solution

The desired parameters and associated values are identified in the figure below.

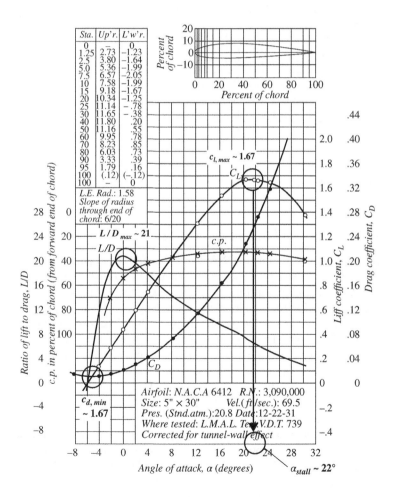

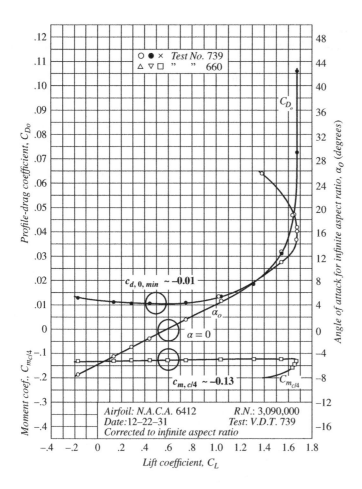

3.8.7 Comparison of Symmetric and Cambered Airfoils

In this section, the predicted airfoil data of a symmetric, NACA 0012 airfoil and a cambered, NACA 2412 airfoil are compared. These airfoils have been and are used in many aircraft designs for lifting surfaces. These are both NACA four-digit series airfoils with 12% maximum thickness. The NACA 2412 has 2% maximum camber located at four-tenths of the chord length ($0.4c$) aft of the airfoil leading edge. The airfoil profiles are shown and compared in Figure 3.102. The aerodynamic predictions were made using an airfoil design and analysis computer software tool [26]. The ability to design and analyze existing airfoil shapes or create new airfoil profiles has become commonplace with modern computer software tools, such as was used here. The present predictions were performed for each airfoil in an incompressible, high Reynolds number flow. Both the pressure drag and the skin friction drag were calculated as a function of angle-of-attack.

The predicted lift curves are compared in Figure 3.103. The lift curve of the symmetric NACA 0012 airfoil is symmetric for positive and negative angles-of-attack, with a zero-lift angle-of-attack of zero. The cambered NACA 2412 airfoil has a positive lift coefficient at zero angle-of-attack and a zero-lift angle-of-attack of negative 2.25°. The linear lift range and stall behaviors of both airfoils are predicted for positive and negative angles-of-attack. The two airfoils have nearly identical lift curve slopes, an indication that this is driven by the airfoil thickness. Due to its camber, the NACA 2412 airfoil has a higher maximum lift coefficient than the 0012 airfoil.

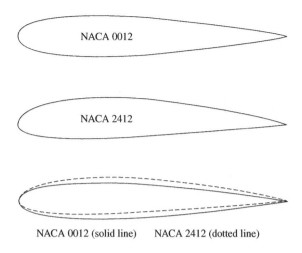

NACA 0012 (solid line)   NACA 2412 (dotted line)

Figure 3.102   NACA 0012 and NACA 2412 airfoils.

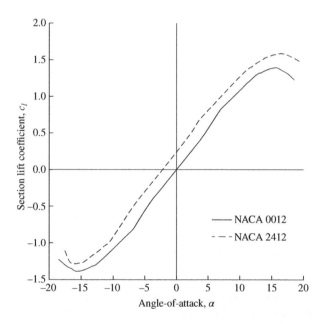

Figure 3.103   Lift curves for symmetric (NACA 0012) and cambered (NACA 2412) airfoils.

The predicted drag curves are shown in Figure 3.104. The NACA 0012 drag curve is symmetric about the zero lift coefficient line as expected. The NACA 2412 drag curve is shifted in the direction of positive lift coefficient (positive angle-of-attack), due to its camber. The minimum drag coefficients of the airfoils are very similar. The NACA 0012 airfoil has a minimum drag coefficient, at a lift coefficient of zero (zero angle-of-attack) of 0.0054. The minimum drag coefficient of the NACA 2412 airfoil is 0.00547 at a lift coefficient of 0.347 (angle-of-attack of 1.0°). The minimum drag coefficient is dominated by the skin friction drag, which is a function of the wetted surface of

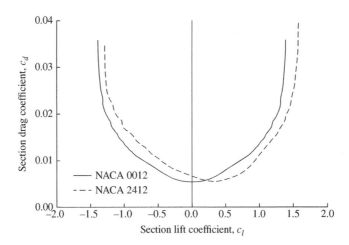

Figure 3.104   Drag curves for symmetric (NACA 0012) and cambered (NACA 2412) airfoils.

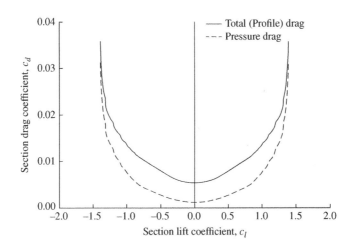

Figure 3.105   Total drag and pressure drag for NACA 0012 airfoil.

the airfoil (linear distance for a two-dimensional shape). The linear distance profiles of the two air-foils are similar, as shown in Figure 3.102, although the cambered airfoil has slightly more surface distance and thus, a slightly higher skin friction and profile drag coefficient.

An interesting feature of performing the airfoil calculations with a software tool is the ability to separate the types of drag making up the total or profile drag of the airfoil. Figure 3.105 shows the pressure drag separated from the profile drag for the NACA 0012 airfoil. The difference between the profile and pressure drag coefficient lines is the skin friction drag coefficient of the airfoil. At small angles-of-attack, the skin friction drag makes up the majority of the profile drag, while the pressure drag dominates at high angles-of-attack. At zero angle-of-attack (lift coefficient of zero), the skin friction drag is 79% of the profile drag, while at an angle-of-attack of 18.5° (lift coefficient of 1.23) the skin friction is only 3% of the total drag.

The pitching moment curves are compared in Figure 3.106. Both curves are fairly linear through the linear lift range with a change in the slope direction in the stall regions. The NACA 0012 airfoil

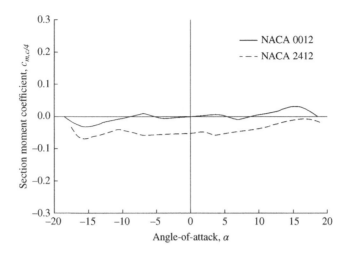

Figure 3.106   Moment curves for symmetric (NACA 0012) and cambered (NACA 2412) airfoils.

has a zero moment coefficient at a lift coefficient of zero (or zero angle-of-attack) and a symmetric moment curve about this zero point. The moment coefficient for the cambered NACA 2412 airfoil is negative through the range of angles-of-attack. A negative pitching moment coefficient indicates a nose-down pitching moment of the airfoil, which is typical of a cambered airfoil or wing alone.

3.9   Three-Dimensional Aerodynamics: Wings

Our aerodynamic discussions, thus far, have focused on two-dimensional flows. There are flow situations and geometries where two-dimensional flow is a good approximation, but it is still an approximation. We live in a three-dimensional world and all real flows over vehicles are three-dimensional in nature. The addition of a third dimension not only adds complexity to the mathematics, but also adds new physical flow phenomena that must be properly accounted for. We have discussed the aerodynamics of two-dimensional bodies, such as cylinders and airfoils. These two-dimensional shapes may be thought of as bodies with infinite width or span. Airfoils are sometimes referred to as *infinite wings*. We start our discussion about three-dimensional aerodynamics with wings that have a finite wingspan or *finite wings*.

3.9.1   Finite Wings

In this section, definitions of finite wing geometry and nomenclature are presented. The aerodynamics of wingtip vortices and induced drag are then discussed.

3.9.1.1   Wing Geometry and Nomenclature

The *planform* of a finite wing is shown in Figure 3.107. Although some wings have simple rectangular planform shapes, most wings have trapezoidal planforms, with a wingspan, b, root chord length, c_r, tip chord length, c_t, and a leading edge sweep angle, Λ. The root and tip chords are of equal length for a rectangular wing. The wing shape shown is the semi-span, $b/2$, of the full span wing. The wing planform area, S, is usually used as the wing reference area for the definition of

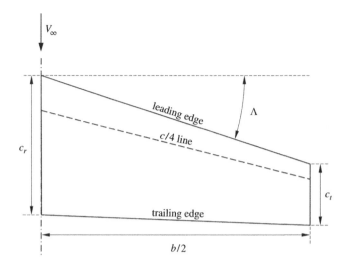

Figure 3.107 Wing planform geometry and nomenclature.

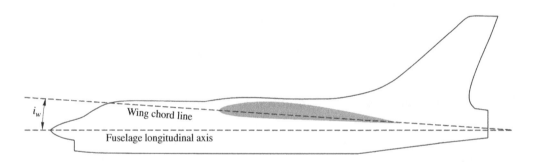

Figure 3.108 Wing incidence angle, i_w.

aerodynamic coefficients. The quarter-chord, $c/4$, line of the wing is the line connecting the points that are one fourth of the chord length aft of the leading edge for each chord section along the span. The wing cross-section is formed by airfoil sections along the wingspan. The type of airfoil section, along the wingspan, may be the same or it may vary along the span.

The angle between the chord line of the wing root airfoil section and the fuselage longitudinal axis is known as the *wing incidence angle*, usually given the symbol i_w, as shown in Figure 3.108. This angle is sometimes referred to as the wing setting angle or mounting angle, since it is the angle at which the wing is mounted to the fuselage. The angle of incidence may be set so that the wing or fuselage generates minimum drag in a cruise flight condition. The incidence angle of a wing is usually a small positive angle, typically about 0–4°. The horizontal tail is also usually set at an incidence angle, i_t, for stability and control considerations, to be discussed in Chapter 6. The tail incidence angle is usually negative for a conventional, aft-mounted tail configuration and positive for a canard, forward-mounted tail configuration.

In almost all cases, the wing incidence angle is a fixed angle that cannot be changed in flight. Of course, there is always the exception in aircraft design. Despite the mechanical difficulties of changing the wing incidence angle in flight, the Voight F-8 *Crusader* was a successful aircraft design with a variable incidence wing. The F-8 wing could pivot upward, from the top of the

Figure 3.109   Voight F-8 *Crusader* with variable incidence wing raised in landing position. (Source: *NASA*.)

fuselage, to increase the wing incidence angle by 7°, as shown in Figure 3.109. Thus, the wing angle-of-attack could be increased at low speeds, for takeoff and landing, by increasing the wing incidence, rather than by raising the aircraft's nose attitude, which reduces forward visibility. The variable incidence wing was one of several innovations in the design of the F-8 *Crusader*, including fuselage area ruling, an all-moving horizontal stabilizer, and titanium structure, which led to the award of the 1956 Collier Trophy to the Voight design team.

The wing taper ratio, λ, is defined as

$$\lambda \equiv \frac{c_t}{c_r} \tag{3.269}$$

Obviously, a rectangular wing has a taper ratio of one. Taper ratios of about 0.4–0.5 are common for subsonic aircraft. Most high-speed, swept wing aircraft have a taper ratio of about 0.2–0.3.

A wing may also have *twist*, where the wing section aerodynamic characteristics vary along the wingspan. The wing may incorporate *aerodynamic twist* where different airfoil sections are used along the span, such that the zero-lift line of the airfoil sections are different from at the wing root. The wing may have *geometric twist* where the angle of incidence of the airfoil sections varies along the span. The geometric twist angle is usually measured as the angle between the airfoil section chord line and the wing root chord line. If the airfoil sections at the wingtip are set at a negative geometric twist angle, with respect to the root chord, the wing has *washout*. If the wingtip sections are set at a positive geometric twist angle, the wing has *washin*. Both aerodynamic and geometric twist serve to change the local angle-of-attack "seen" by the airfoil sections along the wingspan, to tailor the lift distribution and aerodynamic characteristics of the wing. Most wings have washout, of about −3 to −4°, so that the wingtip airfoil sections "see" a lower angle-of-attack than the root sections to make the wing root sections aerodynamically stall before the wingtip sections. Having the stall occur at the wing root, rather than at the wingtip, provides several benefits. First, on many conventional aircraft configurations, the stalled, turbulent, separated flow at the wing root flows aft and strikes the aircraft horizontal tail, resulting in buffeting that is felt by the pilot as a "stall warning". Secondly, since the wingtips remain unstalled, when the wing root is stalled, aileron control is retained in the stall so that the pilot can make roll inputs to keep the aircraft level.

The wing *aspect ratio*, *AR*, is defined as

$$AR \equiv \frac{b^2}{S} \tag{3.270}$$

Since the wing area of a rectangular wing is equal to the span times the constant chord length, the aspect ratio of a rectangular wing, AR_{rect}, is simply

$$AR_{rect} \equiv \frac{b^2}{S} = \frac{b^2}{bc} = \frac{b}{c} \tag{3.271}$$

A wing with high aspect ratio tends to be long in span and short in chord. A low aspect ratio wing has a short span and a longer chord. In the limit of an infinitely long span, a wing geometry approaches that of a two-dimensional airfoil. The lift-to-drag ratio increases with increasing aspect ratio. A general idea of the wing aspect ratios and lift-to-drag ratios for different types of selected aircraft is given in Table 3.13. The aspect ratio is an important wing efficiency parameter.

As we have discussed, many of aviation's early airplane designers attempted to copy nature in designing their airplane wings. Just as an airplane designer chooses the appropriate wing shape to suit the desired mission, nature has given birds different wing shapes that are suited to their type of flight. The Irish botanist D.B.O. Savile [60] describes the function and evolution of bird wings as follows.

> A bird wing is an airfoil combining the functions of an aircraft wing and propeller blade to give lift and thrust. It is radically modified from the vertebrate arm for strength and lightness …
>
> Wing evolution has been affected by the habitats to which birds have adapted (e.g., the open ocean, cliff tops or the closed environment of forests) and by the need to reduce drag, or air resistance. Wing shape and size have been modified to reduce drag and to enable the bird to achieve the most advantageous kind of flight in its usual habitat.
>
> The relationship between the wingspan and the average width of the wing, called the "aspect ratio," also varies. It is calculated by dividing wingspan (tip to tip) by the average width of wing. The long, narrow wings of oceanic soarers have high aspect ratios, enabling these birds to sustain flight over long distances without flapping, and consequently reducing energy expenditure.
>
> The tail is not a rudder, but a combined landing flap and elevator. It is spread and lowered at landing, the wing being pulled forward to keep the center of pressure above

Table 3.13 Wing aspect ratio and lift-to-drag ratios of selected aircraft.

Aircraft	Aspect Ratio, AR	Lift-to-Drag Ratio, L/D
Rockwell Space Shuttle	2.3	4.5 (subsonic)
Boeing F/A-18B	3.5	10.3 ($M_\infty = 0.6$)
Wright *Flyer I*	6.0	8.3
Cessna 150 (general aviation trainer)	6.7	7 (cruise)
Boeing 747 commercial airliner	7.4	17 (cruise)
Lockheed U-2 high-altitude airplane	10.6	28 (cruise)
Schempp-Hirth *Ventus C* sailplane	23.7	43

the center of gravity. In turning, the bird banks through unequal lift on the wings, and raises the tail.

Birds soar (without flapping), hover (with wings beating backward and forward in configurations resembling the figure 8), or achieve fast, level flight (wings beating rhythmically up and down). Wing forms correlate closely with these types of flight, and 4 slightly overlapping wing types can be identified.

Savile identified four types of wings, shown in Figure 3.110, the high aspect ratio wing, the high-speed (pointed, swept) wing, the slotted high-lift wing, and the elliptical wing.

High aspect ratio wings are found on soaring seabirds, such as the albatross and gulls that spend significant amounts of time soaring over the open ocean. The aspect ratio of the albatross wing is about 15 to 18, while the seagull wing aspect ratio is about 8. Similar to a glider, these soaring wings are long, narrow, and pointed. They are designed for high-speed flight and dynamic soaring with low energy expenditure.

The high-speed (pointed, swept) wing has moderate to high aspect ratio, low camber, a tapered, narrow elliptical tip, and is often swept back. The wing trailing edge blends smoothly into the body at the wing root, which reduces flow separation and turbulent drag, much like a wing fillet on an airplane. High-speed wings are flapped constantly with a rapid beat. This wing form is found on birds that make long migratory flights or that feed in the air. Falcons, ducks, swallows, swifts, and hummingbirds possess high-speed wings.

Slotted soaring or high-lift wings have moderate aspect ratios of 6 to 7, deep camber, and wingtip slots, similar in function to leading edge slots in an aircraft wing. The slots enhance the wing's low-speed, high-lift capability, allowing birds to soar at low speed and to takeoff and land in confined areas. This wing type is found on soaring birds of prey, such as buzzards, eagles, hawks, vultures, and owls. The high-lift wing design enables birds to carry heavy loads, such as large prey.

Elliptical wings tend to be found on birds that fly in dense forests, thick shrubbery, or heavy woodland habitats, where maneuverability in tight spaces is desirable. These bird types include

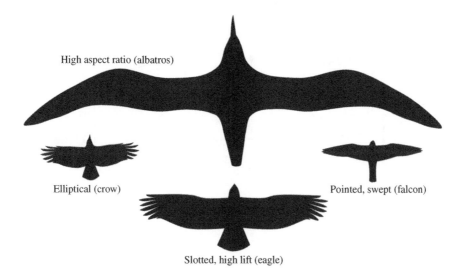

High aspect ratio (albatros)

Elliptical (crow)

Pointed, swept (falcon)

Slotted, high lift (eagle)

Figure 3.110 Savile's four types of bird wings. (Source: *L. Shyamal, "Flight Silhouettes" https:// en.wikipedia.org/wiki/File:FlightSilhouettes.svg, CC-BY-SA-2.5. License at https://creativecommons.org/ licenses/by-sa/2.5/deed.en.*)

sparrows, woodpeckers, doves, and crows. They are also the wing form found on most bats. The wings have elliptical planform shapes, but they are usually short in length with low aspect ratios of about 4 to 5. The wings also have slotting to enhance low-speed lift. High wing-beat frequency is required for rapid takeoff, quick acceleration, and maneuverability.

For two-dimensional airfoils, the aerodynamic center is defined as the chordwise location where the airfoil pitching moment is independent of angle-of-attack. For airfoils in subsonic flow, the aerodynamic center is typically at the quarter chord point. Similar to this concept, an aerodynamic center can be defined for a wing. The wing aerodynamic center is referenced to a type of "average" chord length for the wing, called the *mean aerodynamic chord*, often abbreviated as the *MAC* and denoted by the symbol $\bar{c}$. The mean aerodynamic chord is used as a reference length for various aerodynamic and stability and control analyses. The MAC passes through the center of area or centroid of the wing planform. For a rectangular wing with a constant chord length, determination of the wing center of area is simple and the MAC is simply the constant chord length. For a non-rectangular wing, the determination of the MAC is a bit more complicated, but not too difficult. The MAC may be determined analytically or using a graphical method.

The determination of the mean aerodynamic chord using the graphical method is as follows (Figure 3.111). At the wing root, lines, with lengths equal to the tip chord length, c_t, are drawn above and below the wing root chord. Similarly, lines of length equal to the root chords are drawn above and below the tip chord. Two diagonal lines are drawn, connecting the lines at the top and bottom of the root and tip chords. The intersection of the two diagonal lines is the location of the wing centroid. The MAC passes through the centroid, and has a length $\bar{c}$ equal to the distance from

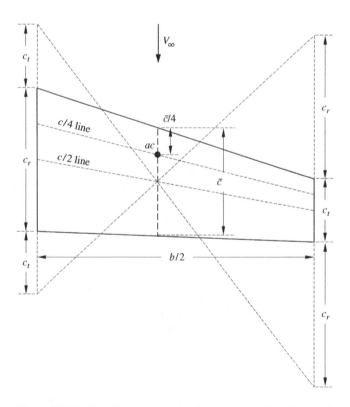

Figure 3.111   Graphical construction for mean aerodynamic chord.

the leading edge to the trailing edge at this location. The MAC quarter chord point is the location of the wing aerodynamic center.

3.9.1.2 Wingtip Vortices, the Wing Vortex System, and Wing Lift

The major difference between a two-dimensional airfoil or infinite wing and a finite wing is the wingtips of the finite wing. The flow over a wing is fundamentally three-dimensional versus the two-dimensional flow over an airfoil. Because of this, the aerodynamics of a wing, including the lift and drag as a function of angle-of-attack, are different from that of an airfoil. For a lift-producing wing, there is high pressure on the lower surface and low pressure on the upper surface. The high pressure air from underneath the wing flows around the wingtip towards the low pressure, as shown in Figure 3.112. This wingtip flow creates a tightly wound, horizontal spiral of air called a *wingtip vortex*. The wingtip vortex leaves the wingtip and flows downstream of the wing, increasing in diameter. The large diameter of a wingtip vortex, visualized with smoke, behind a general aviation airplane is shown in Figure 3.113.

Wingtip vortex trails can extend behind an aircraft for many miles, and may remain intact in the atmosphere for many minutes after the passage of the aircraft. These invisible, horizontal "tornados" can be a serious hazard to aircraft that inadvertently passes through them, with the possibility of upsetting the aircraft or even sending it out of control. The strength of the vortices is related to the lift, and thus the weight, of the aircraft generating them. Since the size and strength of the vortex are proportional to the lift that is generated, a large, heavyweight airliner generates much more powerful wingtip vortices than a small, lightweight general aviation airplane. The state of the atmosphere, whether it is stable and calm or turbulent, plays a major role in the decay of the trailing vortices. Eventually, the vortices are dissipated due to the viscosity of the air.

The wingtip vortices are part of a complete wing vortex system, as shown in Figure 3.114. The vortex system forms a closed circuit, with a *bound vortex* modeled at the wing and a *starting vortex* that trails downstream. The vortices have a circulation, Γ, which is directly related to the wing lift. (Recall that the circulation is mathematically defined as the negative of the line integral around a closed curve in the flow. It is a kinematic property of the flow that is a function of the velocity field and the choice of the closed curve. In our case, the closed curve surrounds the vortex filament.) For our aerodynamic purposes, the circulation is a construct or tool that allows us to quantify the lift.

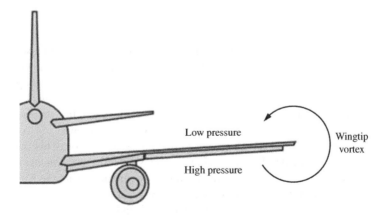

Figure 3.112 Wingtip vortex (view looking forward from aircraft tail).

Figure 3.113   Wingtip vortex made visible with smoke. (Source: *NASA*.)

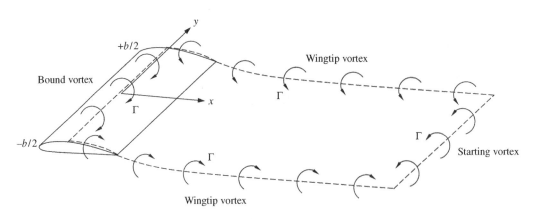

Figure 3.114   Vortex system created by lifting wing.

To properly model the wing lift distribution, the circulation strength must be varied along the span of the wing, such that $\Gamma = \Gamma(y)$, where y is from the left wingtip to the right wingtip, $-b/2 \leq y \leq +b/2$ (Figure 3.114). This model of the wing circulation on a line along the wingspan, called a *lifting line*, was first theorized by the aerodynamicist Ludwig Prandtl. Developed by Prandtl in the early 1900s, the *classical lifting line theory* was one of the first practical applications of aerodynamic theory in calculating the lift of a finite wing. Using lifting line theory, the lift can be calculated from the circulation distribution along the lifting line as follows.

Imagine that you are looking down the wingspan, at an airfoil section. This two-dimensional "slice" of the wing and flow field is the same as Figure 3.35, showing a two-dimensional airfoil with a vortex of circulation, Γ, and a starting vortex trailing downstream. Using the Kutta–Joukowsky theorem for the two-dimensional lift of an airfoil, Equation (3.213), the lift per unit span, L', of the three-dimensional wing, at a y location along the span, is given by

$$L'(y) = \rho_\infty V_\infty \Gamma(y) \tag{3.272}$$

The three-dimensional lift of the wing, L, is obtained by integrating the lift per unit span, L', along the span of the wing, from the left wingtip $(-b/2)$ to the right wingtip $(+b/2)$, as given by

$$L = \int_{-b/2}^{+b/2} L'(y)dy = \rho_\infty V_\infty \int_{-b/2}^{+b/2} \Gamma(y)dy \tag{3.273}$$

where ρ_∞ and V_∞ are the freestream density and velocity, respectively. To obtain the wing lift, the circulation distribution, $\Gamma(y)$, must be known. Prandtl calculated analytical solutions of the wing lift for different circulation distributions, including the specific circulation for an elliptical planform wing and a generalized circulation distribution, applicable to other wing planform shapes. From this, he was able to calculate the drag due to the lift of a finite wing, to be discussed next.

3.9.1.3 Downwash and Induced Drag

The wingtip vortices affect the flow field at the aircraft, as well as downstream of the aircraft. The vortices induce an *upwash* and a *downwash* in the flow field, as shown in Figure 3.115. The downwash is also felt at the wing itself, changing the local flow over the wing.

The spanwise distributions of the lift, lift coefficient, and downwash are qualitatively shown in Figure 3.116 for three different wing planforms: an elliptical planform, a rectangular planform, and a tapered wing with pointed wingtips. For all of the wings, the lift is a maximum at the center of the wing and decreases to zero at the wingtips. The lift distribution for an elliptical wing also has an elliptical shape, while the downwash is uniform across the span. Since the spanwise downwash is constant, the induced and effective angles-of-attack are constant across the span (assuming there is no wing twist), leading to a constant spanwise lift coefficient. The rectangular wing has uniform lift and a uniform lift coefficient over the center-span, decreasing rapidly to zero at the wingtips. The downwash is uniform over the center, increasing rapidly at the wingtips. The tapered planform has non-uniform lift and lift coefficient distributions with the lift at a maximum peak and the lift coefficient at a minimum peak, in the center. The tapered wing downwash is non-uniform across the span.

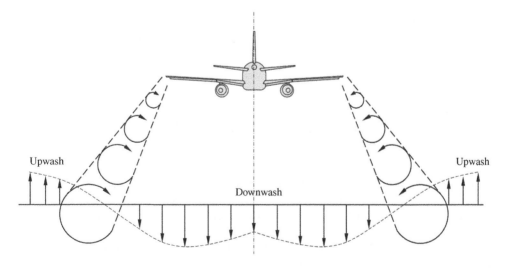

Figure 3.115   Upwash and downwash induced by wingtip vortices.

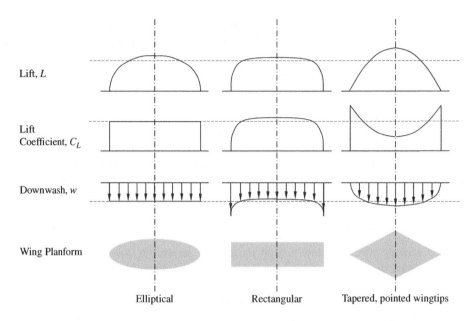

Figure 3.116   Spanwise distributions of lift, lift coefficient, and downwash for several wing planforms.

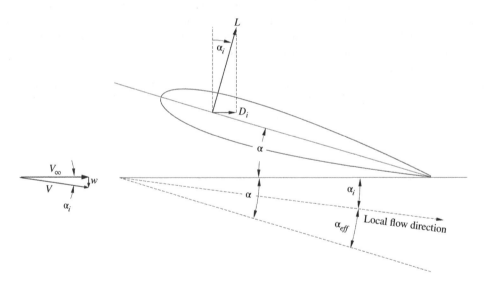

Figure 3.117   Geometry for induced drag.

The effect of the downwash at a local airfoil section of the wing is shown in Figure 3.117. The downwash velocity, w, adds to the freestream velocity, V_∞, resulting in a local flow velocity, V, that is rotated downward by the angle α_i from the freestream direction. The angle α_i is known as the *induced angle-of-attack*, as it is induced by the downwash flow. The downwash velocity is much smaller than the freestream velocity, $w \ll V_\infty$, hence the induced angle-of-attack is a small angle, less than a few degrees (which has been made larger in Figure 3.117 for clarity).

Because of this, the angle-of-attack "seen" by the airfoil section is not the *geometric angle-of-attack*, α, between the chord line and the freestream velocity, rather it is the smaller *effective angle-of-attack*, α_{eff}, between the chord line and the relative velocity. The effective angle-of-attack is defined as

$$\alpha_{eff} = \alpha - \alpha_i \tag{3.274}$$

An outcome of Prandtl's lifting line theory is the solution of the induced angle-of-attack for a generalized lift distribution, given by

$$\alpha_i = \frac{C_L}{\pi e AR} \tag{3.275}$$

where C_L is the wing lift coefficient, AR is the wing aspect ratio, and e is the *Oswald efficiency factor*, or simply the *span efficiency factor*. As introduced in Section 3.7.2, the span efficiency factor is a parameter related to the efficiency of the wing planform shape where an elliptical planform is the most efficient. All other wing planform shapes are less efficient, with span efficiency factors less than one, typically between about 0.85 and 0.95. The induced angle-of-attack, given by Equation (3.275), has units of radians.

The lift vector, L, which is perpendicular to the relative wind, is canted aft by the angle α_i, thus contributing a force in the drag direction, called the induced drag, D_i, given by

$$D_i = L \sin \alpha_i \cong L \alpha_i \tag{3.276}$$

where $\sin \alpha_i \cong \alpha_i$, for small angles. Based on this explanation of the origin of this drag term, the induced drag is also referred to as *drag due to lift*. Alternatively, the induced drag may be explained from an energy perspective, where the wingtip vortices are robbing energy from the aircraft, such that more power must be expended to overcome this drag.

Inserting Equation (3.275), for the induced angle-of-attack, into Equation (3.276), we have for the induced drag

$$D_i = L \frac{C_L}{\pi e AR} \tag{3.277}$$

Dividing both sides of Equation (3.277) by $q_\infty S$, we have

$$\frac{D_i}{q_\infty S} = \left(\frac{L}{q_\infty S} \right) \left(\frac{C_L}{\pi e AR} \right) \tag{3.278}$$

Defining the induced drag coefficient, $C_{D,i}$, we have

$$C_{D,i} = \frac{C_L^2}{\pi e AR} \tag{3.279}$$

Equation (3.279) provides the induced drag coefficient, or drag due to lift, for a finite wing with a generalized lift distribution. This equation is more accurate for wings with larger aspect ratio. As the wing aspect ratio decreases, the lifting line model of the wing is less valid and the accuracy of the lifting line-based relationships decreases.

For the special case of a wing with an elliptical planform shape, Prandtl's solution for the induced drag is given by

$$C_{D,i} = \frac{C_L^2}{\pi AR} \tag{3.280}$$

Comparing Equation (3.280) with (3.279), we see that the elliptical wing is a special case of the generalized solution for the induced drag with a span efficiency factor of one. The span efficiency factor for a non-elliptical wing planform shape is less than one, therefore the induced drag of a non-elliptical wing is greater than that for an elliptical planform. Thus, the elliptical wing has the minimum induced drag of any wing planform shape.

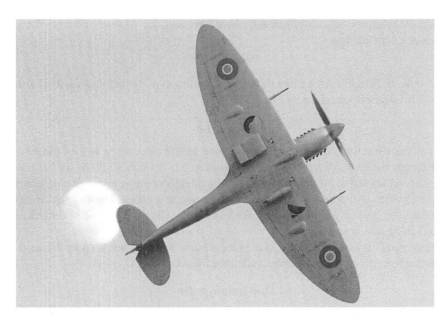

Figure 3.118 The WWII-era Supermarine *Spitfire* with an elliptical planform wing. (Source: *Cpl. Neil Conde, OGL v1.0.*)

This fact was known to early airplane designers who sought to make their wing designs as efficient as possible to increase performance. One of the most famous aircraft designed with an elliptical wing is the World War II-era British Supermarine *Spitfire*[5], shown in Figure 3.118. While the elliptical wing is beneficial from an aerodynamics and performance standpoint, it is not advantageous from a manufacturing perspective. The added complexity of the elliptical wing shape greatly increases the time and expense of fabrication over a simpler rectangular or straight, tapered wing.

The span efficiency factor may also be expressed in terms of the induced drag factor, δ, as

$$e = \frac{1}{1 + \delta} \tag{3.281}$$

Inserting Equation (3.281) into (3.279), the induced drag coefficient may be written as

$$C_{D,i} = \frac{C_L^2}{\pi AR}(1 + \delta) \tag{3.282}$$

The induced drag factor is a constant for a given planform shape. The induced drag is equal to that of the optimum elliptical wing for an induced drag factor of zero. Therefore, Equation (3.282) provides the fractional increase in the induced drag for a non-elliptical planform wing. The value of δ may be calculated using Prandtl's classical lifting line theory. The variation of the span efficiency factor, e, and the induced drag factor, δ, as a function of the wing taper ratio, $\lambda = c_t/c_r$, and aspect ratio, AR, are shown in Figure 3.119 and Figure 3.120, respectively.

At high aspect ratio, the maximum span efficiency factor and the minimum induced drag factor are at a taper ratio approaching 0.4. The induced drag, corresponding to this taper ratio, is less than about 1% greater than that for an elliptical wing. The planform shape of the straight, tapered wing

[5] The initial design of the Spitfire wing was a straight, tapered wing, rather than an elliptical wing. The final, elliptical wing design was selected to increase volume for the mounting of eight wing-mounted machine guns, rather than primarily for aerodynamic efficiency.

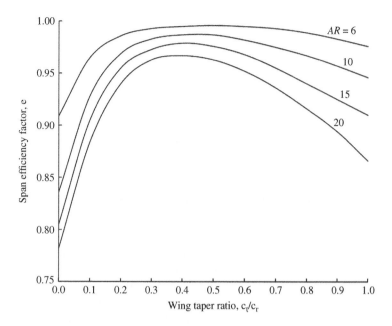

Figure 3.119 Span efficiency factor for unswept, tapered wings. (Source: *R.F. Anderson, "Determination of the Characteristics of Tapered Wings," NACA Report No. 572, 1940.*)

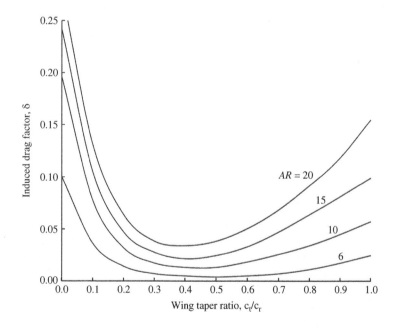

Figure 3.120 Induced drag factor for unswept, tapered wings.

with λ of 0.4 is compared to an elliptical wing in Figure 3.121. A simple rectangular wing, with a taper ratio of one and high aspect ratio, has an induced drag that is about 6% greater than an elliptical wing. Overall, the induced drag varies by up to about 10% over the range of taper ratios from zero to one. In contrast, the effect of aspect ratio on the induced drag is more significant. Since

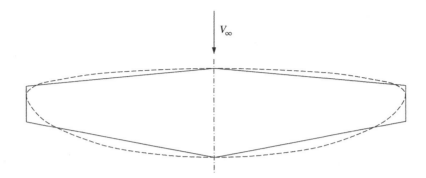

Figure 3.121 Comparison of planform shape of straight wing with a taper ratio, λ, of 0.4 (solid line) and an elliptical wing (dashed line).

the induced drag varies inversely with the aspect ratio, doubling the aspect ratio reduces the induced drag by a factor of two. Prandtl used his lifting line theory to show this strong effect of aspect ratio on induced drag. He later verified this by measuring the drag of seven rectangular wings, with varying aspect ratios, in a wind tunnel. Therefore, the usual strategy for decreasing induced drag of finite wings is to increase the aspect ratio, rather than to build a wing with a taper ratio that matches an elliptical wing lift distribution. Typically, the wingspan is lengthened to increase the aspect ratio, which has the physical effect of moving the influence of the wingtip vortices further outboard from the center of the wing and making the flow over the main portion of the wing more two-dimensional. Rectangular wings, with as high an aspect ratio as practical, are often used on aircraft, especially general aviation aircraft, because they are easier and less costly to fabricate than tapered wings.

Having completed our development of the induced drag, we can recall Equation (3.219), and write the total drag, C_D, of a finite wing (excluding wave drag) as

$$C_D = c_d + C_{D,i} = c_d + \frac{C_L^2}{\pi e AR} \tag{3.283}$$

where c_d is the profile drag.

3.9.2 Lift and Drag Curves of Finite Wings

We now examine the lift and drag curves of finite wings, in contrast to those of infinite wings or two-dimensional airfoils. Of course, the infinite wing may be considered a limiting case of a finite wing with an infinite span or aspect ratio. In the last section, we discovered that the three-dimensional flow effects of a finite wing result in effects that degrade the aerodynamic characteristics of an infinite wing, including the "tilting back" of the lift vector and the addition of induced drag. Thus, we may expect that the characteristics of the finite wing lift and drag curves are similarly adversely affected.

3.9.2.1 Finite Wing Lift Curve

Let us assume that we have a finite wing and an infinite wing at the same lift coefficient. Earlier, it was determined that the wingtips of a finite wing induce a downwash which reduces the geometric angle-of-attack of a finite wing by the induced angle-of-attack, such that the finite wing "sees" an effective angle-of-attack, given by Equation (3.274). Therefore, for a finite wing and an infinite wing at the same lift coefficient, the effective angle-of-attack of the finite wing, α_{eff}, must equal the geometric angle-of-attack of the infinite wing, α_{2D}, and, from Equation (3.274), is given by

$$\alpha_{eff} = \alpha_{2D} = \alpha - \alpha_i \tag{3.284}$$

where α and α_i are the geometric and the induced angles-of-attack of the finite wing, respectively. Synonymous with this is the fact that the finite wing geometric angle-of-attack must be greater than the infinite wing geometric angle-of-attack by the amount equal to the induced angle-of-attack, to produce the same lift coefficient.

$$\alpha = \alpha_{2D} + \alpha_i \tag{3.285}$$

The infinite wing lift curve slope, a_0, is given by

$$a_0 = \frac{dc_l}{d\alpha_{2D}} = \frac{dc_l}{d\alpha_{eff}} = \frac{dc_l}{d(\alpha - \alpha_i)} \tag{3.286}$$

Integrating Equation (3.286), we have

$$\int dc_l = a_0 \int d(\alpha - \alpha_i) \tag{3.287}$$

$$c_l = a_0(\alpha - \alpha_i) + k \tag{3.288}$$

where k is a constant. Inserting Equation (3.275) for the induced angle-of-attack, α_i, into (3.288), we have

$$c_l = a_0 \left(\alpha - \frac{C_L}{\pi eAR} \right) + k \tag{3.289}$$

where C_L is the lift coefficient of the finite wing. Recall that we are assuming that the infinite wing and the finite wing are at the same lift coefficient, $c_l = C_L$, so that Equation (3.290) is

$$C_L = a_0 \left(\alpha - \frac{C_L}{\pi eAR} \right) + k \tag{3.290}$$

Solving for the lift coefficient, we have

$$C_L + a_0 \frac{C_L}{\pi eAR} = C_L \left(1 + \frac{a_0}{\pi eAR} \right) = a_0 \alpha + k \tag{3.291}$$

$$C_L = \left(\frac{a_0}{1 + \frac{a_0}{\pi eAR}} \right) \alpha + \frac{k}{1 + \frac{a_0}{\pi eAR}} = \left(\frac{a_0}{1 + \frac{a_0}{\pi eAR}} \right) \alpha + k' \tag{3.292}$$

where k' is a new constant. The lift coefficient of the finite wing can be expressed as

$$C_L = a\alpha - \alpha_{L=0} \tag{3.293}$$

where a is the lift curve slope of the finite wing and $\alpha_{L=0}$ is the zero-lift angle-of-attack of the finite wing. Setting Equation (3.292) equal to (3.293), we have

$$C_L = \left(\frac{a_0}{1 + \frac{a_0}{\pi eAR}} \right) \alpha + k' = a\alpha - \alpha_{L=0} \tag{3.294}$$

Comparing the left and right sides of Equation (3.294), we see that the finite wing lift curve slope is given by

$$a = \frac{a_0}{1 + \frac{a_0}{\pi e AR}} \tag{3.295}$$

From Equation (3.295), we conclude that for any positive value of the infinite wing lift curve slope, a_0, the lift curve slope of a finite wing, a, is less than that for an infinite wing, or $a < a_0$.

Equation (3.295) also provides some insight into the effect of the wing aspect ratio, AR, on the lift curve slope. Firstly, for an infinite wing, with an infinite aspect ratio, Equation (3.295) simply reduces to $a = a_0$. As the aspect ratio decreases, corresponding to decreasing the wingspan for a given chord length, the lift curve slope decreases. As the wingspan decreases, the wingtip effects become more significant over the span of the wing, resulting in increased downwash, more induced drag, and less lift.

Now, consider the case of finite and infinite wings, both at zero lift. Since there is no lift, the finite wing has no downwash and no induced angle-of-attack. Therefore, at zero lift coefficient, the finite wing has the same zero-lift geometric angle-of-attack, $\alpha_{L=0}$, as the infinite wing.

Let us now consider the other end of the lift curve, where the angle-of-attack is large. We have already determined that, due to downwash, the finite wings "sees" a lower angle-of-attack than its geometric angle-of-attack. Thus, when the finite wing is at the stall geometric angle-of-attack, the wing is "seeing" a lower angle-of-attack and remains unstalled. It is not until the finite wing effective angle-of-attack reaches the stall angle-of-attack, that the wing stalls. From this, we can conclude that the finite wing is at a higher geometric angle-of-attack than the infinite wing stall geometric angle-of-attack, before it reaches stall.

Now we ask, how does the maximum lift coefficient at the stall angle-of-attack compare between the finite and infinite wings? To answer this, return to the spanwise lift distributions shown in Figure 3.116. At the wingtips of the finite wing, the lift distribution must go to zero. The integrated lift force for the finite wing, and the associated wing lift coefficient, are less than is obtainable with an infinite wing. Thus, the finite wing maximum lift coefficient is less than that of the infinite wing. As seen in Figure 3.116, the elliptical wing has the optimum lift distribution for a finite wing to achieve lift coefficients close to an infinite wing. With an elliptical lift distribution, it may be possible for a finite wing to achieve approximately 90% of the two-dimensional maximum lift coefficient, at the same Reynolds number.

Based on the observations that we have made about the lift curve slope, the zero lift angle-of-attack, the stall angle-of-attack, and the maximum lift coefficient, we draw the lift curves of a finite wing and an infinite wing as shown in Figure 3.122. This figure summarizes the comparisons that we have made of the finite wing and infinite wing lift curves, as follows: (1) the zero-lift angles-of-attack are the same, (2) the finite wing lift slope is less than the infinite wing lift slope, (3) to obtain the same lift coefficient, the finite wing geometric angle-of-attack is greater than the infinite wing geometric angle-of-attack, (4) the finite wing stall geometric angle-of-attack is greater than the infinite wing stall geometric angle-of-attack, and (5) the finite wing maximum lift coefficient is less than the infinite wing maximum lift coefficient. We have also made the additional observation that the finite wing lift curve slope decreases with decreasing aspect ratio.

3.9.2.2 Finite Wing Drag Curve

The total drag coefficient, C_D, of a finite wing (ignoring wave drag) was given earlier by Equation (3.283), repeated below.

$$C_D = c_d + C_{D,i} = c_d + \frac{C_L^2}{\pi e AR} \tag{3.296}$$

where c_d is the airfoil profile drag coefficient and $C_{D,i}$ is the induced drag coefficient.

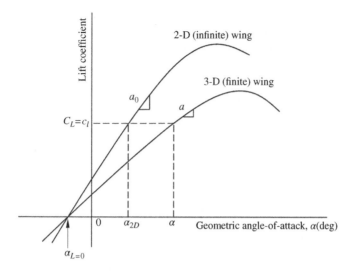

Figure 3.122   Comparison of lift curves for infinite and finite wings.

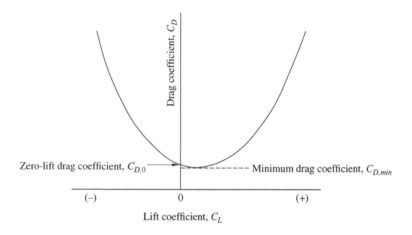

Figure 3.123   Drag polar of a finite wing.

A plot of the variation of the total drag coefficient, C_D, with lift coefficient, C_L, is known as a *drag polar*, shown in Figure 3.123, for a finite wing. The drag polar has a parabolic shape since the drag coefficient varies with the square of the lift coefficient in Equation (3.296). The drag polar has a minimum drag coefficient, $C_{D,min}$, which may be at a lift coefficient greater than zero if the wing airfoil sections have camber. The drag coefficient corresponding to zero lift is the zero-lift drag coefficient, $C_{D,0}$. For a wing with a symmetric airfoil, the minimum drag coefficient and the zero-lift coefficients are the same.

3.9.3   High-Lift Devices

Typically, an airfoil and wing are designed to provide desirable aerodynamic characteristics at cruise flight conditions, where most aircraft spend much of their operational time. However, all

aircraft must take off and land, which occurs at the low speed portion of the aircraft's flight envelope. A wing that has been aerodynamically optimized for higher-speed operation does not usually
have the desired low-speed aerodynamic characteristics for takeoff and landing. Therefore, high-lift
devices are often added to a wing to improve its low-speed aerodynamics without compromising
the high-speed characteristics.

Consider an aircraft flying in steady flight at low speed, near its maximum lift coefficient, $C_{L,max}$.
The minimum airspeed associated with flying at maximum lift coefficient is the stall speed, V_s,
given by Equation (2.48), repeated below.

$$V_s = \sqrt{\frac{2W}{\rho_\infty S C_{L,max}}} \tag{3.297}$$

By examining Equation (3.63), we see that there are two practical ways of decreasing the stall
speed, by increasing the maximum lift coefficient, $C_{L,max}$, or by increasing the wing area, S. This
is usually accomplished using mechanical high-lift devices that are applied to the wing leading or
trailing edges. In this section, we briefly discuss three kinds of high-lift devices, the trailing edge
flap, leading edge slots and slats, and the use of boundary layer control.

3.9.3.1 Flaps

The flap is the most commonly used type of high-lift device used on wings. In its simplest form,
the *plain flap* (see Figure 3.124), a portion of the wing, typically 15 to 25% of the chord length, is
hinged near the trailing edge, allowing it to deflect downward. This downward deflection changes
the shape of the airfoil section, increasing its camber. The primary effects of flap deflection are:

1. an increase in the maximum lift coefficient
2. a change in the zero-lift angle-of-attack to a more negative value
3. a decrease in the stall angle-of-attack
4. a significant increase in the drag.

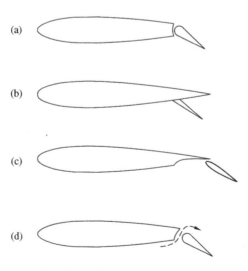

Figure 3.124   Various types of wing flaps, (a) plain, (b) split, (c) Fowler, and (d) slotted.

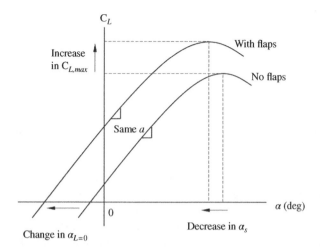

Figure 3.125   Effect of flaps on lift.

The drag increase is not necessarily a disadvantage, as it permits a steeper approach angle without an increase in airspeed. The lift curves, for a wing with and without flap deflection, are shown in Figure 3.125. Flap deflection shifts the lift curve upward, without changing the lift curve slope, a, similar to what is seen in Figure 3.103 with the addition of camber. The maximum lift coefficient can be increased by about 50% with a plain flap and by about 100% with more complex types of flaps, to be discussed next.

Several other types of flaps are shown in Figure 3.124. Typically, the aerodynamic benefits of these other types of flaps come at the cost of increased complexity. For the *split flap*, a portion of the undersurface of the wing is "split off" and deflected downward. A slightly higher maximum lift coefficient can be obtained with a split flap over a plain flap, but there is a greater increase in drag due to the turbulent wake produced.

The *slotted flap* is similar to a plain flap, except that its deflection creates a gap between the wing and the flap leading edge. The gap allows high-energy air, from underneath the wing, to flow over the flap upper surface, which energizes the boundary layer and delays flow separation at high lift coefficients. The slotted flap provides a greater increase in the maximum lift coefficient than the plain or split flap, without as a large a drag increase as the split flap. A wing may have multiple slotted flaps that extend beyond each other to form a row of flaps and slots.

Similar to the slotted flap, the *Fowler flap* creates a slot, but also moves aft along tracks when deflected. This aft movement increases the chord length, so that both the maximum lift coefficient and wing area are increased. The Fowler flap produces the largest increase in maximum lift coefficient over the other types of flaps discussed, with a minimal drag increase.

3.9.3.2   Leading Edge Devices

High-lift leading edge devices include the use of slots and slats, as shown in Figure 3.126. The *fixed slot* operates in a manner similar to the trailing edge slotted flap. High-energy air, from underneath the wing, flows through the slot to energize the upper wing surface boundary layer, delaying separation and stall. The increase in the maximum lift coefficient is due to this delay in the stall to a higher angle-of-attack, as the wing camber is unchanged by the slot. The fixed slot can provide an increase of the maximum lift coefficient by about 0.1–0.2. A disadvantage of the fixed slot is the excessive drag it creates at high flight speeds.

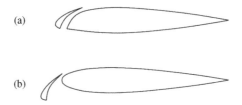

Figure 3.126   High-lift leading edge devices, (a) fixed slot and (b) movable slot–slat.

The *movable or automatic slot* has a *leading edge slat* that moves freely on tracks. At low angle-of-attack, the slat is held against the wing by the local air pressure distribution. At high angle-of-attack, at or near stall, the wing leading edge suction "automatically" lifts the slat away from the wing, forming a slot. The slot–slat arrangement can increase the maximum lift coefficient by as much as 100% for some airfoil sections. Since the slat is stowed at high airspeed, it does not have the drag penalty of the fixed slot, but this is at the cost of increased complexity and weight over the fixed slot.

The slot–slat is particularly useful for thin, high-speed wings that have sharp leading edges, which are susceptible to leading edge flow separation. The slot–slat can delay this leading edge flow separation at low speeds and significantly increase the lift coefficient. On highly swept wings, the slot–slat is effective in reducing spanwise flow on the wing upper surface, which is detrimental to lift, at low airspeed and high angle-of-attack.

The effect of the slot–slat on the wing lift curve is shown in Figure 3.127. As summarized by this figure, the slot–slat increases the maximum lift coefficient by extending the lift curve to a higher stall angle-of-attack, while leaving the zero lift angle-of-attack unaltered.

An increase in the angle-of-attack is required to achieve higher lift coefficients, with a slot–slat. If the wing only has leading edge devices, this may lead to excessively high angles-of-attack during takeoff and landing. Therefore, trailing edge flaps are usually used in conjunction with high-lift leading edge devices. The configuration of the high-lift devices varies depending on the phase of

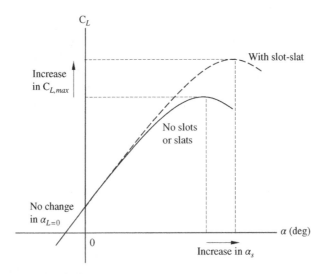

Figure 3.127   Effect of leading edge slats on lift.

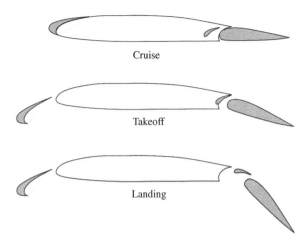

Figure 3.128 High-lift devices during different phases of flight.

flight and the desired flight speed. Typical deployment of the leading and trailing edge devices, during different phases of flight, is shown in Figure 3.128. The leading edge slat is shown stowed and the trailing edge flaps retracted at the top of Figure 3.128. The takeoff configuration is shown in the middle, with the slot–slat extended and the first slotted flap partially deflected. The lower picture shows the landing configuration, with the slot–slat extended and a double-slotted flap fully extended.

3.9.3.3 Boundary Layer Control

In addition to flap-like devices that change the airfoil shape, another means of increasing the maximum lift coefficient is through the use of boundary layer control. At high angle-of-attack, the boundary layer on the wing upper surface may separate, resulting in wing stall and loss of lift. There are two ways to prevent or delay the boundary layer separation and loss of lift, by applying suction to the wing upper surface to remove the low-energy boundary layer flow or by injecting a high-energy flow into the boundary layer. The slots discussed for leading and trailing edge devices are a form of boundary layer control. Both of these techniques energize the boundary layer and enable it to remain attached to a higher angle-of-attack, thus increasing the maximum lift coefficient.

Suction may be applied through slots or small holes on the wing surface. The low-energy flow is replaced by the higher kinetic energy-flow outside the boundary layer. The installation of surface suction systems over large areas of the wing can be complex, typically requiring the installation of vacuum pump systems. The clogging of the small suction holes by dirt and other debris must also be considered.

The installation of surface blowing systems may be more practical than suction systems, as they can utilize high-pressure air from a jet engine compressor. The high-pressure, high kinetic energy air is injected into the low energy boundary layer through slots or small holes, similar to the suction systems, although clogging of the ports may be less of an issue with blowing.

Surface blowing systems are often included with flaps, as the increase in the stall angle-of-attack, due to blowing, improves the decrease in stall angle-of-attack with flaps. Flap systems combined with surface blowing are known as *blown flaps*. In the extreme, the mechanical flap system can be replaced with a high-speed jet at the wing trailing edge, known as a *jet flap*. Many different types of surface blowing systems have been flight tested in the past to improve lift capabilities, especially for short-takeoff-and-landing (STOL) aircraft. Several modern aircraft, such as the Boeing C-17 *Globemaster III* military cargo aircraft, employ *powered lift* systems, where the high-energy jet engine exhaust flow is blown over trailing edge slotted flaps, which increases the maximum lift coefficient and also vectors the exhaust flow to directly augment lift.

3.9.3.4   Spoilers

We have discussed several types of high-lift devices designed to increase lift, but in this section we conclude with a brief description of *spoilers*, devices that are used to decrease lift. Spoilers are flat plate-type flaps, located on the wing upper surface, that are deflected up into the flow, as shown in Figure 3.129, to reduce or "spoil" the lift. The raised spoiler separates the flow on the wing upper surface, resulting in a loss of lift and an increase in drag.

Typically, spoilers are used by commercial and military transports during descents and after landing. *Flight spoilers* are deployed to increase the rate of descent, to slow down, or to do both. *Ground spoilers* are automatically raised upon landing, decreasing lift to prevent an aircraft "bouncing" after landing and to increase the weight on the wheels, which maximizes braking effectiveness. The increased aerodynamic drag also reduces the ground landing roll.

By placing spoilers on the outboard sections of the wing, they can also be used asymmetrically for auxiliary roll control. For example, by raising a spoiler on the left wing, the lift on the left wing is diminished, causing the aircraft to roll to the left. The spoiler drag on the left wing also creates a favorable yawing moment to the left in concert with the left roll. Spoilers are used on most commercial transports to augment the primary aileron roll control. Spoilers have been used on aircraft in the past for primary roll control when landing flaps are incorporated in the full span of the wing trailing edge, leaving no room for conventional ailerons. Spoilers produce less wing twist than conventional ailerons, so they have also been used for primary roll control on very flexible wing structures to minimize wing twist.

Since the operation of spoilers is based upon the non-linear physics of separated flows, spoiler characteristics tend to be non-linear, making their use as a primary flight control device more difficult. Spoilers may exhibit undesirable, non-linear effects upon opening or at high angles-of-attack, where their effectiveness may be reduced or even reversed. The control forces associated with spoilers may also be non-linear, making the control system design more difficult.

As a final note, a very early use of spoilers for primary roll control was on the Northrop P-61 *Black Widow* airplane, that was used to tow our P-51 *Mustang* up to altitude for the glide flight test technique in Section 3.7.6. Unlike the flat plate-type spoiler that is raised up into the flow, the P-61 used a *plug-type spoiler*, which was a curved panel that retracted and extended vertically in and out of the wing. Extensive testing and modification were required to make this roll control perform satisfactorily due to its non-linear effectiveness and non-linear control forces.

Figure 3.129   Extended spoiler separates flow and "spoils" lift.

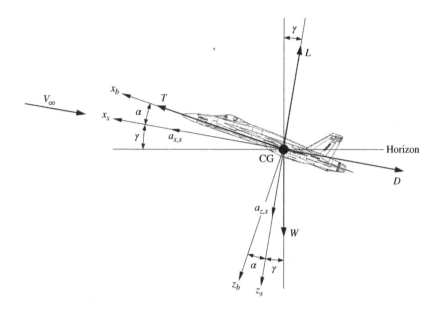

Figure 3.130 Resolution of forces and accelerations on an aircraft in flight.

3.9.4 FTT: Aeromodeling

As discussed earlier, wind tunnel testing is used to obtain aerodynamic coefficients of sub-scale models and in some cases of full-scale vehicles. These ground tests have their limitations and difficulties, including matching the actual flight conditions. The present section discusses how the lift and drag coefficients can be obtained for a full-scale aircraft in flight, using flight test techniques known as aerodynamic modeling, or simply *aeromodeling*.

Consider an aircraft at a flight condition with a velocity V_∞, fligh path angle, γ, and angle-of-attack, α, as shown in Figure 3.130. The flight path angle is the angle between the horizon and the velocity vector or relative wind. The angle-of-attack is measured between the relative wind and the x_b body axis of the aircraft. The forces acting on the aircraft are the lift, L, drag, D, thrust, T, and weight, W. The lift and drag are perpendicular and parallel to the velocity vector, respectively, hence they are aligned with the z_s and x_s stability axes, respectively. The thrust is parallel to the x_b body axis of the aircraft, making an angle α with respect to the velocity. The accelerations of the aircraft center of mass, $a_{x,s}$ and $a_{z,s}$, are in the directions parallel and perpendicular to the flight path, respectively, where the subscript "s" denotes their alignment with the stability axes.

Applying Newton's second law perpendicular to the velocity (parallel to the z_s stability axis) we have

$$W\cos\gamma - L - T\sin\alpha = ma_{z,s} = mg\left(\frac{a_{z,s}}{g}\right) = Wn_{z,s} \qquad (3.298)$$

where m is the mass of the aircraft, g is the acceleration due to gravity, $a_{z,s}$ is the acceleration perpendicular to the velocity, and $n_{z,s}$ is the load factor perpendicular to the velocity.

Solving for the lift, we have

$$L = W\cos\gamma - T\sin\alpha - Wn_{z,s} = W(\cos\gamma - n_{z,s}) - T\sin\alpha \qquad (3.299)$$

Expressing the lift in terms of the lift coefficient, C_L, we have

$$C_L = \frac{W(\cos \gamma - n_{z,s}) - T \sin \alpha}{q_\infty S} \tag{3.300}$$

where q_∞ is the freestream dynamic pressure and S is the wing reference area.

Similarly, applying Newton's second law parallel to the freestream velocity (parallel to the x_s stability axis) we have

$$T \cos \alpha - D - W \sin \gamma = m a_{x,s} = mg \left(\frac{a_{x,s}}{g} \right) = W n_{x,s} \tag{3.301}$$

where $a_{x,s}$ is the acceleration parallel to the velocity, and $n_{x,s}$ is the load factor parallel to the freestream velocity.

Solving for the drag, we have

$$D = T \cos \alpha - W \sin \gamma - W n_{x,s} = T \cos \alpha - W(\sin \gamma + n_{x,s}) \tag{3.302}$$

Expressing the drag in terms of the drag coefficient, C_D, we have

$$C_D = \frac{T \cos \alpha - W(\sin \gamma + n_{x,s})}{q_\infty S} \tag{3.303}$$

Equations (3.300) and (3.303) provide the lift and drag coefficients, respectively, of the aircraft. All of the quantities on the right-hand side of these equations are known or can be measured in flight. The wing reference area, S, is known and is generally constant. The dynamic pressure, q_∞, is known for a given altitude and velocity. The angle-of-attack, α, and flight path angle, γ, can be measured in flight. The weight, W, is obtained from the known aircraft weight and by measuring the fuel weight in flight. To obtain the thrust, T, a thrust model of the engines is required. Typically, aircraft propulsion data, such as engine rpm, engine pressure ratio, or other parameters, are measured in flight and are input into the thrust model, along with the flight conditions. The load factors can be measured using accelerometers. However, accelerometers are typically mounted in a fixed orientation, aligned with the aircraft body axes. Therefore, the accelerometer data must be transformed from the body axes to the stability axes for use in Equations (3.300) and (3.303). The stability axes components of the body axes $n_{x,b}$ and $n_{z,b}$ accelerometer load factors are depicted in Figure 3.131.

The stability axes $n_{x,s}$ and $n_{z,s}$ load factors are calculated from the body axes $n_{x,b}$ and $n_{z,b}$ load factors using the equations below.

$$n_{x,s} = n_{x,b} \cos \alpha - n_{z,b} \sin \alpha \tag{3.304}$$

$$n_{z,s} = n_{x,b} \sin \alpha + n_{z,b} \cos \alpha \tag{3.305}$$

The lift and drag coefficients, calculated using Equations (3.300) and (3.303), respectively, are for one flight condition. Ideally, we would like to calculate these aerodynamic coefficients for a range of angles-of-attack so that we can generate the aircraft lift curve and drag polar. There are two fundamentally different aeromodeling flight test techniques that can be applied to accomplish this. In the *stabilized* aeromodeling methods, data is obtained using a series of trim shots, where the aircraft is stabilized at constant airspeed, altitude, angle-of-attack, and load factor. The airspeed or load factor is changed to obtain data at a different trim shot. Since the aircraft state is not changing, the data can be manually recorded, using hand-held data recording, without the need for a data acquisition system.

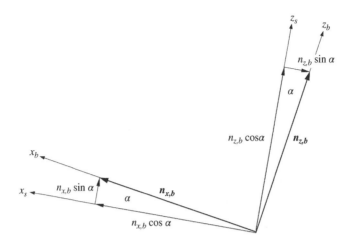

Figure 3.131   Transformation of accelerometer load factor from body axes to stability axes.

In the *dynamic* aeromodeling methods, the aircraft is maneuvered such that the load factor and angle-of-attack are rapidly changing, while maintaining the airspeed and altitude approximately constant. The maneuvers are flown within a predetermined altitude band, where the atmospheric properties do not change significantly, and the airspeed is controlled to within a tight tolerance. The dynamic methods provide much more data per maneuver than the stabilized methods, over a wide range of angle-of-attack. However, the aircraft parameters are changing too rapidly to record the data by hand, so a data acquisition system is required.

Obtaining aeromodeling flight data using stabilized methods is a much slower process than using the dynamic methods. Stabilized methods are slower in terms of the time it takes to collect data during the maneuver and in terms of the amount of data obtained per maneuver. The generation of a lift curve or drag polar can be a time-consuming process using this method, since data for only a single angle-of-attack can be obtained at each trim shot. We investigate both of these types of aeromodeling methods in flying this FTT.

To demonstrate the aeromodeling FTTs to obtain the lift curve and drag polar, you will be flying the Boeing F/A-18E *Super Hornet* supersonic jet aircraft, shown in Figure 3.132. The F/A-18E is a single-seat, twin-engine, supersonic fighter jet aircraft, designed for launching and landing on an aircraft carrier. The F/A-18E/F *Super Hornet* is a larger and more advanced design, evolved from the F/A-18 *Hornet*, described in Section 1.1.2. The first flight of the *Super Hornet* was on 29 November 1995. Selected specifications of the Super Hornet are given in Table 3.14.

3.9.4.1   Stabilized Aeromodeling Methods

You have climbed up to an altitude of 35,000 ft (10,700 m) and the aircraft is in wings-level, 1 *g* flight at Mach 0.8. You must take your time in setting up the trim shot for the data collection, as tight tolerances have been established to obtain high quality data. You are attempting to hold the steady state Mach number to within a tolerance of ±0.005 for three minutes. After a little more time to allow the engines to stabilize, the trim shot is stable at Mach 0.8, 35,000 ft, and a load factor of one. You have just set up the first stabilized aeromodeling method, called the *constant altitude aeromodeling* FTT. You hand-record the required data, including the velocity, pressure altitude, load factor, angle-of-attack, fuel weight, and engine parameters.

Figure 3.132 Boeing F/A-18 F *Super Hornet*, two-seat version of F/A-18E. (Source: *US Navy*.)

Table 3.14 Selected specifications of the Boeing F/A-18 *Super Hornet*.

Item	Specification
Primary function	All weather, supersonic fighter/attack jet aircraft
Manufacturer	Boeing, Seattle, Washington
First flight	29 November 1995
Crew	1 pilot
Powerplant	2 × F414-GE-400 afterburning turbofan engine
Thrust, MIL (ea. engine)	13,000 (62,000 N), military power
Thrust, MAX (ea. engine)	22,000 lb (98,000 N), maximum afterburner
Empty weight	~32,000 lb (14,500 kg)
Maximum takeoff weight	66,000 lb (30,000 kg)
Length	60 ft 1.25 in (18.31 m)
Height	16 ft (4.9 m)
Wingspan	44 ft 8.5 in (13.63 m)
Wing area	500 ft^2 (46 m^2)
Maximum speed	1190 mph (1915 km/h), Mach 1.7+
Service ceiling	>50,000 ft (>15,000 m)
Load factor limits	+7.5 g, −3.0 g

You now set up a steady state test point at a load factor greater than one. You roll the F-18E into approximately a 45° bank left turn and pull back slightly on the stick. You are really looking to set up a desired load factor, rather than a bank angle. Adding a little more left roll, you establish a load factor of 1.49, which is close to your 1.5 *g* target. The bank angle corresponding to the 1.49 g load factor is 46.8°. (The relationship between bank angle and load factor is discussed in

Chapter 5.) You adjust the power, as required, to stabilize at this elevated load factor test point. Once stable for several minutes, you record the data.

If you were taking data at a lower Mach number, it would not be possible to maintain constant altitude at higher load factors. You would then utilize the *descending turn aeromodeling* FTT, where you set up the 1 g trim point and then perform a constant power, descending turn at the higher load factor, while maintaining a constant Mach number. Here, you are simply trading potential energy (altitude) for kinetic energy (velocity) to maintain a constant Mach number at the desired load factor.

After completing several test points at Mach 0.8, you obtain data at Mach 0.9 and 1.2. This is a time consuming process, as it takes several minutes to obtain data at one flight condition. It is not practical to build up the entire lift curve and drag polar using these stabilized methods. You have collected steady state data at several different load factors, corresponding to several different angles-of-attack, at several Mach numbers. The data from the stabilized maneuvers is used to validate or *anchor* the dynamic aeromodeling data, which you collect next.

3.9.4.2 Dynamic Aeromodeling Methods

In setting up for flying the dynamic aeromodeling FTTs, you ensure that the data acquisition system is running, as you know that the parameters will be changing too fast to record by hand. The system is on and functioning properly (in fact, you had it on for the stabilized FTTs also, to collect that data). You will fly the dynamic maneuvers in an altitude data band of 1000 ft (300 m) and attempt to maintain a tolerance of ±0.01 on the Mach number. Since these are dynamic maneuvers, the tolerances are larger than what can be held for the stabilized FTTs.

You climb up, setting your altitude data band between 35,500 ft (10,800 m) and 34,500 ft (10,500 m), so that the heart of your 1000 ft data band is at 35,000 ft. You establish a wings-level, trim shot at an altitude of 35,000 ft, Mach 0.8, and a load factor of 1. The first dynamic FTT that you perform is called the *rollercoaster* or *pushover-pull-up* (*POPU*), which is descriptive of the aircraft trajectory during the maneuver, as shown in Figure 3.133a. Starting from steady, level flight, you slowly push over from 1 g to 0 g, followed by a slow pull-up to 2 g and then relax your back pressure on the stick, returning to 1 g, level flight. The angle-of-attack and lift coefficient decrease during the pushover, with decreasing load factor, and increase during the pull-up, with increasing load factor. You had no problem remaining within the desired altitude data band. Even though this is a dynamic maneuver, the duration is 8–16 s from start to finish with a 0.25–0.50 g/second onset rate. You fly the maneuver as smoothly as you can, with a sinusoidal variation of load factor, while holding constant power and Mach number. You ensure that there are no steps, ramps, or pulses in the trajectory of the aircraft. You are satisfied with your POPU, although you know that you can improve on the maneuver quality with a little practice.

Next, you set up for the *wind-up turn* dynamic aeromodeling FTT. Starting from a Mach 0.8 wings-level, trim shot at 35,500 ft, the top of the data band, you smoothly roll into a right turn, gradually increasing the bank angle and load factor, while maintaining constant power and Mach number, as shown in Figure 3.133b. To hold the Mach number constant, with increasing load factor, you let the aircraft descend, trading altitude for airspeed. You increase the load factor by pulling at a maximum onset rate of 1 g/second. As the maneuver progresses, the load factor builds and you begin to strain under the high g load. You have to lower the aircraft nose more and more, as the bank angle gets steeper and steeper. Finally, you are pulling 7 g, near the maximum 7.4 g load factor of the aircraft, with the aircraft pointing almost straight down, in a near vertical attitude. You knock it off and call the maneuver complete. Using the single *wind-up turn* FTT, you have collected aerodynamic data over a wide range of angle-of-attack, from the trim angle-of-attack, α_{trim}, up to

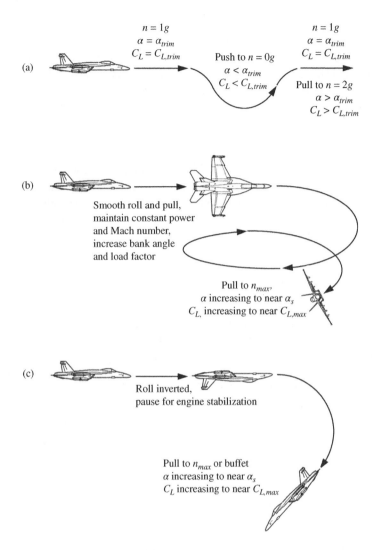

Figure 3.133 Dynamic aeromodeling flight test techniques, (a) pushover pull-up (POPU) or rollercoaster, (b) wind-up turn, and (c) inverted pull-up or split-S.

near the stall angle-of-attack, α_s, and from the trim lift coefficient, $C_{L,trim}$, to near the maximum lift coefficient, $C_{L,max}$.

The final dynamic aeromodeling FTT that you fly is the *inverted pull-up* or *split-S* FTT. This maneuver has higher dynamic effects than the POPU or wind-up turn, because it is completed much faster, in about 2–3 s. You again trim the aircraft for wings-level flight at Mach 0.8 and 35,500 ft. You then roll the F-18E inverted and pause for a moment, hanging in your seat straps. You then smoothly pull the stick back, pulling the aircraft nose down through the horizon, increasing the load factor to the 7.4 g limit, as shown in Figure 3.133c. You vary the g onset rate, by how slow or fast you are pulling back, to maintain a constant Mach number. Once you have reached the load factor limit, you recover the F-18E from the dive to level flight. Similar to the wind-up turn FTT, aerodynamic data was obtained over a wide range of angle-of-attack, up to near the stall angle-of-attack, α_s, and the maximum lift coefficient, $C_{L,max}$. Since the split-S aeromodeling FTT is an aerobatic maneuver,

Lift Coefficient ~ CL

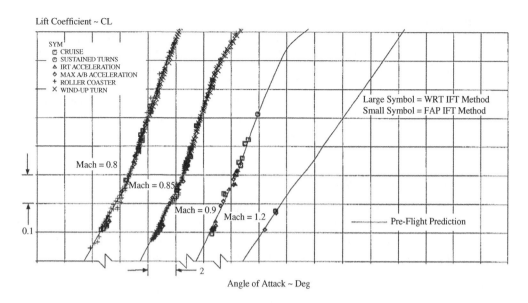

Figure 3.134   Boeing F/A-18E lift curve from flight test data. (Source: *Niewald and Parker, [54].*)

Lift Coefficient ~ CL

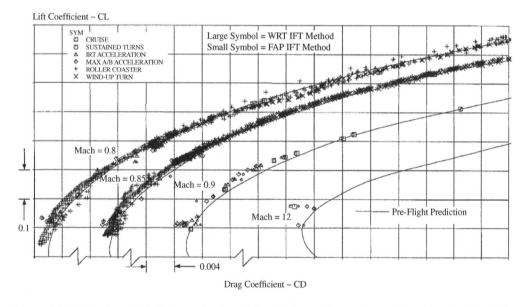

Figure 3.135   Boeing F/A-18E drag polar derived from flight test data. (Source: *Niewald and Parker, [54].*)

it is not appropriate for use in a non-aerobatic aircraft, such as a commercial airliner. However, the POPU and wind-up turn are used for non-aerobatic aircraft, within their load limits.

After your flight, the aerodynamic data that you collected using the stabilized and dynamic aeromodeling FTTs is used to plot the lift curve and drag polar for the F/A-18E, as shown in Figure 3.134 and Figure 3.135, respectively. In these figures, data from the 1 g, constant altitude FTT is labeled as "cruise" and labeled as "sustained turns" for the constant altitude or descending turn FTTs. The split-S data is not shown.

The lift and drag data was obtained for Mach 0.8, 08.5, 0.9, and 1.2. The data at higher angles-of-attack come from the dynamic FTTs, the rollercoaster, and the wind-up turn, since these FTTs are performed up to the maximum load factor. The stabilized FTT data is used to anchor the dynamic FTT data, which is seen to be in good agreement. A pre-flight prediction is seen to match favorably with the flight data. Despite the dynamic nature of the rollercoaster and wind-up turn FTTs, the data obtained is smooth with a small amount of scatter. The data scatter increases with increasing Mach number, as it becomes more difficult to maintain constant Mach number test point conditions at near transonic speeds.

3.9.5 Wings in Ground Effect

Aircraft must fly close to the ground when taking off and landing. When an aircraft is at or below a height above the ground of about one wingspan for an airplane, or about one rotor diameter for a rotorcraft, there is a favorable *ground effect* on the aircraft aerodynamics. In a fundamental sense, ground effect alters the pressure distribution over the aircraft wing, tail, and fuselage, which changes the lift and drag on these surfaces. The primary effect on the main wing is a reduction in the induced drag and an increase in the lift. During takeoff, a pilot can keep the aircraft close to the ground to take advantage of the increased lift and decreased drag in ground effect, allowing the aircraft to accelerate more efficiently to a safe flying speed. When landing, a pilot can sometimes have the feeling that the aircraft is "floating" in ground effect, due to the increase in lift and decrease in drag.

As an aircraft gets closer to the ground, the flow around the aircraft is altered due to the interaction of the flow with the ground. The upwash and downwash, ahead of and behind the aircraft, respectively, are reduced. The trailing vortices, flowing downstream from the aircraft wingtips, strike the ground and are reduced in strength, resulting in decreased induced drag and reduced downwash. The decrease in induced drag results in higher lift, an increase in airspeed, and reduced thrust required to maintain a constant airspeed. The lower strength wingtip vortices have the effect of making the wing act like a wing with a higher aspect ratio. If we assume that the aircraft maintains a constant lift coefficient and constant airspeed as it descends into ground effect, the wing is at a lower angle-of-attack in ground effect due to the decrease in the induced drag. A smaller angle-of-attack is required in ground effect to produce the same lift coefficient as out of ground effect. Conversely, if the angle-of-attack and airspeed are constant while entering ground effect, a higher lift coefficient is obtained, hence producing the effect of "floating" during landing. This may also be stated as: the in-ground effect lift curve slope is steeper than the out of ground effect slope. These effects are shown in Figure 3.136 where the lift and induced drag coefficients for a 4° swept wing, in and out of ground effect, are plotted versus angle-of-attack.

Ground effect also has an impact on the aircraft stability and control. The reduced downwash from the wing increases the local angle-of-attack seen by the tail, resulting in a nose-down pitching moment. These effects on stability and control are discussed in Chapter 6. The changes to the local flow field around the aircraft, mentioned above, may also affect the static pressure measurement, causing errors in the airspeed indication. (We discuss how airspeed is calculated, using the aircraft static pressure measurement, in Chapter 5.) Other factors that impact the intensity of ground effect include the aircraft wing-mounting configuration and the type of ground surface. An airplane with a low-mounted wing configuration is influenced more by ground effect than a high-wing airplane. This is simply because the low-mounted wing is closer to the ground during takeoff and landing. The type of ground surface also affects the magnitude of the ground effect on the aircraft. A smooth, hard surface, such as a concrete runway, has a greater impact than a rough, soft, or unprepared surface, such as water or grass.

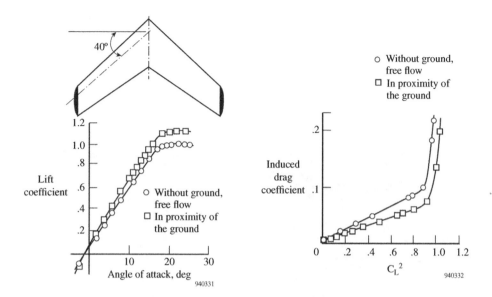

Figure 3.136   Change in lift and induced drag coefficients due to ground effect. (Source: *Corda, NASA TM 4604, 1994, [24]*.)

Early aviators realized the benefits of flying in ground effect. As shown in Figure 3.137, Charles Lindbergh was known to fly low in ground effect to take advantage of improved aircraft performance and range. He commented on this in his autobiographical account of his historic crossing of the Atlantic Ocean in the New York-to-Paris flight.

> As the fog cleared I dropped down closer to the water, sometimes flying within ten feet of the waves and seldom higher than two hundred. There is a cushion of air close to the ground or water through which a plane flies with less effort than when at a higher altitude, and for hours at a time I took advantage of this factor.[6]

Of course, this may not be a prudent operational strategy in today's aviation airspace, as flying in ground effect for long distances presents additional hazards, including colliding with high-rising terrain or manmade obstacles.

It is interesting that the advantages of aerodynamic ground effect are found in nature. Pelicans, sea gulls, cormorants, black skimmers, sea ducks, and sandpipers are among the many birds that take advantage of flight in ground effect. By flying close to the surface of the water, often well below the height of their wingspan, sea birds are able to glide a hundred feet or more without flapping their wings. Often, birds fly in ground effect over water as part of their foraging technique.

The brown pelican (*Pelecanus occidentalis*), a common coastal bird found in the western and southern USA, often flies in ground effect over the water when foraging. Although a clumsy land animal, brown pelicans are graceful in the air. With a wingspan of about 6–8 ft (1.8–2.4 m), they often fly in groups, just above the water surface, in a "V" formation or single line. Upon seeing their prey, the brown pelican dives into the water, bill first, scooping up a fish or crustacean.

The black skimmer (*Rliyncops nigra*), a tern-like seabird found in North and South America, has a foraging habit that takes advantage of ground effect, which is very different from that of

---

[6] Charles A. Lindbergh, *We* (New York: G.P. Putnam's Sons, 1955), pp. 206.

Figure 3.137 Charles Lindbergh flying the *Spirit of St Louis* in ground effect over Yellowstone Lake in Wyoming. (Source: *Courtesy of the National Air and Space Museum, Smithsonian Institution, with permission.*)

the brown pelican. With high aspect ratio wings, skimmers fly in large flocks, low over the water surface, with their lower bill or mandible skimming the water, as shown in Figure 3.138. Using this technique, the black skimmer scoops up small fish, insects, crustaceans, and other prey. By flying in ground effect, the black skimmer flies more efficiently and conserves energy as it feeds.

Over the years, there have been several attempts to build vehicles that operate solely in ground effect. These *wing-in-ground effect* (WIG) vehicles are designed so that their wings take advantage of ground effect, lifting them off the ground into ground effect, but unable to fly in free flight above ground effect. One can debate whether these WIG vehicles are "low-flying" aircraft or surface ships. One of the largest WIG vehicles ever built was the Russian Alexeyev A-90 ekranoplan ground effect vehicle, shown in Figure 3.139 (mounted on pylons), operated by the Russian navy in the 1980s and 1990s. The A-90 vehicle could fly in ground effect above the water at a height of up to about 3000 m (9,840 ft). This huge ground effect vehicle was about 58 m (190 ft) in length with a wingspan of about 31.5 m (103 ft) and had a maximum takeoff weight of 140,000 kg (309,000 lb). Powered by two turbofan jet engines and a tail-mounted turboprop engine, the vehicle could fly in ground effect at a cruise speed of about 215 knots (400 km/h, 249 mph). Designed as a troop transport, the A-90 had a crew of 6 and could transport 150 personnel.

Let us now examine the quantitative changes to the induced drag due to ground effect. The change in the induced drag due to ground effect, $\Delta C_{D_{i,GE}}$, can be expressed as

$$\Delta C_{D_{i,GE}} = C_{D_{i,GE}} - C_{D_i} = -\sigma \, C_{D_i} \tag{3.306}$$

where σ is defined as the ground influence coefficient and C_{D_i} is the induced drag coefficient out of ground effect. The ground influence coefficient is typically a function of the ratio of the height

Figure 3.138 Black skimmer (*Rliyncops nigra*) flying in ground effect, skimming the water for food. (Source: *Dick Daniels, "Black Skimmer" https://commons.wikimedia.org/wiki/File:Black_Skimmer_RWD3 .jpg, CC-BY-SA-3.0. License at https://creativecommons.org/licenses/by-sa/3.0/legalcode.*)

Figure 3.139 Russian Lun-class ekranoplan ground effect vehicle, Alexeyev A-90 *Orlyonok*. (Source: *Alan Wilson, "Alexeyev A-90 Orlyonok" https://commons.wikimedia.org/wiki/File:Alexeyev_A-90_Orlyonok_ (26_white)_(8435214563).jpg, CC-BY-SA-2.0. License at https://creativecommons.org/licenses/by-sa/2.0/ legalcode.*)

above the ground to the wingspan, h/b. Using Equation (3.306), the induced drag coefficient in ground effect, $C_{D_{i,GE}}$, is given by

$$C_{D_{i,GE}} = C_{D_i} + \Delta C_{D_{i,GE}} = C_{D_i} - \sigma C_{D_i} = (1 - \sigma) C_{D_i} \tag{3.307}$$

From Equation (3.307), the ratio of the induced drag in ground effect to the induced drag out of ground effect is

$$\frac{C_{D_{i,GE}}}{C_{D_i}} = 1 - \sigma \tag{3.308}$$

There are several different ground influence coefficients that have been proposed by different researchers, based on either empirical flight data or theoretical foundations. Wetmore and Turner defined a ground influence coefficient as an exponential function of the height-to-wingspan ratio, based on data from towed glider flight tests ([73]), given by

$$\sigma_{Wetmore\ \&\ Turner} = e^{\left[-2.48\left(\frac{2h}{b}\right)^{0.768}\right]} \tag{3.309}$$

Lan and Roskam cite an empirically based ground influence coefficient, which is defined for the region $0.033 < (h/b) < 0.25$ ([48]), as follows.

$$\sigma_{Lan\ \&\ Roskam} = \frac{1 - 1.32\left(\frac{h}{b}\right)}{1.05 + 7.4\left(\frac{h}{b}\right)} \tag{3.310}$$

A theoretically based ground influence coefficient is given by McCormick ([51]), as follows.

$$\sigma_{McCormick} = 1 - \frac{\left(16\frac{h}{b}\right)^2}{1 + \left(16\frac{h}{b}\right)^2} \tag{3.311}$$

Equation (3.311) is based on the Biot–Savart law, providing a mathematical expression for the velocity field associated with a vortex.

The ratios of the induced drag in and out of ground effect, using the three different ground influence coefficients given in Equations (3.309), (3.310), and (3.311), are compared in Figure 3.140. All three ratios are in good agreement very close to the ground, below an h/b of about 0.05, predicting that the induced drag in ground effect is only about 30% of the out-of-ground effect induced drag. Above this height, Westmore and Turner predicts a greater reduction in induced drag due to ground effect than McCormick. The Lan and Roskam predictions are nearly identical with Westmore and Turner, over the height range that the Lan and Roskam coefficient is defined. As the height approaches one wingspan, ground effect is predicted to have a small influence of about one percent or less decrease of the induced drag.

Ground effect data may be obtained from wind tunnel or flight tests. Typically, *static ground effects* data is obtained in a wind tunnel, where measurements are taken on a stationary aircraft model, mounted at fixed heights above the tunnel floor. This type of data simulates the aircraft flying at a constant height above the ground, rather than the real, dynamic situation where the height above the ground is constantly changing, as the aircraft is ascending or descending near the ground. This real situation of *dynamic ground effect* can be simulated in the wind tunnel by using a model that is moved toward or away from a ground plane. Flight tests have also been conducted to obtain dynamic ground effects data on a variety of aircraft. Figure 3.141 shows the change in the

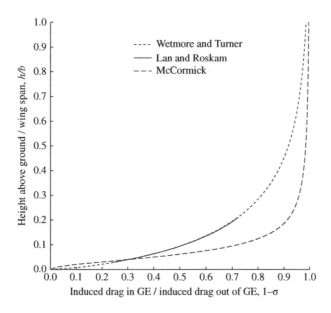

Figure 3.140   Ratios of induced drag, in and out of ground effect (GE).

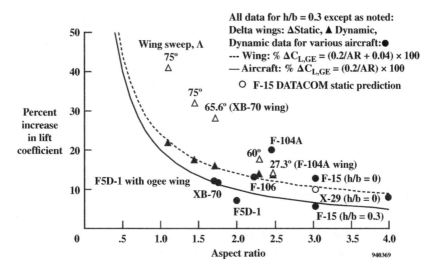

Figure 3.141   Change in lift coefficient due to ground effect for various wings and for various aircraft. (Source: *Corda, NASA TM 4604, 1994, [24].*)

lift coefficient for various wings and for various aircraft, due to ground effect. The percent change in the lift coefficient due to ground effect, %ΔC_{LGE}, has been correlated with aspect ratio, *AR*, for wings alone and full aircraft configurations.

3.10   Compressible, Subsonic and Transonic Flows

For most of our aerodynamic discussions thus far, we have assumed that the flow is subsonic, inviscid, and incompressible. This laid our foundation of aerodynamic theory for low-speed flight, similar to the historical development of aerodynamics as applied to aircraft design. Incompressible flow theory was adequate for application to World War I aircraft with their top speeds of about 120–140 mph (190–220 km/h, Mach < 0.2). During World War II, the speeds of propeller-driven aircraft were well over 400 mph (640 km/h, Mach ~0.5) and the new jet-propelled aircraft would reach close to 600 mph (970 km/h, Mach ~0.8). The higher aircraft speeds also affected propellers, where propeller blade tip speeds were approaching the speed of sound. At these higher speeds, incompressible theory was unable to predict the flight physics accurately.

Now in the realm of *compressible flow*, the air density could not be considered constant, complicating the theoretical aerodynamic picture. These *compressibility effects* result in an increase in the drag coefficient (at a given lift coefficient or angle-of-attack), a decrease in the lift coefficient, and a change in the pitching moment coefficient. These compressibility effects are not restricted to airfoils, as they impact other aircraft components with thickness or which produce lift. The fuselage, horizontal and vertical tail surfaces, engine nacelles, and canopy are examples of the components affected by compressibility. The pressure distributions and hence the forces and moments on these surfaces are changed as Mach one is approached. There are also other aerospace applications where compressibility effects occur and this is an important design consideration. Several other types of rotating aerodynamics, such as the rotating tips of helicopter blades or the blade tips of rotating jet engine machinery, are examples of this. The rotating nature of these types of flows make them even more complex, but the fundamental physics of compressible flow is the same as found on a non-rotating airplane wing.

In this section, we discuss several aspects of high-speed flight, approaching the speed of sound or Mach one, where compressibility effects become important. As the speed of sound is approached, there are drastic changes to the flow field, most notably the appearance of shock waves and regions of separated flow. The pressure distribution over an aircraft is significantly changed due to these phenomena, resulting in increasingly large changes to the forces and moments on the body as the airspeed is increased. We first discuss more details about the nature of the speed of sound, then discuss the changes to the flow field as Mach one is approached, followed by some of the ways that incompressible flow theory can be corrected to predict compressible flows. Finally, we discuss the "sound barrier" and blaze a path towards supersonic flight.

3.10.1   The Speed of Sound

The speed of sound plays an increasingly important role in aerodynamics as the flight speed increases. We already know that the Mach number is an important similarity parameter relating the velocity to the speed of sound. In Chapter 2, it was stated that the speed of sound is proportional to the square root of the static temperature. In this section, we prove this and learn more about the speed of sound.

We hear sound because of the small pressure changes associated with sound waves that vibrate our eardrum. Thus, if we consider a sound wave, moving through the room at the speed of sound, the flow properties are changed by a small amount as they pass through the wave. It is easier to think about this from a frame of reference where the wave is stationary and the flow upstream of the wave has a velocity, u, equal to the speed of sound, a, as shown in Figure 3.142. The flow upstream of the sound wave is at a pressure, p, density, ρ, and temperature, T. After passing through the sound wave, these flow properties, including the flow velocity, are changed by a small amount, given by da, dp, $d\rho$, and dT. The velocity, pressure, density, and temperature downstream of the sound wave are given by $a + da$, $p + dp$, $\rho + d\rho$, and $T + dT$, respectively. We are assuming that the flow is compressible, as the density of the flow changes as it passes through the sound wave.

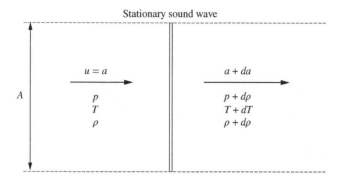

Figure 3.142 Stationary sound wave in a compressible flow.

We assume that the flow is uniform and bounded by a streamtube of constant area A, such that the mass flow in and out of the streamtube, through the sound wave is constant. Applying the continuity equation for conservation of mass, Equation (3.143), to the flow upstream and downstream of the sound wave, we have

$$\rho a A = (\rho + d\rho)(a + da)A \tag{3.312}$$

Simplifying and expanding terms, we have

$$\rho a = (\rho + d\rho)(a + da) = \rho a + a d\rho + \rho da + d\rho da \tag{3.313}$$

The product of the small change in density, $d\rho$, and the small change in the sound velocity, da, is much smaller than the other terms in Equation (3.313) and is therefore neglected. Thus, Equation (3.313) is

$$\rho a = \rho a + a d\rho + \rho da \tag{3.314}$$

$$a d\rho = -\rho da \tag{3.315}$$

Solving for the speed of sound, we have

$$a = -\rho \frac{da}{d\rho} \tag{3.316}$$

Now, using Euler's equation, the momentum equation for a compressible flow, Equation (3.165), we have an expression relating the pressure change, dp, to the change in the speed of sound, da.

$$dp = -\rho a da \tag{3.317}$$

Solving for the change in the speed of sound, we have

$$da = -\frac{dp}{\rho a} \tag{3.318}$$

Inserting Equation (3.318) into (3.316), we have

$$a = -\rho \frac{(-dp/\rho a)}{d\rho} = \frac{1}{a} \left(\frac{dp}{d\rho} \right) \tag{3.319}$$

Solving for the speed of sound, we have

$$a = \sqrt{\frac{dp}{d\rho}} \tag{3.320}$$

The flow through the sound wave does not involve any heat transfer, thus the flow is adiabatic. The changes or gradients of the flow properties are small; in fact, they may be considered infinitesimal changes, so that there are no dissipative phenomena due to gradients in the flow properties. Recall that gradients in the velocity result in viscous losses and gradients in the temperature result in thermal losses, as discussed in Section 3.2.6. Therefore, the flow may be considered adiabatic and reversible, or isentropic. Thus, the isentropic relation, Equation (3.138), is valid, repeated below.

$$\frac{p}{\rho^\gamma} = \text{constant} \equiv c \tag{3.321}$$

or, solving for the pressure

$$p = c\rho^\gamma \tag{3.322}$$

Taking the derivative of the pressure with respect to the density, we have

$$\frac{dp}{d\rho} = \frac{d}{d\rho}(c\rho^\gamma) = c\gamma\rho^{\gamma-1} \tag{3.323}$$

Inserting Equation (3.321) into (3.323) for the constant c, we have

$$\frac{dp}{d\rho} = \frac{p}{\rho^\gamma}(\gamma\rho^{\gamma-1}) = \frac{\gamma p}{\rho} \tag{3.324}$$

Substituting Equation (3.324) into (3.320), the speed of sound is given by

$$a = \sqrt{\frac{\gamma p}{\rho}} \tag{3.325}$$

Using the perfect gas equation of state, Equation (3.61), we can rewrite Equation (3.325) as

$$a = \sqrt{\gamma R T} \tag{3.326}$$

where the ratio of specific heats, γ, and the specific gas constant, R, are properties of the gas, and T is the gas static temperature. Thus, we see that for a calorically perfect gas, with constant specific heats, the speed of sound is only a function of the temperature. Based on this equation, the speed of sound, in air at standard sea level conditions, is calculated as

$$a_{SL} = \sqrt{\gamma R T} = \sqrt{1.4 \left(1716 \frac{\text{ft} \cdot \text{lb}}{\text{slug} \cdot \,^\circ\text{R}}\right)(519\,^\circ\text{R})} = 1116.6\,\text{ft/s} \tag{3.327}$$

Perhaps the earliest known prediction of the speed of sound in air was made by Isaac Newton in his *Principia*, where he calculated the speed of sound at sea level to be 882 mph, about 16% higher than the currently accepted value of 760 mph (1120 ft/s, 340 m/s, 1220 km/h) at standard, sea level conditions. Many early predictions of the speed of sound were based on calculating the time it took for the sound of a gunshot or cannon blast to cover the known distance between the location where the shot was fired and the observer. The flash of the firing was first observed and then the time it took for the sound to reach the observer was measured. The speed of sound was obtained by dividing the known distance by this time. Amazingly, several of these early calculations were within about 1% of the correct value. Deficiencies with this type of measurement include the

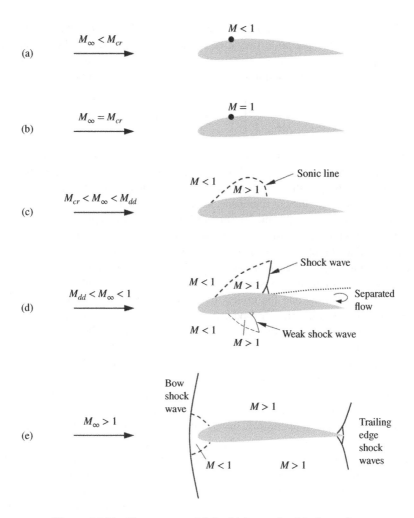

Figure 3.143 Flow over an airfoil with increasing Mach number.

changes in the properties of the air, such as the density and temperature, over the long distance and the inaccuracy of the time measurement.

3.10.2   The Critical Mach Number and Drag Divergence

Consider the subsonic flow over a cambered airfoil, where the freestream Mach number, M_∞, is low, as shown in Figure 3.143a. The flow accelerates over the top surface of the airfoil, such that the local Mach number, M, at a point on the top surface is higher than the freestream value, but it remains subsonic, below Mach one. As the freestream Mach number is increased, the local Mach number at the point on the top surface increases. At some subsonic freestream Mach number, the local Mach number at this point just reaches sonic conditions, as shown in Figure 3.143b. This sonic point is also the location of minimum pressure and maximum velocity on the airfoil surface.

The pressure coefficient corresponding to this minimum pressure point is called the *critical pressure coefficient*, $C_{p,cr}$. Since the pressure coefficient is based on the difference between the local pressure and the freestream pressure, $p - p_\infty$, the pressure coefficient has its maximum *negative* value at this minimum pressure point (negative since the local pressure is below the freestream pressure). An expression for the critical pressure coefficient, as a function of the freestream Mach number, can be obtained from the definition of the pressure coefficient, with the result

$$C_{p,cr} = \frac{2}{\gamma M_\infty^2} \left\{ \left[\frac{2 + (\gamma - 1) M_\infty^2}{\gamma + 1} \right]^{\gamma/(\gamma-1)} - 1 \right\} \qquad (3.328)$$

The critical pressure coefficient is independent of the geometry or angle-of-attack of the airfoil or body. Based on this expression, the critical pressure coefficient decreases in magnitude (becomes a smaller negative number) as the freestream Mach number increases. This makes physical sense since the critical pressure coefficient represents the change in the local pressure, $p - p_\infty$, required to accelerate the flow from the subsonic freestream Mach number to sonic conditions. The higher the freestream Mach number, the smaller the pressure difference required to achieve Mach one.

The freestream Mach number, corresponding to the first occurrence of sonic flow anywhere on the airfoil, is called the *critical Mach number*, M_{cr}. The critical Mach number can be considered a boundary between subsonic and transonic flight, since beyond this Mach number, there is mixed subsonic and supersonic flow and the effects of compressibility begin. The actual value of the critical Mach number, for an airfoil or body, is dependent on the geometry and angle-of-attack of the particular airfoil or body. In general, a thin airfoil or slender body has a higher critical Mach number than a thick airfoil or non-slender body. The critical Mach number is higher at lower angle-of-attack, regardless of geometry, as this represents a smaller disturbance to the flow, similar to the contrast between a thin and thick airfoil.

Let us continue to increase the freestream Mach number beyond the critical Mach number. The sonic point spreads to a small region or "bubble" of supersonic flow on the airfoil upper surface, as shown in Figure 3.143c. The transitions between subsonic and supersonic flow are smooth across the "bubble", bounded by a weak sonic line. The surface pressure distribution, aerodynamic forces, and aerodynamic moments are not strongly affected by this small bubble of supersonic flow.

As the freestream Mach number is further increased, the flow continues to accelerate over the airfoil, creating "pockets" of supersonic flow on the upper and lower surfaces, as shown in Figure 3.143d. The transition from the subsonic freestream to supersonic flow is still smooth, but the transition back from supersonic to subsonic flow is now discontinuous, through a shock wave. The supersonic pocket is larger and the shock wave is stronger on the airfoil upper surface, as compared to the lower surface, due to the greater camber on the upper surface. There is a large pressure increase across the shock wave, creating an adverse pressure gradient along the surface, which separates the boundary layer. In addition to the losses created by the shock waves themselves, there is a large increase in the pressure drag due to the boundary layer separation. The freestream Mach number at which the drag increases dramatically is called the *drag divergence Mach number*, M_{dd}. The significantly altered pressure distribution also results in large changes of the lift and moment coefficients. The drag divergence Mach number can be significantly higher than the critical Mach number, by several tenths of a Mach, but it is less than one. Thus, it can be stated that

$$M_{cr} < M_{dd} < 1 \qquad (3.329)$$

When the freestream flow reaches low supersonic speed, a bow shock wave forms in front of the blunt nose of the airfoil, as shown in Figure 3.143e. There is a subsonic pocket behind the normal part of the shock wave, at the airfoil nose, while the flow remains supersonic downstream of the oblique sections of the shock wave. There are shock waves at the airfoil trailing edge, which change the direction of the flow, above and below the airfoil, to make it parallel with the freestream flow.

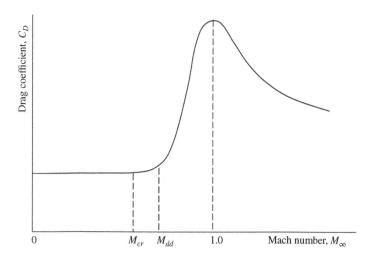

Figure 3.144 Drag coefficient versus Mach number, highlighting the critical and drag divergence Mach numbers.

A generic plot of the drag coefficient versus Mach number, from low subsonic speeds to beyond Mach one, is shown in Figure 3.144. The drag coefficient is relatively constant, from low subsonic, incompressible Mach numbers through the critical Mach number. At the drag divergence Mach number, the drag coefficient increases rapidly to a maximum value at or near Mach one. The maximum drag coefficient does not necessarily reach a maximum at *exactly* Mach one, as it may be slightly less or greater. The drag coefficient then decreases for increasing supersonic Mach numbers. Later in this chapter, we discuss several ways to decrease transonic drag, or delay the onset of drag divergence.

Previously we defined the transonic flow regime as about Mach 0.8–1.2. A more precise definition of the transonic flow boundaries is the range between the critical Mach number, when supersonic flow first appears somewhere on the surface, and the higher Mach number, when the flow is predominantly supersonic, as depicted by Figure 3.143b–e. Between these Mach number boundaries, the flow is a mix of subsonic and supersonic flow.

Figure 3.145 shows the transonic flow over an airfoil in a wind tunnel using a flow visualization technique called schlieren photography, an optical technique that makes the density gradients in the flow visible. The flow properties, including the density, pressure, and temperature, increase discontinuously across a shock wave, so schlieren imaging makes these waves visible. The shock waves on the upper and lower surfaces are clearly evident, as are the separated flow regions (white areas with the appearance of smoke on the top and bottom). This photograph also shows the complex structure of transonic flow.

3.10.3 Compressibility Corrections

Given the large amount of theoretical work and experimental data that existed about incompressible flow in the 1920s, the first approach to address the effects of compressibility was to apply *compressibility corrections* to the existing incompressible theory and experimental database to match the compressible, high-speed flight data. The first compressibility correction to be introduced was the *Prandtl–Glauert rule*, developed independently by Ludwig Prandtl in Germany and Hermann Glauert in England. Glauert published a derivation of his compressibility correction in 1927, while

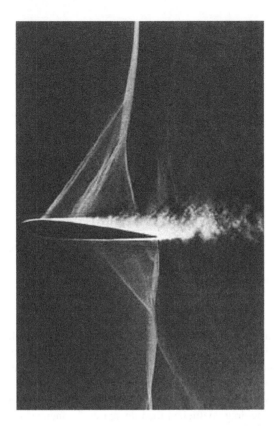

Figure 3.145 Schlieren photograph of transonic flow over an airfoil in a wind tunnel. (Source: *Baals and Corliss, NASA SP-440, 1981, [14].*)

it is purported that Prandtl discussed his similar results in lectures at Gottingen in Germany in the early 1920s. The Prandtl–Glauert rule is derived by assuming that the flow disturbances caused by the airfoil or body are small (the *small disturbance* assumption) and by approximating the non-linear, inviscid equations of fluid flow, the Euler equations, by linear equations. These simplifying assumptions limit the validity of the Prandtl–Glauert rule to thin airfoils or slender bodies at small to moderate angle-of-attack and Mach numbers below the critical Mach number.

The Prandtl–Glauert rule is an amazingly simple correction to incompressible flow, with the compressible pressure coefficient, C_p, given by

$$C_p = \frac{C_{p,0}}{\beta} \tag{3.330}$$

where $C_{p,0}$ is the incompressible pressure coefficient. β is defined as the Prandtl–Glauert compressibility correction factor, given by

$$\beta \equiv \sqrt{1 - M_\infty^2} \tag{3.331}$$

where M_∞ is the freestream Mach number.

Prandtl–Glauert compressibility corrections for the lift coefficient, c_l, and the moment coefficient, c_m, are given by

$$c_l = \frac{c_{l,0}}{\beta} \tag{3.332}$$

$$c_m = \frac{c_{m,0}}{\beta} \tag{3.333}$$

where $c_{l,0}$ and $c_{m,0}$ are the incompressible values of the lift and moment coefficients, respectively. The incompressible values of the lift and moment coefficients are a function of angle-of-attack. Based on thin airfoil theory, the incompressible lift coefficient of a thin, symmetric airfoil is equal to $2\pi\alpha$, as given by Equation (3.267), so the Prandtl–Glauert rule for the lift coefficient of a thin, symmetric airfoil is

$$c_l = \frac{2\pi\alpha}{\beta} \tag{3.334}$$

The compressible lift curve slope, $dc_l/d\alpha$, based on the Prandtl–Glauert rule, is then given by

$$\frac{dc_l}{d\alpha} = \frac{2\pi}{\beta} \text{ per radian} = \frac{2\pi}{57.3\beta} \text{ per degree} \tag{3.335}$$

The compressible lift coefficient, calculated using the Prandtl–Glauert rule, is compared with wind tunnel data for a NACA 0012 airfoil at 2° angle-of-attack in Figure 3.146. The compressible lift coefficient is calculated in two ways: (1) using Equation (3.332) with the incompressible coefficient, $c_{l,0}$, equal to 0.151, obtained from the experimental data, and (2) using symmetric, thin airfoil theory, Equation (3.334), where the incompressible lift coefficient is equal to $2\pi\alpha$. The critical Mach number, M_{cr}, is plotted as a function of lift coefficient. The wind tunnel data is shown for angles-of-attack 0–6°, in 1° increments.

The Prandtl–Glauert compressibility corrections, shown in Figure 3.146, indicate an increase in the lift coefficient with increasing Mach number. This is in general agreement with the wind tunnel

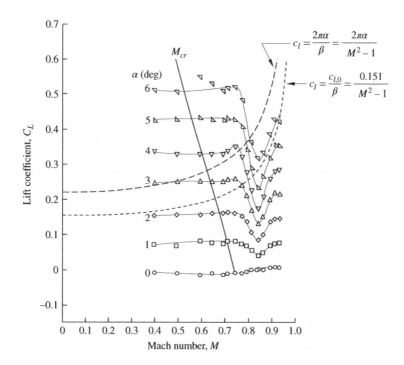

Figure 3.146 Comparison of Prandtl–Glauert rule (2° angle-of-attack) with wind tunnel data for NACA 0012 airfoil. (Source: *Adapted from Ferri, NACA Report L5E21, 1945, [27]*.)

data up to about Mach 0.75, which is near the critical Mach number, although the increase is not as great. Recall that the Prandtl–Glauert approximation is valid only up to the critical Mach number, which is predicted to be about Mach 0.66, based on the correction using Equation (3.328). The wind tunnel data indicates a slightly higher critical Mach number of about 0.68. Thus, it is not surprising that the lift coefficients using the Prandtl–Glauert corrections do not match the wind tunnel data trends above the critical Mach number. Also, the thin airfoil theory-based Prandtl–Glauert prediction is farther from the wind tunnel data than the prediction using the wind tunnel incompressible lift coefficient. This may be expected since the thin airfoil theory calculation is generally applicable to any thin airfoil at any "small" angle-of-attack.

It is worthwhile to pause and digest the wind tunnel data for the change in the lift coefficient, as the Mach number approaches one. The lift coefficient is relatively constant at low Mach numbers, where it can be assumed that the values at Mach 0.4 represent the incompressible values. The lift coefficient is constant up to about Mach 0.5–0.6, then it starts to increase slightly with increasing Mach number. At about Mach 0.75, there is a rapid loss of lift, with the lift coefficient decreasing until about Mach 0.85. The lift coefficient then increases and levels off, at least for the smaller angles-of-attack, at about Mach 0.9 to 0.92, but at a lower magnitude than the incompressible value.

Improved compressibility corrections have been developed, which are slightly more complicated. These include the Karman–Tsien rule, given by

$$C_p = \frac{C_{p,0}}{\beta + \frac{1}{2}\left(\frac{M_\infty^2}{1+\beta}\right)C_{p,0}} \tag{3.336}$$

and Laitone's rule, given by

$$C_p = \frac{C_{p,0}}{\beta + \frac{1}{2}M_\infty^2\left[\frac{1+\left(\frac{\gamma-1}{2}\right)M_\infty^2}{\beta}\right]C_{p,0}} \tag{3.337}$$

The three compressibility corrections are compared in Figure 3.147, assuming an incompressible pressure coefficient, $C_{p,0}$, of -1.0. The Laitone correction predicts the greatest absolute magnitude increase of the pressure coefficient due to compressibility, followed by the Karman–Tsien and the Prandtl–Glauert corrections. The critical pressure coefficient, as a function of Mach number, from Equation (3.328), is also plotted in Figure 3.147. The critical Mach number is obtained at the intersection of this critical pressure coefficient curve with the pressure coefficient curve for the particular compressibility correction (shown by the arrows from the curve intersections to the Mach number horizontal axis in Figure 3.147). The lowest critical Mach number of about 0.56 is predicted by the Laitone correction, followed by a critical Mach numbers of about 0.585 and 0.605 from the Karman–Tsien and Prandtl–Glauert corrections, respectively. While the exact values of the critical Mach number are dependent on the hypothetical incompressible pressure coefficient of -1.0 selected for this example, the trends of the relative magnitudes of the critical Mach number for the different corrections remains the same.

As a reminder, there are limitations to the compressibility corrections. Since they are based on linear theory, they are valid only up to the critical Mach number, before sonic or supersonic flow appears on the airfoil. The corrections are not valid for transonic flow, between about Mach 0.8–1.2, where the flow is a highly non-linear mix of subsonic and supersonic regions. All of these compressibility corrections assume an inviscid flow, where there is no accounting for drag due to skin friction or pressure drag due to flow separation. Thus, there are no compressibility corrections for the drag coefficient. In the past, the identification of the compressibility effects on the drag

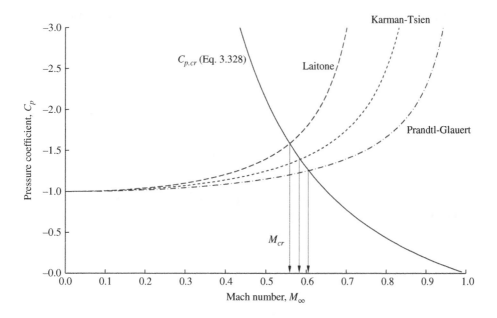

Figure 3.147   Comparison of compressibility corrections.

coefficient relied on experimental data collected in the wind tunnel. This was a difficult task, as there was considerable uncertainty about the accuracy of the wind tunnel data collected at high subsonic speeds due to tunnel interference effects.

Example 3.14   Prandtl–Glauert Compressibility Correction for an Airfoil   *Calculate the lift coefficient and moment coefficient, about the quarter chord point, for a NACA 2412 airfoil in a Mach 0.74 flow at an angle-of-attack of 3°.*

Solution

From Figure 3.100, the low-speed, incompressible lift coefficient, $c_{l,0}$, and moment coefficient, about the quarter chord point, $c_{m,c/4,0}$, of the NACA 2412 airfoil at 3° angle-of-attack, are 0.36 and −0.080, respectively.

From Equation (3.331), the Prandtl–Glauert compressibility correction factor is

$$\beta \equiv \sqrt{1 - M_\infty^2} = \sqrt{1 - (0.74)^2} = 0.6726$$

Using Equation (3.332), the lift coefficient at Mach 0.74 is

$$c_l = \frac{c_{l,0}}{\beta} = \frac{0.36}{0.6726} = 0.535$$

Using Equation (3.333), the moment coefficient at Mach 0.74 is

$$c_m = \frac{c_{m,0}}{\beta} = \frac{-0.080}{0.6726} = -0.119$$

Example 3.15 Compressibility Correction for an Airfoil Based on Thin Airfoil Theory *Using thin airfoil theory, calculate the lift coefficient and the lift curve slope for a thin airfoil in a Mach 0.74 flow at an angle-of-attack of 3°.*

Solution

Based on thin airfoil theory, the incompressible lift coefficient is given by

$$c_{l,0} = 2\pi\alpha = 2\pi \left(3 \, deg \times \frac{\pi}{180} \right) = 0.329$$

From Example 3.14, the Prandtl–Glauert compressibility correction factor, β, at Mach 0.74 is 0.6726. Based on thin airfoil theory, the Prandtl–Glauert rule for the lift coefficient is given by Equation (3.334). The lift coefficient at Mach 0.74 is

$$c_l = \frac{2\pi\alpha}{\beta} = \frac{0.329}{0.6726} = 0.489$$

Based on thin airfoil theory, the incompressible slope of the lift curve is given by

$$\frac{dc_l}{d\alpha} = 2\pi = 6.28 \, per \, radian = 0.109 \, per \, deg$$

Using Equation (3.335), the lift curve slope at Mach 0.74, based on thin airfoil theory, is

$$\frac{dc_l}{d\alpha} = \frac{2\pi}{\beta} = \frac{2\pi}{0.6726} = 9.34 \, per \, radian = 0.163 \, per \, deg$$

3.10.4 The "Sound Barrier"

To fear the unknown is, perhaps, human nature. In the early 1940s, a major unknown in aviation was whether an aircraft could fly faster than the speed of sound. The fear associated with the unknown of supersonic flight was manifested in the more frequent occurrences of airplanes having controllability problems and catastrophic crashes, as they flew faster and faster, approaching the speed of sound. Some believed that there was an impenetrable "sound barrier" that prevented flight beyond the speed of sound. The story of the British de Havilland *Swallow*, high-speed research airplane is one example that fueled the controversy over this "sound barrier".

The DH 108 *Swallow* was an experimental, research aircraft, designed and built by the British de Havilland Aircraft Company, to explore the high-speed flight regime near the speed of sound (Figure 3.148). The *Swallow* was a jet engine powered airplane with a highly swept wing, a single vertical tail, and no horizontal tail surfaces. De Havilland built three DH 108 test airplanes, in 1946 and 1947. On 27 September 1946, de Havilland Chief Test Pilot Geoffrey de Havilland, Jr was performing high-speed dive testing of the *Swallow*. Diving at Mach 0.9, at an altitude of 10,000 ft (3050 m), the airplane suffered a catastrophic structural failure, resulting in the disintegration of the airplane and loss of the pilot. The post-crash accident investigation determined that the *Swallow* had experienced an uncommanded, transonic pitch oscillation, which resulted in aerodynamic loads that exceeded the structural limits of the wings, shearing them off the airplane at the wing roots. The crash of the *Swallow*, at near the speed of sound, further entrenched the notion of the impenetrable "sound barrier".

A total of 480 test flights were completed by the three de Havilland Swallow research aircraft, collecting a wealth of technical data about high-speed flight. The *Swallow* did ultimately fly faster

Figure 3.148   The de Havilland DH 108 *Swallow*. (Source: *US Navy*.)

than the speed of sound, becoming the first British aircraft to break the sound barrier on 6 September 1950. Piloted by British test pilot John Derry, the DH 108 became the third airplane in the world to fly supersonically, with Derry being the seventh pilot to do so. Sadly, the remaining two *Swallow* test aircraft were ultimately lost, along with their test pilots, in subsequent crashes in 1950. One Swallow crashed, again performing transonic dive testing from high altitude, but it is unknown whether the aircraft broke up in a dive or whether the pilot became incapacitated due to a faulty oxygen system. The other *Swallow* crashed during stall testing, when the airplane departed and entered an inverted spin.

In addition to the catastrophic consequences of airplanes attempting to fly supersonically, such as the de Havilland *Swallow*, there was some technical basis for the idea of a sound barrier in the late 1930s. At that time, aerodynamic data or theories were available for the variation of drag in the subsonic and supersonic flow regimes. Subsonic experimental data from wind tunnel tests indicated that the aerodynamic drag coefficient increased with increasing subsonic Mach number, with this increase becoming very steep approaching Mach one. Supersonic linear theory and experimental data also predicted a steep increase in the drag coefficient as Mach one is approached from higher supersonic Mach numbers. As shown in Figure 3.149, the aerodynamic state-of-the-art, in the 1930s, was a harbinger that the aerodynamic drag rises to an infinite value, in the transonic flow region, creating a "drag wall" or "sound barrier". In hindsight, this was an unwarranted conclusion, since it was known at the time that projectiles, such as bullets and artillery shells, traveled at supersonic speeds. The photograph of the shock waves from a bullet traveling at supersonic speed, as shown in Figure 2.21, is evidence of this.

3.10.5   Breaking the Sound Barrier: the Bell X-1 and the Miles M.52

On 14 October 1947, high over the Mojave Desert in Southern California, a B-29 *Superfortress* bomber climbs into the sky carrying a unique payload attached to its belly, a bright saffron-colored (yellow-orange), experimental airplane with the words *Glamorous Glennis* painted on its nose, the name of the pilot's wife. This unique aircraft has a fuselage that is shaped like a .50-caliber bullet and a windscreen that is faired in with the fuselage to preserve its bullet shape. As discussed in the last section, prior to the first flight of a supersonic airplane, it was known that projectiles and bullets traveled at supersonic speeds. This is the reason that this aircraft is shaped like a bullet.

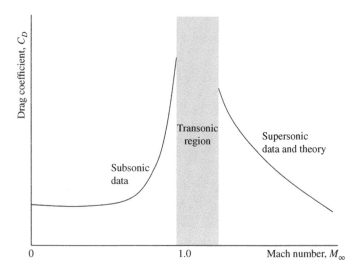

Figure 3.149 State of subsonic and supersonic drag predictions prior to the first supersonic flight. (Source: *Adapted from P.E. Mack, ed., "From Engineering Science to Big Science: The NACA and NASA Collier Trophy Research Project Winner," NASA SP-4219, 1998.*)

It is designed to explore the then unknown regions of transonic and supersonic flight. It has very thin wings with only an 8% thickness (relative to the wing chord). The aircraft propulsion is a Reaction Motors, Inc., XLR11-RM3 four-chamber rocket engine, providing about 6000 lb (27,000 N) of sea level thrust. The 12,250-pound (5560 kg) rocket-powered airplane is fueled with freezing cold liquid oxygen and diluted ethyl alcohol rocket fuel. This rocket-powered aircraft is the experimental Bell X-1, manufactured by the Bell Aircraft Company, Buffalo, New York. The National Advisory Committee for Aeronautics (NACA), the forerunner of NASA, has played a significant role in the design of this research aircraft. Sitting in the single-place cockpit of this first-ever "X-plane" is Captain Charles E. "Chuck" Yeager, a 24-year-old US Air Force test pilot from West Virginia.

This is the 50th flight of the X-1 program and the 9th flight of the X-1 under rocket power. As the B-29 "mothership" approaches an altitude of 20,000 ft (6100 m), we listen in on the flight test radio transmissions, just prior to the drop of the X-1 rocket plane from the B-29. ("Tower" is the Muroc Tower at the Muroc Army Air Field, "Cardenas" is Maj. Robert Cardenas, the B-29 pilot, "Ridley" is Capt. Jackie "Jack" Ridley, the X-1 project engineer, and "Yeager" is Capt. Charles "Chuck" Yeager, the X-1 test pilot. Comments in italics are added for clarity.)

Tower:	Muroc Tower to Air Force (B-29) Eight Zero Zero, clear to drop.
Cardenas:	Roger.
Ridley:	Yeager, this is Ridley. You all set?
Yeager:	Hell, yes, let's get this over with.
Ridley:	Remember those stabilizer settings.
	(The X-1 had an all-moving horizontal stabilizer, a key design feature that helped make supersonic flight possible.)
Yeager:	Roger.
Cardenas:	Eight Zero Zero. Here is your countdown, 10–9–8–7–6–5–3–2–1–Drop.
	(Cardenas had omitted "4" in the countdown. The X-1 release occurred at 10:26 a.m. as the B-29 was flying at 20,000 ft and an indicated airspeed of 250 mph.)

Yeager: Firing Four *(rocket chamber #4)* ... Four fired okay ... will fire Two ... Two on ... will cut off Four ... Four off ... will fire Three ... Three burning now ... will shut off Two and fire One ... One on ... will fire Two again ... Two on ... will fire Four. *(The Bell X-1 was powered by the Reaction Motors, Inc. XLR11 liquid rocket engine, burning ethyl alcohol and liquid oxygen. The XLR11 had four separate combustion chambers and nozzles, each producing 1500 lb (6700 N) of thrust, for a total, combined thrust of 6000 lb (27,000 N). The thrust of each chamber could not be varied or throttled, rather the thrust was controlled by turning the individual rocket chambers on and off.)*

Ridley: How much of a drop *(in the rocket chamber pressure)*?

Yeager: About forty psi ... got a rich mixture ... chamber pressure down ... now going up again ... pressures all normal ... will fire Three again ... Three on ... acceleration good ... have had mild buffet ... usual instability. Say Ridley, make a note here. Elevator effectiveness regained. *(It was known, at the time, that there would be a decrease in elevator effectiveness at transonic Mach numbers. The all-moving stabilizer was moved in small increments during the acceleration to maintain longitudinal stability and proved very effective. Yeager's comment, that elevator effectiveness was regained, was made as the X-1 accelerated through an indicated Mach number of 0.96).*

Ridley: Roger. Noted.

Yeager: Ridley! Make another note. There's something wrong with this Mach meter. It's gone screwy! *(In his post-flight, written test report, Yeager states that the needle of the Mach meter fluctuated momentarily at about Mach 0.98, then went off the scale. Based on radar tracking data, the Bell X-1 had achieved Mach 1.06.)*

Ridley: If it is, we'll fix it. Personally, I think you're seeing things.

Yeager: I guess I am, Jack ... will shut down again ... am shutting off ... shut off ... still going upstairs like a bat ... have jettisoned fuel and LOX ... about thirty percent of each remaining ... still going up ... have shut down now. *(LOX is liquid oxygen.)*

Despite the disbelief and surprise of Yeager and Ridley, the Bell X-1 had indeed become the first manned aircraft to exceed the speed of sound. Figure 3.150 shows the Bell X-1 in powered flight along with some of the flight data from the first manned supersonic flight. The static and total pressure data traces are shown versus time, at the bottom of Figure 3.150, where time, in seconds, is annotated as the numbers on the topmost part of the data traces. The total and static pressure data were obtained from sensors that were connected to pressure orifices, most likely on a Pitot tube and a static pressure source on the tube or fuselage, respectively. The total and static pressure traces, at about 145 s, represent the values of the freestream flow in subsonic flight. When the Bell X-1 went supersonic, shock waves formed on the aircraft in front of the Pitot tube and static pressure orifice. The total and static pressures being measured are then the values behind a shock wave, very probably a normal shock wave. As discussed in a later section, the total pressure rapidly decreases and the static pressure rapidly increases behind a shock wave. This is exactly what is seen in the data traces at the point marked "Mach jump" in Figure 3.150, when the Bell X-1 goes supersonic. The transition from subsonic to supersonic flight is sometimes called a "Mach jump", indicating the step-like change or "jump" in the Mach number, or other parameters such as the total and static pressure. The Bell X-1 flies supersonically for about 17 s and then decelerates

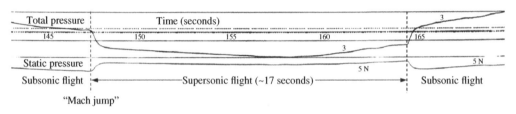

Figure 3.150 The Bell X-1 aircraft and data showing the first "Mach jump" on 14 October 1947. (Source: *Courtesy of NASA with annotations added by author.*)

to subsonic speed, accompanied by the reverse step-wise changes in the pressure traces, where the total pressure increases and the static pressure decreases.

There were 157 test flights of the Bell X-1, flown by 19 test pilots, with the final flight occurring on 23 October 1951. During the course of its test flights, the X-1 reached a maximum Mach number of 1.45 and a maximum altitude of 69,000 ft (21,000 m). It was the first supersonic airplane and the first in a long series of flight research "X-planes". (Jack Ridley also flew the Bell X-1 rocket airplane four times. He broke the sound barrier on his first X-1 flight, reaching Mach 1.23 on 11 March 1949, joining a small and exclusive club of supersonic pilots.) Note in the conversations above, the emphasis on the aircraft's horizontal stabilizer. The *all-moving horizontal stabilizer*, versus the traditional fixed stabilizer and moving elevator, was an innovative flight control design feature that helped make supersonic flight a reality. The all-moving horizontal stabilizer would be incorporated into future supersonic aircraft, including modern-day supersonic aircraft.

The quest for supersonic flight was not limited to efforts in the USA. In 1942, the UK initiated a top secret project to develop a supersonic airplane. The result of these efforts was the Miles Aircraft M.52 turbojet-powered aircraft (Figure 3.151), designed to reach 1000 mph (1600 km/h) in level flight and climb to an altitude of 36,000 ft (11,000 m) in 1.5 minutes. The M.52 had several innovative design features that would be incorporated in future supersonic aircraft.

Unlike the jet aircraft of the era, which had round noses, thick wings, and conventional horizontal stabilizers with hinged elevators, the Miles M.52 had a pointed conical nose, thin wings, and an all-moving horizontal stabilizer. The wing had a biconvex airfoil shape, formed by opposing

Figure 3.151 Miles M.52 supersonic research aircraft. (Source: *UK Government, PD-UKGov.*)

circular arcs (see Figure 3.86). The thin wing, with its sharp leading and trailing edges, significantly reduced the supersonic wave drag as opposed to the blunt-nosed, thick shapes of subsonic airfoils. These wing design features are common characteristics of modern supersonic aircraft. Miles Aircraft modified a light airplane with the M.52 wing and performed low-speed flight tests in 1944 to obtain data for their supersonic wing design.

The M.52 design had an all-moving stabilizer, the advantages of which have already been discussed in association with the Bell X-1. Early design versions of the X-1 had a conventional fixed stabilizer with a moving elevator. After design data for the Miles M.52 was given to Bell Aircraft, the X-1's conventional horizontal tail was changed to an all-moving stabilizer. Miles Aircraft mounted the all-moving tail on a Supermarine *Spitfire* and obtained flight test data up to about Mach 0.86.

The propulsion system of the M.52 was an air-breathing jet engine, but existing jet engines could not produce sufficient thrust to reach supersonic speeds in level flight. To solve this problem, the M.52 jet engine had a novel *reheat* jet pipe, where additional fuel was mixed with unburned fuel in the engine tailpipe, producing a significant increase in thrust. Today, this propulsion component, known as an *afterburner*, is used in all supersonic jet engines. (The term *reheat* is still commonly used by the British, when referring to an *afterburner*.) Another propulsion system innovation of the M.52 was the design of the engine inlet, which had to efficiently decelerate the supersonic flow to subsonic speeds for ingestion into the turbojet engine. The inlet opening was an annular ring between the fuselage outer diameter and a conical centerbody. The supersonic flow entering the inlet is decelerated more efficiently through the oblique, conical shock waves from the centerbody rather than through stronger, normal shocks without a centerbody. This concept of using a centerbody for efficient supersonic flow deceleration is used in many modern supersonic aircraft. In the M.52, the cockpit and pilot seat were located in the conical centerbody with the windscreen faired into the conical nose shape.

The design data for the Miles M.52 was shared with the NACA and Bell Aircraft during the design of the X-1. In addition to the incorporation of the all-moving horizontal stabilizer on the Bell X-1, the M.52 design information may have been of some use in the final design of the Bell

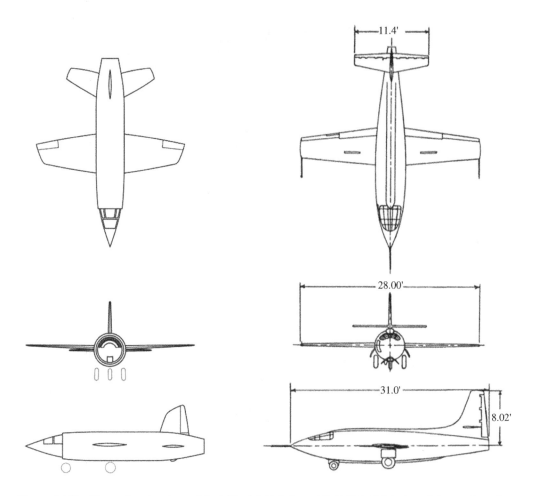

Figure 3.152   Three-view drawings of the Miles M.52 (left) and Bell X-1 (right) aircraft (approximately to scale).   (Source: *Miles M.52 courtesy M.L. Watts, CC0-1.0, Bell X-1 courtesy NASA.*)

X-1. A three-view drawing comparison of the Miles M.52 and Bell X-1 is shown in Figure 3.152. In 1946, the Miles M.52 manned, jet engine-powered supersonic airplane design was changed to an unmanned, rocket-powered missile. Test flights were conducted of the Miles supersonic missile and on 10 October 1948 it reached Mach 1.38 in level, supersonic flight.

It should be mentioned that the Soviet Union also initiated a supersonic flight research program after World War II. They designed and built several prototype research aircraft, including the swept wing La-176 jet engine powered airplane, designed by the Lavochkin design bureau. The La-176 broke the sound barrier in a shallow dive on 26 December 1948, piloted by Soviet test pilot I.E. Fedorov.

The NACA played a major role in attaining supersonic flight in the USA. They performed extensive wind tunnel testing in an attempt to understand the effects of compressibility, but they were limited by the inability to duplicate transonic Mach numbers in the tunnels. NACA aerodynamicist John Stack was prominent in advocating for high-speed research aircraft to explore the aerodynamics, stability, control, and other fundamental issues associated with transonic and supersonic flight. This ultimately led to the Bell X-1 program, where the NACA developed the basic aircraft design specifications, planned the overall flight research test program, and was responsible for the flight data instrumentation, collection, and analyses. The NACA pioneered many of the techniques in obtaining high quality flight test data. Their flight planning also emphasized safety, using the flight test build-up approach, where the flight data was used to verify it was safe to continue to the next test points in the flight envelope.

The X-1 flight tests were conducted at the Muroc Army Air Field in the Mojave Desert of Southern California. The NACA created the Muroc Flight Test Unit on 7 September 1947 to support the Bell X-1 flight tests, staffed by a total of 27 personnel. This small flight research station in the middle of the desert would steadily grow in size, eventually becoming the current NASA Armstrong Flight Research Center on Edwards Air Force Base, California. They would play a major role in the research and flight testing of many of the X-planes that have flown.

The transonic and supersonic flights of the Bell X-1 proved that there was no "sound barrier" and paved the way for future supersonic aircraft. Flight data from the Bell X-1 flights are shown in Figure 3.153, where drag coefficient is plotted versus flight Mach number. The flight data indicates a drag divergence Mach number of about 0.9, where the drag coefficient increases rapidly. The drag coefficient reaches a maximum at about Mach 1.1 and then starts to decrease as the Mach number increases. Based on the Bell X-1 flight results we can return to Figure 3.149 and fill in the behavior of the drag coefficient in the transonic region, as shown in Figure 3.154. The drag coefficient rises in the transonic region to a finite peak and then decreases with increasing supersonic Mach number.

3.11   Supersonic Flow

We have made our way through the sound barrier, beyond transonic flow, and have arrived in the supersonic flow regime. Compressibility, where the density is not constant, is now the rule, rather than the exception as it is for subsonic flow. However, some of the simplicity of linear physics

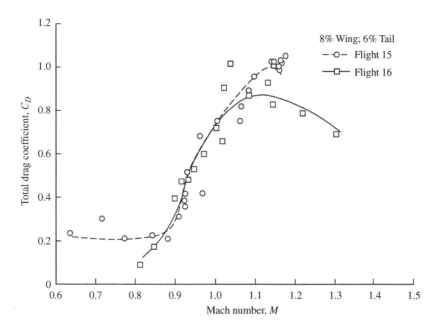

Figure 3.153   Bell X-1 drag coefficient versus Mach number. (Source: *Gardner, NACA RM L8K05, 1948, [32].*)

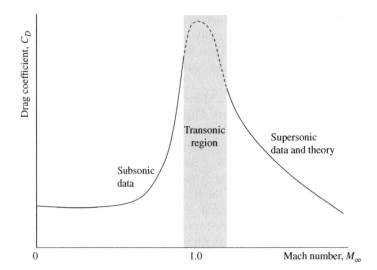

Figure 3.154 Variation of drag coefficient through the transonic flow regime. (Source: *Adapted from P.E. Mack, ed., "From Engineering Science to Big Science: The NACA and NASA Collier Trophy Research Project Winner," NASA SP-4219, 1998.*)

returns with fully supersonic flow as compared with the non-linear characteristics of transonic flow. Earlier, in Chapter 2, the supersonic flow regime was defined as being from Mach 1.2 to 5 and several aspects of supersonic flow were introduced, including shock and expansion waves. In this section, we discuss the details of these supersonic phenomena and their impacts on supersonic flight, including the impact that is literally felt and heard on the ground, known as the sonic boom. We also discuss some of the design innovations developed to defeat, or at least reduce drag for supersonic flight, including the *swept wing* and fuselage *area ruling*. First, we introduce the isentropic flow relations, which relate the static and total properties for an isentropic flow of a calorically perfect gas to the flow Mach number. These relations are used frequently in the analysis of supersonic flows.

3.11.1 Isentropic Flow Relations

As discussed in Section 3.5.7, the isentropic flow assumption, where the flow is adiabatic (no heat transfer) and reversible (no viscous or thermal dissipation or other losses), is applicable to many flow situations, even many supersonic flows. The losses incurred in viscous flows can be confined to the boundary layer, while the inviscid flow, outside the boundary layer, may often be treated as isentropic. Similarly, the losses associated with shock waves in supersonic flows, can be confined to the thin region of the shock wave itself, allowing the flows upstream and downstream of the shock to be considered isentropic. There is a wide range of applicability of the isentropic flow relations to supersonic flows. These include external supersonic flows over compression surfaces, such as wings or propulsive inlet ramps, and internal supersonic flows through ducts and nozzles.

We seek to obtain an expression for the ratio of the total to the static temperature as a function of Mach number. We start with the energy equation for a calorically perfect gas, Equation (3.190), repeated below.

$$c_p T + \frac{V^2}{2} = c_p T_t = \text{constant} \qquad (3.338)$$

Dividing this equation by $c_p T$, we have the ratio of the total temperature, T_t, to the static temperature, T.

$$\frac{T_t}{T} = 1 + \frac{V^2}{2c_p T} \tag{3.339}$$

Using Equation (3.124), relating the specific heat at constant pressure, c_p, to the specific gas constant, R, and the ratio of specific heats, γ, and the definition of the speed of sound, a, from Equation (3.326), we have

$$\frac{T_t}{T} = 1 + \frac{V^2}{2\left(\frac{\gamma R}{\gamma - 1}\right)T} = 1 + \frac{V^2}{\frac{2}{\gamma - 1}(\gamma R T)} = 1 + \frac{V^2}{\frac{2}{\gamma - 1}a^2} \tag{3.340}$$

Using the definition of the Mach number, we have an equation relating the total-to-static temperature ratio to the Mach number.

$$\frac{T_t}{T} = \left[1 + \left(\frac{\gamma - 1}{2}\right)M^2\right] \tag{3.341}$$

Since we have assumed isentropic flow, we can utilize Equation (3.139), relating the pressure, density, and temperature at two states for an isentropic process. Here, the two states are the static flow condition and the stagnation or total state. Thus, we can write Equation (3.139) as

$$\frac{p_t}{p} = \left(\frac{\rho_t}{\rho}\right)^\gamma = \left(\frac{T_t}{T}\right)^{\gamma/(\gamma - 1)} \tag{3.342}$$

Inserting Equation (3.341) into (3.342), we obtain isentropic relations for the total-to-static ratios of the pressure and density.

$$\frac{p_t}{p} = \left[1 + \left(\frac{\gamma - 1}{2}\right)M^2\right]^{\gamma/(\gamma - 1)} \tag{3.343}$$

$$\frac{\rho_t}{\rho} = \left[1 + \left(\frac{\gamma - 1}{2}\right)M^2\right]^{1/(\gamma - 1)} \tag{3.344}$$

The isentropic relations, given by Equations (3.341), (3.343), and (3.344), relate the total-to-static ratio of the temperature, pressure, and density, respectively, to the Mach number for the flow of a calorically perfect gas. These relations provide a straightforward means of calculating the stagnation conditions of a flow given the Mach number and static conditions.

If we take the inverse of Equations (3.341), (3.343), and (3.344), we have the isentropic relations as the static-to-total ratios of the properties.

$$\frac{T}{T_t} = \left[1 + \left(\frac{\gamma - 1}{2}\right)M^2\right]^{-1} \tag{3.345}$$

$$\frac{p}{p_t} = \left[1 + \left(\frac{\gamma - 1}{2}\right)M^2\right]^{-\gamma/(\gamma - 1)} \tag{3.346}$$

$$\frac{\rho}{\rho_t} = \left[1 + \left(\frac{\gamma - 1}{2}\right)M^2\right]^{-1/(\gamma - 1)} \tag{3.347}$$

Each isentropic relation has a maximum value of one at a Mach number of zero, meaning that the static and stagnation values are equal. This makes perfect sense, as the definition of the stagnation

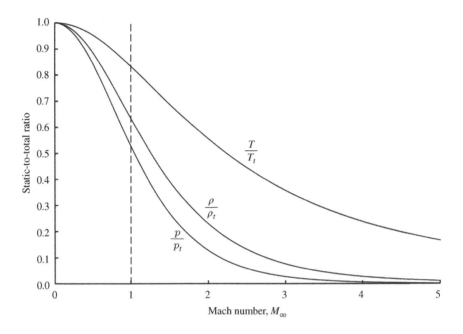

Figure 3.155   Isentropic flow properties versus Mach number.

state is the value obtained by isentropically bringing the flow to zero velocity. The fractional change in the density at a given Mach number from the density at Mach zero is given by

$$\frac{\rho - \rho_{M=0}}{\rho_{M=0}} = \frac{\frac{\rho}{\rho_t} - \left(\frac{\rho}{\rho_t}\right)_{M=0}}{\left(\frac{\rho}{\rho_t}\right)_{M=0}} = \frac{\frac{\rho}{\rho_t} - 1}{1} = \frac{\rho}{\rho_t} - 1 \qquad (3.348)$$

This isentropic equation for the change in the density as a function of Mach number was used earlier to generate the plot shown in Figure 3.9.

The isentropic relations, given by Equations (3.341), (3.343), and (3.344), are plotted in Figure 3.155, assuming normal air with $\gamma = 1.4$. As just mentioned, the static-to-total ratios are equal to one at Mach zero. Each ratio decreases from one with increasing Mach number. Since these relations are used quite often in aerodynamic analyses, tables of calculated values from these equations as a function of Mach number are quite common.

The values of the isentropic ratios are of particular interest at Mach one. Inserting a Mach number equal to one into Equations (3.341), (3.343), and (3.344), we have, for $\gamma = 1.4$, the following values.

$$\frac{T}{T_t} = \left[1 + \left(\frac{\gamma - 1}{2}\right)\right]^{-1} = \left[1 + \left(\frac{1.4 - 1}{2}\right)\right]^{-1} = (1.2)^{-1} = 0.833 \qquad (3.349)$$

$$\frac{p}{p_t} = \left[1 + \left(\frac{\gamma - 1}{2}\right)\right]^{-\gamma/(\gamma-1)} = (1.2)^{-3.5} = 0.528 \qquad (3.350)$$

$$\frac{\rho}{\rho_t} = \left[1 + \left(\frac{\gamma - 1}{2}\right)\right]^{-1/(\gamma-1)} = (1.2)^{-2.5} = 0.634 \qquad (3.351)$$

These values for sonic conditions are useful in many applications, for instance, in analyzing the subsonic and supersonic flows through nozzles.

3.11.2   Shock and Expansion Waves

We have mentioned shock waves several times in the past when discussing transonic and supersonic flows. We know that they start to appear when the flow exceeds sonic velocity or Mach one. The presence of shock waves results in an increase in the drag, called wave drag. We now seek to obtain a deeper understanding of shock waves and some other supersonic wave phenomena, such as Mach waves and expansion waves.

3.11.2.1   Mach Waves

In Section 2.3.7, we introduced the concept of the "bunching up" of acoustic waves (see Figure 2.20), coalescing into a shock wave as the flight speed approached and exceeded the speed of sound, or Mach one. To help visualize this concept further, think about a boat traveling in the water. If the boat is traveling at a speed less than the wave speed (the speed that the waves travel on the surface of the water), then the surface of the water is smooth. When the boat's speed is greater than the wave speed, the waves "bunch up" in front of the boat and create a bow wave. This is analogous to the "bunching up" of the sound waves ahead of a body moving at supersonic speed, creating a shock wave. (Shock wave shapes are sometimes visualized using a *water table*, where the analogy we have just mentioned is used.)

Imagine an infinitely thin flat plate, with a razor sharp leading edge, in a subsonic flow. The pressure disturbance caused by the plate is communicated to the flow through acoustic or sound waves. At subsonic Mach numbers, well below the speed of sound, the presence of the plate is theoretically communicated everywhere in the fluid. As the flow Mach number increases to Mach one, the acoustic waves communicating the plate's pressure disturbance "bunch up" to form an infinitesimally weak compression wave, known as a *Mach wave*. As the flow Mach number further increases beyond Mach one, the acoustic waves cannot move as far upstream and the Mach wave makes an angle, μ, called the *Mach angle*, with respect to the flow direction, as shown on the left in Figure 3.156.

The magnitude of the Mach angle can be obtained as follows. The flat plate is continuously sending "signals" of its presence in the flow via sound waves that propagate at the speed of sound in all directions. These sound waves expand in radius and move downstream with time. At a time t, a sound wave has moved a distance downstream equal to Vt, that is dependent on the flow velocity, as shown on the right in Figure 3.156. At the time t, the sound wave has grown to a radius equal

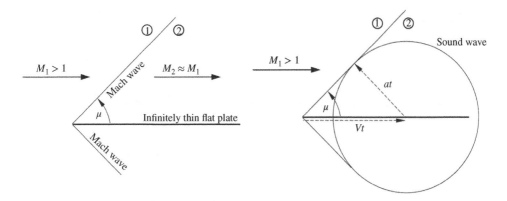

Figure 3.156   Mach wave.

to *at*. From the geometry shown in Figure 3.156, we have

$$\sin \mu = \frac{at}{VT} = \frac{a}{V} = \frac{1}{M} \tag{3.352}$$

The Mach angle, μ, is given by

$$\mu = \sin^{-1} \frac{1}{M} \tag{3.353}$$

The magnitude of the Mach angle is inversely proportional to the sine of the Mach number. Thus, the Mach angle decreases with increasing Mach number, such that the Mach wave gets closer to the body generating the wave. At Mach one, the Mach wave angle is 90° or vertical, which coincides with the acoustic wave pictures we have developed in Figure 2.20.

In our example, the infinitely thin, razor sharp flat plate creates a planar Mach wave in two-dimensions. Three-dimensional Mach waves are also generated by three-dimensional bodies. A needle-like body in a supersonic flow would generate a three-dimensional *Mach cone*, with a Mach angle given by Equation (3.353). Whether two- or three-dimensional, the Mach wave communicates the pressure disturbance of a body in a supersonic flow. The wave is an infinitesimal compression wave, so that there is an infinitesimal change of the properties across the wave. The process is an isentropic compression, so that the total pressure and total temperature remain constant across the Mach wave.

Many of the features of sound waves and Mach waves that we have discussed are seen in Figure 3.157. Here, flow visualization is used to photograph the wave system around a sphere traveling at Mach 3. The sphere is passing over a perforated flat plate. As the sphere passes over the holes in the plate, the bow shock wave sends out weak pressure disturbances, seen as the series of circles below the plate in the photograph. The weak pressure disturbances coalesce into a Mach wave. The waves above the plate are *finite compression waves* or shock waves, which we discuss next.

Figure 3.157   Flow visualization of the wave system around a sphere traveling at Mach 3. (Source: *Courtesy of the US Army Ballistics Research Laboratory, Aberdeen, Maryland.*)

3.11.2.2 Normal Shock Waves

When there is a finite, rather than an infinitesimal, deflection of the flow, caused by a body with finite, rather than infinitesimal thickness, the waves created by the disturbance of the flow by the body, coalesce into a finite compression wave, called a *shock wave*. The change in the flow properties across a shock wave are finite and may be quite large in magnitude for high Mach numbers. The flow may be considered isentropic upstream and downstream of the shock wave, but the flow across the shock wave is non-isentropic. There are viscous and heat transfer losses within the shock wave resulting in an increase of entropy across the wave.

We first look at the *normal shock wave*, where the shock wave is at a 90° angle to the flow. This is the case for the portion of a bow shock wave, directly upstream of the stagnation point of a blunt body in a supersonic flow, as shown on the left side of Figure 3.158. In contrast, the *oblique shock wave* makes an angle less than 90° with respect to the flow direction, as shown on the right side of Figure 3.158, for the supersonic flow over a sharp-nosed wedge. The flow visualization technique used to obtain the images in Figure 3.158 captures the density gradients in the flow as the dark bands, thus, we see that the shock wave is a very thin region with high gradients in density. The gradients in velocity, pressure, and temperature are also very high through the shock. The normal shock wave is quite thin with a nominal thickness of only 10^{-5} in (2.5×10^{-5} cm). Interior to the shock wave, there are dissipative phenomena due to viscosity and heat transfer.

The idealized model of a normal shock wave is an infinitely thin, vertical line, as shown in Figure 3.159, through which the flow properties change discontinuously. For practical engineering problems, we are not interested in the complex viscous and heat transfer physics inside the shock wave. Rather, we are interested in what happens to the flow properties after passing through the shock. Therefore, the idealized model of the normal shock wave works well in this regard. The flow upstream of the shock wave (region 1) is a uniform supersonic flow with Mach number, M_1, velocity, V_1, static pressure, p_1, static density, ρ_1, static temperature, T_1, total pressure, $p_{t,1}$, and total temperature $T_{t,1}$. The flow downstream of the shock wave (region 2) is uniform, but most of the flow properties have changed. A normal shock wave always discontinuously decelerates a supersonic

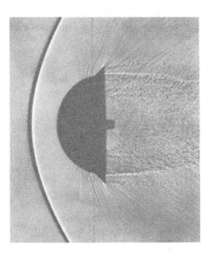

Figure 3.158 A blunt body (left) and sharp-nosed wedge (right) in supersonic flow, where the flow is from left to right. The bow shock wave is near normal upstream of the blunt body stagnation point. The wedge body has oblique shock waves attached to its sharp nose. (Source: *E.P. Hartman, "Adventures in Research: A History of the Ames Research Center, 1940–1965," NASA SP-4302, 1970.*)

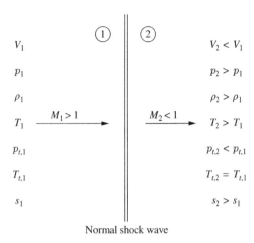

Normal shock wave

Figure 3.159   Flow through a normal shock wave.

flow to a subsonic Mach number, so that $M_2 < 1$. The velocity decreases and the static pressure, static density, and static temperature all increase discontinuously across the normal shock. There is a total pressure loss across the shock, $p_{t,2} < p_{t,1}$, due to the entropy increase, $s_2 > s_1$, across the shock. The entropy increase is due to the dissipative viscous and heat transfer phenomena inside the shock. While there are dissipative phenomena within the shock wave, there is no heat being added or taken away from the shock wave, thus the flow through the normal shock wave is adiabatic making the total temperature constant through the shock, $T_{t,2} = T_{t,1}$.

By applying the fundamental equations of fluid dynamics, the continuity, momentum, and energy equations (see Section 3.6) to the flow upstream and downstream of the normal shock wave, the relationships for the flow properties across the shock may be obtained. This derivation is beyond the scope of the text, thus we simply present some of the results. Equations (3.354), (3.355), and (3.356), below, provide the Mach number, M_2, static pressure ratio, p_2/p_1, and static temperature ratio, T_2/T_1, across a normal shock wave.

$$M_2^2 = \frac{(\gamma - 1)M_1^2 + 2}{2\gamma M_1^2 - (\gamma - 1)} \tag{3.354}$$

$$\frac{p_2}{p_1} = \frac{2\gamma M_1^2 - (\gamma - 1)}{\gamma + 1} \tag{3.355}$$

$$\frac{T_2}{T_1} = \frac{a_2^2}{a_1^2} = \frac{[2\gamma M_1^2 - (\gamma - 1)][(\gamma - 1)M_1^2 + 2]}{(\gamma + 1)^2 M_1^2} \tag{3.356}$$

These normal shock properties are only a function of the upstream Mach number, M_1, and the ratio of specific heats, γ, a property of the type of gas. We often deal with the flow of "normal" air, where the ratio of specific heats is constant and equal to 1.4. Inserting $\gamma = 1.4$ into Equations (3.354), (3.355), and (3.356), we have

$$M_2^2 = \frac{M_1^2 + 5}{7M_1^2 - 1} \tag{3.357}$$

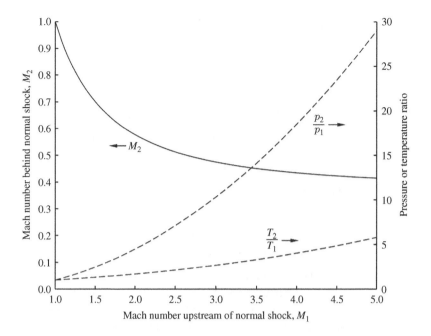

Figure 3.160 Flow properties behind a normal shock wave.

$$\frac{p_2}{p_1} = \frac{7M_1^2 - 1}{6} \qquad (3.358)$$

$$\frac{T_2}{T_1} = \frac{a_2^2}{a_1^2} = \frac{(7M_1^2 - 1)(M_1^2 + 5)}{36M_1^2} \qquad (3.359)$$

Equations (3.354), (3.355), and (3.356) are plotted in Figure 3.160, showing the variation of the downstream Mach number, pressure ratio, and temperature ratio across a normal shock wave, as a function of the upstream Mach number.

Example 3.16 Properties Behind a Normal Shock wave *For the sphere traveling at Mach 3 in Figure 3.157, the bow shock wave is a normal shock at the forward stagnation point. Calculate the Mach number, pressure, and temperature behind the normal shock wave, assuming the sphere is flying through the air at sea level.*

Solution

Using Equation (3.357), the Mach number behind the normal shock is

$$M_2^2 = \frac{M_1^2 + 5}{7M_1^2 - 1} = \frac{(3)^2 + 5}{7(3)^2 - 1} = \frac{14}{62} = 0.2258$$

Using Equations (3.358) and (3.359), the pressure and temperature ratios across the normal shock wave are

$$\frac{p_2}{p_1} = \frac{p_2}{p_{SSL}} = \frac{7M_1^2 - 1}{6} = \frac{7(3)^2 - 1}{6} = \frac{62}{6} = 10.333$$

$$\frac{T_2}{T_1} = \frac{T_2}{T_{SSL}} = \frac{(7M_1^2 - 1)(M_1^2 + 5)}{36M_1^2} = \frac{[7(3)^2 - 1][(3)^2 + 5]}{36(3)^2} = \frac{868}{324} = 2.679$$

The sea level values of pressure and temperature are 2116 lb/ft² and 459.67°R, respectively. Inserting these values into the pressure and temperature ratios and solving for the pressure and temperature behind the normal shock wave, we have

$$p_2 = 9.333 p_{SSL} = 10.333 \left(2,116 \frac{\text{lb}}{\text{ft}^2} \right) = 21,865 \frac{\text{lb}}{\text{ft}^2}$$

$$T_2 = 2.679 T_{SSL} = 2.679(459.67°R) = 1,231.5°R$$

3.11.2.3 Oblique Shock and Expansion Waves

When a supersonic flow is made to turn into itself, such as was shown for the supersonic flow over a wedge in Figure 3.158, an *oblique shock* is formed. An idealized oblique shock wave, formed by the supersonic flow over a wedge of angle, θ, is depicted in Figure 3.161. The flow is uniformly turned through the deflection angle, θ, as it passes through the oblique shock. All of the flow streamlines, downstream of the shock wave, are parallel to the wedge surface and to each other. The oblique shock wave is attached to the apex of the wedge with a shock wave angle, β, with respect to the flow.

The flow properties change discontinuously across the oblique shock wave, in the same manner as with a normal shock. A major difference between the oblique and normal shock waves is that the Mach number behind the oblique shock, M_2, remains supersonic. The other flow properties behind the oblique shock have the same trends as for the normal shock case. However, for the same given upstream Mach number, M_1, the strength of the oblique shock wave is less than the normal shock, thus the magnitude of the changes in the flow properties are less. All of these changes in the flow properties occur discontinuously across the oblique shock wave, as was the case for a normal shock.

The oblique shock wave strength and shock wave angle, β, increase as the flow deflection angle, θ, increases. At some large deflection angle, the oblique shock wave becomes detached and forms a normal shock wave in front of the body. In fact, the normal shock wave is a limiting case of the oblique shock wave, where the shock wave angle is 90°.

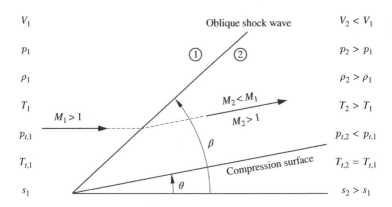

Figure 3.161 Supersonic flow through an oblique shock wave.

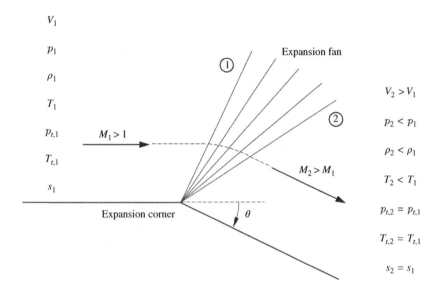

Figure 3.162   Supersonic flow through an expansion wave.

While an oblique shock wave is formed when a supersonic flow is turned *into* itself, an *expansion wave* is created when a supersonic flow is turned *away* from itself, as depicted in Figure 3.162 for a supersonic flow with Mach number, M_1, that is turned through an expansion angle, θ. The expansion wave is composed of a series of Mach waves that are centered at the corner of the turn. The expansion fan continuously turns the upstream supersonic flow through the expansion fan so that the flow is parallel to deflected surface downstream of the expansion. Since the upstream flow is passing through a series of Mach waves, the changes in the flow properties are smooth, continuous, and isentropic. The Mach number and velocity increase through the expansion wave, while the static pressure, static density, and static temperature all decrease. Since the expansion wave process is isentropic, the total pressure, total temperature, and entropy remain constant. In many respects, the expansion wave is the "opposite" of a shock wave.

3.11.3   FTT: Visualizing Shock Waves in Flight

We have seen a few examples of the visualization of shock waves, such as Ernst Mach's photograph of shock waves on a bullet traveling at supersonic speed in Figure 2.21 and the photograph of shock waves on a transonic airfoil in a wind tunnel in Figure 3.145. These examples used optical systems, which make the density gradients in the flow visible. The density increases sharply across shock waves, making them visible using these methods. The ability to visualize the shock waves around bodies traveling at transonic and supersonic speeds can provide information about the aerodynamics of the flow. These types of optical techniques have traditionally been used in ground test facilities due to the mechanical complexities and sizes of the systems. The present flight test technique discusses the application of these types of flow visualization techniques to aircraft in supersonic flight, where shock waves on or around the aircraft are made visible.

As early as the 1940s, pilots were noticing the appearance of gossamer-like filaments "dancing" on the surface of their aircraft wings, when they were in high-speed dives. They were seeing shock waves on their wings as their aircraft touched transonic speeds. The first, perhaps official, reports of these shock wave sightings were by test pilot Major Frederick A. Borsodi of the Army Air Force. In July 1944, Borsodi was conducting flight tests, in the North American P-51D *Mustang*, to assess the effects of compressibility on aircraft handling. He was performing a series of full-power dives, when at Mach 0.86, the aircraft experienced severe buffeting of the empennage. At the time of the buffeting, Borsodi reported seeing shock waves on his wing, extending from the wing root to the wingtip, at the wing's point of maximum thickness. After this incident, he set up a camera system in the aircraft, where he could photograph the shock waves in flight. He successfully captured photographs of the shock waves and shared them with other pilots and engineers, including the eminent aerodynamicist, Theodore von Karman.

Borsodi was indeed seeing shock waves on his wings, which were made visible by the refraction of sunlight through the density gradient of the shock wave. The buffeting he was experiencing was due to the oscillation of the shock wave on the wing surface, which caused an unsteady, cyclic separation of the flow and a turbulent wake that impinged on the aircraft empennage. NACA test pilot, George E. Cooper (later to be known for his flying qualities rating scale, to be discussed in Chapter 6), also saw these shock waves in his P-51 dive test flights. He too photographed these shock waves in flight and, in 1948, published the first report detailing a method for visualizing shock waves in flight [23].

The method is based on the same physics as the *shadowgraph* flow visualization technique that is used in wind tunnels. Whereas the wind tunnel shadowgraph uses an artificial light source, the in-flight technique relies on sunlight to produce *natural shadowgraph* images of shock waves. As shown in Figure 3.163, the in-flight shadowgraph is produced by the refraction of the parallel rays of sunlight through the density discontinuity of the shock wave, where the air density increases from a "low density medium" to a "high density medium". The density change through the shock wave varies with vertical distance from the wing surface, being greatest nearest the surface, thus the amount of light refraction varies, resulting in a dark band (shadow) directly behind the shock wave, followed by a light band (light), as shown in Figure 3.163. The success of the shadowgraph method was dependent on the proper orientation of the sun's rays relative to the wing surface.

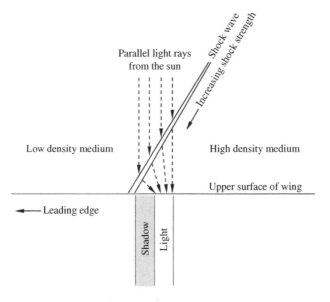

Figure 3.163 Diagram of shadowgraph on a wing in flight. (Source: *Adapted from Cooper and Rathert, NACA RM A8C25, 1948, [23].*)

Figure 3.164   Sunlight shadowgraph of shock waves on a wing in flight. (Source: *Cooper and Rathert, NACA RM A8C25, 1948, [23].*)

Cooper's report was the first to specify the best sun angles to obtain shadowgraph images of shock waves in flight.

Shadowgraphs from Cooper's P-51 flights showed the back-and-forth oscillation of the shock waves, with an amplitude of about 2–3 in (5–8 cm), at transonic Mach numbers, resulting in tail buffeting. The photographs also showed the shocks moving aft, toward the wing trailing edge, as the Mach number increased. Figure 3.164 shows a sunlight shadowgraph from Cooper's report, where the shock waves from the aircraft canopy and wing are visible.

More recently, NASA researchers have taken more precise measurements, during transonic flights in a Lockheed L-1011 transport aircraft, to improve the methods for the in-flight visualization of shock waves using sunlight shadowgraphy [29]. Figure 3.165 shows a natural shadowgraph of a normal shock wave standing near the wingtip of the Lockheed L-1011 aircraft in flight. Today, with the Sun at just the right angle, it is not uncommon for a passenger to see a shock wave dancing on the wing of a commercial airliner. This is the same natural shadowgraph flow visualization that was experienced by test pilots performing high-speed compressibility dives in the 1940s. We now also understand what you saw on the wing of the F-18 during your familiarization flight. The light and dark lines were the natural shadowgraphs of shock waves that became visible on the wing as the aircraft accelerated through the transonic flow regime.

An even more ambitious visualization technique is attempting to capture the shock waves around a complete aircraft in transonic or supersonic flight. The ability to visualize the shock locations, and their relative strengths, over a complete aircraft can provide an understanding of the aerodynamics of different supersonic configurations. Using a more sophisticated technique than the shadowgraph, called *schlieren* imaging, these techniques are still based on the shock wave density gradients and

Figure 3.165   Shock wave on the wingtip of a Lockheed L-1011 aircraft (flow is from right to left). (Source: *Courtesy of NASA/Carla Thomas.*)

the refraction of sunlight. Ground-based schlieren systems that use the Sun as their light source have captured images of shock waves on supersonic aircraft in flight. However, these systems use complex, long-range imaging optics that do not provide the detailed spatial resolution of the shock wave structures.

Newly developed synthetic schlieren techniques apply computer imaging processing to obtain higher resolution images of shock waves in flight. One such method, background oriented schlieren, collects images of a background pattern with and without the supersonic aircraft in the image. The shock waves are captured as distortions in the background pattern, caused by the shock wave's density gradients, by comparing the images with and without the supersonic aircraft. The background pattern can be a natural ground pattern, so that images of the aircraft must be captured from above its flight path as it overflies the background. Figure 3.166 shows a background oriented

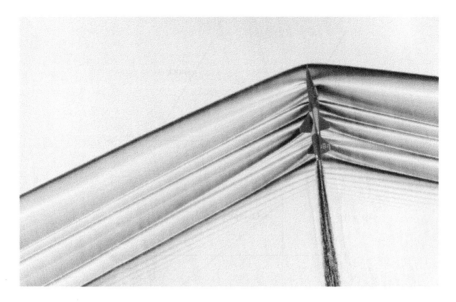

Figure 3.166 Background oriented schlieren image of a T-38 in supersonic flight. (Source: *NASA*.)

schlieren image of a supersonic T-38 aircraft. The T-38 flew over a natural desert background with the images being taken from a NASA Beechcraft *King Air* aircraft, positioned several thousand feet above the T-38.

3.11.4 Sonic Boom

As has been discussed, the supersonic flight of an aircraft produces shock waves, a "piling up" of pressure waves resulting in a discontinuous change in the flow properties such as Mach number, temperature, and pressure. For an aircraft flying at supersonic speeds, there are two primary shock wave systems emanating from the aircraft, a *bow shock wave* off the aircraft nose, and a *tail shock wave* off the aircraft empennage, as shown in Figure 3.167. Shock waves are generated off other aircraft components, such as the wing, engines, and canopy. These other shock waves trail from the aircraft, merging with one of the two primary shock waves in the far field, some distance from the aircraft. Although depicted as two-dimensional straight lines in Figure 3.167, the bow and tail shock waves are three-dimensional, cone-shaped wave structures. These conical shock waves expand outward as they trail downstream of the aircraft, with the base diameter of the conical shock continuously increasing. The angle of the conical shock wave, relative to the freestream flow or flight direction, is set by the freestream Mach number.

The local pressure increases discontinuously above the freestream static pressure, through the bow shock wave at the nose of the aircraft. The pressure then decreases continuously, along the length of the aircraft, to below the freestream static pressure. The flow then passes through the tail shock wave, and the pressure increases discontinuously once again, recovering back to freestream pressure. The pressure distribution just described resembles the shape of a letter "N"; hence it is called an *N-wave*. A simplified diagram of the bow and tail shock waves produced by a supersonic airplane and the N-wave is shown in Figure 3.167. The shock wave *overpressure* is a rapid, almost discontinuous change in the air pressure at the front of the N-wave. In reality, the pressure field around the airplane due to the shock waves is much more complex, with near, mid, and far-field

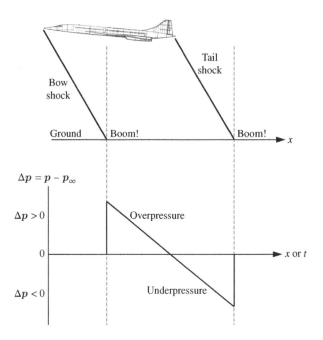

Figure 3.167   Pressure profile through bow and tail shock waves, the N-wave.

pressure profiles that are distinctly different. This is in part due to the fact that the shock waves emanating from the aircraft are not simply two distinct waves, rather there are multiple shock waves of different strengths and geometries that coalesce as they trail from the aircraft. As shown in the figure, the sudden changes in pressure, at the front and back of the N-wave, are heard, and sometimes felt, on the ground as the distinctive double sonic booms of a supersonic aircraft. This double-boom occurs in a time span of about a tenth of a second. Since the aircraft is traveling faster than the waves it is creating, it passes over an observer on the ground before the sonic boom is heard by the ground observer. The sonic boom is not heard inside the aircraft, as the aircraft is moving with the pressure change, rather than having the pressure change sweep by it.

As long as the aircraft is in supersonic flight, its conical sonic boom trails from the aircraft and sweeps across the ground, creating a continuous sonic boom *carpet* along its ground path. The width of the carpet on the ground is approximately one mile (1.6 km) for every 1000 ft (300 m) of altitude of the aircraft. Thus, if the aircraft is flying supersonically at an altitude of 45,000 ft (13,700 m), the width of the sonic boom carpet, sweeping across the ground, is about 45 miles (72 km). The actual movement of the aircraft's shock cone is affected by many atmospheric factors, such as the wind direction, wind speed, and atmospheric turbulence. The strongest sonic boom is felt directly underneath the flight path of the aircraft, its intensity decreasing with lateral distance from this centerline.

The intensity of a sonic boom is a function of many variables. A primary driver is the aircraft's altitude; the higher the aircraft, the less intense the sonic boom, since the shock waves are attenuated with distance. Shock waves, emanating from the aircraft, must travel through the atmosphere, so the sonic boom intensity is affected by atmospheric properties, such as temperature, pressure, humidity, pollution, wind, and turbulence. The sonic boom intensity increases with increasing supersonic Mach number, as would be expected by the increase in shock wave strength with increasing Mach number. The sonic boom overpressure profile is also affected by whether or not the aircraft is maneuvering, which changes its flight path and its wing loading.

Table 3.15 Sonic boom overpressure of various aircraft.

Aircraft	Mach number	Altitude	Overpressure
Lockheed SR-71	3.0	80,000 ft (24,400 m)	0.9 lb/ft^2 (43.1 Pa)
Concorde SST	2.0	52,000 ft (15,800 m)	1.94 lb/ft^2 (92.9 Pa)
Lockheed F-104	1.93	48,000 ft (14,600 m)	0.8 lb/ft^2 (38.3 Pa)
Space Shuttle (approach to land)	1.5	60,000 ft (18,300 m)	1.25 lb/ft^2 (59.9 Pa)

(Source: *Data from NASA Armstrong Fact Sheet: Sonic Booms.*)

The geometry, size, weight, and shape of the aircraft are all factors affecting the sonic boom intensity. The shock wave strength and sonic boom intensity are lower for slender aircraft geometries, that is, for lower values of the ratio of the aircraft's maximum cross-sectional area to its length. The boom intensity increases with the aircraft weight, since flying at higher weights require a larger lift coefficient and larger angle-of-attack, increasing the effective shock deflection angle. A lighter, slender aircraft produces a less intense sonic boom than a heavier, non-slender aircraft. This is seen in Table 3.15 by comparing the sonic boom overpressure of the *Concorde* and Lockheed F-104 *Starfighter* (Figure 3.178), flying at about the same Mach number and altitude. The lighter, more slender F-104 produces an overpressure of less than half that of the heavier, less slender *Concorde*. (The aircraft cross-sectional area includes the wing, where the *Concorde* has a much larger wing than the F-104.)

A measure of the strength of a sonic boom is its *overpressure*, the increase in the pressure above sea level atmospheric pressure (14.7 lb/in^2, 2116 lb/ft^2, 101,300 Pa) at the front of the N-wave. This change in air pressure is typically only a few pounds per square inch. The sonic boom overpressures for various aircraft, flying at supersonic speed at a given altitude, are shown in Table 3.15. The overpressure can have an effect on persons and objects on the ground. At less than 1 lb/ft^2 (0.007 lb/in^2, 48 Pa), the overpressure is audible as a sonic boom, but no harm or damage is expected to persons, structures, or other objects on the ground. However, even at these lower levels where no damage is done, some still find sonic booms objectionable or a nuisance. At an overpressure of about 2–5 lb/ft^2 (0.014–0.035 lb/in^2, 96–240 Pa), minor damage may be incurred by structures on the ground. At much higher overpressures, on the order of about 1000 lb/ft^2 (7 lb/in^2, 48,000 Pa), there may be harm to human eardrums, or possibly internal organs, such as lungs.

The public concern over sonic boom has affected the development of commercial supersonic transport aircraft, often called the SST. Overland, commercial supersonic flight is banned in the USA, due to the issues with sonic boom. Military aircraft are permitted to fly supersonically over land in the USA, in designated supersonic corridors. Currently, there are no industry standards for acceptable sonic boom characteristics of new aircraft designs. Two commercial SSTs have been developed in the past; the Soviet Tupolev Tu-144 (Figure 3.168) and the Aerospatiale-British Aircraft Corporation *Concorde* (Figure 3.169).

The Russians developed the first commercial, supersonic transport, the Tupolev Tu-144, which was capable of Mach 2.15 flight at 66,000 ft (20,000 m). The first flight of the prototype Tu-144 was on 31 December 1968. The Tu-144 had a long slender fuselage, large double-delta wing, two small, retractable canards surfaces on the forward fuselage, and a nose that could "droop" down for increased visibility for takeoff and landing. The supersonic aircraft was powered by four Kolesov turbofan engines with afterburners. The Tu-144 entered commercial service in 1977, but it only flew 55 passenger flights before ending this service in 1978. The aircraft continued to fly commercial cargo flights until 1983. Sixteen Tu-144 supersonic transports were built.

Concorde was a Mach 2 supersonic transport that was developed by a joint venture between the British Aircraft Corporation and the French Aerospatiale. The *Concorde* SST was in commercial,

Figure 3.168   Russian Tupolev Tu-144 commercial SST. (Source: *NASA*.)

Figure 3.169   Aerospatiale-British Aircraft Corporation *Concorde* commercial SST. (Source: *Eduard Marmet*, *"British Airways Concorde G-BOAC" https://en.wikipedia.org/wiki/File:British_Airways_ Concorde_G-BOAC_03.jpg, CC-BY-SA-3.0. License at https://creativecommons.org/licenses/by-sa/3.0/ legalcode*.)

passenger service from 1976 to 2003. Restricted from flying supersonically over land, the *Concorde* flew supersonic transatlantic flights from London, England and Paris, France to New York and Washington DC in the USA. Powered by four Rolls Royce/Snecma Olympus afterburning, turbojet engines, the aircraft could fly up to 128 passengers at up to Mach 2.04 at 60,000 ft (18,000 m). The aircraft had a long slender fuselage, large double-delta wing, and a nose capable of "drooping" for increased visibility during takeoff and landing. Twenty *Concorde* SSTs were built, six being prototype or test aircraft and fourteen aircraft entering commercial service. Although it was never a significant economic success, the *Concorde* demonstrated the technical feasibility of commercial supersonic transport.

Although there are no supersonic airliners in service today, research on commercial supersonic transport airplanes, including addressing the key issue of sonic boom reduction, has continued over the years. Recent attention has been placed on the development of a smaller, commercial supersonic business jet. Some significant sonic boom-related flight tests have been completed, where data has been collected to better understand sonic booms and to evaluate concepts for sonic boom reduction. Two such flight tests are described below.

Flight tests of a highly modified Northrop F-5 aircraft were performed in 2003 and 2004, to evaluate the ability to shape aircraft for significant reduction in sonic boom intensity. The lower fuselage of an F-5 supersonic jet aircraft was modified by attaching a shaped fairing, from the aircraft nose to the engine inlets, as shown in Figure 3.170. A joint program of NASA, DARPA, and Northrop Grumman, 21 flight tests were conducted over a two-year period. The feasibility of reducing the sonic boom intensity by properly shaping the aircraft geometry was demonstrated.

In 2006, Gulfstream Aerospace, Savanah, Georgia, teamed with NASA to flight test their patented *Quiet Spike* concept for sonic boom reduction. Envisioned as part of their supersonic

Figure 3.170   Northrop F-5 Shaped Supersonic Boom Demonstration (SSBD) aircraft. (Source: *NASA.*)

Figure 3.171   NASA F-15B *Quiet Spike* flight test. (Source: *NASA*.)

business jet design, the Gulfstream concept used a slender, telescoping boom on the aircraft nose
to weaken the bow shock wave in supersonic flight. The boom is extended for supersonic flight
and stowed, in a retracted position, during subsonic flight, including takeoff and landing. The
long boom is intended to weaken the intensity of the bow shock wave, much like a longer, more
slender aircraft fuselage. Over 50 test flights of the *Quiet Spike* boom, attached to the nose of a
NASA F-15B aircraft, were performed (Figure 3.171) with reductions measured in the sonic boom
intensity.

3.11.5   Lift and Drag of Supersonic Airfoils

In learning about supersonic flow, we see that it is fundamentally different from subsonic flow.
The appearance of shock waves and expansion waves significantly changes the flow patterns and
pressure distributions on bodies in supersonic flow. Based on this, it is obvious that the forces
and moments on a body, such as an airfoil, derived from these pressure distributions, are different
between subsonic and supersonic flows.

Consider the supersonic flow over a thin, supersonic airfoil, approximated by a thin, flat plate,
as shown in Figure 3.172. Recall from Section 3.11.2, that the strength of a shock wave increases
with larger flow deflection angles. In supersonic flow, a thick airfoil with a large nose radius has
a strong shock wave, perhaps a detached shock wave, at its leading edge, resulting in high wave
drag. Hence, supersonic airfoils are typically thin with sharp, rather than blunt, leading edges, to
reduce wave drag.

The flat plate, in Figure 3.172, is at an angle-of-attack α in a supersonic flow with Mach number
M_∞ and freestream pressure p_∞. Upon reaching the plate leading edge, the supersonic flow is
turned into itself on the bottom of the plate, resulting in an oblique shock wave emanating from

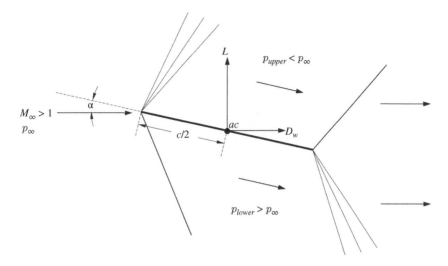

Figure 3.172   Supersonic flow over a thin, flat plate.

the bottom of the plate leading edge. In contrast, the flow is turned away from itself on the top of the plate, resulting in an expansion fan at the upper surface at the plate leading edge. At the plate trailing edge, the flow adjacent to the lower surface is turned back towards the freestream direction through an expansion fan and the upper surface flow is turned back through an oblique shock wave.

In passing through the oblique shock wave on the plate lower surface, the freestream flow pressure is increased, such that $p_{lower} > p_\infty$. This pressure is uniform over the entire region behind the shock wave, so that this is the pressure that is felt at the plate surface. The freestream flow, over the plate upper surface, experiences a pressure decrease in going through the expansion fan at the leading edge, such that $p_{upper} < p_\infty$. This lower pressure is felt on the plate upper surface. The pressure on the plate lower surface is greater in magnitude than on the upper surface, hence this pressure differential produces a resultant force, normal to the flat plate, that can be resolved into a lift, L, and wave drag, D_w, parallel and perpendicular to the freestream flow direction, respectively.

To obtain the actual values of the pressures, on the upper and lower surfaces of the plate, we would need to calculate the flow properties through the oblique shock and expansion waves. This can be done using *shock-expansion theory*, which is beyond the scope of the book. Using this method, we could calculate the actual values of the pressure on the flat plate surface, or the surface of more complex geometries. We would then need to integrate these pressure values over the surface of the body. This integration is simple for a simple geometry, such as the flat plate, but it becomes more complex with geometries that are more complex.

Another approach is based on linearizing the governing equations for supersonic flow, leading to closed form, approximate expressions for aerodynamic quantities that are only a function of the freestream Mach number and angle-of-attack. This *linearized supersonic theory* is also beyond the scope of the text, but the results from this theory are provided below. Based on this theory, the supersonic lift and wave drag coefficients, c_l and $c_{d,w}$, respectively, are given by

$$c_l = \frac{4\alpha}{\sqrt{M_\infty^2 - 1}} \tag{3.360}$$

$$c_{d,w} = \frac{4\alpha^2}{\sqrt{M_\infty^2 - 1}} = c_l \alpha \tag{3.361}$$

where M_∞ is the freestream, supersonic Mach number and α is the angle-of-attack of the body. These coefficients are only a function of the freestream Mach number and angle-of-attack; there are no terms related to the geometry of the body. Thus, these equations are applicable to an airfoil that can be approximated by a thin, flat plate. For an airfoil with thickness and camber, further terms can be added to the wave drag coefficient term, using linearized supersonic theory with slightly more complicated derivations, as

$$c_{d,w} = \frac{4}{\sqrt{M_\infty^2 - 1}}\alpha^2 + f\left(\frac{t}{c}\right) + g(C) \tag{3.362}$$

where $f(t/c)$ and $g(C)$ are functions related to the airfoil thickness and camber, respectively.

In supersonic flow, the aerodynamic center of the airfoil moves aft, relative to subsonic flow. The supersonic aerodynamic center is at the mid-chord of the airfoil, thus

$$x_{ac,supersonic} = \frac{c}{2} \tag{3.363}$$

By taking the derivative of Equation (3.360) with respect to angle-of-attack, we obtain the lift curve slope of an airfoil in supersonic flow, as

$$a_{supersonic} = c_{l_\alpha} = \frac{\partial c_l}{\partial \alpha} = \frac{4}{\sqrt{M_\infty^2 - 1}} \tag{3.364}$$

Here, the supersonic lift curve slope is only a function of the freestream Mach number.

Dividing Equation (3.360) by (3.361), we obtain the inviscid lift-to-drag ratio of a supersonic airfoil as

$$\left(\frac{L}{D}\right)_{supersonic} = \frac{c_l}{c_{d,w}} = \frac{1}{\alpha} \tag{3.365}$$

This is the inviscid lift-to-drag ratio since only the pressure-based wave drag is considered; there are no viscous considerations. The supersonic lift-to-drag ratio is inversely proportional to the angle-of-attack.

The supersonic lift and drag coefficients, as given by linearized supersonic theory in Equations (3.360) and (3.361), are quantitatively compared with values obtained in other speed regimes in Section 3.14.

Example 3.17 Aerodynamics of a Supersonic Airfoil *The wing of the supersonic Lockheed F-16 has a NACA 64A204 airfoil section with a thickness ratio, t/c, of 4%. Calculate the lift coefficient, wave drag coefficient, lift curve slope, and lift-to-drag ratio for this airfoil at a Mach number of 1.4 and angle-of-attack of 4.6°.*

Solution

The F-16 wing may be considered as "thin", since the section has a thickness of only 4% of the wing chord length. Based on this, the wing section may be approximated by a thin flat plate where the equations obtained from linearized supersonic theory are applied.

First, we convert the angle-of-attack from degrees to radians.

$$\alpha = 4.6 \deg \times \frac{\pi}{180} = 0.08029$$

Using Equations (3.360), (3.361), (3.364), and (3.365), we have

$$c_l = \frac{4\alpha}{\sqrt{M_\infty^2 - 1}} = \frac{4(0.08029)}{\sqrt{(1.4)^2 - 1}} = 0.3278$$

$$c_{d,w} = c_l \alpha = (0.3278)(0.08029) = 0.002113$$

$$c_{l_\alpha} = \frac{4}{\sqrt{M_\infty^2 - 1}} = \frac{4}{\sqrt{(1.4)^2 - 1}} = 4.082\,\text{rad}^{-1} = 0.07125\,\text{deg}^{-1}$$

$$\left(\frac{L}{D}\right)_{supersonic} = \frac{1}{\alpha} = \frac{1}{0.08029} = 12.45$$

3.11.6 Supercritical Airfoils

As has been discussed, transonic drag divergence limits the maximum subsonic cruise speed of an aircraft. The drag rise is due to the acceleration of the flow, over the top surface of the airfoil, to locally, supersonic speed and the creation of a wave drag-producing, strong shock wave. Therefore, one might think that an airfoil's transonic drag characteristics could be favorably changed by proper shaping of the airfoil upper surface. This is indeed true, and we can design specific airfoil shapes, called *supercritical* airfoils, that delay the onset and severity of the transonic drag rise, resulting in an increase in the subsonic cruise speed. Supercritical airfoils shapes have reduced curvature in the mid-chord region of their upper surface, resulting in a much flatter upper surface than a conventional airfoil. The camber near the trailing edge of a supercritical airfoil is also greater than a conventional airfoil. The shape of a supercritical airfoil is compared with that of a conventional airfoil in Figure 3.173.

The transonic flow patterns and the associated surface pressure distributions for a supercritical airfoil are compared with those for a conventional airfoil in Figure 3.174. The shock wave on the upper surface of a supercritical airfoil is further aft and weaker than on the conventional airfoil. This further aft, weaker shock wave results in less boundary layer separation on the airfoil surface and therefore less loss of lift and less drag than the conventional airfoil shape. The flatter upper surface of the supercritical airfoil gives a near-constant surface pressure over the top surface, creating more lift towards the aft end of the section.

The increase in the drag divergence Mach number with a supercritical airfoil is shown in Figure 3.175. The drag divergence characteristics of a supercritical airfoil is compared with that of conventional subsonic airfoil (NACA 64-212). The NACA 64-212 is a laminar flow, low-camber airfoil with 12% thickness. This airfoil shape has been used on some general aviation airplanes and early jet aircraft. The drag divergence of each airfoil is evident in Figure 3.175, by the rapid increase in the drag coefficient at a high subsonic Mach number. The drag divergence Mach number of the NACA 64-212 airfoil is seen to be about 0.68, while it is almost Mach 0.8 for the supercritical airfoil. Thus, we see for this comparison, the supercritical airfoil offers a significant

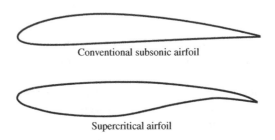

Conventional subsonic airfoil

Supercritical airfoil

Figure 3.173 Conventional and supercritical airfoil shapes.

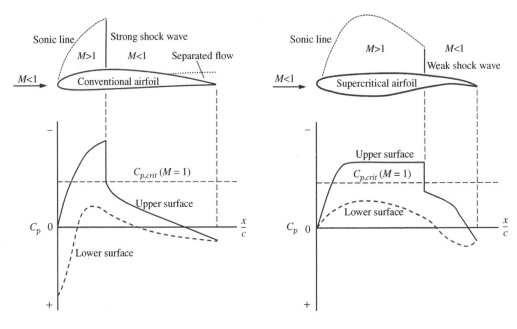

Figure 3.174   Transonic flow patterns and pressure distributions over conventional and supercritical airfoils.

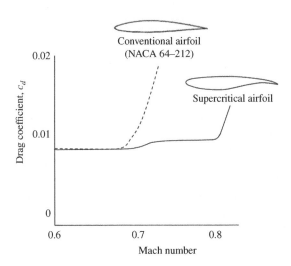

Figure 3.175   Increase in the drag divergence Mach number with a supercritical airfoil.

increase in the drag divergence Mach number: about 0.12. The increase in the drag divergence Mach number enables an aircraft with a supercritical wing to cruise at higher subsonic Mach numbers.

In addition to the transonic aerodynamic benefits, the use of a supercritical airfoil can result in reduced structural weight of the wing. The thicker cross-section of the supercritical airfoil provides more internal volume for a more efficient internal structure. It may also be possible to design a wing with less wing sweep, since the drag divergence Mach number is greater for a supercritical airfoil, leading to a lighter wing structure.

Figure 3.176   NASA TF-8A supercritical wing (SCW) research aircraft. (Source: *NASA*.)

The supercritical airfoil was designed by aerodynamicist Richard Whitcomb (1921–2009), Chief of the Transonic Aerodynamics Branch at NASA Langley Research Center in the 1960s. Whitcomb performed a series of tests in the NASA Langley 8-foot transonic wind tunnel, experimentally verifying the viability of the supercritical airfoil. The first flight tests of a supercritical airfoil concept were on a North American T-2C *Buckeye* with a modified supercritical wing. NASA later modified a Voight TF-8A *Crusader* as a supercritical wing (SCW) research aircraft (Figure 3.176). Flight tests of the SCW proved that a supercritical wing increase the cruise speed, fuel efficiency, and range of a high-subsonic speed aircraft. Most modern aircraft that cruise at high subsonic Mach numbers, such as commercial airliners, incorporate a supercritical airfoil shape in the wing design.

3.11.7   Wings for Supersonic Flight

As discussed in Section 3.10.2, shock waves start to form on the wings of a high-speed aircraft at the critical Mach number, leading to drag divergence, high wave drag, and possible stability and control issues. In this and the next several sections, we discuss several ways to reduce transonic and supersonic drag or to delay the onset of drag divergence. We discuss several design features that make aircraft wing or fuselage geometries optimum for flight at transonic and supersonic speeds. In this section, we focus on the wing in supersonic flight.

By the 1940s, high-performance, propeller-driven airplanes and new jet-powered airplanes were flying near the speed of sound. These airplanes were designed with straight wings, typically having thickness ratios of about 14–18%. At the time, decreasing the airfoil section thickness was the only known way of increasing the critical Mach number. Even the first airplane to break the speed of sound, the Bell X-1, had a thin, straight wing. Further improvements in delaying drag divergence focused on the design of the wing planform shape. We discuss three different approaches to the supersonic wing planform design, the low-aspect ratio straight wing, the swept wing, and the delta wing. In some respects, the delta wing may be considered a subset of the swept wing.

3.11.7.1 Thin, Low-Aspect Ratio, Straight Wings

In Section 3.11.5, the wave drag of a thin, flat plate airfoil, $c_{d,w}$, based on linearized supersonic theory, was given as

$$c_{d,w} = \frac{4\alpha^2}{\sqrt{M_\infty^2 - 1}} \qquad (3.366)$$

and for an airfoil with thickness and camber, the wave drag was given as

$$c_{d,w} = \frac{4}{\sqrt{M_\infty^2 - 1}}\alpha^2 + f\left(\frac{t}{c}\right) + g(C) \qquad (3.367)$$

Thus, the wave drag of a supersonic airfoil varies with the angle-of-attack, α, thickness-to-chord ratio, t/c, and camber. Lower values of wave drag are obtained for thin airfoils with no camber at low angles-of-attack.

The wave drag of a thin, flat plate with a finite span or finite aspect ratio, $C_{D,w}$, is given in [12] as

$$C_{D,w} = c_{d,w}\left(1 - \frac{1}{2AR\sqrt{M_\infty^2 - 1}}\right) \qquad (3.368)$$

The qualitative trend of this equation is that the wave drag coefficient of a (thin flat plate) finite wing decreases with decreasing aspect ratio. It should be noted that the calculation of the wave drag of an actual finite wing in supersonic flow is quite a bit more complicated than for subsonic flow, where simple engineering relationships are available, at least for approximate predictions. Often, computational fluid dynamic solutions are applied for the prediction of lift and drag of a supersonic wing. Approximate, engineering solutions can be obtained using relationships, based on supersonic linear theory, and empirically based constants in these relationships. For our purposes, we take it for granted that the wave drag coefficient decreases with decreasing aspect ratio. Based on this, early supersonic airplanes tended to have very thin, low aspect ratio wings. While low aspect ratio is advantageous for supersonic flight, it is not beneficial for low speed flight, where a large aspect ratio wing is desired. These low aspect ratio wings were getting so thin that they created other non-aerodynamic issues, such as structural problems with carrying flight loads and packaging issues with landing gear, fuel, or actuators for ailerons.

The Douglas X-3 *Stiletto* research aircraft, shown in Figure 3.177, was specifically designed to investigate sustained, high Mach supersonic flight for airplanes with thin, low aspect ratio wings. The X-3 had an extremely slender fuselage, with a very long, pointed nose ahead of the single-place cockpit. Its 66.75 ft (20.35 m) long fuselage was almost three times longer than its stubby, 22.7 ft (6.92 m) wingspan. With a maximum takeoff weight of 22,100 lb (10,800 kg) and a wing area of 166.5 ft^2 (15.47 m^2), the X-3 had a high wing loading of 132.7 lb/ft^2 (647.9 kg/m^2). Because of its extremely small wing, the X-3 had a high takeoff speed of 260 knots (299 mph, 482 km/h). The X-3 was the first airplane to utilize titanium for major structural components, including the wings, which were fabricated from a solid piece of titanium.

The X-3 wing epitomized "thin" and "low aspect ratio", with a hexagonal-shaped airfoil section, a maximum thickness ratio of 4.5%, and a wing aspect ratio of 3.09. The wing had zero incidence,

Figure 3.177 Douglas X-3 *Stiletto* with thin, low-aspect ratio, straight wing. (Source: *NASA*.)

zero dihedral, and zero sweep at the wing ¾-chord line. (We learn about the significance of wing dihedral in Chapter 6). The wing leading edges were so thin and sharp that ground personnel had to be wary of the potential hazard of being cut. The X-3 fuselage and wing designs were focused on minimizing wave drag.

The X-3 was powered by two Westinghouse J34 afterburning turbojet engines, lower-thrust replacements for the originally planned, higher-thrust General Electric J79 turbojets, which were not ready at the time of its first flight. Designed to fly at sustained, cruise speeds up to Mach 2, the X-3 was underpowered and incapable of going supersonic in level flight. The first flight of the X-3 was on 20 October 1952. Later in its 51-flight test program, the X-3 ultimately reached a top speed of Mach 1.2, albeit in a 30° dive.

While the X-3 did not achieve its high Mach flight goals, it did advance the state-of-the-art in many other areas of supersonic aircraft design. The data about its thin, low aspect ratio wing was used in the design of several future supersonic airplanes, most notably the successful Lockheed F-104 *Starfighter*, shown in Figure 2.34. A three-view drawing of the F-104 is shown in Figure 3.178. Similar to the X-3, the F-104 has a small, low aspect ratio wing, with a thickness ratio of 3.36% and an aspect ratio of 2.45. The X-3 also obtained important stability and control data about supersonic aircraft configurations with most of their mass in their fuselage, and little mass in their wings, so-called *fuselage-loaded* configurations.

3.11.7.2 Swept Wings

The idea of using wing sweep to increase the critical Mach number was invented independently by German aerodynamicist Adolf Busemann (1901–1986) in 1935 and later, in 1945, by American

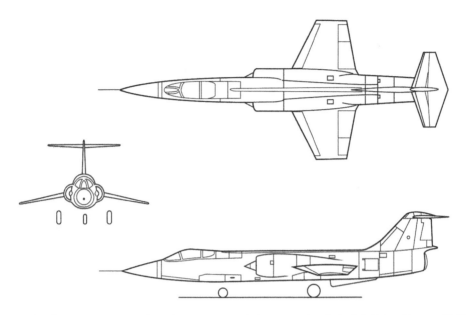

Figure 3.178   Three-view drawing of the F-104 *Starfighter*, capable of Mach 2 flight. (Source: *NASA*.)

aerodynamicist Robert T. Jones (1910–1999). Busemann presented his ideas about the swept wing for high-speed flight, at the 5th Volta Congress on "High Velocities in Aviation" in Rome, Italy in September 1935. Despite the many, world-class aerodynamicists and engineers in attendance, Busemann's presentation about the potential drag reduction offered by swept wings at supersonic speeds went almost unnoticed. This did not stop the German supersonics research on swept wings under Busemann at the Aeronautics Research Laboratory in Braunschweig, Germany. It would not be until after World War II that the USA discovered the large amount of technical data that Germany had about the swept wing. Adolf Busemann assisted with the data transfer and eventually emigrated to the USA, continuing his research at the NACA's Langley Memorial Laboratory. This occurred at about the same time that NACA aerodynamicist Robert T. Jones independently conceived of the use of swept wings for supersonic flight.

The swept wing was actually not new to aviation prior to 1935. Swept wing aircraft had already been flying, but these were subsonic airplanes without tails. As we learn in Chapter 6, there are stability and control advantages of a swept wing design for a tailless airplane, a configuration being used in glider and powered airplane designs in the 1930s. The idea of applying the swept wing to supersonic flight was new in 1935.

In late 1939, the first wind tunnel measurements of a swept wing were made by Hubert Ludwieg at Braunschweig, where Busemann was now the head of the German Institute of Aviation Research. Figure 3.179 shows drag polar data from Ludwieg's testing, where an unswept wing ($\varphi = 0$) and a wing, with a sweep angle of 45° ($\varphi = 45°$), were tested at Mach 0.7 and 0.9 and a Reynolds number of 450,000, based on a center-chord length, l_i, of 23 mm (0.91 in). The swept wing drag polar indicates significantly lower drag coefficients as compared with the unswept wing at the same lift coefficients. For example, comparing the Mach 0.9 data at zero lift ($C_L = 0$), the drag coefficient, C_D, for the unswept wing is about 0.1 and less than half that, about 0.04, for the swept wing. At a higher lift coefficient of 0.7, the Mach 0.9 drag coefficient is about 0.26 for the unswept wing and about 0.18 for the swept wing. This wind tunnel data was the first experimental verification of Busemann's theoretical predictions about wave drag reduction by using wing sweep.

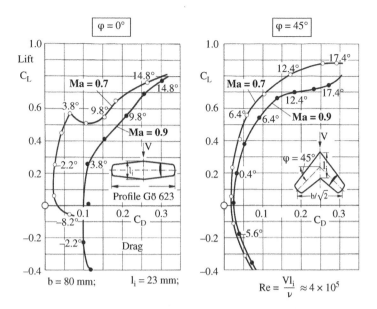

Figure 3.179   First wind tunnel data of a swept wing, Hubert Ludweig, 1939. (Source: *Reproduced with permission of the German Aerospace Center, DLR.*)

The German research on swept wings did not remain in the laboratory. They designed and flew the first swept wing, jet-powered aircraft, the Junkers Ju-287, on 16 August 1944. The Ju-287 was an experimental prototype of a high-speed heavy bomber. Interestingly, this first swept-wing jet airplane had *forward-swept* wings rather than aft swept wings. The Ju-287 had a 66-foot (20 m) wingspan with a forward sweep angle of 25°. Aerodynamically, in terms of wave drag reduction, the flow does not care whether the wing is swept forward or aft. However, there are other issues with forward-swept wings, such as aeroelastic effects at high speeds, that must be addressed.

The forward sweep of the Ju-287 wing was selected based on other considerations. Wind tunnel tests had indicated that an aft-swept wing would stall first at the wingtips, leading to loss of aileron control and stability issues. The forward-swept wing was selected because it stalled at the wing root first, which led to a more benign stall behavior. Another factor concerned the wing structure and location of the bomb bay. With an aft sweep wing design, the passage of the wing spar through the fuselage interfered with the placement of the bomb bay. The wing spar in a forward sweep design passed behind the desired location of the bomb bay.

The Ju-287 was powered by four Jumo 004B turbojet engines, with two mounted under the wings and two mounted on the sides of the fuselage, near the aircraft nose. The Ju-287 prototype completed 17 test flights, reaching a maximum speed of 340 mph (550 km/h). This prototype was destroyed in a bombing raid in 1945, before the end of World War II. Two other prototype aircraft were in fabrication, but were never completed before the war ended and the unfinished aircraft were captured by the Russian Army. Although the Ju-287 never flew to transonic Mach numbers, wind tunnel testing had proven the feasibility of its swept wings to increase the critical Mach number and reduce wave drag. As a final note, the German designer of the Ju-287, Hans Wocke, went on to design the HFB 320 *Hansa Jet* in the 1960s, a business jet with a forward-swept wing. The *Hansa Jet* is the only forward-swept wing jet to ever enter production, with 47 copies built.

Let us now investigate the aerodynamics of how wing sweep increases the critical Mach number. The first explanation is a bit of an oversimplification, but it provides an upper bound to

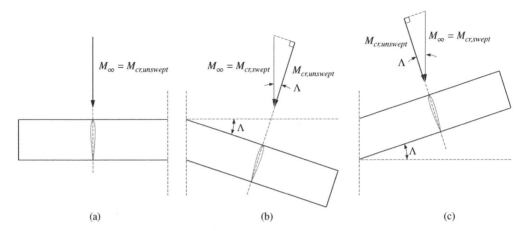

Figure 3.180 Effect of wing sweep relative to Mach number normal to wing leading edge for (a) straight, unswept wing, (b) aft-swept wing, and (c) forward-swept wing.

the potential increase in the critical Mach number. Consider a straight, unswept wing in a flow with a freestream Mach number equal to the critical Mach number of the wing's airfoil section, $M_\infty = M_{cr,unswept}$, as shown in Figure 3.180a. Now, assume that the wing is swept at an angle Λ, as shown in Figure 3.180b. The airfoil section of the swept wing is the same as the straight wing, so that it has a critical Mach number $M_{cr,unswept}$, but the swept-wing section "sees" this critical Mach number at an angle Λ with respect to the freestream flow direction. Therefore, the critical Mach number "seen" by the swept wing is given by

$$M_{cr,swept} = \frac{M_{cr,unswept}}{\cos \Lambda} \tag{3.369}$$

Since $\cos \Lambda$ is less than one, $M_{cr,swept} > M_{cr,unswept}$. Thus, the critical Mach number of the swept wing is increased by the factor $1/\cos \Lambda$. This increase in the critical Mach number also applies to a forward-swept wing, as shown in Figure 3.180c.

For example, if the straight wing has a critical Mach number, $M_{cr,unswept}$, of 0.75, a wing with 20° of sweep has an increased critical Mach number, according to Equation (3.369), of

$$M_{cr,swept} = \frac{M_{cr,unswept}}{\cos \Lambda} = \frac{0.75}{\cos(20°)} = 0.80 \tag{3.370}$$

The swept wing critical Mach number, given by Equation (3.369), represents an upper bound to the potential increase due to wing sweep. This equation assumes that the flow over the swept wing can be modeled as a simple two-dimensional flow over individual airfoil sections, when the swept wing flow is really highly three-dimensional due to spanwise flow.

Another perspective of wing sweep is that it makes the airfoil section "look" thinner to the freestream flow. Consider the straight, unswept wing shown on the left in Figure 3.181. The freestream flow "sees" an airfoil section with a chord length $c_{unswept}$, as shown. For the swept wing, shown on the right of Figure 3.181, the freestream flow "sees" an airfoil section with the same thickness, t, as the unswept wing, but with a longer chord length c_{swept}. Since the thickness-to-chord, t/c, ratio of the swept wing airfoil is smaller than for the unswept wing, that is, $t/c_{unswept} < t/c_{swept}$, the flow "sees" a thinner wing when the wing is swept. As discussed previously, the wave drag is lower and the critical Mach number is greater for a thinner airfoil section.

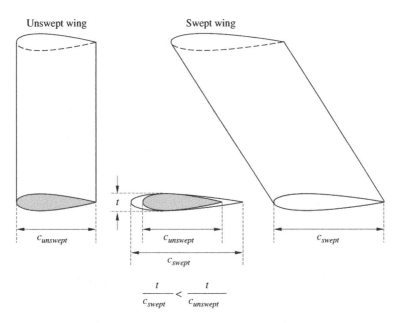

Figure 3.181   Reduction of effective thickness-to-chord ratio using wing sweep. (Source: *Adapted from Talay, NASA SP 367, 1975, [65].*)

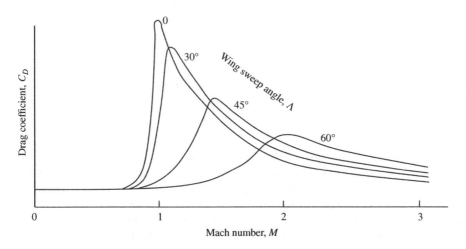

Figure 3.182   Effect of wing sweep on the drag coefficient. (Source: *Hurt, US Navy NAVWEPS 00–80 T-80, 1965, [39].*)

The effect of wing sweep on the drag coefficient as a function of freestream Mach number is shown in Figure 3.182. The increases in the critical Mach number and the drag divergence Mach number, with increasing wing sweep, is evident. The peak in the transonic drag coefficient also decreases with increasing wing sweep.

While there is a reduction in the drag coefficient at transonic and supersonic speeds, wing sweep generally reduces the wing lift. The effect of wing sweep on the subsonic lift coefficient is shown in Figure 3.183, where the lift curve of a swept wing is compared with that of a straight, unswept

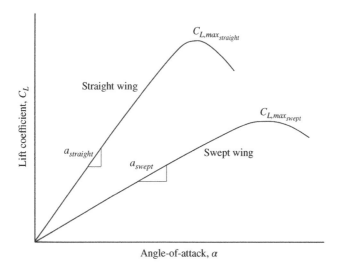

Figure 3.183   Effect of wing sweep on the subsonic lift coefficient.

wing. For a wing of given aspect ratio at subsonic speed, wing sweep decreases the lift curve slope, $dC_L/d\alpha$, decreases the maximum lift coefficient, $C_{L,max}$, and increases the stall angle-of-attack, α_s. The decrease in the lift coefficient also results in a decrease in the wing lift-to-drag ratio, L/D. For a swept wing aircraft, the decrease in the lift coefficient at low speeds results in higher takeoff and landing speeds and higher angles-of-attack, which may reduce cockpit visibility. High-lift devices are required to reduce takeoff and landing speeds and angles-of-attack to acceptable levels. Aerodynamic stall tends to occur outboard, near the wingtips of swept wing, where control surfaces are located, which may lead to controllability issues. There are other stability and control effects, due to wing sweep, which are discussed in Chapter 6.

3.11.7.3   Delta Wings

The limiting case of a swept wing, as the wing taper ratio, $\lambda = c_t/c_r$, goes to zero, is the *delta wing*. The simple delta wing has a triangular planform shape, but there are several variations of this basic shape, as shown in Figure 3.184. The delta wing configuration has found wide application on a variety of supersonic aircraft designs. The delta wing design was pioneered by German aerodynamicist Alexander M. Lippisch (1894–1976) during the 1930s. Lippisch designed several delta wing aircraft and gliders, but with thick wing sections. The British built upon Lippisch's work and developed several delta wing jet aircraft in the 1950s, including the Avro *Vulcan* bomber and the Gloster *Javelin* fighter. In the USA, several successful, supersonic delta wing aircraft were built during the 1950s, using aerodynamic theories for thin delta wings developed by NACA aerodynamicist Robert T. Jones. Many of these early delta wing aircraft were designed and built by the Convair aircraft company, including the first US delta wing aircraft, the XF-92 *Dart*, shown in Figure 3.185. In addition to its delta wing, the *Dart* also had a large, triangular, delta-shaped vertical tail. Convair followed the *Dart* with other successful delta wing aircraft, including the F-102 *Delta Dagger*, F-106 *Delta Dart* (Figure 3.201) and B-58 *Hustler*. The delta wing is used in several modern supersonic aircraft, in a variety of configurations, such as those shown in Figure 3.184. Notable among these are the double-delta configuration used on the Space Shuttle (see Figure 1.79) and the ogival delta used on the *Concorde* SST (Figure 3.169).

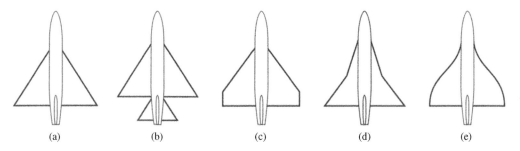

Figure 3.184 Delta wing configurations, (a) simple delta, tailless (b) simple delta with tail, (c) cropped delta, (d) double delta, and (e) ogival delta.

Figure 3.185 Convair XF-92A *Dart*, the first US delta wing aircraft. (Source: *NASA*.)

The aerodynamics of delta wings are complex, three-dimensional flows, due to the non-linear effects associated with a highly swept wing leading edge. However, for small angle-of-attack and small aspect ratio, a linear theory, known as slender wing theory, can be applied to delta wings, where the flow is assumed to be two-dimensional in the cross-flow planes cutting through the wing, perpendicular to the flow direction. Using this linear theory, the lift coefficient, C_L, and the induced drag coefficient, $C_{D,i}$, for a low aspect ratio delta wing at small angle-of-attack are given by

$$C_L = \frac{\pi AR}{2}\alpha \tag{3.371}$$

$$C_{D,i} = \frac{C_L^2}{\pi AR} = \frac{\pi AR}{4}\alpha^2 \tag{3.372}$$

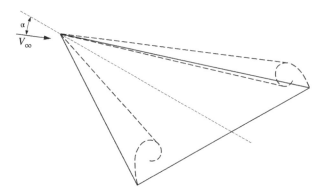

Figure 3.186   Subsonic flow over delta wing at angle-of-attack.

where AR is the delta wing aspect ratio and α is the absolute angle-of-attack. The aspect ratio, AR, of a delta wing is defined as

$$AR = \frac{2b}{c_0} \tag{3.373}$$

where b is the wingspan at the trailing edge and c_0 is the centerline chord length. Equations (3.371) and (3.372) are independent of Mach number and within the limits of the linear theory, used to derive these relationships, they are valid for subsonic and supersonic speeds, albeit only for small angle-of-attack and small aspect ratio.

While we have been focused on the supersonic aspects of swept wings, these wings must also have adequate aerodynamic performance at subsonic speeds. Before we discuss other supersonic characteristics of delta wings, it is worthwhile to obtain an appreciation for the subsonic aerodynamics of these kinds of wings and the tradeoffs that are needed for satisfactory operation in both flight regimes.

For angles-of-attack, which may not be considered "small", the subsonic flow over the delta wing is dominated by a pair of vortices along the swept leading edge of the upper surface, as shown in Figure 3.186. These vortices are created by the high-pressure air, underneath the wing, flowing around the leading edge to the low-pressure region on the upper surface. In attempting to curl around the sharp leading edges, the flow separates, generating the spiraling primary vortices, which "stand off" from the surface, just inboard of the leading edges. The primary vortices are high velocity, horizontal "tornados" that create low pressure, suction regions along the wing leading edges. At higher angle-of-attack, the leading edge suction significantly increases the lift. This additional lift, called *vortex lift*, extends to much higher angles-of-attack than for straight wings. The stall angle-of-attack of a delta wing can be as high as 25–35°. Although the lift curve of a delta wing extends to very high angle-of-attack, the lift coefficient reaches a maximum of about 1.3 because the lift curve slope of a delta wing is small, about 0.05 degree^{-1}, as compared to a straight wing. This leads to very high angle-of-attack during landing, which impedes forward visibility from the cockpit, and calls for creative engineering solutions. For instance, the delta wing supersonic transports, the Russian Tupolev Tu-144 and Angelo-French *Concorde*, droop their noses during landing to enhance forward visibility, as shown in Figure 3.168 and Figure 3.169, respectively.

The leading edge suction produces a resultant force, which not only increases the lift, but also has a component in the drag direction. Despite the large vortex lift increase, the drag increase is substantial enough that the delta wing subsonic lift-to-drag ratio is lower than a straight wing.

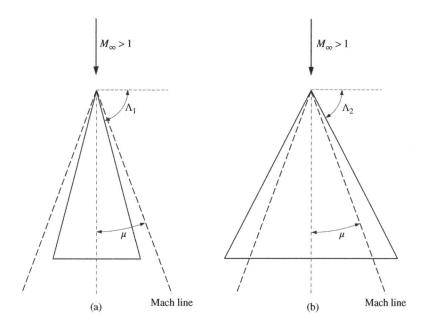

Figure 3.187   Relationship between wing sweep and Mach cone, (a) wing leading edge inside Mach cone, (b) wing leading edge outside Mach cone.

The delta wing lift-to-drag ratio could be increased by minimizing the flow separation around the leading edges, which could be achieved with rounded rather than sharp leading edges. While this is aerodynamically beneficial at subsonic speed, rounded leading edges result in high wave drag at supersonic speed, similar to the drag increase one would expect for a blunt versus a sharp-nosed body in supersonic flow. Since the primary purpose of using a highly swept, delta wing configuration is for supersonic flight, the design trade favors maintaining sharp leading edges.

Consider the wing sweep of a commercial airliner and a military fighter jet. The wing sweep angles are significantly different for these two different types of aircraft. What determines the sweep angle for an aircraft? To answer this, consider the supersonic flow over two wings with different sweep angles, as shown in Figure 3.187. Since the flow is supersonic, there is a Mach cone emanating from the vertex of the wing. (The flow is three-dimensional, so there is a Mach cone, rather than just a Mach wave, as described in Section 3.11.2.1.) The supersonic Mach number, M_∞, is the same for both wings, so the Mach cone angle, μ, is the same for both. Recall that the Mach angle is given by Equation (3.353), as

$$\mu = \sin^{-1} \frac{1}{M_\infty} \tag{3.374}$$

The wing on the left has a sweep angle Λ_1, which is less than the sweep angle Λ_2 of the wing on the right. In Figure 3.187a, the wing leading edge is inside the Mach cone, thus the component of the freestream Mach number normal to the leading edge is subsonic. Even though the flow over the wing is supersonic, this is called a *subsonic leading edge*. In Figure 3.187b, the leading edge is outside the Mach cone, thus the component of the freestream Mach number normal to the leading edge is supersonic. For this *supersonic leading edge*, a shock wave is formed ahead of the leading edge, resulting in additional wave drag, greater than for the subsonic leading edge case.

Therefore, to reduce the wave drag, a wing sweep angle should be selected such that the leading edges are inside the Mach cone. If the wing sweep is selected based on the maximum or cruise

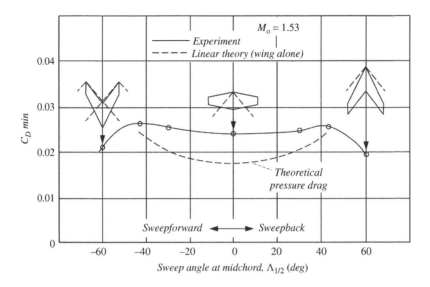

Figure 3.188   Effect of wing sweep on the minimum drag. (Source: *Vincenti, NACA TR-1033, 1951, [71].*)

flight Mach number of the aircraft, the wing leading edges are inside the Mach cone for all Mach numbers less than this selected Mach number, since the Mach cone angle increases with decreasing Mach number. Therefore, for a commercial airliner with a maximum or cruise Mach number that is near sonic, the Mach cone angle is large, near 90°, so that the wing sweep angle can be small. For a military jet with a maximum Mach number of 2.5, the Mach cone angle is 23.6°, requiring a wing sweep angle of no less than 66.4°.

The effect of the wing sweep angle on the minimum drag, at a freestream Mach number of 1.53, is shown in Figure 3.188. Positive sweep angles correspond to aft-swept wings and negative sweep angles correspond to forward-swept wings. The Mach cone angle corresponding to the freestream Mach number is 40.8°. The minimum drag is approximately constant for wing sweep angles of about ±43°. These sweep angles correspond to wings with supersonic leading edges, where the sweep angle is greater (in absolute magnitude) than the Mach cone angle. The drag coefficient decreases dramatically for wing sweep angles greater (in absolute magnitude) than about 43°, where the wing leading edges are subsonic.

3.11.7.4 Variable Sweep Wings

We have seen that it is possible to design a low aspect ratio swept wing that has improved aerodynamic performance and efficiency at supersonic speeds. However, the swept wing may have poor subsonic aerodynamic characteristics, such as high induced drag, low lift-to-drag ratio, or high angles-of-attack to achieve maximum lift. One solution to this problem is the use of a *variable sweep wing*, where the sweep angle of the wing is varied in flight. The advantages of this concept are shown in Figure 3.189, where the maximum lift-to-drag ratio, $(L/D)_{max}$, is plotted versus Mach number for three wing configurations, an optimum straight wing, an optimum swept wing, and a variable sweep wing. The straight wing has a high $(L/D)_{max}$ at subsonic speeds, but it has poor aerodynamic performance at supersonic speeds, perhaps being incapable of supersonic flight at all. In contrast, the swept wing has a poor $(L/D)_{max}$ at subsonic speeds and a higher $(L/D)_{max}$ at supersonic speeds than the optimum straight wing. The variable sweep wing has the aerodynamic

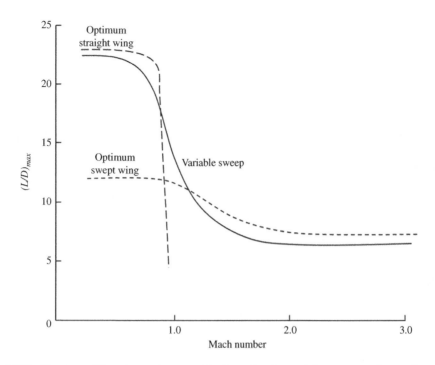

Figure 3.189   Maximum lift-to-drag ratio versus Mach number for straight, unswept wing, optimum swept wing, and variable-sweep wing. (Source: *Talay, NASA SP-367, 1975, [65].)*

benefits of both the straight and the swept wings, nearly matching their maximum lift-to-drag performance at subsonic and supersonic speeds. The downside of incorporating variable sweep is the additional weight and mechanical complexity involved.

The variable wing sweep can be set to a near-straight wing angle, optimum for subsonic flight conditions, such as takeoff, landing, and climb, and to a highly swept angle for transonic and supersonic flight. An intermediate sweep angle may be used for cruise at high subsonic speeds. The variation of wing sweep is shown in the wing sweep sequence of the General Dynamics F-111 *Aardvark*, from a near straight wing to a highly swept wing, in Figure 3.190. The F-111 wing sweep can be varied from 16°, at the full forward or wings spread position, to a maximum aft sweep of 72.5°. By moving the wings from the fully swept to the spread position, the wingspan increases from 32 ft (9.75 m) to 63 ft (19.2 m). This increase in wingspan increases the wing aspect ratio from 1.95 to 7.56, which provides a maximum lift-to-drag ratio of 15.8 at subsonic speeds. As shown in Figure 3.190, the wings and horizontal tail form a delta wing configuration when the wings are swept fully aft.

The rotation or pivot point for the variable sweep wing, such as on the F-111 and other modern aircraft, is outboard of the fuselage centerline. If the pivot point is located on the centerline, the aft movement of the weight of the wings results in a significant aft shift of the aircraft center of gravity (CG). This aft CG shift increases a longitudinal stability parameter, called the static margin (to be discussed in Chapter 6), which increases the aircraft longitudinal stability. This increased stability is not necessarily a good thing, as it decreases the maneuverability of the aircraft. The aft CGshift also requires more longitudinal trim, hence increasing the trim drag. This CG shift with a centerline pivot point can be eliminated by translating the wing forward, when it is rotated, but this greatly increases the mechanical complexity. The alternative solution, developed by NASA research, is to use an outboard pivot point, which does not result in the aft CG shift with wing sweep.

Figure 3.190   Wing sweep sequence for General Dynamics F-111 *Aardvark*. (Source: *US Air Force*.)

The first variable sweep aircraft, where the wing sweep could be changed in flight, was the Bell X-5 research aircraft, shown in Figure 3.191. The design was similar to the German Messerschmitt P.1011 prototype fighter aircraft, which had variable sweep wings, but the sweep could only be changed manually on the ground. The Messerschmitt P.1011 never flew and was captured by the USA after World War II. The German variable-sweep aircraft was shipped to the Bell factory in the United States, where it was studied by the Bell engineers prior to the design of the X-5. However, unlike the P.1011, the Bell X-5 could change its wing sweep in flight to three sweep positions of 20, 40, or 60°.

The first flight of the Bell X-5 was on 20 June 1951. Two X-5 aircraft were built, completing test flights up to Mach 0.9. Valuable aerodynamic, stability, and control data were obtained for variable sweep aircraft at transonic speeds. Unfortunately, the X-5 had poor spin characteristics and one aircraft was lost due to an unrecoverable spin, with the wings in the 60° sweep position.

The effect of wing sweep on aerodynamic performance is shown for the Bell X-5 in Figure 3.192. The lift coefficient, C_L, drag coefficient, C_D, and excess thrust, $F_n - D$, are plotted versus Mach number for wing sweep angles of 20 and 59°. The excess thrust is simply the thrust force, F_n, minus the drag, D, which provides an indication of the capability of the aircraft to accelerate or climb. If the excess thrust is positive, the aircraft has more thrust than drag and can accelerate to higher speed or climb to higher altitude. If the excess thrust is negative, the thrust is less than the drag, and the aircraft cannot accelerate or climb. (We discuss more about excess thrust in Chapter 5.)

Figure 3.191   Bell X-5 variable-sweep research aircraft. (Source: *NASA.*)

At lower subsonic Mach numbers, the 20° sweep angle has a higher lift coefficient and a much lower drag coefficient than the 59° sweep angle. At these low speeds, the excess thrust is positive with 20° of wing sweep, but it is negative with 59° of sweep, due to the high drag. As the aircraft approaches transonic speeds of about Mach 0.81, there is a reversal of these trends as the higher wing sweep shows benefits over the lower sweep angle. While the lift remains about the same for both sweep angles, the drag of the higher swept wing is less at transonic speeds beyond about Mach 0.81. The drag coefficient of the 59° swept wing decreases as the Mach number increases, while the drag coefficient increases with the 20° swept wing. This results in a positive excess thrust with the 59° swept wing in the transonic region, while the excess thrust is negative with the 20° swept wing. As discussed earlier and shown by Figure 3.188, the benefits of wing sweep are not realized until higher sweep angles are used.

The X-5 successfully demonstrated the technological feasibility and aerodynamic benefits of a variable sweep aircraft. Variable sweep wings have been used on a variety of production aircraft, including the General Dynamics F-111 *Aardvark*, Grumman F-14 *Tomcat*, Rockwell B-1 *Lancer*, Panavia *Tornado*, and Mikoyan MiG-23 *Flogger*.

The variable-sweep wing is not restricted to aft sweep. Several variable-sweep aircraft, with forward sweep, have been proposed, although none have been built or flown. Another innovative, variable sweep concept is the *oblique wing*, where a single wing is pivoted about its center point. Hence, one wingtip is move forward, while the other is moved aft, creating an aircraft configuration with a forward-swept wing on one side of the fuselage and an aft-swept wing on the other side.

The oblique wing concept was investigated by the Germans during World War II, with the Blohm and Voss BV P.202 airplane design, with a single oblique wing, and the Messerschmitt Me P.1109 design, which had two oblique wings, one on the top and one on the bottom of the fuselage. The

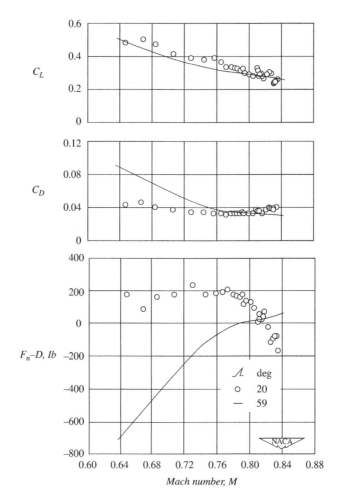

Figure 3.192 Lift coefficient, drag coefficient, and excess thrust for Bell X-5 with 20° and 59° wing sweep, unaccelerated flight at an altitude of approximately 42,000 ft. (Source: *Bellman, NACA RM L53A09c, 1953, [16]*).

two oblique wings of the Me P.1109 rotated in opposite directions, creating an unusual-looking, swept wing biplane airplane. Neither of these oblique airplane designs was ever built.

In the 1970s, the oblique wing concept was revived by NASA aerodynamicist Robert T. Jones, who led analytical and wind tunnel studies of a Mach 1.4 transport-sized, oblique wing airplane. The NASA studies showed there was potential benefit to the oblique wing concept, enough so, that the AD-1 demonstrator research aircraft was designed and built. The NASA AD-1, shown in Figure 3.193, had an oblique wing, which could pivot from a straight wing configuration, with zero sweep, to an oblique sweep of 60°. At zero sweep, the AD-1 had a wingspan of 35 ft 4 in (9.8 m). At 60° of sweep, the wingspan was reduced to 16 ft 2 in (4.9 m). The AD-1 was a small, subsonic, single-place aircraft, powered by two small jet engines, with a maximum speed of 200 mph (320 km/h, 290 ft/s). The objectives of the AD-1 flight research program were to evaluate the low-speed aerodynamics and flying qualities of the oblique wing concept, rather than to explore the transonic aerodynamic characteristics of the oblique wing. The AD-1 first flight was on 21 December 1979. Between 1979 and 1982, the AD-1 flew 79 research flights,

Figure 3.193 NASA AD-1 oblique wing research aircraft showing multiple wing positions, 1980. (Source: *NASA.*)

including flights with the oblique wing at its full sweep of 60°. Unfortunately, the AD-1 exhibited poor handling qualities and aeroelastic effects at sweep angles greater than about 45°. After the AD-1 test program, plans were made for the flight test of an oblique wing installed on a NASA F-8 aircraft, but this was never undertaken. Much work has been performed on the oblique wing concept, including subsonic, transonic, and supersonic wind tunnel tests, computational fluid dynamic studies, and flight tests of several oblique wing unmanned aerial vehicles.

3.11.8 Transonic and Supersonic Area Rule

As we have discussed, flight at transonic and supersonic Mach numbers results in the formation of shock waves with the penalty of significant wave drag. We have seen that a sharp-nosed body with thin swept wings is beneficial in reducing wave drag. Armed with this same knowledge, engineers in the 1950s were designing new jet-powered aircraft with sleek fuselages and thin wings to fly at transonic and supersonic speeds.

A classic example of this was the design of the Convair F-102 *Delta Dagger*, a new jet-powered aircraft with a large delta wing. Convair built two YF-102 prototypes in the early 1950s, preparing to enter flight test, prior to starting production of the aircraft for the US Air Force. However, wind tunnel testing of the F-102 design uncovered a potentially serious aerodynamic issue. The data indicated that the F-102 had much higher transonic drag than expected, so high that it was unknown whether the aircraft would have positive excess thrust to accelerate to supersonic speed. In the subsequent flight tests of the YF-102 prototype in 1954, the wind tunnel data was confirmed, as the aircraft could not accelerate to supersonic speed in level flight.

The NACA had been working on the YF-102 transonic drag issue with Convair, collecting valuable data in the Langley 8-foot high-speed wind tunnel, with slotted walls for transonic operation. (See Section 3.7.4.5 for a discussion about the slotted-wall wind tunnel.) NACA aerodynamicist, Richard T. Whitcomb, made a breakthrough in the understanding of the transonic drag problem and, better yet, came up with a design solution to the problem.

First, Whitcomb made the crucial connection that the transonic wave drag for a low aspect ratio wing–body combination is the same as that for a body of revolution having the same longitudinal, cross-sectional area distribution. In other words, if one measured the cross-sectional area at each axial location along the longitudinal axis of the wing–body and created an axisymmetric body, with the same area versus length, this *equivalent body* has the same wave drag as the wing–body. Then Whitcomb deduced that the transonic drag rise is primarily dependent on this axial distribution of the cross-sectional area normal to the freestream direction.

This equivalent body concept is shown in Figure 3.194a, where the cross-sectional area of the wing–body Section A-A is equal to the area of the equivalent body Section B-B. The equivalent body of the wing–body configuration has a pronounced "bump" where the cross-sectional area increases due to the wing. The drag coefficient versus Mach number for this equivalent body is shown in Figure 3.194b, where the drag coefficient increases to a high level at transonic Mach numbers. The presence of the bump in the area distribution creates strong shock waves and a significant increase in the wave drag. If the equivalent body shape could be smoothed out, such that there were no abrupt increases in area or bumps, strong shock waves are avoided and the drag coefficient is much lower, as shown in the figure for the "smooth body". This is the heart of the *transonic area rule*: to minimize the transonic wave drag, the cross-sectional area, normal to the freestream direction, should change smoothly with axial distance.

How is this smooth area variation accomplished for an aircraft configuration with wings, a tail, and other non-smooth contributions to the area? Consider the two wing–body configurations shown in Figure 3.195. The wing–body in Figure 3.195a is composed of a constant diameter fuselage with a wing attached. The cross-sectional area of the fuselage and wing are plotted beneath the wing–body. For this wing–body shape, there is no area ruling, as the equivalent body has an abrupt discontinuity or bump in the area distribution at the location of the wing, which results in high wave drag.

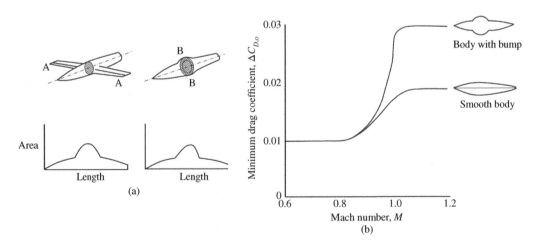

Figure 3.194 Transonic area rule, (a) equivalent body concept and (b) variation of drag coefficient for two bodies. (Source: *Loftin, NASA SP-468, 1985, [50]*.)

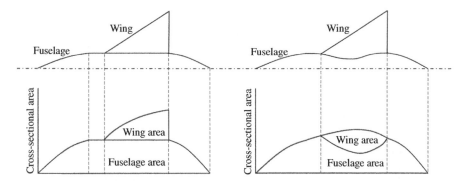

Figure 3.195 Cross-sectional area distribution for a wing–body without area ruling (left) and a wing–body with area ruling (right).

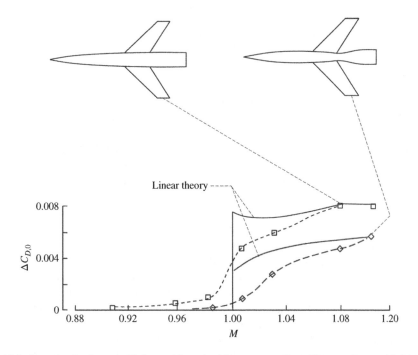

Figure 3.196 Increase in drag coefficient with and without area ruling. (Source: *Jones, NACA TR-1284, 1953, [43].*)

In Figure 3.195b, area ruling is applied to the wing–body shape by reducing the fuselage cross-sectional area at the wing location, resulting in the smooth area distribution shown. This narrowing of the fuselage "waist" gives the body a unique appearance, which is sometimes called a "Coke bottle" fuselage, due to its resemblance to the soft drink bottle shape. To achieve the smooth area distribution, adjustments of the fuselage area are required in other locations, such as the tail and cockpit area. These area adjustments are not always a *reduction* in area. Sometimes an *increase* in body area may be required to produce a smooth area distribution.

The variations of the drag coefficient, versus Mach number, for a non-area ruled wing–body and an area-ruled wing–body, are shown in Figure 3.196. The increase in the drag coefficient

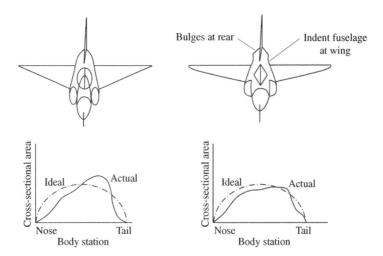

Figure 3.197 Comparison of cross-sectional area distributions for YF-102, without area ruling (left) and with area ruling (right). (Source: *Talay, NASA SP-367, 1975, [65]*.)

for the area-ruled wing–body is about half that of the non-area-ruled wing–body. Also shown in Figure 3.196 is the predicted drag coefficient variations for both types of wing-bodies, using aerodynamic linear theory.

Returning to the transonic drag issues of the YF-102, Convair modified the aircraft to incorporate the transonic area rule concept, adjusting the fuselage area to make the axial distribution of cross-sectional area as smooth as practicable. In addition to "pinching in" the fuselage in certain locations to accommodate the wing area, modifications were made to the nose and canopy, and bulges were installed at the aft end of the aircraft, which *increased* the total cross-sectional area, but made the area distribution smoother in this region. The cross-sectional area distributions of the YF-102, before and after the area rule modifications, are shown in Figure 3.197. Note how the area distribution was made smoother by increasing the cross-sectional areas at the nose and tail, and decreasing the area in the region of the wing. Photographs of the unmodified YF-102 prototype and the production version F-102A, with area ruling modifications, are shown in Figure 3.198. The differences in the nose, mid-fuselage, and tail region are apparent between the YF-102, without area ruling, and the F-102A, with area ruling.

The area-rule-modified YF-102 took to the skies on 20 December 1954 and accelerated to supersonic speed while still climbing to altitude. The significant decrease in the drag coefficient with the modified aircraft is shown in Figure 3.199. The drag coefficient of the unmodified YF-102 prototype is shown as the solid curve in the figure. The total cross-sectional area distribution of the prototype, including the contributions of the various components, is shown in the upper left of the figure. This area distribution of the unmodified aircraft is seen to have several abrupt changes in area with axial distance. The "revised" aircraft, with area rule modifications, is shown as a dashed curve (long and short dashes) in Figure 3.199. The cross-sectional area distribution for this area-ruled aircraft, shown as the dashed curve in the lower right of the figure, is smooth without abrupt changes in area. The drag coefficient, of the area ruled "revised" configuration, is considerably lower than the prototype configuration without area ruling. The "improve nose" configuration adds area rule modifications to the nose of the aircraft, resulting in even further decreases in the drag coefficient.

The Convair F-102 was the first aircraft to which the transonic area rule was applied, and it was a resounding success. NACA aerodynamicist Richard Whitcomb was awarded the 1954 Collier

Figure 3.198 YF-102 prototype without area ruling (left) and F-102A with area ruling (right). (Source: *NASA*.)

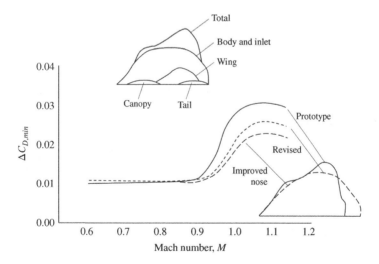

Figure 3.199 Drag coefficient versus Mach number for YF-102 prototype with and without area rule modifications. (Source: *Loftin, NASA SP-468, 1985, [50]*.)

Trophy in recognition of his work. (The area rule concept was classified for several years, so Whitcomb's work was not recognized until 1954 when it was declassified.) The citation for his Collier Trophy read, "For discovery and experimental verification of the area rule, a contribution to base knowledge yielding significantly higher airplane speed and greater range with the same power."

The area rule story does not end here. The transonic area rule concept was extended to supersonic speeds by NACA aerodynamicist Robert T. Jones in 1953, leading to the *supersonic area rule*.

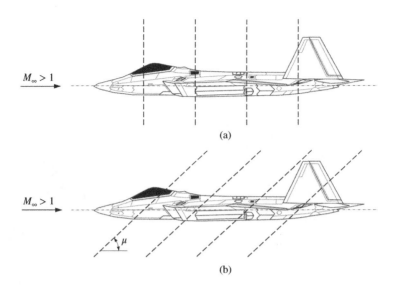

Figure 3.200 Cross-sectional area planes for area ruling, (a) transonic, planes normal to freestream, (b) supersonic, oblique planes parallel to Mach lines.

Recall that the transonic area rule is applied to the body cross-sectional area in planes that are *normal* to the freestream direction, as shown in Figure 3.200a. In contrast, the supersonic area rule is applied to the cross-section area in the oblique planes that are parallel to the Mach lines of the freestream flow. These Mach line planes are at an angle $\mu = \sin^{-1}(1/M_\infty)$, as shown in Figure 3.200b. The angle of these oblique planes change with Mach angle and with the freestream Mach number, therefore changing the relevant cross-sectional area as a function of Mach number. This must be considered when designing an aircraft to operate efficiently over a range of supersonic Mach numbers.

While the Convair F-102 final design was modified to incorporate area ruling, Convair's follow-on to the F-102, the F-106 *Delta Dart*, incorporated the area rule concept from the start. The fuselage area ruling is clearly evident in the top view of the F-106 (first flight on 26 December 1956) shown in Figure 3.201. The area rule concept was applied to many other transonic and supersonic aircraft to follow (see for example the three-view drawing of the Northrop T-38 in Figure 5.23). The area rule is not restricted to military aircraft, as it is applicable to any other aerospace vehicles that fly in the transonic and supersonic flight regimes, including expendable missiles, rocket boosters, and commercial airliners.

3.11.9 Internal Supersonic Flows

We have been focusing on the supersonic flow over wings and fuselages, or the *external* supersonic flow over bodies. There are also many instances of *internal* supersonic flow, where the flow is confined on all sides by solid boundaries. These include the supersonic flow through wind tunnels, jet engine nozzles, and rocket nozzles. These flows may also involve chemical reactions and high temperature effects, but we focus more broadly on the supersonic flow through varying area ducts, where the flow is assumed to be inviscid and isentropic.

Consider the inviscid, isentropic, compressible flow through a varying area duct, as shown in Figure 3.202. At a location where the duct area is A, the flow has a velocity V, pressure p, and

Figure 3.201   Area rule applied to the Convair F-106 *Delta Dart*. (Source: *NASA*.)

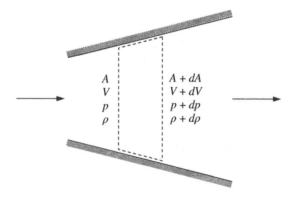

Figure 3.202   Inviscid, isentropic, compressible flow through a varying area duct.

density ρ. A small distance downstream from this location, the duct area has changed by a small amount dA, so that the duct area at this location is $A + dA$. The flow properties have also changed by small amounts dp, dV, and $d\rho$, so that, at this location, the velocity is $V + dV$, the pressure is $p + dp$, and the density is $\rho + d\rho$. This flow is indeed compressible, as the density is variable in the flow.

Since there is no flow through the duct walls, the mass flow rate through the area A is equal to that through the area $A + dA$. This is simply a statement of conservation of mass, which is expressed by the continuity equation, Equation (3.143), as

$$\rho A V = \text{constant}$$

Applying the continuity equation to the mass flow rates through area A and $A + dA$, we have

$$\rho A V = (\rho + d\rho)(A + dA)(V + dV) \tag{3.375}$$

Expanding the right side of the equation, we have

$$\rho A V = (\rho A + \rho dA + A d\rho + d\rho dA)(V + dV)$$
$$\rho A V = \rho A V + \rho V dA + A V d\rho + V d\rho dA + \rho A dV + \rho dA dV + A d\rho dV + d\rho dA dV \tag{3.376}$$

Neglecting the products of the small changes, we have

$$0 = \rho V dA + A V d\rho + \rho A dV \tag{3.377}$$

Dividing through by $\rho V A$, we have

$$0 = \frac{dA}{A} + \frac{d\rho}{\rho} + \frac{dV}{V} \tag{3.378}$$

The change in pressure can be expressed in terms of the density and the change in velocity, by Euler's equation, Equation (3.165), as

$$dp = -\rho V dV$$

Solving for the density, we have

$$\rho = -\frac{dp}{V dV} \tag{3.379}$$

Substituting Equation (3.379) into (3.378), we have

$$0 = \frac{dA}{A} - \frac{d\rho V dV}{dp} + \frac{dV}{V} \tag{3.380}$$

From Equation (3.320), the speed of sound in an isentropic flow is given by

$$a^2 = \frac{dp}{d\rho} \tag{3.381}$$

Substituting Equation (3.381) into (3.380), we have

$$\frac{dA}{A} - \frac{V dV}{a^2} + \frac{dV}{V} = 0 \tag{3.382}$$

Rearranging, we have

$$\frac{dA}{A} = \frac{V dV}{a^2} - \frac{dV}{V} = \frac{V}{V}\left(\frac{V dV}{a^2} - \frac{dV}{V}\right) = \left(\frac{V^2}{a^2} - 1\right)\frac{dV}{V} \tag{3.383}$$

or

$$\frac{dA}{A} = (M^2 - 1)\frac{dV}{V} \tag{3.384}$$

Equation (3.384), known as the *area–velocity relation*, relates the change in velocity to the change in area for the inviscid, isentropic, compressible flow of a fluid in a varying area internal duct. This relationship is strongly influenced by the Mach number of the flow. We can more easily see the effect of changing area on the change in the velocity by rearranging Equation (3.384) as

$$\frac{dV}{V} = \frac{1}{(M^2 - 1)}\frac{dA}{A} \tag{3.385}$$

If the flow is subsonic, with a Mach number less than one, the $(M^2 - 1)$ term is negative, so that a duct with increasing area $(dA > 0)$ results in a decreasing velocity $(dV < 0)$. The opposite is true if the duct area decreases $(dA < 0)$: Equation (3.385) states that the flow velocity increases $(dV > 0)$. For supersonic flow, with a Mach number greater than one, the $(M^2 - 1)$ term is positive, so the flow velocity increases $(dV > 0)$ with increasing duct area $(dA > 0)$. For supersonic flow with a decreasing area $(dA < 0)$, Equation (3.385) states that the flow velocity decreases $(dV < 0)$.

For the trivial case of no area change $(dA = 0)$, Equation (3.385) states that the velocity remains constant $(dV = 0)$. Recall that we are assuming inviscid, isentropic flow, so there is no physical mechanism for the flow to accelerate or decelerate in the constant area duct, regardless of how high a pressure differential may exist between the entrance and exit. With viscosity or heat transfer, there are physical mechanisms for the flow to change velocity in a constant area duct. These types of flow, called Fanno flow and Rayleigh flow, which concern the effects of friction and heat transfer, respectively, are beyond the scope of the text.

The changes in the pressure due to the changes in velocity are given by Euler's equation, Equation (3.165). The changes in the velocity and pressure, with changes in area, are graphically depicted for subsonic flow in Figure 3.203 and for supersonic flow in Figure 3.204.

According to Equation (3.385), an interesting situation occurs at sonic conditions $(M = 1)$, where we have

$$\frac{dV}{V} = \frac{1}{0}\frac{dA}{A} \tag{3.386}$$

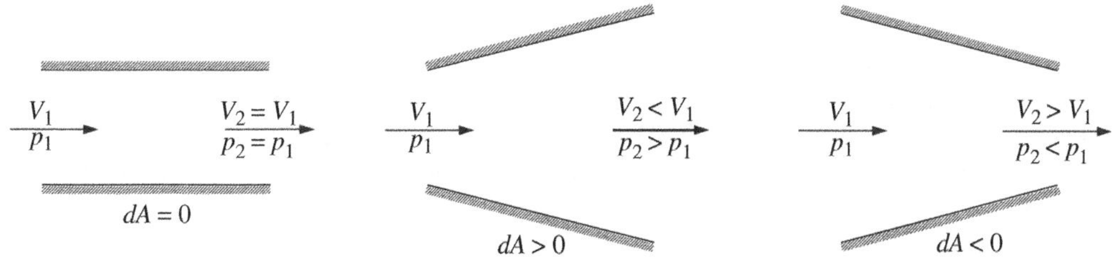

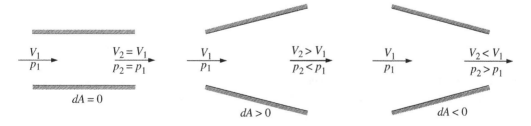

Figure 3.203   Subsonic, isentropic flow through varying area ducts.

Figure 3.204   Supersonic, isentropic flow through varying area ducts.

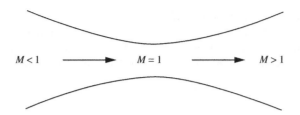

Figure 3.205   Flow through a convergent-divergent supersonic duct or nozzle.

which seems to indicate that the velocity change is infinite. Returning to Equation (3.384), the sonic condition yields

$$\frac{dA}{A} = 0 \tag{3.387}$$

which resolves this dilemma, because Equation (3.386) then becomes an indeterminate form of zero divided by zero, which by L'Hospital's rule of calculus may be a finite number. The derivative of the area equal to zero, in Equation (3.387), has special significance, as this indicates a minimum value of the area in the duct.

Sonic conditions occur at this minimum area, called a *throat*. The conditions at the throat are typically denoted with an asterisk subscript, so that the throat area is denoted by A^*. If a subsonic flow is accelerated in a decreasing area duct, the maximum possible velocity is the sonic velocity at the minimum area. The flow cannot be accelerated to supersonic speed in a converging duct, regardless of the pressure difference between the exit and entrance of the duct. To accelerate the flow from sonic to supersonic conditions, the duct area must be increased downstream of the throat. Thus, to obtain supersonic flow in a duct, starting from a reservoir at near zero speed, a converging-diverging duct must be used, as shown in Figure 3.205.

Imagine now an internal supersonic flow, where sonic flow exists at the throat, and the area varies upstream and downstream of the throat. Using the continuity equation, the mass flow rate at any area A of the duct can be related to the throat conditions, as

$$\rho A V = \rho^* A^* V^* \tag{3.388}$$

where ρ and V are the density and velocity, respectively, at the area A, and ρ^* and V^* are the density and velocity, respectively, at the throat, A^*. Since the Mach number at the throat, M^*, is one, the velocity at the throat is equal to the speed of sound at the throat, a^*. Using this, Equation (3.388) becomes

$$\rho A V = \rho^* A^* a^* \tag{3.389}$$

Solving for the ratio of the arbitrary area, in the duct, to the throat area, we have

$$\frac{A}{A^*} = \frac{\rho^* a^*}{\rho V} \tag{3.390}$$

or

$$\frac{A}{A^*} = \left(\frac{\rho^*}{\rho_t}\right)\left(\frac{\rho_t}{\rho}\right)\left(\frac{a^*}{a}\right)\left(\frac{a}{V}\right) = \left(\frac{\rho^*}{\rho_t}\right)\left(\frac{\rho_t}{\rho}\right)\left(\frac{a^*}{a}\right)\left(\frac{1}{M}\right) \tag{3.391}$$

where ρ_t is the total density, a is the speed of sound at area A, and M is the Mach number at area A.

Using the definition of the speed of sound, given by Equation (3.326), we have

$$\frac{A}{A^*} = \left(\frac{\rho^*}{\rho_t}\right)\left(\frac{\rho_t}{\rho}\right)\frac{\sqrt{\gamma R T^*}}{\sqrt{\gamma R T}}\left(\frac{1}{M}\right) = \left(\frac{\rho^*}{\rho_t}\right)\left(\frac{\rho_t}{\rho}\right)\sqrt{\frac{T^*}{T}}\left(\frac{1}{M}\right) \tag{3.392}$$

or

$$\frac{A}{A^*} = \left(\frac{\rho^*}{\rho_t}\right)\left(\frac{\rho_t}{\rho}\right)\sqrt{\left(\frac{T^*}{T_t}\right)\left(\frac{T_t}{T}\right)}\left(\frac{1}{M}\right) \tag{3.393}$$

where T_t is the total temperature.

Since the flow through the duct is isentropic, the static density is related to the total density by the isentropic relation for the density, Equation (3.347). Therefore, the density ratio at area A is given by

$$\frac{\rho}{\rho_t} = \left[1 + \frac{(\gamma - 1)}{2}M^2\right]^{-1/(\gamma-1)} \tag{3.394}$$

The density ratio at the throat is given by

$$\frac{\rho^*}{\rho_t} = \left[1 + \frac{(\gamma - 1)}{2}(M^*)^2\right]^{-1/(\gamma-1)} = \left(\frac{\gamma + 1}{2}\right)^{-1/(\gamma-1)} \tag{3.395}$$

where M^* is one at the throat.

Similarly, using the isentropic relation for temperature, Equation (3.345), the temperature ratios are given by

$$\frac{T}{T_t} = \left[1 + \frac{(\gamma - 1)}{2}M^2\right]^{-1} \tag{3.396}$$

$$\frac{T^*}{T_t} = \left(\frac{\gamma + 1}{2}\right)^{-1} \tag{3.397}$$

Using the isentropic relations for pressure, Equation (3.346), the pressure ratios are given by

$$\frac{p}{p_t} = \left[1 + \left(\frac{\gamma - 1}{2}\right)M^2\right]^{-\gamma/(\gamma-1)} \tag{3.398}$$

$$\frac{p^*}{p_t} = \left(\frac{\gamma + 1}{2}\right)^{-\gamma/(\gamma-1)} \tag{3.399}$$

Substituting Equations (3.394), (3.395), (3.396), and (3.397) into (3.393), we have

$$\frac{A}{A^*} = \left(\frac{\gamma + 1}{2}\right)^{-1/(\gamma-1)}\left[1 + \frac{(\gamma - 1)}{2}M^2\right]^{1/(\gamma-1)}\sqrt{\left(\frac{\gamma + 1}{2}\right)^{-1}\left[1 + \frac{(\gamma - 1)}{2}M^2\right]}\left(\frac{1}{M}\right) \tag{3.400}$$

Squaring this equation, we have

$$\left(\frac{A}{A^*}\right)^2 = \left(\frac{\gamma + 1}{2}\right)^{-2/(\gamma-1)}\left[1 + \frac{(\gamma - 1)}{2}M^2\right]^{2/(\gamma-1)}\left(\frac{\gamma + 1}{2}\right)^{-1}\left[1 + \frac{(\gamma - 1)}{2}M^2\right]\left(\frac{1}{M^2}\right)$$

$$\left(\frac{A}{A^*}\right)^2 = \left[\frac{2}{\gamma + 1}\left(1 + \frac{(\gamma - 1)}{2}M^2\right)\right]^{(\gamma+1)/(\gamma-1)}\left(\frac{1}{M^2}\right) \tag{3.401}$$

This equation, known as the *Mach-area relation*, provides the Mach number as a function of the area ratio, A/A^* through the duct. The area ratio can be solved for directly, given the Mach number, but an iterative solution is required for the solution of the Mach number for a given area ratio. Tabulated values of the Mach number and area ratio can be found in many textbooks or other sources. The Mach-area relation is plotted in Figure 3.206. There are two Mach numbers

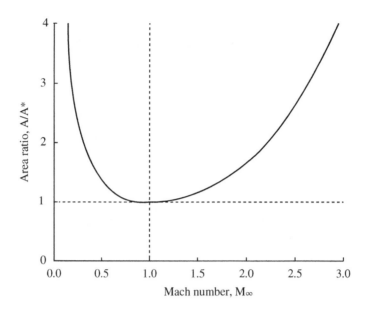

Figure 3.206 The Mach-area relation.

corresponding to each area ratio, a subsonic and a supersonic Mach number. The Mach number is one at an area ratio of one, as expected.

Using the Mach-area relation, Equation (3.401), and the isentropic flow relations, Equations (3.345) to (3.347), the flow properties through a convergent-divergent duct or nozzle may be easily calculated. Assuming that the nozzle area distribution and stagnation conditions are known, the Mach number can be obtained from the Mach-area relation. After the Mach number is known for each area, the pressure, temperature, and density may be calculated, through the nozzle, using the isentropic relations.

Example 3.18 Supersonic Nozzle *A convergent-divergent nozzle, shown below, has a throat diameter, d^*, of 1.2 m and an exit diameter, d_e, of 2.0 m. The stagnation pressure, p_t, and temperature, T_t, are 12.25 Pa and 1000 K, respectively. Calculate the Mach number, pressure, and temperature at the nozzle throat and nozzle exit. Assume isentropic flow of air through the nozzle with supersonic exit conditions.*

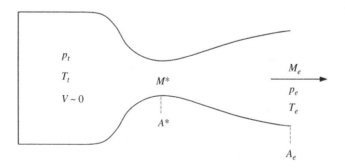

Solution

The Mach number at the nozzle throat, M^*, is one. Using Equations (3.397) and (3.399), respectively, the temperature, T^*, and pressure, p^*, at the throat are given by

$$\frac{T^*}{T_t} = \left(\frac{\gamma + 1}{2}\right)^{-1} = \left(\frac{1.4 + 1}{2}\right)^{-1} = 0.8333$$

$$T^* = 0.8333\,T_t = (0.8333)(1000\,\text{K}) = 833.3\,\text{K}$$

$$\frac{p^*}{p_t} = \left(\frac{\gamma + 1}{2}\right)^{-\gamma/(\gamma-1)} = \left(\frac{1.4 + 1}{2}\right)^{-1.4/(1.4-1)} = 0.5283$$

$$p^* = 0.5283\,T_t = (0.5283)\left(12.25\,\frac{\text{N}}{\text{m}^2}\right) = 6.472\,\frac{\text{N}}{\text{m}^2}$$

The throat area, A^*, and nozzle exit area, A_e, are

$$A^* = \frac{\pi}{4}(d^*)^2 = \frac{\pi}{4}(1.2\,\text{m})^2 = 1.13\,\text{m}^2$$

$$A_e = \frac{\pi}{4}(d_e)^2 = \frac{\pi}{4}(2.0\,\text{m})^2 = 3.14\,\text{m}^2$$

The nozzle exit area ratio is

$$\frac{A_e}{A^*} = \frac{3.14\,\text{m}^2}{1.13\,\text{m}^2} = 2.78$$

The nozzle exit area ratio is related to the exit Mach number through the Mach-area relation, Equation (3.401), given by

$$\left(\frac{A_e}{A^*}\right)^2 = \left[\frac{2}{\gamma + 1}\left(1 + \frac{(\gamma - 1)}{2}M_e^2\right)\right]^{(\gamma+1)/(\gamma-1)}\left(\frac{1}{M_e^2}\right)$$

With a known area ratio, an iterative process is used to solve for the Mach number. (Alternatively, there are tabulated values of the Mach-area relation that can be used.) There are two solutions to the Mach-area relation for a nozzle exit area of 2.78, a subsonic Mach number of 0.2140 and a supersonic Mach number of 2.557. We are interested in the supersonic solution.

Using a nozzle exit Mach number, $M_e = 2.557$, and Equations (3.396) and (3.398), the nozzle exit temperature, T_e, and pressure, p_e, are given by

$$\frac{T_e}{T_t} = \left[1 + \frac{(\gamma - 1)}{2}M_e^2\right]^{-1} = \left[1 + \frac{(1.4 - 1)}{2}(2.557)^2\right]^{-1} = 0.4333$$

$$T_e = 0.4328\,T_t = (0.4333)(1000\,\text{K}) = 433.3\,\text{K}$$

$$\frac{p_e}{p_t} = \left[1 + \left(\frac{\gamma - 1}{2}\right)M_e^2\right]^{-\gamma/(\gamma-1)} = \left[1 + \left(\frac{1.4 - 1}{2}\right)(2.557)^2\right]^{-1.4/(1.4-1)} = 0.0535$$

$$p_e = 0.05332\,p_t = (0.05357)\left(12.25\,\frac{\text{N}}{\text{m}^2}\right) = 0.6562\,\frac{\text{N}}{\text{m}^2}$$

3.12   Viscous Flow

As discussed in Section 3.2.6, a viscous flow is defined as a flow where the transport of mass, momentum, or heat is important. Each of these transport phenomena can be related to a gradient in a flow property. A flow with a gradient in chemical species results in transport of the species mass through diffusion. The transport of heat through conduction occurs in a flow with a temperature gradient. For most of the flows that we discuss, these two transport phenomena are not significant, so when we discuss viscous flow we are primarily concerned with viscosity, which is related to the transport of momentum.

In this context, the words "viscous flow" may conjure up thoughts of the flow of a thick, syrupy fluid, but this is not a necessarily so. While this type of fluid flow is certainly viscous, so is the seemingly "slippery" flow of air over an airplane wing. Viscous flow is simply fluid motion with friction. We are being a bit cavalier with the use of the term "simple" to describe viscous flow, as the complex non-linear physics associated with viscous flow is not quite so simple. The complex physics of viscous flow is why much of theoretical aerodynamics is based upon the assumption of inviscid flow. However, difficulties arise when dealing with aerodynamic drag. With the assumption of inviscid, incompressible flow, we are led to d'Alembert's paradox of zero drag. Here, we see that the presence of friction is an absolute requirement to obtain a finite drag. (We have made the distinction that the inviscid flow must be *incompressible* to arrive at d'Alembert's paradox, because a *compressible* flow may have finite wave drag, due to shock waves, even though the flow is assumed to be inviscid.) For air flowing over a solid surface, we have discussed how the effects of friction can be restricted to a thin region next to the body, called the boundary layer. We have seen that the effects of this thin, boundary layer on aerodynamic drag are profound, leading to the profile drag of airfoils and the zero-lift parasite drag of complete aircraft. Now, we seek to understand the nature of flows with friction in a bit more detail, including how the skin friction drag can actually be calculated.

3.12.1 Skin Friction and Shearing Stress

Consider the flow of air with a freestream velocity, U_∞, over the top of a flat plate, assuming inviscid and viscous flow, as shown in Figure 3.207. The variation of the velocity, U, in the vertical or y-direction is called a *velocity profile*. For the inviscid flow, the velocity profile is uniform and equal to the freestream velocity, U_∞, even at the surface of the plate. There is a relative velocity difference between the freestream velocity, directly above the plate, and the zero velocity of the plate. This relative velocity difference between the inviscid fluid flow and a solid surface is called *slip*. There are no tangential forces (shearing stresses) exerted on the fluid due to this slip condition. Normal forces can be exerted through the fluid layers, so that the plate can "feel" the pressure from the freestream flow.

In reality, there are intermolecular attractions between the fluid molecules and the molecules of the plate surface, causing the fluid molecules to adhere to the surface. This intermolecular attraction or *skin friction* results in a *no slip* condition at the plate surface, where the fluid velocity at the surface is zero. The no slip condition results in tangential friction forces (shearing stresses) being exerted on the fluid. Normal forces (pressure) can still be "felt" through the fluid layers, as for inviscid flow. The no slip condition and the existence of the friction forces are the two primary physical features of viscous flow that make it different from inviscid flow.

Let us consider the nature of the tangential friction force per unit area or shearing stress a bit more closely. If we apply a shearing stress to a solid and a fluid, the solid can resist the shearing stress and not change its shape while the fluid cannot. At some magnitude of the force (per unit area), the solid starts to change its shape or *deform* and the resulting shearing stress, τ_{solid}, is proportional to the deformation per unit length, $\Delta x/L$, or strain, ε, as given by Hooke's law.

$$\tau_{solid} = E\frac{\Delta x}{L} = E\varepsilon \tag{3.402}$$

where the constant of proportionality, E, is Young's modulus of elasticity.

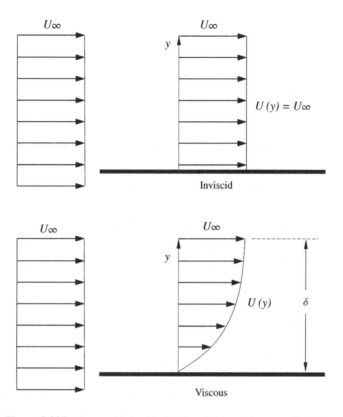

Figure 3.207   Comparison of inviscid and viscous flow on a flat plate.

For a fluid, the shearing stress, τ_{fluid}, is proportional to the *rate of change* of the deformation per unit length, $(\Delta x/L)/\Delta t$, or strain rate, as given by Newton's law of friction.

$$\tau_{fluid} = \mu \frac{(\Delta x/L)}{\Delta t} = \mu \Delta(u/L) \tag{3.403}$$

where the constant of proportionality for a fluid, μ, is the coefficient of viscosity, and $\Delta(u/L)$ is the velocity gradient. The coefficient of viscosity was introduced in Section 3.2.6 as a (momentum) transport property of the fluid. Applying Equation (3.403) to the general velocity profile in Figure 3.207, we have

$$\tau_{fluid} = \mu \frac{d}{dy}[U(y)] = \mu \frac{dU}{dy} \tag{3.404}$$

Thus, we see that, for a fluid, the shearing stress or friction force per unit area is linearly proportional to the fluid viscosity, μ, and the velocity gradient, dU/dy. Fluids that obey Equation (3.404) are called *Newtonian fluids*.

3.12.2   Boundary Layers

Consider a uniform, viscous flow of air over a flat plate. The no slip condition, caused by skin friction, at the plate surface results in a tangential shearing stress that slows the layers of fluid. The slowing of the fluid progresses vertically upward, in the y-direction, into the freestream flow with increasing downstream distance, x, from the leading edge of the plate. Figure 3.208 shows

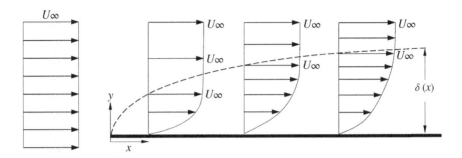

Figure 3.208   Growth of the boundary layer on a flat plate.

velocity profiles at three x-locations along the plate. At each of these locations, the velocity is zero at the plate surface and recovers to the freestream velocity, U_∞, at a height, $\delta(x)$, which is a function of the downstream distance. This thin region of viscous flow, next to the plate surface, is called a *boundary layer* and the height $\delta(x)$ is the *boundary layer thickness*, which "grows" in height with distance from the plate leading edge, as shown in Figure 3.208. As discussed at the beginning of the chapter, the boundary layer concept was conceived by Ludwig Prandtl in 1904. The edge of the boundary layer is defined as the line connecting the boundary layer height at each x-location along the plate, as depicted by the dashed line in Figure 3.208.

The external flow sees the boundary layer as an *effective body*, which is thicker or "fatter" than the actual shape. So, does the viscous flow result in a surface pressure distribution over the body that is different from that for an inviscid flow? The answer is "no", as long as the viscous effects do not cause flow separation, which significantly alters the pressure distribution. While a viscous flow "feels" the tangential frictional forces, the normal forces (pressures) are said to be *impressed* through a boundary layer. Another way of stating this is that there is no pressure gradient from the edge of the boundary layer to the body surface. Therefore, one can perform an inviscid analysis of the flow over a body and the results for the pressure distribution, and hence the lift and drag forces due to the pressure distribution, are applicable to the viscous flow case. This was a significant result from Prandtl's boundary layer theory, as the performance of an inviscid analysis is much simpler than for a viscous flow.

Boundary layers may be *laminar*, *turbulent*, or *transitional* from laminar to turbulent. Just as in the pipe flow experiments of Osborne Reynolds, described in Section 3.2.6, the nature of the boundary layer flow correlates with the Reynolds number. The boundary layer is laminar when the Reynolds number is small and turbulent when it is large. The criteria defined in Equations (3.33), (3.34), and (3.35) are valid, where the flow is laminar below a Reynolds number of about 100,000 and turbulent for a Reynolds number above about 500,000. We use these criteria to predict the nature of the boundary layer on a body, such as an aircraft fuselage or wing. The boundary layer starts at the fuselage nose or wing leading edge and grows in height with downstream distance. The *local Reynolds number*, $Re_{x,\infty}$, at a location x downstream of its starting point, is calculated as

$$Re_{x,\infty} = \frac{\rho_\infty U_\infty x}{\mu_\infty} \tag{3.405}$$

where ρ_∞, U_∞, and μ_∞ are the freestream density, velocity, and coefficient of viscosity, respectively. This value of the local Reynolds number determines whether the boundary layer is laminar or turbulent up to that location.

As discussed earlier, there is a velocity profile in the boundary layer due to viscous effects. The shape of this velocity profile is distinctly different for laminar and turbulent flows. Laminar flows are smooth and regular, while turbulent flows are chaotic and random. The chaotic nature of turbulence causes increased mixing of the high and low velocity fluid elements near the surface of

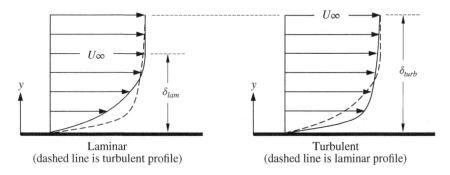

Figure 3.209   Comparison of laminar and turbulent boundary layers on a flat plate.

a body. This results in a "fatter" or fuller turbulent velocity profile, where the average velocity is greater near the surface, than for laminar flow, as shown in Figure 3.209.

Let us see if the assumptions that we made about laminar and turbulent boundary layer velocity profiles makes sense. Returning to Figure 3.209, we calculate the mass flow of fluid between the plate surface up to the vertical height, δ_{turb}, using the two velocity profiles. The mass flow of fluid (per unit width) for the laminar velocity profile is the sum of the mass flow through the distance δ_{lam} and the mass flow through the distance ($\delta_{turb} - \delta_{lam}$), as given by

$$\dot{m} = \rho_\infty \int_0^{\delta_{lam}} U_{lam}(y)\,dy + \rho_\infty U_\infty (\delta_{turb} - \delta_{lam}) \tag{3.406}$$

The mass flow of fluid (per unit width) for the turbulent velocity profile is the mass flow through the distance δ_{turb}, as given by

$$\dot{m} = \rho_\infty \int_0^{\delta_{turb}} U_{turb}(y)\,dy \tag{3.407}$$

Since the top boundary is a streamline of the flow, no mass is passing through this boundary. Therefore, the mass flow through the two velocity profiles are equal. Setting Equation (3.406) equal to (3.407), we have

$$\rho_\infty \int_0^{\delta_{lam}} U_{lam}(y)\,dy + \rho_\infty U_\infty (\delta_{turb} - \delta_{lam}) = \rho_\infty \int_0^{\delta_{turb}} U_{turb}(y)\,dy \tag{3.408}$$

Assuming that the integral of each velocity profile is the product of an average velocity and the boundary layer height, we have

$$\rho_\infty (U_{lam})_{avg}\, \delta_{lam} + \rho_\infty U_\infty (\delta_{turb} - \delta_{lam}) = \rho_\infty (U_{turb})_{avg}\, \delta_{turb} \tag{3.409}$$

where $(U_{lam})_{avg}$ and $(U_{turb})_{avg}$ are the average values of the laminar and turbulent velocities, respectively. Solving for the ratio of the laminar to the turbulent boundary layer heights, we have

$$\frac{\delta_{lam}}{\delta_{turb}} = \frac{U_\infty - (U_{turb})_{avg}}{U_\infty - (U_{lam})_{avg}} < 1 \tag{3.410}$$

The ratio is less than one since the turbulent boundary layer velocity profile is "fatter" or fuller than the laminar profile, making the average velocity in the turbulent boundary layer greater than that for the laminar boundary layer. Equation (3.410) simply states that the laminar boundary layer thickness is less than for the turbulent boundary layer.

$$\delta_{lam} < \delta_{turb} \tag{3.411}$$

This defines the relative magnitudes of the boundary layer heights and verifies that the turbulent boundary layer velocity profile is fuller than the laminar velocity profile, which can be expressed mathematically as

$$\left(\frac{dV}{dy}\right)_{y=0,turb} > \left(\frac{dV}{dy}\right)_{y=0,lam} \tag{3.412}$$

where $(dV/dy)_{y=0,lam}$ and $(dV/dy)_{y=0,turb}$ are the reciprocal of the slopes of the laminar and turbulent velocity profiles at the flat plate surface ($y = 0$), respectively.

Quantitative values for the laminar and turbulent boundary layer heights on a flat plate are calculated using rather simple equations. The laminar boundary layer thickness, δ_{lam}, is obtained from theoretical foundations as

$$\delta_{lam} = \frac{5.2x}{\sqrt{Re_x}} \tag{3.413}$$

where x is the distance from the leading edge of the plate and Re_x is the local Reynolds number, equal to $\rho V x/\mu$. Experimental measurements of the laminar boundary layer height are in agreement with this equation. According to Equation (3.413), the laminar boundary layer height grows with $x/\sqrt{x}$ or the square root of the distance from the plate leading edge, $x^{1/2}$.

For turbulent flow, the equation for the turbulent boundary layer thickness cannot be obtained theoretically. Although there are still no theories of turbulence that adequately explain and predict its nature, there are several empirically based models that allow us to make predictions of turbulent flow and to calculate the associated forces and moments. Thus, even though our theoretical understanding of turbulence is still lacking, our empirical knowledge of turbulent flows allows us to design and fly aerospace vehicles with boundary layers that are often turbulent. Based on empirical data, the flat plate turbulent boundary layer thickness is given by

$$\delta_{turb} = \frac{0.37x}{Re_x^{0.2}} \tag{3.414}$$

where x is the distance from the leading edge of the plate and Re_x is the local Reynolds number equal to $\rho V x/\mu$. The turbulent boundary layer height grows with $x/x^{1/5}$ or $x^{4/5}$, which means it grows faster than the laminar boundary layer, which scales with $x^{1/2}$. Both the laminar and turbulent boundary layer heights are a function of the Reynolds number, which is consistent with our previous discussions.

Now we ask the question, what is the significance of the fuller velocity profile of a turbulent boundary layer? As we found above, the average velocity in the turbulent boundary layer is greater than in the laminar boundary layer. At a macroscopic level, this higher average velocity translates into a greater average kinetic energy in the turbulent boundary layer. This has a significant impact on the ability of the boundary layer to remain attached to a surface. Fluid motion is the result of pressure differences, where a fluid flows in the direction of decreasing pressure, known as a *favorable pressure gradient*. Viscosity dissipates kinetic energy, which slows the flow in the boundary layer, but as long as the favorable pressure gradient is sufficient to overcome this friction, the flow continues to move in the same direction. If the flow is such that the pressure increases with distance, known as an *adverse pressure gradient*, the fluid motion is retarded by both the pressure increase and friction. If these retarding forces overcome the momentum of the fluid molecules, the flow reverses direction and the flow separates. If the boundary layer is turbulent, the flow has a higher mean kinetic energy to resist the adverse pressure gradient and delay flow separation, as compared with a laminar boundary layer.

Think of a fluid molecule as a ball at the top of a hill of height, h, as shown in Figure 3.210. The descending slope of the hill represents the favorable pressure gradient, and the uphill portion the adverse pressure gradient. If there were no friction, the ball would roll down the hill and then roll back up the hill, ending at the same height as it started. (We assume that the ball again reaches a

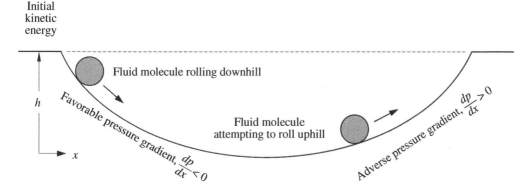

Figure 3.210 Flow separation analogy of a fluid molecule "ball" rolling down and up a hill, in the presence of favorable and adverse pressure gradients, respectively.

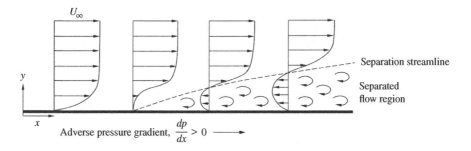

Figure 3.211 Reversed flow in the presence of an adverse pressure gradient.

plateau at the height, h, where it stops.) With friction, the motion of the ball is slowed as it rolls downhill, just as the friction in the boundary layer takes energy from the fluid molecule and slows it down. With friction, the ball rolls uphill and stops at a height lower than its starting height and then rolls backwards, down the slope. Similarly, a fluid molecule is slowed by an adverse pressure gradient and if it has insufficient energy its motion stops and reverses. Fluid molecules in a turbulent boundary layer have more kinetic energy, than in a laminar boundary layer, so that they can move farther downstream, or using our ball analogy, roll further uphill.

Consider the flow over a simple flat plate, as shown in Figure 3.211, where there is an adverse pressure gradient in the x-direction. The fluid closest to the plate surface is slowed by viscosity, due to the action of skin friction on the fluid molecules. If the flow has insufficient energy to resist the adverse pressure gradient, the fluid slows, comes to a stop, and reverses direction, forming a separated flow region. The *separation streamline* divides the separated flow from the unseparated flow. The ability of the turbulent boundary layer to delay flow separation can result in significantly lower pressure drag, due to flow separation, on a body.

To illustrate this, examine Figure 3.212, which shows the laminar and turbulent boundary layer flow over a sphere. The freestream Reynolds number is low, such that the flow over the sphere is laminar. In the upper photo of Figure 3.212, the laminar boundary layer separates just past the top and bottom of the sphere, creating a large separated wake region behind the sphere with an attendant large pressure drag due to flow separation.

In the lower photo, a boundary layer "trip"[7] wire is attached to the front face of the sphere, as annotated in the figure. The trip wire causes the laminar boundary layer to transition to a turbulent

---

[7] Boundary layer trip wires are often used in wind tunnel testing to ensure that the boundary layer over a sub-scale model is turbulent, which is representative of the flow over the full-scale flight vehicle.

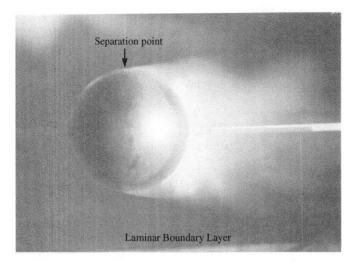

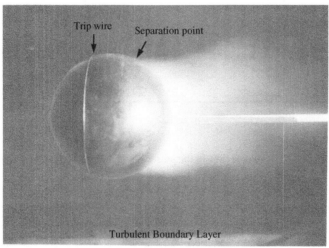

Figure 3.212   Comparison of laminar and turbulent boundary layer flow over a sphere (flow is from left to right). (Source: *Images by permission of the German Aerospace Center, DLR Archive, Gottingen, 1914, annotations by the author.*)

boundary layer. Since the turbulent boundary layer has a higher average kinetic energy than the laminar boundary layer, the flow is able to remain attached well past the laminar separation point, finally separating on the aft portion of the sphere. Since the turbulent boundary layer delayed the flow separation to a further aft position around the sphere, the size of the separated wake is much smaller than for the laminar boundary layer. Thus, the pressure drag due to flow separation is dramatically reduced by the turbulent boundary layer. This significant reduction in the drag due to a turbulent boundary layer is used advantageously in the design of golf balls, where dimples or other surface irregularities are placed on the surface to trip the boundary layer, promoting turbulent flow and delaying flow separation.

A similar result is seen by looking at the pressure distribution over a circular cylinder, as shown in Figure 3.213. The surface pressure coefficient is plotted around the top and bottom of a circular cylinder, starting from the stagnation point ($\theta = 0$) to the aft-most point of the cylinder ($\theta = \pm 180°$). (Note that the pressure coefficient is given the symbol P, rather than C_p, in Figure 3.213.) If the flow is assumed inviscid, the pressure distribution is symmetrical about the

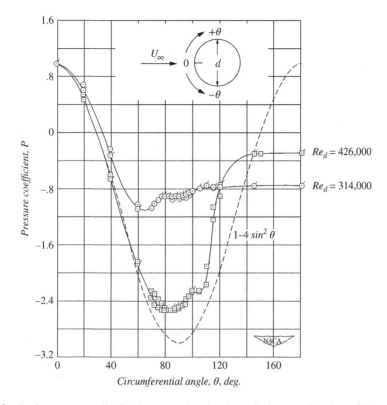

Figure 3.213 Surface pressure distribution around a circular cylinder as a function of Reynolds number (flagged symbols are for negative θ). (Source: *Adapted from F.E. Gowen and E.W. Perkins, "Drag of Circular Cylinders for a Wide Range of Reynolds Numbers and Mach Numbers," NACA Report RM A52C20, Fig. 5a, 19 June 1952.*)

forward and rearward halves of the cylinder. It can be shown that the pressure coefficient around a circular cylinder in an inviscid flow is given by

$$C_p(\theta) = 1 - 4\sin^2\theta \qquad (3.415)$$

The inviscid pressure coefficient is equal to 1 at the front stagnation point ($\theta = 0$), -3 at the top ($\theta = 90°$) of the cylinder, and again equal to 1 at the aft stagnation point ($\theta = 180°$), as shown in Figure 3.213. Since the inviscid pressure distribution is symmetric about a vertical line through the center of the cylinder, the pressure force on the forward half is equal to the pressure force on the rearward half, and there is no net force in the flow direction, or drag, on the cylinder. This zero drag result is d'Alembert's paradox, discussed earlier in the chapter, and is a result of the inviscid flow assumption.

The other pressure distributions in Figure 3.213 are for a viscous flow with increasing Reynolds number, based on the cylinder diameter, Re_d. At a Reynolds number of 314,000, the boundary layer is laminar with the flow separating at a location about 80° around from the forward stagnation point. The pressure coefficient remains constant, with a value of approximately -0.8, in the separated region at the aft end of the cylinder. At a Reynolds number of 426,000, the boundary layer is turbulent, thus the flow remains attached further downstream of the laminar separation point, until about 130° around from the stagnation point. The pressure coefficient is approximately constant, with a value of about -0.35, in the separated flow region. These results for the pressure coefficient distributions are consistent with the visualization of the flow over the sphere for laminar and turbulent boundary layers, shown in Figure 3.212. Since the separated region is smaller and the pressure coefficient is higher, in the separated region, the pressure drag due to flow separation is much less for the turbulent versus the laminar boundary layer.

Example 3.19 Calculation of Boundary Layer Thickness *The Airbus 380 is the world's largest passenger airliner, with a length of 238.6 ft, wingspan of 261.7 ft, and a maximum takeoff weight of 1.27 million pounds. Assume the wing chord is approximated by a flat plate with a length of 35 ft. If the aircraft is flying at 160 knots at an altitude of 8000 ft, calculate the thickness of the boundary layer at the wing trailing edge assuming laminar and turbulent flows.*

Solution

Using Appendix C, the freestream density and temperature at an altitude of 8000 ft are

$$\rho_\infty = \sigma \rho_{SSL} = (0.78609)\left(0.002377\,\frac{\text{slug}}{\text{ft}^3}\right) = 0.001869\,\frac{\text{slug}}{\text{ft}^3}$$

$$T_\infty = \theta T_{SSL} = (0.94502)(519°\text{R}) = 490.5°\text{R}$$

Using Sutherland's Law, Equation (3.16), the freestream coefficient of viscosity is given by

$$\mu_\infty = \left(\frac{T_\infty}{T_{ref}}\right)^{3/2} = \left(\frac{T_\infty}{T_{ref}}\right)^{3/2}\left(\frac{T_{ref}+S}{T_\infty+S}\right)\mu_{ref}$$

$$\mu_\infty = \left(\frac{490.5°\text{R}}{491.6°\text{R}}\right)^{3/2}\left(\frac{491.6°\text{R}+199°\text{R}}{490.5°\text{R}+199°\text{R}}\right)\left(3.584\times10^{-7}\,\frac{\text{slug}}{\text{ft}\cdot\text{s}}\right)$$

$$\mu_\infty = (0.9982)\left(3.584\times10^{-7}\,\frac{\text{slug}}{\text{ft}\cdot\text{s}}\right) = 3.578\times10^{-7}\,\frac{\text{slug}}{\text{ft}\cdot\text{s}}$$

Converting the velocity to consistent units, we have

$$V_\infty = 160\,\text{knots}\times\frac{6076\,\text{ft}}{1\,\text{nm}}\times\frac{1\,\text{hr}}{3600\,\text{s}} = 270.0\,\frac{\text{ft}}{s}$$

The Reynolds number based on the chord length, c, is

$$Re_c = \frac{\rho_\infty V_\infty c}{\mu_\infty} = \frac{\left(0.001869\,\frac{\text{slug}}{\text{ft}^3}\right)\left(270.0\,\frac{\text{ft}}{s}\right)(35\,\text{ft})}{3.578\times10^{-7}\,\frac{\text{slug}}{\text{ft}\cdot\text{s}}} = 4.936\times10^7$$

Using Equation (3.413), the laminar boundary layer thickness at the wing trailing edge is

$$\delta_{lam} = \frac{5.2c}{\sqrt{Re_c}} = \frac{5.2(35\,\text{ft})}{\sqrt{4.936\times10^7}} = 0.02591\,\text{ft} = 0.3109\,\text{in}$$

Using Equation (3.414), the turbulent boundary layer thickness is

$$\delta_{turb} = \frac{0.37c}{Re_c^{0.2}} = \frac{0.37(35\,\text{ft})}{(4.936\times10^7)^{0.2}} = 0.3746\,\text{ft} = 4.495\,\text{in}$$

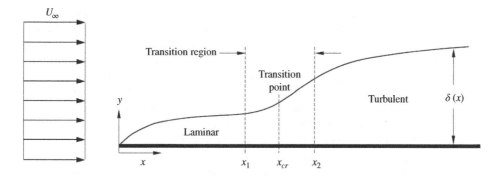

Figure 3.214   Laminar to turbulent boundary layer transition.

The laminar boundary layer thickness is 0.3109 in at the wing trailing edge, while the turbulent thickness is 4.495 in. The ratio of the turbulent-to-laminar thickness is given by

$$\frac{\delta_{turb}}{\delta_{lam}} = \frac{4.495 \, \text{in}}{0.3109 \, \text{in}} = 14.46$$

Hence, the turbulent boundary layer thickness is about 14.5 times larger than the laminar thickness.

3.12.2.1   Boundary Layer Transition

Consider the flow over a flat plate as shown in Figure 3.214, where x is the distance downstream from the leading edge and y is the vertical height above the plate surface. The flow upstream of the flat plate is uniform with a freestream velocity of U_∞. The local Reynolds number can be calculated at any location x, downstream of the plate leading edge, using Equation (3.405). The boundary layer on the plate starts out as laminar and the local Reynolds number is less than about 100,000 in this region. At some downstream distance x_1, the local Reynolds number reaches about 100,000, and the laminar boundary layer starts its transition to a turbulent boundary layer. The boundary layer transition is complete at a distance x_2 downstream of the plate leading edge. The transition occurs over a small but finite distance, known as the boundary layer *transition region*, between points x_1 and x_2. Thus, transition can be more precisely defined as the change, over space and time, of a laminar flow to a turbulent flow, occurring over a certain range of Reynolds number.

For simplicity, the transition region is often modeled as a single point, denoted as the *transition point*, and its distance from the leading edge is called the *critical point* or *critical distance*, x_{cr}. The local Reynolds number corresponding to the critical distance is called the *critical Reynolds number*, Re_{cr}, given by

$$Re_{cr} = \frac{\rho_\infty U_\infty x_{cr}}{\mu_\infty} \tag{3.416}$$

The critical Reynolds number represents the value at which the boundary layer transitions from laminar to turbulent flow.

So, what is physically happening to make the boundary layer transition from laminar to turbulent flow? The answer to this question is often approached from a *stability* perspective, specifically, the stability of the laminar boundary layer and its behavior when disturbed from its equilibrium state. Using this stability theory, it can be proven that laminar flow is unstable when the Reynolds number is above a certain value, but it cannot be analytically proven that the flow then becomes turbulent. We know that the flow becomes turbulent based on experimental evidence. There are some empirically based methods to predict transition, but much like the lack of a theory of turbulence, a theory of transition may elude us for quite some time, if not forever.

The boundary layer can be *tripped* to transition from laminar to turbulent flow, as with the trip wire around the sphere in Figure 3.212. While boundary layer tripping is often used in wind tunnel testing to obtain a turbulent boundary layer, this often occurs in real flight situations also. The surface of a real aircraft is usually not smooth, as there are protuberances such as rivet heads, edges, and gaps that disturb the surface smoothness. Flight through the atmosphere usually results in bugs, dirt, and other debris sticking to the aircraft surfaces, such as the wing leading edges and aircraft nose. Thus, the flight of a real aircraft, within the sensible atmosphere, involves high Reynolds number flow over a rough surface or a surface with many boundary layer trips, which results in turbulent flow. In fact, the existence of laminar flow over the surface of a full-scale aircraft in flight is difficult to obtain.

3.12.3   Skin Friction Drag

We started our discussion about viscous flows by defining shearing stress for a fluid as the product of the coefficient of viscosity, μ, and the velocity gradient. We then discussed the details of the boundary layer velocity profiles for laminar and turbulent flows. Based on this, we now define the shearing stress at the surface of a flat plate or wall, τ_w, as

$$\tau_w = \mu \left(\frac{dV}{dy} \right)_{y=0} \tag{3.417}$$

where μ is the coefficient of viscosity and $(dV/dy)_{y=0}$ is the boundary layer velocity gradient at the wall. Earlier we determined that the turbulent boundary layer velocity profile is fuller than the laminar velocity profile, so that from Equation (3.412), we have

$$\mu \left(\frac{dV}{dy} \right)_{y=0,turb} > \mu \left(\frac{dV}{dy} \right)_{y=0,lam} \tag{3.418}$$

which simply leads to the conclusion that the wall shear stress is greater for a turbulent boundary layer than for a laminar boundary layer, or

$$\tau_{w,turb} > \tau_{w,lam} \tag{3.419}$$

The *local skin friction coefficient*, c_{f_x}, can be defined as

$$c_{f_x} \equiv \frac{\tau_w(x)}{\frac{1}{2}\rho_\infty V_\infty^2} \equiv \frac{\tau_w(x)}{q_\infty} \tag{3.420}$$

where $\tau_w(x)$ is the local wall shear stress, which is a function of the local distance x, and q_∞ is the freestream dynamic pressure. Equation (3.420) defines a dimensionless coefficient that is dependent on the local distance x.

Similar to our earlier discussion about the laminar boundary layer thickness, laminar boundary layer theory can be used to obtain an equation for the laminar skin friction coefficient, $c_{f_{x,lam}}$.

$$c_{f_{x,lam}} = \frac{0.664}{\sqrt{Re_x}} \tag{3.421}$$

The laminar skin friction coefficient is inversely proportional to the square root of the Reynolds number, as is the laminar boundary layer thickness. Inserting Equation (3.421) into (3.420), the laminar wall shear stress, $\tau_{w,lam}$, is given by

$$\tau_{w,lam} = c_{f x, lam} q_\infty = 0.664 \frac{q_\infty}{\sqrt{Re_x}} \tag{3.422}$$

The laminar wall shear stress is a function of the distance along the plate, varying with $x^{-1/2}$.

The flat plate skin friction drag, D_f', is obtained by integrating the shear stress, τ_w, over the length of the plate, L.

$$D_f' = \int_0^L \tau_w dx \tag{3.423}$$

This is the skin friction drag *per unit span*, D_f', denoted by a prime on the drag term. Inserting Equation (3.422) into (3.423), we obtain the flat plate laminar skin friction drag, per unit span.

$$D_{f,lam}' = 0.664 q_\infty \int_0^L \frac{dx}{\sqrt{Re_x}} = \frac{0.664 q_\infty}{\sqrt{\rho_\infty V_\infty / \mu_\infty}} \int_0^L \frac{dx}{\sqrt{x}} = \frac{0.664 q_\infty}{\sqrt{\rho_\infty V_\infty / \mu_\infty}} \left[2 x^{1/2} \right]_0^L \tag{3.424}$$

Evaluating the integral, we have

$$D_{f,lam}' = \frac{0.664 q_\infty}{\sqrt{\rho_\infty V_\infty / \mu_\infty}} \left(2\sqrt{L} \right) = \frac{1.328 q_\infty L}{\sqrt{\rho_\infty V_\infty L / \mu_\infty}} = 1.328 \frac{q_\infty L}{\sqrt{Re_L}} \tag{3.425}$$

where the Reynolds number, Re_L, is now based on the total plate length, L. The flat plate laminar skin friction drag is a function of the freestream dynamic pressure, freestream Reynolds number, and length of the plate. The skin friction drag is greater at higher speed (larger dynamic pressure), for a longer plate, or for lower Reynolds number.

The total skin friction coefficient for the flat plate, C_f, is defined as

$$C_f \equiv \frac{D_f}{q_\infty S_w} \tag{3.426}$$

where S_w is the wetted surface area that is exposed to the flow. (The prime has been dropped on the drag term in Equation (3.426) as this definition of the total skin friction coefficient is valid for the drag of a plate of arbitrary span, not just a unit span.) Inserting Equation (3.425) into (3.426), we have the total laminar skin friction coefficient for the flat plate.

$$C_{f,lam} = \frac{D_{f,lam}'}{q_\infty L(1)} = 1.328 \frac{q_\infty L}{\sqrt{Re_L}} \frac{1}{q_\infty L(1)} = \frac{1.328}{\sqrt{Re_L}} \tag{3.427}$$

For a turbulent boundary layer, an equation for the turbulent skin friction coefficient, $c_{f x, turb}$, can be empirically obtained as

$$c_{f x, turb} = \frac{0.0592}{Re_x^{0.2}} \tag{3.428}$$

Inserting Equation (3.428) into (3.420), the turbulent wall shear stress, $\tau_{w,turb}$, is given by

$$\tau_{w,turb} = c_{f x, turb} q_\infty = 0.0592 \frac{q_\infty}{Re_x^{0.2}} \tag{3.429}$$

Inserting Equation (3.429) into (3.423), the flat plate turbulent skin friction drag, per unit span, is

$$D'_{f,turb} = 0.0592 q_\infty \int_0^L \frac{dx}{Re_x^{0.2}} = \frac{0.0592 q_\infty}{(\rho_\infty V_\infty / \mu_\infty)^{0.2}} \int_0^L \frac{dx}{x^{0.2}}$$

$$D'_{f,turb} = \frac{0.0592 q_\infty}{(\rho_\infty V_\infty / \mu_\infty)^{0.2}} \left[\frac{5}{4} x^{4/5} \right]_0^L = 0.074 \frac{q_\infty L^{4/5}}{(\rho_\infty V_\infty / \mu_\infty)^{0.2}} = 0.074 \frac{q_\infty L}{Re_L^{0.2}} \tag{3.430}$$

Inserting Equation (3.425) into (3.426), we have the total turbulent skin friction coefficient for the flat plate.

$$C_{f,turb} = \frac{D'_{f,turb}}{q_\infty L(1)} = 0.074 \frac{q_\infty L}{Re_L^{0.2}} \frac{1}{q_\infty L(1)} = \frac{0.074}{Re_L^{0.2}} \tag{3.431}$$

Comparing Equation (3.431) with (3.427), we see that the flat plate skin friction coefficient for laminar flow varies with one over the square root of the plate length, $L^{-1/2}$, while it varies with $L^{-1/5}$ for a turbulent flow.

The variations of the laminar and turbulent flat plate skin friction coefficients with Reynolds number based on plate length, given by Equations (3.427) and (3.431), respectively, are shown in Figure 3.215.

Some other empirically derived equations for the skin friction coefficients are given in [53] for turbulent flow at high Reynolds numbers and for a transitional boundary layer. For turbulent boundary layers at high Reynolds numbers, a slightly more complex equation than Equation (3.431), provides another turbulent flat plate skin friction coefficient, $C_{f,turb\,high\,Re}$, given by

$$C_{f,turb\,high\,Re} = \frac{0.455}{(\log_{10} Re_L)^{2.58}} \tag{3.432}$$

This equation may be more accurate than Equation (3.431) at Reynolds numbers greater than about 10 million.

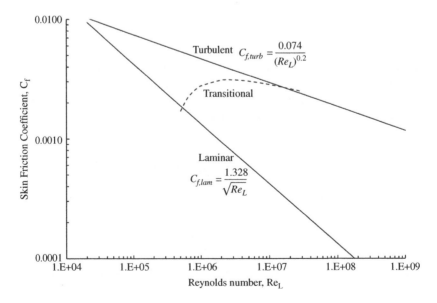

Figure 3.215   Variation of flat plate skin friction coefficient with Reynolds number.

An experimentally based skin friction coefficient for the transitional boundary layer, $C_{f,trans}$, is given by

$$C_{f,trans} = \frac{0.455}{(\log_{10} Re_L)^{2.58}} - \frac{1,700}{Re_L} \tag{3.433}$$

This transitional skin friction coefficient may be used between Reynolds numbers from about 500,000 to about 30 million, although there are many other factors that determine the state of the boundary layer.

The flat plate skin friction equations shown in Figure 3.215 apply to low speed, incompressible flows where the density is assumed to be constant. In fact, everything that we have discussed thus far about viscous flows pertains to *incompressible*, viscous flows. If we were to consider *compressible*, viscous flow, we would need to consider the density as a variable. The pressure is still constant through a compressible boundary layer, as it is for incompressible flow. The analysis of compressible boundary layers is beyond the scope of this text, being suitable for an advanced course in viscous flows.

The incompressible boundary layer skin friction coefficients are functions of the Reynolds number. For compressible boundary layers, they become functions of Reynolds number and Mach number. The calculation of the compressible flat plate skin friction coefficients for laminar and turbulent flow require numerical solutions. The compressible skin friction coefficient decreases with increasing freestream Mach number, assuming the Reynolds number remains constant. This decrease is proportionally larger, in terms of the change from the incompressible value, for a compressible turbulent boundary layer than for a compressible laminar boundary layer.

Example 3.20   Calculation of Skin Friction Drag *For the Airbus 380 in Example 3.19, calculate the skin friction drag on a 100 ft wide span of the wing, assuming laminar, turbulent, and transitional flow.*

Solution

From Example 3.19, the freestream density is 0.001869 slug/ft³, the freestream velocity is 270.0 ft/s and the Reynolds number based on the 35 ft long chord length is 4.931×10⁷. The freestream dynamic pressure is

$$q_\infty = \frac{1}{2} \rho_\infty V_\infty^2 = \frac{1}{2} \left(0.001869 \, \frac{slug}{ft^3} \right) \left(270.0 \, \frac{ft}{s} \right)^2 = 68.13 \, \frac{lb}{ft^2}$$

Using Equation (3.427), the laminar skin friction coefficient is

$$C_{f,lam} = \frac{1.328}{\sqrt{Re_c}} = \frac{1.328}{\sqrt{4.931 \times 10^7}} = 0.0001891$$

The laminar skin friction drag is given by

$$D_{f,lam} = C_{f,lam} q_\infty S_w = (0.0001891) \left(68.13 \, \frac{lb}{ft^2} \right) (35 \, ft \times 100 \, ft) = 45.09 \, lb$$

Using Equation (3.431), the turbulent skin friction coefficient is

$$C_{f,turb} = \frac{0.074}{Re_c^{0.2}} = \frac{0.074}{(4.931 \times 10^7)^{0.2}} = 0.002141$$

The turbulent skin friction drag is given by

$$D_{f,turb} = C_{f,turb} q_\infty S_w = (0.002141) \left(68.13 \frac{\text{lb}}{\text{ft}^2} \right) (35 \, \text{ft} \times 100 \, \text{ft}) = 510.5 \, \text{lb}$$

Using an alternate equation for the turbulent skin friction coefficient at high Reynolds number, Equation (3.432), we have

$$C_{f,turb \, high \, Re} = \frac{0.455}{(\log_{10} 4.931 \times 10^7)^{2.58}} = 0.002354$$

The turbulent skin friction drag is given by

$$D_{f,turb \, high \, Re} = C_{f,turb \, high \, Re} q_\infty S_w = (0.002354) \left(68.13 \frac{\text{lb}}{\text{ft}^2} \right) (35 \, \text{ft} \times 100 \, \text{ft}) = 561.3 \, \text{lb}$$

Thus, we have estimates for the turbulent skin friction drag of 510.5 lb and 561.3 lb. Both of these turbulent drag predictions are an order of magnitude greater than for the laminar boundary layer, highlighting the significant drag reduction with laminar flow.

Using Equation (3.342), the transitional skin friction coefficient is

$$C_{f,trans} = \frac{0.455}{(\log_{10} Re_L)^{2.58}} - \frac{1700}{Re_L}$$

$$C_{f,trans} = \frac{0.455}{(\log_{10} 4.931 \times 10^7)^{2.58}} - \frac{1700}{4.931 \times 10^7} = 0.002320$$

The transitional skin friction drag is given by

$$D_{f,trans} = C_{f,trans} q_\infty S_w = (0.002320) \left(68.13 \frac{\text{lb}}{\text{ft}^2} \right) (35 \, \text{ft} \times 100 \, \text{ft}) = 553.2 \, \text{lb}$$

The skin friction drag prediction for a transitional boundary layer is higher than one of the turbulent skin friction drag predictions and less than the other. Since the transitional boundary layer is composed of a run of a laminar boundary layer, followed by a turbulent boundary layer, we would expect the transitional drag to be in between the values for the laminar and turbulent boundary layers.

3.12.4 Aerodynamic Stall and Departure

In this section, we explore the low-speed boundary of the flight envelope, which was defined in Chapter 2 as the *aerodynamic lift boundary* or *aerodynamic stall* (see Figure 2.22). At this aerodynamic stall boundary, there is a loss of lift due to flow separation on the wing. If left uncorrected, the stall may lead to *departure* of the aircraft from controlled flight. The *stall airspeed* associated with the aerodynamic stall is the minimum airspeed at which the aircraft can maintain steady, level flight. While we primarily discuss the aerodynamics of this type of stall, there are several other definitions of the stall and the stall speed that are discussed.

In flight, the consequences of an inadvertent aerodynamic stall, followed by a departure, can be catastrophic. Aeronautical pioneer, Otto Lilienthal perished when the glider he was piloting, stalled and crashed in 1896. The Wright Brothers first encountered inadvertent stalls when flying their 1901 manned glider. Based on their experience with stalls, and their knowledge of Lilienthal's demise because of a stall, the Wright Brothers chose a canard configuration for their aircraft, which can have more benign stalling characteristics than a conventional aft-tail configuration. The forward canard is designed to stall before the main wing, causing the aircraft nose to drop and reducing the angle-of-attack. Even in modern times, there are a significant number of aviation accidents due to stalls and departures. Since all aircraft must operate near the low-speed boundary of their flight envelope, usually for takeoff, landing, and other low-speed operations, it is important to understand the characteristics of the aircraft along this boundary. There are several ways to define or categorize stalls and departures. After we discuss several of these, we explore the flight test techniques associated with stalls and departures.

3.12.4.1 Stall Definitions

There are several ways to define stall and stall speed. They may be defined based on the aerodynamic lift limit and loss of lift or by other limiting factors, such as uncommanded aircraft motions, high sink rates, control effectiveness issues, or intolerable buffet. The stall definition is different for different types of aircraft, due to differences in geometric configuration, wing planform, airfoil shape, and other considerations. For a given aircraft type, the stall characteristics may change depending on the aircraft configuration (position of flaps, landing gear, speed brake, etc.), center of gravity location, rate of change of angle-of-attack, or other factors. For instance, the stall characteristics and stall speeds are usually different depending on whether or not the flaps are extended. The airfoil and wing geometries play an important role in the aircraft stall characteristics.

The aerodynamic stall is the result of the wing exceeding its *critical* or *stall angle-of-attack*, α_s. The flow over the wing cannot remain attached and there is massive flow separation, leading to significant loss of lift. The large areas of separated flow are shown on a stalled wing, from a wind tunnel test, in Figure 3.216. The flow is seen to reverse direction and flow spanwise along the wing,

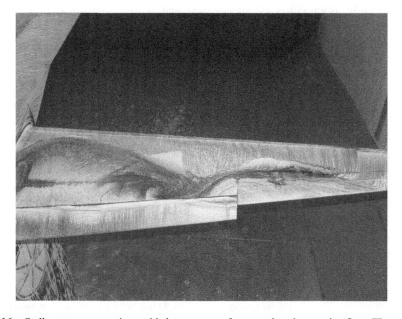

Figure 3.216 Stall pattern on a wing, with large areas of reversed and spanwise flow. The wing is in a wind tunnel where the surface flow is made visible using China clay. (Flow direction is from top to bottom.) (Source: *University of Washington Aeronautical Laboratory, "Wing Airflow Pattern" https://en.wikipedia.org/ wiki/File:Wing_air_flow_pattern.jpg, CC-BY-SA-3.0. License at https://creativecommons.org/licenses/by-sa/ 3.0/legalcode.*)

forming large separation "bubbles", which destroys the lift. The surface flow is made visible using China clay, a mixture of kerosene, clay powder, and fluorescent pigment. The mixture is applied to the wing surface, prior to turning the wind tunnel flow on. When air flows over the wing surface, the kerosene evaporates, leaving streaks of clay powder, marking the flow pattern on the surface.

Consider an aircraft in steady, 1 g flight at constant altitude. The aircraft lift, L, equals the weight, W, so that

$$L = W = q_\infty S C_L = \frac{1}{2} \rho_\infty V_\infty^2 S C_L \qquad (3.434)$$

According to Equation (3.434), if the aircraft slows down (decreasing V_∞), the angle-of-attack must be increased to generate a higher lift coefficient, C_L, to maintain steady, constant altitude (constant ρ_∞) flight. This also assumes that the wing area, S, is constant. If the aircraft slows to the stall speed, V_s, it is at the stall angle-of-attack, α_s, and the maximum lift coefficient, $C_{L,max}$. The stall speed, V_s, for 1 g, steady flight, where the lift equals the weight, was given in Section 2.3.7.1 as Equation (2.48). Rearranging this equation, we have

$$V_s = \sqrt{\frac{2W}{\rho_\infty S C_{L,max}}} = \sqrt{\left(\frac{2}{\rho_\infty C_{L,max}}\right)\left(\frac{W}{S}\right)} \qquad (3.435)$$

where the weight divided by the wing reference area, W/S, is called the *wing loading*. As shown by Equation (3.435), the aerodynamic stall speed is directly proportional to the square root of the aircraft wing loading, W/S, and inversely proportional to the square root of the maximum lift coefficient, $C_{L,max}$.

The wing loading is an important parameter affecting stall speed and other aircraft performance characteristics, such as climb rate, turn performance, and takeoff and landing distances. For a given weight, an aircraft with a high wing loading has a smaller wing than one with a low wing loading. Historically, as aircraft cruise speeds increased, wing loadings got larger as wing sizes decreased to enhance high-speed performance. The larger wing loadings led to higher stall speeds for higher speed aircraft. The wing loading and stall speed of different types of aircraft are shown in Table 3.16. (In SI units, the wing loading is typically expressed by inconsistent units of kg_f/m^2.) The stall speeds are seen to increase proportionally with the increase in the wing loadings.

It is advantageous for an aircraft to have a low stall speed, since this enables low speed takeoff and landing operations. As discussed previously, the maximum lift coefficient can be further increased with the use of high-lift devices, such as flaps, which can further decrease the stall speed by increasing the maximum lift coefficient, $C_{L,max}$. Some high-lift devices also increase the wing area, S, which results in an additional stall speed decrease.

For some types of aircraft or aircraft configurations, the definition of the stall speed based on the maximum lift coefficient is not appropriate. Other limiting factors may be more critical at an airspeed higher than the aerodynamic stall speed. These other factors may be uncommanded aircraft

Table 3.16 Wing loadings and stall speeds of selected aircraft.

Aircraft	Primary function	Wing loading, W/S	Stall speed, V_s
Wright *Flyer I*	1st airplane	1.47 lb/ft² (7.18 kg$_f$/m²)	22 mph (35 km/h)
Schweitzer 2–33	Glider trainer	4.74 lb/ft² (23.1 kg$_f$/m²)	36 mph (58 km/h)
Beechcraft A36 *Bonanza*	General aviation	20.2 lb/ft² (98.6 kg$_f$/m²)	59 mph (95 km/h)
North American P-51 *Mustang*	WW II fighter	41.2 lb/ft² (201 kg$_f$/m²)	95.4 mph (154 km/h)
Northrop T-38 *Talon*	Military jet trainer	69.5 lb/ft² (339 kg$_f$/m²)	146 mph (235 km/h)
Boeing 777–300	Commercial airliner	143 lb/ft² (698 kg$_f$/m²)	165 mph (265 km/h)

motions, high sink rates, control effectiveness issues, or intolerable buffet. Prior to the aerodynamic stall, an aircraft may exhibit undesirable, uncommanded motions in the pitch, roll, or yaw axes, or combinations of these axes. An *uncommanded motion* is any motion of the aircraft, in any of the three axes, that is not a direct result of a control input applied by the pilot. These uncommanded motions can range from a small, uncommanded change in aircraft attitude, such as a small pitch down or wing roll, to a large excursion, such as the aircraft nose dropping through the horizon to a near vertical attitude. Aircraft with low aspect ratio, highly swept, slender wings or delta wings are prone to a dynamic *wing rocking* motion at high angles-of-attack. This oscillatory motion is caused by unsteady vortex shedding from the slender wing or slender fuselage forebody. For aircraft with this type of high angle-of-attack behavior, the stall may be defined by a specific amplitude of the wing rock oscillations, such as an oscillating bank angle of $\pm 20°$.

At a high angle-of-attack, a large lift coefficient results in a large induced drag. Recall that the induced drag is proportional to the lift coefficient squared, as given by Equation (3.279). The large drag increase may result in unacceptably high sink rates, especially when the aircraft is at low altitude, close to the ground, on a landing approach. Even though an unacceptably high sink rate may take the form of a stabilized descent, it may still be a hazardous flight situation. The sink rate may be so high that it would be extremely difficult or impossible to arrest the descent rate to make a safe landing. In this case, the stall may be defined by a minimum speed at the maximum allowable sink rate.

Control effectiveness or lack of control authority issues may also define the stall, which typically involves the pitch control that is being used to increase the angle-of-attack. As the pilot pulls back on the control wheel or stick, the aircraft nose normally pitches up. In this type of stall indication, the nose pitch-up stops prior to aerodynamic stall, despite further aft wheel or stick input. Alternatively, the pitch control limit may be reached prior to the aerodynamic stall, where the control stick is at its full aft position and the angle-of-attack cannot be increased further.

Finally, a stall may be defined by intolerable buffet or vibration that is felt at a high angle-of-attack. This buffet or vibration may be the result of separated flow from the wing impinging on the aircraft tail. Again, this buffet or vibration may occur at an angle-of-attack lower than the aerodynamic stall angle-of-attack.

3.12.4.2 Aerodynamics of Stall

In Section 3.8, the airfoil lift curve was introduced as a plot of the lift coefficient versus angle-of-attack (see Figure 3.90). Up to now, our attention has been focused on the linear lift range, where the lift coefficient varies linearly with the angle-of-attack. This part of the lift curve is used for the majority of a typical flight. The non-linear portion of the lift curve starts near the aerodynamic stall, where the lift coefficient is at a maximum, $C_{L,max}$. This has been shown previously to be a peak in the lift coefficient at the stall angle-of-attack, α_s, followed by a decrease or drop in the lift.

To be clear, the stall is due to exceeding this critical angle-of-attack and is not dependent on the air speed. For example, if we placed an airfoil at $0°$ angle-of-attack in a wind tunnel with a flow velocity below the airfoil's stall speed, the flow over the airfoil would not be separated and there would be no aerodynamic stall. If we now set the airfoil at its stall angle-of-attack, in a flow with a velocity above its stall speed, the flow over airfoil would separate and the airfoil would be stalled. So why can't an airplane fly below stall speed? If we set the angle-of-attack of the airplane's wing to below its stall angle-of-attack, with a speed below its stall speed, the lift produced from the resulting lift coefficient and dynamic pressure would be less than the aircraft weight. The stall speed is associated with the stall in so far as the angle-of-attack is continuously increased to the

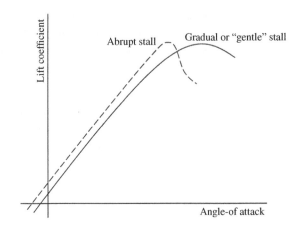

Figure 3.217   Gradual and abrupt stalls.

stall angle-of-attack as the airspeed is decreased, in attempting to maintain the lift equal to the weight.

For conventional airfoils without high-lift devices, the maximum lift coefficient is about 1.8–1.9 and the stall angle-of-attack is usually about 15–18°. However, the nature of the stall, in terms of the rate of change of the lift coefficient with angle-of-attack, can be quite different, depending on several factors. This can be seen in the shape of the lift curve after the stall angle-of-attack, as shown in Figure 3.217. The stall may be "gentle", where the lift coefficient decreases gradually with increasing angle-of-attack, or it may be abrupt, where the lift coefficient drops rapidly after the stall. The difference in the stall behavior may have an impact on flight safety, because a more abrupt stall, with its higher rate of onset, may present less warning to a pilot, in terms of airframe buffet or vibration due to the flow separation, than a gradual stall.

One of the primary factors that affects the type of stall, whether gradual or abrupt, is the shape of the airfoil. A moderate or thick airfoil section, with a large leading edge radius, tends to have a gradual or gentle stall behavior. A thin airfoil, with a small leading edge radius, has a more abrupt stall. This stall behavior is related to the flow separation characteristics of the airfoil. As shown in Figure 3.218, a thick airfoil with a well-rounded leading edge tends to delay flow separation until the trailing edge. Here, a "thick" airfoil is defined as one with a thickness ratio of about 13% or greater. The flow separation starts at the airfoil trailing edge and moves forward as the angle-of-attack is increased. The loss of lift is gradual as the separated region grows and the angle-of-attack is increased. Typically, the pitching moment change due to the stall is small. Separation starts at the leading edge of a thinner airfoil, but usually at a lower angle-of-attack than for the thick airfoil. Here, a "thin" airfoil is defined as one with a thickness ratio of about 6–13%. The stall starts as a separation bubble at the leading edge, which spreads quickly over the length of the airfoil. This type of stall results in an abrupt loss of lift and an abrupt, large change in the pitching moment.

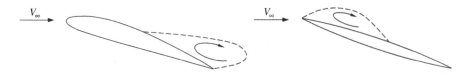

Figure 3.218   Airfoil stall with trailing edge separation (left) and leading edge separation (right).

For a very thin airfoil, with a thickness ratio less than about 6% and a sharp leading edge, the stall also starts with a separation bubble at the leading edge, but the stall progression is different. As the angle-of-attack is increased, the separated flow spreads over the length of the airfoil. At this point, with a separated flow bubble stretching from the leading to the trailing edge, the airfoil achieves its maximum lift. If the angle-of-attack is increased further, the airfoil stalls with a smooth loss of lift, but with a large pitching moment change.

Another primary factor affecting the separation point is the Reynolds number. Flow separation is delayed by a turbulent boundary layer at larger Reynolds number, as discussed previously for the flow over a sphere. Higher Reynolds numbers tend to delay flow separation and move the separation point further aft, towards the airfoil trailing edge.

The flow separation and stall characteristics that we have discussed so far assume a steady-state angle-of-attack or an angle-of-attack that is slowly increasing. If the angle-of-attack is increased rapidly, the *dynamic stall* behavior may be quite different from the steady-state case, due to the unsteady, non-linear aerodynamics of flow separation. The *dynamic stall* angle-of-attack may be larger than the *static stall* angle-of-attack. When the angle-of-attack is rapidly increased, a vortex is created at the airfoil leading edge, which flows over the airfoil upper surface, energizing the boundary layer flow and delaying flow separation. Thus, the angle-of-attack can be dynamically increased to a significantly higher angle before stall occurs and the dynamic stall angle-of-attack may be significantly higher than the static stall angle-of-attack. A higher maximum lift coefficient may also be obtained in the dynamic situation. However, when the dynamic stall does occur, the loss of lift tends to be much more abrupt than the static stall.

We now consider flow separation related to stall for the three-dimensional wing. The two primary factors affecting stall related to the wing geometry are the wing aspect ratio and the planform shape. The effect of aspect ratio was discussed in the section dealing with finite wings, where it was shown that the maximum lift coefficient is increased and the stall angle-of-attack is decreased with increasing aspect ratio.

The effect of the finite wing planform shape on the progression of separated flow, leading to stall, is shown in Figure 3.219. At high angle-of-attack near stall, the flow separation on the rectangular planform starts at the wing trailing edge near the wing root. As the angle-of-attack increases, the separated flow progresses outwards towards the wing leading edge and wingtip. For the tapered planform wing, the separated flow starts more uniformly along the entire trailing edge and progress forward. The separated flow starts at the wingtip on the swept wing planform and moves inboard towards the leading edge and wing root.

Based on the stall progression patterns shown in Figure 3.219, the rectangular planform is advantageous from the standpoint of maintaining attached flow, outboard on the wing, where the ailerons are located. The attached flow over the ailerons enables roll control at higher angles-of-attack. The swept wing planform is seen to be the worst in terms of roll control at high angle-of-attack, with the flow separation starting at the wingtip, making the ailerons ineffective at high angle-of-attack. Another benefit of having the separation start at the wing root is that the turbulent wake from the flow separation impinges on the horizontal tail, providing buffet or vibration, warning the pilot of an impending stall. In this way, the pilot is aware of the impending stall, before the flow separation spreads over the entire wing.

We must keep in mind that the separated flow patterns and stall progression over the left and right wings of an aircraft may not be symmetric. This asymmetry may be caused by a non-uniform freestream flow, aircraft sideslip, or slight differences in the left and right wing geometries. Therefore, the loss of lift due to the flow separation may also not be symmetric, resulting in an asymmetric aircraft motion, such as a roll-off or wing "drop".

Since the topic of stall warning has been mentioned, we conclude this section with some insights into the types of "fixes" that can be incorporated into the wing design to prevent, postpone, warn,

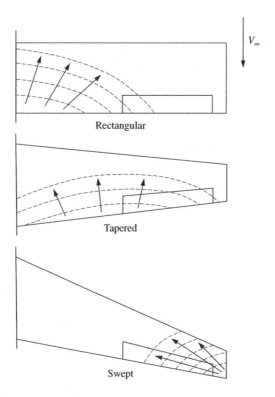

Figure 3.219   Aerodynamic stall progression on different wing planform shapes.

or lessen the severity of aerodynamic stall. Keeping with the notion that root stall is preferred; the wing can be designed with twist, to ensure that the wing root airfoil section stalls prior to the wingtip section. This can be accomplished with geometric twist or washout, where the root airfoil section is set at a higher incidence angle than the tip, or with aerodynamic twist, where the airfoil shape at the root is designed to stall at a lower angle-of-attack than a different airfoil shape at the wingtip.

In addition to the various types of leading edge high-lift devices, such as slots and slats, other types of wing devices or attachments may also be used to improve stall characteristics. A small, sharp-edged *stall strip* may be attached to the wing leading edge, typically near the wing root. The sharp strip makes the stall start where it is attached, resulting in a wing root stall. Other devices placed along the wing leading edge include small, fin-like devices called *vortex generators*. At high angle-of-attack, these small fins create vortices that energize the wing boundary layer, delaying flow separation. In another vortex flow type design, a notch or droop may be made in wing leading edge, called a drooped leading edge, to produce a vortex flow that energizes the boundary layer. Finally, a flat plate-type device, called a stall fence, may be placed in the chordwise direction along the wing to prevent the spanwise spread of flow separation.

Addressing the stall from a warning perspective, there are several types of devices that are based on awareness or monitoring of the stall angle-of-attack. Perhaps the simplest of these is the visible stall warning light or audible stall warning horn, which is activated when the stall angle-of-attack is approached. These warning devices often work by sensing the pressure at the wing leading edge. Many modern aircraft have angle-of-attack indicators, which display angle-of-attack data or information to the pilot.

Other types of mechanical devices serve to prevent the aircraft from stalling. *Stick pushers* are mechanical devices that push the elevator control forward, to reduce the angle-of-attack, as the stall is approached. Mechanical devices, appropriately called *stick shakers*, may also shake or vibrate the flight controls or stick to provide tactile warning of a stall. These types of devices usually rely on sensors that measure angle-of-attack, for their function. In fly-by-wire flight control systems, where the aircraft controls are not mechanically connected to the control surfaces, flight control computers can limit the angle-of-attack that is requested by the pilot. These *angle-of-attack limiters* can provide stall protection throughout the flight envelope.

3.12.4.3 Post-Stall Aerodynamics

In this section, we discuss some of the aerodynamics and aircraft behavior associated with flight beyond the stall, also called the *post-stall* region. Often, flight in this post-stall region is synonymous with *high-angle-of-attack* flight. Up to the stall, the aircraft aerodynamics are assumed linear. A dominant characteristic of post-stall aerodynamics is the non-linear nature of the aerodynamics, which makes it difficult to accurately predict post-stall aerodynamics and hence difficult to predict post-stall aircraft motions.

There are few situations where it is intentionally desired to operate an aircraft in the post-stall region. Flight training may be conducted in this region or some aerobatic aircraft may routinely operate in this region. With the advent of thrust vectoring aircraft, maneuvering in the post-stall region may provide some advantages for military aircraft. Sustained post-stall maneuvering is sometimes termed *supermaneuvrability*. Post-stall maneuvering at very high angle-of-attack has been demonstrated by thrust vectoring aircraft, such as the Rockwell X-31, which was capable of controlled flight at angles-of-attack as high as 70°.

Often, the post-stall region is entered unintentionally and depending on several factors, including the type of aircraft, entry dynamics, pilot inputs, or other factors, the aircraft may depart controlled flight. The *departure* of an aircraft is a transient event that follows loss of aircraft control, which leads to recovery back to controlled flight or to sustained out-of-control behaviors, such as a post-departure gyration, deep stall, or spin. A *post-departure gyration* is a non-steady, out-of-control aircraft motion, such as tumbling about one or more axes of the aircraft. A *deep stall* is a semi-steady state, out-of-control flight condition in which the aircraft is sustained at an angle-of-attack beyond the stall with negligible rotational motions. The aircraft may exhibit some low rate oscillatory motions in yaw, roll, or pitch while in a deep stall. A *spin* is a sustained aircraft rotation in yaw at angles-of-attack above the stall. The rotational motions in the spin may be oscillatory in pitch, roll, and yaw. We discuss spins further in a later section.

The aircraft behavior during a departure includes motions that are uncommanded, of large amplitude, and divergent. *Uncommanded motion* is defined as aircraft motion that is not in response to the pilot inputs, in the normal sense. For instance, with a right stick input, one expects the aircraft to roll to the right, in the normal sense. An aircraft pitch up with a right stick input is an uncommanded motion.

To be clear, an aircraft departure need not be preceded by a stall, as there are other circumstances that can cause a departure. For instance, poor directional stability can lead to a yaw divergence and departure from controlled flight. At high angles-of-attack, asymmetric vortices, shed from the fuselage forebody, can lead to high side forces and yawing moments, which cause a "nose slice" departure. Several military aircraft have had serious problems with nose slice departures, including the McDonnell Douglas F-4 *Phantom*.

The post-stall lift curve of a NACA 0012 airfoil section is shown in Figure 3.220, for angles-of-attack from zero to 180°. This unique aerodynamic data set was obtained from wind tunnel testing of various airfoil sections for use in Darrieus wind turbines, a type of vertical axis wind turbine (see [62] for further details). The Reynolds number based on the airfoil chord length, Re_c, is 10 million. The lift curve up to about 15° angle-of-attack looks like the familiar airfoil lift

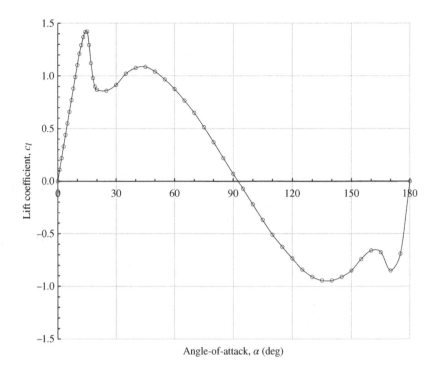

Figure 3.220 Lift curve of NACA 0012 airfoil from zero to 180° angle-of-attack for Reynolds number, based on chord length, of 10 million. (Source: *Data from* [62].)

curves that we are familiar with. The lift coefficient is zero at zero angle-of-attack, as expected for a symmetric airfoil such as the NACA 0012. The lift coefficient is linear with angle-of-attack up to near the stall angle-of-attack of about 15°, where the maximum lift coefficient is about 1.4. Beyond the stall angle-of-attack, the lift coefficient drops dramatically, with an almost vertical slope, reaching a minimum of about 0.86. Beyond an angle-of-attack of about 25°, the lift coefficient again increases with angle-of-attack, with a lift slope that is less than at the lower angle-of-attack range. The lift coefficient reaches a second peak of about 1.1 at an angle-of-attack of about 45°. The second stall of the airfoil at 45° angle-of-attack is more gradual and smoother than the first stall at 15°. The lift coefficient decreases smoothly to zero at about 92° angle-of-attack, where the airfoil is in a near vertical attitude. Beyond 90° angle-of-attack, the airfoil is "backwards", with the trailing edge pointed into the freestream flow. The lift coefficients for angles-of-attack greater than 92° are negative because they correspond to negative angles-of-attack, as the airfoil was rotated continuously from zero to 180° in the wind tunnel. The trends of the "backwards" airfoil are similar to the normal orientation of the airfoil, but the magnitudes of the lift coefficient are lower for corresponding angles-of-attack. The lift returns to zero at 180° angle-of-attack, where it presents a "backwards", yet symmetric, profile to the freestream flow.

3.12.4.4 Spins

A spin is one of the most complex motions of an aircraft. It is a *coupled* motion, a motion about two or more aircraft axes that interact with each other. A spin can be entered from any flight attitude, intentionally or unintentionally. The intentional spin was used in World War I as an

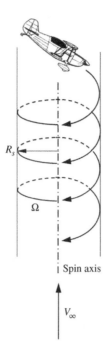

Figure 3.221   Spin geometry.

escape maneuver during combat or as a means of rapidly losing altitude. A departure may or may not lead to an aircraft spin, but a spin must be preceded by a departure. Two necessary ingredients for an aircraft to spin are aerodynamic stall and yawing motion. One or both wings must be aerodynamically stalled for an aircraft to spin, and yawing motion must be present. The yawing motion may be the result of the lift and drag imbalance between the wings in a stall or due to side force in another type of departure, such as a nose slice.

After entering a spin, the aircraft auto-rotates in yaw about a vertical spin axis with a high rate of descent and rapid loss of altitude, as depicted in Figure 3.221. In addition to the yawing motion, there may be oscillatory motions in all three aircraft axes. In the developed spin, the spin characteristics are repeatable from turn to turn, such that the spin may be defined as a steady, helical rotation with angular rate Ω and radius R_s, which may be less than the wingspan. An erect or upright spin is depicted in Figure 3.221, but the spin may also be with the aircraft inverted.

A free-body diagram for an aircraft in a spin is shown in Figure 3.222. The aircraft is in a nose down attitude, making an angle θ with respect to the vertical axis, spinning with an angular rate Ω about the vertical spin axis, and descending at a velocity V_∞. The aircraft is not spinning about its center of gravity, so that the spin has a radius R_s, from the aircraft center of gravity to the center of turn on the vertical spin axis. Even though the wing is stalled, there is still a resultant aerodynamic force, F_R, being produced by the wing. This resultant force can be resolved into a drag, D, parallel to the velocity vector, and a lift, L, perpendicular to the velocity. Therefore, in the spin, the forces acting on the aircraft, at its center of gravity, are the lift L, drag D, and weight W.

Looking at the free-body diagram, one might be tempted to add another horizontal force to balance the lift and "keep the aircraft in equilibrium" or "hold the aircraft in the spin", but the aircraft is in *accelerated* flight and is *not in equilibrium*. Adding this force, commonly known as the *centrifugal force* (from the Greek for "fleeing from the center"), is incorrect. Since the aircraft is descending and turning in yaw, the total velocity vector is the vector sum of the vertical component,

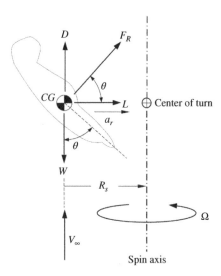

Figure 3.222   Spin free-body diagram.

V_∞, and a yaw component (not shown in the figure). The yaw velocity component is continually changing in direction, which changes the direction of the total velocity vector. Since the velocity vector is changing with time, there is an acceleration. If there were another force that balanced the lift, then the aircraft would be in equilibrium and it would not turn at all. The radial lift is needed to cause the aircraft to yaw.

Applying Newton's second law in the vertical, z-direction, we have

$$\sum F_z = W - D = mg - D = 0 \tag{3.436}$$

where m is the aircraft mass and g is the acceleration due to gravity. Equation (3.436) simply states that the aircraft is in equilibrium in the vertical direction, with the weight balanced by the drag. Solving for the aircraft mass, we have

$$m = \frac{D}{g} \tag{3.437}$$

Applying Newton's second law in the radial direction, along the spin radius, we have

$$\sum F_r = L = ma_r \tag{3.438}$$

where m is the aircraft mass and a_r is the radial acceleration. Assuming that the spin flight path is a circle, the radial acceleration, a_r, is given by

$$a_r = \frac{V_r^2}{R_s} = \frac{(\Omega R_s)^2}{R_s} = \Omega^2 R_s \tag{3.439}$$

where V_∞ is the radial velocity, Ω is the angular velocity in radians per second, and R_s is the spin radius. Substituting Equation (3.334) into (3.333), we have

$$L = m\Omega^2 R_s \tag{3.440}$$

Solving for the spin radius, R_s, and inserting Equation (3.437), we have

$$R_s = \frac{L}{m\Omega^2} = \left(\frac{L}{D}\right)\frac{g}{\Omega^2} \tag{3.441}$$

Thus, the spin radius is directly proportional to the aircraft lift-to-drag ratio, L/D, and inversely proportional to the angular rate, Ω.

Using the geometry in Figure 3.222, the spin radius can be related to the nose up angle from the vertical, θ, as

$$R_s = \frac{g}{\Omega^2} \tan \theta \tag{3.442}$$

As the spin gets flatter (θ increasing), the spin radius decreases and the rotation rate increases. According to [51], for a typical general aviation airplane, the spin radius is of the order of 0.2 of the wingspan with a θ of 45°, decreasing to 0.06 of the wingspan in a flatter spin, with a θ equal to 60°.

The spin rotation rate is considered slow for rotation rates less than 60 deg/s, fast for rates of 60–120 deg/s, and extremely fast for rates greater than 120 deg/s. The British Sopwith *Camel* (Figure 3.223), one of the most successful World War I fighter biplanes, had dangerous spin characteristics that killed many pilots. In a spin, the *Camel* had dizzyingly rapid rotation rates of 180–240 deg/s, while losing 100–200 ft of altitude with each turn of the spin.

Recoveries from departures, stalls, or spins are important considerations in the design of an aircraft configuration and a flight control system. Prior to 1916, before any recovery methods had been developed for spins, an inadvertent spin was most likely fatal for pilot and airplane. The first documented and witnessed spin recovery was by Lt. Wilfred Parke (1889–1912) of the British Royal Navy, flying an Avro Type G, two-seat biplane (Figure 3.224) on 25 August 1912. After a three-hour evaluation flight, Parke was descending to land from an altitude of about 600–700 ft (180–210 m), in a steep, descending, left turn with the engine at idle. Assessing that his glide was too steep, he pulled the nose up and inadvertently entered a left-turning spin. Parke attempted to recover by adding full power, then by pulling the control stick full back and applying full left

Figure 3.223 British Sopwith *Camel*, known for its notorious spin characteristics. (Source: *Unknown RAF photographer, PD-UKGov.*)

Figure 3.224   Avro G biplane. (Source: *Unknown author, 1912, San Diego Air & Space, Museum, San Diego Air & Space Museum Archives, no known copyright restrictions.*)

rudder, which increased the spin rotation rate. Both of these attempts failed to recover the airplane and he was now at an altitude of only about 50 ft (15 m). He then applied hard right rudder; the spin rotation stopped almost immediately and the airplane was back under control. Parke estimated that he had recovered at a height of about 5–6 ft (1.5–1.8 m). This record-setting spin recovery became known as "Parke's dive". Parke's fortunate discovery of a spin recovery technique became known in the aviation community, but it was not formally taught in flight training, as the risk involved in intentionally spinning an aircraft was considered too high.

The first systematic investigation of spins was by the British physicist Frederick A. Lindemann (1886–1957) of the Royal Aircraft Factory, Farnborough, England, some time after 1915. Lindeman analyzed the spin and developed a theoretical spin recovery technique. Unfortunately, at that time, intentional spins were considered too hazardous to actually test his theory. Undaunted, Lindemann requested that he be given flight training so that he could test his theory himself. He was granted permission and, in the summer of 1917, he flew several flights in a Royal Air Factory B.E.2E biplane, entering and successfully recovering from spins, proving his spin recovery method.

In addition, Lindemann was the first to collect flight data about spins. During a spin, he recorded airspeed, normal acceleration, rate of descent, and aircraft attitude using instruments installed in the cockpit. Realizing that the normal aircraft airspeed indicator was inaccurate at the angles-of-attack encountering during a spin, Lindemann installed additional Pitot tubes, oriented to obtain more accurate readings. A spring-type accelerometer was mounted to the pilot's seat, to measure the normal acceleration. He also tried to measure angle-of-attack in the spin with a type of streamer that aligned itself with the flow, indicating the flow angle against a measurement scale behind it. Using this device, the angle-of-attack could be measured to only within 2–3°, due to the fluttering of the streamer. The spins were flown directly above a camera obscura, a type of pinhole camera, on the ground, from which observers could estimate the spin rotation rate and radius. Lindemann reported that a typical spin in the BE2 had a period of about four seconds per turn and a turn radius of about 20 ft (6 m), The spins were quite steep, as he reported a pitch attitude (angle between the aircraft fuselage and the vertical), during the spins, of about 20°. In addition to validating his spin recovery technique, Lindemann deduced several things about the aerodynamics of a spin. He concluded that the flow was steady, since the rates of descent and rotation rates were constant. He also deduced that one wing must be at an angle-of-attack above that for the maximum lift coefficient, that is, one wing was stalled, to sustain the steady spin.

The spin recovery characteristics are strongly influenced by several factors, including the center of gravity position, the location of the tail, and the size of the rudder. Spin susceptibility increases, and ease of spin recovery decreases, as the center of gravity moves aft. At too far an aft center of gravity position, the aircraft may be unrecoverable from a spin. Spin recovery involves reducing the angle-of-attack below the stall angle-of-attack and stopping the rotation in yaw. The primary controls for spin recovery are the elevator, to reduce angle-of-attack, and the rudder, to stop the yawing motion. At post-stall angles-of-attack, it is important that these tail surfaces are not located where they are blanked in the wake of the fuselage or wing, such that there is a significant reduction, or complete loss, of control effectiveness. From this aerodynamic perspective, the inverted spin may be easier to recover from than an erect or upright spin, because the vertical tail is in clean air, which increases its effectiveness. From a piloting standpoint, however, the inverted spin may be more disorienting and uncomfortable than an upright spin, making recovery more difficult.

The exact control inputs required for spin recovery are dependent on the aircraft type, more specifically, on the mass distribution of the aircraft. For aircraft types similar to early airplanes (single-engine, light airplanes), the use of down-elevator and rudder opposite the spin direction is a typical recovery technique. For airplanes with more mass distributed in the wings, called *wing-loaded* airplanes, such as a light twin-engine airplane, down elevator input is often most important, over rudder input. For *fuselage-loaded* airplanes, where most of the mass is in the fuselage – as found in military fighter aircraft with fuselage-buried engines and short aspect ratio wings – aileron input, rolling *into* the spin, along with anti-spin rudder may be required.

Aircraft departure and spin data is obtained from a variety of methods. Specialized vertical spin wind tunnels are used to obtain data on dynamically scaled, sub-scale models, as described in Section 3.7.4.5. A conventional wind tunnel can also be used to collect a database of aerodynamic data for an aircraft configuration over a range of angles-of-attack and orientations. Radio-controlled models have also been used with success. Due to the difficulty in predicting post-stall aerodynamics, manned flight testing to obtain departure and spin data is hazardous, but with proper preparation and training, the flight test risks can be made manageable. In the next section, several flight test techniques are discussed to obtain stall, departure, and spin data on full-scale aircraft in flight.

Example 3.21   Calculation of Spin Rate *Calculate the spin angular velocity of a Christen Eagle II biplane, with a wingspan of 19 ft, 11 in, for a 45° nose-down spin and a flatter, 30° nose-down spin.*

Solution

Using Equation (3.442), the spin angular velocity (in radians per second) is obtained.

$$R_s = \frac{g}{\Omega^2} \tan \theta$$

$$\Omega = \sqrt{\frac{g}{R_s} \tan \theta}$$

The angle θ is measured from the vertical axis, therefore the 45° nose-down spin corresponds to a θ of 45° and the 30° nose-down spin corresponds to a θ of 60°. The spin radius is equal to 0.2b for a θ of 45° and 0.06b for θ of 60°. For the steeper spin, we have

$$\Omega = \sqrt{\frac{g}{0.2b}\tan(45°)} = \sqrt{\frac{\left(32.2\,\frac{\text{ft}}{\text{s}^2}\right)\tan(45°)}{0.2(19.92\,\text{ft})}} = 2.84\,\frac{\text{rad}}{\text{s}} = 163\,\frac{\text{deg}}{\text{s}}$$

For the flatter spin, we have

$$\Omega = \sqrt{\frac{g}{0.06b}\tan(60°)} = \sqrt{\frac{\left(32.2\,\frac{\text{ft}}{\text{s}^2}\right)\tan(60°)}{0.06(19.92\,\text{ft})}} = 6.83\,\frac{\text{rad}}{\text{s}} = 391\,\frac{\text{deg}}{\text{s}}$$

3.12.5   FTT: Stall, Departure, and Spin Flight Testing

Stall, departure, and spin flight testing explore the lower airspeed boundaries of an aircraft's flight envelope. Stall flight testing is required for all airplanes, but departure and spin testing are typically only required for aircraft that are designed for aerobatic flight or maneuvering near the stall angle-of-attack. Stall, departure, and spin flight testing is usually considered high risk, due to the inherent nature of high angle-of-attack aerodynamics, which is non-linear and difficult to predict.

There are many reasons to perform stall, departure, or spin flight testing. Stall testing defines the low-speed boundary of the flight envelope. The aircraft behavior and flight characteristics during these events are obtained, including the aircraft's susceptibility to departure or spinning and any associated warning signs. Testing determines the best recovery methods to be used for stalls, departures, and spins. Flight data may be used to validate or improve the aerodynamic models of the aircraft. The effects on other aircraft systems are determined, such as impacts to the structure, avionics, or air data system.

To learn about stall and departure flight test techniques, you will be flying the Christen *Eagle II* aerobatic biplane (Figure 3.225). The Christen *Eagle* is a high-performance, single-engine, two-place biplane designed for sport aerobatics. It is a *homebuilt airplane*, sold as a kit to be built and assembled by individuals. The aircraft has staggered biplane wings (the top wing is forward of the bottom wing) with equal wingspan, aft-mounted horizontal stabilizers with movable elevators, and a single vertical tail with a movable rudder. The wing structure is wood with a fabric covering. The fuselage and tail have a chromoly steel welded tube structure, with the forward fuselage covered by aluminum panels and the rear covered by fabric. The fuselage cockpit seats two crew in a tandem arrangement, with the pilot in the aft seat and the passenger in the forward seat, underneath a bubble canopy. The aircraft has a tailwheel landing gear configuration with aluminum spring main gear and a steerable tailwheel. The aircraft is powered by a single Lycoming AEIO-360-A1D normally aspirated, air-cooled, horizontally opposed, four-cylinder piston engine producing 200 bhp (149 kW). The engine is designed for aerobatic flight with an oil system capable of supplying engine oil in sustained inverted flight. The first flight of the Christen *Eagle* was in February 1977. A three-view drawing and selected specifications of the Christen *Eagle* are given in Figure 3.226 and Table 3.17, respectively.

Stall and spin testing is typically performed at the aircraft gross weight with various configurations, including different center of gravity positions and various flap, slat, or landing gear positions. The most forward and furthest aft center of gravity positions are typically investigated, which can have a significant effect on the stall and spin characteristics. For aircraft that can carry stores,

Figure 3.225   Christen *Eagle II* aerobatic biplane. (Source: *Courtesy of the author.*)

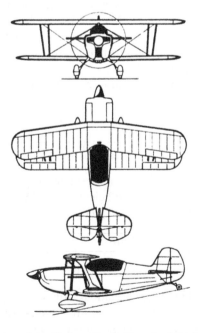

Figure 3.226   Three-view drawing of the Christen *Eagle II* aerobatic biplane. (Source: *Courtesy of Richard Ferriere, with permission.*)

Table 3.17 Selected specifications of the Christen *Eagle II*.

Item	Specification
Primary function	General aviation, sport aerobatics
Manufacturer	Homebuilt (kits by Aviat Aircraft, Afton, Wyoming)
First flight	February 1977
Crew	1 pilot + 1 passenger
Powerplant	Lycoming AEIO-360-A1D four-cylinder engine
Engine power	200 hp (149 kW)
Empty weight	1025 lb (464.9 kg)
Maximum gross weight	1578 lb (715.8 kg)
Length	17 ft 11 in (5.46 m)
Height	6 ft 6 in (1.98 m)
Wingspan	19 ft 11 in (6.07 m)
Wing area	125 ft² (11.6 m²)
Wing loading	12.62 lb/ft² (61.62 kg/m²)
Airfoil	Symmetric, 15% thickness
Maximum speed	184 mph (296 km/h)
Service ceiling	17,000 ft (5200 m)
Load factor limits	+6.0 g, −4.0 g

such as fuel tanks, weapons, pods, or other equipment, the testing is performed with various stores loadings, including no stores, symmetric store loadings, and maximum asymmetric stores loading. Thrust effects may also be investigated, such as power off or on. Stalls at elevated load factors may be of interest, where the stall is performed in constant altitude turns or wings level pullouts from dives. Upright and inverted stalls may be evaluated. You will be flight testing the Christen *Eagle* at maximum gross weight, mid-center of gravity position, and with the engine at idle power. Since the *Eagle* has no flaps and fixed landing gear, there is a single aircraft configuration. You will focus on 1 g, wings level, power-off, upright stalls.

Based on the information in Table 3.17 and using Equation (2.48), the stall speed, V_s, of the Christen *Eagle* is given by

$$V_s = \sqrt{\frac{2W}{\rho S C_{L,max}}} = \sqrt{\frac{2W_{max\,gross}}{\rho_{SSL} S C_{L,max}}} \tag{3.443}$$

The stall speed is calculated at the maximum gross weight, $W_{max\,gross}$, and standard sea level conditions of density, ρ_{SSL}. The maximum lift coefficient of the NACA 0015 airfoil is approximately 1.48. Inserting values into Equation (3.443), the predicted stall speed is

$$V_s = \sqrt{\frac{2(1578\,\text{lb})}{\left(0.002377\,\frac{\text{slug}}{\text{ft}^3}\right)(125\,\text{ft}^2)(1.48)}} = 84.4\,\frac{\text{ft}}{\text{s}} = 57.6\,\frac{\text{mi}}{\text{h}} \tag{3.444}$$

Stall, departure, and spin flight tests are hazardous. Appropriate preparations must be made to reduce the risk as much as possible. A major hazard is the possibility that the aircraft departs controlled flight and will not recover. To mitigate this risk, the testing is always performed at as high an altitude as practical to provide more time to recover, and if a recovery is not possible, to bail out or eject from the aircraft. An altitude limit is specified to bailout or eject in the event the aircraft is unrecoverable. In the Christen *Eagle*, you will wear a backpack emergency parachute.

You will perform a manual bail out in the event the aircraft is out of control and descends below your designated bailout altitude. You will be flying from an airport near sea level, so you will climb up to a minimum altitude of 5000 ft (1500 m) to conduct the testing. Your bailout altitude is 1500 ft (460 m); if the airplane is not recovered to controlled flight by this altitude, then you will bail out.

In larger, more complex aircraft, an emergency recovery device, called a spin recovery parachute or simply, a spin chute, is often mounted on the aircraft to aid in recovery, if needed. The spin chute is a small parachute, stowed in a cylindrical container that is anchored to the aft end of the aircraft. The spin chute is ballistically deployed behind the aircraft to recover the departed or spinning aircraft. A guillotine device cuts the parachute loose after the aircraft returns to controlled flight, allowing the aircraft to fly normally, without the risk of the parachute or lines getting tangled in the tail.

Other test preparations may include special instrumentation or equipment, such as an air data boom with angle-of-attack and angle-of-sideslip vanes. The data may be recorded at higher sample rates to capture the dynamic physics of the stall, departure, or spin. Several cameras may be mounted on the exterior and interior cockpit of the aircraft to capture several different views of the events. Care must be taken to ensure that the addition of the weights, such as cameras, at certain locations on the aircraft, especially at the wingtips, does not significantly alter the spin characteristics. In the past, aircraft and pilots have been lost because of added equipment that degraded spin recovery or made recovery impossible. In the cockpit, additional measurements may be displayed, such as angle-of-attack, angle-of-sideslip, yaw rate, yaw direction, or rudder position. From a human factors perspective, the pilot restraint system must keep the pilot secured in his or her seat, under high positive or negative load factors and in inverted flight. Critical switches and levers, such as for a spin chute deployment, must be checked for accessibility by the pilot. The exterior of the aircraft may be painted to make it highly visible, for ground observers or cameras, especially for when it is in spinning, gyrating, or tumbling attitudes. The wings may be painted different colors, or wide stripes may be painted along the wingspan or fuselage, to make the aircraft attitude more distinct. After these aircraft modifications have been made, it must be checked that the test aircraft is still representative of the production aircraft, otherwise the test data may be valid only for the modified test aircraft and not the aircraft that will be flown in operational service.

Having completed your preparations, you are now ready to go fly the Christen *Eagle* and conduct stall, departure, and spin tests. You strap on the backpack emergency parachute, open the side-hinged bubble canopy from the left side of the airplane, and climb into the aft cockpit seat. The pilot flies from the aft cockpit for weight and balance considerations, to maintain the aircraft center of gravity limits within the forward and aft limits. You start the engine, taxi to the runway, and complete your pre-takeoff checks. Ready to takeoff, you push the throttle full forward, take off, and start your climb. On reaching your test altitude of 5000 ft, you level off and prepare for your first stall.

You will do different types of stall entries to systematically characterize the aircraft's stall behavior. In accordance with the normal flight test buildup approach, you will start with the most benign, lowest risk stall and progress to the more aggressive stalls with potentially higher risk. You will follow the guidelines developed by the US Air Force, as given in Table 3.18. For the most benign stall, the Phase A stall, you will approach the stall with normal control inputs and recover at the first indication of the stall. This stall indication may be a nose drop, uncommanded motion in any axis, sustained, intolerable buffet, or reaching the aft stick stop, such that the control stick cannot be pulled any further aft and the angle-of-attack is not increasing. If the aircraft departs or spins from a Phase A stall, the aircraft is considered extremely susceptible to departure or spins. If the aircraft does not depart or spin, you will move on to the Phase B stall, which is approached with normal control inputs until the stall is reached, whereupon you will apply *aggravated* control inputs for 1 s. Aggravated inputs are misapplied control inputs, such as cross controlling, where a

Table 3.18 Types of stall entries used in flight test (adapted from [63]).

Stall type	Pilot inputs at stall	Departure/spin susceptibility if aircraft departs or spins
Phase A	No aggravation at stall, recover at first indication of stall	Extremely susceptible
Phase B	Aggravate for 1 sec	Susceptible
Phase C	Aggravate for up to 3 sec	Resistant, if it departs or spins, extremely resistant, if it does not depart or spin
Phase D	Aggravate for 15 sec or allow three turns of spin	NA

left rudder input is applied with right roll (control stick to the right), or vice versa. If the aircraft departs or spins from a Phase B stall, the aircraft is considered susceptible to departure or spins. If the aircraft does not depart or spin, you will perform the Phase C stall, which is similar to the Phase B stall, except that the aggravated controls are applied at the stall for a longer duration of up to 3 s. If the aggravated inputs of the Phase C stall are required for the aircraft to depart or spin, the aircraft is considered resistant to departure or spins. If the aircraft does not depart or spin after the aggravated Phase C stall inputs, it is considered extremely resistant to departure or spins. The most aggressive type of stall entry is the Phase D stall, where the aggravated control inputs are applied for up to 15 s or until three turns of a spin are completed.

You reduce power and trim for level flight at an airspeed that is 20% faster than the predicted stall speed from Equation (3.444), or $1.2V_s$, equal to 70 mph (113 km/h). Your first stall entry is a Phase A stall. Maintaining neutral rudders, you reduce power to idle and slowly pull back on the control stick, increasing the angle-of-attack and decreasing the airspeed, at an airspeed *bleed rate* of about 1 mph per second to avoid any dynamic effects in approaching the stall. Your pitch attitude is slowly increasing and you watch the airspeed indicator, as it is winding down. At 59 mph (95 km/h), you feel a slight amount of buffet, right before the nose drops down, accompanied by a slight left wing drop. You glance at the altimeter to see that you have lost only about 200 ft (60 m) of altitude. You quickly release your backpressure on the control stick, lowering the nose, and add full power. The airspeed quickly builds up and you are in controlled flight. From this test, you conclude that the aircraft is not extremely susceptible to departure or spins. The stall speed that you obtained is slightly higher than the prediction, but the prediction is based on the maximum lift coefficient of a two-dimensional airfoil rather than a three-dimensional wing, resulting in a lower predicted stall speed.

Climbing back to an altitude of 5000 ft, you prepare to perform a Phase B stall, where you will aggravate the controls for 1 s. If the aircraft departs and spins, you think about the typical descriptive elements of the spin that you will note. You will be alert to note the spin entry characteristics, the altitude lost per turn of the spin, and the airspeed indications. During the spin, you will use the *SARO spin mode modifiers*, shown in Table 3.19, to describe the spin character. The acronym SARO stands for sense, attitude, rate, and oscillation. The *sense* descriptor specifies whether you are in an erect (upright) or inverted spin. The *attitude* describes the steepness of the aircraft pitch attitude, as quantified by the angle-of-attack ranges given in Table 3.19. The *rate* describes the spin rotation rate, as quantified by the angular rate ranges given in Table 3.19. Finally, the character of the spin oscillations, from mild to violent, are subjectively assessed by the *oscillation* descriptor. There may also be positive and negative *g*-forces that are felt during

Table 3.19 SARO spin mode modifiers [63].

Sense	Attitude	Rate	Oscillations
Erect or upright (positive α)	Extremely steep ($\alpha_s < \alpha < 35°$)	Slow (up to 60°/sec)	Smooth
Inverted (negative α)	Steep ($35° < \alpha < 70°$)	Fast (60–120°/sec)	Mildly oscillatory
	Flat ($\alpha \geq 35°$)	Slow (>120°/sec)	Oscillatory
			Highly oscillatory
			Violently oscillatory

the spin oscillations. For the spin recovery, you will note the effect of the controls, the number of turns to recover, the recovery attitude, and the altitude lost in the spin.

You level off once again at an altitude of 5000 ft and trim the aircraft to $1.2V_s$ or 70 mph. After reducing the power to idle, you slowly and steadily apply aft stick, in the approach to a Phase B stall. At the stall, there is the buffeting as in the previous stall, followed by the nose drop, and a slight left wing drop. To apply Phase B aggravated inputs, you step on the left rudder, holding the stick in the same aft position, with no roll input. The aircraft rolls hard to the left, almost to inverted it seems, the nose drops and then starts yawing rapidly to the left. You look straight down the top of the fuselage, towards the nose, and sight a landmark on the ground, a small house on a hill, as the spin starts. As the ground is whizzing by in a blur, you mentally note the SARO descriptors; the spin sense is upright, the attitude seems flat, the rate is fast, and the oscillation is mildly oscillatory. You glance at the airspeed indicator and it is steady at the stall speed, just below 60 mph. The landmark that you picked out has whizzed by two times now and is coming around for the third time, which is when you will start your spin recovery.

The house on the hill comes into sight and you apply spin recovery control inputs, full opposite (right) rudder followed by forward stick, to reduce the angle-of-attack. After another full turn, the airplane is still spinning! You confirm that you have full right rudder in and add more forward stick, which makes the spin yaw rate increase. The stick is now at the forward stop and the airplane is still spinning. Looking at the altimeter, you are losing altitude at a brisk rate. Your mind is racing as you try to remain calm and think logically about the spin. You are sure that you are upright and spinning to the left, therefore you are applying the correct opposite rudder. Thinking about your SARO observations, you see that the spin is flat, which is surprising since you were expecting a more nose down attitude spin with power off. You recheck the throttle lever, and sure enough, it is not pulled back fully to idle. Your spin has been flattened by the gyroscopic forces of the spinning propeller under power, which raises the nose in a spin. You pull the throttle back hard to its stop and the airplane nose drops slightly. You are still applying the spin recovery right rudder and forward stick inputs. After one more turn, the nose pitches down, the spin rotation stops, and you are under control in a steep dive. You add full power and pull out of the dive, noting that you are just above your bailout altitude. You are thankful that you had the good judgement to perform your testing at a high altitude. Since the aircraft departed and spun after the Phase B stall, you classify it as susceptible to departure and spins. Enough spin testing for today, time to return to the airport and digest the lessons learned.

3.13 Hypersonic Flow

Now is a good time to pause and review the "big picture" concerning aerodynamic flow regimes. Return to Table 2.6, and examine the various aerodynamic flow regimes as a function of Mach number. We have discussed subsonic flow, with a Mach number less than one, transonic flow, with a Mach number of about 0.8–1.2, and supersonic flow, with a Mach number of about 1.2–5. We now discuss the hypersonic flow regime, with a Mach number ranging from about 5 to a theoretically infinite value. As was emphasized in Chapter 2, these Mach number divisions are somewhat arbitrary, in that the changes in the flow phenomena that characterize these different flow regimes occur gradually, rather than as a discontinuous, step function. The predominant changes are often not only a function of the Mach number, they may also depend on the geometry of the body. The same is true for the transition from supersonic to hypersonic flow. The physics that characterizes hypersonic flow becomes more apparent and dominant as the Mach number increases.

In addition to large Mach numbers, hypersonic flight is characterized by dramatic increases in temperature. Quite simply, hypersonic flows are high temperature flows. Just as supersonic flight was met with the dramatic transonic drag rise that was the "sound barrier", one could say that hypersonic flight is confronted with a "thermal barrier" where aerodynamic heating can literally melt steel at very high Mach number. The effects of high Mach number and high temperature make hypersonic flight one of the most challenging flight regimes in which to fly safely.

In the following discussions, we separate the effects of hypersonic flow into those due to high Mach number and those due to high temperature. By doing so, we separate the fluid dynamic effects of high Mach number from the thermochemical effects of high temperature. Both of these phenomena can be explained in the context of an *inviscid* hypersonic flow. Viscous hypersonic flow is briefly discussed, from the perspective of the impact of high Mach number and high temperature on the boundary layer next to the surface of a body. Hypersonic flight is often at very high altitudes, where the air density becomes so low that the assumption of continuum flow may no longer be realistic. This area of aerodynamics, called *rarefied gas dynamics*, associated with hypersonic flow, is briefly discussed. There are several real-world examples of these hypersonic flow phenomena in the upcoming hypersonic flight test technique. We start our discussions about hypersonic flow with a brief discussion about the different types of vehicles associated with hypersonic flight.

3.13.1   Hypersonic Vehicles

After World War II, continual advances in rocket propulsion enabled flight at higher and higher Mach numbers. As discussed in Section 1.3.5.3, the *Bumper-WAC* rocket was the first manmade object to fly at hypersonic speed, on 24 February 1949 reaching a Mach number greater than five. The first person to fly at hypersonic speed was Soviet astronaut Yuri Gagarin, entering the Earth's atmosphere in his *Vostok* spacecraft at over 17,000 mph (27,000 km/h) or Mach 25. The astronauts, returning from the Moon in their *Apollo* space capsule, entered the Earth's atmosphere at Mach 36, the fastest that a manned aerospace vehicle has ever flown.

The aerospace vehicles used for these milestone flights can be classified as different types of hypersonic vehicles. The *Bumper-WAC* rocket and other hypersonic rockets and missiles are *hypersonic accelerator vehicles*, designed to accelerate to hypersonic speeds as quickly as possible. This is in contrast to a *hypersonic cruise vehicle* that is designed to reach hypersonic speeds and cruise at these high Mach numbers for a sustained period. While rocket propulsion may be used for a hypersonic accelerator, more efficient air-breathing, hypersonic propulsion systems are required for a hypersonic cruiser. We discuss some of these potential air-breathing, hypersonic propulsion systems in Chapter 4. The *Vostok*, *Apollo*, and other space vehicles returning to Earth from orbit or beyond are *hypersonic entry vehicles*, which are designed to decelerate from hypersonic to low subsonic speed.

Aerodynamically, these three types of hypersonic vehicles are quite different. The hypersonic entry vehicle is designed to produce high drag to decelerate the vehicle from orbital or higher velocities. In contrast, low drag is desired for the hypersonic accelerator and cruise vehicles. A

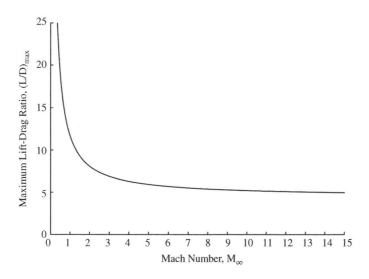

Figure 3.227   Maximum lift-to-drag ratio versus Mach number.

high hypersonic lift-to-drag ratio is advantageous for a cruiser, but may not be of benefit to an accelerator, where one wants to climb out of the thick atmosphere as fast as possible. However, the overall aerodynamic efficiency of a hypersonic vehicle, embodied by the lift-to-drag ratio, is usually low. The variation of the maximum lift-to-drag ratio, $(L/D)_{max}$, with freestream Mach number, M_∞, based on a correlation developed in [47], is given by

$$\left(\frac{L}{D}\right)_{max} = \frac{4.5(M_\infty + 1.6)}{M_\infty} \qquad (3.445)$$

Equation (3.445) is plotted in Figure 3.227. The maximum lift-to-drag ratio is seen to decrease significantly with increasing Mach number. The value of the maximum hypersonic lift-to-drag ratio becomes *Mach number independent* with increasing Mach number (to be discussed shortly). Equation (3.445) and Figure 3.227 indicate that the predicted hypersonic lift-to-drag ratio has a maximum of about five. In practice, the actual lift-to-drag ratio of hypersonic vehicles may be much lower, due to the real world physics and design compromises.

Some types of hypersonic vehicle designs, that take advantage of *flow-containment*, promise a much higher hypersonic lift-to-drag ratio. The *caret wing*, conceived by Nonweiller [56], is a simple shape that uses flow-containment, as shown in Figure 3.228. This shape is derived from the known two-dimensional, supersonic or hypersonic flow over a wedge, as shown in Figure 3.228 for a wedge of angle θ. The supersonic or hypersonic flow over the wedge creates a planar shock wave of angle β, depicted as the shaded plane in the figure. The caret wing is constructed by forming a center wedge with an angle equal to the two-dimensional wedge angle θ and drawing straight leading edges, from this center wedge, to the planar shock wave. The shock wave is fully attached to the leading edges, and the caret wing appears to be "riding" atop the planar shock wave; hence these types of flow-containment vehicles are called *waveriders*. The high pressure beneath the wing, equal to the pressure behind the oblique shock wave, is fully contained, resulting in a high lift-to-drag ratio.

The caret wing is a "point design" shape, designed for a single freestream Mach number, such that at this on-design condition, the shock wave is attached to the leading edges. At other, off-design Mach numbers, the shock wave angle is different, so that the shock wave is no longer attached to the

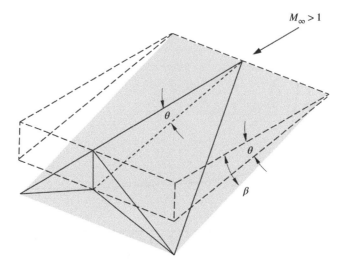

Figure 3.228   Hypersonic caret wing waverider.

wing leading edges. This results in flow spillage of the high pressure air from underneath the wing, which reduces its efficiency. In addition to its off-design performance, there are other design issues with the waverider concept, such as cooling of sharp leading edges and packaging of payloads.

Just as the caret wing was designed from the known flow field over a wedge, waverider configurations can also be constructed from other known flow fields, such as the supersonic or hypersonic flow over a cone or other body. Once this known flow field and shock wave shape are defined, the waverider geometry can be "carved out" such that its leading edges are along the shock wave. There are several advanced designs of the waverider concept, with promising aerodynamic performance, and which address some of the practical design issues that have been mentioned.

Finally, there are natural objects that travel at hypersonic speeds through our atmosphere, such as meteoroids and asteroids that enter the earth's atmosphere from space. A meteoroid or asteroid that becomes visible, as it vaporizes from the intense heat of atmospheric entry, is called a *meteor* or, more colloquially, a "shooting star". If it survives as a solid piece of matter that impacts the surface of the earth, it is then called a *meteorite*. It is estimated that tens of thousands of meteorites, with a mass greater than about 10 grams (0.35 ounces), hit the earth's surface every year. However, most meteors are the size of a grain of sand or a small pebble, with masses of less than a few grams, and burn up in the atmosphere. Meteors may be composed of dense stony or metallic material, if they originate from an asteroid, or they may be a conglomerate of low-density materials, commonly called a "dust ball", if they come from a comet.

Meteors enter the earth's atmosphere at hypersonic speeds of about 11–72 km/s (25,000–160,000 mph). The large difference in the entry speed is due to the earth's rotation, since a meteor's trajectory can be in the same direction as, or opposite to, the earth's rotational speed of about 30 km/s (67,000 mph). The huge kinetic energy associated with a meteor's hypersonic speed *ionizes* the air around the meteor, resulting in its brilliant appearance and bright streak across the sky. *Ionization* is the process whereby electrons are stripped from the oxygen and nitrogen atoms in the air, resulting in "free" electrons that make the air an electronically conducting plasma.

A recent example of a natural object reaching hypersonic speeds within the earth's atmosphere is the Chelyabinsk meteor, which penetrated the atmosphere at an estimated Mach 60 (68,000 km/s, 42,500 miles/s) on 15 February 2013 over the city of Chelyabinsk, Russia. This huge, 10,000 ton (9×10^6 kg) meteor, measuring almost 20 m (66 ft) in diameter, exploded in the air at an altitude of about 30 km (18.6 miles, 98,000 ft), producing a flash of light that was brighter than the sun and sending a shock wave, literally, around the world. The exploding meteor disintegrated into hundreds of fragments, many surviving as meteorites that impacted the ground, some weighing as much as half a ton.

3.13.2   Effects of High Mach Number

Hypersonic flight is synonymous with flight at very high Mach numbers. We have seen that there are fundamental differences in the flow physics between low Mach number, subsonic flow, and flow at supersonic Mach numbers. Once the flight speed approaches the speed of sound, shock waves appear in the flow, a phenomenon not found in subsonic flow. The location and strength of shock waves changes significantly as the Mach number increases.

For a slender body, such as the 7.5°, two-dimensional wedge shown in Figure 3.229, the shock wave is attached to the body, in this case attached to the apex of the wedge. The planar shock wave angle, β, is a function of the freestream Mach number, M_1 (the shock wave angle is depicted for a single freestream Mach number in Figure 3.229). At low supersonic Mach numbers, the shock wave angle is large. The shock wave angle decreases, getting closer to the wedge surface, with increasing Mach number, as shown in Figure 3.229. Numerical values for the shock wave angle are shown in Table 3.20. At Mach 2, the shock wave angle is 36.7°. At Mach 10, the shock wave angle has decreased to 11.9°. At very high Mach numbers of 50 and 100, the shock wave angles are 9.16° and 9.05°, respectively. The shock wave angle is seen to approach a constant value with increasing Mach number, converging to 9.01° at a theoretically infinite Mach number. Thus, we see that the shock wave angle becomes constant or independent of Mach number at very high Mach numbers. This characteristic of hypersonic flows, where certain quantities become independent of Mach number, is known as *Mach number independence*. We expand on this principle later in this section.

Table 3.20 also shows that the strength of the shock wave increases dramatically with increasing Mach number. The flow properties behind the shock wave, in region 2, are compared to the freestream values, in region 1. The ratios of the pressure and temperature behind the shock wave with respect to their freestream values, p_2/p_1 and T_2/T_1, respectively, are given as a function of Mach number. At a low supersonic Mach number of 2, the pressure behind the shock wave

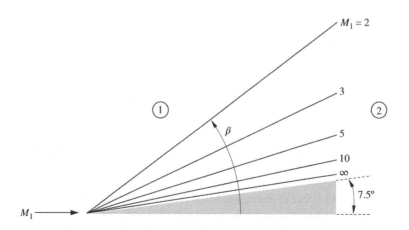

Figure 3.229   Effect of increasing Mach number on the shock wave angle on a slender wedge.

Table 3.20 Shock wave angle, β, and flow properties behind the shock wave for a 7.5° wedge as a function of the freestream Mach number, M_1.

M_1	β	$\dfrac{p_2}{p_1}$	$\dfrac{T_2}{T_1}$
2	36.7°	1.50	1.13
3	25.2°	1.74	1.18
5	17.1°	2.37	1.30
10	11.9°	4.77	1.73
20	9.87°	13.5	3.22
50	9.16°	73.8	13.3
100	9.05°	288	49.1
∞	9.01°	∞	∞

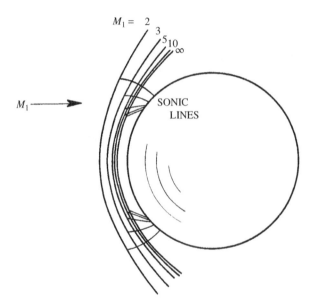

Figure 3.230 Effect of increasing Mach number on the shock wave standoff distance for a sphere. (Source: *Reprinted from* [25], *Fig. 8.*)

is 1.5 times greater than the freestream value. At a hypersonic Mach number of 5, the pressure ratio is 2.37. At very large hypersonic Mach numbers, the pressure and temperature behind the shock wave are orders of magnitude greater than their respective freestream values. Mach number independence does not apply to the pressure or temperature ratios, as they both approach infinity as the Mach number approaches infinity. These values increase with the square of the Mach number normal to the shock wave.

Figure 3.230 shows the effect of increasing Mach number for a blunt body. Here, the bow shock wave ahead of a simple sphere is depicted as a function of freestream Mach number, M_1. The bow shock wave is detached from the sphere and is positioned upstream by a distance known as the *shock detachment distance*, measured along the centerline of the body, parallel to the freestream direction.

The shock detachment distance decreases with increasing Mach number, with the bow shock wave approaching a limiting distance from the front of the sphere as the Mach number approaches infinity. Thus, the shock detachment distance demonstrates Mach number independence for the hypersonic blunt body, similar to the limiting shock wave angle for the slender hypersonic body.

It is also interesting to compare the shock wave shapes between a blunt and slender body. Unlike the straight, planar shock wave on the slender wedge, the bow shock wave on the sphere is highly curved. The shock wave curvature results in transverse (perpendicular to the freestream flow direction) gradients in the flow properties. The portion of the bow shock directly in front of the sphere is perpendicular or normal to the freestream flow direction, thus the flow is decelerated from supersonic or hypersonic speeds to subsonic speed through a normal shock wave. The flow then accelerates around the front of the sphere and reaches supersonic speed, with the dividing line between the subsonic and supersonic flow shown as the *sonic line*, for the different Mach numbers, in Figure 3.230. Moving away from the centerline of the flow, the shock wave changes for a normal shock to an oblique shock with a decreasing shock wave angle.

We now return to the topic of Mach number independence. As we have seen for the flow over a wedge and a sphere, there are certain quantities that become invariant as the Mach number increases to hypersonic values. This phenomenon can be verified analytically and is observed experimentally. If the governing equations of fluid flow, as given in Section 3.6.4, are applied with the appropriate boundary conditions, to the limiting case of very large Mach number, they reduce to forms that are independent of Mach number. The Mach number independence principle is evident in the experimental data shown in Figure 3.231, where the drag coefficient for a sphere and a cone–cylinder are plotted as a function of Mach number. While the drag coefficient changes considerably through the transonic and low supersonic speed regimes, it reaches a near constant value for hypersonic Mach numbers.

Mach number independence applies to certain geometric flow field parameters, such as the shock wave angle, shock detachment distance, and shock wave shape, and to certain non-dimensional

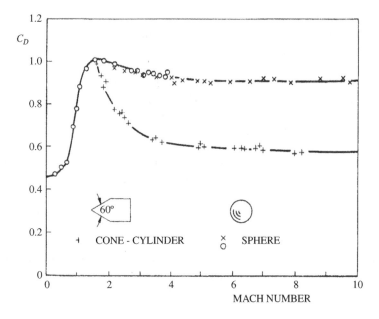

Figure 3.231 Mach number independence of the hypersonic drag coefficient. (Source: *Reprinted from Introduction to Hypersonic Flow, G.G. Chernyi, Fig. 1.8, p. 49, (1961)*, [20], *with permission from Elsevier.*)

aerodynamic variables, such as the pressure coefficient, C_p. Since the lift and wave drag coefficient, C_L and C_{D_w}, respectively, can be derived from the pressure coefficient, they also become Mach number independent.

The drag coefficient for the sphere reaches Mach number independence at about Mach 5 while this Mach number is about 6 or 7 for the cone-cylinder. This is because Mach number independence is a function of the square of the Mach number *normal* to the shock wave, $(M_1^2\sin^2\beta)$, where M_1 is the freestream Mach number and β is the shock wave angle, rather than a function of the freestream Mach number alone. Thus, Mach number independence is achieved when the Mach number normal to the shock wave is very large, as given by

$$M_1^2\sin^2\beta \gg 1 \qquad\qquad (3.446)$$

Since the shock wave angle, β, is dependent on the geometry of the body, the freestream Mach number at which the flow becomes independent of Mach number is also dependent on the body geometry. At the stagnation point of a blunt body, such as the sphere, the shock wave is near normal with β equal to 90° and $(\sin\beta)$ equal to one. For a more slender body, such as the cone–cylinder, the shock wave is more oblique with β less that 90° and $(\sin\beta)$ less than one. Therefore, the square of the normal Mach number, as given by Equation (3.446), is much larger, at the same freestream Mach number, for a blunt body versus a slender body, and Mach number independence is reached at a lower freestream Mach number.

3.13.3   Effects of High Temperature

As has been stated previously, hypersonic flows are high-temperature flows. Consider a hypersonic vehicle flying at Mach 10 and an altitude of 150,000 ft (45,700 m). The static temperature corresponding to this altitude is 479.1 °R (19.43 °F, 266.2 K). Assume that the vehicle has a blunt nose, such that there is a normal shock wave in front of the nose. At Mach 10, the static temperature ratio across the normal shock wave is 20.4, that is, the temperature in the shock layer behind the shock is 20.4 times greater than the freestream static temperature, or 9774 °R (9314 °F, 5430 K)! In addition to shock waves, viscous hypersonic flows experience high temperatures due to skin friction in boundary layers, where the high kinetic energy of the flow is converted into heat.

The temperature calculated above for a normal shock wave assumes that the air is an ideal gas with constant specific heats. In reality, the gas no longer acts like an ideal gas with constant specific heats at hypersonic speeds and the predicted temperature is not accurate. The temperature behind the normal shock is still very high, but it is different from the ideal gas prediction. Previously, it was stated that the temperature of a gas, such as air, is representative of the random motion of the gas molecules. At the non-hypersonic flight conditions that we have discussed up to now, this is a valid statement. However, at hypersonic speeds, the energy in the flow excites much more than just the random motion of the gas molecules.

Let us first examine what is happening at a molecular level as the air temperature increases. Air is primarily composed of nitrogen and oxygen molecules, formed by two atoms of nitrogen or oxygen, respectively. Let us model a diatomic molecule as a "dumbbell" shape, with two spherical atoms connected by a link or rod, as shown in Figure 3.232. The diatomic molecules in air have several modes of freedom that are excited as the temperature is increased. At normal, non-hypersonic conditions of pressure and temperature, a molecule has translational and rotational energy. A diatomic molecule has three *translational energy modes*, where the molecule can linearly translate in three-dimensional space (Figure 3.232a). It is important to note here, that the temperature is a measure of this translational energy only.

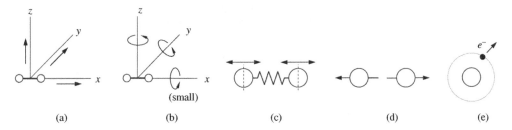

Figure 3.232 Effects of increasing temperature on a diatomic molecule, (a) translational motion, (b) rotational energy excitation, (c) vibrational excitation, (d) dissociation, and (e) ionization.

The molecule has two *rotational energy modes*, where the molecule can rotate or spin about its axes (Figure 3.232b). The energy associated with the rotation of the molecule about its intermolecular axis, the axis connecting the two atoms, is very small compared with the other axes, and is usually neglected, so a diatomic molecule is considered to have two rotational energy modes.

With an increase in temperature, a single *vibrational energy mode* of a diatomic molecule is excited, depicted as the linear, "back-and-forth" vibration of the atoms connected by a spring (Figure 3.232c). As the temperature is increased further, *dissociation* of the molecule occurs, where the bond holding the molecule together is broken, splitting the atoms apart (Figure 3.232d). With the dissociation of the air molecules, the air becomes a *chemically reacting gas*. Finally, at a very high temperature, electrons are stripped from the dissociated atoms to become free electrons in the air (Figure 3.232e). This *ionization* process makes the air an electrically conducting, ionized plasma, which absorbs radio waves, leading to "radio blackout" during certain phases of spacecraft atmospheric entry. The vibrational excitation, dissociation, and ionization of a gas due to increasing temperature are termed *high temperature effects*.

The effects of increasing temperature on the air molecules, shown in Figure 3.232, consume a portion of the flow's kinetic energy. If these high temperature effects are not accounted for, the hypersonic flow temperature is over predicted, since more of the flow's kinetic energy is assumed to be converted into the translational energy, which is the measure of temperature. For the normal shock wave example given earlier, the actual temperature ratio across the shock with the high temperature effects included, is 11.85, rather than 20.4 as predicted using the ideal gas assumption. The actual temperature, behind the normal shock, is 5677 °R (5217 °F, 3154 K), still a very high temperature, but quite a bit less than the ideal gas prediction of 9774 °R (9314.3 °F, 5430 K). (To add some perspective, the melting point of steel is about 3310 °R, 2850 °F, 1839 K). At these elevated temperatures, the assumption of an ideal gas is no longer valid and the ratio of specific heats is not a constant. The gas must be treated as a non-ideal gas, which is more difficult to analyze than an ideal gas. There are tabulations of the data from the analysis of hypersonic, high temperature flows that can be useful for the design and analysis of hypersonic flows. For instance, [78] provides a tabulation of the normal shock properties for hypersonic, high temperature flows, similar to the ideal gas normal shock tables found in the NACA Report 1135 [6].

The flow regimes and high temperature effects associated with hypersonic flows are shown in Figure 3.233 for air at a pressure of 1 atm (101,300 N/m^2, 2116 lb/ft^2). The progression of the flow regimes, from subsonic to hypersonic, and the associated Mach numbers are shown on the left side of the figure. A scale of increasing stagnation temperature is shown on the right side. Recall that the stagnation temperature is the temperature that is obtained if a flow is isentropically brought to rest. Thus, as the flow Mach number increases, there is more energy in the flow to increase the stagnation temperature when the flow is brought to rest.

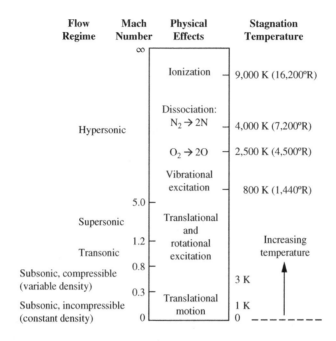

Figure 3.233 Flow regimes and ranges of high temperature effects on air.

At very low temperatures, below about 3 K (−454 °F, 914 °R), the oxygen and nitrogen molecules in the air have only translational energy. As the temperature increases to normal conditions, up to about 800 K (980 °F, 1440 °R), the molecules have both translational and rotational energy. This is the flow regime of subsonic, transonic, and low supersonic flight. Entering the hypersonic flow regime, above about 800 K, the molecules become vibrationally excited. Dissociation of the oxygen molecules into oxygen atoms starts at a temperature of about 2500 K (4000 °F, 4500 °R) and is essentially complete by about 4000 K (6700 °F, 7200 °R). The nitrogen molecules start dissociating at about 4000 K, with dissociation complete by about 9000 K (15,700 °F, 16,200 °R). Ionization of atomic oxygen and nitrogen starts at about 9000 K.

Thus, we see that the high temperatures associated with hypersonic flight can change the "simple" air flowing around a vehicle into a chemically reacting, possibly ionized, gas. So, how do these high temperature effects impact the flight of a hypersonic vehicle? It has already been mentioned that ionization of the air can result in the atmospheric entry "radio blackout" phenomenon for spacecraft. In addition, there are other important aerodynamic impacts to a hypersonic vehicle.

The pressure is not strongly influenced by the high temperature effects of a chemically reacting gas. This is because pressure is a "mechanical" variable that is not strongly affected by the chemistry of the flow. However, the integrated effect of small changes in the surface pressure, due to high temperature effects, can result in more significant impacts to the vehicle's lift, drag, or moments.

Perhaps the most significant impact of the high temperatures associated with hypersonic flight is the increase in aerodynamic heating. The extremely hot shock layer, behind the normal shock, that we have been discussing, radiates its heat energy into the vehicle. This radiative heating from shock waves can be a significant percentage of the total heat input into a hypersonic vehicle. For the *Apollo* space capsule, returning from the Moon and entering the Earth's atmosphere at Mach 36, the radiative heat transfer was about 30% of the total heat load into the vehicle. In addition to heating from high temperature shock layers, there is significant convective heat transfer into a vehicle due to viscous dissipation of the hypersonic kinetic energy in the boundary layers. Unlike lower speed vehicles, the high heat transfer rates of hypersonic flight is a dominant consideration in the design of hypersonic vehicles.

3.13.4   Viscous Hypersonic Flow

Prandtl's boundary layer theory assumes that the effects of viscosity and heat conduction can be confined to a thin layer close to the surface of a body. This enables the separation of analyses for inviscid and viscous flows, which greatly simplifies many aerodynamic problems. At subsonic and low supersonic speeds, the viscous effects of the boundary layer, skin friction, and heat conduction can often be ignored. At high supersonic and hypersonic speeds, the effects of skin friction and heat conduction become significant. Some of the high temperature effects, due to the dissipation of the high kinetic energy in the viscous boundary layer, were discussed in the previous section.

In addition to the skin friction and heat transfer effects, the boundary layer in hypersonic flows may result in a *displacement effect*, which must be considered. The hypersonic boundary layer thickness can grow significantly larger than at lower speeds, such that it adds an effective thickness to a body and displaces the flow outside the boundary layer. This interaction between the hypersonic viscous boundary layer and the inviscid flow, outside the boundary layer, is called *hypersonic viscous interaction*.

Hypersonic flight is often flight at very high altitudes, where the air density is very low. Thus, hypersonic flight can be at relatively low Reynolds numbers, where the boundary layers tend to be laminar. Given this, let us start with a laminar boundary layer, where its thickness, δ, is inversely proportional to the square root of the Reynolds number, as given by Equation (3.413).

$$\delta \propto \frac{1}{\sqrt{Re_x}} = \frac{1}{\sqrt{\frac{\rho V x}{\mu}}} = \sqrt{\frac{\mu}{\rho V x}} \tag{3.447}$$

where Re_x is the Reynolds number based on the length, x, along a flat plate or other surface.

If the density, ρ, and viscosity, μ, in the Reynolds number is evaluated at the surface or wall temperature, T_w, Equation (3.447) becomes

$$\delta \propto \frac{1}{\sqrt{Re_{x,w}}} = \sqrt{\frac{\mu_w}{\rho_w V_\infty x}} \tag{3.448}$$

where V_∞ is the freestream velocity outside the boundary layer and $Re_{x,w}$ is the Reynolds number evaluated at the wall temperature. Multiplying by $\sqrt{\mu_\infty/\mu_\infty}$ and $\sqrt{\rho_\infty/\rho_\infty}$, we have

$$\delta \propto \sqrt{\frac{\mu_\infty}{\rho_\infty V_\infty x}} \sqrt{\frac{\mu_w}{\mu_\infty}} \sqrt{\frac{\rho_\infty}{\rho_w}} = \frac{1}{\sqrt{Re_{x,\infty}}} \sqrt{\frac{\mu_w}{\mu_\infty}} \sqrt{\frac{\rho_\infty}{\rho_w}} \tag{3.449}$$

where $Re_{x,\infty}$ is the Reynolds number evaluated at freestream conditions.

Let us assume that the coefficient of viscosity is linearly related to the wall-to-freestream temperature ratio, T_w/T_∞. (This is a simpler assumption for the viscosity than Sutherland's law given by Equation (3.16), and this makes this development a bit easier.) Thus, we assume

$$\frac{\mu_w}{\mu_\infty} \propto \frac{T_w}{T_\infty} \tag{3.450}$$

Using the perfect gas equation of state and the fact that the pressure is constant through the boundary layer, such that the wall pressure equals the freestream pressure, the ratio of the freestream density to the density corresponding to the wall temperature is given by

$$\frac{\rho_\infty}{\rho_w} \propto \left(\frac{p_\infty}{RT_\infty}\right)\left(\frac{RT_w}{p_\infty}\right) = \frac{T_w}{T_\infty} \tag{3.451}$$

which shows that this density ratio is also proportional to the wall-to-freestream temperature ratio. Inserting Equations (3.450) and (3.451) into Equation (3.449), we have

$$\delta \propto \frac{1}{\sqrt{Re}}\sqrt{\frac{T_w}{T_\infty}}\sqrt{\frac{T_w}{T_\infty}} = \frac{1}{\sqrt{Re}}\left(\frac{T_w}{T_\infty}\right) \tag{3.452}$$

where the subscript on the Reynolds number has been dropped.

Assuming isentropic flow, the wall temperature, T_w, is equal to the flow total temperature, T_t, so that the wall-to-freestream temperature ratio is given by the isentropic relation for the total-to-freestream temperature, Equation (3.341).

$$\frac{T_w}{T_\infty} = \frac{T_t}{T_\infty} = 1 + \frac{\gamma - 1}{2}M_\infty^2 \tag{3.453}$$

Inserting Equation (3.453) into (3.452), we have

$$\delta \propto \frac{1}{\sqrt{Re}}\left(1 + \frac{\gamma - 1}{2}M_\infty^2\right) \tag{3.454}$$

At hypersonic speeds, $M_\infty^2 \gg 1$, so that Equation (3.454) becomes

$$\delta \propto \left(\frac{\gamma - 1}{2}\right)\frac{M_\infty^2}{\sqrt{Re}} \tag{3.455}$$

Thus, we see that the boundary layer thickness grows with the square of the Mach number, such that, at the same Reynolds number, the boundary layer thickness can be orders of magnitude larger at hypersonic speeds than at lower speeds.

The thick hypersonic boundary layer adds an effective thickness, called the *displacement thickness*, to a flat plate or other body surface, making the plate or body "look" thicker to the freestream flow. The inviscid flow, external to the boundary layer, must go around the thick boundary layer, just as if it were a real body shape obstructing the flow. As shown in Figure 3.234, the supersonic flow over a simple flat plate does not create a shock wave (assuming the plate leading edge is very sharp) and the surface pressure, p_w, on the plate equals the freestream pressure, p_∞. In a hypersonic flow, the thick boundary layer effectively blunts the leading edge of the plate, deflecting the hypersonic flow, creating a shock wave, and increasing the surface pressure above freestream pressure, $p_w > p_\infty$. The skin friction drag and heat transfer are also increased due to the thick boundary layer.

In some cases, it is still possible to use an inviscid flow assumption to analyze the hypersonic flow over a body with a thick boundary layer. The boundary layer displacement thickness is added to the original body geometry, as shown in Figure 3.235, to form a new, effective body that can be analyzed using inviscid flow techniques.

As discussed in Section 3.13.2, the shock wave moves closer to the body with increasing Mach number. The thick hypersonic boundary layer occupies proportionally more of the flow between the body and the shock wave, than at lower speeds where the shock wave is not as close and the

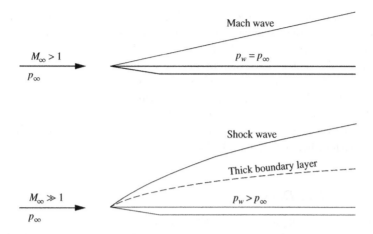

Figure 3.234   Flat plate in supersonic flow (top) and hypersonic flow (bottom).

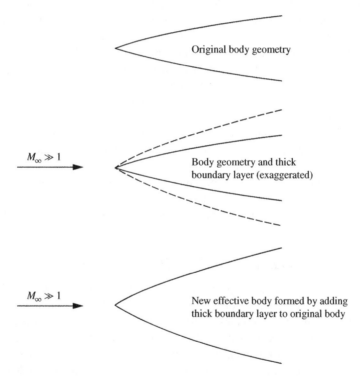

Figure 3.235   Effective body formed by adding thick boundary layer to original body (shock waves not shown).

boundary layer is not as thick. This creates a merged inviscid and viscous *shock layer* between the shock wave and body surface, where the inviscid and viscous effects cannot be separated in an analysis.

In addition to shock waves that are created by boundary layers, there may be other shock waves that impinge on the boundary layer. This *shock wave–boundary layer interaction* is not unique to hypersonic flows, as it is found at lower supersonic Mach numbers, but it is exacerbated at hypersonic speeds by the stronger shock waves, which result in higher pressure increases across the shock, and the thicker, laminar boundary layers, which are less able to withstand adverse pressure gradients. Thus, another viscous-related issue is this shock-induced flow separation, which can significantly impact the vehicle aerodynamics and heat transfer. In addition to an increase in the pressure drag due to flow separation, the pressure distribution over a body can be significantly altered by the separated flow regions. When this flow separation occurs on or near the vehicle's flight control surfaces, it can impact the vehicle stability and control. The location of the shock impingement on the boundary layer is often a point of intense local heating, which can greatly impact the aerodynamic heating on the vehicle.

3.13.5   Effects of Low Density

Hypersonic flight is often flight at very high altitudes, where the density is very small. For hypersonic vehicles, low density effects may become important above about 200,000 ft (61,000 m) or so. Since hypersonic vehicles are flying through the high altitude, *rarified gas* regions of the atmosphere, the study of the low density flows over the vehicles is called *rarefied gas dynamics*.

As discussed in Section 3.3.1, the assumption that the air is a continuum may not be valid at high altitudes. The density may be so low that the motion of the individual gas molecules must be considered. The Knudsen number, *Kn*, defined as the ratio of the mean free path between molecular collisions to the vehicle characteristic length, was identified as the parameter that determines whether continuum or non-continuum flow should be assumed. The dividing line between continuum and non-continuum flow is a Knudsen number of unity, with non-continuum flow assumed for $Kn \geq 1$.

In this non-continuum flow region, the fundamental equations of fluid dynamics are no longer valid and cannot be used. The low densities affect the aerodynamics and heat transfer of a hypersonic vehicle and the models that are used to represent the flows over the vehicle. At these low density flow conditions, shock waves become "smeared" and merge with the flow field close to a body. In addition to the Mach number and Reynolds number, the vehicle aerodynamics and heat transfer become a strong function of the Knudsen number for low density flows.

3.13.6   Approximate Analyses of Inviscid Hypersonic Flow

As with any other speed regime, the hypersonic aerodynamicist is usually interested in predicting the aerodynamic forces and moments on a vehicle, which involves the prediction of the surface pressure distributions. Given the complex physics of hypersonic flows, the prediction of surface pressures on a hypersonic body are expected to be quite difficult. While this is generally true, there are some hypersonic flow problems, which are amenable to very simple types of analyses. In particular, several simple formulas provide excellent approximations for inviscid hypersonic flows. In this section, two of these methods are introduced, one based on supersonic shock-expansion theory and the other based on a 17th century theory of Sir Isaac Newton.

Consider the hypersonic flow over a flat plate at an angle-of-attack, α. There is a system of shock and expansions waves emanating from the plate, as shown in Figure 3.236a. The hypersonic flow expands over the top of the plate and is then turned and recompressed through a shock wave at the trailing edge. The flow is turned and compressed through a shock wave on the plate lower surface and then expanded through an expansion wave at the trailing edge, to join the flow from the upper surface. The flow field properties through the system of shock waves and expansion waves can be calculated using shock-expansion theory, as discussed for supersonic flows.

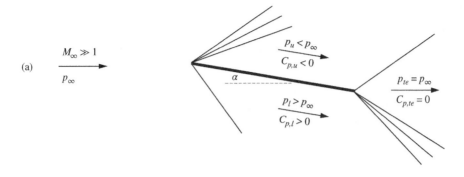

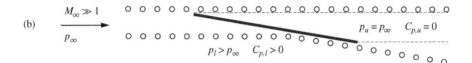

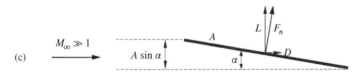

Figure 3.236 Hypersonic flow over a flat plate, (a) shock-expansion theory, (b) Newtonian impact theory, and (c) Newtonian theory flat plate geometry.

If certain assumptions are made about a hypersonic flow, these exact shock-expansion wave equations can be greatly simplified to provide approximate formulas for hypersonic flows. To obtain these simplified equations, it is assumed that the freestream Mach number has a very large, but finite value. The angle-of-attack and the velocity changes or *perturbations* in the flow are assumed small. The *small perturbations* of the flow require that the body be *slender*. In the derivation of these *limiting forms* of the exact shock-expansion equations, applicable to hypersonic flow, a new *hypersonic similarity parameter*, K, is obtained, defined as

$$K \equiv M_\infty \theta = M_\infty \alpha \tag{3.456}$$

where θ is the flow delection angle (in radians), which is equal to the angle-of-attack, α (in radians), for the flat plate example. For hypersonic flows, the hypersonic similarity parameter, $M_\infty \theta$, takes on the role of a governing parameter, essentially replacing the supersonic similarity parameter of the Mach number alone, M_∞.

While the derivation of the approximate forms of the shock-expansion equations is beyond the scope of this text, some of the relevant results are presented. With the assumption of a high, but finite hypersonic Mach number, small angles, and small perturbations, the approximate equation for the pressure coefficient, $C_{p,shock}$, behind a shock wave is given by

$$C_{p,shock} = \left[\frac{\gamma+1}{2} + \sqrt{\left(\frac{\gamma+1}{2}\right)^2 + \frac{4}{K^2}} \right] \alpha^2 \tag{3.457}$$

Similarly, the approximate equation for the pressure coefficient, $C_{p,\,expansion}$, behind an expansion wave is given by

$$C_{p,\,expansion} = \frac{2}{\gamma K^2}\left[1 - \left(1 - \frac{\gamma-1}{2}K\right)^{2\gamma/(\gamma-1)}\right]\alpha^2 \qquad (3.458)$$

The pressure coefficients, in Equations (3.457) and (3.458), are functions of only the hypersonic similarity parameter, K, the angle-of-attack, α, and the specific gas constant, γ.

Returning to the flat plate example, the lift and drag coefficients, C_L and C_D, respectively, of the plate are given by

$$C_L = (C_{p,l} - C_{p,u})\cos\alpha \approx C_{p,l} - C_{p,u} \qquad (3.459)$$

$$C_D = (C_{p,l} - C_{p,u})\sin\alpha \approx (C_{p,l} - C_{p,u})\alpha = C_L\alpha \qquad (3.460)$$

where $C_{p,l}$ and $C_{p,u}$ are the pressure coefficients on the lower and upper surfaces of the plate, respectively, and $\cos\alpha \to 1$ and $\sin\alpha \to \alpha$ for small angle-of-attack.

Since the plate lower surface undergoes a shock wave compression, the lower surface pressure coefficient is given by Equation (3.457). The pressure coefficient for the expansion wave flow on the upper surface is given by Equation (3.458). Inserting Equations (3.457) and (3.458), for $C_{p,l}$ and $C_{p,u}$, respectively, into Equation (3.459), the lift coefficient is

$$C_L = \left\{\frac{\gamma+1}{2} + \sqrt{\left(\frac{\gamma+1}{2}\right)^2 + \frac{4}{K^2}} + \frac{2}{\gamma K^2}\left[1 - \left(1 - \frac{\gamma-1}{2}K\right)^{2\gamma/(\gamma-1)}\right]\right\}\alpha^2 \qquad (3.461)$$

Equation (3.461) provides an approximation of the lift coefficient on a flat plate for a known hypersonic Mach number, angle-of-attack, and specific gas constant. The drag coefficient is the lift coefficient multiplied by the angle-of-attack, as given by Equation (3.460).

As a final simplifying assumption, it can be assumed that the pressure on the upper surface of the flat plate is equal to the freestream pressure. In this case, $C_{p,u} = 0$, so that Equation (3.461) becomes

$$C_L = \left\{\frac{\gamma+1}{2} + \sqrt{\left(\frac{\gamma+1}{2}\right)^2 + \frac{4}{K^2}}\right\}\alpha^2 \qquad (3.462)$$

Next, we consider the model developed by Newton, as postulated in the second volume on fluid mechanics in his 1687 three-volume work, *Philosophiae Naturalis Principia Mathematica* (Latin for *Mathematical Principles of Natural Philosophy*), often simply referred to as *Principia*. Newton considered a fluid flow as a uniform stream of identical, non-interacting particles. He assumed that when the fluid particles impacted a body, such as the flat plate shown in Figure 3.236b, their normal momentum is transferred to the body, resulting in a pressure force on the body. In this *Newtonian theory*, the particles are assumed to "slide" along the body surface, preserving their tangential momentum, as shown in Figure 3.236b. The fluid particles do not affect the surface shielded from the flow. In this aerodynamic "shadow" of the flow, the pressure is assumed to equal the freestream pressure, making the pressure coefficient equal to zero. Newtonian theory was developed for the prediction of low-speed fluid dynamic flows, but it proved to be quite inaccurate for this application. Newtonian theory has found a place in the prediction of inviscid hypersonic flows, where its accuracy is much better.

Consider again the Newtonian model of the hypersonic flow, shown in Figure 3.236b, along with the geometry and force definitions in Figure 3.236c. According to Newton's second law, the normal force, F_n, on the plate is equal to the time rate of change of the flow momentum normal to the plate. Applying Newton's second law for a steady flow velocity, V_∞, we have

$$F_n = \frac{d}{dt}(mV_\infty)_n = \dot{m}_\infty V_{\infty,n} \tag{3.463}$$

where $\dot{m}_\infty$ is the mass flow rate through the cross-sectional area $A \sin \alpha$, and $V_{\infty,n}$ is the velocity normal to the plate. The mass flow rate, through the area $A \sin \alpha$, is given by

$$\dot{m}_\infty = \rho_\infty V_\infty A \sin \alpha \tag{3.464}$$

and the normal velocity is

$$V_{\infty,n} = V_\infty \sin \alpha \tag{3.465}$$

Inserting Equations (3.464) and (3.465) into (3.463), we have

$$F_n = (\rho_\infty V_\infty A \sin \alpha)(V_\infty \sin \alpha) = \rho_\infty V_\infty^2 A \sin^2\alpha \tag{3.466}$$

The pressure force depends only on the body orientation with respect to the freestream flow direction.

Assuming the pressure on the upper surface of the plate is p_∞, we have

$$\frac{F_n}{A} = p - p_\infty = \rho_\infty V_\infty^2 \sin^2\alpha \tag{3.467}$$

Dividing Equation (3.467) by the freestream dynamic pressure, the pressure coefficient, C_p, from Newtonian theory is given by

$$C_p = \frac{p - p_\infty}{\frac{1}{2}\rho_\infty V_\infty^2} = \frac{\rho_\infty V_\infty^2 \sin^2\alpha}{\frac{1}{2}\rho_\infty V_\infty^2} = 2\sin^2\alpha \tag{3.468}$$

Equation (3.468) is Newton's *sine-squared law*, which predicts the pressure coefficient on a surface at an angle α to the freestream hypersonic flow direction.

Inspecting Figure 3.236c, the lift, L, and drag, D, on the flat plate are given by

$$L = F_n \cos \alpha = \rho_\infty V_\infty^2 A \sin^2\alpha \cos \alpha \tag{3.469}$$

$$D = F_n \sin \alpha = \rho_\infty V_\infty^2 A \sin^3\alpha \tag{3.470}$$

The lift-to-drag ratio is given by

$$\frac{L}{D} = \frac{\rho_\infty V_\infty A \sin^2\alpha \cos \alpha}{\rho_\infty V_\infty^2 A \sin^3\alpha} = \frac{\cos \alpha}{\sin \alpha} = \cot \alpha \tag{3.471}$$

The lift and drag coefficients, C_L and C_D, respectively, from Newtonian theory, are

$$C_L = \frac{L}{\frac{1}{2}\rho_\infty V_\infty^2 A} = \frac{\rho_\infty V_\infty^2 A \sin^2\theta \cos \alpha}{\frac{1}{2}\rho_\infty V_\infty^2 A} = 2\sin^2\alpha \cos \alpha \tag{3.472}$$

$$C_D = \frac{D}{\frac{1}{2}\rho_\infty V_\infty^2 A} = \frac{\rho_\infty V_\infty^2 A \sin^3\alpha}{\frac{1}{2}\rho_\infty V_\infty^2 A} = 2\sin^3\alpha \tag{3.473}$$

The lift, drag, and pressure coefficients are a function of the surface angle relative to the freestream direction only and independent of the Mach number or other flow properties. For our

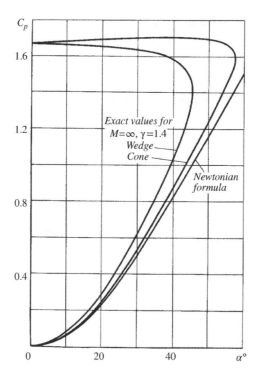

Figure 3.237 Surface pressure coefficients on a wedge and a cone from Newtonian theory and exact solutions for $M_\infty = \infty$ and $\gamma = 1.4$. (Source: *Reprinted from Introduction to Hypersonic Flow, G.G. Chernyi, Fig. 3.2a, p. 100, (1961), [20], with permission from Elsevier.*)

flat plate example, the surface angle relative to the freestream is the angle-of-attack, but this is not always the case. For instance, the surface pressures, and subsequent lift and drag, predicted by Newtonian theory are the same for a flat plate at an angle-of-attack α and a right-circular cone of semi-apex angle α at zero angle-of-attack. Newtonian theory does not differentiate between different types of bodies or geometries, as in the comparsion just given of the flat plate and cone. (We have used uppercase letters for the lift and drag coefficients derived from Newtonian theory just for this reason. The coefficients could be applied to a two-dimensional surface, where lowercase letters are used, or to a three-dimensional body, such as a cone, where uppercase letters are justified.)

Predictions from Newtonian theory for the pressure coefficient on the surface of a two-dimensional wedge and an axisymmetric cone are compared with exact solutions in Figure 3.237. The accuracy of Newtonian theory increases with increasing Mach number, hence this comparison is being made for $M_\infty = \infty$. Newtonian theory is in better agreement with the exact solutions at the lower angles-of-attack, yet still provides a reasonable agreement with the cone solution at higher angles-of-attack. Newtonian theory is in better agreement with the three-dimensional cone flow rather than the two-dimensional wedge flow.

Example 3.22 Calculation of Hypersonic Lift and Drag *The wing of a hypersonic airplane can be approximated by a flat plate with an area of $200\,ft^2$. The airplane is in Mach 8 flight with the wing at an angle-of-attack of $12°$. Calculate the lift coefficient, drag coefficient, and lift-to-drag ratio of the wing, assuming hypersonic small disturbance theory and Newtonian theory.*

Solution

First, we convert the angle-of-attack from degrees to radians.

$$\alpha = 12 \ \text{deg} \times \frac{\pi}{180} = 0.2094$$

From Equation (3.456), the hypersonic similarity parameter is

$$K = M_\infty \alpha = (8)(0.2094) = 1.675$$

Using hypersonic small disturbance theory, the wing lift coefficient is given by Equation (3.462) as

$$C_L = \left\{ \frac{\gamma + 1}{2} + \sqrt{\left(\frac{\gamma + 1}{2} \right)^2 + \frac{4}{K^2}} \right\} \alpha^2$$

$$= \left\{ \frac{1.4 + 1}{2} + \sqrt{\left(\frac{1.4 + 1}{2} \right)^2 + \frac{4}{(1.6752)^2}} \right\} (0.2094)^2 = 0.1268$$

From Equation (3.460), the wing drag coefficient is

$$C_D = C_L \alpha = (0.1268)(0.2094) = 0.02655$$

Dividing the lift coefficient by the drag coefficient, the lift-to-drag ratio of the wing, obtained from hypersonic small disturbance theory, is

$$\frac{L}{D} = \frac{C_L}{C_D} = \frac{0.1268}{0.02655} = 4.776$$

Using Newtonian theory, the wing lift coefficient, given by Equation (3.472), is

$$C_L = 2 \sin^2 \alpha \cos \alpha = 2 \sin^2(12°) \cos(12°) = 0.08457$$

From Equation (3.473), the wing drag coefficient from Newtonian theory is

$$C_D = 2 \sin^3 \alpha = 2 \sin^3(12°) = 0.01797$$

The lift-to-drag ratio of the wing, from Newtonian theory, is given by Equation (3.471) as

$$\frac{L}{D} = \cot \alpha = \cot(12°) = 4.705$$

3.13.7   Aerodynamic Heating

High-speed flows are high kinetic energy flows. This high kinetic energy is dissipated as heat in the boundary layer, adjacent to a body, and in the air around the body. The heat is generated primarily by friction in the boundary layer. Hence, friction is responsible for both skin friction drag and heat transfer in the boundary layer. Recall from Section 3.2.6 that the shear stress is related to the transport of momentum, and heat transfer is related to the transport of heat. If we assume that the mechanisms for these two transport processes are the same (the derivation of this assumption is beyond the scope of the text), then these two viscous effects can be related through the *Reynolds analogy*, given by

$$C_H \approx \frac{C_f}{2} \tag{3.474}$$

where C_f is the skin friction coefficient, defined in Section 3.12.3, and C_H is the Stanton number, defined in Section 3.4.7. Equation (3.474) matches experimental data for Prandtl numbers close to one, which is sufficient for the types of flows that we are considering. Thus, if we are able to predict the skin friction coefficient, Equation (3.474) provides a means of calculating the Stanton number, which proves useful in obtaining predictions of aerodynamic heating.

From basic physics, the three mechanisms of heat transfer are *conduction*, *convection*, and *radiation*. Conduction is heat transfer within a substance or between two substances. The substance may be a solid material or a fluid, such as air. The heat or energy transfer is due to the molecular collisions between particles, where higher energy particles transfer energy to lower energy particles. Conduction does not rely on any motion of the substance as a whole. Thermal conduction was introduced in Section 3.2.6, where the transport of heat is governed by Fourier's Law, given by Equation (3.15).

Convective heat transfer depends on the *mass motion* of the fluid. For aerodynamic heating related to the high-speed flight, we are concerned with *forced convection*, as described in Section 3.4.7, where the mass motion of the fluid is induced by the vehicle motion. If the vehicle is at rest, the primary mechanism for heat transfer is conduction, due to the temperature gradients normal to the vehicle surface. If the vehicle is in motion, heat is transferred due to the temperature gradients and due to the mass movement of the fluid. Hence, convective heat transfer includes molecular conduction heat transfer and gross fluid movement.

Consider the surface of a vehicle in a high-speed flow of velocity, U_∞, as shown in Figure 3.238. Assuming a viscous flow, there is a boundary layer adjacent to the surface with a thickness δ. The velocity at the surface is zero and the velocity profile in the boundary layer is given by $U = U(y)$. The velocity at the edge of the boundary layer is equal to the freestream velocity. The skin friction coefficient, c_f, is related to the shear stress at the surface, τ_w, and the velocity gradient, dU/dy, using Equations (3.14) and (3.420), which gives

$$c_f = \frac{\tau_w}{q_\infty} = \frac{\mu}{q_\infty}\left(\frac{dU}{dy}\right) \tag{3.475}$$

Similar to the velocity boundary layer, the temperature profile, $T = T(y)$, describes a *thermal boundary layer* with a thickness δ_T, as shown in Figure 3.238. The relative thickness of the thermal boundary layer, as compared with the velocity boundary layer, is governed by the Prandtl number.

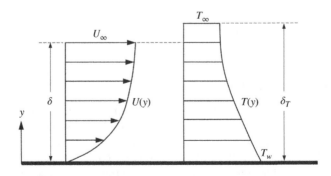

Figure 3.238   Velocity and thermal boundary layers.

Recall, from Equation (3.41), that the Prandtl number is the ratio of the diffusion of momentum to the thermal diffusion. For a Prandtl number of one, the thermal and velocity boundary layer heights are equal. For aerodynamic flows with air, where the Prandtl number is 0.71, the diffusion of momentum is slower than the thermal diffusion, so that the velocity boundary layer is contained within the thermal boundary layer, or $\delta < \delta_T$.

In the thin boundary layer, close to the surface, the velocity is near zero such that the heat transfer is by conduction. The heat flux (rate of heat flow per unit area) at the wall, $\dot{q}_w$, is given by Fourier's Law, Equation (3.15), as

$$\dot{q}_w = k \left(\frac{dT}{dy} \right)_w = k \frac{(T_0 - T_w)}{\delta'} \tag{3.476}$$

where the minus sign has been omitted on the assumption that the direction of the heat transfer is obvious (from high to low temperature) and T_0 is the total temperature.

The value of the thickness, δ', is not a property of the fluid, rather it is a function of the flow velocity (Reynolds number), the pressure gradient in the flow, the Mach number, and the roughness of the wall surface. It is customary to define the convective heat transfer coefficient, h_c, as

$$h_c \equiv \frac{k}{\delta'} \tag{3.477}$$

where values of h_c are often determined experimentally. The convective heat transfer coefficient is a function of the type of fluid and the flow properties, such as the velocity and viscosity. Hence, the heat flux at the wall may be written in terms of the convective heat transfer coefficient as

$$\dot{q}_{c,w} = h_c (T_0 - T_w) \tag{3.478}$$

The Stanton number may be defined in terms of the convective heat transfer coefficient as

$$C_H = \frac{h_c}{\rho_\infty V_\infty c_p} \tag{3.479}$$

Substituting the convective heat transfer coefficient from Equation (3.479) into (3.478), we have

$$\dot{q}_{c,w} = \rho_\infty V_\infty C_H c_p (T_0 - T_w) = \rho_\infty V_\infty C_H (h_0 - h_w) \tag{3.480}$$

where h_0 and h_w are the total enthalpy and enthalpy at the wall, respectively.

For a high-speed flow, the temperature associated with the total enthalpy is much greater than that for the wall enthalpy, so that we can assume $h_0 \gg h_w$, such that $h_0 - h_w \approx h_0$. Hence, Equation (3.480) becomes

$$\dot{q}_c = \rho_\infty V_\infty C_H h_0 \tag{3.481}$$

Similarly, the total enthalpy is assumed to be much greater than the static enthalpy, h_∞, so that

$$h_0 = h_\infty + \frac{V_\infty^2}{2} \approx \frac{V_\infty^2}{2} \tag{3.482}$$

Substituting Equation (3.482) into (3.481), we have

$$\dot{q}_{c,w} = \rho_\infty V_\infty C_H \frac{V_\infty^2}{2} = \frac{1}{2} \rho_\infty V_\infty^3 C_H \tag{3.483}$$

Equation (3.483) gives the heat flux at the wall as a function of the freestream density, freestream velocity, and Stanton number. The heat flux varies with the cube of the velocity, which results in very high heating rates for high speed flows. The increase in heat flux is greater than that for aerodynamic drag, which varies with the velocity squared.

While the heating *rate*, $\dot{q}$, is an important consideration, the total, accumulated heat input, Q, to a vehicle is also important. A relatively low heating rate may not be a problem for a vehicle, but if this low heating rate is applied for a long duration, the total heat input to the vehicle may not be acceptable. It can be shown that the total heat input to a body, with mass m, is given by

$$Q = \frac{1}{4}\left(\frac{C_f}{C_D}\right)(mV_\infty^2) \tag{3.484}$$

where C_f is the body skin friction coefficient and C_D is the body drag coefficient. The total heat input increases with the kinetic energy of the body (mV_∞^2) and is proportional to the ratio of the skin friction drag to the total drag (C_f/C_D). It makes perfect sense that more heat energy goes into a body as the kinetic energy of the body increases.

The ratio of the drag terms leads to an important conclusion concerning the *shape* of a body to minimize the heat load. Recall that the total drag of a body is the sum of the pressure drag and the skin friction drag. To minimize the heating, it is desirable to minimize the skin friction drag relative to the total drag. A body of this type is *blunt*, having high pressure drag and low skin friction drag. In contrast, a *sharp, slender* body has low pressure drag and high skin friction drag. Therefore, a blunt body, or to be more precise, a *blunt-nosed* body results in lower heat load than a sharp-nosed body. It can be shown that the convective heat transfer to the stagnation point of a body, $\dot{q}_{c,stag}$, is inversely proportional to the square root of the nose radius, R_{nose}, of the body. Hence, the large nose radius of a blunt-nosed body results in low stagnation point heat transfer.

The third type of heat transfer is that due to electromagnetic radiation, such as from visible light, infrared, or ultraviolet radiation. Warmth from the Sun or an open fire is an example of radiative heat transfer. As discussed earlier, flight at hypersonic speeds results in strong shock waves with large increases in temperature. These extreme temperatures can make the air, behind the shock, a radiating plasma. This plasma radiates electromagnetic energy, leading to a radiative heat flux on the vehicle. The radiative heat flux, $\dot{q}_R$, (heat transfer rate per unit area) is proportional to a high power of the freestream velocity, ranging from V_∞^5 to V_∞^{12}. Hence, high hypersonic speeds can result in significant radiative heat transfer. The stagnation point heating due to radiative heat transfer is directly proportional to the nose radius, so a blunt-nosed body absorbs more radiative heat energy than a slender-nosed body.

Example 3.23   Calculation of Aerodynamic Heating *The X-30 hypersonic aerospace plane (see Figure 1.83) is flying at Mach 10 and an altitude of 30 km. The undersurface of the vehicle compresses the flow ahead of the supersonic combustion ramjet engine. Assuming this undersurface forebody is 32 m in length, calculate the heat flux to the forebody surface.*

Solution

Using Appendix C, the freestream density and temperature at an altitude of 30,000 m are

$$\rho_\infty = \sigma\rho_{SSL} = (0.01503)\left(1.225\frac{\text{kg}}{\text{m}^3}\right) = 0.01841\ \frac{\text{kg}}{\text{m}^3}$$

$$T_\infty = \theta T_{SSL} = (0.78608)(288\ \text{K}) = 226.6\ \text{K}$$

Using Sutherland's Law, Equation (3.16), the freestream coefficient of viscosity is given by

$$\mu_\infty = \left(\frac{T_\infty}{T_{ref}}\right)^{3/2} = \left(\frac{T_\infty}{T_{ref}}\right)^{3/2} \left(\frac{T_{ref} + S}{T_\infty + S}\right) \mu_{ref}$$

$$\mu_\infty = \left(\frac{226.6\,\text{K}}{273.15\,\text{K}}\right)^{3/2} \left(\frac{273.15\,\text{K} + 110.6\,\text{K}}{226.6\,\text{K} + 110.6\,\text{K}}\right) \left(17.16 \times 10^{-6}\,\frac{\text{kg}}{\text{m}\cdot\text{s}}\right)$$

$$\mu_\infty = (0.8599)\left(17.16 \times 10^{-6}\,\frac{\text{kg}}{\text{m}\cdot\text{s}}\right) = 1.476 \times 10^{-5}\,\frac{\text{kg}}{\text{m}\cdot\text{s}}$$

The freestream velocity is obtained from the Mach number and speed of sound as

$$V_\infty = M_\infty a_\infty = M_\infty \sqrt{\gamma R T_\infty} = (10)\sqrt{1.4\left(287\,\frac{\text{J}}{\text{kg}\cdot\text{K}}\right)(226.6\,\text{K})} = 3017\,\frac{\text{m}}{\text{s}}$$

The Reynolds number based on the forebody length is

$$Re_L = \frac{\rho_\infty V_\infty L}{\mu_\infty} = \frac{\left(0.01841\,\frac{\text{kg}}{\text{m}^3}\right)\left(3017\,\frac{\text{m}}{\text{s}}\right)(32\,\text{m})}{1.476 \times 10^{-5}\,\frac{\text{kg}}{\text{m}\cdot\text{s}}} = 1.204 \times 10^8$$

From Equation (3.431), the turbulent skin friction coefficient is given by

$$C_{f,turb} = \frac{0.074}{Re_L^{0.2}} = \frac{0.074}{(1.204 \times 10^8)^{0.2}} = 0.001791$$

Using Reynolds analogy, Equation (3.474), the Stanton number is

$$C_H \approx \frac{C_f}{2} = \frac{0.001791}{2} = 8.955 \times 10^{-4}$$

Using Equation (3.483), the heat flux at the wall is given by

$$\dot{q}_{c,w} = \frac{1}{2}\rho_\infty V_\infty^3 C_H$$

$$\dot{q}_{c,w} = \frac{1}{2}\left(0.01841\,\frac{\text{kg}}{\text{m}^3}\right)\left(3{,}017\,\frac{\text{m}}{\text{s}}\right)^3 (8.955 \times 10^{-4}) = 2.264 \times 10^5\,\frac{\text{J/s}}{\text{m}^2}$$

3.13.8   FTT: Hypersonic Flight Testing

Hypersonic flight testing is some of the most difficult and complex testing that can be attempted. The hypersonic thermal environment is extreme, placing large heat loads on vehicle structures and systems. Hypersonic testing usually involves flight at very high altitudes, often requiring additional, non-aerodynamic flight control systems. Perhaps the first difficulty in performing hypersonic flight testing is getting the vehicle to the desired hypersonic flight conditions. This is often accomplished using rocket power, as the dream of an air-breathing hypersonic engine is still not a practical reality. To learn about many of the factors involved with hypersonic flight testing, you will take a hypersonic flight in the North American X-15 rocket-powered, hypersonic research aircraft (Figure 3.239).

The X-15 hypersonic research program was a joint venture between NASA, the US Air Force, the US Navy, and North American Aviation during the 1950s and 60s. The X-15 research aircraft

Figure 3.239   North American X-15 hypersonic research airplane. (Source: *NASA.*)

was designed to explore many aspects of flight at hypersonic speeds and at the edge of space. The initial design goals of the X-15 included the capability to fly at hypersonic speeds, up to Mach 6, and to reach the near-space environment, up to an altitude of 250,000 ft (76,000 m). The initial flight test program had four specific objectives: (1) obtain flight data for verification of hypersonic aerodynamic and heat transfer theory and wind tunnel data, (2) investigate high temperature aircraft structures at high flight loads, up to 1200 °F (920 K, 1660 °R), (3) investigate hypersonic stability and control problems associated with atmospheric exit and entry, and (4) investigate the physiological effects of flight in the near-space environment, including the effects of weightlessness and high-g loads on pilot tasks and pilot performance. By the end of the X-15 test program, these objectives were met or exceeded.

The X-15 hypersonic aircraft was designed and built by North American Aviation (NAA), Los Angeles, California, with technical help from the NACA. NAA had a rich heritage as an airplane company, having designed and built very successful aircraft such as the T-6 *Texan* pilot trainer, the P-51 *Mustang* fighter, the B-25 *Mitchell* bomber, and the F-86 *Sabre* jet fighter. They would go on to design and build more famous aerospace vehicles, including the XB-70 *Valkyrie* triple-sonic bomber, the second stage of the *Saturn V* moon rocket, and the *Apollo* command and service modules, which would fly to the Moon. In later years, North American Aviation merged with Rockwell International, making them part of the team that designed and built the Space Shuttle Orbiter vehicle.

The X-15 is a single-seat, mid-wing rocket-powered airplane with a length of 49.5 ft (15.1 m) and a wingspan of 22.36 ft (6.815 m). A three-view drawing of the X-15 is shown in Figure 3.240 and selected specifications are given in Table 3.21. The X-15 powerplant is a Reaction Motors XLR-99 throttlable, liquid rocket engine, burning anhydrous ammonia and liquid oxygen (LOX). The LOX tank in the aircraft holds 1003 gallons (3797 liters) of liquid oxygen and the fuel tank holds 1445 gallons (5470 liters) of anhydrous ammonia, providing a rocket engine burn time of 85 s at maximum thrust, if all of the propellants are consumed. The X-15 has a heat-sink structure,

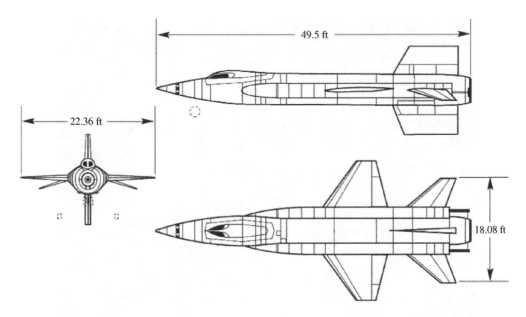

Figure 3.240   Three-view drawing of the X-15 hypersonic research aircraft. (Source: *NASA*.)

Table 3.21   Selected specifications of the X-15 hypersonic research aircraft.

Item	Specification
Primary function	Hypersonic flight research
Manufacturer	North American Aviation, Los Angeles, California
First flight	8 June 1959
Crew	1 pilot
Powerplant	Reaction Motors XLR99 rocket engine
Thrust, maximum	57,000 lb (253,000 N)
Thrust, minimum	28,000 lb (125,000 N)
Launch weight	31,275 lb (14,186 kg)
Burnout weight	12,295 lb (5577 kg)
Length	49.5 ft (15.1 m)
Wing span	22.36 ft (6.815 m)
Wing area	200 ft^2 (18.6 m^2)
Wing loading	170 lb/ft^2 (830 kg$_f$/m^2)
Wing aspect ratio	2.50
Wing airfoil section	NACA 65-005 (modified)

where the outer skin of the vehicle absorbs the tremendous friction heating from hypersonic flight. The outer skin of the aircraft is made of Inconel-X, a nickel-chrome alloy, while the cockpit structure is made of aluminum; thermally isolating it from the skin.

To reach hypersonic speeds, a two-stage aircraft system is used, with a NASA B-52 *Stratofortress* bomber as the first stage and the X-15 as the second stage. The X-15 is hung from a pylon underneath the right wing of the massive B-52 bomber. There are connections from the B-52 "mothership" to the X-15 that provide power and pressurization to the rocket plane during the mated

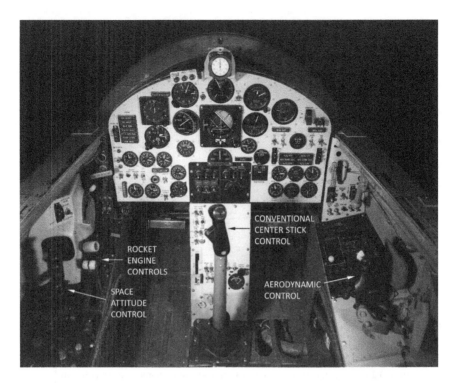

Figure 3.241 X-15 cockpit layout. (Source: *NASA with annotations added.*)

climb to the drop altitude. The B-52 also carries liquid oxygen on board to replenish the X-15 liquid oxygen tank in flight. Unlike one of its X-plane predecessors, the Bell X-1, you must be inside the X-15 cockpit from takeoff, as there is no way to enter the cockpit from the B-52 carrier aircraft.

You don a full pressure suit for your X-15 flight, as the cockpit is unpressurized below 35,000 ft (10,700 m). It also provides your body with protection in the event of a cockpit depressurization at extremely high altitudes. You climb into the X-15 cockpit and strap yourself into the ejection seat. The ejection seat is designed to provide escape capability for the "low altitude, low speed" portion of the flight envelope. Of course, for the X-15 flight envelope, "low altitude" is up to 120,000 ft (36,600 m) and "low speed" is below Mach 4. Outside of these limits, the X-15 airframe and cockpit are designed to protect the pilot.

Once seated, you notice that the cockpit design looks quite conventional, similar to military fighter airplanes of the era, with some significant exceptions (Figure 3.241). There is a conventional center control stick, which moves the all-moving horizontal tail. The left and right horizontal tails deflect symmetrically up and down, like a conventional elevator, for pitch control, while for roll control, the tails deflect differentially. (This all-moving, rolling tail was another advanced design feature of the X-15, as it was not proven technology at the time.) Left and right pedals, at your feet, move the rudder surfaces on the vertical stabilizers for yaw control. However, in addition to the center stick, there are two additional control sticks in the cockpit, left and right sidestick controllers.

The right sidestick controller is mechanically linked to the center control stick, so that it also commands the aerodynamic control surfaces. You will use the right sidestick during powered flight and atmospheric entry, when your arms and the rest of your body will experience high g loads, making the operation of the center stick more difficult. Your arm will be stabilized by an armrest, with the sidestick controlled by hand movements, giving you more precise aircraft control under high g loads. Although its use was very successful in the X-15, the sidestick controller would not find its way into a high g aircraft for 20 years, until the design of the General Dynamics (now Lockheed-Martin) F-16.

You will use the left sidestick controller when the X-15 is outside the atmosphere, where the air is too thin for the operation of conventional aerodynamic controls. The left sidestick activates the reaction control system (RCS) thrusters, which expel high-pressure gas from the decomposition of hydrogen peroxide. RCS nozzles on the aircraft nose provide pitch and yaw control, while roll is controlled by RCS thrusters on the wings. The X-15 RCS was leading the technological path for future spacecraft applications. In later X-15 modifications, an adaptive flight control system automatically blended the RCS with the aerodynamic controls.

The rocket engine control lever is on the left side of the cockpit, located where a jet engine throttle would normally be. The XLR-99 rocket engine is throttlable, allowing you to set the thrust from 40% to 100%, generating 28,000– 57,000 lb (125,000–254,000 N) of thrust. You can also shut down the rocket engine using the control lever.

Your flight is conducted from Edwards Air Force Base in California. The B-52 takes off, with the X-15 hanging on its wing, as shown in Figure 3.242. During the climb to the drop altitude, the B-52 "tops off" your LOX tank to replace the liquid oxygen that is used to pre-cool the rocket engine and that is lost to boil off. You think about how ironic it is that your aircraft skin must survive the over 1000° of high temperature, hypersonic flight, yet the structure and internal components must also be protected from being frozen by the super-cold liquid oxygen and liquid nitrogen on board, stored at about −240 °F to −300 °F (220–160 °R, 122–89 K).

The B-52 levels off at the drop altitude of 45,000 ft (13,700 m) and stabilizes at the drop speed of Mach 0.8. You feel your heart racing in anticipation of the release. The X-15 program was the first to collect biomedical data on pilots under high stress over a range of speeds and g forces, including the weightless environment. This type of monitoring would become routine for manned spaceflight. The measured heart rates of X-15 pilots ranged from about 145 to 185 beats per minute, at several critical phases of the flight, such as release from the B-52, engine shutdown, pullout

Figure 3.242 X-15 carried aloft underneath the wing of the B-52 mothership. (Source: *NASA*.)

Figure 3.243   X-15 release from the B-52 mothership. (Source: *NASA*.)

during atmospheric entry, and landing. This is quite a bit higher than a normal resting rate of about 60–70 beats per minute. While these were high levels, it was determined that this was normal and acceptable, as pilot performance was not affected or degraded. This type of medical data proved invaluable in assessing human performance for spaceflight.

You complete your pre-release checks, including a check of the flight controls and control surface trim settings. There is a final countdown to the release and then you and the X-15 are dropped from the B-52 mothership, as shown in Figure 3.243. The release is a free fall drop, so you become "light in your seat". You are not concerned about re-contact with the carrier aircraft, as the free fall maneuver has been studied in detail by the engineers, including the conduct of wind tunnel tests to understand the separation characteristics. You slide the rocket engine throttle lever sideways, towards you, from OFF to START, and then push it forward towards MAX. You feel the 57,000 lb (254,000 N) of thrust push you back into your seat, as the rocket plane accelerates. At your release weight of 31,275 lb (14,186 kg), the longitudinal acceleration pushing you back into your seat is about 2 g. You use the right sidestick controller to pull the X-15's nose up, setting a flight path angle (angle between the aircraft longitudinal axis and the horizon) of about 40°.

There were two different flight profiles typically flown by the X-15, the maximum altitude profile and the maximum speed profile, as shown in Figure 3.244. For the speed profile, the pilot pulled up into a climb after release and at an altitude of 75,000 ft (22,900 m), performed a negative 2 g pushover to level flight. After leveling off, the rocket engine was shut down and constant altitude was then maintained until starting the glide to landing. The speed profile resulted in the maximum Mach number and the maximum aerodynamic heating of the vehicle.

You are flying the high altitude profile for your flight, as shown in Figure 3.245. You maintain the climb attitude for the entire rocket engine burn. As your propellants are consumed, the longitudinal acceleration increases to a maximum of about 4 g at your burnout weight of 12,295 lb (5577 kg). With a propellant flow rate of 13,000 lb/min (5900 kg/min), the 18,000 lb (8200 kg) of ammonia and LOX are consumed by the rocket engine in a little more than 80 s. The X-15 is accelerating at a tremendous rate to hypersonic speeds. At these high speeds, the skin friction of the air is rapidly

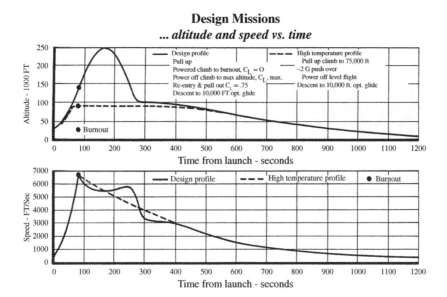

Figure 3.244 The two X-15 mission profiles, the maximum altitude ("design") profile and the maximum speed ("high temperature") profile. (Source: *NASA*.)

Typical mission

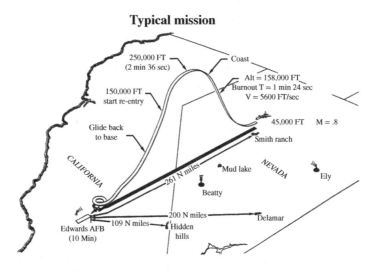

Figure 3.245 X-15 maximum altitude mission profile. (Source: *NASA*.)

heating up the aircraft. You can actually hear the "crackling" of the aircraft due to buckling of the aircraft skin from the heating. You shut the rocket engine down after an 84 s burn. You are at 158,000 ft (48,200 m), traveling at a speed of over 3800 mph (6100 km/h, 5600 ft/s) or a hypersonic Mach number of 5.2. The X-15 is now "coasting" on a ballistic arc, still climbing at an amazing rate. At the apogee of the arc, just 2.5 minutes after your release, you reach a maximum altitude of 250,000 ft (76,000 m), over 47 miles above the surface of the earth, where you can see the earth's curvature. You are just shy of the 50 mile (80 km) altitude limit to earn your US Air Force astronaut wings. You experience weightlessness, collecting more information about the capability of humans

Figure 3.246 Comparison of shock waves on X-15 model at Mach 3.5 (left) and Mach 6 (right). (Source: *NASA.*)

to function in the near-space environment. The impact of weightlessness on humans was a serious concern prior to the advent of spaceflight and the X-15 provided valuable data on this subject.

At this extreme altitude, your aerodynamic control surfaces are ineffective due to the lack of atmosphere, so you switch from the right sidestick to the left sidestick controller, to operate the hydrogen peroxide reaction control thrusters. You use these thrusters to maintain the desired aircraft attitude, which is critical as you start descending into the thicker region of the atmosphere. As the X-15 descends into the thicker atmosphere, the aerodynamic drag starts to increase, slowing the vehicle. If we could see the shock waves on the X-15 vehicle as it descends at a hypersonic Mach number and at a lower supersonic Mach number, they would look as shown in the schlieren photographs in Figure 3.246. These are from tests in the NACA hypervelocity free-flight facility. In this unique type of ground testing, 3–4 in long (7.6–10 cm) X-15 models were fired from a special gun that accelerates them to very high Mach numbers. The models "free fly" past an observing station, where photographs of the shock patterns were captured. Compare the shock wave patterns on the X-15 at the different Mach numbers in Figure 3.246. The shock wave angles are seen to decrease dramatically from Mach 3.5 to 6. The smaller shock wave angles at Mach 6 cause the shock waves from the vehicle nose to impinge on the outboard portions of the main wing, where the aileron control surfaces are located. This may be of concern if the shock impingement separates the flow on the ailerons, making them less effective.

After the flight, engineers analyze the supersonic and hypersonic aerodynamic data for your hypersonic, power-off glide, producing a drag polar, as shown in Figure 3.247, and a plot of maximum lift-to-drag ratio versus Mach number, as shown in Figure 3.248. The drag polars at Mach 3 and 5 are parabolic in shape, as they are for subsonic flight. At a given lift coefficient, the drag coefficient at Mach 5 is less than at Mach 3. The Mach 5 zero-lift drag coefficient is about 0.04, significantly less than the (approximately) 0.062 value at Mach 3. This is in agreement with the trend shown for the decrease in the zero-lift drag coefficient with increasing Mach number, shown in Figure 3.45. The data in Figure 3.248 confirms that the maximum lift-to-drag ratio decreases with increasing Mach number. It is also emphasizes the fact that hypersonic lift-to-drag ratios are typically small, with the X-15 having a maximum lift-to-drag ratio of only about 2.4 at Mach 5. The X-15 wind tunnel data compared well with the flight data, but the supersonic and hypersonic theories over predicted the lift-to-drag ratios.

You know that the drag can be dependent on the nature of the boundary layers, where the drag due to skin friction is much higher with a turbulent boundary layer than a laminar one. Later analyses would show that the X-15 had mostly turbulent boundary layer flow over much of its surfaces at

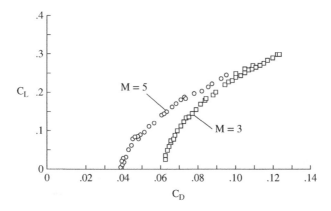

Figure 3.247 X-15 drag polar from power-off flight data. (Source: *Hopkins, et al., NASA TM-X-713, 1962,* [38].)

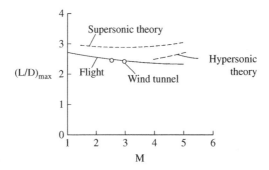

Figure 3.248 X-15 maximum lift-to-drag ratio for power-off flight. (Source: *Hopkins, et al., NASA TM-X-713, 1962,* [38].)

supersonic and hypersonic speeds. This turbulent flow was due to roughness and irregularities in the aircraft skin, which tripped the boundary layers.

You and the aircraft are subjected to considerable g forces, up to about 6 g, during the entry pullout. You continue to descend and decelerate, setting up for the final, subsonic glide to landing on Rogers dry lakebed on Edwards Air Force Base. You enter a 360° circular or overhead pattern at an altitude of 29,000 ft (8800 m) and an airspeed of 300 knots indicated airspeed (KIAS) (560 km/h). At an altitude of 5500 ft (1680 m), which is 3300 ft (700 m) above the ground (the lakebed is at an elevation of 2200 ft (670 m)), you jettison the lower portion of the vertical tail, to provide ground clearance. You start your landing flare at 3000 ft (900 m), 800 ft (240 m) above ground level, at an airspeed of 260 KIAS (480 km/h). At 250 ft (76 m) above the ground and 230 KIAS (430 km/h), you lower the main landing gear, consisting of two narrow skis at the aft end of the fuselage and a nosewheel landing gear (Figure 3.249). You touchdown on the lakebed at 184 KIAS (340 km/h) and an angle-of-attack of about 8°. Your X-15 rocket-powered flight to the near-space altitude of 250,000 ft and back to earth covered a distance of almost 300 miles (480 km), from the release point to landing, with a total flight time of 12 minutes.

After landing, you climb out of the X-15 and look the aircraft over. You look at the left wing and the left horizontal tail, where temperature sensitive paint was applied prior to the flight (Figure 3.250). The paint shows a pattern of contrasting colors, indicating the different surface temperatures obtained in hypersonic flight. This was one of the test techniques used to collect

Figure 3.249   X-15 landing on the dry lakebed. (Source: *NASA*.)

Figure 3.250   Temperature-sensitive paint on the wing and horizontal tail of the X-15. (Source: *NASA*.)

flight data to validate the various existing heat transfer prediction models. Thermocouple instrumentation was also installed on the right side of the aircraft to collect quantitative heat transfer data. The flight data shows surface temperatures greater than 1300 °F (1760 °R, 980 K) near the wing leading edges and temperature above 1000 °F (1460 °R, 810 K) over much of the aircraft structure. In comparing with the existing heat transfer models of the time, the first-of-its-kind

X-15 flight data showed that the models over predicted the heat transfer by 30–40%. The X-15 flight data contributed to the improvement of these heat transfer models, which would later be applied to the design of future manned spacecraft.

The X-15 was the world's first hypersonic airplane. The first flight of the X-15 was an unpowered, glide flight on 8 June 1959, piloted by North American test pilot Scott Crossfield. The 199th and last flight of the X-15 was flown by NASA test pilot William H. Dana on 24 October 1968. A total of 12 test pilots from the US Air Force, US Navy, and NASA flew the X-15, including Neil Armstrong, the first man to set foot on the Moon. The X-15 program is perhaps the most successful manned hypersonic flight research program ever. Three X-15 aircraft flew 199 hypersonic research flights over a nearly 10-year span, setting unofficial altitude and speed records of 354,200 ft (107,960 m, 67.1 mi) on Flight 91 in 1963 and 4520 mph (7274 km/h) or Mach 6.70 on Flight 188 in 1967, respectively. The X-15 explored many areas of manned hypersonic flight in the real flight environment. It taught us *how to fly* a hypersonic airplane, and as important, how to *flight test* a hypersonic airplane. The X-15 program collected a wealth of data about hypersonic aerodynamics, stability and control, structures, materials, and many other areas important to hypersonic vehicle design. Hypersonic data from the X-15 program contributed to the design and development of many future space vehicles that had to fly through the hypersonic flow regime, including the *Mercury*, *Gemini*, *Apollo*, and Space Shuttle vehicles.

In October 1956, the NACA Research Airplane Committee met to review the development of the X-15 program. Noted American aerodynamicist, Hugh L. Dryden spoke of the program goal, "to realize flights of a man-carrying aircraft at hypersonic speeds and high altitudes as soon as possible for explorations to separate the real from the imagined problems and to make known the overlooked and the unexpected problems". [8] Hugh Dryden's statement goes far beyond the hypersonic flights of the X-15, serving as a mantra for the pursuit of flight research and scientific exploration.

3.14   Summary of Lift and Drag Theories

In this chapter, theories have been presented for the predication of lift and drag for incompressible flow ($M_\infty = 0$), subsonic, compressible flow ($0.8 < M_\infty < 1.2$), supersonic flow ($1.2 < M_\infty < 5$), and hypersonic flow ($M_\infty > 5$). A summary of theoretical predictions for the aerodynamics of an airfoil are given in Table 3.22. All of the lift and drag predictions are a function of angle-of-attack. The predictions for incompressible and hypersonic flow are independent of Mach number, while those for subsonic, compressible, and supersonic flows are a function of the Mach number squared. The incompressible, subsonic, compressible, and supersonic predictions are linear with respect to angle-of-attack, while the hypersonic theory is non-linear with respect to angle-of-attack, consistent with the physical nature of these different types of flows.

The incompressible pressure coefficient is simply given as $C_{p,0}$ in Table 3.22, which is usually obtained from experimental data or numerical techniques. For subsonic, compressible flow, this pressure coefficient is adjusted using the Prandtl–Glauert rule. Supersonic linear theory and hypersonic Newtonian theory provide predictions for the pressure coefficient as shown.

The lift coefficients, based on these different theories, are shown in Figure 3.251, for the case of a body flying at an angle-of-attack of 4°. The incompressible result is for a freestream Mach number of zero. The subsonic, compressible curve was calculated for Mach numbers from zero to 1. Supersonic results are for Mach numbers 1–5. Finally, the hypersonic prediction is independent of Mach number, but it is shown only for a Mach number of 5.

---

[8] Gorn, M.H., *Expanding the Envelope: Flight Research at NACA and NASA*, University Press of Kentucky, Lexington, Kentucky, 2001, pp. 3.

Table 3.22 Summary of airfoil lift and drag predictions for different flow regimes.

Parameter	Incompressible	Subsonic, compressible	Supersonic	Hypersonic (Newtonian theory)
Pressure coefficient, C_p	$C_{p,0}$	$\dfrac{C_{p,0}}{\sqrt{1-M_\infty^2}}$	$\dfrac{2\alpha}{\sqrt{M_\infty^2-1}}$	$2\sin^2\alpha$
Lift coefficient, c_l	$2\pi\alpha^*$	$\dfrac{2\pi\alpha}{\sqrt{1-M_\infty^2}}$	$\dfrac{4\alpha}{\sqrt{M_\infty^2-1}}$	$2\sin^2\alpha\cos\alpha$
Drag coefficient, c_d	0	0	$\dfrac{4\alpha^2}{\sqrt{M_\infty^2-1}}$	$2\sin^3\alpha$
Lift curve slope, c_{l_α}	2π	$\dfrac{2\pi}{\sqrt{1-M_\infty^2}}$	$\dfrac{4}{\sqrt{M_\infty^2-1}}$	$4\sin\alpha-6\sin^3\alpha$
Lift-to-drag ratio, L/D	∞	∞	$\dfrac{1}{\alpha}$	$\cot\alpha$
Aerodynamic center, x_{ac}	$\dfrac{c}{4}{}^*$	$\dfrac{c}{4}$	$\dfrac{c}{2}$	—

$*$ Prediction for symmetric airfoil.

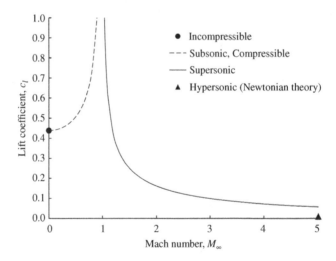

Figure 3.251 Theoretical lift coefficient versus Mach number (4° angle-of-attack).

The predicted incompressible lift coefficient is 0.44, which then increases asymptotically to infinity as the Mach number approaches one. Supersonic theory predicts an infinite lift coefficient at Mach 1 also, followed with the coefficient asymptotically approaching zero at infinite Mach number. Hypersonic Newtonian theory predicts a lift coefficient of 0.0097.

The drag coefficients, predicted using the formulas in Table 3.22, are plotted in Figure 3.252. The incompressible and subsonic, compressible theories predict zero drag. In reality, we know that the subsonic drag coefficient increases asymptotically from a low number to a maximum around Mach one. The supersonic theory predicts an infinite drag coefficient at Mach one, asymptotically approaching zero at infinite Mach number. The hypersonic Newtonian theory predicts a very small drag coefficient equal to 0.00068. The Newtonian theory does not include any viscous effects, which would significantly increase this drag coefficient.

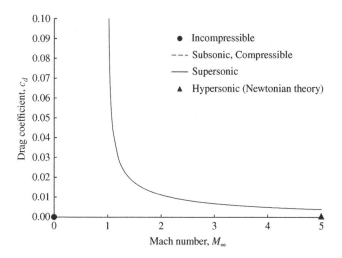

Figure 3.252 Theoretical drag coefficient versus Mach number (4° angle-of-attack).

Finally, for the lift-to-drag ratio, L/D, the incompressible and subsonic, compressible theories predict an infinite value, since they predict zero drag. The supersonic and hypersonic theories predict an L/D of 14.3. The hypersonic value is very close to the supersonic value, since for small angles-of-attack, the Newtonian theory prediction of $\cot \alpha = 1/\tan \alpha \approx 1/\alpha$, which is the same as the supersonic formula. Both of these predictions for L/D are too high, since neither theory takes into account any viscous effects.

In summary, the various theories that have been presented provide some simple equations to predict the aerodynamics of a body from low-speed, incompressible flow to hypersonic speeds. These provide back-of-the-envelope aerodynamic predictions that may be useful in qualitative analyses or in "sanity checks" of more complicated, higher fidelity methods.

References

1. Abbott, I.H. and von Doenhoff, A.E., *Theory of Wing Sections*, Dover, New York, 1959.
2. Abbott, I.H., von Doenhoff, A.E., and Stivers , "Summary of Airfoil Data," NASA Technical Report 824, 1945.
3. Abzug, M.J. and Larrabee, E.E., *Airplane Stability and Control: A History of the Technologies That Made Aviation Possible*, Cambridge University Press, Cambridge, United Kingdom, 2nd edition, 2005.
4. *Aerodynamics*, USAF Test Pilot School, Edwards AFB, California, January 2000.
5. *Aircraft Performance*, USAF Test Pilot School, Edwards AFB, California, January 2000.
6. Ames Research Staff, "Equations, Tables, and Charts for Compressible Flow," NACA Report 1135, Ames Aeronautical Laboratory, Moffett Field, California, 1953.
7. Anderson, D.A., Tannehill, J.C., and Pletcher, R.H., *Computational Fluid Mechanics and Heat Transfer*, Hemisphere Publishing, McGraw-Hill, New York, 1984.
8. Anderson, J.D., Jr, *A History of Aerodynamics*, Cambridge University Press, New York, 1998.
9. Anderson, J.D., Jr, *Computational Fluid Dynamics*, 1st edition, McGraw-Hill, New York, 1995.
10. Anderson, J.D., Jr, *Fundamentals of Aerodynamics*, 3rd edition, McGraw-Hill, New York, 2001.
11. Anderson, J.D., Jr, *Hypersonic and High-Temperature Gas Dynamics*, 2nd edition, AIAA Education Series, Reston, Virginia, 2006.
12. Anderson, J.D., Jr, *Introduction to Flight*, 4th edition, McGraw-Hill, Boston, Massachusetts, 2000.
13. Ashley, H. and Landahl, M., *Aerodynamics of Wings and Bodies*, Dover Publications, Inc., New York, 1965.
14. Baals, D.D. and Corliss, W.R, *Wind Tunnels of NASA*, NASA SP-440, US Government Printing Office, Washington, DC, 1981.

15. Barlow, J.B., Rae, W.H., and Pope, A., *Low-Speed Wind Tunnel Testing*, 3rd edition, John Wiley & Sons, Inc., New York, 1999.

16. Bellman, D.R., "Lift and Drag Characteristics of the Bell X-5 Research Airplane at 59° Sweepback for Mach Numbers from 0.6 to 1.03," NACA RM L53A09c, February 17, 1953.

17. Beggs, G., *Spins in the Pitts Special,* self-published, Odessa, Texas, March 2001.

18. Brinkworth, B.J., "On the Early History of Spinning and Spin Research in the UK, Part 1: The Period 1901–1929," *Journal of Aeronautical History*, Paper No. 2014/03, 2014.

19. Campbell, J.F. and Chambers, J.R., *Patterns in the Sky: Natural Flow Visualization of Aircraft Flow Fields*, NASA SP-54, January 1994.

20. Chernyi, G.G., *Introduction to Hypersonic Flow*, Academic Press, New York, 1961.

21. Chung, T.J., *Computational Fluid Dynamics*, Cambridge University Press, Cambridge, UK, 2002.

22. Coe, P.L., "Review of Drag Cleanup Tests in Langley Full-Scale Tunnel (From 1935 to 1945) Applicable to Current General Aviation Airplanes," NASA TN D-8206, June 1976.

23. Cooper, G.E. and Rathert, G.A., Jr, "Visual Observations of the Shock Wave in Flight," NACA RM A8C25, May 24, 1948.

24. Corda, S., Stephenson, M.T., Burcham, F.W., and Curry, R.E., "Dynamic Ground Effects Flight Test of an F-15 Aircraft," NASA TM 4604, September 1994.

25. Cox, R.N. and Crabtree, L.F., *Elements of Hypersonic Aerodynamics*, The English Universities Press, Ltd., London, Great Britain, 1965.

26. Drela, M., "XFOIL: An Analysis and Design System for Low Reynolds Number Airfoils," in: T.J. Mueller (Ed.), *Low Reynolds Number Aerodynamics, in: Lecture Notes in Engineering*, Springer-Verlag, New York, 1989.

27. Ferri, A., "Completed Tabulation in the United States of Tests of 24 Airfoils at High Mach Numbers," NACA Report L5E21, June 1945.

28. Fillipone, A., *Aerospace Engineering Desk Reference*, Butterworth-Heinemann, Elsevier Inc., Oxford, UK, 1st edition, 2009.

29. Fisher, D.F., Haering, E.A., Noffz, G.K., and Aguilar, J.I., "Determination of Sun Angles for Observations of Shock Waves on a Transport Aircraft," NASA TM-1988-206551, September 1998.

30. Fisher, D.F., Del Frate, J.H., and Zuniga, F.A., "Summary of In-Flight Flow Visualization Obtained From the NASA High Alpha Research Vehicle," NASA TM-101734, January 1991.

31. Gallagher, G.L., Higgins, L.B., Khinoo, L.A., and Pierce, P.W., *Fixed Wing Performance*, USNTPS-FTM-NO. 108, US Naval Test Pilot School, Patuxent River, Maryland, September 30, 1992.

32. Gardner, John J., "Drag Measurements in Flight on the 10-Percent-Thick and 8-Percent-Thick Wing X-1 Airplanes," NACA RM L8K05, 1948.

33. Gibbs-Smith, C.H., *Sir George Cayley's Aeronautics, 1796–1855*, Her Majesty's Stationary Office, London, England, 1962.

34. Hansen, J.R., Editor, *The Wind and Beyond: A Documentary Journey into the History of Aerodynamics in America*, Vol. 1: The Ascent of the Airplane, National Aeronautics and Space Administration, NASA SP2003-4409, Washington, DC, 2003.

35. Hassell, J.L., Jr, Newsom, W.A., Jr, and Yip, L.P., "Full-Scale Wind Tunnel Investigation of the Advanced Technology Light Twin-Engine Airplane (ATLIT)," NASA TP 1591, May 1980.

36. Hoerner, S.F. and Borst, H.V., *Fluid Dynamic Lift*, self-published, Brick Town, New Jersey, 1965.

37. Hoerner, S.F., *Fluid Dynamic Drag*, 2nd edition, self-published, Midland Park, New Jersey, 1985.

38. Hopkins, E.J., Fetterman, D.E., Jr, and Saltzman, E.J., "Comparison of Full-Scale Lift and Drag Characteristics of the X-15 Airplane with Wind-Tunnel Results and Theory," NASA TM-X-713, March 1962.

39. Hurt, H.H., Jr, *Aerodynamics for Naval Aviators*, US Navy NAVWEPS 00-80 T-80, US Government Printing Office, Washington, DC, January 1965.

40. Jacobs, E.N., Ward, K.E., and Pinkerton, R.M., "The Characteristics of 78 Related Airfoil Sections From Tests in the Variable-Density Wind Tunnel," National Advisory Committee for Aeronautics, NACA Report No. 460, 1935.

41. Jenkins, D.R., "Hypersonics Before the Shuttle: A Concise History of the X-15 Research Airplane," Monographs in Aerospace History, NASA Publication SP-2000-4518, Washington, DC, June 2000.

42. John, J.E.A., *Gas Dynamics*, Allyn and Bacon, Inc., Boston, Massachusetts, 1969.

43. Jones, R.T., "Theory of Wing-Body Drag at Supersonic Speeds," NACA Technical Report 1284, July 8, 1953.

44. Karamcheti, K., *Principles of Ideal-Fluid Aerodynamics*, 2nd edition, Robert E. Krieger Publishing Company, Huntington, New York, 1980.

45. Kays, W.M., *Convective Heat and Mass Transfer*, McGraw-Hill Book Company, New York, 1966.

46. Kuethe, A.M. and Chow, C.Y., *Foundations of Aerodynamics: Bases of Aerodynamic Design*, 5th edition, John Wiley & Sons, Inc., New York, 1998.

47. Kuchemann, F.R.S., *The Aerodynamic Design of Aircraft*, 1st edition, Pergamon Press, Oxford, U.K., 1978.

48. Lan, C.T.E. and Roskam, J., *Airplane Aerodynamics and Performance*, Design, Analysis, and Research Corporation (DARcorporation), Lawrence, Kansas, 2003.

49. Lilienthal, O., *Birdflight as the Basis of Aviation*, unabridged facsimile of original work, first published in 1889, translated by A.W. Isenthal, Markowski International, Hummelstown, Pennsylvania, 2001.

50. Loftin, L.K., *Quest for Performance: The Evolution of Modern Aircraft*, NASA SP-468, US Government Printing Office, Washington, DC, 1985.

51. McCormick, B.W., *Aerodynamics, Aeronautics, and Flight Mechanics*, John Wiley & Sons, New York, 1979.

52. Moulden, T.H., *Fundamentals of Transonic Flow*, Krieger Publishing Company, Malabar, Florida, 1991.

53. Nicolai, L.M., *Fundamentals of Aircraft Design*, self-published, METS, Inc., San Jose, California, 1984.

54. Niewald, P.W., and Parker, S.L, "Flight-Test Techniques Employed to Successfully Verify F/A-18E In-Flight Lift and Drag," *Journal of Aircraft*, Vol. 37, No. 2, March-April 2000, pp. 194–200.

55. Nissen, J.M., Burnett, L.G., and Hamilton, W.T., "Correlation of the Drag Characteristics of a P-51B Airplane Obtained from High-Speed Wind Tunnel and Flight Tests," NACA ACR No. 4 K02, February 1945.

56. Nonweiller, G.T., "Aerodynamic Problems of Manned Space Vehicles," *Journal of the Royal Aeronautical Society*, Vol. 63, 1959, pp. 521–528.

57. Oswald, W.B., "General Formulas and Charts for the Calculation of Airplane Performance," NACA-TR-408, 1932.

58. Prandtl, L. and Tietjens, O.G., *Applied Hydro and Aeromechanics*, Dover Publications, Inc., New York, 1934.

59. Raymer, D.P., *Aircraft Design: A Conceptual Approach*, AIAA Education Series, American Institute of Aeronautics and Astronautics, Washington, DC, 2nd edition, 1992.

60. Savile, D.B.O., "Adaptive Evolution of the Avian Wing," *Evolution*, Vol. 11, 1957, pp. 212–224.

61. Shapiro, A.H., *The Dynamics and Thermodynamics of Compressible Fluid Flow*, Vol. 1, John Wiley & Sons, New York, 1953.

62. Sheldahl, R.E. and Klimas, P.C., "Aerodynamic Characteristics of Seven Airfoil Sections Through 180-Degree Angle-of-attack for Use in Aerodynamic Analysis of Vertical Axis Wind Turbines," Sandia National Laboratories Energy Report SAND80-2114, March 1981.

63. "Stall/Post-Stall/Spin Flight Test Demonstration Requirements for Airplanes," US Air Force, MIL-S-83691A, April 15, 1972.

64. Stillwell, W.H., "X-15 Research Results," NASA SP-60, US Government Printing Office, Washington, DC, 1965.

65. Talay, T.A., *Introduction to the Aerodynamics of Flight*, NASA SP-367, US Government Printing Office, Washington, DC, 1975.

66. Tennekes, H., *The Simple Science of Flight: from Insects to Jumbo Jets*, The MIT Press, Cambridge, Massachusetts, 2009.

67. Truitt, R.W., *Hypersonic Aerodynamics*, The Ronald Press Company, New York, 1959.

68. Tsien, H.S., "Similarity Laws for Hypersonic Flows", *Journal of Mathematics and Physics*, vol. 25, 1946, pp. 247–251.

69. Van Ness, H.C., *Understanding Thermodynamics*, Dover Publications, New York, 1983.

70. Van Wylen, G.J. and Sonntag, R.E., *Fundamentals of Classical Thermodynamics*, 2nd edition, John Wiley and Sons, Inc., New York, 1978.

71. Vincenti, W.G., "Comparison Between Theory and Experiment for Wings at Supersonic Speeds," NACA TR-1033, Washington, DC, 1951.

72. Von Karman, T., *Aerodynamics: Selected Topics in the Light of Their Historical Development*, Dover Publications, Inc., New York, 2004.

73. Wetmore, J.W. and Turner, L.I., Jr, "Determination of Ground Effect from Tests of a Glider in Towed Flight," National Advisory Committee for Aeronautics, NACA Report No. 695, 1940.

74. Whitcomb, R.T., "Recent Results Pertaining to the Application of the Area Rule," NACA RM L53I15a, October 28, 1953.

75. Whitcomb, R.T., "Review of NASA Supercritical Airfoils," ICAS Paper No. 74-10, presented at the 9th Congress of the International Council of Aeronautical Sciences, Haifa, Israel, August 25–30, 1974.

76. Whitcomb, R.T., "A Study of the Zero-Lift Drag-Rise Characteristics of Wing-Body Combinations Near the Speed of Sound," National Advisory Committee for Aeronautics, NACA Report 1273, 1956.

77. White, F.M., *Viscous Fluid Flow*, McGraw-Hill, Inc., New York, 1974.

78. Whittliff, C.E. and Curtis, J.T., "Normal Shock Wave Parameters in Equilibrium Air," Cornell Aeronautical Laboratory, Inc., Report No. CAL-111, Cornell University, Buffalo, New York, November 1961.

79. Young, H.D. and Freedman, R.A., *University Physics*, 11th edition, Addison Wesley, San Francisco, California, 2004.

Problems

1. A Lockheed SR-71 *Blackbird* aircraft is flying at Mach 3.0 at an altitude of 80,000 ft. Calculate the Reynolds number per foot for this flight condition. Use Sutherland's Law to calculate the coefficient of viscosity.

2. The wing of a high altitude glider is instrumented to measure the surface pressure and temperature while in flight. At a point on the wing, the pressure and temperature are measured to be 2.58×10^4 N/m^2 and 217.5 K, respectively. Calculate the air density at this point on the wing.

3. Assume that the high altitude glider in Problem 2 is flying over Mars, where the atmosphere is primarily carbon dioxide. (The molecular weight of carbon dioxide is 44 kg/kg·mol.) Assume that the same measurements are obtained for the pressure and temperature at a point on the wing. Calculate the air density at this point on the wing in the Martian atmosphere, assuming the gas in the atmosphere is an ideal gas.

4. A cylindrical fuel tank, with a diameter of 3.20 ft and length of 6.58 ft, is filled with gaseous hydrogen at a pressure and temperature of 2504 psi and 62 °F, respectively. Calculate the number of moles in the tank.

5. A fuel tank is filled with hydrogen gas at a pressure of 2100 lb/ft^2 and temperature of 53 °F. Use the Van der Waals equation of state to determine if the use of the ideal gas equation of state is appropriate for hydrogen at this condition. The Van der Waals constants for hydrogen are

$$a = 25 \times 10^3 \, \frac{\text{N} \cdot \text{m}^4}{(\text{kg} \cdot \text{mol})^2}$$

$$b = 2.66 \times 10^{-2} \, \frac{\text{m}^3}{\text{kg} \cdot \text{mol}}$$

6. The Space Shuttle main landing gear tires were designed for landing speeds up to about 260 mph. The main landing gear tires were used for only a single Space Shuttle landing. The tires were filled with nitrogen gas and were inflated to a maximum pressure of 340 psi. The tires had to survive a wide temperature environment range, from -40 °F in space up to 130 °F at landing. Each tire had a diameter of 44.9 in and width of 16 in. The wheel, upon which the tire was mounted, had a diameter of 21 in and width of 16 in. Calculate the mass and weight of nitrogen gas in a tire at a pressure of 340 psi and temperature of 130 °F.

7. A straight rectangular wing is mounted in a subsonic wind tunnel at an angle-of-attack of 4°. The wingspan of 3 ft extends across the width of the tunnel test section. The wing has a NACA 0012 airfoil section with 9 in chord length. The tunnel test conditions are a freestream velocity of 200 mph, freestream pressure of 14.55 lb/in^2, and freestream temperature of 50 °F. A tunnel force balance measures the normal and axial forces on the wing. (The normal and axial forces are defined as the forces that are perpendicular and parallel to the chord line, respectively, with positive normal force pointing through the upper surface of the wing and positive axial force pointing towards the wing trailing edge.) At the test condition, the force balance measures a normal force of 104.14 lb and an axial force of -5.726 lb. Calculate the lift and drag coefficients at the test condition.

8. For the wind tunnel test conditions described in Problem 7, calculate the Reynolds number based on the wing chord. Would you expect the boundary layer on the wing to be laminar or turbulent? Calculate the boundary layer thickness at the trailing edge of the wing. Since the airfoil section is thin, assume that the distance on the wing upper surface, from the leading edge to the trailing edge, is equal to the chord length.

9. An airplane is flying at an altitude where the freestream pressure, p_∞, density, ρ_∞, and temperature, T_∞, are 4.372 lb/in^2, 0.02868 lb$_m$/ft^2, and -46.9 °F, respectively. The pressure, at

a point on the wing, p_{wing}, is measured to be 1132 lb/ft^2. Assuming isentropic flow, calculate the density and temperature at this point on the wing.

10. A rocket nozzle has an exit diameter, d, of 27.3 cm. The exhaust flow exiting the nozzle has a Mach number, M_e, of 2.3, temperature, T_e, of 787 K, and pressure, p_e, of 101,325 N/m^2. Assuming the exhaust flow is an ideal gas, calculate the nozzle exit mass flow rate.

11. A 1.75 in diameter fuel line supplies gasoline at a mass flow rate of 6.024 lb$_m$/s. Assuming a gasoline density of 0.026 lb$_m$/in^3, calculate the flow velocity in the fuel line.

12. A nozzle has an entrance area, A_i, of 1.50 m^2 and an exit area, A_e, of 4.25 m^2. The velocity at the nozzle entrance, V_i, is 12.3 m/s and the pressure at the exit, p_e, is 1.1 atm. Assuming incompressible flow of air at standard sea level conditions, calculate the nozzle exit velocity and the entrance pressure.

13. The velocity and total temperature in an inviscid, adiabatic flow are 238 m/s and 314 K, respectively. Calculate the static temperature and Mach number of the flow.

14. The wing of the North American AT-6A *Texan*, a World War II era advanced pilot trainer aircraft, has a NACA 2215 airfoil shape at the wing root and a NACA 4412 airfoil shape at the wingtip. "Decode" these NACA airfoil designations.

15. Using the coordinates given in the upper left corner of Figure 3.100, make a plot of the NACA 2412 airfoil shape approximately to scale. The x- and y-axes should be in units of percent of the airfoil chord.

16. Consider the data for the NACA 2412 airfoil section, shown in Figure 3.100. Identify: (1) the maximum lift-to-drag ratio, $(L/D)_{max}$, (2) the maximum lift coefficient, $C_{L,max}$, (3) the stall angle-of-attack, α_{stall}, (4) the minimum drag coefficient, $C_{D,min}$, (5) the minimum profile drag, $C_{D,0,min}$, and (6) the pitching moment coefficient, $C_{M,c/4}$, at zero degrees angle-of-attack.

17. A Lockheed U-2 reconnaissance aircraft has a wingspan of 103 ft. Calculate the percent reduction in the induced drag of the U-2 when it is flying 30 ft above the ground.

18. Calculate the lift coefficient and the moment coefficient, about the quarter chord point, for a NACA 4412 airfoil in a Mach 0.71 flow at an angle-of-attack of 8°.

19. Based on the Prandtl–Glauert compressibility correction, calculate the Mach number that results in a 10% change in the lift coefficient.

20. A common raven is soaring over the Southern California desert at a speed of 18.02 mph and an altitude of 500 ft. The raven's wing has an elliptical shape, span of 4.430 ft, and area of 3.150 ft^2. Calculate the lift coefficient and the induced drag coefficient (in drag counts) of a 2.020 lb raven at this flight condition. (The freestream density at an altitude of 500 ft is 0.0023423 slugs/ft^3.)

21. Calculate the percentage change in wing area that would be required to decrease the stall speed of an aircraft by 5%.

22. The maximum lift coefficient for a Northrop T-38A is 0.86 with flaps up and 1.03 with flaps full down. Calculate the stall speed of the T-38A, for these two flap configurations, at sea level for a weight of 12,000 lb. The T-38A wing reference area is 170 ft^2. Calculate the stall speed in units of ft/s and knots. Calculate the change in the stall speed due to the flaps, in knots and as a percentage change.

23. A skydiver is falling at a terminal velocity of 120 mph. Calculate the pressure, density, and temperature at the stagnation point of the skydiver at an altitude of 2000 m. (At an altitude of 2000 m, the freestream density, pressure, and temperature are 1.0066 kg/m^3 , 79,501 N/m^2, and 275.16 K, respectively.)

24. An aircraft performs an aeromodeling POPU maneuver to obtain aerodynamic data. At one point in the maneuver, the airspeed is 489.1 kt, the altitude is 27,000 ft, the flight path angle is 8.72°, the angle-of-attack is 5.26°, and the aircraft weight is 10,923 lb. An onboard accelerometer, aligned with the aircraft body axes, measures the axial and normal load factors as 0.07471

and -2.231, respectively. A thrust model is used to obtain a thrust of 1227.1 lb. The aircraft has a wing area of 220 ft^2. Calculate the lift and drag coefficients of the aircraft at this flight condition. (Assume a freestream density of 0.0009932 slug/ft^3).

25. An airplane is flying at an altitude of 1800 m and a Mach number 0.13. The pressure, measured at a point on the wing surface, is 79,887 N/m^2. Assuming inviscid flow, what is the Mach number at this point on the wing? If the flow were viscous, what would be the pressure and Mach number at this point on the surface of the wing? (Assume that the freestream pressure at an altitude of 1800 m, is 81,494 N/m^2.)

26. An aircraft is in supersonic flight at a high altitude, where the total temperature, $T_{t,\infty}$, is 867.1 °R. The aircraft's engine inlet has a circular cross-section with an entrance diameter of 3.1 ft. A normal shock wave is located at the inlet entrance. The Mach number upstream of the inlet, M_∞, is 2.47. The static pressure measured downstream of the normal shock wave, p_2, is 11.78 lb/in^2. Assuming an ideal gas, calculate the mass flow rate through the inlet, m, and the freestream dynamic pressure, q_∞.

27. A model of a hypersonic aircraft is being tested in a wind tunnel that uses helium as the test gas ($\gamma = 1.667$). At Mach 3, there is a normal shock wave in front of the blunt nose of the model. For this Mach 3 test condition, calculate the Mach number, pressure, and temperature behind the normal shock wave, assuming a sea level pressure and temperature of 101,325 N/m^2 and 288 K, respectively.

28. An airplane is flying in the Martian atmosphere at a Mach number of 3.2 and an altitude of 50,000 m. At this Mach number, there is a normal shock in front of the airplane's nose. Calculate the Mach number, pressure, and temperature behind the normal shock wave. Assume that the Martian atmosphere is a perfect gas composed of carbon dioxide (constants at 1 atm and 293.15 K may be used). The pressure and temperature of the Martian atmosphere can be modeled using the following equations

$$p(h) = 0.699e^{-0.00009h}$$

$$T(h) = -0.00222h + 249.75$$

where h is the altitude in meters, p is the pressure in kN/m^2, and T is the temperature in K.

29. The Lockheed NF-104A was a modified F-104 with a small liquid-fueled rocket engine installed to enable high altitude, high Mach number flight. Assume that the NF-104A wing is approximated by a thin, flat plate, with a reference area of 212.8 ft^2. The NF-104A is flying in level flight at an altitude of 52,000 ft, where the atmospheric pressure is 221.38 lb/ft^2. If the NF-104A has a weight of 19,000 lb, calculate and plot the aircraft angle-of-attack as a function of Mach number, for Mach numbers between 1.0 and 2.5. Assume that the wing lift coefficient has the same value as that for the airfoil section. (The actual NF-104A wing had a biconvex airfoil shape and a maximum Mach number of 2.2.)

30. An airplane has a wing with a sweepback angle, Λ, of 20°. If the airplane is flying at sea level and an airspeed of 285.5 m/s, what is the Mach number perpendicular to the leading edge of the wing? Comment on the flow over the wing at this flight condition.

31. A supersonic wind tunnel is designed to produce a Mach 2.5 flow at the test section entrance. The wind tunnel nozzle has an exit diameter, d_e, of 18 in. The wind tunnel reservoir pressure, p_r, and temperature, T_r, are 3788.5 lb/ft^2 and 877.5 °R, respectively. Calculate the velocity at the test section entrance (nozzle exit), the mass flow through the nozzle, and the nozzle throat diameter. Assume a perfect gas and isentropic flow throughout the wind tunnel.

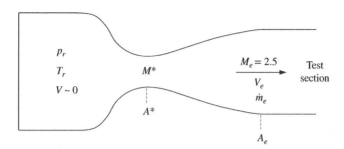

32. Helium gas is flowing through a 2.25 in diameter tube with a mass flow rate of 0.378 lb_m/s. The pressure and temperature of the flowing helium is 80 psi and 52 °F, respectively. Calculate the velocity of the flowing helium gas.

33. An automobile is driving at a velocity of 70 mph at sea level. The roof of a car is 6 ft long. Plot the laminar and turbulent boundary layer thicknesses (in inches) on the car roof, assuming the roof is approximated by a flat plate. Calculate the laminar and turbulent boundary layer thicknesses at the trailing edge of the roof.

34. A sailplane has a wingspan of 75 ft and an aspect ratio of 25.5. Calculate the skin friction drag on the wing, assuming it is a rectangular flat plate, at an airspeed of 100 knots at sea level and at an altitude of 100,000 ft ($\rho = 3.318 \times 10^{-5}$ slug/ft^3, $T = 408.8$ °R).

35. The nose of a hypersonic missile has the shape of a slender, right-circular cone, with a semi-apex angle of 2.5°. The missile is in Mach 8 flight, at 80,000 ft (p = 58.51 lb/ft^2), and an angle-of-attack of 15°, as measured between the freestream flow direction and the centerline of the missile. Using Newtonian theory, calculate the lift, drag, and lift-to-drag ratio of the missile nose, assuming the reference area of the nose cone is 65 ft^2.

36. The Lockheed SR-71 *Blackbird* has a fuselage length of 102.25 ft. Assuming a flat plate length equal to this fuselage length, calculate and plot the heat flux on the plate at 80,000 ft ($\rho = 8.571 \times 10^{-5}$ slug/ft^3, $T = 397.9$ °R), for Mach numbers from 1 to 3. Calculate and plot the heat flux in units of Btu/(s·ft^2).

37. The North American X-15 hypersonic, rocket-powered airplane had a simple wedge-shaped upper and lower vertical tail. The wedge tail had a half-angle of 3° and a chord length of 10.25 ft. Assume that the X-15 is flying at an altitude of 100,000 ft and a dynamic pressure of 1500 lb/ft^2. Using simple Newtonian flow, calculate the coefficient of pressure on the left and right sides of the wedge tail, if the flow upstream of the tail has an angle-of-sideslip of 2° (the flow is coming into the pilot's left ear). Write out the equations for the pressure coefficients before solving for numerical values. Also, first calculate the freestream Mach number at this flight condition to verify that Newtonian flow is applicable. (Assume that the freestream pressure, p_∞, at 100,000 ft is 23.085 lb/ft^2.)

4

Propulsion

Artist's concept of the first flight of the X-43A hypersonic, air-breathing scramjet-powered aircraft.[1] (Source: *NASA*.)

4.1 Introduction

Many of the significant advancements in aerospace vehicles have been paced by advancements in propulsion. As propulsive systems have enabled aerospace vehicles to fly faster and higher, new vehicle configurations and technologies are required to fly at these expanded boundaries of the flight envelope. From this perspective, the breakthroughs in propulsion have driven the design evolution of aerospace vehicles. For aircraft, this is illustrated by the advancement from propeller-driven aircraft to jet powered aircraft, where the great increases in airspeed have led to significant aircraft design advances, such as the swept wing and the all-moving horizontal tail. Advancements in rocket

---

[1] The NASA X-43A was an unmanned, hypersonic research vehicle powered by a hydrogen-fueled, air-breathing, supersonic combustion ramjet or scramjet engine. Three X-43A vehicles were constructed with two reaching hypersonic speeds under air-breathing scramjet power. The second vehicle achieved Mach 6.8 on 27 March 2004 and the third vehicle reached Mach 9.6 on 16 November 2004. The development of air-breathing scramjet propulsion has pushed the boundaries of high-speed flight in the sensible atmosphere, with the promise of airplane-like access to space.

Introduction to Aerospace Engineering with a Flight Test Perspective, First Edition. Stephen Corda.
© 2017 John Wiley & Sons Ltd. Published 2017 by John Wiley & Sons Ltd.
Companion Website: www.wiley.com/go/corda/aerospace_engg_flight_test_persp

propulsion have enabled us to launch vehicles into Earth orbit and beyond, to the Moon and planets. Novel space propulsion, such as solar and ion propulsion, may enable long duration journeys into deep space. In addition to the performance benefits, advancements in propulsion have also led to significant increases in efficiency, improvements in reliability, and reductions in aircraft noise. Advancements in propulsion have turned aircraft into "time machines", with the ability to shrink the time it takes to travel across the globe. In the 1800s, it might have taken a year or more for a covered wagon to cross the United States. Now, we can routinely fly from coast to coast in five or six hours.

The study of propulsion relies heavily on the fundamental areas of aerodynamics, thermodynamics, mechanics, and chemistry. Many propulsion systems involve the flow of gases or liquids through ducts with area variation. These *internal aerodynamic flows* are often complicated by the effects of friction (boundary layers) and heat transfer. The internal aerodynamic flows may be subsonic or supersonic, which may involve compressibility effects and shock waves. In most propulsion concepts, there is a conversion of internal energy of a fuel source to produce thrust, conversions that involve chemical reactions, so propulsive flows may include the effects of chemistry.

In this chapter, we investigate several different types of propulsive devices, which are separated into *air-breathing* and *non-air-breathing* engines. Examples of air-breathing engines include the internal combustion engine, ramjet, turboprop, turbojet, and turbofan engines. These types of engines ingest or "breathe" the atmospheric air, thereby supplying the oxidizer to combust with a chemical fuel. Different types of rockets, including solid and liquid fuel, electric, solar, and nuclear rockets, are examples of non-air-breathing propulsion. These devices do not rely on atmospheric air to produce thrust. The solid and liquid rockets, which rely on chemical combustion, carry their oxidizer and fuel with them. Since non-air-breathing engines do not need atmospheric air to produce thrust, they are able to operate outside of the atmosphere, including in outer space. Each propulsion type has its own suitable application or "niche" in the flight envelope, along with its own particular limitations.

4.1.1   The Concept of Propulsive Thrust

This section serves as a brief, somewhat qualitative, introduction to the concept of propulsive thrust. While touching on the laws of Newton's mechanics and the basics of classical thermodynamics, it is left for later sections to derive the mathematical and quantitative details of propulsive thrust.

The concept of propulsive thrust was certainly *demonstrated* well before the underlying physics was *understood*. One example of this was the *aeolipile*, shown in figure 4.1, sometimes called a *hero engine* after its inventor, hero of alexandria, a 1st century AD Greek mathematician and engineer. The aeolipile was composed of a hollow sphere mounted above a cauldron of heated water. The steam flowed upward into the sphere and was exhausted through L-shaped tubes, causing the sphere to spin. This rudimentary steam-turbine device relied on the force reaction concept of propulsion, but the underlying physics of how it worked was not understood in the time of Hero.

Let us now explore the concept of propulsive thrust using a simple example. Consider a rigid sphere of constant volume, as shown on the left in Figure 4.2, filled with a gas at a pressure, p_0, that is greater than the external, ambient pressure, p_a. Since the forces are balanced in all directions around the sphere, it is in equilibrium and remains stationary. If we were to open a hole in the right side of the sphere, as shown on the right in Figure 4.2, the high pressure gas would start to escape. Let us assume that a gas supply tube is connected to the sphere, such that it fills the sphere with gas as fast as the gas is escaping from the hole. It is also assumed that the connection of the gas supply

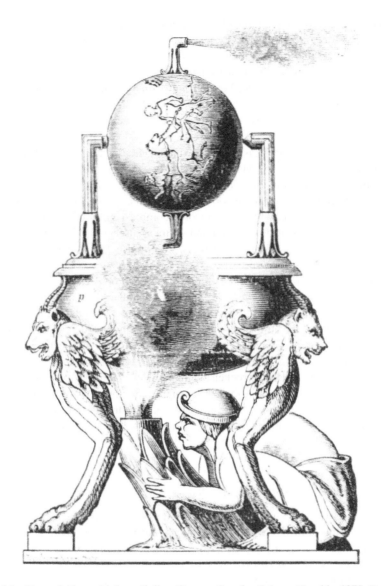

Figure 4.1   Hero of Alexandria's aeolipile. (Source: *Popular Science Monthly, 1878, D-old-70.*)

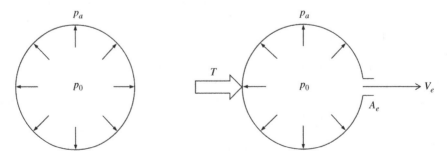

Figure 4.2   Example of fluid momentum and reactive force, a rigid sphere filled with a high pressure gas in equilibrium (left) and a rigid sphere with escaping high pressure gas and steady thrust (right).

line does not interfere with the movement of the sphere. By adding gas to the sphere to make up for the escaping gas, the gas pressure inside the sphere is maintained at a constant pressure, p_0. In this steady-state condition, the gas escapes from the hole with a constant velocity, V_e.

With the high pressure gas escaping from the exit hole in the sphere, the force, T, required to keep the sphere from moving to the left is

$$T = (p_0 - p_a)A_e \tag{4.1}$$

where A_e is the area of the exit hole. Hence, the force or thrust, T, is equal to the pressure difference multiplied by the area over which the pressure difference acts.

Using Newton's second law, this force is equal to the rate of change of the momentum of the escaping gas, such that

$$T = (p_0 - p_a)A_e = \frac{d}{dt}(mV)_e = \left(\frac{dm_e}{dt}\right)V_e = \dot{m}_e V_e \tag{4.2}$$

where m_e is the mass of gas escaping and $\dot{m}_e$ is the mass flow rate of escaping gas. The velocity of the escaping gas, V_e, is constant, so that the time derivative acts only on the mass term in Equation (4.2). Hence, the force or thrust, T, generated by the escaping gas is equal to the mass flow rate of the escaping gas, $\dot{m}_e$, multiplied by the exhaust velocity, V_e.

Fundamentally, both air-breathing and non-air-breathing engines are based on this same physical principle of propulsion, that is, a force is imparted to the propulsive device, and hence, the vehicle, due to the momentum of the fluid or matter that is exhausted or ejected from the engine. The force is the result of the integrated pressure and shear stress distributions in the direction of the vehicle motion and is called the *thrust*. Usually, it is not practical to derive the thrust by literally integrating the distributions of pressure and shear stress inside and outside an engine, due to the geometrical complexity. This force-momentum perspective of thrust proves to be very useful in deriving the engine thrust.

Figure 4.3 shows the application of Equation (4.2) to three fundamentally different types of propulsive devices: the propeller, the air-breathing ducted engine (ramjet or turbojet), and the non-air-breathing rocket engine. In each case, the propulsive device increases the momentum of

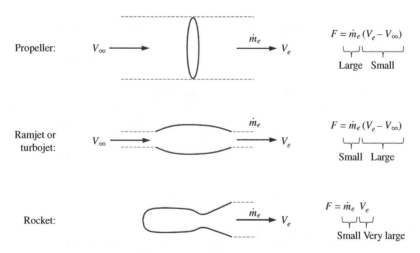

Figure 4.3 Fluid momentum-reactive force principle applied to different types of propulsion. (Source: *Adapted from J.L. Sloop, "Liquid Hydrogen as a Propulsion Fuel, 1945–1959" NASA SP-4404, 1978.*)

Table 4.1 Characteristics of different types of chemical propulsion.

Propulsion type	Engine type	Working fluid	Chemical energy source Chemical combustion of:
Air-breathing	Internal combustion engine + propeller	air	air + fuel
	Turboprop	air	air + fuel
	Turbojet and turbofan	air	air + fuel
	Ramjet and scramjet	air + fuel	air + fuel
	Air-breathing rocket	air + liquid or solid fuel	air + liquid or solid fuel
Non-air-breathing	Liquid rocket	liquid propellants	liquid propellants
	Solid rocket	solid propellant	solid propellant
	Hybrid rocket	hybrid propellant	hybrid propellant

a fluid, resulting in a propulsive force. The fluid is air in the case of the propeller, air mixed with combustion gases for the ramjet or turbojet, and only combustion gases for the rocket. For the air-breathing devices, the momentum involves the change in velocity between the exhaust velocity and the freestream velocity, V_e-V_∞.

The relative magnitudes of the fluid mass flow rate and the fluid exhaust velocity are different for each type of propulsion. A large streamtube of air, with a freestream velocity, V_∞, enters the propeller area and exits with a slightly increased velocity, V_e, hence the air mass flow rate, $\dot{m}_e$, is large and the velocity change, $(V_e$-$V_\infty)$, imparted to the air in the streamtube is small. For the ramjet or turbojet, the air mass flow rate, $\dot{m}_e$, is small while the velocity change, $(V_e$-$V_\infty)$, is large. The rocket has a small mass flow rate, $\dot{m}_e$, and a very large exhaust velocity, V_e.

The propulsive device increases the momentum of a fluid, either through mechanical or thermo-chemical means. The fluid may be the surrounding air going through a propeller or ingested into a ramjet or turbojet. The fluid may be the propellants carried onboard a rocket that are burned and expelled. The energy stored in the propellants is converted to the momentum of the exhaust gases. From a thermodynamic perspective, the engine does work on the fluid, hence it is call the *working fluid* of the propulsive device. Energy is added to the working fluid either mechanically, as in the case of the propeller, or thermally through the release of chemical energy, as in the cases of the ramjet, turbojet, or rocket. Table 4.1 provides a summary of the characteristics of some of the different types of chemical propulsion, defining the working fluid and the chemical energy source.

In the analysis of propulsive flows, the use of inconsistent English units, such as the pound mass, lb_m, the British thermal unit, Btu, and horsepower, hp, is still quite commonplace. Some of these inconsistent units and their conversion into consistent units are discussed below.

Work is defined as a force multiplied by a distance. The SI unit of work, the joule, J, is defined as a force of one newton multiplied by a distance of one meter. The joule is defined in terms of SI base units as

$$1\,\mathrm{J} = 1\,\mathrm{N} \cdot \mathrm{m} = 1\,\frac{\mathrm{kg} \cdot \mathrm{m}^2}{\mathrm{s}^2} \qquad (4.3)$$

In English units, work is defined as a force of one pound multiplied by a distance of one foot. Energy and work have the same units, hence, the units of energy and work are sometimes expressed in the inconsistent English units of the British thermal unit, Btu. The Btu is defined as the amount of energy required to raise the temperature of one pound of water by one degree Fahrenheit. The unit conversions for the Btu are as follows.

$$1\,\text{Btu} = 778\,\text{ft} \cdot \text{lb} = 1055\,\text{J} \tag{4.4}$$

Power is defined as the rate of doing work. The SI unit of power, the watt, W, is defined as one joule per second. In terms of SI base units, the watt is given by

$$1\,\text{W} = 1\,\frac{\text{J}}{\text{s}} = 1\,\frac{\text{N} \cdot \text{m}}{\text{s}} = 1\,\frac{\text{kg} \cdot \text{m}^2}{\text{s}^3} \tag{4.5}$$

Be careful not to confuse the watt, W, with the symbol for work, W.

In the English system, the inconsistent units of horsepower, hp, and British thermal units per hour, Btu/h, are often used. The conversions for power in these units is given as

$$1\,\text{hp} = 550\,\frac{\text{ft} \cdot \text{lb}}{\text{s}} = 2546.7\,\frac{\text{Btu}}{\text{h}} = 745.7\,\text{W} \tag{4.6}$$

4.1.2 Engine Station Numbering

In propulsion, a numbering system has been developed to identify *stations* in the flow direction through an engine. There are some generally accepted conventions for station numbering, but one should always ensure that the numbering scheme being applied is understood. Typical station numbering is shown in Figure 4.4 for air-breathing and non-air-breathing engines.

The station number designates the flow exit or entrance to a specific engine component. We first describe the station numbering for air-breathing engines as shown in Figure 4.4a, b, and c. The inlet or diffuser entrance and exit are designated as stations 1 and 2, respectively. For engines with turbomachinery, such as the turbojet, station 2 is also the entrance to the compressor section and station 3 is the exit to the compressor, as shown in Figure 4.4b and c. For engines without turbomachinery, such as the ramjet, station 2 designates the location where fuel is injected and station 3 is the flameholder location, as shown in Figure 4.4a. The entrance and exit to the combustor are denoted as stations 3 and 4, respectively, for both types of air-breathing engines. For the ramjet, station 4 is the entrance to the nozzle, station 8 is the nozzle throat, and station 9 is the nozzle exit. For the turbojet, stations 4 and 5 are the entrance and exit to the turbine section, respectively. As shown in Figure 4.4b, if the turbojet does not have an afterburner, the flow exits the turbine and enters the convergent nozzle at station 5 and exits the nozzle at station 9. Several station numbers have been omitted in this case so that the nozzle exit station is the same for all of the air-breathing engines. As shown in Figure 4.4c, the turbojet with an afterburner has fuel injection and flameholding sections between stations 5 and 6. The afterburner combustion zone is between stations 6 and 7. Station 7 is the entrance to the convergent-divergent nozzle, station 8 is the nozzle throat, and station 9 is the nozzle exit.

The station numbering scheme for the non-air-breathing rocket is shown in Figure 4.4d. The rocket engine propellants are injected into the combustion chamber at station i. The combustion chamber, including the entrance to the convergent-divergent nozzle is station c. The nozzle throat is station th and the nozzle exit is station e.

Station numbers are also used as subscripts to identify the local flow properties throughout the engine. For example, T_3 is the temperature at the exit of a turbojet compressor or at the entrance to its combustor and p_c is the pressure in a rocket engine combustor. In the text, we follow the engine station numbering as given in Figure 4.4 for engine locations and for designating flow properties through an engine.

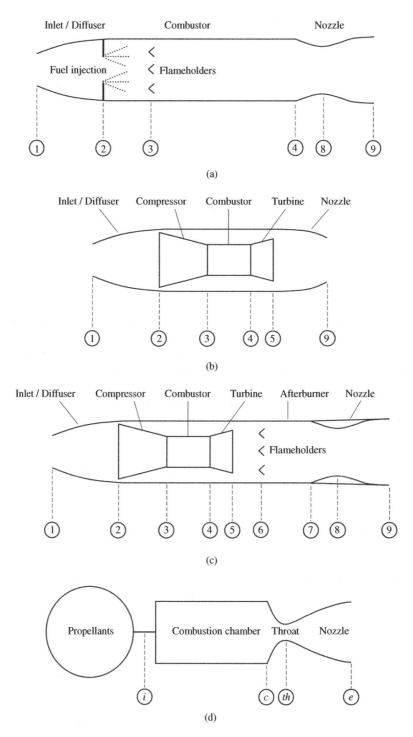

Figure 4.4 Engine station numbering; (a) ramjet, (b) turbojet engine without an afterburner, (c) turbojet with an afterburner, and (d) rocket.

4.2 Propulsive Flows with Heat Addition and Work

In Chapter 3, we derived the governing equations of fluid motion, obtaining the continuity, momentum, and energy equations. These fundamental equations apply to the flow through propulsive devices, such as jet engines and rockets. In this section, we examine the application of the continuity and energy equations to the flow through a propulsive device. In the next section, the thrust produced by various propulsive devices is obtained through the application of Newton's second law.

Consider the propulsive device shown in Figure 4.5, comprising an air inlet, a fuel inflow, and an exhaust exit. A mass flow rate of air, $\dot{m}_i$, enters the device through the area A_i. A mass flow rate of fuel, $\dot{m}_f$, is injected into the device, mixed with the air, and burned. The mass flow rate of combustion products, $\dot{m}_e$, is exhausted from the device through the area A_e.

Applying the principle of conservation of mass to this propulsive device, the mass flow rates into and out of the device are given by

$$\dot{m}_i + \dot{m}_f = \dot{m}_e \tag{4.7}$$

$$\rho_i V_i A_i + \dot{m}_f = \rho_e V_e A_e \tag{4.8}$$

where ρ_i and V_i are the inflow density and velocity, respectively, and ρ_e and V_e are the exhaust density and velocity, respectively.

Consider now the propulsive device as a thermodynamic system, as shown in Figure 4.6. The flow enters through the boundaries of the control system (shown as dotted lines in the figure) at velocity V_i and temperature T_i. The combustion products exit at velocity V_e and temperature T_e. Heat per unit mass, δq, is added to the device by the burning of fuel. If there is a turbine (or propeller) in the device, an amount of work per unit mass, δw_t, is done by the propulsive system.

Applying the first law of thermodynamics, in the form of Equation (3.94), to the system, we have

$$\delta q + \delta w_t = dh - v dp \tag{4.9}$$

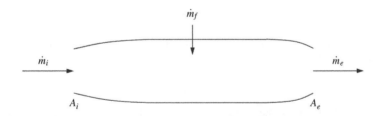

Figure 4.5 Conservation of mass applied to a propulsive device.

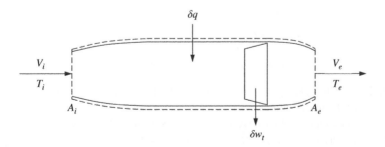

Figure 4.6 Conservation of energy applied to a propulsive device.

Substituting Euler's equation, Equation (3.165), for the change in pressure, dp, we have

$$\delta q + \delta w_t = dh - v(-\rho V dV) = dh - \frac{1}{\rho}(-\rho V dV) = dh + V dV \qquad (4.10)$$

Integrating from the inlet to the exit of the device, we have

$$\int_i^e \delta q + \int_i^e \delta w_t = \int_i^e dh + \int_i^i V dV \qquad (4.11)$$

$$q + w_t = (h_e - h_i) + \frac{1}{2}(V_e^2 - V_i^2) \qquad (4.12)$$

$$q + w_t = \left(h + \frac{V^2}{2}\right)_e - \left(h + \frac{V^2}{2}\right)_i = h_{t,e} - h_{t,i} \qquad (4.13)$$

where q is the total heat added per unit mass through the device, and w is the total work per unit mass delivered by the device. Equation (4.13) states that the heat addition and work done by the propulsive device are due to the difference in the total enthalpies of the flow exiting and entering the device.

Assuming constant specific heats, the enthalpy terms in Equation (4.13) can be rewritten in terms of temperature, yielding

$$q + w_t = \left(c_p T + \frac{V^2}{2}\right)_e - \left(c_p T + \frac{V^2}{2}\right)_i = c_p(T_{t,e} - T_{t,i}) \qquad (4.14)$$

Hence, we see that the heat addition and work done by the propulsive device are due to the difference in the total temperatures of the flow exiting and entering the device. Equation (4.14) is the same energy equation that was derived for fluid flow, Equation (3.191), with the addition of a heat term for a non-adiabatic process and a work term for a work-producing turbine or propeller.

Let us examine the work delivered by the turbine or propeller. The work per unit mass is simply

$$w_t = \frac{W}{m} \qquad (4.15)$$

where W is the work delivered by the turbine or propeller and m is the mass. If we divide the work and mass by time, t, we have

$$w_t = \frac{W}{m} = \frac{W/t}{m/t} = \frac{P}{\dot{m}} \qquad (4.16)$$

where $\dot{m}$ is the mass flow rate through the turbine or propeller and P is the power supplied by the turbine or propeller associated with the work, W, performed. Using Equations (4.13) and (4.14), we can write the power as

$$P = \dot{m}(h_{t,e} - h_{t,i}) - q = \dot{m}c_p(T_{t,e} - T_{t,i}) - q \qquad (4.17)$$

Assuming an adiabatic process ($q = 0$), this becomes

$$P = \dot{m}(h_{t,e} - h_{t,i}) = \dot{m}c_p(T_{t,e} - T_{t,i}) \qquad (4.18)$$

Hence, the turbine or propeller power is a function of the mass flow rate and the difference in the total enthalpies or total temperatures.

The concepts and equations that have been developed for flow through a complete propulsive device, such as a jet engine, are applicable for the flow through a component of the engine. Hence, these equations may be applied to the flow through a component, such as a compressor, combustor, or turbine, as illustrated in the example below.

Example 4.1 Energy Equation Applied to a Jet Engine Combustor *Air enters the combustor of a jet engine at a velocity of 326 ft/s and a temperature of 1155°R. Fuel is burned in the combustor, adding heat per unit mass of 252.7 Btu/lb$_m$ to the flow. The combustion products exit the combustor at a temperature of 2205°R. Calculate the velocity of the combustion products exiting the combustor, assuming a constant specific heat of 6020 ft·lb/(slug·°R).*

Solution

Convert the heat per unit mass added into consistent English units.

$$q = 252.7 \frac{Btu}{lb_m} \times \frac{778 ft \cdot lb}{1 Btu} \times \frac{32.2 lb_m}{1 slug} = 6.331 \times 10^6 \frac{ft \cdot lb}{slug}$$

Using Equation (4.14), we have

$$q + w_t = \left(c_p T + \frac{V^2}{2} \right)_e - \left(c_p T + \frac{V^2}{2} \right)_i$$

Solving for the exit velocity, we have

$$V_e = \sqrt{2 \left[q + w_t + c_p \left(T_i - T_e \right) + \frac{V_i^2}{2} \right]}$$

$$V_e = \sqrt{2 \left[6.331 \times 10^6 \frac{ft \cdot lb}{slug} + 0 + \left(6020 \frac{ft \cdot lb}{slug \cdot °R} \right) (1155°R - 2205°R) + \frac{\left(326 \frac{ft}{s} \right)^2}{2} \right]} = 355 \frac{ft}{s}$$

4.3 Derivation of the Thrust Equations

In this section, the thrust equations are derived for a rocket engine, an air-breathing engine, and a propeller. The derivation for the air-breathing engine applies to a ramjet and turbojet engine, as the development does not differentiate as to what is inside the air-breathing engine, that is, whether there is turbomachinery or other mechanical components inside the engine. These fundamental equations identify the main contributors to the thrust for each type of propulsion and provide insight into the differences between the different types of propulsion.

The rocket and air-breathing engine are modeled as mounted on a rigid pylon in a freestream flow. This model could represent the engine mounted on an aircraft wing pylon in flight or mounted on a sting support in a wind tunnel flow. In deriving the thrust equation, the pylon provides a means of extracting the thrust reaction force.

A control volume approach is used to derive the thrust equations, where a control volume surrounds the engine. The momentum equation of fluid flow is applied in the thrust direction and the forces and momentum fluxes are identified. Steady flow is assumed, where flow properties are constant with time, although they may vary from point to point in space. The time derivative terms in the momentum equation can therefore be neglected with the assumption of steady flow. The force terms can be surface forces or body forces. The surface forces are those due to pressures and viscous shear stresses that act over the control surface, surrounding the control volume. The viscous forces are small compared with the pressure forces, hence, they are ignored. The body forces act over the mass or volume of the fluid and may be due to gravitational or electromagnetic sources. Since the contribution of the body forces is usually negligible for the propulsive flows that are of interest to us, they are also ignored. The momentum flux terms in the momentum equation reduce to terms where a constant mass flow rate is multiplied by a velocity, due to the basic assumption of one-dimensional flow, where the fluid density and velocity are constant over a given cross-sectional area.

The fundamental momentum equation for steady, inviscid flow was discussed in Section 3.6. Here, we introduce an integral form of the momentum equation, which is still an expression of Newton's second law. The x-component of this form of the momentum equation for steady, inviscid flow is given by

$$\sum F_x = \int_S \rho u_x (\vec{u} \cdot \hat{n})\, dS \qquad (4.19)$$

where F_x is the force component in the x-direction, ρ is the fluid density, $\vec{u}$ is the fluid velocity, u_x is the x-component of the fluid velocity, and $\hat{n}$ is the unit normal to the control surface, S. The left-hand side of Equation (4.19) is the sum of the forces in the x-direction and the right-hand side of Equation (4.19) is the net rate of flow of momentum or momentum flux, in the x-direction, out through the control surface.

The approach described above derives the *uninstalled thrust* of the engine, that is, the thrust of the engine without any *installation losses* due to the engine being installed on a vehicle. These installed losses are specific to the type of engine installation, for example, for a wing-mounted, podded engine or for an engine buried in an aircraft fuselage. The uninstalled thrust is the same for any engine whereas the installed thrust varies with the particular engine installation.

4.3.1   Force Accounting

In addition to the thrust force generated by a propulsive device, a drag force acts in a direction opposing the thrust. Similar to the thrust, the drag is the result of the integrated pressure and shear stress distributions in the direction opposite to the vehicle motion. If the thrust is greater than the drag, the vehicle accelerates. If the thrust is equal to the drag, the vehicle is in an equilibrium state and in motion at a constant velocity. If the thrust is less than the drag, the vehicle decelerates. The difference between the thrust and drag plays a critical role in defining the vehicle performance.

The distinction between thrust and drag is not always so clear. It may not suffice to say that the thrust is the resultant of the forces due to the flow *inside* an engine and that the drag is the resultant of the forces due to the flow *external* to the engine and vehicle. There are instances when the propulsive flow acts on the same surfaces that are acted upon by the external flow, so that the distinction between thrust and drag may be unclear. This issue is exacerbated when the propulsion system is highly *integrated* with the vehicle external shape, such as the case with hypersonic vehicles.

In principle, it does not really matter whether a force is called "thrust" or "drag", as long as it is accounted for properly in summing up all of the forces. By defining a *force accounting* system, we can be sure to properly account for all of the forces and ensure that this is done in an accurate and consistent manner. Usually, during the design of a new vehicle, a company establishes a force accounting or thrust-drag "bookkeeping" system. This process also usually specifies whether the aerodynamicist or propulsion engineer is responsible for the bookkeeping of particular forces in performing analyses. This ensures that forces are not counted twice, for instance, by both the aerodynamicist and the propulsion engineer. A force accounting system is essential to correlating results from theoretical design analyses, wind tunnel testing, and flight testing. A defined force accounting system is also required when comparing the performance of different aircraft designs.

The actual force accounting system to be used is usually tailored to the specific aircraft configuration, but there may be some commonality to previous aircraft designs with similar aerodynamic and propulsion configurations. A discriminator for determining whether a force should be bookkept as a thrust or a drag may be based on whether the force changes when the power setting is changed. For instance in an afterburning jet engine, the nozzle opens when afterburner power is selected, so the change in the aerodynamic force, due to exposing more of the external nozzle area to the freestream flow, may be counted as a change in thrust. However, exposing more of the nozzle area to the freestream also creates more drag, so it is certainly acceptable to bookkeep this as a drag force.

There are also distinct differences in handling the force accounting for different aircraft configurations. For example, the force accounting for aircraft with the propulsion system integrated or buried in the fuselage, such as for an F-18, is very different from aircraft with a podded propulsion system, with the engine mounted in a nacelle underneath a wing, such as for a commercial airliner like a Boeing 777. This particular difference is due to how the aerodynamic and propulsive flows through the engine inlet and nozzle interact, impacting how the force accounting should be handled.

Ultimately, a force accounting system should result in the determination of the total force imbalance in the direction of motion, ΔF, between the installed net propulsive force, $F_{n,\,installed}$, and the airframe system drag, D.

$$\Delta F = F_{n,\,installed} - D \tag{4.20}$$

The installed net propulsive force is obtained from the net thrust of the engine as installed in the airframe and may include other force contributions to the engine thrust, such as those due to the effects of power setting or control surface trim settings. The drag term is composed of the external aerodynamic forces in the flight direction and other contributions, including those due to the trim settings of control surfaces. The determination of the drag usually involves a *drag buildup* where the contributions from all of the aircraft components, such as wings, landing gear, tails, and protuberances, are summed up.

4.3.2   Uninstalled Thrust for the Rocket Engine

Consider a rocket engine mounted on a rigid pylon in a freestream flow of velocity V_∞ at an altitude where the static pressure is p_∞, as shown in Figure 4.7. Propellants are burned in the rocket engine combustion chamber and the combustion products are expelled through an exhaust nozzle with an exit area A_e. The exhaust velocity and pressure, u_e and p_e, respectively, are assumed constant across the nozzle exit plane. The rocket engine is positioned such that the exhaust flow is in the x-direction only. The freestream conditions, propellant mass flow rates, and nozzle exit conditions are assumed steady. The thrust of the rocket engine is in the negative x-direction, opposite in direction to the rocket exhaust velocity.

To obtain the rocket motor thrust equation, a control volume is drawn around the rocket engine, bounded by the control surface shown as the dashed line in Figure 4.7. The left side of the control volume is far from the rocket engine such that the x-velocity and pressure on the area A along this boundary are the freestream values, V_∞ and p_∞, respectively. The right side of the control surface is coincident with the exit plane of the rocket nozzle, such that the x-velocity and pressure on the nozzle exit area, A_e, are u_e and p_e, respectively. The x-velocity and pressure along the right side boundary, outside of the nozzle exit area, are the freestream values. The upper and lower boundaries of control surface are far from the engine and parallel to the freestream flow, so there is no flow perpendicular to these boundaries. The upper boundary cuts through the pylon supporting the rocket engine, such that the reaction force on the control surface is the thrust reaction force, T. A mass flow rate of propellants, $\dot{m}_p$, is fed to the engine through the pylon, parallel to the y-axis, which is the sum of the mass flow rates of the oxidizer, $\dot{m}_{ox}$, and the fuel, $\dot{m}_f$.

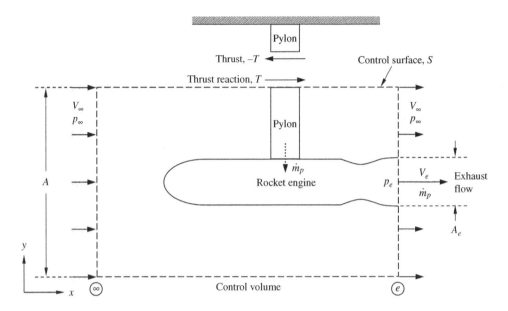

Figure 4.7   Rocket engine and control volume for determination of thrust.

Applying the x-momentum equation for steady, inviscid flow, given by Equation (4.19), to the control volume, the sum of the forces in the x-direction is given by

$$\sum F_x = T + p_\infty A - p_\infty(A - A_e) - p_e A_e = T + (p_\infty - p_e)A_e \qquad (4.21)$$

Since the pressure around the entire control surface has a constant value of p_∞, except at the nozzle exit area, there is a force contribution over this area. The thrust term is in the positive x-direction, but this is the pylon *reaction* force to the thrust force. The rocket engine is producing a thrust force in the negative x-direction that propels the rocket in this direction.

The only contribution to the momentum term in Equation (4.19) is from the flow through the nozzle exit, so that the right side of this equation is

$$\int_{CS} \rho u_x(\vec{u} \cdot \hat{n})\,dS = \rho_e V_e(u_e) A_e = \dot{m}_e V_e \qquad (4.22)$$

where $\rho_e V_e A_e$ is the mass flow rate of the exhaust flow, $\dot{m}_e$, through the nozzle exit area, A_e. The propellant mass flow rate, $\dot{m}_p$, does not contribute to the x-momentum equation since it is perpendicular to the x-axis. The exhaust mass flow rate, $\dot{m}_e$, multiplied by the exhaust velocity, u_e, is the net rate of flow of momentum or momentum flux through the nozzle exit area, A_e. This term is a positive number since the exhaust velocity, V_e, and the normal vector, $\hat{n}$, are both in the positive x-direction, making their dot product a positive number. Using this convention, an outflow from the control surface results in a positive term and an inflow results in a negative term.

Inserting Equations (4.21) and (4.22) into Equation (4.19) and solving for the rocket thrust, we have

$$T = \dot{m}_e V_e + (p_e - p_\infty)A_e \qquad (4.23)$$

The mass flow rate out of the nozzle, $\dot{m}_e$, is equal to the mass flow rate of the propellants, $\dot{m}_p$, into the engine. Using this in Equation (4.23), the thrust of a rocket engine is given by

$$T = F_{n,unistalled} = \dot{m}_p V_e + (p_e - p_\infty)A_e \qquad (4.24)$$

where $F_{n,uninstalled}$ has also been used to emphasize that the thrust, T, is the net uninstalled thrust of the rocket engine. Equation (4.24) states that the rocket engine thrust is a function of the propellant mass flow rate, the nozzle exhaust flow velocity, the nozzle exit area, and the difference between the nozzle exit pressure and the freestream pressure. The rocket thrust is independent of the flight velocity, V_∞.

The thrust is the sum of a *momentum flux term*, $\dot{m}_p V_e$, and a *pressure-area term*, $(p_e - p_\infty)A_e$. The momentum flux term contributes positively to the thrust. The pressure-area term may have a positive or a negative contribution to the thrust, depending on the magnitude of the nozzle exit plane pressure relative to the freestream pressure. The nozzle exhaust flow is established based on the difference between the nozzle exit plane pressure and the ambient pressure. The nozzle exhaust flow may be *underexpanded* ($p_e > p_\infty$), *overexpanded* ($p_e < p_\infty$), or *perfectly expanded* ($p_e = p_\infty$). One might ask, which one of these nozzle expansion cases results in the maximum thrust?

To answer this question, consider supersonic flow through a nozzle, where at a specific nozzle exit area, that we call $A_{e,optimum}$, the nozzle is perfectly expanded such that the nozzle exit pressure equals the ambient pressure, $p_e = p_\infty$. If we make the nozzle a little longer, the supersonic flow expands further to a larger nozzle exit area, $A_{e,longer}$, and the exit pressure decreases below freestream pressure. The pressure-area term, $(p_e - p_\infty)A_e$, in Equation (4.24) has a negative contribution, decreasing the thrust. If we make the nozzle a little shorter than the perfectly expanded nozzle, the flow expands to a smaller nozzle exit area, $A_{e,shorter}$, and the exit pressure is greater than freestream pressure. However, since the nozzle exit area is smaller than for the matched pressure case, $A_{e,shorter} < A_{e,optimum}$, the pressure-area term is also smaller, and despite the increase in exit pressure, the thrust is decreased. Therefore, the maximum thrust is obtained when the nozzle is perfectly expanded where the exit pressure matches the freestream pressure, $p_e = p_\infty$. Setting $p_e = p_\infty$ in Equation (4.24), we obtain the rocket engine thrust equation for a perfectly expanded nozzle, given simply as

$$T = \dot{m}_p V_e \tag{4.25}$$

where the thrust is only a function of the propellant mass flow rate and the nozzle exhaust flow velocity. Recalling that $\dot{m}_e = \dot{m}_p$, Equation (4.25) states that the thrust is dependent only on the conditions at the nozzle exhaust plane for a perfectly expanded nozzle.

Example 4.2 Calculation of Uninstalled Rocket Thrust *A rocket engine has a nozzle with an exit area of 1.2 m². The nozzle exhaust velocity is 2350 m/s and the exhaust flow is perfectly expanded. Calculate the uninstalled rocket engine thrust for a propellant mass flow rate of 1.823 kg/s.*

Solution

The uninstalled rocket engine thrust is given by Equation (4.24) as

$$T = F_{n,uninstalled} = \dot{m}_p V_e + (p_e - p_\infty)A_e$$

For a perfectly expanded nozzle, $p_e = p_\infty$, so that the uninstalled thrust is

$$T = \dot{m}_p V_e = \left(1.823 \frac{kg}{s}\right)\left(2350 \frac{m}{s}\right) = 4284 \, N$$

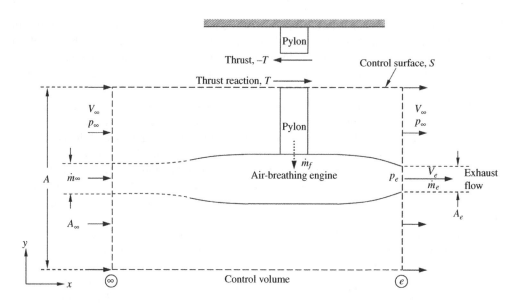

Figure 4.8   Air-breathing engine and control volume for determination of thrust.

4.3.3   Uninstalled Thrust for the Ramjet and Turbojet

The derivation of the thrust equation for an air-breathing ramjet or turbojet engine follows the same procedure as used for the non-air-breathing, rocket engine in the last section. A control volume surrounds the engine in a freestream flow with velocity and pressure, V_∞ and p_∞, respectively, as shown in Figure 4.8. The control surfaces are set up as was done for the rocket engine control volume. The left side control surface is far from the engine, the right side surface is aligned with the nozzle exit plane, and the upper and lower surfaces are parallel to the freestream flow. The major difference from the rocket engine case is the addition of the air inlet for the air-breathing engine, which ingests a mass flow rate of freestream air, $\dot{m}_\infty$, through an area, A_∞, on the left boundary surface. Fuel enters the engine with a mass flow rate $\dot{m}_f$. The air and fuel are burned in the engine and exit as combustion products, through the nozzle exhaust exit area, A_e, with a velocity u_e and a mass flow rate $\dot{m}_e = \dot{m}_\infty + \dot{m}_f$. The pressure at the nozzle exit plane is p_e, which may be different from the freestream pressure, p_∞.

Applying the x-momentum equation for steady, inviscid flow, given by Equation (4.19), to the control volume, the sum of the forces in the x-direction is given by

$$\sum F_x = T + p_\infty A - p_\infty(A - A_e) - p_e A_e = T + (p_\infty - p_e)A_e \tag{4.26}$$

Since the pressure around the entire control surface has a constant value of p_∞, except at the nozzle exit area, there is a force contribution over this area. The thrust term is in the positive x-direction, but recall that this is the pylon *reaction* force to the thrust force. The engine is producing a thrust force in the negative x-direction that propels the engine in this direction.

The momentum term in Equation (4.19) is from the flow through the air inlet and the nozzle exit, so that the right side of this equation is given by

$$\int_{CS} \rho u_x(\vec{u} \cdot \hat{n})dS = \rho_e V_e(V_e)A_e - \rho_\infty V_\infty(V_\infty)A_\infty = \dot{m}_e u_e - \dot{m}_\infty V_\infty \tag{4.27}$$

where $\rho_\infty V_\infty A_\infty$ is the mass flow rate of freestream air, $\dot{m}_\infty$, that is captured by the inlet, and $\rho_e V_e A_e$ is the mass flow rate of the exhaust flow, $\dot{m}_e$, through the nozzle exit area, A_e. The fuel mass flow rate, $\dot{m}_f$, does not contribute to the x-momentum equation since it is perpendicular to the x-axis. For the inlet and exit flows, the mass flow rate multiplied by the velocity is the net rate of flow of momentum or momentum flux. Flow out through the control surface results in a positive term while flow in through the control surface results in a negative term.

Inserting Equations (4.26) and (4.27) into Equation (4.19) and solving for the air-breathing engine thrust, we have

$$T = \dot{m}_e V_e - \dot{m}_\infty V_\infty + (p_e - p_\infty)A_e \qquad (4.28)$$

Comparing this thrust equation for an air-breathing engine with Equation (4.23) for a non-air-breathing, rocket engine, they are the same except for the addition of the air inlet momentum term, $\dot{m}_\infty V_\infty$, for the air-breather.

The mass flow rate out of the nozzle, $\dot{m}_e$, is equal to the sum of mass flow rates of the fuel, $\dot{m}_f$, and the freestream air, $\dot{m}_\infty$. Using this in Equation (4.28), the thrust of the air-breathing engine is given by

$$T = F_{n,uninstalled} = (\dot{m}_\infty + \dot{m}_f)V_e - \dot{m}_\infty V_\infty + (p_e - p_\infty)A_e \qquad (4.29)$$

where $F_{n,uninstalled}$ is used to emphasize that the thrust, T, is the net uninstalled thrust of the engine. Equation (4.29) states that the air-breathing engine thrust is a function of the fuel and air mass flow rates, the velocities of the nozzle exhaust and freestream flows, the nozzle exit area, and the difference between the nozzle exit pressure and the freestream pressure.

The thrust in Equation (4.29) is the sum of three terms: the *momentum thrust*, $(\dot{m}_\infty + \dot{m}_f)V_e$ or $\dot{m}_e V_e$, the *ram drag*, $\dot{m}_\infty V_\infty$, and the *pressure thrust*, $A_e(p_e - p_\infty)$. The *momentum thrust* is the exhaust flow momentum flux through the nozzle, which contributes positively to the thrust. The ram drag is the time rate of change of the engine inlet flow momentum, which is a negative contribution to the thrust (a drag term). The ram drag is the drag penalty of decelerating the freestream flow at the inlet. The pressure thrust is a pressure–area force which acts over the nozzle exit area, A_e, in exactly the same manner as for the rocket engine nozzle. This term may be a positive or a negative contribution to the thrust depending on the magnitude of the exit pressure relative to the freestream pressure. The maximum thrust is obtained for a perfectly expanded nozzle, as was explained for a rocket nozzle in the previous section. The air-breathing engine thrust equation, for the case of a perfectly expanded nozzle, is given by

$$T = (\dot{m}_\infty + \dot{m}_f)V_e - \dot{m}_\infty V_\infty \qquad (4.30)$$

Assuming that the fuel mass flow rate is much smaller than the air mass flow rate, $\dot{m}_f \ll \dot{m}_\infty$, Equation (4.30) becomes

$$T = \dot{m}_\infty V_e - \dot{m}_\infty V_\infty = \dot{m}_\infty (V_e - V_\infty) \qquad (4.31)$$

Equation (4.31) states that the thrust is directly proportional to the air mass flow rate entering the engine inlet and the difference between the flow velocities exiting and entering the engine. The thrust is increased with a larger engine inlet opening that can ingest a larger mass flow of air or by increasing the difference between the exhaust flow velocity and the flight velocity.

The nozzle exhaust flow momentum flux and the pressure-area force on the nozzle exit area are both nozzle related contributions to the thrust. These terms are defined as the *gross thrust*, F_g, given by

$$F_g \equiv (\dot{m}_\infty + \dot{m}_f)V_e + A_e(p_e - p_\infty) \qquad (4.32)$$

The gross thrust is the thrust that the engine nozzle produces, assuming there are no losses.

The *ram drag*, D_{ram}, is defined as

$$D_{ram} \equiv \dot{m}_\infty V_\infty \tag{4.33}$$

Inserting the definitions in Equations (4.32) and (4.33) into Equation (4.29), the uninstalled thrust for an air-breathing engine can be written simply as the difference between the gross thrust and the ram drag, given by

$$T = F_{n,uninstalled} = F_g - D_{ram} \tag{4.34}$$

Since the ram drag decreases the gross thrust, we are justified in calling the uninstalled thrust a *net* thrust. The thrust equation for an air-breathing engine physically states that the net thrust produced by the engine, assuming no installation losses, is equal to the gross thrust produced by the flow through the engine nozzle minus the inlet ram drag.

4.3.4   Installed Thrust for an Air-Breathing Engine

Consider an air-breathing engine that has been installed in a wing-mounted nacelle or cowling, as shown in Figure 4.9. The net uninstalled thrust of the engine alone, $F_{n,uninstalled}$, is given by Equation (4.34), as the gross thrust, F_g, minus the inlet ram drag, D_{ram}. The installation of the engine in the nacelle introduces several force contributions, which are added to the net uninstalled thrust to yield the net installed thrust, $F_{n,installed}$, as shown in Figure 4.9.

The freestream air entering the inlet forms a captured streamtube of air, depicted by the dashed lines entering the inlet in Figure 4.9. The shape of this captured streamtube varies depending on the flight speed and the mass flow requirements of the engine. The pressure and shear stress acting on the surface of this captured streamtube result in a drag force, called the *pre-entry* or *additive drag*, D_{add}. It can be shown that the additive drag is a negative contribution to the thrust, regardless of the shape of the captured streamtube.

The air flowing over the cowling results in pressure and shear stress distributions over its surface, which, when integrated, result in forces that can be resolved in the thrust direction. The viscous drag on the nacelle due to skin friction is lumped into a *friction drag*, $D_{friction}$, that contributes negatively to the uninstalled thrust.

The air flow over the cowling leading edge or lip is much like the flow over the leading edge and upper surface of an airfoil. The integration of the pressure distribution over the cowl lip results in a suction force with a component in the thrust direction called the lip thrust, F_{lip}, which contributes positively to the uninstalled thrust. The difference between the additive drag and the lip thrust is called the spillage drag, $D_{spillage}$, defined as

$$D_{spillage} \equiv D_{add} - F_{lip} \tag{4.35}$$

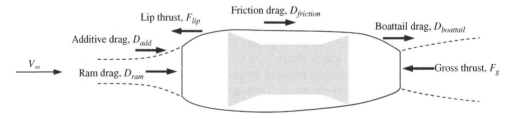

Figure 4.9   Contributions to air-breathing engine installed thrust. (Source: *Adapted from Aircraft Propulsion, S. Farokhi, Fig. 3.13, p. 125, (2014), [3], with permission from John Wiley & Sons, Inc.*)

Similarly, the air flowing over the aft end of the cowling creates a pressure distribution that, when integrated, results in a drag force called the *boattail drag*, $D_{boattail}$.

In summary, the net installed thrust, $F_{n,installed}$, is given by

$$F_{n,installed} = F_g - D_{ram} - D_{add} + F_{lip} - D_{friction} - D_{boattail} \tag{4.36}$$

In terms of the uninstalled thrust, Equation (4.36) may be written as

$$F_{n,installed} = F_{n,uninstalled} - D_{spillage} - D_{friction} - D_{boattail} \tag{4.37}$$

4.3.5 Thrust Equation for a Propeller

In this section, the thrust equation for a propeller-driven aircraft is developed. The effects of the aircraft fuselage, engine, cowling, etc. are ignored, to obtain the thrust generated solely by the *unducted* propeller. An *unducted* propeller is entirely in the airstream, in contrast to a *ducted* propeller, which is surrounded by a mechanical duct. The propeller is approximated by an infinitely thin, circular actuator disk, which can be thought of as a propeller with an infinite number of propeller blades. The actuator disk has the same diameter as the propeller and has a cross-sectional area, A_p. The loading on the propeller actuator disk is assumed uniform. The propeller is attached to an engine, which is supported by a vertical rigid pylon; however, it is assumed that there are no effects on the flow due to the engine or pylon. The engine and pylon are included as a means of obtaining a thrust force reaction.

A streamtube of incompressible air flows through the propeller actuator disk, as shown in Figure 4.10. The sides of the streamtube are streamlines of the flow, such that no flow passes through these side boundaries. The inflow and outflow boundaries of the streamtube have circular cross-sectional areas of A_i and A_e, respectively. The inflow and exit boundaries are far enough away from the propeller flow so that the boundary static pressure is equal to the freestream static pressure, p_∞. The inflow and exit boundary velocities are V_∞ and V_e, respectively. Along the streamtube wall boundaries, the velocity and pressure vary and are not equal to the freestream values. Since the flow is incompressible, the entire flow field has a constant density, ρ_∞. The streamtube cross-sectional area decreases continuously as the air accelerates, from area A_i to area A_e, in accordance with Bernoulli's equation for an incompressible flow.

A cylindrical control volume surrounds the propeller streamtube. The left and right sides of the control volume have equal areas, A_∞. The left and right boundaries are coincident with the streamtube entrance and exit planes, respectively. The velocity and pressure at the left and right side boundaries are equal to the freestream velocity and pressure, V_∞ and p_∞, respectively. The cylindrical surfaces of the control volume (the upper and lower boundaries in Figure 4.10) are parallel to the freestream flow streamlines. The velocity and pressure along this surface are equal to the freestream values, V_∞ and p_∞, respectively.

Stations 1 and 2 are just upstream and downstream of the propeller actuator disk, respectively, such that $A_1 = A_2 = A_p$. The flow velocity at station 1, just upstream of the propeller, is greater than the freestream velocity by an amount ΔV_p, such that

$$V_1 = V_\infty + \Delta V_p \tag{4.38}$$

The flow is allowed to pass through the propeller disk, with a constant uniform velocity, such that $V_1 = V_2$. If the flow velocity were to increase through the propeller such that $V_2 > V_1$, this would imply an infinite flow acceleration through the infinitely thin disk. Any rotation of the flow, induced by a rotating propeller, is ignored.

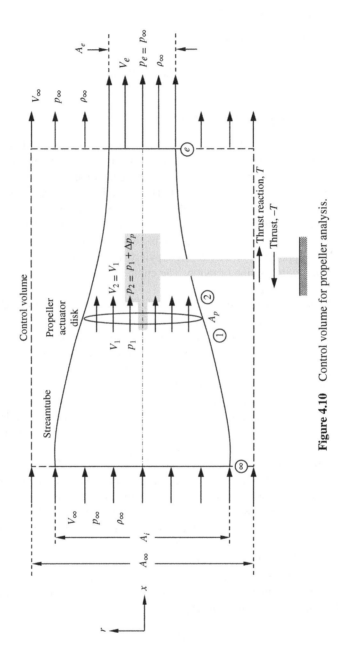

Figure 4.10 Control volume for propeller analysis.

Unlike the velocity, the static pressure increases discontinuously by an amount Δp_p through the propeller, increasing from a uniform pressure, p_1, upstream of the propeller, to a uniform pressure, $p_2 = p_1 + \Delta p_p$, downstream of the propeller. Physically, this pressure increase is due to the energy added by the propeller to the flow passing through the disk. With the assumption of constant pressure, upstream and downstream of the propeller, it is not necessary to know the detailed pressure distributions on individual propeller blades.

At the far downstream exit plane (station e), the flow is uniform and the streamlines are parallel to the freestream flow. The exit flow pressure is equal to the freestream pressure, $p_e = p_\infty$, and the exit flow velocity is greater than the freestream velocity by an amount ΔV, such that $V_e = V_\infty + \Delta V$.

We now apply Bernoulli's equation to the incompressible flow, upstream (station 1) and downstream (station 2) of the propeller. Bernoulli's equation cannot be applied to the flow *through* the propeller disk, since the propeller is adding energy to the flow at this location. For the flow upstream of the disk, we have

$$p_\infty + \frac{1}{2}\rho_\infty V_\infty^2 = p_1 + \frac{1}{2}\rho_\infty V_1^2 \tag{4.39}$$

Downstream of the disk, we have

$$p_2 + \frac{1}{2}\rho_\infty V_2^2 = p_\infty + \frac{1}{2}\rho_\infty V_e^2 \tag{4.40}$$

Using Equations (4.39) and (4.40) to solve for the pressure difference across the propeller disk, we have

$$p_2 - p_1 = \Delta p_p = \left(p_\infty + \frac{1}{2}\rho_\infty V_e^2 - \frac{1}{2}\rho_\infty V_2^2\right) - \left(p_\infty + \frac{1}{2}\rho_\infty V_\infty^2 - \frac{1}{2}\rho_\infty V_1^2\right) \tag{4.41}$$

Recall that, through the propeller disk, the velocity is *continuous* ($V_1 = V_2$), while the pressure is *discontinuous*, ($p_1 \neq p_2$). Therefore, Equation (4.41) simplifies to

$$\Delta p_p = \frac{1}{2}\rho_\infty V_e^2 - \frac{1}{2}\rho_\infty V_\infty^2 = \frac{1}{2}\rho_\infty(V_e^2 - V_\infty^2) \tag{4.42}$$

The propeller thrust, T, is equal to the force on the actuator disk, which is equal to the pressure difference across the disk, $(p_2 - p_1)$, multiplied by the disk area, A_p, or

$$T = (p_2 - p_1)A_p = [(p_1 + \Delta p_p) - p_1]A_p = \Delta p_p A_p \tag{4.43}$$

which states that the propeller thrust is simply equal to the pressure rise across the propeller disk, Δp_p, multiplied by the disk area. Inserting Equation (4.42) into (4.43), we have

$$T = \Delta p_p A_p = \frac{1}{2}\rho_\infty(V_e^2 - V_\infty^2)A_p = \frac{1}{2}\rho_\infty(V_e + V_\infty)(V_e - V_\infty)A_p \tag{4.44}$$

We now apply the x-component of the momentum equation, Equation (4.19), to the control volume around the propeller flow streamtube.

$$\sum F_x = T + p_\infty A_\infty - p_\infty A_\infty = \int_S \rho u_x(\vec{u} \cdot \hat{n})\, dS = \rho_\infty V_e^2 A_e - \rho_\infty V_\infty^2 A_i \tag{4.45}$$

where $\rho_\infty V_e^2 A_e$ is the momentum flux exiting the streamtube control volume through the exit area, A_e, and $\rho_\infty V_\infty^2 A_i$ is the momentum flux entering the streamtube control volume through the entrance area, A_i. Simplifying, we have

$$T = \rho_\infty V_e^2 A_e - \rho_\infty V_\infty^2 A_i = (\rho_\infty V_e A_e)V_e - (\rho_\infty V_\infty A_i)V_\infty \tag{4.46}$$

The mass flow rate, $\dot{m}$, through the propeller streamtube is constant, such that

$$\dot{m} = \rho_\infty V_\infty A_i = \rho_\infty V_e A_e = \rho_\infty V_1 A_p \qquad (4.47)$$

where $(\rho_\infty V_\infty A_i)$, $(\rho_\infty V_e A_e)$, and $(\rho_\infty V_1 A_p)$ are the mass flow rates through the streamtube entrance, exit, and propeller, respectively. Using Equation (4.47) in (4.46), we have

$$T = \dot{m}(V_e - V_\infty) = (\rho_\infty V_1 A_p)(V_e - V_\infty) \qquad (4.48)$$

Setting Equation (4.48) equal to (4.44), we have

$$T = (\rho_\infty V_1 A_p)(V_e - V_\infty) = \frac{1}{2}\rho_\infty(V_e + V_\infty)(V_e - V_\infty)A_p \qquad (4.49)$$

Solving Equation (4.49) for the velocity through the propeller, V_1, we have

$$V_1 = \frac{V_e + V_\infty}{2} \qquad (4.50)$$

Equation (4.50) states that the velocity through the propeller is equal to the average of the velocity far upstream and far downstream of the propeller.

Solving Equation (4.50) for the velocity at the streamtube exit, we have

$$V_e = 2V_1 - V_\infty \qquad (4.51)$$

Substituting Equation (4.51) into (4.48), we have

$$T = \dot{m}(V_e - V_\infty) = \dot{m}(2V_1 - V_\infty - V_\infty) = 2\dot{m}(V_1 - V_\infty) \qquad (4.52)$$

Recalling that the velocity through the propeller is greater than the freestream velocity, as given by Equation (4.38), we have

$$T = 2\dot{m}(V_\infty + \Delta V_p - V_\infty) = 2\dot{m}_p \Delta V_p \qquad (4.53)$$

Thus, we see that the propeller thrust is directly proportional to the mass flow rate through the propeller, $\dot{m}_p$ and the increase in velocity through the propeller, ΔV_p.

Example 4.3 Calculation of Propeller Thrust *A propeller-driven airplane is flying at an airspeed of 170 mph and an altitude of 3000 ft (air density of 0.002175 slug/ft³). The velocity of the streamtube of air captured by the 5 foot diameter propeller, far downstream of the propeller, is 320 ft/s. Calculate the thrust of the propeller and the velocity at the propeller disk.*

Solution

Convert the airspeed to consistent units.

$$V_\infty = 170\frac{mi}{h} \times \frac{5280\,ft}{1\,mi} \times \frac{1\,h}{3600\,s} = 249.3\frac{ft}{s}$$

Calculate the area of the propeller disk.

$$A_p = \frac{\pi}{4}(5\,ft)^2 = 19.6\,ft^2$$

Using Equation (4.44), the propeller thrust is

$$T = \frac{1}{2}\rho_\infty(V_e^2 - V_\infty^2)A_p = \frac{1}{2}\left(0.002175\,\frac{slug}{ft^3}\right)\left[\left(320\frac{ft}{s}\right)^2 - \left(249.3\frac{ft}{s}\right)^2\right](19.6\,ft^2) = 857.9\,lb$$

From Equation (4.50), the velocity at the propeller disk is

$$V_1 = \frac{V_e + V_\infty}{2} = \frac{320\frac{ft}{s} + 249.3\frac{ft}{s}}{2} = 284.7\frac{ft}{s}$$

4.4 Thrust and Power Curves for Propeller-Driven and Jet Engines

Now that we have derived equations for the thrust produced by various types of propulsion, let us examine how the thrust and power vary as a function of velocity for these propulsion types. The thrust equations for rocket, jet, and propeller-driven propulsion are summarized in Table 4.2.

While thrust is commonly used to describe the propulsive output of jet engines, power is typically used for propeller-driven piston engines. Recall that power, P, is defined as work per unit time, t. Work, W, is defined as a force, F, acting through a distance, x, where the displacement is in the same direction as the force. The force is the thrust, T, generated by the propulsive device. Thus, the power may be expressed as

$$P = \frac{\Delta W}{\Delta t} = \frac{T\Delta x}{\Delta t} = TV_\infty \tag{4.54}$$

where the $\Delta x/\Delta t$ is the flight velocity, V_∞, of the aircraft. Recall that the units of power are typically given as horsepower in English units and as watts in SI units. The unit conversions between these systems are given by Equations (4.5) and (4.6).

Typical thrust and power curves for rocket, jet, and propeller-driven, piston engine propulsion are shown in Figure 4.11. The non-air-breathing rocket thrust is constant and independent of flight velocity. Therefore, the power calculated from the rocket thrust is simply the constant thrust multiplied by the flight velocity, as given by Equation (4.54), resulting in a linear increasing power with increasing velocity.

The thrust available from a jet engine is nearly constant with flight velocity. The thrust decreases slightly up to about Mach 0.3. At low speed, the ram drag, D_{ram}, increases more than the gross thrust, causing a net decrease in the thrust available. As the speed increases beyond about Mach 0.3, there is an increase of the density and pressure in the inlet, resulting in *ram recovery*, which increases the gross thrust much more than the decrease due to ram drag. Therefore, the thrust available increases with increasing speed above about Mach 0.3. For practical engineering purposes, these thrust changes are relatively small, such that the thrust available from a jet engine is often assumed to be constant with velocity.

Table 4.2 Summary of thrust equations for selected types of propulsion.

Propulsion type	Thrust equation
Rocket	$T = \dot{m}_p V_e$
Jet	$T = F_g - D_{ram} = \dot{m}_e V_e - \dot{m}_\infty V_\infty \cong \dot{m}_\infty(V_e - V_\infty)$
Propeller-driven	$T = 2\dot{m}_p\Delta V_p$

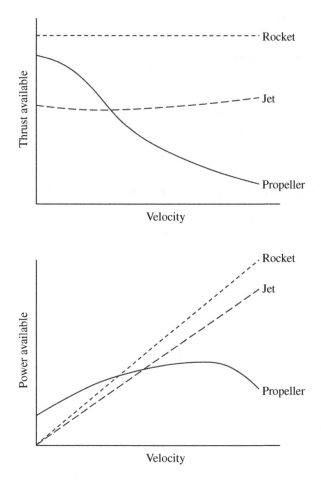

Figure 4.11   Thrust and power versus velocity for selected types of propulsion.

The power available, P_A, from a propeller-piston engine combination is given by

$$P_A = T_A V_\infty = \eta_P P = \eta_P T V_\infty \tag{4.55}$$

where P is the power output from the engine to the propeller shaft and η_P is the propeller efficiency. The propeller efficiency decreases the power output from the engine, varying as shown in Figure 4.11. (The propeller efficiency is discussed in Section 4.5.8.4.) The thrust available from the propeller varies as shown in Figure 4.11, according to the specified power available and Equation (4.54). The thrust of the propeller-driven, piston engine decreases with forward flight speed, due to the losses incurred by the propeller at high speed.

4.4.1   FTT: In-Flight Thrust Measurement

Aircraft performance is driven by the thrust produced by the propulsion system, and the aerodynamic drag due to the aircraft airframe. Therefore, to assess performance, the thrust and drag must be accurately known. Ideally, we would like to separate the thrust from the drag, which requires separate measurements of these forces. A critical requirement to accomplish this is the use of a force accounting system, where the thrust-drag accounting or bookkeeping is clearly defined, as discussed in Section 4.3.1.

In the previous chapter, we discussed several ways to predict the aerodynamic drag, including using various analytical methods, such as computational fluid dynamics, or ground techniques, such as wind tunnel testing. Typically, flight test data is required to validate the drag predictions for the complete aircraft. We discussed certain flight test techniques to obtain the aerodynamic drag, but unless the aircraft is unpowered, these techniques require a model of the propulsion system. Therefore, in-flight thrust determination is critical in accurately predicting aerodynamic drag and in validating the thrust models. After the aerodynamic and propulsion models have been flight-validated, the accuracy of performance predictions is greatly improved. In the present flight test technique, we discuss three methods for in-flight thrust determination: *direct measurement* using strain gauges, the *gas generator method*, and the exhaust nozzle *traversing rake method*.

For these in-flight thrust flight test techniques, you will be flying the Convair F-106 *Delta Dart*, as shown in Figure 4.12. The *Delta Dart* was an all-weather, single-engine, delta wing, supersonic interceptor aircraft for the US Air Force. During the 1950s, aircraft, such as the F-102, F-104, and F-106, were designed with a dedicated mission of intercepting enemy high-altitude nuclear bombers. They had the performance capability to accelerate to high supersonic speed and climb to high altitudes to intercept the enemy aircraft. The F-106 was one the Century Series of jet aircraft, which included the F-100, F-101, F-102, F-104, and F-105. These aircraft were the first series of US fighter-bomber and interceptor aircraft, which had supersonic performance capability. The first flight of the F-106 was on 26 December 1956.

The F-106 epitomizes many of the aerodynamic characteristics of a supersonic aircraft, discussed in Chapter 3, including a thin, 60° swept-back delta wing and a slender, area-ruled fuselage, as shown in the three-view drawing of Figure 4.13 (see also Figure 3.201). Powered by a single Pratt & Whitney J75-17 axial flow turbojet engine with afterburner, the F-106 is capable of Mach 2.3

Figure 4.12   Convair F-106B *Delta Dart* with underwing engine nacelles.   (Source: *NASA*.)

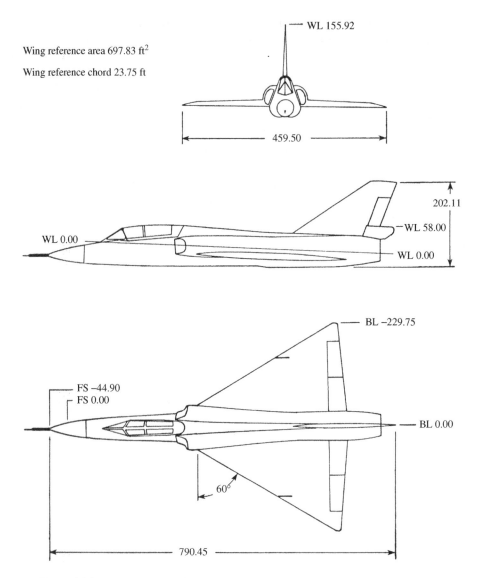

Wing reference area 697.83 ft^2

Wing reference chord 23.75 ft

WL 155.92

459.50

202.11

WL 0.00

WL 58.00

WL 0.00

BL −229.75

FS −44.90

FS 0.00

BL 0.00

60°

790.45

Figure 4.13   Three-view drawing of the Convair F-106B *Delta Dart*. (Source: *NASA*.)

flight at high altitudes. The F-106B is the two-seat variant of the single-seat version, the F-106A. Selected specifications of the Convair F-106B *Delta Dart* are given in Table 4.3.

The aircraft you will be flying is an F-106B aircraft that was modified by NASA for propulsion flight research in the 1970s. The aircraft was modified to carry two General Electric J85-GE-13 afterburning turbojet engines, housed in nacelles, mounted underneath the wing, as shown in Figure 4.12. Other aircraft modifications included the installation of a fuel supply system for the underwing J85 engines and J85 engine throttle controls and engine instruments in the aft cockpit (Figure 4.14), J85 engine sensors, and a data acquisition system. Further details about this modified F-106 aircraft and the underwing engine installation are given in [6].

The J85 engine nacelles are located 6.11 ft (1.86 m) outboard of the aircraft centerline. The nacelles are inclined downward at a 4.5° angle, with respect to the wing chord line, such that

Table 4.3 Selected specifications of the Convair F-106B *Delta Dart*.

Item	Specification
Primary function	All weather, supersonic interceptor and trainer
Manufacturer	Convair, San Diego, California
First flight	26 December 1956
Crew	1 pilot + 1 instructor pilot or flight test engineer
Powerplant	J75-17 afterburning turbojet engine
Thrust, MIL	17,000 lb (75.6 kN), military power
Thrust, MAX	24,500 lb (109 kN), maximum afterburner
Empty weight	25,140 lb (11,400 kg)
Maximum takeoff weight	40,080 lb (18,180 kg)
Length	65.87 ft (20.08 m)
Height	20.3 ft (6.19 m)
Wingspan	38.29 ft (11.67 m)
Wing area	697.83 ft^2 (64.83 m^2)
Wing aspect ratio	2.10
Wing sweep	60°
Airfoil	NACA 0004-75 modified
Maximum speed	1525 mph (2455 km/h), Mach 2.3 at 40,000 ft
Service ceiling	57,000 ft (17,400 m)

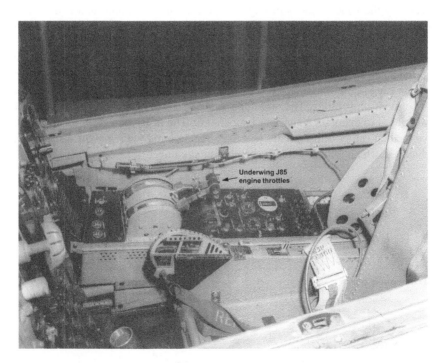

Figure 4.14 Convair F-106B aft cockpit with throttle controls for underwing J85 engines. (Source: *NASA with annotation added.*)

the aft end of the nacelles are tangent to the wing lower surface at its trailing edge. Each nacelle is attached to the wing by three bearing-mounted links, at forward, center, and aft locations along the nacelle, which allow the engines to translate freely in the axial thrust direction. The forward and aft links transfer all of the loads acting on the nacelle directly to the wing, except for the loads acting in the engine thrust direction. The center link transfers the axial direction loads to the wing through a load cell. The axial force on the nacelle is measured using strain gauges mounted on the load cell.

Assuming a steady-state condition (zero axial acceleration), the load cell provides a direct measurement of the engine gross thrust minus the total drag, as shown schematically in Figure 4.15 (the 4.5° incline angle of the nacelle is ignored). The J85 underwing engine is a podded engine, hence the total drag includes the nacelle drag, and the load cell is measuring an installed thrust, as described in Section 4.3.4. For a buried engine where the thrust measuring strain gauges are on the engine mounting structure, the direct force measurement is similar to the net thrust, as given by Equation (4.34). There may be other potential axial forces on a buried engine, from seals, cabling and plumbing lines, etc., which must be accounted for or reduced to zero when using the direct thrust measurement technique. Reference [1] may be consulted for details of direct thrust measurements made for buried turbofan engines in an F-15, using strain gauges mounted to the engine support structure, at speeds up to Mach 2.

Conceptually, the direct measurement technique is perhaps the most straightforward method for obtaining in-flight thrust. However, the way that the thrust forces produced by the engine are transmitted to the airframe can be extremely complex. In the past, direct thrust measurements have provided less than desired accuracy due to difficulties with the mounting and calibration of the load-sensing strain gauges and the inability to account for secondary load paths. This difficulty is exacerbated by the wide varieties of engine types, installations, and inlet and nozzle geometries. In general, the strain gauge mounting and secondary load path issues are less problematic for podded engines, such as for the underwing J85 engines, as compared with buried engines. Strain gauge issues are often associated with temperature variations, such as those due to altitude changes, supersonic flight, or afterburner use. Temperature-related errors can be reduced by performing thorough strain gauge calibrations as a function of temperature or by keeping the thrust measuring system at a constant temperature. This was done for the F-106 system, where heaters and insulation blankets were used to maintain a constant temperature. The instrumentation requirements for the direct force method can be substantially less than that required with analytical in-flight determination techniques, which typically require a substantial number of sensors distributed throughout

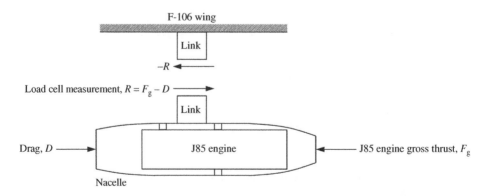

Figure 4.15 F-016 underwing engine load cell measurement of thrust minus drag.

an engine. With careful planning, installation, and calibration of instrumentation, it is possible to obtain high accuracy with a direct force measurement.

After takeoff, your pilot levels off at an altitude of 15,000 ft (4570 m) and trims the F-106 for steady, level flight at Mach 0.6. You are on condition for your first propulsion test point. You have control of the J85 underwing engines, using the engine controls on the right side panel of your cockpit, as shown in Figure 4.14. You control the power settings for the two J85 engines, which changes the trim condition. The pilot, who has control of the F-106 J75 engine, adjusts the J75 engine power, as required, to obtain the desired trim Mach number for the test point. You advance the two J85 engine throttles to partial power. The additional thrust from the J85 engines causes the test point Mach number to increase. The pilot reduces the power of the F-106 J75 engine, as required, to obtain a trim shot at Mach 0.6. The data system records the J85 engine data and the nacelle load cell data for the test point. As discussed earlier, the load cell obtains a *direct measurement* of the nacelle thrust minus drag.

The data acquisition system records in-flight measurements of flow properties, at various stations in the engine gas path. In the *gas generator method*, these in-flight measurements are related to similar measurements made in ground tests to calculate the thrust. For example, in-flight measurements of the compressor fan pressure ratio and fan speed may be used with a ground test derived calibration curve to obtain the engine air mass flow rate. The exact parameters to be measured in flight are dependent on the ground test models or calibrations curves that are applied. The ground test database may include analytical predictions and ground test data from sea level test stands or altitude test cells. If the ground test data is collected from static tests at sea level, the data must be extrapolated to the airspeeds and altitudes at the flight condition, which can lead to inaccuracies.

The final method that you employ to calculate the in-flight thrust is the traversing or *swinging rake method*. You activate the system that moves an instrumented rake through the exhaust flow of the J85 engine in flight. The rake is populated with sensors that measure the static and total pressure, total temperature, and flow direction in the nozzle exhaust plane. From these measurements, the nozzle exhaust mass flow and engine gross thrust can be calculated. Survivability and warping of the traversing rake in the hot engine exhaust flow are issues for these systems. The rake completes its automated traverse of the engine exhaust plane and you are ready to move on to the next test point at a higher power setting.

You subsequently complete the Mach 0.6 test points at different J85 power settings, up to maximum thrust with full afterburner. After the Mach 0.6 test points, you obtain in-flight thrust data up to Mach 1.3, at an altitude of 25,000 ft (7620 m), using the same process of setting the desired J85 thrust, following by adjustment of the J75 thrust, to obtain the desired test point Mach number. Hence, you are able to obtain in-flight thrust data for engine thrust levels from a low thrust setting to maximum thrust, over a range of Mach numbers from subsonic to low supersonic speeds.

4.5 Air-Breathing Propulsion

All of the engines discussed in this section "breathe" air from the atmosphere, and this air is used as the oxidizer that is mixed with various types of fuel. This limits the operation of vehicles powered by air-breathing engines to within the confines of the sensible atmosphere. In this section, we discuss five different air-breathing engines: the internal combustion engine, the ramjet, the turbojet, the turbofan, and the turboprop. These five types of engines may be categorized in terms of three types of air-breathing propulsion: the internal combustion engine–propeller combination,

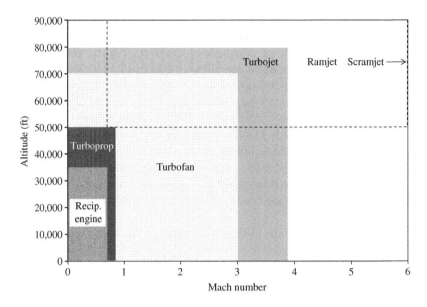

Figure 4.16   Approximate flight envelopes of various types of air-breathing propulsion.

the air-breathing engine with no moving parts (ramjet), and the air-breathing engine that utilizes turbomachinery (turbojet, turbofan, and turboprop).

The approximate flight envelopes of the various types of air-breathing engines are shown on an altitude–Mach number plot in Figure 4.16. Reciprocating engines generally operate from zero airspeed to high subsonic Mach numbers and altitudes from sea level to about 35,000 ft (11,000 m). Turboprop engines operate up to slightly higher subsonic Mach numbers and altitudes up to about 50,000 ft (15,000 m). Supersonic flight is possible with turbofan and turbojet engines, which can operate up to Mach numbers of about 3 and 4, respectively, and altitudes up to about 70,000 ft (21,000 m) and 80,000 ft (24,000 m), respectively. The internal combustion engine, turbojet, turbofan, and turboprops are capable of *static operation*, that is, operation at zero airspeed. The subsonic ramjet and supersonic combustion ramjet or *scramjet*, cannot operate at zero airspeed, as they rely on the forward vehicle motion to provide the air mass flow ingestion and compression for the engine. The ramjet can operate at high subsonic Mach numbers, albeit with poor efficiency. Ramjet operation is optimal at Mach numbers of about 3–5, and high altitudes above about 50,000 ft (15,000 m). High altitude, high Mach number supersonic flight, at or above about Mach 6, is required for scramjet operation. The upper Mach number limit of scramjet engines is still to be determined.

4.5.1   Air-Breathing Propulsion Performance Parameters

This section defines selected parameters and relations that are important in the evaluation of air-breathing engine performance. Figure 4.17 depicts an air-breathing engine with an inlet/diffuser, combustor, and nozzle. The relationships that are developed from this general description of an air-breathing engine are applicable to the various types of air-breathing engines that are discussed in later sections, including ramjets, turbojets, and turbofans. The station numbers and flow properties used in this section are also given in Figure 4.17.

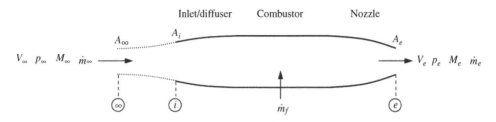

Figure 4.17   Air-breathing engine flow properties.

A streamtube of freestream air with velocity, V_∞, pressure, p_∞, Mach number, M_∞, and mass flow rate, $\dot{m}_\infty$, enters the engine inlet. The area of the streamtube, A_∞, of freestream air may be different from the inlet area, A_i, depending on the flight speed and the mass flow requirements of the engine. The diffuser decelerates the flow and increases the pressure and temperature of the flow to the optimum conditions for entrance into the combustor. A mass flow of fuel, $\dot{m}_f$, is added in the combustor, which mixes and burns with the mass flow of air. The products of combustion are exhausted through a nozzle with area, A_e, with a velocity, V_e, pressure, p_e, Mach number, M_e, and mass flow rate, $\dot{m}_e$. From Section 4.3.3, the uninstalled thrust of an air-breathing engine with a perfectly expanded nozzle is given by

$$T = \dot{m}_e V_e - \dot{m}_\infty V_\infty \tag{4.56}$$

4.5.1.1   Thrust-to-Weight Ratio

In addition to the thrust, the *weight* of the engine or propulsive system is an important propulsion parameter. It is obviously better for the engine or propulsive system to be lighter rather than heavier. However, the individual magnitude of the thrust or weight is not sufficient to describe the complete propulsive performance, and it may not be meaningful in comparing different engines or propulsion types. Thus, we combine these two parameters to form the dimensionless *thrust-to-weight ratio*, T/W. Typically, the thrust, T is the maximum *static* (zero velocity) *thrust at sea level* and the weight, W, is the *maximum gross vehicle weight*. The thrust can vary due to changes in the power setting or flight condition (altitude and airspeed). The weight can vary due to fuel or propellant consumption or different payload configurations. Since it is desired to maximize the thrust and minimize weight, the overall goal is to maximize the thrust-to-weight ratio. As we pursue further in Chapter 5, the thrust-to-weight ratio plays an important role in vehicle performance. We see that the higher the thrust-to-weight ratio, the greater the capability of the vehicle to accelerate or climb.

To get a feel for the thrust-to-weight ratio, Table 4.4 provides T/W for selected aerospace vehicles, where values for the maximum sea-level static thrust and the maximum gross weight are used, unless otherwise specified. Typically, the thrust-to-weight ratio of jet transport aircraft is about 0.15–0.25. Business-class jet aircraft typically have slightly higher thrust-to-weight ratios, as shown by the Learjet 85 in Table 4.4. Very high altitude jet aircraft have thrust-to-weight ratios of about 0.5, as shown by the Lockheed U-2 in Table 4.4. Jet fighter aircraft have typical thrust-to-weight ratios of about 0.5–0.9, with the possibility of values slightly greater than one at light weights. This is shown in Table 4.4 for the F-16, where the thrust-to-weight ratio is 0.615 at maximum gross weight and 1.05 at a light weight. A thrust-to-weight ratio greater than one means that the vehicle can *accelerate* in a vertical climb. This is a requirement for rockets where the typical thrust-to-weight ratios are greater than about 1.2, as given by the Space Shuttle and Atlas V rocket in Table 4.4.

Table 4.4 Thrust-to-weight ratios of selected aerospace vehicles.

Vehicle	Thrust, T^*	Weight, W^{**}	Thrust-to-weight ratio, T/W
Boeing 767	63,300 lb (282 kN)	412,000 lb (186,880 kg)	0.154
Bombardier Learjet 85	10,526 lb (46.82 kN)	33,500 lb (15,195 kg)	0.314
Lockheed U-2	19,000 lb (84.5 kN)	40,000 lb (18,144 kg)	0.475
Lockheed SR-71	68,000 lb (302.5 kN)	143,000 lb (64,864 kg)	0.476
Northrop T-38A	5800 lb (25.8 kN)	12,093 lb (5485 kg)	0.480
Lockheed F-16	29,500 lb (131 kN)	48,000 lb (21,772 kg)	0.615
Lockheed F-16, lightweight	29,500 lb (131 kN)	28,000 lb (12,701 kg)	1.05
Space Shuttle	6780 klb (30,159 kN)	4400 klb (1,996,000 kg)	1.54
Atlas V rocket	1931 klb (8590 kN)	1205 klb (546,578 kg)	1.60

*Maximum static, sea level thrust.
**Maximum gross weight, unless otherwise specified.

4.5.1.2 Specific Impulse

Ideally, a propulsion system produces the maximum amount of thrust with the minimum consumption of fuel. This measure of propulsive efficiency is embodied by the *specific impulse*, I_{sp}, defined as the thrust, T, per unit weight flow rate of fuel, $\dot{W}_f$, as given by

$$I_{sp} \equiv \frac{T}{\dot{W}_f} = \frac{T}{\dot{m}_f g_0} \qquad (4.57)$$

where $\dot{m}_f$ is the fuel mass flow rate and g_0 is the acceleration due to gravity at sea level. Assuming that a consistent set of units is used, the specific impulse has units of *seconds*. Maximizing the specific impulse corresponds to maximizing the thrust or minimizing the consumption of fuel.

A comparison of specific impulse for several different propulsion cycles is shown in Figure 4.18, where specific impulse is plotted against the flight Mach number. Generally, the specific impulse decreases significantly with increasing Mach number. The exception to this is the non-air-breathing rocket engine, which maintains a near constant specific impulse of about 400 s, independent of flight Mach number. Recall, from Equation (4.25), that the thrust of a rocket engine is independent of the flight velocity. Since a rocket engine does not ingest freestream air, it does not suffer from Mach number related air ingestion losses.

Turbojet and turbofan engines provide the highest specific impulses, on the order of many thousands of seconds, as high as 7000–8000 s, at low, subsonic Mach numbers up to low supersonic Mach numbers. The turbofan has a higher specific impulse than the turbojet. Adding an afterburner to the turbojet or turbofan engine results in maximum specific impulses of about 4000 s, at low supersonic Mach numbers, decreasing to about 2000 s at about Mach 3. Ramjets are more efficient from about Mach 3 to 5, with specific impulses of about 1000–2000 s. Above about Mach 5 to 6, the supersonic combustion ramjet or scramjet promises to provide the best specific impulse, although many of the practical aspects of this cycle are still to be proven, especially for very high Mach number.

4.5.1.3 Specific Fuel Consumption

The fuel consumption of an engine is the rate at which the fuel is burned, typically quantified as a mass or weight flow rate of fuel. The *specific fuel consumption* is the rate of fuel consumption

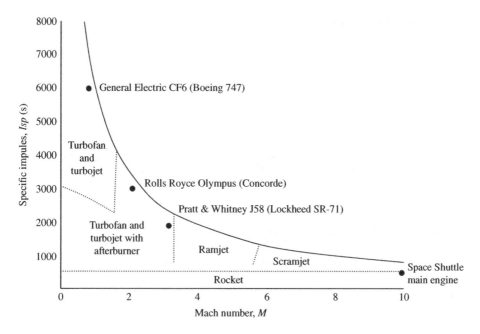

Figure 4.18   Specific impulse comparison of different propulsion types.

divided by the thrust or power produced. The specific fuel consumption is a measure of the fuel efficiency of the engine per unit thrust or power produced.

For jet engines, the specific fuel consumption is expressed as the *thrust specific fuel consumption* (TSFC), defined as the weight flow rate of fuel, $\dot{W}_f$, per unit thrust, T, produced by the engine.

$$TSFC \equiv \frac{\dot{W}_f}{T} \tag{4.58}$$

In the English system, the units of TSFC are commonly given as pounds of fuel per hour per pound of thrust, (lb/h)/lb. In the SI system, the units of TSFC are often given in units of grams of fuel per second per newton of thrust, (g/s)/N. These units should be converted to consistent SI units of N/(N·s) when performing calculations, as illustrated in the Example problems below. In both systems of units, TSFC has units of 1/time, so that its numerical value is the same in either system. Care should be exercised when dealing with TSFC as it is quoted in a wide variety of units. The TSFC of turbojet engines is typically 0.75–1.1 (lb/h)/lb. Turbofans engines have typical TSFC values of about 0.3–0.75 (lb/h)/lb.

Reciprocating piston engine fuel consumption is typically given in terms of specific fuel consumption, c, defined as the weight of fuel consumed per unit of power per unit of time, as given by

$$c \equiv \frac{\dot{W}_f}{P} \tag{4.59}$$

where P is the engine power. The units of c are often given in inconsistent English units of pounds of fuel per hour per horsepower, (lb/h)/hp. Consistent units for c, in English and SI units, are

$$[c] = \frac{\text{lb of fuel}}{(\text{ft} \cdot \text{lb/s})/\text{s}} \quad \text{or} \quad \frac{\text{N of fuel}}{\text{J/s}} \tag{4.60}$$

Example 4.4 Specific Impulse *The Space Shuttle used two Thiokol (now ATK) solid rocket boosters, each producing 2.8 million pounds of thrust at sea level. If the mass flow rate of propellant is 11,814 lb$_m$/s, calculate the specific impulse.*

Solution

Using Equation (4.57), the specific impulse is

$$I_{sp} \equiv \frac{T}{\dot{W}_f} = \frac{T}{\dot{m}_f g_0} = \frac{2,800,000 \, lb}{\left(11,814 \dfrac{lb_m}{s} \times \dfrac{1 \, slug}{32.2 \, lb_m}\right)\left(32.2 \dfrac{ft}{s^2}\right)} = 237.0 \, s$$

Example 4.5a Thrust Specific Fuel Consumption (English units) *A General Electric F404 jet engine, used in the McDonnell Douglas F-18 Hornet, has a military thrust of 11,000 lb (maximum, non-afterburning thrust) at a fuel flow of 8965 lb/h and maximum thrust of 17,700 lb (with full afterburner) at a fuel flow of 30,798 lb/h. Calculate the thrust specific fuel consumption for these conditions.*

Solution

Using Equation (4.58), the thrust specific fuel consumption at military thrust is

$$TSFC \equiv \frac{\dot{W}_f}{T} = \frac{8965 \, lb/h}{11,000 \, lb} = 0.815 \frac{lb}{lb \cdot h}$$

The thrust specific fuel consumption at maximum thrust is

$$TSFC \equiv \frac{\dot{W}_f}{T} = \frac{30,798 \, lb/h}{17,700 \, lb} = 1.74 \frac{lb}{lb \cdot h}$$

Example 4.5b Thrust Specific Fuel Consumption (SI Units) *The Boeing 747 Jumbo Jet has a thrust specific fuel consumption of 17.1 g/(kN · s). Convert this TSFC to consistent SI units.*

Solution

Converting TSFC to consistent SI units, we have

$$TSFC = 17.1 \frac{g}{kN \cdot s} \times \frac{1 \, kg}{1000 \, g} \times 9.81 \frac{m}{s^2} \times \frac{1 \, kN}{1000 \, N} = 1.678 \times 10^{-4} \frac{N}{N \cdot s}$$

4.5.1.4 Propulsive Efficiency

A propulsion system converts the *engine power* into *thrust power* that propels the aircraft. The efficiency with which the propulsion system is able to perform this conversion is defined as the propulsive efficiency, η_p, given by

$$\eta_p \equiv \frac{\text{Thrust power}}{\text{Engine power}} \qquad (4.61)$$

The thrust power is defined as the product of the thrust, T, and freestream velocity, V_∞.

$$\text{Thrust power} \equiv TV_\infty \tag{4.62}$$

In Figure 4.17, an air-breathing engine was defined as a device that ingests a mass flow rate of freestream air, $\dot{m}_\infty$, at a freestream velocity, V_∞, increases its kinetic energy by the combustion of fuel, and expels the products at a mass flow rate, $\dot{m}_e$, and a velocity, V_e. The engine power can be defined as the difference in the time rate of change of the kinetic energy of the fluid exiting, KE_e, and entering the engine, KE_∞.

$$\text{Engine power} \equiv \frac{d}{dt}(KE_e - KE_\infty) \tag{4.63}$$

Substituting Equations (4.62) and (4.63) into the definition of the propulsive efficiency, Equation (4.61), we have

$$\eta_p = \frac{TV_\infty}{\dfrac{d}{dt}(KE_e - KE_\infty)} \tag{4.64}$$

The kinetic energies can be expressed in terms of the mass flow rates and velocities for the freestream and the exhaust flows.

$$\frac{d}{dt}(KE_e - KE_\infty) = \frac{d}{dt}\left(\frac{1}{2}\dot{m}_e V_e^2 - \frac{1}{2}\dot{m}_\infty V_\infty^2\right) \tag{4.65}$$

Substituting for the thrust from Equation (4.56), the thrust power can be written as

$$TV_\infty = (\dot{m}_e V_e - \dot{m}_\infty V_\infty)V_\infty \tag{4.66}$$

Substituting Equations (4.65) and (4.66) into Equation (4.64), we have

$$\eta_p = \frac{(\dot{m}_e V_e - \dot{m}_\infty V_\infty)V_\infty}{\dfrac{1}{2}\dot{m}_e V_e^2 - \dfrac{1}{2}\dot{m}_\infty V_\infty^2} \tag{4.67}$$

If the fuel mass flow rate, $\dot{m}_f$, is assumed to be small with respect to the freestream air mass flow rate, $\dot{m}_\infty$, that is, $\dot{m}_f \ll \dot{m}_\infty$, we have

$$\dot{m}_e = \dot{m}_\infty + \dot{m}_f \approx \dot{m}_\infty \tag{4.68}$$

Substituting Equation (4.68) into (4.67), the propulsive efficiency is given by

$$\eta_p = \frac{(\dot{m}_\infty V_e - \dot{m}_\infty V_\infty)V_\infty}{\dfrac{1}{2}\dot{m}_\infty V_e^2 - \dfrac{1}{2}\dot{m}_\infty V_\infty^2} = \frac{\dot{m}_\infty(V_e - V_\infty)V_\infty}{\dfrac{1}{2}\dot{m}_\infty(V_e^2 - V_\infty^2)} = \frac{2V_\infty}{V_e + V_\infty} = \frac{2}{\dfrac{V_e}{V_\infty} + 1} \tag{4.69}$$

This equation gives the propulsive efficiency as a function of the ratio of the engine exhaust velocity to the freestream velocity, V_e/V_∞. According to Equation (4.69), a propulsive efficiency of one, or 100%, is obtained when the exhaust velocity, V_e, equals the freestream velocity, V_∞, or when the velocity ratio, V_e/V_∞, is equal to one. Of course, an engine with an exhaust velocity equal to the flight velocity produces zero thrust. In fact, our thrust equation states that the thrust increases with increasing exhaust velocity relative to the freestream or flight velocity. The propulsive efficiency is plotted versus the exhaust-to-flight velocity, V_e/V_∞, in Figure 4.19. Also plotted is a non-dimensional specific thrust, defined as

$$\frac{T}{\dot{m}_\infty V_\infty} = \frac{V_e}{V_\infty} - 1 \tag{4.70}$$

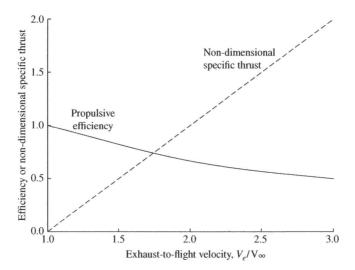

Figure 4.19 Comparison of propulsive efficiency and thrust.

which is simply Equation (4.56) rearranged. It probably comes as no surprise, that a lower thrust engine has a higher propulsive efficiency than a higher thrust device and vice versa. Hence, a propulsion device that takes a large mass of air and imparts a small velocity increase to the freestream, such as a propeller, is more efficient than one that takes a smaller mass of air and increases the freestream velocity by a large amount, such as a jet engine or rocket motor.

4.5.2 The Ramjet

The ramjet is perhaps the simplest air-breathing engine in both concept and mechanical construction. A ramjet may have no moving parts at all, in contrast to the complex moving machinery inside an internal combustion engine or the rapidly rotating turbomachinery in a jet engine. The ramjet is a high-speed propulsion device, providing optimum performance at about Mach 3–5.

The concept of ramjet propulsion was envisioned in the early 1900s, with the first patent of a subsonic ramjet cycle in the United States by Lake in 1909. This was followed by the patent of a ram compression-based jet engine in 1913 by the Frenchman Rene Lorin (1877–1933). Neither Lake nor Lorin were able to construct an engine based on their concepts, due to the many technological limitations of their time, notably the lack of suitable materials to handle the high temperatures involved with the operation of a ramjet engine.

The primary components of an axially symmetric ramjet engine are shown in Figure 4.20. The flow properties through the engine are depicted. Air enters the ramjet inlet, which can be of several different designs. An inlet with a sharp-nosed centerbody is shown in Figure 4.20. Let us examine the flow through the ramjet in supersonic flight.

The supersonic freestream air strikes the inlet centerbody, generating an oblique shock wave. The freestream air passes through this oblique shock wave and is compressed, increasing the static pressure and temperature and decreasing the total pressure. The flow is still supersonic downstream of the oblique shock (OS) wave, but at a lower Mach number than the freestream Mach number. The inlet compression system terminates in a normal shock (NS) wave, which decelerates the flow to subsonic speed. The static pressure and temperature are increased further, and the total pressure decreased further by the normal shock wave. The flow enters the diffuser, a duct with increasing

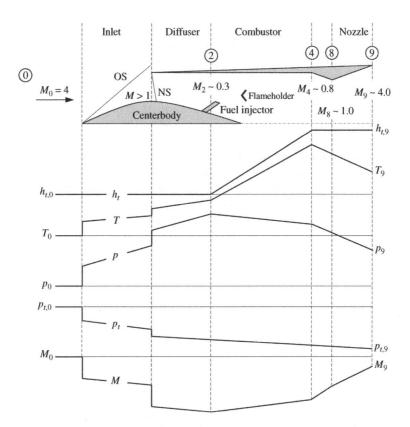

Figure 4.20 Ramjet engine components and internal flow parameters. (Source: *Adapted from Aircraft Propulsion, S. Farokhi, Fig. 12.43, p. 886, (2014),* [3], *with permission from John Wiley & Sons, Inc.*)

area, which further decreases the Mach number, static pressure, and static temperature. The air flow enters the combustor at about Mach 0.3 where fuel is added from fuel injectors. The fuel and air are mixed and burned in the combustor, where the combustion flames are stabilized by flameholders. The heat addition from the combustion increases the static temperature and the Mach number, while the static and total pressures are decreased through the combustor. The combustion products exit the combustor and enter the convergent-divergent nozzle at about Mach 0.8. The subsonic flow accelerates in the convergent part of the nozzle, reaching Mach 1 at the nozzle throat. In the divergent section of the nozzle, the flow expands and accelerates to about Mach 5 at the nozzle exit. The static pressure, static temperature, and total pressure decrease through the nozzle.

Despite its simplicity and advantages, the ramjet has its limitations and disadvantages. If we had a ramjet on a table in front of us, we could not "turn it on" and have air flowing through the engine. There is no mechanism to suck in air, compress the air, and produce thrust in this static condition; in other words, the ramjet cannot produce static thrust. The ramjet must be in *motion* to produce thrust, and while it can operate at subsonic flight conditions, its performance and efficiency are poor and it may not be capable of producing positive thrust (thrust greater than drag) at these low speeds. The ramjet's performance is best at supersonic speeds, starting at about Mach 2–3, where ram compression of the air is accomplished through oblique and normal shock waves.

As the flight Mach number increases, the temperature of the flow entering the ramjet combustor increases dramatically. At about Mach 6, the temperature of the flow is so high that the molecules of oxygen and nitrogen, in the air entering the combustor, are broken apart into individual atoms of oxygen and nitrogen. This chemical reaction of the air, called *dissociation*, absorbs much of the chemical energy in the high temperature flow, resulting in poor combustion and low thrust. Hence, the ramjet has an upper limit of about Mach 6 based on this high temperature effect on combustion. This problem can be solved by performing the combustion at supersonic speeds, in an engine called a supersonic combustion ramjet, or *scramjet*, to be discussed in a later section.

Ramjet propulsion has found wide application in a variety of missile designs, but only limited success for manned aircraft. Since the ramjet cannot produce static thrust, a first stage propulsion system is needed to accelerate the device to operating speed, typically about Mach 3. This first stage could be a solid rocket motor or a jet engine.

An example of a ramjet-powered missile is the US Navy *Talos* missile, shown in Figure 4.21. Developed in the mid-1950s, the *Talos* missile was the first operational, ramjet system in the US Navy. A first-stage, solid rocket booster was used to accelerate the missile to about Mach 2.2 and was then jettisoned. The ramjet engine operated from Mach 2.2 to a cruising Mach number of about 2.7, at an altitude of 70,000 ft (21,000 m). The *Talos* missile had an inlet with a centerbody spike that compressed the supersonic freestream air through a conical shock wave. A cutaway view of the *Talos* ramjet is shown in Figure 4.22, depicting the relative simplicity of the ramjet engine design.

Figure 4.21   Ramjet-powered *Talos* missile with first stage rocket motor ignited. (Source: *US Navy.*)

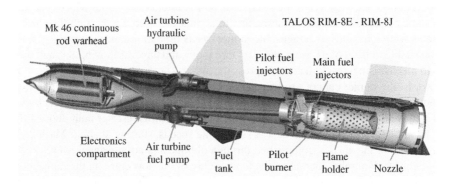

Figure 4.22   Cutaway view of *Talos* ramjet missile. (Source: *Courtesy of Phillip R. Hays, with permission.*)

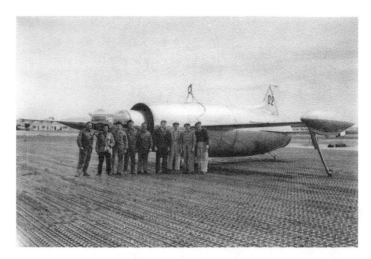

Figure 4.23 French Leduc 020 ramjet-powered research airplane. (Source: *User: Alain31-commonswiki, "Leduc 020"* https://en.wikipedia.org/wiki/File:Leduc020.jpg, *CC-BY-SA-3.0. License at https://creative commons.org/licenses/by-sa/3.0/legalcode.*)

There have been few examples of manned ramjet-powered aircraft. In the 1950s, the French engineer Rene Leduc designed and built several different models of a manned ramjet-powered research aircraft. One example, the Leduc 020 aircraft, shown in Figure 4.23, resembled a "flying ramjet engine" with wings. The nose of the aircraft was the centerbody compression spike of the ramjet engine. The pilot sat in this centerbody, which had clear windscreen surfaces for the pilot to look through. To solve the problem of the ramjet's inability to produce static thrust, Leduc's ramjet aircraft were carried on top of a specially modified, four-engine transport aircraft and then released at a subsonic airspeed where the ramjet could operate. The first flight of a Leduc ramjet-powered aircraft was on 21 April 1949, when test pilot Jean Gonord released from the carrier aircraft and climbed away on ramjet power. The flight testing of Leduc's ramjet-powered aircraft continued for several years, amassing almost 250 free-flights of the Leduc 021 alone. Ultimately, the higher performance and efficiency of turbojet propulsion was the demise of Leduc's ramjet-powered aircraft.

The application of ramjet propulsion has not been limited to missiles and airplanes. The Hiller Aircraft Company designed and built several prototype ramjet-powered helicopters in the 1950s. One example, the Hiller YH-32 *Hornet*, shown in Figure 4.24, used two Hiller 8RJ2B ramjet engines, mounted on the tips of the rotor blades. Each small ramjet engine weighed 13 lb (5.9 kg) and had a thrust of about 40 lb (178 N), producing a total equivalent power of about 90 hp (67 W). Since the ramjet engines could not produce static thrust, a small motor started the rotor spinning at 50 rpm, which started the thrust-producing flow through the ramjets. Once operating, the ramjet engines spun the main rotor at about 550 rpm.

An advantage of the rotor-tip propulsion was that there was no torque developed from the main rotor. In a conventional helicopter, the main rotor blades are spun by an engine attached to the helicopter, so that the helicopter tries to spin in the opposite direction to the rotor. The ramjet helicopter did not produce this torque, so a conventional anti-torque tail rotor was not required. This made the mechanical design of the ramjet helicopter simpler than a conventionally powered helicopter. The rotor tip speeds still needed to remain subsonic, as in a conventional helicopter, which meant that the ramjet engines were operating at subsonic speeds where their performance and efficiency were poor. This resulted in poor overall vehicle performance, including high fuel consumption, low range, and a low maximum airspeed of about 80 mph (129 km/h). The ramjet

Figure 4.24   Hiller YH-32 *Hornet* ramjet-powered helicopter. (Source: *US Air Force photo courtesy of The Ray Watkins Collection.*)

engines and their exhaust also proved to be extremely noisy and highly visible. While the exhaust flames coming out of the ramjets were probably a spectacular sight, the noise and high visibility were not advantageous from a flight operations perspective. Although military versions of the Hiller ramjet-powered helicopter were evaluated by the US Army and US Navy, the concept of ramjet-powered helicopters was abandoned due to their deficiencies and the improvements in more conventional, turbine-powered helicopters.

Example 4.6 Ramjet Performance *A ramjet-powered missile is flying at Mach 4.5 and an altitude of 55,000 ft. The ramjet engine is producing 1200 lb of thrust with a fuel flow rate of 0.654 lb/s. Calculate the specific impulse, and the thrust specific fuel consumption of the ramjet.*

Solution

Using Equation (4.57), the specific impulse is

$$I_{sp} \equiv \frac{T}{\dot{W}_f} = \frac{1200 \, lb}{0.654 \, \frac{lb}{s}} = 1835 \, s$$

Using Equation (4.58), the thrust specific fuel consumption is

$$TSFC \equiv \frac{\dot{W}_f}{T} = \frac{0.654 \, \frac{lb}{s} \times \frac{3600 \, s}{h}}{1200 \, lb} = 1.962 \frac{lb}{lb \cdot h}$$

4.5.3   The Gas Generator

The *gas turbine generator* – often simply called a *gas turbine* or *gas generator* – is the core building block for many air-breathing engines that use turbomachinery. The turbojet, turbofan, and turboprop all have a gas generator at their core, around which other necessary components are added. The gas generator can be considered an internal combustion engine, since the combustion occurring inside the gas generator changes the composition of the working fluid, from air and fuel to the combustion gas products of the fuel–air mixture. The purpose of the gas generator in an aircraft engine is to produce a supply of high pressure, high temperature gas for use by the engine to produce thrust.

The invention of the gas turbine engine was far ahead of its time. In 1791, an English inventor, John Barber (1734–1801) patented the gas turbine engine shown in Figure 4.25. Wood, coal, oil, or other combustible material was burned to create a hot gas, which was collected in a receiver and cooled. The gas was then compressed and pumped into a combustion chamber, which Barber called an "exploder", and ignited. The hot high-pressure combustion products were exhausted onto paddle wheel vanes, producing motive power. Unfortunately, the materials and manufacturing technology of the late 18th century was insufficient for Barber to build his engine. Barber's gas turbine concept was fundamentally sound, as evidenced by the fabrication of several functioning devices in modern times.

As shown in Figure 4.26, the gas generator is composed of three primary components, a *compressor*, a *burner* or *combustor*, and a *turbine*. The gas generator station numbering starts with station 2 at the compressor entrance or *compressor face*. This is done to be consistent with future station

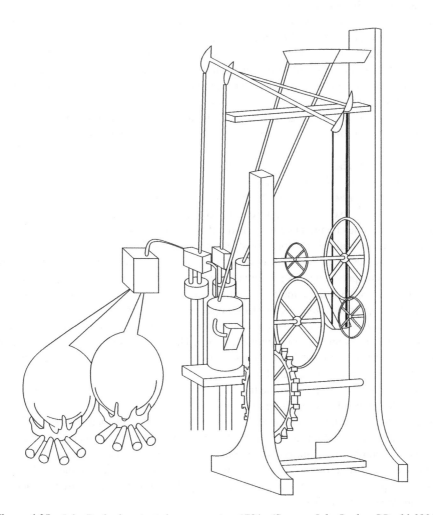

Figure 4.25 John Barber's patented gas generator, 1791. (Source: *John Barber, PD-old-100.*)

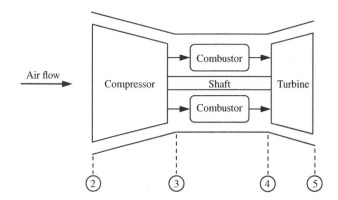

Figure 4.26   Components of a gas generator.

numbering for air-breathing engines with a gas generator, where additional engine components are located upstream of the compressor.

A mass flow of air is supplied to the compressor (station 2 in Figure 4.26) through an air inlet. The diameter of the compressor circular inlet area or *engine face* is usually sized by this air mass flow rate. The function of the compressor is to increase the pressure and temperature of the air to conditions that are optimum for efficient combustion. The compressor "squeezes" the air to a smaller volume, allowing the combustion to occur at a reduced volume. The compressor is driven by the turbine, which is connected to the compressor by a shaft or *spool*.

The compressor may be of the axial flow or centrifugal flow type. In the axial flow compressor, the direction of the air flow is parallel to the axis of rotation, axially through the compressor. The axial compressor is composed of alternating rows of rotating blades, called *rotors*, and stationary blades, called *stators*. The compressor blades have airfoil type cross-sections and very low aspect ratio. The air is *mechanically compressed* by each set of rotating and stationary blades, collectively called a *compressor stage*, incrementally increasing the pressure and temperature of the air. The diameter of the compressor decreases through each stage with the decrease in the cross-sectional area proportional to the change in density, as the flow is compressed.

Typically, there are multiple compressor stages, as many as 20 in modern jet engines. Each compressor stage increases the pressure of the air by only a small amount, perhaps as little as 10–15%, which makes the compression process more efficient. To obtain even higher efficiencies, there may be dual compressors, with their multiple, individual stages, that are rotated at different speeds. The upstream *low-pressure compressor* (LPC) operates at a lower pressure than the downstream, *high-pressure compressor* (HPC). In these dual compressor configurations, there are also dual, concentric spools that are connected to separate turbines. The compressor in a modern jet engine may have an overall pressure ratio of 15 or 20, meaning that the pressure exiting the compressor is 15 or 20 times higher than when it entered.

In the centrifugal flow compressor, the flow is turned 90° from the rotational axis in the compression process. Air enters the compressor at the center of an impeller, as shown in Figure 4.27, and is compressed through rotation and acceleration of the flow. The air then enters a diffuser where the velocity decreases and the pressure increases. The flow is fed into a manifold where it is then dumped into the combustor. Compared to the axial compressor, the centrifugal compressor is less efficient and has a larger cross-sectional area, which results in higher aerodynamic drag. Multiple centrifugal compressors are used to increase efficiency. The maximum pressure ratio across a centrifugal compressor is about 5:1.

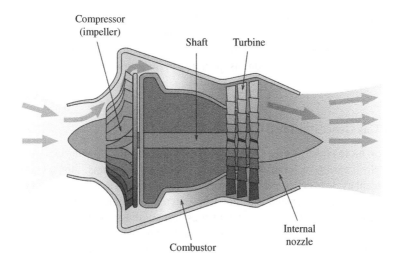

Compressor
(impeller)

Shaft   Turbine

Internal
nozzle

Combustor

Figure 4.27 Centrifugal flow compressor. (Source: *Adapted from User: Tachymeter, "Turbojet Operation Centrifugal Flow" https://en.wikipedia.org/wiki/File:Turbojet_operation-centrifugal_flow-en.svg, CC-BY-SA-3.0. License at https://creativecommons.org/licenses/by-sa/3.0/legalcode.*)

The air exits the compressor and enters the combustor (station 3 in Figure 4.26) at a high pressure, high temperature, and low velocity. The flow entering the combustor typically has a Mach number of about 0.3 or less. In the combustor, the air is mixed with fuel, ignited, and burned at near constant pressure (stations 3 to 4 in Figure 4.26). About half of the air that enters the combustor is not mixed with fuel and flows along the burner surfaces to provide cooling. This air, called *secondary air*, is heated and later mixed with the *primary air* that is mixed and burned with fuel, before exiting the combustor. The hot combustion gases are designed to exit the combustor at a uniform temperature, not to exceed the material limits of the turbine.

The combustion products enter the turbine (station 4 in Figure 4.26), another set of rotating and stationary blades, at maximum temperatures of between about 1750–2000 K (2700–3100 °F). This is beyond the material temperature limit of turbine blade materials, typically a nickel-based superalloy. Several cooling strategies are used to enable the turbine blades to survive the extremely high temperature environment. The turbine blades are actively cooled inside and out, by circulating cooling air through passages internal to the blades and by flowing a cooling film of air over their surfaces. Cooling air is drawn from the compressor for active cooling in the turbine. The turbine blades may also have ceramic or other high temperature coatings for thermal protection. The *turbine inlet temperature* is usually the limiting factor in operation and maximum performance of the engine. Since the combustor and turbine both handle the hot combustion gases, they are collectively called the "hot section" of the engine.

The high-pressure, high-temperature combustion gas products are *expanded* through the turbine, decreasing the gas pressure and temperature. Unlike the compressor, which *does work on the working fluid (air)*, the turbine *extracts work from the working fluid (combustion gas products)*. The rotating turbine is connected to the compressor through a shaft, or *spool*, making the compressor rotate. Approximately 75% of the energy derived from combustion is used by the turbine to drive the compressor.

There are fewer turbine stages than compression stages since the expansion process can be performed in larger pressure increments than the compression process. The pressure increases in the compression process result in an adverse pressure gradient for the boundary layers, making them more susceptible to flow separation. The pressure decrease through the turbine results in a favorable pressure gradient, which is less susceptible to flow separation. The efficiency of turbines is typically higher than compressors due to these favorable pressure gradients. The exhaust gas exits the turbine (station 5 in Figure 4.26) and is available to the engine to extract additional work and produce thrust.

4.5.3.1   The Gas Generator Ideal Cycle: the Brayton Cycle

Let us now look more closely at the thermodynamic details of the flow through the gas generator with the aid of the pressure–volume and temperature–entropy, diagrams, shown in Figure 4.28. The gas generator ideal cycle is called the Brayton cycle, named after George Brayton (1830–1892), an American mechanical engineer and inventor who patented a single-cylinder internal combustion engine in 1872, which operated with constant pressure combustion. The Brayton cycle assumes that the flows through the compressor and turbine are isentropic, that is, adiabatic and reversible.

The ideal gas generator cycle starts at the thermodynamic state labeled as point 2, where air enters the compressor with a static pressure and temperature of p_2 and T_2, respectively. The pressure is increased from state 2 to 3, with work entering the system to drive the compressor. Since the compression process is assumed to be isentropic, the temperature increases along a constant entropy line, or *isentrope*, from state 2 to state 3, exiting the compressor at a temperature T_3.

The combustion process occurs between states 3 and 4. The combustion is assumed to occur along a constant pressure line, or *isobar*, such that the pressure at states 3 and 4 are equal ($p_3 = p_4$). The entropy increases in the combustion process ($s_4 > s_3$) due to wall friction and turbulent mixing losses. Significant heat is added to the gas, increasing the gas total temperature from $T_{t,3}$ to $T_{t,4}$, where $T_{t,4}$ is a limit temperature, typically set by the turbine. The maximum turbine inlet temperature is usually set by the temperature limits of the turbine blade materials, factoring in the benefits obtained from thermal barrier coatings and active blade cooling. Current, state-of-the-art, maximum turbine inlet temperatures are about 2000 K (3600 °R). The maximum flight Mach number of an aircraft powered by jet engines can be limited by the maximum turbine inlet temperature.

For the ideal turbine process, from state 4 to 5, there are no losses, so the entropy of the gas remains constant ($s_4 = s_5$) and the temperature decreases from T_4 to T_5 along an isentrope. Work is extracted from the flow, reducing the gas pressure from p_4 to p_5. In the Brayton cycle, the gas returns to its initial state 2 to complete the ideal thermodynamic cycle. In an actual gas generator process, there is no closed thermodynamic cycle, as the working fluid exits the turbine as combustion products.

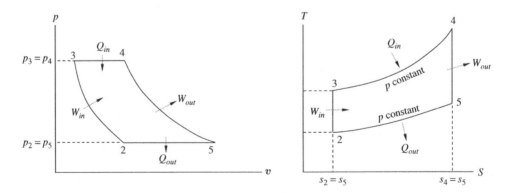

Figure 4.28   Gas generator ideal Brayton cycle.

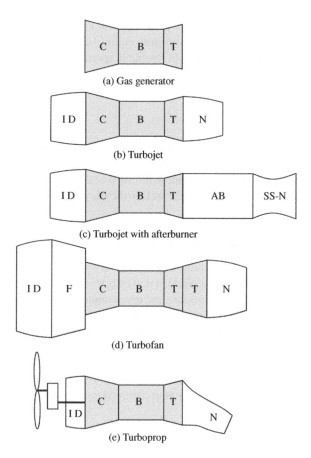

Figure 4.29 Air-breathing turbine engines based on the gas generator (I D = inlet-diffuser, C = compressor, B = burner, T = turbine, N = nozzle, SS-N = supersonic nozzle, AB = afterburner, F = fan).

4.5.3.2 Air-Breathing Engines Based on the Gas Generator

Several different air-breathing, turbine engines are built around a basic gas generator at their core, as shown in Figure 4.29. These are *ducted engines*, where the momentum of the working fluid (air) is increased as it flows through a duct. The working fluid is mechanically compressed, combusted with a fuel, expanded through a turbine to drive the compressor, and expanded through a nozzle to produce thrust. A significant advantage of these turbine-based engines over the ramjet is their capability to generate static thrust. Their capability to mechanically compress the freestream air allows these engines to generate thrust at zero airspeed. This capability comes at a price, since power must be extracted from the flow by the turbine to drive the compressor.

The basic *gas generator*, composed of a compressor (C), burner (B), and turbine (T), is shown in Figure 4.29a. The *turbojet* engine is composed of the basic gas generator with an *air inlet and diffuser* (I D) upstream of the compressor and an *exhaust nozzle* (N) downstream of the turbine, as shown in Figure 4.29b. An *afterburner* can be added downstream of the turbine in the turbojet (or turbofan), as shown in Figure 4.29c, allowing for additional fuel injection and burning, resulting in a significant thrust increase. The basic *turbofan* engine, shown in Figure 4.29d, is created by adding a *fan* (F) between the inlet and compressor with an additional turbine (T) to drive the fan. Some of the inlet air passing through the fan bypasses the compressor and is exhausted around the engine, acting much like a ducted propeller. The *turboprop* engine is composed of the basic gas generator with a turbine added to drive a propeller, as shown in Figure 4.29e. The rotational speed of the gas generator shaft is reduced through a gearbox to rotational speeds appropriate for the propeller. The turbojet, turbofan, and turboprop engines are discussed in more detail in the following sections.

4.5.4 The Turbojet Engine

The basic turbojet engine is constructed by adding an inlet and an exhaust system to the gas generator. The exhaust system may consist of a nozzle or may include an afterburner duct, where additional fuel is added to generate substantially more thrust. We start by looking at how the thermodynamic cycle is changed by adding an inlet and nozzle to the gas generator, to form the ideal turbojet.

4.5.4.1 Ideal Turbojet Thermodynamic Cycle

The thermodynamic cycle of the ideal turbojet is shown in Figure 4.30, where an inlet is added upstream of the compressor and a nozzle is added downstream of the turbine. The air flow entering the inlet is compressed, increasing the pressure and temperature and decreasing the volume (states 0 to 2). The inlet flow is assumed to be isentropic, following the same isentrope as the compressor flow process, on the T–s diagram, and the same line of constant pv^γ as the compressor flow on the p–v diagram. The constant pressure combustion process from state 3 to 4 is the same as the gas generator. After being expanded through the gas generator turbine (state 4 to 5), the flow is expanded further in the nozzle (state 5 to 9), decreasing the pressure and temperature and increasing the volume. Similar to the inlet, the nozzle flow is assumed to be isentropic in this ideal case.

4.5.4.2 Turbojet Flow Properties and Thrust

The components of a turbojet engine, without an afterburner section, are shown in Figure 4.31. The gas generator forms the core of this engine. The function of the air inlet is to capture the required mass flow of air and efficiently decelerate the flow to a subsonic speed for entrance to the compressor. In decelerating the flow, the inlet also provides some compression of the air, increasing its pressure. The inlet must operate efficiently through a wide range of speeds, from zero to possibly high supersonic speeds. The inlet geometry is shaped to minimize flow losses for the given speed regime.

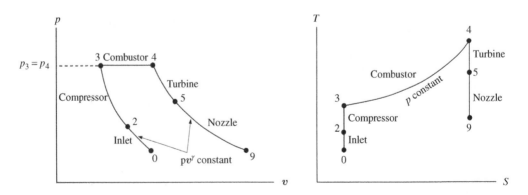

Figure 4.30 Ideal turbojet cycle.

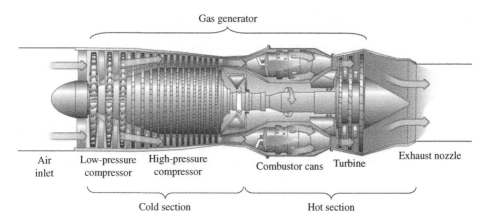

Figure 4.31 Components of a turbojet engine. (Source: *Adapted from Fig. 14-1, Airplane Flying Handbook, Federal Aviation Administration, FAA-H-8083-3A, 2004.*)

Inlets designed for subsonic flight only can be *fixed geometry*, divergent ducts (recall that a subsonic flow decelerates in a diverging duct). The inlet cross-sectional area must be sized to capture the required mass flow of air for the desired thrust level. As the flight Mach number increases to supersonic speeds, shock waves form in front of a fixed geometry, subsonic inlet, reducing its efficiency. At Mach numbers greater than about 1.6–1.8, an air inlet requires *variable geometry* for efficient operation, where the geometry of the inlet can be changed with Mach number. In the design of a supersonic inlet, the location and strength of shock waves are a major consideration in obtaining acceptable inlet efficiency and performance. The details of subsonic and supersonic inlets are discussed in Section 4.5.7.

The workings of the compressor, combustor, and turbine, as described for the gas generator, are the same in the turbojet. The compressor accepts the inlet air and increases its pressure and temperature for the combustor. Typically, there is a low-pressure compressor section and a high-pressure section, as described earlier for the gas generator. The compressor section operates at much lower temperatures than the combustor and turbine, hence, it is termed the *cold section* of the engine. Air extracted from the compressor section, called *bleed air*, is used for various purposes in the aircraft, including turbine cooling, cabin pressurization, and heating for engine inlet anti-icing. Fuel is mixed with the air from the compressor and burned in the combustor. The turbine extracts work from the expanding combustion gases to drive the compressor and other aircraft accessories, such as fuel, oil, and hydraulic pumps. The *hot section* refers to the combustor and turbine, which operate at much higher temperatures than the compressor sections.

After exiting the turbine, the flow enters the nozzle where it is further expanded, reducing the pressure and increasing the velocity. The function of the exhaust nozzle is to accelerate the flow exiting the turbine to a high velocity and high momentum, generating thrust to propel the aircraft. For low thrust, subsonic aircraft, the exhaust nozzle can be a simple converging duct. The velocity increases and the pressure decreases in the diverging area. A supersonic aircraft requires a convergent-divergent nozzle, where the subsonic flow, exiting the turbine, accelerates in the converging section to Mach 1 at the nozzle throat and then expands to supersonic speeds in the diverging section. The flow expansion is controlled by the pressure at the turbine exit or nozzle entrance and the ambient pressure at the nozzle exit. Most convergent-divergent nozzles are variable geometry, where the nozzle throat and exit areas can be changed to maximize the thrust.

The afterburner (not shown in Figure 4.31) is a long duct, added to the back of some turbojet engines, where additional fuel is injected directly into the exhaust stream. The burning of this additional fuel results in a 50–80% thrust increase, but with a significant increase in fuel consumption. Afterburners are used by military aircraft for acceleration and maneuvering. The *Concorde* supersonic airliner used afterburners to sustain Mach 2 cruise flight. The afterburner duct appears almost "empty" as compared with the turbomachinery-filled gas generator. The length of the afterburner is required to provide adequate time for the injection, mixing, and burning of the fuel into the high-speed air stream exiting the turbine. Spray bars are used to inject and distribute the fuel over the cross-sectional area of the afterburner duct. The combustion in the afterburner is stabilized by flameholders, small bluff bodies that create pockets of recirculating, low velocity flow that "hold" the flame. The operation of the afterburner relies on the availability of excess oxygen, in the flow exiting the turbine, to burn with the fuel being injected. This *fuel lean* condition means that less fuel was added in the combustor than could burn with the available oxygen in the air. Thus, the combustion products exiting the turbine have unburned oxygen that can be burned in the afterburner.

A cutaway view of the General Electric J85 non-afterburning, turbojet engine is shown in Figure 4.32. This turbojet powers the Cessna A-37 attack aircraft, while the afterburning models are used in the Northrop T-38 and F-5. Addition of the afterburner duct approximately doubles the length of the engine.

The axial distributions of the flow properties through a turbojet engine are shown in Figure 4.33. The values of static pressure, static temperature, and velocity, averaged over the cross-sectional area, are plotted versus the axial distance through the engine. The pressure rises through the compressor, increasing to over ten times its entrance value due to the work done by the compressor on the air. The temperature also increases through the compressor. The velocity is constant through the compressor by design, since a constant velocity compression process is more efficient. Combustion occurs at near constant pressure in the combustor, with the temperature rising to a high value. The flow is expanded through the turbine, decreasing the pressure and temperature. The pressure drops significantly, as work is extracted from the flow by the turbine to drive the compressor. The flow is further expanded through the exhaust nozzle, significantly increasing the exhaust velocity. Two temperature distributions are shown for the exhaust, with and without afterburner. Without afterburner, the temperature remains about constant through the nozzle expansion. With afterburner in operation, the temperature increases to a very high value.

Figure 4.32   Cutaway view of a non-afterburning, General Electric J85 turbojet engine. (Source: *Sanjay Acharya, "J85-GE-17A Turbojet Engine"* https://en.wikipedia.org/wiki/File:J85_ge_17a_turbojet_engine. jpg, CC-BY-SA-3.0. *License at https://creativecommons.org/licenses/by-sa/3.0/legalcode.*)

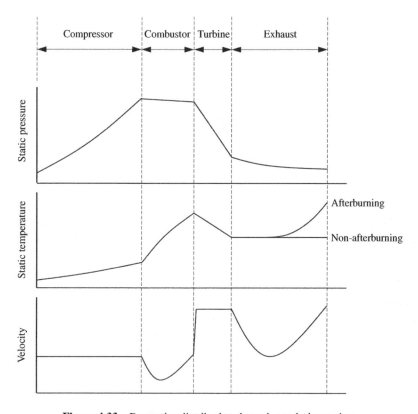

Figure 4.33 Properties distribution through a turbojet engine.

4.5.4.3 Birth of the Turbojet Engine

The invention of a gas turbine powered jet aircraft was far ahead of its time, with the first patent filed by the Frenchman Maxime Guillaume in 1921. Guillaume's invention was an amazingly modern version of the jet engine. It was an axial-flow engine with an axial-flow compressor that fed a combustor, the compressor being driven by an axial-flow turbine. Guillaume's engine was never constructed, as the technology of his time was not sufficient to build such a device to power an aircraft.

In the development of the jet engine, there were a myriad detailed engineering, manufacturing, and materials problems that had to be solved. The internal engine flow path environment exposed the various engine components to high stresses, high pressures, and high temperatures. The rotating components had to provide adequate performance and efficiency, while holding tight mechanical tolerances and clearances from the outer engine case. These problems were even more challenging when the engine had to be integrated into an aircraft, where weight is so critical.

The birth of the practical jet engine was the result of the near simultaneous, yet independent, efforts of two men, Frank Whittle (1907–1996) in England and Hans von Ohain (1911–1998) in Germany. Von Ohain designed and built an axial flow turbojet engine that was successfully ground tested in September 1937. Interestingly, this first jet engine used gaseous hydrogen fuel, because of its wide combustibility range as compared with hydrocarbon fuels and to avoid the then unsolved combustion problems with liquid hydrocarbon fuels. Von Ohain's later jet engines used liquid hydrocarbon fuels. Prior to these engine ground tests, von Ohain was already working with the German aircraft manufacturer, Ernst Heinkel, who had a passion for developing higher-speed, higher-altitude aircraft. Their combined efforts culminated in the first flight of a jet-powered

Figure 4.34   Heinkel He 178, the first jet airplane. (Source: *US Air Force, PD-US.*)

airplane, the Heinkel He 178 with a von Ohain designed HeS 3b turbojet engine (Figure 4.34) on 27 August 1939. Using gasoline for fuel, the HeS 3b turbojet engine produced about 1000 lb (4448 N) of thrust.

In January 1930, over nine years before the first flight of the He 178, a Royal Air Force officer, Frank Whittle, filed a patent for a centrifugal flow turbojet engine. Whittle tried for years to obtain government support to develop and build his jet engine, without success. Finally, in 1935, he received private financial support to build his first engine. Whittle's engine was a centrifugal flow engine, in contrast to von Ohain's axial flow engine. In April 1937, a Whittle turbojet engine, with a single-stage centrifugal compressor and a single-stage turbine, was successfully ground tested. On 15 May 1941, the Gloster E.28 *Pioneer* experimental aircraft, powered by a Whittle turbojet engine, made its first flight (Figure 4.35).

A comparison of the world's first two jet aircraft and their engines is given in Table 4.5. The He 178 was about the same length as the E.28, but it weighed almost 700 lb (300 kg) more and had a much smaller wing. In addition to the smaller wing, the He 178 had a more slender fuselage than the E.28, due to the smaller diameter of its axial flow jet engine as compared with the *Pioneer*'s

Figure 4.35   Gloster E.28 *Pioneer* jet engine powered airplane. (Source: *S.A. Devon, Royal Air Force Official Photographer, PD-UKGov.*)

Table 4.5 Comparison of the first two jet-powered aircraft.

Item	Heinkel He-178	Gloster E.28
Jet engine inventor	Hans von Ohain	Frank Whittle
Aircraft designer	Ernst Heinkel	George Carter
First flight	27 August 1939	15 May 1941
Length	24.5 ft (7.48 m)	25.3 ft (7.74 m)
Wingspan	23.25 ft (7.20 m)	29 ft (8.84 m)
Wing area	85 ft^2 (7.9 m^2)	146 ft^2 (13.6 m^2)
Empty weight	3572 lb (1620 kg)	2886 lb (1309 kg)
Gross weight	4406 lb (1998 kg)	3750 lb (1701 kg)
Powerplant	Heinkel HeS 3b turbojet	Power Jets W.1 turbojet
Thrust	992 lb (4413 N)	860 lb (13,760 N)
Thrust-to-weight	0.225	0.229
Maximum speed	435 mph (700 km/h)	338 mph (544 km/h)

centrifugal flow engine. The Heinkel HeS 3b jet engine produced over 130 lb (580 N) more thrust than the Power Jets W.1 engine. The maximum speed of the He-178 was about 100 mph (160 km/h) faster than the E.28. The aircraft performance difference may be due to the aerodynamic impacts of the different fuselage and wing sizes. The thrust-to-weight ratio of the two aircraft are nearly identical.

The USA was also eager to enter the jet age, signing an agreement with the British, which gave them a Whittle jet engine and a set of engine drawings. This sharing of hardware and information led to the first ground test of a jet engine in the USA on 18 March 1942, a General Electric (GE) copy of the Whittle engine. The GE engine had a weight of about 1000 lb (450 kg) and developed 1250 lb (5560 N) of static thrust. The first flight of a jet-powered airplane in the USA occurred on 3 October 1942, with the flight of the Bell XP-59A *Aircomet*, powered by two General Electric GE 1-A turbojet engines (Figure 4.36).

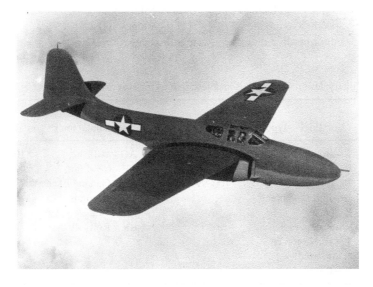

Figure 4.36 Bell XP-59A *Aircomet*, the first US jet airplane. (Source: *US Air Force.*)

4.5.4.4 Jet Engine Designations

There is an "alphabet soup" of letters and numbers for the designation of jet engines. Commercial engine manufacturers use their own designation systems. Below are a few examples of engine designations from some major commercial engine companies.

The General Electric CF6-50E2 high-bypass turbofan engine

GE	company	General Electric
CF6	model or "family"	commercial fan 6
50	specific engine series	originally, representative of takeoff thrust rating
E2	specific aircraft type	version for Boeing 747

The Rolls Royce RB.211 *Trent* high-bypass turbofan engine

RB	company	Rolls Barnoldswick (Rolls Royce, Ltd.)
211	engine family	numeric designation used during development
Trent	British river	Rolls Royce assigns the name of a British river to their jet engines after the engine is put into service

The US military have their own engine designation system, which includes renaming of commercial engines used in military aircraft. The US military designator symbols and descriptions for air-breathing engines is given in Table 4.6. In addition to the designations in the table, there may be an additional suffix letter at the end of the engine designation signifying a minor modification. These are then illustrated by the following examples of the US military designation system.

The Pratt & Whitney YF119-PW-100 L turbofan engine, used in the YF-22 and YF-23 prototype advanced tactical fighter (ATF) aircraft, is decoded as follows.

Y	status prefix	prototype engine for YF-22 and YF-23 prototype ATF aircraft
F	engine type	turbofan
119	model number	model 119
PW	engine manufacturer	Pratt & Whitney
100	specific engine model	specific model 100
L	minor modification	modified engine

Table 4.6 US Military designator symbols and descriptions for air-breathing engines.

Status Prefix (optional)	Engine type	Engine model	Engine manufacturer*	Specific model
X – Experimental	C – Rotating combustion	100–399 Air Force	AD – Allison	100–399 Air Force
Y – Prototype	F – Turbofan	400–699 Navy	CF – CFM Intl.	400–699 Navy
	J – Turbojet	700–999 Army	GE – General Electric	700–999 Army
	O – Piston, opposed		LD – Lycoming	
	P – Other		PW – Pratt & Whitney	
	R – Piston, radial		RR – Rolls Royce	
	T – Turboprop or turboshaft			
	V – Piston			

*Listing does not include all possible engine manufacturers.

The General Electric F404-GE-402 turbofan engine, used in the Boeing F/A-18C/D *Hornet*, is decoded as follows.

F	engine type	turbofan
404	model number	model 404
GE	engine manufacturer	General Electric
402	specific engine model	specific model 402

The General Electric F404-GE-IN20 turbofan engine, used in India's light combat aircraft, *Tejas*, is decoded as follows.

F	engine type	turbofan
404	model number	model 404
GE	engine manufacturer	General Electric
IN20	specific engine model	F404 engine for the Indian Air Force

4.5.5   The Turbofan Engine

As discussed in Section 4.5.1.4, the propulsive efficiency of the turbojet is low because it takes a relatively small mass flow rate of air and imparts a large velocity increase to the exhaust flow. The efficiency can be improved by increasing the inlet air mass flow rate and decreasing the exhaust flow velocity. The turbofan engine does this by adding a large ducted fan in front of the compressor, which increases the inlet mass flow rate and extracts energy from the exhaust stream. Some of the kinetic energy in the exhaust flow is used to increase the inlet mass flow rate and decrease the exit velocity, which increases the net propulsive efficiency of the engine. Most modern commercial and military jet aircraft use turbofan engines due to the significant increase in efficiency over turbojets, without a lack of performance.

The turbofan engine has the same components as a turbojet engine, plus an additional turbine stage to drive the large fan in front of the compressor. The basic components of a high bypass ratio, twin-spool turbofan engine are shown in Figure 4.37. The separate, low-pressure and high-pressure turbine stages are operated at different pressures and rotate at different speeds. They also drive separate, concentric shafts or *spools*. The additional low-pressure turbine stage may also drive an additional low-pressure compressor stage, upstream of the high-pressure compressor. The fan flow exhausts through a bypass nozzle and may generate a significant percentage of the total engine thrust.

The *bypass ratio* is defined as the ratio of the air mass flow rate through the fan, $\dot{m}_{fan}$, to the air mass flow rate through the gas generator or *core*, $\dot{m}_{core}$. The bypass ratio is given the symbol α (not to be confused with angle-of-attack) and is given by

$$\alpha = \frac{\dot{m}_{fan}}{\dot{m}_{core}} \tag{4.71}$$

The total mass flow rate of air entering the engine, $\dot{m}_{\infty}$, is the sum of the mass flow rates of the core and the fan, and may be written in terms of bypass ratio as

$$\dot{m}_{\infty} = \dot{m}_{core} + \dot{m}_{fan} = (1 + \alpha)\dot{m}_{core} \tag{4.72}$$

There is a tradeoff between engine efficiency and performance concerning bypass ratio. The engine efficiency improves with increasing bypass ratio, but the fan diameter also gets larger, which

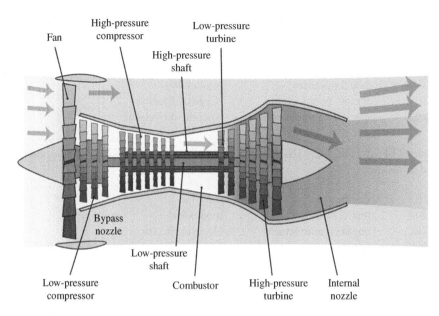

Figure 4.37 Components of a high-bypass ratio, twin-spool turbofan engine. (Source: *Adapted from K. Aainsqatsi, "Turbofan Operation" https://en.wikipedia.org/wiki/File:Turbofan_operation.svg, CC-BY-SA-3.0. License at https://creativecommons.org/licenses/by-sa/3.0/legalcode.*)

increases the aircraft frontal area and hence the aerodynamic drag. At one extreme of bypass ratio is the pure turbojet with a bypass ratio of zero, since there is no fan flow. Early turbofan engines had low bypass ratios of about 1 to 2. Modern military aircraft have low bypass ratios of less than one, as a compromise between fuel efficiency and performance. Modern commercial airliners have high bypass ratios of between about 5 and 10. Future aircraft may have *ultra-high bypass* ratios of between 10 and 20, with potentially higher efficiency.

The bypass ratio and other selected specifications of a few turbofan engines are given in Table 4.7. The turbofan engines used in the military fighter aircraft have low bypass ratios of around 0.3:1. The turbofan engines used in commercial and military transport aircraft (Boeing 757, 777, and C-17) have high bypass ratios of about 6:1. The modern turbofan engines used in the Boeing 787 have even higher bypass ratios of 10:1. The increase in thrust specific fuel consumption (TSFC) and fan size, with increasing bypass ratio, is evident in the table.

Table 4.7 Selected specifications of turbofan engines.

Engine	Aircraft	Fan diameter	Thrust specific fuel consumption, TSFC	Bypass ratio, α	Fan pressure ratio
General Electric F404	F-18	31 in	0.81 lb/(lb-h)*	0.34:1	3.9:1
Pratt & Whitney F100	F-15, F-16	35 in	0.76 lb/(lb-h)*	0.36:1	3.0:1
Pratt & Whitney PW2000	B757, C-17	78.5 in	0.33 lb/(lb-h)	5.9:1	1.74:1
Pratt & Whitney PW4000	B777	112 in	–	6.4:1	1.7:1
Rolls Royce *Trent 1000*	B787	112 in	–	10:1	–

*Military power (maximum, non-afterburning)

In a high bypass ratio turbofan engine, the thrust from the fan flow is usually much greater than the thrust from the core flow. The fan flow thrust may be as high as 75% of the total engine thrust. In some respects, the fan is similar to a large propeller with a duct around it. However, unlike a conventional propeller, a fan is composed of a large number of *fan blades*, perhaps as many as 50, surrounded by a *shroud*. The fan functions somewhat like a single-stage compressor, increasing the pressure of the flow through the fan. This is usually expressed as the fan *pressure ratio*, which is given in Table 4.7. The military turbofan engines increase the pressure of the fan flow about two- or three-fold, while the commercial engines increase the pressure by about 70%.

In the high bypass ratio turbofan engine, shown in Figure 4.37, the fan exhaust flow is separate from the core exhaust. In some turbofan engines, such as the low bypass ratio engine shown in Figure 4.38, the fan shroud is extended to the back of the engine to form a common nozzle with the engine core. This provides for mixing of the colder, lower-velocity fan exhaust flow with the hotter, higher-velocity core exhaust flow. This mixing of the fan and core flows, prior to expansion in the nozzle, can provide additional thrust, although there is a weight penalty due to the longer fan duct.

Example 4.7 Bypass Ratio and Mass Flow *If the total air mass flow rate, $\dot{m}_{\infty}$, through a Rolls Royce Trent 1000 high bypass ratio turbofan engine, is 2610 lb/s, calculate the mass flow rate through the engine core and the fan.*

Solution

The total air mass flow rate is related to the mass flow rate through the core by Equation (4.72) as

$$\dot{m}_{\infty} = (1 + \alpha)\dot{m}_{core}$$

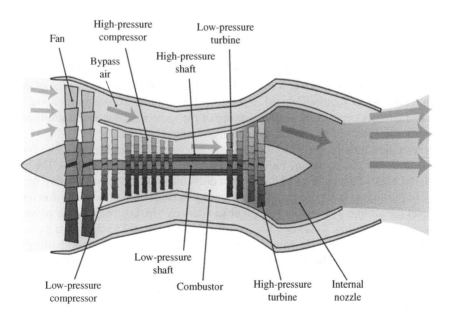

Figure 4.38 Components of a low-bypass ratio, twin-spool turbofan engine. (Source: *Adapted from K. Aainsqatsi, "Turbofan Operation" https://en.wikipedia.org/wiki/File:Turbofan_operation_lbp.svg, CC-BY-SA-3.0. License at https://creativecommons.org/licenses/by-sa/3.0/legalcode.*)

From Table 4.7, the Trent 1000 *has a bypass ratio of 10. Solving for the mass flow rate through the engine core, we have*

$$\dot{m}_{core} = \frac{\dot{m}_\infty}{1 + \alpha} = \frac{2610 \frac{kg}{s}}{1 + 10} = 237.3 \frac{kg}{s}$$

The bypass ratio is defined by Equation (4.71) as

$$\alpha = \frac{\dot{m}_{fan}}{\dot{m}_{core}}$$

Solving for the fan mass flow rate, we have

$$\dot{m}_{fan} = \alpha \dot{m}_{core} = (10) \left(237.3 \frac{kg}{s} \right) = 2373 \frac{kg}{s}$$

4.5.6 The Turboprop and Turboshaft Engines

The turboprop engine utilizes a propeller to generate thrust, where the propeller is driven by a gas generator. Almost all of the power output from the gas generator turbine is used to drive the propeller. The high velocity exhaust gases from the gas generator contribute about 10% of the total thrust. The basic turboprop engine is composed of two major assemblies, a *gas generator section*, and a *power section*. The gas generator section contains the basic components of a gas generator, a compressor, combustor, and turbine. The power section contains the power turbine, the reduction gearbox, and propeller driveshaft. There are inlet and exhaust ducts that direct air into the compressor and out of the turbine, respectively. Some of the high velocity, exhaust gas is expanded further through a nozzle to generate thrust.

In the *free power turbine* turboprop configuration, the propeller is driven by a *power turbine* that is independent or "free" of the gas generator turbine that drives the compressor. As such, the power turbine can rotate at a different speed from the gas generator turbine, which avoids the need for special transmissions. The power turbine turns a propeller of much larger diameter, so a *reduction gearbox* is required to prevent rotating the propeller at too high a speed, which would overstress the propeller. The reduction gearbox converts the low torque, high rotational speed of the turbine into a high torque, lower rotational speed of the propeller. A turboprop engine is about 1.5 times heavier than a turbojet engine with a gas generator of the same size, due to the additional weight of the power turbine, reduction gearbox, propeller, and propeller controls.

In the *reverse-flow* turboprop engine, the gas generator is placed "backwards". The components of a reverse flow, free power turbine turboprop engine are shown in Figure 4.39. The air intakes are towards the back of the engine and the exhaust ducts are at the front end. Placing the air inlets at the back of the engine is advantageous in preventing the ingestion of debris or foreign objects. Screens cover the air inlets to prevent foreign object damage (FOD). The inlet air passes through an axial compressor section, followed by a centrifugal compressor. The flow exits the centrifugal compressor, still going towards the front of the engine. The flow enters the combustors or burner cans and is burned with fuel. The combustion gases exit in the reverse direction, flowing towards the back of the engine. The flow is turned 180° and enters the turbine. The flow spins the compressor turbine, which drives the compressor, and the power turbine, which drives the propeller. After exiting the turbine, the exhaust gas is turned 180° once again, to exit the engine through an exhaust pipe (not shown in the figure), generating a small amount of thrust. The flow reversals, interior to the gas generator, make the engine more compact (shorter) than an axial-flow engine, albeit with a larger diameter owing to the centrifugal rotating machinery.

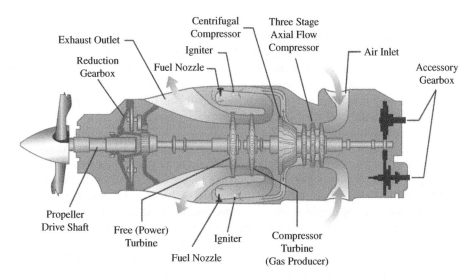

Figure 4.39 Components of a reverse flow, free power turbine turboprop engine. (Source: *Fig. 14-5, Airplane Flying Handbook, Federal Aviation Administration, FAA-H-8083-3A, 2004.*)

Figure 4.40 US Air Force C-12 *Huron* powered by two Pratt & Whitney PT-6A turboprop engines. (Source: *US Air Force.*)

Turboprop-powered aircraft have the same airspeed limitations of any propeller-driven aircraft, due to compressibility effects on the propellers near the speed of sound. Generally, turboprop aircraft are most efficient below flight speeds of about 450 mph (720 km/h). An example of a turboprop-powered airplane is the twin-engine Beechcraft C-12 *Huron*, shown in Figure 4.40. The C-12 is used by the US military for a variety of flight operations, including passenger, medical, and cargo transport and reconnaissance roles. The C-12 is powered by two Pratt & Whitney PT-6A turboprop engines with 850 shp (635 kW) (shaft horsepower) each.

The turboshaft engine is similar to a turboprop, but instead of turning a propeller, the turboshaft engine power output is used to turn a driveshaft. The driveshaft may turn a propeller or rotor, as in a rotorcraft. Unlike the turboprop engine, the hot exhaust gases of a turboshaft engine are expanded further to lower pressures in the turbine, thus extracting more work, which is added to the shaft power. Thus, the exhaust gases from a turboshaft engine contribute little to the total thrust. Because of their high power output, light weight, and small size, turboshaft engines have a variety of applications, including being used in rotorcraft, auxiliary power units, ships, tanks, and other industrial power generation equipment.

A unique application of a turboshaft engine is found in the Lockheed-Martin F-35 *Lightning II*, a *short takeoff and vertical landing* or *STOVL* supersonic fighter aircraft, shown in Figure 4.41. The STOVL propulsion system of the F-35 is shown in Figure 4.42. The F-35 has a single Pratt & Whitney F135-PW-600, low bypass ratio turbofan engine, with a swivel nozzle that can swivel 90° downward and a horizontally mounted *lift fan*, located just aft of the cockpit. In forward flight, the powerplant operates like a conventional turbofan engine. In hover mode, the powerplant functions as both a turbofan and a turboshaft engine. Some of the engine power is used to turn a shaft, which drives the lift fan, blowing unheated air downward, generating about 20,000 lb (89,000 N) of thrust. The turbofan is still operating, providing about 18,000 lb (80,000 N) of vertical thrust from the jet exhaust through the swivel nozzle at the rear of the aircraft. About 10% of the engine thrust can also be diverted to two roll posts under the wings for attitude control.

4.5.7   More about Inlets and Nozzles for Air-Breathing Engines

The inlet and nozzle are essential components of any ducted, air-breathing engine, such as ramjets, turbojets, and turbofans. The inlet must provide an efficient means for the engine to ingest the required mass flow of air. The nozzle must efficiently expand the combustion products of the engine to a high velocity to generate thrust. Unlike the other internal engine components that have been discussed, the inlet and nozzle are exposed to both an internal flow through the engine and an *external* air flow that affects their operation and performance. The flow through the inlet experiences an

Figure 4.41   Lockheed-Martin F-35 *Lightning II* short takeoff and vertical landing (STOVL) aircraft. (Source: *US Air Force.*)

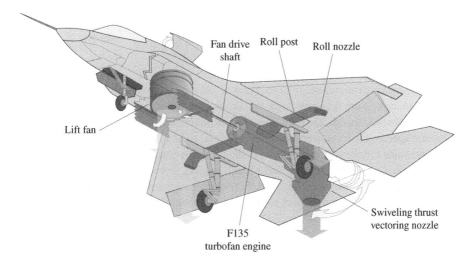

Figure 4.42 STOVL propulsion system of the Lockheed-Martin F-35 aircraft.   (Source: *Tosaka,* *"F-35B Joint Strike Fighter (Thrust Vectoring Nozzle and Lift Fan"* https://en.wikipedia.org/wiki/File: F-35B_Joint_Strike_Fighter_(thrust_vectoring_nozzle_and_lift_fan).PNG, *CC-BY-SA-3.0. License at* https://creativecommons.org/licenses/by-sa/3.0/legalcode.)

increasing pressure with downstream distance, while the nozzle flow is expanding with the pressure decreasing. The inlet flow sees an adverse pressure gradient, while the nozzle flow has a favorable pressure gradient. The potential for boundary layer separation of the inlet flow is exacerbated due to the adverse pressure gradient. In this section, more details about the inlet and nozzle are provided, in addition to defining some parameters that quantify their efficiency.

4.5.7.1   Inlet Requirements and Total Pressure Recovery

The inlet must deliver the required mass flow of air to the compressor or fan entrance or *face*, but there are several other critical requirements as to how this mass of air is delivered. The freestream flow must be accelerated or decelerated, depending on the phase of flight, to the appropriate subsonic Mach number acceptable to the compressor. Typically, this Mach number is about 0.4–0.6. If the flight speed is below these Mach numbers, such as during takeoff and slow-speed flight, the freestream flow must be accelerated by the inlet. At higher Mach flight, typically during cruise and descent, the inlet must decelerate the freestream flow. The delivered mass flow of air must have as uniform a velocity profile as possible at the compressor or fan face. Flow non-uniformities or *flow distortion* can have extremely adverse effects on the operation and performance of the compressor or fan, resulting in loss of thrust or vibrations that could result in blade failures. The inlet must be as insensitive as possible to aircraft attitude, including angle-of-attack or angle-of-sideslip, and atmospheric disturbance and turbulence.

There are several ways of characterizing the efficiency and performance of an inlet. One measure of the efficiency of the flow processes through any engine component is the change in total pressure of the flow between the entrance and exit of the component. Recall that for an isentropic process, where there are no losses due to friction or shock waves, the total pressure remains constant. A total pressure-based efficiency is captured as a ratio of the total pressure of the flow exiting the component to the flow total pressure entering it. For the inlet, this is defined as the inlet total pressure recovery, π_d, given as

$$\pi_d \equiv \frac{p_{t_2}}{p_{t_\infty}} \tag{4.73}$$

where p_{t_2} is the total pressure at the inlet exit and p_{t_∞} is the freestream total pressure entering the inlet. The inlet total pressure recovery has a maximum value of one, corresponding to an isentropic process. At low subsonic speeds, the inlet flow approaches an isentropic process. At higher subsonic and supersonic speeds, the inlet flow is usually assumed to be a non-isentropic, adiabatic process, so the total pressure recovery is typically less than one for the flow through the inlet. The total pressure ratio decreases with increasing flight Mach number. For subsonic inlets, the total pressure losses are primarily due to viscous effects, while for supersonic inlets, the losses are primarily due to shock waves. For modern jet transport aircraft, the inlet total pressure recovery is typically high, with values of 0.97 or higher.

4.5.7.2 Subsonic Inlets

Subsonic inlets are typically of the *fixed geometry* type, usually sized for a cruise flight condition, which may be about Mach 0.8–0.9 for a subsonic transport aircraft. Since the inlet geometry cannot change, the freestream flow adjusts to the flight speed and the mass flow demands of the engine, as shown in Figure 4.43. At static (zero airspeed) or low airspeed conditions, such as during takeoff, the engine demands a larger mass flow of air than the cruise design condition to produce high thrust. Thus, the fixed geometry inlet accelerates a large streamtube of freestream air, with a cross-sectional area, A_∞, greater than the inlet entrance area, A_1, into the inlet, as shown in Figure 4.43a. At its cruise design condition, shown in Figure 4.43b, the inlet ingests a mass flow of freestream air with the same cross-sectional area as the inlet entrance area ($A_\infty = A_1$). At an airspeed greater than the cruise design condition, the freestream airspeed is high and must be decelerated going into the inlet, as shown in Figure 4.43c. A smaller mass flow of air is required for this lower thrust condition ($A_\infty < A_1$), so the freestream air that does not enter the inlet is *spilled* around the inlet opening.

Note the shape of the leading edges of the inlet entrance, or the *inlet lips*, shown in Figure 4.43. If the inlet is on a podded engine, such as on a commercial airliner, the inlet leading edges are referred to as *cowl lips*, since the podded engine is covered by an engine cowl. Subsonic inlets have rounded inlet lips with a radius of curvature much like the leading edge of an airfoil. This leading edge shape promotes the smooth flow of subsonic air and avoids boundary layer separation around the lip and into the inlet duct. This shape is also best for reducing the external aerodynamic drag around the inlet. If designed properly, the flow around the rounded inlet lips leads to an aerodynamic suction force, similar to that on the upper surface of an airfoil. This suction force has a component in the thrust direction, resulting in an inlet *lip thrust*. The rounded shape of subsonic inlet lips is not beneficial for supersonic inlets, as the blunt leading edge would lead to detached shock waves and large total pressure losses. Supersonic inlets have sharp lips, much like a supersonic airfoil leading edge.

4.5.7.3 Supersonic Inlets

At supersonic speeds, the inlet must decelerate the flow to subsonic speeds as efficiently as possible. (The exception to this is hypersonic inlets designed for a supersonic combustion ramjet, or *scramjet* engine, where the flow entering the combustor is supersonic. Scramjet propulsion is discussed in

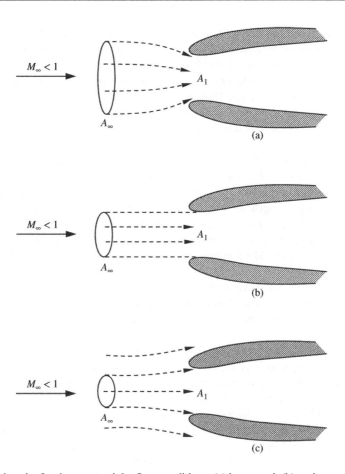

Figure 4.43 Subsonic, fixed geometry inlet flow conditions, (a) low speed, (b) cruise, and (c) high speed.

Section 4.8.1.) As discussed previously, the total pressure recovery decreases sharply at supersonic flight speeds due to shock wave losses.

The simplest type of supersonic inlet is the fixed geometry, *Pitot* or *normal shock inlet*, shown in Figure 4.44. The inlet entrance geometry is typically a rounded opening that may or may not be

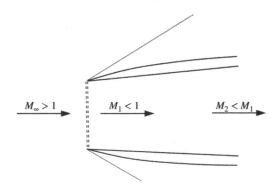

Figure 4.44 Normal shock inlet.

axisymmetric. At supersonic speed, a single normal shock wave is established at the inlet entrance, decelerating the supersonic freestream flow to subsonic speed behind the shock. At the design Mach number of the inlet, the normal shock wave is attached to the inlet lips and the freestream flow capture area is equal to the inlet entrance area. The inlet lips are as sharp as practical, so that the normal shock can remain attached. There may be weak oblique shock waves that emanate from the inlet lips, due to the finite angle described by the inlet lips. At off-design Mach numbers, the normal shock wave may be detached from the inlet lips and freestream flow spills around the inlet opening.

As we learned in Chapter 3, the total pressure loss across an oblique shock wave is less than across a normal shock wave. Therefore, for a supersonic inlet to have a higher pressure recovery than a normal shock inlet, it must have a compression system based on oblique rather than normal shocks. Figure 4.45 depicts such an inlet, a multiple oblique shock, supersonic inlet where the oblique shock waves are generated from two-dimensional or axisymmetric ramps that deflect the freestream flow. The freestream flow is compressed and decelerated through two oblique shocks and a normal shock. Since the Mach number behind an oblique shock wave is always supersonic, the inlet compression system must terminate in a normal shock to obtain a subsonic inlet exit flow. Even though the flow must eventually pass through a normal shock wave, the Mach number upstream of this normal shock wave has been decreased through the oblique shocks, so the total pressure loss is less than the freestream flow passing through a single normal shock wave at a higher Mach number.

It can be shown that the total pressure recovery is improved by increasing the number of oblique shock waves in the compression process. The maximum total pressure recovery, using multiple oblique shocks terminating in a single normal shock, is obtained when the oblique shocks are of equal strength, and the total pressure loss across each of these oblique shock waves is the same. This result, obtained by the Austrian physicist Klaus Oswatisch in 1944, is shown in Figure 4.46, where the maximum total pressure recovery, p_{t_2}/p_{t_0}, is plotted versus the flight Mach number as a function of the number of oblique shock waves, n. (n is the number of oblique shock waves, so that $n = 2$ corresponds to two oblique shock waves, followed by a single normal shock, and $n = 0$ corresponds to no oblique shocks and a single normal shock.) For example, at a freestream Mach number of 4, approximately 62% of the freestream total pressure is recovered ($p_{t_2}/p_{t_0} \cong 0.62$) with three oblique shocks followed by a single normal shock, while the total pressure recovery is only about 14% with a single normal shock ($p_{t_2}/p_{t_0} \cong 0.14$).

In the limit, the highest total pressure recovery is obtained with an infinite number of oblique shocks ($n = \infty$). This can be obtained with the *isentropic compression ramp*, shown in Figure 4.47, where the flow is turned in infinitesimal increments, generating very weak waves that isentropically compress the flow. This type of flow-efficient turning has been used on two-dimensional inlet ramps and as the contour for *isentropic compression spikes* that are the centerbody of an axisymmetric inlet. While the total pressure recovery is excellent, issues with the isentropic ramp or spike include

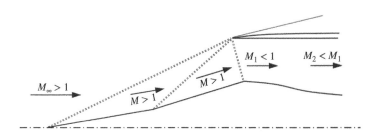

Figure 4.45   Supersonic inlet with multiple oblique shocks terminating in a normal shock.

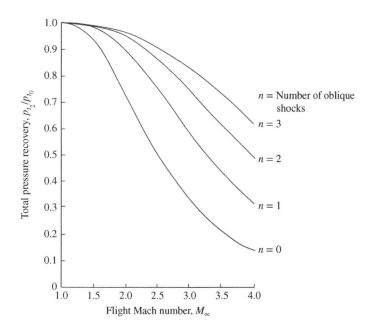

Figure 4.46 Optimum total pressure recovery of multiple oblique shocks and a single normal shock. (Source: *Adapted from Oswatisch*, [15].)

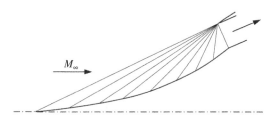

Figure 4.47   Isentropic compression ramp.

the mechanical and structural complexity in its fabrication as compared with a straight surface. In addition, at the end of the isentropic turn, the flow has been turned far away from the axial centerline of the engine. The requirement to turn the flow back to the centerline adds length and weight to the inlet.

The Lockheed-Martin F-16 *Fighting Falcon* has a fixed geometry, normal shock type inlet with two fixed ramps, as shown in Figure 4.48. After entering the F-16 inlet, the flow is turned by a 6° ramp, followed by a 6.67° isentropic compression ramp, turning the flow a total of 12.67°. The total pressure recovery of the fixed geometry inlet of the F-16 decreases dramatically at speeds greater than about Mach 1.4. The poor total pressure recovery of the fixed geometry, normal shock inlet at higher supersonic Mach numbers limits its use to below about Mach 1.8.

While multiple inlet ramps significantly increase the total pressure recovery, the inlet efficiency is improved at only one Mach number (or a small range of Mach numbers) if the inlet ramp geometry is fixed. To make the inlet efficient over a broad range of subsonic and supersonic Mach numbers, the inlet ramp geometry is varied with Mach numbers. This is accomplished with the *variable geometry supersonic inlet*, where the inlet ramps move to change their angle

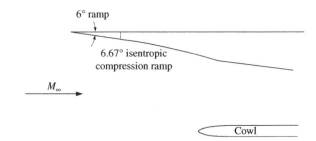

Figure 4.48   Fixed double-ramp inlet on the F-16 *Fighting Falcon*.

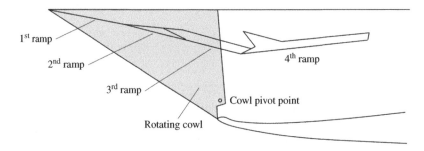

Figure 4.49   Variable geometry, multiple-ramp inlet on the F-15 *Eagle*. (Source: *Adapted from Fig. 10, F.W. Burcham, T.A. Maine, C.G. Fullerton, and L.D. Webb, "Development and Flight Evaluation of an Emergency Digital Flight Control System Using Only Engine Thrust on an F-15 Airplane" NASA TP-3627, September 1996.*)

with respect to the flow and to change the area variation in the duct. For inlets with rectangular cross-sections, such as on the Boeing F-15 *Eagle* or Grumman F-14 *Tomcat*, the movable ramps are planar surfaces that are rotated about a hinge line. The multiple-ramp, variable geometry inlet on the F-15 is shown in Figure 4.49. The variable geometry inlets of the F-15 and F-14 provide significantly higher inlet total pressure recovery than the fixed geometry F-16 inlet at comparable Mach numbers. The variable geometry inlets also enable these aircraft to fly at much higher Mach numbers.

4.5.7.4   Nozzle Requirements and Types

The primary function of a propulsion system nozzle is to expand the exhaust flow to a high velocity, thereby generating thrust. As with the inlet, the nozzle is exposed to both an internal and external flow, which interact and influence each other. The external aerodynamic drag of the nozzle must be considered in the design and installation of the nozzle in the vehicle. The nozzle is also exposed to a high temperature flow of combustion products, which may require cooling of the nozzle.

The favorable pressure gradient in the nozzle flow makes the design of the nozzle somewhat easier, at least from a fluid dynamic perspective, than the inlet with its adverse pressure gradient. However, the mechanical design of the nozzle can be quite complex, especially if variable geometry is required. An even more complex mechanical design problem is the requirement for thrust vectoring, where the nozzle is rotated or vanes are deflected to point the thrust force in different directions.

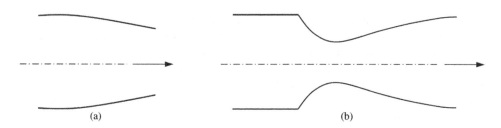

Figure 4.50 Types of jet engine nozzles, (a) convergent and (b) convergent-divergent.

There are two basic types of nozzles for jet engines, the convergent nozzle and the *convergent-divergent* or *C–D nozzle*, as shown in Figure 4.50. The *C–D* nozzle is mechanically more complex and heavier than a simple converging nozzle. Generally used in subsonic aircraft, the convergent nozzle is a simple converging area duct. Convergent-divergent nozzles are used on supersonic aircraft with afterburning engines, where the nozzle area ratio variation is required for optimum engine performance. The *C–D* nozzle is a duct that converges in area, and then diverges. The nozzle *throat* is the location of the minimum area in the *C–D* nozzle.

4.5.7.5 Nozzle Efficiency and Performance Parameters

As with the nozzle flows discussed in Chapter 3, the isentropic flow assumption can often be applied to a propulsive nozzle, where the viscous and thermal losses are assumed to be zero. Similar to the inlet flow, a measure of the losses or irreversibilities in the nozzle flow can be quantified by the nozzle total pressure ratio, π_n, defined as

$$\pi_n \equiv \frac{p_{t_9}}{p_{t_7}} \tag{4.74}$$

where p_{t_7} and p_{t_9} are the total pressures at the entrance and exit of the nozzle, respectively. A nozzle total pressure ratio of one corresponds to isentropic nozzle flow.

An important nozzle performance figure of merit, is the *nozzle pressure ratio* (*NPR*) defined as

$$NPR \equiv \frac{p_{t_7}}{p_\infty} \tag{4.75}$$

where p_{t_7} is the total pressure of the flow entering the nozzle and p_∞ is the freestream ambient static pressure. These parameters are directly analogous to a total pressure, p_{t_7}, in a reservoir or tank, connected to a nozzle that is exhausting into an ambient atmosphere with a static pressure, p_∞. As explained in Section 4.3.2, the thrust is maximized for a perfectly expanded nozzle where the nozzle exit pressure equals the ambient pressure. Thus, the *NPR* determines the required nozzle exit area to obtain a perfectly expanded nozzle and maximum thrust. (Recall that the total to static pressure ratio is used to determine the Mach number, from which the area ratio can be obtained.) The ambient static pressure is a function of altitude; hence, the *NPR* and the required nozzle exit area for maximum thrust vary with altitude. Hence, the *NPR* determines the flow characteristics and performance of the nozzle. The nozzle pressure ratio is used as a metric to help decide when it is best to use a convergent versus a *C–D* nozzle. By performing an analysis of the gross thrust produced by these different nozzle types, it can be determined that the *C–D* produces significantly more thrust than the convergent nozzle when the *NPR* is greater than about 5 or 6.

When the supersonic flow expands through a diverging nozzle, the Mach number increases and the static pressure decreases. We have already defined the perfectly expanded case, where the nozzle exit pressure equals the ambient pressure. We now define the other two possibilities for the nozzle expansion. A nozzle flow is *underexpanded* if the nozzle exit pressure is greater than the freestream ambient pressure. Remember, that supersonic flow expansion decreases the pressure, so that this underexpanded flow may be expanded further to decrease the pressure to ambient pressure. A nozzle flow is *overexpanded* if the nozzle exit pressure is less than the freestream ambient pressure. Here, the flow has been expanded "too much" such that the nozzle exit pressure has decreased to below the ambient pressure.

The three nozzle exhaust flow cases are shown in Figure 4.51, where p_{t_7} is the total pressure at the nozzle entrance, p_9 is the nozzle exit static pressure, and p_∞ is the freestream ambient static pressure. If there is a mismatch in static pressure between the nozzle exit plane pressure and the freestream ambient pressure, the nozzle exit flow adjusts itself, through shock waves or expansion waves, to match the freestream pressure. For the perfectly expanded nozzle ($p_9 = p_\infty$) in Figure 4.51b, there is no pressure mismatch, so the flow exits the nozzle with no adjustments.

For the underexpanded nozzle ($p_9 > p_\infty$) in Figure 4.51a, the nozzle exit pressure is higher than the ambient pressure, so an expansion wave emanates from the exit to decrease the pressure. The flow goes through the first expansion and matches the ambient pressure (region 1), but the expansion has turned the flow such that it is not parallel to the centerline axis of the nozzle. The flow adjusts again, by passing through another expansion fan, which aligns the flow direction with the centerline (region 2). However, this additional expansion has decreased the pressure to below ambient pressure. The flow adjusts once again by passing through a shock wave to increase the pressure (region 3). The pressure matches the ambient pressure, but the flow is once again misaligned. The flow is realigned with the centerline by passing through another shock wave (region 4), with another

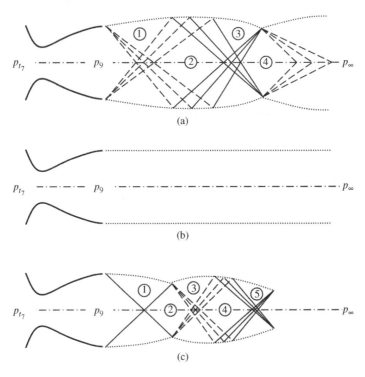

(a)

(b)

(c)

Figure 4.51 Nozzle exhaust flows, (a) underexpanded ($p_9 > p_\infty$), (b) perfectly expanded ($p_9 = p_\infty$), and (c) overexpanded ($p_9 < p_\infty$).

Figure 4.52   Overexpanded nozzle exhaust flow in afterburning exhaust flow of the Lockheed SR-71 *Blackbird* during takeoff. (Source: *NASA.*)

overshoot in pressure. This process of alternating expansion and compression waves continues, with the mismatches in pressure decreasing and the waves becoming weaker and weaker, until the nozzle exit flow pressure equilibrates with the ambient pressure. The situation is similar for the overexpanded nozzle in Figure 4.51c, except that the nozzle exit pressure starts out lower than ambient, requiring a shock wave to increase the pressure.

A photograph of an overexpanded nozzle flow is shown in Figure 4.52. The exhaust gas is made brightly visible by the temperature rises that accompany the pressure rises, in the alternating shock wave patterns of the exhaust flow adjusting to the ambient air pressure. These types of flow patterns are often visible in the afterburning jet exhaust of military aircraft.

4.5.7.6   Thrust Vectoring Nozzles

Thrust vectoring of the nozzle exhaust flow can provide advantages in aircraft performance and maneuverability. It can give an aircraft the capability to take off and land vertically or can significantly reduce an aircraft's takeoff and landing distances. Thrust vectoring has changed the role of aircraft jet propulsion, from only providing propulsive thrust, to being integrated with the aircraft flight control system. The concept of aircraft *supermaneuvrability* has been realized by using thrust vectoring as a low-speed and high angle-of-attack flight control, when conventional, aerodynamic flight controls are ineffective. Thrust vectoring has enabled the flight of *tailless aircraft*, where the aircraft's vertical tail has been significantly reduced in size or eliminated.

The inclusion of thrust vectoring makes the nozzle mechanical design much more difficult. Various schemes have been used in the past, including deflection of vanes or paddles in the nozzle exhaust stream or rotation of the nozzle as a unit. The thrust vectoring scheme on the previously mentioned Lockheed-Martin F-35 is able to rotate the complete nozzle unit by 90° (Figure 4.42).

The first operational thrust vectoring nozzles were used on the British Aerospace AV-8 *Harrier* VTOL military aircraft, which entered service in 1968. The *Harrier* propulsion system ducts engine compressor air and turbine exhaust flow through four swiveling nozzles, mounted forward and aft on either sides of the fuselage. By swiveling the nozzles and adjusting the proportion of thrust through each nozzle, the *Harrier* can take off and land vertically, hover, and fly forwards or backwards.

The Lockheed-Martin F-22 *Raptor* fighter aircraft incorporates thrust vectoring nozzles on its two Pratt & Whitney F119-PW-100 afterburning, turbofan engines. Movable horizontal flap surfaces, located above and below the nozzle, move up and down, providing thrust vectoring in the engine pitch axis. The engine thrust can be vectored up or down by 20°. The engine thrust vectoring is integrated into the aircraft flight control system so that the vectored thrust is used with the conventional flight control system, which enhances the maneuvering capability of the aircraft. By vectoring the nozzles in the same direction, a pitch input can be commanded. Since there are two engines, differential thrust vectoring provides roll inputs also.

4.5.8 The Reciprocating, Piston Engine–Propeller Combination

For much of early aviation, the primary type of propulsion was the gasoline-fueled, reciprocating, internal combustion, piston engine driving a propeller. With the advent of steam power during this era, there were some early attempts to use the steam engine in a heavier-than-air aircraft, but these engines proved to be much too heavy, especially in relation to their low power output. Aeronautical propulsion for heavier-than-air vehicles requires powerplants with a high *power-to-weight ratio*, meaning that they produce as high a power output with as light a weight as possible.

Many of the early airplane designers realized that the powerplants available to them, most of which were designed for industrial or automotive applications, were not suitable for heavier-than-air flying machines. Several of these airplane designers initiated their own efforts to build an engine specifically designed for an airplane. Perhaps the most successful of these early engine designers was Charles Manley, who was assisting Samuel Pierpoint Langley, secretary of the prestigious Smithsonian Institution, in his efforts to fly the first manned, heavier-than-air airplane. Manley designed and built a water-cooled, five-cylinder, radial engine that produced 52 hp (39 kW) and weighed 208 lb (94.3 kg), giving a power-to-weight ratio of 0.25 hp/lb (0.41 kW/kg). Manley's radial engine was the most advanced aircraft engine in the world at the time. The Wright brothers also built their own aircraft engines with the help of machinist Charlie Taylor. They built a water-cooled, four-cylinder, in-line engine that produced 12 hp (8.9 kW) and weighed about 200 lb (91 kg), with a power-to-weight ratio of 0.06 hp/lb (0.1 kW/kg), about one-fourth of Manley's engine.

The power output of reciprocating piston engines increased dramatically over the next few decades. For example, the North American XP-51, discussed in Section 3.7.6, had a liquid-cooled, 12-cylinder piston engine weighing 1645 lb (746 kg), that produced almost 2000 hp (1490 kW) at takeoff. This is a power-to-weight ratio of about 1.21 hp/lb (1.99 kW/kg), over 20 times greater than the Wright brothers' engine. The reciprocating, piston engine-propeller combination is still the best option today for aircraft that fly at relatively low airspeed, below about 250 mph (400 km/h), and low altitude, below about 20,000 ft (600 m). This is the airspeed–altitude regime of many general aviation aircraft and some unmanned aerial vehicles.

4.5.8.1 The Reciprocating Piston Internal Combustion Engine

In discussing the early aircraft engines, it was stated that Charles Manley's engine was a *radial* design and the Wright brothers was an *in-line* engine. The distinction between radial and in-line engines has to do with the arrangement of the engine cylinders, the structures that house the moving

Figure 4.53 Air cooled, radial, reciprocating engine–propeller combination on a Boeing Stearman aircraft. (Source: *User: Groman123, "Boeing Stearman" https://commons.wikimedia.org/wiki/File:Boeing_Stearman_(20285733933).jpg, CC-BY-SA-2.0. License at https://creativecommons.org/licenses/by-sa/2.0/legalcode.*)

Figure 4.54 Lycoming IO-540 air-cooled, horizontally opposed, reciprocating engine–propeller combination. (Source: *Photo courtesy of Lycoming Engines, a Division of Avco Corporation. All rights reserved.*)

pistons in the engine. In a radial engine, the cylinders are located along radial lines from the center of the engine, as shown in Figure 4.53. The radial engine was commonly used in early aircraft designs and is usually found installed on vintage aircraft today, such as the Stearman biplane shown in Figure 4.53. For an in-line engine, the cylinders are arranged in the same plane or in line with each other, as shown in Figure 4.54. In the horizontally opposed engine shown in the figure, the cylinders are horizontally mounted, with the left and right cylinders facing each other, or *opposed* to each other. The horizontally opposed engine arrangement is in common use today on most general aviation aircraft.

As is evident when comparing pictures of these engine types, the radial engine presents a much larger cross-section to the air stream than the in-line engine, resulting in higher aerodynamic drag. The impact of the higher aerodynamic drag on aircraft performance is one reason the in-line engine became more commonly accepted for use in small aircraft. For both engine types, an aerodynamic covering, the *engine cowling*, is fitted around the engine cylinders. Early airplanes with radial engines did not have cowlings, and suffered from significant aerodynamic drag due to engine cylinders that were exposed to the air stream. The Stearman biplane, shown in Figure 4.53, is an example of an airplane with an uncowled radial engine.

In the 1920s and 30s, the NACA conducted numerous wind tunnel investigations with the goal of significantly reducing aerodynamic drag by using engine cowlings. It was discovered that a radial engine enclosed in what became known as the "NACA cowling" had less than one-fifth of the drag of the engine without a cowling. This was a breakthrough in aerodynamic efficiency for aircraft designs. The NACA was awarded the prestigious 1929 Collier Trophy for the development of a low-drag cowling for radial, air-cooled engines.

Both of these engine types are typically *air-cooled*, where the engine is cooled by the flow of the air stream over the engine cylinders and other engine components. Each cylinder has an array of cooling fins around it, to maximize the cooling surface area for convective heat transfer between the hot engine and the cooler air stream. The cowling plays an important role in engine cooling. While the air flow around the outside of the cowling is important to the aerodynamic drag, the air flowing inside the cowling must be properly directed to cool the engine cylinders. In fact, a well-designed cowling can actually improve the cooling of an engine with no cowling. While air-cooled aircraft engines are quite common, closed circuit cooling, using a fluid coolant and radiator system, as found in automobiles, is also sometimes used.

A schematic of a horizontally opposed, reciprocating piston, four-cylinder aircraft engine is shown in Figure 4.55. The cylinders are mounted to the central engine case. Each cylinder houses a piston that moves in a back-and-forth or *reciprocating* motion within the cylinder. The pistons are

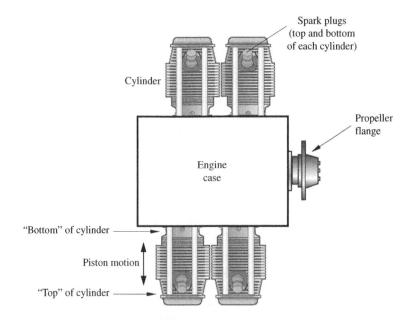

Figure 4.55   Horizontally opposed, reciprocating piston aircraft engine.

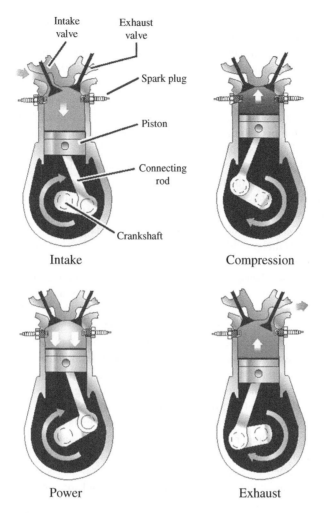

Intake valve   Exhaust valve

Spark plug

Piston

Connecting rod

Crankshaft

Intake   Compression

Power   Exhaust

Figure 4.56 Four-stroke internal combustion engine cycle. (Source: *Adapted from Fig. 6–5, US Federal Aviation Administration*, [4].)

connected to a central crankshaft in the engine case by a connecting rod. Even though the cylinders are mounted horizontally, it is normal to refer to the "top" and "bottom" of the cylinder, especially when referring to the motion of the piston. The bottom of the cylinder is mounted to the engine case and the top of the cylinder is the furthest from the engine case. The linear back-and-forth motion of the pistons is converted to a rotating motion in the crankshaft, as shown in Figure 4.56. The rotating motion of the crankshaft spins the propeller, which is attached to the propeller flange. When the propeller is connected directly to the crankshaft of the engine, the propeller spins at the same rotational speed, or rpm (revolutions per minute), as the engine crankshaft. This type of arrangement is called a *direct-drive engine*. Some engines have reduction gears between the engine and the propeller to reduce the propeller rpm below that of the engine, so-called *geared-engines*.

Other components and accessories attached to the engine usually include a starter, the magnetos and ignition system wiring, alternators or generators to produce electrical power, and vacuum pumps to power cockpit instruments. A lubrication system circulates oil to the moving parts of the engine, including the pistons and crankshaft.

Many, if not most, reciprocating internal combustion aircraft engines burn aviation gasoline, or *avgas*, a special blend of gasoline specifically designed for aircraft use. Some aircraft engines have been certified for use with automotive gasoline. Significant recent effort has been devoted by general aviation engine manufacturers to develop aircraft diesel engines that can operate on kerosene-based fuels.

The reciprocating internal combustion engine operates on a four-stroke process, composed of an intake stroke, compression stroke, power stroke, and exhaust, as depicted in Figure 4.56. On the intake stroke, the piston moves to the bottom of the cylinder and the intake valve opens at the top of the cylinder. Fuel and air are supplied to the cylinder by either a *carbureted* or *fuel-injection* system. A carbureted system uses a carburetor to provide a blended fuel–air mixture to each cylinder through the open intake valve. In the fuel injection system, fuel is injected directly into the cylinder and air is drawn into the cylinder through the intake valve. Air is supplied to the engine using an *air induction system*, which usually includes some sort of air filter to remove contaminants from the ingested freestream air. To obtain increased power, especially at higher altitudes, some aircraft engines have components that provide additional compression of the ingested air. A *supercharger* is an engine-driven air compression system, while a *turbocharger* is a turbine-driven system powered by the engine exhaust gases. Once the fuel–air mixture is in the cylinder, the intake valve closes.

The piston now moves towards the top of the cylinder, compressing the fuel–air charge in the compression stroke. After the piston has reached the limit of its travel near the top of the cylinder, the fuel–air mixture is ignited with an electrical spark from spark plugs. Each cylinder has two spark plugs to improve combustion efficiency and provide redundancy. The spark plug is powered by a magneto, a type of engine-driven electricity generator. The combustion of the fuel–air charge creates a high pressure in the cylinder, which moves the piston back towards the bottom of the cylinder. In this power stroke, the fuel–air combustion is producing the power to turn the crankshaft and the propeller. At the end of the power stroke, the exhaust valve at the top of the cylinder opens and the combustion products are expelled through the exhaust system. At the completion of this exhaust stroke, the cylinder is ready to accept another fuel–air charge, and the cycle starts again with the intake stroke. Since the engine has multiple cylinders, typically four or more, the four-stroke process is timed between the various cylinders so that each cylinder is sequentially producing power. This engine timing results in a continuous and smooth production of power.

4.5.8.2 Gasoline-Fueled Internal Combustion Engine Ideal Cycle: the Otto Cycle

The operation of a gasoline-fueled, reciprocating, internal combustion engine is approximated by the *ideal Otto cycle*, named after the German engineer, Nikolaus Otto (1832–1891), the designer of the first practical, four-stroke internal combustion engine. The Otto cycle is a constant volume combustion process, in contrast to the constant pressure combustion of the Brayton cycle, which approximates the gas generator cycle. The four-stroke Otto cycle is shown for a single cylinder-piston of an internal combustion engine on a pressure–volume, or *p–V*, diagram in Figure 4.57.

During the intake stroke (state 0 to 1), the volume increases as the piston moves from the top to the bottom of the cylinder and the fuel–air mixture enters the cylinder. In the ideal Otto cycle, the volume increases at just the right rate such that the pressure remains constant during the intake stroke. During the compression stroke (state 1 to 2), the piston moves from the bottom to the top of the cylinder, compressing the fuel–air mixture and decreasing the volume. The compression process is assumed to be isentropic, so that from Equation (3.139), we have

$$\frac{p_1}{\rho_1^\gamma} = \frac{p_2}{\rho_2^\gamma} = C \tag{4.76}$$

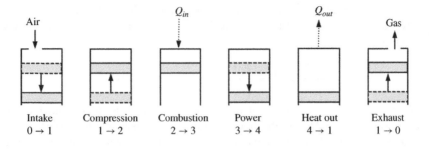

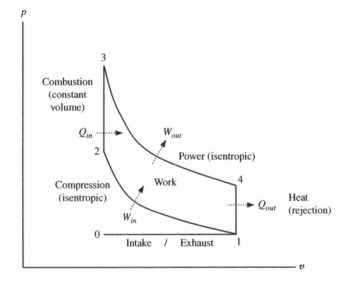

Figure 4.57 Internal combustion engine ideal Otto cycle.

where C is a constant and γ is the ratio of specific heats of the gaseous fuel–air mixture. Using the specific volume, $v = 1/\rho$, we have

$$p_1 v_1^{\gamma} = p_2 v_2^{\gamma} = C \tag{4.77}$$

Since the mass is constant, we have

$$p_1 \mathcal{V}_1^{\gamma} = p_2 \mathcal{V}_2^{\gamma} = C \tag{4.78}$$

where $\mathcal{V}$ is the volume of the fuel–air mixture. Therefore, pressure, p, at any point during the isentropic compression can be expressed in terms of the volume, $\mathcal{V}$, as

$$p = C\mathcal{V}^{-\gamma} \tag{4.79}$$

Using Equation (4.79), the work done by the piston on the system (the fuel and air in the cylinder), W_{in}, is given by

$$W_{in} = -\int_{\mathcal{V}_1}^{\mathcal{V}_2} p d\mathcal{V} = -C \int_{\mathcal{V}_1}^{\mathcal{V}_2} \mathcal{V}^{-\gamma} d\mathcal{V} = -\frac{C}{\gamma - 1}(\mathcal{V}_2^{1-\gamma} - \mathcal{V}_1^{1-\gamma}) \tag{4.80}$$

Inserting Equation (4.78) into (4.80), we have an expression for the work done by the piston on the gas in the cylinder as a function of the initial and final states (states 1 and 2, respectively) in the compression process.

$$W_{in} = -\frac{p_2 \mathcal{V}_2^\gamma}{\gamma - 1}(\mathcal{V}_2^{1-\gamma}) + \frac{p_1 \mathcal{V}_1^\gamma}{\gamma - 1}(\mathcal{V}_1^{1-\gamma}) = -\frac{1}{\gamma - 1}(p_2 \mathcal{V}_2 - p_1 \mathcal{V}_1) \tag{4.81}$$

After the compression process is complete, the fuel–air mixture is spark-ignited, and the combustion occurs very rapidly. The combustion is essentially at constant volume (state 2 to 3), since the piston has not yet started moving down. Using the first law of thermodynamics, we can relate the incremental heat added, δq, to the incremental temperature change, dT, for a constant volume process ($dv = 0$) as

$$\delta q = de + pdv = de = c_v dT \tag{4.82}$$

Using Equation (4.82), the heat added to the system by the combustion process, Q_{in}, is given by

$$Q_{in} = c_v(T_3 - T_2) \tag{4.83}$$

where T_2 and T_3 are the temperatures at the beginning and end of the combustion process.

During the power stroke (state 3 to 4), the combustion gases do work on the piston, moving it down in the cylinder. This expansion is assumed to take place isentropically, so that we can use the results of Equation (4.81) to give the work being produced, W_{out}, as

$$W_{out} = \int_{\mathcal{V}_3}^{\mathcal{V}_4} pd\mathcal{V} = \frac{1}{\gamma - 1}(p_3 \mathcal{V}_3 - p_4 \mathcal{V}_4) \tag{4.84}$$

The integral for the work in Equation (4.84) is now positive, since the gas (the system) is doing work on the piston, rather than the piston doing work on the system.

Heat leaves the cylinder when the exhaust valve opens (state 4 to 1). This occurs in a constant volume process, so that the heat loss, Q_{out}, can be written as

$$Q_{out} = c_v(T_1 - T_4) \tag{4.85}$$

The exhaust stroke is completed as the piston moves to the bottom of the cylinder (state 1 to 0) and the cylinder is ready to restart the cycle.

The net work done by the system, ΔW, is equal to the difference between the heat added to the system, Q_{in}, and the heat that leaves the system, Q_{out}.

$$\Delta W = W_{out} - W_{in} = Q_{in} - Q_{out} \tag{4.86}$$

This net work is represented by the area enclosed by the cycle on the p–V diagram, as shown in Figure 4.57.

4.5.8.3   Diesel-Fueled Internal Combustion Engine Ideal Cycle: the Diesel Cycle

An internal combustion engine can be operated on a variety of fuels. The gasoline-fueled engine found widespread use in early aviation and is still widely used in modern-day general aviation. Diesel-fueled aircraft engines have not found wide acceptance, even though diesel engines have some advantages over gasoline engines, including their higher specific fuel consumption. Diesel fuel also has a safety advantage over gasoline since its vapors do not ignite or explode as easily as gasoline vapors. However, the lower power-to-weight ratio of diesel engines, has historically made them heavier than gasoline engines, a significant disadvantage for aircraft applications.

Recent issues with the high cost and future availability of aviation gasoline have brought about renewed interest in the diesel-fueled aircraft engine. Advancements in diesel engine technology have resulted in diesel engines with higher power-to-weight ratios, making the aircraft diesel engine more feasible. Several aircraft manufacturers have developed or are developing aircraft with diesel power, capable of using kerosene-based jet fuels or automotive diesel fuel.

In 1897, German engineer Rudolph Diesel proposed the ideal thermodynamic cycle that bears his name. The ideal Diesel engine cycle is shown on a p–V diagram in Figure 4.58. Air enters the engine cylinder during the intake stroke (state 0 to 1) and is isentropically compressed during the compression stroke (state 1 to 2). Unlike the Otto cycle, only air is compressed, rather than fuel and air, in the compression process. This allows the Diesel cycle to operate at higher compression ratios than the Otto cycle, since there is no risk of auto-ignition during compression. This is advantageous, since the efficiency of the Diesel cycle is lower than that of the Otto cycle, if they are both operated at the same compression ratio. However, by operating the Diesel engine at a higher compression ratio than the Otto engine, the Diesel engine's efficiency can exceed that of the Otto engine. After the compression process, fuel is injected into the cylinder and is ignited by the heat generated from the compression. The Diesel engine utilizes *compression-ignition* of the fuel–air mixture, rather than *spark-ignition*, as in the Otto engine. In the ideal Diesel cycle, the combustion of the fuel–air mixture occurs at constant pressure (state 2 to 3), in contrast to the constant volume combustion process in the ideal Otto cycle. During the power stroke, the combustion gases expand in an isentropic expansion (state 3 to 4). At the end of the power stroke, the cylinder exhaust valve opens and the heat is transferred out of the cylinder in a constant volume or isochoric process (state 4 to 1). Finally, the exhaust stroke is completed as a constant pressure, decreasing volume process (state 1 to 0), returning the cycle to its starting point.

Let us examine the Diesel cycle from a thermodynamic perspective. Work is being done by the piston on the system (the air in the cylinder) during the compression process (state 1 to 2), denoted by W_{in} in Figure 4.58. Since this compression is occurring isentropically, there is no heat loss or gain during this process. During the constant pressure combustion process (state 2 to 3), heat is added to the system, Q_{in}, due to the combustion of the fuel and air. Work is being done by the system on the piston, W_{out}, during the power stroke (state 3 to 4). This is an isentropic expansion, so there is no heat loss or gain. When the exhaust valve opens and the exhaust products exit the

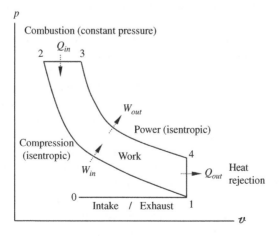

Figure 4.58   Internal combustion engine ideal Diesel cycle.

cylinder, heat leaves the system, Q_{out}, in an isochoric process (state 4 to 1). As in the Otto cycle, the net work produced by the Diesel cycle is equal to the difference between the heat added and the heat lost by the system during the cycle, as represented by the area enclosed by the cycle in the p–V diagram.

The thermal efficiency, η_{diesel}, of the ideal Diesel cycle is given by

$$\eta_{diesel} = 1 + \frac{Q_{out}}{Q_{in}} = 1 + \frac{c_v(T_1 - T_4)}{c_v(T_3 - T_2)} = 1 - \frac{1}{\gamma}\left[\frac{T_1\left(\dfrac{T_4}{T_1} - 1\right)}{T_2\left(\dfrac{T_3}{T_2} - 1\right)}\right] \tag{4.87}$$

4.5.8.4   The Propeller

The power produced by the internal combustion engine is used to rotate the engine crankshaft, which is connected to the propeller. The power delivered by the engine to the crankshaft is called the *brake horsepower* (BHP). The engine power delivered to the propeller shaft is the called the *shaft horsepower* (SHP). In a non-geared engine, found in most general aviation aircraft, the crankshaft is connected directly to the propeller, so that the shaft horsepower is equal to the brake horsepower. In a geared engine, as in a turboprop aircraft, there are reduction gears between the engine and the propeller, which reduce the engine rpm to a lower propeller rpm. In this case, the shaft horsepower is less than the brake horsepower due to the losses in the reduction gears. In either case, the propeller converts the engine shaft horsepower into *propulsive horsepower*.

The propeller converts the engine power into a thrust force. The thrust generated by the propeller is proportional to the rate of mass flow of air that passes through the propeller and the velocity increase that the propeller imparts to this mass of air. However, unlike a jet engine, which imparts a large velocity change to a small volume of air, a propeller has a large diameter and imparts a small velocity change to a large volume of air.

In many respects, a propeller can be thought of as a rotating wing, although the rotation of the propeller make its aerodynamics even more complicated. Similar to the airfoil compressibility issues that were discussed in Chapter 3, a propeller can experience compressibility effects as the propeller tip speeds approach high subsonic or near sonic speeds. These compressibility issues severely reduce the effectiveness and efficiency of the propeller and this is a factor that limits the maximum speed of propeller-driven aircraft.

In addition to producing thrust, the spinning propeller has other effects on an aircraft. In a single-engine airplane, the spinning propeller creates a spiraling flow or air, called a *slipstream*, that flows around the aircraft fuselage and over the tail. Since the propeller is adding energy to the freestream flow, the slipstream velocity is greater than the freestream velocity, which results in increased lift or drag on the aircraft surfaces that it flows over. (Recall that the aerodynamic force is proportional to the square of the velocity.) The increased magnitude of the forces can change the moments on the aircraft, thus affecting the stability and control. For instance, when the spiraling flow reaches the aircraft's tail, it may strike the vertical stabilizer, imparting a yawing moment on the aircraft. The slipstream effects are more significant at high power settings, such as during takeoff and climb, when the propeller is turning at maximum rpm.

The propeller may be attached to the aircraft in a *tractor* configuration, where the thrust force acts to *pull* the aircraft through the air, or in a *pusher* configuration, where the thrust forces acts to *push* the aircraft. In the tractor configuration, the propeller is facing forward on the aircraft, while it is facing aft in the pusher configuration. The Beechcraft *Bonanza* (Figure 2.23) is an example of

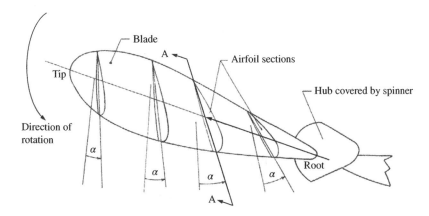

Figure 4.59 Propeller components and airfoil sections. (Source: *From Talay, NASA SP 367*, [17].)

an aircraft with a tractor propeller configuration, while the Wright Flyer (see the photograph at the beginning of Chapter 1) had a pusher configuration. Some aircraft, such as the Cessna *Skymaster* (Figure 1.19), have both tractor and pusher propellers.

As shown in Figure 4.59, an aircraft propeller assembly often consists of the propeller *blades*, the center *hub* to which the blades are attached, and a *spinner* that fits over the hub to reduce aerodynamic drag. In its simplest form, the propeller assembly may be a two-bladed propeller, constructed as a single piece of wood or metal, which is bolted to the engine. The propeller may have two or more blades, with some propellers having as many as eight blades. If we slice a propeller blade into cross-sections, as shown in Figure 4.59, we see that it is made up of a series of different airfoil sections along the length of the blade. The orientation of the airfoil shapes varies from the tip to the root of the blade, such that the different airfoil sections along the blade "see" a slightly different, local angle-of-attack, α. As it turns out, it is desirable for these local angles-of-attack along the blade length to be about equal. We discuss this in further detail after we learn a little more about the propeller geometry and flow.

Consider the aerodynamic force produced by each airfoil section due to the relative wind at the local angle-of-attack. This aerodynamic force can be resolved into components, one of which is the contribution to the thrust force of the propeller. Ideally, the thrust force is constant along the blade length, such that if the propeller was a solid circular disk, with a radius equal to the propeller blade radius, the propeller disk has a constant *disk loading*. With the proper choices of the blade airfoil sections and the blade orientations, the local angles-of-attack of the blade sections can be made to be about equal, to provide a constant disk loading.

Imagine that we cut through Section A-A of the propeller blade in Figure 4.59, so that we are looking towards the spinner, at a cambered airfoil section of the propeller that is a distance r from the center of rotation, as shown in Figure 4.60. The propeller is in a freestream flow of velocity V_∞ and is spinning at a constant angular velocity $r\omega$. The angle between the plane of rotation of the propeller and the blade section chord line is defined as the *blade angle*, β (not to be confused with the angle-of-sideslip, β).

The propeller can be thought of as a small, rotating wing, "flying" in a relative wind that is the vector sum of the freestream velocity, $\vec{V}_\infty$, and flow velocity due to its rotation, $\vec{V}_{rot}$, as shown in Figure 4.60. The relative velocity "seen" by the propeller, $\vec{V}_{rel}$, is given by

$$\vec{V}_{rel} = \vec{V}_\infty + \vec{V}_{rot} = \vec{V}_\infty + \vec{r} \times \vec{\omega} \tag{4.88}$$

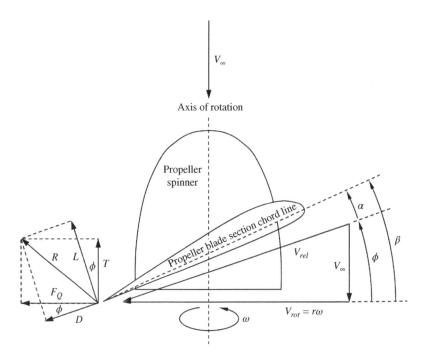

Figure 4.60 Propeller airfoil section local flow velocities, flow angles, and forces (section A-A of Figure 4.59).

where the velocity due to the propeller or tangential velocity, $\vec{V}_{rot}$, is equal to the vector cross product of the radial vector to the blade section, $\vec{r}$, and the angular velocity of the spinning propeller, $\vec{\omega}$. Equation (4.88) tells us that highest relative velocity is seen at the propeller tips, where $\vec{r}$ has its largest value.

The angular velocity, ω, is related to the propeller rotational speed by

$$\omega = 2\pi n \tag{4.89}$$

where n is the rotational speed of the propeller or the number of propeller revolutions per second.

The angle between the relative velocity, V_{rel}, and the chord line of the blade airfoil section is the local *angle-of-attack*, α. The angle between the relative velocity and the propeller plane of rotation is defined as the propeller *helix angle* or *angle of advance*, ϕ. The propeller blade angle is the sum of the helix angle and the blade section angle-of-attack, given by

$$\beta = \phi + \alpha \tag{4.90}$$

The helix angle may be expressed in terms of the freestream velocity and the tangential velocity, as

$$\phi = \tan^{-1}\left(\frac{V_\infty}{V_{rot}}\right) = \tan^{-1}\left(\frac{V_\infty}{r\omega}\right) \tag{4.91}$$

Using Equation (4.89), Equation (4.91) can be written as

$$\phi = \tan^{-1}\left(\frac{V_\infty}{r2\pi n}\right) = \tan^{-1}\left(\frac{V_\infty}{\pi n D}\right) = \tan^{-1}\left(\frac{J}{\pi}\right) \tag{4.92}$$

where D is the propeller diameter and J is defined as the dimensionless propeller *advance ratio*, given by

$$J \equiv \frac{V_\infty}{nD} \tag{4.93}$$

Since the blade section is at an angle-of-attack, α, in a relative wind, V_{rel}, there is a resultant force acting on the section, R, as shown in Figure 4.60. The resultant aerodynamic force, R, may be resolved into a thrust force, T, parallel to the freestream velocity and a torque force, F_Q, in the propeller plane of rotation. The thrust is the force that moves the aircraft forward through the air. The torque force must be overcome by the engine in turning the propeller. Thus not all of the engine's power is available to convert to thrust, as some of the power must be used to counter the torque force.

Alternatively, the resultant force could be resolved into a lift, L, and a drag, D, perpendicular and parallel to the relative wind, V_{rel}, respectively. The blade section thrust may be written in terms of the lift and drag as

$$T = L \cos \phi - D \sin \phi \tag{4.94}$$

Similarly, the torque force may be written as

$$F_Q = D \cos \phi - L \sin \phi \tag{4.95}$$

To obtain the total thrust and torque forces, the contributions from each blade section are summed over the length of each propeller blade (recall that the aircraft propeller may have two or more blades).

Recall that we are looking at a local section of the propeller blade, located at a distance r from the center of the propeller. Thus, the blade angle-of-attack, α, blade angle, β, and helix angle, ϕ, vary from the propeller root to the tip, that is, $\alpha = \alpha(r)$, $\beta = \beta(r)$, and $\phi = \phi(r)$. Although the blade angle varies along the blade length, a representative blade angle, measured at a distance of 75% of the blade length from the hub, is sometimes specified.

As discussed earlier, it is desired to have an approximately constant local angle-of-attack along the blade length to obtain a constant propeller disk loading. Since the propeller is spinning, the tangential velocity, V_{rot}, increases with distance from the center of the propeller. Therefore, the helix angle varies from a large angle at the propeller root to a smaller angle at the tip, as shown in Figure 4.61. To maintain a constant local angle-of-attack along the blade length, the local blade angle increases from root to tip, as does the helix angle, according to Equation (4.90). Thus, the propeller blade is twisted from the tip to the root, with the blade and helix angles increasing from the tip to the root, as shown in Figure 4.61. In summary, to obtain a constant angle-of-attack along the blade length, and hence a constant propeller disk loading, the propeller blade is twisted to account for the change in the tangential velocity along the blade length.

For some propellers, called *fixed pitch propellers*, the blade angle is fixed along its length and cannot be changed. (Here, we are introducing the term *pitch* to be synonymous with *blade angle*, which is in commonplace usage.[2]) Since a fixed pitch setting is not optimum for all airspeeds and engine power settings, the propeller blade angles are set for efficient operation at a single flight condition, usually cruise or climb. In other types of propellers, called *controllable pitch propellers*, the blade angle can be adjusted to a desired angle in flight, thereby operating the propeller in an optimum setting for the given airspeed and power setting. In a *constant speed propeller*, a propeller

---

[2] To be technically precise, the propeller pitch is not the same as the blade angle. The *pitch* or *geometric pitch* is the distance, in inches, that the propeller advances in one revolution, analogous to the distance a wood screw advances in a piece of wood in a 360° turn with a screwdriver.

governor is used to automatically control the blade angle to provide a desired constant engine rpm, regardless of changes in the power setting or airspeed. For example, during takeoff the propeller is set to a small blade angle or *fine pitch* to obtain maximum thrust. The small blade angle corresponds to a small blade angle-of-attack, which reduces the torque force, the force that the engine must overcome to turn the propeller. The lower torque force allows the engine to spin the propeller at higher rpm, which imparts a higher velocity to the air going through the propeller, thus increasing the thrust. The opposite situation applies to cruising flight, where the propeller is set to a large blade angle or *coarse pitch*, to allow the engine to operate at lower rpm, which is more fuel-efficient.

It should come as no surprise that the blade section angle-of-attack is a critical parameter governing the propeller performance. This is analogous to the importance of angle-of-attack for airfoil sections on a wing. Referring back to Figure 4.60 and Equation (4.90), we see that the blade angle and the helix angle are the two important parameters in setting the blade angle-of-attack. Based on this and using Equation (4.93), we can choose to use the blade angle, β, and the advance ratio, J, to assess propeller performance. Figure 4.62 shows a typical propeller performance chart, where the propeller efficiency, η_p, is plotted against the propeller advance ratio, J, as a function of the blade angle, β. The propeller efficiency, η_p, (not to be confused with the propulsive efficiency) is defined as

$$\eta_p = \frac{P_A}{P} = \frac{T_A V_\infty}{P} \tag{4.96}$$

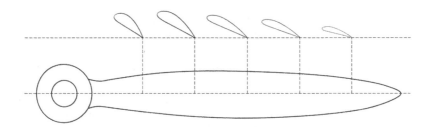

Figure 4.61   Change in propeller blade angle from root to tip.

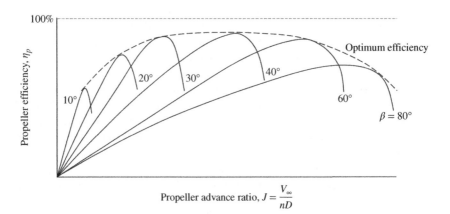

Figure 4.62   Propeller efficiency chart.

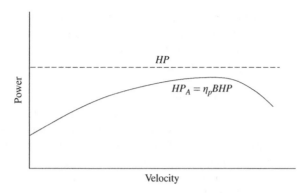

Figure 4.63 Typical propeller-driven engine power available.

where P_A is the propeller power available, P is the engine shaft horsepower, and T_A is the thrust available. The propeller efficiency is a measure of the percentage of engine shaft power that is converted into power available to spin the propeller. An efficiency of 1.0 would mean that 100% of the engine shaft power is being used to turn the propeller. In reality, there are frictional and other losses in the propeller system, such that the propeller efficiency is always less than one. Propeller efficiencies of about 0.85–0.88 are achievable. Based on the envelope of maximum efficiency for the propeller efficiency, shown in Figure 4.62, the typical trend of horsepower available from a propeller-driven engine is as shown in Figure 4.63. There is a loss in horsepower, over the velocity range, due to the propeller efficiency.

For a fixed blade angle, the propeller efficiency increases with increasing advance ratio to a maximum, then decreases. Hence, a fixed pitch propeller may have a narrow range of efficiency as a function of advance ratio, which embodies forward airspeed and propeller rotational speed or engine power setting. If a line is drawn, connecting the peaks in the propeller efficiency for the different blade angles, as shown by the dashed line in Figure 4.62, the blade angles providing the best or optimum efficiency can be found as a function of the advance ratio. This provides the needed information for selecting the proper blade angle of a controllable pitch propeller as a function of the airspeed and power setting.

As a final note about variable-pitch propellers, some have the capability to *feather* in flight, where the blade angle is set to a high angle so that the propeller is not producing forward thrust. The feather setting is used when an engine has stopped producing power in flight and it is desired to get the propeller in a minimum drag configuration to enhance engine-out maneuverability or gliding. Usually, the propeller stops spinning when placed in the feather position. Another feature of some variable-pitch propellers is the ability to set *reversible pitch* with the blades, where the blades are at a large negative angle-of-attack. Reversible pitch results in a negative thrust force, which is used after landing touchdown, to reduce landing ground roll distance.

4.5.8.5 The Electric Motor-Propeller Combination

This section makes brief mention of aircraft propulsion using an electrical motor, instead of an internal combustion engine, to drive a propeller. The source of electrical power may be batteries, fuel cells, solar cells, or power beaming. To date, most of the applications of this type of propulsion have been for light aircraft, such as powered gliders, ultralights, small general aviation aircraft, or unmanned aerial vehicles.

The first flight of a manned electric airplane was of a modified Brditschka HB-3 motorglider in Austria on 23 October 1973. The motorglider was modified with a 10 kW (13 hp) electric motor that spun a single pusher propeller. Using nickel-cadmium batteries to power the electric motor, the electric aircraft could only fly for about 12 minutes and climb to a maximum altitude of about 400 m (1300 ft).

Technical advantages of electric aircraft propulsion, over hydrocarbon-fueled propulsion, may include lower cooling drag, constant power output independent of altitude, and less overall system complexity and size. Operational benefits may include reduced operational costs, extremely quiet operation, and no vehicle emissions. Some of the issues with electric aircraft propulsion include energy storage weight, reliability and safety of some types of batteries, and low specific power (power per unit weight) output. With current technology, electric motors for aviation have power outputs of less than about 100 kW (130 hp).

An example of an electric aircraft in development is the prototype Airbus *E-Fan*, shown in Figure 4.64. The *E-Fan* is a two-seat, twin-motor, low-wing monoplane of all composite construction. The aircraft is propelled by two ducted, variable-pitch fans powered by two electric motors. The aircraft has a length of 6.67 m (22 ft), wingspan of 9.50 m (31 ft), and a maximum takeoff weight of 550 kg (1200 lb). The propulsion system delivers a total of 1.5 kN (430 lb) of static thrust, resulting in a cruise speed of about 160 km/h (100 mph).

The electric power is provided by 250-volt lithium polymer battery packs, with a total weight of 167 kg (368 lb), mounted in the inboard section of the wings. The electric batteries give the aircraft an endurance of about 1 hour. The system has a backup battery that provides power for an emergency landing if the primary batteries fail or are discharged. The electric power drives two motors, each providing 30 kW (40 hp) of power to an eight-bladed ducted fan.

A unique feature of this electric aircraft design is the motorized landing gear. The aircraft has fore and aft, retractable main landing gear and a small, outrigger wheel under each wing. The main wheels are powered by a 6 kW (8 hp) electric motor, making the aircraft an electric-powered vehicle on the ground. The *E-Fan* can taxi and accelerate to 60 km/h (37 mph) using the motorized wheels, without use of the ducted fan propulsion.

Figure 4.64 Airbus *E-Fan* prototype electric airplane, 2015. (Source: *Dick Schwarz, "E Fan Airbus" https://de.wikipedia.org/wiki/Datei:E_FAN_Airbus_DS_20140524_1237.jpg, CC-BY-SA-3.0. License at https://creativecommons.org/licenses/by-sa/3.0/legalcode.*)

Table 4.8 Summary of air-breathing propulsion thermodynamic cycles and processes.

Type of air-breathing engine	Thermodynamic cycle	Combustion process
Ramjet, turbojet, turbofan	Brayton cycle	Constant pressure
Internal combustion engine (gasoline)	Otto cycle	Constant volume
Internal combustion engine (diesel)	Diesel cycle	Constant pressure
Intermittent combustion engine (PDE)	Humphrey cycle	Constant volume

Airbus has targeted the prototype *E-Fan* design for the pilot training market. It has other designs for electric aircraft for the commercial regional airline transport market. There are still issues for use of electric aircraft in these markets, as aircraft regulatory and certification requirements do not yet exist for electric aircraft.

4.5.9   Summary of Thermodynamic Cycles for Air-Breathing Engines

The thermodynamic cycles and processes of the various air-breathing engines that have been discussed are summarized in this section to allow for a convenient comparison. The air-breathing engines that have been discussed include the ramjet, turbojet, turbofan, turboprop, and internal combustion engine. The operation of these various air-breathing engines can be approximated with an ideal thermodynamic cycle. Table 4.8 summarizes the thermodynamic cycle and process associated with each type of air-breathing engine.

4.5.10   GTT: the Engine Test Cell and Test Stand

The present GTT describes some of the techniques used to test propulsion systems in specialized ground facilities. These include indoor and outdoor test cells and test stands, where complete engines can be operated at up to full power. Ground testing of propulsion systems can significantly reduce the number of flight tests required for propulsion testing and may significantly reduce flight test risks.

In the early development of the jet engine, it was recognized that there was a need for ground facilities and ground test techniques to test these engines. One of the earliest instances of this was the creation of the Bavarian Motor Works (BMW) jet engine test facility by the Germans in 1944. After World War II, the BMW facility was dismantled, and much of the equipment was brought to the USA to become the US Air Force Arnold Engineering Development Center (AEDC) Engine Test Facility (ETF) in Tullahoma, Tennessee. The first turbojet engine test at AEDC occurred on 3 May 1954. In addition to military propulsion test facilities such as AEDC, there are several civilian test facilities operated by civilian commercial engine companies and by government research organizations, such as NASA.

Testing is performed on a variety of propulsion systems, including air-breathing jet engines and non-air-breathing rocket engines. The tests may be of flightworthy, complete engines or of heavier, non-flyable "boilerplate hardware" that is perhaps less complex, less costly, and more survivable than the flight hardware. Complete engines may be tested, as can components of engines, such as inlet or nozzle systems. Figure 4.65 shows a complete jet engine being tested in an AEDC test cell. The hot engine exhaust is ducted out of the cell through a special exhaust system. A large, conically shaped screen is placed over the engine inlet to prevent the ingestion of debris or foreign objects.

Figure 4.65   Jet engine test cell, Arnold Engineering Development Center (AEDC). (Source: *US Air Force*.)

There are a wide variety of tests that are conducted in the development and operation of a propulsion system. Tests may be conducted to collect information about engine performance, operability, or durability. The ground tests may simulate the entire life of an engine in a shorter time, by running the engine under special test conditions. The objectives of propulsion ground testing may be for pure research and development. Experimental engines or propulsion systems may be tested on the ground, simulating conditions that are too difficult or too costly to obtain in flight. The test engines are typically heavily instrumented to collect the desired data.

Some indoor engine test cells have the capability to simulate flight conditions, including high Mach number and high altitude. In addition to normal, sea level test conditions, some test cells can simulate flight at up to altitudes of 100,000 ft (30,500 m). There are several facilities that can test air-breathing engines at low hypersonic Mach numbers of less than about five, and some very specialized facilities that can provide test conditions at much higher Mach numbers. A critical aspect of simulating flight at these hypersonic speeds is the matching of the total temperature or total enthalpy. Figure 4.66 shows the components of a hypersonic engine in an AEDC test cell.

Many commercial jet engines are tested on outdoor test stands or indoor test facilities, as shown in Figure 4.67. The turbofan engine shown is mounted with provisions for supplying power, fuel, instrumentation connections, and other operating requirements. Since the test is conducted at zero airspeed, a fixture, called a *bell mouth*, is placed on the engine inlet, to smoothly turn the air into the engine. Outdoor engine test stands may utilize special equipment to simulate different environmental conditions. A wind generator, resembling a small wind tunnel, may be used to blow air across the engine inlet to simulate crosswinds. A large, spherically shaped structure, called a turbulence control sphere, may be placed over the engine inlet to smooth out atmospheric disturbances from wind and turbulence. Other types of specialized tests include testing of engines under distressed conditions, including the ingestion of golf ball-sized hail or monsoon rain-levels of water. Engine fan blades are rigged to separate with the engine running at full power to evaluate the effect on the fan and on the engine ingestion of a fan blade. Perhaps one of the most interesting engine ingestion tests is the use of pneumatic air cannons to shoot bird carcasses into engines, simulating a bird strike.

Figure 4.66   Components of a hypersonic propulsion system in an engine test cell, Arnold Engineering Development Center (AEDC).   (Source: *US Air Force.*)

Figure 4.67   Turbofan engine, with a bellmouth, mounted in an indoor test facility.   (Source: *Cherry Salvesen, "Test Facility" https://commons.wikimedia.org/wiki/File:Test_Facility.jpg, CC-BY-SA-3.0. License at https://creativecommons.org/licenses/by-sa/3.0/legalcode.*)

4.5.11 FTT: Flying Engine Testbeds

In the last section, ground test techniques for testing full-size engines in indoor test cells or on outdoor test stands were discussed. The present FTT discusses the flying engine testbed, aircraft that are configured to carry full-size air-breathing engines that can be tested in flight. While some ground facilities can duplicate some parts of a flight envelope, there is often no substitute for the real flight environment with its real-world atmospheric characteristics, including turbulence, wind shear, temperature variations, and other characteristics. The flying engine testbed can also fly the test engine through a range of operational flight profiles, such as takeoff, climbs, descents, landings, and other maneuvers that may be difficult to simulate in real time in a ground facility. Several examples of flying engine testbed aircraft are given below.

During the 1960s, a Lockheed *Constellation* was converted to a flying engine testbed, as shown in Figure 4.68. The *Constellation* was a propeller-driven, four-engine airliner, designed and manufactured by Lockheed during the 1940s and 1950s. The aircraft had a unique triple-vertical tail configuration and a dolphin-shaped fuselage. The engine testbed *Constellation* had a support structure added to the upper fuselage, where test engines could be mounted, as shown.

The Northrop P-61 *Black Widow* was designed specifically as a nighttime attack aircraft during World War II. (The *Black Widow* was also used to tow the XP-51 in the aerodynamics flight test technique in Chapter 3.) The twin engine P-61 had a twin-boom configuration with a single horizontal tail, which joined the two vertical tails. A modified P-61 was used by the NACA, after World War II, as a subsonic flight testbed for ramjet engines. The ramjet was mounted underneath the fuselage, as shown in Figure 4.69.

A North American AJ-2 *Savage* was modified by AVCO Lycoming as an engine testbed in the 1970s. The AJ-2 was a US Navy carrier-borne aircraft, designed and built by North American Aviation after World War II. The aircraft had two wing-mounted piston engines and a single turbojet engine in the fuselage. The modified AJ-2 was flown by Lycoming to test its turbofan jet engine, which was mounted underneath the fuselage, as shown in Figure 4.70. The test engine could be retracted into the fuselage bomb bay of the aircraft.

Current flying engine testbed aircraft that are used to test subsonic jet engines are typically large, four-engine transport aircraft. Many different types of aircraft have been used in the past. Current engine testbeds are operated by several commercial engine manufacturers using a variety of aircraft,

Figure 4.68 Lockheed L-749 *Constellation flying engine testbed.* (Source: *Courtesy of Michel Gilland, "Lockheed L-749, CEV" https://commons.wikimedia.org/wiki/File:Lockheed_L-749_Constellation,_CEV_-_ Centre_d%27essais_en_ Vol_AN0665578.jpg, GFDL-1.2. License at https://commons.wikimedia.org/wiki/ Commons:GNU_Free_ Documentation_License,_version_1.2.*)

Figure 4.69   Northrop P-61 *Black Widow* in flight with operating ramjet test engine. (Source: *NASA*.)

Figure 4.70   North American AJ-2 *Savage* configured to flight test an AVCO Lycoming turbofan engine. (Source: *US Navy*.)

including the Boeing 747, Boeing 757, and several Airbus models. By using a four-engine aircraft, one of the engines can be replaced by a test engine without sacrificing safety, as the aircraft can fly safely on its three remaining standard production engines. Often the aircraft cabin has adequate space to house data acquisition equipment and real-time test monitoring stations for engineers. In this way, the aircraft cabin becomes an in-flight control room that can be staffed by engineers and technicians to monitor the engine testing during flight. This provides the utmost flexibility in the location of testing, as the control room and data acquisition are always with the aircraft.

4.6   Rocket Propulsion

Rocket propulsion is perhaps the earliest form of flight propulsion, dating back to the 11th or 12th century, when the Chinese launched gunpowder-filled bamboo tubes as a form of fireworks. Some time later, they attached these rudimentary solid rocket motors to arrows, creating a rocket-powered weapon of sorts. The arrow shaft provided some flight stability, although the accuracy of these rocket-powered arrows was probably questionable. Much later, in the early 19th century, solid propellants rockets were developed in England by William Congreve. These rockets supposedly had a range of 3000 yards (2700 m), probably with poor accuracy. Congreve's rockets were used by the British in the War of 1812, including being fired at American troops in Baltimore's Fort McHenry, as immortalized by "the rockets' red glare" in the American national anthem, the *Star Spangled Banner*. Although, rockets were developed for military uses in the 18th and 19th centuries, the most significant advancements in rocket propulsion took place in the 20th century. Great strides in rocket propulsion were made during the 20th century, most notably in Russia, Germany, and the USA. In Chapter 1, the pioneering achievements in rocket propulsion of Robert Goddard (1882–1945) were discussed.

One of the earliest visionaries of rocket propulsion, especially of its potential application to space travel, was Konstantin Tsiolkovsky (1857–1935), a Russian schoolteacher. In 1903, Tsiolkovsky published a paper entitled "Exploration of Space with Reaction Devices", where he discussed the use of rockets to escape the orbit of the Earth. He was the first to publish the *rocket equation*, which relates the burnout velocity of a rocket to the exhaust velocity and initial-to-final *mass ratio* of the rocket (the rocket equation and these relevant terms are defined later in this section). In 1929, Tsiolkovsky published the idea of multi-stage rockets, a concept that makes spaceflight into earth orbit and beyond realizable. Tsiolkovsky's rocket concepts included the use of liquid oxygen and liquid hydrogen as propellants. In addition to his significant achievements in rocket science, Tsiolkovsky contributed to early aviation, including designing a monoplane in 1894 and constructing the first Russian wind tunnel in 1897.

Tsiolkovsky's ideas about using rocket propulsion for space travel have become reality. As he theorized, rockets carry all of their fuel and oxidizer onboard the vehicle, making it possible for them to operate inside or outside of the atmosphere, including in the vacuum of space or under water. Rocket propulsion is used today for a variety of applications. Rocket propulsion is used as the *primary propulsion for launch vehicles* that lift payloads from the surface of the earth into earth orbit and into space beyond earth orbit. It is used for *in-space propulsion* of orbiting spacecraft, including on-orbit maneuvering, station keeping (maintaining the spacecraft in its proper orbit), and attitude control. In-space propulsion also includes the rocket engines that send spacecraft to other celestial bodies and into deep space.

In the most common type of rocket propulsion, called *chemical rocket propulsion*, the fuel and oxidizer are chemicals that may be in liquid, solid, or gaseous form. The energy stored in the propellants is converted by *chemical combustion* into combustion products that are ejected at a high velocity from the rocket, creating thrust. The maximum energy available to accelerate a chemical rocket is limited by the energy in the chemical propellants, whether they are liquid, solid, or gas. This fundamentally limits the performance of chemical rocket propulsion.

We discuss two types of chemical rocket propulsion, the *liquid-propellant rocket engine* and the *solid-propellant rocket motor*. Other types of rocket propulsion are possible that do not utilize chemical combustion. In these other types of rocket propulsion, a thrust-producing, high kinetic energy ejectant is created by heat addition from other energy sources, such as nuclear reactions or solar radiation. The ejected matter can be a fluid, a gas, an electrically activated gas called a plasma, or even packets of energy. We briefly describe these more exotic forms of rocket propulsion, to include nuclear, electric, and solar propulsion. Before we discuss the various types of rocket propulsion, we start by discussing the thermodynamics of rocket propulsion.

4.6.1   Thrust Chamber Thermodynamics

The basic components of a chemical rocket propulsion system, shown in Figure 4.71, are the *propellant system*, the *combustion chamber*, and the *exhaust nozzle*. The propellant system is composed

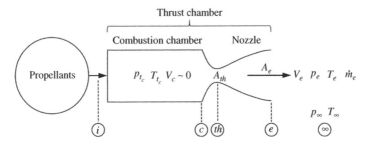

Figure 4.71 Rocket thrust chamber nomenclature and flow properties.

of propellant storage and a feed system to deliver the propellants to the combustion chamber. The propellant may be stored in one or more tanks, as in a liquid-propellant system, or they may be an integral part of the motor, as in the solid-propellant system. The propellant feed system may include additional tanks with inert gases, pumps, valves, feed plumbing lines, or other components. The propellants are burned in the combustion chamber and the combustion products are accelerated through an exhaust nozzle to generate thrust. The combination of the combustion chamber and the nozzle is called the *thrust chamber*.

The station numbers and flow properties are specified in Figure 4.71. The propellants are injected into the combustion chamber at station i. The propellants are burned at a total pressure, p_{t_c}, and total temperature, T_{t_c}, in the combustion chamber. The combustion chamber is connected to a convergent-divergent, supersonic nozzle. The nozzle has a sonic throat with a cross-sectional area A_{th} (station th) and an exit cross-sectional area A_e. The flow properties at the nozzle exit (station e) are the velocity, V_e, pressure, p_e, temperature, T_e, and mass flow rate, $\dot{m}_e$. The nozzle exit mass flow rate is equal to the mass flow rate of the propellants, $\dot{m}_p$. The nozzle flow exhausts into the freestream air (station ∞), which is at a pressure and temperature of p_∞ and T_∞, respectively.

The flow through the rocket engine thrust chamber is shown on a temperature–entropy diagram in Figure 4.72. The propellants enter the combustion chamber at a static temperature and static entropy of T_i and s_i, respectively. An amount of heat is added in the combustion chamber by the burning of the propellants, which occurs at a constant pressure, p_{t_c}. The constant-pressure heat addition raises the static temperature to T_c and increases the static entropy to s_c. The heat added (per unit mass) is equal to the heating value of the propellants, Q_R. The increase in the total temperature of the propellants from state i to c can be expressed as

$$T_{t_c} = T_{t_i} + \frac{Q_R}{c_p} \qquad (4.97)$$

Assuming an adiabatic reaction, the total temperature, T_{t_c}, is called the *adiabatic flame temperature*. Since the velocity in the combustion chamber is approximately zero ($V_c \cong 0$), the static and total temperatures are equal, so that the heat addition, Q_R, results in an increase of static temperature from T_i to T_c equal to Q_R/c_p, as shown in Figure 4.72. The static and total pressures are also equal in the combustion chamber, as shown in Figure 4.72 by the $p = p_{t_c}$ line from state i to state c.

The combustion products enter the supersonic exhaust nozzle (state c), are accelerated to sonic conditions at the nozzle throat (state $*$), and exit the nozzle (state e) with velocity, V_e, pressure, p_e, and temperature, T_e. The flow through the nozzle is assumed to be isentropic (constant entropy), so that $s_c = s^* = s_e$. Since the flow through the thrust chamber is isentropic, the total pressure through the nozzle is simply equal to that in the combustion chamber, that is, $p_{t_c} = p_t^* = p_{t_e}$.

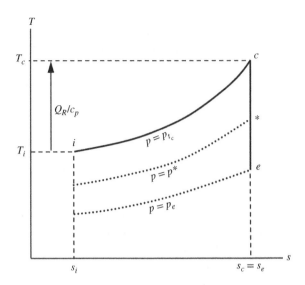

Figure 4.72   Rocket thrust chamber T–s diagram.

4.6.2   Rocket Propulsion Performance Parameters

This section defines selected parameters and relations that are important in the evaluation of rocket engine performance. The nomenclature for the rocket thrust chamber, shown in Figure 4.71, is followed for the parameters and relations given below. The flow through the thrust chamber is assumed to be isentropic, so the total pressure and total temperature are constant. In this section, we simply use p_t and T_t to denote the total pressure and total temperature, respectively, in the thrust chamber. In addition to assuming that the flow through the thrust chamber is isentropic, it is assumed that the combustion products (the working fluid) are a perfect gas with constant composition. Thus, the perfect gas equation of state is valid, and the gas is assumed to have constant specific heats. The following definitions and relations apply to the rocket engine either at static (zero velocity) conditions or in flight at a freestream velocity, V_∞.

4.6.2.1   Thrust

The thrust force generated by the rocket engine is an obvious performance parameter. From Equation (4.24), the thrust of a rocket engine, T, is given by

$$T = \dot{m}_p V_e + (p_e - p_\infty)A_e \tag{4.98}$$

One fundamental difference between a non-air-breathing rocket engine and an air-breathing engine is that the rocket engine can operate in an airless environment, such as in the vacuum of space, where the ambient pressure, p_∞, is zero. At this condition, the rocket thrust is given by

$$T_{vac} = \dot{m}_p V_e + p_e A_e \tag{4.99}$$

where T_{vac} is *vacuum thrust*. The vacuum thrust can be 10–30% higher than the sea level thrust. The maximum thrust for a given nozzle condition is in vacuum where $p_\infty = 0$. Do not confuse this with the maximum thrust condition for a perfectly expanded nozzle, as the thrust is highest in vacuum if the nozzle flow is perfectly expanded to $p_e = p_\infty = 0$ (see Section 4.3.2).

For the case of a perfectly expanded nozzle, the rocket engine thrust is simply

$$T = \dot{m}_p V_e \tag{4.100}$$

The rocket thrust in Equations (4.98) and (4.100) is independent of the flight speed of the rocket. Higher thrust is obtained by increasing the propellant mass flow rate or increasing the exhaust velocity.

4.6.2.2 Exhaust Velocity

From the last section, it was found that the exhaust velocity is an important parameter affecting the rocket thrust. Let us look at the exhaust velocity more closely. Since the exhaust flow through the nozzle is assumed to be isentropic, the total enthalpy, h_{t_c}, is constant in the thrust chamber and can be written in terms of the nozzle exit static enthalpy, h_e, and velocity, V_e, as

$$h_{t_c} = h_e + \frac{V_e^2}{2} \tag{4.101}$$

Using the definition of enthalpy, Equation (4.101) becomes

$$c_p T_t = c_p T_e + \frac{V_e^2}{2} \tag{4.102}$$

where c_p is the specific heat at constant pressure, and T_t is the chamber or nozzle total temperature, which is constant throughout the thrust chamber. Solving for the exhaust velocity, V_e, we have

$$V_e = \sqrt{2c_p(T_t - T_e)} = \sqrt{2c_p T_t \left(1 - \frac{T_e}{T_t}\right)} \tag{4.103}$$

In reality, the exhaust velocity is not uniform across the nozzle exit plane, but rather has a non-uniform velocity profile. The exhaust velocity in Equation (4.103) represents an *effective exhaust velocity* that is an average equivalent exhaust velocity.

Using Equation (3.124), the specific heat can be expressed in terms of γ and the specific gas constant, R. Since the flow is isentropic, $T_{t,e} = T_t$, so that the temperature ratio can be expressed in terms of the pressure ratio, p_e/p_t, using the isentropic relation, Equation (3.139), and Equation (4.103) becomes

$$V_e = \left\{ \frac{2\gamma R T_t}{\gamma - 1} \left[1 - \left(\frac{p_e}{p_t}\right)^{(\gamma-1)/\gamma} \right] \right\}^{1/2} \tag{4.104}$$

The specific gas constant, R, is equal to the universal gas constant, $\mathcal{R}$, divided by the gas molecular weight, $\mathcal{M}$, given by Equation (3.57), so Equation (4.105) is

$$V_e = \left\{ \frac{2\gamma \mathcal{R} T_t}{(\gamma - 1)\mathcal{M}} \left[1 - \left(\frac{p_e}{p_t}\right)^{(\gamma-1)/\gamma} \right] \right\}^{1/2} \tag{4.105}$$

Equation (4.105) states that a gas with a low molecular weight increases the exhaust velocity, which increases the thrust. This explains why hydrogen, the lowest molecular weight substance, is a preferred working fluid for many types of rocket engines.

Since the exhaust velocity is an important parameter in assessing the performance of a rocket engine, how can one obtain a value of this parameter from an engine test? Looking at

Equation (4.105), one would have to measure the total temperature and total pressure in the combustion chamber and the nozzle exit pressure. This might be possible, but would require specialized instrumentation that had the ability to survive the harsh environments in the combustion chamber and nozzle flow. In addition, multiple measurements may be required to adequately characterize the non-uniform flows. Rearranging Equation (4.100), we have an expression for the exhaust velocity that may provide a simpler solution.

$$V_e = \frac{T}{\dot{m}_p} \tag{4.106}$$

The exhaust velocity can be obtained from Equation (4.106) by measuring the thrust and the propellant flow rates, which may be easier to measure than the inflow pressures and temperatures. The thrust of the rocket engine can be measured on a test stand with a load cell and the propellant flow rates can be measured with various types of flow meters.

4.6.2.3   Thrust Chamber Mass Flow Rate

It can be shown that the mass flow rate through a supersonic nozzle, $\dot{m}_e$, is given by

$$\dot{m}_e = \frac{A^* p_t}{\sqrt{RT_t}} \sqrt{\gamma \left(\frac{2}{\gamma + 1} \right)^{(\gamma+1)/(\gamma-1)}} = \dot{m}_p \tag{4.107}$$

where, in our rocket engine application, p_t and T_t are the combustion chamber total pressure and total temperature, respectively, A^* is nozzle throat area, R is the specific gas constant, and γ is the ratio of specific heats. Since the nozzle exhaust and propellant mass flow rates are equal ($\dot{m}_e = \dot{m}_p$), Equation (4.107) provides the propellant mass flow rate in terms of the combustion chamber properties and the nozzle throat area.

4.6.2.4   Specific Impulse

Similar to an air-breathing engine, the specific impulse, I_{sp}, can be defined for a non-air-breathing rocket engine as

$$I_{sp} \equiv \frac{T}{\dot{w}_p} = \frac{T}{\dot{m}_p g_0} \tag{4.108}$$

where T is the rocket thrust, $\dot{w}_p$ is the propellant weight flow rate, $\dot{m}_p$ is the propellant mass flow rate, and g_0 is the acceleration due to gravity at sea level. For the rocket specific impulse, the weight or mass flow rates in Equation (4.108) include both the fuel and oxidizer, whereas only the fuel mass flow rate is used for the air-breathing specific impulse. Assuming that a consistent set of units is used, the specific impulse has units of seconds. Maximizing the specific impulse corresponds to maximizing the thrust or minimizing the consumption of propellant.

Substituting the expression for the thrust of a perfectly expanded nozzle from Equation (4.100) into (4.108), we have

$$I_{sp} = \frac{T}{g_0 \dot{m}_p} = \frac{\dot{m}_e V_e}{g_0 \dot{m}_p} = \frac{V_e}{g_0} \tag{4.109}$$

Now, inserting the expression for the nozzle exit velocity, Equation (4.104), into (4.109), we have

$$I_{sp} = \frac{1}{g_0} \left\{ \frac{2\gamma RT_t}{\gamma - 1} \left[1 - \left(\frac{p_e}{p_t} \right)^{(\gamma-1)/\gamma} \right] \right\}^{1/2} \tag{4.110}$$

Table 4.9 Theoretical performance of selected rocket engine liquid propellants.

Oxidizer	Fuel	Molecular weight, $\mathcal{M}$ (kg/mol)	Adiabatic flame temperature, T_{t_c} (K)	Characteristic velocity, c^* (m/s)	Specific impulse, I_{sp} (s)
Fluorine	Hydrogen	8.9	3080	2530	390
Fluorine	Hydrazine	18.5	4550	2130	340
Oxygen	Hydrogen	8.9	2960	2430	300
Oxygen	Methane	20.6	3530	1835	295
Oxygen	RP-1 (kerosene)	21.9	3570	1770	285

Equations (4.109) and (4.110) provide some insight into how to maximize the rocket specific impulse. Equation (4.109) clearly shows that specific impulse is directly proportional to the exhaust velocity, V_e. It states that the higher the exhaust velocity, the higher the specific impulse. This is also captured in Equation (4.110), through the exit-pressure-to-total-pressure ratio term, p_e/p_t. Higher exhaust velocities correspond to lower values of this pressure ratio, which give higher values of specific impulse in Equation (4.110).

Equation (4.110) also indicates that a higher specific impulse can be obtained with a higher combustion chamber total temperature, T_t. According to Equation (4.97), the rise in the combustion chamber total temperature depends on the heat of reaction, Q_R, of the propellants. More energetic, highly reactive propellants have a higher heat of reaction and result in a higher chamber total temperature. Table 4.9 provides values of molecular weight, adiabatic flame temperature, and specific impulse for selected rocket engine propellants. As expected, the more energetic, lower molecular weight propellants deliver higher specific impulses.

Lastly, the specific impulse is seen to be directly proportional to the specific gas constant, R. Recalling the definition of the specific gas constant as the universal gas constant, $\mathcal{R}$, divided by the gas molecular weight, $\mathcal{M}$, a larger value for the specific gas constant is obtained with lower values of the molecular weight. Thus, the specific impulse is increased by using a low molecular weight or light gas, such as hydrogen, for the rocket engine working fluid. To summarize, the specific impulse of a liquid rocket engine can be increased with (1) higher exhaust velocity, (2) higher combustion chamber temperature, and (3) use of a low molecular weight gas. Rocket engine specific impulse versus Mach number was shown in Figure 4.18. The rocket engine specific impulse is independent of Mach number with a maximum limit of about 400–450 s.

4.6.2.5 Characteristic Exhaust Velocity

The performance of the combustion chamber and propellants can be characterized by a parameter, with units of velocity, called the *characteristic velocity*, c^* (pronounced "c-star"), defined as

$$c^* \equiv \frac{p_t A^*}{\dot{m}_p} \tag{4.111}$$

where p_t is the combustion chamber total pressure, A^* is the nozzle throat area, and $\dot{m}_p$ is the propellant mass flow rate. These three quantities are readily obtained experimentally for a given thrust chamber, allowing the performance comparison of different thrust chambers by comparison of the characteristic velocity.

Inserting Equation (4.107) for the propellant mass flow rate into Equation (4.111), an equation for the characteristic velocity, independent of the nozzle geometry, is given by

$$c^* = \sqrt{\dfrac{RT_t}{\gamma\left(\dfrac{2}{\gamma+1}\right)^{(\gamma+1)/(\gamma-1)}}} \tag{4.112}$$

The combustion chamber temperature, the ratio of specific heats, and the specific gas constant depend on the choice of propellants. Thus, the characteristic velocity is useful for comparing combustion chamber designs and propellant combinations. Values of the characteristic velocity for selected propellant combinations are given in Table 4.9.

A c^* *efficiency* can be defined, which is equal to an experimentally measured c^*, from Equation (4.111), divided by a theoretical maximum c^* from Equation (4.112). It is a measure of the degree of completeness of the chemical energy release in the combustion chamber and the combustion chamber's efficiency in converting the propellant chemical energy into a high-pressure, high-temperature gas. The c^* *efficiency* is typically 92–99.5%.

4.6.2.6   Thrust Coefficient

Substituting Equations (4.105) and (4.107), for the mass flow rate and the exhaust velocity, respectively, into Equation (4.98) for the rocket thrust, we obtain the *ideal thrust equation*, given by

$$T = p_t A^* \sqrt{\frac{2\gamma^2}{\gamma-1}\left(\frac{2}{\gamma+1}\right)^{(\gamma+1)/(\gamma-1)}\left[1-\left(\frac{p_e}{p_t}\right)^{(\gamma-1)/\gamma}\right]} + \left(\frac{p_e}{p_t}-\frac{p_\infty}{p_t}\right)\frac{A_e}{A^*} \tag{4.113}$$

Dividing Equation (4.113) by $p_t A^*$, the rocket nozzle thrust coefficient, C_T, is defined as

$$C_T \equiv \frac{T}{p_t A^*} = \sqrt{\frac{2\gamma^2}{\gamma-1}\left(\frac{2}{\gamma+1}\right)^{(\gamma+1)/(\gamma-1)}\left[1-\left(\frac{p_e}{p_t}\right)^{(\gamma-1)/\gamma}\right]} + \left(\frac{p_e}{p_t}-\frac{p_\infty}{p_t}\right)\frac{A_e}{A^*} \tag{4.114}$$

Equation (4.113) shows that the thrust coefficient is independent of the combustion chamber temperature, and thus independent of the propellant choice, and is a function of the nozzle geometry only. Even though the pressure terms, in Equation (4.113), relate to the nozzle flow also, they are a function of the nozzle area ratio. Therefore, the thrust coefficient can be treated as strictly a nozzle parameter. At first thought, it may seem unusual that the thrust coefficient is independent of the chamber combustion temperature and the gas molecular weight, since these parameters play such an important role for thrust production through the exhaust velocity equation. However, given that we are simply dealing with isentropic, supersonic flow through a nozzle, it is appropriate that the thrust coefficient is only a function of the nozzle geometry and pressure distribution.

Using the definition of the thrust coefficient in Equation (4.114) and the characteristic velocity in Equation (4.111), the thrust can be written as

$$T = C_T p_t A^* = C_T \dot{m}_p c^* \tag{4.115}$$

Equation (4.115) is a simple expression for the rocket engine thrust, which characterizes it in terms of the propellant mass flow rate, $\dot{m}_p$, the performance of the combustion chamber, through the characteristic velocity, c^*, and the nozzle performance, through the thrust coefficient, C_T.

Example 4.8 Calculation of Rocket Engine Performance Parameters *A rocket engine uses liquid methane and liquid oxygen for its fuel and oxidizer, respectively. The pressure and temperature in the rocket engine combustion chamber are 45 atm and 3480 K, respectively. The ratio of specific heats of the combustion products is 1.22 and the nozzle throat area is 0.180 m². Assuming the rocket nozzle is perfectly expanded to a pressure of 1 atm, calculate the exhaust velocity, exhaust mass flow rate, specific impulse, characteristic velocity, and thrust coefficient.*

Solution

From Table 4.9, the molecular weight of the oxygen–methane propellant mixture is 20.6 kg/kg·mol. The specific gas constant for the mixture is

$$R = \frac{\mathcal{R}}{\mathcal{M}} = \frac{8314 \frac{J}{kg \cdot mol \cdot K}}{20.6 \frac{kg}{kg \cdot mol}} = 403.6 \frac{J}{kg \cdot K}$$

Using Equation (4.104), the exhaust velocity is

$$V_e = \left\{ \frac{2\gamma R T_t}{\gamma - 1} \left[1 - \left(\frac{p_e}{p_t} \right)^{(\gamma-1)/\gamma} \right] \right\}^{1/2}$$

$$V_e = \left\{ \frac{2(1.22)\left(403.6 \frac{J}{kg \cdot K}\right)(3480\,K)}{1.22 - 1} \left[1 - \left(\frac{1\,atm}{45\,atm} \right)^{(1.22-1)/1.22} \right] \right\}^{1/2} = 2781 \frac{m}{s}$$

Using Equation (4.107), the exhaust mass flow rate is

$$\dot{m}_e = \frac{A^* p_t}{\sqrt{R T_t}} \sqrt{\gamma \left(\frac{2}{\gamma + 1} \right)^{(\gamma+1)/(\gamma-1)}}$$

$$\dot{m}_e = \frac{(0.180)\left(45\,atm \times \frac{101{,}325\,N/m^2}{1\,atm}\right)}{\sqrt{\left(403.6 \frac{J}{kg \cdot K}\right)(3480\,K)}} \sqrt{1.22\left(\frac{2}{2.22}\right)^{(2.22/0.22)}} = 451.8 \frac{kg}{s}$$

Using Equation (4.109), the specific impulse is

$$I_{sp} = \frac{V_e}{g_0} = \frac{2781 \frac{m}{s}}{9.8 \frac{m}{s^2}} = 283.8\,s$$

Using Equation (4.112), the characteristic velocity is

$$c^* = \sqrt{\frac{R T_t}{\gamma \left(\frac{2}{\gamma + 1} \right)^{(\gamma+1)/(\gamma-1)}}} = \sqrt{\frac{\left(403.6 \frac{J}{kg \cdot K}\right)(3{,}480\,K)}{1.22\left(\frac{2}{2.22}\right)^{(2.22/0.22)}}} = 1{,}817 \frac{m}{s}$$

Using Equation (4.114), the thrust coefficient is

$$C_T = \sqrt{\frac{2\gamma^2}{\gamma-1}\left(\frac{2}{\gamma+1}\right)^{(\gamma+1)/(\gamma-1)}\left[1-\left(\frac{p_e}{p_t}\right)^{(\gamma-1)/\gamma}\right]} + \left(\frac{p_e}{p_t} - \frac{p_\infty}{p_t}\right)\frac{A_e}{A^*}$$

Since the nozzle is perfectly expanded, $p_e = p_\infty$, and the thrust coefficient is

$$C_T = \sqrt{\frac{2(1.22)^2}{0.22}\left(\frac{2}{2.22}\right)^{(2.22/0.22)}\left[1-\left(\frac{1\,atm}{45\,atm}\right)^{(0.22/1.22)}\right]} = 1.531$$

4.6.2.7 The Rocket Equation

In this section, we develop the *rocket equation*, which relates the velocity change, ΔV, imparted to a rocket vehicle, to the engine specific impulse, I_{sp}, and the decrease in propellant mass. This relationship was first published by Konstantin Tsiolkovsky in the early 1900s.

Consider a rocket at a point in its flight trajectory, where it has a velocity, V, and a flight path angle, θ, as shown in Figure 4.73. The rocket is assumed to be a point mass with a total mass, m, that includes the mass of the structure, payload, and propellants. The forces acting on the rocket are the engine thrust, T, the aerodynamic drag, D, and its weight, W, which is equal to the total mass multiplied by acceleration due to gravity, mg. The thrust and drag act in a direction parallel with the velocity vector. The weight acts vertical downward.

Applying Newton's second law parallel to the velocity vector, we have

$$\sum F_{\parallel V} = m\frac{dV}{dt} \tag{4.116}$$

$$T - D - W\sin\theta = T - D - mg\sin\theta = m\frac{dV}{dt} \tag{4.117}$$

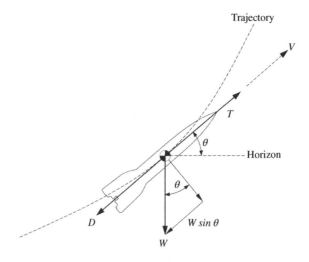

Figure 4.73 Free-body diagram of rocket in flight.

The thrust is related to the specific impulse and the mass flow rate of propellant, $\dot{m}_p$, through Equation (4.108). Substituting this relation into Equation (4.117) for the thrust, we have

$$\dot{m}_p g_0 I_{sp} - D - mg \sin \theta = m \frac{dV}{dt} \tag{4.118}$$

The propellant mass is consumed by the engine as the rocket ascends along its trajectory, so that the propellant mass flow rate is the decrease in the vehicle total mass as given by

$$\dot{m}_p = -\frac{dm}{dt} \tag{4.119}$$

Substituting Equation (4.119) into (4.118), we have

$$\left(-\frac{dm}{dt}\right) g_0 I_{sp} - D - mg \sin \theta = m \frac{dV}{dt} \tag{4.120}$$

Multiplying through by dt, dividing by m, and rearranging, we have

$$dV = -g_0 I_{sp} \frac{dm}{m} - \frac{D}{m} dt - g \sin \theta dt \tag{4.121}$$

We now have an equation that may be integrated to obtain the change in velocity as a function of the decreasing mass, aerodynamic drag, and gravity effects. Let us first look at the simplest case, where the aerodynamic drag and gravity effects are negligible compared with the thrust (recall that the first term on the right-hand side of Equation (4.121) is the thrust in terms of the specific impulse and mass). This case might represent the rocket traveling in space where there is no atmosphere and the effects of gravity are small compared with the thrust. For this case, Equation (4.121) becomes

$$dV = -g_0 I_{sp} \frac{dm}{m} \tag{4.122}$$

Integrating Equation (4.123) from an initial state, where the velocity and mass are V_1 and m_1, respectively, to a final state, where the velocity and mass are V_2 and m_2, respectively, we have

$$\int_{V_1}^{V_2} dV = -g_0 I_{sp} \int_{m_1}^{m_2} \frac{dm}{m} \tag{4.123}$$

$$V_2 - V_1 \equiv \Delta V = g_0 I_{sp} \ln \left(\frac{m_1}{m_2}\right) \tag{4.124}$$

Equation (4.124) is the *rocket equation*, which gives the velocity change, ΔV, imparted to the rocket, by burning an amount of propellant mass, $(m_2 - m_1)$, in a rocket engine with a specific impulse, I_{sp}, from an initial state to a final state in the trajectory.

Let us assume that the initial state is at lift-off, where the velocity is zero and the mass is m_i, and the final state is at burnout, where the velocity is V_b and the mass is m_f. Assuming that all of the propellants are consumed, Equation (4.124) becomes

$$V_b = g_0 I_{sp} \ln \left(\frac{m_i}{m_f}\right) = g_0 I_{sp} \ln \left(\frac{1}{MR}\right) \tag{4.125}$$

where the mass ratio, MR, has been defined as

$$MR \equiv \frac{m_f}{m_i} \tag{4.126}$$

The propellant mass, m_p, which has been consumed, is given by $m_p = m_i - m_f$. Equation (4.125) gives the final, burnout velocity of a rocket after consuming all of its propellant after lift-off.

Let us now consider the case where the effect of gravity is not negligible. For this case, Equation (4.121) becomes

$$dV = -g_0 I_{sp} \frac{dm}{m} - g \sin \theta dt \tag{4.127}$$

Integrating Equation (4.127) from an initial state at time t_1, where the velocity and mass are V_1 and m_1, respectively, to a final state at time t_2, where the velocity and mass are V_2 and m_2, respectively, we have

$$\int_{V_1}^{V_2} dV = -g_0 I_{sp} \int_{m_1}^{m_2} \frac{dm}{m} - \int_{t_1}^{t_2} g \sin \theta dt \tag{4.128}$$

$$V_2 - V_1 = g_0 I_{sp} \ln \left(\frac{m_1}{m_2} \right) - (g \sin \theta)_{av} (t_2 - t_1) \tag{4.129}$$

$$\Delta V = g_0 I_{sp} \ln \left(\frac{m_1}{m_2} \right) - (g \sin \theta)_{av} \Delta t \tag{4.130}$$

where $(g \sin \theta)_{av}$ is the time-averaged value of the acceleration due to gravity and the flight path angle, and Δt is the time span of the propellant burn, equal to $(t_2 - t_1)$. Obtaining expressions for the time-averaged values is beyond the scope of this text; however, assuming these quantities are known, Equation (4.130) provides a means of obtaining the velocity change, ΔV, including gravity effects. The gravity term in Equation (4.128) may also be obtained using numerical integration on a digital computer, for a known flight trajectory, where the altitude (from which g can be calculated) and the flight path angle, θ, are defined as a function of time.

Similarly, integrating Equation (4.121) to account for the effects of aerodynamic drag is beyond the scope of the text. Simplifying assumptions can be made to obtain a closed form solution or, as in the case of the gravity term, numerical integration can be applied to obtain a solution.

Example 4.9 V-2 Rocket Burnout Velocity *The V-2 rocket (see Figure 1.76) had a total launch mass of 12,500 kg, with 3800 kg of fuel and 4900 kg of liquid oxygen. The V-2 liquid rocket engine had a specific impulse of 200 s. Calculate the burnout velocity assuming 90% of the propellants are consumed, neglecting gravity effects and aerodynamic drag.*

Solution

The total propellant mass, m_p, is

$$m_p = 3800 \, kg + 4900 \, kg = 8700 \, kg$$

The final, burnout mass, m_f, assuming 90% of the propellants are consumed, is

$$m_f = m_i - 0.9 m_p = 12{,}500 \, kg - 0.9(8700 \, kg) = 4670 \, kg$$

Using Equation (4.125), the burnout velocity, V_b, neglecting gravity effects and aerodynamic drag, is

$$V_b = g_0 I_{sp} \ln \left(\frac{m_i}{m_f} \right) = \left(9.81 \frac{m}{s^2} \right) (200 \, s) \ln \left(\frac{12{,}500 \, kg}{4670 \, kg} \right) = 1931.7 \frac{m}{s}$$

$$V_b = 1931.7 \frac{m}{s} = 6889.3 \frac{km}{h} = 4280.8 \frac{mi}{h}$$

4.6.3   *Liquid-Propellant Rocket Propulsion*

A liquid rocket engine is so-named because it uses liquid propellants. The propellants comprise an oxidizer and a fuel, which are stored in one or more thin-walled tanks at low pressures. The propellants may be liquid *bipropellants*, composed of a liquid oxidizer that is separated from a liquid fuel, or a *monopropellant*, where the oxidizer and fuel are chemically combined in a single liquid. Bipropellants are stored in separate tanks, while a monopropellant is stored in a single tank. The liquid propellants typically constitute about 25% of the total launch weight of a liquid-propellant rocket. The basic components of a liquid propellant rocket engine system are the propellant storage tanks, a combustion chamber, associated feed system plumbing, valves, and regulators to move propellants from the tanks to the combustion chamber, a propellant injection system, an ignition system, and an exhaust nozzle.

The fuel and oxidizer are injected into the combustion chamber using an injector system, which typically consists of a series of small holes, arranged in a pattern, to optimize the mixing of the fuel and oxidizer. The fuel and oxidizer injection may be coaxial, such that the oxidizer is coaxially injected around a central fuel injection port. Types of injectors include the showerhead, impinging, and swirl injectors. In the showerhead injector, the fuel and oxidizer are injected through an injector faceplate, perforated with many small holes, analogous to a bath showerhead. In the impinging injector, the injected fuel and oxidizer are aimed so that the two streams impinge on each other, a short distance downstream of the injector face, which improves mixing. The propellants are injected in a swirling manner in the swirl injector, which may also improve mixing.

Once the fuel and oxidizer are mixed, an ignition source is required to initiate combustion. Pyrotechnic, electric spark, or chemical ignition may be used. The timing of the ignition is critical in liquid propellant rocket engines. If excessive amounts of propellants are allowed to accumulate in the combustion chamber prior to ignition, a *hard start* or explosive ignition of the propellants may occur, which may over pressurize or destroy the chamber.

Combustion may also be initiated using *hypergolic* propellants, which ignite spontaneously with each other. After combustion has been established, the propellants are switched to the non-hypergolic, primary fuel and oxidizer. Some liquid rocket engines operate solely on hypergolic propellants. Hydrazine and nitrogen tetroxide are a common hypergolic fuel and oxidizer combination. Advantages of hypergolic propellant systems include their simplicity, reliability, and their restart capability, since the fuel and oxidizer need only be mixed to initiate burning. Disadvantages include difficulties in handling, due to their toxicity and corrosiveness, and their lower specific impulse. Hypergolic propellant systems are typically used for spacecraft maneuvering systems and upper stages of launch vehicles.

The combustion chamber and nozzle are exposed to the extremely high temperatures of the combustion gases. Thermal protection may be provided by passive or active cooling techniques. Passive thermal protection may include thermal barrier or ablative coatings on the walls of the combustion chamber or nozzle. Active cooling involves the circulation of fuel or other fluid in or on the walls of the combustion chamber or nozzle. *Regenerative cooling* refers to the circulation of fuel for active cooling, which is then burned in the engine.

The propellant feed system may be a *pressure-fed* or *pump-fed* system. In a pressure-fed system, the propellants are moved by a high-pressure, inert gas, such as nitrogen. Turbopumps and other types of turbomachinery are utilized in a pump-fed system to move propellants. In both types of systems, a series of pressure regulators, check valves, and precision flow valves are usually required. The hot combustion gases are accelerated and exhausted through a supersonic nozzle to produce thrust. A liquid propellant rocket engine can be started, stopped, and restarted by opening and closing the appropriate propellant valves. The thrust of a liquid propellant engine can be varied or *throttled* by controlling the propellant flow rates.

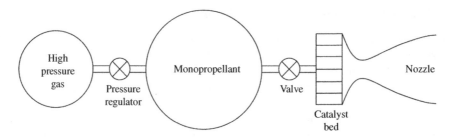

Figure 4.74   Liquid monopropellant rocket engine.

Monopropellant liquid rocket engines are relatively simple propulsive devices with low thrust and low specific impulse. They are commonly used for spacecraft attitude control. The basic components of a monopropellant rocket engine are shown in Figure 4.74. The propellant storage tank is filled with a monopropellant, such as hydrazine or hydrogen peroxide. Many monopropellant systems utilize a gas pressure-fed system where a high-pressure inert gas is used to move the monopropellant out of its storage tank. The monopropellant is passed through a catalyst bed, such as a platinum mesh, which decomposes it into a hot combustion gas. The decomposed products are accelerated and exhausted through a supersonic nozzle to develop thrust. The monopropellant thruster may also simply use a high-pressure, *cold gas propellant*, such as nitrogen gas, which is exhausted through a nozzle. The monopropellant rocket can be turned on and off, making it ideal for use as an attitude control motor.

Large thrust rocket engines are generally of the bipropellant type. Examples of liquid bipropellant oxidizers include liquid oxygen and nitric acid. Kerosene, gasoline, alcohol, and liquid hydrogen are examples of liquid bipropellant fuels. The propellants are often liquefied gas or *cryogenic propellants*, which require storage at very low temperatures. Liquid oxygen and liquid hydrogen storage are at about $-183\,°C$ ($90.2\,K$, $-297\,°F$) and $-253\,°C$ ($20.2\,K$, $423\,°F$), respectively. Cryogenic storage tanks must be vented to release the pressure rise from vaporization of the liquid propellants. The propellants may be fed to the thrust chamber using a pressure-fed or pump-fed system.

A pressure-fed, bipropellant rocket system is shown in Figure 4.75. The feed system is composed of propellant storage tanks, high-pressure gas tanks, typically filled with an inert gas such as helium, and plumbing, valves, and regulators between the tanks and thrust chamber. Operation of the system consists of opening the valves upstream and downstream of the propellant tanks, which allows the high-pressure gas to "blow-down" into the propellant tanks, pushing the propellants into the

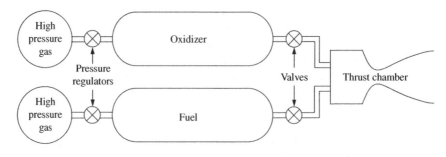

Figure 4.75   Pressure-fed liquid-bipropellant rocket engine.

thrust chamber. Hence, this type of feed system is sometimes called a *blow-down* system. The pressure of the gas entering the propellant tanks is controlled by the pressure regulators, which maintain a desired, constant set pressure until the pressure in the high-pressure tanks decreases below the regulator set pressure. An advantage of the pressure-fed system is its simplicity and reliability. There are no mechanical pumps or other machinery needed to move the propellants, as the motive power for moving the propellants is simply the blow-down of the high-pressure gas into the propellant tanks. A disadvantage of the pressure-fed system is the heavier, thicker-walled tanks needed to handle the high system pressures, which may operate at hundreds or even thousands of pounds per square inch.

A pump-fed, bipropellant liquid rocket system is shown in Figure 4.76. As in the pressure-fed system, the oxidizer and fuel are stored in separate tanks, but these propellants are delivered to the thrust chamber using pumps rather than a high-pressure gas. There are separate oxidizer and fuel pumps that are driven by a hot gas turbine. The turbine is powered by a gas generator, essentially another combustion device that may burn the same propellants as the rocket engine. Since the turbine is driven by a gas generator, this type of pump-fed engine is called a *gas-generator cycle* rocket engine. In some rocket engines, the gas generator may use other propellants, such as hydrogen peroxide. For large rocket engines, the propellant flow rates in a pump-fed system can be very high, perhaps hundreds of gallons of propellant per second, requiring large and complex turbopump and gas generator systems.

An example of a large, pump-fed, bipropellant rocket engine is the Rocketdyne F-1, shown in Figure 4.77. The F-1 rocket engine was developed in the 1950s and used in the *Saturn V* rockets that were flown to the Moon in the 1960s and 1970s (see Figure 1.72). There were five F-1 engines in the S-IC first stage of the *Saturn V* rocket. Each F-1 engine produced 1.5 million lb (6.7 MN) of thrust at sea level with a specific impulse of 260 s. The F-1 is still the most powerful liquid-fueled rocket engine, with a single thrust chamber and single nozzle, that has ever flown. The F-1 was a large engine by any standards, with a length of 19 ft (5.8 m), a nozzle exit diameter of 11 ft 7 in (3.53 m), and a flight weight of 18,500 lb (8390 kg).

The F-1 propellants were RP-1 kerosene and liquid oxygen. The propellants were fed to the combustion chamber by separate fuel and oxidizer pumps, driven by a hot gas turbine. The turbopump system was exposed to extremes in temperature, ranging from the 1465 °F (1069 K, 796 °C) hot gas entering the turbine to the −300 °F (89 K, −184 °C) liquid oxygen flowing through the pump. The F-1 propellant flow rates achieved by the turbopumps were staggering. The oxidizer flow rate was 24,811 gal/min (3945 lb/s, 1789 kg/s) and the fuel flow rate was 15,471 gal/min (1738 lb/s, 788.3 kg/s). There were five F-1 engines in the *Saturn V* first stage, with a total propellant flow rate of 204,410 gal/min or 3357 gal/s (473.6 lb/s, 214.8 kg/s)! The rated burn duration of each engine

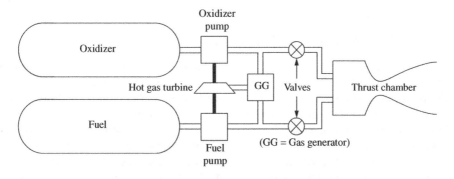

Figure 4.76 Pump-fed liquid-bipropellant rocket engine.

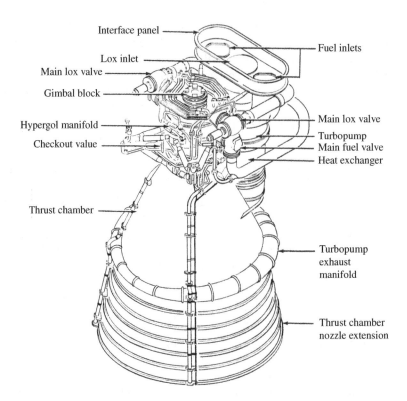

Figure 4.77   The F-1 turbopump-fed, liquid bipropellant rocket engine. (Source: *NASA*.)

was about 2.5 min (150 s). After the first stage burn, the *Saturn V* rocket had a speed of about 6200 mph (9980 km/h) at an altitude of 220,000 ft (67,000 m, 41.7 miles).

The propellants were injected into the thrust chamber, where the combustion temperature was 5970 °F (3572 K, 3299 °C) and the chamber pressure was 965 psi (678 kN/m^2). Combustion was initiated using hypergolic propellants, which were then switched to the primary fuel and oxidizer after combustion was established.

The thrust chamber had a nozzle extension, which increased the expansion ratio (ratio of the nozzle exit area to the throat area) of the nozzle from 10:1 to 16:1. The turbopump exhaust manifold was wrapped around the thrust chamber. The cooler turbopump exhaust gases were injected along the walls of the nozzle to provide film cooling from the higher temperature nozzle flow.

4.6.4   Solid-Propellant Rocket Propulsion

Unlike the liquid-propellant rocket, where the propellants are separated, the fuel and oxidizer in a solid-propellant rocket are mixed together in a combined, solid propellant. As discussed earlier, solid-propellant rockets date back to the earliest form of rocket propulsion. Solid-propellant rocket motors have reached a high degree of technical advancement, capable of providing reliable, high-thrust performance for relatively low cost. Solid propellant rockets are capable of launching small payloads, i.e. less than about 2000 kg (4400 lb), into low earth orbit (LEO) or payloads of about 500 kg (1100 lb) beyond earth orbit. Solid rocket first stage motors can have specific impulses of as high as about 280 s, with values of 175–250 s being common. This compares to specific impulses of over 450 s for hydrogen–oxygen liquid propellant rocket engines. The lower

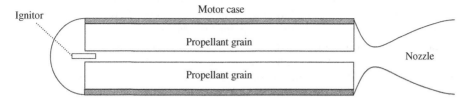

Figure 4.78   Solid rocket motor.

performance of solid propellant rocket motors, relative to liquid propellant rocket engines, do not make them suitable for use as the primary propulsion for larger space launch vehicles. Solid propellant "strap-on" boosters are attached to liquid propellant rockets to increase their launch weight capability, as for the Space Shuttle. They are also used for the final stage propulsion of booster systems that place satellites into earth orbit.

The propellant in a solid rocket motor is called the *propellant grain*. The propellant grain is bonded to the inside of a metal or composite cylinder, called the rocket motor *case*, as shown in Figure 4.78. The propellant grain must possess structural mechanical properties that resist cracking during ground handling of the rocket and in flight. Cracks in the grain could result in a catastrophic explosion of the propellant. The motor case is a pressure vessel that is designed for the high pressures and temperatures of combustion. The motor case may be fabricated from a variety of materials, ranging from cardboard for simple, black powder hobby rocketry motors to steel as used in the Space Shuttle solid rocket boosters. High strength-to-weight, composite materials, such as carbon fiber, are also used for solid rocket motor cases. The motor case may be lined with an insulating material, to protect it from the high temperatures of the burning propellant.

There is a hole in the center of the grain, called the *perforation*, which can have various cross-sectional shapes. Since the grain burns from the inside surface of the perforation outward to the case, the perforation pattern is designed to obtain the desired propellant burn rate, and hence the desired thrust profile of the motor. Various solid rocket motor perforation shapes are shown in Figure 4.79, along with their associated thrust curves. The type of thrust profile is given at the top

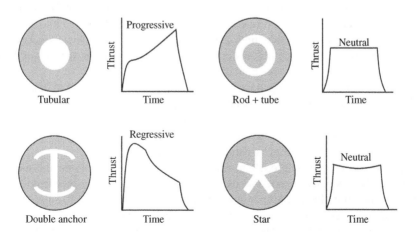

Figure 4.79   Various solid rocket motor perforation shapes and thrust curves.

of each thrust curve. For example, a progressive thrust profile provides an increasing thrust level over time, while the thrust decreases during the burn with a regressive profile.

The combustion process consumes the solid propellant and generates a high pressure, high temperature gas that is exhausted through the nozzle to produce thrust. The thrust produced by the solid motor is directly proportional to the combustion gas mass flow rate. The rate of propellant consumption is equal to the mass flow rate (mass per unit time) of the combustion gases, m_g, and is given by

$$m_g = \rho_g A_b r \tag{4.131}$$

where ρ_g is the density of the solid propellant, A_b is the propellant surface area that is burning, and r is the recession rate of the burning surface or the burning rate (linear distance of recession per unit time). The recession rate, r, can be obtained using empirical equations that are generally a function of the propellant type, propellant temperature, combustion pressure, and other factors. Since larger mass flow rates of combustion gases result in higher thrust, we see from Equation (4.131) that higher thrust is obtained by increasing the solid propellant density, the surface burn area, or propellant recession rate.

An ignitor is mounted at the top or head end of the motor, in the center of the perforation. After ignition, combustion of the solid propellant occurs on the exposed, interior surfaces of the grain, consuming the propellant from the central area of the case to its outer diameter. The hot combustion gases from the burning of the solid fuel exit through the nozzle at the base end of the motor. Unlike a liquid rocket engine, no propellant feed systems are required in a solid rocket motor, so there are no valves, plumbing lines, pressure regulators, or other feed system components. Lacking valves to close off the flow of propellants, thrust termination in solid rockets may be accomplished by blowing off the rocket nozzle or opening vents in the walls of the combustion chamber. Both of these methods cause the combustion chamber pressure to drop dramatically, extinguishing the combustion process. The nozzle geometry or chamber venting can be controlled in advanced solid rockets, enabling cut-off, restart, or throttling capabilities.

The two primary types of solid propellants are *homogeneous* and *heterogeneous* or *composite*. The distinction between these types of propellants is based on how the fuel and oxidizer are combined to make up the propellant. In a homogeneous propellant, the fuel and oxidizer are combined at a molecular level, that is, the molecules of the propellant contain both the fuel and oxidizer. A common type of homogeneous solid propellant is a combination of two monopropellants, nitroglycerin and nitrocellulose, also called a double-base propellant. The molecular formulas for nitroglycerin $[C_3H_5(NO_2)_3]$ and nitrocellulose $[C_6H_7O_2(NO_2)_3]$ contain both hydrocarbon-based fuel and oxygen atoms. In this double-base, solid propellant, the nitroglycerin is dissolved into a nitrocellulose gel. Since nitroglycerin is an unstable, high-energy monopropellant, the lower-energy nitrocellulose propellant serves as a stabilizer for the combined solid propellant. These types of double-base propellants have a specific impulse of about 235 s.

The composition of the solid propellant usually contains other additives, to delay decomposition, increase performance, improve the mechanical properties, or for other purposes. For instance, metallic powders, such as aluminum, magnesium boron, and beryllium, are sometimes added to increase the propellant specific impulse and fuel density. The specific impulse of a double-base propellant is increased to about 250 s by using a metallic additive.

A heterogeneous or composite propellant consists of an oxidizer mixed into and suspended in a plastic-like or rubber-like *fuel binder*. The oxidizer, in a ground crystal or powdered form, is usually an ammonium nitrate-based (AN) or ammonium perchlorate-based (AP) substance, although potassium nitrates, potassium chlorates, nitronium perchlorate, and other substances are also used. The fuel binder may be a synthetic rubber or a common plastic, such as hydroxyl terminated polybutadiene (HTPB), polybutadiene acrylonitrile (PBAN), or polyurethane. Powdered metals may also

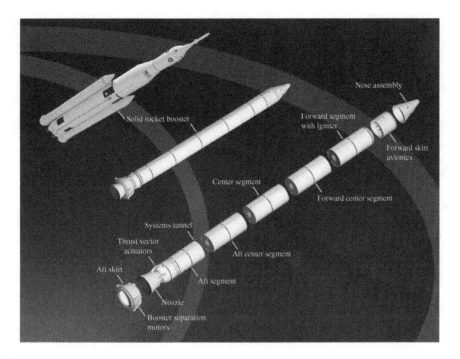

Figure 4.80 Orbital ATK Space Launch System (SLS) five-segment booster. (Source: *NASA*.)

be combined into the binder for added performance. Specific impulses near 300 s can be achieved with composite propellants.

The world's largest solid rocket motor ever built, the Orbital ATK Space Launch System (SLS) five-segment booster (FSB), is shown in Figure 4.80. The NASA SLS is the heavy-lift rocket system designed to replace the Space Transportation System or Space Shuttle. Derived from the Space Shuttle, four-segment solid rocket booster, the FSB is composed of five composite propellant segments, each using ammonium perchlorate oxidizer and aluminum fuel mixed into a polybutadiene acrylonitrile (PBAN) fuel binder. The SLS utilizes two of these 12 ft (3.7 m) diameter, 177 ft (53.9 m) long solid rocket motors, with a total propellant weight of about 1.4 million lb (0.64 million kg). The two solid rocket boosters generate about 7.2 million lb (32 million N) of thrust, providing about 75% of the total vehicle thrust at launch. The FSBs burn for about two minutes, consuming about 5.5 tons (11,000 lb, 5000 kg) of propellant per second. A ground firing of the SLS five-segment booster is shown in Figure 4.81.

4.6.5 Hybrid-Propellant Rocket Propulsion

As the name implies, a hybrid rocket engine combines aspects of a liquid and a solid rocket engine. The propellants of a hybrid rocket are composed of a liquid or gaseous component and a solid component. Better performance and operation are usually obtained with a liquid or gaseous oxidizer and a solid fuel, although hybrid rockets using liquid or gaseous fuels and solid oxidizers are possible and have been tested in the past.

Common hybrid oxidizers include liquid or gaseous oxygen, nitrous oxide, and hydrogen peroxide. Many solid fuels used in hybrid rockets are ubiquitous, non-hazardous substances that one would not normally associate with rocket fuel. These common hybrid rocket solid fuels include

Figure 4.81 Ground firing of the largest solid rocket motor ever built, the SLS five-segment booster. (Source: *Photo courtesy of Orbital ATK, by permission.*)

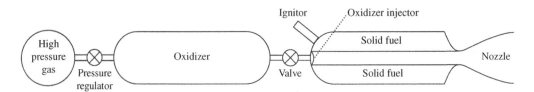

Figure 4.82 Hybrid rocket engine.

Plexiglas, (polymethylmethacrylate), paraffin wax, HTPB synthetic rubber, and various other types of synthetic plastics. To increase performance, high-energy additives, such as magnesium, aluminum, beryllium, and lithium, can be mixed into the solid fuel, which then acts as a binder for these additives.

A typical, liquid-oxidizer, solid-fuel hybrid rocket system is shown in Figure 4.82. The basic components of the system are a liquid oxidizer-filled tank and a solid fuel chamber with an attached exhaust nozzle. In the simple, pressure-fed system shown in Figure 4.82, a high pressure inert gas, such as helium, is used to force the oxidizer into the solid fuel chamber. A pump-fed system can also be used, with turbopumps taking the place of a high pressure gas source, to flow the oxidizer into the combustion chamber. A series of valves and regulators are required to control the flow of the pressurant gas and liquid oxidizer.

An ignition system provides a source of heat, which gasifies the solid fuel at the head end of the motor. This can be accomplished by injecting a pyrophoric substance (a substance that spontaneously ignites when exposed to air), such as a mixture triethyl aluminum (TEA) and triethyl borane (TEB), into the combustion chamber or by using a propane or hydrogen-fueled ignition system. The pressurized oxidizer is injected into the combustion chamber and reacts with the vaporized fuel. An electric spark ignitor system can also be used, with gaseous oxidizers, to ignite the gaseous

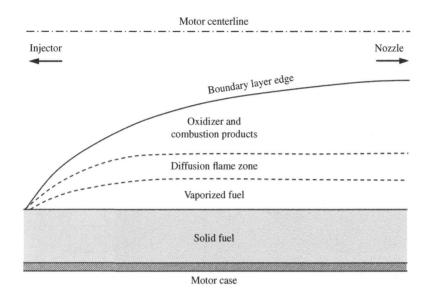

Figure 4.83   Hybrid rocket combustion of fuel and oxidizer in a diffusion flame zone.

fuel–oxidizer mixture. The oxidizer flow also spreads the combustion flame, from the head end of the motor, down the length of the solid fuel.

The combustion of the oxidizer and the vaporized fuel occurs in a diffusion flame zone within the boundary layer above the surface of the solid fuel, as shown in Figure 4.83. This type of combustion is a slow process, relative to combustion in solid and liquid rocket motors, since it is dependent on the vaporization of the solid fuel at its surface, followed by the mixing of the vaporized fuel with the oxidizer in the narrow flame zone. The resulting rate at which the fuel surface is vaporized or recedes during the burn, called the fuel regression rate, is low. The fuel regression rate of a hybrid can be an order of magnitude smaller than for a solid fuel motor. For example, if the fuel regression rate of a solid motor is about 1 cm/s, the hybrid may have a much smaller regression rate of about 0.1 cm/s.

The simplest fuel grain geometry has a single hole or port through the center of the motor. The use of multiple combustion ports in the fuel grain, as shown by the "wagon wheel" multi-port design in Figure 4.84, provides increased fuel surface area, which increases the regression rate, fuel flow, combustion efficiency, and thrust of the motor. The multi-port design also increases the turbulence of the flow exiting the fuel grain and entering the mixing chamber upstream of the nozzle, improving the mixing and combustion of unreacted fuel and oxidizer. But the multi-port design has some drawbacks, including poor volumetric efficiency as compared to a solid motor (a given hybrid motor cylindrical volume contains less fuel due to the numerous holes or ports), structural integrity issues, and difficulty in getting all of the fuel segments to burn at the same uniform rate.

Some hybrid motors have a pre-combustion or vaporization chamber, located at the head end of the motor upstream of the solid fuel grain, to help initiate combustion. Some hybrid motors also have an aft mixing chamber, located downstream of the solid fuel grain and upstream of the nozzle, to burn any residual fuel and oxidizer prior to being exhausted from the nozzle.

One disadvantage of hybrid combustion is its tendency towards instability. These combustion instabilities are felt as large pressure oscillations with a frequency near the natural frequency of the propellant feed system or combustion chamber volume. This can lead to the failure of the rocket motor due to excessively high pressures or high heat transfer rates. This is not to imply that

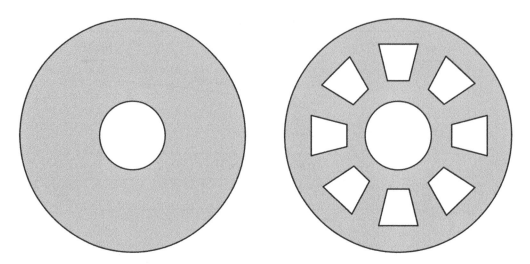

Figure 4.84   Single port (left) and multiple-port, wagon wheel (right) fuel grain.

combustion instability is limited to hybrid rockets, as it is also sometimes found in liquid and solid propellant rocket engines. The physics of combustion instability is still not well understood and "fixes" are often left to trial-and-error experimental testing. Exacerbating the problem is the need to conduct testing on full-scale hardware, as the instabilities tend not to scale down, which would allow testing using less expensive, sub-scale hardware.

Combustion instabilities in liquid and hybrid propellant rockets can be categorized based on the frequency of the oscillations. Low frequency oscillations of less than about 100 Hz (100 cycles per second), known as *chugging*, are due to coupling between the natural frequencies of the combustion process and the propellant feed system. High frequency oscillations on the order of 1000 Hz, known as *screaming*, are related to the acoustic vibration modes of the combustion chamber volume. Chugging can usually be mitigated by proper design of the propellant injectors. Screaming is not well understood and fixes can be elusive, as there are many different combustion chamber vibration modes, which tend to be sensitive to small changes in the system design.

Hybrid rocket propulsion can provide advantages over liquid and solid rocket propulsion in the areas of simplicity, safety, cost, performance, and operation. With far fewer valves and plumbing, the mechanical simplicity of the hybrid rocket engine is an obvious advantage over liquid rocket engines, but less so in comparison to a pure solid rocket. The high-density hybrid solid fuels reduce the overall system volume, again more of an advantage over liquid rather than solid rockets.

In terms of safety, the fact that the hybrid rocket fuel is often a non-volatile substance, such as a rubber or plastic, is an inherent safety benefit over the typically volatile, corrosive, or toxic fuels used in liquid and solid rockets. The fabrication, storage, and transport of hybrid rocket fuels are also much easier and safer. The fuel and oxidizer of a hybrid rocket are stored in different states of matter, making it less likely that unintentional mixing will result in an explosion, unlike the more likely explosion hazard with mixing two liquid propellants in a liquid rocket engine. The separation of propellants in a hybrid also makes the accidental ignition, detonation, and firing of the motor less likely than for a solid rocket motor. The relative mechanical simplicity and advantages in safety, which require less complexity in handling, translate into potentially lower costs for the hybrid propulsion.

The specific impulse of hybrid rocket motors can be superior to that of solid motors and some bipropellant liquid engines. The bulk densities of the hybrid propellants are comparable with that of a solid motor, but the more energetic liquid oxidizers, used in a hybrid, increase the hybrid's performance over the solid. The incorporation of high-energy additives into the solid fuel can further increase hybrid rocket performance.

The hybrid rocket motor has many beneficial operational similarities to a liquid rocket engine. By controlling the flow of the liquid propellant, a hybrid rocket motor can be started, stopped, restarted, and throttled, capabilities similar to a liquid rocket engine, but not possible with a solid motor.

The technology, testing, and flight experience associated with hybrid rocket propulsion are much less mature than for solid or liquid rocket propulsion. Hybrid rocket propulsion has been used for sounding rockets, but there has been limited development, testing, and flights of large-scale hybrid-powered rockets. The largest hybrid rocket motors tested to date were 250,000 lb (1.1 kN) thrust-class motors that were ground fired during a NASA and DARPA program in the 1990s. These large-scale hybrid motor tests were of limited success as there were significant issues with combustion instabilities and non-uniform burning in the combustion ports. Smaller-scale hybrid rocket motors have been flown in the Scaled Composites Spaceship One and Spaceship Two sub-orbital vehicles.

4.6.6   Types of Rocket Nozzles

In previous chapters, we have discussed the (internal) aerodynamics and propulsion related to convergent-divergent nozzles. Assuming there are no viscous and thermal losses, the rocket nozzle flow can be assumed isentropic. If we also assume that the flow through the nozzle is one-dimensional, the Mach-area relation, Equation (3.401), and the isentropic flow relations, Equations (3.345) to (3.347), then provide the Mach number and flow properties through a rocket nozzle of defined area distribution. The thrust of the rocket nozzle is given by Equation (4.24), where the maximum thrust is obtained for an ideal nozzle with an expansion ratio (nozzle exit area-to-throat area ratio) chosen to perfectly expand the flow to ambient pressure. Thus, we have already developed the relationships that define the aerodynamic and propulsion aspects of the rocket nozzle.

One critical difference between rocket nozzles and those designed for air-breathing engines concerns their range of operation, in particular, the ambient conditions at the nozzle exit. A rocket may operate from sea level to the vacuum of space, where the nozzle exit ambient back pressure varies from sea level pressure to near zero. The nozzle pressure ratio, or NPR, (nozzle total pressure divided by the nozzle exit ambient static pressure) for a rocket operating over this range of ambient conditions, may vary from about 50, at sea level, to infinity in space. To expand the flow to very high nozzle pressure ratios, a very large nozzle expansion ratio is required. A rocket nozzle, optimized for operation in near-space, may have an expansion ratio of 100. A fixed geometry nozzle, with a fixed expansion ratio, is optimum for a single altitude, typically a very high altitude. Therefore, the nozzle is operating sub-optimally at all lower altitudes. There are several mechanical schemes to change the nozzle expansion ratio during flight, such as with an extendable nozzle extension, but these add weight and complexity to the nozzle design. There are also some aerodynamic ways to have the nozzle operate more optimally at varying altitudes, as we discuss shortly.

We now wish to elaborate on the various types of nozzle geometries that are in use and have been developed, highlighting several of their advantages and disadvantages. There are three primary categories of convergent-divergent rocket nozzles: *conical*, *bell*, and *annular* (also called plug or altitude-compensating nozzles), as shown in Figure 4.85. The expansion-deflection, aerospike, and truncated aerospike, shown as Figure 4.85c, d, and e, respectively, are examples of annular nozzles with a center body or plug inside the nozzle. The annular nozzles provide altitude compensation,

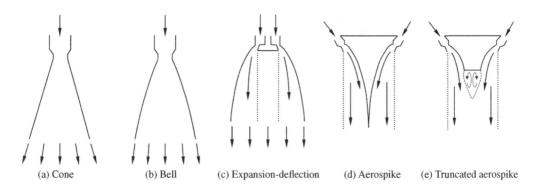

(a) Cone   (b) Bell   (c) Expansion-deflection   (d) Aerospike   (e) Truncated aerospike

Figure 4.85   Types of rocket engine nozzles.

as discussed shortly. The most widely used rocket nozzle in use today is the bell shape, while the annular is probably the least used due to its complexity even though its theoretical performance is higher than the other nozzle types.

Typically, all of these types of rocket nozzles have circular cross-sections, although there are exceptions. They all have converging-diverging sections, with a minimum area or throat between these two sections. The flow is subsonic in the converging section, sonic at the throat, and supersonic in the diverging section. Of these three areas, the design of the diverging section is the most critical to nozzle performance. The large, favorable pressure gradients, in the converging section and throat, keep the flow attached so that these areas can tolerate a wide range of geometries without serious losses. The favorable pressure gradients are much smaller in the diverging, supersonic section and the flow is more susceptible to boundary layer separation, which incurs large performance losses. In addition, improper contouring of the diverging section can also result in loss-producing shock waves. This is the reason why the primary difference in the nozzle types is in the design of the diverging section.

As its name implies, the conical nozzle has a diverging section with a simple cone-shape, as shown in Figure 4.85a. The conical nozzle is the simplest and most economical to fabricate. The walls of the supersonic section diverge from the throat at a constant angle. This constant wall angle leads to *divergence losses*: losses in flow momentum and thrust because the flow exiting the nozzle is not uniform and parallel with the axial component of the velocity. While a smaller divergence angle produces more thrust by maximizing the axial component of the exit velocity, it results in a longer, heavier nozzle as compared with a larger divergence angle, which is shorter and lighter, albeit with low performance. The optimum conical nozzle half angle is 12–18°, which is a compromise between length/weight and performance. Conical nozzles are still used today for smaller applications.

The bell-shaped rocket nozzle, shown in Figure 4.85b, has many advantages over the conical nozzle, making it the most commonly used nozzle for rocket engines today. The bell nozzle is shorter, lighter, and has higher performance than the conical nozzle. The bell nozzle contour diverges at a large angle, immediately downstream of the throat, and then has a gradual divergence that becomes small at the nozzle exit. This results in efficient expansion of the supersonic flow and small divergence losses since the exit flow is more uniform and parallel.

All fixed expansion ratio nozzles are perfectly expanded at a single altitude. At all other altitudes, the nozzle is over- or underexpanded and the performance is less than ideal. Typically, the nozzle is overexpanded at lift-off and underexpanded at high altitude. A solution to this issue is offered by the *annular nozzle*.

The annular nozzle may be of the *expansion-deflection* type (Figure 4.85c) or *aerospike* type (Figures 4.85d and e). The expansion-deflection nozzle has an annular throat that surrounds the centerbody or plug at the base of the nozzle. The aerospike nozzle has several, individual or modular thrust chambers, each with a small nozzle, arranged around an axisymmetric, central plug or spike centerbody. The most efficient shape for the spike contour is an isentropic expansion ramp, but this tends to be prohibitively long and heavy. The truncated aerospike nozzle is much shorter, but a secondary "bleed" flow must be injected into the blunt base to alleviate flow separation and base drag. The cooling of the plug, centerbody, or base is an issue that adds some complexity to these types of nozzles.

Both types of annular nozzles have a hot gas boundary that adjusts to the nozzle exit ambient back pressure, which changes with altitude, so they are called *altitude-compensating* nozzles. The hot gas boundary is inside the diverging section of the expansion-deflection nozzle and exterior to the aerospike nozzle, as shown by the dotted vertical lines in Figure 4.85c, d, and e. The hot gas boundary acts as a self-adjusting aerodynamic inner (for the expansion-deflection nozzle) or outer (for the aerospike nozzle) wall of the nozzle. As the altitude increases and the back pressure decreases, the gas boundary expands and the hot gas flow fills more of the interior of the nozzle, changing the pressure distribution on the nozzle walls. As the altitude increases and the back pressure decreases, the hot gas boundary expands outward, changing the pressure distribution on the spike. The altitude-compensation makes the off-design performance of these nozzles superior to the cone or bell-shaped nozzles.

The aerospike nozzle offers some interesting advantages over conventional shaped nozzles. The smaller, modular combustion chambers of the aerospike may be easier and less costly to develop and test than the larger device for a conventional nozzle. The thrust of the modular combustion chambers can be controlled individually, providing a thrust vectoring capability without the weight and complexity of actuators and gimbals to swivel a conventional nozzle. The advantages of the aerospike nozzle have led to some ground tests of large-scale hardware, but there has been no significant flight experience with these types of nozzles. The proposed Lockheed-Martin X-33, a single-stage-to-orbit reusable launch vehicle design, incorporated a linear aerospike rocket engine, where the modular combustion chambers are in a linear, rather than annular, arrangement (Figure 4.86).

4.7   Other Types of Non-Air-Breathing Propulsion

In this section, several types of non-air-breathing rocket propulsion are described that do not rely on chemical propellants as the source of energy. Fundamentally, in chemical rocket propulsion, the energy available to accelerate a vehicle is limited by the chemical energy of the propellants. This limits the maximum specific impulse of chemical rockets to about 400–450 s. Other types of non-chemical rocket propulsion do not have this limitation and can theoretically deliver much higher impulses to accelerate a rocket. These *thermal rocket propulsion* systems use other types of energy sources, such as nuclear, electric, or solar power, to heat a low molecular weight working fluid, often hydrogen. By expanding, accelerating, and exhausting the hot working fluid through a supersonic nozzle, extremely high specific impulses, perhaps over 1000 s, are theorized with thermal propulsion.

Some of these types of thermal propulsion could be used to launch rockets into space from the earth's surface, while some are suitable only for use in space, for operations such as orbit transfer or stabilization. These kinds of in-space propulsion are typically of low thrust, but this low thrust can be applied for long durations, making it possible to accelerate a space vehicle to a high velocity

Figure 4.86 Lockheed-Martin X-33 single-stage-to-orbit reusable launch vehicle with aerospike nozzles. (Source: *NASA*.)

over a long period of time. As such, this type of propulsion may be beneficial for long duration space missions, such as interplanetary missions or journeys into deep space.

4.7.1 Nuclear Rocket Propulsion

Theoretically, the fuel energy density of a nuclear energy-derived propellant is perhaps a hundred times greater than for a chemical propellant. The resulting exhaust velocities are predicted to be twice as high for a nuclear rocket, resulting in half the propellant launch mass of a chemical rocket. Specific impulses of about 500–1200 s have been predicted using nuclear rocket engines, which far exceeds the best obtainable values of about 450 s for conventional chemical rocket propulsion. With this potential, significant research and development of nuclear rocket propulsion has been conducted in the past, although no nuclear-powered rocket has ever been flown. From the mid-1950s through the early 1970s, several different nuclear rocket designs were built and ground tested in the USA.

All nuclear rocket propulsion concepts have serious environmental and safety issues that must be addressed. Shielding must be used to protect both equipment and personnel from the damaging effects of radiation from the nuclear reactions. While nuclear rocket engines have an inherent radiation exposure risk from the propulsion system itself, their performance increase may significantly decrease the travel time to distant worlds, such as Mars, thereby significantly decreasing the exposure of astronauts from space radiation. There are also hazards associated with an accident or crash of a nuclear-powered rocket, which could contaminate areas with radioactive material and debris. Mitigating these issues typically involves heavy shielding or containment vessels, which can dramatically increase the vehicle weight.

Fundamentally, nuclear rocket propulsion is based upon the addition of heat from a nuclear reaction to a working fluid, such as hydrogen. Unlike chemical rocket propulsion, there is no combustion of propellants in nuclear rocket propulsion. According to Equation (4.104), the highest exhaust velocity is obtained with the lowest molecular weight working fluid at a given temperature. Liquid hydrogen is an optimum choice for a working fluid since it has the lowest molecular weight of the elements. Ammonia has also been proposed as a working fluid, due to its higher density and ease of handling, although it provides only about half the specific impulse of hydrogen. The hot propellant is then expanded through a rocket nozzle to generate thrust. These types of nuclear energy-based propulsion systems are sometimes called *nuclear thermal rocket engines*.

The transfer of heat to a working fluid using nuclear energy can be accomplished in three ways: using *radioactive decay*, nuclear *fusion*, or nuclear *fission*. Radioactive decay-based nuclear propulsion relies on the decay of radioactive isotopes, which generates heat. This technique has been used successfully to generate electrical power in space vehicles, satellites, and deep-space probes, but has not been successfully applied as a means of rocket propulsion. Nuclear rocket propulsion using fusion has been investigated, but no practical concepts have been advanced due to our lack of fundamental understanding of the physics. Fission-based thermal rocket engines generate heat through fission of a radioactive material, such as uranium. The working fluid is passed through a nuclear reactor to transfer the heat from the fission reactions to the fluid. The nuclear reactor may operate at temperatures above 2500 K (4000 °F), providing the capability to add substantial energy to the working fluid. However, this also presents challenges in the design of the hardware that must operate at these extreme temperatures.

The solid-core, hydrogen-cooled, fission reactor, is at the heart of the nuclear thermal rocket engine shown in Figure 4.87. Liquid hydrogen is pumped from a storage tank into a cooling circuit around the reactor and rocket nozzle. The hydrogen is then injected into the reactor, absorbs the heat from the nuclear reactions, and exits the reactor as high temperature, hydrogen gas. The hot hydrogen gas is accelerated through a supersonic nozzle to a high exhaust velocity, producing the thrust to propel the space vehicle. A small amount of hot hydrogen gas is diverted from the reactor to the turbine, which powers the pump. A radiation shield protects the components and people from the reactor's radiation.

Perhaps the most intensive research and development of nuclear rocket propulsion was the Nuclear Engine for Rocket Vehicle Application (NERVA) program conducted by the US Atomic Energy Commission, NASA, and the Los Alamos National Laboratory, in the 1960s and 1970s. The NERVA program objective was to demonstrate the feasibility of nuclear thermal rocket engines for

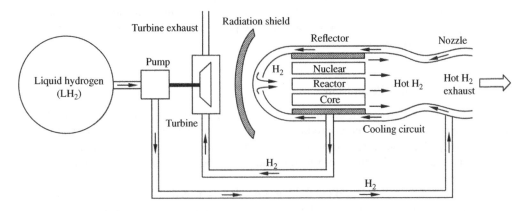

Figure 4.87 Nuclear thermal rocket engine.

space propulsion. Several different nuclear rocket engines were built and tested during the program, many with flight-rated components. Ground tests of nuclear thermal rocket engines achieved thrust levels of about 980 kN (210,000 lb) with a specific impulse of about 850 s. The hydrogen working fluid in these tests was heated to about 2500 K (4000 °F). By the end of the program, some believed that the nuclear rocket propulsion technology was in hand for use in space travel. However, funding was cut for the development of nuclear rocket propulsion in 1973, with the waning of the US manned space exploration program. There has been some renewed interest in the use of nuclear rocket propulsion for manned, deep space missions by NASA and other organizations.

4.7.2 Electric Spacecraft Propulsion

Electrical power is the basis of electric spacecraft propulsion, although the electrical power does not directly generate the thrust force. The electrical power may be provided by a variety of means, including nuclear, solar, battery, or other sources. The size, mass, and efficiency of the electrical power sources are issues with making this type of propulsion viable. Electric rocket engines or thrusters are typically low-thrust devices, with thrust levels much less than 1 N (0.2 lb), but they possess very high specific impulses. Given their very low thrust levels, the electric rocket must be run for a very long time, perhaps weeks or even months, to impart a significant velocity to a space vehicle. As such, this type of propulsion may be suitable for very long duration missions into deep space. Electric propulsion has also been used for in-space attitude control for orbiting vehicles, where low thrust is acceptable.

Three types of electric spacecraft propulsion are discussed: electrothermal propulsion, electrostatic, and electromagnetic propulsion. Electrothermal propulsion is conceptually similar to chemical rocket propulsion, where a working fluid is heated and then expanded and accelerated through an exhaust nozzle to generate thrust. However, the electrothermal thrusters use electrical power to heat the propellants, rather than chemical combustion. Electrostatic and electromagnetic thrusters are departures from the concept of a thermal rocket engine where a working fluid is expanded through a nozzle to produce thrust. Both are based instead on the principles of magnetohydrodynamics, where ionized gases are acted upon by either electric or magnetic fields to generate thrust.

Electrostatic propulsion is based on ionizing a propellant and accelerating it in a static electric field. In electromagnetic propulsion, the propellant is converted to an electrically conducting plasma that is accelerated by the interaction of an electric current and a magnetic field. Electrostatic and electromagnetic thrusters operates only in the vacuum of space.

The range of specific impulse obtainable with these types of electric propulsion is shown in Figure 4.88. Also shown in this figure are the electrical power requirements for the various types of electric propulsion, which may be significant. The power may be supplied by spacecraft solar panels or nuclear power sources.

4.7.2.1 Electrothermal Propulsion

Electrothermal thrusters resemble chemical propellant rocket engines, in that a propellant is heated and then expanded through a thrust-producing nozzle. The propellant is heated electrically, typically by flowing electrical current through resistors or creating an electric arc discharge. A variety of propellants may be used in these thrusters, including hydrogen, nitrogen, ammonium, or decomposed hydrazine.

In the *resistojet thruster*, the working fluid or propellant is heated by flowing it over electrically heated resistors, such as wire coils or other metal surfaces. After expansion through a nozzle, the resistojet can generate about 200–300 mN (0.04–0.07 lb) of thrust with specific impulses of about

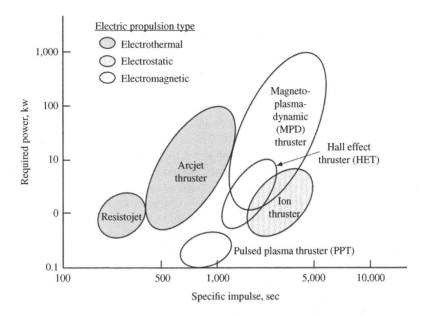

Figure 4.88 Power requirements and specific impulse of various types of electric propulsion. (Source: *Adapted from Rocket Propulsion Elements, G.P. Sutton and O. Biblarz, Fig. 19.1, (2001)*, [16], *with permission from John Wiley & Sons, Inc.*)

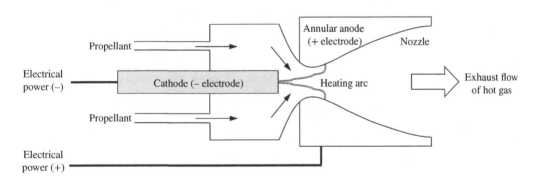

Figure 4.89 Arcjet thruster.

200–400 s. Resistojet thrusters are used on a variety of satellites for attitude control and station keeping.

A schematic of an *arcjet thruster* is shown in Figure 4.89. Electrical power is supplied to a cathode, located in the center of the chamber, and an annular anode that is upstream of the cathode. A very high temperature electric arc, perhaps as hot as 15,000 K (26,000 °F), bridges the gap between the cathode and anode. Propellants are supplied to the thruster chamber, where they flow through the hot electric discharge, reaching extremely high temperatures, perhaps as high as 20,000 K (35,000 °F) in localized areas. The hot propellant gas is expanded through a nozzle, where exhaust velocities can reach 1000–5000 m/s (3000–16,000 ft/s). The thrust levels of arc-heated, electrothermal thrusters are low, ranging from about 200 mN (0.04 lb) to 1 N (0.2 lb), with specific impulses of about 400–1200 s. The higher thrust devices require large amounts of power, approximately 100 kW or more. Arcjet thrusters are operational on a variety of satellites. The ranges of specific impulse and the required power of the resistojet and the arcjet thruster are shown in Figure 4.88.

4.7.2.2 Electrostatic Propulsion

Electrostatic propulsion is based upon using an electrostatic field to accelerate an ionized gas propellant to very high velocities of up to 60,000 m/s (200,000 ft/s). The propellant particles are accelerated by an electrostatic force or Coulomb force, the attraction or repulsion of particles due to their electric charge. This does not use thermodynamic expansion and acceleration of the gas through a rocket nozzle to generate thrust. An ionized gas is one in which electrons have been stripped from the atoms in the gas, creating positively charged ions and giving the gas a positive charge. These positively charged particles are accelerated in the same direction to high velocities, yielding a high momentum, thrust-generating beam of particles. To obtain the highest momentum, it is desirable to use propellants with high molecular mass. This is opposite to the use of the lowest molecular weight propellant, to obtain high thrust, in a thermal rocket engine.

How the charged particles are created is a discriminator for the different types of electrostatic propulsion. An *electron bombardment ion thruster* creates positively charged ions by bombarding a monatomic gas, such as xenon or mercury with electrons emitted from a heated cathode. A cesium propellant vapor is passed through a hot, porous tungsten contact ionizer in the *ion contact thruster*. In the *colloid electrostatic thruster*, droplets of propellant are passed through an electric field to give them a positive or negative charge.

The components of an electron bombardment ion thruster are shown in Figure 4.90. A gaseous propellant, such as xenon or mercury, is injected into the ionization chamber. Electrons are emitted from an electrically heated cathode and are attracted to the anode. The electrons collide with the atoms of the propellant gas, splitting off electrons and creating positively charged ions. A magnetic field, created by a coil around the ionization chamber, increases the ionization efficiency by causing the electrons, emitted by the cathode, to spiral in the chamber, which increases the number of collisions between electrons and propellant atoms. The positively charged ionized plasma is moved towards an electrostatic accelerator grid, a porous electrode with a positive charge on one side and a negative charge on the other. This particle acceleration is analogous to how electrons are accelerated in a television picture tube. The propellant plasma exits the chamber as a beam of accelerated ions. To prevent a buildup of a negative charge on the ionization chamber and the spacecraft, which

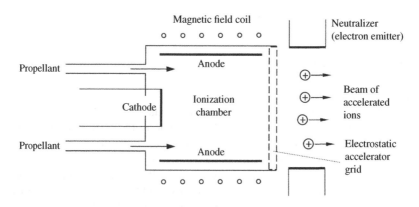

Figure 4.90   Electron bombardment ion thruster.

would also retard the exiting ion beam, the positively charged beam is electrically neutralized after it has exited the chamber by injecting electrons into the beam.

The ion thruster has a thrust range of about 0.01–200 mN (2×10^{-6} to 0.04 lb) with specific impulses of about 1500–5000 s. The range of specific impulse and the power requirements of the ion thruster is shown in Figure 4.88.

4.7.2.3   Electromagnetic Propulsion

Similar to electrostatic propulsion, electromagnetic propulsion does not rely on the thermodynamic expansion and acceleration of the propellant working fluid through a rocket nozzle to generate thrust. Electromagnetic propulsion is based on the fundamental physics of electromagnetic theory, dealing with the interaction of electrical currents and magnetic fields. In electromagnetic propulsion, heat is added to the propellant, converting it to a *plasma*, an energized hot gas consisting of a mixture of electrons, positive ions, and neutral particles that is electrically conducting at high temperatures. An electric field is applied to the plasma, which creates a high current within the plasma. When the current interacts with a perpendicular magnetic field, a force is created at right angles to the current and magnetic field, called the Lorentz force. This force accelerates the propellant to very high velocities of about 1000–50,000 m/s (3000–160,000 ft/s), generating a thrust force. In this section, we briefly discuss three types of electromagnetic propulsive devices, the pulsed plasma thruster (PPT), the Hall effect thruster (HET), and the magnetoplasmadynamic (MPD) thruster. The power requirements and specific impulses of these thrusters are shown in Figure 4.88.

Pulsed Plasma Thruster (PPT)
The *pulsed plasma thruster* (PPT) is perhaps the simplest form of an electromagnetic propulsive device. A schematic of a PPT is shown in Figure 4.91. Conceptually, the device consists of two charged electrode plates, a cathode plate, an anode plate, and a capacitor in an electrical circuit connected to a power source.

The operation of the PPT is as follows. The capacitor in the electrical circuit is first charged by a power source. When the capacitor is discharged, a plasma arc bridges the cathode and anode plates. A small amount of propellant is injected, and the plasma arc vaporizes the propellant into

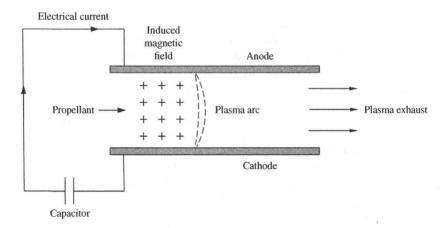

Figure 4.91   Pulsed plasma thruster.

a charged plasma gas cloud. Solid propellants are typically used in PPTs, although PPTs have been designed to use gaseous or liquid propellants. The synthetic fluorocarbon solid, polytetrafluoroethylene, commonly known by its brand name Teflon, is the most commonly used PPT solid propellant. The Teflon solid is fed into the PPT, ablating away as the plasma arc vaporizes the material.

The charged plasma gas completes the electrical circuit between the anode and cathode plates allowing electrical current to flow through the plasma cloud. The current-carrying plasma induces a magnetic field, perpendicular to the current direction (the magnetic field is shown as going into the page in Figure 4.91). The propellant plasma is accelerated by the Lorentz force created by the interaction of the electrical current and the magnetic field. This force is in a direction at right angles to the electrical current and the magnetic field and parallel to the charged plates. Thus the small propellant plasma cloud is exhausted from the PPT at a high velocity, creating a small, but finite thrust pulse, which ends when the capacitor is discharged. The capacitor is recharged by the power source to restart the cycle. Thus, the plasma and the thrust are generated in pulses as the capacitor charges and discharges. In practice, the pulses can be fast enough so that the thrust is seemingly continuous and smooth. PPTs are designed to operate reliably for millions of pulse cycles.

Figure 4.92 shows a pulsed plasma thruster that is used on the Earth Observing 1 (EO-1) spacecraft, launched in 2000, for pitch axis attitude control. There are two thrusters, pointing in opposite directions, at the top of the PPT pictured in the figure. The EO-1 PPT uses solid Teflon for propellant and delivers thrust levels between 0.05–10 mN (1×10^{-5} to 2×10^{-3} lb) with high specific impulses of 900–1200 s, with a power consumption of 1–100 W (3.4 to 340 Btu/h, 0.74–74 lb-ft/s).

Hall Effect Thruster (HET)

The Hall effect thruster ionizes a propellant and accelerates the ionized gas to generate thrust. A diagram of a cylindrical HET is shown in Figure 4.93. The main body of the HET is a cylinder with a cylindrical cavity wrapped around a centerbody. The electrical power supply is attached to a hollow cathode, located near the cavity opening, and an anode ring at the base of the cylindrical cavity. Powerful electromagnets at the inner and outer ring of the cavity create a radial magnetic field.

The operation of the HET is as follows. Electrons are generated and discharged from the negatively charged cathode and are attracted towards the positively charged anode. As these electrons accelerate towards the anode, the powerful magnetic field traps the electrons near the entrance to the cavity. The electrons are acted on by the *Hall effect*, the production of a voltage difference transverse to the electrical current (the flow of electrons from the cathode to the anode) and the magnetic field that is perpendicular to the current. The voltage difference results in a current flow called the *Hall current*, where the electrons move in a circular path around the cavity, spiraling down towards the anode. Propellant, typically inert high molecular weight gases, such as xenon or krypton, is injected into the cavity, at the anode. The trapped spiraling electrons collide with the propellant atoms, knocking off other electrons from the atoms, and creating an ionized propellant gas. The positively charged ions are attracted towards the negatively charged cathode and accelerate out of the cavity, creating an ion beam. The ions are much more massive than the electrons, so they are not affected by the magnetic field. Some of the electrons from the cathode are attracted to the ions that are exiting the thruster, thus electrically neutralizing the beam. To summarize, the Hall effect thruster traps electrons in a magnetic field, which enhances their interaction with, and ionization of, a propellant gas, which is then accelerated out of the thruster by an electrical field to generate a thrust force.

The power requirements and specific impulses obtainable with Hall thrusters are shown in Figure 4.88. HETs have a thrust range of about 0.01–2000 mN (2×10^{-6} to 0.4 lb) with specific

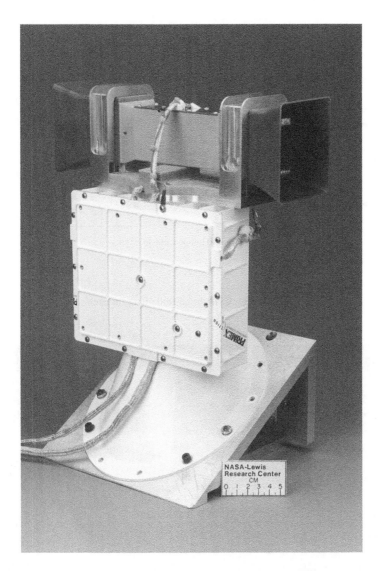

Figure 4.92 Pulsed plasma thruster (PPT) used on Earth Observing 1 spacecraft. (Source: *NASA*.)

impulses of about 1500–3000 s. High power Hall effect thrusters have generated up to 3 N (0.2 lb) of thrust in the laboratory. Hall effect thrusters are used routinely on commercial communications satellites for orbit insertion and station keeping.

A 6 kW laboratory model of a xenon Hall thruster is shown Figure 4.94. The throttling range (input power) for this thruster is approximately 1–10 kW (3400–34,000 Btu/h, 740–7400 lb-ft/s), with a specific impulse between approximately 1000–3000 s.

Magnetoplasmadynamic (MPD) Thruster

The *magentoplasmadynamic* (MPD) *thruster* is similar in configuration to the electrothermal arcjet thruster shown in Figure 4.89. In fact, the MPD thruster is sometimes referred to as the MPD arcjet. One major difference is that the MPD propellant exhaust is not thermodynamically

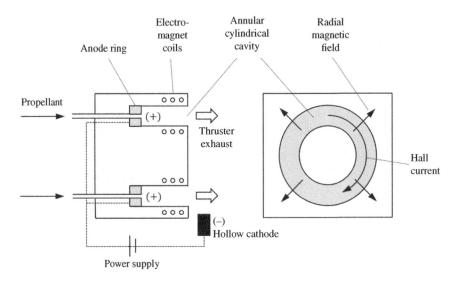

Figure 4.93   Cylindrical Hall effect thruster.

Figure 4.94   6 kW xenon Hall thruster. (Source: *NASA*.)

expanded and accelerated through the exhaust nozzle. The propellant enters the chamber and is ionized by the heating arc. A wide variety of propellants have been used, including xenon, hydrogen, argon, and lithium. The electrically conducting plasma allows current to flow between the anode and cathode, which sets up an induced magnetic field, similar to the effect in the pulsed plasma thruster. The propellant plasma is accelerated by the Lorentz force due to the interaction of the electric current and the magnetic field and is exhausted from the MPD thruster at a high velocity.

The magnetic field in the MPD thruster may be *self-induced*, as just described, or it may be an *applied magnetic field* by locating magnet rings (permanent magnets or electromagnets) around the exhaust chamber. To obtain the desired propellant acceleration, the electric and magnetic fields in the MPD thruster are much stronger than in the electrothermal arcjet thruster.

The power requirements and specific impulses of MPD thrusters are shown in Figure 4.88. MPD thrusters may have significantly higher power demands than electrothermal or electrostatic thrusters. They provide much higher specific impulses than electrothermal thrusters and comparable values relative to ion thrusters.

4.7.3   Solar Propulsion

Solar propulsion is based on using sunlight or solar radiation as the major component of the propulsion system. Typically, very large sunlight concentrators, or collectors, are required for these types of systems, making them practical only for in-space uses, such as in earth orbit or deep space missions. These are low thrust systems as compared with conventional chemical rocket propulsion. Two quite different solar power approaches – the *solar thermal rocket* and the *solar sail* – are discussed below.

4.7.3.1   The Solar Thermal Rocket

This concept uses solar radiation to heat and expand a propellant through a conventional rocket nozzle, thereby producing thrust (Figure 4.95). The propellant is a low molecular weight working fluid, such as hydrogen or ammonia. Specific impulses of about 700–1000 s are predicted, using hydrogen as the propellant working fluid. A large solar collector, such as a parabolic mirror, focuses sunlight to heat the propellant in a direct or indirect manner. The solar radiation is focused directly on the propellant in the direct heating method. The propellant is pumped through a heat exchanger that is heated by sunlight in the indirect method. Only small-scale hardware ground test evaluations of the solar thermal rocket concept have been completed to date.

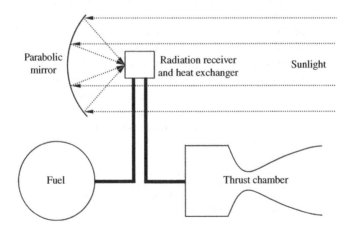

Figure 4.95   The solar thermal rocket.

Figure 4.96 The Age of Discovery caravel, with triangular lateen sails, designed for exploration. The caravel shown is the *Vera Cruz*, a replica of one of the ships in which Pedro Cabral discovered Brazil in 1500. (Source: *User: jad99, "Schiff" https://commons.wikimedia.org/wiki/File:Schiff_(14009000674).jpg, CC-BY-SA-2.0. License at https://creativecommons.org/licenses/by-sa/2.0/legalcode.*)

4.7.3.2   The Solar Sail

During the Age of Discovery,[3] European sailing ships left the coastal waters of the "Old World" and set sail across the vast expanses of the Indian and Atlantic Oceans, in search of new trade routes, new commercial goods, and the undiscovered lands of the "New World". Many of these ocean voyages were in a relatively small ship, designed specifically for exploration, called a *caravel*, shown in Figure 4.96. With a typical length of about 75 ft (23 m), the caravel was a broad-beamed ship with a shallow draft,[4] enhancing its ability to sail in uncharted waters of unknown depth. The caravel had a top speed of about 8 knots (9 mph, 15 km/h) and an average speed of about 4 knots (4.6 mph, 7.4 km/h), giving it a sailing range of about 100 miles (160 km) per day. With a small crew, the caravel had sufficient cargo space for ocean voyages of up to about a year.

The ocean-going, exploration caravel was propelled by the wind, which filled its triangular *lateen sails*, made of woven linen or cotton. Unlike the ancient square sail, the triangular lateen sail allowed ships to sail against the wind, greatly increasing the capability of ships to explore the oceans. The lateen sail functions much like an airplane wing, where the air flow across the sail creates a pressure differential between the concave and convex surfaces of the sail, resulting in an

---

[3] The Age of Discovery (also called the Age of Exploration) was a period of global exploration by the Europeans in the 15th and 16th centuries. Explorers from Portugal, Spain, England, France, and the Netherlands journeyed to the coasts of Africa, the archipelagoes in the Atlantic, and the Americas. This period included the voyage of Vasco de Gama to India, the discovery of the Americas by Christopher Columbus, the exploration of Brazil by Pedro Alvares Cabral, and Ferdinand Magellan's attempted circumnavigation of the globe.

[4] The *beam* is the width of the ship at its widest point. A larger beam, relative to its length, makes the ship slower, but more stable. *Draft* is the vertical distance from the ship's waterline to the bottom of the hull.

aerodynamic force that propels the ship forward. The period of great exploration, during the Age of Discovery, was made possible in part by the technological advancement of the lateen sail.

Just as the lateen sail enabled the caravel to explore distant lands, the *solar sail* (also called a *light sail* or *photon sail*) is a low-thrust, in-space propulsion concept that may allow spacecraft to explore deep space. The solar sail is a large, ultra-thin, lightweight surface that is attached to the spacecraft body or payload. The sail is pushed by the solar radiation pressure from stars, much as the wind pushes against the sail of a ship. All vehicles in space are affected by solar pressure and its effects on a spacecraft's orbit, attitude, or trajectory must be taken into account. Solar pressure is used beneficially by spacecraft to perform fine attitude adjustments or to allow them to remain stationary at a fixed point in space. Unlike conventional chemical rockets that must carry all of their propellants with them, the sunlight propellant source for the solar sail is inexhaustible. The propellant working material of the solar sail is the solar energy in photons of light. Alternatively, concepts have been proposed where the radiation energy is provided by large lasers rather than from sunlight, but strictly speaking, this is (laser) *beam sailing* rather than solar sailing.

Three of the basic types of solar sails are the square sail, the heliogyro, and the spinning disk sail, as shown in Figure 4.97. With the square sail, the sail material is attached to a rigid frame structure, somewhat similar to an ordinary kite. Both the heliogyro and spinning disk solar sails are spun to provide stabilization and maintain the desired orientation. The spinning motion is accomplished by using control vanes, which are miniature solar sails, or by offsetting the spacecraft's center of mass from the sail's center of solar pressure. For all of these solar sails, the structural frames are typically composed of stiff lightweight composite tubes that weigh less than an ounce per foot. The sail material is folded or stowed during rocket launch and unfurled once in space. For the heliogyro sail, the spinning motion serves to extend the sail material from a central hub along structural spokes. Solar sail designs have few moving parts, other than the expandable structures and associated mechanisms required to unfurl a sail in space.

Sail materials are typically made of very thin, lightweight material with reflective coatings giving them mirror-like finishes. The material thickness can be 1/100th the thickness of a piece of paper. Current technology sail materials include polyester films, such as aluminized Mylar, or space-rated insulating materials. Advanced sail materials being investigated include composite, carbon fiber meshes. The trajectory of solar sails must be planned such that they do not get too close to the Sun, as solar radiation may increase the sail material temperature beyond its limits. The sail temperature is a function of the distance from the Sun, the sail's angle relative to the incident sunlight, and the sail's reflectivity and emissivity.

Because of the very low thrust developed from solar pressure, solar sail designs tend to be very large, with sail dimensions about tens or thousands of meters. Let us calculate the solar pressure force that is obtainable from a very large solar sail. Assume a very large, square solar sail

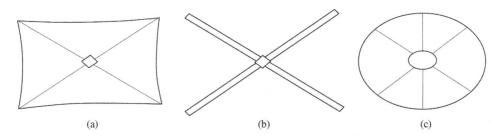

(a)   (b)   (c)

Figure 4.97 Types of solar sails, (a) square, (b) heliogyro, and (c) spinning disk sails. (Source: *Adapted from NASA.*)

of side dimension 1000 m (1 km, 0.6 mi, 3300 ft) by 1000 m, giving a sail area, A_{sail}, of 10^6 m^2 (1.0×10^7 ft^2). The solar sail concept is based on the fact that photons of light have momentum. Einstein's equation for the equivalence of energy, E, and mass, m, can be written in terms of the momentum of a photon, p, as

$$E = mc^2 = pc \qquad (4.132)$$

where c is the speed of light. Solving for the momentum of a photon, we have

$$p = \frac{E}{c} \qquad (4.133)$$

Using Newton's second law and Equation (4.133), the solar pressure force, F, on the sail, imparted by the momentum of the photons, is given by

$$F = \frac{d}{dt}(mV) = \frac{d}{dt}(mc) = \frac{d}{dt}(p) = \frac{d}{dt}\left(\frac{E}{c}\right) = \frac{1}{c}\frac{dE}{dt} = \frac{P}{c} \qquad (4.134)$$

where P is the solar power imparted by the photons. The solar power per unit area, $\frac{P}{A}$, is known as the solar irradiance, and has a value of 1360 W/m^2 at a distance of one AU (astronomical unit), the distance from the Earth to the Sun. Therefore, the solar pressure force is given by

$$F = \frac{(P/A)}{c}(A_{sail}) = \left(\frac{1360\,\text{W/m}^2}{3 \times 10^8\,\text{m/s}}\right)(1000\,\text{m})^2 = 4.53\,\text{N} \qquad (4.135)$$

This result assumes perfect reflection of the photons from the sail, when in reality there are losses due to absorption of the solar radiation, sail curvature, wrinkles in the sail material, and other factors. Even without these losses, we see that the solar pressure force on a 1×1 km, sail is very small, only about 4.5 N (~1 lb).

Since the solar radiation fuel for this propulsion system is inexhaustible, the very small force can act on the sail for a very long time, continuously accelerating the sail by a very small amount, allowing it to reach high velocities over time. Thus, it may be advantageous to use solar sail propulsion on long space voyages. Another advantage is that the solar sail propulsion system has few, if any, moving parts, making it a reliable system for long space journeys.

The first practical solar sail spacecraft was the *IKAROS* (Interplanetary Kite-craft Accelerated by Radiation Of the Sun), launched by the Japan Aerospace Exploration Agency (JAXA) on 21 May 2010 (Figure 4.98). The 315 kg (694 lb) spacecraft is the first to use a solar sail as its primary in-space propulsion. In June 2010, *IKAROS* deployed its sail, using a combination of a spinning motion and the unfurling of the sail's support frame. In the spin-deployment of the sail, small 0.5 kg (1.1 lb) tip masses, located in the corners of the square sail, helped to pull the sail outward. The spacecraft had two small cameras that could be ejected, so that they could take photographs of the deployed sail. The square solar sail was 14 m (46 ft) on each side, giving it a surface area of 196 m^2 (2110 ft^2). Made of a 7.5 μm (0.00030 in) thick, polyimide film, with an evaporated aluminum coating, the total mass of the sail was only 2 kg (4.4 lb), not including the tip masses. The spacecraft's attitude was controlled using LCD (liquid crystal diode) panels, located along the sail's square perimeter. The LCD panels could be turned on and off, changing their reflectivity and the resulting momentum transfer from the solar pressure. When an LCD panel was turned on, it diffused light and reduced the momentum transfer in that area of the sail. When the LCD panel was turned off, the sail reflected more light in that area and more momentum was transferred. The circular, centerbody of the spacecraft was located at the center of the sail (Figure 4.98), containing several of the scientific instruments. Thin-film solar cells, that were embedded in the sail, provided the spacecraft power. In July 2010, it was confirmed that solar sail propulsion was successfully accelerating the spacecraft. *IKAROS* sailed past the planet Venus on 8 December 2010. By August

Figure 4.98   JAXA *IKAROS*.   (Source: *Mirecki, https://en.wikipedia.org/wiki/File:IKAROS.jpg, CC-BY-SA-3.0. License at https://creativecommons.org/licenses/by-sa/3.0.*)

2013, the spacecraft had increased its speed by a total of about 400 m/s (1300 ft/s, 900 mph) due to solar propulsion, as it entered an orbit around the Sun.

4.8   Other Types of Air-Breathing Propulsion

In this section, several types of advanced air-breathing propulsion are described, other than the ramjet, turbojet, turbofan, and internal combustion engine. Most of these propulsion types are still being developed, with the promise of new propulsion capabilities and applications. Several of these have applications to very high-speed flight at hypersonic speeds at very high altitudes. They all rely on the earth's atmosphere for oxygen to burn with fuel, hence, they are confined to the limits of the sensible atmosphere.

4.8.1   The Scramjet

In a ramjet engine, the supersonic freestream flow through the inlet terminates in a normal shock wave, resulting in subsonic flow entering the combustor, and combustion occurs at subsonic speeds. The operation of the ramjet engine is limited to a maximum flight speed of about Mach 5, due to large total pressure losses and high temperature increases across the terminal normal shock wave. The high temperatures result in thrust losses from dissociation of the air and issues with survivability of the structure. To operate at flight speeds above Mach 5, the terminal normal shock must be avoided, leading to the concept of the supersonic combustion ramjet, or *scramjet*.

In the scramjet engine, the hypersonic freestream flow is decelerated, but not to subsonic speed. Since there is no terminal normal shock wave, the large normal shock total pressure losses are avoided and static temperature increases are lower, resulting in less dissociation losses. Heat transfer to the engine structure is very high, since the stagnation temperature is very high due to the large freestream Mach number. Combustion in the scramjet occurs at supersonic speed, with the flow entering the combustor at about Mach 2–3, depending on the freestream Mach number. Contrast this with the Mach 0.3 flow entering a turbojet or turbofan combustion chamber. The injection, mixing, and efficient combustion of fuel in the supersonic combustor flow are some of the very difficult design problems in the design of scramjet engines.

The first patent of a supersonic combustion ramjet was submitted in 1965 by Frederick Billig and Gordon Dugger of The Johns Hopkins University Applied Physics Laboratory, Laurel, Maryland. The patent was for a scramjet-powered missile, as shown in Figure 4.99. The upper drawing in the figure shows the scramjet-powered missile, while the lower drawing includes the non-air-breathing rocket booster attached to the scramjet, required to accelerate the scramjet to hypersonic Mach numbers where it can operate. The scramjet engine shown has an axisymmetric geometry with a conical spike inlet centerbody.

A simplified schematic of an *airframe-integrated* scramjet engine is shown in Figure 4.100. Unlike subsonic turbojet or turbofan engines, which tend to be podded engines hung underneath a wing, the scramjet engine is integrated with the entire airframe of the vehicle. The lower surface of the vehicle is part of the hypersonic propulsion system, with the vehicle forebody contributing to the freestream flow compression for the inlet and the vehicle afterbody expanding the exhaust flow as part of the nozzle. The hypersonic freestream flow is compressed by the forebody and inlet through a series of oblique shock waves, decreasing the local Mach number and increasing the pressure and temperature. The flow is supersonic as it exits the inlet and enters the *isolator* section. The isolator serves to isolate the inlet from pre-combustion shock waves that are generated by the downstream combustor. If this pre-combustion shock system were allowed to move upstream into the inlet, the inlet could unstart and disrupt the flow entering the engine. At lower hypersonic flight

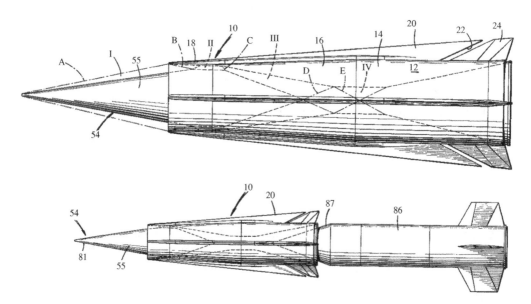

Figure 4.99 Scramjet-powered missile patented by Billig and Dugger. (Source: *US Design Patent 4,291,533 A, US Patent and Trademark Office, December 30, 1965.*)

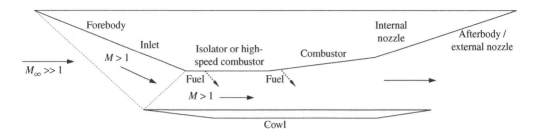

Figure 4.100   Simplified schematic of an airframe-integrated scramjet engine.

Mach numbers, where the supersonic Mach number exiting the inlet is lower, fuel is injected in the downstream location of the combustor and no fuel is injected at the isolator location. At higher hypersonic flight speeds, the isolator is no longer required for shock wave isolation and serves as additional length for mixing and burning of fuel in the combustor. At these higher speeds, fuel is injected further upstream in the isolator, which now serves as part of the high-speed combustor. The combustion products exit the combustor and are expanded in the nozzle, composed of an internal nozzle, which includes the cowl, and an external, open or scarfed nozzle formed by the vehicle afterbody.

Scramjet engine design and development are still at the cutting edge of research and testing. Several successful flight demonstrations of scramjet engines in unmanned vehicles, including the first flights of a scramjet-powered aircraft, the X-43A, mentioned at the start of this chapter, have paved the way for future practical applications. There is still much room for innovation and technological advancement in the design of these hypersonic engines. The dream of airplane-like hypersonic flight within the atmosphere or a spaceplane that can takeoff and fly into earth orbit may depend on the development of the scramjet engine.

4.8.2   Combined Cycle Propulsion

There are several types of propulsion systems that combine different types of propulsive cycles. These *combined cycle propulsion* systems may use aspects of ramjets, turbojets, rockets, or other propulsion types. Each one of these propulsive cycles is optimum for certain Mach numbers, such that combining them may provide a more efficient propulsion system for a wider Mach number range. Often, the difficulty in combining different types of propulsion cycles lies in the efficient integration of the components and systems required for each type of propulsion, due to aerodynamic, weight, and other constraints that may outweigh the benefits of combining cycles.

Hypersonic air-breathing propulsion concepts, such as the scramjet, are not capable of generating static thrust. This limitation makes it necessary to accelerate a hypersonic vehicle to sufficient speed to enable the ramjet cycle of the scramjet engine to produce positive thrust; that is, thrust greater than drag. One solution to the problem is to integrate the scramjet engine with a low-speed propulsion system that is capable of accelerating the vehicle to the ramjet take-over speed, typically around Mach 3. (We are calling the propulsion system that accelerates the vehicle to Mach 3, a "low-speed" system, as this is a low speed relative to the normal scramjet operation speeds.)

Potential candidates for this low-speed propulsion include non-air-breathing rockets, air-breathing rockets, and turbine-based engines. These types of propulsion systems that combine different propulsive cycles are called *combined cycle propulsion* systems. Integration of turbine engines with a hypersonic, air-breathing propulsion system is termed a *turbine-based combined cycle* (TBCC) propulsion system, as shown in Figure 4.101a. Using a rocket for the

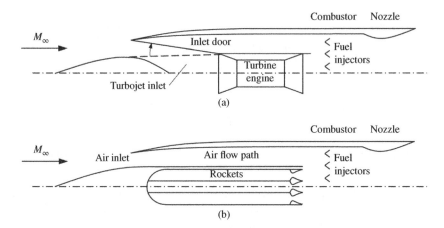

Figure 4.101 Combined cycle propulsion (a) turbine-based combined cycle propulsion (TBCC). (b) rocket-based combined cycle (RBCC).

low-speed system, integrated with the high-speed, hypersonic system is called a *rocket-based combined cycle* (RBCC) propulsion system, as shown in Figure 4.101b. The TBCC has a dual flow path configuration, where the air entering the engine can go through the turbine system or the high-speed scramjet engine. An inlet door is positioned to direct the flow either into the turbine engine or the scramjet. The RBCC has a single flow path, since the rocket is non-air-breathing. It may be desirable to close off the air inlet at low speeds when only the rocket is operating, but at higher speeds, it may be beneficial to have the fuel-rich rocket exhaust mix with the air to provide additional combustion and thrust. These types of integrated low- and high-speed propulsion systems have the promise of powering a hypersonic vehicle that can take off and land like a conventional airplane, cruise at subsonic through hypersonic Mach numbers, and perhaps even fly into orbit.

This single-stage-to-orbit or SSTO concept has been pursued several times in the past, the most recent being the X-30 National Aerospace Plane design effort in the late 1980s and early 1990s (see Figure 1.83). It is still a much sought after aerospace dream to have a true spaceplane that can takeoff from a conventional runway, fly into space, and return for a conventional landing.

4.8.3 Unsteady Wave Propulsion

In this section, we discuss the *pulsejet* and the *pulse detonation engine* (PDE), two types of propulsive devices that generate *intermittent thrust*. The combustion of fuel and air in a jet engine may be considered a steady process. In the jet engine, air is ingested and compressed at a steady rate, then mixed and burned with fuel in a combustion chamber at a steady rate, and the combustion products are exhausted through a nozzle, which generates a constant thrust. The propulsive devices discussed in the present section operate based on *unsteady or intermittent combustion*. After the combustion of the fuel–air charge, the combustion chamber is emptied of its combustion products, exiting through a nozzle and generating a pulse of thrust. The combustion chamber is refilled with a new fuel–air charge to start the process once again. In this unsteady combustion process, the thrust is generated intermittently in pulses, albeit the frequency of the pulses may be quite high.

This unsteady process has similarities with the force-generating process of an individual cylinder of an internal combustion engine. Each cylinder goes through a cycle of intake, compression, combustion, and exhaust, generating a force "pulse" that rotates the propeller. However, the internal combustion engine, as a whole, delivers a substantially constant thrust by utilizing many cylinders, where the combustion in each cylinder is sequenced to deliver constant power. Another similarity between the internal combustion engine and the unsteady engines is that combustion occurs in a constant volume process, in contrast to the constant pressure process of the jet engine.

4.8.3.1 The Pulsejet

The pulse jet engine, or *pulsejet*, is a form of jet propulsion based on *intermittent combustion*. Combustion and thrust production occur in cyclic pulses, producing a pulsating exhaust jet, hence its name. While the pulsejet has a high thrust-to-weight ratio, it has a low thrust specific fuel consumption. The pulsejet can produce static thrust and is limited to subsonic flight speeds up to about 600 mph (960 km/h, Mach 0.8) due to limitations of the air intake system. The pulsejet is one of the simplest types of jet propulsion devices, with no, or very few, moving parts. There are two main types of pulsejet engines, the *valved pulsejet* and the *valve-less pulsejet*.

The major components of a valved pulsejet are shown in Figure 4.102. The device is a simple tube, typically constructed of steel, with an inlet/diffuser, a one-way inlet valve, a combustor, and an exhaust tube terminating in a nozzle. The one-way, mechanical inlet valve only allows air to come into the engine. There are one or more fuel injectors in the combustor and an electric ignitor for starting, typically a spark plug.

Pulsejet operation is best described by starting with the first fuel–air charge that is in the combustion chamber. This fuel–air mixture is ignited by the electric spark, and combustion occurs rapidly at constant volume, with a large increase in pressure. Propane is often used for fuel in the pulsejet, although almost any fuel can be used. Non-conventional particulate fuels, such as sawdust and coal powder, have been used in the past.

The combustion frequency of the device is dependent on the length of the acoustically resonant exhaust tube, similar to the frequencies of an organ pipe. The frequencies increase with tube length. Short tubes of less than about a foot may have frequencies of several hundred hertz (cycles per second), while a 5–6 ft long (1.5 to 1.8 m) tube may have a frequency of about 50 Hz. When operating, the pulsejet engine makes a distinctive "buzzing" sound with the pitch related to the tube frequency. The pulsejet is known for the high intensity noise that it can generate, which can be a detriment to commercial applications.

The high pressure in the combustion chamber forces the one-way inlet valves to close, and the hot, high-pressure combustion gases escape out the exhaust tube and through the nozzle, generating

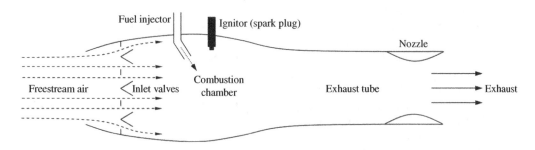

Figure 4.102 Pulsejet components.

a pulse of thrust. The high-speed exhaust flow lowers the combustion chamber pressure to below the ambient pressure, which allows the inlet valve to open, and a new mass of fresh air is ingested into the combustion chamber. In this manner, the alternating high and low pressures in the combustion chamber control the closing and opening, respectively, of the mechanical inlet valve and the mass flow of air into and out of the chamber.

There is a variety of configurations of the inlet one-way valve. Two of the most common types of inlet valves are the daisy valve and the rectangular valve grid. As its name implies, the daisy valve looks like the flower with petals that get wider from its center. Made of thin sheet metal, the daisy valve is placed against the downstream cross-section of a circular disk in the center of the tube, with each petal covering an opening in the disk that allows air into the combustor. The center of the daisy valve is attached to the disk, so that each petal can fold back towards the combustor, like a "flapper", allowing air to enter when the downstream pressure is lowered. The rectangular valve grid operates in a similar fashion, with flapper-type pieces of sheet metal that cover openings in a rectangular grid. These types of flapper valves are prone to metal fatigue and breakage due to their high rate of cycling. The flapper-type valve also limits the maximum flight speeds of the pulsejet, since above a certain speed, the flapper valve may not open enough to allow a sufficient mass flow of air to be ingested for adequate performance.

When the fuel is again injected into the combustion chamber, it also flows into the exhaust tube, where it meets the escaping hot exhaust gases. The fuel in the exhaust tube ignites and flashes back to the combustion chamber, thus starting the combustion process once again. After the initial use of the spark plug to start the combustion process, it is no longer required, as the ignition process is self-sustaining.

The *valve-less pulsejet* has no moving parts. It operates on the same cycle as the valved pulsejet, but the geometry of the valve-less pulsejet acts as an "aerodynamic valve" that controls the flow of gases in and out of the engine. The valve-less pulse jet is a U-shaped metal tube, open at both ends, with the intake tube and combustion chamber located on one side of the U-tube. The intake tube faces backwards, in the same direction as the exhaust opening, since exhaust gases are expelled through both during the cycle. The backwards-facing intake tube thus contributes to the thrust.

Operation of the valve-less pulsejet is as follows. Air enters the intake tube and burns with fuel in the combustor. The hot combustion products are expelled from the intake and exhaust openings. The high-speed flow exiting both openings creates a low pressure in the combustor, which draws in fresh air from the intake tube and pulls back some of the hot gases from the exhaust. The U-tube geometry must be properly "tuned" with the combustion cycle to create a self-sustaining combustion cycle.

The pulsejet has not seen wide acceptance as a means of propulsion for aerospace vehicles. However, pulsejets have found a niche in industrial applications, being used as high-output heaters, cyclone filters, and industrial dryers. Similar to the ramjet, the pulsejet has been tried as a means of propulsion on several different types of aerospace vehicles, including being attached to the tips of helicopter rotor blades. Perhaps the most noteworthy aerospace application of the pulsejet was as the primary propulsion for cruise missiles.

During World War II, the Germans developed the V-1 cruise missile, powered by an Argus As 014 valved pulsejet. The engine operated at a frequency of about 45 Hz, making a distinctive low frequency, buzzing sound; hence, the V-1 was dubbed the "buzz bomb". Germany produced more than 30,000 V-1 pulsejet-powered missiles between 1944 and 1945.

The V-1 missile was 27.1 ft (8.26 m) in length, with a wingspan of 17.7 ft (5.39 m), a gross weight of 5023 lb (2278 kg), and could carry a payload of 2100 lb (953 kg) of explosives. The pulsejet engine had a length of 12 ft (3.7 m), a maximum diameter of 22 in (0.56 m), and a weight of 344 lb (153 kg). The tailpipe of the pulsejet was 69 in long (1.75 m) with a diameter of 15 in (0.38 m). The engine delivered 500 lb (2224 N) of static thrust and a maximum thrust of about 750 lb (3300 N) in flight. The V-1 could fly at about 400 mph (644 km/h) at an altitude of about 4000 ft (1200 m)

Figure 4.103 US Navy JB-2 *Loon* pulsejet missile launch, Point Mugu, California, 1947. The pulsejet engine is on top of the missile, and the solid rocket motor used for launch is on bottom. (Source: *US Navy.*)

with a range of about 150 miles (240 km). The vehicle was steam catapult-launched from a 200 ft (60 m) long inclined ramp, accelerating it to about 250 mph (400 km/h).

In 1944, the USA reverse-engineered the German V-1 and the Argus pulsejet from several captured missiles, building about 1000 copies, designated the JB-2 *Loon* (Figure 4.103). The JB-2 airframes were built by Republic Aviation and the pulsejet engines were built by the Ford Motor Company. The JB-2 gave the USA valuable engineering experience with pulsejet propulsion and advanced the development of cruise missiles.

4.8.3.2 The Pulse Detonation Engine (PDE)

The *pulse detonation engine* (PDE), sometimes called a pulse detonation wave engine (PDWE), is another form of jet propulsion based on intermittent combustion. Similar to the pulsejet, the PDE is a relatively simple mechanical device. The PDE can produce static thrust and, theoretically, can operate from subsonic to low hypersonic flight speeds.

The pulse detonation engine is based on the *constant volume combustion, Humphrey cycle*, in contrast to the *constant pressure combustion, Brayton cycle*, that is the basis of the ideal turbojet engine. The Humphrey and Brayton cycles are compared in pressure–volume and temperature–entropy diagrams in Figure 4.104. The Humphrey cycle is depicted as states 1-2-3-4-1 and the Brayton cycle is states 1-2-5-6-1. The combustion in the Humphrey cycle (state 2 to 3) is at constant volume, while it is at constant pressure in the Brayton cycle (state 2 to 5). The Humphrey cycle, constant volume combustion results in a pressure gain in the combustor $(p_3 > p_2)$, which enhances the cycle efficiency. As seen in the temperature–entropy plot, the same temperature rise is achieved during combustion with the Humphrey cycle as with the Brayton cycle $(T_3 = T_5)$, but with a lower increase in entropy $(s_3 < s_5)$. The predicted thrust specific fuel consumption of the PDE is comparable to afterburning turbojet engines and exceeds that of ramjets.

PDE combustion is fundamentally different from combustion in other jet engines, including turbojets and pulsejets. In these other forms of propulsion, combustion is based on *deflagration*,

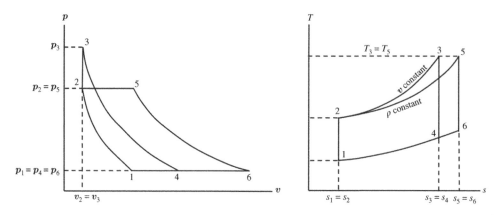

Figure 4.104 Comparison of Brayton and Humphrey thermodynamic cycles. (Source: *Adapted from Aircraft Propulsion, S. Farokhi, Fig. 1.23, p. 13, (2014), [3], with permission from John Wiley & Sons, Inc.*)

where the burning of the fuel–air mixture is a steady process occurring at subsonic speed, with a small increase in pressure. In the ideal Brayton cycle, this pressure increase is assumed to be so small that the constant pressure combustion may be assumed. PDE combustion is based upon *detonation*, where a supersonic detonation wave is used to compress and combust the fuel–air mixture, creating a large increase in pressure. The combustion occurs so rapidly that the fuel–air mixture does not have time to expand, so that the process occurs at near constant volume.

Similar to the pulsejet, there are also two main types of PDEs, the valved and valve-less PDE. The basic cyclic operation of the PDE is also similar to the pulsejet. Consider a PDE tube-type device, with the combustion chamber filled with a fuel–air mixture. The mixture is detonated, which requires a considerable amount of energy. There are several different schemes to initiate the detonation, but this is still an active area of research. The detonation wave speeds through the combustion chamber, significantly increasing the pressure and temperature. The wave continues down the exhaust tube and exits through the nozzle. The high-speed wave, exiting the tube, reduces the combustion chamber pressure to below ambient, which then opens the inlet valve system to let a fresh charge of air into the tube and restart the detonation combustion cycle. The frequency of the detonations can be quite high, on the order of 60 Hz (60 detonations/sec). Given this very high combustion cycle frequency, a very fast acting mechanical valve is required. There has been some success with the use of a rotary-type valve, which spins to cover and uncover openings on an intake disk. The unsteady combustion also results in high noise levels and significant vibration issues, which must be addressed for successful integration into a flight vehicle. Research and development has continued with PDE propulsion, with design and testing of experimental PDE engines.

References

1. Conners, T.R. and Sims, R.L., "Full Flight Envelope Direct Thrust Measurement on a Supersonic Aircraft," NASA TM-1998-206560, July 1998.
2. Covert, E.E., editor, *Thrust and Drag: Its Prediction and Verification*, Progress in Aeronautics and Astronautics Vol. **98**, American Institute of Aeronautics and Astronautics, New York, New York, 1985.
3. Farokhi, S., *Aircraft Propulsion*, 2nd edition, John Wiley & Sons, New York, 2014.
4. Federal Aviation Administration, US Department of Transportation, *Pilot's Handbook of Aeronautical Knowledge*, FAA-H-8083-25A, Oklahoma City, Oklahoma, 2008.
5. Foa, J.V., *Elements of Flight Propulsion*, John Wiley & Sons, New York, 1960.

6. Groth, H.W., Samanich, N.E., and Blumenthal, P.Z., "In-flight Thrust Measuring System for Underwing Nacelles Installed on a Modified F-106 Aircraft," NASA TM X-2356, August, 1971.

7. *Handbook of Aircraft Fuel Properties*, CRC Report No. 635, 3rd edition, Coordinating Research Council, Inc., Alpharetta, Georgia, 2004.

8. Hill, P.G. and Peterson, C.R., *Mechanics and Thermodynamics of Propulsion*, 2nd edition, Addison-Wesley, Reading, Massachusetts, 1992.

9. Kerrebrock, J.L., *Aircraft Engines and Gas Turbines*, 1st edition, MIT Press, Cambridge, Massachusetts, 1981.

10. Loftin, L.K., *Quest for Performance: the Evolution of Modern Aircraft*, NASA SP-468, US Government Printing Office, Washington, DC, 1985.

11. Lucian of Samosata, *True History*, translated by Francis Hicks, A.H. Bullen, London, England, 1894.

12. Mattingly, J.D., *Elements of Gas Turbine Propulsion*, 1st edition, McGraw-Hill, Inc., New York, 1996.

13. MIDAP Study Group, "Guide to In-Flight Thrust Measurement of Turbojet and Fan Engines," Advisory Group for Aerospace Research and Development, AGARDograph No. 237, January 1979.

14. Moore, M.D. and Fredericks, B., "Misconceptions of Electric Propulsion Aircraft and their Emerging Aviation Markets," 52nd Aerospace Sciences Meeting, National Harbor, Maryland, 13–17 January 2014, pp. 52–58.

15. Oswatisch, K., "Pressure Recovery for Missiles with Reaction Propulsion at High Supersonic Speeds (The Efficiency of Shock Diffusers)," NACA TM 1140, January 1944.

16. Sutton, G.P. and Biblarz, O., *Rocket Propulsion Elements*, 7th edition, John Wiley and Sons, Inc., New York, 2001.

17. Talay, T.A., *Introduction to the Aerodynamics of Flight*, NASA SP-367, US Government Printing Office, Washington, DC, 1975.

18. Van Wylen, G.J. and Sonntag, R.E., *Fundamentals of Classical Thermodynamics*, 2nd edition, John Wiley and Sons, Inc., New York, 1978.

19. Waltrup, P.J., White, M.E., Zarlingo, F., and Gravlin, E.S., "History of Ramjet and Scramjet Propulsion Development for US Navy Missiles," *The Johns Hopkins APL Technical Digest*, Vol. **18**, No. 21, 1997, pp. 234–243.

20. Yechout, T.R., *Introduction to Aircraft Flight Mechanics*, 2nd edition, American Institute of Aeronautics and Astronautics, Inc., Reston, Virginia, 2014.

Problems

1. The air enters a combustor of a jet engine with a velocity of 290 ft/s and exits at 195 ft/s. If 770.8 Btu/lb_m of heat per unit mass is added, calculate the temperature rise in the combustor. Assume a constant specific heat of 6020 ft·lb/(slug·°R).

2. Air enters an engine at a velocity of 160 m/s with an enthalpy of 290,000 J/kg. Fuel is burned in the engine, adding heat per unit mass of 54,000 J/kg to the flow. The flow exits the engine at a velocity of 300 m/s with an enthalpy of 283,000 J/kg. Calculate the work per unit mass delivered by the engine.

3. A 1.70 m diameter propeller produces a thrust of 6140 N while in flight at an airspeed of 311.0 km/h and an altitude of 3000 m (air density of 0.9092 kg/m^3). Calculate the velocity at the propeller disk and the mass flow rate of air going through the propeller.

4. The thrust and specific impulse of a rocket engine are often specified in terms of the performance at sea level and in the vacuum of space. The Space Shuttle Main Engine (SSME) had a sea level thrust of 1859 kN and a vacuum thrust of 2279 kN. If the sea level propellant mass flow rate is 163.4 kg/s and the propellant mass flow rate in vacuum is 156.6 kg/s, calculate the SSME specific impulse at sea level and in vacuum.

5. A rocket engine produces a thrust of 9900 lb and a specific impulse of 303 s. Calculate the propellant mass flow rate corresponding to these conditions.

6. The Lockheed SR-71 *Blackbird* is powered by two Pratt & Whitney J58 air-turboramjet engines. At Mach 3.2, each J58 produces 32,500 lb of thrust. If the J58 fuel flow rate is 9200 gal/h, calculate the thrust specific fuel consumption. (The J58 burned JP-7 jet fuel with a density of 6.67 lb/gal.)

7. An aircraft has a turbojet engine (Engine 1) where the exhaust exits the engine at 2400 ft/s. Another aircraft has a turbofan jet engine (Engine 2) where the exhaust speed is 800 ft/s. Assuming that both aircraft are flying at the same airspeed of 350.0 mph at sea level (density of

$0.002377 \, \text{slugs/ft}^3$), calculate the propulsive efficiency of each engine. How much larger does the Engine 2 inlet area have to be to have the same thrust as Engine 1? Compare the mass flow rates of each engine, assuming equivalent thrust.

8. The General Electric GE90 is a high bypass ratio turbofan engine with a bypass ratio of 8.4 and a fan diameter of 3.124 m. If the total air mass flow rate, $\dot{m}_\infty$, is 1350 kg/s, calculate the flow velocity through the fan at sea level conditions.

9. A ramjet engine is being tested in a direct-connect test facility, where only the internal flow through the engine is simulated. Air enters the ramjet inlet at a velocity of 510 mph and a mass flow rate of 31.7 lb_m/s. Fuel is injected into the engine at a mass flow rate of 1.44 lb_m/s. At this test condition, a thrust force of 830 lb is measured. If the flow is perfectly expanded at the exit of the ramjet engine nozzle, calculate the ram drag and the velocity of the exhaust gas.

10. A rocket engine produces 3.885×10^6 N of thrust with a specific impulse of 310 s. If the cylindrical propellant tank has a diameter of 3.78 m, calculate the tank length required for a 241 second burn using RP-1 fuel with a density 820 kg/m^3.

11. To calculate the takeoff thrust of an aircraft, the takeoff airspeed is assumed to be small, therefore the ram drag can be neglected. Starting with the complete thrust equation for an air-breathing engine given by

$$T = (\dot{m}_a + \dot{m}_f)u_e - \dot{m}_a V_\infty + A_e(p_e - p_\infty)$$

obtain an equation for the takeoff thrust. Assume a perfectly expanded nozzle and that the air mass flow rate is much greater than the fuel mass flow rate. Calculate the takeoff thrust if the air mass flow rate is 112 kg/s and the engine exhaust flow velocity is 887 m/s. If the fuel mass flow rate is 2.5 kg/s, calculate the takeoff thrust if this fuel flow rate is included. Also, calculate the percent difference in the takeoff thrust calculation, with and without the fuel mass flow rate included.

12. An oxygen–hydrogen propellant rocket engine has the specifications listed in Table 4.9. Calculate the rocket nozzle exit velocity and the ratio of the nozzle exit pressure to the combustion chamber pressure. The ratio of specific heats of the combustion gases is 1.21.

13. Five F-1 rocket engines were used on the first stage of the Apollo *Saturn V* launch vehicle, which took man to the moon. The F-1 remains the largest single-chamber, liquid-fueled rocket engine ever built, with each engine producing about 1.5 million lb (6.67 MN) of thrust. For the F-1 rocket engine, the mass flow rates of RP-1 kerosene fuel and liquid oxygen, entering the combustion chamber, are 1738 lb_m/s and 3945 lb_m/s, respectively. Assuming that the pressure, temperature, and density of the exhaust gas flow at the exit plane of the rocket engine nozzle are 877 lb/ft^2, 2462 °R, and 1.671×10^{-4} slug/ft^3, calculate the Mach number at the nozzle exit. Assume that the flow is uniform at the nozzle exit and that the exhaust gas is an ideal gas with a ratio of specific heat, γ, of 1.23. The nozzle exit has a diameter of 11 ft 7 in.

14. A rocket is launched vertically and maintains a vertical trajectory until burnout, 41 s later. Assuming a specific impulse of 250 s and a mass ratio of 0.410, calculate the burnout velocity without gravity effects. Assuming an average gravitational acceleration of 9.69 m/s^2 over the trajectory, calculate the burnout velocity with gravity effects. What is the burnout velocity with gravity effects, if the trajectory is horizontal?

5

Performance

Charles Lindbergh lifts off on a test flight in the Ryan NYP *Spirit of St Louis,* prior to being the first person to fly non-stop from New York to Paris on 20–21 May 1927. (Source: *National Air and Space Museum, Smithsonian Institution.*)

Introduction to Aerospace Engineering with a Flight Test Perspective, First Edition. Stephen Corda.
© 2017 John Wiley & Sons Ltd. Published 2017 by John Wiley & Sons Ltd.
Companion Website: www.wiley.com/go/corda/aerospace_engg_flight_test_persp

A few days later the plane was completely assembled in its hangar, and on April 28, or sixty days after the order had been placed, I gave the "Spirit of St. Louis" her test flight. The actual performance was above the theoretical. The plane was off the ground in six and one-eight seconds, or in 165 ft, and was carrying over 400 lbs. in extra gas tanks and equipment. The maximum airspeed meter reading was 126 M.P.H. and the climb was excellent.

Charles Lindbergh writing about the first flight performance of the *Spirit of St Louis*[1]

5.1 Introduction

Historically, the development of aircraft has often focused on the desire to fly faster, higher, farther, or for a longer time. The long range, long endurance transatlantic flight in 1927 by Charles Lindbergh certainly exemplifies this quest. In the past two chapters, the desire to fly faster and higher has been a motivator for many developments in aerodynamics and propulsion. The historical increases in aircraft speed and altitude capabilities are shown for a wide variety of aircraft in Figure 5.1.

Aircraft performance can be considered a subset of the larger study of aircraft *flight mechanics*, which includes the disciplines of performance, stability and control, and aeroelasticity. Flight mechanics is an *applied* engineering subject, rather than a fundamental discipline, such as aerodynamics and thermodynamics. Aircraft performance is discussed in the present chapter and stability and control in Chapter 6. Performance is an engineering discipline that relies on inputs from several other engineering areas, especially aerodynamics and propulsion. The aerodynamic and propulsive characteristics of an aircraft typically bound its performance capabilities. The aerodynamic lift and drag characteristics of an aircraft are embodied by its airframe. The propulsive thrust and drag characteristics are a result of the aircraft propulsion system. The previous two chapters sought to explain the individual concepts of lift, drag, and thrust, and ways to predict or model them. The present chapter combines these individual predictions or models to allow us to predict aircraft performance.

Performance predictions and testing seek to answer questions about an aircraft's capabilities, such as the following.

- How fast can the aircraft fly?
- How high can the aircraft fly?
- How far can the aircraft fly?
- How long can the aircraft remain airborne?
- How much payload can the aircraft carry?
- How long a runway is required for takeoff or landing?
- How fast can the aircraft climb?
- How maneuverable is the aircraft?

The breadth of performance analyses generally encompasses an aircraft's flight profile, as shown in Figure 5.2. This usually includes the takeoff, climb, cruise, maneuvers such as turns in the horizontal and vertical planes, descent, and landing. We seek to answer many of the performance questions posed at these various flight conditions.

The data from performance evaluations may be used for various purposes, some more technically driven than others. Technical objectives of performance evaluations may include specification

---

[1] Charles A. Lindbergh, *We* (New York: G.P. Putnam's Sons, 1955), pp. 206.

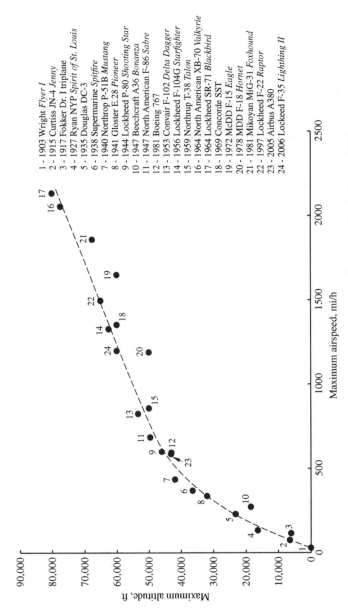

Figure 5.1 Historical increase in aircraft airspeed and altitude capabilities.

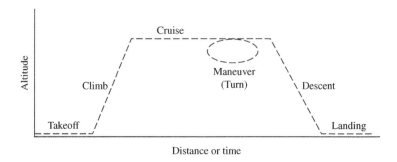

Figure 5.2   Typical flight conditions considered in performance analyses.

compliance, aircraft flight manual development, aircraft flight envelope determination, or mission suitability determination. Non-technical purposes may be related to marketing of the aircraft by a manufacturer to prospective customers. An aircraft manufacturer may be able to sell more aircraft because its aircraft can fly faster or farther than the competition. As the adage goes, "performance sells airplanes."

Performance deals primarily with the *forces* acting on the aircraft and the effects of these forces on the aircraft's flight path. The mathematics of performance analysis involves primarily the force equation, embodied by Newton's second law of motion. The assumption is often made that the aircraft is a point mass, where the total mass of the aircraft is approximated as a single point located at the center of gravity. The typical forces acting on this point mass aircraft are the four forces discussed in Section 2.3.5, the lift, drag, thrust, and weight. Using this simple model, we are able to construct free-body diagrams of the aircraft, with the forces acting on it, apply Newton's second law, and obtain reasonable performance estimates. The accuracy of the performance prediction is highly dependent on the fidelity of the force calculations, most notably how the lift, drag, and thrust are modeled or predicted.

In analyzing aircraft performance, it is the aircraft energy state versus the aircraft motion that is important. Performance involves the capability of an aircraft to increase or decrease its energy states. The aircraft performance can often be determined by simply analyzing the initial and final energy states of the aircraft, ignoring the actual trajectory or motion of the aircraft between these two states. The energy state can be uniquely defined by specifying the aircraft's potential and kinetic energies, or more simply, the aircraft's altitude and velocity. This perspective allows us to evaluate several performance characteristics using an energy concepts approach.

It is insightful to think about the time scales associated with aircraft performance problems. Here, we mean the actual timescales over which the actual phenomena that we are interested in analyzing are occurring. For example, if we consider an aircraft takeoff or climb, these events usually occur over several minutes. A cruise flight condition may occur over several hours. As these examples illustrate, the timescales for aircraft performance are relatively long, on the order of minutes or even hours. This implies that performance deals with steady state situations, where things are in equilibrium or perhaps changing rather slowly with time. The timescales associated with performance problems of interest are shown in Table 5.1. They are compared with the timescales associated with problems of interest in the disciplines of stability and control and aeroelasticity. The timescales of problems in stability and control are small, on the order of seconds, while aeroelastic problems have even smaller timescales of less than a second. The understanding of the timescales involved with the physics of a problem is important in the mathematical formulation of the problem and the assumptions that can be made. From a more pragmatic perspective, it is critical to know the

Table 5.1 Comparison of performance, stability and control, and aeroelasticity timescales for problems of interest.

Discipline	Problems of interest	Timescale of problems
Performance	Maximum speed	Long
	Ceiling	(minutes or hours)
	Rate of climb	
	Range	
	Endurance	
	Flight path optimization	
Stability and control	Stability	Small
	Control	(seconds)
	Maneuver	
	Flying qualities	
Aeroelasticity	Interactions between	Very small
	inertial, elastic, and	(less than a second)
	aerodynamic forces	
	(control reversal,	
	divergence, flutter, etc.)	

appropriate timescale if one is to make meaningful measurements of the phenomenon, either in the laboratory or in flight.

Aircraft operate in the sensible atmosphere, deriving their lift, drag, and thrust due to their interaction with the air. Since the properties of the atmosphere, such as the pressure and temperature, can vary significantly with altitude, this can greatly affect the aerodynamic and propulsion characteristics of an aircraft, and thus affect the aircraft's performance. The understanding and quantitative definition of the atmosphere and its properties are critical when evaluating aircraft performance. Before we can start our discussions about the atmosphere, we must first provide some definitions of altitude.

5.2 Altitude Definitions

We have been loosely using the term *altitude* in our previous discussions, but we now get a little more precise with our terminology. Suppose that you are flying in an aircraft or spacecraft in the atmosphere above the surface of the earth. There are several ways that we can define the vehicle's vertical location above the earth.

The *geometric* altitude, h_g, is the physical, linear distance measured from mean sea level (MSL), defined as the average height of the ocean's surface, to the vehicle. The geometric altitude is sometimes referred to as the *tapeline* altitude, since this is the altitude that would be obtained if a tape measure were used to measure the distance from sea level to the vehicle position. The units of geometric altitude are usually given as "ft MSL" or "m MSL".

We are often interested in knowing the height of the aircraft above the local terrain or above ground level (AGL), h_{AGL}. Since the local ground level or ground elevation, h_e, may be different from sea level elevation, the height above ground level is different from the geometric altitude. If the ground elevation is known, the geometric altitude may be related to the height above ground level by

$$h_g = h_{AGL} + h_e \tag{5.1}$$

If the center of the earth is used as the reference, rather than mean seal level, we can define the *absolute altitude*, h_a, as the sum of the geometric altitude, h_g, and the radius of the earth, R_E.

$$h_a = h_g + R_E \qquad (5.2)$$

Typically, absolute altitude is used for spaceflight applications, and geometric altitude is used for aircraft flight applications. When we are dealing with the very large distances associated with spaceflight, we must take into account the variation of the acceleration due to gravity with distance. This is usually not critical for aircraft flying in the sensible atmosphere. Let us look at how the acceleration due to gravity varies with altitude in the example problem below.

Example 5.1 Acceleration Due to Gravity at Altitude *Assume that you are flying in an aircraft at a geometric altitude of 50,000 ft (15,200 m). What is the actual acceleration due to gravity at this altitude versus the assumption of a constant, sea level value of 32.174 ft/s² (9.8066 m/s²)?*

Solution

Newton's universal law of gravitation states that two bodies of masses m_1 and m_2 are attracted to each other by a gravitational force, F, in a manner that is directly proportional to the product of their masses and inversely proportional to the square of the distance, r, between them. The gravitational law is expressed as

$$F = \frac{Gm_1 m_2}{r^2} \qquad (5.3)$$

where, G is the gravitational constant (not to be confused with the acceleration due to gravity, g).

Now, assume that one body is the aircraft, with mass $m_{aircraft}$, and the other is the earth, with mass m_E. For our aircraft at 50,000 ft, the distance, r, between our aircraft and the center of the earth is the absolute altitude, h_a. The gravitational force felt by our aircraft, $F_{g,50,000\,ft}$, is simply the mass of the aircraft multiplied by the local acceleration due to gravity at our absolute altitude, g. Therefore, Equation (5.3) applied to our aircraft at 50,000 ft is given by

$$F_{g,50,000\,ft} = m_{aircraft} g = \frac{Gm_E m_{aircraft}}{h_a^2}$$

Solving for the acceleration due to gravity at our absolute altitude, we have

$$g = \frac{Gm_E}{h_a^2} \qquad (5.4)$$

Now, let us assume that our aircraft is sitting on the ground at sea level, so that $r = R_E$. The gravitational force acting on our aircraft at sea level, $F_{g,SL}$, is given by

$$F_{g,SL} = m_{aircraft} g_0 = \frac{Gm_E m_{aircraft}}{R_E^2}$$

where we have designated the acceleration due to gravity at sea level as g_0. Solving for the acceleration due to gravity at sea level, we obtain

$$g_0 = \frac{Gm_E}{R_E^2} \qquad (5.5)$$

Solving Equations (5.4) and (5.5) for common terms, we obtain

$$Gm_E = g_0 R_E^2 = g h_a^2$$

Finally, solving for the local acceleration due to gravity at a given absolute altitude, we have

$$g = g_0 \left(\frac{R_E}{h_a}\right)^2 = g_0 \left(\frac{R_E}{R_E + h_g}\right)^2 \tag{5.6}$$

This equation relates the local acceleration due to gravity to the gravitational acceleration at sea level. Inserting values for our geometric altitude, the acceleration due to gravity at sea level, and the earth's mean radius ($R_E = 3959\ miles = 6371.4\ km = 20{,}903{,}520\ ft$), the acceleration due to gravity at 50,000 ft is

$$g = g_0 \left(\frac{R_E}{R_E + h_g}\right)^2 = 32.174\ ft/sec^2 \left(\frac{20{,}903{,}520\ ft}{20{,}903{,}520\ ft + 50{,}000\ ft}\right)^2 = 32.021\ ft/s^2 \tag{5.7}$$

Thus we see that the acceleration due to gravity at 50,000 ft is only 0.476% smaller than the typically assumed constant, sea level value. This difference would get even smaller as the altitude of our aircraft decreased. So it appears that for aircraft related applications, we can ignore the change in the acceleration due to gravity with altitude and use the constant, sea level value of 32.174 ft/s^2 (9.8066 m/s^2) for most cases.

(It is worth noting that we have made a simplifying assumption in our calculation. The acceleration due to gravity at sea level varies slightly with your location, because the earth is not a perfect sphere, rather is it an oblate spheroid that bulges out at the equator. The earth's radius varies from about 3950 miles (6360 km) at the poles to about 3963 miles (6378 km) at the equator. These differences in the value of R_E would introduce a negligible difference in the value for the acceleration due to gravity at sea level, which is small enough for us to ignore.)

We can now define another altitude, the *geopotential altitude*, h, as the altitude created by assuming that the acceleration due to gravity is the constant, sea level value, g_0. Let us relate this new geopotential altitude to the geometric altitude. Consider an aircraft flying, with a weight mg, at a given geometric altitude, h_g, that corresponds to a different geopotential altitude, h. If the altitude of the aircraft is changed by a small amount, its potential energy is changed by a small amount. This change in potential energy is the same, whether measured in terms of the geometric or geopotential altitudes. Equating the change in potential energy in terms of the geopotential altitude, h, and the constant acceleration of gravity, g_0, and the geometric altitude, h_g, and the variable acceleration due to gravity, g, we have

$$mg_0 dh = mg\,dh_g \tag{5.8}$$

Rearranging, we have

$$dh = \frac{g}{g_0} dh_g \tag{5.9}$$

Substituting Equation (5.6) into (5.9), we have

$$dh = \left(\frac{R_E}{R_E + h_g}\right)^2 dh_g \tag{5.10}$$

Integrating Equation (5.10) from sea level to a given altitude, we have

$$\int_0^h dh = \int_0^{h_g} \left(\frac{R_E}{R_E + h_g}\right)^2 dh_g = \left[-\left(\frac{R_E^2}{R_E + h_g}\right)\right]_0^{h_g} \tag{5.11}$$

Table 5.2   Definitions of altitude.

Type of altitude	Symbol	Definition
Geometric (tapeline) altitude	h_g	Height measured above mean sea level
Height above ground level	h_{AGL}	Height measured above local elevation
Absolute altitude	h_a	Height measured from center of earth
Geopotential altitude	h	Altitude based on constant acceleration due to gravity
Pressure altitude	h_p	Altitude corresponding to standard day pressure
Temperature altitude	h_T	Altitude corresponding to standard day temperature
Density altitude	h_ρ	Altitude corresponding to standard day density

$$h = -\left(\frac{R_E^2}{R_E + h_g}\right) + \left(\frac{R_E^2}{R_E}\right) = \frac{-R_E^3 + R_E^2(R_E + h_g)}{(R_E + h_g)R_E} = \frac{-R_E^3 + R_E^3 + R_E^2 h_g}{(R_E + h_g)R_E} \tag{5.12}$$

$$h = \frac{R_E h_g}{R_E + h_g} \tag{5.13}$$

Equation (5.13) provides the desired relationship between the geopotential altitude, h, and the geometric altitude, h_g.

We soon define a model of a *standard atmosphere* that provides values of the pressure, temperature, and density as a function of altitude. Using this standard atmosphere model, we can define several additional types of altitudes. The *pressure altitude*, h_p, is based on the measurement of the air pressure. It is defined as the altitude from the standard atmosphere model, assuming that the measured pressure equals the standard day value. Similarly, the temperature and density altitudes, h_T and h_ρ, respectively, are defined as the altitudes obtained from the standard atmosphere model, assuming that the measured temperature and density, respectively, are equal to the standard day values. We discuss these altitudes, based on the standard atmosphere, in more detail shortly.

The different definitions of altitude that have been discussed are summarized in Table 5.2.

5.3   Physical Description of the Atmosphere

The earth's atmosphere plays a critical role in the operation and performance of almost all aerospace vehicles. All aircraft must operate within the atmosphere to create the aerodynamic lift required to sustain their flight and often to enable their propulsion. Space vehicles, or at least their rocket-propelled boosters, must fly through the atmosphere to get into space. The aerodynamics and performance of aerospace vehicles are highly dependent on the nature of the atmosphere.

The earth's atmosphere is a highly dynamic system, with properties that are constantly changing in three-dimensional space and time. Consider how the weather around you changes from place to place and from one moment to the next. However, it is not practical to develop and use a truly dynamic atmospheric model that describes how the atmospheric properties, such as pressure, density, and temperature, vary in three-dimensional space and time. Instead, we develop atmospheric models that are constant or static with respect to time and vary only with linear distance from the surface of the earth. However, before we develop our atmospheric model, let us get to know our earth's atmosphere a little better.

The atmosphere is a thin layer of gas that surrounds and protects the earth. It provides breathable air for the earth's inhabitants and shields them from the damaging effects of space radiation. When viewed from space, the earth's atmosphere appears to be quite thin and fragile, as shown in

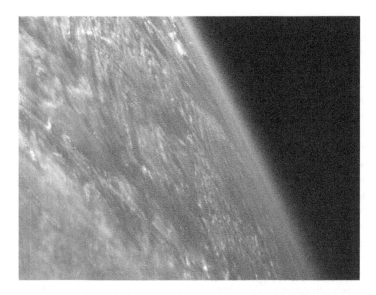

Figure 5.3 The earth's atmosphere as viewed from space. (Source: *NASA.*)

Figure 5.3. To offer some perspective, the diameter of the Earth is about 7918 miles (12,743 km) and the thickness of the atmosphere is only about 2% of this diameter or about 155 miles (250 km). We could also define the thickness of the atmosphere as the boundary between the realm of aircraft and spacecraft. Below this boundary, aerodynamic forces, such as lift and drag and aero-heating are significant. Above this boundary, aerodynamics plays little to no role in the motion and control of an aerospace vehicle. One definition of this boundary, called the *von Karman line*, is 100 km (62 miles) above the surface of the earth.

5.3.1 Chemical Composition of the Atmosphere

The atmosphere is mostly composed of molecular nitrogen and oxygen. The chemical composition, in volume percent, of gases in the atmosphere is approximately 78% nitrogen, 21% oxygen, 1% argon, 0–5% water vapor, 0.035% carbon dioxide, and other smaller proportions of trace species. We call this mixture of gases, in the atmosphere, *air*. Nitrogen is an inert gas and is not directly required to sustain life, while of course, oxygen is essential for life. Water vapor is the gas phase of water and is invisible. The percent concentration of water vapor is variable, however, and rarely exceeds 5% even for very humid conditions. The percent concentration of water vapor should not be confused with relative humidity. Relative humidity is the percentage amount of water vapor in the air divided by the maximum amount of water vapor that the air can hold. When water vapor in the atmosphere condenses to the liquid or solid phase, many of the visible signs of weather are produced, such as clouds, rain, snow, and ice. One of the trace gas species, ozone, is of particular importance in aerospace propulsion. Ozone, with a chemical formula O_3, is composed of three oxygen atoms per molecule, versus oxygen's two atoms (O_2). The ozone layer in the atmosphere is critical in filtering out most of the sun's harmful ultraviolet radiation.

The total quantity of atmospheric gases decreases with increasing altitude, but their relative proportions remain about the same up to high altitudes of about 90 km (56 mi). The concentrations of nitrogen, oxygen, argon, carbon dioxide, and other long-lived chemical gas species are uniform throughout the atmosphere due to turbulent mixing. Water vapor is found mostly in the lower atmosphere, well below about 10,000 ft (3000 m), as it condenses and changes phase when air is lifted, producing clouds, rain, snow, and ice. Highly reactive gas species, such as ozone, have short lives in the atmosphere and therefore do not have a chance to mix uniformly throughout the atmosphere.

The percent concentrations of nitrogen and oxygen remain constant up to an altitude of about 50 miles (264,000 ft, 80 km). While this percent concentration remains constant with increasing altitude, the atmospheric pressure decreases and results in a decrease in the nitrogen and oxygen gas pressures with increasing altitude. Focusing on oxygen, it is this decreased *driving* pressure, needed for moving oxygen from our lungs into our bloodstream, which makes it impossible to breathe and survive at high altitudes above about 8000 m (26,000 ft). Therefore, to fly safely at these high altitudes, we must either increase the oxygen pressure or increase the quantity of oxygen.

Increasing the oxygen pressure is accomplished through aircraft cabin pressurization, typically used on large commercial and business class aircraft. The pressurization pushes more oxygen from the lungs into the blood, but does not change the nominal 21% oxygen level. Cabin pressurization requires an aircraft fuselage or cockpit area that is structurally designed to handle the stresses imposed by the pressure differential between the internal, pressurized area and the outside, low pressure atmosphere.

Increasing the oxygen quantity does increase the percent oxygen level above 21%, but does not necessarily raise the reduced pressure environment. Since more oxygen is made available to the lungs, less pressure is required to move it into the bloodstream. This type of supplemental aircraft oxygen system is viable for use on aircraft up to altitudes of about 40,000 ft (12,000 m), without cabin pressurization. Some aircraft, such as military aircraft, use a combination of cabin pressurization and supplemental oxygen. Supplemental oxygen systems are also used as redundant or backup systems for pressurized cabin systems.

5.3.2   Layers of the Atmosphere

Based on chemical composition, the atmosphere can be divided into two layers, the *homosphere* below 90 km (56 mi) and the *heterosphere* above 90 km. The chemical composition in the homosphere is constant, composed of the same relative proportion of gas species, as described in the previous section. In the heterosphere, the gas species separate out into layers based on their densities. The higher density gases, such as oxygen and helium, settle out into lower layers, while the lightest gas, hydrogen, is in the upper layers.

Typically, the layers of the atmosphere are distinguished based on temperature. Five distinct layers of the atmosphere can be distinguished, based on the variation of temperature with altitude within each layer. The five layers of the atmosphere, in ascending order from sea level, are the troposphere, stratosphere, mesosphere, thermosphere, and exosphere, as shown in Figure 5.4. The upper limit of each layer is given the suffix *pause*, such as the tropopause, stratopause, mesopause, and thermopause. The boundaries of each atmospheric layer are imprecise, as they can vary with seasonal variations of temperature and with geographical latitude on the earth. Nevertheless, general boundaries may be defined to distinguish between the various layers.

5.3.2.1   Troposphere

The *troposphere* (from the Greek *trope* for *turning, changing*) extends from sea level to an average altitude of about 38,000 ft (11.6 km). This is an average height above the earth's surface, since it varies from about 30,000 ft (9 km) over the poles to as high as about 60,000 ft (18 km) at the equator. Most of the earth's weather, including clouds, rain, and snow, occurs within the troposphere.

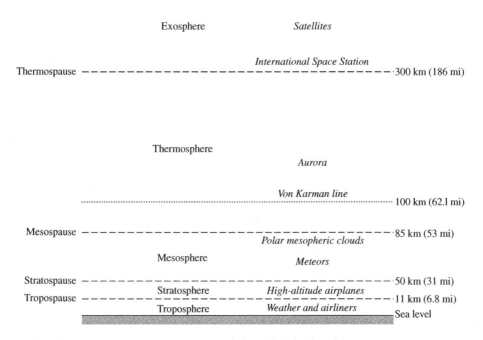

Figure 5.4   Layers of the atmosphere (not to scale).

The troposphere is often considered the realm of the *sensible* atmosphere. Approximately 80% of the mass of the atmosphere is in the troposphere. Much of civilian aviation, including commercial airliner and general aviation flight, takes place in the troposphere.

The temperature in the troposphere decreases with altitude at an average linear rate of about –3.5 °F per 1000 ft or about 6.5 °C per 1000 m. This decreasing rate of change of vertical temperature is defined as the *lapse rate*, Γ. Therefore, the temperature in the troposphere decreases from about 59 °F (15 °C) at sea level to about –69 °F (–56.1 °C) at 38,000 ft (11.6 km). Above this, in the tropopause, the temperature is a relatively constant value of –69 °F (–56.1 °C) up to about 60,000 ft (18 km).

While the temperature generally decreases with altitude in the troposphere, there are isolated, thin layers within the troposphere where the temperature increases with altitude, known as temperature inversions. Unlike the rest of the troposphere, there is not much atmospheric mixing in these temperature inversions, leading to conditions such as freezing rain and summertime smog. The lack of vertical mixing in the tropopause also leads to the horizontal spreading or stratification of cloud tops, resulting in the characteristic anvil shapes of thunderstorms clouds.

5.3.2.2   Stratosphere

Above the tropopause, the *stratosphere* (from the Greek *strato* for *layer, sheet*) extends to as high as about 165,000 ft (50 km). Due to the variation in the height of the troposphere between the earth's poles and equator, the bottom of the stratosphere varies between about 30,000 ft (9 km) and about 60,000 ft (18 km). Since this layer is composed of very dry air and has significant vertical stability, i.e. little vertical mixing of air, there is little cloud formation in the stratosphere. An exception to this is polar stratospheric clouds, iridescent clouds that occur in the winter over the poles at altitudes of about 50,000–80,000 ft (15–24 km).

Some civilian, commercial flights and many military flights occur in the lower boundaries of the stratosphere. Some military reconnaissance aircraft are able to routinely fly higher in the stratosphere, such as the Lockheed U-2 with a cruising altitude of over 70,000 ft (21 km) and the Lockheed SR-71 (which is no longer in service) with a cruising altitude of 80,000 ft (24 km).

The ozone layer, which protects the Earth from ultraviolet radiation, lies within the stratosphere. Absorption of ultraviolet radiation by the ozone molecules results in heating of the stratosphere to a maximum temperature at the top of the stratosphere and start of the stratopause. Due to this solar radiative heating, the lapse rate in the stratosphere is negative, that is, the temperature increases with altitude. The temperature increases from about –69 °F (–56.1 °C) at 60,000 ft (18 km) to about 26.3 °F (–3.2 °C) at 165,000 ft (50 km).

There is no distinct boundary that defines the end of the earth's atmosphere and the start of space. Rather, the atmosphere continuously and gradually thins with increasing altitude. It is safe to say that somewhere above the stratosphere marks the edge of space. The US Air Force awards astronaut wings for flight above 50 miles (80 km), while other organizations accept the von Karman line, an altitude of 100 km (62 miles), as the beginning of space.

5.3.2.3 Mesosphere

The *mesosphere* (from the Greek *mesos* for *middle, intermediate*), extends from about 165,000 ft (50 km) to about 280,000 ft (85 km) above the surface of the earth. The temperature decreases with altitude, from 26.3 °F (–3.2 °C) at 165,000 ft (50 km) to –118 °F (–83.3 °C) at 280,000 ft (85 km). The mesopause, the constant temperature layer at the top of the mesosphere, is the coldest region of the atmosphere. The mesosphere absorbs harmful radiation, including cosmic radiation and the sun's ultraviolet and X-ray radiation. There is no atmosphere to scatter sunlight, so the sky appears black.

The highest clouds in our atmosphere are the polar mesospheric clouds, also called *noctilucent* or *night-shining* clouds, due to their appearance as delicate, shining threads, as shown in Figure 5.5. These clouds are at the edge of space, forming at altitudes of between 47 and 53 miles (75.6 and 85.3 km). Noctilucent clouds have been increasing in frequency and intensity and may serve as

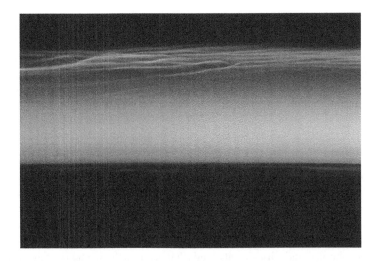

Figure 5.5 Polar mesospheric clouds as viewed from the International Space Station. (Source: *NASA*.)

sensitive indicators of changes in our high atmosphere. How these clouds are formed is still uncertain, even as to whether they are due to extraterrestrial sources, such as meteor dust, or terrestrial sources, such as rocket exhaust particles.

5.3.2.4   Thermosphere

The *thermosphere* (from the Greek *thermo* for *heat*) is the thickest atmospheric layer and extends from 53 miles (85 km) to 186 miles (300 km) above the earth's surface. Absorption of solar ultraviolet radiation causes the temperature to dramatically increase within the thermosphere, from −118 °F (−83.3 °C) at 53 miles (85 km) to 1340 °F (1000 K) at 186 miles (300 km). Increases in solar activity, such as from solar flares and the 7-year solar cycle maximum, result in larger temperature rises within the thermosphere.

Above an altitude of about 62 miles (100 km), aerodynamic forces, acting on an aerospace vehicle, may be considered negligible. This 100 km demarcation is denoted as the Von Karman line. Above about 100 km (328,000 ft), the gas particles are spaced quite far apart, such that the distance between particle collisions, called the *mean free path*, is greater than about 3.3 ft (1 m). Contrast this with the atmosphere at sea level, where the mean free path is 2.36×10^{-6} in (6×10^{-6} cm). Despite this large gas particle spacing in the thermosphere, a spacecraft orbiting the earth, in the thermosphere, experiences an atmospheric drag due to gas particle impacts. This drag is significantly smaller than the aerodynamic drag in the troposphere, but it does affect the orbital mechanics of the spacecraft. In fact, it is probably not practical for a spacecraft or satellite to orbit the earth below about 125 miles (200 km), as the orbit would degrade in just a few days and the space vehicle would fall out of its orbit, back to earth.

5.3.2.5   Exosphere and Hard Space

The *exosphere* (from the Greek *exo* for *outside*) is the outermost layer of the earth's atmosphere, starting at 186 miles (300 km) and eventually merging with the interplanetary medium. The temperature in the exosphere is constant at 1340 °F (1000 K), although this varies with solar cycle. Ultraviolet radiation dissociates most of the molecular oxygen into atomic oxygen, which can be highly reactive with spacecraft materials.

The region above the exosphere is called *hard space*, which is filled with electromagnetic radiation and cosmic particles. For example, the particle density at a height of 1243 miles (2000 km) is about 10 billion particles per cubic meter, so hard space is not empty space.

Hard space is the realm of orbital space vehicles. For instance, the typical Space Shuttle orbit was at about 200 miles (322 km) above the earth. The International Space Station's orbit is between 205 miles (330 km) and 270 miles (435 km) above the earth. Both of these spacecraft reside in what's called low earth orbit (LEO).

5.3.3   GTT: Cabin Pressurization Test

This ground test technique discusses several areas related to the pressurization of aircraft and spacecraft cabins. While it is related to aircraft structures, it also pertains to our discussions about the atmosphere, especially at high altitudes. To fly at high altitudes or in space, aircraft or spacecraft cabins must be pressurized to maintain a breathable and comfortable atmosphere for occupants. The vehicle cabin must be structurally designed for the stresses imposed by the pressure difference between the higher pressure in the cabin interior and the low pressure of the external atmosphere.

Often, a test cabin is built and instrumented for pressurization tests on the ground. The ground testing is performed to verify predictions and to validate the structural integrity of the cabin pressure vessel. The cabin is typically tested for tens of thousands of pressure cycles, where it is pressurized and depressurized, to simulate the pressurization cycles that the cabin experiences during its lifetime in flight operations. Some cabin pressurization tests are performed using a water tank, where the cabin is immersed in a pool of water and pressurized.

Before entering commercial service and at normal maintenance intervals, aircraft cabins of commercial airliners are pressure tested on the ground to check for any significant leaks in the cabin pressure vessel. The cabin may be pressurized with high pressure air from one of the jet engines, an auxiliary power unit, or from a ground unit. The proper operation of the cabin pressurization system is checked, and any significant leaks are sealed.

In normal operation, aircraft or spacecraft cabins are pressurized with compressed air from various sources. High-pressure air is typically "bled" off from the compressor stage of a jet engine to pressurize an aircraft cabin. The pressure to which a cabin is pressurized, corresponding to a cabin altitude, varies with the cruising altitude of the aircraft and with the type of aircraft. A Boeing 767 airliner maintains a cabin altitude of about 6900 ft (2100 m), corresponding to a cabin pressure of 11.4 psi (78.6 kPa), when at a cruising altitude of 39,000 ft (11,900 m). Typical differential pressures, between the cabin interior and the exterior atmosphere, are about 8–9 psi (55–62 kPa).

For spacecraft cabin pressurization, high-pressure gas is supplied from storage tanks. Since 1961, Russian spacecraft maintain a near-sea-level cabin altitude (cabin pressure of 14.7 psi or 102 kPa) at all times, using a nitrogen–oxygen gas mixture. The USA has pressurized its spacecraft to higher cabin altitudes, around 25,000 ft (7600 m) (cabin pressure of 5.5 psi or 37.7 kPa) for the *Mercury* and *Gemini* spacecraft and about 27,000 ft (8230 m) (cabin pressure of 5.0 psi or 34.5 kPa) for the *Apollo* spacecraft. The lower cabin pressure allows for a lighter structure, since the pressure differential between the cabin interior and the exterior space environment is much smaller. US spacecraft used a pure oxygen environment with a slightly greater than sea level cabin pressure before launch. This high-pressure pure oxygen cabin environment was abandoned after the *Apollo 1* capsule fire, during a 1967 ground test, which killed the three-man crew. Since then, the US has used a nitrogen–oxygen gas mixture with a sea level cabin altitude at launch, transitioning to a low-pressure, pure oxygen environment in space.

Important lessons about cabin pressurization tests and the design of pressurized aircraft fuselages were learned with the advent of the world's first commercial jet airliner, the British de Havilland *Comet*. The *Comet* was the first passenger airliner designed to fly at high altitudes, above 30,000 ft (9000 m), with a pressurized passenger cabin. The prototype of the *Comet* airliner first flew on 27 July 1949. The speed and comfort of the passenger service offered by the *Comet* were a great success at first. The design of the *Comet* was advanced with its sleek aerodynamic airframe and modern jet propulsion. A unique feature of the Comet fuselage design was the large square windows in the passenger cabin, as shown in Figure 5.6.

In 1954, there were several catastrophic, in-flight break-ups of the *Comet*, which were eventually attributed to the design and installation of these square windows and an inadequate understanding of metal fatigue due to repeated cabin pressurization cycles. The shape of the square windows led to stress concentrations in their small radius corners, which was exacerbated by the type of rivet holes that were used to install the windows. The punch riveting used to install the windows resulted in a more ragged, imperfect hole that led to fatigue cracks, which could propagate from the hole.

Although portions of the fuselage of the *Comet* were subjected to pressure cycle testing during its development, the techniques used were not adequate to identify the failure modes that ultimately led to the in-flight break-ups. After the accidents, more thorough tests of the *Comet* were conducted, including a pressure test of a complete fuselage, submerged in a large water tank. After repeated

Figure 5.6   DeHavilland *Comet* with square cabin windows. (Source: *United Kingdom Government, British Official Photographer, 1949, PD-UKGov.*)

pressure cycles, simulating over 3000 flight pressurization cycles, the fuselage structure failed at a corner of one of the square windows. The lessons learned from the *Comet* were incorporated into the design of all future aircraft with pressurized cabins, including the use of oval windows in the fuselage, which eliminated stress concentrations at the small radius corners.

5.4   Equation of Fluid Statics: The Hydrostatic Equation

We seek to develop a standard model of a static atmosphere that can be used for engineering analyses. This static model provides atmospheric properties, such as pressure, density, and temperature, as a function of distance above the surface of the earth. As a first step in the development of this model, we derive an expression that quantifies the variation of atmospheric pressure with altitude.

Consider a cylindrical fluid element as shown in Figure 5.7. The fluid element has a base area dA and a height, dh_g. The weight, W, of the fluid element is given by

$$W = \rho g \, dA \, dh_g \tag{5.14}$$

where ρ is the fluid element density, g is the acceleration due to gravity, and $dAdh_g$ is the fluid element volume. The pressure at the base of the fluid is p acting over the base area, dA. The force on the base area is given by

$$F_{base} = p \, dA \tag{5.15}$$

The pressure decreases with the height, h_g, changing by a small amount, dp, over the small height change, dh_g, of the fluid element. The force on the top of the fluid element is therefore given by

$$F_{top} = (p + dp)dA \tag{5.16}$$

The fluid element is in equilibrium, not moving up or down, so that the sum of the forces on the element is zero. The forces acting on the fluid element are its weight and the forces on the top and base areas. Summing these forces, we have

$$\sum F = 0 = F_{base} - F_{top} - W \tag{5.17}$$

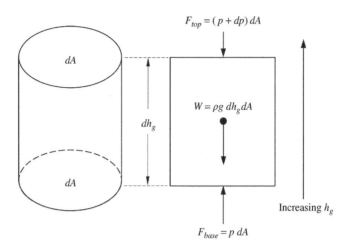

Figure 5.7   Free-body diagram of force balance on a fluid element.

Substituting Equations (5.14), (5.15), and (5.16) into Equation (5.17) we have

$$\sum F = 0 = pdA - (p + dp)dA - \rho gdAdh_g \tag{5.18}$$

Solving for the change in pressure, dp, we obtain

$$dp = -\rho gdh_g \tag{5.19}$$

This is the desired relationship, relating the change in pressure, dp, to the change in distance, dh_g. An increase in altitude (positive dh_g) results in a decrease in pressure (negative dp). Equation (5.19) is the known as the *hydrostatic equation* and relates the pressure change with distance for any fluid of density ρ in static equilibrium. This equation is also applicable to the change in pressure as a function of depth in water or any other fluid.

We can write the hydrostatic equation in terms of the geopotential altitude, h, and the constant acceleration due to gravity, g_0. Using Equation (5.9) in (5.19), the hydrostatic equation becomes

$$dp = -\rho g_0 dh \tag{5.20}$$

Let us integrate Equation (5.20) from sea level ($h = 0$), where the pressure is p_{SL}, to a height above sea level, h, where the pressure is p.

$$\int_{p_{SL}}^{p} dp = -\rho g_0 \int_{0}^{h} dh \tag{5.21}$$

$$p - p_{SL} = -\rho g_0 h \tag{5.22}$$

Solving for the pressure at the height above sea level, we have

$$p = p_{SL} - \rho g_0 h \tag{5.23}$$

Equation (5.23) shows that the pressure decreases from sea level pressure with increasing height above sea level.

Example 5.2 Weight of a Column of Air *Use the hydrostatic equation to calculate the weight of a cylindrical column of air, with a cross-sectional area of one square inch that stretches from*

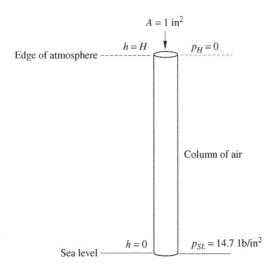

Figure 5.8   Cylindrical column of air stretching from sea level to edge of the atmosphere.

the surface of the earth at sea level to the edge of atmosphere. Assume that the static pressure is 14.7 lb/in² at sea level and zero at the edge of the atmosphere. Also, assume that the density and acceleration due to gravity are constants.

Solution

The cylindrical column of air is shown in Figure 5.8.

Integrating the hydrostatic equation, Equation (5.20), from sea level to the edge of the atmosphere.

$$\int_{p_{SL}}^{0} dp = -\rho g_0 \int_{0}^{H} dh$$

$$0 - p_{SL} = -\rho g_0 (H - 0)$$

$$p_{SL} = \rho g_0 H = \frac{W}{A}$$

where the quantity ρgH is the weight per unit cross-sectional area, W/A, of the air column. The weight of the air column is

$$W = p_{SL} A = \left(14.7 \ \frac{lb}{in^2} \right) (1 \ in^2) = 14.7 \ lb$$

Thus, we see that the sea level air pressure of 14.7 lb/in² may be interpreted as the weight of the entire atmosphere in a cylindrical column, with a cross-sectional area of one square inch, that stretches from sea level to the edge of the atmosphere. (We could have also obtained this same result by substituting the boundary conditions directly into Equation (5.23).)

Example 5.3 The U-Tube Manometer *The U-tube manometer is a simple device that is used to measure pressures in a laboratory setting, such as a wind tunnel. The device is a U-shaped*

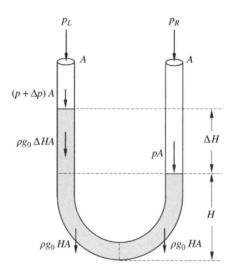

Figure 5.9   Forces on the static fluid inside a U-tube manometer.

glass tube that is filled with a liquid whose density is known, such as mercury or water. One end of the tube is connected to a small hole or port, where the pressure is to be measured. The other end is connected to a reference pressure port or simply left open to sense the ambient, atmospheric pressure. The manometer measures the pressure difference between the two ends. Derive a formula for measuring pressure using a U-tube manometer.

Solution

Consider a U-tube manometer with a tube cross-section area, A, filled with a liquid of known density ρ. Assume that there is a pressure difference, Δp, between the two ends of the manometer such that the liquid is pushed down on one side and drawn up on the other side. The pressures on the right and left sides are p and $p + \Delta p$, respectively. The difference in heights of liquid on the left and right sides is ΔH. The forces on the static liquid inside the manometer are shown in Figure 5.9.

The liquid in the left and right sides of the lower part of the manometer, up to a height H, have equal volumes and equal weights of $\rho g_0 HA$. The weight of the liquid on the left side of the manometer, with a height ΔH, is equal to $\rho g_0 \Delta HA$. The pressure force on the left side of the liquid column is $(p + \Delta p)A$. The liquid column on the right side has a pressure force of pA. Since the liquid is stationary, the forces on the left side are balanced with the forces on the right side, as shown in Figure 5.9. Thus, we have

$$(p + \Delta p)A + \rho g_0 \Delta HA + \rho g_0 HA = pA + \rho g_0 HA \tag{5.24}$$

or

$$\Delta p = -\rho g_0 \Delta H \tag{5.25}$$

Equation (5.25) relates the pressure change measured by the manometer, Δp, to the difference in height of the manometer liquid, ΔH. If the left side of the manometer senses a pressure p_L and the right side senses a pressure p_R, the manometer measures the pressure difference given by

$$\Delta p = p_R - p_L = \rho g_0 \Delta H \tag{5.26}$$

If one of these pressure is a known, reference pressure, the other pressure may be determined. For instance, if the right side of the manometer is open to the atmosphere, it senses the known ambient pressure, p_a, and the manometer can be used to measure the pressure p_L, given by

$$p_L = p_a - \rho g \Delta H \tag{5.27}$$

Hence, the pressure at the port location can be obtained from the height difference of the liquid in the manometer and the known ambient or reference pressure. Since manometers were often filled with mercury, "inches of mercury" or "in Hg" were used as a unit of pressure. This unit system is still used today as a unit of pressure, in various applications, such as in meteorology and in aircraft altimeters.

As a numerical example, let us assume that a mercury-filled U-tube manometer is used to measure the pressure on the surface of a wing model in a wind tunnel. The left side of the manometer is connected to a port on the surface of the wing and the right side is open to the atmosphere at sea level. The sea level atmospheric pressure and the density of mercury are known (the density of mercury is 13.534 g/cm³). When the wind tunnel is turned on, the manometer measures a pressure difference, Δp, of 35.6 cm.

In consistent units, the density of mercury is

$$\rho_{Hg} = 13.534 \, \frac{g}{cm^3} \times \frac{1 \, kg}{1000 \, g} \times \left(\frac{100 \, cm}{1 \, m} \right)^3 = 13{,}534 \, \frac{kg}{m^3}$$

Using Equation (5.27), the pressure on the wing surface is

$$p_L = p_a - \rho_{Hg} g \Delta H = 101{,}325 \, \frac{N}{m^2} - \left(13{,}534 \, \frac{kg}{m^3} \right) \left(9.81 \, \frac{m}{s^2} \right) (0.356 \, m) = 54{,}059 \, \frac{N}{m^2}$$

5.5 The Standard Atmosphere

The earth's atmosphere is in a constant state of change, where the atmospheric properties, such as pressure, density, and temperature, vary with time and location. For example, if we measure the temperature at 20,000 ft (6096 m) on a summer versus a winter day, we would expect the values to differ. Similarly, if we take the same temperature measurement at two different locations, such as Alaska and Florida, we would expect a variation due to the different geographic locations. Why does this matter? Well, if we conduct flight tests of an airplane on two different days or two different locations, we may get quite different performance results for takeoff, climb, cruise, or other tasks, because the atmospheric conditions are different. It would be impossible to define the performance of the airplane objectively. However, if we *standardize* the data from these different tests to a common reference atmosphere, then we could objectively define the performance. To accomplish this, a *standard atmosphere* has been defined.

The standard atmosphere is an idealized model of a static atmosphere that does not change with time or location. The model provides values of the pressure, temperature, and density as a function of vertical distance. The air in the standard atmosphere is assumed to be a perfect gas that obeys the perfect gas equation of state. It is assumed that there is no wind or turbulence in the atmosphere, that is, the air is at rest with respect to the earth. The air is assumed to be dry, with no water vapor present. The property values in the standard atmosphere model represent averages over long periods of time.

The first standard atmosphere models were developed in the USA and in Europe in the 1920s. In 1952, the International Civil Aviation Organization (ICAO) introduced an internationally accepted atmosphere model. There have been subsequent improvements and updates to atmospheric models over the years. Today, there are standard atmosphere models available, below 120 km (74.6 mi), for different latitudes and different seasons of the year. Above 120 km, there are different models for variations in solar (sunspot) activity. There are also cold, hot, polar, and tropical models, developed by the US Military. NASA has recently developed an atmospheric model, which includes wind, in addition to pressure, temperature, and density. This model was developed for trajectory simulations, such as the reentry of the Space Shuttle external fuel tank.

We use the 1976 US Standard Atmosphere [28], in this text, which is typically used in engineering, along with the 1962 US Standard Atmosphere. The date refers to the year that the model was developed. The 1976 and 1962 atmosphere models are identical below 50 km (31 mi), but above this altitude, the 1976 model has improved models and replaces the 1962 model. The 1976 Standard Atmosphere model represents idealized, year-round, mean conditions at a mid-latitude location on the earth, for moderate solar activity. The model extends from 5 km (3.1 mi) below mean sea level to an altitude of 1000 km (621 mi).

5.5.1 Development of the Standard Atmosphere Model

We wish to develop an idealized model of the atmosphere, where the properties of the air are defined as a function of vertical distance from the earth's surface (mean sea level). In our previous analyses of a gas, such as air, we usually defined the state of the air in terms of its thermodynamic variables, such as the pressure, temperature, and density, and the flow velocity. Therefore, there are four unknowns (pressure, temperature, density, and velocity), which requires four equations or relationships to define these unknowns. We are assuming a static atmosphere, so velocity is zero. It is also assumed that the air is a perfect gas, so we have the perfect gas equation of state, relating the pressure, density, and temperature. We have also developed a relationship relating the change in pressure to the density and altitude for a static fluid, the hydrostatic equation. The altitude is the independent variable, so it is not an additional unknown. We now have two equations (perfect gas equation of state and the hydrostatic equation) for the solution of three unknown dependent variables (pressure, temperature, and density) with altitude as the independent variable. We need an additional relationship to solve for the properties in the atmospheric model. For this final relationship, we define the temperature as a function of altitude, based on empirical data from many measurements from ground stations, balloons, sounding rockets, satellites, and other means. This temperature profile for the development of the standard atmosphere is shown up to an altitude of about 110 km (68.4 mi) in Figure 5.10.

The standard temperature profile is composed of linear segments that represent either isothermal regions, where the temperature is constant with altitude, or gradient regions, where the temperature varies linearly with altitude. The standard temperature and geopotential altitude of the reference points, denoted by numbers in Figure 5.10, connecting the linear segments are given in Table 5.3. The change of temperature with altitude, known as the *lapse rate*, is also listed for the gradient regions. The calculation of the standard pressure and density, corresponding to the defined standard temperature profile, is discussed next.

Consider the hydrostatic equation, Equation (5.20), given by

$$dp = -\rho g_0 dh$$

Dividing by the perfect gas equation of state, Equation (3.61), we have

$$\frac{dp}{p} = -\frac{\rho g_0 dh}{\rho RT} = -\frac{g_0}{RT} dh \tag{5.28}$$

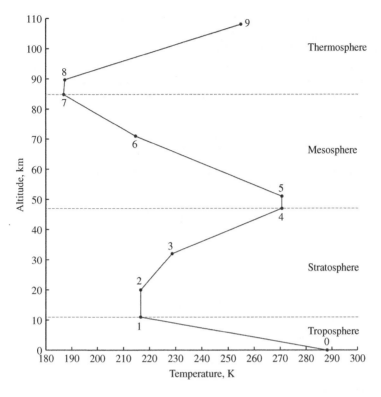

Figure 5.10   Temperature distribution in 1976 US Standard Atmosphere.

Table 5.3   Reference altitudes, temperature lapse rates, and properties for 1976 Standard Atmosphere.

Reference point (Figure 5.10)	Layer of atmosphere	Geopotential altitude		Temperature lapse rate		Standard temperature	Pressure	Density
		(km)	(ft)	(K/km)	(K/kft)	(K)	(N/m^2)	(kg/m^3)
0	Troposphere	0	0	−6.5	−1.9812	288.15	101,325	1.2250
1	Tropopause	11	36,089	0	0	216.65	22,632.1	0.36392
2	Stratosphere	20	65,617	1.0	0.3048	216.65	54,74.89	8.8035×10^{-2}
3	Stratosphere	32	104,9867	02.8	0.85344	228.65	868.019	1.3550×10^{-2}
4	Stratopause	47	154199	0	0	270.65	110.906	1.4275×10^{-3}
5	Mesosphere	51	167,323	−2.8	−0.85344	270.65	66.9389	8.6160×10^{-4}
6	Mesosphere	71	232,940	−2.0	−0.6096	214.65	3.95642	6.4211×10^{-5}
7	Mesopause	84.852	278,386	—	0	186.95	—	—
8	Thermosphere	89.716	294,344	—	—	187.36	—	—
9	Thermosphere	108	354,754	—	—	254.93	—	—

Equation (5.28) is a differential equation that relates the change in pressure to the change in altitude, as a function of the temperature profile. This equation can be integrated over the isothermal or gradient regions of the standard atmosphere to provide the desired variation of pressure with altitude of the standard atmosphere. Once the variation of pressure is found, the equation of state is used to obtain the variation of the density in the standard atmosphere. The integrations for the isothermal layer and the gradient layer are presented below.

For the isothermal region, Equation (5.28) is integrated from the bottom or base of the isothermal region, with a geopotential altitude, h_n, and pressure, p_n, to a point above this base, with a geopotential altitude, h, and pressure, p. The subscript n denotes the reference point at the base of the isothermal region, where n is reference point 1, 4, or 7 in the standard atmosphere shown in Figure 5.10. Thus, we have

$$\int_{p_n}^{p} \frac{dp}{p} = -\frac{g_0}{RT} \int_{h_n}^{h} dh \tag{5.29}$$

where the temperature, T, is a constant since we are integrating over the isothermal layer. Completing the integration, we have

$$\ln \frac{p}{p_n} = -\frac{g_0}{RT}(h - h_n) \tag{5.30}$$

Solving for the pressure, we have

$$p = p(h) = p_n \, e^{-\frac{g_0}{RT}(h-h_n)} \tag{5.31}$$

Equation (5.31) is the desired result, providing an equation defining the pressure as a function of altitude and temperature for an isothermal region of the standard atmosphere.

Substituting for the pressure in Equation (5.31), using the equation of state, we have

$$\rho RT = (\rho_n RT_n) \, e^{-\frac{g_0}{RT}(h-h_n)} \tag{5.32}$$

Since $T = T_n$ in the isothermal region, we have

$$\rho = \rho(h) = \rho_n \, e^{-\frac{g_0}{RT}(h-h_n)} \tag{5.33}$$

Equation (5.33) specifies the density as a function of altitude and temperature for an isothermal region of the standard atmosphere.

Now, consider the gradient regions of the standard atmosphere. The constant lapse rate in each of the gradient regions, a_n, is defined as

$$a_n \equiv \frac{dT}{dh} = \text{constant} \tag{5.34}$$

where dT is the change in temperature, dh is the change in altitude, and n denotes the base of the gradient region, which may be reference point 0, 2, 3, 5, 6, or 8, in Figure 5.10. Solving for the change in temperature, we have

$$dT = a_n dh \tag{5.35}$$

Inserting Equation (5.35) into (5.28), we have a differential equation for the change in pressure as a function of the change in temperature in the gradient region.

$$\frac{dp}{p} = -\frac{g_0}{RT} dh = -\frac{g_0}{RT} \left(\frac{dT}{a_n} \right) \tag{5.36}$$

Integrating from the base of the gradient region, where the pressure and temperature are, p_n and T_n, respectively, to a point above this base, where the pressure and temperature are, p and T, respectively, we have

$$\int_{p_n}^{p} \frac{dp}{p} = -\frac{g_0}{a_n R} \int_{T_n}^{T} \frac{dT}{T}$$

$$\ln\left(\frac{p}{p_n}\right) = -\frac{g_0}{a_n R}\ln\left(\frac{T}{T_n}\right)$$

$$p = p_n\left(\frac{T}{T_n}\right)^{-\frac{g_0}{a_n R}} \tag{5.37}$$

Equation (5.37) provides the pressure as a function of the temperature in the gradient region of the standard atmosphere. To obtain pressure as a function of the altitude, we integrate Equation (5.35) from the base of the gradient region to a geopotential altitude, h.

$$\int_{T_n}^{T} dT = a_n\int_{h_n}^{h} dh$$

$$T - T_n = a_n(h - h_n) \tag{5.38}$$

or

$$\frac{T}{T_n} = 1 + \frac{a_n}{T_n}(h - h_n) \tag{5.39}$$

Equation (5.39) provides the temperature as a function of altitude for the gradients regions of the standard atmosphere. Inserting Equation (5.39) into (5.37), we have

$$p = p(h) = p_n\left[1 + \frac{a_n}{T_n}\left(h - h_n\right)\right]^{-\frac{g_0}{a_n R}} \tag{5.40}$$

Equation (5.40) provides the pressure as a function of altitude for the gradient regions of the standard atmosphere.

Using the equation of state, we can obtain the standard density for the gradient regions, as

$$\rho = \rho(h) = \rho_n\left[1 + \frac{a_n}{T_n}\left(h - h_n\right)\right]^{-\left(\frac{g_0}{a_n R} + 1\right)} \tag{5.41}$$

We are often interested in the properties of air at standard atmospheric conditions at sea level. These standard sea level values, denoted by the subscript "SSL", for pressure, temperature, and density are given below for various unit systems. These standard conditions were introduced in Chapter 3 and are listed in Table 3.1.

$$p_{SSL} = 29.92 \text{ in Hg} = 1 \text{ atm} = 2{,}116 \text{ lb/ft}^2 = 101{,}320 \text{ N/m}^2 \tag{5.42}$$

$$T_{SSL} = 15°C = 59°F = 288.15 \text{ K} = 459.67°R \tag{5.43}$$

$$\rho_{SSL} = 1.225 \text{ kg/m}^3 = 0.002377 \text{ slugs/ft}^3 \tag{5.44}$$

The variations of the temperature, pressure, and density with altitude, for the standard atmosphere, are shown in Figure 5.11, from sea level to an altitude of 20 km (65,600 ft, 12.4 mi). This altitude range covers the flight envelopes of most aircraft. The temperature decreases linearly with altitude, at a rate of –6.5 K/km (–1.98 K/1000 ft), up to the top of the troposphere, at an altitude of 11 km (36,100 ft). Above the tropopause, the standard temperature is constant at 216.65 K (–69.696 °F), up to an altitude of 20 km. The pressure and density decrease continuously, from their sea level values, up to 20 km. Values for the standard atmosphere are tabulated and can be referenced from various sources (including Appendix C).

Example 5.4 Calculation of the Standard Atmosphere *Calculate the standard pressure, density, and temperature at an altitude of 6050 m.*

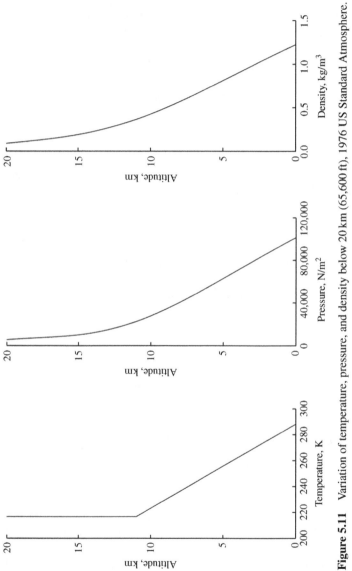

Figure 5.11 Variation of temperature, pressure, and density below 20 km (65,600 ft), 1976 US Standard Atmosphere.

Solution

An altitude of 6050 m is in the troposphere, a gradient region of the standard atmosphere. From Table 5.3 the pressure, temperature, density, and lapse rate at the base of the gradient region, where n = 0, are

$$p_0 = 101{,}325 \frac{N}{m^2}$$

$$T_0 = 288.15 \ K$$

$$\rho_0 = 1.2250 \frac{kg}{m^3}$$

$$a_0 = -0.0065 \frac{K}{m}$$

Inserting values into Equation (5.31), we have

$$p = p(h) = 101{,}325 \frac{N}{m^2} \left[1 + \frac{-0.0065 \frac{K}{m}}{288.15 \ K} (h - 0) \right]^{-\frac{9.81 \frac{m}{s^2}}{\left(-0.0065 \frac{K}{m}\right)\left(287 \frac{N \cdot m}{kg \cdot K}\right)}}$$

Solving this equation at an altitude of 6050 m, we have

$$p = p(6050 \ m) = \left(101{,}325 \frac{N}{m^2} \right) (0.86353)^{5.2586} = 46{,}839 \frac{N}{m^2}$$

The density at 6050 m is given by Equation (5.41) as

$$\rho = \rho(h) = 1.2250 \frac{kg}{m^3} \left[1 + \frac{-0.0065 \frac{K}{m}}{288.15 \ K} (h - 0) \right]^{-\left[\frac{9.81 \frac{m}{s^2}}{\left(-0.0065 \frac{K}{m}\right)\left(287 \frac{N \cdot m}{kg \cdot K}\right)} + 1 \right]}$$

$$\rho = \rho(6050 \ m) = \left(1.2250 \frac{kg}{m^3} \right) (0.86353)^{4.2586} = 0.65578 \frac{kg}{m^3}$$

The temperature at 6050 m is given by Equation (5.38) as

$$T = T(h) = 288.15 \ K + \left(-0.0065 \frac{K}{m} \right) h$$

$$T = T(6050 \ m) = 288.15 \ K + \left(-0.0065 \frac{K}{m} \right) 6050 \ m = 248.83 \ K$$

5.5.2 Temperature, Pressure, and Density Ratios

It is often useful to express the temperature, pressure, and density in terms of non-dimensional ratios, where the value is divided by the standard sea level value. Using this convention, we can define the temperature ratio, θ, pressure ratio, δ, and density ratio, σ, as

$$\theta \equiv \frac{T}{T_{SSL}} \tag{5.45}$$

$$\delta \equiv \frac{p}{p_{SSL}} \tag{5.46}$$

$$\sigma \equiv \frac{\rho}{\rho_{SSL}} \tag{5.47}$$

Tabulated values of these ratios are provided in Appendix C, allowing for the calculation of atmospheric properties at any given altitude.

By using the ideal gas equation of state, we can relate any of the ratios in terms of the other ratios. For example, we can express the density ratio, in terms of the pressure and temperature ratios, as

$$\sigma = \frac{\rho}{\rho_{SSL}} = \frac{p/RT}{p_{SSL}/RT_{SSL}} = \frac{p}{p_{SSL}} \frac{T_{SSL}}{T} = \frac{\delta}{\theta} \tag{5.48}$$

Care must be taken when using these ratios, to prevent erroneous calculations of the atmospheric properties. Correct results are always be obtained if consistent English or SI units are used in the ratios. If other units are used, correct results are obtained if the conversion factor involved is a multiple of the consistent unit, since these multiplications cancel out. If the conversion requires addition of a constant value, such as in the case of temperature conversions, erroneous results are obtained if consistent units are not used. The example problems below illustrate this issue.

Example 5.5 Temperature, Pressure, and Density Ratios Using Consistent Units *Calculate the pressure, temperature, and density corresponding to a standard altitude of 17,000 ft.*

Solution

From Appendix C, for a standard altitude of 17,000 ft, the temperature, pressure, and density ratios are 0.88321, 0.52060, and 0.58948, respectively. Using Equations (5.45), (5.46), and (5.47), respectively, the temperature, pressure, and density are calculated as

$$T = \theta T_{SSL} = 0.88321(519°R) = 458.4°R$$

$$p = \delta p_{SSL} = 0.52060 \left(2,116 \frac{lb}{ft^2} \right) = 1101.6 \frac{lb}{ft^2}$$

$$\rho = \sigma \rho_{SSL} = 0.58948 \left(0.002377 \frac{slug}{ft^3} \right) = 0.001401 \frac{slug}{ft^3}$$

Example 5.6 Temperature, Pressure, and Density Altitudes Using Inconsistent Units *The results of previous example are used to illustrate the correct and incorrect use of inconsistent units.*

Solution

Using the temperature result from Example 5.5, in consistent units, the temperature ratio is

$$\theta \equiv \frac{T}{T_{SSL}} = \frac{458.4°R}{519°R} = 0.8832$$

which matches the value found in Appendix C. Now, calculating the temperature ratio using inconsistent temperature units of Fahrenheit, we have

$$\theta \equiv \frac{T}{T_{SSL}} = \frac{458.4°R - 459}{519°R - 459} = \frac{-0.6°F}{60°F} = -0.01$$

which produces an erroneous result.

Using the pressure result from Example 5.5, in consistent units, the pressure ratio is

$$\delta \equiv \frac{p}{p_{SSL}} = \frac{1,101.6 \frac{lb}{ft^2}}{2,116 \frac{lb}{ft^2}} = 0.5206$$

which matches the value found in Appendix C. Now, calculating the pressure ratio using inconsistent pressure units of lb/in^2, we have

$$\delta \equiv \frac{p}{p_{SSL}} = \frac{1,101.6 \frac{lb}{ft^2} \times \frac{1\,ft^2}{144\,in^2}}{2,116 \frac{lb}{ft^2} \times \frac{1\,ft^2}{144\,in^2}} = \frac{7.65 \frac{lb}{in^2}}{14.69 \frac{lb}{in^2}} = 0.5208$$

which is a correct result (within round off error), since the conversion involved multiplication that cancels out.

Using the density result from Example 5.5, in consistent units, the density is

$$\sigma \equiv \frac{\rho}{\rho_{SSL}} = \frac{0.001401 \frac{slug}{ft^3}}{0.002377 \frac{slug}{ft^3}} = 0.5894$$

which matches the value found in Appendix C. Now, calculating the density ratio using inconsistent density units of lb$_m$/ft^3, we have

$$\sigma \equiv \frac{\rho}{\rho_{SSL}} = \frac{0.001401 \frac{slug}{ft^3} \times \frac{32.2\,lb_m}{1\,slug}}{0.002377 \frac{slug}{ft^3} \times \frac{32.2\,lb_m}{1\,slug}} = \frac{0.04511 \frac{lb_m}{ft^3}}{0.07654 \frac{lb_m}{ft^3}} = 0.5894$$

which is a correct result (within round off error), since the conversion involved multiplication that cancels out.

5.6 Air Data System Measurements

In assessing aircraft flight performance, *air data measurements* are required to define the flight conditions and attitude of the aircraft. The flight conditions include the aircraft altitude, airspeed, and, for high-speed aircraft, the Mach number. Attitude information includes the aircraft angle-of-attack and angle-of-sideslip. Most of these parameters cannot be measured directly; they must be calculated from measurements of other flow properties. The collection of sensors used to obtain these types of air data parameters is known as an *air data system*. Sophisticated air data systems include *air data computers* that add corrections to the measurements and process the air data for use by other instruments or systems in the aircraft.

The air data system sensors may be mounted at various locations on the aircraft or they may be mounted on a single *air data boom*, as shown in Figure 5.12. Typical sensors on an air data boom include those for measuring pressure, temperatures, and flow angles. The air data boom is used to position the air data sensors in the freestream flow, away from any interference effects of the aircraft. Air data booms are typically several feet in length, but the miniaturization of sensors and electronics allows the construction of much shorter booms. Of course, the boom length must be an appropriate "fit" for the size of the aircraft. The boom is typically mounted on the nose of the aircraft to get it as far into the unobstructed, freestream flow as possible. Air data booms can be used at transonic and supersonic speeds, but the boom must be designed to handle the appropriate steady and unsteady aerodynamic loads that may be experienced at these high speeds.

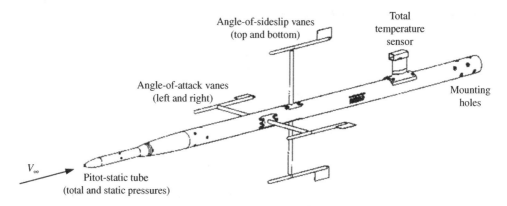

Figure 5.12   Flight test air data boom.

Altitude and airspeed are usually obtained using a *Pitot-static system*, which measures freestream pressures. To obtain Mach number, a temperature measurement is required in addition to these Pitot-static measurements. The flow angles are often measured using moving vane devices, much like a weather vane. In this section, we discuss the details of how these different measurements are obtained in a flow. We discuss how a Pitot-static system works, how altitude and airspeed are measured, the different types of airspeed, and the errors associated with the measurements of altitude and airspeed. We will fly a flight test technique to measure the errors and calibrate a Pitot-static system of a supersonic aircraft.

5.6.1   The Pitot-Static System

The basic parts of an aircraft Pitot-static system are a Pitot tube, a static pressure port, pressure-sensing instruments, and the associated plumbing connecting the different components, as shown in Figure 5.13. In its simplest form, the Pitot tube is a straight tube, placed parallel

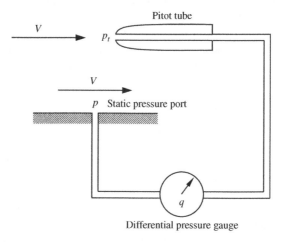

Figure 5.13   Pitot-static system.

with the freestream flow, with one end open and the other end closed. The freestream flow, with velocity V_∞, enters the open end of the tube and is brought to rest at the closed end, thus providing a measurement of the flow total pressure, p_t.

The static port is simply a flush orifice or hole on a surface that is parallel to the flow. Unlike the Pitot tube, which brings the flow to zero velocity, the static port senses only the random motion of the gas molecules of the moving fluid, that is, the flow static pressure, p. Sometimes the static pressure port is combined with the Pitot tube, where the static pressure is measured on the side of the tube, in a device called a Pitot-static probe. In other installations, the static pressure port is located elsewhere on the aircraft, often on the side of the fuselage. The location of the static pressure orifice is selected to minimize the measurement error, but there is always an error that must be corrected. Later, we examine the flight test techniques used to find this static pressure error correction.

The measured total and static pressures are fed to a mechanical or digital system that converts the measurements into an altitude or airspeed. Only the static pressure measurement is required for altitude and both the static and total measurements are required for airspeed. The calculation of the airspeed, from the measured pressures, differs depending on the Mach number regime. Obtaining altitude from static pressure is somewhat more straightforward, although corrections must typically be applied for different Mach number regimes of flight.

5.6.2   Measurement of Altitude

The accurate measurement of altitude is essential for the safe and efficient operation of aircraft during normal flight operations and during flight test operations. Accurate altitude information is required for terrain and obstacle clearance and for vertical separation of aircraft flying in the same airspace. For flight test, accurate altitude data is critical for the analysis and comparison of test data.

Various types of altitude were discussed in Section 5.2 and are summarized in Table 5.2. In this section, we discuss several of the techniques and devices that may be used to measure these different types of altitudes. In particular, we discuss the most common altitude-measuring instrument, the altimeter.

The height above the local elevation or ground level, h_{AGL}, may be measured directly using radar. In its simplest form, a radar altimeter works by measuring the time that it takes for radio signals, transmitted by the aircraft, to reflect from the surface back to the aircraft. By knowing the time and the speed of the radio signals (the speed of light), the distance can be calculated. Modern-day radar altimeters use a more complicated and more accurate method, measuring the frequency shift of the radio waves to deduce the distance. This type of height measurement is used in terrain warning systems, to warn of flying too low to the ground, or in terrain-following systems, where it is desired to fly very close to the ground.

The geometric altitude, h_g, can be obtained from the radar altitude, assuming the local elevation is known. A more direct measurement of geometric altitude is provided by the global positioning system (GPS). GPS is a satellite-based system, which measures the time for radio signals to be received from several satellites to calculate a spatial position. Data must be received by four or more satellites for a GPS receiver to determine a three-dimensional position (latitude, longitude, and altitude). Altitude is determined to an accuracy of within about 3–5 m (10–16 ft) using differential GPS (DGPS), where the satellite GPS data is corrected using ground-based systems.

Since geometric altitude is based on acceleration due to gravity that varies with altitude, one could theoretically measure geometric altitude using a highly sensitive accelerometer. This is not practical for aircraft applications, especially since the aircraft accelerations affect such an instrument.

The pressure, temperature, and density altitudes, h_p, h_T, and h_ρ, respectively, could be obtained by measuring atmospheric properties and using the relationships derived for the standard atmosphere. For example, the temperature altitude could be obtained by measuring the atmospheric temperature and using Equation (5.38) to solve for the altitude. In performance testing, we are interested in the forces acting on the aircraft. The forces are a function of the air density, through the dynamic pressure, so the use of density altitude may seem like the best choice. In reality, any of these properties may be used to determine the altitude, since the perfect gas equation of state is used to obtain the density. Unfortunately, the use of temperature or density altitude leads to inaccurate results due to many factors associated with non-standard conditions, including non-standard lapse rates, temperature inversions, or changes in atmospheric properties due to seasons, the day–night cycle, and geographic location.

The atmospheric pressure is least affected by these many issues and hence the pressure altitude, h_p, is the best choice for use in an altitude-measuring instrument. This instrument, known as an *altimeter*, senses the ambient, atmospheric pressure and uses the standard atmosphere relationships, Equations (5.31) and (5.40), to determine the pressure altitude. The functional components of an altimeter are shown in Figure 5.14. The altimeter instrument case is connected to the static pressure source of the Pitot-static system, so that the case internal pressure is equal to the static pressure, p. Inside the case, there is a metal bellows, or more precisely, a series of aneroid wafers or diaphragms, which are evacuated so that the pressure inside the bellows is near zero. (*Aneroid* is an adjective that refers to a device that operates due to the effect of ambient air pressure on a diaphragm.) The bellows expands or contracts due to the static pressure inside the case. The bellows are connected to mechanical gears that move pointers or needles, due to the expansion or contraction of the bellows, which indicate the altitude on the altimeter indicator face. The mechanical gears inside the altimeter are designed to translate the sensed pressure into an altitude according to the standard atmosphere relations, Equations (5.31) and (5.40).

The base pressure at sea level in Equation (5.40) can be mechanically adjusted in the altimeter, setting the value to the local, ambient sea level barometric pressure of the day. This adjustment is made by turning the barometric scale adjustment knob (Figure 5.14), setting the desired pressure on an indicator called the Kollsman window which shows the atmospheric pressure in units of inches

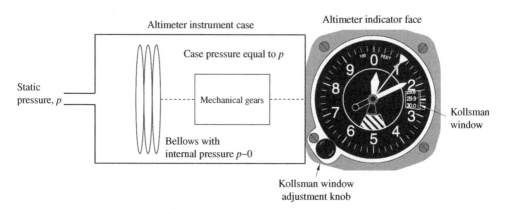

Figure 5.14 Schematic of an altimeter. (The small needle on the indicator face displays thousands of feet and the longer needle shows hundreds of feet. The thin, outer pointer indicates tens of thousands of feet. The altimeter is showing an altitude of 10,180 ft. The Kollsman window is set at 29.92 in Hg.).

of mercury (in Hg). When the pressure in the Kollsman window is set to the local atmospheric pressure, the altimeter reads the local elevation. If the pressure in the Kollsman window is set to a value of 29.92 in Hg (2116.4 lb/ft^2, 101,325 N/m^2), the altimeter indicates the (standard day) pressure altitude.

The altimeters of all aircraft flying at or above an altitude of 18,000 ft (5490 m) are set to 29.92 in Hg, so that they read pressure altitude. In this way, all aircraft, flying at these high altitudes, are using the same basis or datum plane for altitude, which helps to ensure vertical separation between aircraft flying in the same airspace. In addition, aircraft that fly at these high altitudes are typically flying at high speed, and resetting altimeters to local barometric pressures would be required every few minutes, which would be impractical.

For flight test, altimeters are typically set to 29.92 in Hg so that they display pressure altitude. This allows for the standardization of flight test data collected at different atmospheric conditions. Performance flight test data is collected at different geographical locations, at different times of the day and different seasons of the year, under a variety of atmospheric conditions. The test day conditions may be far from what we have defined as standard day atmospheric conditions. The forces acting on the aircraft, such as the drag, are a function of the air density, which can vary considerably from non-standard conditions. In addition, the propulsion system thrust is affected by changes in the air density.

To provide a means of comparing performance results, the test day data is reduced to standard day conditions. For tests at standard or non-standard conditions, the test day pressure corresponds to a unique pressure altitude. Tests performed at different times or different locations can always be flown at the same pressure altitude. Since the performance data is collected at the same pressure altitude, the test day data can then be corrected for non-standard, test day temperature to obtain data at standard pressure and temperature conditions. These non-standard temperature corrections are applied directly to the flight test data, not to the pressure altitude. To relate the pressure altitude to the geopotential altitude would require integration of Equation (5.28) with the test day variation of temperature versus altitude. While a temperature survey of the atmosphere could be obtained, it is not practical to do this every time a test is conducted.

5.6.3 Measurement of Airspeed

The accurate measurement of airspeed is critical for the safe and efficient flight of aircraft in normal operations and in the conduct of flight testing. Accurate knowledge of the airspeed is required at the low and high speed limits of an aircraft, to avoid stall and potential loss of control, at the low speed limit, and to prevent exceeding high speed limits where catastrophic structural damage could occur. In addition to the flight operations aspects, accurate airspeed information is essential for the analysis and comparison of flight test data. The measurement of aircraft airspeed has been an evolution from very simple, approximate techniques to very accurate methods based on the Pitot-static system.

In the very early days of aviation, airplane wings were structurally braced with wires. These wires vibrated in the windstream, producing a sound that varied in pitch as the vibration changed with airspeed. This enabled a crude method of airspeed measurement, where a pilot, in an open cockpit airplane, used his "calibrated ear" to listen to the pitch of the wires to estimate the airspeed. This measurement technique was not very accurate and soon became impractical as aircraft performance increased. In addition, wires for structural support and open cockpits soon disappeared as aircraft became more advanced.

While altitude measurement requires only the sensing of the freestream static pressure, the measurement of airspeed requires the sensing of both the static pressure and the total pressure. Components of the airspeed indicator are similar to the altimeter, with a metal bellows that is

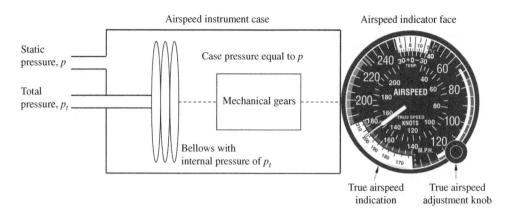

Figure 5.15 Schematic of airspeed indicator. (The instrument is indicating an airspeed of 176 mph or 152 knots, and a true airspeed of 202 knots.)

connected to mechanical gears that move an indicator needle on the instrument face, as shown in Figure 5.15. The static pressure source is connected to the instrument case, so that the inside of the instrument is at the ambient static pressure, p. The total pressure source is plumbed to the bellows, so that the bellows internal pressure is equal to the total pressure, p_t. Hence, the bellows senses the *difference* between the total and static pressure, Δp, given by

$$\Delta p = p_t - p \tag{5.49}$$

The expansion or contraction of the bellows, due to the pressure difference, Δp, results in the airspeed indication through the movement of the mechanical gears and levers. The airspeed displayed on the instrument is simply called the *indicated airspeed*. The airspeed indicator can also provide the *true airspeed*, the airspeed of the aircraft relative to the air mass through which the aircraft is flying. The different types of airspeed are explained in detail in later sections.

Even though the same two pressures are sensed, the calculation of the airspeed differs considerably depending on the flow Mach number, that is, whether the flow is incompressible, subsonic and compressible, or supersonic. We have already laid the groundwork for defining the airspeed in these different flight regimes by developing equations that relate the flow thermodynamic properties (pressure, temperature, or density) to the flow velocity or Mach number. We start with the low speed regime of subsonic, incompressible flow.

5.6.3.1 Subsonic, Incompressible Flow

As discussed in Section 3.6.2, the governing equation for isentropic, incompressible flow is Bernoulli's equation, Equation (3.180). Applying this equation to the Pitot-static system in Figure 5.13, we have

$$p + \frac{1}{2}\rho V^2 = p + q = p_t \tag{5.50}$$

where p is the freestream static pressure, ρ is the freestream density, V is the freestream velocity, q is the freestream dynamic pressure, and p_t is the total pressure. Solving Equation (5.50) for the velocity, we have

$$V = \sqrt{2\left(\frac{p_t - p}{\rho}\right)} = \sqrt{2\left(\frac{\Delta p}{\rho}\right)} \tag{5.51}$$

Using Equation (5.51), the airspeed is calculated for an isentropic, incompressible flow by measuring the pressure difference, Δp, and the density, ρ. An airspeed indicator, as shown in Figure 5.15, with a single bellows unit, can provide the pressure difference measurement.

If the density in Equation (5.51) is the freestream air density, the velocity given by Equation is defined as the *true airspeed*, V_t, given by

$$V_t = \sqrt{2\left(\frac{p_t - p}{\rho_\infty}\right)} = \sqrt{2\left(\frac{\Delta p}{\rho_\infty}\right)} \tag{5.52}$$

where ρ_∞ is used to emphasize that the density is the freestream density. Obtaining the freestream density requires an additional measurement of the freestream temperature, from which the density could be calculated assuming the perfect gas equation of state. This temperature measurement is possible with a separate temperature probe and the addition of another bellows unit in the airspeed instrument case.

The applicability of Equation (5.51) is limited to flows with Mach numbers below about 0.3 or airspeeds less than about 200 mph ($\sim$300 ft/s, $\sim$100 m/s). We now move on to the airspeed measurements for higher speed flows.

5.6.3.2 Subsonic, Compressible Flow

As the flow Mach number increases above about Mach 0.3, the constant density assumption is not valid. The assumption of isentropic flow is still valid, as long as there are no shock waves present. The relationships between static pressure, total pressure, and Mach number were developed in Section 3.11.1, where the total-to-static pressure ratio is

$$\frac{p_t}{p} = \left[1 + \left(\frac{\gamma - 1}{2}\right)M^2\right]^{\gamma/(\gamma-1)} \tag{5.53}$$

Solving Equation (5.53) for the Mach number, we have

$$M = \sqrt{\frac{2}{\gamma - 1}\left[\left(\frac{p_t}{p}\right)^{(\gamma-1)/\gamma} - 1\right]} \tag{5.54}$$

or, in terms of the pressure, difference, Δp, we have

$$M = \sqrt{\frac{2}{\gamma - 1}\left[\left(\frac{\Delta p}{p} + 1\right)^{(\gamma-1)/\gamma} - 1\right]} \tag{5.55}$$

The true airspeed, V_t, can be related to the Mach number by

$$V_t = Ma_\infty = M\sqrt{\gamma R T_\infty} = M\sqrt{\frac{\gamma p}{\rho_\infty}} \tag{5.56}$$

where a_∞, T_∞, and ρ_∞ are the freestream values of the speed of sound, temperature, and density, respectively. Inserting Equation (5.54) into (5.56), we have

$$V_t = a_\infty\sqrt{\frac{2}{\gamma - 1}\left[\left(\frac{p_t}{p}\right)^{(\gamma-1)/\gamma} - 1\right]} = \sqrt{\frac{2\gamma p}{\rho_\infty(\gamma - 1)}\left[\left(\frac{p_t}{p}\right)^{(\gamma-1)/\gamma} - 1\right]} \tag{5.57}$$

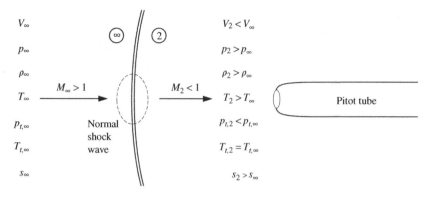

Figure 5.16 Pitot tube in supersonic flow.

Rearranging the total-to-static pressure ratio term, we have

$$V_t = \sqrt{\frac{2\gamma}{(\gamma - 1)} \left(\frac{p}{\rho_\infty}\right) \left[\left(\frac{p_t - p}{p} + 1\right)^{(\gamma-1)/\gamma} - 1\right]}$$

$$V_t = \sqrt{\frac{2\gamma}{(\gamma - 1)} \left(\frac{p}{\rho_\infty}\right) \left[\left(\frac{\Delta p}{p} + 1\right)^{(\gamma-1)/\gamma} - 1\right]} \tag{5.58}$$

The true velocity for an isentropic, compressible flow is a function of the freestream density, ρ_∞, the freestream static pressure, p, and the pressure difference, Δp. An airspeed instrument, as shown in Figure 5.15, can only measure the pressure difference and does not provide a measurement of the static pressure itself. Therefore, to measure the static pressure, a second, independent bellows unit is required inside the instrument. To obtain the freestream density, a third independent bellows unit and a separate temperature probe are required to measure the freestream temperature. Thus, to obtain the true velocity in a subsonic, compressible flow, three independent bellows units and a separate temperature probe are required to measure the pressure difference, the static pressure, and the temperature (to obtain the density).

Equation (5.58) must be used instead of Equation (5.52), when the flow is compressible, at Mach numbers greater than about 0.3. However, Equation (5.58) is based on the assumption of isentropic flow, so it is valid only up to high subsonic Mach numbers before shock waves start to appear in the flow. We now address airspeed measurement in supersonic flow with shock waves.

5.6.3.3 Supersonic, Compressible Flow

When the flow becomes supersonic, shock waves are present in the flow, as discussed in Section 3.11.2. Looking at Equation (5.58), determination of the airspeed requires sensing of the freestream total pressure. However, we know that the total pressure changes dramatically across shock waves. At some supersonic Mach number, a bow shock wave forms in front of the Pitot tube, as shown in Figure 5.16. Thus, the value of the total pressure, sensed by the Pitot tube, is different from the freestream value as used in Equation (5.58).

The bow shock wave ahead of the Pitot tube is a normal shock wave directly in front of the blunt nose of the tube. The change in the flow properties across this normal shock wave is shown in

Figure 5.16. There is a loss of total pressure across the normal shock wave. While the flow properties are changed in crossing the shock wave, due to the non-isentropic processes inside the shock wave, the flows upstream and downstream of the shock may be considered isentropic. Therefore, the ratio of the total-to-static pressure, behind the normal shock wave, in front of the Pitot tube, may be expressed by the isentropic relation given by Equation (3.343), as

$$\frac{p_{t,2}}{p_2} = \left[1 + \left(\frac{\gamma - 1}{2}\right) M_2^2\right]^{\gamma/(\gamma-1)} \tag{5.59}$$

where the subscript "2" denotes the properties behind the normal shock wave. Using the normal shock relations, from Section 3.11.2.2, the Mach number behind the normal shock, M_2, wave is given by

$$M_2^2 = \frac{(\gamma - 1)M_\infty^2 + 2}{2\gamma M_\infty^2 - (\gamma - 1)} \tag{5.60}$$

and the ratio of the static pressure across the normal shock, p_2/p_∞, is given by

$$\frac{p_2}{p_\infty} = \frac{2\gamma M_\infty^2 - (\gamma - 1)}{\gamma + 1} \tag{5.61}$$

Multiplying Equation (5.59) by (5.61), we have an equation for the ratio of the total pressure behind the normal shock, $p_{t,2}$, to the freestream static pressure, p_∞.

$$\frac{p_{t,2}}{p_\infty} = \left(\frac{p_{t,2}}{p_2}\right)\left(\frac{p_2}{p_\infty}\right) = \left[1 + \left(\frac{\gamma - 1}{2}\right) M_2^2\right]^{\gamma/(\gamma-1)} \left[\frac{2\gamma M_\infty^2 - (\gamma - 1)}{\gamma + 1}\right] \tag{5.62}$$

Substituting Equation (5.60) into (5.62), we have

$$\frac{p_{t,2}}{p_\infty} = \left\{1 + \left(\frac{\gamma - 1}{2}\right)\left[\frac{(\gamma - 1)M_\infty^2 + 2}{2\gamma M_\infty^2 - (\gamma - 1)}\right]\right\}^{\gamma/(\gamma-1)} \left[\frac{2\gamma M_\infty^2 - (\gamma - 1)}{\gamma + 1}\right]$$

$$\frac{p_{t,2}}{p_\infty} = \left\{1 + \left[\frac{(\gamma - 1)^2 M_\infty^2 + 2(\gamma - 1)}{4\gamma M_\infty^2 - 2(\gamma - 1)}\right]\right\}^{\gamma/(\gamma-1)} \left(\frac{1 - \gamma + 2\gamma M_\infty^2}{\gamma + 1}\right)$$

$$\frac{p_{t,2}}{p_\infty} = \left[\frac{4\gamma M_\infty^2 - 2(\gamma - 1) + (\gamma - 1)^2 M_\infty^2 + 2(\gamma - 1)}{4\gamma M_\infty^2 - 2(\gamma - 1)}\right]^{\gamma/(\gamma-1)} \left(\frac{1 - \gamma + 2\gamma M_\infty^2}{\gamma + 1}\right)$$

$$\frac{p_{t,2}}{p_\infty} = \left[\frac{4\gamma M_\infty^2 + \left(\gamma^2 - 2\gamma + 1\right) M_\infty^2}{4\gamma M_\infty^2 - 2(\gamma - 1)}\right]^{\gamma/(\gamma-1)} \left(\frac{1 - \gamma + 2\gamma M_\infty^2}{\gamma + 1}\right)$$

$$\frac{p_{t,2}}{p_\infty} = \left[\frac{\left(\gamma^2 + 2\gamma + 1\right) M_\infty^2}{4\gamma M_\infty^2 - 2(\gamma - 1)}\right]^{\gamma/(\gamma-1)} \left(\frac{1 - \gamma + 2\gamma M_\infty^2}{\gamma + 1}\right)$$

$$\frac{p_{t,2}}{p_\infty} = \left[\frac{(\gamma + 1)^2 M_\infty^2}{4\gamma M_\infty^2 - 2(\gamma - 1)}\right]^{\gamma/(\gamma-1)} \left(\frac{1 - \gamma + 2\gamma M_\infty^2}{\gamma + 1}\right) \tag{5.63}$$

Equation (5.63) is the *Rayleigh–Pitot tube formula*, which relates the total pressure behind the normal shock, $p_{t,2}$, to the freestream static pressure, p_∞, in terms of the freestream Mach number,

M_∞. In a supersonic flow, the total pressure behind the normal shock wave is measured by the Pitot tube. The static pressure is also measured at some location on the aircraft and must be corrected to provide the value of the freestream static pressure.

Writing Equation (5.63) in terms of the pressure, difference, Δp, we have

$$\frac{p_{t,2} - p_\infty}{p_\infty} = \frac{\Delta p}{p_\infty} = \left[\frac{(\gamma + 1)^2 M_\infty^2}{4\gamma M_\infty^2 - 2(\gamma - 1)}\right]^{\gamma/(\gamma-1)} \left(\frac{1 - \gamma + 2\gamma M_\infty^2}{\gamma + 1}\right) - 1 \tag{5.64}$$

Thus, to obtain the freestream Mach number, measurements of the pressure difference, Δp, and the freestream static pressure, p_∞, are required. An instrument, known as a *Mach meter*, with two independent bellows units, can display the freestream Mach number using Equation (5.64). However, the Mach number is a function of temperature and there is no temperature term in Equation (5.64). Obtaining the Mach number from Equation (5.64) inherently assumes that the temperature is the standard temperature. Therefore, this *indicated Mach number*, as it is called, is not accurate if the freestream temperature is different from standard. To obtain *true Mach number*, a temperature measurement is required to correct for non-standard temperature errors.

Substituting Equation (5.56) into (5.64), we obtain a relation in terms of the true airspeed.

$$\frac{\Delta p}{p_\infty} = \left[\frac{(\gamma + 1)^2 \left(\frac{V_t}{a_\infty}\right)^2}{4\gamma \left(\frac{V_t}{a_\infty}\right)^2 - 2(\gamma - 1)}\right]^{\gamma/(\gamma-1)} \left[\frac{1 - \gamma + 2\gamma \left(\frac{V_t}{a_\infty}\right)^2}{\gamma + 1}\right] - 1 \tag{5.65}$$

Expressing the freestream speed of sound in terms of the freestream static pressure and density through Equation (5.56), we have

$$\frac{\Delta p}{p_\infty} = \left\{\left[\frac{(\gamma + 1)^2 \left(\frac{V_t}{\sqrt{\gamma p/\rho}}\right)^2}{4\gamma \left(\frac{V_t}{\sqrt{\gamma p/\rho}}\right)^2 - 2(\gamma - 1)}\right]^{\gamma/(\gamma-1)} \left[\frac{1 - \gamma + 2\gamma \left(\frac{V_t}{\sqrt{\gamma p/\rho}}\right)^2}{\gamma + 1}\right]\right\} - 1 \tag{5.66}$$

Equation (5.66) is the desired relationships for determining true airspeed in a supersonic flow. With measured values for the density, ρ, the freestream static pressure, p, and the pressure difference, Δp, this equation can be solved iteratively for the true airspeed. As with subsonic, compressible flow, three independent bellows units, inside the airspeed instrument, and a separate temperature probe are required to obtain the true airspeed in a supersonic flow.

5.6.4 Types of Airspeed

Thus far, we have introduced the true airspeed and discussed its measurement for different flight regimes. In this section, other types of airspeed are defined, which are frequently used in aerospace engineering and flight testing. These other types of airspeed include equivalent, calibrated, and indicated airspeed.

5.6.4.1 True Airspeed

The true airspeed, V_t, relationships that have already been developed, for the various regimes of flight, are summarized in Table 5.4. For all of the flight regimes, the true velocity is a function of the freestream density, ρ, the freestream static pressure, p, and the pressure difference,

Table 5.4 Summary of equations for true velocity.

Flight regime	Equation for true airspeed, V_t
Subsonic, incompressible	$V_t = \sqrt{2\left(\dfrac{p_t - p}{\rho}\right)}$
Subsonic, compressible	$V_t = \sqrt{\dfrac{2\gamma}{(\gamma - 1)}\left(\dfrac{p}{\rho}\right)\left[\left(\dfrac{\Delta p}{p} + 1\right)^{(\gamma-1)/\gamma} - 1\right]}$
Supersonic, compressible	$\dfrac{\Delta p}{p_\infty} = \left\{\left[\dfrac{(\gamma+1)^2\left(\dfrac{V_t}{\sqrt{\gamma p/\rho}}\right)^2}{4\gamma\left(\dfrac{V_t}{\sqrt{\gamma p/\rho}}\right)^2 - 2(\gamma-1)}\right]^{\gamma/(\gamma-1)}\left[\dfrac{1 - \gamma + 2\gamma\left(\dfrac{V_t}{\sqrt{\gamma p/\rho}}\right)^2}{\gamma + 1}\right]\right\} - 1$

$\Delta p = p_t - p$. (For simplicity, the freestream subscript "∞" has been omitted from the parameters in Table 5.4.).

Assuming that these three parameters are measured, a true airspeed indicator can be built, based on these true airspeed equations. Measurements of the static pressure and the pressure difference are obtained from a Pitot-static system. Obtaining the freestream density is more difficult, typically requiring the measurement of the freestream temperature. Despite these difficulties, true airspeed indicators have been fabricated and used in airplanes in the past. However, true airspeed indicators tend to be mechanically complex, difficult to calibrate, and have had reliability and accuracy issues. Fortunately, true airspeed is usually not required in flight, except perhaps as an aid in navigation for the determination of the ground speed. This is even less of an issue today with the easy availability of ground speed from global positioning system (GPS) technology.

5.6.4.2 Equivalent Airspeed

As a first step in simplifying the measurement requirements, the dependence of the true airspeed on the freestream density can be removed by assuming that the density is equal to the constant, standard sea level density, ρ_{SSL}. Applying this to the true velocity equation for subsonic, compressible flow, Equation (5.58), we define a new airspeed, the *equivalent airspeed*, V_e, as

$$V_e = \sqrt{\frac{2\gamma}{(\gamma - 1)}\left(\frac{p}{\rho_{SSL}}\right)\left[\left(\frac{\Delta p}{p} + 1\right)^{(\gamma-1)/\gamma} - 1\right]} \qquad (5.67)$$

By comparing Equations (5.58) and (5.67), the true and the equivalent airspeeds are related by the square root of the density ratio, σ, given by

$$V_t = \sqrt{\frac{\rho_{SSL}}{\rho}}\, V_e = \frac{V_e}{\sqrt{\sigma}} \qquad (5.68)$$

Hence, the true airspeed is equal to the equivalent airspeed corrected for non-standard, sea level density. Since the density ratio is usually less than one, the equivalent airspeed is usually less than the true airspeed. At standard sea level conditions, the equivalent and true airspeeds are equal. The obvious advantage in using the equivalent airspeed is that a measurement of density or temperature is no longer required.

Equivalent airspeed is particularly useful when dealing with flight structures, where the structural loads scale with constant dynamic pressure, q. Using Equation (5.68), the dynamic pressure may be written as

$$q = \frac{1}{2}\rho V_t^2 = \frac{1}{2}\rho_{SSL} V_e^2 \tag{5.69}$$

Thus, we see that a constant dynamic pressure equates with a constant equivalent airspeed. Therefore, constant dynamic pressure flight loads data can be obtained by flying at constant equivalent airspeed. According to Equation (5.69), flight at constant dynamic pressure is also independent of a specific altitude.

This altitude independence of equivalent airspeed is useful for many performance-related aircraft speeds, such as stall speed, landing approach speed, and flap limit speeds. Consider an aircraft in steady, equilibrium flight, such that the lift equals the weight, so that, using Equation (5.69), we have

$$L = W = qSC_L = \left(\frac{1}{2}\rho V_t^2\right) SC_L = \left(\frac{1}{2}\rho_{SSL} V_e^2\right) SC_L \tag{5.70}$$

Solving for the equivalent airspeed, we have

$$V_e = \sqrt{\frac{2W}{\rho_{SSL} SC_L}} \tag{5.71}$$

where the weight, W, standard sea level density, ρ_{SSL}, wing reference area, S, and the lift coefficient, C_L, are constants, making the equivalent airspeed constant for this steady, equilibrium flight condition. Therefore, a performance speed, such as the stall speed or landing approach speed, correspond to a constant equivalent airspeed, which does not vary with altitude. This greatly simplifies flying an aircraft, where critical performance speeds are the same, regardless of altitude.

Returning to our discussion about airspeed indicators, an equivalent airspeed indicator still has the disadvantage of being mechanically complex, with associated calibration, reliability, and accuracy issues. Despite this complexity, there are several aircraft that utilize equivalent airspeed indicators, most notably, the Space Shuttle and the triple-sonic SR-71 *Blackbird*. Both of these vehicles operate at higher Mach numbers where the increased accuracy of using equivalent airspeed is required.

5.6.4.3   Calibrated Airspeed

To further simplify the true airspeed equation, we set both the density and the pressure, in Equation (5.58), to standard sea level values, and define a *calibrated airspeed*, V_c, given by

$$V_c = \sqrt{\frac{2\gamma}{(\gamma - 1)}\left(\frac{p_{SSL}}{\rho_{SSL}}\right)\left[\left(\frac{\Delta p}{p_{SSL}} + 1\right)^{(\gamma-1)/\gamma} - 1\right]} \tag{5.72}$$

The only unknown in the calibrated airspeed equation is the pressure difference, Δp. In a mechanical airspeed instrument, a single bellows unit is used to measure the pressure difference Δp, making this a much simpler instrument with fewer calibration, reliability, and accuracy issues. Most airspeed instruments are designed to use the calibrated airspeed given by Equation (5.72).

By comparing Equations (5.67) and (5.72), the calibrated and equivalent airspeeds are related as

$$V_e = \frac{\sqrt{\frac{2\gamma}{(\gamma-1)}\left(\frac{p}{\rho_{SSL}}\right)\left[\left(\frac{\Delta p}{p}+1\right)^{(\gamma-1)/\gamma}-1\right]}}{\sqrt{\frac{2\gamma}{(\gamma-1)}\left(\frac{p_{SSL}}{\rho_{SSL}}\right)\left[\left(\frac{\Delta p}{p_{SSL}}+1\right)^{(\gamma-1)/\gamma}-1\right]}} V_c = fV_c \qquad (5.73)$$

where a *pressure correction factor*, f, has been defined as

$$f \equiv \frac{\sqrt{\frac{2\gamma}{(\gamma-1)}\left(\frac{p}{\rho_{SSL}}\right)\left[\left(\frac{\Delta p}{p}+1\right)^{(\gamma-1)/\gamma}-1\right]}}{\sqrt{\frac{2\gamma}{(\gamma-1)}\left(\frac{p_{SSL}}{\rho_{SSL}}\right)\left[\left(\frac{\Delta p}{p_{SSL}}+1\right)^{(\gamma-1)/\gamma}-1\right]}} = \frac{V_e}{V_c} \qquad (5.74)$$

Hence, the equivalent airspeed is calibrated airspeed, corrected for non-standard, sea level pressure. All of the variables in Equation (5.74) are constant except the static pressure, p, and the pressure difference, Δp. By setting the pressure p to the pressure altitude and using Equation (5.72) to solve for the pressure difference, Δp, a table of pressure correction factors can be created for values of calibrated airspeed as a function of the pressure altitude, as shown in Table 5.5. Thus, the pressure correction factor for a calibrated airspeed, at a given pressure altitude, can be obtained to calculate the equivalent airspeed.

A correction to the calibrated airspeed, ΔV_c, may be presented in tabular or graphical form, as a function of calibrated airspeed and pressure altitude. This correction is then simply added to the calibrated airspeed to obtain equivalent airspeed, as

$$V_e = V_c + \Delta V_c \qquad (5.75)$$

The correction to the calibrated airspeed, to obtain equivalent airspeed, is often referred to as a *compressibility correction*. This correction is independent of the aircraft type, so that the same correction applies to any aircraft.

Since the static pressure, p, is usually less than the standard sea level pressure, p_{SSL}, the pressure correction factor, f, is usually less than one and the calibrated airspeed is usually greater than the equivalent airspeed. When the pressure is equal to the standard sea level pressure, the calibrated and equivalent airspeeds are equal.

Table 5.5 Pressure correction factors, f.

Pressure altitude (ft)	Calibrated airspeed, V_c (knots)			
	100	150	200	250
0 (sea level)	1.000	1.000	1.000	1.000
5000	0.999	0.999	0.998	0.997
10,000	0.999	0.997	0.995	0.992
20,000	0.997	0.993	0.987	0.981
30,000	0.993	0.986	0.975	0.963
40,000	0.988	0.974	0.957	0.937
50,000	0.979	0.957	0.930	0.901

The difference between the calibrated and equivalent airspeeds is small at low airspeeds, such as the stall speed. This justifies the use of calibrated airspeed for performance speeds, such as takeoff and landing speeds, since these speeds are usually defined as multiples of the stall speed.

Using Equation (5.68), the equivalent airspeed can be related to the true airspeed, as

$$V_c = \frac{V_e}{f} = \frac{\sqrt{\sigma}}{f} V_t \tag{5.76}$$

It has already been established that the equivalent airspeed is usually less than the true airspeed and less than the calibrated airspeed. The square root of the density ratio divided by the pressure correction factor is usually less than one, making the calibrated airspeed usually less than the true airspeed.

5.6.4.4   Indicated Airspeed

The *indicated airspeed*, V_i, is the airspeed that is read on the dial of the airspeed indicator. Most airspeed indicators are designed to display the calibrated airspeed, using Equation (5.72). However, the indicated airspeed is not equal to the calibrated airspeed due to errors in the instrument and in the Pitot-static measurements.

Instrument errors include those due to the mechanical or electrical workings of the instrument. The instrument may have inherent errors due to manufacturing discrepancies, imperfect mechanization, magnetic fields, friction, inertia, hysteresis, and scale error. Periodic laboratory calibration of the instrument is required to correct for these errors.

The Pitot-static system errors, discussed in an upcoming section, are those due to errors in the measurements of the total or static pressures. The static pressure measurement is the primary source of error, called the *static pressure position error*. The indicated airspeed is adjusted by an instrument correction, ΔV_{instr}, and a position error correction, ΔV_{pc}, to give the corrected calibrated airspeed as

$$V_c = V_i + \Delta V_{instr} + \Delta V_{pc} \tag{5.77}$$

Ideally, the airspeed instrument is mechanically and electrically designed to produce as little error as possible and the static pressure measurement is made in a location, and using methods, that result in small errors. Airspeed measurement and indication systems are designed to produce as small an airspeed error as possible in the critical flight regimes, such as at low speeds. Typically, tables or graphs are provided in flight manuals that provide indicated airspeed corrections to calibrated airspeed.

5.6.4.5   Airspeed Conversions and Summary

Let us summarize our discussion of the various types of airspeed. If you are flying in an airplane and read the airspeed indicator in the cockpit, you are reading the *indicated airspeed*. If you want to know how fast you are flying, relative to the air mass through which the airplane is flying, you want to know your *true airspeed*. If you want to know how close you are to the stall speed or what airspeed to fly for your landing approach, you may want to use the *equivalent airspeed*, since it does not change with altitude. However, most airspeed indicators are designed to display the *calibrated airspeed*. Fortunately, the calibrated airspeed is usually close to the equivalent airspeed, at low airspeed, critical regimes of flight. A summary of the different types of airspeed is given in Table 5.6.

Table 5.6 Summary of types of airspeed.

Airspeed	Symbol	Description
Indicated airspeed	V_i	Airspeed read off airspeed indicator
Calibrated airspeed	V_c	Indicated airspeed corrected for instrument and position error
Equivalent airspeed	V_e	Calibrated airspeed corrected for non-standard, sea level pressure
True airspeed	V_t	Equivalent airspeed corrected for non-standard, sea level density, airspeed relative to air mass

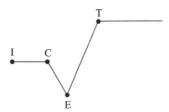

Figure 5.17   ICE-T memory aid for relative magnitudes of types of airspeed.

Figure 5.17 is a memory aid for the relative magnitudes of the different airspeeds and provides a typical order of conversion from one airspeed to another (shown in the following example problem). The acronym "ICE-T" gives the airspeed conversion order for indicated, calibrated, equivalent, and true airspeed. The location of each letter on the square root symbol gives the relative magnitude of each airspeed type. This tells us that the indicated and calibrated airspeeds are of near equal magnitude, that the equivalent airspeed is smaller than the indicated and calibrated airspeeds, and that the true airspeed is larger than the other airspeeds. Finally, the square root symbol is also a reminder that it is necessary to take a square root of the density ratio to convert from equivalent to true airspeed.

Example 5.7 Airspeed Conversions *An F-16 aircraft is flying at an altitude of 10,000 ft with an indicated airspeed of 250 KIAS. Assuming standard atmospheric conditions, calculate the calibrated, equivalent, and true airspeeds. Assume that the instrument error, ΔV_{instr}, is −0.25 knots and the position error, ΔV_{pc}, is +0.85 knots.*

Solution

Using Equation (5.77), the calibrated airspeed is

$$V_c = V_i + \Delta V_{instr} + \Delta V_{pc} = 250\,kt - 0.25\,kt + 0.85\,kt = 250.6\,kt$$

From Table 5.5, for a pressure altitude of 10,000 ft and a calibrated airspeed of 250.6 kt ≅ 250 kt, the pressure correction factor, f, is 0.992. Using Equation (5.73), the equivalent airspeed is

$$V_e = fV_c = 0.992(250.6\,kt) = 248.6\,kt$$

From Appendix C, the density ratio, σ, at 10,000 ft is 0.73860. Using Equation (5.76), the true airspeed is

$$V_t = \frac{V_e}{\sqrt{\sigma}} = \frac{248.6\,kt}{\sqrt{0.73860}} = 289.3\,kt$$

5.6.5   Pitot-Static System Errors

The basic Pitot-static system is typically composed of a total pressure probe and a static pressure orifice, which are plumbed to pressure sensing instruments with pneumatic tubing. Ideally, this system accurately measures the freestream total and static pressures. In reality, the measured pressures are different from their freestream values due to errors associated with the Pitot-static system. Sources of Pitot-static system errors include the instruments (*instrument error*), the pressure tubing (*pressure lag*), and the location of the pressure measurements (*position error*).

We have discussed the *instrument error* related to the mechanical and electrical workings of the airspeed instrument. These types of instrument error are also applicable to other Pitot-static instruments, such as the altimeter. Laboratory or on-aircraft ground calibrations are typically used to quantify the instrument error.

Since the instruments are connected to the Pitot and static ports by a finite length of pneumatic tubing, there is an error associated with the lag in pressure propagation from the pressure port to the instrument through the tubing. Physically, this lag is due to the friction and inertia of the air in the tubing, the finite volume of air in the system that must be filled, and the finite speed of propagation of the pressure waves. The *lag error* is more significant during flight conditions with large or fast pressure changes, such as climbing, descending, accelerating, and decelerating flight. Since the lag error is dependent on the specific installation in an aircraft, the lag error is usually obtained from an on-aircraft ground calibration.

Finally, there are *position errors* due to the physical locations and installations of the total and static pressure sensing ports. The *total pressure position error* is usually small and assumed to be zero. The *static pressure position error* is usually the primary source of position error in a Pitot-static system. We discuss both of these types of position error in more detail below. A wealth of information and data related to Pitot-static system related measurements and errors can be found in [9] and [24].

5.6.5.1   Total Pressure Position Error

The total pressure is typically measured using a Pitot tube. In general, it is much less difficult to obtain an accurate measurement of the total pressure than of the freestream static pressure. The primary source of error in measuring the total pressure is due to flow angularity, that is, when the flow is at an angle with respect to the tube. The magnitude of the error is highly dependent on the geometry of the Pitot tube.

If the Pitot tube is aligned with the flow direction in subsonic flow, the total pressure measurement is, for all practical purposes, independent of the probe geometry. Almost any open-ended tube provides an accurate measurement of the total pressure. This assumption is valid as long as the tube is not located in a region of the flow that is not representative of the freestream total pressure, such as the boundary layer, wing wake, propeller wash, or engine exhaust. At high subsonic or transonic speeds, the Pitot tube should not be located in sonic regions where shock waves form. In supersonic flow, the Pitot tube should be located upstream of the bow shock waves formed by the aircraft. There is still a normal shock wave in front of the Pitot tube in supersonic flow, but this can be properly accounted for, as described in Section 5.6.3.3. Usually, it is not difficult to locate a Pitot tube in a location that satisfies all of the requirements for the different flow regimes. Typical locations that provide accurate measurements include ahead of the fuselage, wing, or vertical fin, mounted on a short boom or strut.

If the Pitot tube is not aligned with the flow, the total pressure starts to decrease at some flow angle, which is dependent on the shape of the tube nose and the size of the tube opening relative to the tube frontal area. The range of flow angles through which the Pitot tube measures the total pressure, to some defined level of accuracy (usually 1%), is called the *range of insensitivity*. Early Pitot tubes had hemispherical nose shapes with a small opening. These types of tubes have a range of insensitivity of only about ±5°, when only total pressure is measured.

From 1951 to 1953, the NACA conducted a series of wind tunnel tests on a variety of Pitot tube geometries [10]. The tests were performed in five different wind tunnels, at subsonic, transonic, and supersonic Mach numbers from 0.26 to 2.40 and at angles-of-attack up to 67°. The tests varied several geometric parameters of the Pitot tubes, including the tube opening, relative to the tube frontal area, the tube internal entry shape, and the tube nose shape. From the NACA test results, it is reasonable to assume that practical Pitot tube geometries can be obtained, which provide a range of insensitivity of close to ±20° in angle-of-attack or sideslip. For these flight conditions, the total pressure position error is very small and is usually assumed to be zero.

For high angles-of-attack or sideslip, corrections to the total pressure measurement must be made or special types of Pitot probes are used, designed for high flow angles. The *swivel-head probe* has weather vane type fins that allow it to pivot or swivel, aligning the probe with the flow direction. While this type of probe can be useful for subsonic flight test work, it is not practical for operational use. A non-moving, fixed probe, designed by the German aerodynamicist G. Kiel in 1935, has the total pressure tube placed inside another venturi-like tube, which acts as a shield. The Kiel tube has a range of insensitivity of greater than ±40° at subsonic speeds. It has been used successfully for high angle-of-attack flight testing.

5.6.5.2 Static Pressure Position Error

The static pressure position error is usually the major source of error in the Pitot-static system. This error is primarily due to the location of the static pressure port and secondarily to the orifice size and edge shape of the hole. Usually, the errors due to the secondary factors can be kept very small, assuming appropriate design guidelines are followed. The smallest error is obtained by using a small, round hole with clean, sharp edges, free from burrs, damage, or deformation.

The pressure field around an aircraft in flight varies with the Mach number and the lift coefficient (or angle-of-attack). A typical subsonic pressure distribution on an aircraft fuselage, at zero angle-of-sideslip, is shown in Figure 5.18, with the local pressure coefficient, C_p, defined as

$$C_p = \frac{\Delta p}{q_c} = \frac{p - p_\infty}{p_t - p_\infty} \tag{5.78}$$

where p is the local pressure on the fuselage surface, along a line midway up the side of the fuselage, and q_c is the difference between the total pressure and freestream static pressure, sometimes called the *compressible q*. The parameter q_c is equal to the dynamic pressure, $q = \frac{1}{2}\rho V^2$, for incompressible flow only. It is different from the incompressible dynamic pressure as the Mach number increases and compressibility effects become important. The parameter q_c is always equal to the difference between the total pressure and the freestream static pressure.

The local pressure is greater ($C_p > 1$) or less ($C_p < 1$) than the freestream pressure at different axial locations along the fuselage. There are several locations (marked by a dashed line in Figure 5.18), where the local pressure is equal to the freestream static pressure ($C_p = 0$). These locations of zero static pressure position error are best suited for the location of a static pressure port. Often, static pressure ports are located on the left and right sides of the fuselage and manifolded together to null errors induced by sideslip. Once the static pressure port locations are determined, the static pressure position error is determined using flight test (to be described in an upcoming flight test technique).

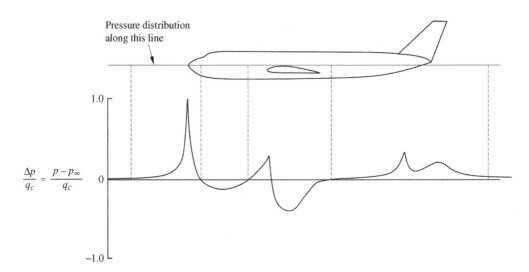

Figure 5.18 Typical subsonic static pressure distribution on aircraft fuselage. (Source: *Adapted from E.A. Haering, "Air Measurement and Calibration," NASA TM-104316, December 1995, Fig. 3.*)

The *static pressure position error*, Δp_{pc}, is defined as

$$\Delta p_{pc} = p_s - p_\infty \tag{5.79}$$

where p_s is the static pressure, measured at the sensing port, and p_∞ is the freestream static pressure.

From the static pressure position error, other position error corrections may be determined. As was given by Equation (5.77), the calibrated airspeed is equal to the indicated airspeed, V_i, corrected for instrument error, ΔV_{instr}, and position error, ΔV_{pc}, as

$$V_c \equiv V_i + \Delta V_{instr} + \Delta V_{pc} = V_{ic} + \Delta V_{pc} \tag{5.80}$$

where V_{ic} is the indicated airspeed corrected for instrument error. To be precise, the position error correction for velocity can also include the total pressure position error correction, but we are assuming that this error is zero. Therefore, the position error correction is composed solely of that due to the static pressure error.

Similarly, position error corrections can be applied to the altitude and Mach number, as

$$h_c = h_{ic} + \Delta h_{pc} \tag{5.81}$$

$$M = M_{ic} + \Delta M_{pc} \tag{5.82}$$

where h_{ic} and M_{ic} are the instrument corrected, indicated altitude and Mach number, respectively, Δh_{pc} and ΔM_{pc} are the position error corrections for the altitude and Mach number, respectively, and h_c and M are the altitude and Mach number, corrected for instrument and position errors, respectively.

Typical altitude and airspeed position error corrections for a supersonic aircraft are shown in Figure 5.19. The charts provide the position error corrections as a function of the indicated airspeed in knots (KIAS). The largest corrections are required in the transonic region, as the aircraft passes from subsonic to supersonic flight. In this speed region, the correction increases discontinuously (in positive or negative magnitude) and then reverses discontinuously with a correction in the opposite direction, of approximately the same absolute magnitude.

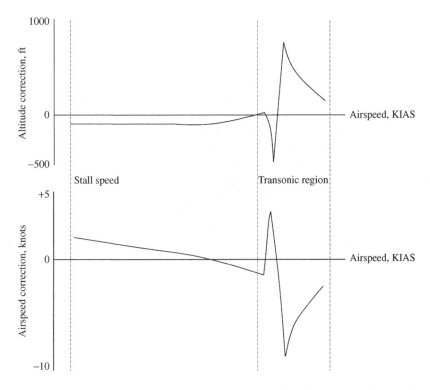

Figure 5.19   Altitude and airspeed position error corrections for a supersonic aircraft.

From Equation (5.79), the *static pressure position error coefficient*, $\Delta p_{pc}/q_C$, is defined as

$$\frac{\Delta p_{pc}}{q_C} = \frac{p_s - p_\infty}{p_t - p_\infty} \tag{5.83}$$

By non-dimensionalizing the static pressure position error by q_C, the position error curves for various altitudes collapse into a single curve.

5.6.6   Other Air Data Measurements

Other types of air data measurements are important, especially for flight testing. These include measurements of the air temperature and the flow direction, typically angle-of-attack and angle-of-sideslip. A brief discussion of these types of measurements is given below.

5.6.6.1   Temperature Measurement

Ideally, a direct measurement of the flow static temperature is desired for the calculation of the true velocity and for other research uses. Recall that the static temperature is the temperature that is measured when moving at the velocity of the flow. At first thought, one might assume that the static temperature could be measured at the wall or surface of a body, similar to the measurement of the static pressure. However, due to the viscous boundary layer, the velocity at the wall is zero, which makes the wall temperature different from the static temperature outside the boundary layer.

The static pressure measurement at the wall is valid since the pressure in the freestream, outside the boundary layer, is impressed through the boundary layer to the wall. Alternatively, the total or stagnation temperature can be readily measured, from which the static temperature can be calculated. Another benefit of measuring the total temperature is that it is not affected by the presence of shock waves. Since a shock wave is an adiabatic process, the total temperature is constant through the shock wave, making the total temperature measurement valid for both subsonic and supersonic flows.

The air total temperature is typically measured using a probe that is mounted on the skin of the aircraft fuselage or on an air data boom, as shown in Figure 5.12. A schematic of a total temperature probe is shown in Figure 5.20. The temperature is typically measured using a calibrated electrical resistance element, where the resistance is a function of temperature. The flow enters the mouth of the probe and turns a 90° angle before being brought to rest at the electrical resistance, sensing element. The flow turning provides protection of the sensing element from impingement of particles, such as dirt, sand, insects, etc. The probe is typically shielded to prevent radiative heat loss from the sensing element. The electrical power to the probe is very low to avoid heat conduction to the sensing element.

The total temperature, T_t, is related to the static temperature, T_∞, through the adiabatic relationship, Equation (3.345), as

$$\frac{T_t}{T_\infty} = 1 + \frac{(\gamma - 1)}{2}M_\infty^2 \tag{5.84}$$

Solving for the static temperature, we have

$$T_\infty = \frac{T_t}{1 + \frac{(\gamma-1)}{2}M_\infty^2} \tag{5.85}$$

where M_∞ is the freestream Mach number. In reality, not all of the flow velocity is converted to temperature, so that the probe measures a temperature, T_r, different from the actual total temperature, T_t. The difference in these temperatures is captured in a *temperature recovery factor*, r, defined as

$$r \equiv \frac{T_r - T_\infty}{T_t - T_\infty} \tag{5.86}$$

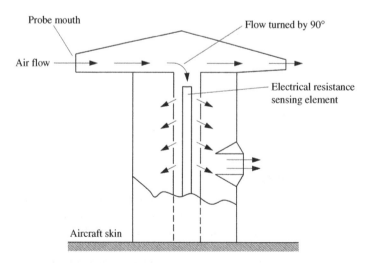

Figure 5.20 Schematic of total temperature probe. (Source: *Adapted from* [2], *Fig. 1–11.*)

For a well-designed total temperature probe, the recovery factor is close to one. Values for the recovery factor are typically obtained through calibration of the total temperature probe in a wind tunnel.

Solving Equation (5.86) for the measured temperature, T_r, we have

$$T_r = T_\infty + (T_t - T_\infty)r = rT_t + (1 - r)T_\infty \tag{5.87}$$

Dividing by the static freestream temperature, we have

$$\frac{T_r}{T_\infty} = r\frac{T_t}{T_\infty} + (1 - r) \tag{5.88}$$

Inserting Equation (5.85) into (5.88), gives

$$\frac{T_r}{T_\infty} = r\left[1 + \frac{(\gamma - 1)}{2}M_\infty^2\right] + (1 - r) = r + r\frac{(\gamma - 1)}{2}M_\infty^2 + 1 - r$$

$$\frac{T_r}{T_\infty} = 1 + r\frac{(\gamma - 1)}{2}M_\infty^2 \tag{5.89}$$

Equation (5.89) is similar to Equation (5.84), with the inclusion of the recovery factor. Solving Equation (5.89) for the static temperature, we have

$$T_\infty = \frac{T_r}{1 + r\frac{(\gamma - 1)}{2}M_\infty^2} \tag{5.90}$$

With a probe-measured value of T_r, a known recovery factor, r, and an indicated Mach number, M_∞, obtained from a Pitot-static measurement, the static temperature, T_∞, is calculated using Equation (5.90).

5.6.6.2 Flow Direction Measurement

The measurement of flow directions, typically angle-of-attack and angle-of-sideslip, are often important in flight operations and flight test. As shown in Figure 5.12, rotating vanes are used to measure the angle-of-attack and angle-of-sideslip. The vanes align themselves with the local flow direction, much like a weather vane on the ground. Vanes mounted in the aircraft x-y plane measure angle-of-attack and angle-of-sideslip is measured with vanes mounted in the x-z plane. In addition to their use on air data booms for flight test, moving vanes are often mounted on the forward fuselage, near the nose of larger aircraft in normal flight operations. Flow angle vanes typically require calibrations that are specific to the mounting location.

Another type of flow direction sensor system uses measurements of several static pressures, typically on the head of a probe or the nose of a vehicle. The pressure orifices are arranged circumferentially around the central axis of the probe or vehicle nose. Since these systems use a series of flush-mounted pressure orifices, they are called *flush air data systems* or *FADS*. Differential pressures are used to determine flow directions. Extensive wind tunnel calibrations are usually required for the particular geometry of the FADS installation.

A research FADS, using 11 pressure orifice measurements (one total pressure and 10 circumferential static pressures), on the nose of a NASA F-18, is shown in Figure 5.21. This system was used to measure the aircraft airspeed, altitude, and freestream flow directions. These types of flush air data systems are especially attractive for vehicles that cannot use conventional probes that protrude into the flow, such as hypersonic vehicles, where high heating rates would destroy conventional probes, or stealth vehicles, where conventional probes would compromise stealth characteristics.

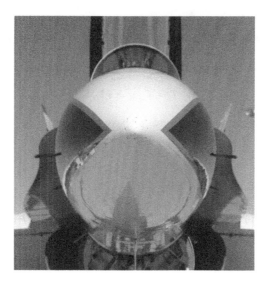

Figure 5.21   Research flush air data system (FADS) on the nose of a NASA F-18 aircraft. (Source: *NASA.*)

5.6.7   FTT: Altitude and Airspeed Calibration

This section discusses the flight test techniques (FTTs) used to obtain the static pressure position error correction. This correction to the static pressure is required for accurate altitude, airspeed, and Mach number information. The airspeed and Mach number are also a function of the total pressure; the error in this measurement is small and may be neglected, at low angles-of-attack, as discussed earlier.

The objective of a position error calibration method is to obtain the static pressure position error, $\Delta p_{pc} = p_s - p_\infty$, from which the static pressure position errors for altitude, Δh_{pc}, velocity, ΔV_{pc}, and Mach number, ΔM_{pc}, can be calculated. In some methods, the static pressure position error, Δp_{pc}, is measured directly, which provides the altitude position error. In other methods, the static pressure position error is derived from direct measurements of the velocity or Mach number position error.

According to Gracey [9], the calibration methods can be categorized, based on one of four parameters, from which the position error is derived: (1) the freestream static pressure, (2) the total temperature, (3) the true airspeed, or (4) the Mach number. For the freestream static pressure method, the static pressure, at the sensing port, p_s, is measured and the freestream static pressure, p_∞, is either measured or calculated, to obtain the static pressure position error, Δp_{pc}. For the temperature method, p_s is measured and p_∞ is derived from measurement of the total temperature, T_t, and a pressure–temperature survey. For the true airspeed method, p_s is measured and p_∞ is derived from measurements of freestream velocity, V_∞, and total temperature, T_t. The freestream Mach number, M_∞, and the sensed static pressure, p_s, are measured, to obtain Δp_{pc}, in the Mach number method.

The freestream static pressure and true airspeed methods are most commonly used in flight test. Brief descriptions of selected FTTs, using these two types of methods, are given in Table 5.7. Some of these FTTs can only be performed at low altitudes, while others are better suited for high altitudes. There may also be airspeed limitations of the various FTTs, depending on the performance capability of the aircraft at the different altitudes. Often, a combination of the FTTs must be applied to obtain a full envelope calibration of the static pressure position error of an aircraft. For some of

Figure 5.22   Northrop T-38A *Talon* supersonic trainer. (Source: *US Air Force.*)

the FTTs, the aircraft must be in steady, level flight, while for others the aircraft can be climbing, descending, accelerating, or decelerating.

The selection of which FTT should be used is a function of several parameters, including the desired accuracy of the correction, the range of altitudes and airspeeds where the correction is required, and the instrumentation available. The level of instrumentation required varies with the selected FTT. Some require modest hand-held instrumentation, while others require more sophisticated instrumentation and data acquisition systems.

For your air data calibration flight, you will fly an FTT using a freestream static pressure method, the tower fly-by FTT, and one using a true airspeed method, the ground speed course FTT. You will fly the Northrop T-38A *Talon*, shown in Figure 5.22, for these FTTs. The Northrop T-38A is a two-place, twin-turbojet, supersonic trainer used by the military. It has a small, low aspect ratio, thin wing, all-moving horizontal tail, and tandem cockpits. The first flight of the T-38A was in 1959 and it is still in military service today throughout the world. A three-view drawing of the T-38A is shown in Figure 5.23. Selected specifications of the Northrop T-38A *Talon* are given in Table 5.8.

After takeoff in the T-38, there is no reason to climb too high, since the first position error calibration FTT that you will perform is the tower fly-by, which is flown close to the ground. Leveling off at 1000 ft (305 m) AGL (above ground level), you trim the aircraft for 300 KIAS (345 mph, 556 km/h) and set 29.92 in Hg in the altimeter, so that it indicates pressure altitude. The fly-by tower is off to your right, sitting in the middle of a large dry lakebed adjacent to a broad, black stripe, "painted" with oil, which stretches for miles across the dry lakebed. You will line up on this black stripe to fly an accurate line past the tower. You descend down to 500 ft (152 m) AGL, holding 250 KIAS. You are on the base leg, perpendicular to the black fly-by line. You turn right to line up on the black fly-by line and descend down to 200 ft (61 m) AGL. When you are established on the run-in line, you descend down to your final tower fly-by altitude of 100 ft (31 m) and push the throttles up to accelerate to the first fly-by airspeed of 300 KIAS. You keep the altitude and airspeed steady, trying to hold the altitude to within ±50 ft (15 m) and the airspeed to within ±5 KIAS (6 mph, 9 km/h). You want to be sure to that you stay at a height of at least one wingspan or more above the ground, to remain out of ground effect.

Table 5.7 Selected air data calibration methods to obtain position error.

Calibration method (FTT)	Description

Freestream static pressure methods (Δp_{pc} from measurement of p_s and measurement or calculation of p_∞):

Pacer aircraft	Uncalibrated aircraft is flown in close formation with calibrated pacer aircraft with p_s and p_∞ directly measured on aircraft. Must be flown far enough apart to prevent interference effects, typically one wing span. Requires precise formation flight. Suitable for any altitudes and airspeeds within compatible capabilities of aircraft.
Trailing bomb or Trailing cone	Aircraft static pressure is compared with freestream static pressure measured on bomb-shaped body or pressure tubing, with cone-shaped drag device, suspended on long length of pressure tubing, trailing behind aircraft with p_s and p_∞ directly measured at altitude of aircraft. Trailing tube or body instabilities may be unpredictable and may limit airspeeds.
Radar altimeter	Aircraft calibrated radar altimeter measures height to obtain altimeter position error in level flight at low altitude. Sensitive pressure altimeter at ground level (aircraft altimeter on ground can be used) provides ground altitude. Radar height is added to ground level pressure altitude to provide true altitude.
Tower fly-by	Aircraft flown past fly-by tower with height above ground measured using known tower height, deviation of aircraft above or below tower reference (by sighting aircraft through an optical grid), and geometry; p_s directly measured on aircraft and p_∞ derived from aircraft height, measured from tower, and pressure gradient. Aircraft indicated altitude is compared with altitude measured from tower. Suitable for subsonic flight only.
Space positioning (tracking radar or ground camera)	Optical or radar tracking system used to measure aircraft altitude and ground speed. Three tracking stations triangulate aircraft linear and angular positions; p_s directly measured on aircraft and p_∞ calculated from measurements of p and T on ground and assumed standard temperature gradient. Complex hardware and software systems required. Suitable for subsonic transonic, and supersonic flight and for climbs, descents, accelerations, and decelerations.

True airspeed methods (Δp_{pc} derived from measurement of V_∞):

Ground speed course	Aircraft ground speed from measurement of time to fly known distance at constant airspeed. Wind effects cancelled by flying in opposite directions of straight-line course or flying triangular course. Compare average ground speed to Pitot-static system derived airspeed to obtain velocity position error. Suitable for subsonic, low altitude only.
All altitude speed course	Differentiated from ground speed course, since suitable for all altitudes. GPS is required. Compare drift correction GPS ground speed to Pitot-static system derived airspeed to obtain velocity position error. Various courses may be flown, including gentle turn and cloverleaf.

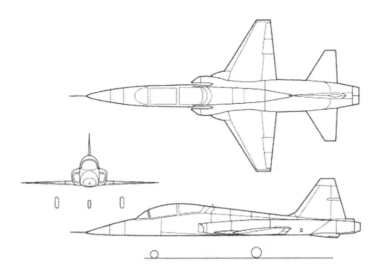

Figure 5.23   Three-view drawing of the Northrop T-38A *Talon*. (Source: *NASA*.)

Table 5.8   Selected specifications of the Northrop T-38A *Talon*.

Item	Specification
Primary function	Advanced supersonic jet trainer
Manufacturer	Northrop Corporation, Los Angeles, California
First flight	10 March 1959
Crew	1 pilot + 1 instructor pilot
Powerplant	$2 \times$ J85-GE-5 afterburning turbojet engine
Thrust, MIL (ea. engine)	2050 lb (9120 N), military power
Thrust, MAX (ea. engine)	2900 lb (12,900 N), maximum afterburner
Empty weight	7200 lb (3270 kg)
Maximum takeoff weight	12,093 lb (5485 kg)
Length	46 ft 4 in (14.1 m)
Height	12 ft 10 in (3.91 m)
Wingspan	25 ft 3 in (7.70 m)
Wing area	170 ft^2 (15.8 m^2)
Airfoil	NACA 65A004.8
Maximum speed	812 mph (1307 km/h), Mach 1.08 at sea level
Service ceiling	$>$50,000 ft ($>$15,000 m)
Load factor limits	$+7.33\ g, -3.0\ g$

You make a final radio call to announce that you are on the fly-by line and state that you are at an airspeed of 300 knots. As you near the fly-by tower, your backseater records the indicated altitude, h_{ic}, indicated velocity, V_{ic}, outside air temperature, T_{ic}, and fuel weight. (It is assumed that these are instrument corrected measurements.) The personnel in the fly-by tower have watched the T-38 turn, descend, and line up on the fly-by line. An engineer in the tower is looking intently through a small eyepiece, sighting across a vertical framework of wires that form a grid, waiting for the T-38 to fly past. The T-38 zooms past the tower at 300 knots, 100 ft above the ground. The engineer

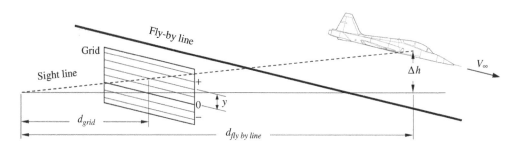

Figure 5.24   Tower fly-by geometry.

obtains the grid reading, the vertical location, y, of the T-38 on the wire grid as it flies past, as shown in Figure 5.24.

Using the geometry shown in Figure 5.24, the "truth source" pressure altitude, h_c, of the T-38 is calculated as

$$h_c = h_{c,tower} + \Delta h = h_{c,tower} + \left(\frac{d_{flyby\,line}}{d_{grid}} \right) y \qquad (5.91)$$

where $h_{c,tower}$ is the known pressure altitude of the fly-by tower, $d_{flyby\,line}$ is the known distance from the sighting device to the fly-by line, d_{grid} is the distance from the sighting device to the grid, and y is the vertical grid reading. The altitude position error correction is then simply given by

$$\Delta h_{pc} = h_c - h_{ic} \qquad (5.92)$$

(A density correction may also be applied, based on the temperature measurements in the aircraft and tower, which we neglected for simplicity.) The static position error correction may be obtained from this altitude correction and thence the velocity and Mach number corrections.

You climb back up to 1000 ft in the T-38 and prepare for the next tower fly-by at a different airspeed. You will perform multiple passes at different airspeeds, ranging from 190 KIAS (220 mph, 350 km/h) to 575 KIAS (660 mph, 1060 km/h), which provides corrections from about Mach 0.29 to 0.87.

After you have completed the tower fly-bys, you set up for another position error calibration FTT, the ground speed course. You will fly a straight-line path along a measured distance course, between two landmarks. You can use the fly-by line as your straight-line path, with a landmark near the edge of the dry lakebed as the start point and the tower as the end point. The distance between these two points is known and you will measure the time it takes to fly between these two points. You again need to fly this course at low altitude, so that the start and end point landmarks can be more accurately sighted.

You again line up on the fly-by line, dropping down to a height of 100 ft AGL and trimming the T-38 for 250 KIAS. The fly-by line is on a magnetic heading of 240° and you check the alignment of the magnetic compass in the cockpit. You maintain this heading and do not correct for any crosswind drift of the aircraft, since the airspeed is measured in the direction that the aircraft is heading, not along its ground track, which is influenced by the wind. The flight path geometry for the ground speed course is shown in Figure 5.25. It is best if the ground speed course is flown with zero wind, but you have a tail wind, V_{TW}, and a crosswind, V_{XW}, when you fly the course.

As the starting landmark approaches, your backseater records the indicated airspeed, V_{ic}, indicated altitude, h_{ic}, and outside air temperature, T_{ic}, then looks out of the right side of the cockpit and starts a stopwatch as the landmark passes. You do your best to fly a constant airspeed with a tight tolerance of ±1 knot (1.15 mph, 1.9 km/h) and a constant heading as the end point

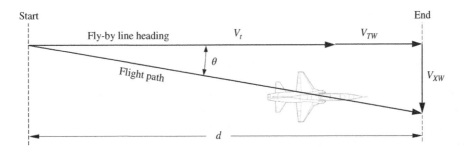

Figure 5.25   Ground speed course.

approaches. When the tower end point passes, the backseater stops the timer and obtains a time t_1 for flying the known distance d. The velocity for this ground course is given by

$$\frac{d}{t_1} = V_t + V_{TW} \tag{5.93}$$

where V_t is the aircraft true airspeed and V_{TW} is the tailwind velocity. After completing this ground course run, you make a 180° turn, so that you can fly the course in the opposite direction. This reciprocal heading technique cancels the effects of the headwind or tailwind. You now set up on a magnetic heading of 060°, the reciprocal of the 240° heading for the first run, where the fly-by tower is now the start point and the landmark at the edge of the lakebed is the end point. You line up on the fly-by line once again and maintain the 060 heading for the second run, being sure not to correct for crosswind drift. This also ensures that the velocity vectors for the two course runs are parallel. After this second ground course run, your backseater obtains a time of t_2. The velocity for this second ground course is given by

$$\frac{d}{t_2} = V_t - V_{TW} \tag{5.94}$$

where the tailwind velocity is now subtracted from the true velocity.

The true velocity is solved for by adding Equations (5.93) and (5.94), where the tailwind velocity is cancelled out, yielding

$$V_t = \frac{1}{2}\left(\frac{d}{t_1} + \frac{d}{t_2}\right) \tag{5.95}$$

Thus, the true velocity is the average of the speeds obtained from the two, reciprocal heading ground course runs.

The calibrated airspeed, V_c, can be calculated using Equation (5.68), assuming that the calibrated airspeed equals the equivalent airspeed, V_e, which is a good assumption at low altitude, giving

$$V_c = V_e = \sqrt{\sigma}\,V_t = \sqrt{\frac{\delta}{\theta}}\,V_t \approx \frac{V_t}{\sqrt{\theta}} \tag{5.96}$$

where σ, δ, and θ are the density, pressure, and temperature ratios, given by Equations (5.45), (5.46), and (5.47), respectively. The pressure and temperature ratios are calculated from the measured values of altitude and temperature. Since the ground course is flown close to the ground, the pressure ratio, δ, may be assumed to be one.

With the calculated calibrated airspeed, the velocity position error is given by

$$\Delta V_{pc} = V_c - V_{ic} \tag{5.97}$$

The static position error correction may be obtained from this velocity correction and thence the altitude and Mach number corrections.

Example 5.8 Ground Speed Course *A T-38 jet uses the ground speed course FTT to obtain a velocity position error correction. The jet flies at a height of 100 ft AGL, between two landmarks that are 6000 ft apart. Using the reciprocal heading technique, the jet flies in one direction and obtains a time, t_1, of 18.27 s and a time, t_2, of 19.68 s, in the opposite direction. The T-38 aircrew record an indicated velocity, V_{ic}, of 190 KIAS and an outside air temperature, T_{ic}, of 71.3°F. Calculate the velocity position error correction, ΔV_{pc}.*

Solution

Using Equation (5.95), the T-38 true airspeed is

$$V_t = \frac{1}{2}\left(\frac{d}{t_1} + \frac{d}{t_2}\right) = \frac{1}{2}\left(\frac{6000\,ft}{18.27\,s} + \frac{6000\,ft}{19.68\,s}\right) = 316.6\frac{ft}{s} = 187.6\,kt$$

Using Equation (5.45), the temperature ratio, θ, is

$$\theta = \frac{T_{ic}}{T_{SSL}} = \frac{(71.3 + 459.69)°R}{518.69°R} = \frac{530.99}{518.69} = 1.0237$$

Using Equation (5.96), the calibrated airspeed is

$$V_c \approx \frac{V_t}{\sqrt{\theta}} = \frac{316.6\frac{ft}{s}}{\sqrt{1.0237}} = 312.9\frac{ft}{s} = 185.4\,kt$$

Using Equation (5.97), the velocity position error correction is

$$\Delta V_{pc} = V_c - V_{ic} = 185.4\,kt - 190\,kt = -4.6\,kt$$

5.7 The Equations of Motion for Unaccelerated Flight

Aircraft performance deals with the translational motion of the vehicle in three-dimensional space. The application of Newton's second law of motion to the vehicle's translational motion yields three force equations, one for each dimension in three-dimensional space. We restrict ourselves to motion in two dimensions, parallel and perpendicular to the direction of motion. The free-body diagram for this case is shown in Figure 5.26.

The aircraft is in a wings level attitude, flying with a velocity V_∞ and angle-of-attack α, measured between the fuselage reference line (*FRL*) and the relative wind (*RW*), along a flight path at an angle γ with respect to the horizon. Although drawn along the flight path direction, the velocity may be changing with time, thus the aircraft may have an acceleration a. The forces acting on the aircraft are the lift, L, weight, W, thrust, T, and drag, D. The lift and drag forces act in directions perpendicular and parallel to the relative wind (*RW*), respectively. The weight acts perpendicularly to the horizon, towards the ground. The thrust acts in a direction opposite to the drag at a thrust angle, α_T, between the thrust vector and the relative wind (*RW*).

Applying Newton's second law, in vector form, to Figure 5.26, we have

$$\sum \vec{F} = m\vec{a} = m\frac{d\vec{V}_\infty}{dt} \tag{5.98}$$

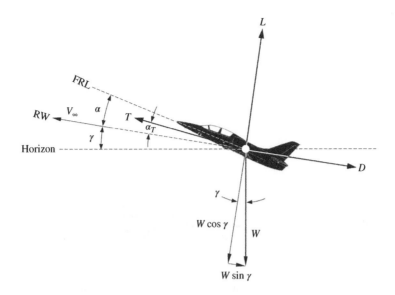

Figure 5.26   Free-body diagram for an aircraft in two-dimensional flight.

Resolving Equation (5.98) into components parallel and perpendicular to the velocity, respectively, gives

$$\sum F_{\parallel} = ma = m\frac{dV_{\infty}}{dt} \tag{5.99}$$

$$\sum F_{\perp} = ma_r = m\frac{V_{\infty}^2}{R} \tag{5.100}$$

where $\sum F_{\parallel}$ and $\sum F_{\perp}$ are the sum of the forces parallel and perpendicular to the flight path direction, respectively, m is the vehicle mass, and a is the acceleration in the flight path direction. The acceleration perpendicular to the flight path direction is the radial or centripetal acceleration, a_r, equal to V_{∞}^2/R, where R is the radius of curvature of the flight path, as shown in Figure 5.27. Equations (5.99) and (5.100) are the general equations of motion for the two-dimensional, accelerated motion of an aircraft.

Inserting the forces from Figure 5.26 into Equations (5.99) and (5.100), we have

$$T\cos\alpha_T - D - W\sin\gamma = \frac{W}{g}\frac{dV_{\infty}}{dt} \tag{5.101}$$

$$L - W\cos\gamma + T\sin\alpha_T = \frac{W}{g}\frac{V_{\infty}^2}{R} \tag{5.102}$$

Our performance problems of interest fall into two categories, unaccelerated motion with a straight-line flight path and accelerated motion with a curved flight path. The aircraft is assumed to be in unaccelerated, straight-line flight during the climb, cruise, and descent segments of the flight profile. The aircraft is assumed to be in accelerated motion during takeoff, landing, and turning flight. The velocity may be changing in magnitude, direction, or both during accelerated flight.

Let us examine unaccelerated motion, where the velocity is constant and the acceleration is zero. Normally, the thrust angle, α_T, is small, such that $\cos\alpha_T \approx 1$ and $\sin\alpha_T \approx 0$. Using these

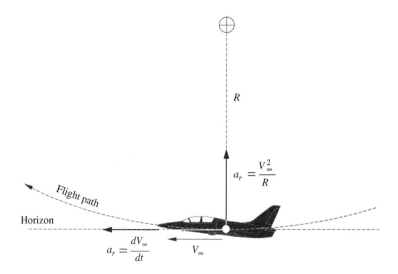

Figure 5.27   Accelerations acting on an aircraft flying curvilinear, two-dimensional flight path.

Table 5.9   Equations of motion for steady, unaccelerated flight.

Flight condition	Flight path angle, γ	$\sum F_z$	$\sum F_x$
Level, unaccelerated flight	$\gamma = 0$	$L = W$	$T = D$
Climbing, unaccelerated flight	$+\gamma$	$L = W \cos\gamma \cong W$	$T = D + W \sin\gamma$
Gliding, unaccelerated flight ($T = 0$)	$-\gamma$	$L = W \cos\gamma$	$D = W \sin(-\gamma)$

assumptions, Equations (5.101) and (5.102) become

$$T = D + W \sin\gamma \tag{5.103}$$

$$L = W \cos\gamma \tag{5.104}$$

Equations (5.103) and (5.104) are the equations of motion for a vehicle in unaccelerated flight in two dimensions. These equations can be applied to unaccelerated level (constant altitude), climbing, or descending flight with the proper choice of the flight path angle, γ. The equations of motion for these three flight conditions were obtained earlier in Chapter 2, when free-body diagrams were introduced. The equations of motion for unaccelerated flight are summarized in Table 5.9, including the associated flight path angle. For climbing flight, we assume small climb angles, such that $\cos\gamma \cong 1$. Therefore, the lift equals the weight in this approximation. Gliding, unaccelerated flight is simply descending, unaccelerated flight with zero thrust. We apply these unaccelerated equations of motion in analyzing cruise, climb, and gliding performance.

5.8   Level Flight Performance

Level, unaccelerated flight is usually associated with the *cruise* segment of a flight profile. The aircraft is assumed to be in a wings level attitude, at constant airspeed and constant altitude. In this section, we are interested in answering performance questions such as, "How far can the aircraft fly and for how long?" These are the topics of range and endurance, respectively, which are important cruise performance metrics. We are also interested in the airspeeds, fuel flows, and altitude ceilings, associated with cruise flight.

Cruise performance is important from the standpoint of the time and distance it takes for an aircraft to get somewhere, whether it is for an airliner getting to a destination or a fighter jet or bomber getting to a combat area and back. Requirements for cruise performance are usually set by a customer, such as an airline or the military, to ensure that the aircraft meets their mission objectives. There are usually no safety-of-flight requirements levied on cruise performance.

As was derived earlier, the aircraft equations of motion for level, unaccelerated flight are such that the four forces are in balance; the lift equals the weight and the thrust equals the drag.

$$L = W \tag{5.105}$$

$$T = D \tag{5.106}$$

These simple equalities are the foundation for evaluating cruise performance.

5.8.1   Thrust Required in Level, Unaccelerated Flight

As given by Equation (5.106), the *thrust required*, T_R, for steady, unaccelerated, level flight is simply equal to the total aircraft drag. Thus, we have

$$T_R = D = D_0 + D_i = T_{R,0} + T_{R,i} \tag{5.107}$$

The total drag is composed of the zero-lift drag, D_0, and the drag due to lift or induced drag, D_i. Equating the thrust required and the drag, the total thrust required is therefore composed of the zero-lift thrust required, $T_{R,0}$, and the lift-induced thrust required, $T_{R,i}$. Thus, for steady, level, unaccelerated flight, a plot of the zero-lift drag, D_0, the lift-induced drag, D_i, and the total drag, D, is a plot of the zero-lift thrust required, $T_{R,L=0}$, lift-induced thrust required, $T_{R,i}$, and total thrust required, T_R, as shown in Figure 5.28.

Dividing Equation (5.106) by (5.105), we have

$$\frac{T}{W} = \frac{D}{L} = \frac{1}{(L/D)} \tag{5.108}$$

As discussed in Section 2.3.5, the thrust-to-weight ratio, T/W, for level, unaccelerated flight varies inversely with the lift-to-drag ratio, L/D. This simple equation highlights the connection between the aircraft propulsion system, embodied by the thrust-to-weight ratio, and the aircraft aerodynamics, in the lift-to-drag ratio. The highest thrust-to-weight ratio, at a given aircraft weight, is obtained at the maximum lift-to-drag ratio, $(L/D)_{max}$.

Solving Equation (5.108) for the thrust, we have

$$T_R = \frac{W}{(L/D)} = \frac{W}{(C_L/C_D)} \tag{5.109}$$

where we have defined the thrust required, T_R, to maintain steady, level, unaccelerated flight. As expected, the thrust required varies inversely with the lift-to-drag ratio, such that the minimum thrust required, $T_{R,min}$, at a given weight, is obtained at the maximum lift-to-drag ratio.

$$T_{R,min} = \frac{W}{(L/D)_{max}} \tag{5.110}$$

Since the lift-to-drag ratio embodies the aerodynamic efficiency of the aircraft, it is appropriate that maximizing the aerodynamic efficiency results in the least thrust required for level, unaccelerated flight.

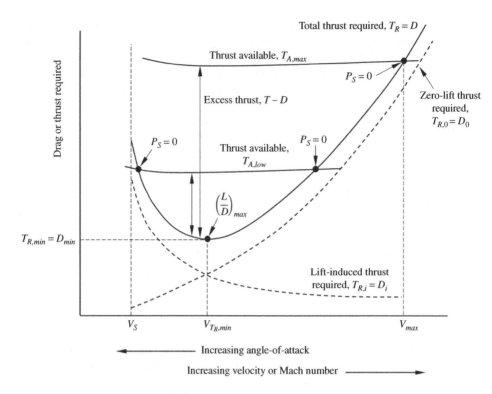

Figure 5.28   Thrust required and thrust available.

The lift, L, in Equation (5.108), is given by

$$L = W = q_\infty S C_L \tag{5.111}$$

Solving for the lift coefficient, we have

$$C_L = \frac{W}{q_\infty S} \tag{5.112}$$

The total drag is given by Equation (3.230), so Equation (5.107) can be expanded as

$$T_R = D = q_\infty S C_D = q_\infty S(C_{D,0} + C_{D,i}) = q_\infty S\left(C_{D,0} + \frac{C_L^2}{\pi e AR} \right) \tag{5.113}$$

where $C_{D,0}$ is the zero-lift drag coefficient and $C_{D,i}$ is the induced drag coefficient. Substituting Equations (5.112) and (5.113) into (5.108), we have

$$\frac{T}{W} = \frac{q_\infty S\left(C_{D,0} + \frac{C_L^2}{\pi e AR} \right)}{W} = \frac{q_\infty S\left[C_{D,0} + \frac{1}{(\pi e AR)}\left(\frac{W}{q_\infty S} \right)^2 \right]}{W}$$

$$\frac{T}{W} = q_\infty C_{D,0}\left(\frac{1}{W/S} \right) + \frac{1}{q_\infty \pi e AR}\left(\frac{W}{S} \right) \tag{5.114}$$

Thus, the thrust-to-weight ratio, T/W, for steady, level, unaccelerated flight is related to the wing loading, W/S, of the aircraft. Multiplying Equation (5.113) by the weight gives

$$T_R = q_\infty S C_{D,0} + \frac{W^2}{q_\infty S \pi e AR} = T_{R,0} + T_{R,i} \tag{5.115}$$

where $T_{R,0}$ and $T_{R,i}$ are defined as the zero-lift thrust required and the lift-induced thrust required, respectively, as given by Equation (5.107) earlier.

The thrust required, given by Equation (5.115), is a function of the dynamic pressure, zero-lift drag coefficient, weight, and other wing-related geometric properties. All of these parameters are related to either the flight condition or the aircraft airframe, which includes its weight, aerodynamics and wing geometry. The thrust required is not dependent on any parameters related to the propulsion system.

Example 5.9 Thrust-to-Weight Ratio versus Wing Loading in Level, Unaccelerated Flight *Calculate and plot the thrust-to-weight ratio, T/W, versus wing loading, W/S, for a Boeing 747 Jumbo Jet airliner flying in steady, level, unaccelerated flight at an airspeed of 900 km/h, using the specifications given in the table below. Assume that the aircraft weight varies from the heavier weight to the lighter weight given in the table.*

Parameter	Specification
Weight, light	1,600,000 N
Weight, heavy	2,830,000 N
Wing span	59.74 m
Wing area	520.2 m²
Zero-lift drag, $C_{D,0}$	0.036
Span efficiency factor, e	0.7
Stall speed, V_s	200 km/h

Solution

A sample calculation is given for a weight of 2,830,000 kg. The results for the range of weights and wing loadings are given in the table below. First, the velocity is converted to meters per second.

$$V_\infty = 900 \, \frac{km}{h} \times 1000 \, \frac{m}{km} \times \frac{1}{3600} \frac{h}{s} = 250.0 \, \frac{m}{s}$$

The wing aspect ratio is given by

$$AR = \frac{b^2}{S} = \frac{(59.74 \, m)^2}{520.2 \, m^2} = 6.861$$

The dynamic pressure corresponding to a velocity of 900 km/s at sea level is

$$q_\infty = \frac{1}{2} \rho_\infty V_\infty^2 = \frac{1}{2} \left(1.225 \, \frac{kg}{m^3} \right) \left(250.0 \, \frac{m}{s} \right)^2 = 38,281.25 \, \frac{N}{m^2}$$

The wing loading for a weight of 2,830,000 N is

$$\frac{W}{S} = \frac{2,830,000 \, N}{520.2 \, m^2} = 5440.2 \, \frac{N}{m^2}$$

The wing loading is typically specified in inconsistent units of kg_f/m^2. Therefore, we have

$$\frac{W}{S} = 5440.2 \ \frac{N}{m^2} \times \frac{1 \ kg_f}{9.81 \ N} = 554.6 \ \frac{kg_f}{m^2}$$

Using Equation (5.114), the thrust-to-weight ratio is

$$\frac{T}{W} = q_\infty C_{D,0} \left(\frac{1}{W/S}\right) + \frac{1}{q_\infty \pi e AR} \left(\frac{W}{S}\right)$$

$$\frac{T}{W} = \frac{\left(38,281.25 \ \frac{N}{m^2}\right)(0.036)}{5440.2 \ \frac{N}{m^2}} + \frac{5440.2 \ \frac{N}{m^2}}{\left(38,281.25 \ \frac{N}{m^2}\right)\pi(0.7)(6.861)} = 0.2627$$

The computations of the thrust-to-weight ratio, for the range of wing loadings, are tabulated in the table and plotted in the figure below.

W (N)	W/S (kg$_f$/m^2)	T/W
2,830,000	554.6	0.2627
2,800,000	548.7	0.2654
2,700,000	529.1	0.2745
2,600,000	509.5	0.2844
2,500,000	489.9	0.2951
2,200,000	470.3	0.3067
2,100,000	450.7	0.3194
2,000,000	431.1	0.3332
1,900,000	411.5	0.3484
1,800,000	391.9	0.3651
1,700,000	372.3	0.3836
1,600,000	352.7	0.4043

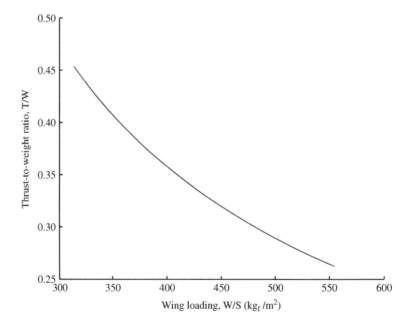

5.8.2 Velocity and Lift Coefficient for Minimum Thrust Required

We now seek to determine the flight conditions corresponding to minimum thrust required, at a given weight, in steady, level, unaccelerated flight. This minimum thrust required point corresponds to the condition for the maximum lift-to-drag ratio, as was given by Equation (5.110).

To obtain the velocity corresponding to the minimum thrust required point, we take the derivate of Equation (5.115) with respect to the velocity, V_∞, and set this to zero.

$$\frac{dT_R}{dV_\infty} = \frac{d}{dV_\infty}\left(q_\infty S C_{D,0} + \frac{W^2}{q_\infty S \pi e AR}\right) = 0 \tag{5.116}$$

Expanding the dynamic pressure, q_∞, in terms of the density, ρ_∞, and velocity, V_∞, we have

$$\frac{dT_R}{dV_\infty} = \frac{d}{dV_\infty}\left(\frac{1}{2}\rho_\infty V_\infty^2 S C_{D,0} + \frac{W^2}{\frac{1}{2}\rho_\infty V_\infty^2 S \pi e AR}\right) = 0$$

$$\frac{dT_R}{dV_\infty} = (\rho_\infty S C_{D,0})V_\infty - \left(\frac{4W^2}{\rho_\infty S \pi e AR}\right)\frac{1}{V_\infty^3} = 0 \tag{5.117}$$

Solving for the velocity gives

$$(\rho_\infty S C_{D,0})V_\infty^4 - \left(\frac{4W^2}{\rho_\infty S \pi e AR}\right) = 0 \tag{5.118}$$

$$V_{T_R,min} = \left(\frac{4W^2}{\rho_\infty^2 S^2 C_{D,0} \pi e AR}\right)^{1/4} \tag{5.119}$$

where $V_{T_R,min}$ is the velocity in steady, level, unaccelerated flight where the thrust required is at a minimum. This is also the velocity corresponding to the minimum total drag, D_{min}, as shown in Figure 5.28.

Substituting the velocity for minimum thrust, Equation (5.119), into Equation (5.111), we have

$$L = W = q_\infty S C_L = \frac{1}{2}\rho_\infty V_\infty^2 S C_L = \frac{1}{2}\rho_\infty \left(\frac{4W^2}{\rho_\infty^2 S^2 C_{D,0} \pi e AR}\right)^{1/2} S C_L$$

$$W = W\left(\frac{1}{C_{D,0}\pi e AR}\right)^{1/2} C_L \tag{5.120}$$

Solving for the lift coefficient,

$$(C_L)_{T_R,min} = \sqrt{C_{D,0}\pi e AR} \tag{5.121}$$

where $(C_L)_{T_R,min}$ is the lift coefficient corresponding to steady, level, unaccelerated flight at the minimum thrust velocity. Solving for the zero-lift drag coefficient gives

$$C_{D,0} = \frac{(C_L)_{T_R,min}^2}{\pi e AR} = C_{D,i} \tag{5.122}$$

At the minimum thrust required or minimum drag point for steady, level, unaccelerated flight, the zero-lift drag equals the induced drag. This is shown graphically, in Figure 5.28, as the point of intersection of the zero-lift thrust required (or drag) curve with the curve for the lift-induced thrust required (or drag). Thus, the total drag coefficient is given by

$$(C_D)_{T_R,min} = C_{D,0} + C_{D,i} = 2C_{D,0} = 2C_{D,i} \tag{5.123}$$

Thus, the minimum drag is equal to either twice the zero-lift drag or twice the lift-induced drag.

Example 5.10 Velocity and Lift Coefficient for Minimum Thrust in Level, Unaccelerated Flight *Using the specifications in Example 5.9, calculate the velocity and lift coefficient for minimum thrust for the Boeing 747* Jumbo Jet *in steady, level, unaccelerated flight at sea level, assuming the aircraft is at the heavy weight listed.*

Solution

Using Equation (5.119), the velocity for minimum thrust is

$$V_{T_R,min} = \left(\frac{4W^2}{\rho_\infty^2 S^2 C_{D,0} \pi e AR} \right)^{1/4}$$

$$V_{T_R,min} = \left[\frac{4(2,830,000\ N)^2}{\left(1.225\ \frac{kg}{m^3}\right)^2 (520.2\ m^2)^2 (0.036) \pi (0.7)(6.861)} \right]^{1/4} = 109.8\ \frac{m}{s} = 395.2 \frac{km}{h}$$

Using Equation (5.121), the lift coefficient for minimum thrust is

$$(C_L)_{T_R,min} = \sqrt{C_{D,0} \pi e AR} = \sqrt{(0.036)\pi(0.7)(6.861)} = 0.7370$$

5.8.3 Thrust Available and Maximum Velocity

We have seen that the thrust required for steady, level, unaccelerated flight is related to the aircraft airframe characteristics and not related to the propulsion system. We now address the *thrust available*, T_A, which is directly related to the propulsion system. The thrust produced by different types of propulsive devices was discussed in Section 4.4, with the variation of thrust as a function of airspeed as shown in Figure 4.11. Regardless of the propulsion type, the thrust available is selected by setting the throttle to a desired setting, from a minimum to a maximum thrust level.

The thrust available for a jet engine, for two different throttle settings, a low thrust throttle setting, $T_{A,low}$, and a maximum thrust setting, $T_{A,max}$, are shown in Figure 5.28. At both throttle settings, the thrust available is nearly constant with velocity, indicative of a jet engine, as discussed in Section 4.4.

Let us first focus on the thrust available curve for the low thrust throttle setting. The thrust available curve intercepts the thrust required curve at two points, at a low velocity or Mach number and a high velocity or Mach number. At the intersection of these curves, the thrust available is equal to the thrust required or drag, thus the aircraft is in equilibrium or in steady, level, unaccelerated flight. From an energy perspective, a quantity called the specific excess power at these two points is zero, $P_s = 0$. The specific excess power is directly proportional to the *excess thrust*, defined as the difference between the thrust available and thrust required.

$$\text{Excess thrust} \equiv T_A - T_R \tag{5.124}$$

At the lower velocity point, the angle-of-attack is greater than at the higher velocity point. The higher speed equilibrium point represents the *maximum velocity* obtainable by the aircraft in level flight, *at the selected throttle setting*, and at a given weight.

In between these two equilibrium flight conditions, the thrust available is greater than the thrust required or the excess thrust is positive. If the excess thrust is greater than zero, the aircraft can accelerate, climb, or accelerate and climb. The excess thrust, at the two equilibrium points, is zero. The concept of excess thrust is important for level and climb performance. The larger the excess thrust, the greater the aircraft's level and climb performance.

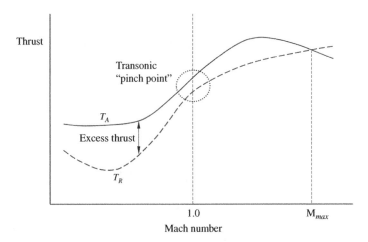

Figure 5.29   Transonic "pinch point" for a supersonic aircraft.

Now, assume that the engine throttle is advanced to its maximum thrust setting. The thrust available is now shown by the $T_{A,max}$ curve in Figure 5.28. The aircraft now has positive excess thrust, the difference between the maximum thrust available curve and the thrust required curve, so that it accelerates and/or climbs. This acceleration or climb continues until the vehicle reaches equilibrium, such that the maximum thrust available equals the thrust required. This occurs at the high speed intersection of the maximum thrust available curve and the thrust required curve, where again $P_s = 0$. Since the thrust is at its maximum, this point represents the *maximum velocity* obtainable by the aircraft in steady, level flight. At the higher thrust settings, there may not be a low-speed equilibrium point. This is because the stall speed, V_s, may make this low-speed point unobtainable.

The thrust available and thrust required curves for a supersonic aircraft are shown in Figure 5.29. The rapid increase of transonic drag results in a large increase of the thrust required and a significant decrease of excess thrust in this region. The excess thrust can become very small, go to zero, or even become negative for some aircraft, depending on the aerodynamic configuration, such as the carriage of external stores, which can add significant drag. The atmospheric temperature can also affect the thrust available, with higher temperatures resulting in a decrease in thrust. These factors affect the capability of an aircraft to accelerate through the transonic region. The flight region, where the excess thrust is significantly reduced, such that the thrust available and thrust required curves are "pinched" closer together, is called the transonic "pinch point". A supersonic aircraft must be designed and configured properly to maintain an acceptable excess thrust capability through this region.

Example 5.11 Thrust Required, Thrust Available, and Maximum Velocity in Level, Unaccelerated Flight *Calculate and plot the thrust available and thrust required, at sea level, for a Boeing 747 Jumbo Jet airliner, from the stall speed to a speed of 900 km/h, using the specifications given in Example 5.9. Assume the aircraft is at the heavy weight in the table. From the plot, estimate the maximum velocity.*

The aircraft is powered by four Pratt & Whitney JT9D-7A high bypass ratio, turbofan engines. Each engine produces a static sea level thrust of 205,063 N. The sea level thrust of each engine, $T_{eng,SL}$, varies with forward velocity, V_∞, according to the equation

$$T_{eng,SL} = 205{,}063 \text{ N} - \left(681.53 \ \frac{\text{kg}}{\text{s}}\right) V_\infty + \left(2.236 \ \frac{\text{kg}}{\text{m}}\right) V_\infty^2$$

where V_∞ has units of m/s.

Solution

A sample calculation is given for an airspeed of 400 km/h. The results for the range of velocities are given in the table below. First, the velocity is converted to meters per second.

$$V_\infty = 400 \frac{km}{h} \times 1000 \frac{m}{km} \times \frac{1}{3600} \frac{h}{s} = 111.1 \frac{m}{s}$$

The wing aspect ratio is given by

$$AR = \frac{b^2}{S} = \frac{(59.74 \ m)^2}{520.2 \ m^2} = 6.861$$

From Equation (5.105) and the definition of lift, we have

$$W = L = \frac{1}{2}\rho_\infty V_\infty^2 S C_L$$

Solving for the lift coefficient, in steady, level, unaccelerated flight gives

$$C_L = \frac{W}{\frac{1}{2}\rho_\infty V_\infty^2 S} = \frac{2,830,000 \ N}{\frac{1}{2}\left(1.225 \frac{kg}{m^3}\right)\left(111.1 \frac{m}{s}\right)^2 (520.2 \ m^2)} = 0.7192$$

From Equation (3.230), the total aircraft drag is given by

$$C_D = C_{D,0} + \frac{C_L^2}{\pi e AR} = 0.036 + \frac{(0.7192)^2}{\pi(0.7)(6.861)} = 0.07028$$

The lift-to-drag ratio is

$$\frac{L}{D} = \frac{C_L}{C_D} = \frac{0.7196}{0.07028} = 10.233$$

From Equation (5.109), the thrust required is

$$T_R = \frac{W}{(L/D)} = \frac{W}{(C_L/C_D)} = \frac{2,830,000 \ N}{10.233} = 276,556 \ N$$

The total thrust available from the four engines is given by

$$T_A = 4 \times \left[205,063 \ N - \left(681.53 \frac{kg}{s}\right) V_\infty + \left(2.236 \frac{kg}{m}\right) V_\infty^2\right]$$

$$T_A = 4 \times \left[205,063 \ N - \left(681.53 \frac{kg}{s}\right)\left(111.1 \frac{m}{s}\right) + \left(2.236 \frac{kg}{m}\right)\left(111.1 \frac{m}{s}\right)^2\right]$$

$$T_A = 4 \times (156,944 \ N) = 627,776 \ N$$

The computations of the thrust required and thrust available, for the range of velocities, are tabulated in the table and plotted in the figure below.

The maximum velocity is obtained at the intersection of the thrust available and thrust required curves, as shown in the figure. From the figure, the maximum velocity is estimated to be about 860 km/h. (A more precise value could be obtained by calculating more velocity points.)

V_∞(km/h)	V_∞(m/s)	C_L	C_D	L/D	T_R(N)	T_A(N)
200	55.56	2.878	0.5849	4.920	575,169	696,406
300	83.33	1.279	0.1444	8.856	319,552	655,186
400	111.1	0.7192	0.07028	10.233	276,556	627,776
500	138.8	0.4604	0.05005	9.199	307,628	614,155
600	166.6	0.3198	0.04278	7.475	378,597	614,343
700	194.4	0.2349	0.03966	5.924	477,743	628,334
800	222.2	0.1799	0.03814	4.715	600,175	656,127
900	250.0	0.1421	0.03734	3.806	743,556	697,722

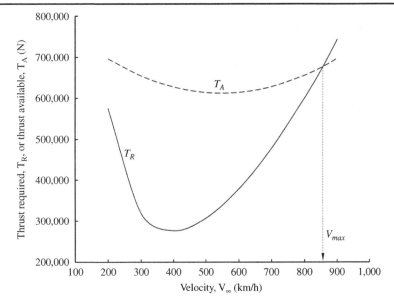

5.8.4 Power Required and Power Available

Thrust is typically used to describe the propulsive output of jet engines, while power is used for propeller-driven engines. Therefore, in the next few sections, we develop the relationships for power relevant to steady, level, unaccelerated flight, similar to the development for thrust. As discussed in Section 4.4, power is defined as thrust multiplied by the flight velocity, thus the power required, P_R, is given by the product of the thrust required, T_R, and the velocity.

$$P_R = T_R V_\infty \tag{5.125}$$

Inserting Equation (5.109) for the thrust required, we have

$$P_R = T_R V_\infty = \left(\frac{W}{L/D}\right) V_\infty = \left(\frac{W}{C_L/C_D}\right) V_\infty \tag{5.126}$$

Using Equation (5.111), equating the lift and weight for steady, level, unaccelerated flight, the velocity is given by

$$V_\infty = \sqrt{\frac{2W}{\rho_\infty S C_L}} \tag{5.127}$$

Inserting Equation (5.127) into (5.126), we have

$$P_R = \left(\frac{W}{C_L/C_D}\right)\sqrt{\frac{2W}{\rho_\infty S C_L}} = \sqrt{\frac{2}{\rho_\infty S}}(W^{3/2})\left(\frac{1}{C_L^{3/2}/C_D}\right) \tag{5.128}$$

Equation (5.128) shows that the power required varies inversely with $C_L^{3/2}/C_D$, whereas the thrust required varied inversely with C_L/C_D, as given by Equation (5.109). The minimum power required, $P_{R,min}$, occurs for $(C_L^{3/2}/C_D)_{max}$, in contrast to $(C_L/C_D)_{max}$ for minimum thrust required, $T_{R,min}$.

Using Equation (5.113), the powered required may also be couched in terms of the total aircraft drag, D, as given by

$$P_R = T_R V_\infty = D V_\infty = q_\infty S C_D V_\infty = q_\infty S \left(C_{D,0} + \frac{C_L^2}{\pi e AR}\right) V_\infty \tag{5.129}$$

Inserting Equation (5.112) for the lift coefficient, we have

$$P_R = q_\infty S \left[C_{D,0} + \left(\frac{1}{\pi e AR}\right)\left(\frac{W}{q_\infty S}\right)^2\right] V_\infty = \left(q_\infty S C_{D,0} + \frac{W^2}{q_\infty S \pi e AR}\right) V_\infty \tag{5.130}$$

Expanding the dynamic pressure, q_∞, in terms of the density, ρ_∞, and velocity, V_∞, gives

$$P_R = \frac{1}{2}\rho_\infty V_\infty^3 S C_{D,0} + \frac{W^2}{\frac{1}{2}\rho_\infty V_\infty S \pi e AR} = P_{R,0} + P_{R,i} \tag{5.131}$$

where $P_{R,0}$ and $P_{R,i}$ are the zero-lift power required and the lift-induced powered required, respectively, similar to the zero-lift thrust required, $T_{R,0}$, and lift-induced thrust required, $T_{R,i}$, given in Equation (5.115). As expected, similar to the thrust required, the power required is dependent on the flight condition (airspeed and altitude), weight, and airframe-related parameters. It is independent of any propulsion-related parameters.

The zero-lift, lift-induced, and total power required curves are shown in Figure 5.30. The minimum power required point corresponds to $(C_L^{3/2}/C_D)_{max}$ while the minimum thrust required point corresponds to $(C_L/C_D)_{max}$. The velocity and lift coefficient for minimum power required are derived in the following section.

The power available, P_A, for jet and propeller-driven, piston engines was developed in Section 4.4. These power available curves are drawn with the power required curves in Figure 5.31. The power required and power available curves are matched with the thrust required and thrust available curves in this figure. The point of intersection of the power available and power required curves yields the maximum velocity, regardless of propulsion type. These maximum velocities, based on the power curves, line up with the thrust curves, as would be expected. These are equilibrium points, where the aircraft is stable in steady, level, unaccelerated flight where the specific excess power is zero.

It is possible for the power required, P_R, and power available, P_A, curves to intersect at two points, at a low speed and a high speed, labeled as points 1 and 2, respectively, in Figure 5.32. This could occur at low power available settings or at high altitudes. These points represents steady, level flight where the power available equals the power required. Point 1 has a trim speed, $V_{trim,1}$, that is slower than the speed for minimum power required, $V_{P_R,min}$, while the trim speed at Point 2, $V_{trim,2}$, is faster than $V_{P_R,min}$.

Let us first consider flight at Point 2, where the power available equals the power required, $P_A = P_R$, and the aircraft is in steady, level flight at an airspeed, $V_{trim,2}$. If the speed of the aircraft

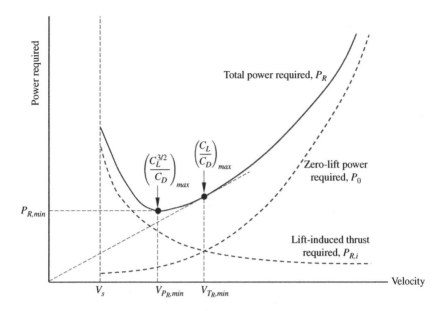

Figure 5.30   Power required curves.

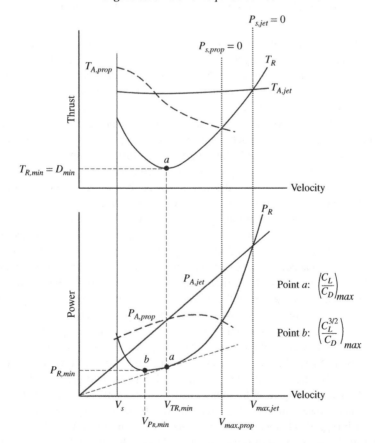

Figure 5.31   Thrust and power curves for propeller-driven and jet-powered aircraft.

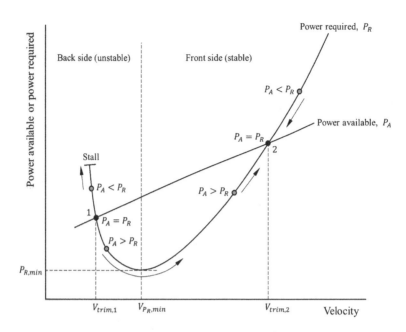

Figure 5.32   Concept of front and back side of the power curve.

increases above $V_{trim,2}$, while maintaining the same power setting, the power available is less than the power required, $P_A < P_R$. This negative excess power causes the aircraft to slow down until it equilibrates back at the Point 2 trim speed. Now, suppose the airspeed decreases from the Point 2 trim speed, again while maintaining the same power setting. At this slower speed, the power available exceeds the power required, $P_A > P_R$, which causes the aircraft to accelerate back to the Point 2 trim speed, where the power available again equals the power required. Thus, we see that the aircraft exhibits a *speed stability*, such that if the aircraft is perturbed from its trim speed, either slower or faster, it moves back *towards* the original, equilibrium, trimmed flight condition. This *stable* region of the power versus velocity curve is called the *front side* of the power curve.

Now, consider the low speed, trimmed Point 1, where the power available equals the power required, $P_A = P_R$, and the aircraft is in steady, level flight at an airspeed, $V_{trim,1}$. If the airspeed increases above, $V_{trim,1}$, while maintaining the same power setting, the power available is greater than the power required, $P_A > P_R$. This excess power causes the aircraft to accelerate to higher airspeed, with an attendant further reduction in the power required. The power required continues to decrease, as the aircraft accelerates, reaching a minimum at $V_{P_R,min}$. The power required starts to increase after this point, but it is still less than the power available, so that the aircraft continues to accelerate until it reaches Point 2, where the power available equals the power required. Returning to the trim condition at Point 1, let us assume that the airspeed decreases from the trim speed, $V_{trim,1}$, again maintaining the same constant power setting. At the slower speed, the power required increases, such that the power available is less than the power required, $P_A < P_R$. This causes the airspeed to decrease further, which then causes a further increase in the power required. The danger with this situation is that the aircraft may continue to decelerate into the stall. The only way that the aircraft can maintain steady, level flight is if the power available is increased. Thus, the aircraft is in a situation where power must be *added* to fly *slower*. In both cases where the airspeed changes from the Point 1 trim speed, whether increasing or decreasing, the aircraft exhibits *speed instability*, moving *away* from the original, equilibrium flight condition. This *unstable* region is called the *back side* of the power curve.

5.8.5 Velocity and Lift Coefficient for Minimum Power Required

The conditions for minimum power required, at a given weight, are different from those for the minimum thrust required. The velocity and lift coefficient, corresponding to minimum power required, are obtained in a similar manner as used for the minimum thrust required condition. Taking the derivate of Equation (5.131) with respect to the velocity, V_∞, and setting this to zero, we have

$$\frac{dP_R}{dV_\infty} = \frac{d}{dV_\infty}\left[\left(\frac{1}{2}\rho_\infty S C_{D,0}\right)V_\infty^3 + \left(\frac{W^2}{\frac{1}{2}\rho_\infty S\pi eAR}\right)\frac{1}{V_\infty}\right] = 0$$

$$\frac{dP_R}{dV_\infty} = \left(\frac{3}{2}\rho_\infty S C_{D,0}\right)V_\infty^2 - \left(\frac{W^2}{\frac{1}{2}\rho_\infty S\pi eAR}\right)\frac{1}{V_\infty^2} = 0$$

$$\left(\frac{3}{2}\rho_\infty S C_{D,0}\right)V_\infty^4 - \left(\frac{W^2}{\frac{1}{2}\rho_\infty S\pi eAR}\right) = 0 \tag{5.132}$$

Solving for the velocity gives

$$V_\infty^4 = \frac{1}{\left(\frac{3}{2}\rho_\infty S C_{D,0}\right)}\left(\frac{W^2}{\frac{1}{2}\rho_\infty S\pi eAR}\right) = \frac{4}{3}\left(\frac{W^2}{\rho_\infty^2 S^2 C_{D,0}\pi eAR}\right) \tag{5.133}$$

$$V_{P_R,min} = \left[\frac{4}{3}\left(\frac{W^2}{\rho_\infty^2 S^2 C_{D,0}\pi eAR}\right)\right]^{1/4} \tag{5.134}$$

where $V_{P_R,min}$ is the velocity in steady, level, unaccelerated flight where the power required is at a minimum. Comparing $V_{P_R,min}$ with $V_{T_R,min}$, from Equation (5.119), we see that

$$V_{P_R,min} = \left(\frac{1}{3}\right)^{1/4}V_{T_R,min} = 0.7598\, V_{T_R,min} \tag{5.135}$$

Thus, the velocity for minimum power required is about 24% lower than the velocity for minimum thrust required or minimum drag, as depicted in Figure 5.30.

To obtain the lift coefficient corresponding to the minimum power required, we substitute Equation (5.134) into Equation (5.111).

$$L = W = q_\infty S C_L = \frac{1}{2}\rho_\infty V_\infty^2 S C_L = \frac{1}{2}\rho_\infty\left[\frac{4}{3}\left(\frac{W^2}{\rho_\infty^2 S^2 C_{D,0}\pi eAR}\right)\right]^{1/2} S C_L$$

$$W = W C_L\sqrt{\frac{1}{3C_{D,0}\pi eAR}}$$

Solving for the lift coefficient, we have

$$(C_L)_{P_R,min} = \sqrt{3C_{D,0}\pi eAR} \tag{5.136}$$

where $(C_L)_{P_R,min}$ is the lift coefficient corresponding to steady, level, unaccelerated flight at the minimum power velocity. Comparing $(C_L)_{P_R,min}$ with $(C_L)_{T_R,min}$, from Equation (5.121), we see that

$$(C_L)_{P_R,min} = \sqrt{3}\,(C_L)_{T_R,min} = 1.7321\,(C_L)_{T_R,min} \tag{5.137}$$

Thus, the lift coefficient corresponding to minimum power required is about 73% greater than the lift coefficient for minimum thrust required or minimum drag.

Solving Equation (5.136) for the zero-lift drag coefficient, we have

$$C_{D,0} = \frac{(C_L)^2_{P_R,min}}{3\pi e AR} = \frac{C_{D,i}}{3} \tag{5.138}$$

At the minimum power required point, the induced drag coefficient is three times greater than the zero-lift drag coefficient. The induced drag coefficient and the zero-lift drag coefficient are equal at the minimum thrust required or minimum drag point.

$$(C_D)_{P_R,min} = C_{D,0} + C_{D,i} = C_{D,0} + 3C_{D,0} = 4C_{D,0} \tag{5.139}$$

Thus, the total drag coefficient at minimum power required is four times the zero-lift drag coefficient. Comparing $(C_D)_{P_R,min}$ with $(C_D)_{T_R,min}$, given by Equation (5.123), we have

$$(C_D)_{P_R,min} = 2(C_D)_{T_R,min} \tag{5.140}$$

Although the drag coefficient is twice as high at the minimum power point as compared with the minimum thrust point, the drag is not this high. This is because the velocity, and hence the dynamic pressure, at the minimum power point is lower than at the minimum thrust point. Consider the ratio of the total drag at minimum power required, $D_{P_R,min}$, to the drag at minimum thrust required, $D_{T_R,min}$, given by

$$\frac{D_{P_R,min}}{D_{T_R,min}} = \frac{(q_\infty)_{P_R,min} S(C_D)_{P_R,min}}{(q_\infty)_{T_R,min} S(C_D)_{T_R,min}} = \frac{\frac{1}{2}\rho_\infty (V_{P_R,min})^2 (C_D)_{P_R,min}}{\frac{1}{2}\rho_\infty (V_{T_R,min})^2 (C_D)_{T_R,min}}$$

$$\frac{D_{P_R,min}}{D_{T_R,min}} = \left(\frac{V_{P_R,min}}{V_{T_R,min}}\right)^2 \frac{(C_D)_{P_R,min}}{(C_D)_{T_R,min}} \tag{5.141}$$

Inserting Equations (5.135) and (5.140) into (5.141), we have

$$\frac{D_{P_R,min}}{D_{T_R,min}} = \left(\frac{0.7598\, V_{T_R,min}}{V_{T_R,min}}\right)^2 \left[\frac{2(C_D)_{T_R,min}}{(C_D)_{T_R,min}}\right] = 2(0.7598)^2 = 1.155 \tag{5.142}$$

Thus, the total drag at minimum power required is only about 16% larger than the total drag at minimum thrust required.

Let us now compare the lift-to-drag ratio at minimum power required, $(L/D)_{P_R,min}$, and at minimum thrust required, $(L/D)_{T_R,min}$, at a given weight, and hence the lift, are the same. The ratio of the lift-to-drag ratios is given by

$$\frac{(L/D)_{P_R,min}}{(L/D)_{T_R,min}} = \frac{1/D_{P_R,min}}{1/D_{T_R,min}} = \frac{D_{T_R,min}}{D_{P_R,min}} \tag{5.143}$$

Inserting Equation (5.142) into (5.143), we have

$$\frac{(L/D)_{P_R,min}}{(L/D)_{T_R,min}} = \frac{D_{T_R,min}}{1.155\, D_{T_R,min}} = \frac{1}{1.155} = 0.8658 \tag{5.144}$$

Since, the lift-to-drag ratio at minimum thrust required is the maximum lift-to-drag ratio, we have

$$\left(\frac{L}{D}\right)_{P_R,min} = 0.8658\left(\frac{L}{D}\right)_{T_R,min} = 0.8658\left(\frac{L}{D}\right)_{max} \tag{5.145}$$

Thus, the lift-to-drag ratio, at minimum power required, is about 86% of the maximum lift-to-drag ratio.

Table 5.10 Summary of thrust required and power required for steady, level, unaccelerated flight.

	Thrust required	Power required	Comparison
	$T_R = q_\infty S C_{D,0} + \dfrac{W^2}{q_\infty S \pi e AR}$	$P_R = q_\infty V_\infty S C_{D,0} + \dfrac{W^2 V_\infty}{q_\infty S \pi e AR}$	$P_R = T_R V_\infty$
V	$V_{T_R,min} = \left(\dfrac{4W^2}{\rho_\infty^2 S^2 C_{D,0} \pi e AR}\right)^{1/4}$	$V_{P_R,min} = \left[\dfrac{4}{3}\left(\dfrac{W^2}{\rho_\infty^2 S^2 C_{D,0} \pi e AR}\right)\right]^{1/4}$	$V_{P_R,min} = 0.7598\, V_{T_R,min}$
C_L	$(C_L)_{T_R,min} = \sqrt{C_{D,0}\pi e AR}$	$(C_L)_{P_R,min} = \sqrt{3 C_{D,0}\pi e AR}$	$(C_L)_{P_R,min} = 1.7321 (C_L)_{T_R,min}$
$C_{D,0}$	$(C_{D,0})_{T_R,min} = C_{D,i}$	$(C_{D,0})_{P_R,min} = \dfrac{C_{D,i}}{3}$	$(C_{D,0})_{P_R,min} = 3(C_{D,0})_{T_R,min}$
C_D	$(C_D)_{T_R,min} = 2C_{D,0}$ or $2C_{D,i}$	$(C_D)_{P_R,min} = 4C_{D,0}$ or $\dfrac{4}{3}C_{D,i}$	$(C_D)_{P_R,min} = 2(C_D)_{T_R,min}$
D	$D_{T_R,min} = 2q_\infty S C_{D,i}$	$D_{P_R,min} = \dfrac{4}{3}q_\infty S C_{D,i}$	$D_{P_R,min} = 1.155 D_{T_R,min}$
$\dfrac{L}{D}$	$\left(\dfrac{L}{D}\right)_{T_R,min} = \left(\dfrac{L}{D}\right)_{max}$	$\left(\dfrac{L}{D}\right)_{P_R,min} = 0.8658\left(\dfrac{L}{D}\right)_{max}$	$\left(\dfrac{L}{D}\right)_{P_R,min} = 0.8658\left(\dfrac{L}{D}\right)_{T_R,min}$

A summary of equations for thrust and power required, including relationships at the minimum thrust or power required condition, for steady, level, unaccelerated flight are given in Table 5.10. A comparison of the thrust required and power required relationships is also provided.

5.8.6 Range and Endurance

Usually, the objective of the cruise segment of a flight profile is to reach a destination. The distance traveled and the time it takes to get there are often important aspects of the flight. In this section, we examine several of these aspects of cruise performance. In particular, we are interested in the *range* and *endurance* associated with cruise performance. The *range*, R, is the total distance, usually measured in air miles, that can be flown for a given fuel load. The units of range are miles or nautical miles in English units and kilometers in SI units. The *endurance* is the time that the aircraft can remain aloft for a given fuel load. The units of endurance are units of time, typically hours.

An aircraft designer would like to maximize both the range and endurance of an aircraft. In maximizing the range, R, we want to maximize the air distance traveled and minimize the weight of fuel consumed, W_f. In maximizing the endurance, E, we want to maximize the time aloft and minimize the weight of fuel consumed, W_f. The range and endurance are affected by the flight conditions and several aircraft-related factors. Flight condition factors include the airspeed, altitude, and ambient temperature. Aircraft-related factors, affecting the range and endurance, include the airframe aerodynamics, the aircraft and fuel weights, and the center of gravity location. Both the range and endurance differ considerably, based on the propulsion type and the associated fuel consumption. We develop equations for range and endurance, specific to propeller-driven and jet-powered aircraft.

In developing equations to quantify the endurance and range, we define the *specific range*, SR, and the *specific endurance*, SE. These specific quantities are defined as the ratio of the range or endurance, respectively, to the weight of fuel consumed, W_f.

The specific range is defined as

$$SR \equiv \frac{\text{range}}{\text{weight of fuel used}} = \frac{R}{W_f} = \frac{V_\infty dt}{W_f} = \frac{V_\infty}{dW_f/dt} = \frac{V_\infty}{\dot{W}_f} \tag{5.146}$$

where V_∞ is the aircraft velocity and $\dot{W}_f$ is the rate of fuel consumption.

The specific endurance is defined as

$$SE \equiv \frac{\text{time aloft}}{\text{weight of fuel used}} = \frac{E}{W_f} = \frac{dt}{dW_f} = \frac{1}{\dot{W}_f} \qquad (5.147)$$

The units of $\dot{W}_f$ are pounds of fuel per hour, lb/h, in English units, or N/h in consistent SI units.

Consider an aircraft with a weight, W_0, composed of the aircraft weight without fuel, W_1, and the fuel weight, W_f, as given by

$$W_0 = W_1 + W_f \qquad (5.148)$$

When an amount of fuel, dW_f, is consumed, the aircraft weight changes by an amount, dW, therefore

$$dW = dW_f \qquad (5.149)$$

The rate of fuel consumption, $\dot{W}_f$, is given by

$$\dot{W}_f = -\frac{dW_f}{dt} = -\frac{dW}{dt} \qquad (5.150)$$

or

$$dt = -\frac{dW}{\dot{W}_f} \qquad (5.151)$$

The range, R, is the integral of the distance increment, ds, from an initial time, t_0, and location, s_0, when the aircraft weight is W_0 to a final time, t_1, and location, s_1, when the aircraft weight is W_1.

$$R = \int_{s_0}^{s_1} ds = \int_{t_0}^{t_1} V_\infty dt = -\int_{W_0}^{W_1} \frac{V_\infty}{\dot{W}_f} dW = -\int_{W_0}^{W_1} SR \, dW \qquad (5.152)$$

Hence, the range is the integral of the specific range from the initial to the final weight.

The endurance, E, is the integral of the time increment, dt, given by Equation (5.151), from an initial time, t_0, when the aircraft weight is W_0 to a final time, t_1, when the aircraft weight is W_1.

$$E = \int_{t_0}^{t_1} dt = -\int_{W_0}^{W_1} \frac{dW}{\dot{W}_f} = -\int_{W_0}^{W_1} \frac{dW}{\dot{W}_f} = -\int_{W_0}^{W_1} SE \, dW \qquad (5.153)$$

Hence, the endurance is the integral of the specific endurance from the initial to the final weight.

We now apply Equations (5.153) and (5.152) to aircraft with specific types of propulsion, the propeller-driven aircraft and the jet-powered aircraft.

5.8.6.1 Range and Endurance for a Propeller-Driven Aircraft

For a propeller-driven aircraft, the change in fuel weight, dW_f, is given by

$$dW_f = dW = -cPdt \qquad (5.154)$$

where c is the specific fuel consumption (see Section 4.5.1.3), P is the engine power, and dt is the increment in time over which the fuel is consumed. Recall that the specific fuel consumption for a piston engine is defined as the weight of fuel consumed per unit of power per unit of time.

Inserting Equation (5.154) into (5.150), the fuel consumption, $\dot{W}_f$, is given by

$$\dot{W}_f = -\frac{dW}{dt} = -\frac{(-cP)}{dt} = \frac{cPdt}{dt} = \frac{cP_A}{\eta_P} = \frac{cT_A V_\infty}{\eta_P} \qquad (5.155)$$

where the power, P, has been replaced in terms of the power available and propeller efficiency, as given by Equation (4.96). The power available is then replaced by the thrust available, T_A, multiplied by the velocity, V_∞, from Equation (4.55).

To obtain an expression for the range, we start by inserting Equation (5.155) into (5.152) to obtain

$$R = -\int_{W_0}^{W_1} \frac{V_\infty}{\dot{W}_f} dW = -\frac{\eta_P}{c} \int_{W_0}^{W_1} \frac{dW}{T_A} \tag{5.156}$$

where the propeller efficiency, η_P, and the specific fuel consumption, c, are assumed to be constants.

Assuming steady, level, unaccelerated flight, the lift is equal to the weight and the thrust available is equal to the drag. Thus, multiplying Equation (5.156) by the lift over the weight, which is equal to unity, and substituting the drag for the thrust available, we have

$$R = -\frac{\eta_P}{c} \int_{W_0}^{W_1} \left(\frac{L}{W}\right) \frac{dW}{D} = -\frac{\eta_P}{c} \left(\frac{C_L}{C_D}\right) \int_{W_0}^{W_1} \frac{dW}{W}$$

$$R = \frac{\eta_P}{c} \left(\frac{C_L}{C_D}\right) \ln\left(\frac{W_0}{W_1}\right) \tag{5.157}$$

Equation (5.157) is the *Breguet range formula*, named after the French airplane designer and aviator, Louis-Charles Breguet. Breguet's airplane company of the early 1900s would eventually become the French airline company, Air France. The Breguet range formula is an expression for the range of a propeller-driven, piston-powered aircraft. This expression tells us that a long-range, propeller-driven aircraft should have a high propeller efficiency, η_P, a large initial-to-final weight, W_0/W_1, maximizing the fuel weight, W_f, and a minimum specific fuel consumption, c. A long-range aircraft should fly at the maximum lift-to-drag ratio, $(L/D)_{max}$, which is the most aerodynamically efficient flight condition.

To obtain an expression for the endurance, we start by inserting Equation (5.155) into (5.153).

$$E = -\int_{W_0}^{W_1} \frac{dW}{\dot{W}_f} = -\frac{\eta_P}{c} \int_{W_0}^{W_1} \frac{dW}{T_A V_\infty} \tag{5.158}$$

where the propeller efficiency, η_P, and the specific fuel consumption, c, are assumed to be constants.

Again assuming steady, level, unaccelerated flight, the lift is equal to the weight and the thrust available is equal to the drag. Thus, multiplying Equation (5.158) by the lift over the weight, which is equal to unity, and substituting the drag for the thrust available, we have

$$E = -\frac{\eta_P}{c} \int_{W_0}^{W_1} \left(\frac{L}{W}\right) \frac{dW}{D V_\infty} = -\frac{\eta_P}{c} \int_{W_0}^{W_1} \left(\frac{L}{D}\right) \frac{dW}{V_\infty W} \tag{5.159}$$

or

$$E = -\frac{\eta_P}{c} \left(\frac{C_L}{C_D}\right) \int_{W_0}^{W_1} \frac{dW}{V_\infty W} \tag{5.160}$$

where the lift-to-drag ratio, C_L/C_D, is assumed to be constant.

Using the definition of the lift, which equals the weight, we obtain an expression for the velocity, as

$$L = W = \frac{1}{2}\rho_\infty V_\infty^2 S C_L$$

$$V_\infty = \sqrt{\frac{2W}{\rho_\infty S C_L}} \tag{5.161}$$

Substituting Equation (5.161) into (5.160), we have

$$E = -\frac{\eta_P}{c} \left(\frac{C_L}{C_D}\right) \int_{W_0}^{W_1} \sqrt{\frac{\rho_\infty S C_L}{2W}} \frac{dW}{W} = -\frac{\eta_P}{c} \left(\frac{C_L}{C_D}\right) \sqrt{\frac{\rho_\infty S C_L}{2}} \int_{W_0}^{W_1} \frac{dW}{W^{3/2}} \tag{5.162}$$

where the density, ρ_∞, is assumed to be constant, which means that the aircraft is at a constant altitude for the evaluation of the endurance. Performing the integration, we have

$$E = 2\frac{\eta_P}{c}\left(\frac{C_L^{3/2}}{C_D}\right)\sqrt{\frac{\rho_\infty S}{2}}[W^{-1/2}]_{W_0}^{W_1}$$

$$E = \frac{\eta_P}{c}\left(\frac{C_L^{3/2}}{C_D}\right)\sqrt{\frac{\rho_\infty S}{2}}\left(\frac{1}{\sqrt{W_1}} - \frac{1}{\sqrt{W_0}}\right) \tag{5.163}$$

Equation (5.163) is the *Breguet endurance formula* for the endurance of a propeller-driven, piston-powered aircraft. This expression tells us that a high endurance, propeller-driven aircraft should have a high propeller efficiency, η_P, a very large wing area, S, maximum fuel weight, W_f (which maximizes W_0 and minimizes W_1), and have a minimum specific fuel consumption, c. It should fly at sea level, to maximize the air density, ρ_∞, and at a flight condition where the quantity $C_L^{3/2}/C_D$ is maximized. The endurance is proportional to $C_L^{3/2}/C_D$, while the range is proportional to C_L/C_D.

5.8.6.2   Range and Endurance for a Jet-Powered Aircraft

For a jet-powered aircraft, the change in fuel weight, dW_f, is given by

$$dW_f = dW = -(TSFC)T_A dt \tag{5.164}$$

where *TSFC* is the thrust specific fuel consumption (see Section 4.5.1.3). Recall that the thrust specific fuel consumption is defined as the weight flow rate of fuel consumed per unit thrust.

Using Equation (5.154), the fuel consumption, $\dot{W}_f$, is given by

$$\dot{W}_f = -\frac{dW}{dt} = -\frac{-(TSFC)T_A dt}{dt} = (TSFC)T_A \tag{5.165}$$

Inserting Equation (5.165) into (5.152), the range is given by

$$R = -\int_{W_0}^{W_1} V_\infty \frac{dW}{\dot{W}_f} = -\int_{W_0}^{W_1} V_\infty \frac{dW}{(TSFC)T_A} \tag{5.166}$$

Assuming steady, level, unaccelerated flight, the lift is equal to the weight and the thrust available is equal to the drag. Thus, multiplying Equation (5.166) by the lift over the weight, which is equal to unity, and substituting the drag for the thrust available, we have

$$R = -\frac{1}{TSFC}\int_{W_0}^{W_1}\left(\frac{L}{W}\right)\frac{V_\infty}{D}dW = -\frac{1}{TSFC}\left(\frac{C_L}{C_D}\right)\int_{W_0}^{W_1}\frac{V_\infty}{W}dW \tag{5.167}$$

where the thrust specific fuel consumption, *TSFC*, and the lift-to-drag ratio, C_L/C_D, are assumed to be constants. Inserting Equation (5.161) for the velocity, we have

$$R = -\frac{1}{TSFC}\left(\frac{C_L}{C_D}\right)\int_{W_0}^{W_1}\sqrt{\frac{2W}{\rho_\infty S C_L}}\frac{dW}{W} = -\frac{1}{TSFC}\left(\frac{\sqrt{C_L}}{C_D}\right)\sqrt{\frac{2}{\rho_\infty S}}\int_{W_0}^{W_1}\frac{dW}{\sqrt{W}} \tag{5.168}$$

where the density, ρ_∞, is assumed to be constant, which means the aircraft is at a constant altitude for the evaluation of the range. Performing the integration, we have

$$R = -\frac{2}{TSFC}\left(\frac{\sqrt{C_L}}{C_D}\right)\sqrt{\frac{2}{\rho_\infty S}}[W^{-1/2}]_{W_0}^{W_1}$$

$$R = \frac{2}{TSFC}\left(\frac{\sqrt{C_L}}{C_D}\right)\sqrt{\frac{2}{\rho_\infty S}}(\sqrt{W_0} - \sqrt{W_1}) \tag{5.169}$$

Equation (5.169) is the range formula for a jet-powered aircraft. This expression tells us that a long-range, jet-powered aircraft should have a small wing area, S, maximum fuel weight, W_f, (maximum difference between W_0 and W_1), and a minimum thrust specific fuel consumption, $TSFC$. It should fly at high altitude, where the air density, ρ_∞, is low and at a flight condition where the quantity $\sqrt{C_L}/C_D$ is maximized.

To obtain an expression for the endurance, we start by inserting Equation (5.165) into (5.152).

$$E = \int_{t_0}^{t_1} dt = -\int_{W_0}^{W_1} \frac{dW}{\dot{W}_f} = -\int_{W_0}^{W_1} \frac{dW}{(TSFC)T_A} \tag{5.170}$$

Assuming steady, level, unaccelerated flight, the lift is equal to the weight and the thrust available is equal to the drag. Thus, multiplying Equation (5.170) by the lift over the weight, which is equal to unity, and substituting the drag for the thrust available, we have

$$E = -\frac{1}{(TSFC)}\int_{W_0}^{W_1}\left(\frac{L}{W}\right)\frac{dW}{D} = -\frac{1}{(TSFC)}\left(\frac{C_L}{C_D}\right)\int_{W_0}^{W_1}\frac{dW}{W} \tag{5.171}$$

where the thrust specific fuel consumption, $TSFC$, and the lift-to-drag ratio, C_L/C_D, are assumed to be constants. Performing the integration, we have

$$E = \left(\frac{1}{TSFC}\right)\left(\frac{C_L}{C_D}\right)\ln\left(\frac{W_0}{W_1}\right) \tag{5.172}$$

Equation (5.172) is the endurance formula for a jet-powered aircraft. This expression tells us that a long-endurance, jet-powered aircraft should have a large initial-to-final weight, W_0/W_1, maximizing the fuel weight, W_f, and a minimum thrust specific fuel consumption, $TSFC$. It should fly at the maximum lift-to-drag ratio, $(L/D)_{max}$, which is the most aerodynamically efficient flight condition.

The range and endurance equations for propeller-driven and jet-powered aircraft are summarized in Table 5.11.

Example 5.12 Calculation of Range and Endurance *Using the specifications and results from Examples 5.9 and 5.11, calculate the range and endurance of the Boeing 747 with an airspeed, V_∞, of 400 km/h at sea level. Assume an initial weight, W_0, of 2,700,000 N, a final weight, W_1, of 1,900,000 N, and a thrust specific fuel consumption, TSFC, of 1.678×10^{-4} N/(N·s).*

Table 5.11 Range and endurance equations for jet-powered and propeller-driven aircraft.

Parameter	Propeller-driven aircraft	Jet-powered aircraft
Range, R	$\dfrac{\eta_P}{c}\left(\dfrac{C_L}{C_D}\right)\ln\left(\dfrac{W_0}{W_1}\right)$	$\dfrac{2}{TSFC}\left(\dfrac{\sqrt{C_L}}{C_D}\right)\sqrt{\dfrac{2}{\rho_\infty S}}(\sqrt{W_0} - \sqrt{W_1})$
Endurance, E	$\dfrac{\eta_P}{c}\left(\dfrac{C_L^{3/2}}{C_D}\right)\sqrt{\dfrac{\rho_\infty S}{2}}\left(\dfrac{1}{\sqrt{W_1}} - \dfrac{1}{\sqrt{W_0}}\right)$	$\left(\dfrac{1}{TSFC}\right)\left(\dfrac{C_L}{C_D}\right)\ln\left(\dfrac{W_0}{W_1}\right)$

Solution

From Example 5.9, the Boeing 747 wing area, S, is 520.2 m². From Example 5.11, the lift and drag coefficients of the Boeing 747 at 400 km/h are 0.7192 and 0.07028, respectively.

Using Equation (5.169), the range is given by

$$R = \frac{2}{TSFC} \left(\frac{\sqrt{C_L}}{C_D} \right) \sqrt{\frac{2}{\rho_\infty S}} (\sqrt{W_0} - \sqrt{W_1})$$

$$R = \frac{2}{\left(1.678 \times 10^{-4} \frac{N}{N \cdot s} \right)} \left(\frac{\sqrt{0.7192}}{0.07028} \right) \sqrt{\frac{2}{\left(1.225 \frac{kg}{m^3} \right) (520.2 \ m^2)}}$$

$$\times (\sqrt{2,700,000 \ N} - \sqrt{1,900,000 \ N})$$

$$R = 2.134 \times 10^6 \ m = 2,133 \ km$$

Using Equation (5.172), the endurance is given by

$$E = \left(\frac{1}{TSFC} \right) \left(\frac{C_L}{C_D} \right) \ln \left(\frac{W_0}{W_1} \right)$$

$$E = \frac{1}{\left(1.678 \times 10^{-4} \frac{N}{N \cdot s} \right)} \left(\frac{0.7192}{0.07028} \right) \ln \left(\frac{2,700,000 \ N}{1,900,000 \ N} \right) = 21,430 \ s = 5.95 \ h$$

5.8.7 FTT: Cruise Performance

Level flight or cruise performance flight testing is conducted to determine the endurance and range of an aircraft. Quantifying and understanding these cruise performance parameters is critical to the operation of all aircraft. The endurance and range are fundamentally functions of the characteristics of the aircraft airframe and its propulsion system. The aircraft's airframe determines the aerodynamics, the lift and drag, and the weight of fuel that may be carried, while the thrust and fuel flow are related to the propulsion system. Cruise performance flight testing is usually straightforward, in terms of the types of maneuvers involved. Data is collected with the aircraft in steady, wings level flight. In this steady-state flight condition, the aircraft lift equals the weight and the thrust equals the drag.

A legendary example of the demonstration of aircraft range and endurance is the transatlantic flight of Charles Lindbergh in 1927. Lindbergh, then a 25-year-old, US Air Mail pilot, was the first person to fly solo, non-stop, across the Atlantic Ocean, from New York to Paris, France. He flew a specially redesigned, single-engine, Ryan M-2 monoplane, designated the Ryan NYP (for New York to Paris) designed and built by the Ryan Airlines, Inc., San Diego, California (see frontispiece photo of Chapter 5 and Figure 3.137). The aircraft was christened the *Spirit of St Louis*, in honor of Lindbergh's financial supporters in St Louis, Missouri. Lindbergh's long-range, long-endurance flight was a testament to aircraft capabilities of the time. Taking off from Roosevelt Airfield in Garden City (Long Island), New York, he landed 33½ hours later at Le Bourget Aerodrome in Paris, France, covering a distance of about 3610 miles (5810 km). It was also a testament to human endurance, as by the time Lindbergh landed in Paris, he had not slept for about 55 hours, due to preparations and anticipation of the flight.

Modifications to the Ryan M-2 included lengthening of the wing by 10 ft (3.05 m) and increasing the fuel capacity to 450 gallons (1700 liters). The 36 in (91 cm) wide, 32 in (81 cm) long, and 51 in

(129 cm) high cockpit was cramped for Lindbergh's 6 ft, 3 in (1.9 m) frame. Fully loaded, the *Spirit of St Louis* had a gross weight of 5135 lb (2330 kg) and an empty weight of 2150 lb (975 kg). The Ryan NYP was powered by an air-cooled, nine-cylinder Wright *Whirlwind* J-5C radial engine, which produced 223 hp (166 kW) at 1800 rpm. A three-view drawing of the Ryan NYP is shown in Figure 5.33 and selected specifications are given in Table 5.12.

You will fly the cruise performance flight test techniques in the *Spirit of St Louis*. You climb into the cramped cockpit and sit in the pilot seat, which to your surprise is simply a wicker seat, as shown in Figure 5.34. There are conventional stick and rudder flight controls and a throttle lever on your left side. The instrument panel is sparse, with just the essential flight instruments. Another surprise is that you have no forward visibility from the cockpit! The cockpit forward windscreen has been sacrificed to accommodate a large fuselage fuel tank. There is a periscope device, with an angled mirror that extends from the left side of the fuselage (the periscope is shown in its extended position on the left side of Figure 5.34.). You can look at the angled mirror through a small, 3×5 in (7.6×12.7 cm) rectangular hole in the cockpit panel, giving you a glimpse of what is in front of you. The sliding horizontal lever, next to the rectangular hole, allows you to retract the periscope to reduce aerodynamic drag.

Below the instrument panel, you see an array of plumbing lines and levers, called a Lunkenheimer distributor, which is connected to all of the fuel tanks. There are a total of five fuel tanks, three in the wings, one in the center fuselage, and one in the forward fuselage. It is possible to pump fuel from any tank to any other using the distributor, which is useful in maintaining the proper aircraft longitudinal and lateral balance. With its 450 gallon (2754 lb, 1249 kg) fuel capacity, the *Spirit of St Louis* has a fuel fraction (fuel weight divided by fully loaded, gross weight) of 52.6%, which was much higher than aircraft of the era. For comparison, typical fuel fractions are about 12–15% for a modern general aviation airplane and about 40% for a modern commercial jet airliner. A few specialized airplanes, designed for very long-endurance flight, are constructed of lightweight composite materials and have had very high fuel fractions of 70–85%.

The *Spirit of St Louis* is fully fueled with 450 gallons of gasoline for your flight. After engine start, you taxi out to the end of a long grass airstrip for takeoff. You are starting your flight early in the morning so that there is a better chance of smooth air, which is needed to obtain high accuracy cruise performance data. Lined up at the end of the grass runway, you push the throttle full forward and the aircraft starts to move forward. As you trundle down the grass runway, you push the control stick forward, allowing the tailwheel to lift off the ground so that you are rolling on the two main gear tires. The aircraft is at its maximum takeoff weight of 5135 lb (2330 kg), so you are expecting a long takeoff ground roll, especially since the coefficient of rolling resistance is some 30–40% greater for the grass compared to a hard surfaced runway (see Table 5.17). You pull back on the stick slightly to coax the aircraft into the air as soon as it will fly. The main tires lift off the grass and you are airborne. You keep the aircraft within a wingspan of the ground, in ground effect, letting the airspeed build. Finally, with ample flying speed, you start your climb to your first cruise performance test point.

Cruise performance flight test techniques are flown in wings level flight, at constant altitude and airspeed. Part of the difficulty of cruise performance flight testing is the large amount of data that is required. Typically, it is desired to characterize the level flight performance of an aircraft over its full flight envelope and for a variety of gross weights, which can result in a large matrix of test points. Usually, data is collected in increments of altitude, typically every 5000 ft (1500 m) or so, over a range of airspeeds, and for selected gross weights. Data may be collected at only the maximum gross weight, which may be the only data presented in a flight manual.

You start by collecting some cruise performance flight data using the *constant pressure altitude* flight test technique. You level off in the *Spirit of St Louis* at a pressure altitude of 4000 ft (1220 m). You advance the throttle full forward with the engine tachometer reading about 1950 rpm. Under

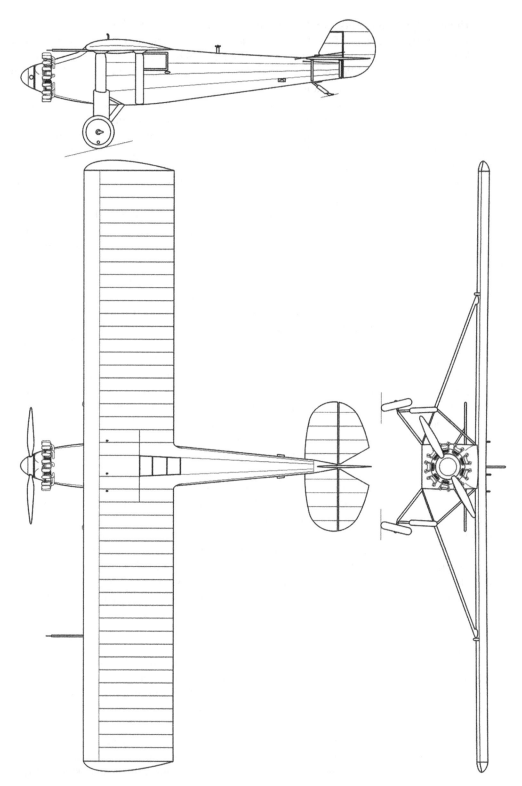

Figure 5.33   Three-view drawing of the Ryan NYP *Spirit of St Louis*. (Source: *Kaboldy, "Ryan NYP" https://en.wikipedia.org/wiki/File:Ryan_NYP.svg, CC-BY-SA-4.0. License at https://creativecommons.org/ licenses/by-sa/4.0/legalcode.*)

Table 5.12   Selected specifications of the Ryan NYP *Spirit of St Louis*.

Item	Specification
Primary function	Long endurance flight, 1st aircraft to cross the Atlantic Ocean
Manufacturer	Ryan Aeronautical Company, San Diego, California
First flight	28 April 1927
Crew	1 pilot
Powerplant	Wright *Whirlwind* J-5C air-cooled, nine-cylinder, radial engine
Engine power	223 hp (166 kW) at 1800 rpm
Fuel capacity	450 gallons (1700 liters) of gasoline
Empty weight	2150 lb (975 kg)
Maximum takeoff weight	5135 lb (2330 kg)
Fuel fraction	0.526
Power-to-weight ratio*	23.6 bhp/lb (10.7 hp/kg)
Length	27 ft 7 in (8.41 m)
Wingspan	46 ft (14 m)
Wing chord	7 ft (2.1 m)
Wing area	319 ft^2 (29.6 m^2)
Wing airfoil	Clark Y
Wing loading*	16.5 lb/ft^2 (80.6 kg$_f$/m^2)
Airfoil	Clark Y
Maximum speed*	120 mph (193 km/h)
Economic speed*	97 mph (156 km/h)
Range	4110 mi (6614 km)
Service ceiling	16,400 ft (5000 m)

*At maximum takeoff weight

full power, the aircraft accelerates to its maximum flight speed at this altitude. Now, you must have patience and hold the trim condition steady to allow the engine and airspeed to stabilize. This is a time-consuming process, but it is critical in obtaining accurate level performance flight data. Typical airspeed stabilization requirements are a change of no more than 1 knot per minute, which equates to a change in the flight path acceleration of about 0.001 g and a change in the drag or fuel flow of approximately 1%. After several minutes, you feel that the aircraft is stable at the test point, with the altitude changing less than ±100 ft (30.5 m) and the airspeed stable at the maximum level airspeed of 120 mph (193 km/h).

Given the steady-state nature of these cruise performance test points, you can hand-record much of the flight data. You record the altitude, airspeed, engine parameters, air temperature, fuel flow, and fuel quantity. This data set is for the aircraft near its maximum gross weight. Often, a high accuracy data system is used to measure the flight path acceleration, fuel flow, and other parameters.

With this test point complete, you reduce the engine power and repeat the above process at another lower stabilized airspeed. You obtain test points at this altitude for airspeeds down through the low speed end of the flight envelope. After the airspeed range has been covered at this altitude, you climb up to several higher altitudes and collect data using the constant altitude FTT. The data collection process is very time-consuming, as each test point requires several minutes to stabilize.

Power available and power required curves, based on flight test data, for the Ryan NYP *Spirit of St Louis* are shown in Figure 5.35. These curves highlight many of the items of level, flight performance that have been discussed. There are three power required curves shown, corresponding to three different aircraft weights of the *Spirit of St Louis*. The power required curves shift up (higher power) and to the right (higher velocity) with increasing weight. The minimum power required, and

Figure 5.34   Cockpit and instrument panel of Ryan NYP *Spirit of St Louis*. (Source: *National Air and Space Museum, Smithsonian Institution.*)

the associated velocity, increase significantly with increasing weight, from about 49 bhp (36.5 kW) and 57 mph (91.7 km/h) at a gross weight of 2415 lb (1095 kg) to about 154 bhp (115 kW) and 83 mph (134 km/h) at a gross weight of 5130 lb (2327 kg). The left uppermost boundary of the power required curve is likely at or very close to the stall speed, which increases significantly with increasing weight. The stall speed at 2415 lb is about 48 mph (77.2 km/h) increasing to about 72 mph (116 km/h) at 5130 lb. The power available curve is a straight line near the top of the chart, showing a maximum power of 237 bhp (177 kW) at 1950 rpm. The excess power, the difference between the power available and power required curves, decreases significantly with increasing weight. The maximum airspeed is found at the high-speed intersection of the power available and power required curves. The maximum airspeed is about 125 mph (201 km/h) at 2415 lb and about 120 mph (193 km/h) at 5130 lb.

Endurance versus distance is plotted for the *Spirit of St Louis* in Figure 5.36. Curves are shown for a no-wind condition and for a tail wind of 10 mph (16 km/h). The endurance is given as 47.5 hours

N.A.C.A. Technical Note No.257

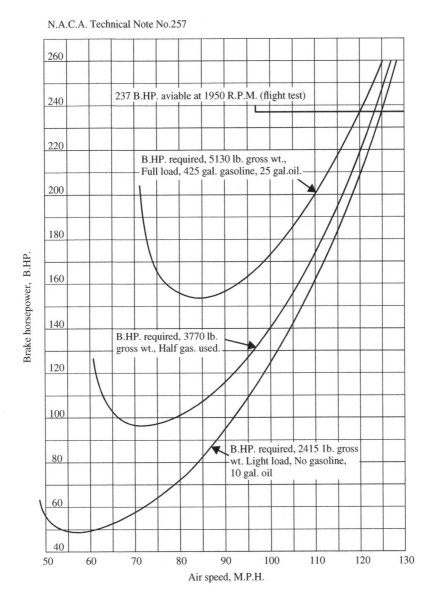

Figure 5.35 Ryan NYP *Spirit of St Louis* power required and power available. (Source: *Hall, NACA TN 257, 1927,* [11].)

for what was termed a "practical speed", which was 95 mph (153 km/h) at the start of the flight at heavy weight and slowed to 75 mph (121 km/h) at the end of the flight at light weight. The range associated with these speeds is given as 4040 miles (6502 km) at a no-wind condition.

There are several schemes to more efficiently evaluate the level flight performance, rather than flying every flight condition in steady, level flight. One such scheme is the *range factor method*. The range factor, *RF*, is defined as

$$RF \equiv SR \times W = \frac{R}{W_f/W} = \frac{V_\infty}{\dot{W}_f}W \qquad (5.173)$$

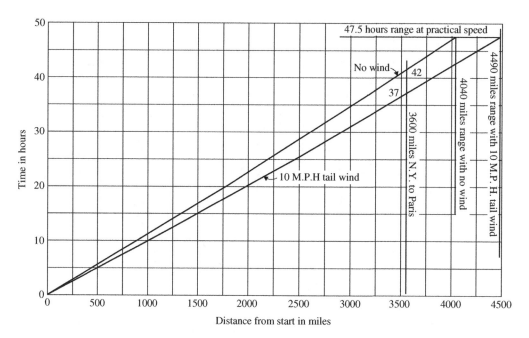

Figure 5.36 Ryan NYP *Spirit of St Louis* endurance versus distance. (Source: *Hall, NACA TN 257, 1927,* [11].)

where SR is the specific range, defined by Equation (5.146), W is the aircraft total weight, and $\dot{W}_f$ is the fuel flow. The range factor is directly proportional to the range, R, and inversely proportional to the fuel fraction, W_f/W. Hence, the maximum range factor corresponds to the maximum range for a given fuel fraction.

Expressing the velocity, V_∞, in terms of the Mach number, M_∞, and speed of sound, a_∞, we have

$$RF = \left(\frac{W}{\dot{W}_f}\right) M_\infty \sqrt{\gamma R T_\infty} \tag{5.174}$$

Dividing the weight and fuel flow by the pressure ratio, $\delta = p_\infty/p_{SL}$, and inserting the temperature ratio, $\theta = T_\infty/T_{SL}$, gives

$$RF = \left(\frac{W/\delta}{\dot{W}_f/\delta}\right) M_\infty \sqrt{\gamma R \theta T_{SL}} = \left[\frac{W/\delta}{\dot{W}_f/(\delta\sqrt{\theta})}\right] M_\infty \sqrt{\gamma R T_{SL}} \tag{5.175}$$

If the fuel flow is assumed constant, the range factor is a function of two parameters for a given altitude, the Mach number and a new parameter, W/δ. Hence, we seek to find the Mach number and value of W/δ, corresponding to the maximum range factor. To determine this, several flights are performed, at different values of W/δ, taking data over the airspeed or Mach range of the aircraft for each value of W/δ. From each of these constant W/δ flights, there is a Mach number where the specific range is a maximum, as shown in Figure 5.37a. The range factor, corresponding to the maximum specific range, for each W/δ flight, is calculated, using Equation (5.175), and plotted versus the flight Mach number, as shown in Figure 5.37b. From this plot, the maximum range factor and the corresponding optimum Mach number are determined. Finally, using this optimum Mach

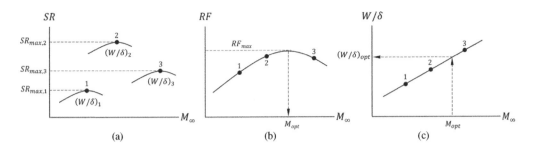

Figure 5.37 Determination of flight Mach number and W/δ for maximum range.

number, the optimum value of W/δ is obtained from a plot of W/δ, corresponding to the maximum specific range, versus Mach number, as shown in Figure 5.37c.

Let us consider what a constant W/δ flight profile might look like for the *Spirit of St Louis*. Let us assume that you have again taken off at a maximum gross weight of 5135 lb and climbed up to an altitude of 4000 ft. You have burned off a little fuel, so that your weight is now 5120 lb (2320 kg). At 4000 ft, the static pressure is 1827.7 lb/ft^2 (87,510 N/m^2), making the pressure ratio equal to

$$\delta = \frac{p_\infty}{p_{SL}} = \frac{1827.7 \text{ lb/ft}^2}{2116.2 \text{ lb/ft}^2} = 0.8637 \tag{5.176}$$

The value of W/δ is

$$\frac{W}{\delta} = \frac{5120 \text{ lb}}{0.8637} = 5927.8 \text{ lb} \tag{5.177}$$

As you continue to fly, fuel is burned and the aircraft weight decreases. To maintain a constant W/δ of 5927.8 lb, the pressure ratio must also decrease. Hence, you must increase altitude or climb to maintain the desired constant W/δ as the weight decreases. The flight profile required to maintain a constant W/δ of 5927.8 lb is shown in Figure 5.38. To fly a constant W/δ profile, these types of

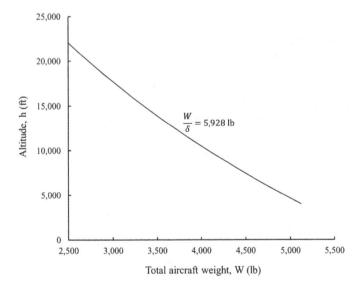

Figure 5.38 Altitude versus weight for constant W/δ.

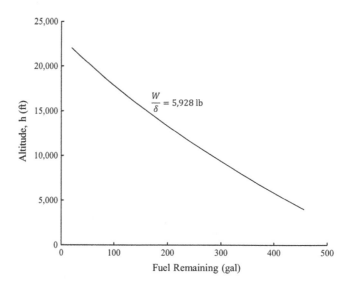

Figure 5.39   Altitude versus fuel remaining for constant W/δ.

plots must be prepared before the flight so that the pilot can adjust the aircraft altitude accordingly as the fuel is burned off. In practice, the W/δ profile is usually prepared as in Figure 5.39, where the altitude is plotted versus the fuel quantity remaining, which is more easily tracked in real-time during a flight.

As an addendum to this FTT, we mention another noteworthy long-endurance transatlantic flight that occurred between 9 and 11 August 2003. On 9 August 2003, *The Spirit of Butts' Farm* model airplane was launched from Cape Spear, near St Johns Newfoundland, and landed, 38 hours, 52 minutes, 19 s later on 11 August 2003, at Mannin Beach, near Clifden, Ireland, becoming the first model airplane to fly across the Atlantic Ocean. The model airplane flew a distance of 1881 miles (3028 km), maintaining an average speed of 48 mph (77 km/h), with a tail wind, and an altitude of approximately 1000 ft (300 m), being controlled by an autopilot for most of the flight. The model airplane consumed 99.2% of its fuel load during the flight, landing with only about 1.5 ounces (44 ml) of fuel remaining.

The Spirit of Butts' Farm was named after R. Beecher Butts, whose farm was used for much of the flight testing, and also in homage of Charles Lindbergh's *Spirit of St Louis*. It was also designated the TAM-5, for Transatlantic Model No. 5. The TAM-5 was the fifth attempt to cross the Atlantic, with four previous TAM airplanes crashing into the ocean due to mechanical or weather-related problems.

The model airplane was designed by record-setting aeromodeler and retired American metallurgist, Maynard Hill. The airplane was designed to be as simple and "low-tech" as possible, to increase reliability, and to reduce weight. A three-view drawing of the TAM-5 is shown in Figure 5.40 and selected specifications are given in Table 5.13. Constructed of balsa wood, with a Mylar covering, the model airplane had only one aileron, no rudder, and no landing gear. The TAM-5 had a length of about 6 ft (1.8 m), a wingspan of about 6 ft, and had a fully fueled, gross weight of about 11 lb (5 kg). The airplane was powered by a 0.61 cubic inch (10 cc), 4-stroke model airplane engine, burning lantern fuel and turning a 14-inch (35.6 cm) diameter propeller at 3800 rpm.

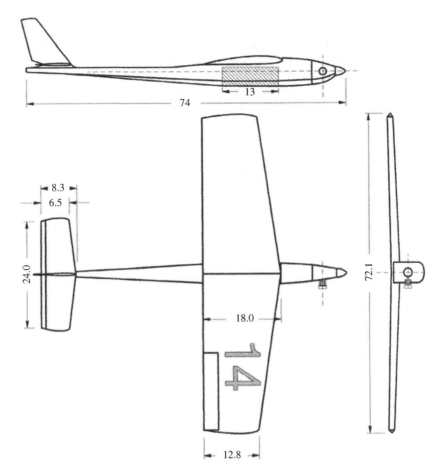

Figure 5.40   Three-view drawing of the TAM-5 model aircraft (dimensions in inches). (Source: *Courtesy of Maynard Hill, drawn by Art Kresse.*)

Table 5.13   Selected specifications of the TAM-5 *Spirit of Butts' Farm.*

Item	Specification
Primary function	Long endurance flight
Designer and manufacturer	Maynard Hill, Silver Spring, Maryland
Crew	Unmanned
Powerplant	OS Engines 0.61 cubic inch (10 cc) four-stroke engine
Fuel capacity	118 oz (1.49 liters) of lantern fuel
Empty weight	5.96 lb (2.70 kg)
Gross weight	10.99 lb (4.987 kg)
Length	74 in (188 cm)
Wingspan	72.1 in (183 cm)
Cruise speed	42 mph (68 km/h)
Range*	1881 miles (3028 km)
Ceiling*	~1000 ft (300 m)

*As flown for transatlantic crossing

5.9   Climb Performance

After takeoff, all aircraft must climb to clear obstacles and reach cruising altitudes. Climb performance is directly linked to propulsive and aerodynamic characteristics of the aircraft. This is embodied in the excess thrust, the difference between the thrust available and the thrust required. Of course, the aircraft weight is a factor in the climb performance also. In this section, we examine several aspects of climb performance, including the rate of climb, the angle of climb, the time to climb, and fuel used during climb.

5.9.1   Maximum Angle and Maximum Rate of Climb

Two important climb performance parameters are the maximum angle of climb and the maximum rate of climb. The maximum angle of climb provides the maximum flight path angle for terrain or obstacle clearance. The maximum rate of climb provides the maximum altitude gain in the shortest time. In addition to obtaining the values for the maximum climb angle and the maximum rate, the velocities for these climbs are important in flying these maximum performance climbs.

Consider an aircraft in a steady constant airspeed climb with a flight path angle, γ, as depicted in Figure 5.41. As with steady, level, unaccelerated flight, it is assumed that the thrust angle and the climb angles are small. Summing the forces parallel and perpendicular to the flight direction, respectively, we have

$$T = D + W \sin \gamma \tag{5.178}$$

$$L = W \cos \gamma \cong W \tag{5.179}$$

where the climb angle, γ, is assumed to be small, such that $\cos \gamma \cong 1$.

By assuming that the climb angle is small, the equation of motion for climbing flight, perpendicular to the flight path, is the same as for level flight. Thus, the level flight results for thrust required and thrust available are assumed valid for climbing flight. These assumptions restrict the following climbing flight analysis to small climb angles less than about 15–20°. Therefore, for a steady, unaccelerated climb, the thrust and drag correspond to the thrust available and thrust required,

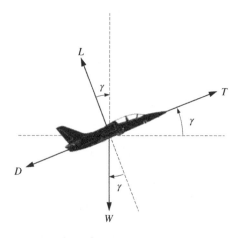

Figure 5.41   Forces on aircraft in steady, constant velocity climb.

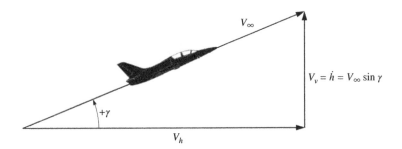

Figure 5.42   Geometry for steady, constant velocity climb.

respectively. Equations (5.178) and (5.179) can be written as

$$T_A = D - W \sin\gamma = T_R - W \sin\gamma \tag{5.180}$$

$$L = W \tag{5.181}$$

In climbing flight, the thrust is less than the drag and the lift is less than the weight. Unlike level flight, the thrust is now supporting a portion of the aircraft weight. The geometry for a normal climb is shown in Figure 5.42. The aircraft is in a steady, unaccelerated climb at a constant velocity, V_∞, and constant climb angle, γ. The horizontal and vertical velocities are V_h and V_v, respectively.

Solving Equation (5.180) for the climb angle, we have

$$\sin\gamma = \frac{T - D}{W} = \frac{T_A - T_R}{W} \tag{5.182}$$

or

$$\gamma = \sin^{-1}\left(\frac{T - D}{W}\right) = \sin^{-1}\left(\frac{T_A - T_R}{W}\right) \tag{5.183}$$

where the angle is proportional to the excess thrust, $T - D$, and inversely proportional to the weight, W. The climb angle increases with increasing excess thrust or decreasing weight. Conversely, less excess thrust of a heavier aircraft result in a lower climb angle.

The excess thrust divided by weight, in Equation (5.183), is defined as the *specific excess thrust*.

$$\text{Specific excess thrust} \equiv \frac{T - D}{W} = \frac{T_A - T_R}{W} \tag{5.184}$$

(Recall that we make a quantity a *specific* quantity, by dividing by the weight.) The climb angle is proportional to the specific excess thrust. The maximum climb angle, γ_{max}, commonly called the *best angle of climb*, is obtained when the specific excess thrust is maximized, as given by

$$\gamma_{max} = \sin^{-1}\left(\frac{T - D}{W}\right)_{max} = \sin^{-1}\left(\frac{T_A - T_R}{W}\right)_{max} \tag{5.185}$$

The maximum excess thrust is shown for a propeller-driven and a jet-powered aircraft, at a given weight, by the thrust curves on the left side of Figure 5.43. The velocity corresponding to the maximum excess thrust is the *best angle of climb speed*, designated as V_x. The curves in Figure 5.43 are generic, therefore it should not be inferred that the best angle of climb speed for a propeller-driven aircraft is necessarily lower than for a jet-powered aircraft.

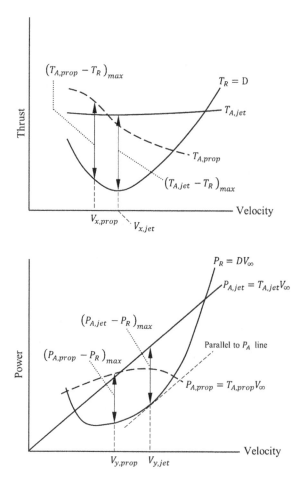

Figure 5.43   Climb performance for propeller-driven and jet-powered aircraft.

The rate of climb is equal to the rate of change of altitude with time, dh/dt, which equals the vertical velocity, V_v. From Figure 5.42, the vertical velocity or rate of climb is given by

$$V_v = \frac{dh}{dt} = \dot{h} = V_\infty \sin \gamma \tag{5.186}$$

Using Equation (5.182), we have

$$\dot{h} = V_\infty \left(\frac{T - D}{W} \right) = V_\infty \left(\frac{T_A - T_R}{W} \right) \tag{5.187}$$

Since power is equal to thrust times velocity, we have

$$\dot{h} = \frac{T_A V_\infty - T_R V_\infty}{W} = \frac{P_A - P_R}{W} \equiv P_s = \text{specific excess power} \tag{5.188}$$

The rate of climb is equal to the excess power, $P_A - P_R$, divided by the weight, a quantity defined as the *specific excess power*, P_s. The maximum rate of climb, $\dot{h}_{max}$, is obtained when the specific excess power is maximized, as given by

$$\dot{h}_{max} = \left(\frac{P_A - P_R}{W} \right)_{max} \tag{5.189}$$

This is shown by the power curves on the right side of Figure 5.43, for a propeller-driven and a jet-powered aircraft. The maximum rate of climb airspeed, designated as V_y, is the airspeed corresponding to the maximum specific excess power, as shown in the figure. Again, these curves are generic, so it should not be inferred that the best rate of climb speed for a propeller-driven aircraft is necessarily lower than for a jet-powered aircraft. The best rate of climb airspeed, denoted as V_y, is also shown for the two propulsion types.

5.9.2   Time to Climb

Another important climb performance metric is the time to climb from one altitude to another. From Equation (5.186), the time increment, dt, to climb an altitude increment, dh, is given by

$$dt = \frac{dh}{\dot{h}} \tag{5.190}$$

Integrating Equation (5.190) from a starting time and altitude, t_0 and h_0, respectively, to a final time and altitude, t_1 and h_1, respectively, gives

$$\int_{t_0}^{t_1} dt = \int_{h_0}^{h_1} \frac{dh}{\dot{h}} \tag{5.191}$$

or

$$t_1 - t_0 = \Delta t = \int_{h_0}^{h_1} \frac{dh}{\dot{h}} \tag{5.192}$$

Thus, if the rate of climb, $\dot{h}$, is known as a function of altitude, h, the time to climb can be obtained by integrating Equation (5.192). If an analytical equation for $\dot{h}$ is not known, the integral can be evaluated numerically or graphically, assuming that numerical data is available for $\dot{h}$ versus altitude, as might be obtained in a flight test. The graphical solution is obtained by plotting $1/\dot{h}$ versus altitude and calculating the area under this curve.

The integration in Equation (5.192) assumes that the weight is constant during the climb. In reality, fuel is consumed during the climb, reducing the aircraft weight with increasing altitude. Thus, Equation (5.192) should be considered an approximation for the actual time to climb.

A closed form solution for the time to climb can be obtained by assuming that the rate of climb is a decreasing linear function of altitude. McCormick [19] suggests a linear function for rate of climb, $\dot{h}$, given by

$$\dot{h} = \left(1 - \frac{h}{h_a} \right) \dot{h}_{SL} \tag{5.193}$$

where h_a is the absolute ceiling and $\dot{h}_{SL}$ is the sea level rate of climb, which is a constant. Using this function, the rate of climb is equal to $\dot{h}_{SL}$ at sea level ($h = 0$) and decreases linearly to zero at the absolute ceiling ($h = h_a$). Inserting Equation (5.193) into (5.192), we have

$$\Delta t = \frac{1}{\dot{h}_{SL}} \int_{h_0}^{h_1} \frac{dh}{\left(1 - \frac{h}{h_a} \right)} = \frac{1}{\dot{h}_{SL}} \left[-h_a \ln \left(1 - \frac{h}{h_a} \right) \right]_{h_0}^{h_1}$$

$$\Delta t = \frac{h_a}{\dot{h}_{SL}} \left[\ln \left(1 - \frac{h_0}{h_a} \right) - \ln \left(1 - \frac{h_1}{h_a} \right) \right] \tag{5.194}$$

Assuming that the climb starts at sea level, $h_0 = 0$, and ends at an altitude, $h_1 = H$, we have

$$\Delta t = -\frac{h_a}{\dot{h}_{SL}} \ln \left(1 - \frac{H}{h_a} \right) \tag{5.195}$$

The integral in Equation (5.195) becomes undefined as the final altitude approaches the absolute altitude.

Example 5.13 Calculation of Time to Climb *A single-engine aircraft has an absolute ceiling of 19,000 ft and sea level rate of climb of 800 ft/min. Calculate the time to climb from sea level to 6000 ft.*

Solution

The time to climb from sea level to an altitude H is given by Equation (5.195) as

$$\Delta t = -\frac{h_a}{\dot{h}_{SL}} \ln \left(1 - \frac{H}{h_a} \right)$$

The time to climb from sea level to an altitude of 6000 ft is

$$\Delta t = -\left(\frac{19,000 ft}{800 \frac{ft}{min}} \right) \ln \left(1 - \frac{6000 ft}{19,000 ft} \right) = 9.01 \, min$$

Example 5.14 Calculation of Time to Climb by Integrating Flight Data *Rate of climb flight data is obtained for a Cessna 172RG Cutlass from sea level to an altitude of 12,000 ft, as shown in the table below. Using this data, calculate the time to climb from sea level to 12,000 ft.*

Altitude h (ft)	Rate of climb $\dot{h}$ (ft/min)
0	853
2000	755
4000	652
6000	561
8000	462
10,000	365
12,000	274

Solution

The time to climb from an altitude h_0 to an altitude h_1 is given by Equation (5.192) as

$$\Delta t = \int_{h_0}^{h_1} \frac{dh}{\dot{h}} = \int_{h_0}^{h_1} \left(\frac{1}{ROC} \right) dh$$

Using the data in the table, one over the rate of climb, (1/ROC), is plotted versus altitude, as shown in the figure below. A curve fit is applied to the data, yielding a cubic equation for (1/ROC) as a function of altitude, h.

$$\frac{1}{ROC} = 1.4976 \times 10^{-15} h^3 - 9.4426 \times 10^{-12} h^2 + 1.0415 \times 10^{-7} h + 0.001637$$

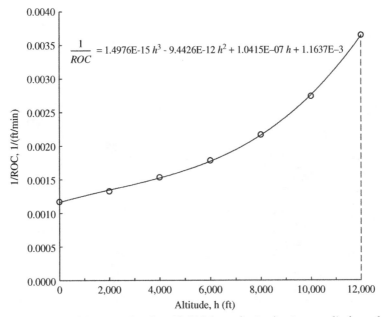

This equation is integrated from sea level to 12,000 ft to obtain the time to climb, as follows.

$$\Delta t = \int_0^{12,000\,ft} (1.4976 \times 10^{-15}h^3 - 9.4426 \times 10^{-12}h^2 + 1.0415 \times 10^{-7}h + 0.001637)dh$$

$$\Delta t = \left[1.4976 \times 10^{-15}\left(\frac{h^4}{4}\right) - 9.4426\left(\frac{h^3}{3}\right) + 1.0415\left(\frac{h^2}{2}\right) + 1.1637h\right]_0^{12,000\,ft}$$

$$\Delta t = (1.4976 \times 10^{-15})\frac{(12,000)^4}{4} - (9.4426 \times 10^{-12})\frac{(12,000)^3}{3}$$

$$+ (1.0415 \times 10^{-7})\frac{(12,000)^2}{2} + (0.001637)(12,000)$$

$$\Delta t = 7.764\,min - 5.439\,min + 7.499\,min + 19.644\,min = 29.47\,min$$

5.9.3   FTT: Climb Performance

Climb performance flight testing seeks to measure the actual aircraft performance for various types of climbs and to determine the *climb schedules* for these various types of climbs. The climb performance may be quantified in terms of the minimum time or minimum fuel used in climbing to an altitude or an energy level. A climb schedule is typically specified in terms of the best airspeeds or Mach numbers to fly for a given type of climb, as a function of altitude. A commonly defined climb schedule specifies the airspeeds to fly, versus altitude, to obtain the best rate of climb. Climb speeds or schedules may also be determined for minimum fuel climbs or for maximum climb angles, to clear an obstacle.

Flight test techniques that are commonly used for climb performance testing are the *level acceleration* FTT and the *sawtooth climb* FTT. You will fly the level acceleration in a later FTT, related to specific excess power, so we focus on the sawtooth climb FTT in the present section. You will fly sawtooth climbs in the single-engine Cessna 172RG *Cutlass*, shown in Figure 5.44.

Figure 5.44   Cessna 172 single-engine, general aviation airplane (fixed-gear version shown), configured for flight testing with two wingtip-mounted air data booms and wing leading edge cuffs. (Source: *NASA*.)

Designed and manufactured by the Cessna Aircraft Company, Wichita, Kansas, the Cessna 172RG is a four-place, single-engine airplane. The Cessna 172RG is the retractable landing gear variant of the fixed-gear Cessna 172. The Cessna 172 line of aircraft has been extremely successful, with over 43,000 172 s built as of 2015, more than any other aircraft. The Cessna 172RG is commonly used for personal general aviation flying and flight training. The airplane has a high-mounted wing, aft-mounted horizontal tail, single vertical tail, and retractable, tricycle landing gear. The Cessna 172RG is powered by a single Lycoming O-360-F1A6 normally aspirated, air-cooled, horizontally opposed, four-cylinder piston engine producing 180 bhp (134 kW) at 2700 rpm. The first flight of the Cessna 172 with a fixed, tricycle landing gear was on 12 June 1955. The first flight of the Cessna 172RG, with retractable landing gear, was in 1980. A three-view drawing of the Cessna 172 RG is shown in Figure 5.45 and selected specifications are given in Table 5.14.

The test plan is to obtain climb performance data at four constant airspeeds, of 70 knots (81 mph, 130 km/h), 80 knots (92 mph, 148 km/h), 90 knots (104 mph, 67 km/h), and 100 knots (115 mph, 185 km/h), and at three pressure altitudes, of 2000 ft (610 m), 4000 ft (1220 m), and 6000 ft (1830 m). This results in a test matrix of 12 climb test points to be flown. To reduce the effects of the wind, you will fly the climbs perpendicularly to the wind direction. In addition, you will fly two climbs for each test point, in opposite directions, to cancel any wind effects that are present. Therefore, your test matrix is doubled to 24 climb test points. For each altitude, the altitude data band is from 500 ft (152 m) below to 500 ft above the target altitude. You will collect data for two minutes while flying through the altitude data band at a constant airspeed. The flight path of the series of climbs and descents that you will fly looks like the teeth of a saw, hence the name *sawtooth climbs*.

Climb performance data is typically presented for an aircraft at its maximum gross weight; hence you take off in the Cessna 172RG with an aircraft weight of 2650 lb (1200 kg). You climb and level off at an altitude of 1000 ft (305 m), which is 500 ft below your first altitude data band of 1500 ft (457 m) to 2500 ft (762 m). You check that the aircraft is in the proper clean configuration, with the flaps and landing gear retracted. You add full throttle and allow the engine to stabilize. You

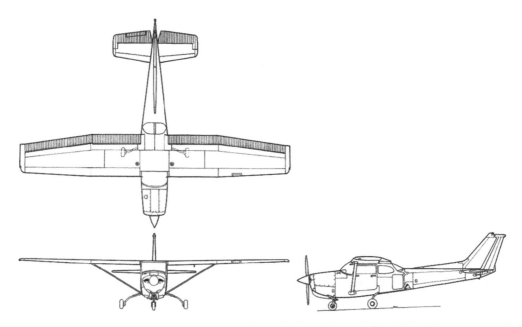

Figure 5.45 Three-view drawing of the Cessna 172RG *Cutlass*. (Source: *Adapted from figure courtesy of Richard Ferriere.*)

Table 5.14 Selected specifications of the Cessna 172RG *Cutlass*.

Item	Specification
Primary function	General aviation aircraft
Manufacturer	Cessna Aircraft Company, Wichita, Kansas
First flight	12 June 1955 (Cessna 172 with fixed landing gear)
Crew	1 pilot + 3 passengers
Powerplant	Lycoming O-360-F1A6 four-cylinder engine
Engine power	180 bhp (134 kW) at 2700 rpm
Empty weight	1555 lb (705.3 kg)
Maximum takeoff weight	2650 lb (1202 kg)
Length	27 ft 5 in (8.36 m)
Height	8 ft 9.5 in (2.68 m)
Wingspan	36 ft 0 in (11.0 m)
Wing area	174 ft^2 (16.2 m^2)
Wing loading	15.2 lb/ft^2 (74.2 kg$_f$/m^2)
Airfoil	NACA 2412
Never exceed speed	164 knots (189 mph, 304 km/h)
Service ceiling	17,000 ft (5200 m)
Load factor limits	+3.8 g, −1.52 g

pull back on the yoke to bleed off the speed to your first test point airspeed of 70 knots. As you are pulling back the aircraft starts to climb, but you have 500 ft to establish the desired test airspeed before you enter the altitude data band at 1500 ft. You've overshot your 70 knot target airspeed by a little bit, but it is steady at 73 knots (84 mph, 135 km/h) as you reach 1500 ft, so you keep this speed.

As you enter the altitude data band at 1500 ft, you start your timer. While maintaining the constant 73 knot climb, you record the altitude at 30 s intervals. You make small, precise pitch corrections,

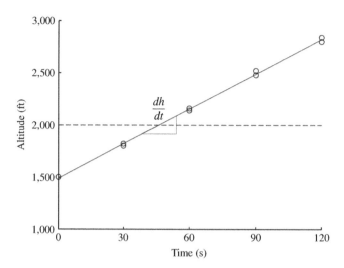

Figure 5.46   Cessna 172RG rate of climb, dh/dt, determination for 70 knot sawtooth climbs in opposite directions for an altitude of 2000 ft.

as required, to keep the airspeed constant. At the two minute mark, you are at 2430 ft (741 m). After you have climbed through the top of the data band, at 2500 ft, you reduce power and descend back down to an altitude of 1000 ft. You make a 180° turn, so that your heading is now in the opposite direction to your first climb direction. You repeat the above sawtooth climb, in the opposite direction, at a constant airspeed of 70 knots through the same altitude data band. The remaining test points take you about an hour to complete. Over this time, your aircraft weight decreases due to the fuel consumed. You will need to standardize all of the climb data to maximum gross weight during the data reduction to account for this weight reduction.

After you land, you have a climb performance data set comprised of 24 constant airspeed climbs. The first task is to determine the rate of climb, dh/dt, for each constant airspeed and altitude. The altitude versus time data is plotted for the two 70 knot sawtooth climbs, performed in opposite directions, for an altitude of 2000 ft, as shown in Figure 5.46. The rate of climb is calculated as the slope of the line, dh/dt, through these data points, at an altitude of 2000 ft, as shown. From this figure, the rate of climb at an altitude of 2000 ft is determined to be 666 ft/min (203 m/min) at an airspeed of 73 knots.

The rates of climb, obtained in this manner, are plotted versus calibrated airspeed for each altitude, as shown in Figure 5.47. The horizontal line, tangent to the curve for each altitude, defines the maximum rate of climb for that altitude. A vertical line, drawn from this point, to the horizontal axis, is the best rate of climb velocity, V_y, for that altitude. This maximum rate of climb point could be better defined if there were more data points in this region. In practice, an altitude versus time plot should be made while the test points are flown, so that additional test point airspeeds can be identified in real time. These new airspeeds are then flown to increase the number of data points near the maximum rate of climb, which more accurately defines this value. A line, drawn from the origin, in Figure 5.47, to the tangent point of an altitude curve, defines the best angle of climb speed, V_x, and the associated rate of climb.

The lines connecting the tangent points define the climb schedules for the best rate and best angle of climbs, as shown. Once the climb schedules are defined, *check climbs* are flown, following these airspeed versus altitude profiles, to verify their accuracy and to evaluate operational considerations, such as forward visibility in the climb, engine cooling in the climb attitude, and other factors.

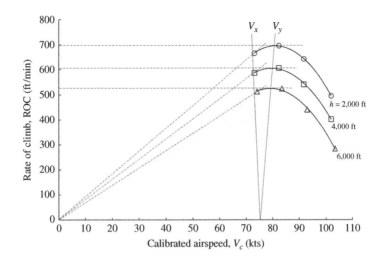

Figure 5.47   Rate of climb versus airspeed from Cessna 172RG sawtooth climb flight data.

5.10   Glide Performance

Steady, constant velocity gliding flight was introduced in Chapter 3, when we flew the North American XP-51 *Mustang* in a flight test technique that was focused on obtaining the aerodynamic lift and drag of the aircraft. Here, we expand on some aspects of gliding performance. The geometry for steady, constant velocity, hence unaccelerated, gliding flight is shown in Figure 5.48. The aircraft is descending at a velocity, V_∞, with a horizontal velocity, V_h, vertical velocity, V_v, and a negative flight path angle, $-\gamma$. The angle, θ, is the magnitude of the negative flight path angle. The forces acting on the aircraft are the lift, drag, and weight, as shown in Figure 5.49. (The drag force is shown displaced from the center of mass for clarity.) The thrust is assumed to be zero for gliding flight.

From Figure 5.49, the sum of the forces perpendicular and parallel to the flight direction for steady, unaccelerated, gliding flight are given by

$$L = W \cos \theta \cong W \tag{5.196}$$

$$D = W \sin \theta \tag{5.197}$$

where, similar to the climbing flight case, the glide angle is assumed to be small, such that $\cos \theta \cong 1$. This assumption allows us to use the level flight results for power required later in the analysis.

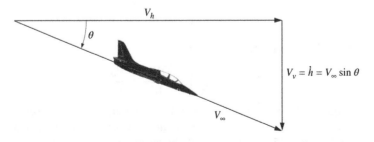

Figure 5.48   Geometry for steady, constant velocity glide.

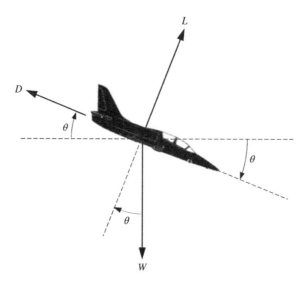

Figure 5.49   Forces on aircraft in steady, constant velocity glide.

Dividing Equation (5.197) by (5.196) and solving for the glide angle, θ, gives

$$\theta = \tan^{-1}\left(\frac{1}{L/D}\right) \tag{5.198}$$

The glide angle, θ, is inversely proportional to the lift-to-drag ratio, L/D, such that the minimum glide angle, θ_{min}, is obtained by flying at the maximum lift-to-drag ratio, $(L/D)_{max}$.

$$\theta_{min} = \tan^{-1}\left[\frac{1}{(L/D)_{max}}\right] \tag{5.199}$$

From Equation (5.197), we can also obtain an expression for the glide angle in terms of the drag and weight.

$$\theta = \sin^{-1}\left(\frac{D}{W}\right) \tag{5.200}$$

Thus, we observe that, for a given aircraft weight, the minimum glide angle also occurs at the minimum drag condition, D_{min}.

$$\theta_{min} = \sin^{-1}\left(\frac{D_{min}}{W}\right) \tag{5.201}$$

From Figure 5.48, the rate of descent or vertical velocity, V_v, is given by

$$V_v = V_\infty \sin\theta \tag{5.202}$$

Inserting Equation (5.197) into (5.202), we have

$$V_v = V_\infty\left(\frac{D}{W}\right) = \frac{DV_\infty}{W} = \frac{P_R}{W} \tag{5.203}$$

where the power required, P_R, is assumed to equal the drag multiplied by the velocity, an assumption that is truly valid only for level, unaccelerated flight, However, by assuming that the glide angle is not too large, say less than 15–20°, the level flight assumption for power required may be used for gliding flight.

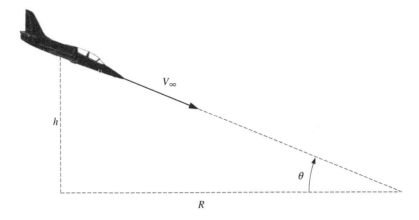

Figure 5.50 Geometry for steady, constant velocity glide range.

The minimum rate of descent or minimum sink, $V_{v,min}$, for a given weight, is obtained at the minimum power required, $P_{R,min}$.

$$V_{v,min} = \frac{P_{R,min}}{W} \qquad (5.204)$$

The geometry for the calculation of the horizontal gliding distance is shown in Figure 5.50. The *horizontal gliding range*, R, is given by

$$R = \frac{h}{\tan \theta} \qquad (5.205)$$

where h is the altitude at the start of the glide.

Inserting Equation (5.198) into (5.204), we have

$$R = h\left(\frac{L}{D}\right) \qquad (5.206)$$

Hence, the horizontal gliding distance is directly proportional to the lift-to-drag ratio and, of course, the starting altitude. The maximum range, R_{max}, is obtained by flying at the maximum lift-to-drag ratio, $(L/D)_{max}$.

$$R_{max} = h\left(\frac{L}{D}\right)_{max} \qquad (5.207)$$

The *glide ratio* is defined as the ratio of the horizontal distance flown relative to the altitude loss, R/h. Thus, we see, from Equation (5.204), that the maximum glide ratio is equal to the maximum lift-to-drag ratio.

$$\left(\frac{R}{h}\right)_{max} = \left(\frac{L}{D}\right)_{max} \qquad (5.208)$$

5.11 The Polar Diagram

The *polar diagram* is an efficient, graphical visualization of the velocities associated with steady, unaccelerated flight, as shown in Figure 5.51. The curve is the locus of points that represent the horizontal, vertical, and total velocities for a given aircraft weight, configuration, altitude, and power setting. The horizontal and vertical axes of the plot are the horizontal and vertical velocities, respectively, in steady, unaccelerated flight.

A line drawn from the origin to the any point on the curve, shown by Point 1, represents a steady climbing flight condition at a total velocity, V, horizontal velocity, V_h, vertical velocity or rate of

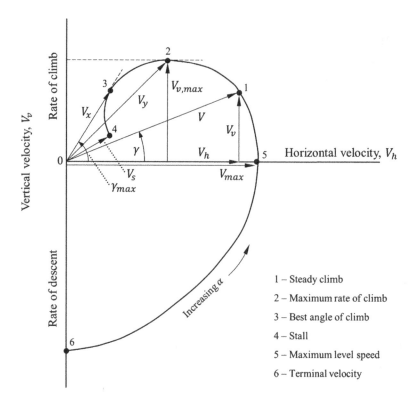

Figure 5.51   Polar diagram for an aircraft at a given weight, configuration, altitude, and power setting.

climb, V_v, and flight path angle, γ. The length of the vector represents the true airspeed at that flight condition. Point 2 represents the maximum rate of climb, with a total velocity, V_y, and the maximum vertical velocity or rate of climb, $V_{v,maxh}$, on the plot. The line extending from the origin to the tangent point of the curve represents the best angle of climb condition, given by Point 3, with a velocity, V_x, and the maximum climb angle, γ_{max}. The best angle of climb speed is less than the best rate of climb speed. Stall is shown by Point 4, with the stall speed, V_s. The maximum speed in level flight, V_{max}, is Point 5, with the longest horizontal velocity component of the curve. Finally, the longest vertical velocity component is shown by Point 6, the terminal velocity. This is the vertical velocity that the aircraft would reach if it were pointed straight down, in a vertical dive with a flight path angle of $\gamma = -90°$, at the given power setting.

The polar diagram also indicates the trend in the aircraft angle-of-attack, with angle-of-attack increasing as one travels counterclockwise along the curve. Thus, the terminal velocity, vertical dive has the lowest angle-of-attack. The angle-of-attack in level flight at the maximum speed is greater. The angle-of-attack for the best angle of climb is greater than that for the best rate of climb. The stall has the greatest angle-of-attack over the velocity range depicted by the curve. These are all relative comparisons of the angle-of-attack. The polar diagram does not provide any quantitative values for the angle-of-attack at any of the flight conditions.

The polar diagram for steady, power-off, gliding flight is shown in Figure 5.52. The gliding flight curve has "shrunk" from the curve corresponding to a high power setting. Point 1 represents gliding at the minimum or shallowest glide angle, γ_{min}. This point corresponds to flight at the maximum lift-to-drag ratio, $(L/D)_{max}$, since flying at the minimum glide angle results in the maximum horizontal distance traveled for a given altitude. The velocity at Point 1 is the maximum L/D airspeed, $V_{(L/D),max}$. Point 2 has the lowest vertical velocity; hence it is the minimum sink

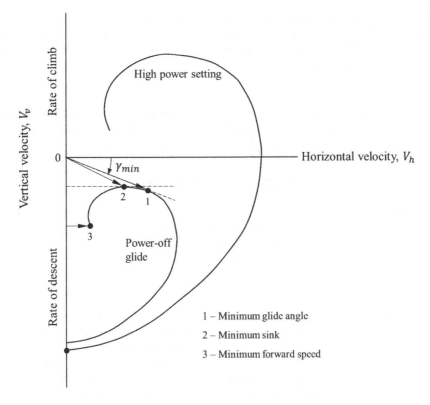

Figure 5.52 Polar diagram for an aircraft in gliding flight.

condition. Flying at the minimum sink airspeed, $V_{min\,sink}$, results in the lowest rate of descent. The velocity $V_{(L/D),max}$ at Point 1 is greater than $V_{min\,sink}$ at Point 2, but the aircraft glides farther at the higher airspeed because the glide angle is shallower. Flying at any airspeed slower than $V_{(L/D),max}$, in an attempt at what is called "stretching the glide", results in a reduced gliding distance.

The fact that the glide angle for minimum sink, at Point 2, is steeper than the minimum glide angle, γ_{min} at Point 1, may seem counterintuitive. From Equation (5.201), the minimum glide angle occurs at the minimum drag condition. This minimum glide angle or minimum drag condition is shown as Point a on the thrust required (or drag) curve in Figure 5.31, where the associated velocity is $V_{T_R,min}$. From Equation (5.203), minimum sink is obtained at minimum power required, $P_{R,min}$. This minimum sink or minimum power condition is shown as Point b on the power required curve in Figure 5.31, where the associated velocity is $V_{P_R,min}$. The drag associated with the minimum sink or minimum power condition (Point b) is higher than at the minimum drag condition (Point a), which requires flying at a steeper glide angle.

Point 3 represents gliding flight at the minimum forward or horizontal speed, $V_{h,min}$, as shown graphically in Figure 5.52. This is not the minimum total velocity, as other points on the curve, result in a lower total velocity. However, these other points have a higher forward or horizontal velocity.

5.12 Energy Concepts

The energy concepts approach quantifies aircraft performance based on the *energy state* of the aircraft. During the 1950s, energy techniques were developed, independently in Germany, the UK, and the USA, to analyze the performance and determine the best climb airspeeds for new jet-powered aircraft. These methods were first used in Germany, by Kaiser [14], for performance and climb analyses for the new Messerschmitt Me-262 *Swallow* jet aircraft. Later, publications in the UK, by Lush [17], and in the USA, by Rutowski [25], described energy techniques applied to aircraft performance and climb trajectories.

In energy state analysis, the aircraft is assumed to be a point mass, located at the aircraft center of gravity. The aircraft is acted upon by *conservative forces*, that is, there are no losses or dissipative phenomena associated with the forces. The *total energy*, E, of an aircraft is the sum of its *potential energy*, PE, due to its height above the ground, and its *kinetic energy*, KE, due to its motion, given by

$$E = PE + KE = mgh + \frac{1}{2}mV^2 \tag{5.209}$$

where m is the aircraft mass, g is the acceleration due to gravity, h is the height above ground level, and V is the aircraft velocity. Dividing Equation (5.209) by the aircraft weight, $W = mg$, the *specific energy* or *energy height*, E_s, is defined as

$$E_s \equiv \frac{E}{W} = h + \frac{1}{2}\frac{V^2}{g} \tag{5.210}$$

The units of the energy height are feet or meters, the same as for an altitude. The energy height defines the *energy state* of the aircraft, in terms of the sum of its potential and kinetic energies, per unit weight. The energy state is given by two variables, the aircraft altitude, h, and velocity, V. Lines of constant energy height can be drawn on an altitude-versus-velocity plot, as shown in Figure 5.53. An aircraft's energy state can be shown on this plot, based on its altitude and velocity.

Consider an aircraft at an altitude of 10,000 ft (3048 m) with zero velocity ($V = 0$), as depicted by Point A in Figure 5.53. From Equation (5.210), the specific energy or energy height, E_s, of this aircraft is simply 10,000 ft. Now imagine this aircraft enters a high-speed dive, trading all of its

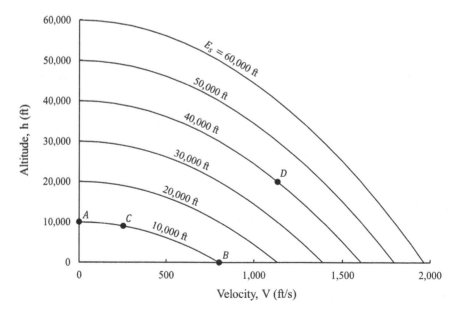

Figure 5.53   Lines of constant energy height, E_s.

altitude (potential energy) for velocity (kinetic energy). Solving Equation (5.210) for the velocity at zero altitude ($h = 0$), we have

$$V = \sqrt{2g(E_s - h)} = \sqrt{2\left(32.2\,\frac{\text{ft}}{\text{s}^2}\right)(10{,}000\,\text{ft} - 0)} = 802.5\,\frac{\text{ft}}{\text{s}} \tag{5.211}$$

which is Point B in Figure 5.53. Now, assume that the aircraft climbs back to 9000 ft (2743 m), its velocity is

$$V = \sqrt{2g(E_s - h)} = \sqrt{2\left(32.2\,\frac{\text{ft}}{\text{s}^2}\right)(10{,}000\,\text{ft} - 9000\,\text{ft})} = 253.8\,\frac{\text{ft}}{\text{s}} \tag{5.212}$$

which is Point C in Figure 5.53. In all of these cases, the aircraft has changed its altitude (potential energy) and velocity (kinetic energy), but its energy state has remained constant, equal to 10,000 ft. The constant E_s line defines all of the flight conditions, specified by an altitude and a velocity, where the aircraft can maintain steady equilibrium flight.

This also assumes that the aircraft maintains a constant thrust setting, constant weight, and constant aerodynamic configuration (position of flaps, landing gear, speed brake, etc.), which fixes the drag. If any of these parameters change, the energy state changes. For example, if the thrust is increased, the energy state increases. This provides insight into how the energy state can be changed. Reaching another energy state, such as the higher state depicted by Point D in Figure 5.53, requires a change in the weight, thrust, drag, or a combination of these. By common sense, getting to a higher energy state requires a weight decrease, thrust increase, or drag decrease. The opposite is true to reach a lower energy state.

The equilibrium flight condition is defined from an energy perspective only, and does not consider other physics associated with the flight condition, such as the aerodynamic stall speed or the structural limit speed. Since we have assumed that the forces acting on the aircraft are conservative, the aircraft can exchange potential energy (altitude) and kinetic energy (velocity), along a constant E_s line, without any losses. This also assumes that these energy exchanges can occur instantaneously, in zero time. In reality, there are irreversible viscous and heat transfer losses, such that the aircraft would not achieve the velocity of Point B, starting from Point A, or reach the altitude of Point A, starting from Point B.

Next, we define the time rate of change of the energy height as the *specific excess power*, P_s, given by

$$P_s \equiv \frac{dE_s}{dt} = \frac{dh}{dt} + \frac{V}{g}\frac{dV}{dt} \tag{5.213}$$

The specific excess power is the sum of the rate of change of the altitude or climb rate, dh/dt, and the rate of change of the velocity or acceleration, dV/dt. The specific excess power has the dimensions of a velocity or rate of climb, that is, ft/s or m/s.

The excess power is equal to the net excess force, F_{excess}, acting on the aircraft, times the velocity. (Recall that power is a force times a velocity.) The excess force, acting on the aircraft, is the thrust minus the drag, $T - D$. Therefore, the specific excess power in Equation (5.213) can also be expressed as

$$P_s = \frac{F_{excess}V}{W} = \frac{(T - D)V}{W} \tag{5.214}$$

Combining Equations (5.213) and (5.214) gives

$$P_s = \frac{dE_s}{dt} = \frac{dh}{dt} + \frac{V}{g}\frac{dV}{dt} = \frac{(T - D)V}{W} \tag{5.215}$$

As embodied by Equation (5.215), the specific excess power characterizes an aircraft's capability to change altitude (dh/dt term), change velocity (dV/dt term), or change energy states, as a result of the difference between the thrust of the engine and the drag of the airframe. This also highlights the earlier comment concerning the energy state's dependence on the weight, thrust, and drag. The rate of change of the energy state of an aircraft, or the specific excess power, is directly proportional to the excess thrust, $T - D$, and the velocity, V, and inversely proportional to the weight, W. If the specific excess power is positive, corresponding to a positive excess thrust, the aircraft can change its energy state by climbing, accelerating, or a combination of these. If the specific excess power is zero, the energy state of the aircraft is constant and the aircraft is in equilibrium.

A specific excess power plot for a subsonic aircraft, the North American F-86 *Sabre* jet aircraft (Figure 5.54), is shown in Figure 5.55. The F-86 was the first US swept-wing fighter jet, which flew for the first time on 1 October 1947. The specific excess power plot, or P_s plot, consists of lines of constant P_s, drawn on an altitude versus velocity or Mach number chart. The plot is valid for a fixed aircraft weight, configuration (drag), thrust, and load factor. Typically, a P_s plot is for the aircraft at its maximum weight, nominal cruise configuration, maximum thrust, and 1 g load factor, although plots can be created for other conditions. Lines of constant specific energy, E_s, are also shown, where the value of the specific energy is equal to the altitude given at zero Mach number.

The $P_s = 0$ line is the dividing line between energy states where the aircraft has positive ($P_s > 0$) or negative ($P_s < 0$) specific excess power. Along the $P_s = 0$ curve, the thrust equals the drag and the aircraft can maintain steady, equilibrium flight. For those altitude–velocity flight conditions where $P_s < 0$, the aircraft cannot sustain steady, level flight. For those altitude–velocity conditions where $P_s > 0$, the aircraft has excess energy available to climb, increase velocity, or both. The uppermost point on the $P_s = 0$ line represents the maximum altitude obtainable in steady equilibrium flight. The furthest point, to the right, on the $P_s = 0$ line along the x-axis, where $h = 0$, represents the maximum Mach number sustainable in steady equilibrium flight at sea level.

Figure 5.54   North American F-86 *Sabre*. (Source: *Maritz, https://en.wikipedia.org/wiki/File:F86-01.jpg, CC–BY-SA-3.0. License at https://creativecommons.org/licenses/by-sa/3.0.*)

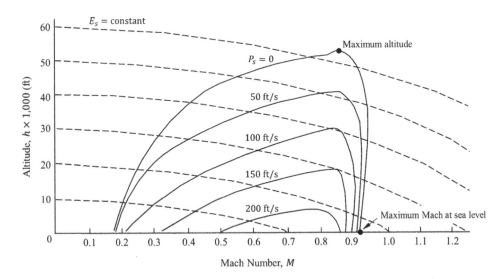

Figure 5.55   Specific excess power plot for a subsonic airplane, the North American F-86 *Sabre*.

Below the $P_s = 0$ line are lines of constant P_s greater than zero. For these positive P_s points, the thrust is greater than the drag and the aircraft can climb or accelerate. In fact, for the constant values of the weight, configuration, thrust, and load factor, upon which the chart is based, the aircraft is not in equilibrium at a point where the P_s is greater than zero. The aircraft must change velocity, altitude, or both, and moves away from this point until it reaches equilibrium with $P_s = 0$. The magnitudes of the P_s greater than zero points indicate the rate at which the aircraft can change altitude or airspeed.

The P_s plot for a supersonic airplane has a different shape than that of a subsonic airplane, as shown in Figure 5.56, where the $P_s = 0$ curves are shown for each. The difference is due to the transonic drag rise experienced by a supersonic aircraft near Mach 1. Since the thrust is constant,

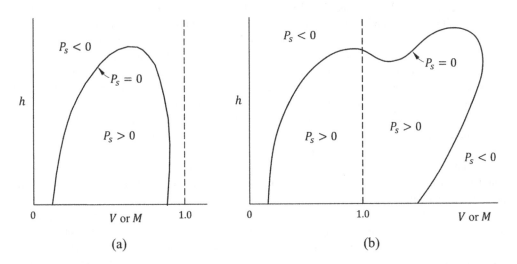

Figure 5.56   Comparison of P_s plots for (a) subsonic and (b) supersonic aircraft.

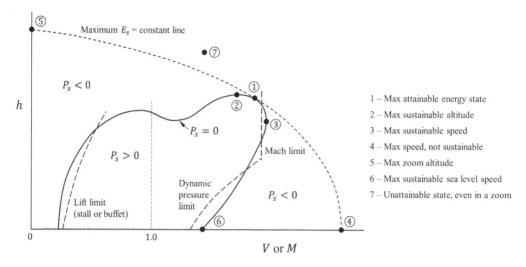

Figure 5.57   Energy state accessibility and flight envelope limits on P_s plot.

the transonic drag increase results in a decrease of the excess thrust and a dip in the P_s curves for a supersonic aircraft.

Several energy-related flight conditions are identified, relative to the $P_s = 0$ curve of a supersonic aircraft, in Figure 5.57. These flight conditions represent the energy states that are accessible, or sometimes not accessible, to the aircraft from an energy perspective. Point 1 is the maximum attainable energy state of the aircraft in steady, equilibrium flight. This point is tangential to the maximum specific energy line. The maximum altitude and maximum velocity, sustainable in steady equilibrium flight, are given by Points 2 and 3, respectively. Point 4 is at the same energy level as Point 1, so this point is theoretically accessible by the aircraft. Point 4 represents the maximum airspeed that can be reached by exchanging all of the aircraft's potential energy for kinetic energy or velocity. Since the specific excess power is less than zero at this point, the aircraft cannot sustain this velocity and must move back to some place on the $P_s = 0$ curve. Similarly, Point 5 is at the same energy level as Points 1 and 4 and is accessible but not sustainable. Point 5 represents the maximum altitude that can be reached in a *zoom climb*, where all of the aircraft's energy is traded for altitude. Again, this altitude can be reached but the aircraft cannot remain there in equilibrium flight. The maximum sustainable velocity, in steady equilibrium flight at sea level is represented by Point 6, which rests on the $P_s = 0$ curve, where the thrust equals the drag. Finally, Point 7 is shown as a flight condition that is not accessible at all, since its energy level is above the maximum energy state of the aircraft at its given weight, thrust, drag, and load factor.

While the $P_s = 0$ curve represents all of the possible airspeed–altitude points where the aircraft can maintain steady, equilibrium flight, all of these points may not be within the flight envelope of the aircraft. The low-speed, lift limit, identified by aerodynamic stall or intolerable low-speed buffet, may be more limiting on the low-speed boundary of the $P_s = 0$ curve. At the high-speed boundaries of the $P_s = 0$ curve, the dynamic pressure limit and the maximum airspeed or Mach number limit may be more restrictive.

The P_s plot for the supersonic Lockheed F-104 *Starfighter* (see Figure 3.178), incorporating flight envelope limits, is shown in Figure 5.58. This P_s plot corresponds to the F-104 at a nominal weight, clean configuration, maximum, full-afterburning thrust, and 1 g load factor. The P_s curves have a dip in the transonic region that is characteristic of supersonic aircraft. Some portions of the P_s curves have been clipped by the lift limit, at low speeds, and the dynamic pressure and

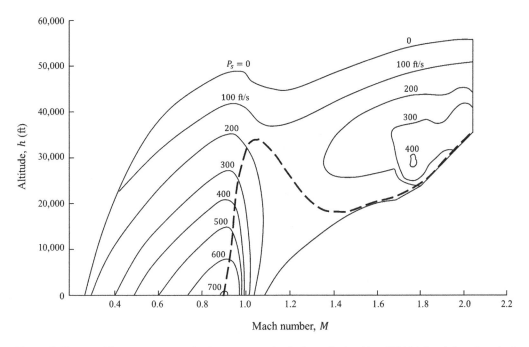

Figure 5.58 Specific excess power plot for a supersonic airplane, the Lockheed F-104 *Starfighter* (nominal weight, clean configuration, maximum thrust, and 1 g). (Source: *US Air Force*, [22].)

Mach limits, at high Mach numbers. The magnitude of the constant P_s values for the F-104 are considerably larger than for the subsonic F-86, in Figure 5.55, indicative of the F-104's higher excess thrust capability. This higher excess thrust can be attributed to the higher thrust engine of the F-104 and its lower aerodynamic drag design for supersonic flight.

The P_s contours change if any of the constant conditions (weight, configuration, thrust, or load factor), upon which the plot is based, are changed. The $P_s = 0$ line shrinks, along with the other P_s contours, if the weight, drag, or load factor is increased or if the thrust is decreased. This is graphically shown in Figure 5.59, for the Lockheed F-104, where the load factor has been increased from 1 to 3 g. At this higher, sustained load factor, the $P_s = 0$ boundary has shrunk, with the aircraft capable of steady, equilibrium flight in a much smaller region of the flight envelope. The aircraft has a small "island" of flight conditions where it can sustain 3 g flight at supersonic speeds, but it must be at a lower load factor to get to these flight conditions. These types of specific power plots at elevated load factors provide insights into the turn performance capability of the aircraft.

Energy techniques were originally developed to determine best climb paths for aircraft. Several types of useful optimum climb paths can be obtained from the specific excess power plot, including the *maximum rate of climb path* and the *optimum energy climb path*.

The maximum rate of climb path is defined as the path that results in the maximum increase in altitude per unit time. This climb path can be graphically constructed by connecting the tangent points between lines of constant altitude and the peaks in the specific excess power, P_s, contours, as shown in Figure 5.60. The maximum rate of climb path gives the maximum rate of change of P_s, which is the same as the maximum change in the excess thrust.

The optimum energy climb path is defined as the path that yields the maximum increase in the energy state per unit time. This climb path is graphically constructed by connecting the points

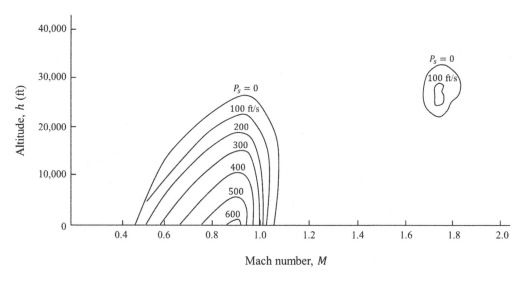

Figure 5.59   Specific excess power plot for the Lockheed F-104 *Starfighter* at 3 g.   (Source: *US Air Force,* [22].)

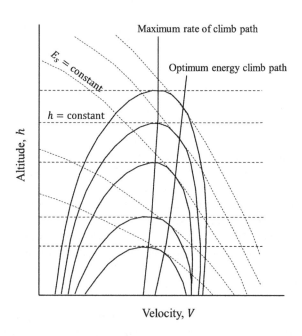

Figure 5.60   Maximum rate of climb and optimum energy climb paths for a subsonic aircraft.

where the lines of constant specific energy, E_s, are tangent to the specific excess power, P_s, lines, as shown in Figure 5.60. The optimum energy path gets the aircraft to the highest energy state in the shortest time, rather than the highest altitude as in the maximum rate of climb path.

The optimum climb paths, shown in Figure 5.60, are for a subsonic aircraft. For a supersonic aircraft, the optimum climb paths in the subsonic region are the same for a subsonic aircraft. In the

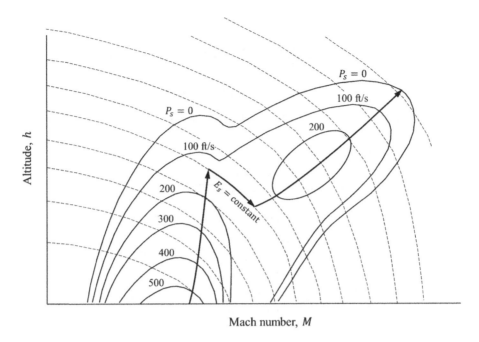

Figure 5.61   Optimum energy climb path for a supersonic aircraft.

supersonic region, the optimum climb paths follow the same rules as in the subsonic region. However, connecting the subsonic and supersonic climb paths, which occurs in the transonic region, is different. As shown in Figure 5.61 for an optimum energy climb, the subsonic and supersonic climb paths are connected with a constant specific energy dive. Theoretically, this constant energy transition takes place instantaneously with no losses. In reality, the transition takes a finite amount of time, including the change in the aircraft attitude, which also cannot occur instantaneously. The optimum energy climb path for the F-104 is shown in Figure 5.58 as the dashed line trajectory. The transitions between the various segments of the climb path occur more smoothly than depicted in Figure 5.61, with "rounded corners" in the trajectory, versus abrupt transitions. At high Mach number, the optimum climb is restricted by the dynamic pressure flight envelope limit.

Example 5.15 Calculation of Kinetic and Potential Energies *A Boeing F-18* Hornet *is flying at an altitude of 500 ft and an airspeed of 450 knots. If the F-18 has a weight of 35,000 lb, calculate its potential energy, kinetic energy, total energy, and specific energy at this flight condition. Also, calculate the potential and kinetic energies as a percentage of the total energy.*

Solution

The potential energy is given by

$$PE = mgh = (35,000\,lb)(500\,ft) = 1.750 \times 10^7\,ft \cdot lb$$

Convert the airspeed into consistent units.

$$V = 450\frac{nm}{h} \times \frac{6076\,ft}{1\,nm} \times \frac{1\,h}{3600\,s} = 759.5\,ft/s$$

The kinetic energy is given by

$$KE = \frac{1}{2}mV^2 = \frac{1}{2}\left(\frac{35{,}000\,lb}{32.174\,ft/s^2}\right)(759.5\,ft/s)^2 = 3.138 \times 10^8\,ft \cdot lb$$

The total energy is the sum of the potential and kinetic energy.

$$E = PE + KE = 1.75 \times 10^7\,ft \cdot lb + 3.1375 \times 10^8\,ft\,lb = 3.313 \times 10^8\,ft \cdot lb$$

The energy height is given by

$$E_s = \frac{E}{W} = \frac{3.313 \times 10^8\,ft \cdot lb}{35{,}000\,lb} = 9464\,ft$$

The potential and kinetic energies as a percentage of the total energy are

$$\frac{PE}{E}\% = \frac{1.750 \times 10^7\,ft \cdot lb}{3.313 \times 10^8\,ft \cdot lb} \times 100\% = 5.28\%$$

$$\frac{KE}{E}\% = \frac{3.138 \times 10^8\,ft \cdot lb}{3.313 \times 10^8\,ft \cdot lb} \times 100\% = 94.72\%$$

Example 5.16 Calculation of Specific Excess Power *A North American F-86* Sabre *jet performs a level acceleration at an altitude of 24,000 ft. At Mach 0.5, the aircraft weight is 14,927 lb and the rate of change of velocity is 5.22 ft/s². Calculate the specific excess power and the excess thrust at this point during the level acceleration.*

Solution

From Appendix C, the temperature at an altitude of 24,000 ft is

$$T = \theta T_{SSL} = (0.83518)(519°R) = 433.5°R$$

The velocity is given by

$$V = Ma = M\sqrt{\gamma RT} = (0.5)\sqrt{(1.4)\left(1716\,\frac{ft \cdot lb}{slug \cdot °R}\right)(433.5°R)} = 510.3\,\frac{ft}{s}$$

From Equation (5.215), the specific excess power is given by

$$P_s = \frac{dh}{dt} + \frac{V}{g}\frac{dV}{dt}$$

Since the aircraft is performing a level acceleration, the rate of change of altitude, dh/dt, is zero, and we have

$$P_s = 0 + \left(\frac{510.3\,\frac{ft}{s}}{32.2\,\frac{ft}{s^2}}\right)\left(5.22\,\frac{ft}{s^2}\right) = 82.7\,\frac{ft}{s}$$

Also, from Equation (5.215), we have

$$P_s = \frac{(T - D)V}{W}$$

Solving for the excess thrust gives

$$T - D = \frac{P_s W}{V} = \frac{\left(82.7\frac{ft}{s}\right)(14,927\,lb)}{510.3\frac{ft}{s}} = 2419\,lb$$

5.12.1   FTT: Specific Excess Power

The flight test techniques, described in this section, are used to collect the data needed to create a specific excess power or P_s plot for an aircraft. The two FTTs that are typically flown are the *sawtooth climb* and the *level acceleration*. Both of these methods are non-steady-state maneuvers, where either the altitude or airspeed is constantly changing. However, they are used to collect a large amount of useful data in a short time. Typically, a data acquisition system is used to record the flight, but hand-held data recording may also be used.

Recall from Equation (5.215), that the specific excess power is given by

$$P_s = \frac{dh}{dt} + \frac{V}{g}\frac{dV}{dt} \tag{5.216}$$

where dh/dt is the rate of climb and dV/dt is the acceleration. The sawtooth climb FTT is flown at a constant airspeed, so that $dV/dt = 0$ and Equation (5.216) becomes

$$P_s = \frac{dh}{dt} \tag{5.217}$$

The level acceleration FTT is flown at a constant altitude, so that $dh/dt = 0$ and Equation (5.216) becomes

$$P_s = \frac{V}{g}\frac{dV}{dt} \tag{5.218}$$

The sawtooth climb FTT is a short, timed climb at constant airspeed. As indicated by Equation (5.217), the specific excess power is obtained by measuring the rate of climb, dh/dt. The series of climbs and descents flown with this technique has the appearance of the series of teeth on a cutting saw, hence its name. The sawtooth climbs are typically performed at airspeeds bracketing the expected best rate of climb speed. This method is generally suited for slow speed aircraft or for those parts of the flight envelope where the specific excess power is small, such as in the landing approach flight configuration. Using this method, the specific excess power contours and the maximum rate of climb airspeed can be obtained.

In the level acceleration method, a constant altitude or level acceleration is flown. As indicated by Equation (5.218), the specific excess power is obtained by measuring the rate of change of velocity or acceleration, dV/dt. The acceleration is performed from slightly higher than the minimum airspeed to near the maximum airspeed. Because of this wide range of airspeeds, the level acceleration FTT provides a large amount of data in one maneuver. This method is suitable for high performance aircraft at subsonic and supersonic speeds. The level acceleration FTT is used to obtain the specific excess power contours, the level flight acceleration time, fuel consumption data, and subsonic and supersonic climb schedules. Since the sawtooth climb FTT was flown previously, to determine climb performance, we focus on the level acceleration FTT in obtaining a specific excess power plot.

You will be flying the level acceleration FTT at 20,000 ft (6096 m) in the Lockheed F-104 *Starfighter* supersonic interceptor, shown in Figure 2.34. A three-view drawing of the F-104 is shown in Figure 3.178. Selected specifications of the *Starfighter* are given in Table 5.15.

Table 5.15 Selected specifications of the Lockheed F-104G *Starfighter*.

Item	Specification
Primary function	All weather, Mach 2 supersonic interceptor
Manufacturer	Lockheed Skunk Works, Burbank, California
First flight	17 February 1959
Crew	1 pilot
Powerplant	J79-GE-11A afterburning turbojet
Thrust, MIL	10,000 lb (44,500 N), military power
Thrust, MAX	15,600 lb (69,400 N), maximum afterburner
Empty weight	14,000 lb (6350 kg)
Maximum takeoff weight	29,027 lb (13,166 kg)
Length	54 ft 8 in (16.7 m)
Height	13 ft 6 in (4.1 m)
Wingspan	21 ft 9 in (6.4 m)
Wing area	196.1 ft^2 (18.22 m^2)
Wing loading	148 lb/ft^2 (723 kg/m^2)
Aspect ratio	2.45
Airfoil	3.36% thick biconvex
Maximum speed	1328 mph (2137 km/h), Mach 2+
Service ceiling	58,000 ft (17,700 m)

Recall that a specific excess power plot pertains to a specific aircraft configuration, weight, power setting, and load factor. Your level acceleration will be with the aircraft in a clean configuration, a weight of 18,000 lb (8160 kg), maximum power (full afterburner), and a load factor of one. After an exhilarating takeoff, you climb up to an altitude of 19,000 ft (5800 m). (With its low aspect ratio thin wing, the F-104 has a high takeoff speed of 190 knots (220 mph, 350 km/h)!)

Level at an altitude of 19,000 ft, you stabilize the F-104 at an indicated airspeed of 200 knots (230 mph, 370 km/h, Mach 0.44), which is 10% faster than your 181 knot (208 mph, 335 km/h, Mach 0.40) stall speed or $1.1V_s$. You look forward to seeing where the horizon cuts through the canopy at this airspeed, making a mental picture of the aircraft's pitch attitude. This is the level flight sight picture reference that you will use to enter the level acceleration at 20,000 ft. You set the horizontal stabilizer to a position where you expect the aircraft to be trimmed at about midway through the acceleration. You set the trim at this mid-band setting because setting it at the low or high end of the airspeed range will result in excessive stick forces at high or low speeds, respectively. You will not re-trim the F-104 during the acceleration, as this will lead to non-smooth or "jerky" pitch motions.

You are ready to start your climbing entry into the level acceleration. You push the throttle forward to FULL, commanding maximum afterburner power. You use your pitch attitude to maintain the 200 knot airspeed, as the engine stabilizes. You make final checks of your engine instruments and confirm that the data acquisition system is on. You let the F-104 climb towards 20,000 ft and just below this target altitude, you pushover to the level flight sight picture that you made a mental note of earlier.

The F-014 is accelerating briskly in maximum afterburner, as you watch the airspeed dial rotating to higher and higher airspeeds. Airspeed and time are the primary parameters of interest, but you are recording many other parameters of interest, including fuel flow, outside air temperature, altitude, and vertical velocity. You strive to be smooth on the controls during the acceleration, using the pitch control to hold altitude and referencing the horizon for your attitude rather than "chasing" your instruments, which have an inherent lag. As you are accelerating, you feel the stick forces change, but you do not change the trim setting that you made before you started, because this would lead to jerky pitch changes of the aircraft. As you approach Mach 1, you prepare for the abrupt changes or "jumps" in the airspeed and altitude indications, due to shock waves affecting the Pitot-static system. As you pass through the transonic speed range, from about Mach 0.9 to 1.1, the altitude and airspeed indications "jump", following the characteristic profiles of the static pressure position error corrections (see Figure 5.19 for example).

The F-104 continues to accelerate to supersonic speeds, past Mach 1.5, 1.6, 1.7, then up to Mach 1.8. Approaching 920 knots or Mach 1.95 (1060 mph, 1700 km/h), the increase in velocity has slowed to about 1 knot per second. You are at the end of the level acceleration, where you *anchor* the end point. You enter a slight descent, losing about 200 ft (60 m) in altitude, increasing the airspeed by about 10 knots (12 mph, 19 km/h). You then reset a level flight attitude and allow the F-104 to accelerate to an end point airspeed of 970 knots or Mach 2.06 (1120 mph, 1800 km/h). Having completed the level acceleration, you pull the throttle back, decelerate and start your descent back to the airport.

Using Equation (5.218), the specific excess power, P_s, is calculated using the velocity versus time data from your level acceleration. A plot of the calculated specific excess power versus calibrated airspeed is shown in Figure 5.62. Starting at near the stall speed (point 1), the specific excess power increases to over 400 ft/s (122 m/s) between points 5 and 6. The P_s decreases dramatically through the transonic range where there is a large increase in drag between points 6 and 7, and then increases slightly at low supersonic speeds. As high supersonic speeds are reached, the excess thrust decreases (points 8 and 9), reducing the specific excess power to zero at the end point (point 10).

Horizontal lines of constant P_s are drawn on Figure 5.62, intersecting the level acceleration curve at two velocities for each value of P_s. If we were to fly a level acceleration at 30,000 ft (9100 m), we would obtain two more velocity points for each constant value of P_s. Hence, by flying a level acceleration at several different altitudes, one can generate a constant P_s curve comprising these altitude–velocity points, as shown in Figure 5.63. The level acceleration, flown at an altitude of 20,000 ft, is shown with the labeled numbers corresponding to those in Figure 5.62.

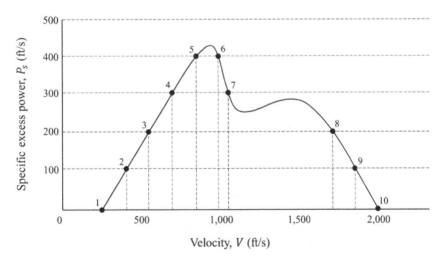

Figure 5.62 Specific excess power versus velocity from Lockheed F-104 level acceleration at 20,000 ft (clean configuration, weight 18,000 lb, maximum power, and load factor one).

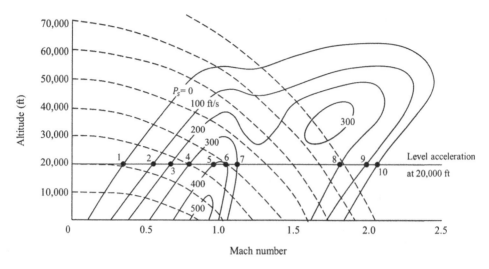

Figure 5.63   Lockheed F-104 level acceleration at 20,000 ft overlaid on specific excess power plot (clean configuration, weight 18,000 lb, maximum power, and load factor one). (Source: *US Air Force, [22].*)

5.13   Turn Performance

We know that an aircraft does not just fly in a straight line from takeoff to landing; it must maneuver, changing its direction and speed during a flight. Up to now, we have assumed the aircraft motion is rectilinear, along a straight line, with zero acceleration, and thus with constant velocity. Remember that velocity is a vector, with a magnitude and a direction. In unaccelerated flight, the magnitude and the direction of the velocity are constant. If the magnitude or direction of the velocity is changed, the result is a linear or radial acceleration, respectively. In this section, we focus on curved-path motion with constant speed and a radial acceleration. We discuss curved flight paths, with constant speed, when the aircraft's motion is either entirely in the horizontal or vertical plane. For these curved paths, we define the turn performance equations, some of the limitations on turn performance, and introduce the turn performance chart. We look first at the turn in the horizontal plane.

5.13.1   The Level Turn

Assume that we are flying in an aircraft at constant velocity and constant altitude, in straight-and-level flight. The linear acceleration is zero since the speed is constant. There is a normal acceleration of 1 g on the aircraft, pointing downward. Sitting in the airplane, we feel a force equal to our mass multiplied by this normal acceleration, or simply our weight. A *turn* is defined as a change from the straight-line flight path to a curved path, in a horizontal, vertical, or oblique plane. If the change in the flight path is only in the horizontal plane, the result is a *level turn*, as shown in Figure 5.64.

For a level turn at constant speed, V_∞, the curved flight path describes a circle, with a constant radius, R, in a horizontal plane at constant altitude, as shown in Figure 5.64. Even at constant speed, there is a *radial acceleration, a_r,* pointing towards the center of the turn. The radial acceleration is due to the constantly changing *direction* of the velocity vector, rather than a change in the velocity magnitude.

We now ask the fundamental question, how do we get the aircraft to turn? Prior to entering the turn, the aircraft is flying in a straight-line, in an equilibrium state. The four forces, acting on the

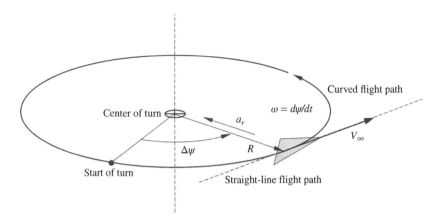

Figure 5.64 Constant airspeed, level turn.

aircraft are in balance: the lift equals the weight and the thrust equals the drag. A force must be changed to generate a radial acceleration to turn the aircraft. Typically, this is done aerodynamically by moving a control surface (ailerons, elevator, or rudder). In a level turn, the ailerons are deflected to roll or bank the aircraft, tilting the lift vector to produce a force and radial acceleration directed towards the center of the turn. The rudder alone could also be used to produce a side force to turn the aircraft. Since there is no aileron input, the bank angle is zero and the lift vector is not tilted. This type of rudder-only turn is sometimes called a "flat turn", since the aircraft remains level or "flat" with zero bank angle during the turn. Deflecting the elevator produces a pitching motion that results in a turn in the vertical plane. A change in the thrust force can also be used to turn the aircraft. The direction of the thrust vector can be changed, using a thrust-vectoring engine nozzle, to produce roll and side forces, resulting in a turn. We now examine the forces in the turn more closely to develop equations for the level turn.

5.13.1.1 Level Turn Performance Equations

In evaluating aircraft turn performance, three of the most important parameters are the *turn rate*, *turn radius*, and *load factor*. The turn rate tells us "how fast or slow" we are turning, the turn radius quantifies "how small or how big" a turn we are making, and the load factor quantifies the acceleration factor or g's that the aircraft structure and the people are subjected to, in the turn. In this section, we develop the equations to quantify these three important turn performance parameters for a level turn.

In performing a turn, we can distinguish between a *sustained* turn and an *instantaneous* turn. In a sustained turn, the aircraft maintains a constant altitude and airspeed and the turn radius, turn rate, and load factor are constant. In an instantaneous turn, the aircraft cannot maintain these constant conditions, rather the turn entry conditions can only be maintained at the instant that the turn is initiated. After turn entry, the airspeed, altitude, or both, decrease in an instantaneous turn. From an energy perspective, the sustained turn is a constant energy maneuver while the instantaneous turn is an energy-losing maneuver. Instantaneous turn performance is important for aircraft such as military fighters, where it may relate to "nose pointing" of the aircraft for combat maneuvering or weapons release. In quantifying the turn radius, turn rate, and load factor, we focus on the steady-state sustained turn where these quantities are constant.

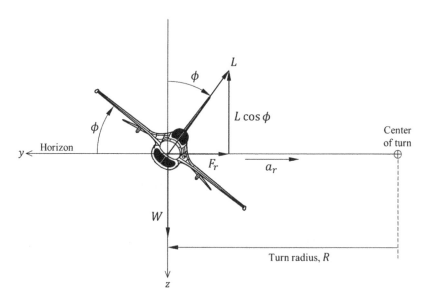

Figure 5.65   Forces on an aircraft in a steady, level turn.

A free-body diagram of an aircraft in a steady, level, sustained turn is shown in Figure 5.65. The aircraft is at a constant airspeed, V_∞, constant altitude, and constant bank angle, ϕ. The aircraft weight, W, acts vertically downward. The lift vector is rotated from the vertical through the bank angle, ϕ. There is a component of the lift in the vertical direction, $L\cos\phi$, which balances the weight, since the aircraft is at a constant altitude. There is also a component of the lift in the radial direction, F_r, pointing towards the center of the turn circle. The radial force gives rise to a radial acceleration, a_r, sometimes called the *centripetal acceleration* (from the Greek, for "seeking the center").

Looking at the free-body diagram, one might be tempted to add another horizontal force to balance the radial force and "keep the aircraft in equilibrium" or "hold the aircraft in the turn". Adding this force, commonly known as the *centrifugal force* (from the Greek for "fleeing from the center"), is incorrect. Although the aircraft is in steady, level flight (constant altitude, not wings-level flight), it is in accelerated flight and is not in equilibrium. The direction of the velocity vector is continuously changing, resulting in a velocity vector that changes with time, and hence an acceleration. If there were another force that balanced the radial force, F_r, then the aircraft would not turn, but would continue in a straight-line path. The radial force is needed to cause the aircraft to turn. This may seem contradictory to your intuition, where you know that when you make a turn in an automobile or an airplane, your body is pushed in the direction opposite to the turn, as though there were a centrifugal force pushing you to the outside of the turn. In reality, you are moving in an accelerating, non-inertial frame of reference, where the vehicle is turning *into* you as you continue in a straight line.

Applying Newton's second law in the vertical, z-direction, we have

$$\sum F_z = W - L\cos\phi = ma_z = 0 \tag{5.219}$$

where the vertical acceleration, a_z, is zero since the aircraft is in steady, constant altitude flight. Rearranging Equation (5.219), we have

$$W = L\cos\phi \tag{5.220}$$

Thus, we see that, to maintain constant altitude flight, the aircraft weight, W, is balanced by the vertical component of the lift, $L \cos \phi$. Rearranging Equation (5.220), we have

$$\frac{L}{W} = \frac{1}{\cos \phi} \tag{5.221}$$

The lift divided by the weight is the load factor, n. Thus, for a level turn, the load factor is given by

$$n = \frac{1}{\cos \phi} \tag{5.222}$$

The units of load factor are given in g's, so that when the aircraft is in steady, wings-level flight, the lift equals the weight and the load factor is 1 g. When the load factor is other than 1 g, say 4 g, the lift is equal to four times the weight.

Equation (5.222) gives the *sustained* load factor since we have assumed that the aircraft is in a *sustained* level turn. This means that the aircraft can maintain this load factor while in the steady turn. According to Equation (5.222), in a level turn, the sustained load factor, n, is only a function of the aircraft bank angle, ϕ. Equation (5.222) is plotted in Figure 5.66, showing that load factor, in a level turn, increases with increasing bank angle, with the load factor going to infinity as the bank angle goes to 90°.

Applying Newton's second law to Figure 5.65 in the y-direction, along the radius of the turn, we have

$$\sum F_y = F_r = L \sin \phi = ma_r \tag{5.223}$$

where a_r is the radial acceleration. The flight path is a circle, so that the radial acceleration is given by

$$a_r = \frac{V_\infty^2}{R} \tag{5.224}$$

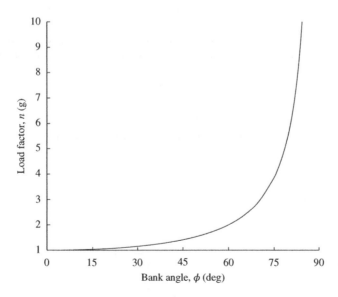

Figure 5.66 Load factor as a function of bank angle in a level turn.

Substituting Equation (5.224) into (5.223), we have

$$L \sin \phi = m \frac{V_\infty^2}{R} \tag{5.225}$$

Solving for the turn radius, R, and using Equation (5.221), we have

$$R = m \frac{V_\infty^2}{L \sin \phi} = \frac{W}{g} \frac{V_\infty^2}{L \sin \phi} = \frac{V_\infty^2}{g} \frac{\cos \phi}{\sin \phi} \tag{5.226}$$

We can relate the bank angle terms in Equation (5.226) to the load factor as follows.

$$\frac{\sin \phi}{\cos \phi} = \sqrt{\frac{\sin^2 \phi}{\cos^2 \phi}} = \sqrt{\frac{1 - \cos^2 \phi}{\cos^2 \phi}} = \sqrt{\frac{1}{\cos^2 \phi} - 1} = \sqrt{n^2 - 1} \tag{5.227}$$

Inserting Equation (5.227) into (5.226), we have an equation for the turn radius in a level turn.

$$R = \frac{V_\infty^2}{g \sqrt{n^2 - 1}} \tag{5.228}$$

Equation (5.228) shows that the turn radius is a function of the velocity and the load factor, increasing with higher velocity or lower load factor. Conversely, the turn radius is minimized with lower velocity or higher load factor.

Rearranging Equation (5.228) as

$$g \sqrt{n^2 - 1} = \frac{V_\infty^2}{R} = a_r \tag{5.229}$$

we see that the term $g \sqrt{n^2 - 1}$ is the radial acceleration, a_r. The radial load factor, n_r, is defined as

$$n_r \equiv \sqrt{n^2 - 1} \tag{5.230}$$

For the level turn, the radial acceleration and radial load factor act entirely in the horizontal plane.

Prior to entering a level turn, the aircraft is pointing towards a specific direction, called the *heading*, ψ. After entering the turn, the *turn rate*, $\dot{\psi}$, is defined as the time rate of change of the aircraft heading, given by

$$\dot{\psi} = \frac{\Delta \psi}{\Delta t} \tag{5.231}$$

where $\Delta \psi$ is the change in the heading, as shown in Figure 5.64, and Δt is the change in time. The turn rate is equivalent to the aircraft angular velocity, ω, which is given by

$$\omega = \frac{V_\infty}{R} = \dot{\psi} \tag{5.232}$$

where V_∞ is the aircraft velocity and R is the turn radius. Inserting Equation (5.228) into (5.232), we have

$$\omega = \frac{V_\infty}{R} = \frac{V_\infty}{\left(\dfrac{V_\infty^2}{g \sqrt{n^2 - 1}} \right)} \tag{5.233}$$

or

$$\omega = \frac{g \sqrt{n^2 - 1}}{V_\infty} \tag{5.234}$$

The turn rate is a function of the load factor and the velocity, increasing with higher load factor or lower velocity. Conversely, the turn rate is minimized with lower load factor or higher airspeed.

The turn performance equations, given by Equations (5.222), (5.228), and (5.234), provide the load factor, turn radius, and turn rate for a level turn, respectively. There are no parameters in these equations that refer to a specific aircraft or type of aircraft. In other words, the turn radius and turn rate are the same, in any aircraft, given the same velocity and load factor, and the load factor is the same in any aircraft at the same bank angle.

The parameters of interest in the turn performance equations are the load factor, n, turn radius, R, turn rate, ω, and velocity, V_∞. If we know any two of these parameters, we can calculate the other two using the turn performance equations. In flight, it is straightforward to measure the velocity and the load factor or the turn rate. We investigate this further in the turn performance flight test technique.

In evaluating turn performance, it is usually desired to determine the *minimum turn radius* and *maximum turn rate*. These limits on turn performance may seem obvious for military aircraft, where a smaller turn radius and larger turn rate could provide a combat advantage against an adversary or in the deployment of weapons. However, these limits are also important for commercial aircraft that need to maneuver efficiently in congested airspace. By inspection of Equation (5.228), we can make the broad observation that the turn radius is minimized by minimizing the velocity and maximizing the load factor. Similarly, by inspection of Equation (5.234), we observe that the turn rate is maximized by minimizing the velocity and maximizing the load factor.

If we were to measure the load factor versus Mach number for a level turn in flight, we would obtain a plot as shown on the upper left in Figure 5.67. From this plot, the Mach number corresponding to the maximum load factor, $M_{n,max}$, can be identified. With the known values of Mach number and load factor, plots for the turn radius and the turn rate are generated, using Equations (5.228) and (5.234), respectively, as shown in the upper right and lower plots in Figure 5.67. From these plots, the Mach number for the minimum radius turn, $M_{R,min}$, and for the maximum rate turn,

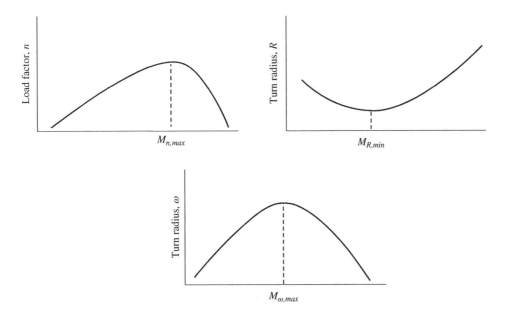

Figure 5.67 Turn performance plots for a level turn.

$M_{\omega,max}$, are obtained. The three Mach number that have been identified are not equal. We discuss their relationship to each other shortly.

Example 5.17 Level Turn Performance *An aircraft enters a level turn at a constant airspeed of 315 mph. It completes a full 360° circle in 27 s, maintaining the entry speed constant throughout the turn. Calculate the turn rate, turn radius, load factor, and bank angle of the level turn.*

Solution

First, we convert the airspeed to consistent units.

$$V_\infty = 315\,\frac{mi}{h} \times \frac{5280\,ft}{1\,mi} \times \frac{1\,h}{3600\,s} = 462.0\frac{ft}{s}$$

The turn rate, ω, is

$$\omega = \frac{360\,deg}{27\,sec} = 13.33\frac{deg}{s} = 0.2327\frac{rad}{s}$$

Equation (5.234) is used to solve to the load factor, n.

$$n = \sqrt{\left(\frac{\omega V_\infty}{g}\right)^2 + 1} = \sqrt{\left[\frac{\left(0.2327\frac{rad}{sec}\right)\left(462.0\frac{ft}{s}\right)}{32.17\frac{ft}{s^2}}\right]^2 + 1} = 3.488\,g$$

Using Equation (5.228), the turn radius, R, is

$$R = \frac{V_\infty^2}{g\sqrt{n^2 - 1}} = \frac{\left(462.0\frac{ft}{s}\right)^2}{32.17\frac{ft}{s^2}\sqrt{(3.488)^2 - 1}} = 1986\,ft$$

Equation (5.222) is used to solve for the bank angle, ϕ.

$$\phi = \cos^{-1}\left(\frac{1}{n}\right) = \cos^{-1}\left(\frac{1}{3.488}\right) = 73.34\,\text{deg}$$

5.13.1.2 The Turning Stall

In Chapter 2, we defined the stall speed for 1 g, wings-level flight. However, does this stall speed change for turning flight? For 1 g, wings-level flight, the aircraft lift equals the weight, so that the stall speed, $V_{s,1g}$, is given by Equation (2.48), repeated below.

$$V_{s,1g} = \sqrt{\frac{2L}{\rho_\infty S C_{L,max}}} = \sqrt{\frac{2W}{\rho_\infty S C_{L,max}}} \tag{5.235}$$

If we now enter a level turn with a load factor n, the aircraft lift in the turn, L_{turn}, is equal to the weight multiplied by the load factor, nW.

$$L_{turn} = nW \tag{5.236}$$

Inserting Equation (5.236) into (5.235), the aircraft stall speed in the turn, $V_{s,turn}$, is given by

$$V_{s,turn} = \sqrt{\frac{2L_{turn}}{\rho_\infty S C_{L,max}}} = \sqrt{\frac{2nW}{\rho_\infty S C_{L,max}}} \qquad (5.237)$$

For 1 g, wings-level flight, where $n = 1$, Equation (5.237) simply reduces to Equation (5.235) and the 1 g stall speed. Comparing Equations (5.235) and (5.237), we have

$$V_{s,turn} = V_{s,1g} \sqrt{n} \qquad (5.238)$$

which states that the stall speed in a level turn increases from the 1 g, wings-level stall speed with the square root of the load factor.

Rearranging Equation (5.238) and using Equation (5.222), we have

$$\frac{V_{s,turn}}{V_{s,1g}} = \sqrt{n} = \sqrt{\frac{1}{\cos \phi}} \qquad (5.239)$$

which states that the aircraft stall speed in a level turn, relative to the wings-level, 1 g stall speed, increases with increasing load factor or increasing bank angle. Equation (5.239) is plotted in Figure 5.68, showing the dramatic increase in stall speed with bank angle. As expected, the stall speed goes to an infinite value as the bank angle approaches 90°, where the wing does not produce vertical lift.

5.13.1.3   The Turn Performance Chart

The turn performance chart is a plot of turn rate versus Mach number as a function of the turn radius, as shown in Figure 5.69. This chart is independent of aircraft type, since it is generated

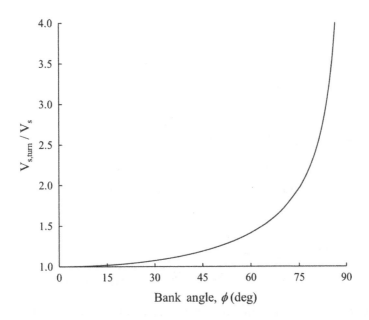

Figure 5.68   Stall speed as a function of bank angle in a level turn.

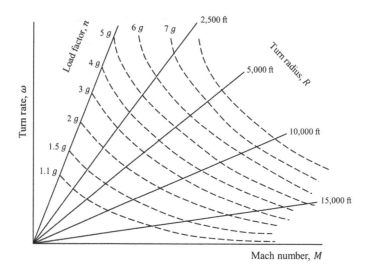

Figure 5.69   Turn performance chart.

using the turn performance equations, which are independent of aircraft type. The plot is valid for only one altitude, since Mach number is used which requires the specification of a temperature corresponding to an altitude. The plot would be valid for all altitudes if velocity was plotted instead of Mach number. The turn performance chart graphically relates the turn radius, turn rate, load factor, and Mach number for a level turn.

By overlaying a specific excess power curve on the turn performance chart, as shown in Figure 5.70, the chart is made aircraft specific, since the P_s curve describes the characteristics of a specific aircraft. The $P_s = 0$ overlay curve is of particular interest because we can identify several

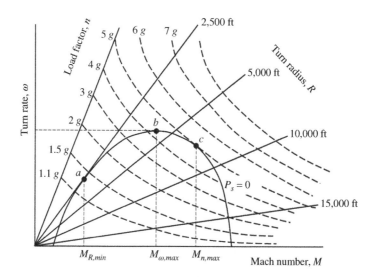

Figure 5.70   The turn performance chart with specific excess power ($P_s = 0$) overlay.

sustained turn performance limits. Recall that the $P_s = 0$ line corresponds to all of the points in the flight envelope (Mach number and altitude) where the aircraft can stabilize in level flight. The point where the $P_s = 0$ curve is tangent to a line of constant turn radius, point a in Figure 5.70, is the minimum sustained turn radius (2500 ft). The top of the $P_s = 0$ curve, point b in Figure 5.70, is the maximum sustained turn rate (ω_{max}). The point where the $P_s = 0$ curve is tangent to a line of constant load factor, point c in Figure 5.70, is the maximum sustained load factor (5 g). The Mach number corresponding to these limiting turn performance values can also be identified. The Mach number for the maximum sustained load factor, $M_{n,max}$, is greater than or equal to the Mach number for the maximum turn rate, $M_{\omega,max}$, which is greater than or equal to the Mach number for the minimum turn radius, $M_{R,min}$. We can write this as

$$M_{n,max} \geq M_{\omega,max} \geq M_{R,min} \tag{5.240}$$

For some aircraft, at certain altitudes, the Mach number corresponding to the maximum sustained load factor is coincident with the Mach number for maximum turn rate.

If we now bound the turn performance chart with the aircraft flight envelope limits, we obtain a plot as shown in Figure 5.71, known as a "doghouse" plot, because of its resemblance to the front of a doghouse. The flight envelope bounds the plot by the lift or stall limit on the left side, the load factor limit on the upper right, and the Mach number or dynamic pressure limit on the right side. The doghouse plot is an aircraft-specific turn performance chart that includes the aircraft's flight envelope boundaries. The overlaid $P_s = 0$ curve is the boundary of *sustained* turn performance. The aircraft has positive P_s below the $P_s = 0$ curve and negative P_s above this curve. An aircraft has its highest turn rate at the intersection of the lift limit and the load factor limit, at the top of the doghouse. This is an instantaneous turn rate, since P_s is negative at this point. The velocity, corresponding to the highest instantaneous turn rate, V^*, is called the *corner velocity*. To turn at this highest rate, an aircraft needs to be flying at the corner velocity.

A doghouse plot for the Lockheed F-104G *Starfighter* is shown in Figure 5.72. The F-104 performance is for a weight of 17,880 lb (8110 kg), maximum power, an altitude of 10,000 ft (3030 m), and a clean configuration (flaps and landing gear up). The data in the chart has been corrected for

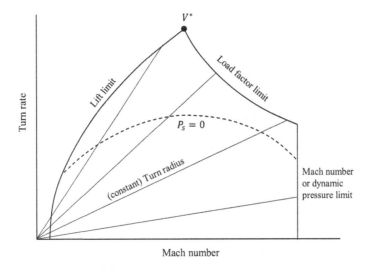

Figure 5.71   The "doghouse" plot.

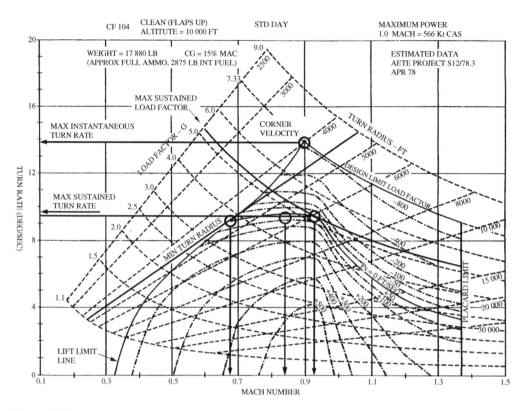

Figure 5.72 Lockheed F-104G *Starfighter* doghouse plot. (Source: *US Air Force, F/RF/TF-104G Flight Manual, T.O. 1 F-104G-1, 31 March 1975.*)

standard day conditions. Thus this turn performance chart is for the F-104 flying in this particular aircraft configuration at these flight conditions.

To obtain the *sustained* turn performance values of the F-104 from the chart, we first find the $P_s = 0$ line (solid line among the many P_s lines). The $P_s = 0$ line is tangent to a load factor curve of about $n = 5.2$ (circled). Thus, the F-104 has a maximum sustained load factor of 5.2 g at a Mach number of about 0.93. The maximum sustained turn rate is 9.7 deg/s at a Mach number of 0.86 (arrow pointing to left axis of chart). The maximum instantaneous turn rate is 13.9 deg/s at a Mach number of 0.90, corresponding to the corner velocity. The minimum sustained turn radius point is located where the $P_s = 0$ line intersects a line of constant turn radius. The minimum turn radius is approximately 4500 ft (1400 m) at a Mach number of about 0.68. Thus, we can write Equation (5.240) for the F-104 sustained turn performance as

$$M_{n,max}(0.93) \geq M_{\omega,max}(0.86) \geq M_{R,min}(0.68) \tag{5.241}$$

5.13.2   Turns in the Vertical Plane

We now look at two curved flight paths in the vertical plane only, the pull-up and pull-down maneuvers. Similar to the level turn in the horizontal plane, it is assumed that the vertical turns are performed at constant airspeed, resulting in a constant radius flight path that describes a vertical circle. The altitude varies in flying the vertical circle.

In the vertical plane turns, the weight contributes directly to the radial force, whereas it did not contribute at all in the horizontal plane, level turn. The weight contribution varies depending on the orientation of the weight vector relative to the lift vector. A general expression for the radial load factor, n_r, in terms of the flight path angle, γ, is given by

$$n_r = n - \cos \gamma \tag{5.242}$$

This equation for radial load factor is consistent with the earlier definition, given by Equation (5.230), as Equation (5.242) equals (5.230) for the case of a level turn where $\gamma = 0$ and $n = 1$. The radial load factor, given by Equation (5.242), acts only in the vertical plane.

5.13.2.1   The Pull-Up Maneuver

The free-body diagram for the pull-up maneuver in the vertical plane is shown in Figure 5.73. Entering the pull-up, the aircraft is in wings-level flight at a constant airspeed, V, with the lift, L, equal to the weight, W. To enter the vertical turn, the elevator is deflected, increasing the angle-of-attack, which increases the lift. A radial force, F_r, is created, equal to the change in lift, ΔL, which curves the flight path since the total lift is now greater than the weight.

Applying Newton's second law to Figure 5.73 along the radius of the vertical turn, we have

$$\sum F_{radial} = F_r + L - W = \Delta L + L - W = L_{turn} - W = ma_r \tag{5.243}$$

where L_{turn} is defined as the increased lift, $L + \Delta L$, that initiated the turn. The increased lift can be expressed in terms of the load factor, as

$$L_{turn} = nW \tag{5.244}$$

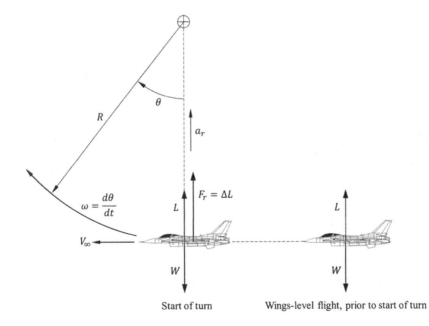

Figure 5.73   Pull-up maneuver in the vertical plane.

Inserting Equation (5.244) into (5.243) and using the definition of the radial acceleration, a_r, given by Equation (5.224), we have

$$L_{turn} - W = nW - W = W(n-1) = m\frac{V_\infty^2}{R} = \frac{WV_\infty^2}{gR} \qquad (5.245)$$

or

$$n - 1 = \frac{V_\infty^2}{gR} \qquad (5.246)$$

Solving for the vertical turn radius, we have

$$R = \frac{V_\infty^2}{g(n-1)} \qquad (5.247)$$

The vertical turn radius is a function of the velocity and load factor, increasing with higher velocity or lower load factor. Conversely, the vertical radius is minimized with lower airspeed or higher load factor.

The turn rate, ω is given by

$$\omega = \frac{d\theta}{dt} = \frac{V_\infty}{R} \qquad (5.248)$$

Inserting Equation (5.247), for the turn radius, into Equation (5.248), we have for the vertical turn rate

$$\omega = \frac{g\sqrt{n^2 - 1}}{V_\infty} \qquad (5.249)$$

The vertical turn rate is a function of the velocity and load factor, increasing with lower velocity or higher load factor. Conversely, the vertical radius is minimized with higher airspeed or lower load factor.

5.13.2.2 The Pull-Down Maneuver

The free-body diagram for the pull-down maneuver in the vertical plane is shown in Figure 5.74. Similar to the pull-up maneuver, the aircraft enters the pull-down maneuver in wings level flight at a constant airspeed, V, except that the aircraft is upside down or inverted. The lift, L, must still equal the weight, W, to be flying at constant altitude. The lift is simply being generated by a negative angle-of-attack of the wing. To enter the vertical turn, the elevator is deflected, increasing the angle-of-attack, which increases the lift in the downward direction. A radial force, F_r, is created, equal to the lift, L_{turn}, which curves the flight path.

Applying Newton's second law to Figure 5.74 along the radius of the vertical turn, we have

$$\sum F_{radial} = F_r + W = L_{turn} + W = ma_r \qquad (5.250)$$

The lift can be expressed in terms of the load factor, as

$$L_{turn} = nW \qquad (5.251)$$

Inserting Equation (5.251) into (5.250) and using the definition of the radial acceleration, a_r, given by Equation (5.224), we have

$$L_{turn} + W = nW + W = W(n+1) = m\frac{V_\infty^2}{R} = \frac{WV_\infty^2}{gR} \qquad (5.252)$$

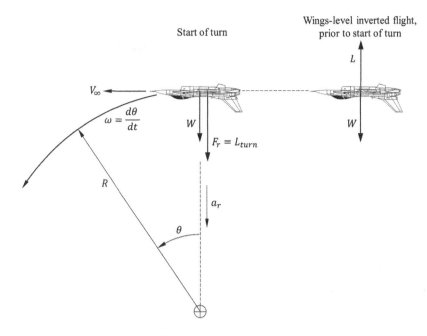

Figure 5.74   Pull-down maneuver in the vertical plane.

or

$$n + 1 = \frac{V_\infty^2}{gR} \tag{5.253}$$

Solving for the vertical turn radius, we have

$$R = \frac{V_\infty^2}{g(n + 1)} \tag{5.254}$$

The vertical turn radius is a function of the velocity and load factor, increasing with higher velocity or lower load factor. Conversely, the vertical radius is minimized with lower airspeed or higher load factor.

The turn rate, ω, is given by

$$\omega = \frac{d\theta}{dt} = \frac{V_\infty}{R} \tag{5.255}$$

Inserting Equation (5.254), for the turn radius, into Equation (5.255), we have for the vertical turn rate

$$\omega = \frac{g\sqrt{n^2 + 1}}{V_\infty} \tag{5.256}$$

The vertical turn rate is a function of the velocity and load factor, increasing with lower velocity or higher load factor. Conversely, the vertical radius is minimized with higher airspeed or lower load factor.

Comparing the turn radius and turn rate equations for the pull-up, Equations (5.247) and (5.249) respectively, with those for the pull-down maneuvers, Equations (5.254) and (5.256) respectively, they differ by the subtraction or addition of one from the load factor. The unity term represents the effect of gravity, or the weight force, on the turn radius or turn rate. For the pull-up maneuver, the aircraft weight increases the turn radius and decreases the turn rate. The opposite is true for the pull-down maneuver, where gravity serves to decrease the turn radius and increase the turn rate.

5.13.3   Turn Performance and the V–n Diagram

As has been mentioned, in assessing turn performance, we are usually interested in the minimum turn radius and the maximum turn rate. For the level turn, we have noted that these limiting values are obtained by minimizing the velocity and maximizing the load factor. The minimum airspeed is set by the stall speed and the maximum load factor is set by aircraft structural considerations.

Equations (5.228) and (5.234), for the turn radius and the turn rate, respectively, can be manipulated to obtain the following relationships for the minimum turn radius, R_{min}, and the maximum turn rate, ω_{max}.

$$R_{min} = \frac{2}{\rho_\infty g C_{L,max}} \left(\frac{W}{S} \right) \tag{5.257}$$

$$\omega_{max} = g \sqrt{\frac{\rho_\infty C_{L,max} n_{max}}{2(W/S)}} \tag{5.258}$$

These parameters are now couched in terms of the freestream density, ρ_∞, the maximum lift coefficient, $C_{L,max}$, wing loading, W/S, and the maximum load factor, n_{max}. These equations indicate that the minimum turn radius and the maximum turn rate are obtained at the highest density, the maximum lift coefficient, and the highest wing loading. We should expect the best turn performance at sea level, as this is where the density is the highest. Wing loading is usually set by aircraft design factors, other than turn performance, such as cruise performance. Therefore, we are usually "stuck" with whatever wing loading the aircraft has, in determining the turn performance. This leaves the maximum lift coefficient and the maximum load factor. Here, there is a problem, as it is not possible for the aircraft to fly at both its maximum lift coefficient and its maximum load factor, throughout the flight envelope. This becomes clear by examining the V–n diagram for the aircraft.

The V–n diagram was introduced in Chapter 2 as a plot that bounds an aircraft flight envelope in terms of its aircraft's aerodynamic lift limits and its structural limits. At lower airspeeds, the aircraft stalls before the maximum load factor is reached, as shown in Figure 2.32. This holds true for both the positive and negative lift and load limits. Below an airspeed denoted as V_A in Figure 2.32, the highest load factor that can be obtained is less than the limit load factor, thus the turn performance is less than if the limit load factor could be attained.

The point where the lift limit line intersects the load limit line is called the *maneuver point*, and the airspeed at this point is called the *corner speed*, V_A. Since the lift coefficient and the load factor are both at their maximum at the maneuver point, the turn radius is at its minimum, and the turn rate is at its maximum at this point. For a given aircraft type, the maximum turn performance is achieved when the aircraft is flying at the corner speed, which is an important speed to know for air combat maneuvering.

This corner speed also has structural implications. Below the corner speed, the aircraft cannot incur structural damage since it stalls before reaching the limit load factor. Above the corner speed, this is not the case, as the aircraft can reach the limit load and beyond, where the structure can be damaged. Thus, the corner speed is also called the *maneuvering speed*, a maximum airspeed to fly in turbulent air to avoid the possibility of structural damage.

An equation for the corner speed, V_A, can be obtained as follows. The maximum load factor, n_{max}, is given by

$$n_{max} = \frac{L}{W} = \frac{\left(\frac{1}{2} \rho_\infty V_\infty^2 \right) S C_{L,max}}{W} \tag{5.259}$$

Figure 5.75   Lockheed-Martin F-16 *Fighting Falcon* supersonic aircraft. (Source: *US Air Force.*)

By definition, the corner speed is the airspeed, V_∞, where the lift coefficient and the load factor are at their maximums. Therefore, solving Equation (5.259) for the corner speed, we have

$$V_A = \sqrt{\frac{2n_{max}}{\rho_\infty C_{L,max}} \left(\frac{W}{S} \right)} \tag{5.260}$$

5.13.4   FTT: Turn Performance

In this section, you will fly the Lockheed Martin F-16 *Fighting Falcon* to learn about the flight test techniques used to evaluate sustained turn performance. We discuss three *stabilized method* turn performance FTTs: the *stabilized load factor* FTT, the *stabilized airspeed* FTT, and the *timed turn* FTT. The four parameters of interest in sustained turn performance are load factor, n, velocity, V, turn radius, R, and turn rate or angular velocity, ω. If we can measure any two of these four, the remaining two parameters may be calculated using Equations (5.228) and (5.234). The load factor, velocity, and turn rate (time to turn) are easy to measure in flight, while the turn radius is not.

You will fly several of the turn performance FTTs in the General Dynamics (now Lockheed Martin) F-16 *Fighting Falcon* (Figure 5.75). Originally designed and built by General Dynamics, Fort Worth, Texas, the F-16 was introduced in the late 1970s as a relatively low cost, lightweight, highly maneuverable, supersonic fighter. The F-16 has a slender fuselage that is blended into its low aspect ratio, swept wing, aft-mounted, all-moving horizontal stabilators, and a single vertical tail. Powered by a single turbofan jet engine with afterburner, capable of generating almost 30,000 lb (133,000 N) of thrust, the F-16 has a top speed near Mach 2. The first flight of the F-16 was on 20 January 1974. A three-view drawing of the F-16 is shown in Figure 5.76 and selected specifications are given in Table 5.16.

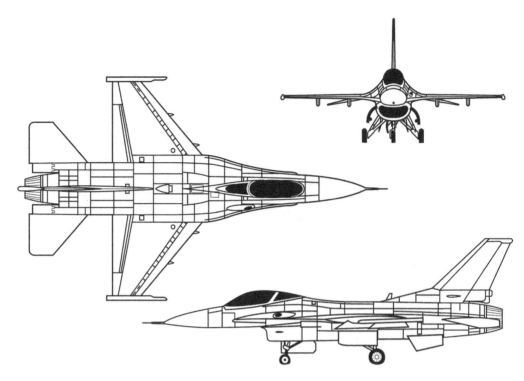

Figure 5.76   Three-view drawing of the Lockheed-Martin F-16 *Fighting Falcon*. (Source: *NASA*.)

Table 5.16   Selected specifications of the Lockheed-Martin F-16 *Fighting Falcon*.

Item	Specification
Primary function	Multirole, supersonic fighter aircraft
Manufacturer	General Dynamics (now Lockheed Martin), Fort Worth, Texas
First flight	20 January 1974
Crew	1 pilot
Powerplant	F110-GE-100/129 afterburning turbofan engine
Thrust, MIL	17,100 (76,100 N), military power
Thrust, MAX	28,600 lb (127,200 N), maximum afterburner
Empty weight	18,900 lb (8570 kg)
Maximum takeoff weight	42,300 lb (19,200 kg)
Length	49 ft 5 in (14.8 m)
Height	16 ft (4.8 m)
Wingspan	32 ft 8 in (9.8 m)
Wing area	300 ft^2 (27.9 m^2)
Airfoil	NACA 64A204
Maximum speed	1320 mph (2120 km/h), Mach 2
Service ceiling	>50,000 ft (>15,240 m)
Positive load factor limit	+9.0 g

After you enter the cockpit of the F-16 and sit on the ejection seat, you notice that the seatback is reclined significantly more than other fighter jets, such as the F-18 that you flew for your familiarization flight. The ejection seat has a tilt-back angle of 30°, compared with the more modest 15–20° found in other fighter jets. The reclined seat is designed to give the pilot increased g-force tolerance, by reducing the vertical distance that the heart has to pump blood up into the brain. This is beneficial in flying the turn performance FTTs, where high load factors are encountered in turning flight. The F-16 structure is designed for higher load factors than other fighter aircraft, with a normal limit load factor of nine. After the single-piece, bubble canopy closes, you appreciate the exceptional field of view around you, especially over the sides of the fuselage and to the rear. There is no center control stick; instead, there is a sidestick controller on your right. There are conventional rudder pedals at your feet and a single throttle lever on your left side. There is a large head-up display (HUD), mounted in front of you that looks like an angled piece of thick glass. Critical flight information is displayed on the HUD without obstructing your view, which allows you to keep your view outside the cockpit during maneuvers.

After takeoff, you climb the F-16 to an altitude of 1000 ft (300 m). Your F-16 is at a gross weight of 20,000 lb (9,000 kg) with no external stores or fuel tanks attached. The general technique for the stabilized load factor FTT and the constant airspeed FTT is to fly at a constant power setting, constant airspeed, and constant load factor. You advance the throttle, selecting maximum afterburner, and the F-16 accelerates to its maximum airspeed. You allow the aircraft to stabilize and this is your first turn performance data point, corresponding to a load factor, n, of one. Your power is set and constant at maximum afterburning thrust, so now you have to choose between trying to hold either the load factor or the airspeed constant.

Before you choose, consider the stabilized turn performance of an aircraft, represented by the load factor versus velocity curve shown in Figure 5.77. The specific excess power, P_s, for this curve is equal to zero, since the aircraft is in equilibrium, with the thrust equal to the drag. This curve also assumes a constant power setting. Your F-16 is currently at the point on the horizontal axis, where the velocity is V_{max} and load factor is one (point 1 in Figure 5.77). Your next test point, point 2, is at a lower velocity and a higher load factor. If you try to maintain a constant velocity at point 2, a small deviation in the velocity results in a large deviation in the load factor, due to the slope of the curve at point 2. In contrast, if you try to maintain a constant load factor at point 2, a small deviation in the load factor results in a small deviation in the velocity. Hence, it is better to use the stabilized load factor technique, as the velocity is decreased from V_{max} at point 1, until reaching a point where the slope of the curve flattens out (between points 5 and 8). At this point, a small deviation in load factor results in a large deviation in the velocity, whereas a small deviation in

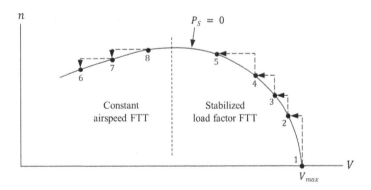

Figure 5.77   Sustained turn performance flight test techniques.

velocity results in a small deviation in the load factor. Hence, when the slope of the curve flattens out, the constant airspeed FTT should be used (points 6, 7, and 8). This can be graphically seen in flight by plotting the load factor versus velocity in real time as the test points are completed. In addition, the pilot finds it progressively more difficult to stabilize the airspeed by holding the load factor constant and this is a signal to transition from the stabilized load factor FTT to the constant airspeed FTT.

Having decided to use the stabilized load factor FTT, your next test point is a load factor of two. You smoothly roll the F-16 into a 60° bank, while applying back pressure to the sidestick, establishing a 2-g turn. The power is still set at maximum afterburning thrust. You allow the airspeed to bleed down and stabilize, while holding the load factor constant. When the airspeed is stable, you record the load factor and velocity, from which the turn radius and turn rate can be calculated (of course, you would need a data system or a backseater to record the data, as your hands are full, maintaining the turning test point). You incrementally increase the load factor, taking data at each successively lower, stabilized airspeed. At some load factor, the airspeed no longer stabilizes and continues to decrease; it is time to transition to the constant airspeed technique.

You reduce engine power and decelerate to set up for these lower airspeed turn performance test points. You stabilize the F-16 at an airspeed that is about 50 knots (58 mph, 93 km/h) slower than your first test point airspeed (Point 6). You then advance the throttle, selecting maximum afterburner, and allow the aircraft to accelerate. As the test airspeed is approached, you smoothly increase the bank angle and sidestick back pressure to stabilize at an airspeed below the test airspeed. You maintain the constant airspeed test point with stick back pressure. You slowly relax your back pressure on the sidestick, which allows the airspeed to increase to the test airspeed. You hold this test airspeed and allow the load factor to stabilize. After the load factor stabilizes, you record the airspeed and load factor. After this test point is complete, you incrementally increase the airspeed and stabilize the load factor, until an airspeed is reached where the load factor does not stabilize. After you land, your turn performance data can be used to produce a doghouse plot.

As may be obvious, the stabilized load factor and constant airspeed turn performance FTTs are best suited for aircraft that can be safely flown at high load factors. Another turn performance FTT, the *timed turn* technique, is suitable for aircraft with load factor limits of two or less. The timed turn is essentially a variation of the constant airspeed method. The technique is normally used at low airspeeds and low load factors or when a measurement of load factor is not available.

The timed turn is flown by setting constant power and then setting either a constant bank angle or a constant airspeed. If a specific, fixed bank angle is selected, then the airspeed is allowed to stabilize to the constant value corresponding to this constant bank angle. If a specific, fixed airspeed is selected, the bank angle is adjusted to the constant value that maintains this constant airspeed. The aircraft is flown in a horizontal circle, starting and ending on the same heading, and the time, Δt, for this 360° turn is recorded. The turn rate, ω, is then simply calculated as $2\pi/\Delta t$. The load factor and turn radius can be calculated from the turn rate and the airspeed.

5.14 Takeoff and Landing Performance

At the start of a flight, an aircraft must take off, and at the end of its flight, it must land. In this section, we examine the performance for horizontal takeoff and landing of a conventional fixed-wing aircraft. In this sense, we are not considering aircraft that are capable of supporting some portion of their weight with thrust. It is beneficial to minimize both the takeoff and landing distances and to perform these maneuvers at as low an airspeed as practical. There are other important issues, such as obstacle clearance and effects of wind, that can affect the takeoff and landing performance. Representative flight profiles for takeoff and landing are shown in Figure 5.78. Both the takeoff

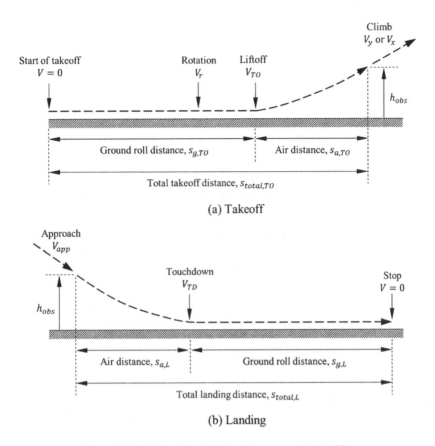

Figure 5.78   Takeoff and landing distances and velocities.

and landing are composed of a ground segment, where the aircraft is rolling on the ground, and an air segment, where the aircraft is airborne.

The takeoff starts at zero velocity with the aircraft in a level attitude. The aircraft accelerates down the runway, in a level attitude, with the airspeed increasing. Upon reaching the rotation airspeed, V_r, the aircraft nose is rotated upward to increase the angle-of-attack. At the liftoff or takeoff speed, V_{TO}, typically no less than about 10% above the stall speed, the aircraft leaves the ground and is airborne. The ground roll segment is the distance, $s_{g,TO}$, between the start of the takeoff roll and the liftoff spot. After liftoff, the aircraft accelerates from liftoff speed to the climb airspeed, typically 20% above the stall speed. The air segment, $s_{a,TO}$, is the distance from liftoff to the distance required to clear a specified height above the ground, typically an obstacle clearance height, h_{obs}. The required obstacle clearance height is 50 ft (15 m) for military and small civilian aircraft and 35 ft (10.7 m) for civilian commercial transport aircraft. The total takeoff distance, $s_{total,TO}$, is the sum of the ground roll distance, $s_{g,TO}$, and the air distance, $s_{a,TO}$.

The landing starts with an air segment, where the aircraft is flying at the approach speed, V_{app}, typically 20–30% above stall speed, at a height above the ground equal to an obstacle clearance height, h_{obs}, of 50 ft (15 m). The air distance for landing, $s_{a,L}$, is the distance from the location of the aircraft at the obstacle distance height to the touchdown spot. After touchdown at velocity, V_{TD}, the aircraft travels a ground roll distance, $s_{g,L}$, to a stop where the velocity is zero. The total landing distance, $s_{total,L}$, is the sum of the ground roll distance, $s_{g,L}$, and the air distance, $s_{a,L}$.

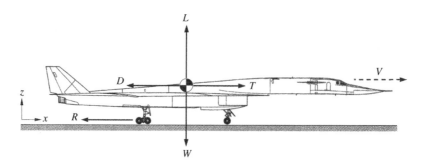

Figure 5.79 Free-body diagram during takeoff and landing.

The forces acting on an aircraft during takeoff or landing are shown in Figure 5.79. There are the same four forces that act on an aircraft in flight, the lift, L, and weight, W, perpendicular to the velocity, V, and the thrust, T, and drag, D, parallel to the velocity. In addition to these four forces, there is a resistance force, R, due to the rolling friction of the landing gear tires on the ground, that acts in a direction opposite to the thrust. The rolling friction force is given by

$$R = \mu(W - L) \tag{5.261}$$

where $(W - L)$ is the net normal force acting on the tires and μ is the coefficient of rolling resistance due to friction. The rolling friction is variable, depending on the runway surface material and whether or not the aircraft wheel brakes are applied. For the minimum takeoff distance, it is assumed that no brakes are applied. For the minimum landing distance, it is assumed that full braking application is used throughout the landing ground roll. Typical values of the coefficient of rolling resistance for various runway surfaces is shown in Table 5.17. The rolling resistance can increase by an order of magnitude with the application of brakes.

The typical variations of axial forces during the takeoff and landing ground rolls are shown in Figure 5.80 and Figure 5.81, respectively. At the start of the takeoff, the velocity is zero, so the lift and drag are also zero. Before the aircraft starts moving, the rolling friction force is also zero. After the thrust is increased, typically to a maximum thrust setting, the aircraft accelerates, and the velocity increases. The lift and drag increase with the square of the velocity. When the aircraft starts to move, the rolling friction force is at a maximum and decreases as the net normal force decreases as the lift increases. The combined retarding force, composed of the drag and the rolling friction force, increases with distance as the drag is continuously increasing. The net acceleration of the

Table 5.17 Typical values of the coefficient of rolling resistance.

Runway surface	Coefficient of rolling resistance, μ	
	Brakes off	Brakes on
Dry concrete/asphalt	0.02–0.05	0.2–0.5
Wet concrete/asphalt	0.05	0.15–0.3
Icy concrete/asphalt	0.02	0.06–0.1
Firm dirt	0.04	0.3
Hard turf	0.05	0.4
Soft turf	0.07	0.2
Wet grass	0.08	0.2

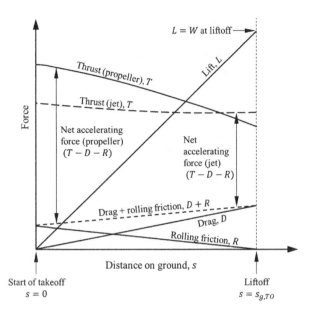

Figure 5.80   Typical variation of axial forces acting on aircraft during takeoff ground roll.

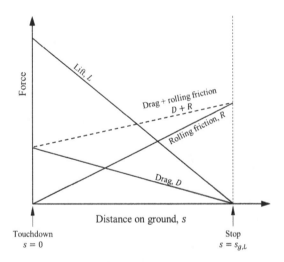

Figure 5.81   Typical variation of axial forces acting on aircraft during landing ground roll.

aircraft is the difference between the thrust and the combined drag and rolling friction force. The thrust variation during the takeoff is dependent on the type of propulsion. For a propeller-driven aircraft, the thrust decreases with increasing velocity, while for a jet-powered aircraft, the thrust is approximately constant. When the lift is greater than the weight, the aircraft lifts off and the rolling friction force goes to zero.

During takeoff and landing, the aircraft experiences an increase in lift and a decrease in drag due to ground effect, as discussed in Section 3.9.5. The ground effect increases the lift when the aircraft is within a height of about a wingspan above the ground. The magnitude of the changes in lift and drag are calculated as discussed in Section 3.9.5.

The forces during the landing ground roll, as shown in Figure 5.81, are the same as for takeoff, except for differences in the magnitude and directions. At touchdown, the lift is at its maximum, equal to the aircraft weight. Typically, lift reducing devices are deployed, such as spoilers, to prevent the aircraft from "bouncing" back into the air and to increase the net normal force for braking action. The rolling friction force is zero at touchdown and increases as the net normal force increases with decreasing lift. The brakes are also applied, increasing the coefficient of rolling friction, as shown in Table 5.17. The engine thrust is assumed to be zero at touchdown. The thrust force can be a decelerating force if thrust reversers are employed. The drag force decreases, during the landing ground roll, with the square of the decreasing velocity. High-speed military aircraft may further increase the drag by deploying a drag parachute. As the aircraft comes to a stop, all of the forces decrease to zero.

We now seek to develop a relationship for the takeoff or landing distance as a function of the forces on the aircraft and the velocity, We apply Newton's second law to the takeoff or landing situation, in the x-direction, parallel to the velocity, as given by

$$\sum F_x = ma_x = m\frac{dV}{dt} \tag{5.262}$$

There is a finite acceleration during the takeoff and a finite deceleration during landing. Inserting the various forces into Equation (5.262), we have

$$T - (D + R) = T - D - \mu(W - L) = m\frac{dV}{dt} = \frac{W}{g}\frac{dV}{dt} \tag{5.263}$$

Dividing by (W/g), multiplying by dt, and rearranging, we have

$$dV = [T - D - \mu(W - L)]\frac{g}{W}dt \tag{5.264}$$

Integrating Equation (5.264) between two arbitrary points along the takeoff or landing ground roll, where constant, average values of thrust, drag, and lift are assumed, we have

$$\int_{V_1}^{V_2} dV = \int_{t_1}^{t_2} [T - D - \mu(W - L)]_{avg}\frac{g}{W}dt \tag{5.265}$$

$$V_2 - V_1 = [T - D - \mu(W - L)]_{avg}\frac{g}{W}(t_2 - t_1) \tag{5.266}$$

The rate of change of velocity can expanded in terms of the ground distance, s, as

$$\frac{dV}{dt} = \frac{dV}{ds}\frac{ds}{dt} = \frac{dV}{ds}V \tag{5.267}$$

The rate of change of velocity can be related to the forces, using Equation (5.266), as

$$\frac{dV}{dt} = \frac{\Delta V}{\Delta t} = \frac{V_2 - V_1}{t_2 - t_1} = [T - D - \mu(W - L)]_{avg}\frac{g}{W} \tag{5.268}$$

Inserting Equation (5.268) into (5.267), we have

$$[T - D - \mu(W - L)]_{avg}\frac{g}{W} = \frac{dV}{ds}V \tag{5.269}$$

or

$$[T - D - \mu(W - L)]_{avg}\frac{g}{W}ds = VdV \tag{5.270}$$

We now wish to integrate Equation (5.270), but the thrust, lift, and drag forces vary with distance and velocity. It can be assumed that the weight remains constant during the takeoff or landing ground roll. To simplify the analysis, we assume that the thrust, drag, and lift are constant, equal to average values during the takeoff or landing ground rolls, as was done in Equation (5.265). Thus, Equation (5.270) is integrated between two points along the ground roll, as

$$[T - D - \mu(W - L)]_{avg} \frac{g}{W} \int_{s_1}^{s_2} ds = \int_{V_1}^{V_2} V dV \tag{5.271}$$

where $[T - D - \mu(W - L)]_{avg}$ is the average value of the forces. The result of the integration is given by

$$s_2 - s_1 = \frac{W}{g[T - D - \mu(W - L)]_{avg}} \left(\frac{V_2^2}{2} - \frac{V_1^2}{2} \right) \tag{5.272}$$

Equation (5.272) relates the distance between two points in the takeoff or landing ground roll with the velocities at these two points. We can now use this relationship to obtain the takeoff and landing ground roll distances.

5.14.1 Takeoff Distance

Equation (5.272) is applied to the takeoff ground roll, between the start of the takeoff roll and the liftoff point, as shown in Figure 5.78. At the start of the takeoff roll, the distance, s_1, is assumed to be zero and the velocity, V_1, is also zero. At liftoff, the distance, s_2, is the takeoff ground roll distance, $s_{g,TO}$, and the velocity, V_2, is the liftoff velocity, V_{TO}. Inserting these values into Equation (5.272), we have

$$s_{g,TO} = \frac{W}{g[T - D - \mu(W - L)]_{avg}} \frac{V_{TO}^2}{2} \tag{5.273}$$

Equation (5.273) provides the ground roll distance for takeoff, $s_{g,TO}$, as a function of the liftoff speed, V_{TO}, and the average forces acting on the aircraft.

A similar type of analysis can be performed to obtain the distance along the ground for the air segment of the takeoff, $s_{a,TO}$ (Figure 5.78). The analysis is performed between the two end points of the air segment, where the initial velocity is the liftoff speed, V_{TO}, and the end velocity is the airspeed at the obstacle clearance height of 50 ft, V_{50}. The result for the takeoff air segment distance is

$$s_{a,TO} = \frac{W}{(T - D)_{avg}} \left[\left(\frac{V_{50}^2 - V_{TO}^2}{2g} \right) + 50 \right] \tag{5.274}$$

The total ground distance for the takeoff is the sum of the ground roll distance and the air segment distance, given by

$$s_{total,TO} = s_{g,TO} + s_{a,TO} \tag{5.275}$$

Examining Equations (5.273) and (5.274), we can assess the impacts of the various terms on the takeoff distance. As may be obvious, the takeoff distance is increased for heavier weight, lower thrust, or higher drag. Delaying the liftoff to a higher airspeed also increases the takeoff ground roll. The type of runway surface changes the coefficient of rolling friction, which affects the takeoff distance. Of course, the takeoff distance is decreased by reducing the weight, decreasing the drag, or increasing the thrust. Weight reductions may be obtained by carrying a lighter load of passengers, cargo, or fuel, for instance. Thrust increases may be achieved by using an afterburner in a jet engine or by using supplemental, strap-on rockets, commonly called JATO (jet-assisted takeoff) bottles. Naval carrier-borne aircraft get an increase in forward speed from the force of a steam-powered catapult launch system. The thrust is also affected by the ambient air density, with decreased thrust available from propeller-driven and jet engines at high densities, corresponding to high altitudes or high temperatures. Thus, the takeoff ground distance can be significantly increased for a takeoff at a high altitude airport on a hot day.

The use of high-lift devices, such as flaps, may also decrease the takeoff distance, although this may not seem evident by inspection of Equation (5.273). Let us dig a little deeper into this. The liftoff speed is typically 10% above the stall speed, to provide a safety margin. Thus, using Equation (2.48) for the stall speed, V_s, the liftoff speed is

$$V_{TO} = 1.1 V_s = 1.1 \sqrt{\frac{2W}{\rho_\infty S C_{L,max}}} \tag{5.276}$$

Inserting Equation (5.276) into (5.273), we have

$$s_{g,TO} = \frac{1.21 W^2}{\rho_\infty S C_{L,max} g [T - D - \mu(W - L)]_{avg}} \tag{5.277}$$

Equation (5.277) shows that the takeoff ground distance can be decreased by using a high-lift device to increase the maximum lift coefficient, $C_{L,max}$, or increase the wing planform area, S. These also increase the lift, which decreases the rolling friction force. Equation (5.277) also emphasizes the impact of the weight on the takeoff distance, as the distance varies with the weight squared. For instance, a 10% increase in the weight results in a 21% increase in the takeoff ground distance.

Other factors that are not included in our takeoff analysis include the effects of wind, runway slope, and pilot technique. A headwind or tailwind can significantly decrease or increase, respectively, the takeoff ground distance. An upward or downward sloping runway can increase or decrease, respectively, the distance by adding a detrimental or favorable weight contribution, respectively, to the axial force. Finally, pilot technique can significantly change the takeoff distance. The pilot's ability to capture the proper liftoff speed and attitude is one of the significant factors affecting the ground distance.

5.14.2 Landing Distance

We now look at the landing ground distance in a similar fashion as was done for the takeoff case. Equation (5.272) is applied to the landing ground roll, between the touchdown point and the full stop of the aircraft. At touchdown, the distance, s_1, is assumed to be zero and the velocity, V_1, is the touchdown speed, V_{TD}. When the aircraft is stopped, the distance, s_2, is the landing ground roll distance, $s_{g,L}$, and the velocity, V_2, is zero. The thrust is assumed to be zero for the landing ground distance. Inserting these values into Equation (5.272), we have

$$0 - s_{g,L} = \frac{W}{g[-D - \mu(W - L)]_{avg}} \left(\frac{V_{TD}^2}{2} - 0 \right) \tag{5.278}$$

or

$$s_{g,L} = \frac{W}{g[D + \mu(W - L)]_{avg}} \frac{V_{TD}^2}{2} \tag{5.279}$$

Equation (5.279) provides the ground roll distance for landing, $s_{g,L}$, as a function of the touchdown speed, V_{TD}, and the average forces acting on the aircraft.

A similar type of analysis can be performed to obtain the distance along the ground for the air segment of the landing, $s_{a,L}$ (Figure 5.78). The analysis is performed between the two end points of the air segment, where the initial velocity is the approach speed, V_{app}, and the end velocity is the touchdown airspeed, V_{TD}. The result for the landing air segment distance is

$$s_{a,L} = \frac{W}{(T-D)_{avg}}\left[\left(\frac{V_{TD}^2 - V_{50}^2}{2g}\right) - 50\right] \tag{5.280}$$

The total ground distance for the landing is the sum of the ground roll distance and the air segment distance, given by

$$s_{total,TO} = s_{g,TO} + s_{a,TO} \tag{5.281}$$

Examining Equations (5.279) and (5.280), we can assess the impacts of the various terms on the landing distance. The landing distance is increased for heavier weight, lower drag, or touching down at a higher airspeed. The type of runway surface changes the coefficient of rolling friction, with a higher coefficient decreasing the distance. The landing distance is decreased by reducing the weight or increasing the drag. The landing distance can be decreased by reducing the lift or increasing the drag. Reducing the lift, typically by deploying wing spoilers or retracting landing flaps, increases the net normal force, which increases the rolling friction force. Drag can be maximized by deploying drag devices such as wing spoilers, speedbrakes (flap-like surfaces that are raised up into the flow), or drag parachutes. The aircraft pitch attitude may be maintained at a high angle, in a technique called aerobraking, to produce additional drag. Reverse thrust may also be applied during the ground roll to provide an additional force decelerating the aircraft. The landing ground distance, including reverse thrust, is given by

$$s_{g,L} = \frac{W^2}{\rho_\infty SC_{L,max}g[-T - D - \mu(W - L)]_{avg}} \tag{5.282}$$

Similar to the takeoff situation, there are several factors that affect the landing distance, including wind, runway slope, air density, and pilot technique.

Example 5.18 Calculation of Takeoff Performance *The North American XB-70A Valkyrie (Figure 5.82) takes off on a hard surface, dry concrete runway at sea level, with a weight of 519,000 lb. During the takeoff roll, each engine is producing an average thrust of 25,000 lb. Assume that the coefficient of rolling resistance is 0.03, and that the average lift and drag during the takeoff roll are 210,000 lb and 8000 lb, respectively. The aircraft liftoff airspeed is 211.0 kt. The XB-70A has a wing area of 6298 ft². Calculate the aircraft lift coefficient at liftoff and the takeoff ground distance.*

5.14.3 Solution

First, convert the liftoff speed, V_{TO}, into consistent units.

$$V_{TO} = 211.0\,\frac{nmi}{h} \times \frac{6076\,ft}{1\,nmi} \times \frac{1\,h}{3600\,s} = 356.1\,\frac{ft}{s}$$

The liftoff speed is given by

$$V_{TO} = \sqrt{\frac{2W}{\rho_\infty SC_L}}$$

Figure 5.82 North American XB-70A *Valkyrie* in flight with wingtips drooped. (Source: *NASA*.)

Solving for the lift coefficient, we have

$$C_L = \frac{2W}{\rho_\infty S V_{TO}^2} = \frac{2(519,000\,lb)}{\left(0.002377\,\frac{slug}{ft^3}\right)(6298\,ft^2)\left(356.1\,\frac{ft}{s}\right)^2} = 0.547$$

The total average thrust produced by the six engines during the takeoff roll is

$$T = 6 \times 25,000\,lb = 150,000\,lb$$

The takeoff ground distance is given by

$$s_{g,TO} = \frac{W}{g[T - D - \mu(W - L)]_{avg}} \frac{V_{TO}^2}{2}$$

$$s_{g,TO} = \frac{519,000\,lb}{32.2\,\frac{ft}{s^2}[150,000\,lb - 8000\,lb - 0.03(519,000\,lb - 210,000\,lb)]} \frac{\left(356.1\,\frac{ft}{s}\right)^2}{2} = 7700\,ft$$

5.14.4 FTT: Takeoff Performance

Takeoff and landing performance flight tests are typically performed to develop takeoff and landing procedures and to verify performance predictions. Techniques are developed for normal, short, and soft field takeoff and landings, engine-out situations, rejected takeoffs, abused takeoffs, where controls are misapplied, and other special or abnormal situations. Data is obtained for a variety of runway conditions, including dry, wet, icy, and soft or rough surfaces. In the present flight test technique, we focus on takeoff performance.

Typical takeoff performance data that is collected include the ground roll distance, the air distance to clear a 50 foot obstacle, and the optimum rotation speeds for different gross weights and aircraft configurations. The takeoff performance results are influenced by a large number of variables, including individual pilot technique, runway surface condition and slope, wind, and aircraft weight. Some of these variables are difficult to measure accurately or compensate for, so it is usually only possible to estimate the takeoff characteristics of the aircraft within broad limits, relying on statistical averages of a large number of takeoffs to reduce errors.

Variables related to pilot technique include brake release and power application procedures, directional control inputs (nose wheel steering, differential braking, rudder), aileron and elevator positions during acceleration, and rotation technique (airspeed, pitch rate, angle-of-attack at liftoff, and climb out angle). To reduce the effects of pilot technique, a repeatable, well-defined takeoff procedure is followed as closely as possible. For example, consistent procedures should be developed for throttle setting prior to brake release, throttle technique at and immediately after brake release, control position during acceleration, selecting rotation airspeed and rate, aircraft attitude at liftoff, and landing gear and flap retraction points.

To learn about the takeoff performance flight test technique, you will be flying the North American XB-70A *Valkyrie*, triple-sonic bomber prototype, as shown in Figure 5.82. Designed in the late 1950s by North American Aviation, the XB-70A was the prototype for the planned B-70 high-altitude bomber, capable of cruise flight at Mach 3. As shown by the three-view drawing in Figure 5.83, the *Valkyrie* was a very large aircraft, almost 200 ft (61 m) in length, with a takeoff

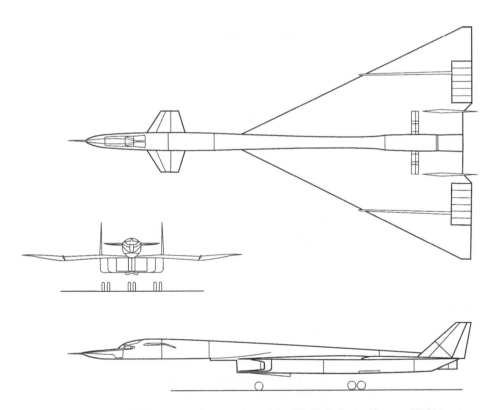

Figure 5.83 Three-view drawing of the XB-70 *Valkyrie*. (Source: *NASA*.)

weight over half a million pounds. It had a large low aspect ratio delta wing, with a sweep of 65°, a wingspan of 105 ft (32 m), twin movable vertical stabilizers, and two large, fuselage-mounted canard surfaces. The movable canards were used to aid in trimming the aircraft and they had trailing edge flaps. Movable elevon control surfaces were located at the trailing edge of the delta wing, providing both pitch and roll control. The XB-70 was powered by six General Electric YJ93-GE-3 afterburning turbojet engines, each producing 30,000 lb (133,400 N) of thrust in full afterburner. The aircraft was fabricated primarily of stainless steel and titanium, to handle the high temperatures associated with Mach 3.1 flight at 73,000 ft (22,000 m). The nose of the aircraft had a variable-geometry ramp that could be faired into the cockpit windscreen for high-Mach flight or lowered for increased visibility during landing. Selected specifications of the North American XB-70A are provided in Table 5.18.

The large delta wing had the unique feature of providing compression lift in supersonic flight, by drooping the outer wing sections by as much as 65°, as shown in Figure 5.82. This compression lift is similar to the waverider concept described in Section 3.13, where the high pressure flow behind the shock wave is contained under the wing. The downward deflected wingtips also provided additional vertical surface area, which increased the directional stability at high speed. The delta wing of the No. 1 XB-70 had zero dihedral, which resulted in poor lateral–directional stability at Mach 3. The No. 1 XB-70 only flew to Mach 3 once, being limited to Mach 2.5 after it demonstrated poor stability at higher Mach numbers. The No. 2 XB-70 incorporated 5° of wing dihedral, which improved its lateral–directional stability for flight beyond Mach 2.5.

Only two XB-70A prototypes were built, with the first flight of the No. 1 on 21 September 1964 and the first flight of the No. 2 prototype on 17 July 1965. Although production of the B-70 bomber did not materialize, the two XB-70s aircraft performed valuable research in support of the design of supersonic transport (SST) aircraft. The No. 1 XB-70 completed a total of 83 flights and is now on display at the US Air Force Museum at Wright-Patterson Air Force Base, Dayton, Ohio. The No. 2 XB-70 completed 46 flights, before it crashed following a mid-air collision with a NASA F-104 chase aircraft during a flight to obtain photographs of a formation of General Electric-powered aircraft on 8 June 1966.

Table 5.18   Selected specifications of the North American XB-70A *Valkyrie*.

Item	Specification
Primary function	Supersonic, high altitude bomber prototype
Manufacturer	North American Aviation, Los Angeles, California
First flight	21 September 1964
Crew	1 pilot + 1 co-pilot
Powerplant	6 × General Electric YJ93-GE-3 afterburning turbojet
Thrust (ea. engine)	30,000 lb each (133,400 N), maximum afterburner
Empty weight	253,600 lb (115,030 kg)
Takeoff weight	542,000 lb (246,000 kg)
Length	193.4 ft (58.9 m)
Height	30.75 ft (9.37 m)
Wing span	105 ft (32.0 m)
Wing area	6298 ft^2 (585.1 m^2)
Wing aspect ratio	1.751
Airfoil	Hexagonal section
Maximum speed	2056 mph (3309 km/h), Mach 3.1 at 73,000 ft (22,250 m)
Range	4288 mi (6900 km)
Service ceiling	77,350 ft (23,600 m)

You climb aboard the massive XB-70 and slide into the left pilot seat in the cockpit. Your co-pilot straps into the right seat. The pilot flight controls consist of a conventional yoke and rudder pedals. There are six throttle levers in the center console to control the thrust of the six YJ93-GE-3 after-burning turbojet engines, with engine gauges laid out above these. You have a nice view from the cockpit, perched about 20 ft (6 m) above the ground, with good forward visibility through the wind-screen since the variable-geometry nose ramp is down. You are sitting far forward of the engine inlets and the nose landing gear, which are located at about the middle of the fuselage.

After engine start-up, you taxi to the end of the runway and perform your pre-flight checks and configure the aircraft for the takeoff performance tests. For normal takeoffs, you set the forepart of the canard at 0°, with the canard flap set to 20°. You move the yoke back and forward to check that the elevons on the trailing edge of the delta wing are moving up and down, which provides pitch control during takeoff. Then you move the yoke from left to right to check the differential movement of the elevons, which provides roll control. Your crew chief on the ground communicates to you that the elevons are moving properly. You check the wingtip fold setting, located on the left-side panel above the throttles next to the landing gear handle, and verify that they are undeflected for takeoff.

Prior to taking the runway, you record the fuel weight, so that a takeoff weight may be calculated. Your co-pilot will use a hand-held data technique, called the "eyes right" method, during the takeoff, where he will visually estimate the ground roll distance by looking out the right cockpit window and referencing markers on the side of the runway. This is not highly accurate, but it provides another piece of data.

The aircraft is highly instrumented so that the takeoff parameters are recorded for later data analysis. The runway conditions, including the runway slope and composition, have already been recorded by the test team. For each takeoff test, the wind speed, pressure altitude, and temperature will be recorded. After the tests, the takeoff performance data will be standardized to correct for wind, runway slope, thrust, weight, and air density.

In addition, the ground team will use a cinetheodolite tracking system, a combination camera and surveying instrument, which will precisely record the aircraft's takeoff trajectory. Modern data collection also includes the use of global positioning system (GPS) instruments, as discussed previously.

Taxing onto the runway, you line up at a predetermined point, ensure that the aircraft nose wheel is straight, and hold the brakes. As was briefed prior to the flight, you will follow the defined takeoff procedure as closely as possible, to reduce any effects of individual pilot techniques. While holding the brakes, you advance the six throttle levers to set the engines to minimum afterburner power. After checking that the engine instruments are stabilized you release the brakes, the co-pilot starts a stopwatch, and you advance the throttles to maximum afterburner power. You accelerate to 20 knots below your intended liftoff speed, and then pull back on the yoke, rotating the aircraft to a 10° nose up pitch attitude, which you establish by using the visual reference of placing the upper surface of the fuselage on the horizon. You use nosewheel and rudder steering, as required, to maintain the runway heading, but you are careful not to use any differential braking, which would increase the takeoff distance. You hold this attitude until the aircraft lifts off. You note the indicated airspeeds for nose wheel rotation, nose wheel liftoff, and main wheel liftoff. You maintain this takeoff attitude and configuration until the aircraft is 50 ft above the ground.

After completing many more takeoffs at a variety of weights, it is time to look at the data. The test ground roll distance, $s_{g,t}$, to liftoff is plotted versus the average weight during the ground roll, $\bar{W}_t$, in Figure 5.84a. The aircraft weights range from about 400,000 lb (181,000 kg) to about 540,000 lb (245,000 kg). The test ground roll distance data is scattered and does not correlate well with aircraft weight. For example, the dispersion in the distance is about 3000 ft (900 m) for a typical weight of 520,000 lb (236,000 kg). It is likely that the takeoff data is affected by other factors, in addition to the takeoff weight. The takeoff data is corrected to zero wind, a constant lift coefficient at liftoff,

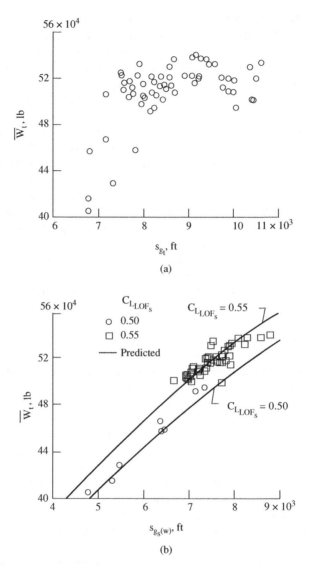

Figure 5.84   North American XB-70 ground roll landing data, (a) uncorrected, (b) corrected.   (Source: *Larson and Schweikhard, NASA TM-X-2215, 1971, [17].*)

standard thrust, and standard air density and plotted in Figure 5.84b. The lighter takeoff weights, less than 500,000 lb (227,000 kg), were corrected to a lift coefficient at liftoff, $C_{L,LOF,s}$, of 0.50. For heavier weights, above 500,000 lb, a lift coefficient of 0.55 was used. The standardized takeoff data correlates much better with the test weight and with the pre-test predictions. At the lightest weight, the takeoff distance is about 4800 ft (1500 m) and about 8900 ft (2700 m) at the heaviest weights.

References

1. *Aerodynamics*, USAF Test Pilot School, Edwards AFB, California, January 2000.
2. *Aircraft Performance*, USAF Test Pilot School, Edwards AFB, California, January 2000.

3. Anderson, J.D., Jr, *Aircraft Performance and Design*, WCB/McGraw-Hill, Boston, Massachusetts, 1999.
4. Corda, S. Stephenson, M.T., Burcham, F.W., and Curry, R.E., "Dynamic Ground Effects Flight Test of an F-15 Aircraft," *NASA TM-4604*, September 1994.
5. Diehl, Walter S., "Standard Atmosphere – Tables and Data," *NACA Report* **218**, 1925.
6. Federal Aviation Administration, US Department of Transportation, *Aviation Maintenance Technician Handbook – Airframe*, Vol. **1**, Oklahoma City, Oklahoma, 2012.
7. Fillipone, A., *Aerospace Engineering Desk Reference*, Butterworth-Heinemann, Elsevier Inc., Oxford, UK, 1st edition, 2009.
8. Gallagher, G.L., Higgins, L.B., Khinoo, L.A., and Pierce, P.W., *Fixed Wing Performance*, USNTPS-FTM-NO. 108, US Naval Test Pilot School, Patuxent River, Maryland, September 30, 1992.
9. Gracey, W., "Measurement of Aircraft Speed and Altitude," *NASA Reference Publication* **1046**, 1980.
10. Gracey, W., "Wind Tunnel Investigation of a Number of Total Pressure Tubes at High Angles of Attack – Subsonic, Transonic, and Supersonic Speeds," *NACA Report* **1303**, 1957.
11. Hall, D.A., "Technical Preparation of the Airplane *Spirit of St. Louis*," NACA Technical Note No. 257, July, 1927.
12. Hurt, H.H., Jr *Aerodynamics for Naval Aviators*, US Navy NAVWEPS 00–80 T-80, US Government Printing Office, Washington, DC, January 1965.
13. Jackson, P. (ed.), *Jane's All the World's Aircraft: 2002–2003*, Jane's Information Group Limited, Coulsdon, Surrey, United Kingdom, 2002.
14. John, J.E.A., *Gas Dynamics*, Allyn and Bacon, Inc., Boston, Massachusetts, 1969.
15. Kaiser, F., "Der Steigflug mit Strahlflugzeugen; Teilbericht 1: Bahngeschwindigkeit fur Besten Steigens (The Climb with Jet Airplane; Speeds for Best Climb)," Versuch-Bericht Nr 262 O2 L 44, Messerschmitt A.G., Lechfield, May 1, 1944, Translated into English as R.T.P./T.I.B. Translation No. G.DC/15/148 T, Ministry of Supply, United Kingdom.
16. Khurana, I., *Medical Physiology for Undergraduate Students*, 1st edition, Elsevier, New Delhi, India, 2012.
17. Larson, T.J. and Schweikhard, W.G., "Verification of Takeoff Performance Predictions for the XB-70 Airplane," *NASA TM X-2215*, March 1971.
18. Lush, K.J., "A Review of the Problem of Choosing a Climb Technique with Proposals for a New Climb Technique for High Performance Aircraft," Aeronautical Research Council Report, Memo. 2557, 1951.
19. McCormick, B.W., *Aerodynamics, Aeronautics, and Flight Mechanics*, John Wiley & Sons, New York, 1979.
20. Newman, D., *Interactive Aerospace Engineering and Design*, McGraw-Hill, New York, 2002.
21. Olson, W.M., *Aircraft Performance Flight Testing*, US Air Force Flight Test Center, Edwards, California, AFFTC-TIH-99-01, September 2000.
22. *Performance Flight Test Phase*, Vol. **I**, US Air Force Test Pilot School, Edwards AFB, California, August 1991.
23. "Record Setting Transatlantic Flight," *Model Airplane News*, **88**, January 2004.
24. Rosemount Engineering Company, "Flight Calibration of Aircraft Static Pressure Systems," Report No. RD-66-3, REC Report 76431, February 1966.
25. Rutowski, E.S., "Energy Approach to the General Aircraft Performance Problem," *Journal of the Aeronautical Sciences*, vol. **21**, no. 23, March 1954, pp.187–195.
26. Talay, T.A., *Introduction to the Aerodynamics of Flight*, NASA SP-367, US Government Printing Office, Washington, DC, 1975.
27. Wallace, J.M. and Hobbs, P.V., *Atmospheric Science: An Introductory Survey*, 2nd edition, Academic Press, Burlington, Massachusetts, 2006.
28. *US Standard Atmosphere, 1976*, US Government Printing Office, Washington, DC, October 1976.
29. Yechout, T.R., *Introduction to Aircraft Flight Mechanics*, 2nd edition, American Institute of Aeronautics and Astronautics, Inc., Reston, Virginia, 2014.
30. Young, H.D. and Freedman, R.A., *University Physics*, 11th edition, Addison Wesley, San Francisco, California, 2004.

Problems

1. A mountain climber has a mass of 100 kg. Calculate the mountain climber's weight at the surface of the earth and on the top of Mt Everest, 29,029 ft above the surface of the earth. What is the percent change in the mountain climber's weight, on the top of Mt Everest, as compared to on the surface of the earth? (The radius of the earth is 6371.4 km and the mass of the earth is 5.98×10^{24} kg.)

2. Assume that the U-tube manometer, used to measure the pressure on the wing in Example 5.3, is filled with water (density of 1 g/cm^3) rather than mercury. Calculate the height difference

measured by the water-filled manometer to obtain the same measured pressure of 54,059 N/m^2, as in the example. What is the issue with using water in the manometer rather than mercury?

3. Calculate the temperature, pressure, and density at an altitude of 6223 m (interpolate as required).

4. You are flying a Northrop T-38 *Talon* at an altitude of 15,000 ft and your indicated airspeed is 225 KIAS. Assuming standard atmospheric conditions, calculate the calibrated, equivalent, and true airspeeds. Assume that the instrument error, ΔV_{instr}, is 0.10 knots and the position error, ΔV_{pc}, is −0.55 knots.

5. An aircraft is flying at Mach 0.670. A total temperature probe, mounted on the aircraft, measures a temperature of 475.5 °R. Assuming that the probe has a recovery factor of 0.98, calculate the freestream static temperature (in degrees Fahrenheit). Calculate the altitude of the aircraft, assuming a standard atmosphere.

6. You are in a fly-by tower, preparing to collect data for the Pitot-static system calibration of an F-16 aircraft. The distance from your sighting device in the tower to the fly-by line is 850 ft. The distance from your sighting device to the grid is 2.25 ft. You measure an atmospheric pressure and temperature in the tower of 13.171 lb/in^2 and 59 °F, respectively. You sight through the grid as the F-16 performs the tower fly-by and measure a vertical grid reading of 7.4 in. The instrument corrected pressure altitude reading in the cockpit of the F-16 is 3160 ft. Calculate the pressure altitude and altitude position error correction for the F-16.

7. The Cessna 310 is a light, twin-engine, general aviation aircraft with a low-wing configuration, powered by two six-cylinder, horizontally opposed, piston engines. The aircraft has a wing span of 36.9 ft, wing area of 179 ft^2, zero-lift drag coefficient of 0.0267, and span efficiency factor of 0.810. The Cessna 310 is flying at an altitude of 8000 ft and an airspeed of 190 mi/h. Calculate the wing loading and thrust-to-weight ratio for the Cessna 310 at a weight of 5500 lb.

8. For the Cessna 310 aircraft in Problem 7, calculate the velocity, lift coefficient, and lift corresponding to minimum power required.

9. For the Cessna 310 in Problem 7, calculate the lift-to-drag ratio and the power required at the specified flight condition. Convert the power required to units of horsepower.

10. A jet aircraft is flying in steady, level flight at an altitude of 37,000 ft. The aircraft lift coefficient at this flight condition is 0.540. The aircraft has a wing area of 450 ft^2 and a thrust specific fuel consumption, TSFC, of 1.60×10^{-4} lb/(lb·s). The final weight, W_1, of the aircraft is 33,800 lb. If the aircraft consumes 8350 lb of fuel, plot the range and endurance as a function of the lift-to-drag ratio, for values of L/D from 5 to 12. Perform a hand calculation of the range and endurance for a lift-to-drag ratio of 6.

11. Rate of climb flight data is obtained for an aircraft from sea level to an altitude of 16,000 ft, as shown in the table below. Using this data, calculate the time to climb from 1500 ft to 15,000 ft.

Altitude h (ft)	Rate of climb $\dot{h}$ (ft/min)
0	1,220
2,000	1,134
4,000	1,040
6,000	956
8,000	861
10,000	770
12,000	689
14,000	595
16,000	512

12. A glider is flying at an airspeed of 42.0 mph with a lift-to-drag ratio of 22. Calculate the glide angle, the rate of descent, and the horizontal glide distance from an altitude of 2500 m.

13. A Lockheed U-2 is flying at an altitude and airspeed of 70,400 ft and 95 knots, respectively. If the U-2 has a weight of 17,000 lb, calculate its potential energy, kinetic energy, total energy, and specific energy at this flight condition. Also, calculate the potential and kinetic energies as a percentage of the total energy.

14. A North American *Sabre* jet performs a sawtooth climb at a constant airspeed of 300 kt. At a point during the climb, the rate of climb is 3820 ft/min. Calculate the specific excess power and the excess thrust at this point in the climb. Assume that the aircraft has a constant weight of 13,700 lb during the climb.

15. An aerobatic Pitts *Special* biplane enters a split-S maneuver at an airspeed of 95 mph and a load factor of 4 g. At what altitude must the pilot initiate the split-S so as to complete the maneuver at 500 ft above the ground?

16. You are flying an Extra 300 aerobatic aircraft with a 1 g, wings-level stall speed of 60 knots. At an airspeed of 150 knots, you roll the aircraft into a constant bank angle, level turn and steadily increase the load factor by pulling back on the control stick. You watch the airspeed decrease as the load factor increases, and the aircraft stalls at an airspeed of 105 knots. What was the load factor and bank angle of the turn when the aircraft stalled?

17. A Lockheed F-16 is performing a 9 g turn at an airspeed of 530 knots. Calculate the turn radius (in feet and statute miles) and turn rate (in degrees per second). How long will it take for the F-16 to complete a 180° heading change?

18. For a weight of 10,200 lb and an altitude of 25,000 ft, the corner velocity of a high-performance jet aircraft is at a Mach number of 0.93 and a maximum load factor of 7.33. If the wing reference area of the aircraft is 205 ft^2, calculate the maximum lift coefficient.

19. The North American XB-70 *Valkyrie* touches down on a hard surface, dry concrete runway at sea level, with a weight of 380,000 lb and a speed of 193 KIAS. Moderate braking is applied, yielding a coefficient of rolling resistance of 0.3. Calculate the landing ground distance for the XB-70, assuming that the lift is zero at touchdown and the zero-lift drag coefficient is 0.007. The XB-70 has a wing area of 6298 ft^2.

6

Stability and Control

Restored Northrop N-9 M prototype flying wing. (Source: *Courtesy of Bernardo Malfitano, UnderstandingAirplanes.com.*)

Introduction to Aerospace Engineering with a Flight Test Perspective, First Edition. Stephen Corda.
© 2017 John Wiley & Sons Ltd. Published 2017 by John Wiley & Sons Ltd.
Companion Website: www.wiley.com/go/corda/aerospace_engg_flight_test_persp

A surprisingly large number of people, both within and without the aircraft industry, still appear to question the economic reasons for going to all the trouble to build an all-wing airplane. "Sure," they say, "after a lot of practice people can learn to walk on their hands, but it's most uncomfortable and unnatural, so why do it when nothing is gained thereby?" Actually, there are startling gains to be made in the aerodynamic and structural efficiency of an all-wing type, provided that certain basic requirements can be fulfilled by the type under question. These requirements can be simply stated as follows:

First, the airplane must be large enough so that the all-wing principle can be fully utilized. This is a matter closely related to the density of the elements comprising the weight empty and the useful load to be carried within the wing ...

The second basic requirement is that the all-wing airplane be designed to have sufficient stability and controllability for practical operation as a military or commercial airplane. We believe this requirement has been fully met by hundreds of flights completed with this type, and we are fully convinced of its practicability after having built a dozen different airplanes embodying scores of different configurations incorporating the all-wing principle.

John K. "Jack" Northrop, 35th Wilbur Wright Memorial Lecture, 1947[1]

6.1 Introduction

Jack Northrop was a visionary aircraft designer, who pioneered the flying wing concept in the 1940s. His company designed, built, and flight tested several prototype flying wing designs in the 1940s, which eventually culminated in the Northrop B-2 flying wing bomber (see Figure 1.27). In many ways, the flying wing represents an optimum design from an aerodynamic perspective. However, an aircraft is considered a poor design if it is unstable and uncontrollable in flight. If the flying wing is disturbed from steady, level flight by atmospheric turbulence or a pilot input, will the aircraft return to its equilibrium flight condition or go out of control? Will the aircraft fly smoothly through this turbulence or will it be a very uncomfortable "ride" for the pilot? What flight control surfaces are required for the pilot to adequately control the aircraft? How much should these control surfaces move or deflect? How much force does the pilot have to use to move the surfaces? These are the types of stability and control questions and issues that are examined in this chapter.

Aircraft fly in three-dimensional space. They have six degrees of freedom, three translational or linear motions (up, down, and sideways) and three rotational motions (pitch, roll, and yaw). Aircraft performance, discussed in the previous chapter, dealt with the translational motions caused by the forces acting on the aircraft. Stability and control deals with the aircraft rotational motions, about the aircraft center of gravity, as a response to the *moments* acting on the aircraft. Since we must now consider the moments, the equations of motion for stability and control include the moment equations. In the most general case of three-dimensional aircraft motion, we have to consider all six degrees of freedom, embodied by three force equations and three moment equations. We are not able to find closed form solutions to this set of six coupled differential equations, but we find that, by making certain simplifying assumptions, we can reduce the number of equations and obtain approximate solutions that provide valuable insight into the aircraft motion.

Two important characteristics of the aircraft motion are its *instantaneous* response and its response *over time*. In *static* stability, we are focused on the aircraft's instantaneous response to an

---

[1] John K. Northrop, "The Development of the All-Wing Aircraft", 35th Wilbur Wright Memorial Lecture, The Royal Aeronautical Society Journal, Vol. 51, pp. 481–510, 1947.

input. In *dynamic* stability, we are interested in the response or motion of the aircraft over time. In evaluating static stability, we consider the different *equilibrium states* of the aircraft, rather than the dynamics of its motion. Dynamic stability involves the study of the aircraft motion over time.

Let us think again about the timescales that are characteristic of the physical phenomena of interest in various engineering disciplines, as was given in Table 5.1. In the study of aircraft performance, we determined that the timescales were long, on the order of minutes or hours. For stability and control, we are often interested in more dynamic situations, where the timescale is much smaller, on the order of seconds. This small timescale makes the job of obtaining flight data a bit more difficult. We may need more complicated data acquisition equipment to record the aircraft responses at a high data rate.

Rather than trying to immediately understand the complex, three-dimensional motion of an aircraft, we separate our study of stability and control into simpler pieces. We are able to decouple the two-dimensional, *longitudinal (pitch)* motion of the aircraft from the more complicated *lateral-directional (roll-yaw)* motions. The longitudinal motions are *symmetric*, wings-level movement of the aircraft center of gravity in the vertical plane of the aircraft. The lateral motions are *asymmetric*, involving rolling and yawing motions, where the velocity vector is not in the vertical plane of symmetry.

When learning any new discipline, there is new terminology, definitions, and nomenclature. We start with some stability and control definitions and nomenclature.

6.2 Aircraft Stability

In assessing aircraft stability, we start with the aircraft in an equilibrium condition, disturb this condition, and then observe the aircraft response. A body is defined to be in a state of equilibrium when it is at rest or in steady, unaccelerated motion. For the aircraft to be in an *equilibrium* or *trim state*, the resultant, external force and moment about the aircraft center of gravity must be zero. (You had experience with this trim state by setting up the trim shot when flying the Extra 300 aircraft flight test technique of Section 2.3.6.) This may be expressed in equation form as

$$\sum \vec{F}_{CG} = 0 \tag{6.1}$$

$$\sum \vec{M}_{CG} = 0 \tag{6.2}$$

where $\vec{F}_{CG}$ and $\vec{M}_{CG}$ are the forces and moments, respectively, about the aircraft center of gravity, *CG*. If the forces and moments are non-zero, there are translational (due to the forces) or rotational (due to the moments) accelerations acting on the aircraft. The translational and rotational accelerations are zero in equilibrium or trimmed flight. The equilibrium or trimmed state is not restricted to level, constant altitude flight, as the aircraft could be in a steady, trimmed climb, descent, or other flight condition.

Stability refers to the *tendency* of a body to return to its equilibrium state after it has been disturbed. For an aircraft, this disturbance or upset may be due to a pilot input, atmospheric turbulence, a wind gust, or other event. An aircraft must have adequate stability to maintain equilibrium or trimmed flight, over a wide range of airspeeds and altitudes in the flight envelope, and in a variety of flight conditions, such as level, climbing, or descending flight. Aircraft stability also serves to reduce the workload of the pilot, such that constant attention is not required to fly the aircraft. The stability may be positive, negative, or neutral.

Many aircraft are designed to be *inherently stable*, which is a property of the basic aircraft configuration or airframe. For an aircraft that is not stable, *artificial stability* may be required, which can be provided by some type of automatic flight control system. Artificial stability can stabilize an

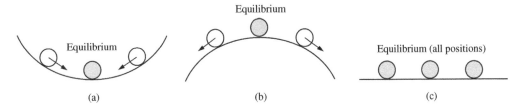

Figure 6.1 Definitions of static stability, (a) positive static stability, (b) negative static stability, and (c) neutral static stability.

inherently unstable aircraft. A *stability augmentation system* (SAS) may also be used to improve the stability characteristics of an aircraft. Many modern aircraft are designed with *relaxed stability*, which can greatly enhance maneuverability. Digital computers, coupled with mechanical actuators connected to control surfaces, are used in "fly-by-wire" flight control systems to provide artificial or augmented stability. Even without a fly-by-wire system, it may be acceptable for an aircraft to have some degree of instability that is not corrected, as long as it can be controlled with a manageable workload. We discuss some of these stability enhancements in future sections.

We separate aircraft stability into the *static* and *dynamic* stability, related to the initial and long-term response of the aircraft to a disturbance. *Static stability* deals with the *initial* tendency of a body to return to equilibrium after being disturbed, while *dynamic stability* is concerned with the tendency or *response over time* of a body to return to equilibrium after being disturbed. We examine the *time history of the response* in assessing dynamic stability.

6.2.1 Static Stability

The concept of static stability is often visualized by considering a ball in the three situations shown in Figure 6.1. The ball is in an equilibrium state, resting at the bottom of a bowl, in Figure 6.1a. If the ball is displaced from its equilibrium position, by moving the ball up the concave surface, and then released, gravity causes the ball to roll back towards its equilibrium position at the bottom of the bowl. After being disturbed from its equilibrium state, the ball's *initial* tendency is to return to the equilibrium state; hence, the ball demonstrates *positive static stability*, or is *statically stable*.

In Figure 6.1b, the ball is in an equilibrium state at the top of an upside down bowl. If the ball is disturbed from its equilibrium state, gravity causes the ball to roll down the side of the bowl, away from the equilibrium position. After being disturbed from its equilibrium state, the ball's initial tendency is to move away from the equilibrium state, hence the ball demonstrates *negative static stability*, or is *statically unstable*.

There can be degrees of positive and negative static stability. If the sides of the bowl are made shallower or steeper, the stability of the ball is made more or less stable, respectively, in its motion. The degree of stability directly impacts the controllability of an aircraft.

Finally, if the ball is placed on a flat surface, as shown in Figure 6.1c, it is in an equilibrium state at any position on the surface. If displaced, the ball remains in equilibrium at its new displaced position, hence the ball demonstrates *neutral static stability*.

6.2.2 Dynamic Stability

Dynamic stability concerns the motion over time of the aircraft after the initial disturbance from equilibrium. Imagine an aircraft in an equilibrium state at a cruise flight condition with a constant

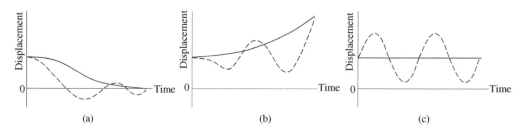

Figure 6.2 Dynamic stability, non-oscillatory (solid), oscillatory (dashed), (a) positive dynamic stability, (b) negative dynamic stability, and (c) neutral dynamic stability.

airspeed, constant altitude, and constant angle-of-attack. If the aircraft encounters a turbulent gust that disturbs the equilibrium condition and raises the nose, what will the subsequent aircraft motion look like? Assuming that we restrict ourselves to the pitching motion of the aircraft, the nose of the aircraft may prescribe an up or down motion over time, as shown in Figure 6.2. The aircraft nose is displaced (upward on the vertical axis) from the equilibrium position (horizontal axis) and the motion of the nose, over time, is given by the various curves.

If the aircraft nose drops and returns to the equilibrium position over time, as shown by the solid curve in Figure 6.2a, the motion has *positive dynamic stability* or is *dynamically stable*. The motion may be aperiodic (non-oscillatory) or oscillatory, as shown by the solid and dashed curves, respectively. In both cases, the amplitude of the displacement is getting smaller with time, exhibiting *positive damping* of the disturbance. After the initial disturbance, the initial tendency is for the nose to return to the equilibrium position, thus these motions also have positive static stability. The curves in Figure 6.2a exhibit both *positive static stability* and *positive dynamic stability*.

If the aircraft nose continues to rise after the initial disturbance, diverging from the equilibrium position, the motion has *negative dynamic stability* or is *dynamically unstable*, as shown in Figure 6.2b. Again, the motion may be aperiodic or oscillatory, as shown by the solid and dashed curves, respectively. These motions exhibit *negative damping*, where the amplitude of the displacement grows larger over time. For the non-oscillatory case (solid curve), the initial tendency after release is to diverge from equilibrium, thus this represents negative static stability. For the oscillatory motion (dashed curve), the initial tendency after release is for the nose to return to equilibrium, thus this motion has positive static stability.

We conclude from this that positive static stability does not guarantee positive dynamic stability. In other words, static stability is a necessary, but not a sufficient, condition for dynamic stability. On the other hand, negative static stability results in negative dynamic stability.

Figure 6.2c shows non-oscillatory and oscillatory curves for neutral dynamic stability. Here, when the aircraft nose is displaced from its equilibrium position, it either remains in the displaced position (solid curve) or it oscillates about the displace position (dashed curve). The solid curve displays neutral static stability, as well as neutral dynamic stability. The dashed curve displays negative static stability and neutral dynamic stability.

Although we have been considering the displacement of the aircraft in pitch, we could have considered a displacement in roll, yaw, or a combination of these. For instance, an aircraft could be dynamically stable or unstable in roll or yaw. The classification of the dynamic behavior is the same in all cases, with some complications as more modes of motion are added.

In quantitatively assessing dynamic stability, we are usually interested in the time it takes for a disturbance to damp to half its initial amplitude, if the motion is dynamically stable or convergent, or the time it takes to double in amplitude, if the motion is dynamically unstable or divergent. If the motion is oscillatory in nature, the frequency and period of the motion are important parameters.

6.3   Aircraft Control

While an aircraft must have an adequate amount of stability, whether inherent or artificial, it must also be capable of maneuvering and changing its equilibrium flight condition. An aircraft must be able to change speed, altitude, heading, angle of climb, and maneuver. Aircraft *control* relates to the ability of the vehicle to respond to control inputs, typically from deflections of aerodynamic control surfaces. In our study of aircraft control, we are interested in these control surface displacements and the forces associated with these deflections. Aircraft controls must be powerful enough to maintain and change the equilibrium states of the aircraft throughout the range of airspeeds and altitudes of its flight envelope. For example, it is not acceptable for the controls to be adequate during level, cruising flight, but be incapable of controlling the aircraft during takeoff and landing.

Many early airplane designers attempted to design inherently stable aircraft, with the misconception that inherent stability would alleviate much of the requirements for aircraft control. As some of these designers learned, aircraft designed with too much inherent stability and lack of controllability were difficult, if not impossible, to fly safely.

An opposite approach was taken by the Wright brothers. The aircraft that they designed and flew were unstable in all three axes, pitch, roll, and yaw. They believed that an unstable aircraft would enhance maneuverability in flight, but their aircraft were very difficult to control in flight. The Wright brothers compensated for this by devoting many hours to flight training in their gliders, learning how to control their aircraft.

Both of these stability and control approaches were lacking in obtaining satisfactory aircraft flying and handling qualities for mass-produced aircraft. There is a balance that must be struck between the degree of stability and the amount of controllability. Adequate stability does not necessarily translate into adequate controllability. In fact, a high degree of stability tends to reduce controllability, as it is more difficult to change the aircraft's equilibrium state. The upper limits in the aircraft stability are set by the lower limits of the available controllability.

6.3.1   Flight Controls

The *flight control surfaces* for a conventional aircraft were introduced in Section 1.2.2.2. The primary control surfaces are composed of movable elevators, ailerons, and rudder for pitch, roll, and yaw control, respectively. Secondary control surfaces may include flaps, spoilers, slats, or speed brakes. These are aerodynamic control surfaces as they generate forces and moments due to the flow of air over their surfaces. Other types of non-aerodynamic control surfaces are possible, such as thrust vectoring or weight shifting, as used in a hang glider. The control surfaces are connected to the pilot controllers, such as a center stick, control wheel, or rudder pedals, through either mechanical, hydro-mechanical, or electrical linkages. The *flight control system* translates the inputs, typically from a pilot or computer, to the flight control surfaces. This may happen through direct mechanical connection, hydro-mechanical devices, computers, or other devices. The flight control system may be of a *reversible* or *non-reversible* type.

In a reversible flight control system, the cockpit controls are connected directly to the flight control surfaces through mechanical linkages, which include pushrods, cables, pulleys, and sometimes chains, as shown in Figure 6.3a. The aerodynamic forces and moments on the control surfaces are fed back to the cockpit controls (center stick, control wheel, rudder pedals, etc.). For example, if the pilot moves the control stick forward or aft, this input causes the elevator to move down or up, respectively. If the trailing edge of the elevator is moved up or down by hand, this causes the control stick to move aft or forward, respectively. (This assumes a conventional aircraft configuration with the tail mounted aft of the wing.) Since movement of the cockpit controls cause the control surface to move and vice versa, the system is dubbed "reversible". The pilot must provide all of the

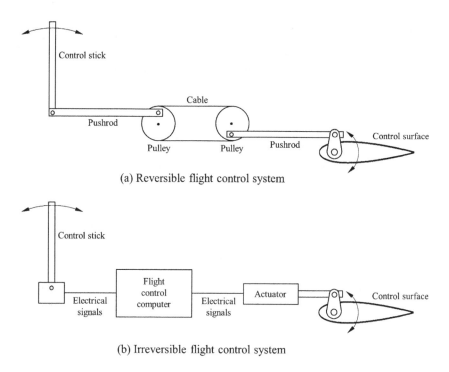

(a) Reversible flight control system

(b) Irreversible flight control system

Figure 6.3   Types of flight control systems.

input power to move the aerodynamic flight control surfaces. Reversible flight control systems are relatively simple and are normally used on smaller, lower speed aircraft, where the air loads on the control surfaces do not generate intolerable cockpit control forces. Aircraft designed up until about the 1940s had reversible flight controls. As aircraft became faster and larger, the control forces grew to be greater than a pilot's muscular capability. Hydraulically boosted, irreversible flight controls were invented to handle these higher forces and moments, associated with larger aircraft and the increasing airspeed capabilities of jet aircraft.

In an irreversible flight control system, the cockpit controls are electronically connected to a controller, typically some type of computer, which translates the pilot input into a commanded position of a control surface, as shown in Figure 6.3b. These computer-controlled flight control systems, where the cockpit controls are connected by electrical wires to the flight control computers, are sometimes called "fly-by-wire" systems. The "muscle" in an irreversible flight control system is no longer due to the pilot, rather the control surface is moved using a hydraulic or electromechanical actuator. There is no feedback of the aerodynamic forces and moments on the control surfaces to the pilot controllers. If the control surface is moved by hand at the surface, the cockpit control does not move, hence, the name "irreversible". Since there is no feedback from the surface to the controller, some type of artificial "feel" system is required to provide cockpit control forces for the pilot. This artificial feel is often accomplished by a combination of springs and bob weights on the control stick and springs on the rudder pedals.

6.3.2   Stick-Fixed and Stick-Free Stability

The stability of an aircraft differs depending on whether the control surfaces are in a fixed or "frozen" position or are free to move or "float", after the aircraft has been disturbed from its trim condition. The aircraft stability may increase or decrease depending on whether the controls are fixed or free.

In *stick-fixed stability*, all of the control surfaces are assumed to be fixed and do not move with changes of the aerodynamic forces and moments on the surfaces. The term *stick-fixed* was derived from the fact that, for a reversible flight control system, the control surfaces are kept in a fixed position by holding or fixing the pilot controllers, such as a center stick or rudder pedals, in a stationary position. Stick-fixed stability is a measure of the *free response* of the aircraft.

In *stick-free stability*, the control surface is assumed to be free to move or "float", after the disturbance of the aircraft from its trim state. The forces and moments on the surface change the control surface positions over time, until it reaches an equilibrium position where the forces are balanced and the moment acting to rotate the surface, at its hinge point, is zero. In a reversible flight control system, the pilot controllers are assumed to be released after the disturbance, so that they are free to "float" with the movement of the control surfaces. It is assumed that there are no forces applied to the pilot controllers by the pilot.

Even with a rigid aircraft structure, a human pilot cannot hold the control surface of a reversible flight control system in a perfectly fixed position. The stick-fixed assumption is more closely approached with an irreversible flight control system, where hydro-mechanical or electromechanical systems hold the surface fixed in place. On the other side of the spectrum, stick-free controls are also an idealization, since friction in the flight control system makes it so the controls are not perfectly free. Thus, the stick-fixed and stick-free assumptions provide the idealized limits in the movement of the control surfaces and bounds the stability and control of the vehicle.

So why is the stability different if the controls are fixed or free? To answer this question, consider an aircraft flying at a steady, trim condition with an angle-of-attack, α. The horizontal tail generates an aerodynamic force, which acts through the moment arm from the tail to the aircraft center of gravity. This tail-generated aerodynamic moment is a major contributor to the longitudinal stability of the aircraft. Now, assume that a gust upsets the trim condition and increases the aircraft angle-of-attack. This new flight attitude changes the air flowing over the horizontal tail, thus changing the aerodynamic force that it is generating. If the pilot holds the control stick position fixed, the elevator position also remains fixed (assuming a reversible flight control system). Let us assume that this fixed elevator position is such that the elevator is approximately in line with, or faired with, the horizontal stabilizer (recall that the horizontal tail is composed of the fixed horizontal stabilizer and the moving elevator). With this fixed elevator position, the combined horizontal stabilizer and elevator generate an aerodynamic force and resulting moment, which we call F_h and M_h, respectively. Now, suppose that the pilot releases the control stick after the gust upset, so that the elevator is free to float. The flow over the tail moves the elevator to a new position, rotating the elevator trailing edge up from its faired position, for example. The flow now sees the horizontal tail as the fixed horizontal stabilizer and a deflected elevator, which generates an aerodynamic force and moment, F_h' and M_h', which are different from F_h and M_h. Since the force and moment, created by the horizontal tail, is different depending on whether the controls are fixed or free, the aircraft stability is different.

6.4 Aircraft Body Axes, Sign Conventions, and Nomenclature

In our discussions of aircraft performance, we were concerned with the translational motion of the aircraft center of mass in relation to a fixed, inertial coordinate system attached to the Earth. In discussing stability and control, we are concerned with the rotational motions of the aircraft about its own center of mass; hence, we adopt a *body fixed coordinate system*, with its origin at the aircraft center of mass, as shown in Figure 6.4. The body axes are a right-handed coordinate

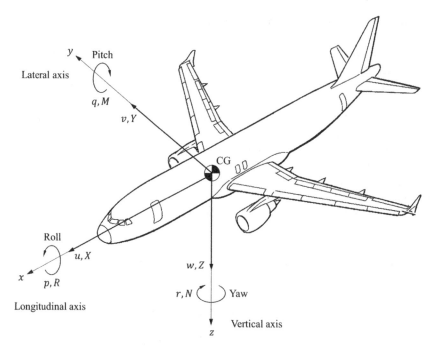

Figure 6.4 Body axis coordinate system nomenclature.

system that obeys the right-hand rule. The x or *longitudinal axis* is along the aircraft fuselage and is positive pointing out the aircraft nose. The y or *lateral axis* is along the aircraft wingspan and is positive pointing out the right wing. The z or *vertical axis* is perpendicular to the x–y plane and is positive pointing towards the Earth.

The aircraft translational motion is described by a total velocity, V_∞, with components u, v, and w, in the x-, y-, and z-axes, respectively. The aircraft vector velocity, $\vec{V}_\infty$, is defined as the sum of the component velocities, as given by

$$\vec{V}_\infty = \vec{u} + \vec{v} + \vec{w} \tag{6.3}$$

The magnitude of the velocity is given by

$$V_\infty = \sqrt{u^2 + v^2 + w^2} \tag{6.4}$$

The aircraft rotational motion is described by angular rates, the roll rate, p, the pitch rate, q, and the yaw rate, r, about the x-, y-, and z-axes, respectively. The positive direction of the angular rates follow the right-hand rule.

The x-, y-, and z-components of the resultant aerodynamic force are given by the axial force, X, the side force, Y, and the normal force, Z, respectively. Other forces which may act on the aircraft include those due to thrust or gravity. Usually, the propulsive forces in the y and z directions and the gravity forces in the x and y directions are assumed to be zero.

R, M, and N are the rolling, pitching, and yawing moments, respectively about the x-, y-, and z-axes, respectively. The moments may be due to the aerodynamic forces or due to the thrust force not acting through the center of mass. R is the rolling moment about the x or roll axis, M is the pitching moment about the y or pitch axis, and N is the yawing moment about the z or yaw axis. (Normally, the symbol L is used to denote the rolling moment, but the symbol R is adopted in the text for the rolling moment, to avoid confusion with using L for both lift and rolling moment.) The positive direction of the moments follow the right-hand rule.

We are usually dealing with the forces and moments in non-dimensional, coefficient form. The axial, side, and normal force coefficients, C_X, C_Y, and C_Z, respectively, are defined as

$$C_X = \frac{X}{q_\infty S}, \quad C_Y = \frac{Y}{q_\infty S}, \quad C_Z = \frac{Z}{q_\infty S} \tag{6.5}$$

where q_∞ is the freestream dynamic pressure and S is the wing planform area. The rolling, pitching, and yawing moment coefficients, C_R, C_M, and C_N, respectively, are defined as

$$C_R = \frac{R}{q_\infty S l}, \quad C_M = \frac{M}{q_\infty S l}, \quad C_N = \frac{N}{q_\infty S l} \tag{6.6}$$

where the additional characteristic length term, l, is required to non-dimensionalize the moments. The characteristic length is usually the wingspan, b, for the rolling and yawing moments and the wing chord length, c, for the pitching moment. These aerodynamic forces and moments are typically a function of the Mach number, Reynolds number, angle-of-attack, and angle-of-sideslip. By convention, we use uppercase letters for three-dimensional forces and moments, such as for wings and the complete aircraft, and lowercase letters for two-dimensional forces and moments, such as for airfoil sections.

In our study of stability and control, we are often interested in the change of an aerodynamic coefficient with a change in the direction of the relative wind or a change in a control surface position. A *stability derivative* is defined as a change in the coefficient with respect to a change in the angle-of-attack, α, for longitudinal motion or with respect to a change in the angle-of-sideslip, β, for lateral or directional motion. For example, the following stability derivatives

$$C_{M_\alpha} = \frac{\partial C_M}{\partial \alpha}, \quad C_{N_\beta} = \frac{\partial C_N}{\partial \beta} \tag{6.7}$$

define the longitudinal static stability or pitch stability, C_{M_α}, as the change in the pitching moment coefficient, C_M, with angle-of-attack, α, and the directional static stability or weathercock stability, C_{N_β}, as the change in the yawing moment coefficient, C_N, with angle-of-sideslip, β. The sign of the stability derivative is important in determining the vehicle stability.

A *control derivative* or *control power* is defined as a change in the coefficient due to a change in the control surface deflection. For example, the following control powers

$$C_{M_{\delta_e}} = \frac{\partial C_M}{\partial \delta_e}, \quad C_{R_{\delta_a}} = \frac{\partial C_R}{\partial \delta_a} \tag{6.8}$$

define the elevator or longitudinal control power, $C_{M_{\delta_e}}$, as the change in the pitching moment, C_M, due to a change in the elevator deflection, δ_e, and the aileron or lateral control power, $C_{R_{\delta_a}}$, as the change in the rolling moment coefficient, C_R, due to a change in the aileron deflection, δ_a. The larger the absolute magnitude of the control power, the larger the moment that is generated by the control surface deflection.

A summary of the component terms in the aircraft body axes is given in Table 6.1. Some of the terms are still to be defined in upcoming discussions. As is probably obvious by now, careful attention to nomenclature is very important in stability and control.

The aircraft angle-of-attack, α, and angle-of-sideslip, β, were defined in Section 2.3.2, in terms of the velocity components as (see Figure 2.12)

$$\alpha = \tan^{-1}\frac{w}{u} \cong \frac{w}{u} \tag{6.9}$$

Table 6.1 Definition of components terms in body axis coordinate system.

Parameter	x	y	z
Translational velocities	u	v	w
Angular rates (roll, pitch, yaw)	p	q	r
Aerodynamic forces (axial, side, normal)	X	Y	Z
Propulsive forces	T_x	T_y	T_z
Gravitational forces	W_x	W_y	W_z
Moments (rolling, pitching, yawing)	R	M	N
Moment of inertias	I_x	I_y	I_z
Product of inertias	I_{yz}	I_{xz}	I_{xy}
Control forces (aileron, elevator, rudder)	F_a	F_e	F_r
Control deflections (aileron, elevator, rudder)	δ_a	δ_e	δ_r

$$\beta = \sin^{-1}\frac{v}{V_\infty} \cong \frac{v}{V_\infty} \qquad (6.10)$$

The aircraft has principal moments of inertia, I_x, I_y, and I_z, and products of inertia, I_{yz}, I_{xz}, and I_{xy}, about the x-, y-, and z-axes, respectively. The inertias are a function of the shape and mass distribution of the aircraft. The larger the moments of inertia, the greater the resistance of the body to rotation. Since the inertias are referenced to the fixed body axis system, they remain constant with rotation of the aircraft.

The deflection of the primary flight control surfaces, composed of the ailerons, elevator, and rudder, generates aerodynamic forces and moments that change the aircraft stability. The control forces generated by the ailerons, elevator, and rudder are defined as F_a, F_e, and F_r, respectively. The aileron, elevator, and rudder control surface deflections are defined as δ_a, δ_e, and δ_r, respectively. The sign conventions for the control surface deflections are given in Table 6.2. A positive elevator deflection, $+\delta_e$, is defined as the elevator trailing edge moving downward, or simply stated as trailing edge down (TED). A positive rudder deflection, $+\delta_r$, is defined as the rudder trailing edge left (TEL). The sign convention for the aileron deflection is a little more complicated. A positive aileron deflection, $+\delta_a$, is defined as trailing edge down, for either the left or the right aileron. The total aileron deflection, δ_a, is defined as the difference between the left and right aileron deflections, as given by

$$\delta_a = \frac{1}{2}(\delta_{a,left} - \delta_{a,right}) \qquad (6.11)$$

The control surface deflections result in incremental changes of the moments acting on the aircraft and incremental changes in the aircraft attitude. As a word of caution, the sign conventions used for deflections of control surfaces and pilot controllers (stick, wheel, rudder pedals, etc.) varies

Table 6.2 Sign convention for positive control surface deflection.

Control surface	Symbol	Direction	Result
Elevator	$+\delta_e$	TED	$+\Delta L_t, -\Delta M, -\Delta \alpha$
Aileron	$+\delta_a$	TED*	$+\Delta R, +\Delta \phi$
Rudder	$+\delta_r$	TEL	$+\Delta Y_r, -\Delta N, +\Delta \beta$

*Applies to left and right aileron.

within the aerospace industry and government organizations. Care must be taken to be sure which conventions are being used. The present text adopts the sign conventions used by the Air Force Flight Test Center, Edwards, California. The sign conventions for the incremental changes are given in Table 6.2.

A positive (trailing edge down) elevator deflection, $+\delta_e$, results in an increase of the tail lift $+\Delta L_t$, a negative (nose down) increment of the pitching moment, $-\Delta M$, and a decrease of the angle-of-attack, $-\Delta \alpha$. A positive total aileron deflection, $+\delta_a$, results in a positive (right wing down) increment of the rolling moment, $+\Delta R$, and a positive (right wing down) change of the bank angle, $+\Delta \phi$. A positive (trailing edge left) rudder deflection, $+\delta_r$, results in an increase of the rudder side force, $+\Delta Y_r$, a negative (nose left) increment of the yawing moment, $+\Delta N$, and a positive (wind in right ear) change of the angle-of-sideslip, $+\Delta \beta$. (Recall that a positive sideslip angle is referred to as "wind in the right ear" as this is what the pilot would feel in an open cockpit airplane with the nose pointing left relative to the velocity vector.) The sign convention conforms to the right-hand-rule applied at the control surface hinge line, with the right thumb pointing in the positive axis direction. For example, if the right thumb points along the hinge line of the elevator or ailerons, in the direction of the positive y-axis, the fingers curl in the direction of making the control surface rotate trailing edge down. If the right thumb points along the rudder hinge line, in the direction of the positive z-axis, the fingers curl to make the rudder rotate its trailing edge to the left.

6.5   Longitudinal Static Stability

In this section, we focus on the longitudinal static stability of an aircraft. In analyzing the longitudinal stability of an aircraft, we are interested in the pitching moment about the aircraft center of gravity. This pitching moment may be stabilizing, destabilizing, or neutral, in terms of the aircraft's longitudinal stability. *Longitudinal balance* is defined as the condition where the net pitching moment acting at the aircraft center of gravity is zero. If the net pitching moment is not zero, the aircraft has a rotational acceleration in the direction of the unbalanced moment. Longitudinal static stability is quantified in terms of the relative locations of the aerodynamic center and center of gravity of an aircraft. We determine a particular center of gravity position, called the *neutral point*, where the aircraft has neutral longitudinal static stability. Hence, the neutral point represents a boundary between stability and instability.

6.5.1   The Pitching Moment Curve

Consider an aircraft in steady, unaccelerated flight at a constant velocity, V_∞, and an absolute angle-of-attack, α_a, as shown in Figure 6.5. Recall that the absolute angle-of-attack is the angle between the freestream velocity and the zero-lift line, in this case, the zero-lift line of the complete airplane (see Section 3.8.5.1). The longitudinal stability of the aircraft is dictated by the pitching moment about the center of gravity, M_{CG}, as shown. Moments acting on the aircraft are due to the contributions of the forces and moments of the aircraft's mass distribution (weight), propulsion system (thrust), and aerodynamics (lift, drag, and moment), as shown in Figure 6.6. Since we are taking the moment about the center of gravity (CG), the weight, W, does not contribute to this moment since it acts through the CG. We assume that the thrust, T, is acting through the CG, so it also does not contribute to the pitching moment about the center of gravity.

The aerodynamic contributions to the pitching moment are fundamentally due to the integrated pressure and shear stress distributions on the complete aircraft. The aerodynamic contributions to the pitching moment can be separated into those due to the various component parts of the aircraft, such as the fuselage, wing, tail, etc. For our quantitative analysis of longitudinal stability, we only

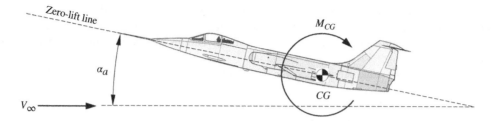

Figure 6.5 Pitching moment about the center of gravity.

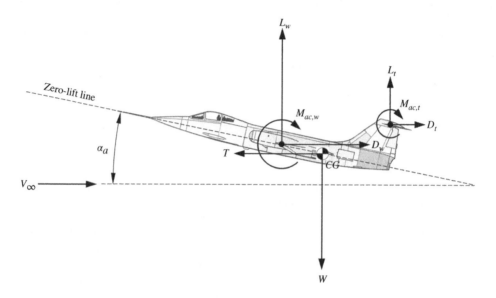

Figure 6.6 Contributions to the pitching moment about the center of gravity.

include the contributions due to the wing and horizontal tail, since these are the major contributors. We assess the effects of other components on the longitudinal stability in a qualitative manner. The aerodynamic contributions of the wing are due to the lift, L_w, drag, D_w, and moment, $M_{ac,w}$, respectively, which act at the aerodynamic center (ac) of the wing. Recall from Section 3.8.3 that the moment about the aerodynamic center is independent of angle-of-attack and may be translated anywhere on the body. Similarly, the lift, L_t, drag, D_t, and moment, $M_{ac,t}$, about the aerodynamic center of the horizontal tail contribute to the pitching moment about the center of gravity. (For the Lockheed F-104 *Starfighter* aircraft, depicted in Figure 6.6, the horizontal tail is a "T-tail", high-mounted on the vertical stabilizer, which has a larger moment arm, relative to the CG, as compared with a conventional, fuselage-mounted horizontal tail.)

Let us now consider the pitching moment about the center of gravity as a function of the angle-of-attack for three different airplanes, represented by curves 1, 2, and 3 in Figure 6.7. This figure shows curves of the non-dimensional, pitching moment coefficient about the center of gravity, $C_{M,CG}$, versus the absolute angle-of-attack, α_a. The pitching moment coefficient about the center of gravity is defined as

$$C_{M,CG} = \frac{M_{CG}}{q_\infty S \bar{c}} \qquad (6.12)$$

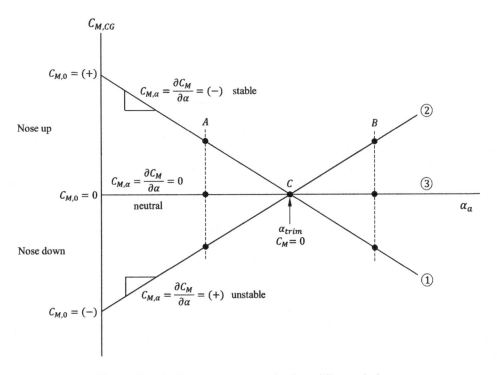

Figure 6.7 Pitching moment curves for three different airplanes.

where M_{CG} is the dimensional pitching moment about the center of gravity, q_∞ is the freestream dynamic pressure, S is the wing reference planform area, and $\bar{c}$ is the mean aerodynamic chord of the wing (see Section 3.9.1.1). Positive values of $C_{M,CG}$ on the vertical axis correspond to a nose up pitching moment and negative $C_{M,CG}$ correspond to a nose down moment. By using the absolute angle-of-attack, the lift is zero at $\alpha_a = 0$. The three curves in Figure 6.7 correspond to a fixed elevator position and a constant airspeed. The curves are linear because they correspond to the linear region of the lift curves (lift versus angle-of-attack). All three airplanes are longitudinally balanced or trimmed at a positive trim angle-of-attack, α_{trim} (Point C), where the pitching moment is zero, $C_{M,CG} = 0$ (CG subscripts omitted in Figure 6.7 for simplicity).

Consider airplane 1 with the pitching moment versus angle-of-attack given by curve 1 in Figure 6.7. Assume that the airplane is at point C, in steady, trimmed flight. The pitching moment, about the center of gravity, at zero lift, $C_{M,0}$, is positive, giving this airplane a pitching moment curve with a negative slope, that is, $\partial C_{M,CG}/\partial \alpha_a < 0$. Now, assume that airplane 1 is disturbed by a wind gust, so that its angle-of-attack decreases to point A. At an angle-of-attack less than the trim angle-of-attack, the pitching moment is positive, creating a nose up pitching moment. The nose-up pitching moment tends to move the airplane's nose up and increase its angle-of-attack. Thus, after being disturbed from its trimmed, equilibrium position, to a lower angle-of-attack, the initial tendency of the airplane is to return to the higher, trimmed angle-of-attack. Now, assume the wind gust disturbs airplane 1 from point C to point B, where the angle-of-attack is increased. At the higher angle-of-attack along curve 1, the pitching moment is negative with a nose-down pitching moment. The nose-down moment decreases the angle-of-attack, tending to return the airplane to its trim angle-of-attack. Thus, airplane 1, with a pitching moment curve as depicted by curve 1, has positive longitudinal static stability. From this, we conclude that positive longitudinal static stability corresponds to a pitching moment curve with a negative slope. The magnitude of the slope changes the degree of the stability. If the pitching moment slope is steeper, the static stability is stronger or, in other words, the *pitch stiffness* is greater. The opposite is obviously true, where a shallower slope results in less pitch stiffness.

Now consider airplane 2 with the pitching moment curve 2 in Figure 6.7. Curve 2 has a positive slope with a negative value of the pitching moment at zero lift. We start again in steady, trimmed flight at point C. If the wind gust disturbance reduces the angle-of-attack to point A on curve 2, the pitching moment is negative with a nose-down pitching moment. The nose-down moment tends to decrease the angle-of-attack, moving it further away from its trim value. If the wind gust disturbance increases the angle-of-attack to point B on curve 2, the pitching moment is positive and the resulting nose up pitching moment tends to increase the angle-of-attack further, thus moving away from the trim position. Thus, airplane 2, with pitching moment curve 2, has negative longitudinal stability. From this, we conclude that negative longitudinal static stability corresponds to a pitching moment curve with a positive slope.

Airplane 3, with pitching moment curve 3, has neutral longitudinal static stability, as the pitching moment remains zero for any displacement from the trimmed position. Thus, we conclude that a pitching moment curve such as curve 3, with zero slope, corresponds to neutral longitudinal stability.

Consider the pitching moment curve for another airplane, in Figure 6.8. Airplane 4 has positive longitudinal static stability, but there is no angle-of-attack for which the airplane can be trimmed or longitudinally balanced, such that $C_{M,CG} = 0$. The airplane cannot be trimmed for steady, equilibrium flight at any positive angle-of-attack, which is not desirable. The condition that caused this dilemma is the fact that the zero-lift moment coefficient, $C_{M,0}$, for airplane 4 is negative. From this, we conclude that, in addition to positive static stability, an aircraft must have a positive zero-lift pitching moment to enable it to fly in steady, trimmed flight at a usable, positive angle-of-attack.

Summarizing, the requirements for longitudinal static stability and balance are that the slope of the pitching moment curve must be negative and the pitching moment at zero lift must be positive, so that the aircraft can be trimmed at a positive, usable angle-of-attack. These requirements may be expressed as

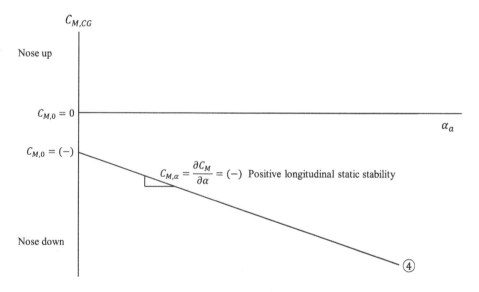

Figure 6.8   Pitching moment curves for an unbalanced airplane.

$$\frac{\partial C_{M,CG}}{\partial \alpha_a} < 0 \tag{6.13}$$

$$C_{M,0} > 0 \tag{6.14}$$

6.5.2 Configurations with Longitudinal Static Stability and Balance

The criteria, defined in the last section, are the basis for designing aircraft configurations that have longitudinal static stability and longitudinal balance. It is possible to satisfy both of these requirements with a wing alone or a wing–tail combination, with a horizontal tail in front of or behind the wing.

Let us first consider the longitudinal balance of a wing alone. For a wing with a symmetrical airfoil section, the wing is at zero angle-of-attack at zero lift. Therefore, the pitching moment at zero lift, $C_{M,0}$, is zero for a wing with a symmetric section. If the wing has a section with positive camber, the pitching moment at zero lift is negative, as can be verified by inspection of airfoil section data. The pitching moment at zero lift is positive for a wing with a negative camber airfoil section. These observations are summarized in Figure 6.9.

Assuming that it is possible to obtain longitudinal static stability for a wing alone, say by suitable placement of the center of gravity, we have the following conclusions concerning the different wing sections. A wing with a symmetric section is able to fly in steady, trimmed flight only at zero lift or zero angle-of-attack. A wing, with positive camber, does not have longitudinal balance and cannot fly in steady, trimmed flight at any positive angle-of-attack. Steady, trimmed flight is possible, at any positive angle-of-attack for the wing with negative camber. Hence, it is possible to design a flying wing that has longitudinal static stability and balance, if the wing section has negative camber.

Let us address the wing with positive camber or zero camber. To obtain longitudinal balance, another lifting surface must be used to provide a positive pitching moment when the wing is at zero lift. This additional surface is a horizontal tail, which may be located forward or aft of the wing, as shown in Figure 6.10. If the tail is mounted aft of the wing, it must be set at a negative incidence angle to generate a negative lift and a nose up pitching moment when the wing is at zero lift. In a canard configuration, where the tail is mounted forward of the wing, the tail must have a positive incidence angle, to provide a positive lift and a positive pitching moment, when the wing lift is zero.

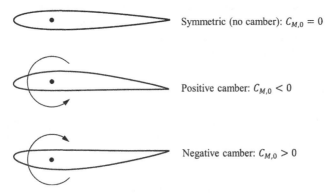

Symmetric (no camber): $C_{M,0} = 0$

Positive camber: $C_{M,0} < 0$

Negative camber: $C_{M,0} > 0$

Figure 6.9 Pitching moment at zero lift, $C_{M,0}$, for different airfoil sections. (Source: *Adapted from Dynamics of Flight: Stability and Control, B. Etkin and L.D. Reid, Fig. 2.5, p. 22, (1996), [7], with permission from John Wiley & Sons, Inc.*)

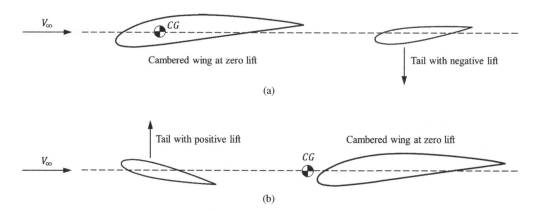

Figure 6.10 Wing–tail arrangements with positive $C_{M,0}$, (a) conventional, aft-mounted tail and (b) forward-mounted tail or canard configuration. (Source: *Adapted from Dynamics of Flight: Stability and Control, B. Etkin and L.D. Reid, Fig. 2.6, p. 22, (1996), [7], with permission from John Wiley & Sons, Inc.*)

The canard offers the advantage of adding to the overall positive lift of the vehicle, whereas the conventional, aft-mounted tail produces negative lift, which must be counteracted by increased wing lift. The canard, with its forward placement, is not affected by flow interference downwash from the wing. The canard configuration has also been used to provide aerodynamic stall protection. By setting the canard incidence angle such that it stalls prior to the wing, at high angle-of-attack, the loss of canard lift causes the aircraft nose to rotate down, reducing the angle-of-attack and preventing the wing from ever reaching its stall angle-of-attack. While good for longitudinal balance, the canard does contribute negatively to the longitudinal static stability. With an increase or decrease in the angle-of-attack, the canard generates a positive or negative lift, respectively, which drives the airplane away from equilibrium. This deficiency is usually easily corrected with proper placement of the center of gravity.

Perhaps the first understanding of longitudinal static stability was by Sir George Cayley in the early 1800s. In 1804, Cayley designed a fixed-wing, monoplane glider with a main wing and an aft-mounted tail for longitudinal stability. This simple model glider was the first airplane design with a "conventional" configuration, as we know it today. As shown in Figure 6.11, the model glider had a stick-like fuselage, about a meter in length, with a main, kite-shaped wing, mounted with an angle of incidence to the flight path, and an adjustable, cruciform tail. The center of gravity of the glider could be adjusted with a movable weight. Cayley's many flights with this glider were perhaps the first longitudinally stable flights of a fixed-wing airplane.

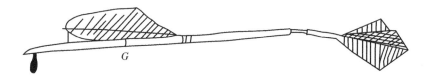

Figure 6.11 Sir George Cayley's fixed-wing, monoplane glider with an adjustable cruciform tail, 1804. (Source: *Hodgson, John Edmund, Aeronautical and Miscellaneous Notebook (ca. 1799–1826) of Sir George Cayley, with an Appendix Comprising a List of the Cayley Papers, W. Heffer & Sons, Ltd., Cambridge, 1933, Newcomen Society Extra Publication No. 3.*)

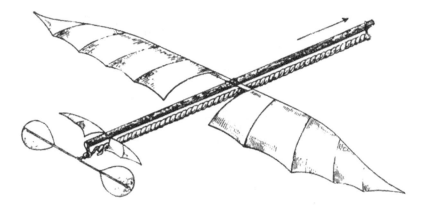

Figure 6.12 Alphonse Penaud's rubber-band-powered model airplane, 1871. (Source: *Alphonse Penaud, 1871, PD-old-70.*)

In 1871, the Frenchman, Alphonse Penaud, designed and built a statically stable model airplane with a modern airplane configuration, which he called the "Planophore". As shown in Figure 6.12, the model had a stick fuselage, single wing, and an aft-mounted tail, which acted as both a vertical and horizontal stabilizer, somewhat like the modern V-tail. The model was 20 in (50.4 cm) long with a wingspan of 18 in (46 cm). It was rubber-band-powered with a pusher propeller. The model had longitudinal and directional static stability due to the tail and lateral static stability due to slight upward curvature of the wingtips, called wing dihedral (to be discussed in a later section). In 1871, Penaud flew his model airplane in a circular path over a distance of about 131 ft (40 m), staying aloft for about 11 s. This flight of Penaud's Planophore was perhaps the first flight of a statically stable, powered, fixed-wing airplane.

Let us return to the possible configuration of a wing alone with negative camber. The planform shape of this negative camber wing is important in designing the tailless airplane. If the negative camber wing is straight (no sweep), it tends to have poor aerodynamic characteristics, including high drag and low maximum lift coefficient. While the longitudinal static stability is satisfactory, the negative camber, straight wing tends to have unsatisfactory dynamic stability characteristics. Lastly, the center of gravity range that provides longitudinal static stability, for this geometry, is too small to be of practical use.

If the flying wing has a swept planform shape with twist at the wingtips, longitudinal balance can be obtained with positive camber. As shown in Figure 6.13, when the swept-wing is at its zero lift angle-of-attack, the forward portion of the wing is at a positive angle-of-attack and the portions near the wingtips see a negative angle-of-attack. The forward part of the swept-wing produces a positive lift and the wingtip area produces a negative lift. The net lift of the wing is zero, but the positive and negative lift forces can produce a positive pitching moment, resulting in longitudinal balance. The wing twist could be geometric, or it could be accomplished aerodynamically, such that the airfoil sections in the center part of the wing have positive camber and the wingtip parts have negative camber.

Some say that the tailless, flying wing was inspired by nature, in the form of the Zanonia seed, found in the tropical forests of Java, Indonesia. Released by the hundreds from large gourds suspended in the Alsomitra vines, the Zanonia seed has a set of paper-thin wings, as shown in Figure 6.14, allowing it to glide over long distances to spread the seeds in the forest. With a wingspan of about 13 cm (5.1 in), the seed has a flying wing configuration that has longitudinal static stability, due to its swept-back wing and twisted tips. The seeds have a graceful, phugoid-like oscillating motion in flight (phugoid motion is discussed in Section 6.9.1), accelerating downward

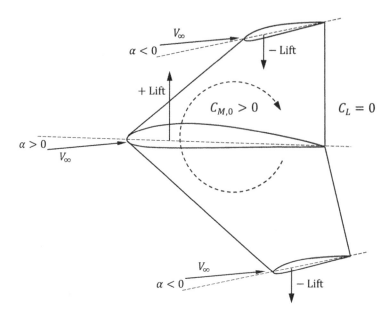

Figure 6.13 Swept-back wing with twisted tips – a flying wing with longitudinal static stability. (Source: *Adapted from* <u>*Dynamics of Flight: Stability and Control,*</u> *B. Etkin and L.D. Reid, Fig. 2.7, p. 23, (1996),* [7], *with permission from John Wiley & Sons, Inc.*)

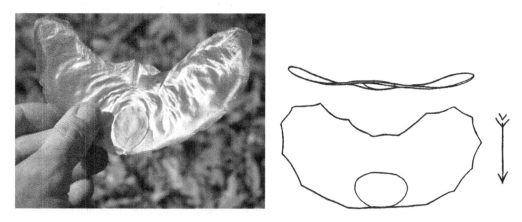

Figure 6.14 The Zanonia seed "flying wing" with wing sweep and twisted tips. (Source: *Left photo courtesy of Scott Zona, "Alsomitra Macrocarpa Seed" https://en.wikipedia.org/wiki/File:Alsomitra_macrocarpa_ seed_(syn._Zanonia_macrocarpa).jpg, CC-BY-SA-2.0. License at https://creativecommons.org/licenses/by-sa/2.0/legalcode. Right diagram from Alfried Gymnich,* <u>*Der Gleit – und Segelflugzeubau,*</u> *Richard Carl Schmidt & Co., Berlin, Germany, 1925.*)

then slowly pitching up to near-aerodynamic stall, then pitching over to again accelerate towards the ground.

The gliding flight of the Zanonia seed inspired the dreams of several early aviation pioneers. In the early 1900s, Austrian Ignaz "Igo" Etrich (1879–1967) designed and flew several Zanonia seed-inspired gliders and airplanes, one of which is shown in Figure 6.15. Etrich's *Taube* series of airplanes were somewhat successful, with variants of the design used by the Germans in World War I. British soldier and engineer, John Dunne (1875–1949) studied the Zanonia seed for his tailless,

Figure 6.15 Zanonia seed-inspired glider design of Igo Etrich, ca. 1904. (Source: *Alfried Gymnich, Der Gleit – und Segelflugzeubau, Richard Carl Schmidt & Co., Berlin, Germany, 1925.*)

swept wing designs. Dunne designed and successfully flew several tailless flying wing gliders and powered airplanes, which demonstrated inherent stability, patterned after the Zanonia seed. The German inventor, Karl Jatho (1873–1933) was inspired by the Zanonia seed for the design of his airplane wings. Between August and November 1903, Jatho made a series of powered hops in Hanover, Germany, with the longest hop covering a distance of 60 m (197 ft).

6.5.3 Contributions of Aircraft Components to the Pitching Moment

The total pitching moment about the aircraft center of gravity may be obtained by combining the contributions from the various parts of the aircraft, such as the wing, horizontal tail, fuselage, and propulsion system components. A more accurate result is obtained by including the mutual aerodynamic interference between these various parts. In this section, the contributions of various airplane components, to the pitching moment about the center of gravity, are evaluated. The contributions of the wing and horizontal tail are quantitatively determined, while the contributions of the fuselage and propulsion system are qualitatively evaluated. In general, the pitching moment about the center of gravity, $C_{M,CG}$, can be expressed as

$$C_{M,CG} = C_{M,CG,0} + \left(\frac{\partial C_{M,CG}}{\partial \alpha_a} \right) \alpha_a = C_{M,CG,0} + C_{M,CG_a}\alpha_a \tag{6.15}$$

where $C_{M,CG,0}$ is the zero-lift pitching moment, $\partial C_{M,CG}/\partial \alpha_a$ or C_{M,CG_a} is the slope of the pitching moment curve, and α_a is the absolute angle-of-attack. These two pitching moment parameters play an important role in the longitudinal balance or trim and longitudinal static stability of the aircraft.

In all cases, the aircraft is considered a rigid body, where the structure does not deform due to applied loads. Under certain conditions, such as flight at high dynamic pressure, aeroelastic effects may cause deformation of the structure, such as wing bending and twisting, which can significantly affect the forces and moments. These effects are beyond the scope of the present discussion and are left to more advanced treatments.

We start with a major contributor to the aircraft pitching moment, that of the wing.

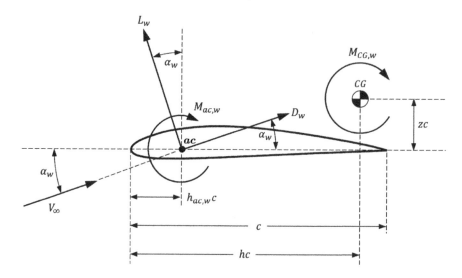

Figure 6.16 Wing forces and moments that contribute to the pitching moment.

6.5.3.1 Wing Contribution to the Pitching Moment

Consider the aerodynamic forces and moment on a wing in a flow with a freestream velocity, V_∞, at an angle-of-attack, α_w, relative to the wing chord line, c. The aerodynamic forces and moments on the wing are the lift, L_w, drag, D_w, and moment about the wing aerodynamic center, $M_{ac,w}$, as shown in Figure 6.16. The wing aerodynamic center, ac, is located a distance $h_{ac,w}c$ from the wing leading edge. The center of gravity (CG) is located a distance hc from the wing leading edge and a vertical height zc above the chord line.

The pitching moment about the center of gravity due to the wing, $M_{CG,w}$, is given by

$$M_{CG,w} = M_{ac,w} + (L_w \cos\alpha_w)(hc - h_{ac,w}c)$$
$$+ (D_w \sin\alpha_w)(hc - h_{ac,w}c) + (L_w \sin\alpha_w)(zc) - (D_w \cos\alpha_w)(zc) \qquad (6.16)$$

The components of the lift, $L_w \cos\alpha_w$ and $L_w \sin\alpha_w$, perpendicular and parallel to the chord line, respectively, contribute a positive (nose up) pitching moment about the center of gravity. The drag component perpendicular to the chord line, $D_w \sin\alpha_w$, contributes a positive moment, while the parallel drag component, $D_w \cos\alpha_w$, contributes a negative (nose down) moment about the CG.

If the angle-of-attack is assumed to be small, such that $\cos\alpha_w \approx 1$ and $\sin\alpha_w \approx \alpha_w$, Equation (6.16) becomes

$$M_{CG,w} = M_{ac,w} + (L_w + D_w\alpha_w)(h - h_{ac,w})c + (L_w\alpha_w - D_w)(zc) \qquad (6.17)$$

Dividing through by $q_\infty Sc$, we obtain Equation (6.17) in coefficient form.

$$C_{M,CG,w} = C_{M,ac,w} + (C_{L,w} + C_{D,w}\alpha_w)(h - h_{ac,w}) + (C_{L,w}\alpha_w - C_{D,w})z \qquad (6.18)$$

We can make other simplifying assumptions based on the mass properties and aerodynamics of most airplanes. The vertical location of the center of gravity is usually close to the chord line, such that $z \approx 0$ may be assumed. The lift is usually much greater than the drag, so that $C_{L,w} \gg C_{D,w}$. If the wing angle-of-attack is small, α_w in radians is much less than one, so that $C_{D,w}\alpha_w \ll C_{L,w}$.

Using these simplifying assumptions in Equation (6.18), we have

$$C_{M,CG,w} = C_{M,ac,w} + C_{L,w}(h - h_{ac,w}) \qquad (6.19)$$

(Remember that h and $h_{ac,w}$ are fractions of the wing chord length, c.)

In general, the wing lift coefficient, $C_{L,w}$, is given by

$$C_{L,w} = C_{L,0,w} + \left(\frac{dC_{L,w}}{d\alpha_w}\right)\alpha_w = C_{L,0,w} + a_w\alpha_w \qquad (6.20)$$

where $C_{L,0,w}$ is the wing lift coefficient at zero angle-of-attack and a_w is the lift curve slope of the wing. Inserting Equation (6.20) into (6.19), we have

$$C_{M,CG,w} = C_{M,ac,w} + (C_{L,0,w} + a_w\alpha_w)(h - h_{ac,w}) \qquad (6.21)$$

If we assume that the wing angle-of-attack, α_w, is the absolute angle-of-attack, then $C_{L,0,w}$ equals zero, by definition. By assuming that the wing angle-of-attack is the same as the absolute angle-of-attack, we are assuming that the chord line and the zero-lift line are nearly coincident, which for most wings is a valid assumption. Therefore, Equation (6.21) becomes

$$C_{M,CG,w} = C_{M,ac,w} + [a_w(h - h_{ac,w})]\alpha_w \qquad (6.22)$$

where α_w is now the absolute angle-of-attack of the wing.

Equation (6.22) gives the pitching moment coefficient, $C_{M,CG,w}$, about the aircraft center of gravity due to the aerodynamic forces and moment on the wing. Within the simplifying assumptions that were made, the wing contributions to the pitching moment are due to the moment about the wing aerodynamic center and that due to the wing lift acting through the moment arm between the wing aerodynamic center and the center of gravity. Equation (6.22) has the form of Equation (6.15), given by

$$C_{M,CG,w} = C_{M,CG,0,w} + \left(\frac{\partial C_{M,CG,w}}{\partial \alpha_w}\right)\alpha_w \qquad (6.23)$$

where

$$C_{M,CG,0,w} = C_{M,ac,w} \qquad (6.24)$$

and

$$\frac{\partial C_{M,CG,w}}{\partial \alpha_w} = a_w(h - h_{ac,w}) \qquad (6.25)$$

Let us apply the conditions for longitudinal static stability and balance – Equations (6.13) and (6.14), respectively – to the pitching moment equation due to the wing, Equation (6.22). For the static stability condition, the slope of the pitching moment curve must be negative. Taking the derivative of Equation (6.22) with respect to the absolute angle-of-attack, we have

$$\frac{\partial C_{M,CG,w}}{\partial \alpha_a} = \frac{\partial C_{M,CG,w}}{\partial \alpha_w} = a_w(h - h_{ac,w}) < 0 \qquad (6.26)$$

For longitudinal balance, the pitching moment coefficient, at zero lift, must be greater than zero. Evaluating Equation (6.22) at zero lift, where by definition, the absolute angle-of-attack is zero, $\alpha_w = 0$, we have

$$C_{M,CG,0,w} = C_{M,ac,w} > 0 \qquad (6.27)$$

where $C_{M,0,CG,w}$ is the pitching moment coefficient about the center of gravity, at zero lift, due to the wing. (Equations (6.26) and (6.27) could have also been obtained directly from Equations (6.25) and (6.24), respectively.)

Equations (6.26) and (6.27) are the requirements for longitudinal static stability and balance for a wing alone, or a flying wing. For static stability, Equation (6.26) dictates that $h_{ac,w} > h$, which means that the wing aerodynamic center must be aft of the center of gravity. For most airplanes, the aerodynamic center is usually slightly forward of the center of gravity, hence, the wing alone in these cases would not have longitudinal static stability. For longitudinal balance, according to Equation (6.27), the wing pitching moment coefficient about the aerodynamic center, $C_{M,ac,w}$, must be positive. As discussed in Section 6.5.2 and shown in Figure 6.9, this can be obtained with an airfoil section with negative camber. Most airplanes use wings with positive camber, which would be unbalanced for a wing alone, unless the sweep and twist "fixes" are applied as discussed in Section 6.5.2. In general, we can conclude that a wing alone is usually destabilizing and may be unbalanced, assuming a positively cambered wing is desired. Given this situation for most airplanes, another lifting surface must be used to provide stability and balance, so we now discuss the contribution to the pitching moment of a horizontal tail.

Example 6.1 Calculation of the Wing Pitching Moment *A rectangular wing, with a chord length of 5.10 ft, is at an absolute angle-of-attack of 4.20°. The aerodynamic center of the wing is located 0.230 ft forward of the aircraft center of gravity. The lift slope of the wing is 0.0912 deg⁻¹, and the wing pitching moment about the aerodynamic center is −0.108. Calculate the pitching moment coefficient of the wing about the aircraft center of gravity.*

Solution

Using Equation (6.22), the pitching moment coefficient of the wing about the aircraft center of gravity is given by

$$C_{M,CG,w} = C_{M,ac,w} + [a_w(h - h_{ac,w})]\alpha_w$$

$$C_{M,CG,w} = -0.108 + \left[(0.0912 \text{ deg}^{-1})\left(\frac{0.230\,\text{ft}}{5.10\,\text{ft}}\right)\right]\alpha_w$$

$$C_{M,CG,w} = -0.108 + (0.004113 \text{ deg}^{-1})\alpha_w$$

where the wing absolute angle-of-attack, α_w, is in degrees. At a wing absolute angle-of-attack of 4.20°, the pitching moment due to the wing is

$$C_{M,CG,w} = -0.108 + (0.004113 \text{ deg}^{-1})(4.2 \text{ deg}) = -0.09073$$

Hence, the wing alone produces a nose down pitching moment.

6.5.3.2   Tail Contribution to the Pitching Moment

The contributions of the horizontal tail alone to the pitching moment about the center of gravity are the same as for the wing alone. However, the tail contributions are altered by the aerodynamic influences of the wing, whether the tail is mounted forward or aft of the wing. If the tail is mounted forward of the wing, it is affected by the upwash of the wing. We quantify the contributions to the pitching moment for a conventional, aft-mounted horizontal tail, which is affected by the downwash

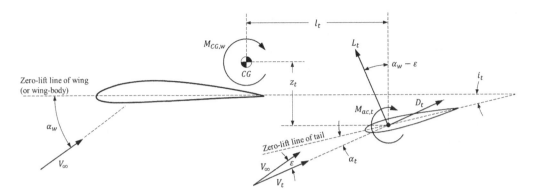

Figure 6.17   Horizontal tail forces and moments that contribute to the pitching moment.

from the wing. The analysis for the canard follows the same process, with different geometry to account for upwash.

Consider the horizontal tail located aft of a wing, as shown in Figure 6.17. The wing and tail locations are referenced to the zero-lift line of the wing, so we assume that the absolute angle-of-attack is being referenced also. The zero-lift line could also be that of the wing mounted on a fuselage, called a wing–body combination, but for simplicity, we reference the wing alone. (Fuselage effects are considered in the following section.) The wing is located slightly forward of the center of gravity (CG) and the horizontal tail aerodynamic center is located at a horizontal distance l_t and a vertical distance, z_t from the CG. The horizontal tail is composed of the stationary horizontal stabilizer and the movable elevator. It is assumed that the elevator is in a fixed position, faired with the stabilizer. The horizontal tail is set at an incidence angle, i_t, relative to the zero-lift line, where positive incidence angle is measured upward. Thus, for the nose-down incidence setting of the tail, shown in Figure 6.17, the incidence angle is negative.

The wing, forward of the tail, sees the freestream flow at a velocity V_∞ and angle-of-attack, α_w. At the tail, the flow is deflected downward by the downwash of the finite wing, as discussed in Section 3.9.1.3. The amount of downwash is a function of the tail location relative to the wing. Immediately behind the wing, the downwash angle is theoretically equal to the wing angle-of-attack. The downwash angle decreases with distance behind the wing, approaching an angle equal to about half the wing angle-of-attack, at the tail.

The velocity of the flow has also been slowed due to the drag over the wing. Thus, the tail sees a relative wind, V_t, that is deflected downward by the angle ε and decreased in velocity from V_∞. The angle-of-attack seen by the tail, α_t, is given by

$$\alpha_t = \alpha_w - \varepsilon + i_t \tag{6.28}$$

Since $V_t < V_\infty$, the dynamic pressure at the tail, q_t, is less than the freestream dynamic pressure, q_∞. The ratio of the dynamic pressures is defined as the *horizontal tail efficiency*, η_t, given by

$$\eta_t \equiv \frac{q_t}{q_\infty} = \frac{\frac{1}{2}\rho_\infty V_t^2}{\frac{1}{2}\rho_\infty V_\infty^2} = \left(\frac{V_t}{V_\infty}\right)^2 \tag{6.29}$$

The tail efficiency is less than one for an aft-mounted tail and greater than one for a canard configuration. Typical values of the tail efficiency are about 0.8–1.2.

The forces and moment, on the horizontal tail, which contribute to the pitching moment about the center of gravity are the tail lift, L_t, tail drag, D_t, and the moment about the tail aerodynamic center,

$D_{ac,t}$. While the components of the wing lift and drag, L_w and D_w, respectively, are perpendicular and parallel, respectively, to the freestream velocity, V_∞, the tail lift and drag are perpendicular and parallel, respectively, to the local velocity at the tail, V_t. Therefore, taking the total lift, L, as perpendicular to V_∞, as is normally the case, the total lift is given by

$$L = L_w + (L_t \cos \varepsilon - D_t \sin \varepsilon) \tag{6.30}$$

where $(L_t \cos \varepsilon - D_t \sin \varepsilon)$ is the lift component of the tail, perpendicular to V_∞. Assuming the downwash angle, ε, is very small, we have $\cos \varepsilon \approx 1$ and $\sin \varepsilon \approx 0$. Therefore, the total lift simply becomes

$$L = L_w + L_t \tag{6.31}$$

Dividing by the freestream dynamic pressure, q_∞, and the wing planform reference area, S, we have

$$\frac{L}{q_\infty S} = \frac{L_w}{q_\infty S} + \frac{L_t}{q_\infty S} = \frac{L_w}{q_\infty S} + \left(\frac{q_t}{q_t}\right)\left(\frac{S_t}{S_t}\right)\frac{L_t}{q_\infty S} = \frac{L_w}{q_\infty S} + \left(\frac{q_t}{q_\infty}\right)\left(\frac{S_t}{S}\right)\frac{L_t}{q_t S_t} \tag{6.32}$$

where q_t is the local dynamic pressure at the tail and S_t is the planform reference area of the horizontal tail. The total lift coefficient is defined as

$$C_L \equiv \frac{L}{q_\infty S} \tag{6.33}$$

The wing lift coefficient, $C_{L,w}$, is defined as

$$C_{L,w} \equiv \frac{L_w}{q_\infty S} \tag{6.34}$$

The tail lift coefficient, $C_{L,t}$, is defined as

$$C_{L,t} \equiv \frac{L_t}{q_t S_t} \tag{6.35}$$

Inserting Equations (6.29), (6.33), (6.34), and (6.35) into (6.32), we have

$$C_L = C_{L,w} + \left(\frac{q_t}{q_\infty}\right)\left(\frac{S_t}{S}\right)C_{L,t} = C_{L,w} + \eta_t\left(\frac{S_t}{S}\right)C_{L,t} \tag{6.36}$$

Thus, the total lift coefficient is the sum of the wing lift coefficient plus the tail lift coefficient, which has been adjusted for upwash or downwash and the reference area.

The pitching moment about the center of gravity due to the forces and moment on the tail, $M_{CG,t}$, is given by

$$M_{CG,t} = M_{ac,t} - l_t[L_t \cos(\alpha_w - \varepsilon) + D_t \sin(\alpha_w - \varepsilon)] + L_t \sin(\alpha_w - \varepsilon)z_t - D_t \cos(\alpha_w - \varepsilon)\, z_t \tag{6.37}$$

The rather lengthy Equation (6.37) is shortened by making several assumptions that are valid for most airplanes, as follows. For most airplanes, the vertical distance between the horizontal tail aerodynamic center and the center of gravity, z_t, is much smaller than the horizontal distance, l_t, so that $z_t \ll l_t$. Assuming that the wing angle-of-attack, α_w, and the downwash angle, ε, are small, then the angle $(\alpha_w - \varepsilon)$ is also small, such that $\cos(\alpha_w - \varepsilon) \approx 1$ and $\sin(\alpha_w - \varepsilon) \approx 0$. Finally, for most airplanes, flying at low angles-of-attack, the tail drag is much smaller than the tail lift, $D_t \ll L_t$, and the tail moment is small, $M_{ac,t} \sim 0$. Applying all of these assumptions to Equation (6.37), the tail contribution to the pitching moment about the center of gravity is reduced to simply

$$M_{CG,t} = -l_t L_t \tag{6.38}$$

which is the tail lift, L_t, multiplied by the horizontal distance between the tail aerodynamic center and the center of gravity, l_t.

Dividing Equation (6.38) by $q_\infty S c$ to obtain the moment coefficient about the center of gravity due to the tail, $C_{M,CG,t}$, we have

$$C_{M,CG,t} = \frac{M_{CG,t}}{q_\infty S c} = -\left(\frac{l_t}{c}\right)\left(\frac{S_t}{S_t}\right)\left(\frac{L_t}{q_\infty S}\right) = -\left(\frac{l_t}{c}\right)\left(\frac{S_t}{S}\right)\left(\frac{q_\infty}{q_t}\right)\left(\frac{L_t}{q_\infty S_t}\right) \tag{6.39}$$

Using the definitions of the tail efficiency and the tail lift coefficient, Equations (6.29) and (6.35), respectively, we have

$$C_{M,CG,t} = -\left(\frac{l_t}{c}\frac{S_t}{S}\right)\eta_t C_{L,t} \tag{6.40}$$

The term in parentheses on the right-hand side of Equation (6.40) is defined as the *horizontal tail volume ratio*, $\mathcal{V}_H$, given by

$$\mathcal{V}_H \equiv \frac{l_t S_t}{cS} \tag{6.41}$$

Typical values of the horizontal tail volume ratio are about 0.5–0.7 for a single-engine, general aviation airplane, about 0.4 for a military jet fighter aircraft, about 0.7 for a military jet trainer aircraft, and about 1.0 for a commercial jet transport [14].

Thus, the horizontal tail contribution to the pitching moment coefficient about the center of gravity is given by

$$C_{M,CG,t} = -\mathcal{V}_H \eta_t C_{L,t} \tag{6.42}$$

We wish to apply the criteria for longitudinal static stability to the moment due to the tail contribution, so we need to rewrite Equation (6.42) in terms of the angle-of-attack, so that we can apply Equation (6.13).

The tail lift coefficient, $C_{L,t}$, is given by

$$C_{L,t} = a_t \alpha_t = a_t(\alpha_w - \varepsilon + i_t) \tag{6.43}$$

where a_t is the lift curve slope of the tail and the tail angle-of-attack, α_t, has been replaced with Equation (6.28).

The downwash angle, ε, is usually an experimentally derived quantity, often from the results of wind tunnel testing. The downwash can be approximated as

$$\varepsilon = \varepsilon_0 + \frac{\partial \varepsilon}{\partial \alpha}\alpha_w \tag{6.44}$$

where ε_0 is the downwash angle when the wing (or later, the combined wing and fuselage) is at zero lift (or zero absolute angle-of-attack) and $\partial \varepsilon / \partial \alpha$ is the change in the downwash angle with angle-of-attack. Even at zero lift, there is a downwash ε_0 due to wing twist or due to the induced velocity field of a fuselage. The downwash derivative term, $\partial \varepsilon / \partial \alpha$, derives from the wing trailing vortex system. Theoretically, the downwash, ε, (in radians) for a wing with an elliptical lift distribution, is given in [13] as

$$\varepsilon = \left(\frac{2}{\pi AR}\right)C_{L,w} \tag{6.45}$$

where $C_{L,w}$ is the wing lift coefficient and AR is the wing aspect ratio. This expression states that the downwash decreases with increasing wing aspect ratio, going to zero for an infinite (2D) wing, as expected. It also states that the downwash increases with increasing lift, which makes sense, since the strength of the trailing vortices increase with increasing lift. Taking the derivative of Equation (6.45) with respect to angle-of-attack, an expression for the downwash derivative term is obtained as

$$\frac{\partial \varepsilon}{\partial \alpha} = \left(\frac{2}{\pi AR} \right) \frac{\partial C_{L,w}}{\partial \alpha} = \left(\frac{2}{\pi AR} \right) a_w \tag{6.46}$$

Inserting Equation (6.44) into (6.43), we have

$$C_{L,t} = a_t \left(\alpha_w - \varepsilon_0 - \frac{\partial \varepsilon}{\partial \alpha} \alpha_w + i_t \right) = -a_t(\varepsilon_0 - i_t) + a_t \alpha_w \left(1 - \frac{\partial \varepsilon}{\partial \alpha} \right) \tag{6.47}$$

Inserting Equation (6.47), for the tail lift coefficient, into (6.42), we have

$$C_{M,CG,t} = V_H \eta_t a_t (\varepsilon_0 - i_t) - \left[V_H \eta_t a_t \left(1 - \frac{\partial \varepsilon}{\partial \alpha} \right) \right] \alpha_w \tag{6.48}$$

Equation (6.48) is the desired result, providing the horizontal tail contribution to the pitching moment about the center of gravity, as a function of the angle-of-attack. Equation (6.48) has the form of Equation (6.15), given by

$$C_{M,CG,t} = C_{M,CG,0,t} + \left(\frac{\partial C_{M,CG,t}}{\partial \alpha_w} \right) \alpha_w \tag{6.49}$$

where

$$C_{M,CG,0,t} = V_H \eta_t a_t (\varepsilon_0 - i_t) \tag{6.50}$$

and

$$\frac{\partial C_{M,CG,t}}{\partial \alpha_w} = - \left[V_H \eta_t a_t \left(1 - \frac{\partial \varepsilon}{\partial \alpha} \right) \right] \tag{6.51}$$

We assess the conditions for longitudinal static stability and longitudinal balance, due to the tail contributions to the pitching moment, in the next section.

Example 6.2 Calculation of the Horizontal Tail Pitching Moment *An aircraft has a rectangular wing and an aft-mounted horizontal tail with the specifications given in the table below.*

Parameter	Value
Wing area, S	$193 \, \text{ft}^2$
Wing chord, c	$5.10 \, \text{ft}$
Wing span, b	$36.3 \, \text{ft}$
Wing lift curve slope, a_w	$0.0912 \, \text{deg}^{-1}$
Horizontal tail area, S_t	$34.1 \, \text{ft}^2$
Horizontal tail lift curve slope, a_t	$0.0940 \, \text{deg}^{-1}$
Horizontal tail incidence angle, i_t	$-2.7°$
Horizontal tail efficiency, η_t	0.960
CG to horizontal tail ac distance, l_t	$15.3 \, \text{ft}$

If the wing is at an absolute angle-of-attack of $4.20°$, calculate the horizontal tail angle-of-attack, the tail lift coefficient, and the contribution of the tail to the pitching moment coefficient. Assume that the downwash at zero lift, ε_0, is zero.

Solution

Using Equation (6.41), the horizontal tail volume ratio is

$$V_H = \frac{l_t S_t}{cS} = \frac{(15.3\,\text{ft})(34.1\,\text{ft}^2)}{(5.10\,\text{ft})(193\,\text{ft}^2)} = 0.530$$

The wing aspect ratio of the rectangular wing is

$$AR = \frac{b^2}{S} = \frac{b^2}{bc} = \frac{b}{c} = \frac{36.3\,\text{ft}}{5.10\,\text{ft}} = 7.12$$

Using Equation (6.46), the downwash term is given by

$$\frac{\partial \varepsilon}{\partial \alpha} = \left(\frac{2}{\pi AR}\right) a_w = \left[\frac{2}{\pi\,(7.12)}\right]\left(0.0912\,\frac{1}{\text{deg}} \times \frac{180}{\pi}\right) = 0.467$$

Using Equation (6.28), the horizontal tail angle-of-attack is

$$\alpha_t = \alpha_w - \varepsilon + i_t = \alpha_w - \left(\varepsilon_0 + \frac{\partial \varepsilon}{\partial \alpha}\alpha_w\right) + i_t = 4.2 - 0 - (0.467)4.2 - 2.7 = -0.461\,\text{deg}$$

Using Equation (6.43), the horizontal tail lift coefficient is

$$C_{L,t} = a_t \alpha_t = (0.0940\,\text{deg}^{-1})(-0.461\,\text{deg}) = -0.0434$$

From Equation (6.48), the contribution of the horizontal tail to the pitching moment coefficient about the aircraft center of gravity is given by

$$C_{M,CG,t} = V_H \eta_t a_t(\varepsilon_0 - i_t) - \left[V_H \eta_t a_t\left(1 - \frac{\partial \varepsilon}{\partial \alpha}\right)\right]\alpha_w$$

Inserting values into the expression for the tail moment coefficient, we have

$$C_{M,CG,t} = V_H \eta_t a_t(\varepsilon_0 - i_t) - \left[V_H \eta_t a_t\left(1 - \frac{\partial \varepsilon}{\partial \alpha}\right)\right]\alpha_w$$

$$C_{M,CG,t} = (0.530)(0.960)(0.0940\,\text{deg}^{-1})(0 + 2.7\,\text{deg})$$
$$- [(0.530)(0.960)(0.0940\,\text{deg}^{-1})(1 - 0.467)]\alpha_w$$

$$C_{M,CG,t} = 0.1291 - (0.02549\,\text{deg}^{-1})\alpha_w$$

where the wing absolute angle-of-attack, α_w, is in degrees. At a wing angle-of-attack of 4.20 degrees, the pitching moment due to the tail is

$$C_{M,CG,t} = 0.1291 - (0.02549\,\text{deg}^{-1})(4.2\,\text{deg}) = 0.0220$$

The horizontal tail produces a nose up pitching moment.

6.5.3.3   Combined Contributions of the Wing and Tail to the Pitching Moment

We now combine the results that have been obtained for the contributions of the wing and the horizontal tail to the pitching moment about the center of gravity, $C_{M,CG,wt}$, which may be written as

$$C_{M,CG,wt} = C_{M,CG,w} + C_{M,CG,t} \tag{6.52}$$

Inserting Equation (6.19), for the wing contribution in terms of the wing lift coefficient, and Equation (6.42), for the tail contribution in terms of the tail lift coefficient, we have

$$C_{M,CG,wt} = C_{M,ac,w} + C_{L,w}(h - h_{ac,w}) - \mathcal{V}_H \eta_t C_{L,t} \tag{6.53}$$

Inserting Equations (6.22) and (6.48) into (6.52), the pitching moment coefficient about the center of gravity due to the wing and tail, in terms of the angle-of-attack, is given by

$$C_{M,CG,wt} = C_{M,ac,w} + a_w \alpha_w (h - h_{ac,w}) + \mathcal{V}_H \eta_t a_t (\varepsilon_0 - i_t) - \left[\mathcal{V}_H \eta_t a_t \left(1 - \frac{\partial \varepsilon}{\partial \alpha} \right) \right] \alpha_w$$

$$C_{M,CG,wt} = C_{M,ac,w} + \mathcal{V}_H \eta_t a_t (\varepsilon_0 - i_t) + a_w \left[h - h_{ac,w} - \mathcal{V}_H \eta_t \frac{a_t}{a_w} \left(1 - \frac{\partial \varepsilon}{\partial \alpha} \right) \right] \alpha_w \tag{6.54}$$

Equations (6.53) or (6.54) provide the total pitching moment coefficient about the center of gravity, due to the contributions of the wing and horizontal tail. Equation (6.54) has the form of Equation (6.15), given by

$$C_{M,CG,wt} = C_{M,CG,0,wt} + \left(\frac{\partial C_{M,CG,wt}}{\partial \alpha_w} \right) \alpha_w \tag{6.55}$$

where

$$C_{M,CG,0,wt} = C_{M,ac,w} + \mathcal{V}_H \eta_t a_t (\varepsilon_0 - i_t) \tag{6.56}$$

and

$$\frac{\partial C_{M,CG,wt}}{\partial \alpha_w} = a_w \left[h - h_{ac,w} - \mathcal{V}_H \eta_t \frac{a_t}{a_w} \left(1 - \frac{\partial \varepsilon}{\partial \alpha} \right) \right] \tag{6.57}$$

Let us now apply the conditions for longitudinal static stability and balance, Equations (6.13) and (6.14), respectively, to the total pitching moment, Equation (6.54). For longitudinal balance, the pitching moment coefficient, at zero lift, must be greater than zero. Evaluating Equation (6.54) at zero lift, where by definition, the absolute angle-of-attack is zero, $\alpha_w = 0$, we have

$$C_{M,CG,0,wt} = C_{M,ac,w} + \mathcal{V}_H \eta_t a_t (\varepsilon_0 - i_t) > 0 \tag{6.58}$$

where $C_{M,0,CG}$ is the pitching moment coefficient about the center of gravity, at zero lift. From Equation (6.58), the criterion for longitudinal balance may be written as

$$\mathcal{V}_H \eta_t a_t (\varepsilon_0 - i_t) > -C_{M,ac,w} \tag{6.59}$$

For most conventional aircraft, the wing moment coefficient about the aerodynamic center is a negative quantity, hence $a_t \mathcal{V}_H (\varepsilon_0 - i_t)$ in Equation (6.59) must be a positive number for longitudinal balance. The tail lift slope, a_t, tail volume, $\mathcal{V}_H$, and downwash, ε_0, are all positive numbers. Therefore, the horizontal tail should be mounted at a sufficiently large, negative incidence angle, i_t, to satisfy Equation (6.59).

For the static stability condition, the slope of the pitching moment curve must be negative. Taking the derivative of Equation (6.54) with respect to the absolute angle-of-attack, we have

$$\frac{\partial C_{M,CG,wt}}{\partial \alpha_a} = \frac{\partial C_{M,CG,wt}}{\partial \alpha_w} = a_w \left[h - h_{ac,w} - V_H \eta_t \frac{a_t}{a_w} \left(1 - \frac{\partial \varepsilon}{\partial \alpha} \right) \right] < 0 \qquad (6.60)$$

All of the terms on the right-hand side of Equation (6.60) are "knobs that can be turned" in designing an aircraft with longitudinal static stability. The two most influential parameters are the location of the center of gravity, h, and the tail volume, V_H. The derivative $\partial C_{M,CG}/\partial \alpha_a$ can almost always be made negative by the proper choice of the center of gravity location. Increasing the horizontal tail volume ratio, which essentially means having a larger horizontal tail, also increases the pitch stiffness.

Example 6.3 Calculation of the Pitching Moments of the Wing and Tail *Using the specifications for the wing and tail, given in Examples 6.1 and 6.2, respectively, calculate the pitching moment, about the aircraft center of gravity, due to the wing, the horizontal tail, and the combined wing and tail, as a function of absolute angle-of-attack of the wing.*

Solution

From Example 6.1, the pitching moment due to the wing is given by

$$C_{M,CG,w} = C_{M,CG,w} = -0.108 + (0.004113 \text{ deg}^{-1})\alpha_w$$

From Example 6.2, the pitching moment due to the horizontal tail is given by

$$C_{M,CG,t} = 0.1291 - (0.02549 \text{ deg}^{-1})\alpha_w$$

The combined pitching moment is the sum of the wing and tail contributions.

$$C_{M,CG,wt} = C_{M,CG,w} + C_{M,CG,t}$$
$$C_{M,CG,wt} = -0.108 + (0.004113 \text{ deg}^{-1})\alpha_w + 0.1291 - (0.02549 \text{ deg}^{-1})\alpha_w$$
$$C_{M,CG,wt} = 0.0211 - (0.02138 \text{ deg}^{-1})\alpha_w$$

Numerical values for these pitching moment coefficients are given in the table below. The pitching moment due to the wing alone, horizontal tail alone, and combined wing and tail are detailed as a function of the absolute angle-of-attack. As expected, the wing-alone pitching moment has negative longitudinal static stability, while the horizontal tail provides the stabilizing influence to give the wing–tail combination positive longitudinal static stability.

Absolute angle-of-attack, α (deg)	$C_{M,CG,bw}$	$C_{M,CG,t}$	$C_{M,CG}$
0	−0.1080	0.1291	0.02110
4.2 (Example 6.2)	−0.09073	0.02204	−0.06870
5	−0.08744	0.001674	−0.08580
10	−0.06687	−0.1258	−0.1927
15	−0.04631	−0.2532	−0.2996

6.5.3.4   Fuselage Contribution to the Pitching Moment

The fuselage of most airplanes is a long, slender cylindrical body. The aerodynamics of this type of body alone can be quantified as a function of Mach number and angle-of-attack, so that a lift, drag, and moment, acting at an aerodynamic center could be obtained. Once a wing is attached, there are mutual interference effects induced by the wing and body. Therefore, the lift, drag, and moment of the wing–body combination are not equal to the linear summation of the lift, drag, and moment of the wing and the body separately.

With the addition of a fuselage (body), the equations for the pitching moment about the center of gravity, due to the contributions of a wing–body and tail, have the same form as Equations (6.53) and (6.54), for the wing and tail contributions. Hence, the equations for the total pitching moment about the center of gravity, due to the wing–body and the tail, $C_{M,CG}$, are

$$C_{M,CG} = C_{M,ac,wb} + C_{L,wb}(h - h_{ac,wb}) - \mathcal{V}_H \eta_t C_{L,t} \tag{6.61}$$

$$C_{M,CG} = C_{M,ac,wb} + \mathcal{V}_H \eta_t a_t(\varepsilon_0 - i_t) + a_{wb}\left[h - h_{ac,wb} - \mathcal{V}_H \eta_t \frac{a_t}{a_{wb}}\left(1 - \frac{\partial\varepsilon}{\partial\alpha}\right)\right]\alpha_{wb} \tag{6.62}$$

where the aerodynamic and geometric parameters are now referenced to the wing–body (wb) rather than the wing (w). Equation (6.62) has the form of Equation (6.15), given by

$$C_{M,CG} = C_{M,CG,0} + \left(\frac{\partial C_{M,CG}}{\partial\alpha_{wb}}\right)\alpha_{wb} \tag{6.63}$$

where

$$C_{M,CG,0} = C_{M,ac,wb} + \mathcal{V}_H \eta_t a_t(\varepsilon_0 - i_t) \tag{6.64}$$

and

$$\frac{\partial C_{M,CG}}{\partial\alpha_{wb}} = a_{wb}\left[h - h_{ac,wb} - \mathcal{V}_H \eta_t \frac{a_t}{a_{wb}}\left(1 - \frac{\partial\varepsilon}{\partial\alpha}\right)\right] \tag{6.65}$$

The addition of a fuselage to a wing results in a forward shift in the aerodynamic center and an increase in the lift curve slope. The wing interference typically results in a fuselage flow field that produces a positive pitching moment, which increases with increasing angle-of-attack. Hence, the fuselage contribution to the pitching moment about the center of gravity contributes a destabilizing moment to the total pitching moment.

The longitudinal static stability of the fuselage, wing, and tail alone and the complete aircraft is shown in Figure 6.18, where the pitching moment about the center of gravity, $C_{M,CG}$, is plotted versus absolute angle-of-attack, α_a. (We have shown the fuselage alone to illustrate its destabilizing contribution to the pitching moment. Its contribution should, of course, be properly calculated by including the wing interference effects, as discussed earlier.) The wing alone and the fuselage alone are both destabilizing and unbalanced, with a positive pitching moment slope and negative $C_{M,0}$. The tail alone is stabilizing, with a negative pitching moment slope and positive $C_{M,0}$. The combined contributions of the fuselage, wing, and tail result in a complete aircraft with longitudinal static stability and balance.

6.5.3.5   Propulsion System Contribution to the Pitching Moment

The propulsion system can have a significant effect on the aircraft longitudinal static stability and balance, but the evaluation of the propulsion system contribution can be complex. This is made more difficult because of the many different types of propulsion systems, including propeller-driven, jet, and rocket engines, and the wide variety of propulsion system installations,

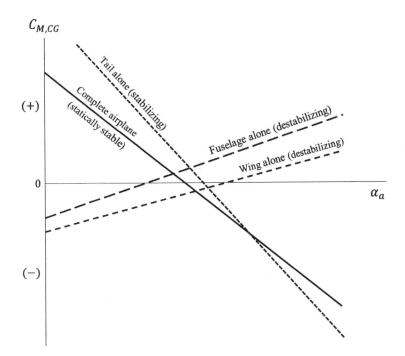

Figure 6.18   Longitudinal static stability contributions of aircraft components. (Source: *Talay, NASA SP 367, 1975,* [16].)

including underwing pod-mounted, and fuselage-buried. Modern aircraft with thrust-vectoring capabilities can use this capability to enhance stability, which may be taken into consideration for longitudinal control. Since the propulsion contributions are often difficult to predict analytically, they can be obtained through wind tunnel tests of powered models that can duplicate the operation of the propulsion system.

The propulsion system contributions are due to direct and indirect effects of the propulsive unit. The direct effects are due to the forces acting on the propulsion unit itself. The thrust force, if vertically offset from the center of gravity, contributes directly to the pitching moment. For a propeller-driven aircraft, there is also a force normal to the plane of the propeller rotation at angle-of-attack, which contributes directly to the pitching moment. For jet-powered aircraft, there can also be a normal force on the air inlet at angle-of-attack. The contribution to the pitching moment about the center of gravity, due to the direct effects of the propulsion system, $C_{M,CG,p}$, can be expressed in the form of Equation (6.15), as

$$C_{M,CG,p} = C_{M,CG,0,p} + \left(\frac{\partial C_{M,CG,p}}{\partial \alpha_a} \right) \alpha_a \qquad (6.66)$$

where $C_{M,CG,0,p}$ is the zero-lift pitching moment due to the propulsion system and $\partial C_{M,CG,p}/\partial \alpha_a$ is the slope of the pitching moment curve due to the propulsion system. The moment created by a thrust offset contributes directly to the $C_{M,CG,0,p}$ term while the moment due to the normal force varies with angle-of-attack, contributing to the $\partial C_{M,CG,p}/\partial \alpha_a$ term.

Indirect effects of the propulsion system involve the interaction of the flow created or induced by the propulsion unit on the wing–body or tail. For a propeller-driven aircraft, there is a propeller slipstream that affects the flow over the wing, the wing downwash, and the tail efficiency. If the propeller is located in the proper location ahead of the wing, the high velocity slipstream over the top of the wing can significantly increase the lift. The propeller slipstream can also increase the local velocity over the tail, increasing the tail efficiency. The exhaust of a jet-powered aircraft entrains the flow around it, inducing a flow towards the center of the exhaust jet. If the horizontal tail is located near this induced flow field, the local tail angle-of-attack may be changed. These interference effects may be included in the calculation of the wing–body and tail contributions to the pitching moment.

6.5.4   Neutral Point and Static Margin

Consider the pitching moment coefficient curve versus angle-of-attack for a complete aircraft that possesses longitudinal static stability and longitudinal trim, as shown by curve 1 in Figure 6.7. The zero-lift moment coefficient, $C_{M,CG,0}$, is positive, and the slope of the moment curve, $\partial C_{M,CG}/\partial \alpha$, is negative. The degree of static stability or amount of pitch stiffness is determined by the moment curve slope. The steeper the (negative) slope, the more stable the aircraft or the higher the pitch stiffness. If we were to, somehow, gradually reduce the static stability of the aircraft, the slope of the moment curve would become less and less steep. A limiting stability point is reached when the slope is zero, as shown by curve 3 in Figure 6.7, and the aircraft becomes neutrally stable. Further reduction in the aircraft stability makes the aircraft statically unstable and the slope of the pitching moment curve positive, as given by curve 2 in Figure 6.7. Hence, curve 3 and neutral stability is a boundary between stability and instability.

Consider the equation for the slope of the pitching moment curve for the complete aircraft, given by Equation (6.65). The stability of the aircraft can be changed by moving the location of the center of gravity, h. We now wish to determine the one location of the center of gravity, defined as the *neutral point*, h_n, where the aircraft has neutral stability. We can determine the neutral point by setting the slope of the pitching moment curve for the complete aircraft, Equation (6.65), to zero. Thus, we have

$$\frac{\partial C_{M,CG}}{\partial \alpha_a} = a_{wb} \left[h - h_{ac,wb} - \mathcal{V}_H \eta_t \frac{a_t}{a_{wb}} \left(1 - \frac{\partial \varepsilon}{\partial \alpha} \right) \right] = 0 \qquad (6.67)$$

Solving for the center of gravity position, we have

$$h_n = h_{ac,wb} + \mathcal{V}_H \eta_t \frac{a_t}{a_{wb}} \left(1 - \frac{\partial \varepsilon}{\partial \alpha} \right) \qquad (6.68)$$

All of the quantities on the right-hand side of this equation are determined and fixed for a given aircraft configuration. Therefore, the neutral point is at a fixed location in the airplane, for a given configuration, that does not vary during flight. Remember that h is not a dimensional distance, rather it is a fraction of the wing chord length c, measured from the wing leading edge. The dimensional distance to the neutral point is therefore equal to $h_n c$. Recall that the pitching moment equation that was used, Equation (6.67), was developed for an elevator fixed condition, hence, Equation (6.68) is the *stick-fixed neutral point*. A corresponding *stick-free neutral point*, h'_n, can also be derived. The stick-free neutral point is typically forward of the stick-fixed neutral point.

The neutral point represents the boundary between positive and negative static stability. At the center of gravity position equal to the neutral point, $h_C = h_n$, the slope of the pitching moment curve is zero, $C_{M_\alpha} = 0$, and the aircraft has neutral stability, as shown by Point C in Figure 6.19. At the neutral point, the pitching moment about the center of gravity, $C_{M,CG}$, is independent of angle-of-attack. Therefore, similar to our previous definition of the aerodynamic center of a wing, the neutral point may be considered the aerodynamic center of the complete aircraft. Hence, the lift, drag, and angle-of-attack-independent pitching moment of the complete aircraft can be represented as acting at the neutral point.

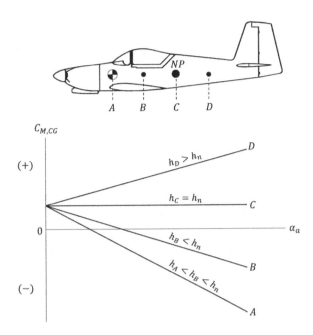

Figure 6.19   Effect of center of gravity position on static stability for fixed elevator position.

Assume that the neutral point is at location C on the aircraft pictured in Figure 6.19. If the center of gravity is moved forward of the neutral point, to position B, such that $h_B < h_n$, the slope of the pitching moment curve is negative, $C_{M_\alpha} < 0$, and the aircraft is statically stable. Moving the center of gravity even further forward to position A, such that $h_A < h_B < h_n$, increases the static stability or pitch stiffness. On the other hand, if the center of gravity is moved aft of the neutral point to position D, such that $h_D > h_n$, the slope of the pitching moment curve is positive, $C_{M_\alpha} > 0$, and the aircraft is statically unstable. Thus, the neutral point represents the *aft center of gravity limit* for the aircraft. Either the stick-fixed or the stick-free neutral point sets the aft CG limit, with the limit set by whichever one is further forward. Pilots must be careful in loading an aircraft, to ensure that the center of gravity location does not fall aft of the aft limit, since the aircraft will have negative static stability and may be unsafe to fly.

One might conclude that it would always be best to fly with the center of gravity as far forward as possible, to increase the stability of the aircraft. This is not the case, as there are other issues with a center of gravity that is too far forward. From a stability and control perspective, "too much" stability may be an issue in that the aircraft will be "heavier" on the controls to maneuver in pitch, making rotation of the nose for takeoff or landing difficult or even impossible, for example. There are also cruise performance impacts with a forward center of gravity position. Moving the center of gravity forward increases the nose down pitching moment since the moment arm between the CG and the neutral point or aircraft aerodynamic center increases. To trim the aircraft in level flight with a larger nose down moment requires more up elevator deflection, to increase the nose up moment created by the horizontal tail. This increases the negative lift of the tail, which requires more lift from the wing. The higher wing lift is produced with a higher angle-of-attack, which increases the drag and results in lower cruise speed performance. An aft center of gravity position unloads the horizontal tail, decreasing the lift required from the wing. This decreases the drag and increases the cruise speed. Theoretically, the best cruise speed is obtained with the center of gravity located at the neutral point, but this results in a neutrally stable aircraft, which would be difficult to control.

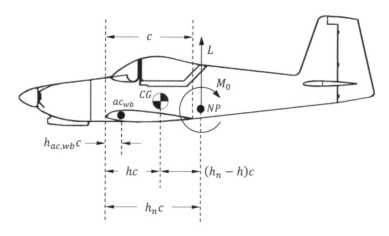

Figure 6.20   Geometry for neutral point and static margin.

The difference between the neutral point and the actual location of the center of gravity is defined as the static margin. The *stick-fixed static margin*, SM, is defined as

$$SM \equiv h_n - h \qquad (6.69)$$

The geometry for the static margin, neutral point, and wing–body aerodynamic center are shown in Figure 6.20.

Since there is a stick-free neutral point, there is also a *stick-free static margin*, SM', defined as

$$SM' \equiv h'_n - h \qquad (6.70)$$

Since the stick-free neutral point is typically forward of the stick-fixed neutral point, the stick-free static margin is typically less than the stick-fixed static margin. Hence, the stick-free longitudinal static stability is typically less than for the stick-fixed case. It is usually required that an aircraft have a minimum stick-fixed or stick-free static margin of at least 5% of the mean aerodynamic chord (MAC). This minimum static margin often sets the location of the aft-most center of gravity position.

The static margin can be related to the longitudinal static stability as follows. Solving Equation (6.68) for the position of the wing–body aerodynamic center, $h_{ac,wb}$, we have

$$h_{ac,wb} = h_n - \mathcal{V}_H \frac{a_t}{a_{wb}} \left(1 - \frac{\partial \varepsilon}{\partial \alpha} \right) \qquad (6.71)$$

Substituting this into Equation (6.65), we simply have

$$C_{M_\alpha} = \frac{\partial C_{M,CG}}{\partial \alpha_a} = -a_{wb}(h_n - h) = -a_{wb} \times SM \qquad (6.72)$$

Equation (6.72) states that the longitudinal static stability is directly proportional to the static margin. The static stability or pitch stiffness increases with increasing static margin, which is simply another way of saying that the stability increases the further forward the CG is positioned. Of course, the opposite is true, that the stability is degraded with decreasing static margin.

Recall that the aerodynamic center of a wing is located at about the quarter-chord point in subsonic flight and at mid-chord in supersonic flight. The aerodynamic center of the complete aircraft or the neutral point also moves aft in going from subsonic to supersonic flight. Hence, the static margin increases from subsonic to supersonic flight, increasing the longitudinal static stability of an aircraft flying supersonically.

Another flight condition can significantly affect the neutral point and static margin. When the aircraft is close to the ground, usually during takeoff and landing, the aircraft aerodynamics is significantly altered, which affects the stability and trim. This *ground effect*, discussed in Section 3.9.5, causes a reduction in the downwash angle at the tail, ε, and increases the wing–body and tail lift slopes, a_{wb} and a_t, respectively. Referring back to the equation for the determination of the neutral point, Equation (6.68), the ratio of the lift slopes, a_t/a_{wb}, and the downwash derivative, $\partial \varepsilon/\partial \alpha$, are decreased in ground effect. The decrease in the lift slopes tends to move the neutral point forward, while the decrease in the downwash derivative has the opposite effect. The decrease in $\partial \varepsilon/\partial \alpha$ usually dominates, resulting in a large rearward shift of the neutral point due to ground effect. Thus, ground effect results in an increase in the static margin and an increase in the longitudinal static stability.

Finally, Equation (6.72) provides a way to determine the neutral point from aerodynamic test data of the lift coefficient and moment coefficient versus angle-of-attack. If the lift curve slope of the wing–body, $a_{wb} = C_{L_\alpha}$, and the pitching moment slope, C_{M_α}, are obtained from test data for a given center of gravity position, h, the neural point may be simply calculated as

$$h_n = -\frac{C_{M_\alpha}}{a_{wb}} + h = -\frac{C_{M_\alpha}}{C_{L_\alpha}} + h \tag{6.73}$$

Example 6.4 Calculation of the Neutral point and Static Margin *Assuming an aircraft has a wing and horizontal tail with the specifications given in Example 6.2, calculate the location of the neutral point. Assume that the aerodynamic center of the wing–body is located at 0.242c and the aircraft center of gravity is at 0.380c.*

Solution

The location of the neutral point is calculated using Equation (6.68) as

$$h_n = h_{ac,wb} + \mathcal{V}_H \eta_t \frac{a_t}{a_{wb}} \left(1 - \frac{\partial \varepsilon}{\partial \alpha}\right)$$

$$h_n = 0.242 + (0.530)(0.960) \left(\frac{0.0940 \text{ deg}^{-1}}{0.0912 \text{ deg}^{-1}}\right)(1 - 0.467) = 0.522$$

The center of gravity position is forward of the neutral point, therefore the wing–body is statically stable. Using Equation (6.69), the stick-fixed static margin is given by

$$SM \equiv h_n - h = 0.522 - 0.380 = 0.142$$

The static margin is positive, also indicating positive static stability.

6.6 Longitudinal Control

For a statically stable, balanced aircraft, with a constant thrust setting, constant weight (constant center of gravity position), and fixed elevator position, there is a single trim condition, where the aircraft is flying at a trimmed angle-of-attack. Recall that our static stability analysis assumed that the aircraft is at a constant velocity, thus this trim condition also corresponds to a single airspeed.

(This can also be easily proven by recalling that the lift equals the weight in steady, level flight, where the trim angle-of-attack corresponds to a trim lift coefficient.) An aircraft that can only be trimmed at a single airspeed and angle-of-attack is not very useful operationally and is probably difficult or unsafe to fly at other speeds and angles-of-attack. We would like the ability to trim an aircraft for steady flight at other airspeeds and angles-of-attack. This is the focus of the present section. We investigate *longitudinal control* of an aircraft in the context of changing from one steady, trim condition to another. In this sense, we are evaluating the *static control* of an aircraft versus dynamic control or maneuvering.

We have already discussed one way that the trim condition is changed, by changing the center of gravity position, as shown in Figure 6.19. Examining curves A and B, in this figure, we see that we can obtain different trim angles-of-attack by changing the center of gravity position. Note from these curves that changing the center of gravity position does not change the moment coefficient at zero lift, $C_{M,0}$, rather it results in a change in the pitch stability, C_{M_α}. This may not be desirable for an aircraft, especially since the stability or pitch stiffness decreases with decreasing airspeed. It should be mentioned that this type of aircraft control has been used in the past, most notably for flying vehicles that were predecessors of the modern hang glider, which use weight shifting of a person's body position for control.

Almost all aircraft are controlled longitudinally by producing an aerodynamic force to change the pitching moment. Modern aircraft may also use a propulsive force, in the form of thrust vectoring, to accomplish or enhance longitudinal control. Typically, aerodynamic control surfaces, which may be mounted forward (canard) or aft on the aircraft, are deflected to produce an incremental lift force, resulting in a pitching moment about the aircraft center of gravity. All of the control surface may be deflected, as in an all-moving horizontal stabilizer, or part of the surface may be used, as in an elevator.

In evaluating aircraft control, we are often interested in the *control effectiveness* and the *control forces*. The *longitudinal control effectiveness* is a measure of how effective the longitudinal control is in producing the desired pitching moment. The *longitudinal control force* is a measure of how much force is required to move the longitudinal control surface, by the pilot or actuator, to obtain the desired pitching moment. The longitudinal control force is related to the longitudinal control surface *hinge moments*, the aerodynamic moments, which resist the rotation of the control surface. For our discussions, we focus on an aft-mounted horizontal tail where the control surface is an elevator.

6.6.1 Elevator Effectiveness and Control Power

Recall that the static stability analysis that we performed assumed that the elevator position was fixed. We now assume that the elevator can be deflected up or down, creating a force that generates a moment about the center of gravity. Deflecting the elevator on the horizontal stabilizer is analogous to deflecting a flap on a wing. As discussed in Section 3.9.3, the deflection of a flap shifts the lift curve, without changing the lift slope, as shown in Figure 3.125.

The same is true for an elevator deflection, as shown in Figure 6.21, where the horizontal tail lift coefficient, $C_{L,t}$, is plotted versus the local angle-of-attack at the tail, α_t. Point 1 in this figure represents the tail at an angle-of-attack $\alpha_{t,1}$ with the elevator at zero deflection, $\delta_e = 0$. If the elevator is deflected trailing edge down (TED), this is analogous to lowering a flap on a wing. Recall, from Section 6.4, that deflecting the elevator trailing edge down is positive in sign, that is, $\delta_e > 0$. The positive elevator deflection causes the lift curve to move up and to the left, maintaining the same tail lift curve slope, a_t. If the angle-of-attack were to remain the same as the undeflected elevator angle-of-attack, the lift coefficient is increased by an amount $+\Delta C_{L,t}$, shown as point 2 in

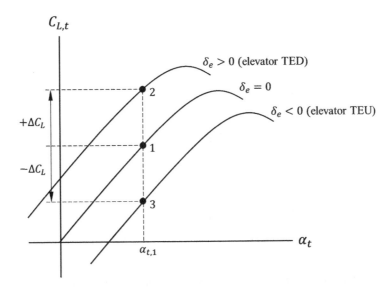

Figure 6.21 Effect of elevator deflection on the tail lift coefficient.

Figure 6.21. The opposite effect occurs with a negative elevator deflection, that is, deflecting the elevator trailing edge up (TEU). The lift curve shifts down and to the right, again, keeps the tail lift curve slope constant. At the same tail angle-of-attack $\alpha_{t,1}$, the tail lift is decreased by an increment, $-\Delta C_{L,t}$, as depicted by point 3.

Now, consider the change in the aircraft pitching moment about the center of gravity, $C_{M,CG}$, due to elevator deflection, as shown in Figure 6.22. Point 1 represents the steady, trimmed condition with $C_{M,CG} = 0$ and zero elevator deflection, $\delta_e = 0$. A positive elevator deflection results in an increase in the tail lift (point 2 in Figure 6.21), which tends to rotate the aircraft nose down relative to the center of gravity, yielding a negative increment in the pitching moment, $-\Delta C_{M,CG}$, as shown by point 2 in Figure 6.22. A negative elevator deflection results in a decrease in tail lift (point 3 in Figure 6.21), which tends to rotate the aircraft nose up relative to the center of gravity, yielding a positive increment in the pitching moment, $+\Delta C_{M,CG}$, as shown by point 3 in Figure 6.22. The elevator deflection changes the value of the pitching moment at zero lift, $C_{M,0}$, but does not change the static stability or pitch stiffness, as shown by the constant slope of the moment curve with elevator deflection.

The process of obtaining a new trim point is as follows. Consider an aircraft in steady, level flight at a trim angle-of-attack, $\alpha_{trim,1}$, with zero elevator deflection, $\delta_e = 0$, as shown in Figure 6.23. The aircraft is at a lift coefficient, $C_{L,1}$, indicated by point 1 on the lift curve, and the moment is zero since the aircraft is trimmed, as indicated by point 1 on the moment curve. The elevator is deflected trailing edge down ($\delta_e > 0$), increasing the tail lift and resulting in a shift of the entire lift curve by an amount $+\Delta C_L$. The increased tail lift produces a nose down pitching moment, shifting the moment curve by an amount $-\Delta C_M$. The aircraft retains the same static stability or pitch stiffness since the slope of the moment curve is unchanged. The new moment curve intercepts the $C_M = 0$ line at a new trim angle-of-attack, $\alpha_{trim,2}$, indicated by point 2 in Figure 6.23. This new trim angle-of-attack is less than the original angle-of-attack, $\alpha_{trim,2} < \alpha_{trim,1}$, so that the new lift coefficient is less than the original lift coefficient, $C_{L,2} < C_{L,1}$, as indicated by point 2 on the lift curve. Since the lift coefficient has decreased, the velocity must increase to maintain level flight at the new trim point, so that $V_2 > V_1$.

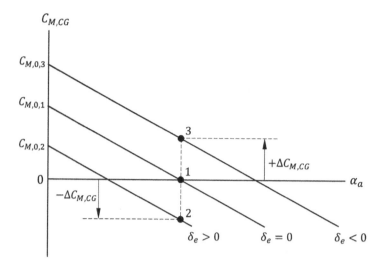

Figure 6.22   Effect of elevator deflection on the pitching moment.

Clearly, the aircraft lift and pitching moment coefficients are a function of the aircraft angle-of-attack. We have just shown that they are also a function of the elevator deflection. The aircraft lift coefficient and pitching moment about the center of gravity (CG), as a function of absolute angle-of-attack and elevator deflection may be expressed as

$$C_L = \left(\frac{\partial C_L}{\partial \alpha}\right) \alpha + \left(\frac{\partial C_L}{\partial \delta_e}\right) \delta_e = C_{L_\alpha} \alpha + C_{L_{\delta_e}} \delta_e = a\alpha + C_{L_{\delta_e}} \delta_e \tag{6.74}$$

$$C_M = C_{M,0} + \left(\frac{\partial C_M}{\partial \alpha}\right) \alpha + \left(\frac{\partial C_M}{\partial \delta_e}\right) \delta_e = C_{M,0} + C_{M_\alpha} \alpha + C_{M_{\delta_e}} \delta_e \tag{6.75}$$

where the subscripts for the absolute angle-of-attack and for the center of gravity have been omitted for simplicity. In Equation (6.74), a is the lift curve slope and $C_{L_{\delta_e}}$ is the change in the lift coefficient due to elevator deflection. Similarly, C_{M_α} is the pitching moment curve slope or pitch stiffness and $C_{M_{\delta_e}}$ is the change in the moment coefficient due to elevator deflection.

From Equation (6.36), the total lift coefficient may be expressed in terms of the wing–body lift coefficient, $C_{L,wb}$, and the tail lift coefficient, $C_{L,t}$, as

$$C_L = C_{L,wb} + \eta_t \left(\frac{S_t}{S}\right) C_{L,t} \tag{6.76}$$

where η_t is the tail efficiency, S_t is the horizontal tail planform area, and S is the wing planform area. Taking the derivative of Equation (6.76) with respect to the elevator deflection, δ_e, we have

$$C_{L_{\delta_e}} = \frac{\partial C_L}{\partial \delta_e} = \frac{\partial C_{L,wb}}{\partial \delta_e} + \eta_t \left(\frac{S_t}{S}\right) \frac{\partial C_{L,t}}{\partial \delta_e} \tag{6.77}$$

The change in the wing–body lift coefficient due to elevator deflection, $\partial C_{L,wb}/\partial \delta_e$, in Equation (6.77), is typically small for aircraft with a tail, whether aft-mounted or canard, and can usually be neglected. This term is not small for tailless aircraft. Hence, for a conventional aircraft with a tail, we have

$$C_{L_{\delta_e}} = \eta_t \left(\frac{S_t}{S}\right) \frac{\partial C_{L,t}}{\partial \delta_e} = \eta_t \left(\frac{S_t}{S}\right) C_{L,t_{\delta_e}} = \eta_t \left(\frac{S_t}{S}\right) a_e \tag{6.78}$$

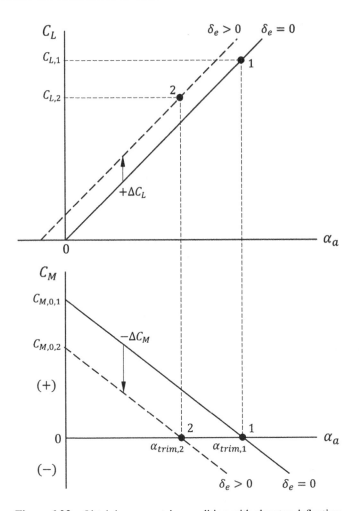

Figure 6.23   Obtaining a new trim condition with elevator deflection.

where the change in the tail lift coefficient due to elevator deflection, $C_{L,t_{\delta_e}}$, is a control derivative defined as the *elevator lift effectiveness*, or simply the *elevator effectiveness*, a_e. It is applicable only to a tailed aircraft with an elevator or all-moving stabilizer. The elevator effectiveness for an aircraft with a tail is proportional to the size of the elevator or an all-moving stabilizer relative to the wing area, S_t/S. The elevator effectiveness is a measure of the lift-producing capability of the elevator as a function of deflection. A higher elevator effectiveness indicates that the elevator is more effective at producing lift, and hence a pitching moment. From [12], the elevator effectiveness may be estimated from

$$C_{L,t_{\delta_e}} = a_e = \left(\frac{\partial C_{L,t}}{\partial \alpha_t} \right) \left(\frac{\partial \alpha_t}{\partial \delta_e} \right) = a_t \tau \tag{6.79}$$

where a_t is the horizontal tail lift curve slope and τ is the *elevator effectiveness parameter*, which is a function S_t/S. The elevator effectiveness parameter ranges from zero to about 0.8, for S_t/S from 0 to 0.7. Since the tail lift curve slope and the elevator effectiveness parameter are positive quantities, the elevator effectiveness is also always a positive number.

Inserting Equation (6.78) into (6.74), the aircraft lift may be written as

$$C_L = C_{L_\alpha} \alpha + C_{L_{\delta_e}} \delta_e = a\alpha + \eta_t \left(\frac{S_t}{S} \right) a_e \delta_e \qquad (6.80)$$

Using the definition of the elevator effectiveness, the lift coefficient of the tail alone, $C_{L,t}$, is given by

$$C_{L,t} = C_{L,t_\alpha} \alpha_t + C_{L,t_{\delta_e}} \delta_e = a_t \alpha_t + a_e \delta_e \qquad (6.81)$$

Substituting Equation (6.81) into (6.61), the aircraft pitching moment, about the center of gravity, can be written as

$$C_{M,CG} = C_{M,ac,wb} + C_{L,wb}(h - h_{ac,wb}) - V_H \eta_t (a_t \alpha_t + a_e \delta_e) \qquad (6.82)$$

Equation (6.82) provides the aircraft pitching moment as a function of the elevator deflection, δ_e. Taking the derivative of the moment with respect to the elevator deflection, we have

$$C_{M_{\delta_e}} = \frac{\partial C_M}{\partial \delta_e} = \frac{\partial C_{L,wb}}{\partial \delta_e}(h - h_{ac,wb}) - V_H \eta_t a_e \qquad (6.83)$$

The deflection of the elevator results in a significant change in the tail lift, but it does not cause a large change in the lift of the wing–body. Therefore, the change in the wing–body lift due to elevator deflection, $\partial C_{L,wb} / \partial \delta_e$, may be considered negligible, reducing Equation (6.83) to

$$C_{M_{\delta_e}} = -V_H \eta_t a_e \qquad (6.84)$$

The parameter $C_{M_{\delta_e}}$ is a control derivative known as the *elevator control power*, which is a measure of the pitching moment produced as a function of elevator deflection. A larger elevator control power indicates the capability to produce a larger pitching moment. The elevator control power is a function of the horizontal tail volume ratio, defined by Equation (6.41). Thus, the control power can be increased by increasing the horizontal tail planform area, S_t, or by increasing the moment arm to the tail, l_t.

Consider all of the quantities on the right-hand side of Equation (6.84). They are all fixed, constant values, set by the aircraft configuration. Hence, the change in the pitching moment due to an elevator deflection, δ_e, is given by

$$\Delta C_{M,CG} = \left(\frac{\partial C_M}{\partial \delta_e} \right) \delta_e = -V_H \eta_t a_e \delta_e \qquad (6.85)$$

Equation (6.85) provides the numerical value of the increment in the pitching moment due to an elevator deflection, ΔC_M, as shown in Figure 6.23.

Example 6.5 Calculation of Elevator Effectiveness and Control Power *Using the specifications for the horizontal tail given in Example 6.2, calculate the ratio of the horizontal tail area to the wing area, the elevator effectiveness, the change in the total lift coefficient due to elevator deflection, and the elevator control power. Assume a value of 0.78 for the elevator effectiveness parameter. If the aircraft lift coefficient is 1.10 and the elevator is deflected 2°, calculate the new lift coefficient.*

Solution

The ratio of the horizontal tail area to the wing area is given by

$$\frac{S_t}{S} = \frac{34.1 \, ft^2}{193 \, ft^2} = 0.177$$

Using Equation (6.79), the elevator effectiveness, a_e, is

$$a_e = a_t \tau = (0.0940 \, \text{deg}^{-1})(0.78) = 0.0733 \, \text{deg}^{-1}$$

Using Equation (6.78), the change in the total lift coefficient due to elevator deflection is given by

$$C_{L_{\delta_e}} = \eta_t \left(\frac{S_t}{S} \right) a_e = (0.960)(0.177)(0.0733 \, \text{deg}^{-1}) = 0.0125 \, \text{deg}^{-1}$$

The new lift coefficient, $C_{L,new}$, due to a 2° deflection of the elevator is given by

$$C_{L,new} = C_L + C_{L_{\delta_e}} \delta_e$$

$$C_{L,new} = 1.10 + (0.0125 \, \text{deg}^{-1})(2 \, \text{deg}) = 1.125$$

Using Equation (6.84), the elevator control power is

$$C_{M_{\delta_e}} = -\mathcal{V}_H \eta_t a_e = -(0.530)(0.960)(0.0733 \, \text{deg}^{-1}) = -0.0373 \, \text{deg}^{-1}$$

6.6.2   Calculation of New Trim Conditions Due to Elevator Deflection

In the previous section, we showed that a new trim condition, with a new trim angle-of-attack and a new trim velocity, could be obtained for a given elevator deflection, but we did not show how any of the new trim conditions could be quantified. In this section, we develop the equations to calculate the conditions at the new trim point, including the elevator deflection angle to trim and the new trim angle-of-attack and airspeed.

The pitching moment for a given angle-of-attack and elevator deflection is given by Equation (6.75). The definition of the trimmed state is the point on the pitching moment curve where the moment coefficient, C_M, is equal to zero. Setting the pitching moment to zero in this equation, we have

$$0 = C_{M,0} + C_{M_\alpha} \alpha_{trim} + C_{M_{\delta_e}} \delta_{e,trim} \tag{6.86}$$

where, now, the angle-of-attack and elevator deflection correspond to the trimmed condition, α_{trim} and $\delta_{e,trim}$, respectively.

Solving for the elevator deflection required to trim, we have

$$\delta_{e,trim} = -\left(\frac{C_{M,0} + C_{M_\alpha} \alpha_{trim}}{C_{M_{\delta_e}}} \right) \tag{6.87}$$

This equation provides the elevator deflection required to trim at a given absolute angle-of-attack, α_{trim}. The pitching moment at zero lift, $C_{M,0}$, the static stability, C_{M_α}, and the elevator control power, $C_{M_{\delta_e}}$, are typically known quantities, obtained from ground test or computational fluid dynamic analyses.

If the elevator trim deflection is known, the trim angle-of-attack can be calculated from Equation (6.86) as

$$\alpha_{trim} = -\left(\frac{C_{M,0} + C_{M_{\delta_e}}\delta_{e,trim}}{C_{M_\alpha}}\right) \tag{6.88}$$

Using Equation (6.80), the trim coefficient of lift, $C_{L,trim}$, may be calculated as

$$C_{L,trim} = C_{L_\alpha}\alpha_{trim} + C_{L_{\delta_e}}\delta_{e,trim} = a\alpha_{trim} + \eta_t\left(\frac{S_t}{S}\right)a_e\delta_{e,trim} \tag{6.89}$$

where the wing lift curve slope, a, the tail efficiency, η_t, the horizontal tail planform area, S_t, the wing planform area, S, and the elevator effectiveness, a_e, are assumed to be known from testing or analysis.

The trim velocity may be found from equating the weight and lift for steady, level flight at the trim point, as

$$W = L = q_\infty S C_{L,trim} = \frac{1}{2}\rho_\infty V_{trim}^2 S C_{L,trim} \tag{6.90}$$

Solving for the trim velocity, we have

$$V_{trim} = \sqrt{\frac{2W}{\rho_\infty S C_{L,trim}}} \tag{6.91}$$

Equation (6.89) may also be used to obtain the trim angle-of-attack in terms of the lift-based coefficients and the trim elevator deflection, as

$$\alpha_{trim} = \frac{C_{L,trim} - C_{L_{\delta_e}}\delta_{e,trim}}{C_{L_\alpha}} \tag{6.92}$$

This trim angle-of-attack may be substituted into Equation (6.75) to obtain the elevator deflection, required for trim, in terms of stability and control derivatives only.

$$\delta_{e,trim} = -\frac{C_{M,0}}{C_{M_{\delta_e}}} - \frac{C_{M_\alpha}}{C_{M_{\delta_e}}}\left(\frac{C_{L,trim} - C_{L_{\delta_e}}\delta_{e,trim}}{C_{L_\alpha}}\right)$$

$$\delta_{e,trim} = -\frac{C_{M,0}C_{L_\alpha}}{C_{M_{\delta_e}}C_{L_\alpha}} - \frac{C_{M_\alpha}C_{L,trim}}{C_{M_{\delta_e}}C_{L_\alpha}} + \frac{C_{M_\alpha}C_{L_{\delta_e}}}{C_{M_{\delta_e}}C_{L_\alpha}}\delta_{e,trim}$$

$$C_{M_{\delta_e}}C_{L_\alpha}\delta_{e,trim} - C_{M_\alpha}C_{L_{\delta_e}}\delta_{e,trim} = -C_{M,0}C_{L_\alpha} - C_{M_\alpha}C_{L,trim}$$

$$\delta_{e,trim} = -\left(\frac{C_{M,0}C_{L_\alpha} + C_{M_\alpha}C_{L,trim}}{C_{M_{\delta_e}}C_{L_\alpha} - C_{M_\alpha}C_{L_{\delta_e}}}\right) \tag{6.93}$$

Example 6.6 Elevator Deflection and Trim Angle-of-Attack *The aircraft in Example 6.5 is flying in steady, level flight at a trim airspeed of 190 knots and an altitude of 24,000 ft. The aircraft has a weight of 4315 lb and a trimmed elevator deflection of 4.70 deg. The lift curve slope of the aircraft is 0.098 deg^{-1}. Calculate the angle-of-attack at this flight condition.*

Solution

The lift coefficient at the trim condition is obtained using Equation (6.91).

$$V_{trim} = \sqrt{\frac{2W}{\rho_\infty S C_{L,trim}}}$$

Convert the trim airspeed into consistent units

$$V_{trim} = 190 \, kt \times \frac{6076 \, ft}{1 \, kt} \times \frac{1 \, h}{3600 \, s} = 321 \, \frac{ft}{s}$$

Using Appendix C, the density at 24,000 ft is

$$\rho_\infty = \sigma \rho_{SSL} = (0.46462)\left(0.002377 \, \frac{slug}{ft^3}\right) = 0.001104 \, \frac{slug}{ft^3}$$

Solving for the trim lift coefficient, we have

$$C_{L,trim} = \frac{2W}{\rho_\infty S V_{trim}^2}$$

$$C_{L,trim} = \frac{2(4315 \, lb)}{\left(0.001104 \, \frac{slug}{ft^3}\right)(193 \, ft^2)\left(321 \, \frac{ft}{s}\right)^2} = 0.393$$

From Equation (6.92), the trim angle-of-attack is given by

$$\alpha_{trim} = \frac{C_{L,trim} - C_{L_{\delta_e}} \delta_{e,trim}}{C_{L_\alpha}}$$

From Example 6.5, the change in the lift coefficient due to elevator deflection, was calculated as 0.0125 deg^{-1}. Solving for the trim angle-of-attack, we have

$$\alpha_{trim} = \frac{0.393 - (0.0125 \text{ deg}^{-1})(4.70 \text{ deg})}{0.098 \text{ deg}^{-1}} = 3.41 \text{ deg}$$

6.6.3 Elevator Hinge Moment

The three aerodynamic control surfaces typically found on an aircraft, the elevator, ailerons, and rudder, are often flap-type surfaces that are attached to a lifting or force-generating surface by a mechanical hinge. Hinges attach the elevator to the horizontal stabilizer, the ailerons to the wing, and the rudder to the vertical stabilizer. The horizontal tail is composed of the non-moving horizontal stabilizer and the moving elevator. A typical geometric configuration for a horizontal stabilizer, elevator, and elevator hinge is shown in Figure 6.24a, where c_t is the chord length of the horizontal tail and c_e is the elevator chord length, defined as the distance between the hinge line and the elevator trailing edge. A simplified model of the horizontal stabilizer and elevator is shown in Figure 6.24b, where the horizontal stabilizer and elevator are represented by flat plates, with the elevator rotating about the hinge line. This simplified model is typically used for stability and control analyses.

For a conventional airplane configuration, with the horizontal tail aft of the main wing, the air approaches the tail at a local tail angle-of-attack, α_t, as shown in Figure 6.25. The downwash from

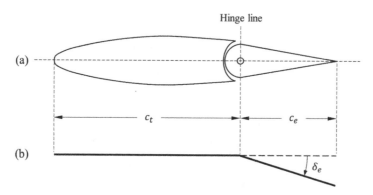

Figure 6.24 The horizontal tail, (a) geometry of horizontal stabilizer and elevator and (b) simplified horizontal tail model.

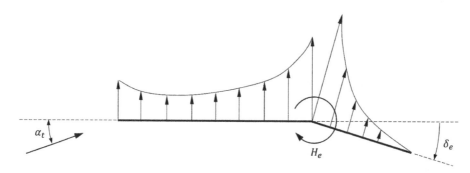

Figure 6.25 Pressure distribution on the horizontal stabilizer and elevator due to tail angle-of-attack, α_t, and elevator deflection, δ_e.

the wing makes the tail angle-of-attack different from the wing angle-of-attack. The elevator may be deflected by an angle δ_e, where δ_e is considered positive for a downward deflection of the elevator trailing edge. The aerodynamic forces and moments on the horizontal stabilizer and elevator are due to the pressure distribution over these surfaces, which is a function of the tail angle-of-attack and elevator deflection. Since the elevator is hinged, the pressure distribution on its surface may result in an aerodynamic *hinge moment*, H_e, about the hinge line, as shown in Figure 6.25. A positive hinge moment is defined as one that tends to cause a positive deflection of the elevator.

To rotate the elevator, the aerodynamic hinge moment must be overcome by applying a force to the control stick. The force may by supplied by a pilot through a mechanical connection between the control stick and the elevator, or it may be supplied by an actuator that is controlled by the pilot or a computer. The magnitude of the hinge moment must be known to properly design the flight control system and to ensure that it is acceptable throughout the flight envelope.

The non-dimensional elevator hinge moment, $C_{h,e}$, can be defined as

$$C_{h,e} = \frac{H_e}{qS_e c_e} = \frac{H_e}{\frac{1}{2}\rho V^2 S_e c_e} \tag{6.94}$$

where H_e is the dimensional elevator hinge moment, S_e is the elevator planform area that is aft of the hinge line, c_e is the elevator chord length as defined in Figure 6.24, and q, ρ, and V are the local flow properties seen by the horizontal tail.

In practice, the analytical prediction of the aerodynamic hinge moment is quite difficult, due to the complexity of the many geometrical variables that affect the hinge moment, including the elevator chord ratio, c_e/c_t, the hinge location, the elevator nose radius and trailing edge angle, the elevator planform, and the gap between the aft end of the horizontal stabilizer and the elevator leading edge. Another complicating factor in the prediction of the elevator hinge moment is the sensitivity of the moment to the type of boundary layer on the elevator. The hinge moment can be measured in flight, using load cells or other devices, but care must be taken to obtain an accurate measurement.

As explained above, the elevator hinge moment is a function of the tail angle-of-attack, α_t, and the elevator deflection angle, δ_e. Therefore, the elevator hinge moment, $C_{h,e}$, can be written as

$$C_{h,e} = \frac{\partial C_{h,e}}{\partial \alpha_t}\alpha_t + \frac{\partial C_{h,e}}{\partial \delta_e}\delta_e = C_{h,e_{\alpha_t}}\alpha_t + C_{h,e_{\delta_e}}\delta_e \qquad (6.95)$$

where $C_{h,e_{\alpha_t}}$ and $C_{h,e_{\delta_e}}$ are the derivatives of the elevator hinge moment with respect to the tail angle-of-attack and the elevator deflection angle, respectively. Equation (6.95) assumes that the variation of the hinge moment, with either tail angle-of-attack or elevator deflection angle, is linear. This assumption is valid for subsonic and supersonic flow, but not accurate for transonic flow, making transonic elevator hinge moment predictions especially difficult.

6.6.4 Stick-Free Longitudinal Static Stability

The longitudinal stability and control discussions thus far have assumed that the elevator is in a fixed position, so-called stick-fixed stability. Even the discussion concerning deflection of the elevator to establish a new trim condition assumed that the elevator is deflected and then remains in a new, fixed position. We now embark on a discussion of stick-free static longitudinal stability and control, where the elevator is free to float. Aerodynamic forces and moments act on the free-moving elevator until the control surface reaches an equilibrium position where the forces are balanced and the hinge moment is zero. A primary effect of the stick-free assumption is a change in the horizontal tail lift curve slope.

Typically, the stick-free stability is less than stick-fixed. With the stick-free assumption, the control surface is moved to a new position by the aerodynamic forces and moment, whereas this movement is resisted in the stick-fixed assumption. In this sense, the stick-fixed system is "stiffer" in its resistance to being disturbed from a trim state. Ideally, it is desirable to design an aircraft where the difference between the stick-fixed and stick-free stability is small.

For a free elevator control surface, the hinge moment, as given by Equation (6.95), is zero, so that we have

$$0 = C_{h,e_{\alpha_t}}\alpha_t + C_{h,e_{\delta_e}}\delta_{e,free} \qquad (6.96)$$

where $\delta_{e,free}$ is the stick-free elevator deflection. Solving for $\delta_{e,free}$, we have

$$\delta_{e,free} = -\left(\frac{C_{h,e_{\alpha_t}}}{C_{h,e_{\delta_e}}}\right)\alpha_t \qquad (6.97)$$

The changes in the elevator hinge moment due to the tail angle-of-attack, $C_{h,e_{\alpha_t}}$, and due to elevator deflection, $C_{h,e_{\delta_e}}$, are usually negative quantities, hence the stick-free elevator deflection, $\delta_{e,free}$, is negative (TEU) for positive tail angle-of-attack.

If the elevator is free to move, it moves to a new position after the aircraft has been disturbed from an equilibrium position. Hence the tail lift is different between a free-moving elevator and a fixed elevator. This change in the tail lift results in a change in the longitudinal static stability. Based on Equation (6.81), the tail lift coefficient for a free-moving elevator, $C'_{L,t}$, may be written as

$$C'_{L,t} = a_t \alpha_t + a_e \delta_{e,free} \tag{6.98}$$

A prime on a stability or control parameter is typically used to differentiate between stick-fixed (unprimed) and stick-free (primed). Inserting Equation (6.97) into (6.98), we have

$$C'_{L,t} = a_t \alpha_t + a_e \left[-\left(\frac{C_{h,e_{\alpha_t}}}{C_{h,e_{\delta_e}}} \right) \alpha_t \right] = \left[1 - \frac{a_e}{a_t} \left(\frac{C_{h,e_{\alpha_t}}}{C_{h,e_{\delta_e}}} \right) \right] a_t \alpha_t = F a_t \alpha_t \tag{6.99}$$

where the *free elevator factor*, F, is defined as

$$F = 1 - \frac{a_e}{a_t} \left(\frac{C_{h,e_{\alpha_t}}}{C_{h,e_{\delta_e}}} \right) \tag{6.100}$$

Comparing the tail lift with a free elevator, Equation (6.99), with the tail lift with a fixed elevator, Equation (6.43), we see that they simply differ by the free elevator factor, F. This factor is a measure of the reductions in the tail lift and of the tail lift curve slope, and hence the reduction in the longitudinal static stability, due to a free elevator. This makes perfect sense, as the deflection of the elevator, when it is free to move, changes the effective camber of the horizontal tail, thus changing the lift characteristics. If $F = 1$, the stick-free tail lift and tail lift curve slope are equal to the stick-fixed tail values. Therefore, to represent a free elevator, F must be less than one. For typical, conventional airplanes, F is about 0.7–0.8.

The equations for stick-free longitudinal static stability can be obtained using the same methodology as was used for the stick-fixed case. The stick-free stability and control parameters are compared with the stick-fixed parameters in Table 6.3. It is evident that a primary effect of the stick-free assumption on the longitudinal static stability is the reduction of the horizontal tail lift curve slope by the free elevator factor (emphasized by bold lettering in the table). Since the free elevator factor is less than one, the stick-free neutral point is less than that for the stick-fixed case. Hence, the stick-free static margin is less than the stick-fixed static margin, resulting in reduced longitudinal static stability for a free elevator.

6.6.5 Longitudinal Control Forces

To control an aircraft in three-dimensional flight, forces must be applied to deflect the aerodynamic control surfaces, which produce moments in pitch, roll, and yaw. For longitudinal and lateral control, forces must be applied to the control wheel or stick, commonly called *stick forces*, to deflect

Table 6.3 Comparison of stick-fixed and stick-free stability parameters.

Parameter	Stick-fixed	Stick-free
Tail lift	$C_{L,t} = a_t \alpha_t$	$C'_{L,t} = F a_t \alpha_t$
Moment at zero lift	$C_{M,0} = C_{M,ac,wb} + a_t \mathcal{V}_H (\varepsilon_0 - i_t)$	$C'_{M,0} = C_{M,ac,wb} + F a_t \mathcal{V}_H (\varepsilon_0 - i_t)$
Neutral point	$h_n = h_{ac,w} - \mathcal{V}_H \eta_t \dfrac{a_t}{a_{wb}} \left(1 - \dfrac{\partial \varepsilon}{\partial \alpha} \right)$	$h'_n = h_{ac,wb} + \mathcal{V}_H \eta_t \dfrac{F a_t}{a_{wb}} \left(1 - \dfrac{\partial \varepsilon}{\partial \alpha} \right)$
Pitch stability	$C_{M_\alpha} = -a_{wb}(h_n - h)$	$C'_{M_\alpha} = -a_{wb}(h'_n - h)$

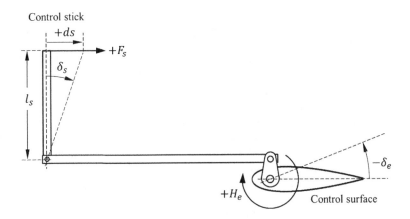

Figure 6.26   Geometry for stick force and hinge moment.

the elevator and ailerons, respectively. For directional control, forces must be applied to the rudder pedals, commonly called *pedal forces*, to deflect the rudder. In this section, we develop relationships for the elevator stick force. Similar expressions can be obtained for the aileron stick and rudder pedal forces.

By convention, for longitudinal control, a push force on the control stick should rotate the aircraft nose down and the airspeed should increase, while a pull force should rotate the nose up and the airspeed should decrease. The control forces must be within acceptable limits throughout the flight envelope, with the maximum limits set by human muscular capabilities. The control forces cannot be so high that the pilot cannot move the controls at high speed or high load factors that are within the aircraft's flight envelope. On the other hand, there are minimum control force limits also. If the control forces are too "light", it may be easy for the pilot to over control the aircraft, which can lead to over stressing the aircraft structure. The control force should be indicative of the severity of the motion that results from the control input.

For a reversible flight control system, such as the simple elevator control shown in Figure 6.26, the stick force is directly proportional to the elevator hinge moment. To see this, we can apply the principle of conservation of energy, where the change in the energy of our simple elevator control system is the sum of the heat added and work done by the system. We assume that there is no heat added to the system and that there are no losses, such as friction, in the linkages of the system, so that the change in energy is zero. The energy equation thus reduces to the work performed at the control stick and the elevator. Let us assume that the control stick, of length l_s, is pulled aft by a positive force, F_s, displacing the stick by a small, positive angle, δ_s. (By convention, a positive stick force is a pull force and a negative stick force is a push force.) The aft stick input causes the elevator to deflect the trailing edge up, to an angle $-\delta_e$ (recall that a TEU elevator deflection has a negative sign) with a positive hinge moment, H_e. The work done by moving the control stick is the force, F_s multiplied by the linear displacement, ds. The work done by the elevator is the hinge moment, H_e, multiplied by the angular displacement, $-\delta_e$. Thus, the energy equation for the work performed is given by

$$0 = F_s ds + H_e(-\delta_e) = F_s l_s \sin \delta_s - H_e \delta_e \tag{6.101}$$

Solving for the stick force, we have

$$F_s = \frac{\delta_e}{l_s \delta_s} H_e = G H_e \tag{6.102}$$

where G is gearing ratio, defined as

$$G \equiv \frac{\delta_e}{l_s \delta_s} \tag{6.103}$$

As expected, Equation (6.102) shows that stick force is directly proportional to the elevator hinge moment, with the gearing ratio as the constant of proportionality. The gearing ratio is a measure of the mechanical advantage in the control system. The larger the gearing ratio, the larger the elevator deflection, δ_e, for a given displacement of the control stick, δ_s. The gearing ratio is inversely proportional to the length of the control stick, l_s. A longer control stick has to be displaced further to obtain the same elevator deflection as a shorter stick. It may be desirable to make the gearing ratio constant as a function of elevator deflection or it may be necessary to have it vary in some way, using mechanical linkages and devices.

Substituting Equation (6.94), for the hinge moment, into Equation (6.102), we have

$$F_s = GC_{h,e} q_\infty S_e c_e \tag{6.104}$$

Equation (6.104) shows that the stick force is also proportional to the freestream dynamic pressure and the size of the elevator. Hence, the stick force increases with the square of the freestream velocity. The stick force increases with the size of the control surface; in fact, it increases dramatically with the cube of the aircraft size, being proportional to the product $S_e c_e$.

In addition to the magnitude of the stick force, the variation of the stick force with velocity, or the *stick force gradient*, around the trim point is important. The stick force gradient, $\partial F_s / \partial V$, is a measure of the change in the stick force required to produce a given change in the airspeed. A typical variation of the stick force with velocity is shown in Figure 6.27, where the slope of the stick force versus velocity is negative, $\partial F_s / \partial V < 0$. This curve provides an indication of the *speed stability* of an aircraft. Consider an aircraft in steady, trimmed flight at the trim velocity, V_{trim}, as indicated in Figure 6.27. Now assume that a disturbance causes the aircraft to slow down, such that the velocity is less than the trim velocity, $V < V_{trim}$, as indicated by point A. At the lower velocity, a positive (pull) stick force is required to maintain level flight, or the nose will pitch down and the velocity will increase back to the trim point. If a disturbance results in an airspeed greater than the trim speed, $V > V_{trim}$, (Point B), a negative (push) stick force is required to maintain level flight, or the nose will pitch up and the velocity will decrease back to the trim point. Hence, an aircraft with a stick force gradient curve versus velocity, as shown in Figure 6.27, demonstrates positive speed stability, where the aircraft tends to return to its trim point after being disturbed from its trim velocity. Thus, an aircraft with a negative stick force gradient, $\partial F_s / \partial V < 0$, at its trim point, has positive speed stability. An aircraft is more resistant to disturbances in airspeed or more *speed stable* with a steeper stick force gradient. The opposite is true, that an aircraft with a positive stick force gradient, $\partial F_s / \partial V > 0$, at its trim point, has negative speed stability.

It can be shown that the stick force gradient is proportional to the following parameters.

$$\frac{\partial F_s}{\partial V} \propto G \frac{WS_e c_e}{V_{trim}} (h - h'_n) \tag{6.105}$$

Thus, the stick force gradient is proportional to the gearing ratio, wing loading (through the aircraft weight), and the stick free static margin. It is inversely proportional to the trim speed, increasing with decreasing trim speed. The gradient is very sensitive to aircraft size, increasing with the cube of the size.

Another issue related to the control forces has to do with the forward center of gravity position. As discussed in Chapter 2, the forward and aft center of gravity limits are important safety-related limits for an aircraft. The aft center of gravity limit is set by the neutral point and the minimum static margin, as discussed in Section 6.5.4. The forward limit is related to the control forces, as discussed below.

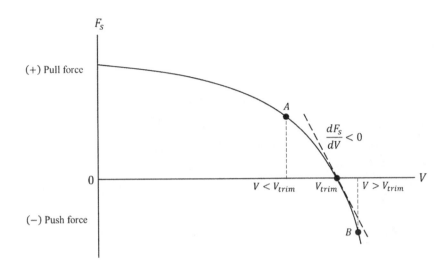

Figure 6.27 Stick force versus velocity.

The longitudinal static stability of an aircraft increases as the CG is moved forward. As the aircraft stability increases, the longitudinal control forces become heavier and larger control surface deflections are required. While this may be acceptable, perhaps even advantageous, for some types of aircraft and missions, the control forces become excessive at some forward CG position, such that maneuvering is not practical or the control surface deflections exceed mechanical limits. Hence, the forward CG limit is constrained by limits on the longitudinal control forces or the elevator deflection.

With the CG at its forward limit, the longitudinal control force must not exceed a limiting value, which may be defined by an FAA or military specification. As this applies throughout the aircraft flight (V–n) envelope, the control force requirement at the forward CG location may be specified as a control force per g or as a control force gradient in a trimmed flight condition, anywhere in the envelope. This is especially important during landing, where the control forces must not exceed the specified maximum at the trimmed approach speed down to landing.

The other possible limiting factor for the forward CG location is the maximum angle that the elevator can be deflected. This maximum angle is usually set by a mechanical stop in the elevator system. During takeoff and landing, the elevator must be deflected to rotate the aircraft's nose up. For takeoff, the nose up rotation is required to lift the nose wheel off the ground. For landing, the nose is usually rotated up in a landing *flare*, to increase the angle-of-attack as the airspeed decreases and to avoid touching down on the nose wheel first and "wheel barrowing" the landing, which may damage the nose landing gear or cause directional stability problems during the landing roll out. As the CG is moved forward, a larger elevator deflection is required to obtain a given nose rotation. At some forward CG location, the full or maximum elevator travel is reached. At this forward CG location, the elevator angle required for takeoff or landing cannot exceed the maximum possible elevator angle.

6.6.6 FTT: Longitudinal Static Stability

The flight test techniques used to determine the longitudinal static stability and neutral point of an aircraft are introduced in the present section. You will fly the Piper PA32 *Saratoga*, shown

Figure 6.28   Piper PA32 *Saratoga* six-place, single-engine airplane. (Source: *Courtesy of the author.*)

in Figure 6.28, to learn about these FTTs. The Piper *Saratoga* is a six-place, single-engine air-plane, with a low-mounted wing, conventional aft-mounted tail, and fixed landing gear. It is pow-ered by a single Lycoming IO-540-K1G5D, normally aspirated, air-cooled, horizontally opposed, six-cylinder, 300 hp (224 kW) piston engine, turning a three-bladed propeller. The *Saratoga* is used as a personal general aviation airplane and as a utility aircraft for commercial passenger and cargo transportation. Selected specifications of the Piper PA32 *Saratoga* are provided in Table 6.4. Designed and manufactured by the Piper Aircraft Corporation, Vero Beach, Florida, the first flight of the Piper PA32 *Saratoga* was on 6 December 1963.

There are two fundamentally different methods that can be used to obtain longitudinal static stability data in flight, the *stabilized method* and the *acceleration-deceleration method*. For the stabilized FTTs, data is collected at constant power, constant altitude, stabilized test points, at airspeeds above and below a selected trim airspeed. Given the stabilized nature of the test points, the technique is suitable for hand-held data collection. For the acceleration-deceleration FTT, a constant altitude acceleration or deceleration is performed to collect the test data. The aircraft is never in complete equilibrium while data is being collected, making this method less accurate that the stabilized method. However, the acceleration-deceleration FTT is more efficient in obtaining data over a large flight envelope. A data acquisition system is required to record the flight data, since the flight conditions are changing too quickly for hand recording. You will use the stabilized method to collect longitudinal static stability data in the Piper *Saratoga*.

To determine the longitudinal static stability and neutral point, data is required at different center of gravity (CG) positions. Conceivably, you could fly the airplane at progressively aft CG positions and assess the airplane's stability. However, this is a dangerous approach, since at some aft CG position the airplane will be neutrally stable or unstable, making it difficult or impossible to fly safely. You will follow a much safer approach, where you will obtain data at two CG positions,

Table 6.4   Selected specifications of the Piper PA32 *Saratoga*.

Item	Specification
Primary function	General aviation and commercial utility aircraft
Manufacturer	Piper Aircraft Corporation, Vero Beach, Florida
First flight	6 December 1963
Crew	1 pilot + 5 passengers
Powerplant	Lycoming IO-540-K1G5D six-cylinder engine
Engine power	300 hp (224 kW)
Empty weight	1920 lb (870.9 kg)
Maximum gross weight	3600 lb (1633 kg)
Length	27.7 ft (8.44 m)
Height	7.9 ft (2.4 m)
Wingspan	36.2 ft (11.0 m)
Wing area	174.5 ft^2 (16.21 m^2)
Wing loading	20.2 lb/ft^2 (98.6 kg$_f$/m^2)
Airfoil	NACA 65–415
Cruise speed	146 knots (168 mph, 272 km/h)
Service ceiling	17,000 ft (5200 m)
Load factor limits	+3.8 g, no inverted maneuvers approved

far from the neutral stability boundary. The CG position will be changed in flight by moving the location of a flight test engineer from the forward to the aft row of seats in the airplane cabin. You start with the engineer in the forward seat, resulting in a forward CG position.

The longitudinal static stability is usually obtained for the airplane in at least two configurations, the clean configuration, with the flaps and landing gear retracted, and in the approach configuration, with the flaps and landing gear extended. You will obtain data for the Piper *Saratoga* in the clean configuration, but since the airplane has fixed landing gear, this simply means with the flaps retracted.

After takeoff in the *Saratoga*, you climb to an altitude of 4500 ft (1370 m). The first task is to stabilize the aircraft at a selected trim airspeed of 100 knots (115 mph, 185 km/h). This trim airspeed is selected to be in the middle of the airspeed range where data will be collected. In other words, stabilized test points will be flown at airspeeds above and below this selected trim airspeed. The static stability is measured from this trim airspeed condition. After a few small power adjustments, the airplane is stable at an airspeed of 102 knots (117 mph, 189 km/h). It is not critical to have an airspeed of exactly 100 knots, as long as the airspeed is close to this value and stable. You take your hands off the control wheel and the airspeed remains at 102 knots for ten seconds, confirming that you have a solid trim point. You record the data at this trim point, including the airspeed, altitude, fuel quantity, angle-of-attack, elevator deflection, and longitudinal stick force, parameters that are part of the flight test instrumentation on this aircraft.

You complete this test point and prepare to set up a test point at a lower airspeed. To reach the lower airspeed, you do not want to touch the throttle, as you want to maintain a constant power setting. To decrease the airspeed, you pull back a little on the control wheel or yoke, increasing the angle-of-attack and increasing the drag. The lift also increases slightly, which causes the airplane to climb a little, but the test point stays within a typical altitude data band of ±1000 ft (300 m). Your target airspeed is about 10% slower than the trim airspeed. By holding backpressure on the control wheel, you stabilize the airplane at 92 knots (106 mph, 170 km/h).

There is inherent friction in the reversible flight control system, due to all of the cables, pulleys, and other mechanical interferences, which affects the stick force data. The friction varies, depending on whether you are pulling or pushing on the control wheel. To stabilize the test point, you have had to push and pull the control wheel back and forth a little bit. Therefore to ensure that the stick force data is consistent, you release the controls and let the airplane nose drop slightly, then pull back to reset the stabilized condition. You will do this for all of the slower than trim speed test points, so that they are all set with the controls being pulled back. For the faster than trim speed test points, you will release the controls until the nose rises, then push forward so that all of these test points are set with the controls pushed forward.

You are stabilized at an airspeed of 92 knots, having to hold a little backpressure on the yoke to maintain the constant airspeed. You record the test point data as before. Now, you are ready to proceed to a test point at a higher airspeed than the 102 knot trim airspeed. You push forward on the control wheel and allow the airplane to accelerate. Your target airspeed is about 10% faster than the trim airspeed. Holding some forward stick force, you stabilize the airplane at 111 knots (128 mph, 206 km/h) and again collect the test point data. You collect data for several more test points, down to an airspeed of 74 knots (85 mph, 137 km/h) and up to 121 knots (139 mph, 224 km/h). After collecting this data, you are complete with the test points at this forward CG position. Now, you have the flight test engineer change seats, from the front row to the aft, moving the CG aft. You repeat all of the test points that were flown for the forward CG position at the aft CG position. You are fortunate, perhaps skillful, in setting up the trim airspeed for this new, aft CG position at the same 102 knots as the forward CG position. After collecting the data for the test points at this aft CG position, you fly back to the airport and land.

A plot of the elevator deflection angle required to stabilize or trim the airplane, $\delta_{e,trim}$, at the different calibrated airspeeds, V_c, are shown in Figure 6.29, for the two different center of gravity positions. The 102 knot trim airspeed condition is marked by the dotted line in the figure. The required trim elevator deflection at 102 knots is $-3.46°$ for the forward CG position (86 in) and $-3.05°$ for the aft CG position (88 in). These negative values of elevator deflection correspond to the elevator being deflected trailing edge up (TEU) and the control wheel being pulled back. For the test points at speeds slower than the trim airspeed, more trailing edge up elevator deflection is required to stabilize. For the test points at speeds faster than the trim airspeed, less trailing edge

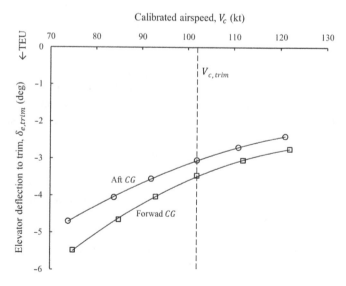

Figure 6.29 Elevator deflection versus airspeed.(Source: *Figure created by author based on data trends in* [6], *with permission of Matthew DiMaiolo.*)

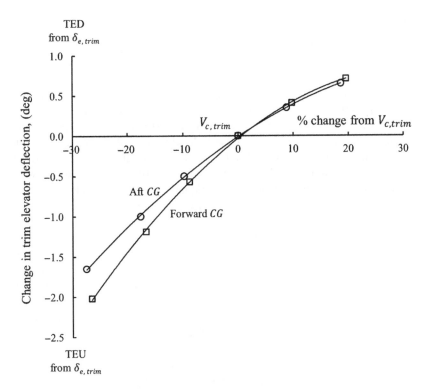

Figure 6.30 Elevator deflection versus airspeed from initial trimmed flight condition. (Source: *Figure created by author based on data trends in* [6], *with permission of Matthew DiMaiolo.*)

up elevator deflection is required to stabilize. These results are also plotted in Figure 6.30, in a different format. In this figure, the change in the elevator deflection from the initial trim condition is plotted versus the percent change from the initial trim calibrated airspeed.

These figures represent the stick-fixed longitudinal static stability of the airplane, similar to the moment coefficient versus angle-of-attack plots discussed earlier. As the airspeed is decreased from the initial trim or equilibrium point (102 knots), the elevator deflection and control wheel backpressure must be increased to counter a nose down pitching moment to stabilize the airplane. Thus, the tendency of the airplane is to pitch nose down as the airspeed is decreased, which tends to increase the airspeed and return the airplane to the initial equilibrium condition. The opposite is true when the airspeed is increased from the initial trim or equilibrium point. At higher airspeeds, the elevator deflection and the control wheel backpressure must be decreased because the airplane has a nose up pitching moment. Thus, the tendency of the airplane is to pitch nose up as the airspeed in increased, tending to decrease the airspeed and return the airplane to the initial equilibrium condition. Based on these observations, it can be concluded that a plot of elevator deflection versus airspeed with a negative slope, as shown in Figure 6.29, indicates stick-fixed longitudinal static stability. If plotted as the change in the elevator deflection from the initial equilibrium condition, the slope is positive as shown in Figure 6.30. The slope is steeper for the forward CG position, indicating higher positive static stability than for the aft CG position. This figure also verifies that the airplane can be trimmed at a useable airspeed.

To determine the stick-fixed neutral point, we start by taking the derivative of the trim elevator deflection, Equation (6.93), with respect to the lift coefficient at the initial trimmed condition,

$C_{L,trim}$, yielding

$$\frac{d\delta_{e,trim}}{dC_{L,trim}} = -\left(\frac{C_{M_\alpha}}{C_{M_{\delta_e}}C_{L_\alpha} - C_{M_\alpha}C_{L_{\delta_e}}}\right) \tag{6.106}$$

At the neutral point, the slope of the pitching moment coefficient with respect to angle-of-attack is equal to zero, $C_{M_\alpha} = 0$ (see also Figure 6.19). For $C_{M_\alpha} = 0$, Equation (6.106) equals zero. Therefore, if we plot the derivative, $d\delta_{e,trim}/dC_{L,trim}$, versus CG position, the neutral point is identified as the CG position where this derivative is zero. First, the elevator deflection angle is plotted versus lift coefficient, as shown in Figure 6.31. The lift coefficient is calculated for steady level flight, using the flight condition (altitude and airspeed), wing planform area, and airplane weight corresponding to the measured fuel weight, as given by Equation (5.112). From Figure 6.31, values of the derivative, $d\delta_{e,trim}/dC_{L,trim}$, are obtained, for the two center of gravity positions, as the slope of the data.

The values of the derivative, $d\delta_{e,trim}/dC_{L,trim}$, are plotted versus the center of gravity position in Figure 6.32. The stick-fixed neutral point, h_n, is determined by drawing a straight line through the two data points, intercepting the horizontal axis where the derivate, $d\delta_{e,trim}/dC_{L,trim}$, is zero. From Figure 6.32, the flight-determined, stick-fixed neutral point is at a center of gravity position of 101.1 in (256.8 cm).

A similar data reduction process is followed for the determination of the *stick-free* longitudinal static stability and *stick-free* neutral point, h'_n. The major difference is that the elevator stick force, F_e, is the relevant parameter instead of the elevator deflection, δ_e. By developing a plot of the derivate of the stick force with respect to the lift coefficient versus CG position, similar to Figure 6.32, the stick-free neutral point can be obtained from flight data.

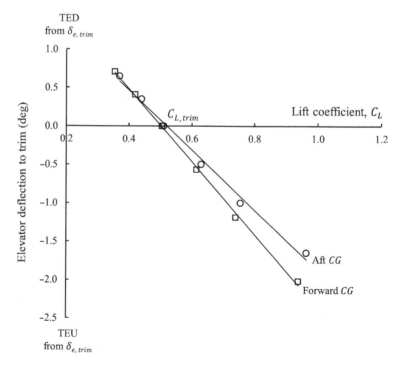

Figure 6.31 Elevator deflection versus lift coefficient. (Source: *Figure created by author based on data trends in* [6], *with permission of Matthew DiMaiolo.*)

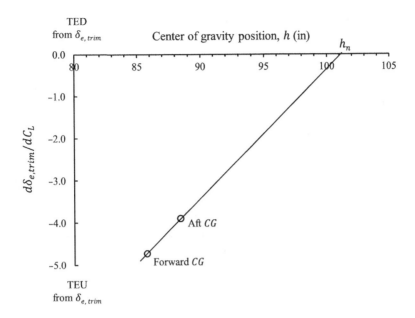

Figure 6.32   Determination of stick-fixed neutral point. (Source: *Figure created by author based on data trends in* [6], *with permission of Matthew DiMaiolo.*)

6.7   Lateral-Directional Static Stability and Control

Longitudinal static stability concerns the pitching motion of the aircraft in its plane of symmetry. Referring back to the aircraft coordinate system in Figure 6.4, the pitching motion was about the y-axis in the x–z plane. In analyzing longitudinal static stability, it was assumed that the aircraft was in steady flight at a velocity, V_∞, angle-of-attack, α, and a dynamic pressure, q_∞. The forces and moment acting on the aircraft were the lift, drag, thrust, and pitching moment.

We now examine the rolling and yawing motions of the aircraft, rotations about the x- and z-axes, respectively. We evaluate the behavior of the aircraft after being displace in yaw, β, and roll, ϕ, rather than angle-of-attack, α. The side force, Y, yawing moment, N, and rolling moment, R, act on the aircraft in lateral-directional motion. Since the center of gravity lies in the aircraft's plane of symmetry (the x–z plane), it is not a dominant parameter for lateral-directional motion, as it was for longitudinal motion. The vertical location of the center of gravity can have an effect on the aircraft lateral-directional motion, but this is usually small.

The lateral-directional motions are more complicated than the longitudinal motion of the aircraft. For one thing, the roll and yaw motions tend to be *cross-coupled*, that is, rotation about one axis results in motions in both axes. A roll rate, p, not only produces a rolling moment, R, but also produces a yawing moment, N. A yaw displacement, β, or a yaw rate, r, results in both a yawing moment, N, and a rolling moment, R. The lateral-directional controls, the aileron and rudder, are also cross-coupled. Deflecting the ailerons produce both roll and yaw. Similarly, a rudder deflection results in yaw and roll.

The force and moment coefficients of interest for lateral-direction motion are the side force, yawing moment, and rolling moment coefficients. The side force coefficient, C_Y, is defined as

$$C_Y \equiv \frac{Y}{qS} \tag{6.107}$$

The yawing moment coefficient, C_N, is defined as

$$C_N \equiv \frac{N}{qSb} \tag{6.108}$$

One must be careful not to confuse the yawing moment coefficient with the normal force coefficient. The rolling moment coefficient, C_R, is defined as

$$C_R \equiv \frac{R}{qSb} \tag{6.109}$$

where q is the dynamic pressure, S is the wing planform area, and b is the wingspan. The characteristic length used to non-dimensionalized the lateral-directional moments is the wingspan, b, rather than the chord length, as for the pitching moment.

Keep in mind that we are still discussing *static* stability, but for lateral-directional motions rather than longitudinal. Hence, we are still interested in the *tendency* of the aircraft to return to or diverge from equilibrium, after being disturbed from its trim point.

6.7.1   Directional Static Stability

Directional stability is sometimes called *weathercock* stability, because of the similarity between an aircraft aligning itself with the relative wind and a weathervane (in the shape of a rooster). Directional static stability deals with stability about the vertical or z-axis of the aircraft. Since we are interested in *static* stability, we evaluate the *tendency* of the aircraft to return to or diverge from equilibrium. As such, we are not interested in the motion of the aircraft over time, where there may be coupling of motions in different axes. Because of this, the directional static stability can be isolated to yawing motion in the x–y plane only. From this perspective, directional static stability is similar to longitudinal static stability, with the rotation occurring around a different axis and a different plane. Many of the relationships are similar between these two motions.

Consider the aircraft in the x–y plane, shown in Figure 6.33. It is originally flying in steady, equilibrium flight with its longitudinal (x-axis) aligned with the freestream velocity, V_∞. A disturbance causes the aircraft to yaw, moving the nose left. This "wind in the right ear" rotation corresponds to a positive sideslip angle, $+\beta$ (see sign conventions in Section 6.4), as shown in Figure 6.33. The sideslip angle sets the vertical tail at an angle-of-attack relative to the freestream velocity, creating a lift force, L_v, perpendicular to the velocity vector. (The subscript v is used to reference the vertical tail fin.) The lift on the vertical tail results in a restoring moment, N, around the aircraft center

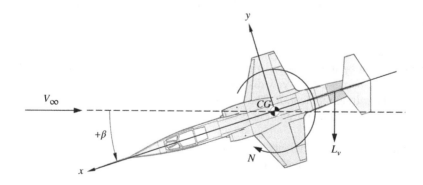

Figure 6.33   Aircraft with positive sideslip and restoring yawing moment.

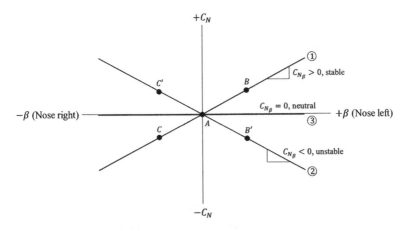

Figure 6.34 Directional static stability.

of gravity (CG), which tends to "straighten" the aircraft and return it to its original equilibrium position. The vertical tail is acting like the horizontal stabilizer, for longitudinal static stability, in creating a moment about the center of gravity. However, unlike the horizontal tail, the vertical tail is set at a zero incidence angle relative to the fuselage longitudinal axis, so that the aircraft remains aligned with the velocity vector in steady, trimmed flight. (There is a slight exception to this for propeller-driven aircraft, due to the propeller slipstream, which is discussed for lateral stability.) If the aircraft is disturbed in the opposite yaw direction, corresponding to "wind in the left ear" and negative β, the lift force and the restoring moment are in the opposite direction, which tends to restore equilibrium. Hence, the aircraft has directional static stability with the tendency to return to its equilibrium position after being disturbed in the yaw direction.

Consider the yawing moment coefficient, C_N, versus angle-of-sideslip, β, for three different aircraft, as shown in Figure 6.34. Aircraft 1 is in steady, trimmed flight, with the aircraft longitudinal (x-axis) aligned with the freestream velocity. As depicted by point A in the figure, the yawing moment and the angle-of-sideslip are zero at the trim point. A disturbance causes the nose of the aircraft to yaw left, such that the sideslip angle is $+\beta$ and the yawing moment is positive, $+C_N$, as depicted by point B. The positive yawing moment rotates the nose to the right, tending to restore the aircraft to equilibrium. If the disturbance yaws the nose right, to a sideslip angle $-\beta$, a negative yawing moment, $-C_N$, is created, shown by point C, which rotates the nose left, again, tending to restore equilibrium. Hence, aircraft 1 demonstrates directional static stability and to do so, it must have a yawing moment versus angle-of-sideslip variation as shown in Figure 6.34.

The slope of the yawing moment curve is called the *directional static stability* or the *yaw stiffness*, C_{N_β}. For positive directional static stability, the slope must be positive, so that

$$C_{N_\beta} = \frac{\partial C_N}{\partial \beta} > 0 \tag{6.110}$$

Now consider aircraft 2, with a yawing moment curve as shown in Figure 6.34. The aircraft is originally at its trim condition at point A. A disturbance causes the nose to yaw left to a positive sideslip angle, $+\beta$, creating a negative yawing moment, $-C_N$, as depicted by point B'. The negative yawing moment tends to rotate the aircraft nose to a larger sideslip angle, further away from its original trim point. When the nose is yawed to the right $(-\beta)$, to point C', the positive yawing moment, $+C_N$, tends to rotate the nose to a larger negative angle-of-sideslip, away from the trim point. Hence, aircraft 2 does not have directional static stability and we conclude that a negative yawing moment slope, $\partial C_N / \partial \beta < 0$, is destabilizing, in terms of directional static stability.

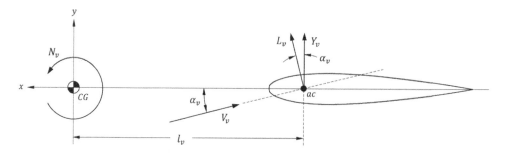

Figure 6.35   Vertical tail geometry.

Aircraft 3 has neutral directional static stability with $\partial C_N/\partial \beta = 0$. Displacement in yaw from its trim condition results in no yawing moment, and the aircraft remains at its displaced angle-of-sideslip.

We now seek to develop a quantitative relationship for the directional static stability or yaw stiffness, C_{N_β}, similar to what was developed for longitudinal static stability or pitch stiffness, C_{M_α}. Consider the vertical tail that is mounted aft of the aircraft center of gravity, as shown from above in Figure 6.35. The x-axis points along the aircraft fuselage, out to the nose, and the y-axis points out the right wing. The vertical tail is aligned with the x-axis, having zero incidence angle. The vertical tail has a symmetric airfoil section so that the zero-lift line and the chord line are the same. The vertical tail is composed of the stationary vertical stabilizer or fin and the movable rudder. It is assumed that the rudder is in a fixed position, faired with the vertical stabilizer. The aerodynamic center of the vertical tail is located a distance l_v behind the center of gravity.

Now assume that the aircraft is yawed to the right, such that the "wind is in the left ear" for someone sitting on top of the vertical tail. The local flow velocity seen by the vertical tail, V_v, is different from the freestream velocity due to the influence of the fuselage and wing. The vertical tail velocity, V_v, is at an angle-of-attack, α_v, relative to the zero-lift line or chord line of the vertical tail, given by

$$\alpha_v = -\beta + \sigma \tag{6.111}$$

where the sideslip angle is negative in sign since the nose was yawed to the right. The vertical tail angle-of-attack is different from the sideslip angle due to *sidewash*, the disruption of the flow by the fuselage and wings, which changes the local flow direction. The change in the flow direction at the vertical tail, due to sidewash, is analogous to the effect of downwash on the horizontal tail. The sidewash angle, σ, is added to the sideslip angle to obtain the local angle-of-attack at the vertical tail, α_v. The sidewash angle is positive when it corresponds to flow in the positive y-direction, which tends to increase the local angle-of-attack.

For small local angles-of-attack, α_v, the side force, Y_v, on the vertical tail is equal to the lift, L_v, so that

$$Y_v = L_v \cos \alpha_v \cong L_v = q_v S_v C_{L,v} \tag{6.112}$$

where q_v is the local dynamic pressure, S_v is the planform area of the vertical tail, and $C_{L,v}$ is the vertical tail lift coefficient. The side force and lift act at the aerodynamic center of the vertical tail.

The vertical tail lift coefficient may be written as

$$C_{L,v} = \frac{\partial C_{L,v}}{\partial \alpha} \alpha_v = a_v \alpha_v = a_v(-\beta + \sigma) \tag{6.113}$$

where a_v is the lift curve slope of the vertical tail.

Inserting Equation (6.113) into (6.112), the side force is given by

$$Y_v = q_v S_v a_v (-\beta + \sigma) \tag{6.114}$$

The yawing moment, about the center of gravity, N_v, created by the vertical tail side force is given by

$$N_v = -Y_v l_v = -q_v S_v l_v a_v (-\beta + \sigma) \tag{6.115}$$

where l_v is the distance between the aircraft center of gravity and the vertical tail aerodynamic center. The sign of the yawing moment is negative.

The vertical tail yawing moment coefficient, $C_{N,v}$, is defined as

$$C_{N,v} \equiv \frac{N_v}{q_{wb} S b} = -\frac{q_v S_v l_v}{q_{wb} S b} a_v (-\beta + \sigma) \tag{6.116}$$

where q_{wb} is the dynamic pressure of the flow over the wing–body, S is the wing planform area, and b is the wingspan.

Analogous to the horizontal tail efficiency, η_t, the *vertical tail efficiency*, η_v, is defined as

$$\eta_v \equiv \frac{q_v}{q_{wb}} = \left(\frac{V_v}{V}\right)^2 \tag{6.117}$$

Similar to the horizontal tail efficiency, the magnitude of the vertical tail efficiency is less than one. Analogous to the horizontal tail volume ratio, $\mathcal{V}_H$, the *vertical tail volume ratio*, $\mathcal{V}_v$, is defined as

$$\mathcal{V}_v \equiv \frac{S_v l_v}{S b} \tag{6.118}$$

Typical values of the vertical tail volume ratio are about 0.04 for a single-engine general aviation airplane, about 0.07 for a military jet fighter aircraft, about 0.06 for a military jet trainer aircraft, and about 0.09 for a commercial jet transport [14]. The vertical tail volume ratios are about an order of magnitude smaller than the horizontal ratios, due to the non-dimensionalizing, characteristic length of wingspan instead of chord.

Substituting Equations (6.117) and (6.118) into (6.116), we have

$$C_{N,v} = -\eta_v \mathcal{V}_v a_v (-\beta + \sigma) \tag{6.119}$$

The directional static stability is obtained by taking the derivative of the yawing moment coefficient, Equation (6.119), with respect to the angle-of-sideslip, β, given by

$$C_{N,v_\beta} = \frac{\partial C_{N,v}}{\partial \beta} = \eta_v \mathcal{V}_v a_v \left(1 - \frac{\partial \sigma}{\partial \beta}\right) \tag{6.120}$$

Equation (6.120) is the contribution to the directional static stability by the vertical tail. It states that directional static stability is directly proportional to the vertical tail efficiency, the vertical tail volume ratio, and the vertical tail lift curve slope. As may seem obvious, increasing the vertical tail volume ratio by increasing the size of the vertical tail increases the directional static stability. This equation for directional static stability is analogous to the equation for longitudinal static stability, Equation (6.51).

The contributions of other aircraft components are compared qualitatively in Figure 6.36, in the form of curves for the yawing moment coefficient versus sideslip angle. Positive slopes indicate

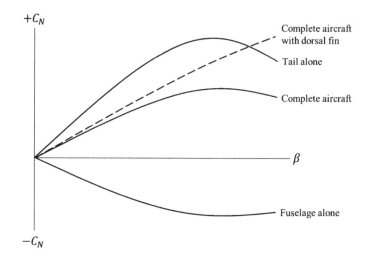

Figure 6.36 Directional static stability contributions of various aircraft components. (Source: *Talay, NASA SP 367, 1975*, [16].)

a positive contribution to the directional static stability, while negative slopes indicate that the component is destabilizing. The side area of the fuselage contributes negatively to the directional static stability, as shown. The fuselage, and engine nacelles, are typically destabilizing influences to directional stability. Usually, the contribution of the wing is small, in comparison to the fuselage, except at high aircraft angles-of-attack. The vertical tail has the largest effect in providing positive directional static stability for the complete aircraft. At some sideslip angle, the flow over the vertical tail aerodynamically stalls, drastically reducing the vertical tail lift. The loss of side force results in loss of the restoring yawing moment, required for stabilization. At these higher sideslip angles, the complete aircraft can become directionally unstable. The addition of ventral fins, low aspect ratio vertical fins that are typically attached to the upper fuselage, forward of the vertical fin, or to the underside of the fuselage, may provide additional directional stability through this higher sideslip angle range.

Example 6.7 Contribution of the Vertical Tail to Directional Static Stability *An aircraft has a wing and vertical tail with specifications given in the table below. Calculate the contribution of the vertical tail to the aircraft directional static stability. Assume that the change in the sidewash angle with sideslip, $\partial\sigma/\partial\beta$, is zero.*

Parameter	Value
Wing area, S	193 ft^2
Wing span, b	36.3 ft
Vertical tail area, S_v	22.3 ft^2
Vertical tail lift curve slope, a_v	0.0985 deg^{-1}
Vertical tail efficiency, η_v	0.980
CG to vertical tail ac distance, l_v	16.1 ft

Solution

Using Equation (6.118), the vertical tail volume ratio is given by

$$\mathcal{V}_v = \frac{S_v l_v}{Sb} = \frac{(22.3 \, ft^2)(16.1 \, ft)}{(193 \, ft^2)(36.3 \, ft)} = 0.0512$$

The directional static stability is given by Equation (6.120) as

$$C_{N,v_\beta} = \frac{\partial C_{N,v}}{\partial \beta} = \eta_v \mathcal{V}_v a_v \left(1 - \frac{\partial \sigma}{\partial \beta}\right)$$

The vertical lift curve slope is converted from per degrees to per radian.

$$a_v = (0.0985 \, \deg^{-1}) \times \frac{180}{\pi} = 5.64$$

Inserting values into the equation for the directional static stability, we have

$$C_{N,v_\beta} = (0.980)(0.0512)(5.64)(1 - 0) = 0.283$$

As expected, the vertical tail contributes positively to the directional static stability.

6.7.2 Directional Control

Usually, flight at zero sideslip is desired. Assuming an aircraft has a symmetric configuration and positive directional static stability, it tends to fly "straight" with zero sideslip. However, with the inevitability of configuration asymmetries, atmospheric upsets, asymmetric thrust, propeller slipstream effects, asymmetric flows of turning or maneuvering flight, and other factors, yawing moments act on the aircraft and the angle-of-sideslip is non-zero. There may also be flight conditions where it is desired to intentionally fly with sideslip. These conditions may include the forward slip for crosswind control during landing or the sideslip to increase drag and the glide path angle. The aircraft control, to counter undesired yawing moments and to add intentional sideslip, is the rudder. As described in Section 1.2.2.2, the rudder is a vertical, hinged flap, attached to the aft part of the vertical stabilizer. Rudder deflection changes the lift force created by the vertical tail, much like the wing flap or the elevator on the horizontal tail.

Consider the vertical tail at zero sideslip, as shown in Figure 6.37. A positive deflection of the rudder, $+\delta_r$, creates a positive side force, $+Y_v$, which results in a negative yawing moment about

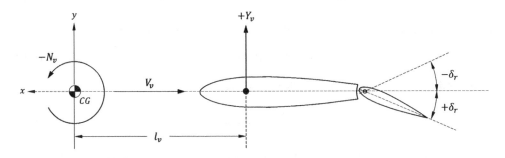

Figure 6.37 Rudder geometry and side force at zero sideslip.

the center of gravity, $-N_v$, given by

$$N_v = -Y_v l_v = -q_v S_v C_{L,v} l_v \tag{6.121}$$

where $C_{L,v}$ is the lift coefficient of the vertical tail with the rudder deflected.

The vertical tail yawing moment coefficient is given by

$$C_{N,v} = -\frac{N_v}{q_{wb} S b} = \frac{-q_v S_v l_v}{q_{wb} S b} C_{L,v} = -\eta_v \mathcal{V}_V C_{L,v} \tag{6.122}$$

where the vertical tail efficiency and the vertical tail volume ratio, Equations (6.117) and (6.118), respectively, have been inserted.

The vertical tail lift coefficient (assuming zero sidewash) may be written as

$$C_{L,v} = \left(\frac{\partial C_{L,v}}{\partial \beta}\right) \beta + \left(\frac{\partial C_{L,v}}{\partial \delta_r}\right) \delta_r = C_{L,v_\beta} \beta + C_{L,v_{\delta_r}} \delta_r \tag{6.123}$$

Since the sideslip angle is zero ($\beta = 0$), this becomes

$$C_{L,v} = \left(\frac{d C_{L,v}}{d \delta_r}\right) \delta_r = C_{L,v_{\delta_r}} \delta_r \tag{6.124}$$

Inserting Equation (6.124) into (6.122), we have

$$C_{N,v} = -\eta_v \mathcal{V}_V C_{L,v_{\delta_r}} \delta_r \tag{6.125}$$

Taking the derivative of Equation (6.125), with respect to rudder deflection, we have

$$C_{N,v_{\delta_r}} = \frac{\partial C_{N,v}}{\partial \delta_r} = -\eta_v \mathcal{V}_V C_{L,v_{\delta_r}} \tag{6.126}$$

The derivative of the yawing moment with respect to rudder deflection is defined as the *rudder effectiveness* or *rudder control power*. The larger the value of $C_{N,v_{\delta_r}}$, the larger the effectiveness or power of the rudder in producing a yawing moment for a given rudder deflection. In our analysis, it was assumed that the rudder was deflected and then held fixed at the deflected position. Thus, Equation (6.126) is the stick-fixed rudder effectiveness.

As discussed earlier, there are flight conditions when it is desired to maintain a steady sideslip angle. We now seek to determine a relationship for the steady sideslip angle that can be obtained for a given rudder deflection. The aircraft yawing moment, about the center of gravity, in a steady sideslip (assuming zero sidewash) is given by

$$C_{N,v} = \left(\frac{\partial C_{N,v}}{\partial \beta}\right) \beta + \left(\frac{\partial C_{N,v}}{\partial \delta_r}\right) \delta_r = C_{N,v_\beta} \beta + C_{N,v_{\delta_r}} \delta_r \tag{6.127}$$

For a steady, equilibrium flight condition, the moment must equal zero, so we have

$$0 = C_{N,v_\beta} \beta + C_{N,v_{\delta_r}} \delta_r \tag{6.128}$$

Solving for the sideslip angle per rudder deflection, we have

$$\frac{\beta}{\delta_r} = -\frac{C_{N,v_{\delta_r}}}{C_{N,v_\beta}} \tag{6.129}$$

Equation (6.129) states that the steady sideslip that can be obtained for a given rudder deflection is proportional to the rudder control power, $C_{N,v_{\delta_r}}$, and inversely proportional to the directional static stability or yaw stiffness, C_{N,v_β}. If the control power is large, then it is possible to obtain a higher sideslip angle than if the control power was smaller. If the aircraft is more directionally stable or "stiffer" in yaw, less sideslip can be obtained for a given rudder deflection than if the aircraft has less yaw stiffness.

6.7.3   Lateral Static Stability

We now consider static stability of the aircraft about the x-axis, referred to as lateral or roll stability. Similar to longitudinal and directional static stability, lateral static stability concerns the tendency of the aircraft to return to or diverge from a steady, wings-level, trimmed condition after being disturbed, in this case, by a roll upset. If a restoring moment is generated after a roll upset from wings-level flight, then the aircraft has positive lateral stability. If the generated moment increases the roll, away from wings-level flight, then the aircraft has negative lateral stability.

The criteria for determining longitudinal or directional static stability is based on the change in the moment with respect to a displacement, in the plane of motion. Longitudinal static stability is evaluated based on the pitch stiffness, the change in the pitching moment with respect to a change in the angle-of-attack, $\partial C_M / \partial \alpha$. Directional static stability is evaluated based on the yaw stiffness, the change in the yawing moment with respect to a change in the angle-of-sideslip, $\partial C_N / \partial \beta$. Following this line of reasoning, one might surmise that lateral static stability should be based on the roll stiffness, the change in the rolling moment with respect to a change in the bank angle, $\partial C_R / \partial \phi$. Let us consider this premise.

Consider an aircraft in steady, wings-level flight that is constrained to a single degree-of-freedom, that being rotation or roll about the x-axis. If we assume that the angle-of-attack is small, we can assume that the freestream velocity vector is parallel to the x-axis. The aerodynamic lift and drag, generated by the freestream flow over the aircraft, are in the aircraft's plane of symmetry. Now, if the aircraft rolls to a bank angle, ϕ, the flow is still symmetric with respect to the aircraft's plane of symmetry, therefore the lift and drag are unchanged. Hence, the change in the rolling moment with a change in the bank angle, $\partial C_R / \partial \phi$, or the roll stiffness is zero. This argument can be extended to an aircraft at an angle-of-attack, where the aircraft rolls about the velocity vector. In this more general case, no restoring moment is generated by an aircraft rolling about its velocity vector. However, it can be shown that if an aircraft, at angle-of-attack, rolls about its x-axis, a sideslip develops which creates a rolling moment. This is characteristic of aircraft with slender fuselages and small aspect ratio wings.

Thus, the rolling moment required for lateral static stability is generated due to sideslip, rather than roll. When an aircraft rolls, a sideslip is generated, resulting in a rolling moment. Hence, the critical stability derivative for lateral static stability is the derivative of the rolling moment with respect to sideslip, $C_{R_\beta} = \partial C_R / \partial \beta$, rather than roll.

Consider three aircraft with variations of rolling moment, C_R, versus sideslip, β, as shown in Figure 6.38. All of the aircraft are initially in steady, wings-level, trimmed flight with zero sideslip ($\beta = 0$) and zero rolling moment ($C_R = 0$), as depicted by point A. Aircraft 1 experiences a roll upset, which drops the right wing and rolls the airplane to the right. The roll tilts the lift vector, moving the airplane sideways to the right, resulting in a positive sideslip ($+\beta$) with "wind in the right ear". The positive sideslip produces a negative rolling moment, $-C_R$, as depicted by point B in Figure 6.38. The negative rolling moment tends to roll the aircraft to the left, back towards a wings-level attitude. If the upset rolls the airplane to the left, a negative sideslip ($-\beta$) develops, producing a positive rolling moment, $+C_R$, as depicted by point C. The positive rolling moment tends to roll the aircraft to the right, back towards wings-level flight. Hence, we conclude that

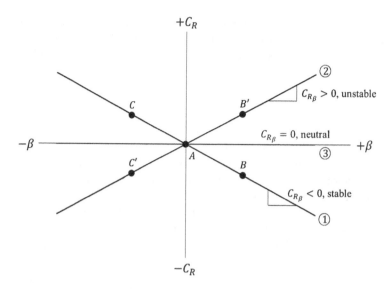

Figure 6.38 Lateral static stability.

aircraft 1 has positive lateral static stability with the requirement for stability being

$$C_{R_\beta} = \frac{\partial C_R}{\partial \beta} < 0 \tag{6.130}$$

For aircraft 2, an upset that rolls the airplane to the right causes a sideslip to the left $(+\beta)$ and a positive rolling moment, $+C_R$, as depicted by point B'. The positive rolling moment tends to roll the aircraft to the right, driving the airplane further from wings-level flight. If the upset rolls the airplane to the left, a negative sideslip $(-\beta)$ develops, producing a negative rolling moment, $-C_R$, as depicted by point C'. The negative rolling moment produces more left roll, tending to roll the airplane further from equilibrium. Hence, we conclude that aircraft 2 has negative lateral static stability with $C_{R_\beta} > 0$.

Aircraft 3 has neutral lateral static stability with $C_{R_\beta} = 0$. Displacement in yaw, from its trim condition, results in no rolling moment, and the aircraft remains at its displaced angle-of-sideslip.

The roll created by sideslip is influenced by several factors, including the wing dihedral, the wing sweep, and the position of the wing on the fuselage. The dominant contributor is the *wing dihedral*, hence C_{R_β} is sometimes referred to as *dihedral effect*. The wing dihedral is defined as the spanwise inclination of the wing with respect to the horizontal or y-axis. The dihedral angle, Γ, is positive when the wingtip is above the y-axis and negative when it is below the horizontal. Negative dihedral is also called *anhedral*. Examples of wing dihedral for different aircraft are shown in Figure 6.39.

The generation of the rolling moment, C_R, due to dihedral effect is explained with the aid of Figure 6.40. Consider an aircraft flying in steady wings-level flight at a velocity V_∞. If the aircraft enters a left sideslip (due to right roll), with a sideslip angle β, there is a relative wind seen by the side of the airplane, $V_{sideslip}$, given by

$$V_{sideslip} = V_\infty \sin \beta \cong V_\infty \beta \tag{6.131}$$

where $\sin \beta \cong \beta$ for small sideslip angles. For an airplane, with wing dihedral Γ, the sideways velocity has an upward component on the right wing and a downward component on the left wing, as shown in Figure 6.40. The magnitude of this normal component is given by

$$V_{normal} = V_\infty \beta \sin \Gamma \cong V_\infty \beta \Gamma \tag{6.132}$$

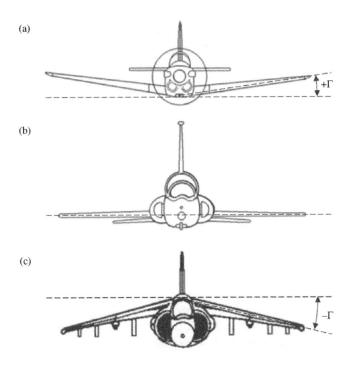

Figure 6.39 Examples of wing dihedral, (a) Beechcraft T-34C *Turbo Mentor* with positive dihedral, (b) Northrop T-38 *Talon* with zero dihedral, and (c) British Aerospace AV-8B *Harrier* with negative dihedral or anhedral.

where $\sin \Gamma \cong \Gamma$, assuming the dihedral angle is small. This normal velocity vector component adds to the freestream velocity vector as indicated in the lower portion of Figure 6.42. On the right wing, the upward normal velocity adds to the freestream velocity vector to form a local velocity, V_R, and increases the local angle-of-attack by an amount $\Delta \alpha_R$, given by

$$\Delta \alpha_R \cong \tan(\Delta \alpha_R) = \frac{V_\infty \beta \Gamma}{V_\infty} = \beta \Gamma \tag{6.133}$$

The left wing sees a local velocity, V_L, where the local angle-of-attack is decreased by an amount, $\Delta \alpha_L$, equal to $-\beta \Gamma$. The increase in the local angle-of-attack results in a lift increase on the right wing. The local angle-of-attack decrease on the left wing results in decreased lift on that wing. The difference in the lift between the right and left wings, $L_R - L_L$, results in a rolling moment, C_R, which rolls the airplane to the left (negative rolling moment), tending to restore wings-level flight. Hence, we see that positive wing dihedral contributes to the positive lateral static stability of the aircraft.

From [11], the dihedral effect, C_{R_β}, may be approximated as

$$C_{R_\beta} = \frac{a}{6} \left(\frac{1 + 2\lambda}{1 + \lambda} \right) \Gamma \tag{6.134}$$

where a is the wing lift curve slope and λ is the wing taper ratio, given by Equation (3.269). This equation loses accuracy with decreasing wing aspect ratio and should not be used for low aspect

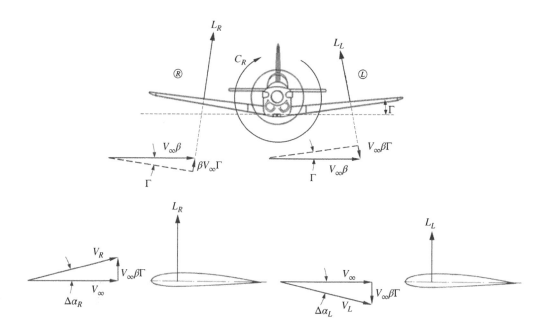

Figure 6.40   Rolling moment due to dihedral effect.

ratio wings. The dihedral effect is directly proportional to the wing lift curve slope and the wing dihedral angle. For a straight wing ($\lambda = 1$), the dihedral effect is simply

$$C_{R_{\beta},straight\ wing} = \frac{1}{4}a\Gamma \tag{6.135}$$

Wing sweep can also be a major contributor to lateral static stability. Consider the swept wing at a positive sideslip angle, as shown in Figure 6.41. We assume that the lift of the swept wing is a function of the velocity normal to the wing. Due to the sideslip, the windward (right) side of the swept wing sees a larger normal velocity than the leeward (left) side. Therefore, the lift is greater on the windward side and less on the leeward side of the wing. The difference in lift creates a moment rolling the wing to the left, countering the right roll, which produced the sideslip. Hence, wing sweep adds to the positive lateral static stability of an aircraft. For highly swept wings, the contribution to lateral stability may become excessive, making the aircraft too stable. In these cases, anhedral may be added to the wing to make it less stable.

The effect of the wing placement on the lateral static stability is shown for an aircraft with a sideslip flow, in Figure 6.42. On the windward side of the fuselage, the flow turns up over the top of the fuselage and turns down at the bottom. On the leeward side of the fuselage, the flow turns down at the top and turns up at the bottom. These local flow direction changes around the fuselage, due to the sideslip, affect the angle-of-attack seen by the wing sections near the fuselage. For a high-mounted wing, the local wing angle-of-attack is increased on the windward side of the fuselage and decreased on the leeward side. Thus, the windward side of the wing experiences an increase in lift and the leeward side a decrease in lift. The resulting difference in lift produces a restoring moment, which tends to roll the airplane to a wings-level attitude. The opposite is true for a low-mounted wing, where the local wing angle-of-attack is decreased on the windward side of the fuselage and increased on the leeward side. This results in a lift difference that produces a destabilizing rolling moment. Thus, a high-mounted wing provides a positive contribution to the

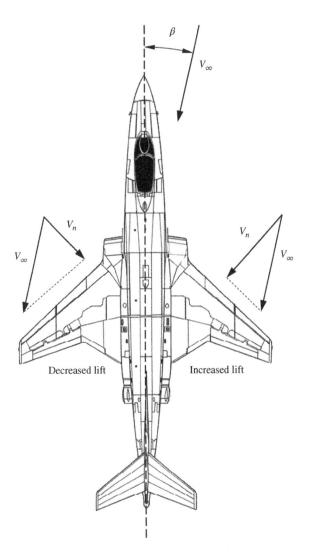

Figure 6.41   Effect of wing sweep on lateral static stability.

lateral static stability, while a low-mounted wing provides a negative contribution. Because of this, a low-wing airplane requires more stabilizing dihedral than a high-wing airplane.

6.7.4   Roll Control

Roll control is typically accomplished using ailerons or spoilers. These have been described previously, with the sign convention given in Section 6.4 and the aileron deflection angle being defined by Equation (6.11). Both types of roll control devices work by modifying the spanwise lift distribution of the wing, which creates a rolling moment. Differential deflection of the horizontal stabilizer is also used in most modern fighter aircraft to generate rolling moments.

When an airplane is rolled to the right using ailerons, the aileron on the left wing deflects downward and the aileron on the right wing deflects upward. The downward deflected aileron increases

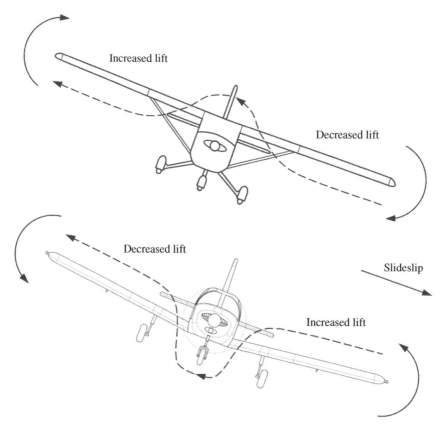

Figure 6.42   Effect of wing location on lateral static stability, high wing is stabilizing, low wing is destabilizing.

the lift of the left wing, while the downward deflected aileron decreases the lift of the right wing. This lift difference produces a moment, rolling the airplane to the right. The difference in lift also results in a drag difference, with the higher lift left wing producing more induced drag than the right wing. This drag difference results in a yawing moment, turning the airplane nose to the left. This left turn is opposite in direction to the intended roll to the right, hence it is termed *adverse yaw*. Typically, the rudder control is used to correct for the adverse yaw, but ailerons can be designed to minimize the adverse yaw. Spoilers do not suffer from adverse yaw, since they operate by decreasing lift and increasing drag on the side of the wing where the spoiler is raised, in the direction of the turn. Since the yaw due to drag, for a spoiler, is in the direction of the turn, it is termed *proverse yaw*.

Aileron roll control is fundamentally different from pitch control using the elevator or directional control using the rudder. The elevator and rudder are *displacement controls*: a constant deflection of the control produces a constant angular displacement of the aircraft. For instance, if the elevator is set at a constant deflection angle, the airplane's nose pitches up to a constant pitch attitude. If the rudder is set at a constant deflection angle, the aircraft's nose yaws to a constant yaw attitude. In contrast, the ailerons are *rate controls*: a constant aileron deflection results in a constant *rate of roll* rather than a constant angular displacement in roll or bank angle. Thus, if the ailerons are set to a constant deflection angle, the aircraft rolls at a constant rate.

Two control derivatives associated with roll control are the *aileron control power*, defined as

$$C_{R_{\delta_a}} = \frac{\partial C_R}{\partial \delta_a} \qquad (6.136)$$

and the *yawing moment due to aileron deflection*, defined as

$$C_{N_{\delta_a}} = \frac{\partial C_N}{\partial \delta_a} \qquad (6.137)$$

As discussed earlier, the yaw due to aileron deflection may be adverse or proverse, depending on the type of roll control surface used. The aileron control power is the change in the rolling moment due to aileron deflection. The higher the aileron control power, the larger the rolling moment that can be produced for a given aileron deflection. The aileron control power is a function of the size, location, and amount of deflection of the ailerons. In general, higher aileron control power can be obtained with a larger chord or span of the ailerons, relative to the wing. The spanwise location of the ailerons determines the rolling moment arm, such that ailerons located further outboard on the wing produce a larger $C_{R_{\delta_a}}$. Larger aileron deflection angles can also increase the aileron control power, but aileron deflections greater than about 20° may cause the surface to aerodynamically stall, decreasing the aileron power.

6.7.5 FTT: Lateral-Directional Static Stability

In this section, we discuss the flight test techniques used to evaluate the lateral-directional static stability of an aircraft. We are focused on the rolling and yawing motions of the aircraft and the aircraft's tendency to return to equilibrium after being disturbed from its trim condition. For this FTT, you will fly a unique aircraft with no wings and no engine, the NASA M2-F1 lifting body, shown in Figure 6.43. You will apply the lateral-directional static stability FTTs to the subsonic, gliding flight of the M2-F1. The lateral-directional static stability characteristics of this unconventional configuration are also probably not conventional, which you will determine in your test flight.

After the *Apollo* moon landings, NASA and the US space industry were investigating concepts for the next generation of US spacecraft. All of the US manned spacecraft flown to date, in the *Mercury*, *Gemini*, and *Apollo* programs, had been capsule vehicles, which launched vertically using a rocket booster and landed vertically using a parachute descent. There was much interest in developing a new, reusable spacecraft, which still launched vertically using a rocket, but landed horizontally like an airplane. The configuration that was eventually selected was the Space Shuttle Orbiter with a low aspect ratio, highly swept, double-delta wing (see Figure 1.79). However, during the 1960s, an alternate innovative wingless vehicle configuration was explored, called the *lifting body*. Even though the lifting body configuration was not selected, the data obtained from extensive analyses, ground tests, and flight tests of the lifting body shape was still useful in the development of the Space Shuttle Orbiter. Recently, there has been renewed interest in lifting body shapes as a reusable space vehicle.

Conceived by engineers at NASA in the mid-1950s, the lifting body concept was of a wingless, blunt-shaped body, which was well suited for the high-temperature, hypersonic flight associated with entry from earth orbit. Without wings, the lifting body generates aerodynamic lift from its body shape only; hence, they tended to have low lift-to-drag ratios. However, this lift-to-drag ratio was enough to give the lifting body a much larger landing footprint and cross range than a ballistic capsule. Extensive flight tests were performed on a variety of different lifting body vehicles with different shapes, seeking to understand the aerodynamics, stability, and control of these unconventionally shaped vehicles. These tests were conducted throughout the lifting body flight envelope, from subsonic to hypersonic speeds.

Figure 6.43   NASA M2-F1 "flying bathtub" lifting body being towed to altitude for a test flight. (Source: *NASA.*)

The first flight of the M2-F1 occurred on 16 August 1963. By the end of the flight test program in August 1966, the M2-F1 had successfully completed 77 gliding test flights from altitudes as high as 12,000 ft (3700 m). The success of the lightweight M2-F1 subsonic flights led to rocket-powered, supersonic flight tests of the heavyweight, Northrop-built M2-F2 lifting body and other lifting body vehicles, including the Northrop HL-10 and US Air Force X-24.

The M2-F1 was the first manned lifting body, designed to explore the subsonic glide characteristics of the lifting body shape. The M2-F1 was wingless, with a blunt, highly rounded underside, a flat upper surface, and a blunt nose. The body was a blunt 13° half-cone with a tapered afterbody. It had short, twin vertical fins with movable rudder surfaces, all-moving horizontal elevons extending outward from each vertical, and a large, trailing edge flap at the back of the body. The elevons and trailing edge flap were deflected up and down for pitch control; roll control was achieved by differential deflection of the elevons, and directional control by deflection of the rudders.

The vehicle had an internal welded-steel tube fuselage structure, covered with a 3/32″ (2.4 mm) thick mahogany plywood shell. The vertical fins, rudders, and stabilators were of aluminum sheet construction, while the trailing edge elevator flap was made of aluminum tubing, covered by fabric. The M2-F1 had fixed, tricycle landing gear, modified from a Cessna general aviation airplane. The pilot sat in a small cockpit in the middle of the vehicle, underneath a modified sailplane canopy of molded Plexiglas. Additional Plexiglas windows were placed in the nose and sides of the vehicle to provide enhanced cockpit visibility during landing. With its resemblance to a bathtub, the M2-F1 was referred to as the "flying bathtub". A three-view drawing of the M2-F1 is shown in Figure 6.44 and selected specifications are given in Table 6.5.

Prior to the first flight of the manned M2-F1 lifting body, test data was obtained from various sources. The full-scale M2-F1 lifting body was tested in the NASA Ames 40 × 80 ft wind tunnel, collecting subsonic data up to 85 knots (98 mph, 157 km/h) and for angles-of-attack from zero to 22°. To be precise, these were tests of the manned vehicle, since a pilot sat in the M2-F1 to move the flight controls during the wind tunnel runs. Sub-scale flight tests of the M2-F1 were performed by dropping a radio-controlled, scale model of the vehicle from a radio-controlled mothership, carrier model airplane.

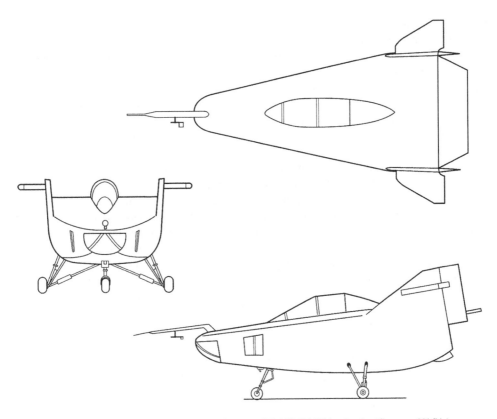

Figure 6.44 Three-view drawing of the NASA M2-F1 lifting body. (Source: *NASA.*)

Table 6.5 Selected specifications of the M2-F1 lifting body.

Item	Specification
Primary function	Wingless lifting body flight research
Manufacturer	Briegleb Glider Company, El Mirage, California
First flight	16 August 1963
Crew	1 pilot
Powerplant	Unpowered
Empty weight	1000 lb (454 kg)
Maximum takeoff weight	1250 lb (567 kg)
Length	20 ft (6.1 m)
Height	9 ft 6 in (2.89 m)
Span of lifting body	14 ft 2 in (4.32 m)
Planform area	139 ft^2 (12.9 m^2)
"Wing" loading	9 lb/ft^2 (43.9 kg/m^2)
Maximum speed	130 knots (150 mph, 240 km/h)

Since the M2-F1 is unpowered, it has to be towed into the air, similar to a sailplane or glider. For your first flight, the M2-F1 will be towed behind a ground vehicle. You climb into the cockpit of the lifting body and strap yourself into the seat. The cockpit is sparse, with just the essential flight instruments. There are conventional flight controls, a center stick that controls pitch and roll, and rudder pedals for yaw control.

One end of a towline is connected to the M2-F1 and the other end is connected to a "souped up" 1963 Pontiac convertible. The Pontiac has a modified engine and racing slick tires to enable it to tow the M2-F1 to as high a speed as possible. The M2-F1 and the Pontiac are lined up at one end of a long dry lakebed, providing several miles of unobstructed flat surface. There are two people sitting in the Pontiac, the driver in the left front seat and an observer, sitting in a backwards-facing seat, so that he can watch you in the M2-F1. The driver starts up the Pontiac and moves forward to take the slack out of the towline. The observer signals to you and you signal back that you are ready to go. The driver mashes down the accelerator of the Pontiac and roars forward, dragging you behind it.

At about 75 knots (86 mph, 139 km/h), the M2-F1 lifts off the lakebed and is airborne. The tow car continues to accelerate. Looking at your airspeed indicator, you reach an airspeed of 95 knots (109 mph, 176 km/h) and climb to an altitude of about 20 ft (6.1 m). You then release the towline and are gliding in free flight. Your glide lasts about 20 s before you touch down on the dry lakebed. You fly several more car tow flights to gain more flight experience. When the day's flying is done, the car-tow glide flights have provided you with valuable experience about the flying and landing characteristics of the M2-F1 vehicle, as they did for the earlier NACA pilots.

The following week, the M2-F1 is repositioned at the edge of the lakebed for another tow flight, but this time the vehicle will be towed to a much higher altitude by a Doulas C-47 *Sky-train* multi-engine transport airplane (the military variant of the Douglas DC-3 *Dakota*). For these air-launched flights, the M2-F1 is equipped with a modified Cessna T-37 ejection seat. The vehicle also incorporates a small 180 lb (800 N) thrust, solid fuel, landing-assist rocket motor in the tail. Due to the vehicle's low lift-to-drag ratio and poor cockpit visibility during the landing flare, the landing-assist rocket is installed to provide additional maneuvering time, if needed. The motor fires for approximately 11 s, increasing the vehicle lift-to-drag ratio from 2.8 to 4.5. The landing-assist rocket can be fired during the landing flare to extend the flight for those 11 seconds, if you need it.

You climb into the M2-F1 once again and strap into the ejection seat. The 1000 ft (300 m) long towline is connected between the M2-F1 and the C-47. The C-47 starts its takeoff roll on the dry lakebed with you in tow. Once airborne, you maneuver to keep the M2-F1 out of the turbulent wake of the much larger C-47 transport. You find that the best location is a high tow position, approximately 20 ft (6 m) above the C-47. This also provides good forward visibility for you, by looking through the Plexiglas nose of the M2-F1. Your tow speed is about 87 knots (100 mph, 161 km/h) as you continue your climb above the desert landscape. The first thing that you notice about the lateral flying qualities of the M2-F1 is that it is difficult to maintain perfectly wings-level flight. Later analysis shows that this is due to the flexibility of the flight control system, which results in small differential deflections of the elevons. You also notice that there is a large dihedral effect: when the vehicle is upset by a little bit of wind shear, or turbulence, from the C-47, causing even the slightest sideslip, the M2-F1 responds with a rapid roll rate.

At an altitude of 12,000 ft (3700 m), you pull the towline release and you are in free flight. The dihedral effect that you noticed while being towed is worse in free flight. The towline provided some directional stability, but that is now gone. The vehicle rolls to some moderate bank angles before you can counteract the motion. You are at 95 knots (109 mph, 176 km/h), with a significant descent rate of 3600 ft/min (1100 m/min). At this large descent rate, you do not have much time to complete your test points. Normally, the test point set up requires a trim shot in steady, level flight, but in your case you attempt to maintain a constant descent rate for the trim condition.

The first flight test technique that you use is a *rudder pulse* or *singlet*, a short, step-like rudder input in one direction only. You step on the left rudder, yawing the nose of the M2-F1 to the left, then step on the right rudder, centering the rudder pedals. You look out of the left side of the cockpit to observe the vehicle response to your input. The rolling motion appears to be much greater than the yaw motion, such that the ϕ/β ratio is large (see Section 6.9.3). The rudder deflection, δ_r, due to the rudder pulse is shown in Figure 6.45. The maneuver appears to start with 1° of positive sideslip, β, present. After the rudder pulse, the sideslip angle increases then decreases, overshooting the starting sideslip angle, then increasing again, appearing to oscillate about the starting condition. The rudder input results in a significant left (negative) roll rate, p, which is much greater than the yaw rate, r, which confirms your observation of a large ϕ/β ratio.

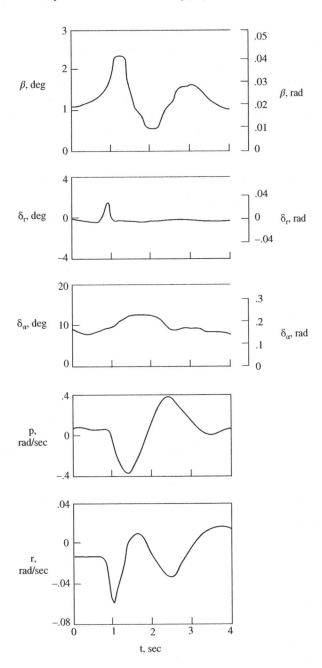

Figure 6.45 Time history plot from rudder pulse in M2-F1 lifting body. (Source: *Smith, NASA TN D-3022, 1965*, [15].)

You set up for the next lateral-directional FTT. You are at 95 knots, with an angle-of-attack of 2°. Keeping the rudders centered and fixed, you push the control stick fully over to the left, performing an *aileron roll*. The horizon rotates around you at a lazy rate, with a slow rate of roll. Your left roll input causes significant adverse yaw, with the vehicle nose moving to the right. Due to the strong dihedral effect, the right yaw results in right roll, which acts to counter your left roll and results in a sluggish roll rate.

Next, you set up to perform a *steady-heading sideslip* maneuver, where you will yaw the vehicle while maintaining a wings-level attitude by applying lateral stick inputs, as required. There are two steady-heading sideslip techniques that you can use, the *stabilized method* and the *slowly varying method*. For the stabilized method, you stabilize the vehicle in three to four sideslip increments, up to the maximum sideslip angle that is obtainable. For the slowly varying method, the rudder input is applied continuously, at a slow, steady, and smooth rate. Again, lateral stick is used as required to keep the wings level. You will start with small rudder inputs, since you do not know how the vehicle will respond. A large rudder input could result in a vertical fin stall and a serious yaw or roll departure.

In setting up for the steady-heading sideslip, you pull back on the control stick to slow the M2-F1 to 80 knots (92 mph, 148 km/h) with an angle-of-attack of 12°. Using the stabilized method, you apply a small left rudder input of about a quarter of the full rudder travel, yawing the nose to the left. The dihedral effect causes the M2-F1 to roll to the left. You counter the left roll with the application of right aileron. You repeat the steady-heading sideslip in the opposite direction and obtain similar results. Your small rudder input requires a large aileron input to maintain the steady-heading sideslip.

The lateral static stability, $C_{l,\beta}$, (recall that we are using $C_{R,\beta}$ for this term) and the directional static stability, $C_{n,\beta}$, of the M2-F1, from the wind tunnel tests and the flight test, are plotted versus angle-of-attack in Figure 6.46. The flight data confirms the wind tunnel prediction of positive directional static stability (positive $C_{n,\beta}$) and positive lateral static stability (negative $C_{l,\beta}$). The flight data indicates that the M2-F1 may possess greater lateral stability at higher angles-of-attack, than predicted by the wind tunnel. The flight data also suggests that the M2-F1 directional stability is greater than predicted by the wind tunnel, at all angles-of-attack.

After the test points are done, it is time to set up for landing. At 1000 ft (300 m) above the ground, you lower the nose of the M2-F1, increasing your speed to 150 mph (240 km/h). At 200 ft (61 m), you pull back on the control stick to arrest the 20° dive angle and start the landing flare. You feel like you are flaring a bit too high, so you fire the small tail rocket to buy a little more time to straighten things out. You do not feel any significant changes to the handling qualities of the M2-F1 with the rocket firing. You adjust your pitch attitude, reduce your vertical velocity, and let the M2-F1 settle back onto the lakebed for a nice landing. Your free flight lasted only two minutes, but you learned a lot about lateral-directional flight test techniques in that short time.

6.8   Summary of Static Stability and Control Derivatives

The various static stability derivatives and control powers, discussed in the previous sections, are summarized in Table 6.6. The lift curve slope, C_{L_α}, is neither of these, but is included in the table for completeness of important parameters that have been discussed. A comprehensive collection of prediction methods to calculate various aerodynamic stability and control derivatives can be found in the *US Air Force Stability and Control DATCOM* [9].

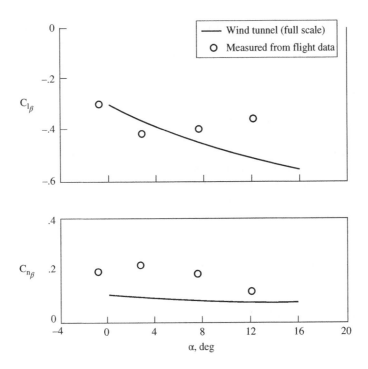

Figure 6.46   Lateral and directional static stability results from M2-F1 flight. (Source: *Smith, NASA TN D-3022, 1965*, [15].)

Table 6.6   Summary of static stability and control derivatives.

Parameter	Symbol	Comment
Lift curve slope	C_{L_α}	
Longitudinal static stability (pitch stiffness)	C_{M_α}	Slope is negative for positive stability
Directional static stability (weathercock stability)	C_{N_β}	Slope is positive for positive stability
Lateral static stability (dihedral effect)	C_{R_β}	Slope is negative for positive stability
Elevator effectiveness	$C_{L_{\delta_e}}$	
Elevator control power	$C_{M_{\delta_e}}$	
Rudder control power	$C_{N_{\delta_r}}$	
Aileron control power	$C_{R_{\delta_a}}$	
Yawing moment due to aileron deflection	$C_{N_{\delta_a}}$	

6.9   Dynamic Stability

We now examine the dynamic stability, where we are concerned with the motion of the aircraft over time, after the initial disturbance from equilibrium. Static stability dealt with the *initial* response of the aircraft after being disturbed from equilibrium. The forces and moments acting on the aircraft are not in equilibrium during the dynamic motion.

Dynamic stability deals with the *time history* and *final tendency* to return to an equilibrium state. As was shown in Figure 6.2, the dynamic stability may be positive, negative, or neutral with either non-oscillatory or oscillatory motion over time. Usually, the time history of the dynamic response deals with the amplitude of the displacement, such as the altitude, with time. The degree of dynamic stability is usually quantified in terms of the time for the amplitude to decrease by half for a converging oscillation, called the *time to half*, or the time for the amplitude to double for a diverging oscillation, called the *time to double*.

Similar to our study of static stability, we separate or decouple the aircraft dynamic motion into longitudinal and lateral-direction motions. Longitudinal dynamic motion is symmetric, wings-level motion of the aircraft center of gravity in the *x–z* or vertical plane. Longitudinal dynamic stability is concerned with the time history of the motion after the aircraft has been disturbed, from its equilibrium or trim condition, in airspeed or angle-of-attack, which may be caused by turbulence, a gust upset, or a control input. Lateral-directional motion is asymmetric with rolling, yawing, and sideslipping motions. Lateral-directional dynamic stability is concerned with the time history of the motion after the aircraft has been disturbed, from its equilibrium or trim condition, in yaw or roll, which may be caused by turbulence, a gust upset, or a control input.

There are five classic modes of dynamic motion, two longitudinal modes and three lateral-directional modes. The longitudinal modes are the *long period* or *phugoid* and the *short period*. The lateral-directional modes are the *Dutch roll*, the *spiral mode*, and the *roll mode*.

6.9.1 Long Period or Phugoid

The *long period* or *phugoid* [2] is a longitudinal dynamic motion characterized by an alternating climbing and descending motion, as shown in Figure 6.47. As its name implies, the motion has a long period, typically about 30–90 s, where the period, *T*, is defined as the time required to complete one cycle of the oscillating motion. The motion has a low frequency, ω, the frequency being inversely proportional to the period. As depicted in Figure 6.47, there is a slow oscillation of airspeed and altitude about an equilibrium or trim airspeed and altitude, as the aircraft climbs and descends. This exchange of altitude and airspeed may be considered a continual exchange of potential and kinetic energies about an equilibrium energy point. The airspeed decreases to below the trim speed in the climb and increases to above the trim speed in the dive, reaching a minimum speed at the top of the climb and a maximum speed at the bottom of the dive. The aircraft pitch attitude continually changes, while the angle-of-attack remains nearly constant. (Recall that the angle-of-attack is the angle between the velocity vector and the aircraft reference line, usually the wing chord line, and the pitch angle is the angle between the horizon and the aircraft reference line.) The oscillation is lightly damped or may be slightly divergent, but it is usually controllable and correctible, even if it is unstable, due to its long period and low frequency.

Starting at the top of a cycle, indicated as point *A* in Figure 6.47, the airplane is at its maximum altitude and minimum airspeed, with the airplane pitch attitude near level or at the attitude of the original trim condition. The loss of airspeed results in a loss of lift, since the angle-of-attack remains constant. The lift is less than the weight, so that the airplane noses over, increasing the (negative) pitch attitude, and starts to descend. The airspeed increases in the dive, increasing the lift, but the drag also increases. The drag acts to *dampen* the motion, by decreasing the airspeed. The aircraft overshoots and dives through the original trim airspeed and altitude, point *B*. Increasing the aerodynamic drag of the aircraft adds damping to the phugoid. At the bottom of the cycle, indicated as point *C* in Figure 6.47, the airspeed is at a maximum and the altitude at a minimum, with the airplane attitude again near level pitch attitude or at the original trim condition pitch

---

[2] The word phugoid was first used by English engineer Frederick W. Lanchester (1868–1946), in his 1907 book *Aerodynamics*, to describe the long period motion. Lanchester seems to have misinterpreted the word to mean "to fly", as in the flight of a bird or an airplane, when it actually derives from the Greek word "to flee", as in flight (or running away) from danger.

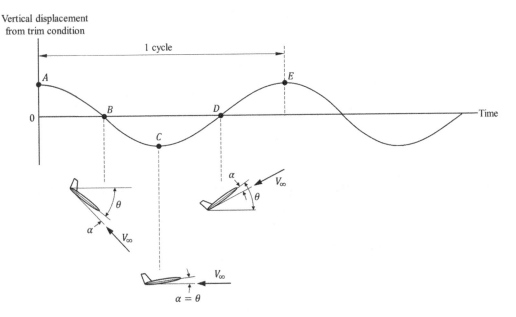

Figure 6.47   Long period or phugoid mode.

attitude. Due to the increase in airspeed over the trim airspeed, the lift is greater than the weight, and the airplane starts to climb in a nose up attitude. The airspeed continues to decrease as the airplane again climbs through the horizon, point D, until it reaches the minimum at the top of the oscillation, point E, and the cycle begins again. If the phugoid is damped, the overshoots of the trim point become smaller with time and the amplitude of the oscillations decrease. The oscillations grow in amplitude for a divergent phugoid and remain the same for a neutrally damped phugoid.

Consider the perspective of someone who is flying next to an aircraft, observing its phugoid motion. Assume that the observer is flying at the other aircraft's trim airspeed and altitude, so that this other aircraft is stationary, with respect to the observer, except for the excursions in two-dimensional space from the trim conditions. The observer sees the other aircraft move up and down, forward and backwards, depending on the changes in the airspeed and altitude due to the phugoid. If the phugoid has zero damping, the observer, at point O, sees the other aircraft motion as shown by the solid, elliptical curve in Figure 6.48. Since there is no damping, the motion is symmetric about the observer's position. The points labeled A, B, C, and D correspond to the positions in the phugoid shown in Figure 6.47. If the phugoid has positive damping, the elliptical curve spirals in, as the oscillations get smaller and smaller, as shown by the dashed, spiral curve in Figure 6.48.

Typically, the difference between the stick-fixed and stick-free phugoid is small. However, the characteristics of the phugoid may be dramatically different depending on whether the aircraft is in a clean, cruise configuration (flaps and landing gear retracted) or in a landing configuration (flaps and landing gear down). The difference in the phugoid is caused by the change in the aircraft lift and drag, due to the configuration changes.

The characteristics of the dynamic motion are often expressed in terms of the *damping ratio* and the *natural* or *undamped frequency* of the motion. The damping ratio, ζ, is defined as the period, T, divided by the time for the oscillation to subside. The natural or undamped frequency, ω_n, is the oscillation frequency of the motion if the damping were zero. This is the highest frequency possible for the motion. Damping always decreases the natural frequency.

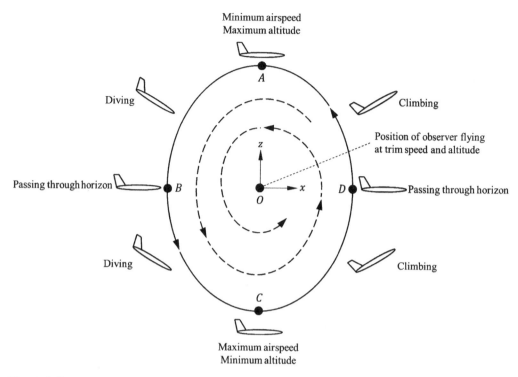

Figure 6.48 Phugoid motion as seen by observer flying at trim airspeed and altitude. Sold curve is phugoid with zero damping, dashed curve is with positive damping.

The damping ratio of the phugoid mode, ζ_{ph}, is inversely proportional to the square root of the lift-to-drag ratio, L/D, as given by

$$\zeta_{ph} = \frac{1}{\sqrt{2}} \frac{1}{(L/D)} \tag{6.138}$$

Thus, at higher L/D, such as in a cruise configuration, the phugoid damping is less than at lower L/D, as in the landing configuration. This is beneficial, as there is more damping when approaching to land, with the flaps and landing gear extended.

The undamped natural frequency of the phugoid, $\omega_{n,ph}$, is inversely proportional to the freestream, trim velocity, V_∞, given by

$$\omega_{n,ph} = \frac{\sqrt{2}g}{V_\infty} \tag{6.139}$$

where g is the acceleration due to gravity. The units of the natural frequency are radians per second.

The period of the undamped phugoid, T_{ph}, is

$$T_{ph} = \frac{2\pi}{\omega_{n,ph}} = \frac{2\pi}{\sqrt{2}} \frac{V_\infty}{g} \tag{6.140}$$

Thus, at higher airspeed, the phugoid frequency is lower and the period is longer, than at lower airspeed.

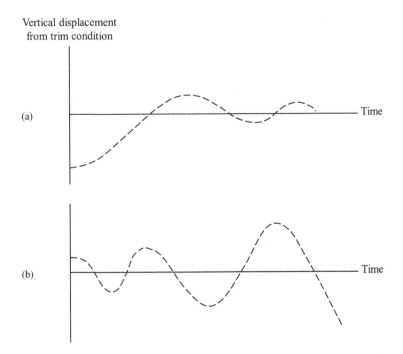

Vertical displacement
from trim condition

(a)   Time

(b)   Time

Figure 6.49   Short period mode, (a) stable (convergent) and (b) unstable (divergent).

The phugoid is also affected by the aircraft center of gravity (CG) location. As the CG moves aft, the phugoid period increases and the damping decreases. This is consistent with the fact that as the CG moves aft, the static margin decreases, with an attendant decrease in stability.

6.9.2   Short Period

The *short period* is a longitudinal dynamic motion that has a 1–3 s period and a high frequency. The airspeed, altitude, and flight path angle remain nearly constant, with rapid changes in the angle-of-attack and pitch attitude, as shown in Figure 6.49. The short period motion may be characterized as a rapid and abrupt porpoising motion, with the aircraft pitching up and down about its center of gravity. A stable short period mode is characterized by the time it takes for the amplitude of the oscillation to reduce to half, called the *time to half*. In contrast, when the short period mode is unstable, the *time to double* is the time it takes for the oscillation amplitude to increase twofold or to double. It is usually desirable for the short period mode to be heavily damped, where the angle-of-attack returns to its trim condition quickly with few overshoots.

The short period mode is more critical to safe flight than the phugoid. A low frequency, lightly damped or divergent short period mode can be difficult to control and easy to over control, leading to pilot induced oscillations (discussed in Section 6.10). These can quickly lead to severe oscillations that overstress the aircraft structure. It is also very important for many piloting tasks, including takeoff and landing. The short period motion is especially important for piloting tasks in military aircraft, such as air-to-air gunnery, air-to-ground bombing, aerial refueling, and formation flying.

The short period mode can be excited by a change in the angle-of-attack, such as due to turbulence, gust upset, or control input. The time to half, after an upset or control input, increases with decreasing damping. Typically, the horizontal tail provides most of the damping for the short period motion. Hence, there is reduced aerodynamic damping at high altitude, where the decrease in air density results in a decrease in the aerodynamic force and moment produced by the tail. In general for the short period mode, if the aircraft has low static stability, the period is increased, the frequency is decreased, and the time to half increases.

As with the phugoid, an aft shift of the center of gravity tends to increase the period and decrease the damping of the short period mode. As the aircraft center of gravity moves aft, the short period becomes less stable and may become unstable. Since it is desirable to have a further aft center of gravity to reduce trim drag at cruise, stability augmentation may be required to dampen the short period motion.

6.9.3   Dutch Roll

The *Dutch roll*[3] mode is a dynamic lateral-directional motion, where roll and yaw are coupled with the same frequency, but out of phase with each other. The motion is a coupled yaw and roll in one direction, followed by coupled yaw and roll in the opposite direction that overshoots the equilibrium position and continues to oscillate back and forth. The changes in sideslip and bank angle are relatively rapid. The motion is usually dynamically stable, but it is lightly damped, resulting in a motion that may be objectionable and uncomfortable. The period of the oscillation is about 3–15 s for a light aircraft and may be up to 60 s in a heavy aircraft.

The Dutch roll oscillation is shown in Figure 6.50. The motion starts with the right wing up, at its maximum upward bank angle, ϕ, and the aircraft yawing to the right (right wing moving aft), as shown in Figure 6.50a. As the aircraft is yawing, it is rolling towards a wings-level attitude. When the aircraft reaches the maximum yaw angle, β, the wings are rolling through a level attitude, as shown in Figure 6.50b. The aircraft starts to yaw to the left and continues to roll to the right (right wing moving forward and down). When the aircraft passes through zero yaw, the right wing has reached it maximum downward bank angle, ϕ, as shown in Figure 6.50c. The aircraft continues to yaw left and starts to roll to the left (right wing moving forward and up). When the aircraft reaches the maximum yaw angle, β, the wings are again rolling through a level attitude, as shown in Figure 6.50d. The aircraft then starts yawing to the right and continues to roll to the left (right wing moving aft and up), returning to the position shown in Figure 6.50a and the cycle begins again.

The right wingtip traces out an ellipse during the Dutch roll oscillation, as shown in Figure 6.51, where the letters correspond to the various times in one cycle of the oscillation shown in Figure 6.50. The ellipse defines a key descriptive parameter of the Dutch roll, the bank-to-sideslip ratio, or ϕ/β ratio. If the ϕ/β ratio is low, the Dutch roll is dominated by yawing motion and the major axis of the ellipse is horizontal. If the ϕ/β ratio is high, rolling motion dominates and the major axis of the ellipse is vertical.

The Dutch roll mode characteristics depend on the relative degree of static directional stability, $C_{N,\beta}$, as compared with the degree of static lateral stability or dihedral effect, $C_{R,\beta}$. When the dihedral effect is strong, in comparison with the static directional stability, the Dutch roll mode has weak damping and is more objectionable. The Dutch roll motion is heavily damped and not objectionable when the static directional stability is stronger than the directional stability.

As discussed in Section 6.7.3, wing sweep can make a significant contribution to the dihedral effect. The strength of the dihedral effect is a function of the lift being produced by the wing, hence, a function of the lift coefficient. At low speed, when the lift coefficient is large, the dihedral effect

---

[3] Although unclear, the term Dutch roll may have derived from the aircraft motion resembling that of a Dutch skater, who weaves from side-to-side as the skater's weight shifts from one foot to the other.

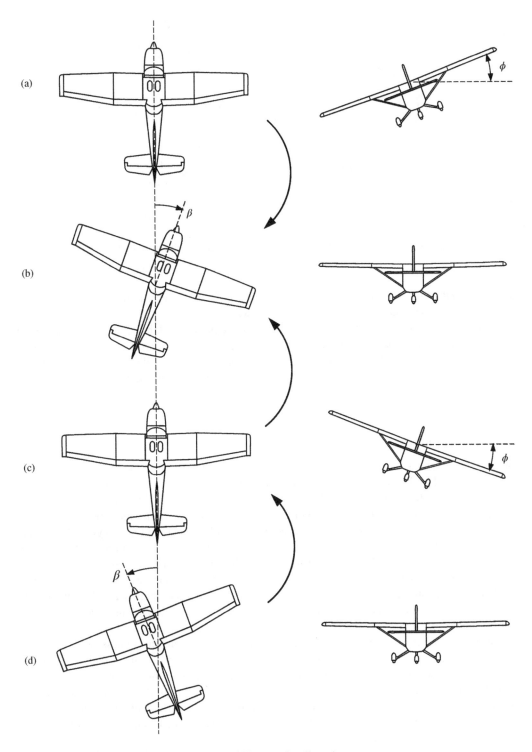

Figure 6.50 Dutch roll mode.

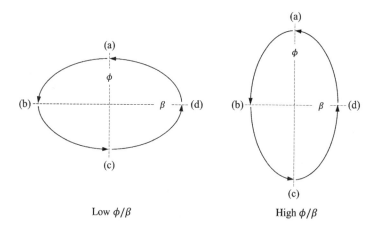

Low ϕ/β High ϕ/β

Figure 6.51   Dutch roll ϕ/β ratio (letters correspond to positions in Figure 6.50).

is strong and the Dutch roll motion is increased. The Dutch roll is decreased at high speed, when the lift coefficient is small and the dihedral effect is weak.

The Dutch roll mode is usually easily controlled by the pilot, but stability augmentation, in the form of a *yaw damper*, may be required, if the damping is low or the frequency is high. The yaw damper automatically applies rudder inputs to damp out the yaw oscillations. If the Dutch roll oscillations are allowed to become excessive, the structural limits of the tail may be exceeded, with catastrophic results.

6.9.4   Spiral Mode

The *spiral mode* is a slow, combined yawing and rolling motion of the aircraft, which may be stable, neutral, or unstable. The spiral mode may be excited by an upset in roll or yaw. If the spiral mode is stable, the aircraft returns to wings-level flight after the disturbance. If the aircraft remains in a turn with a constant bank angle, the spiral mode has neutral stability. If the spiral mode is unstable, the aircraft motion is a non-oscillatory descending turn with steepening bank angle, which leads to spiral divergence, as shown in Figure 6.52. After the initial upset from equilibrium, the changes in attitude occur relatively slowly, occurring over 15–30 s or more. The aircraft nose down pitch attitude and bank angle steepen, with a continuous increase in the airspeed and load factor and rapid decrease in altitude. Although the aircraft motion describes a descending, helical path, with decreasing radius, the motion is not a spin, as the wings are not stalled and the airspeed is increasing to high values.

The spiral mode is easily controllable when the pilot has situational awareness of the aircraft's attitude relative to the horizon. This is straightforward when the pilot can visually acquire the horizon. When the aircraft is in the clouds or at night, awareness of the aircraft attitude may be more difficult, especially without proper instrument training. Because of the increasing normal load factor, the pilot may not realize that the aircraft is turning. If the unusual attitude of the aircraft is not recognized and corrected, the aircraft will exceed its never-exceed airspeed and ultimate load factor, with catastrophic results.

The spiral mode characteristics depend on the relative degree of static directional stability, $C_{N,\beta}$, as compared with the degree of static lateral stability or dihedral effect, $C_{R,\beta}$. If the aircraft has strong directional stability, the aircraft nose tends to align itself with the wind after a disturbance,

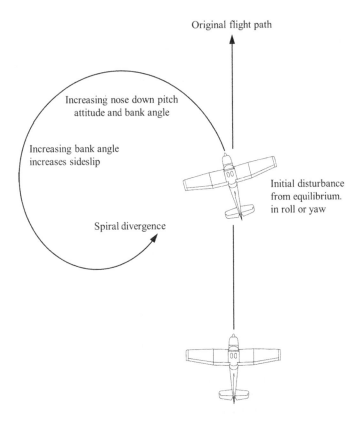

Figure 6.52   Spiral divergence.

while the weak dihedral effect does not roll the aircraft to wings-level flight; hence, the mode is unstable. The continually increasing bank angle causes the sideslip angle to increase, which leads to a tighter and tighter spiral turn. If the dihedral effect is strong, as compared with directional stability, the aircraft tends to roll wings level after an upset, hence it has a stable spiral mode. Since the Dutch roll mode increases with stronger dihedral effect, there is a tradeoff between having spiral stability and objectionable Dutch roll. Typically, an aircraft is designed to have the minimum necessary dihedral effect to reduce the Dutch roll, as a weak spiral mode is easily manageable, while an objectionable Dutch roll is not desired.

6.9.5   Roll Mode

The *roll mode* is the aircraft roll response to a roll upset or roll command. The aircraft motion is a roll acceleration, characterized by an exponentially decaying rise in the roll rate, which damps to a steady-state roll rate. The roll mode is usually stable at low and moderate angles-of-attack, but it can become unstable at high angles-of-attack. The time from an initial roll input to the final steady-state roll response is typically a few seconds. If the time is too long, the aircraft feels sluggish in roll, taking too long to reach a desired roll rate. If the time is too short, the aircraft feels too "loose" in roll, responding quickly to every roll upset, such as in turbulence.

The roll mode response to aileron deflection is shown in Figure 6.53. If the mode is pure rolling motion, the roll rate increases exponentially, until reaching a steady-state roll rate, as shown by the solid line in the figure. The dashed curves show the effects of the relative degree of static directional

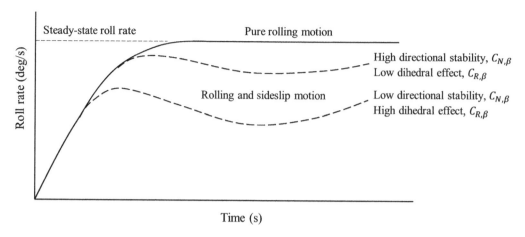

Figure 6.53 Roll response to aileron deflection. (Source: *Adapted from Hurt, US Navy NAVWEPS 00–80 T-80, 1965*, [10].)

stability, $C_{N,\beta}$, as compared with the static lateral stability or dihedral effect, $C_{R,\beta}$. As discussed in Section 6.7.4, aileron deflection produces adverse yaw, in a direction opposite to the commanded roll. The sideslip and yaw, due to adverse yaw, are resisted by the directional stability of the aircraft. If sideslip is produced by the adverse yaw, the dihedral effect results in a rolling moment that opposes the roll response and reduces the roll rate. Hence, if the aircraft has high directional stability and low dihedral effect, this provides better roll performance than if the directional stability is low and the dihedral effect is high, as shown in Figure 6.53.

The roll mode should not be confused with another dynamic motion of swept-wing aircraft at high angle-of-attack, known as *wing rock*. Wing rock is a sustaining roll oscillation, caused by the unsteady shedding of vortices from highly swept wings or slender fuselage forebodies at high angles-of-attack. This roll oscillation is an undesirable flying quality for aircraft that may need to accurately maneuver and point at high angles-of-attack.

6.9.6 FTT: Longitudinal Dynamic Stability

In this flight test technique, you will evaluate the longitudinal dynamic stability of an aircraft. By applying longitudinal control inputs, you will excite the long period or phugoid mode and the short period mode. After application of the control input, you will observe and measure the aircraft response over time. You will assess the dynamic motion for stick-fixed and stick-free conditions. Your longitudinal dynamic flight test will be flown in the Piper PA31 *Navajo* twin-engine airplane, shown in Figure 6.54.

The Piper PA31 *Navajo* is a cabin-class, twin-engine aircraft, designed and manufactured by Piper Aircraft Company, Vero Beach, Florida. The *Navajo* is used as a commuter transport, cargo carrier, and personal general aviation aircraft. The *Navajo* has a conventional configuration, with a low-mounted wing, aft-mounted horizontal tail, and retractable tricycle landing gear. Powered by two Lycoming TIO-540 turbocharged air-cooled horizontally opposed 6-cylinder piston engines, each with 310 hp (231 kW), the *Navajo* has a maximum cruising speed of 227 knots (261 mph, 420 km/h). The aircraft can accommodate two pilots in the cockpit and four passengers in the main cabin. The first flight of the Piper PA31 *Navajo* was on 30 September 1964. Selected specifications of the PA31 are given in Table 6.7.

Figure 6.54   Piper PA31 *Navajo* twin-engine, cabin-class aircraft. (Source: *Courtesy of the author.*)

Table 6.7   Selected specifications of the Piper PA31 *Navajo*.

Item	Specification
Primary function	General aviation and commercial utility aircraft
Manufacturer	Piper Aircraft, Vero Beach, Florida
First flight	30 September 1964
Crew	2 pilots + 4 passengers
Powerplant	2 × Lycoming TIO-540 turbocharged, 6 cylinder, piston engine
Engine power (each engine)	310 hp (231 kW)
Empty weight	3842 lb (1740 kg)
Maximum gross weight	6500 lb (2950 kg)
Length	32 ft 7 in (9.94 m)
Height	13 ft 0 in (3.96 m)
Wingspan	40 ft 8 in (12.40 m)
Wing area	229 ft^2 (21.3 m^2)
Airfoil	NACA 63A415 at root, NACA 63A212 at tip
Stall speed	63 kt (73 mph, 118 km/h) with full flaps
Maximum cruise speed	227 knots (261 mph, 420 km/h)
Service ceiling	26,300 ft (8015 m)

The PA31 *Navajo* is ready for your flight, with the weight near the maximum gross weight limit and an aft center of gravity. You take off and climb to an altitude of 9900 ft (3020 m), a safe altitude to perform the longitudinal dynamic flight test techniques. You first set up for the long period or phugoid evaluations. You set the power of the two engines to give you a low-cruise trim airspeed of 140 knots (161 mph, 259 km/h). You allow the engines and the flight condition to stabilize, as a solid trim shot is essential before you start the long period FTT maneuver, to obtain good quality data. To excite the long period dynamic motion, you apply a *singlet* of long duration. First, you smoothly pull back on the yoke, decreasing the airspeed by 10 knots (12 mph, 19 km/h) and increasing the altitude by about 100 ft (30 m). Then you rapidly reset the yoke position back to where it was at the trim airspeed, returning the elevator to its trim position. If the pitch control is not returned to its original trim position, the motion may have a slight climb or descent superimposed on the dynamic data. Finally, you release the controls, to obtain the stick-free response of the aircraft. You make very small roll inputs to keep the wings level, but you do not make any pitch inputs during the dynamic response.

The airspeed continues to decrease to about 132 knots (152 mph, 244 km/h) and the altitude increases to about 10,020 ft (3050 m), as shown by your flight data in Figure 6.55. The aircraft's nose then smoothly pitches down, with the airspeed increasing and the altitude decreasing. The flight path of the *Navajo* "bottoms out" at about 9750 ft (2970 m) and an airspeed of about 152 knots (175 mph, 282 km/h). The aircraft continues this oscillatory motion as shown, with relatively slow changes in airspeed and altitude. The oscillation is lightly damped, as the peaks and valleys in the airspeed and altitude are getting slightly smaller with each cycle. You have an angle-of-attack indicator in the cockpit and you observe that the angle-of-attack remains nearly constant during these oscillations.

Using a stopwatch, you measure the period of the phugoid oscillation. You wait for the short period oscillation to damp out, which typically occurs within a half cycle after the excitation input. You start timing when the vertical velocity indicator or VVI passes through zero, indicating a descent. When the bottom of the oscillation is reached, the VVI reverses direction and then passes through zero, indicating a climb. When the VVI indicator again passes through zero, indicating a descent once again, a complete phugoid cycle has been completed and you stop timing. You measure the period of the stick-free phugoid oscillation as 40 s. The amplitude of the oscillations is slowly getting smaller. The slow up and down, oscillating motion of the aircraft is a little uncomfortable, somewhat reminiscent of the uncomfortable motion of a boat slowly rocking up and down in ocean swells. After several minutes, you take control of the aircraft and stop the oscillating motion.

You now set up a 140 knot trim shot for a stick-fixed phugoid test point. After stabilizing at 140 knots and 9900 ft, you again apply a decreasing airspeed input to excite the dynamic motion. This time, when you return the yoke to its trim position, you keep the controls fixed to obtain the stick-fixed aircraft response. The airspeed and altitude oscillations are still lightly damped, but the amplitudes of the oscillations are much smaller than for the stick-free case, as shown in Figure 6.55. After several minutes, you again take control of the aircraft, stop the oscillation, and consider the flight data.

First, by examining the altitude versus time plots for the stick-free and stick-fixed cases it is evident that you did not return the pitch control exactly to the trim position, as there is a slight descent during the stick-free response and a slight climb during the stick-fixed response. Despite this, the flight data looks good enough for further analysis.

By measuring the time between peaks in the amplitude of the flight data in Figure 6.55, the periods of the stick-fixed phugoid, $T_{ph,fixed}$, and stick-free phugoid, $T_{ph,free}$, are approximately 42 s and 39 s, respectively. As expected, the period of the stick-fixed phugoid is greater than for the stick-free oscillation.

From the flight data, the stick-fixed and stick-free damped natural frequencies, $\omega_{d,ph,fixed}$ and $\omega_{d,ph,free}$, respectively, are calculated as

$$\omega_{d,ph,fixed} = \frac{2\pi}{T_{ph,fixed}} = \frac{2\pi}{42\,s} = 0.150\,rad/s \tag{6.141}$$

$$\omega_{d,ph,free} = \frac{2\pi}{T_{ph,free}} = \frac{2\pi}{39\,s} = 0.161\,rad/s \tag{6.142}$$

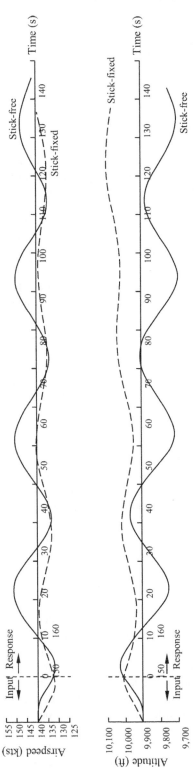

Figure 6.55 Stick-fixed and stick-free phugoid motion of the Piper PA31 *Navajo*. (Source: *Figure created by author based on data trends in* [6], *with permission of Matthew DiMaiolo.*)

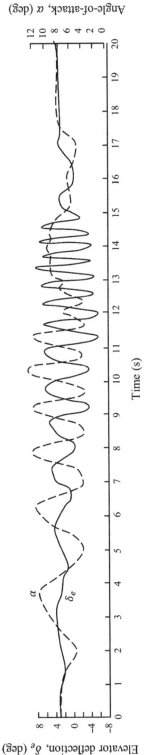

Figure 6.56 Frequency sweep to identify short period frequency of the Piper PA31 *Navajo*. (Source: *Figure created by author based on data trends in* [6], *with permission of Matthew DiMaiolo.*)

Using Equations (6.139) and (6.140), the undamped natural frequency, $\omega_{n,ph}$, and the undamped period, T_{ph}, of the phugoid, respectively, are calculated as

$$\omega_{n,ph} = \frac{\sqrt{2}g}{V_\infty} = \frac{\sqrt{2}(32.2\,\text{ft/s}^2)}{140\,\text{kt} \times \frac{6076\,\text{ft}}{3600\,\text{s}}} = 0.193\,\text{rad/s} \qquad (6.143)$$

$$T_{ph} = \frac{2\pi}{\omega_{n,ph}} = \frac{2\pi}{0.193\,\text{rad/s}} = 32.6\,\text{s} \qquad (6.144)$$

As expected, the predicted undamped frequency is higher than the predicted damped frequency, and the period of the undamped oscillation is shorter than for the damped oscillation.

Having completed the long period FTT, you are ready to move on to the short period FTTs. We focus on the qualitative aspects of the short period motion, as most of the quantitative aspects are beyond the scope of the text. You again trim the aircraft for an airspeed of 140 knots at an altitude of 9900 ft. The first maneuver that you will perform is called a *frequency sweep*, which provides you with the undamped natural frequency of the short period motion, the frequency that excites the short period motion. This is the *undamped* natural frequency because you will be driving or exciting the motion with pitch inputs. Once you have identified this natural frequency, you can use it to excite the short period in another FTT that you will perform.

You start by slowly moving the yoke back and forth, in a smooth continuous motion. This pitch control movement deflects the elevator up and down, which causes the Navajo's nose to pitch up and down. You move the yoke back and forth at a faster and faster rate. In essence, you are varying the frequency of this periodic input to cover a range of frequencies, searching for the natural frequency of the airplane. You must be careful to use only small inputs, as the input frequency approaches the natural frequency, to avoid overstressing the airframe. As you are pumping the yoke faster and faster, the aircraft nose is responding by moving up and down, relative to the horizon. This motion is not very comfortable as the rate increases. At some fast rate, you notice that the aircraft nose is not responding properly: the nose motion becomes *out of phase* with the pitch input, so that the nose is moving in the opposite direction of your pitch input. Soon, the nose is barely moving, despite the fact that you are moving the yoke back and forth at a very fast rate.

Figure 6.56 shows the frequency sweep that you just performed in the PA31. The elevator deflection, δ_e, (left vertical scale) and angle-of-attack, α, (right vertical scale) are plotted versus time. The frequency of the elevator deflection input increases with time. The angle-of-attack response of the aircraft remains in phase with the input, until about 11.5 s, when the amplitude of the aircraft response starts to diminish and then it stops responding at about 12.5 s. The undamped natural frequency of the short period occurs when the response reaches its maximum amplitude due to the input, which occurs at about 10.5 s. You will use this identified frequency to excite the short period motion in the next FTT that you will perform. In simpler terms, the frequency sweep has identified how fast you should move the yoke (the natural frequency) in applying the input to excite the short period.

You once again trim the aircraft at 140 knots and 9900 ft. You will use an input called a *doublet* to excite the short period, applied at the rate identified by the frequency sweep that you just completed. You push on the yoke, applying a nose down pitch rate, then pull the yoke back, applying a nose up pitch, at the same rate, and then push the yoke forward again, returning to the trim attitude and airspeed. You attempt to apply these pitch rates at the natural frequency rate of stick movement. If applied properly, the doublet excites the short period, but does not excite the long period, because the maneuver starts and ends at the trim attitude and airspeed. After the input, you can hold or release the yoke to obtain the stick-fixed or stick-free response, respectively.

Your doublet input is shown by the elevator deflection plot in Figure 6.57. The aircraft response is heavily damped and convergent for both the stick-fixed and stick-free cases. You watch the nose of

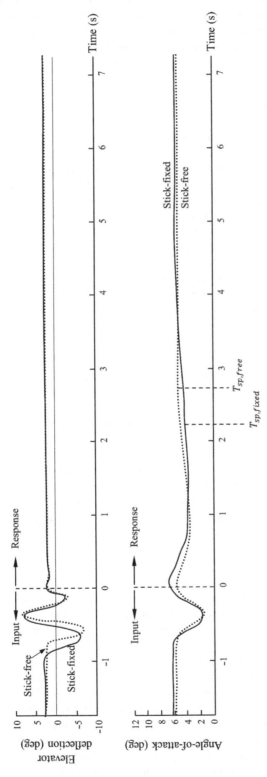

Figure 6.57 Stick-fixed and stick-free short period motion of the Piper PA31 *Navajo*. (Source: *Figure created by author based on data trends in* [6], *with permission of Matthew DiMaiolo.*)

the Navajo and see the rapid changes in the pitch attitude relative to the horizon. You can certainly feel the response in terms of the normal acceleration on your body. You glance at the airspeed and altitude indicators during the responses and see that they remain constant. However, you see that the angle-of-attack is changing rapidly during the responses. From the flight data, the periods of the stick-fixed and stick-free short period motion are about 2.8 s and 2.2 s, respectively.

6.10 Handling Qualities

This chapter has primarily discussed the stability and control characteristics of the aircraft, independent of any inputs from a human pilot or non-human controller, a topic we choose to define as *flying qualities*. This is distinguished from *handling qualities*, which concerns the dynamics, response, and control characteristics of the aircraft with inputs by the pilot or controller. Be mindful that in the literature these terms are sometimes used interchangeably.

Handling qualities seek to understand the response of the *total system*, composed of the aircraft and the pilot or controller. According to [5], "handling qualities refers to those qualities or characteristics of an aircraft that govern the ease and precision with which a pilot is able to perform the tasks required in support of an aircraft role." In some ways, handling qualities evaluations of "how an aircraft flies" may be viewed as subjective, since after all, there is usually human opinion involved. There are techniques to make this assessment as objective as possible.

An example of when a pilot's input may dramatically affect the aircraft flight characteristics is the *pilot induced oscillation* or *PIO*, a sustained or uncontrollable oscillation caused by the pilot's efforts to control the aircraft. The PIO occurs when the aircraft attitude, angular rate, or normal acceleration is 180° out of phase with the pilot's control inputs. This can occur when the characteristic dynamic motion of the aircraft, such as the short period mode, has the same timescale as the lag time in pilot response. As discussed earlier, the short period mode is about 1–3 s, which is about the same as the time it takes for a pilot to apply a control input in response to the short period oscillation. This may lead to the control input being applied out of phase with the oscillation, which can have a reinforcing effect, rather than a damping effect.

Consider an aircraft that is dynamically stable with controls fixed. When disturbed from equilibrium, the aircraft response may be oscillatory, but it returns to equilibrium over time. If we now add a pilot, who attempts to correct the aircraft oscillations with controls inputs that are out of phase with the oscillatory motion, the result may be an unstable, diverging oscillation. This has been seen many times in the past during the landing phase of flight, where the aircraft describes a diverging pitch oscillation due to out-of-phase pilot control inputs, which can be disastrous close to the ground.

In this short section, we discuss handling qualities, where human input is critical in evaluating the aircraft flight characteristics, related to stability and control. There is a broad spectrum of other areas, not related to stability and control, where the interaction of the human with the aerospace vehicle system is important to the safe and efficient operation of the system. This area of *human factors* encompasses human interaction with most things in the cockpit, including the displays, instruments, switches, windscreens or windows, and even the seats. Human factors also deal with *workload* of the aircraft operators, in performing their piloting or other flight related tasks.

The human pilot is the handling qualities "sensor" and handling qualities data is primarily the comments and ratings of the pilot. From this perspective, the data is reliant on human *perception* and *opinion*, which can lead to issues with human subjectivity. Two pilots who fly the same airplane may have vastly different perceptions and opinions about how it flies, dependent on factors such as their experience, skill level, or ability to adapt. Pilots, especially test pilots, tend to be excellent *compensators*, with the ability to make the poor flying characteristics of an aircraft acceptable. Another factor involves communication and terminology. Two pilots may use different adjectives and terms to describe the same flight characteristics, with different interpretations by the analyst,

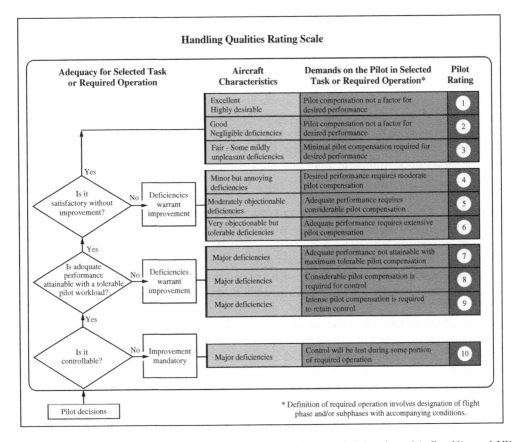

Figure 6.58 Cooper–Harper handling qualities rating scale. (Source: *P.F. Borchers, J.A. Franklin, and J.W. Fletcher, "Flight Research at NASA Ames," NASA SP-3300, 1998.*)

aircraft designer, or other pilots. Many different handling quality rating scales have been proposed over the years, but a generally accepted system is the *Cooper–Harper scale*, published in 1969 [5].

The Cooper–Harper rating scale assigns a numerical rating as an indication of the ease or difficulty of achieving the desired aircraft characteristics, as shown in Figure 6.58. The numerical scale ranges from 1 to 10, with 1 denoting the best aircraft handling characteristics and 10 the worst. The numerical rating is a notation for the aircraft characteristics description; it is not a numerical value with mathematical significance. For instance, an aircraft with a Cooper–Harper rating of 4 is not twice as "bad" as an aircraft with a rating of 2. Since the rating scale relies on pilot perception and opinion, it is still considered subjective. As seen in Figure 6.58, the scale takes the pilot opinion through a flow chart-like rating process, where the aircraft characteristics are based on the demands of the aircraft on the pilot in performing a selected task or required operation to a specific degree of precision.

6.10.1   FTT: Variable-Stability Aircraft

There are special types of aircraft that are designed to simulate the performance, stability, and control characteristics of other aircraft in the real flight environment. These *variable-stability aircraft* have been used in the research and development of many new aircraft designs and flight control systems. This concept has expanded to include aircraft that act as *in-flight simulators*, where pilots and engineers can experience the performance and flying qualities of a variety of different aircraft in the real flight environment.

A variable-stability aircraft may be defined as an aircraft that can simulate the stability and control characteristics of another aircraft, in the real flight environment. The aircraft that is being simulated may be a real aircraft type or it may by a fictitious aircraft, with stability and control characteristics of interest. Typically the variable-stability capability is enabled by a system that senses the aircraft state (airspeed, attitude, angle-of-attack, angle-of-sideslip, load factor, etc.) and moves the existing flight control surfaces, independently of the pilot inputs, to simulate the desired stability and control characteristics. However, these control surface motions are not communicated to the pilot's controls, so that the pilot flies the aircraft with the use of the conventional flight controls.

The existing flight control surfaces may be operated within their normal ranges of deflection and rates of travel or the variable-stability system may have the ability to command different ranges and rates. For instance, flaps, which normally only deflect trailing edge down, may be enabled to deflect upwards, producing a down force on the aircraft. Control surfaces may be made to deflect at higher rates to simulate desired accelerations and angular rates. Additional aerodynamic control surfaces may be installed on the aircraft, such as vertical side force generators on the wings, to enable the variable-stability system to generate additional aerodynamic forces.

The variable-stability system is typically controlled by analog or digital computers, which can be configured or programmed to provide the desired flight characteristics. For the variable-stability system to accurately represent a desired aircraft configuration or flight control system, a mathematical model must be prepared that represents the aircraft's aerodynamic, stability, and control characteristics. This is no small task, which must be completed prior to any flights.

Prior to the use of variable-stability aircraft, the geometry of an aircraft had to be physically altered to assess the impact on its flying qualities in flight. The development of the Ryan FR-1 *Fireball*,[4] a new carrier-borne fighter aircraft design for the US Navy in the 1940s, is an example of this approach. To evaluate the amount of wing dihedral to use in the final design of the aircraft, Ryan built three *Fireball* aircraft, each with a different amount of wing dihedral. Needless to say, this approach is not practical for most aircraft development programs. In the mid-1940s, this inspired NACA engineer William M. Kauffman to conceive of an aircraft where the stability and control characteristics could be varied with a stability augmentation system that used servomotors to drive the control surfaces, independent of pilot input. Kauffman's work led to the first aircraft with a variable-stability system, a NACA-modified Grumman F6F-3 *Hellcat*, shown in Figure 6.59. The effective wing dihedral was changed in the F6F by using a servomotor to deflect the ailerons in response to sensed sideslip. The pilot retained conventional control of the aircraft, but the flight controls were modified so that the pilot's control stick did not move when the variable-stability system moved the ailerons. The variable-stability system in the F6F later included the rudder and the aircraft was used extensively for lateral-directional flying qualities evaluations.

With the advent of high-performance, swept-wing jet aircraft, the NACA modified a North American F-86 *Sabre* and a North American F-100 *Super Sabre* into variable-stability platforms. The F-100 was the first aircraft to have three-axis (pitch, roll, and yaw) variable-stability capability. In the following years, many other variable-stability aircraft were developed by NACA (later

---

[4] The Ryan FR-1 *Fireball* was a single-place fighter design with a low-mounted wing, conventional empennage, and retractable, tricycle landing gear. The propulsion system of the *Fireball* was unconventional, with two types of engines, a radial piston engine and a turbojet engine. The *Fireball* was the US Navy's first jet-powered aircraft. The piston engine compensated for the early jet engine's poor acceleration capability from low to high power. The first flight of the FR-1 *Fireball* was on 25 June 1944, without its jet engine, which was installed soon afterward. Only 66 aircraft were built.

Figure 6.59 The first variable-stability aircraft, a NACA-modified Grumman F6F-3 *Hellcat*. (Source: *San Diego Air & Space Museum Archives, no known copyright restrictions.*)

to become NASA), the Cornell Aeronautical Laboratories (later to become Calspan), Princeton University, Boeing, and other organizations in the USA. Variable-stability aircraft have also been built in other countries, by the Royal Aircraft Establishment in England, the German Aerospace Center, and other organizations in France, Japan, and China. Many different types of aircraft have been modified to serve as variable-stability platforms, including single-engine military jets, single-engine general aviation airplanes, twin engine, piston-powered bombers, small business jets, and rotorcraft. NASA modified the Boeing CH-47 *Chinook* helicopter, shown in Figure 1.35, to serve as a variable-stability rotorcraft.

Another variable-stability aircraft was the Calspan Total In-flight Simulator (TIFS), shown in Figure 6.60, which was a highly modified Convair 131B *Samaritan* transport aircraft. The aircraft modifications included replacement of the piston engines with turboprop motors, installation of vertical side force generators on the wings (one can be seen on the far left of Figure 6.60), and addition of a reconfigurable nose section. The aircraft nose could house the radome and radar systems of another aircraft or it could be configured as a completely separate, simulation cockpit. The aircraft could be flown by pilots in the simulation cockpit, which was configured to represent the simulated aircraft, while safety pilots flew from the normal cockpit.

Variable-stability aircraft have played an important role in the development of a wide range of new aircraft and flight control system designs. This impressive list includes the North American X-15, the Lockheed F-104 *Starfighter*, the North American XB-70 *Valkyrie*, the Convair B-58 *Hustler*, the Lockheed A-12 (forerunner to the SR-71 *Blackbird*), the Lockheed F-117A *Nighthawk*, the Grumman X-29, and the Space Shuttle Orbiter. They have provided valuable in-flight training to pilots, allowing them to acquaint themselves with the flight characteristics of a new aircraft design, before flying the new aircraft for the first time.

Figure 6.60   The Calspan Total In-flight Simulator (TIFS), a highly modified Convair 131B *Samaritan* transport aircraft. (Source: *US Air Force.*)

6.11   FTT: First Flight

With this final flight test technique, we come full circle in our introduction to aerospace engineering, returning to where we started in Chapter 1, with the first flight of an aerospace vehicle. The successful completion of a first flight draws on much of the material that has been discussed up to now. Many of the first flights described in Chapter 1 did not benefit from the aerospace engineering knowledge and flight test lessons learned that have filled this book. The risks involved with many early first flights were high, sometimes leading to catastrophic results. The risk can never be reduced to zero, but with thorough preparation and prudent decision-making, the risk of a first flight can be reduced to an acceptable level.

The first flight may be of a completely new type of vehicle, such as the Northrop *Tacit Blue*, shown in Figure 6.61. This unusual aircraft was designed and built, in secrecy, by Northrop in the early 1980s as a demonstrator for the newly emerging stealth or low observable technology. The *Tacit Blue* first flew in February 1982 and successfully completed 135 test flights over a three-year period. The first flight may also be of an existing vehicle that has been modified, such as of the highly modified Gulfstream 550 business jet, shown in Figure 6.62. The aircraft's outer mold line has been highly modified to integrate radar equipment and other sensors. Internal modifications included changes to various systems, including the electrical and fuel systems. Whether a new vehicle or a modified-existing vehicle, many of the fundamental tenants associated with a first flight are applicable.

Prior to a first flight, significant engineering preparation is required. Extensive analyses and ground testing are completed to predict the aerodynamics, performance, flying qualities, systems performance, and structural integrity of the vehicle. Aerodynamic, stability, and control models are developed, which are used to predict how the vehicle will fly. These models are often applied to a piloted flight simulator, which is used for pilot training and further engineering analyses. Depending

Figure 6.61   The first flight may be of a new vehicle, such as the Northrop *Tacit Blue* technology demonstrator aircraft. (Source: *US Air Force.*)

Figure 6.62   The first flight may be of a highly modified vehicle, such as this Gulfstream 550 business jet modified for an airborne surveillance role. (Source: *User: Alert5, "RSAF G550-AEW" https://commons .wikimedia.org/wiki/File:RSAF_G550-AEW.jpg, CC-BY-SA-4.0. License at https://creativecommons.org/ licenses/by-sa/4.0/legalcode.*)

on the fidelity of the simulation, the pilot can obtain valuable and realistic experience with the flight characteristics of the aircraft, before it is flown for the first time. In addition, the simulator can be used to further reduce risk by evaluating uncertainties in the engineering predictions. The flight characteristics of the vehicle can be evaluated at the limits of the uncertainties in the aerodynamic, stability, or control parameters, to determine the sensitivities to these parameters. For example, if there is a large uncertainty in the longitudinal static stability of the vehicle, the pitch stiffness, $C_{M,\alpha}$, may be varied over its uncertainty range, in the simulator, to evaluate the effects on the flight characteristics. If it turns out that the flying qualities are acceptable at the limits of the uncertainty, then the uncertainty in the longitudinal stability prediction may be acceptable. On the other hand, if the flying qualities are very sensitive to changes in the parameter, then this provides focus for further work to increase the fidelity of the parameter prediction.

Another important area for pre-first flight predictions and testing concerns the vehicle structure. Typically, ground tests are performed to verify the static structural integrity of the vehicle. The vehicle structure, such as the wing, is loaded with weights, often using sand bags, to verify the soundness of the structure at the design limit loads. Sometimes the structure is loaded to failure to verify the ultimate limit loads. Dynamic structural analyses and testing is also performed, to evaluate phenomena such as flutter and other aeroelastic instabilities, involving coupling of aerodynamic, elastic (structural), and inertial forces on the vehicle. These types of instabilities are non-linear and may lead to catastrophic failure of the structure. We have not discussed these areas in this text, but they are a critical area of concern for a first flight.

For your first flight FTT, we will not focus on a particular aircraft; rather we assume that you will be flying a new or highly modified aircraft. Prior to the first flight, you have spent many hours in the flight simulator, becoming thoroughly familiar with the normal and emergency procedures for the aircraft and learning everything that you can about its flight characteristics. You have performed several taxi tests, where you have checked the steering system and other onboard systems, including the test instrumentation system. You have conducted high-speed taxi tests on the runway to test the high-speed ground handling and the braking system. Having completed all of these ground tasks, you are ready to fly the first flight.

After a thorough pre-flight briefing, you proceed out to your aircraft. It is still early in the morning, with the air still and the surface winds calm, as desired for a first flight. There will be another aircraft flying alongside you today, a *safety chase aircraft*, which can visually examine your aircraft in flight to ensure that everything is normal. The safety chase can also provide assistance to you, in the event of an emergency. In addition to providing a visual assessment of your aircraft, it can provide communications, navigation, and other guidance as you may require during an in-flight emergency. You start the engine of your aircraft and check that all systems are nominal. You check communications with the chase aircraft and with the ground control room, where there are engineers watching computer screens, filled with telemetered data from your aircraft. You taxi out to the end of the runway with your chase airplane in trail behind you. After obtaining takeoff clearances, your chase aircraft taxis onto the runway first and takes off. The chase circles back in the airport pattern to join up with you, in an airborne pick-up, when you take off. You taxi onto the runway, line up on the runway centerline, and look down the long 12,000 ft (3700 m) of wide concrete. It is no accident that this runway and airport have been selected for the first flight. The very long runway provides added safety if you have to abort the takeoff or if your landing roll out is very long. The airport has few obstructions nearby, such as building, towers, or trees, which could be impacted if you experience loss of engine power or other problem on climb out. The airport is also well equipped with trained emergency personnel and vehicles to render aid in an emergency.

You take a deep breath, advance the engine throttle, check your engine gauges again, and release the brakes. The aircraft starts it roll down the runway. Your engine gauges are still normal and your airspeed indicator "comes off the peg" from zero and your airspeed is "alive". The aircraft reaches rotation speed; you pull back on the control stick and are airborne. The aircraft accelerates to climb airspeed and everything looks nominal, as your chase aircraft comes aboard on your right wing. Even though you are flying an aircraft with retractable landing gear, you keep the landing gear extended for this first flight as an added measure of safety. Now that you have successfully made the first takeoff in this aircraft, you remember that a primary objective of this first flight is to land safely.

You climb up to 15,000 ft (4600 m), well below the 50,000 ft (15,000 m) ceiling of the aircraft. You want to be at a high enough altitude to provide time to handle a problem, but you do not want to be anywhere near the altitude limits of the aircraft. This generally applies to all of the aircraft limits. You do not want to fly the first flight near any of the limits of the flight envelope, including the altitude, speed, dynamic pressure, or other limits. The first flight should be in the "heart" of the flight envelope, far from any of the limits and well within the linear range of the engineering predictions, where the fidelity is highest. Even though your aircraft is capable of supersonic speed, your first flight will be at low subsonic speeds, far from the non-linear, transonic speed regime.

You level off at an altitude of 15,000 ft and trim the aircraft for steady level flight at a subsonic airspeed. This trim point provides flight data to help validate the aerodynamic predictions and models in the linear range of the aerodynamic coefficients. You perform several trim shots at different speeds, collecting more aerodynamic data in the linear range. The trim shots also provide longitudinal static stability data, as was demonstrated in the longitudinal static stability FTT. You perform several steady-heading sideslips to evaluate static directional stability and stabilized turns to assess lateral static stability, as was shown in the lateral-directional static stability FTT. To evaluate dynamic stability, you apply doublets as were used in the dynamic longitudinal stability FTT. In applying these various different inputs, you are careful to apply low amplitude, low to medium rate inputs, not inputs with full control deflections or maximum rates. You want to avoid large angles-of-attack, large angles-of-sideslip, and extreme attitudes. You also keep the normal load factor within a range of about 0.8–1.5 g maximum. Your safety chase aircraft observes all of your test points from a safe distance so as not to interfere with your maneuvers, but close enough to observe any anomalies with the aircraft.

Your next test points are at your planned approach-to-landing speed. You are following the standard buildup approach, starting at lower risk, higher airspeed test points, before proceeding to the higher risk, lower speed test points. You pull the power back and decelerate to your planned approach-to-landing speed. Again remembering your objective to land safely, you want to evaluate the flight characteristics at your landing approach speed, at a high altitude, before you are close to the ground during the actual landing. In this way, you can familiarize yourself with how the aircraft will fly during the landing approach and identify any potential problems. Your goal is again to validate the aerodynamic models and to evaluate handling qualities at your approach speed. If there are differences between the predictions and the flight test data, the models will be updated after the flight. Of course, significant differences may warrant adjustments to your landing plan, such as increasing your approach airspeed if the handling characteristics are undesirable at lower airspeeds.

You proceed through the various handling qualities evaluations at the approach speed and the aircraft characteristics are matching the predictions fairly well. After completing these maneuvers, you fly a practice approach to an imaginary runway at your altitude, followed by a practice *go-around*, a procedure that you will use to abort the landing, if required, and go around the landing pattern to set up another landing attempt. With this task completed, you start your descent back to the airport. You enter the airport landing pattern and fly the final approach to the runway. As you get closer to the ground, you notice that the motion and visual cues are quite different from those in the flight simulator, which is expected and manageable. You are grateful for that long, 12,000 ft of runway as you are not trying to "spot land" the aircraft at a specific point on the runway; rather you want to make the landing as smooth and steady as possible. As the aircraft enters ground effect, you feel it "float" down the runway until the aircraft tires softly chirp as they touch down on the runway surface. Your first flight of the aircraft is successfully completed.

References

1. Air Force Flight Test Center, *Flying Qualities Testing*, Edwards Air Force Base, California, 20 February 2002.
2. Abzug, M.J. and Larrabee, E.E., *Airplane Stability and Control: A History of the Technologies That Made Aviation Possible*, Cambridge University Press, Cambridge, United Kingdom, 2nd edition, 2005.
3. Anderson, J.D., Jr, *Introduction to Flight*, 4th edition, McGraw-Hill, Boston, Massachusetts, 2000.
4. Borchers, P.F., Franklin, J.A., and Fletcher, J.W., *Flight Research at Ames, NASA SP-3300*, US Government Printing Office, Washington, DC, 1998.
5. Cooper, G.E. and Harper, R.P., Jr, "The Use of Pilot Rating in the Evaluation of Aircraft Handling Qualities," NASA TN D-5153, 1969.
6. DiMaiolo, M.J., "Flight Testing the Piper PA-32 Saratoga and PA-31 Navajo," Master's Thesis, University of Tennessee, 2015.
7. Etkin, B. and Reid, L.D., *Dynamics of Flight: Stability and Control*, John Wiley & Sons, New York, 3rd edition, 1996.
8. Etkin, B., *Dynamics of Atmospheric Flight*, John Wiley & Sons, New York, 1972.
9. Hoak, D.E., et al., "The USAF Stability and Control DATCOM (Data Compendium)," Air Force Wright Aeronautical Laboratories, TR-83-3048, October 1960 (Revised April 1978).
10. Hurt, H.H., Jr, *Aerodynamics for Naval Aviators*, US Navy NAVWEPS 00-80 T-80, US Government Printing Office, Washington, DC, January 1965.
11. McCormick, B.W., *Aerodynamics, Aeronautics, and Flight Mechanics*, John Wiley & Sons, New York, 1979.
12. Nelson, R.C., *Flight Stability and Automatic Control*, McGraw-Hill, Boston, Massachusetts, 2nd edition, 1998.
13. Perkins, C.D. and Hage, R.E., *Airplane Performance Stability and Control*, John Wiley & Sons, New York, 1949.
14. Raymer, D.P., *Aircraft Design: A Conceptual Approach*, AIAA Education Series, American Institute of Aeronautics and Astronautics, Washington, DC, 2nd edition, 1992.
15. Smith, H.J., "Evaluation of the Lateral-Directional Stability and Control Characteristics of the Lightweight M2-F1 Lifting Body and Low Speeds," NASA TN D-3022, September 1965.
16. Talay, T.A., *Introduction to the Aerodynamics of Flight, NASA SP-367*, US Government Printing Office, Washington, DC, 1975.
17. Ward, D.T. and Strganac, T.W., *Introduction to Flight Test Engineering*, 2nd edition, Kendall Hunt Publishing Company, Dubuque, Iowa, 1996.
18. Weingarten, N.C., "History of In-Flight Simulation and Flying Qualities Research at Calspan," *AIAA Journal of Aircraft*, Vol. 42, No. 2, March/April 2005.
19. Yechout, T.R., *Introduction to Aircraft Flight Mechanics*, 2nd edition, American Institute of Aeronautics and Astronautics, Inc., Reston, Virginia, 2014.

Problems

1. At a given velocity, an aircraft has a pitching moment coefficient about its center of gravity, at zero lift, of 0.0621. At an absolute angle-of-attack of 3° and the same velocity, the moment coefficient is 0.0152. Is the aircraft longitudinally statically stable and balanced? What is the trim angle-of-attack at this velocity?

2. At a given velocity, an aircraft has a pitching moment coefficient about its center of gravity, at zero lift, of 0.0411. At an absolute angle-of-attack of 4° and the same velocity, the moment coefficient is 0.0621. Is the aircraft longitudinally statically stable and balanced? What is the trim angle-of-attack at this velocity?

3. A rectangular wing, with a chord length of 7 ft 3 in, has a lift curve slope of $0.0987 \, \text{deg}^{-1}$. The aerodynamic center of the wing is located at the quarter chord point and the center of gravity is 2 ft 4 in aft of the wing leading edge. The pitching moment about the aerodynamic center of the wing is −0.1023. Plot the wing pitching moment coefficient as a function of angle-of-attack, from 0° to 15°.

4. Using the aircraft specifications and wing angle-of-attack from Example 6.2, plot the horizontal tail pitching moment coefficient, about the aircraft center of gravity, as a function of the tail incidence angle, from $-3°$ to $+1°$.

5. An aircraft configuration consists of a fuselage (body), wing, and aft-mounted horizontal tail. Specifications of the configuration are given in the table below.

Parameter	Value
Wing area, S	$182 \, \text{ft}^2$
Wing chord, c	$5.32 \, \text{ft}$
Wing span, b	$34.2 \, \text{ft}$
Horizontal tail area, S_t	$28.1 \, \text{ft}^2$
Horizontal tail lift curve slope, a_t	$0.0981 \, \text{deg}^{-1}$
Horizontal tail incidence angle, i_t	$-2.21°$
Horizontal tail efficiency, η_t	0.980
Distance from CG to tail ac, l_t	$14.3 \, \text{ft}$
Downwash at zero lift, ε_0	$1.02°$

The wing–body (without the tail) is tested in a wind tunnel and the following lift and moment data are obtained as a function of angle-of-attack at a center of gravity location of $0.290c$.

α (deg)	$C_{L,wb}$	$C_{M,wb}$
-1.23	0	-0.1344
0	0.1199	-0.1275
5	0.6074	-0.09929
10	1.095	-0.07111
15	1.582	-0.04293

Plot the pitching moment about the center of gravity due to the wing–body, the horizontal tail, and the complete aircraft for absolute angles-of-attack from $0°$ to $20°$. Is the aircraft longitudinally statically stable and longitudinally balanced?

6. Solve for the horizontal tail area, S_t, required to make the aircraft, in Problem 5, longitudinally balanced with a pitching moment coefficient at zero lift, of the complete aircraft, equal to 0.05. Calculate the trim angle-of-attack. Plot the pitching moment coefficient curves, for the horizontal tail, wing–body, and complete aircraft, versus angle-of-attack for absolute angles-of-attack from $0°$ to $15°$.

7. If the center of gravity of the aircraft in Problem 5 is at $0.29c$, calculate the location of the neutral point and the static margin.

8. An aircraft is in steady, level flight at an airspeed of 425 km/h. The aircraft has a weight of 65,300 N, a wing area of $23.2 \, \text{m}^2$, and a horizontal tail area of $5.3 \, \text{m}^2$. The horizontal tail efficiency is 0.99 and the elevator effectiveness is $3.67 \, \text{radian}^{-1}$. Assuming standard sea level conditions, calculate the change in the lift coefficient if the elevator is deflected $3.1°$.

Appendix A

Constants

A.1 Miscellaneous Constants

Symbol	Description	SI Units	English Units
a	speed of sound (at sea level)	340.2 m/s	1116.6 ft/s
c_p	specific heat at constant pressure for air	1006 J/kg·K	6020.7 ft·lb/slug·°R
c_v	specific heat at constant volume for air	719 J/kg·K	4303.1 ft·lb/slug·°R
g	acceleration due to gravity (at sea level)	9.81 m/s^2	32.17 ft/s^2
$\mathcal{M}$	molecular weight of air	28.96 kg/(kg mol)	28.96 slug/(slug mol)
R	specific gas constant for air	287 J/(kg·K)	1716 ft·lb/(slug·°R)
$\mathcal{R}$	universal gas constant	8314 J/(kg mol·K)	1545 ft·lb/(lb$_m$ mol·°R)
γ	ratio of specific heats for air	1.4	1.4

A.2 Properties of Air at Standard Sea Level Conditions

Property	Symbol	SI Units	English Units
Density	ρ_{SSL}	1.225 kg/m^3	0.002377 slug/ft^3
Pressure	p_{SSL}	101,325 N/m^2	2116 lb/ft^2
Temperature	T_{SSL}	288 K (15°C)	519°R (59°F)
Speed of sound	a_{SSL}	340.2 m/s	1116.6 ft/s
Dynamic viscosity	μ_{SSL}	17.89×10^{-6} kg/(m·s)	0.3737×10^{-6} slug/(ft·s)
Thermal conductivity	k_{SSL}	0.02533 J/(m·s·K)	4.067×10^{-6} Btu/(ft·s·°R)

Introduction to Aerospace Engineering with a Flight Test Perspective, First Edition. Stephen Corda.
© 2017 John Wiley & Sons Ltd. Published 2017 by John Wiley & Sons Ltd.
Companion Website: www.wiley.com/go/corda/aerospace_engg_flight_test_persp

Appendix B

Conversions

B.1 Unit Conversions

$$1\,\text{atm} = 2116\,\frac{\text{lb}}{\text{ft}^2} = 1.01325 \times 10^5\,\frac{\text{N}}{\text{m}^2}$$

$$1\,\text{Btu} = 778\,\text{ft} \cdot \text{lb} = 1055\,\text{J}$$

$$1\,\text{ft} = 0.3048\,\text{m}$$

$$1\,\text{gal} = 3.785\,\text{liters}$$

$$1\,\text{hp} = 550\,\frac{\text{ft} \cdot \text{lb}}{\text{s}} = 2546.1\,\frac{\text{Btu}}{\text{h}} = 745.7\,\text{W}$$

$$1\,\frac{\text{lb}}{\text{in}^2} = 6895.0\,\frac{\text{N}}{\text{m}^2}$$

$$1\,\frac{\text{lb}}{\text{ft}^2} = 47.88\,\frac{\text{N}}{\text{m}^2}$$

$$1\,\text{mi} = 5280\,\text{ft} = 1609\,\text{m}$$

$$1\,\text{nm} = 6076\,\text{ft} = 11{,}852\,\text{m}$$

$$1\,\text{slug} = 32.2\,\text{lb}_m = 14.594\,\text{kg}$$

Introduction to Aerospace Engineering with a Flight Test Perspective, First Edition. Stephen Corda.
© 2017 John Wiley & Sons Ltd. Published 2017 by John Wiley & Sons Ltd.
Companion Website: www.wiley.com/go/corda/aerospace_engg_flight_test_persp

B.2   Temperature Unit Conversions

$$1.8\,°R = 1\,K$$
$$K = °C + 273.15$$
$$°R = °F + 459.67$$
$$°C = \frac{5}{9}(°F - 32)$$
$$°F = \frac{9}{5}(°C + 32)$$

Appendix C

Properties of the 1976 US Standard Atmosphere

C.1 English Units

Property	Symbol	English units
Temperature	T_{SSL}	519°R
Pressure	p_{SSL}	2116 lb/ft^2
Density	ρ_{SSL}	0.002377 slug/ft^3

Standard sea level atmospheric conditions.

Altitude, h_g (ft)	Temperature ratio, $\theta = T/T_{SSL}$	Pressure ratio, $\delta = p/p_{SSL}$	Density ratio, $\sigma = \rho/\rho_{SSL}$
−1,000	1.00688	1.03667	1.02957
0	1.00000	1.00000	1.00000
1,000	0.99312	0.96442	0.97106
2,000	0.98625	0.92983	0.94279
3,000	0.97938	0.89628	0.91511
4,000	0.97250	0.86367	0.88810
5,000	0.96563	0.83210	0.86169
6,000	0.95876	0.80144	0.83590
7,000	0.95189	0.77171	0.81070
8,000	0.94502	0.74289	0.78609
9,000	0.93815	0.71491	0.76207
10,000	0.93128	0.68784	0.73860

Introduction to Aerospace Engineering with a Flight Test Perspective, First Edition. Stephen Corda.
© 2017 John Wiley & Sons Ltd. Published 2017 by John Wiley & Sons Ltd.
Companion Website: www.wiley.com/go/corda/aerospace_engg_flight_test_persp

Altitude, h_g (ft)	Temperature ratio, $\theta = T/T_{SSL}$	Pressure ratio, $\delta = p/p_{SSL}$	Density ratio, $\sigma = \rho/\rho_{SSL}$
11,000	0.92441	0.66161	0.71567
12,000	0.91754	0.63614	0.69334
13,000	0.91067	0.61152	0.67150
14,000	0.90381	0.58766	0.65022
15,000	0.89694	0.56460	0.62944
16,000	0.89007	0.54225	0.60920
17,000	0.88321	0.52060	0.58948
18,000	0.87635	0.49972	0.57021
19,000	0.86948	0.47949	0.55145
20,000	0.86262	0.45991	0.53315
21,000	0.85576	0.44101	0.51535
22,000	0.84890	0.42274	0.49798
23,000	0.84204	0.40509	0.48107
24,000	0.83518	0.38804	0.46462
25,000	0.82832	0.37158	0.44859
26,000	0.82146	0.35569	0.43299
27,000	0.81460	0.34036	0.41782
28,000	0.80774	0.32556	0.40305
29,000	0.80088	0.31130	0.38869
30,000	0.79403	0.29755	0.37473
31,000	0.78717	0.28429	0.36115
32,000	0.78032	0.27151	0.34795
33,000	0.77346	0.25921	0.33513
34,000	0.76661	0.24737	0.32267
35,000	0.75976	0.23597	0.31058
36,000	0.75291	0.22499	0.29883
37,000	0.75187	0.21447	0.28525
38,000	0.75187	0.20444	0.27191
39,000	0.75187	0.19488	0.25920
40,000	0.75187	0.18577	0.24708
41,000	0.75187	0.17709	0.23552
42,000	0.75187	0.16881	0.22452
43,000	0.75187	0.16092	0.21402
44,000	0.75187	0.15340	0.20402
45,000	0.75187	0.14623	0.19449
46,000	0.75187	0.13940	0.18540
47,000	0.75187	0.13288	0.17674
48,000	0.75187	0.12668	0.16848
49,000	0.75187	0.12076	0.16061
50,000	0.75187	0.11512	0.15310
51,000	0.75187	0.10974	0.14596
52,000	0.75187	0.10462	0.13914
53,000	0.75187	0.09973	0.13264
54,000	0.75187	0.09507	0.12645

Altitude, h_g (ft)	Temperature ratio, $\theta = T/T_{SSL}$	Pressure ratio, $\delta = p/p_{SSL}$	Density ratio, $\sigma = \rho/\rho_{SSL}$
55,000	0.75187	0.09063	0.12055
56,000	0.75187	0.08640	0.11492
57,000	0.75187	0.08237	0.10955
58,000	0.75187	0.07853	0.10444
59,000	0.75187	0.07486	0.09957
60,000	0.75187	0.07137	0.09492
61,000	0.75187	0.06804	0.09049
62,000	0.75187	0.06486	0.08627
63,000	0.75187	0.06184	0.08224
64,000	0.75187	0.05895	0.07841
65,000	0.75187	0.05620	0.07475
66,000	0.75205	0.05358	0.07125
67,000	0.75310	0.05108	0.06783
68,000	0.75415	0.04871	0.06458
69,000	0.75520	0.04645	0.06150
70,000	0.75625	0.04429	0.05856
75,000	0.76151	0.03496	0.04592
80,000	0.76676	0.02765	0.03606
85,000	0.77200	0.02190	0.02837
90,000	0.77725	0.01738	0.02236
95,000	0.78249	0.01381	0.01765
100,000	0.78773	0.01100	0.01396

C.2   SI Units

Property	Symbol	SI units
Temperature	T_{SSL}	288 K
Pressure	p_{SSL}	101,325 N/m^2
Density	ρ_{SSL}	1.225 kg/m^3

Standard sea level atmospheric conditions.

Altitude, h_g (m)	Temperature ratio, $\theta = T/T_{SSL}$	Pressure ratio, $\delta = p/p_{SSL}$	Density ratio, $\sigma = \rho/\rho_{SSL}$
−500	1.01128	1.06075	1.04890
0	1.00000	1.00000	1.00000
500	0.98872	0.94213	0.95290
1,000	0.97745	0.88701	0.90751
1,500	0.96617	0.83454	0.86376

Altitude, h_g (m)	Temperature ratio, $\theta = T/T_{SSL}$	Pressure ratio, $\delta = p/p_{SSL}$	Density ratio, $\sigma = \rho/\rho_{SSL}$
2,000	0.95490	0.78461	0.82171
2,500	0.94363	0.73715	0.78118
3,000	0.93236	0.69204	0.74224
3,500	0.92109	0.64920	0.70482
4,000	0.90983	0.60854	0.66886
4,500	0.89856	0.56998	0.63432
5,000	0.88730	0.53341	0.60117
5,500	0.87604	0.49878	0.56936
6,000	0.86478	0.46601	0.53887
6,500	0.85352	0.43499	0.50964
7,000	0.84227	0.40567	0.48165
7,500	0.83102	0.37799	0.45485
8,000	0.81977	0.35186	0.42922
8,500	0.80852	0.32720	0.40470
9,000	0.79727	0.30398	0.38127
10,000	0.78602	0.28211	0.35891
10,500	0.77478	0.26153	0.33756
11,000	0.76353	0.24219	0.31720
11,500	0.75229	0.22403	0.29780
12,000	0.75187	0.20711	0.27545
12,500	0.75187	0.19145	0.25464
13,000	0.75187	0.17699	0.23541
13,500	0.75187	0.16363	0.21763
14,000	0.75187	0.15128	0.20119
14,500	0.75187	0.13985	0.18601
15,000	0.75187	0.12930	0.17197
15,500	0.75187	0.11954	0.15899
16,000	0.75187	0.10218	0.13589
16,500	0.75187	0.09447	0.12564
17,000	0.75187	0.08734	0.11616
17,500	0.75187	0.08075	0.10740
18,000	0.75187	0.07466	0.09931
18,500	0.75187	0.06903	0.09181
19,000	0.75187	0.06383	0.08490
19,500	0.75187	0.05902	0.07850
20,000	0.75187	0.05457	0.07258
20,500	0.75337	0.05046	0.06698
21,000	0.75510	0.04667	0.06181
21,500	0.75682	0.04317	0.05705
22,000	0.75854	0.03995	0.05266
22,500	0.76027	0.03697	0.04862
23,000	0.76199	0.03422	0.04490

Altitude, h_g (m)	Temperature ratio, $\theta = T/T_{SSL}$	Pressure ratio, $\delta = p/p_{SSL}$	Density ratio, $\sigma = \rho/\rho_{SSL}$
23,500	0.76371	0.03168	0.04148
24,000	0.76543	0.02933	0.03832
24,500	0.76716	0.02716	0.03541
25,000	0.76888	0.02516	0.03272
25,500	0.77060	0.02331	0.03025
26,000	0.77232	0.02160	0.02796
26,500	0.77404	0.02002	0.02586
27,000	0.77576	0.01855	0.02392
27,500	0.77748	0.01720	0.02212
28,000	0.77920	0.01595	0.02047
28,500	0.78092	0.01479	0.01894
29,000	0.78264	0.01372	0.01753
29,500	0.78436	0.01273	0.01623
30,000	0.78608	0.01181	0.01503
30,500	0.78780	0.01096	0.01392
31,000	0.78952	0.01018	0.01289
31,500	0.79124	0.00945	0.01194
32,000	0.79295	0.00877	0.01107
32,500	0.79676	0.00815	0.01023
33,000	0.80157	0.00757	0.00945
33,500	0.80638	0.00704	0.00873
34,000	0.81119	0.00655	0.00807
34,500	0.81599	0.00609	0.00747
35,000	0.82080	0.00567	0.00691
35,500	0.82560	0.00528	0.00640
36,000	0.83041	0.00492	0.00592
36,500	0.83521	0.00459	0.00549
37,000	0.84002	0.00428	0.00509
37,500	0.84482	0.00399	0.00472
38,000	0.84962	0.00372	0.00438
38,500	0.85442	0.00347	0.00407
39,000	0.85922	0.00325	0.00378
39,500	0.86402	0.00303	0.00351
40,000	0.86882	0.00283	0.00326

Index

Introduction to Aerospace Engineering with a Flight Test Perspective, First Edition. Stephen Corda.
© 2017 John Wiley & Sons Ltd. Published 2017 by John Wiley & Sons Ltd.
Companion Website: www.wiley.com/go/corda/aerospace_engg_flight_test_persp

Printed and bound by CPI Group (UK) Ltd, Croydon, CR0 4YY

16/04/2025

14658837-0001